GW01406604

Edible Medicinal and Non-Medicinal Plants

T.K. Lim

Edible Medicinal and Non-Medicinal Plants

Volume 3, Fruits

ISBN 978-94-007-2533-1 e-ISBN 978-94-007-2534-8
DOI 10.1007/978-94-007-2534-8
Springer Dordrecht Heidelberg London New York

Library of Congress Control Number: 2011944279

Printed on acid-free paper

Springer is part of Springer Science+Business Media (www.springer.com)

Disclaimer

The author and publisher of this work have checked with sources believed to be reliable in their efforts to confirm the accuracy and completeness of the information presented herein and that the information is in accordance with the standard practices accepted at the time of publication. However, neither the author nor publishers warrant that information is in every aspect accurate and complete and they are not responsible for errors or omissions or for consequences from the application of the information in this work. This book is a work of reference and is not intended to supply nutritive or medical advice to any individual. The information contained in the notes on edibility, uses, nutritive values, medicinal attributes and medicinal uses and suchlike included here are recorded information and do not constitute recommendations. No responsibility will be taken for readers' own actions.

Contents

Introduction

This book continues as volume 3 of a multi-compendium on *Edible Medicinal and Non-Medicinal Plants*. It focuses on edible fruits/seeds used fresh, cooked or processed into other by-products, or as vegetables, spices, stimulant, edible oils and beverages. It covers species from the following families: Ginkgoaceae, Gnetaceae, Juglandaceae, Lauraceae, Lecythidaceae, Magnoliaceae, Malpighiaceae, Malvaceae, Marantaceae, Meliaceae, Moraceae, Moringaceae, Muntigiaceae, Musaceae, Myristicaceae and Myrtaceae. However, not all the edible species in these families are included for want of coloured illustrations. The edible species dealt with in this work include to a larger extent lesser-known, wild and under-utilized crops and also common and widely grown crops.

As in the preceding two volumes, topics covered include: taxonomy (botanical name and synonyms); common English and vernacular names; origin and distribution; agro-ecological requirements; edible plant part and uses; plant botany; nutritive and medicinal/pharmacological properties with up-to-date research findings, traditional medicinal uses other non-edible uses; and selected/cited references for further reading.

Ginkgoaceae is a family of temperate gymnosperms which appeared during the Mesozoic Era, of which the only extant representative and living fossil is *Ginkgo biloba*. *Gingkgo biloba* has both culinary and medicinal uses. Several thousands of scientific papers have been published on the phytochemicals and associated pharmacological and medicinal properties of the aerial plant parts of *G. biloba*. The edible seed is rich in niacin, and vitamin A, phosphorus and potassium. It is a good source of starch and protein, but is low in unsaturated or monounsaturated fats. The seed also contains vitamin B1 (thiamine), B2 (riboflavin), vitamin C and iron, sodium and calcium (USDA 2010). Important bioactive constituents reported to occur in the medicinally used Ginkgo leaves include terpene trilactones, i.e., ginkgolides A, B, C, J and bilobalide, many flavonol glycosides, biflavones, proanthocyanidins, alkylphenols, simple phenolic acids, and polyprenols (van Beek 2002).

Gnetaceae is a representative of tropical gymnosperms. *Gnetum*, a genus of about 30–35 species, is the sole genus in the family Gnetaceae and order Gnetales. They are tropical, evergreen trees, shrubs and lianas and occur in Indomalaysia, tropical parts of West Africa, Fiji and the northern regions of South America. Many *Gnetum* species including *Gnetum gnemon* are edible, with the seeds being roasted, and the foliage used as a leaf vegetable. *Gnetum gnemon* contains bioactive chemicals like flavonostilbenes and stilbenes that play a role in various pharmacological activities. *Gnetum gnemon* is found in Assam, southeast Asia, the Philippines and Papua New Guinea, Fiji, Solomon Islands and Vanuatu.

The large and economically important Juglandaceae, or the walnut and hickory family is a family of deciduous, semi-evergreen, or evergreen, monoecious (rarely dioecious) trees, rarely shrubs in the order Fagales. The family contains 9 genera and 50 or more species, which are distributed mainly in the north temperate zone but extend through Central America along the Andes Mountains to Argentina and, in scattered stands, from temperate Asia to the highlands of Java and New Guinea. The commercially important nut-producing trees include walnut (*Juglans regia*), pecan (*Carya illinoinensis*), and hickory (*Carya* spp). Walnut, hickory, and gaulin (*Alfaroa costaricensis*) are also valuable timber trees. Both Persian walnut, *Juglans regia*, and pecan nut which are covered in this volume, have culinary, nutritive and medicinal attributes.

The Lauraceae or laurel family contains about 55 genera and over 2,000 species world-wide, mostly from warm subtropical or tropical regions, especially Southeast Asia and Brazil. Most are aromatic evergreen trees or shrubs, a few genera are deciduous, and *Cassytha* is a genus of parasitic vines. The Lauraceae are economically important as sources of medicine, timber, nutritious fruits (e.g., *Persea americana*), spices (e.g. *Cinnamomum aromaticum, C. verum, Laurus nobilis* covered in later volumes), and perfumes and essential oils. Avocados are important oil-rich and nutritious fruit with health and medicinal properties, that are now planted in warm climates across the world. *Litsea garciae* is another edible tropical fruit but is lesser-known and under-utilised. The hard wood of several species is a source for timber around the world.

Lecythidaceae, a tropical plant family, is indigenous to South America and Madagascar. It has about 20 genera and 250–300 species of woody plants. Neotropical Lecythidaceae comprises ecologically dominant species in the Amazonian forests and are spectacular plants with showy flowers and large woody fruits. They include the edible and economically important Brazil nut (*Bertholletia excelsa*), and the edible, lesser-known paradise nut or monkey nut (*Lecythis* spp.). Other edible but lesser-known species are the *Barringtoni*a species which are eaten in southeast Asian and the Pacific Island countries. The genus *Barringtonia* is also placed in the family, Barringtoniaceae.

Magnoliaceae comprises about 225 species in 7 genera. Magnoliaceae is better known for its ornamental species and timber species. The bark and flowers from several species are believed to possess medicinal qualities. In this family the edible fruit species that is treated in this volume is *Michelia mediocris*, a highly valued and productive indigenous Vietnamese timber species. The fruit and seeds of this species have good potential as a spice.

Malpighiaceae comprises approximately 75 genera and 1,300 species, all of which are native to the tropics and subtropics. About 80% of the genera and 90% of the species occur in the New World (the Caribbean and the southernmost United States to Argentina) and the rest in the Old World (Africa, Madagascar, and Indomalaysia to New Caledonia and the Philippines). The Malpighiaceae are shrubs, small trees, or woody lianas. Of the two edible genera *Malpighi*a and *Bunchosi*a, the former also has species with pharmacological and medicinal attributes. Acerola (*Malpighia emarginata*) has been reported to have very high vitamin C content, much higher than other fruits like pineapple, araçá (*Eugenia stipitata*), cashew, guava, kiwi, orange, lemon, and strawberry. Acerola has also reported to have carotenoids and bioflavonoids which contribute to its high antioxidant capacity and provide important nutritive and pharmacological values.

The Marantaceae or arrowroot or prayer plant family, is a family of flowering, herbaceous plants under the order Zingiberales. Based on nucleotide sequence variation, 59 species (21 genera) formed the ingroup, and 12 species (12 genera) of other Zingiberales formed the outgroup (Andersson and Chase 2001). There is no support for the traditional subdivision of Marantaceae into a triovulate and a uniovulate tribe or the informal groups previously proposed (Andersson 1981). Based on phylogeny it is concluded that Africa where early diversification of the family took place, in spite of being much poorer in species, is the most likely ancestral area of Marantaceae. The family is found in the lowland tropics of Asia and Africa, mainly (80%) in American tropics, occasionally subtropics, southern United States to northern Argentina. The family is known for its large starchy rhizomes and house-hold ornamental plants. The most significant food plant is *Maranta arundinacea*, cultivated in tropical regions worldwide for arrowroot starch. However, one species, *Thaumatococcus daniellii* produces fruit with edible aril which furnished a natural source of thuamatin, an intensely sweet protein which is about 100,000 times sweeter than sugar on a molar basis and 3,000 times on a weight basis. Thaumatin is used as a sweetener and flavour enhancer for food, desserts, confectionary and beverages.

Malvaceae has been circumscribed to embrace the non-monophyletic families, Bombacaceae, Tiliaceae, and Sterculiaceae, which have always been considered very close to the traditional Malvaceae *sensu stricto,* a very homogeneous and cladistically monophyletic group. Following this circumscription which is based on newer techniques, Malvaceae *sensu lato* now include all of these families so as to have a monophyletic group. The circumscription of the Malvaceae is still controversial. A close relationship between Bombacaceae and Malvaceae has long been recognized but until recently the families have been kept separate in most classification systems, and continue to be separated in many recent references, including the reference work in classification of flowering plants by Heywood et al. (2007) and Takhtajan (2009). However, the Angiosperm Phylogeny Group (2003, 2009) have lumped them together into a larger family Malvaceae *sensu lato*. Heywood et al. (2007) assert "although closely related to Malvaceae, molecular data supports their separation. Only pollen and habit seem to provide a morphological basis for the separation." Contrariwise they say: "One approach is to lump them (the families in the core Malvales, including Bombacaceae) all into a 'super' Malvaceae, recognizing them as subfamilies. The other, taken here, is to recognize each of these ten groups as families". Members of the Bombacaceae have been covered in volume 1. In this volume, members of Sterculiaceae (e.g. kola, cacao, cupuassu) are included together with species belonging to the traditional Malavaceae *sensu stricto* which comprises the mallows, abutilons, cotton, okra, hibiscuses and related plants. Species of Malvaceae *sensu lato* provide sources of fibre, food and beverages, medicines, timber, and in horticulture (ornamental). Also some members are deemed as weeds or invasive species. The species with edible fruits/seeds and medicinal properties covered in this volume include *Grewia asiatica, Abelmoschus esculentus, Scaphium macropodum, Sterculia foetida, Sterculia monosperma and Sterculia parviflora, Theobroma bicolour, T. cacao* and *T. grandiflorum.* Due to their high concentration of catechins and procyanidins, bioactive compounds with distinct properties, cocoa and chocolate products may have beneficial health effects against oxidative stress and chronic inflammation, risk factors for cancer and other chronic diseases (Maskarinec 2009).

The Meliaceae or mahogany family comprises about 50 genera and 550 species, with a pantropical distribution but a weak penetration into the temperate zone. One genus (*Toona*) extends north into temperate China and south into southeast Australia, and another (*Melia*) nearly as far north. The species are evergreen or deciduous trees or tree-lets and rarely shrubs; the bark sometimes with a milky latex. Meliaceae species are very common trees in the understory of lowland primary forest throughout Malesia. Various species are used for vegetable oil, soap-making, insecticides, and highly prized wood mahogany (*Swietenia* spp. and *Aglai*a spp.). Species that provide edible fruits are mainly tropical and include various *Aglaia* spp., the duku, langsat, lonkong (*Lansium domesticum*) and the santol (*Sandoricum koetjape*). The latter two species are popular and widely eaten fruits in southeast Asia and also have several pharmacological properties; various plant have been used in traditional folkloric medicine.

The Moraceae family comprises between 37 and 43 genera and 1,100–1,400 species, widespread in tropical and subtropical areas but less common in temperate areas. They comprise trees, shrubs, vines, frequently with milky or watery latex. Flowers occur usually in heads and are unisexual; ovule is anatropous or campylotropous and united into a more or less fleshy compound fruits. Economically, the most important species are those of *Morus* and *Maclura* associated with the production of silk. Some species in *Broussonetia*, *Maclura*, and *Morus* are important for paper making. Some *Artocarpus* and *Broussonetia* species are used for furniture or timber.

Some species in *Artocarpus*, *Ficus*, *Prainea*, *Treculia* and *Morus* have edible fruit. The common edible tropical *Artocarpus* species include the bread fruit *A. altilis*, the breadnut *A. camansi*, jackfruit *A. heterophyllus*, chempedak *A. integer* and the marang or terap *A. odratissimus*. Many of the edible *Artocapus* species contain bioactive compounds such as the prenylated flavonoids or stilbenoids, and lectins which have significant pharmacological activities. The edible *Ficus* species include the common and popular fig *Ficus carica* and other lesser-known fig trees like the elephant ear fig tree, *F. auricalata*, cluster fig, *F. racemosa*, the creeping ivy fig, *F. pumila* and dinner plate fig tree *F. dammaropsis*. Many of the *Ficus* species have medicinal attributes. *Prainea limpato* is a rare species with unusual stellate, grosteques looking fruit which is edible. The edible *Morus* species include the red (*M. rubra*), white (*M. alba*) and black (*M. nigra*) mulberries, the plant parts of which have bioactive chemicals with pharmacological activities.

Moringaceae or horseradish tree family comprise only one genus with 12 species, found mainly in tropical and subtropical climates. The most widely known species is *Moringa oleifera*, a multi-purpose tree native to the foothills of the Himalayas in north-western India and cultivated pan-tropically. *M. stenopetala*, an African species, is also widely cultivated, but to a much lesser extent than *M. oleifera*. *Moringa oleifera* (horseradish or drumstick tree) has edible fruits and leaves. The seeds provide "ben oil" used in perfumery and light lubricants and the seeds are also used to purify water and removal of industrial pollutants and heavy metals. *Moringa oleifera* oil was found to have potential as acceptable feedstock for biodiesel. The leaves made highly nutritious cattle feed and the roots are also a source of edible condiment. The tree's bark, roots, fruit, flowers, leaves, seeds, and gum are also used medicinally.

Muntigiaceae is indigenous to the neotropics. The small family includes the monotypic genera, *Muntingia*, *Dicraspidia* and *Neotessmania*. They were previously included in Elaeocarpaceae, Tiliaceae or Flacourtiaceae. Muntigiaceae is closely related to the rosid order Mavales (Sterculiaceae, Tiliaceae, Bombaceae and Malvaceae) and several other families but the relationships are still obscure and unresolved. *Muntingia calabura,* the type species, has edible fruits and contains phytochemicals with pharmacological properties.

The genus *Musa* in the family Musaceae is divided into four sections, including members of both seeded and non-seeded (parthenocarpic) types. Two of the sections contain species with a chromosome number of 2n = 20 (*Callimusa* and *Australimusa*) while the other two sections (*Eumusa* and *Rhodochlamys*) have species with a basic chromosome number of 11 (2n = 22). The majority of cultivated bananas arises from the *Eumusa* group of species. This section is the biggest in the genus and the most geographically widespread, with species being found from India, throughout South East Asia to the Pacific Islands.

Linnaeus first classified banana (*Musa*) into two species based on their culinary use, *Musa sapientum* for dessert bananas and *Musa paradisiaca* for plantains. This distinction is entirely semantic and artificial with no botanical basis and no consistent culinary basis. In 1948, Cheesman found that *Musa sapientum* and *Musa paradisiaca*, described by Linnaeus, were actually cultivars and intra and interspecific hybridizations of two wild and seedy species, *Musa acuminata* and *Musa balbisiana,* each contributing the A and B genomes respectively. The identification of *Musa* cultivars has traditionally been based upon various combinations of morphological, phenological and floral criteria. The preponderance of cultivars magnified the taxonomic problems of classifying *Musa* until Simmonds and Shepherd (1955) devised a scoring system based on 15 diagnostic morphological characters to differentiate *M. acuminata* cultivars from *M. balbisiana* cultivars and their hybrids into six genome groups. Generally, modern classifications of banana cultivars follow Simmonds' and Shepherd's system. The accepted names for bananas are *Musa acuminata*, *Musa balbisiana* or *Musa acuminata* × *balbisiana*, depending on their ancestral genome. Examples of the new classification scheme adopted include: *Musa acuminata* (AA group) 'Lakatan', *Musa acuminata* (AAA Group) 'Gros

Michel', *Musa acuminata x balbisiana* (AAB Group) 'Horn Plantain', *Musa acuminata x balbisiana* (AAB Group) 'Pisang Raja' *Musa acuminata x balbisiana* (ABB Group) 'Bluggoe'. Other edible Muss spp covered in this volume are *Musa troglodytarum* (Fei bananas), *Musa velutina* and *Musa zebrina*.

As described above, most banana cultivars are derived from two species, *Musa acuminata* (A genome) and *Musa balbisiana* (B genome). However, Shepherd and Ferreira (1982) found cultivars derived from hybridizations with *M. schizocarpa* (S genome), which was subsequently confirmed by Carreel et al. (1993). Several landraces containing the two genomes *acuminata* and species from the Australimusa section (T genome) and two landraces containing the three genomes, A, B and T have been found in Papua New Guinea and a Philippine clone (Butuhan) is considered to be the result of an ancient hybridization between *M. balbisiana* and *M. textilis* (T genome) (Carreel et al. 1993).

Myristicaceae, the nutmeg family comprises about 20 genera and approximately 500 species of evergreen trees and shrubs found in tropical Asia to the Pacific islands and also in Africa and tropical America. The most well known and widely cultivated species is the spice, *Myristica fragrans*, the nutmeg or mace. Nutmeg has culinary and medicinal uses. Two other edible species covered in this volume are *Myristica fatua Myrtaceae*, and *Horsfeldia australiana.*

Myrtaceae , the myrtle family, placed within the order Myrtales comprises at least 133 genera and 3,800 species of woody shrubs to tall trees. It has centers of diversity in Australia, southeast Asia, and tropical to southern temperate America, but has little representation in Africa. The family is distinguished by a combination of the following features: entire aromatic leaves containing oil glands, flower parts in multiples of four or five, ovary half inferior to inferior, numerous brightly coloured and conspicuous stamens, internal phloem, and vestured pits on the xylem vessels. Until relatively recently, the family has been considered to be naturally divisible into two subfamilies, the fleshy-fruited Myrtoideae and the capsular-fruited Leptospermoideae. This was seriously challenged by Johnson and Briggs (1984) who concluded, from a cladistic analysis based on morphological and anatomical characters, that these subfamilies must be abandoned. Species of the myrtle family provide many valuable products, including timber (e.g. *Eucalyptus*), essential oils and spices (e.g. allspice, cloves), and horticultural plants (e.g. ornamentals such as *Verticordia, Callistemon, Leptospermum*) and edible fruits such as the common guava, strawberry guava, other *Psidium* spp., Feijoa, myrtle, rose myrtle, jaboticaba, *Eugenia* spp. *Myrciaria* spp. *and Syzygium* spp. Many of these myrtaceous plants also have medicinal properties.

Selected References

Andersson L (1981) The neotropical genera of Marantaceae. Circumscription and relationships. Nordic J Bot 1:218–245

Andersson L, Chase MW (2001) Phylogeny and classification of Marantaceae. Bot J Linn Soc 135(3): 275–287

Angiosperm Phylogeny Group II (2003) An update of the Angiosperm Phylogeny Group classification for the orders and families of flowering plants: APG II. Bot J Linn Soc 141:399–436

Angiosperm Phylogeny Group III (2009) An update of the Angiosperm Phylogeny Group classification for the orders and families of flowering plants: APG III. Bot J Linn Soc 161:L105–L121

Baum DA, Dewitt Smith S, Yen A, Alverson WS, Nyffeler R, Whitlock BA, Oldham RL (2004) Phylogenetic relationships of Malvatheca (Bombacoideae and Malvoideae; Malvaceae sensu lato) as inferred from plastid DNA sequences. Am J Bot 91(11): 1863–1871

Bayer C, Chase MW, Fay MF (1998) Muntingiaceae, a new family of dicotyledons with malvalean affinities. Taxon 47:37–42

Bayer C, Fay MF, de Bruijn AY, Salvolainen V, Morton CM, Kubitzki K, Alverson WS, Chase MW (1999) Support for an expanded family concept of Malvaceae within a recircumscribed order Malvales: a combined analysis of plastid atpB and rbcL DNA sequences. Bot J Linn Soc 129(4):267–303

Carlquist S (1996) Wood and bark anatomy of lianoid Indomalesian and Asiatic species of *Gnetum*. Bot J Linn Soc 121:1–24

Carreel F, Fauré S, Gonzalez de Léon D, Lagoda PJL, Perrier X, Bakry F, Tézenas du Montcel H, Lanaud C, Horry J-P (1993) Evaluation de la diversité génétique chez les bananiers diploïdes à l'IRFA-CIRAD. Fruits (numéro spécial):25–40

Cheesman EE (1948) Classification of bananas IIIc *Musa paradisiaca* Linn. and *Musa sapientum* Linn. Kew Bull 2:146–153

Daniells J, Jenny C, Tomekpe K (2001) Musalogue: a catalogue of Musa germplasm. Diversity in the genus Musa (Arnaud E, Sharrock S. compli.). International Network for the Improvement of Banana and Plantain, Montpellier, France. <www.inibap.org/publications/musalogue.pdf>

de Vos AM, Hatada M, van der Wel H, Krabbendam H, Peerdeman AF, Kim SH (1985) Three-dimensional structure of thaumatin I, an intensely sweet protein. Proc Natl Acad Sci USA 82(5):1406–1409

Govaerts R et al (2008) World checklist of Myrtaceae. Royal Botanic Gardens, Kew. xv + 455 pp

Heywood VH, Brummitt RK, Culham A, Seberg O (2007) Flowering plant families of the world. Firefly Books, Richmond Hill, 424 pp

Holttum RE (1951) The Marantaceae of Malaya. Gard Bull Singapore 13:254–296

Hunt D (ed) (1998) *Magnolias* and their allies. International Dendrology Society & Magnolia Society. http://en.wikipedia.org/wiki/Magnoliaceae

Johnson LAS, Briggs BG (1984) Myrtales and Myrtaceae – a phylogenetic analysis. Ann Missouri Bot Gard 71: 700–756

Kostermans AJGH (1957) Lauraceae. Reinwardtia 4(2): 193–256

Li B, Wilson TK (2008) Myristicaceae. In: Wu ZY, Raven PH, Hong DY (eds) Flora of China. Menispermaceae through Capparaceae, vol 7. Science Press/Missouri Botanical Garden Press, Beijing/St. Louis

Liu Y, Xia N, Nooteboom HP (2008) Magnoliaceae. In: Wu ZY, Raven PH, Hong DY (eds) Flora of China. Menispermaceae through Capparaceae, vol 7. Science Press/Missouri Botanical Garden Press, Beijing/St. Louis

Lu A, Stone DE, Grauke LJ (1999) Juglandaceae. In: Wu ZY, Raven PH (eds) Flora of China. Cycadaceae through Fagaceae, vol 4. Science Press/Missouri Botanical Garden Press, Beijing/St. Louis, pp 277–285

Mabberley DJ, Pannell CM (1989) Meliaceae. In: Ng FSP (ed) Tree flora of Malaya. vol 4. Longman, Kuala Lumpur, pp 199–260

Manos PS, Stone DE (2001) Evolution, phylogeny and systematics of the Juglandaceae. Ann Missouri Bot Gard 88:231–269

Maskarinec G (2009) Cancer protective properties of cocoa: a review of the epidemiologic evidence. Nutr Cancer 61(5):573–579

Mori SA, Prance GT (1990a) Lecythidaceae – part II. The zygomorphic-flowered New World genera (*Couroupita, Corythophora, Bertholletia, Couratari, Eschweilera, & Lecythis*). Fl Neotrop 21(2):1–376

Mori SA, Prance GT (1990b) Taxonomy, ecology, and economic botany of the Brazil nut (*Bertholletia excelsa* Humb. and Bonpl.: Lecythidaceae). Adv Econ Bot 8:130–150

Mori SA, Tsou CC, Wu CC, Cronholm B, Anderberg A (2007) Evolution of Lecythidaceae with an emphasis on the circumscription of Neotropical genera: information from combined *ndh*F and *trn*L-F sequence data. Amer J Bot 94(3):289–301

Pennington TD, Styles BT (1975) A generic monograph of the Meliaceae. Blumea 22:419–540

Schultes RE, Raffauf RF (1990) The healing forest: medicinal and toxic plants of the northwest Amazonia. Dioscorides Press, Portland

Sherperd K, Ferreira FR (1982) The PNG biological foundation's banana collection at Laloki, Port Moresby, Papua New Guinea. IBGR/SEAN Newslett 8:28–34

Simmonds NW, Shepherd K (1955) The taxonomy and origins of cultivated bananas. Bot J Linn Soc 55:302–312

Stover RH, Simmonds NW (1987) Bananas, 3rd edn. Longman, London

Takhtajan A (2009) Flowering plants, 2nd edn. Springer, New York, 872 pp

US Department of Agriculture, Agricultural Research Service (USDA) (2010) USDA National Nutrient Database for Standard Reference, Release 23. Nutrient Data Laboratory Home Page. http://www.ars.usda.gov/ba/bhnrc/ndl

van Beek TA (2002) Chemical analysis of *Ginkgo biloba* leaves and extracts. J Chromatogr A 967(1):21–55

Wagner WL, Herbst DR, Sohmer SH (1990) Manual of the flowering plants of Hawai'i, Special publication 83. University of Hawaii Press and Bishop Museum Press, Honolulu, 1854 pp

Watson L, Dallwitz M (1992 onwards) The families of flowering plants: descriptions, illustrations, identification, and information retrieval. Version: 30th June 2010. http://delta-intkey.com

Wilson PG, O'Brien MM, Gadek PA, Quinn CJ (2001) Myrtaceae revisited: a reassessment of infrafamilial groups. Am J Bot 88(11):2013–2025

Wu Z, Zhou ZK, Gilbert MG (2003) Moraceae link. In: Wu ZY, Raven PH, Hong DY (eds) Flora of China. Ulmaceae through Basellaceae, vol 5. Science Press/Missouri Botanical Garden Press, Beijing/St. Louis

Ginkgo biloba

Scientific Name

Ginkgo biloba **L.**

Synonyms

Ginkgo biloba Siebold & Zucc., *Ginkyo biloba* Mayr, *Pterophyllus salisburiensis* Nelson, *Saliburya biloba* Hoffmanns., *Saliburya biloba* Hoffmansegg, *Salisburia adiantifolia* Sm., *Salisburia biloba* (L.) Hoffsgg.

Family

Ginkgoceae

Common/English Names

Common Ginkgo, Duck's Foot Tree, Gingko, Gingko Biloba, Gingko Nuts, Golden Fossil Tree, Kew Tree, Maidenhair Tree

Vernacular Names

Afrikaans: Vrekboom Ginkyo;
Arabic: Gingko, Ginkgoes, Ginkgos, Mabad Ag;
Bohemian: Ginko, Jinan Dvoulaločný;
Brazil: Guincos, Nogueira-Do-Japão;
Chinese: Ginnan, Icho, Paikua Su, Pakgor, Bai Guo, Su, Ya-Chiao;
Croatian: Ginko;
Czech: Jinan Dvoulalocný;
Dutch: Chinese tempelboom, Ginkgo, Japanse Notenboom, Japanse tempelboom, Tempelboom, Waaierboom;
Danish: Ginkgo,Tempeltrć;
Dutch: Chinese Tempelboom, Gingko, Ginkgo Biloba, Japanse Notenboom, Japanse Tempelboom.
Eastonian: Hõlmikpuu;
Finnish: Neidonhiuspuu, Neidonhiuspuut, Temppelipuu;
French: Arbre Aux Mille Écus, Arbre Aux Quarante Écus, Arbre Fossile, Arbre Sacré Des Temples D'asie, Noyer Du Japon Arbre À Noix, Arbre Aux Quarante Ecus, Arbre Des Pagodes, Noyer Du Japon;
German: Fächerblattbaum, Ginkgobaum, Ginko, Goldfruchtbaum, Japanbaum, Japanischer Nussbaum, Mädchenhaarbaum, Silberaprikose, Tempelbaum, Chinesischer Tempelbaum, Elefantenohrbaum, Entenfußbaum, Fächerbaum, Fächerblattbaum, Frauenhaarbaum, Goethebaum, Großvater-Enkel-Baum, Japanischer Nußbaum, Japanischer Tempelbaum, Mädchenhaarbaum, Silberaprikose, Silberpflaume, Tempelbaum, Weiße Frucht;
Greek: Gigko, Gingko, Gkingko;
Hungarian: Ginkgófa, Páfrányfenyő;
Icelandic: Musteristré, Musterisviður;
India: Balkuwari (Hindu);
Italian: Ginko;
Japanese: Ichou, Ginkyo, Ginnan;
Korean: Apgaksu, Baekgwamok, Gongsonsu, Eun-Hang-Na-Moo, Haeng-Ja-Mok, Okgwamok;

T.K. Lim, *Edible Medicinal And Non-Medicinal Plants: Volume 3, Fruits*,
DOI 10.1007/978-94-007-2534-8_1, © Springer Science+Business Media B.V. 2012

Nepali: Bal Kumari;
Norwegian: Ginkgo, Tempeltre;
Polish: Chiński, Miłorząb Dwudzielny, Miłorząb Dwuklapowy, Miłorząb Japoński;
Portuguese: Ginkgo, Nogueira-Do-Japão;
Russian: Ginkgo;
Singapore: Pakgor Su;
Slovašcina: Ginko Dvokrpi;
Slovencina: Ginko Dvojlaločné;
South Africa: Vrekboom;
Spanish: Arbol Sagrado, Árbol De Oro, Árbol De Las Pagodas, Árbol De Los Cuarenta Escudos, Árbol De Los Escudos, Gingo, Ginkgo;
Swedish: Gingko, Kinesiskt Tempelträd, Tempelträd;
Taiwan: Yin Xing;
Thailand: Pae Guay;
Turkish: Fosil Ağacı, Gingko Ağacının, Japon Eriği, Japon Eriği Olarak Bilinir, Mabet Ağacı;
Vietnamese: Bạch Quả, Cây Bạch Quả, Cây lá quat.

Origin/Distribution

Ginkgo is native to Far East Asia – China, Japan and Korea. It is commonly planted in Buddhist and Taoist temples in East Asia. It has been introduced to other temperate areas in both hemispheres.

Agroecology

Ginkgo is a cool temperate species but does not tolerate extreme frost. It grows in lowland broadleaved forests and valleys up to 2,000 m altitude. It tolerates a range of soil types but thrives on acidic, well-watered, well-drained, (pH = 5–5.5), yellow loess. It thrives best in full sun.

Edible Plant Parts and Uses

Peeled nuts (endosperms) are eaten roasted, baked, boiled in soups, porridge, stews, or fried in dishes with meat or other vegetables. Ginkgo endosperms are popular in congee or herbal sweet-dessert soups, commonly serves at birthdays, weddings and the lunar New year as part of the famous vegetarian dish called Buddha's Delight. Japanese used the endosperm in dishes such as *Chawanmushi* and in other dishes. The endosperms are available dried or canned whence it is sold as "White nuts" in many Asian grocery stores. Roasted endosperms are relished in China and Japan and marketed widely in East Asia. They are also used as spice especially for fish dishes. A report says that the seed can be eaten raw whilst another says that large quantities of the seed are toxic. It needs to be heated or cooked before being eaten in order to destroy a mildly acrimonious compound. An edible oil is obtained from the seed.

Botany

A deciduous, resinous, dioecious branched tree, to 40 m tall with light grey or greyish brown bark that is longitudinally fissured especially on old trees and with trunk diameter reaching 1.5 m in old trees (Plates 1 and 2). Male trees show an upright and irregular form, female trees are low and spreading. Branches are stiff, covered with elliptic leaf scars and dimorphic, the elongated bearing alternate leaves and the abbreviated terminated with whorl of leaves surrounding a bud (Plates 1, 2 and 4). Leaves (Plates 2, 3, 4 and 5) are borne on 3–10 cm long petioles which are channelled on the adaxial surface, lamina is

Plate 1 Habit of old Ginkgo tree

Plate 2 Lush foliage of Ginkgo tree

Plate 3 *Ginkgo* leaves

Plate 4 Fruiting Ginkgo branch

fan-shaped, to 13×8–15 cm, mostly 1.5 times wider than long, glossy pale green (resembling those of the maidenhair fern or *Adiantum*), turning bright yellow in Autumn, with irregularly

Plate 5 Close-up Ginkgo leave and fruit

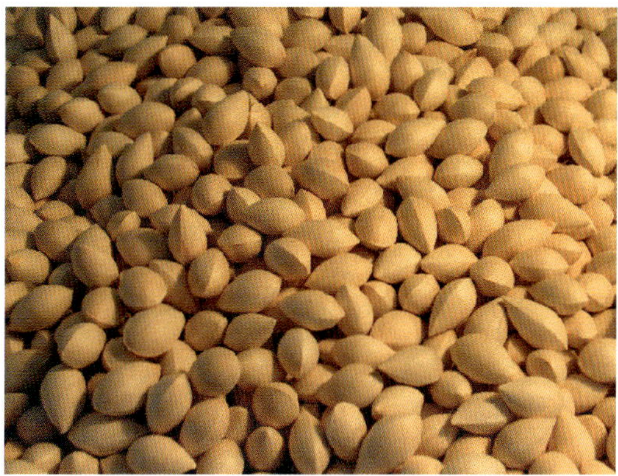

Plate 6 Hard-shelled Ginkgo seeds

toothed or notched upper margins and dichotomously veined. Trees flower after 20–35 years, females exhibiting an abundance of ovules in pairs on stalks each containing an egg cell, initially very green, but later turning greenish-yellow, then orange and brown. The male flowers are ivory-coloured, catkin-like pollen cones (microsporangia), 3–6 on each short shoot containing boat-shaped pollen sacs with widely gaping slit. A single naked ovule ripens into a elliptic, narrowly obovoid, or ovoid, 2.5–3.5×1.6–2.2 cm green fruit (Plates 4 and 5). Seed with yellow, or orange-yellow, glaucous, sarcotesta with acrid, rancid odour when ripe; sclerotesta white, with 2 or 3 longitudinal ridges; endotesta pale reddish brown (Plates 6 and 7). The fleshy-coated seeds are frequently incorrectly designated as fruits or nuts.

Plate 7 Ginkgo seeds intact and de-shelled

Nutritive/Medicinal Properties

The nutrient composition (per 100 g edible portion) of raw ginkgo nuts (exclude 24% shell refuse) was reported as water 55.20 g, energy 182 kcal (761 kJ), protein 4.32 g, total lipid 1.68 g, ash 1.20 g, carbohydrates 37.60 g, Ca 2 mg, Fe 1.00 mg, Mg 27 mg, P 124 mg, K 510 mg, Na 7 mg, Zn 0.34 mg, Cu 0.274 mg, Mn 0.113 mg, vitamin C 15 mg, thiamine 0.220 mg, riboflavin 0.090 mg, niacin 6.00 mg, pantothenic acid 0.160 mg, vitamin B-6 0.328 mg, total folate 54 μg, vitamin A RAE 28 μg, vitamin A 558 IU, total saturated fatty acids 0.319 g, 14:0 (myristic acid) 0.006 g, 16:0 (palmitic acid) 0.288 g, 18:0 (stearic acid) 0.016 g; total monounsaturated fatty acids 0.619 g, 16:1 undifferentiated (palmitoleic acid) 0.079 g, 18:1 undifferentiated (oleic acid) 0.512 g, 20:1 (gadoleic acid) 0.010 g; total polyunsaturated fatty acids 0.618 g, 18:2 undifferentiated (linoleic acid) 0.578 g, 18:3 undifferentiated (linolenic acid) 0.021 g; tryptophan 0.071 g, threonine 0.268 g, isoleucine 0.209 g, leucine 0.316 g, lysine 0.206 g, methionine 0.055 g, cystine 0.023 g, phenylalanine 0.171 g, tyrosine 0.061 g, valine 0.283 g, arginine 0.420 g, histidine 0.102 g, alanine 0.247 g, aspartic acid 0.543 g, glutamic acid 0.836 g, glycine 0.232 g, proline 0.347 g and serine 290 g (U.S. Department of Agriculture and Agricultural Research Service 2010).

The seed is rich in niacin, and vitamin A, phosphorus and potassium. It is a good source of starch and protein, but is low in fats. These fats are mostly unsaturated or monounsaturated. The seed also contain vitamin B1 (thiamine), B2 (riboflavin), vitamin C and iron, sodium and calcium.

Other Phytochemicals

Important bioactive constituents reported to occur in the medicinally used *Ginkgo* leaves include terpene trilactones, i.e., ginkgolides A, B, C, J and bilobalide, many flavonol glycosides, biflavones, proanthocyanidins, alkylphenols, simple phenolic acids, 6-hydroxykynurenic acid, 4-O-methylpyridoxine and polyprenols (van Beek 2002). However, in the commercially important Ginkgo extracts some of these compound classes may not be present. Since 2001 over 3,000 papers on *Ginkgo biloba* had been published, and about 400 of them pertain to chemical analysis in a broad sense and in the same period over 2,500 patents were filed on *Ginkgo* and the very few related to analysis were mentioned as well (van Beek and Montoro 2009). A sharp contrast to the plethora of papers on terpene trilactones, flavonol glycosides, and ginkgolic acids forms the low number of papers on biflavones, proanthocyanidins, simple phenolics, simple acids, and other constituents that make up the remaining 70% of Ginkgo standardised extracts.

Five terpene lactones were determined in ginkgo dry extract (Ekman et al. 2009). The content of bilobalide was found to be in the range of 2.6–3.4% in all samples, whereas the sum of ginkgolides A, B and C was found to be in the range of 3.0–3.6%. Ginkgolide J was found in the range of 0.3–0.6%.

Six flavonoid constituents (quercetin, isorhamnetin, kaempferol, bilobetin, ginkgetin and sciadopitysin) were isolated from *Ginkgo biloba* leaves (Chi et al. 1997) and also the flavonoid, 5, 7, 4′-trihydroxy-flavone from the leaves (Chi and Xu 1998). Wang et al. (2007a) identified and purified genistein from the hydrolysate of *G. biloba* leaf extract. Eight flavonoid compounds, namely rutin, myricetin, quercitrin, quercetin, luteolin, kaempferol, apigenin, and isorhamnetin were determined in *G. biloba* (Tang et al. 2010).

Five types of ginkgolic acid (C13:0, C15:1, C17:2, C15:0 and C17:1) were determined in *Ginkgo biloba* leaves (Yang et al. 2002). The relative percentage content of ginkgolic acids C15:1 and C17:1 was about 85%. Ginkgolic acid C17:2 had not been reported in China. The content of ginkgolic acids in the leaves of *Ginkgo biloba* collected in April, May and June was 1.48%, 1.19% and 1.11% respectively. The average recovery of *Ginkgo biloba* leaves collected in June was 97.0%. Six kinds of ginkgolic acid (C13:0, C15:1, C17:2, C15:0, C17:1 and an unknown compound C17:3 tentatively) were determined in the *Ginkgo biloba* extract (Wu et al. 2003). The relative percentage concentration of ginkgolic acids C13:0, C15:1 and C17:1 was above 94%. The content of ginkgolic acids in one type of *Ginkgo biloba* extract preparations (tablet) was 49.2 μg/g. The average recovery was 98.2%.

An antifungal protein ginkbilobin-2 (Gnk2) was isolated from *Ginkgo biloba* seeds (Miyakawa et al. 2007). It did not exhibit homology to other pathogenesis-related proteins, but did displayed homology to the extracellular domain of plant cysteine-rich receptor-like kinases. *G. biloba* 11S seed storage protein, ginnacin, was purified by sequential anion-exchange and gel-filtration chromatography (Jin et al. 2008). *G. biloba* also contained allergenic and toxic compounds such as ginkgotoxin (Leistner and Drewke 2010).

A number of secondary metabolites comprising terpenoids, polyphenols, allyl phenols, organic acids, carbohydrates, fatty acids and lipids, inorganic salts and amino acids had been isolated from the plant. However, the main bioactive constituents found were terpene trilactones and flavonoid glycosides, considered responsible for the pharmacological activities of its standardized leaf extract (Singh et al. 2008). Leaf extract of *Ginkgo biloba* (GBE) was reported to be increasingly used as an herbal medicine for the treatment of neurodegenerative, cardiovascular and cerebrovascular diseases (Gu et al. 2009). Extracts from the leaves of *Ginkgo biloba* had been used in Chinese medicine for thousands of years (He et al. 2009). To-date, various standardized preparations from *G. biloba* leaf extract had been developed. Products prepared from *Ginkgo biloba* are leading botanical dietary supplements in the United States and top-selling phytopharmaceuticals especially in Europe and (Leistner and Drewke 2010). In European medicine, *G. biloba* medications are employed to improve memory, to treat neuronal disorders such as tinnitus or intermittent claudication, and to ameliorate brain metabolism and peripheral blood flow. During the past 20 years, an estimated two billion daily doses (120 mg) of *Ginkgo biloba* (GB) had been sold (Pérez 2009). French and German agencies consider it to be effective for the treatment of several diseases. *Ginkgo biloba* (EGb 761) extract had been reported to be the most prescribed phytomedicine in Europe for the treatment of cerebral insufficiency and vascular diseases (Vilar et al. 2009). European Regulation 1924/2006 states that all health claims made on foods need to be substantiated scientifically. After evaluation of 35 human intervention studies, the three health claims, namely improvement of blood circulation, improvement of symptoms of old age and improvement of memory of herbal products with *G. biloba* could not be substantiated because of the lack of evidence (Fransen et al. 2010).

Ginkgo biloba leaves were found to contain two major bioactive constituents, flavonoid glycosides (24%) and terpene lactones (6%), along with less than 5 ppm of the allergenic component, ginkgolic acid (Mahadevan and Park 2008). The *Ginkgo* leaf extract had been reported to have neuroprotective, anticancer, cardioprotective, stress alleviating, and memory enhancing effects and possible effects on tinnitus, geriatric complaints, and psychiatric disorders. The therapeutic mechanisms of action of the Ginkgo leaf extract had been suggested to be through its antioxidant, antiplatelet, antihypoxic, antiedemic, hemorrheologic, and microcirculatory actions, where the flavonoid and the terpenoid constituents may act in a complementary manner. Toxicity studies showed Ginkgo leaf extract to be relatively safe for consumption, although a few side effects had been reported, that is, intracerebral hemorrhage, gastrointestinal disturbances, headaches, dizziness, and allergic skin reactions. The use of Ginkgo leaf extract may be promising for treatment of certain conditions, although its long-term use still needs to be evaluated.

A plethora of papers running into several thousands have been published on the phytochemicals and associated pharmacological and medicinal properties, some of which are highlighted below.

Antioxidant Activity

Flavonoids of *Ginkgo biloba* were shown to have some protective effects on the damage of anti-oxidizing system of mice induced by acute alcohol administration (Yao et al. 2005). The *G. biloba* supplement prevented the rise of malondialdehyde level and the decrease of glutathione, glutathione peroxidase and superoxide dismutase caused by acute alcohol intakes. The findings of studies by Brunetti et al. (2006) showed that *Ginkgo biloba* extract containing 24.1% flavonoids and 181% terpene lactones bilobalide (0.542%), ginkgolide A (0.570%), ginkgolide B (0.293%), ginkgolide C (0.263%), and ginkgolide J (0.138%) pretreatment completely reversed both basal and hydrogen peroxide-stimulated isoprostane production. Isoprostanes are prostaglandin (PG) isomers generated from oxygen radical peroxidation of arachidonic acid, which are reliable markers of membrane oxidative damage. Amyloid beta-peptide-induced isoprostane production was also inhibited, both in young and aged rats. This suggested that the oxygen radical scavenging properties of the *Ginkgo biloba* extract were fully effective in young, as well as in old rats, with a greater inhibition of isoprostane production in the latter.

A novel polysaccharide (GBP50S2) with high antioxidant activity was isolated from *Ginkgo biloba* (Yuan et al. 2010). The backbone of GBP50S2 was composed of (1→4)-linked α-d-mannopyranosyl residues which branched at O-3. The three branches consisted of β-l-rhamnopyranosyl residues, (1→4)-linked α-d-galactopyranosyl terminated with β-l-rhamnopyranosyl residues, and (1→3,4)-linked α-d-mannopyranosyl terminated with β-l-rhamnopyranosyl residues, respectively. In the in-vitro antioxidant assay, GBP50S2 was found to possess DPPH (2,2-diphenyl-1-picrylhydrazyl) radical-scavenging activity and hydroxyl radical-scavenging activity with IC_{50} values of 0.412 mg/ml and 0.482 mg/ml, respectively.

The results of studies by Qa'dan et al. (2011) showed that all the isolated compounds from the tannin fraction of *G. biloba* leaves, namely the new trimeric prodelphinidin, epigallocatechin-(4β→8)-epigallocatechin-(4β→8)-catechin, catechin, epigallocatechin, gallocatechin, and three dimeric proanthocyanidins exhibited potent free radical scavenging activities which were higher than that of BHT (butylated hydroxytoluene) (IC_{50}=15.5 μg/ml). This suggested that the condensed tannins from *G. biloba* leaves strongly contributed to the overall antioxidant effects. The dimeric prodelphinidin epigallocatechin-(4β→8)-epigallocatechin exhibited the strongest DPPH radical scavenging activity with IC_{50} value of 1.7 μg/ml followed by the trimeric prodelphinidin, epigallocatechin-(4β→8)-epigallocatechin-(4β→8)-catechin (IC_{50}=2.1 μg/ml). The crude leaf extract exhibited high DPPH radical scavenging activity with IC_{50} of 15.5 μg/ml comparable with that of BHT. Polymeric proanthocyanidins were eluted after the fractions of flavonol glycosides and biflavone glycosides from an extract from *Ginkgo biloba* leaves (Qaâdan et al. 2010). A purified proanthocyanidin polymer accounted for 86.6% of the total proanthocyanidins, and for 37.7% of the total antioxidant activity of the leaf extract.

In a comparative in-vitro antioxidative study of the fluid extract from maidenhair tree (*Ginkgo biloba*), motherwort (*Leonurus cardiaca*) and hawthorn (*Crataegus monogyna*), the radical scavenging capacity in the DPPH* reaction system determined was in the following order: hawthorn (70.37%) < the fluid extract of maidenhair tree (82.63%) < the fluid extract of motherwort (84.89%), while in the ABTS*+(2,2′-azino-bis(3-ethylbenzthiazoline-6-sulphonic acid)) reaction system, the manifestation of the radical scavenging capacity was in the following order: the fluid extract of hawthorn (87.09%) < the fluid extract of motherwort (88.28%) < the fluid extract of maidenhair tree (88.39%) (Bernatoniene et al. 2009). The results showed that in the DPPH* reaction system, fluid extract of motherwort manifested higher antioxidant activity, compared to the fluid extracts of maidenhair tree and hawthorn. By contrast, in the ABTS*+reaction system, higher antioxidant activity was found in the fluid extract of maidenhair tree, compared to the fluid extracts of motherwort and hawthorn. This would suggest that preparations manufactured from these herbal raw materials could be used as effective preventive means and valuable additional remedies in the treatment of diseases caused by oxidative stress.

Studies conducted by Erdogan et al. (2006) found that bleomycin induced oxidative stress in rats could be prevented by *Ginkgo biloba* treatment via high plasma anti-oxidant enzyme (superoxide dismutase, glutathione peroxidase) activities together with decreased radical production from xanthine oxidase. Bleomycin is an antineoplastic agent and its clinical usage is limited by its toxicity in inducing plasma oxidative stress injury. Studies by Liu et al. (2009c) showed that extract of *Ginkgo biloba* (EGb) had stronger antioxidant activities than flavonoids, but terpenoids did not show antioxidant activity. EGb and flavonoids but not terpenoids were demonstrated to significantly induce the antioxidant enzyme glutamate cysteine ligase catalytic subunit (GCLC), directly scavenged O^{2-}, OH^- and inhibited rat erythrocyte hemolysis and lipid peroxidation of rat liver homogenate. The antioxidant activities of the flavonoids were weaker than those of EGb containing similar dose of flavonoids.

Neuroprotective/Alzheimer's Disease/Dementia Activity

Ginkgo biloba extract, EGb 761, had been therapeutically used for several decades to increase peripheral and cerebral blood flow as well as for the treatment of dementia. EGb 761 is currently used as symptomatic treatment for cerebral insufficiency that occurs during normal ageing or which may be due to degenerative dementia, vascular dementia or mixed forms of both, and for neurosensory disturbances (Ahlemeyer and Krieglstein 2003a). EGb 761 was found to be a complex mixture containing flavonoid glycosides, terpene lactones (non-flavone fraction) and various other constituents believed to contribute to its neuroprotective and vasotropic effects (Ahlemeyer and Krieglstein 2003a, b). A profusion of basic and clinical studies, conducted both in-vitro and in-vivo, had shown *Ginkgo biloba* extract EGb 761 to have neuroprotective and "anti-stress" properties.

DeFeudis and Drieu (2000), Ahlemeyer and Krieglstein (2003a) elucidated that EGb 761 had several major actions: cognition enhancement, improvement of blood rheology and tissue metabolism, and counteracting the detrimental effects of ischaemia. Several mechanisms of action were found to be useful in explaining how EGb 761 benefited patients with Alzheimer's disease and other age-related, neurodegenerative disorders. In animals, EGb 761 was found to possess antioxidant and free radical-scavenging activities, it reversed age-related losses in brain alpha 1-adrenergic, 5-HT1A and muscarinic receptors, protected against ischaemic neuronal death, preserved the function of the hippocampal mossy fiber system, increased hippocampal high-affinity choline uptake, suppressed the down-regulation of hippocampal glucocorticoid receptors, enhanced neuronal plasticity, and counteracted the cognitive deficits that follow stress or traumatic brain injury. Identified chemical constituents of EGb 761 had been associated with certain actions. Both flavonoid and ginkgolide constituents were involved in the free radical-scavenging and antioxidant effects of EGb 761 which decreased tissue levels of reactive oxygen species (ROS) and inhibited membrane lipid peroxidation. Regarding EGb 761-induced regulation of cerebral glucose utilization, bilobalide was found to increase the respiratory control ratio of mitochondria by protecting against uncoupling of oxidative phosphorylation, thereby increasing ATP levels, a result supported by the finding that bilobalide increased the expression of the mitochondrial DNA-encoded COX III subunit of cytochrome oxidase. With regard to its "anti-stress" effect, EGb 761 was reported to act via its ginkgolide constituents to decrease the expression of the peripheral benzodiazepine receptor (PBR) of the adrenal cortex.

In-vitro studies

Results of in-vitro studies on hippocampal primary cultured cells suggested that neuroprotective effects of *Ginkgo biloba* extract (EGb 761) were partly associated with its antioxidant properties (Bastianetto et al. 2000a). Co-treatment with EGb 761 dose-dependently (10–100 μg/ml) protected hippocampal neurons against toxicity induced by Abeta fragments, with a maximal and complete protection at the highest concentration tested. EGb 761 (100 μg/ml) was even able to protect hippocampal cells from a pre-exposure to Abeta 25–35 and Abeta 1–40. EGb 761 was also able to both protect and rescue hippocampal cells from toxicity induced by H_2O_2 (50–150 μM), a major peroxide possibly involved in mediating Abeta toxicity. Moreover, EGb 761 (10–100 μg/ml), completely blocked Abeta-induced events, e.g. reactive oxygen species accumulation and apoptosis. An excess of the free radical nitric oxide (NO) was viewed as a deleterious factor involved in various CNS disorders. Results of studies conducted by Bastianetto et al. (2000b) suggested that the protective and rescuing abilities of EGb 761 were not only attributable to the antioxidant properties of its flavonoid constituents but also via their ability as a No scavenger in inhibiting NO-stimulated protein kinase C activity. The terpenoid constituents of EGb 761, known as bilobalide and ginkgolide B, as well as inhibitors of phospholipases A [3-[(4-octadecyl) benzoyl] acrylic acid (OBAA)] and C (U-73122), failed to display any significant effects. The results also highlighted its possible effectiveness in neurodegenerative diseases, e.g. Alzheimer's disease via the inhibition of Abeta-induced toxicity and apoptosis. *Ginkgo biloba* extract EGb 761 exhibited neuroprotective abilities against dysfunction and death of neurons caused by Abeta deposits (Bastianetto and Quirion 2002). Beta-amyloid (Abeta) deposition had been postulated to play a causal role in the lesions that occur in Alzheimer's disease. A co-treatment with EGb 761 dose-dependently protected hippocampal cells against toxicity induced by Abeta fragments. EGb 761, also completely blocked Abeta-induced events, such as reactive oxygen species accumulation and apoptosis. Shi et al. (2009) found that EGb 761 was able to prevent Abeta (1–42)-induced cell apoptosis, reactive oxygen species (ROS) accumulation, mitochondrial dysfunction and activation of c-jun N-terminal kinase (JNK), extracellular signal-regulated kinase 1/2 (ERK1/2) and Akt signalling pathways. Both quercetin and ginkgolide B may be involved in the inhibitory effects of EGb 761 on JNK, ERK1/2 and Akt signalling pathways. Ginkgolide B also ameliorated mitochondrial functions but quercetin failed to show this effect. Additional experiments suggested that the protective effects of EGb 761 against Abeta toxicity may be associated with its antioxidant and platelet activating factor (PAF) antagonist activities. Quercetin but not ginkgolide B was one of the constituents responsible for the antioxidant action of EGb 761. Both quercetin and ginkgolide B may be involved in the PAF antagonist activity of EGb 761.

Results of studies by Luo et al. (2002) suggested that neuronal damage in Alzheimer's disease might be due to two factors: a direct Abeta toxicity and the apoptosis initiated by the mitochondria. The results also indicted that multiple cellular and molecular neuroprotective mechanisms, including attenuation of apoptosis and direct inhibition of Abeta aggregation, underpinned the neuroprotective effects of EGb 761. EGb 761 significantly mitigated mitochondrion-initiated apoptosis and reduced the activity of caspase 3, a key enzyme in the apoptosis cell-signalling cascade. The results of in-vitro studies by Lee et al. (2004) suggested that ginkgolide B but not ginkgolide A may elicit its anti-amnesic effect by minimizing the inhibitory effect of beta-amyloid peptides on cholinergic transmission in hippocampal brain slices. Treatment with EGb 761 and ginkgolide B could protect the rat neurons against glutamate-induced injury (Xu et al. 2010). EGb 761 and ginkgolide B increased cell viability, reduced apoptosis rate and decreased LDH leakage in varying degree. The protective effect of ginkgolide B was superior to EGb 761.

Ahlemeyer and Krieglstein (2003b) examined the neuroprotective and anti-apoptotic ability of the main constituents of the non-flavone fraction, the ginkgolides A, B, C, J and bilobalide. In focal cerebral ischemia models, pre-treatment of bilobalide before ischemia, dose-dependently reduced infarct area and infarct volume in mouse brain. Pre-treatment with ginkgolide A and ginkgolide B reduced the infarct area in the mouse model of focal ischemia. In primary cultures of hippocampal neurons and astrocytes from neonatal rats, ginkgolide B (1 μM) and bilobalide (10 μM) protected the neurons against damage caused by glutamate. Bilobalide (0.1 μM) was able to increase the viability of cultured neurons from chick embryo telencepalon when exposed to cyanide. In addition, bilobalide (100 μM) protected cultured rat hippocampal neurons from apoptosis caused by serum deprivation (24h), whereas ginkgolide B (100 μM) and bilobalide (100 μM) reduced apoptotic damage induced by staurosporine (300 nM). Ginkgolide A failed to affect apoptotic damage neither in serum-deprived nor in staurosporine-treated neurons. The results suggested that some of the components of the non-flavone fraction of EGb 761 possessed neuroprotective and anti-apoptotic capacity, and that bilobalide was the most potent one. Contrariwise, ginkgolic acids (100–500 μM) produced neuronal death, but these constituents were removed from EGb 761 below an amount of 0.0005%. Their findings provided experimental evidence for a neuroprotective effect of EGb 761 that agreed with clinical studies showing the efficacy of an oral treatment in patients with mild and moderate dementia.

Studies by Yao et al. (2001) demonstrated that *Ginkgo biloba* leaf extract (EGb 761) dose-dependently inhibited the formation of beta-amyloid-derived diffusible neurotoxic soluble ligands (ADDLs), suggested to be involved in the pathogenesis of Alzheimer's disease. The results indicated that the terpenoid and flavonoid constituents of EGb 761, were responsible for rescuing the neuronal PC12 cells from Abeta-induced apoptosis and cell death; their mechanism of action being distinct of their antioxidant properties. Bilobalide, a terpene extracted from *G. biloba* leaves, protected neuronal PC12 cells from beta-amyloid peptide fragment 25–35 (A beta 25–35)-induced cytotoxicity (Zhou et al. 2000). Bilobalide also inhibited A-beta 25-induced elevation of lipid peroxidation and decline of antioxidant enzyme activities. Accumulation of amyloid beta (Abeta) in form of senile plaques was postulated to play a central role in the pathogenesis of Alzheimer's disease mediated by oxidative stress and increasing evidence showed that Abeta generates free radicals in-vitro, which mediated the toxicity of this peptide (Eckert et al. 2003). In their study in PC12 cells they found that EGb 761 protected mitochondria from the attack of hydrogen peroxide, antimycin and Abeta. In addition, they found that EGb 761 reduced ROS levels and ROS-induced apoptosis in lymphocytes from aged mice treated orally with EGb 761 for 2 weeks. Their data further emphasized neuroprotective properties of EGb 761, such as protection against Abeta-toxicity, and antiapoptotic properties, which they postulated were probably due to its preventive effects on mitochondria. One of the components of *Ginkgo biloba* leaf extract, ginkgolide B, a potent platelet-activating factor (PAF) antagonist was found to exhibit neuroprotective property (Smith et al. 1996). The neuroprotective activity was postulated to be attributable to the terpene fraction of *Ginkgo biloba*, which contained the ginkgolides, as well as the flavonoid free radical scavengers.

Longpré et al. (2006) demonstrated that EGb 761 could prevent the activation of NF-kappaB, ERK1/2, and JNK pathways induced by Abeta and could also activate SIRT1 on neuroblastoma cell line N2a. EGb 761 and its flavonoid fraction (CP 205) could also prevent the Abeta fibril (fAbeta) formation in-vitro. They showed that Abeta was less toxic to N2a neuroblastoma cells when the peptide was previously incubated with the flavonoid fraction or EGb 761 during the fibril formation period. *Ginkgo biloba* leaf extract, EGb 50, was found to be capable of enhancing the proliferation of Schwann cells cultured in-vitro, which may be one of the important mechanisms to promote peripheral nerve regeneration (Lin et al. 2008).

Results of studies suggested that EGb 761 promoted clearance of Abeta from the brain by regulating the expression of RAGE and LRP-1 during brain ischemia (Yan et al. 2008). EGb 761 significantly reversed chronic hypoxic and hypoglycemic-induced upregulation of RAGE (receptor for advanced end glycation products) expression and downregulation of LRP-1 (low-density lipoprotein receptor-related protein-1) expression. In-vitro studies highlighted the beneficial effect of *G. biloba* extract on the performance of cellular oxidative phosphorylation system and restoration of Abeta-induced mitochondrial dysfunction in energy metabolism (Rhein et al. 2010). Studies in isolated rat hippocampal neurons indicated that the modulatory effects of GBE on N-methyl-D-aspartate (NMDA)-activated currents may contribute to the neuroprotective effects of GBE and the modulatory effect of nanometer GBE on NMDA-activated current was greater than that of mGBE (Li et al. 2011).

A new *Ginkgo biloba* extract P8A (TTL), 70% enriched with terpene trilactones, hindered A beta (1–42) induced inhibition of long-term potentiation in the CA1 region of mouse hippocampal slices (Vitolo et al. 2009). This neuroprotective effect was attributed largely to ginkgolide J that completely replicated the effect of the extract. Ginkgolide J was also capable of inhibiting apoptosis of rodent hippocampal neurons caused by A-beta (1–42). This beneficial and multi-faceted mode of action of the ginkgolide makes it a new and promising lead in designing therapies against Alzheimer's disease. Using human SH-SY5Y neuroblastoma cells and primary hippocampal neurons, Shi et al. (2010b) found that bilobalide prevented Abeta 1–42-, H(2)O(2)- and serum deprivation-induced apoptosis. Bilobalide dose-dependently increased PI3K activity and levels of phosphorylated Akt. The results further suggested that the PI3K/Akt pathway might be involved in the protective effects of bilobalide. Since modern technology allows production of purified bilobalide with high bioavailability, bilobalide may be useful in developing therapy for diseases involving age-associated neurodegeneration.

Studies by Zhao et al. (2011) confirmed that the neuroprotective effect of *Ginkgo biloba* extract EGb 761 was mediated in part by inhibition of cytosolic phospholipase A_2 (cPLA_2), an enzyme that is known to play a key role in mediating secondary pathogenesis after acute spinal cord injury. EGb 761 administration significantly reversed the elevated expression of phosphorylated cPLA_2 (p-cPLA_2), a marker of cPLA_2 activation, and neuronal death caused by insults with glutamate and hydrogen peroxide. They demonstrated that the extracellular signal-regulated kinase 1/2 signaling pathway was involved in EGb 761's modulation of cPLA_2 phosphorylation.

In-vivo studies

Smith and Luo (2003), Luo (2006) found that treatment of neuroblastoma cells or transgenic *Caenorhabditis elegans* both expressing accumulation of Abeta, with *Ginkgo biloba* extract EGb 761 significantly attenuated the basal as well as the induced levels of hydrogen peroxide-related reactive oxygen species (ROS). Further, EGb 761 eased its toxicity in the transgenic *C. elegans*. Wu et al. (2006) found that EGb 761 and one of its components, ginkgolide A, alleviated Abeta-induced pathological behaviors, including paralysis, and reduced chemotaxis behavior and 5-HT hypersensitivity in a transgenic *Caenorhabditis elegans* (Wu et al. 2006). The findings suggested that (1) EGb 761 suppressed Abeta-related pathological behaviors, (2) the protection against Abeta toxicity by EGb 761 was mediated primarily by modulating Abeta oligomeric species, and (3) ginkgolide A had therapeutic potential for prevention and treatment of AD.

The results of focal cerebral ischemia model studies conducted by Ahlemeyer and Krieglstein (2003b) suggested that some of the constituents of the non-flavone fraction of *Ginkgo biloba* extract EGb 761 exhibited neuroprotective and anti-apoptotic capacity, and that bilobalide (5–20 mg/kg, s. c.) was the most potent one. In contrast, ginkgolic acids (100–500 μM) induced neuronal death, which showed features of apoptosis as well as of necrosis, but these constituents were removed from EGb 761 below an amount of 0.0005%. The data provided experimental evidence for a neuroprotective effect of EGb 761 that agreed with clinical studies showing the efficacy of an oral treatment in patients with mild and moderate dementia. Results of separate animal studies indicated that long-term pre-treatment of EGb 761 administered either alone or in combination with drugs such as $MgSO_4$, FK506, or

MK-801 exerted significantly effective neuroprotection on infarct volume in gerbil ischemic brains (Chung et al. 2006). Standardized *Ginkgo biloba* extract EGb 761 was found to exhibit potential beneficial effects to patients with Alzheimer's disease (AD) using a mouse model (Tchantchou et al. 2007). EGb 761 significantly and dose-dependently increased cell proliferation in the hippocampus of both young (6 months) and old (22 months) transgenic mice, and the total number of neuronal precursor cells in-vitro. Administration of EGb 761 reduced Abeta oligomers and restored cAMP response element binding protein (CREB) phosphorylation in the hippocampus of these mice. The present findings suggested that (1) enhanced neurogenesis by EGb 761 may be mediated by activation of CREB, (2) stimulation of neurogenesis by EGb 761 may contribute to its beneficial effects in AD patients and improved cognitive functions in the mouse model of AD, and (3) EGb 761 had therapeutic potential for the prevention and improved treatment of AD.

Studies indicated that orally administered *Ginkgo biloba* extract can protect the brain against beta-amyloid from changes leading to memory deficit through its effect on the cholinergic system in rats (Tang et al. 2002). The extract reversed the decrease in choline actyltransferase activities in the hippocampus. Yao et al. (2004) found *Ginkgo biloba* extract (EGb 761) could improve cognitive function in patients with Alzheimer's disease. In aging rats, EGb 761 treatment decreased free circulating cholesterol and suppressed the production of brain beta-amyloid precursor protein and amyloid beta-peptide. Augustin et al. (2009) in studies with mice transgenic for human APP (Tg2576) found that the potential neuroprotective properties of EGb 761 may partly be related to its amyloid precursor protein lowering activity. Amyloid precursor protein appeared to be an important molecular target of EGb 761 in relation to the duration of the *Ginkgo biloba* treatment and/or the age of the animals.

The results of animal studies by Saleem et al. (2008) demonstrated that EGb 761 could be used as a preventive or therapeutic agent in cerebral ischemia and suggest that heme oxygenase 1 contributed, at least in part, to its neuroprotective effect. Results of animal studies suggested that administration of EGb761, Selenium and its combination with EGb761 exerted significant neuroprotective effects on ischemia/reperfusion (I/R) injury in a rat model of transient global cerebral I/R via suppression of oxidative stress (Erbil et al. 2008). A standardized extract of *Ginkgo biloba*, EGb 761, was shown to exert a neuroprotective effect against permanent and transient focal cerebral ischemia in rats (Koh 2009). EGb 761 decreased the elevated Bad and Bcl-X(L) interaction caused by Ischemic brain injury Ischemic brain injury. The results of animal studies by Koh (2010) confirmed that EGb 761 protected neuronal cells against ischemic brain injury by preventing injury-induced decreases in p70S6 kinase and S6 phosphorylation.

Oral administration of *Ginkgo biloba* extract (EGb 761) to a mouse model of Alzheimer's disease was confirmed to enhance hippocampal neurogenesis (Tchantchou et al. 2009). Bilobalide and quercetin, two of its components, were also found to enhance in a dose-dependent manner, neurogenesis and synaptogenesis in mice brains suggesting a common final signalling pathway mediated by phosphorylation of cyclic-AMP Response Element Binding Protein (CREB) in the hippocampal neurons. Synaptogenesis and neurogenesis in adulthood could serve as a therapeutic target for the prevention and treatment of Alzheimer's disease. Studies showed that EGb761 treatment could accelerate recovery of the pathological synaptic plasticity in vascular dementia model rats with apparent and long-lasting dysfunction of learning and memory (Zhang and Wang 2008). The results suggested that EGb 761 played an important and improving role on learning and memory dysfunction of vascular dementia.

EGb 761, a standardized extract of *Ginkgo biloba* was found to exert protective effects against ischemic brain injury (Cho et al. 2009). The study in adult male rats suggested that EGb 761 prevented cell death due to brain injury and that EGb 761 protection was effected by preventing the injury-induce decrease of Akt phosphorylation. Additionally, EGb 761 inhibited the injury-induced increase of cleaved caspase-3 levels. Shi et al. (2010c) found that *Ginkgo biloba* extract EGb 761 protected against the decrease of cytochrome c oxidase (COX) activity, mitochondrial ATP (adenosine-5′-triphosphate) and mitochondrial glutathione (GSH) levels in both platelets and hippocampi of ovariectomized rats, suggesting its peripheral and central effects against estrogen withdrawal-induced degeneration. In contrast, in sham-operated rats, EGb 761 enhanced mitochondrial GSH content in platelets but not in hippocampi, suggesting that EGb 761 may help to enhance the functional reserve of mitochondria, but this effect was limited to the peripheral nervous system Therefore, while EGb 761's effect may be limited to the outside of the nervous system under normal physiological conditions, EGb 761 may be a potential protective agent against central neurodegeneration in post-menopausal women.

EGb 761 protected against mitochondrial dysfunction in platelets of young and old mice, suggesting a peripheral effect of this herb in the prevention and treatment of age-associated degeneration (Shi et al. 2010a). In contrast, in hippocampi, protective effects of EGb 761 were observed only in the old mice, probably due to an age-associated increase in the permeability of the blood brain barrier (BBB). Therefore, while EGb 761 has a potential anti-aging effect, its central effect can be affected by in-vivo factors such as the BBB permeability. EGb-treatment exhibited remarkable protective effect in the rat's brainstem and cerebellum by restoring the reactive oxygen species-stressed alterations in young, old and hypoxic brain regions via enhancement of the decreased levels of superoxide dismutase (SOD) and glutathione (GSH) levels in the old temporal cortex and brainstem (Martin et al. 2011).

A single dose of 600 mg/kg *Ginkgo biloba* extract EGb 761® resulted in maximum plasma concentrations of 176, 341, and 183 ng/ml for quercetin, kaempferol, and isorhamnetin/tamarixetin, respectively and in maximum brain concentrations of 291 ng/g protein for kaempferol and 161 ng/g protein for isorhamnetin/tamarixetin (Rangel-Ordóñez et al. 2010). Repeated administration of the same dose for 8 days led to an approximate 4.5-fold increase in the plasma concentration for quercetin, 11.5-fold increase for kaempferol, and tenfold increase for isorhamnetin/tamarixetin. In the brain, an approximate twofold increase was observed for kaempferol and isorhamnetin/tamarixetin. About 90% of the determined flavonoids were distributed in the hippocampus, frontal cortex, striatum, and cerebellum, which together represent only 38% of the whole brain. The results showed the ability of the flavonoid constituents to cross the blood-brain barrier in rats.

Clinical trials, reviews

Contrary to earlier studies, in a recent study involving 523 individuals with dementia, *G. biloba* at 120 mg twice a day was not effective in reducing either the overall incidence rate of dementia or Alzheimer's disease incidence in elderly individuals with normal cognition or those with MCI (Mild Cognitive Impairment) (Dekosky et al. 2008). The overall dementia rate was 3.3 per 100 person-years in participants assigned to *G. biloba* and 2.9 per 100 person-years in the placebo group. Ihl et al. (2011) conducted a multi-centre trial of 410 outpatients with mild to moderate dementia (Alzheimer's disease, vascular dementia or mixed form) where patients were randomly allocated to double-blind treatment with 240 mg of EGb 761 or placebo once daily for 24 weeks. EGb 761, 240 mg once-daily, was found significantly superior to placebo in the treatment of patients with dementia with neuropsychiatric symptoms. Weinmann et al. (2010) conducted a meta-analysis on nine 12-week trials that met their criteria to assess the effects of *Ginkgo biloba* in Alzheimer's disease as well as vascular and mixed dementia. They found that *Ginkgo biloba* appeared more effective than placebo.

Kasper and Schubert (2009) reported that the efficacy of EGb 761 in the treatment of dementia (Alzheimer's disease and vascular dementia) had been studied in ten randomised, controlled, double-blind clinical trials. In three of the four large trials conducted in accordance with recent recommendations EGb 761 was significantly superior to placebo with respect to cognitive performance and one or more further (global, functional or behavioural) outcomes demonstrating the clinical relevance of the findings. The findings from the six smaller trials were in line with those of the large trials. One trial was inconclusive, but of questionable external validity due to uncommonly rigorous patient selection. Subgroup analyses of this study together with the findings from the most recent clinical trial suggest that EGb 761 may be most beneficial to patients with neuropsychiatric symptoms, who actually constitute the majority of dementia patients. Delay in symptom progression, rates of clinically significant treatment response and numbers needed to treat (NNT) found for EGb 761 were in the same range as those reported for cholinesterase inhibitors. In an exploratory trial comparing EGb 761 and donepezil, no statistically significant or clinically relevant differences were seen. The *Ginkgo biloba* extract EGb 761 was found to interfere with pathomechanisms relevant to dementia, such as Abeta aggregation, mitochondrial dysfunction, insulin resistance, and hypoperfusion. They thus asserted that EGb 761 has its place in the treatment of dementia.

Napryeyenko et al. (2009) conducted a randomised controlled trial of 359 demetia patients aged 50 years or more for 22 weeks to ascertain treatment effects of *G. biloba* extract EGb 761 on dementia. They found that under EGb 761 treatment the Short Syndrome Test total score improved by −3.03 and −3.4 points in patients with Alzheimer's disease (AD) and vascular dementia (VaD), respectively, whereas the patients on placebo deteriorated by +1.2 and +1.5 points, respectively. Yancheva et al. (2009) conducted a randomised, double-blind exploratory trial to compare treatment effects and tolerability of EGb 761(R), donepezil and combined treatment in patients with AD and neuropsychiatric features. They found no significant difference in the efficiency between EGb 761(R) and donepezil; a combination therapy was more superior to a mono-therapy with fewer side effects under a combination therapy than under mono-therapy with donepezil. Retrospective analyses of data from a 24-week randomized, placebo-controlled, double-blind clinical trial of EGb 761® (240 mg once daily) were performed involving 410 outpatients with mild to moderate AD, VaD or AD with cerebrovascular disease, each associated with neuropsychiatric features (Ihl et al. 2010). The results showed more net benefit from treatment for the EGb 761® gro, with faster deterioration in the placebo group.

Birks and Grimley Evans (2009) found in their review of 36 randomized, double-blind studies on *Ginkgo biloba* for cognitive impairment and dementia that the results from the more recent trials showed inconsistent results for cognition, activities of daily living, mood, depression and carer burden. Of the four most recent trials to report results three found no difference between *Ginkgo biloba* and placebo, and one found very large treatment effects in favour of *Ginkgo biloba*. Many of the early trials used unsatisfactory small methods and thus publication bias could not be excluded. *Ginkgo biloba* appeared to be safe in use with no excess side effects compared with placebo. The evidence that *Ginkgo biloba* had predictable and clinically significant benefit for people with dementia or cognitive impairment was inconsistent and unreliable. Janssen et al. (2010) conducted a systemic review of randomised controlled trials on *G. biloba* in patients with Alzheimer's disease. In six eligible studies, they found high heterogeneity for most outcomes except safety issues. Among studies administering high-dose ginkgo (240 mg), all studies favoured treatment though effects remained heterogeneous. In this subgroup, a benefit of ginkgo was found for activities of daily living. Cognition and accompanying psychopathological symptoms showed an indication of a benefit. A recent systematic review of single Chinese herbs for Alzheimer's disease treatment was conducted by Fu and Li (2011). After a thorough search of various databases they selected seven Chinese herbs including *Ginkgo biloba* and six randomized controlled clinical trials that matched their predefined criteria. They found that the current evidence in support of their use for Alzheimer's disease treatment was inconclusive or inadequate, although no serious adverse events were reported.

The revised (second) consensus statement from the British Association for Psychopharmacology with regards to clinical practice with anti-dementia drugs states that cholinesterase inhibitors (donepezil, rivastigmine, and galantamine) are effective for mild to moderate Alzheimer's disease (A) and memantine for moderate to severe Alzheimer's disease (A) (O'Brien and Burns 2011). However, other drugs including statins, antiinflammatory drugs, vitamin E and *Ginkgo biloba*, cannot be recommended either for the treatment or prevention of Alzheimer's disease until further evidence is available.

Ginkgo and Parkinson's Disease

Excessive N-methyl-D-aspartate (NMDA) receptor activation had been implicated in the pathophysiology of chronic neurodegenerative diseases, such as Parkinson's disease (PD) and Alzheimer's disease (AD) (Koutsilieri and Riederer 2007). NMDA receptors were found to play central roles in a number of physiological processes, including long-term potentiation in the hippocampus, synaptogenesis and synaptic plasticity. In-vivo studies using an animal model of Parkinson's disease induced by 1-methyl-4-phenylpyridinium MPP(+) showed that the protective effect of EGb 761 against MPP(+) neurotoxicity in may be due in part to the regulation of copper homeostasis in the brain (Rojas et al. 2009). EGb 761 pre-treatment of the MPP (+) group prevented changes in the copper content of the striatum, midbrain, and hippocampus of the mice. Studies showed that rats pretreated for a week with EGb 761 exerted neuroprotective effects and reduced the behavioural deficit in 6-hydroxydopamine (6-OHDA) lesions in rat with Parkinson's disease (Kim et al. 2004). The results also indicated a possible role for the extract in the treatment of Parkinson's disease.

The Chinese compound prescription *Ginkgo biloba* Pingchan Recipe (GBPR) was found to have distinct inhibitory effect against the neurotoxicity of NO probably by producing an antioxidative effect through decreasing neuronal nitric oxide synthase synthesis in the brain (Zhang et al. 2009a). Neuronal nitric oxide synthase (nNOS) mRNA expression was detected in the striatum and substantia nigra of the Parkinson disease model mice, and GBPR treatment significantly reduced its expressions.

Cognitive/Nootropic Activity

In-vivo studies

Ginkgo biloba extract EGb 761 was found to independently improve changes in passive avoidance learning and brain membrane fluidity in mice (Stoll et al. 1996). There was a significant improvement in short-term memory, measured by the avoidance latency 60 s after the aversive stimulus, and of membrane fluidity in the aged animals, but no improvement in long-term memory as measured by the avoidance latency 24 hours after shock. Winter (1998) found that chronic post-session administration of EGb 761 at a dose of 50 mg/kg had no effect on continuous learning but the same dose given pre-session resulted in a trend toward fewer sessions to reach criterion performance as well as fewer errors in tested in an eight-arm radial maze. Additionally, it was observed that rats chronically treated with EGb 761 lived significantly longer than vehicle-treated subjects. EGb 761 administered presession produced a dose-related decrease in total, retroactive, and proactive errors. Further, the data suggested that the procedures employed, i.e., continuous learning and delayed non-matching to position tasks in aged rats, were capable of detecting drugs of possible value in the treatment of human cognitive impairment.

Stackman et al. (2003) found that chronic *Ginkgo biloba* treatment could block an age-dependent decline in spatial cognition without altering Abeta levels and without suppressing protein oxidation in a transgenic mouse model of Alzheimer's disease, Tg2576 mice with a mutant form of human Abeta precursor protein exhibiting age-related cognitive deficits, Abeta plaque deposition, and oxidative damage in the brain. *Ginkgo biloba*-treated Tg2576 mice exhibited spatial memory retention comparable to wild type mice during the probe test, while untreated Tg2576 mice exhibited a spatial learning impairment. Studies in rats with kindling-induced epilepsy showed that EGb treatment could improve learning-memory ability in epileptic rats at different developmental phases in a dose-dependent manner, possibly through a reduction of NMDA receptor expression in the rat hippocampus. (Duan and Yuan 2008).

A standard extract of *Ginkgo biloba* (EGb 761) has been used in the treatment of various common geriatric complaints including vertigo, short-term memory loss, hearing loss, lack of attention, or vigilance (Yang et al. 2007) . They demonstrated that acute systemic administration of EGb 761 facilitated the acquisition of conditioned fear in rats. Results suggested that acute EGb 761 administration modulated extinction of conditioned fear by activating extracellular signal-regulated kinase (ERK1/2) phosphorylation. Oliveira et al. (2009) provided new evidence for a role of EGb 761 on memory but also identified molecular changes that underpinned the fear memory consolidation. They observed that rats submitted to acute and subacute EGb 761 treatments had acquisition of fear conditioning. They demonstrated that EGb 761 modulated GAP-43, CREB-1 and GFAP expression in the prefrontal cortex, amygdala and hippocampus.

Results of animal studies demonstrated that repeated daily pre-session ginkgo injection promoted acquisition of a spatial working memory task, but neither acute nor chronic post-training exposure augmented spatial working memory (Satvat and Mallet 2009). Ginkgo-treated rats reached the training criteria significantly faster and made fewer errors and in subsequent experiment, they found that post-training Ginkgo administration did not enhance memory. They concluded that ongoing Ginkgo administration did not offer any continued beneficial effects in an already-learned working memory task.

Studies by Blecharz-Klin et al. (2009) suggested that long-term administration of *Ginkgo biloba* leaf extract could improve spatial memory and motivation with significant changes in the content and metabolism of monoamines and metabolites in several brain regions of aged rats. Studies by Fehske et al. (2009) confirmed that norepinephrine, the serotonin, the dopamine uptake transporters and monoamine oxidases (MAO) activity were inhibited by EGb 761 in-vitro, although rather high concentrations were required for inhibition of MAO-A and MAO-B activity. However, after 14 days of daily oral treatment with 100 mg/kg EGb 761 only norepinephrine uptake was significantly decreased in NMRI mice, while 5-Hydroxytryptophan uptake and MAO activity were not affected. These findings provided an explanation for the enhancement of dopaminergic neurotransmission by EGb 761 seen in animal models, presumably linked to its positive effects on cognition and attention. Results of recent animal studies demonstrated that chronic but not acute treatment with standardized *G. biloba* extract EGb 761 increased dopaminergic transmission in the prefrontal cortex of conscious rats, improving cognitive function (Yoshitake et al. 2010). Treatment with the main components of EGb 761 revealed that the increase in dopamine levels was mostly caused by the flavonol glycosides and ginkgolide fractions, whereas bilobalide treatment was without effect. Gingkoselect (100 mg/kg/day, orally, 21 days) normalized cognitive deficits in rats chronically treated with corticosterone, and improved spatial memory above the control levels in the chronically stressed rats (Walesiuk and Braszko 2010).

Clinical trials/reviews

Grässel (1992) conducted a double-blind, randomized placebo-controlled study of 24 weeks duration with 72 outpatients with cerebral insufficiency to determine the effect of *Ginkgo biloba* extract on mental performance. Statistically significant improvement in the short-term memory after 6 weeks and of the learning rate after 24 weeks were observed in the ginkgo group, but not in the placebo group (longitudinal analysis). The results indicated that treatment *Ginkgo biloba* extract improved mental/mnestic performance.

The findings of a double-blind placebo-controlled design showed that 120 mg of *Ginkgo biloba* had no acute nootropic effects in healthy older humans (Nathan et al. 2002). Dartigues et al. (2007) conducted a prospective community-based cohort study of 3,534 subjects aged 65 and older to ascertain the effects of treatment for memory impairment and the *Ginkgo biloba* extract (EGb 761) on dementia, mortality, and survival without dementia. They found that the initial consumption of *Ginkgo biloba* did not modify the risk of dementia whereas the consumption of other treatments for memory impairment was associated with a higher risk of dementia. Subjects who took *Ginkgo biloba* had a significantly lower risk of mortality in the long term The initial consumption of treatment for memory impairment other than *Ginkgo biloba* did not modify the risk of mortality. Overall, the results suggested that treatment with EGb 761 may increase the probability of survival in the elderly population.

In a placebo-controlled, multi-dose, double-blind, balanced, crossover design involving 20 participants, acute administration of *Ginkgo biloba* was found capable of producing a sustained improvement in attention in healthy young volunteers (Kennedy et al. 2000). Compared with the placebo, administration of ginkgo produced a number of significant changes on the performance measures. The most striking of these was a dose-dependent improvement of the 'speed of attention' factor following both 240 mg and 360 mg of the extract, which was evident at 2.5 h and was still present at 6 hours. Additionally, there were a number of time- and dose-specific changes (both positive and negative) in performance of the other factors. Similar results were obtained in another placebo-controlled double-blind design study where after acute administration of Ginkgo improved performance were found in tests of attention and episodic and working memory, mental flexibility and planning in university students (Elsabagh et al. 2005). However, there were no effects after 6 weeks, suggesting that tolerance develops to the effects in young, healthy participants. Cieza et al. (2003) conducted a month-long, randomized double-blind, mono-center study with parallel groups with 66 healthy subjects of both sexes aged between 50 and 65 with no age-related cognitive impairments to ascertain the effects of *Ginkgo biloba* on healthy elderly subjects. They found a positive effect of EGb 761 on the subjective emotional well-being of healthy elderly persons in variables such as depression, fatigue, anger and Self Rating Depression Scale (SDS).

Bäurle et al. (2009) found that the fresh leaf extract of *Ginkgo biloba* was a safe, effective, adjuvant treatment option for patients with mild cognitive impairments. In their study, about half of all patients experienced an improvement in their memory and their ability to concentrate, as well as a decrease in symptoms of forgetfulness. The majority of investigators and patients judged the treatment to be effective. The tablets were very well tolerated and, as a treatment for their cognitive impairment, highly accepted (90% would take them again). Wang et al. (2010a) conducted a meta-analysis of six randomized placebo-controlled clinical trials to determine the effectiveness of standardized *Ginkgo biloba* extract (GbE) on cognitive symptoms of dementia. Considering baseline risk in the assessment of treatment effect, GbE was found to be effective for cognitive functions in dementia with the treatment of 6 months. In a 12-week, randomised double blind, placebo controlled trial completed by 152 patients, *Ginkgo biloba* extract (EGb –761) treatment significantly decreased the Abnormal Involuntary Movement Scale (AIMS) total score in patients with tardive dyskinesia compared to those who were given a placebo (Zhang et al. 2011). EGb-761 appeared to be an effective treatment for reducing the symptoms of tardive dyskinesia in schizophrenia patients, and improvement may be mediated through the well-known antioxidant activity of this extract.

The results of a 6-week study indicated that ginkgo did not facilitate performance on standard neuropsychological tests of learning, memory, attention, and concentration or naming and verbal fluency in elderly adults without cognitive impairment (Solomon et al. 2002). The ginkgo group also did not differ from the control group in terms of self-reported memory function or global rating by spouses, friends, and relatives. The data suggested that when taken following the manufacturer's instructions, ginkgo provided no measurable benefit in memory or related cognitive function to adults with healthy cognitive function.

Dodge et al. (2008) conducted a randomized, placebo-controlled, double-blind, 42-month pilot study with 118 cognitively intact subjects randomized to standardized GBE or placebo. They found in unadjusted analyses, *Ginkgo biloba* extract (GBE) neither altered the risk of progression from normal to Clinical Dementia Rating (CDR) = 0.5, nor protected against a decline in memory function. Secondary analysis taking into account medication adherence showed a protective effect of GBE on the progression to CDR = 0.5 and memory decline. They suggested that larger prevention trials taking into account medication adherence may clarify the effectiveness of GBE. Snitz et al. (2009) conducted a randomized, double-blind, placebo-controlled clinical trial of 3,069 community-dwelling participants aged 72–96 years in six academic medical centers in the United States between 2000 and 2008, with a median follow-up of 6.1 years to ascertain the effects of *G. biloba* in slowing the rates of global or domain-specific cognitive decline in older adults. They found that compared with placebo, the use of *G. biloba*, 120 mg twice daily, did not result in less cognitive decline in older adults with normal cognition or with mild cognitive impairment. There was no significant effect modification of treatment on rate of decline by age, sex, race, education, APOE*E4 allele, or baseline mild cognitive impairment

The review of 29 randomized controlled group-studies on the specificity of *G. biloba* in neuropsychological improvement by Kaschel (2009) found little specific information could be obtained from trials for treatment of dementia. A pattern of pharmacological actions on cognitive processes were gleaned from studies for mild cognitive impairment (MCI), depression, multiple sclerosis and healthy young and elderly subjects. There was consistent evidence that chronic administration improved selective attention, some executive processes and long-term memory for verbal and non-verbal material. The author asserted that more comprehensive trials were needed to overcome the technical flaws in the selection of tests and the interpretation of their results favouring predominantly beta-errors.

Singh et al. (2010) conducted a meta-analysis on the usage of *Ginkgo* or its derived products in schizophrenia and focused on six studies that fulfilled the selection criteria comprising 466 cases on ginkgo and 362 cases on placebo. They found that ginkgo as an add-on therapy to antipsychotic medication produced statistically significant moderate improvement in total and negative symptoms of chronic schizophrenia. Ginkgo as add-on therapy ameliorated the symptoms of chronic schizophrenia.

Ginkgo and Vertigo

Haguenauer et al. (1986) conducted a 3-month, multicenter double-blind drug vs. placebo study with 70 patients with vertiginous syndrome of recent onset and undetermined origin. The effectiveness of *Ginkgo biloba* extract on the intensity, frequency and duration of the disorder was statistically significant. At the end of the trial, 47% of the patients treated were rid of their symptoms as against 18% of those who received the placebo. In an open, controlled study, 44 patients complaining of vertigo, dizziness, or both, caused by vascular vestibular disorders were randomly treated with extract of *Ginkgo biloba* (EGb 761) 80 mg twice daily or with β-histine dihydrochloride 16 mg twice daily for 3 months (Cesarani et al. 1998). The results suggested that EGb 761 and β-histine dihydrochloride operated at different equilibrium receptor sites and showed that EGb 761 could appreciably improve oculomotor and visuovestibular function. Hamann (2007) conducted a systematic review of randomised, double-blind, placebo controlled clinical examinations and found that EGb 761 was efficacious in the treatment of vertiginous syndromes. The beneficial effect of EGb 761 on vestibular compensation had been demonstrated in preclinical and clinical studies.

Ginkgo and Intermittent Claudication, Peripheral Artery Disease

Drabaek et al. (1996) conducted a double blind cross-over study comparing the effects of the *Ginkgo biloba* extract GB-8 at a dose of 120 mg o.d. with placebo with 18 patients with stable intermittent claudication. The results showed that treatment with the *Ginkgo biloba* extract GB-8 improved some cognitive functions in elderly patients with moderate arterial insufficiency, whereas the extract did not change signs and symptoms of vascular disease in the patients. Peters et al. (1998) conducted a multicentric, randomized, placebo-controlled double-blind study on *Ginkgo biloba* special extract EGb 761 (Tebonin forte) in 111 patients suffering from peripheral occlusive arterial disease (POAD). The results of their study indicated that treatment with EGb 761 in POAD patients with Fontaine stage II b was very safe and produced a significant and therapeutically relevant prolongation of the patients' walking distance. Beneficial effects of *Ginkgo biloba* on peripheral arterial occlusive disease had been repeatedly shown in clinical trials, especially after use of EGb 761, a standardized special extract (Koltermann et al. 2007). One of the underlying mechanisms for the protective cardiovascular properties of EGb 76 was that the extract increased endothelial nitric oxide production by increasing endothelial nitric oxide synthase (eNOS) promoter activity and eNOS expression in-vitro. The extract also induced acute relaxation of isolated aortic rings and NO-dependent reduction of blood pressure in-vivo in rats.

Wang et al. (2007b) conducted A 24-week double-blind, placebo-controlled *Ginkgo biloba* trial with the first 12-week period as a non-exercise control stage and the second 12-week period as an exercise training stage in 22 patients with peripheral arterial disease.They found that supervised exercise training combined with *Ginkgo biloba* treatment did not generate greater beneficial effects than exercise training alone in patients with peripheral arterial disease. In a double-blind, placebo-controlled, parallel design trial with a 4-month duration of 62 adults aged 70±8 years with claudication symptoms of PAD (peripheral arterial disease) *Ginkgo biloba* treatment produced a modest but insignificant increase in maximal treadmill walking time and flow-mediated vasodilation (Gardner et al. 2008). The data did not support the use of *Ginkgo biloba* as an effective therapy for PAD.

In a meta-analysis of eight randomized trials involving 400 patients with intermittent claudication, Pittler and Ernst (2000) found a useful increase in mean pain-free walking distance, and significantly better than placebo, with about 120 mg of ginkgo extract every day. However, in a recent meta-analysis of 11 randomized controlled trials of *Ginkgo biloba* extract versus placebo in people with intermittent claudication found no evidence that *Ginkgo biloba* had a clinically significant benefit for patients with intermittent claudication (Nicolaï et al. 2010).

Ginkgo and Attention-Deficit Hyperactivity Disorder (ADHD)/ Autistic Disorder

A combination herbal product containing American ginseng extract, *Panax quinquefolium* (200 mg) and *Ginkgo biloba extract* (50 mg) treatment may improve symptoms of attention-deficit hyperactivity disorder (ADHD) in 36 children ranging in age from 3 to 17 years (Lyon et al. 2001) After 2 weeks of treatment, the proportion of the subjects exhibiting improvement (i.e., decrease in T-score of at least 5 points) ranged from 31% for the anxious-shy attribute to 67% for the psychosomatic attribute. After 4 weeks of treatment, the proportion of subjects exhibiting improvement ranged from 44% for the social problems attribute to 74% for the Conners' ADHD index and the DSM-IV hyperactive-impulsive attribute. Five (14%) of 36 subjects reported adverse events, only 2 of which were considered related to the study medication.

A preliminary study of six psychiatric outpatients diagnosed with attention-deficit disorder (ADD) indicated that *Ginkgo biloba* might be a beneficial and useful treatment of ADD, with minimal side effects (Niederhofer 2010). During *Ginkgo biloba* treatment, the attention-deficit disorder (ADD) patients' mean scores in Wender Utah ratings improved significantly overall and in hyperactivity, inattention, and immaturity factors.

In a recent double blind, randomized, parallel group comparison trial, administration of

G. biloba (Ginkgo T.D.) was less efficacious than methylphenidate in the treatment of Attention-Deficit/Hyperactivity Disorder (ADHD) (Salehi et al. 2010). *G. biloba* had lower Parent ADHD Rating and Teacher ADHD Rating scores at the endpoint. The difference between the Ginkgo T.D. and methylphenidate groups in the frequency of side effects was not significant except for decreased appetite, headache and insomnia that were observed more frequently in the methylphenidate group.

In an observational study, three patients administered *Ginkgo biloba* EGb 761 for 4 weeks showed some improvement on the Aberrant Behavior and Symptom Checklist in autistic disorder (Niederhofer 2009). Autism is characterised by deficits in reciprocal social interaction, verbal and nonverbal communication, and imaginative activity. From the psychopharmacological point of view, clonidine, metylphenidate and neuroleptics may improve some of these aspects, but with a remarkable risk of adverse side effects. The results suggested that *Ginkgo biloba* might be effective at least as an add-on therapy.

Ginkgo and Migraine, Antihypertensive Activity

Studies indicated a significant difference among *Ginkgo biloba* treatments in all functional benefits (total phenolics content, antioxidant activity, phenolic profile, and the potential in vitro inhibitory effects on α-amylase, α-glucosidase, and Angiotensin I-Converting Enzyme (ACE) enzymes activities) evaluated in the leaf extracts and also found important seasonal variation related to the same functional parameters (Pinto Mda et al. 2009). In general, the aqueous extracts had higher total phenolic content than the ethanolic extracts. Also, no correlation was found between total phenolics and antioxidant activity. In relation to the Angiotensin I-Converting Enzyme (ACE) inhibition in management of hypertension, only ethanolic extracts had inhibitory activity.

Brinkley et al. (2010) determined the effects of *G. biloba* extract (240 mg/day) blood pressure and incidence of hypertension in 3,069 participants (mean age, 79 years; 46% female; 96% white) from the *Ginkgo* Evaluation of Memory (GEM) study. Over a median follow-up of 6.1 years, there were similar changes in blood pressure and pulse pressure in the *G. biloba* and placebo groups. The rate of incident hypertension also did not differ between participants assigned to *G. biloba* versus placebo. Their data indicated that *G. biloba* did not reduce blood pressure or the incidence of hypertension in elderly men and women.

Ginkgolide B, a constituent extract from *Ginkgo biloba* leaf, was found to have efficacy in the prophylactic treatment of migraine with aura (MA) in women (D'Andrea et al. 2009). The study was carried out in a 6-month multicentric, open, preliminary trial using a combination treatment of 60 mg *Ginkgo biloba* terpenes, phytosome, 11 mg coenzyme Q 10, and 8.7 mg vitamin B2 (Migrasoll), administered twice daily to 50 women suffering from migraine with typical aura, or migraine aura without headache. Ginkgolide B was effective in reducing MA frequency and duration. The effect was clearly evident in the first bimester of treatment and was further enhanced during the second. In a study of school-aged children with migraine, 3 months of treatment with ginkgolide B/coenzyme Q10/riboflavin/magnesium complex, the mean frequency per month of migraine was significantly decreased (Esposito and Carotenuto 2011). The findings suggested that in childhood headache management, the use of alternative treatments must be considered not to evoke a placebo effect, but as soft therapy without adverse reaction. Data from an open-labelled prospective trial showed that ginkgolide B appeared to be effective and well-tolerated as preventive treatment in reducing migraine attack frequency and in attenuating the use of symptomatic medication in children (24) with primary headache (Usai et al. 2010).

Antihyperlipidemic/ Antihypercholesterolemic Activity

Ginkgo biloba extract, GBE was found to decrease the total cholesterol content in cultured hepatocytes and inhibited the activity of HMG-CoA (3-hydroxy-3-methylglutaryl-coenzyme A) reductase (Xie et al. 2009b). In addition, GBE decreased cholesterol influx, whereas lovastatin enhanced cholesterol influx. GBE treatment induced significant increases in the expression of cholesterogenic genes and genes involved in cholesterol metabolism, such as SREBF2. The data indicated that the two compounds modulated cholesterol metabolism through distinct mechanisms.

Ginkgo biloba leaf extract, EGB was found to exert multi-directional lipid-lowering effects on the rat metabonome, including limitation of the absorption of cholesterol, inactivation of HMGCoA (3-hydroxy-3-methylglutaryl-coenzyme A) and favourable regulation of profiles of essential polyunsaturated fatty acid (EFA) (Zhang et al. 2009b). EGB lowered total cholesterol and low density lipoprotein cholesterol levels and raised high density lipoprotein cholesterol levels in rat plasma compared with those of the diet-induced hyperlipidemia group. EGB also enhanced the activities of lipoprotein lipase and hepatic lipase and excretion of fecal bile acid in rats from the EGB-prevention and-treatment groups. Further, elevated levels of sorbitol, tyrosine, glutamine and glucose, and decreased levels of citric acid, galactose, palmitic acid, arachidonic acid, acetic acid, cholesterol, butyrate, creatinine, linoleate, ornithine and proline, were observed in the plasma of rats treated with EGB.

Antimutagenic Activity

Ginkgo biloba extract (EGb 761) was found to have antimutagenic activity (Vilar et al. 2009). Using the micronucleus test in mouse bone marrow, the extract was found to exert protective effects against mitomycin C and cyclophosphamide-induced mutagencity. All doses of Egb 761 were significantly effective in reducing the frequency of micronucleated polychromatic erythrocytes, when compared with mitomycin C or cyclophosphamide alone. The results suggested that Egb 761 possessed both direct and indirect antimutagenic potential. Incorporating *Ginkgo biloba* leaf extract (GBE) to cigarettes was found to reduce the mutagenicity and toxicity of cigarettes though elimination of free radicals (Wang et al. 2010b). Hsu et al. (2009) found that EGb conferred protection from oxidative stress-related apoptosis induced by cigarette smoke extract in human pulmonary artery endothelial cells and its therapeutic effects depended on transcriptional upregulation of heme oxygenase-1 by the extract via the MAPKs/Nrf2 pathway.

Anticancer Activity

Studies indicated that treatment dosage may determine the effect of ginkgolide B, a major active component of *Ginkgo biloba* extracts, on ethanol-induced ROS generation and cell apoptosis in human Hepatoma G2 cells; this apoptotic effect was inhibited by low (5–25 μM) doses of ginkgolide B, but enhanced by high (50–100 μM) doses of ginkgolide B (Chan and Hsuuw 2007). Additionally the increased intracellular oxidative stress induced by ethanol was enhanced by high doses of ginkgolide B but decreased following treatment with low concentrations of ginkgolide B. The dose–response effects of ginkgolide B on reactive oxygen species (ROS) generation were directly correlated with cell apoptotic biochemical changes including c-Jun N-terminal kinase (JNK) activation, caspase-3 activation, and DNA fragmentation. This supported the notion that an appropriate dosage of ginkgolide B may aid in decreasing the toxic effects of ethanol.

Treatment of MCF-10A human mammary epithelial cells with noncytotoxic concentrations of *G. biloba* extract (25–300 μg/ml for 24 or 48 h) was found to increase CYP1B1 and CYP1A1 gene expression and aryl hydrocarbon receptor (AhR) activity (Rajaraman et al. 2009). The AhR signalling pathway regulated the production of CYP1B1 and CYP1A1, which catalysed the bioactivation of various procarcinogens. Bilobalide and ginkgolides A, B, C, and J did not participate in the modulation of CYP1B1 and CYP1A1 gene expression or AhR activation by *G. biloba* extract. In contrast, quercetin increased CYP1B1 and CYP1A1 gene expression and activated AhR, whereas kaempferol and isorhamnetin suppressed constitutive CYP1B1 expression and antagonized AhR activation by benzo[a]pyrene. The findings provided an impetus for future investigations on the effect of *G. biloba* extract in CYP1-mediated chemical carcinogenesis.

Jiang et al. (2009) found that *Ginkgo biloba* extract enhanced anti-oxidative activity and inhibited the progression of gastric precancerous lesions induced by N-methyl-N′-nitro-N-nitrosoguanidine (MNNG), via the modulation of cell proliferation and apoptosis. The incidence of mild to severe intestinal metaplasia and dysplasia, ISA (index of structural atypia) and NGI (nucleo-glandular index), were significantly lower in the *G. biloba* extract treated groups than those in the control group. In the extract treated groups, the activity of SOD (superoxide dismutase) was enhanced and the concentration of MDA (malondialdehyde) was reduced. Further, expressions of Bcl-2, c-myc and FasL decreased in the extract treated groups, while the expression of Fas increased. *Ginkgo biloba* extract EGb(R) 761 was found to be a potent anti-angiogenic drug and to have potential for use in anti-angiogenesis based tumour prevention and adjuvant therapy (Koltermann et al. 2009). The underlying mechanism for its anti-angiogenic was via activation of protein tyrosine phosphatases, leading to inhibition of the Raf-MEK-ERK pathway.

Feng et al. (2009) reported that *Ginkgo biloba* extract, EGb 761(containing flavonoid glycosides), and it analogues EGb 761-H (containing mainly flavonoid aglycones and terpene trilactones), and EGb 761-DT-H (containing mainly flavonoid aglycones) displayed cytotoxicity and inhibitory activity with IC_{50} values of 46.36 μM, 10.27 μM, and 14.93 μM in leukemic 1210 (L1210) cell-based assays, respectively. This elicited 41.74%, 60.72%, and 63.76% reductions in tumour weight after 10 days of treatment, respectively. The researchers concluded that the anticancer activity of EGb 761 could be improved by increasing the levels of the aglycone form of the flavonoid. Terpene trilactones could not exert the anticancer effects of flavonoids in-vivo. Raising the levels of the free radical scavenger enzymes superoxide dismutase (SOD), glutathione (GST), and catalase (CAT) was postulated to be one of the involved anticancer mechanisms.

Kaempferol from *G. biloba* extract was found to effectively inhibit pancreatic cancer (MIA PaCa-2 and Panc-1) cell proliferation and induced cancer cell apoptosis, which may sensitize pancreatic tumour cells to chemotherapy (Zhang et al. 2008). Kaempferol treatment also significantly decreased 3H-thymidine incorporation in both MIA PaCa-2 and Panc-1 cells. Combination treatment of low amounts of kaempferol and 5-fluorouracil produced an additive effect on the inhibition of MIA PaCa-2 cell proliferation. Further, kaempferol had significantly less cytotoxicity than 5-fluorouracil in normal human pancreatic ductal epithelial cells. The results suggested that kaempferol may have clinical applications as adjuvant therapy in the treatment of pancreatic cancer.

Studies by Hao et al. (2009b) showed that rats injected with aflatoxin B1(AFB1) and treated with *Ginkgo biloba* extract (EGb 761) had significantly smaller gammaglutamyl transpeptidase-positive hyperplastic cell foci (gamma-GT foci) than those in rats injected with AFB1 without EGB 761. The prevalence of hepatocelluiar carcinoma in EGb treated group (26.92%) was significantly lower than that in untreated rats (76%). The results suggested that EGB 761 effectively inhibited hepatocarcinogenesis induced by aflatoxin B1 in rats, which may be related to its antioxidant activity.

Ginkgolic acid extracted from the seed coat of *Ginkgo biloba* inhibited the growth of tumorogenic Hep-2 and Tac8113 cell lines in a both dose- and time-dependent fashion (Zhou et al. 2010a). The antitumour action of ginkgolic acid was due to inhibition of cell proliferation by retarding the progress of cell cycle at GO/G1 phase and inducing apoptosis by downregulating the expression of anti-apoptotic Bcl-2 protein and upregulating the expression of pro-apoptotic Bax protein, eventually leading to a decrease in the Bcl-2/Bax ratio in tumour cells. Ginkgolic acid may be a candidate for new antitumour drugs.

Ginkgo biloba leaf extract (EGb 761) had been proven to induce caspase-3-dependent apoptosis in oral cavity cancer cells (Kang et al. 2010). Results of further studies suggested that kaempferol and quercetin, two components of EGb 761, effectively induced caspase-3-dependent apoptosis of oral cavity cancer cells (SCC-1483, SCC-25 and SCC-QLL1) and showed cleavage of poly (ADP-ribose) polymerase. Both components could be considered as possible anti-oral cavity cancer agents.

A total of 3,069 Ginkgo Evaluation of Memory (GEM) participants 75+years of age were randomized to twice-daily doses of either 120 mg ginkgo extract (EGb 761) or placebo and followed for a median 6.1 years in a randomized, double-blind, placebo-controlled clinical trial of ginkgo supplementation (Biggs et al. 2010). During the intervention, there were 148 cancer hospitalizations in the placebo group and 162 in the EGb 761 group. Among the site-specific cancers analyzed, they observed an increased risk of breast and colorectal cancer, and a reduced risk of prostate cancer. Overall, the results did not support the hypothesis that regular use of *Ginkgo biloba* reduces the risk of cancer.

Cardiovascular Disease Activity

There is an increasing evidence of the potential role of *Ginkgo biloba extract* GBE in treating cardiovascular diseases (Zhou et al. 2004). Extensive in-vitro cell culture and animal model studies had shown GBE to exert its action through diverse mechanisms. GBE had been reported to have antioxidant properties, to modulate vasomotor function, to reduce adhesion of blood cells to endothelium, to suppress activation of platelets and smooth muscle cells, to affect ion channels, and to change signal transduction. Further, relevant clinical trials with CBE had been carried out, particularly in the treatment of arterial and venous insufficiency and in the prevention of thrombosis. The authors also discussed the controversial clinical findings and the possible adverse interactions between GBE and other drugs.

In-vitro studies

Ginkgo biloba extract increased endothelial progenitor-cell numbers and functional activity (Chen et al. 2004; Wang et al. 2004). Incubation of isolated human mononuclear cells with *Ginkgo biloba* extract increased the number of endothelial progenitor cells in a time- and dose-dependent manner. Moreover, the extract also dose- and time-dependently promoted endothelial progenitor cells proliferative, migratory, adhesive, and in-vitro vasculogenesis capacity. The results of further studies indicated that *Ginkgo biloba* extract retarded the onset of endothelial progenitor cells senescence, which may be related to activation of telomerase through the PI3k/Akt signalling pathway (Dong et al. 2007). The suppression of endothelial progenitor cells senescence by *Ginkgo biloba* extract in-vitro may augment the functional activity of endothelial progenitor cells in a way that is important for potential cell therapy. In-vitro studies showed that *Ginkgo biloba* extract, GBE, might exert its anti-atherogenesis and vascular protective effects by inducing vascular HO-1 expression (Chen et al. 2011). Pre-treatment with GBE reduced TNF-α-induced endothelial adhesiveness and, increased HO-1 expression and enzyme activity in human aortic endothelial cells (HAECs) via the activation of p38 and Nrf-2 (nuclear factor-erythroid 2-related factor 2) pathways, a mechanism in which oxidative stress was not directly involved. Studies showed that *Ginkgo biloba* (EGb) and quercetin could inhibit angiotensin II-induced cardiomyocyte hypertrophy through a reactive oxygen species-dependent pathway, the effect of quercetin might be related to the JNK and c-fos cascade (Wu and Gu 2007). EGb and quercetin were able to enhance the superoxide dismutase activity and reduce the production of malondialdehye in neonatal rat cardiomyocytes. Results of studies suggested the involvement of oxidatively modified low-density lipoprotein in the development of human atherosclerosis through vascular endothelial growth factor (VEGF) induction in monocytes, and that EGb 761 prevented in-vitro atherogenesis, probably by decreasing VEGF expression in human THP-1 and inhibition of monocyte/macrophage-derived foam cell formation (Liu et al. 2009a).

Studies showed that *Ginkgo biloba* extract (EGB) therapy could improve cardiac function and energy metabolism in rats with heart failure induced by adriamycin by increasing the expression and production of ghrelins that promoted positive myocardial energy metabolism (Xu et al. 2009). Hao et al. (2009a) found that ginkgolide B (GB) improved the function of left ventricle from ischemia-reperfusion injury and decreased infarct size and the release of lactate dehydrogenase. The expression of protein Bcl-2 was upregulated by GB and the ratio of Bax to Bcl-2 was decreased by GB. Their results showed that GB could partly prevent ischemia-reperfusion injury in the rat's heart. Ginkgo leaf extract (EGB 50) could prevent animal model of arrhythmia induced by aconitine and ouabain (Wang et al. 2010c). The extract exerted inhibitory effects on delayed afterdepolarizations (DADs) and triggered activity (TA) induced by ouabain and high Ca2+ in guinea pig papillary muscles.

In-vivo studies

Animal studies showed that Ginaton (*G. biloba* extract) preconditioning could prolong the survival time after discordant cardiac xenograft, and significantly alleviated pathological lesion from acute xenograft vascular rejection combined with cyclosporine A (Huang et al. 2006).

Improvement in cardiac function by *Ginkgo biloba* extract (EGb) in rats with acute myocardial injury induced by isoproterenol was significant (Ding et al. 2009). The cardiac parameters of the control group were compromised significantly in both systolic and diastolic function. In control groups, plasma activities of aspartate aminotransferase (AST), Lactate dehydrogenase (LDH), creatinase kinase (CK), α-hydroxybutyrate dehydrogenase (HBDH), CKMB (Creatinase Kinase Muscle and Brain) and ventricular weight index (LV and RV/BW) were elevated significantly. With the treatment with EGb and metoprolol, the enzymes and ventricular weight index were significantly ameliorated. The results indicated that *G. biloba* extract was beneficial to cardiac performance by improving myocardium enzymes and cardiac function in isoproterenol induced myocardial injury in rats. Animal studies showed that *Ginkgo biloba* leaf extract (EGb) exhibited protective and ameliorative effects against myocardial ischemia injury induced by isoproterenol (Xiong et al. 2009). The protective mechanisms of EGb may be related to the elevation of plasma apelin contents and apelin mRNA level in the myocardium.

Studies by Liu et al. (2008) suggested that GBE could reverse the hyperhomocysteine-induced neointima formation in rabbits following balloon injury, and the suppressive effect of GBE on the migration and proliferation of vascular smooth muscle cells (VSMCs) may contribute to its action. Hyperhomocysteinemia is associated with the risk of atherosclerosis and restenosis after angioplasty. *Ginkgo biloba* extract (GBE) was found to possess protective effect on myocardial damage after craniocerebral injury, which was postulated to be associated with lowering of anti-oxidation stress level of myocardial cellular mitochondria (Hao et al. 2010). Pretreatment of GBE resulted in the decrease of electrocardiograph abnormality occurrence, serum CK-MB level, and degree of myocardial damage, as well as the increase of Mn-SOD activity in post-craniocerebral injured rats.

Clinical studies/reviews

In an open prospective study of 20 outpatients with a long history of elevated fibrinogen levels and plasma viscosity, and a variety of underlying diseases, treatment with *Ginkgo biloba* extract (EGb 761) for 12 weeks resulted in significant improvement in the fibrinogen levels and hemorrheological properties (Witte et al. 1992). Deng et al. (2009) found that adding *Ginkgo biloba* extract (Ginaton) to the cardioplegia perfusion in patients during peri-operative period of cardiac surgery induced the production of plasma vascular endothelial growth factor (VEGF), which may be one of the mechanisms for its myocardial protection. Kuller et al. (2010) conducted a double-blind, placebo-controlled trial involving 3,069 participants over 75 years of age with mean follow-up of 6.1 years. They found no evidence that *G. biloba* (120 mg of EGb 761 twice daily) reduced total or cardiovascular disease mortality or cardiovascular disease events. There were more peripheral vascular disease events in the placebo arm.

Siegel et al. (2007) found that after a 2-month therapy with *Ginkgo biloba* extract, atherosclerotic nanoplaque formation and size were significantly reduced in eight patients who had undergone an aortocoronary bypass operation. Further, superoxide dismutase (SOD) activity was enhanced, the quotient oxLDL/LDL lowered and lipoprotein (a) concentration decreased in the patients' blood after the 2-month medication regimen. The concentration of the vasodilating substances cAMP and cGMP was increased. *Ginkgo biloba* extract (EGb 761) exhibited a dose-dependent antivasospasmic effect in experimental subarachnoid hemorrhage animal model (Kotil et al. 2008). When *Ginkgo biloba* extract was administered at a dose of 45 mg/kg per day, it was effective against vasospasm, however, this effect disappeared at a dose of 90 mg/kg per day.

Patel and Hamadeh (2009) in their review found that improved nutritional status was of utmost significance in alleviating the detrimental effects of amyotrophic lateral sclerosis (ALS). They asserted that vitamin E, folic acid, alpha lipoic acid, lyophilized red wine, coenzyme Q10, epigallocatechin gallate, *Ginkgo biloba*, melatonin, Cu chelators, and regular low and moderate intensity exercise, as well as treatments with catalase and l-carnitine, held promise in mitigating the effects of ALS, whereas caloric restriction, malnutrition and high-intensity exercise were contraindicated in their disease model.

Antiplatelet Aggregation Activity

Ryu et al. (2009) showed that in in-vitro assays using freshly isolated human platelets, the combination of cilostazol and *Ginkgo biloba* extract (GB) showed superior inhibition of both the shear and the collagen-induced platelet aggregation to those of each drug alone. Cilostazol and *Ginkgo biloba* extract (GB) are antiplatelet agents that are commonly used remedies for peripheral arterial disease. In various peripheral vascular diseases, thrombosis and thromboembolic occlusions of major and minor blood vessels are a major complication. In accordance with these enhanced in-vitro antiplatelet activities, the combinative therapy showed enhanced anti-thrombotic effects in in-vivo pulmonary embolism model and arterial thrombosis model. In particular, the increase of survival rate in pulmonary embolism model by combination treatment of cilostazol (25 mg/kg) and GB (20 mg/kg) was higher more than twofold of those of the respective drugs. Notably, the combination of cilostazol and GB did not show a significant effect on the bleeding time, prothrombin time (PT) and activated partial thromboplastin time (aPTT) increase, suggesting that GB may potentiate the antiplatelet effect of cilostazol without the prolongation of bleeding time or coagulation time. These studies suggested that combinative therapy of GB and cilostazol might offer enhanced anti-thrombotic efficacies without increasing side-effects.

In an open-label, randomized, two-period, two-treatment, two-sequence, single-dose cross-over study of a small group of healthy Korean male volunteers, Kim et al. (2010) found that the addition of a single dose of *Ginkgo biloba* extract did not prolong the bleeding time and was not associated with additional antiplatelet effects compared with the administration of ticlopidine (antiplatelet agent) alone. The co-administration of *Ginkgo biloba* extract with ticlopidine was not associated with any significant changes in the pharmacokinetic profile of ticlopidine compared with ticlopidine administered alone.

Ginkgo and Osteoporosis/Sarcopenia

Thirty days treatment with *G. biloba* (56 mg/kg/day) (EGB) of rats with mandibular glucocorticoid-induced-osteoporosis restored mesial and distal periodontal bone support and significantly increased the mandibular cortical thickness. (Lucinda et al. 2010a). The results suggested that EGB may be effective in the treatment of osteoporosis. They also found that treatment with EGB (28 and 56 mg/kg) significantly reversed the loss of the alveolar bone of the mandible and of the trabecular bone of the femur in Wistar rats with glucocorticoid-induced-osteoporosis (Lucinda et al. 2010b).

Bidon et al. (2009) found that *Ginkgo biloba* extract, EGb 761 could reactivate a juvenile profile in the skeletal muscle of sarcopenic rats by transcriptional reprogramming and may represent a novel treatment for sarcopenia in being more manageable and less cumbersome than exercise and caloric restriction. EGb 761 induced a gain in muscular mass that was associated with an improvement of the muscular performances. These changes were found to be accompanied by the transcriptional reprogramming of genes related to myogenesis through the TGFbeta signalling pathway and to energy production via fatty acids and glucose oxidation. EGb 761 restored a more juvenile gene expression pattern by regenerating the aged muscle and reversing the age-related metabolic shift from lipids to glucose utilization.

Optic/Ophthalmic-Protective Activity

Ginkgo biloba extract was shown to protect optic nerves from injury. Ginexin (*Ginkgo biloba* extract) was found to increase the survival of the retinal ganglion cell in the rat optic nerve crush injury model (Kang et al. 2003). However, it did not significantly influence the intravitreal glutamate concentration. *Ginkgo biloba* extract (EGb 761) was found to inhibit the apoptosis of retinal ganglion cells (RGC) in guinea pigs after optic nerve transection, thus protect the morphology and function of RGCs (Xie et al. 2009a).

Ma et al. (2009) found that intragastral applications of a *G. biloba* extract applied after an experimental and standardized optic nerve crush in rats were associated with a higher survival rate of retinal ganglion cells in a dosage-dependent manner. Results of studies suggested that intraperitoneal injections of a *Ginkgo biloba* extract given prior to and daily after an experimental and standardized optic nerve crush in rats were associated with a higher survival rate of retinal ganglion cells (Ma et al. 2010).*Ginkgo biloba* extract (GBE) was found to prevent human lens epithelial cells from high glucose-induced apoptosis through suppressing oxidative stress, reducing the ratio of Bax to Bcl-2, and decreasing the activity of caspase-3 (Wu et al. 2008), indicating GBE to have potential protective effect against diabetic cataract formation.

Quaranta et al. (2003) conducted a prospective, randomized, placebo-controlled, double-masked cross-over trial in 27 patients with bilateral visual field damage resulting from normal tension glaucoma to evaluate the effect of *Ginkgo biloba* extract (GBE) on normal tension glaucoma. After GBE treatment, a significant improvement in visual fields indices was observed. No significant changes were found in intraocular pressure, blood pressure, or heart rate after placebo or GBE treatment.

Russo et al. (2009) in clinical studies found that patients with allergic conjunctivitis treated with *Ginkgo biloba* extract plus hyaluronic acid ophthalmic solution (GB-HA) displayed a significant decrease in the appearance of clinical symptoms such as conjunctival hyperemia, conjunctival discharge, and chemosis compared with patients treated with only hyaluronic acid. Further, all patients treated with GB-HA showed a significant improvement of subjective symptoms such as itching, photophobia, stinging, and lacrimation, compared to HA patients. The results suggested that *Ginkgo biloba* extract may exert therapeutic activity in the treatment of seasonal allergic conjunctivitis.

Ginkgo and Tinnitus/ Otoprotective Activity

Holstein (2001) reviewed 19 clinical trials investigating the effects of tinnitus treatment with *Ginkgo biloba* special extract EGb 761. He found that the results of eight controlled studies on tinnitus due to cerebrovascular insufficiency or labyrinthine disorders of varying genesis for the most part showed a statistically significant superiority of treatment with EGb 761 as compared with placebo or reference drugs applied for periods of 1–3 months. He also reported that open studies, some involving large numbers of patients, revealed appreciable improvements under *ginkgo* treatment and the tolerability of *Ginkgo biloba* was excellent.

Smith et al. (2005) in their review of published work on the effects of *Ginkgo biloba* on tinnitus found the paucity of effective animal model and systemic clinical trials using double-blind and placebo-controlled designs. They found that the restricted number of clinical studies yielding positive results were technically flawed and not published in respected peer reviewed journals. Two most systematic clinical trials, both double-blind and placebo controlled, and published in respected peer-reviewed journals, had yielded negative results and suggested that *Ginkgo biloba* extracts were of little more use in the treatment of tinnitus than a placebo. Similarly, the review by Hilton and Stuart (2004) found that there was no reliable evidence that *Ginkgo biloba* was effective for the primary complaint of tinnitus although the incidence of side effects was small. In a more recent paper, Canis et al. (2011) in a retrospective study of 94 patients found that administration of simvastatin or *G. biloba* over 4 months, showed no significant efficacy in treatment of subacute tinnitus.

In a randomized, prospective, double-blind study involving 72 patients, Reisser and Weidauer (2001) found that treatment of sudden deafness with ginkgo special extract EGb 761 was as effective as treatment with pentoxifylline. The two therapeutic schedules were equally well tolerated and exhibited a statistically significant equivalence in improvement or in return to normal of the auditory thresholds in the two patient groups. Further, no differences were found between the treatment groups with regard to improvement in hearing and reduction in tinnitus which arose at the same time as the sudden hearing loss. In another randomized to double-blind, 12 week study of 52 patients, a combination of infusion therapy with ginkgo special extract EGb 761 followed by oral administration of the same appeared to be effective and safe in alleviating the symptoms associated with tinnitus aurium (Morgenstern and Biermann 2002). The results were supported by the secondary outcome measures for efficacy (e.g. decreased hearing loss, improved self-assessment of subjective impairment).

In a double blind, placebo controlled trial using postal questionnaires of 1,121 healthy people (18–70 years) with tinnitus that was comparatively stable, 12 weeks treatment 50 mg *Ginkgo biloba* extract LI 1370 given three times daily was found no more effective than placebo in treating tinnitus (Drew 2001). In a multicentre, randomized, double-blind phase III study of 106 patients, a higher dosage of EGb 761 administered orally appeared to hasten the recovery of acute idiopathic sudden sensorineural hearing loss (ISSHL) patients, which was already observed after 1 week of treatment (Burschka et al. 2001). The researchers found it justifiable to treat patients who had unilateral ISSHL of less than 75 dB and neither tinnitus nor vertigo with 120 mg oral EGb 761 twice daily.

Studies demonstrated that EGb 761 protected against cisplatin-induced ototoxicity (Huang et al. 2007). Cisplatin-induced ototoxicity is a major dose-limiting side effect in anticancer chemotherapy. Cisplatin-induced ototoxicity had been correlated to depletion of the cochlear antioxidant system and increased lipid peroxidation. Rats treated with EGb 761 plus cisplatin did not show significant auditory brainstem response (ABR) threshold shifts, endocochlear potentials (EPs) were decreased less than 20% compared to 50% of cisplatin treated rats, and the outer hair in the basal turn of cochleae remained intact versus severe hair loss in cisplatin-treated rats. Myringosclerosis was significantly more severe in rats in the control and saline groups than in *Ginkgo biloba* treated groups (Kaptan et al. 2008). The levels of nitrite in ginkgo-treated groups were significantly lower than in untreated and saline-treated groups, while glutathione peroxidase levels were significantly higher. The levels of malondialdehyde and superoxide dismutase were lower in ginkgo groups but not significantly. By scavenging free oxygen radicals, ginkgo extract prevented the formation of myringosclerosis. The results of studies by Emir et al. (2009) showed that formation of experimental myringosclerosis was reduced or inhibited and tympanic membranes were thinner after systemic *Ginkgo biloba* extract administration.

The results of studies in guinea pigs indicated that *Ginkgo biloba* extract, EGb 761 exhibited a protective effect against gentamicin ototoxicity through a reduction in the formation of reactive oxygen species (ROS) and nitric oxide and subsequent inhibition of hair cell apoptosis in the cochlea (Yang et al. 2011). Individual EGb 761 components quercetin, bilobalide, ginkgolide A and ginkgolide B, but not kaempferol, significantly prevented gentamicin-induced hair cell damage. EGb 761 treatment was found to prevent significantly aging-related caspase-induced activities within the cochleae in young (4 months) and aged-mature (12 months) rats (Nevado et al. 2010). In the short EGb 761 treatment, young rats showed lower levels of caspase-3/7 than aged-mature rats. Reduced caspase-3/7 activity in presence of EGb 761 correlated with significant improvements of auditory steady-state responses (ASSR) threshold shifts. The data indicated that EGb 761 treatment had a significant benefit with an early and preventive effect, reversing the deleterious effect of aging in the integrity of the rat cochlea, even in the late stage of the rat lifespan.

Olfactory function in patients with postviral olfactory loss was significantly improved by both treatment modalities (prednisolone monotherapy and combination therapy of prednisolone followed by *G. biloba*) (Seo et al. 2009)). Although the treatment response was not statistically different between the monotherapy group and the combination therapy group, the addition of *G. biloba* showed a tendency of greater efficacy in the treatment of postviral olfactory loss.

Antiinflammatory Activity

Ginkgetin, a biflavone from *Ginkgo biloba* leaves, was reported to be a phospholipase A (2) inhibitor and this compound showed the potent antiarthritic activity in rat adjuvant-induced arthritis as well as analgesic activity (Kim et al. 1997). Ginkgetin strongly reduced arthritic inflammation in an animal model of rat adjuvant-induced arthritis via intraperintonel injection. Histological examination of the knee joints confirmed the findings. It also showed a dose-dependent inhibition in an animal model of acetic-acid-induced writhing. All these results indicated that ginkgetin may be a potential antiarthritic agent with analgesic activity. In a separate study, ginkgetin (1–10 μM) and the biflavonoid mixture (10–50 μg/ml), mainly a 1: 1 mixture of ginkgetin and isoginkgetin, from *G. biloba* leaves, inhibited production of prostaglandin E2 from lipopolysaccharide-induced RAW 264.7 cells (Kwak et al. 2002). This inhibition was mediated, at least in part, by down-regulation of COX-2 expression. Further, ginkgetin and the biflavonoid mixture (100–1,000 μg/ear) dose-dependently inhibited skin inflammation of croton oil induced ear edema in mice by topical application. The data suggested that ginkgetin from *G. biloba* leaves down-regulated COX-2 induction in-vivo and this down-regulating potential was associated with an anti-inflammatory activity against skin inflammatory responses. Son et al. (2005) found that ginkgetin inhibited cyclooxygenase-2 (COX-2) dependent phases of prostaglandin D (2) (PGD (2)) generation in bone marrow-derived mast cells (BMMC) in a concentration-dependent manner with IC_{50} values of 0.75 μM. The decrease in quantity of the PGD (2) product was accompanied by a decrease in the COX-2 protein level. Further, ginkgetin inhibited the production of leukotriene C(4) in a dose dependent manner, with an IC_{50}value of 0.33 μM and inhibited degranulation reaction in a dose dependent manner, with an IC_{50} value of 6.52 μM. The studies demonstrated that ginkgetin possessed a dual cyclooxygenase-2/5-lipoxygenase inhibitory activity and might provide a basis for novel anti-inflammatory agents. The data from in-vitro studies suggested that kaempferol and quercetin, essential ingredients in *Ginkgo biloba* extract (GBE), may overcome the dose problem of GBE and play a valuable role, clinically, in controlling mucin hypersecretion in airway inflammation (Kwon et al. 2009). Kaempferol and quercetin dose-dependently suppressed MUC5AC mRNA expression in NCI-H292 cells.

In a subsequent study, ginkgetin (20–80 μg/ear/treatment) inhibited ear edema (22.8–30.5%)

and prostaglandin E2 production (30.2–31.1%) induced by multiple treatment of 12-O-tetradecanoylphorbol-13-acetate (Lim et al. 2006). Ginkgetin was also found to reduce epidermal hyperplasia and suppress the expression of proinflammatory gene, interleukin-1beta. The data suggested ginkgetin may be beneficial against chronic skin inflammatory disorders like atopic dermatitis.

Pre-treatment of laboratory animals with *Ginkgo biloba* extract (EGb) inhibited the in-vivo production of TNF tumour necrosis factor-alpha (TNF-alpha), a pro-inflammatory cytokine, after challenge with bacterial lipopolysaccharide (Wadsworth et al. 2001). EGb diminished LPS-induced NF-kappaB but had no effect on lipopolysaccharide-induced TNF-alpha transcription. Both EGb and quercetin inhibited ERK1/2 phosphorylation and p38 MAPK activity, which were important in the post-transcriptional regulation of TNF-alpha mRNA. Results of studies by Weng et al. (2008) suggested that one of the anti-inflammatory mechanisms of *G. Biloba* leaf extract may be attributed to the stimulation of eosinophils in broncho alveloar lavage fluid of asthmatic mice.

Animal studies showed that *Ginkgo biloba* extract 50 (GBE50) could decrease the content of interleukin-6 (IL-6) and increased the content of IL-4 in the myocardium after ischemia-reperfusion injury (Bao et al. 2010). The data suggested that GBE50 could regulate the inflammatory reaction after ischemia-reperfusion injury by inhibiting inflammatory cytokines and promoting anti-inflammatory cytokines.

EGb 761, astaxanthin and vitamin C were shown to interact summatively to suppress inflammation with efficacy equal to or better than ibuprofen, a widely used non-steroidal anti-inflammatory drug (NSAID) (Haines et al. 2011). The combination treatment decreased in counts of eosinophils, neutrophils, macrophages and enhanced levels of cAMP (Cyclic adenosine monophosphate) and cGMP (cyclic guanosine monophosphate) in asthmatic guinea pigs. The researchers asserted that such combinations of non-toxic phytochemicals constituted powerful tools for the prevention of onset of acute and chronic inflammatory disease if consumed regularly by healthy individuals; and may also augment the effectiveness of therapy for those with established illness.

Antidiabetic Activity

Studies in spontaneously diabetic BioBreeding/Ottawa Karlsburg (BB/OK) rats showed that pre-treatment of diabetic myocardium with *Ginkgo biloba* extract (EGb 761), resulted in an improvement in (ultra)structural alterations of the myocardial nervous network of most of these parameters and alterations of cardiac sympathetic integrity and activity compared to unprotected myocardium (Schneider et al. 2009.). The results suggested that EGb may act as a potent therapeutic adjuvant in diabetics with respect to cardiovascular autonomic neuropathy, which may contribute to the prevention of late complications in diabetes.

Feeding of *Ginkgo biloba* extract improved the glucose metabolism in diabetic rats and reduced the diabetes-induced diaphragm damage (Li et al. 2010a). The action mechanism may be related to stimulation of the mRNA expression of glucose transporter 4 (GLUT4) in the diaphragm and improvement in the uptake and metabolism of blood glucose. Results of studies by Zhou et al. (2011) showed that GBE suppressed glucose uptake under normal conditions, while it remarkably improved glucose tolerance under insulin-resistant conditions. GBE primarily exerted its effects by stimulating IRS-2 transcription. The data indicated the potential of GBE to prevent insulin resistance and to be a promising anti-diabetic drug. Li et al. (2010b) found that GbE could enhance contraction capacity of diaphragm from type 2 diabetic rats by increasing the aerobic oxidation capacity, glycolytic capacity and the function of respiratory chain. GBE also inhibited swelling and degeneration of diaphragm mitochondrions.

Hepatoprotective Activity

Pretreatment of rats undergoing ischaemia-reperfusion with both N-acetylcysteine (NAC) and EGb 761 (*Ginkgo biloba* extract) clearly reduced 8-hydroxydeoxyguanosine formation and lipid peroxidation in the rat liver tissue (Keles et al. 2008). These findings suggested that antioxidant compounds such as NAC and EGb 761 may be useful in preventing postischaemic reperfusion injury in hepatic tissue. Findings of studies in rats by Taki et al. (2009) suggested that bilobalide (30 mg/kg, gavage) markedly induced hepatic cytochrome P450 (CYP) enzyme activity, but the induction could be mitigated due to rapid elimination from the liver. Pretreatment of *Ginkgo biloba extract* (GbE), for 5 days, prevented most of the rat liver damage caused by carbon tetrachloride (CCl4) (Chávez-Morales et al. 2010). GbE significantly reduced serum activities of alanine aminotransferase and aspartate aminotransferase (54% and 65%, respectively), compared to CCl4-treated rats and partially prevented the increase of liver malondialdehyde (55%) and the decrease of albumin concentration to 12%. This pretreatment prevented the down-regulation of tumour necrosis factor alpha and up-regulated the interleukin 6 (IL-6) mRNA steady-state level. Furthermore, the GbE reduced the amount of necrotic areas in the central lobe area, compared to CCl4-treated rat. Zhang et al. (2007) demonstrated that *Ginkgo biloba* extract (EGb) could decrease the portal vein pressure and improve hepatic microcirculation in carbon tetrachloride treated rats. The mechanisms of this effect may involve its inhibition on endothelin, platelet-activating factor, lipid peroxidation, and down regulation of the hepatic iNOS and nitric oxide expressions.

Extract of *Ginkgo biloba* (EGB) was found to have antioxidant and hepatoprotective effects and could inhibit liver fibrosis in rats with diet-induced non-alcoholic steatohepatitis (Zhou et al. 2010b). EGB treated rats had significantly lower levels of serum alanine transaminase aspartate aminotransferase, fibrosis markers, and pathological grading of liver fibrosis and the staining intensity of nuclear factor, NF-κBp65 protein in the liver compared to those in the model group. The serum biochemical, fibrosis markers, superoxide dismutase (SOD), malondialdehyde (MDA), the pathological changes, and the expression levels of nuclear factor NF-κBp65 protein in the liver were seen. In-vivo results of animal studies showed that *G. biloba* extract was a potent protector against uranium-induced toxicity, and its protective role was dose-dependent (Yapar et al. 2010). Treatment with *G. biloba* produced amelioration in oxidative damage and biochemical indices of hepatotoxicity and nephrotoxicity induced by uranium. Its strongest effect was observed at a dose of 150 mg/kg of body weight.

Antiobesity/Hypolipidemic Activity

The findings of studies by Bustanji et al. (2011) demonstrated that the hypolipidaemic effects of *G. biloba* extract could be in part be attributed to the inhibition of pancreatic lipase by terpene trilactones. Ginkgolides A, B, and bilobalide were found to inhibit pancreatic lipase significantly with IC_{50} of 22.9, 90.0, and 60.1 μg/ml, respectively.

Gu et al. (2009) demonstrated the molecular basis for the pleiotropic effects of GBE50, particularly those involved in lipid metabolism in the liver of rats on a high-fat diet. Transcriptome profiling analysis showed that several genes were modulated by GBE50 in liver, including those involved in lipid metabolism, carbohydrate metabolism, vascular constriction, ion transportation, neuronal systems and drug metabolism.

Renal Protective Activity

Zhou et al. (2005) found that pretreatment of rats with *Ginkgo biloba* extract, EGb prior to renal ischemia-reperfusion could significantly prevent the reduction of super oxide dismutase, Na^+-K^+-ATPase, Ca^{2+}-ATPase activity in renal cortex and the increase of malondialdehyde content in renal cortex, decreased the concentration of blood urea nitrogen and creatine in the plasma. The pathological changes of proximal tubular cells in rats kidneys induced by ischemia-reperfusion were also prevented by the pretreatment with EGb.

Ginkgo biloba extract (GBE) was shown to protect against cisplatin-induced renal failure in rats (Gulec et al. 2006). These results indicate that increased renal xanthine oxidase, adenosine deaminase and myeloperoxidas activities, as well as malondialdehyde and nitric oxide levels play a critical role in cisplatin nephrotoxicity. The 8-week treatment with extract from *Ginkgo biloba* leaf extract decreased blood lipid levels and urinic protein in children with primary nephritic syndrome and improved their clinical symptoms and the renal function (Zhong et al. 2007). The levels of urinic protein and blood lipid in *Ginkgo* extract group were significantly lower than those in prednisome plus dipyridamole group, The results suggested Ginkgo extract to have potential as an adjuvant treatment of steroid therapy in such children.

Ginkgo biloba leaf extract was found to suppress the development of renal interstitial fibrosis in rats partially via down-regulation of transglutaminase expression (Wang et al. 2009a). Results of in-vitro studies suggested that *Ginkgo biloba* extract exhibited protective effects against glomerulosclerosis in diabetic nephropathy of mesangial cells (Tang et al. 2009). The extract decreased oxidative stress, expression of Smad2/3, type IV collagen, laminin and TGF-beta (1) (Transforming growth factor beta) mRNA levels and increased Smad7 expression.

Ginkgo and Premenstrual Syndrome

Ozgoli et al. (2009) conducted a single-blind, randomized, placebo-controlled trial in 85 students diagnosed with premenstrual syndrome (PMS), to ascertain the efficacy of *G. biloba* in treating PMS. Treatment with *G. biloba* was found to reduce significantly the severity of PMS symptoms compared to the placebo.

Ginkgo and Testes Protective Activity

G. biloba extract was found to protect against ethanol-induced oxidative injury in rat testes which may be associated with HO-1 activity enhancement, free radicals elimination, lipid per-oxidation reaction inhibition (Li et al. 2005). Testes of EGB treated rats had decreased levels of reactive oxygen species and malondialdehyde but increased levels of glutathione reductase (GST), glutathione peroxidase (GSH-Px), catalase (CAT), superoxide dismutase (SOD), and glutathione (GSH) compared to the ethanol group. EGB also enhanced expression of HO-1 (heme oxygenase-1) mRNA activity remarkably.

Pre-treatment of rats with *Ginkgo biloba* before testicular torsion had a protective effect due to its free radical scavenging property (Akgül et al. 2009). Extract treated rats had significantly decreased apoptotic cells, endothelial (eNOS) and inducible (iNOS) nitric oxide synthases in testes compared to untreated rats with testicular torsion.

Studies by Yeh et al. (2009) demonstrated that *Ginkgo biloba* extract 761 EGb protected against the oxidative and apoptotic actions of doxorubicin on testes. EGb pre-treatment effectively alleviated all of these doxorubicin-induced abnormalities in testes that included impaired spermatogenesis, depleted haploid germ cell subpopulations, increased lipid peroxidation products (malondialdehyde), depressed antioxidant enzyme activities (superoxide dismutase, glutathione peroxidase and glutathione), reduced antioxidant enzyme expression (superoxide dismutase) and elevated apoptotic indexes (pro-apoptotic modulation of Bcl-2 family proteins, intensification of p53 and Apaf-1, release of mitochondrial cytochrome c, activation of caspase-3 and increase of terminal deoxynucleotidyl transferase nick-end labelling/sub-haploid cells). The results suggested that EGb may be a promising adjuvant therapy medicine, potentially ameliorating testicular toxicity of this anti-neoplastic agent in clinical practice.

Epileptic Activity

The extract of *Ginkgo biloba* was found to affect the formation of kindling epilepsy (Ivetic et al. 2008). Studies showed that the process of epileptogenesis was influenced by EGb 761.

Bioelectric activity of the brain was registered throughout the development of kindling with and without standardized extracts from dried ginkgo leaves (EGb 761). It was established that if the animals received EGb 761, significantly weaker electric stimuli was necessary for the development of the epileptogenic focus and the threshold after discharge were longer, while the number of necessary electrostimulations for the appearance of full kindling was less and the latency was shorter.

Antidepressant Activity

Ginkgo biloba leaf extract was found to have antidepressant effect as evaluated by two behavioural models, the forced swimming test in rats and tail suspension test in mice (Sakakibara et al. 2006). In addition, the extract clearly decreased immobility time in the tail suspension test after acute inter-peritoneal treatment at a dosage of 50 and 100 mg/kg body weight.

Ginkgo and Skin Protective Activity

Ginkgo biloba extract especially the flavonoid fractions: quercetin, kaempferol, sciadopitysin, ginkgetin, isoginkgetin, enhanced the proliferation of normal human skin fibroblast in-vitro as measured by MTT (3-[4, 5-dimethylthiazole-2-yl]-2,5-diphenyl-tetrazolium bromide) assay and direct haemocytometer cell count (Kim et al. 1997). The time-course release of lactic dehydrogenase was determined as the cytotoxicity index (%) during 24 h following a high dose UVB irradiation (200 mJ/cm^2). Sciadopitysin, isoginkgetin/ginkgetin treatments lowered cytotoxicity indices to 50.81 and 67.81%, respectively, compared to 95.38% for the untreated control. The antioxidant potential of biflavones of *Ginkgo biloba* was elucidated on the basis of structure-related activity and hydroxy- and methyl-substitutions on the basic structure of these flavonoids. *Ginkgo biloba* extract (GbE) at 1 or 10% concentrations and vitamin E (10–20%) concentrations protected BALB/c hairless mice skin from trichloroethylene induced irritation damage of erythema and edema (Wang et al. 2009b). Both agents also reduced the elevated levels of nitric oxide in the dorsal skin caused by trichloroethylene.

Results of studies suggested that *G. biloba* leaf extract (GBE), alone or in combination with epigallocatechin-3-gallate (EGCG) exerted a potent inhibition on vascular endothelial growth factor (VEGF) and CXCL8/IL-8 levels in normal human keratinocytes (NHKs) activated by tumour necrosis factor alpha (TNFalpha) (Trompezinski et al. 2010). GBE alone or with EGCG may contribute to anti-inflammatory properties in skin diseases associated with angiogenesis. Herbal anti-wrinkle cosmetic formulated from ginkgo (*Ginkgo biloba*) was found to increase skin moisturization (27.88%) and smoothness (4.32%) and reduced roughness (0.4%) and wrinkles (4.63%), whereas the formula containing tea and rooibos (*Camellia sinensis* and *Aspalathus linearis*) showed the best efficacy on wrinkle reduction (9.9%) (Chuarienthong et al. 2010). Dal Belo et al. (2009) using fresh dermatomed human Caucasian skin, found that EGCG and quercetin from green tea and *G. biloba* extracts vehiculated in cosmetic formulations presented good skin penetration and retention, imparting favourable skin effects.

Drug Potentiating/Drug Interaction/ Suppression of Dioxin Toxicity

In in-vitro studies of human primary hepatocyte (HPH), and HepG2 cells, Li et al. (2009) found that the bioactive terpenoids and flavonoids of GBE exhibited differential induction of drug metabolizing enzymes through selective activation of pregnane X receptor (PXR), constitutive androstane receptor (CAR), and aryl hydrocarbon receptor (AhR). Cell-based reporter assays in HepG2 showed that ginkgolide A and ginkgolide B were potent activators of PXR; quercetin and kaempferol activated PXR, CAR, and AhR, whereas bilobalide elicited no effects on these xenobiotic receptors. In separate studies, Yeung et al. (2008) found that *G. biloba* extract activated pregnane X receptor (mPXR and hPXR) in a cell-based reporter gene assay and induced human cytochrome P450 enzymes,CYP3A4, CYP3A5, and ABCB1 gene expression in hPXR-expressing LS180 human colon adenocarcinoma cells. Dioxins enter the body mainly through the diet, bind to the aryl hydrocarbon receptor (AhR), and cause various toxicological effects. Mukai et al. (2009) found that oral administration of kaempferol or *Ginkgo biloba* extract (EGb) containing 24% flavonol at 100 mg/kg body weight suppressed AhR transformation induced by 3-methylcholanthrene in the liver of mice. They found that inhibition of p-glycoprotein enhanced the suppressive effect of kaempferol on transformation of the aryl hydrocarbon receptor through an increase in the intracellular kaempferol concentration.

Results of animal studies by Fontana et al. (2005) found that EGb 761 administration increased the effects of drugs that modified motor behaviour in mice. After repeated treatment with 80 mg/kg EGb 761, a significant increase in the cataleptic effect induced by both haloperidol (a dopamine receptor inhibitory drug) and L G-nitroarginine (L-NOARG, a nitric oxide synthase inhibitor), was observed. High dose of ginkgo extract was found to increase the quantity of rat liver microsome protein and the activity of CYP3A and promote the metabolism of simvastatin (Liu et al. 2009b). Results of studies in human liver and intestinal microsomes by Mohamed and Frye (2010) indicated that *G. biloba* extract or quercetin- and kaempferol-rich supplements may inhibit intestinal and hepatic glucuronidation of mycophenolic acid. Terpene lactones did not show inhibition of MPA glucuronidation. Data from studies by Zuo et al. (2010) suggested that GBE, when taken in normally recommended doses over a 4-week time period, may not affect the pharmacokinetics of diazepam via CYP2C19 and the excretion of one of its metabolite, N-desmethyldiazepam in healthy volunteers. No drug-drug interaction was observed between GBE and diazepam. Studies in healthy male volunteers suggested that long-term use of GBE significantly influenced talinolol disposition in humans, probably by affecting the activity of P-glycoprotein and/or other drug transporters (Fan et al. 2009). Lau and Chang (2009) found that *G. biloba* extract and its flavonol aglycones were naturally occurring inhibitors of in-vitro CYP2B6 catalytic activity and bupropion hydroxylation. Bupropion is an antidepressant and a tobacco use cessation agent. Serious adverse effects of high dosages of bupropion had been reported, including the onset of seizure.

Kim et al. (2009) conducted an open-label, two-period, two-treatment, and single-dose, randomized-sequence crossover study with 24 healthy Korean male volunteers to compare the pharmacokinetic characteristics of ticlopidine in a combined fixed-dose tablet of ticlopidine/ ginkgo extract with the concomitant administration of ticlopidine and ginkgo extract tablets. A total of 24 adverse events were reported in 13 of 24 subjects: nausea (3 cases), diarrhea (3), dizziness (3), epigastric discomfort (2), headache (2), rhinorrhea (2), purulent sputum (2), dyspepsia (1), upper abdominal pain (1), cough (1), pharyngolaryngeal pain (1), oropharyngeal swelling (1), dysphonia (1), and dysphagia (1). All were considered mild or moderate in nature. There was no statistically significant difference adverse event. Administration of a single dose of a combined fixed-dose formulation of ticlopidine 250 mg/ ginkgo extract 80-mg tablets and concomitant administration of ticlopidine and ginkgo extract tablets did not result in statistically significant differences in the pharmacokinetics of ticlopidine in healthy Korean male volunteers.

Results of studies indicated that *Ginkgo biloba*-related drug metabolizing enzymes may cause herb-drug interactions and contribute to hepatotoxicity (Guo et al. 2010). Drug metabolizing genes were significantly altered in response to GBE treatments. Alterations in the expression of genes coding for drug metabolizing enzymes, the NRF2-mediated oxidative stress response pathway, and the Myc gene-centered network named "cell cycle, cellular movement, and cancer" were found.

Ginkgo and Vitiligo

Parsad et al. (2003). conducted a double-blind placebo-controlled trial with 52 patients to evaluate the efficacy of *G. biloba* extract in controlling the activity of the disease process in patients with limited and slow-spreading vitiligo and in inducing repigmentation of vitiliginous areas. A statistically significant cessation of active progression of depigmentation was noted in patients treated with *G. biloba*. Marked to complete repigmentation was seen in ten patients treated with Ginkgo, whereas only two patients in the placebo group showed similar repigmentation. The *G. biloba* extract was well tolerated and appeared to be a simple, safe and fairly effective therapy for arresting the progression of the disease. Szczurko et al. (2011) conducted an open label pilot clinical trial with 12 an open label pilot clinical trial to evaluate the effect of *G. biloba* on vitiligo. Vitiligo is a common hypopigmentation disorder with significant psychological impact if occurring before adulthood. Ingestion of 60 mg of *Ginkgo biloba* was associated with a significant improvement in total Vitiligo Area Scoring Index (VASI) measures and the Vitiligo European Task Force (VETF) spread, and a trend towards improvement on VETF measures of vitiligo lesion area and staging.

Antifertility and Sexual Behaviour Activity

Studies by Shiao and Chan (2009) showed that ginkgolide B extracts (GKB), the major active component of *Ginkgo biloba*, induced a significant reduction in the rate of oocyte maturation, fertilization, and in vitro embryonic development. Treatment of oocytes with 1–6 μM GKB during in vitro maturation (IVM) led to increased resorption of postimplantation embryos and decreased placental and fetal weights. GKB ingestion led to decreased oocyte maturation and in vitro fertilization, as well as early embryo developmental injury, specifically, inhibition of development to the blastocyst stage in-vivo. Earlier studies by the group demonstrated that ginkgolide treatment of mouse blastocysts induced apoptosis, decreased cell number, hindered early postimplantation blastocyst development, and increased early-stage blastocyst death (Chan 2005). It was further revealed that ginkgolide B retarded the proliferation and development of mouse embryonic stem cells and blastocysts in-vitro and caused developmental injury in-vivo (Chan 2006). Treatment with 10 μM ginkgolide B caused resorption of post-implantation blastocysts and foetal weight loss. Further studies showed that treatment of mouse embryonic stem cells with a JNK-specific inhibitor reduced ginkgolide B-induced activation of both JNK and caspase-3, indicating that JNK activity is required for ginkgolide B-induced caspase activation (Hsuuw et al. 2009). Experiments using caspase-3 inhibitors and antisense oligonucleotides against p21-activated protein kinase 2 (PAK2) showed that caspase-3 activation was required for PAK2 activation and both of these activations were required for ginkgolide B-induced apoptosis in mouse embryonic stem cells. Fernandes et al. (2010) found that *Ginkgo biloba* extract, GBE treatment in pregnant Wistar rats, during the tubal transit and implantation period, produced no toxic effect on the maternal organism and did not cause embryonic death, growth retardation, and/or fetal malformations.

The results of studies by Yeh et al. (2008) showed that EGb 761 (especially at the dose of 50 mg/kg) enhanced the copulatory behaviour of male rats and suggested that the dopaminergic system, which regulated prolactin secretion, may be involved in the stimulatory effect of EGb 761. Yeh et al. (2010) found that chronic treatment of adult Long-Evans male rats with of *Ginkgo biloba* leaves (EGb 761) significantly decreased the non-contact erection latency, but increased the number of non contact erections and the dopamine levels in the bed nucleus of the stria terminalis and medial preoptic area in rats compared to the controls. The extract had been reported to significantly facilitate copulation in male rats.

Meston et al. (2008) assessed the long-term effects of *Ginkgo biloba* extract (GBE) on sexual function in 68 sexually dysfunctional women who were randomly assigned to 8 weeks treatment of either (1) GBE (300 mg/daily), (2) placebo, (3) sex therapy which focused on training women to attend to genital sensations, or (4) sex therapy plus GBE. They found that when combined with sex therapy, but not alone, long-term GBE treatment significantly enhanced sexual desire and contentment beyond placebo. Sex therapy alone significantly increased orgasm function compared with placebo. Long-term GBE administration did not significantly enhance arousal responses beyond placebo. Teaching women to focus on genital sensations during sex enhanced certain aspects of women's sexual functioning.

Allergy, Ginkgotoxin and Side Effects

Ginkgotoxin (4′-methoxypyridoxol) is a constituent not only of *G. biloba* seeds but also the leaves (Leistner and Drewke 2010). The seed of *G. biloba* is called "gin-nan" in Japan, and although it is known to be a medication and a food, it is also the cause of "gin-nan sitotoxism", a food poisoning that has been reported about 70 times with 27% lethality. These intoxications occurred between 1930 and 1996 with a rather high incidence during and after World War II when food was limited. At sub-lethal doses, symptoms of poisoning are eleptiformic seizures, unconsciousness, and paralysis of the legs. Higher doses may lead to convulsions and death in animal studies. In addition to the presumed, but not unequivocally proven, beneficial health effects of *G. biloba* products, these preparations also carry a clear potential for adverse effects, particularly in susceptible individuals. It is therefore important that the large number of *G. biloba* product users and their health care providers be made aware of these risks, in order to enable them to make informed decisions about the use of *G. biloba* preparations.

Studies showed that *Ginkgo biloba* extract could be teratogenic when given to pregnant mice (Zehra et al. 2010). Besides decrease in weight and crown-rump length of fetuses, other gross structural malformations included round shaped eye and orbits, syndactyly, malformed pinnae, nostrils, lips and jaws The results obtained substantiated the early finding that *Ginkgo biloba* could be teratogenic when given to pregnant mothers.

One-month combination treatment of extracts of ginseng and *Ginkgo biloba* (Naoweikang) significantly increased the level of acetylcholine in whole brain of amyloid beta-protein (Abeta) treated rats (Liu et al. 2004).

Ginkgo side effects and cautions include: possible increased risk of bleeding, gastrointestinal discomfort, nausea, vomiting, diarrhea, headaches, dizziness, heart palpitations, and restlessness (DeKosky et al. 2008; Snitz et al. 2009).

Results of studies by He et al. (2009) suggested *G. biloba* leaf extract had a dual action, both protective and disruptive, on red blood cells, depending on the presence of an exogenous stress. *G. biloba* leaf extract was found to have a protective role on red blood cells against Abeta- and hypotonic pressure-induced haemolysis, peroxide-induced lipoperoxidation, as well as glutathione consumption and methaemoglobin formation. On the other hand, *G. biloba* leaf extract also exhibited damage to red blood cells by increasing cell fragility, changing cellular morphology and inducing glutathione consumption and methaemoglobin formation, especially when applied at high doses. These anti- and pro-oxidative activities of polyphenolic substances were thought to be involved in the dual function of *G. biloba* leaf extract. The results of their study suggested that high doses of herbal remedies and dietary supplements could be toxic to red blood cells.

Traditional Medicinal Uses

Ginkgo has been used for centuries in traditional oriental medicine since ancient times. Beside the seed, the leaves have recently been found to posses medicinally active and health promoting compounds. In particular, the leaves have been used to stimulate the blood circulation and to tonify the brain, reduce lethargy, improve memory, mental fuzziness, and vertigo and to give an improved sense of well-being. They were also found effective in improving peripheral arterial circulation to brain, heart, limbs, ears and eyes, in reducing cardiovascular risks and in treating hearing disorders such as tinnitus where these result from poor circulation or damage by free radicals. The leaves contain ginkgolides, that inhibit allergic responses and so were used in treating disorders such as asthma, eye disorders cerebral insufficiency, senile dementia, Alzheimer's disease and senility.

Ginkgo seed is antibacterial, antifungal, astringent, anticancer, digestive, expectorant, tonic, sedative, antitussive, astringent and vermifuge. The kernel is macerated in vegetable oil for 100 days and then the pulp is used in the treatment of pulmonary tuberculosis, asthma, bronchitis coughs with thick phlegm urinary incontinence and stabilizes spermatogenesis. It is used to eliminate damp heat, dampness and stops vaginal discharge in yeast infection.

The roots are used as a cure for leucorrhoea and the bark yields tannin.

Other Uses

The Ginkgo tree is sacred to Buddhists and Taoists and is often planted near temples. The tree is also planted as avenue trees or as landscape trees in parks. The wood is light, soft and has insect repelling qualities and is used in furniture making. An oil from the seed is used as a fuel in lighting. A soap substitute is produced by mixing the pulp of the seed with oil or wine. The leaves can be used as pesticides.

Comments

Ginkgo biloba is one of the best known examples of a living fossil.

Selected References

Ahlemeyer B, Krieglstein J (2003a) Neuroprotective effects of *Ginkgo biloba* extract. Cell Mol Life Sci 60(9):1779–1792

Ahlemeyer B, Krieglstein J (2003b) Pharmacological studies supporting the therapeutic use of *Ginkgo biloba* extract for Alzheimer's disease. Pharmacopsychiatry 36(Suppl 1):S8–S14

Akgül T, Karagüzel E, Sürer H, Yağmurdur H, Ayyildiz A, Ustün H, Germiyanoğlu C (2009) *Ginkgo biloba* (EGB 761) affects apoptosis and nitric-oxide synthases in testicular torsion: an experimental study. Int Urol Nephrol 41(3):531–536

Augustin S, Rimbach G, Augustin K, Schliebs R, Wolffram S, Cermak R (2009) Effect of a short- and long-term treatment with *Ginkgo biloba* extract on amyloid precursor protein levels in a transgenic mouse model relevant to Alzheimer's disease. Arch Biochem Biophys 481(2):177–182

Bao YM, Liu AH, Zhang ZX, Li Y, Wang XY (2010) Effects of *Ginkgo biloba* extract 50 preconditioning on contents of inflammation-related cytokines in myocardium of rats with ischemia-reperfusion injury. Zhong Xi Yi Jie He Xue Bao 8(4):373–378, In Chinese

Bastianetto S, Quirion R (2002) EGb 761 is a neuroprotective agent against beta-amyloid toxicity. Cell Mol Biol (Noisy-le-Grand) 48(6):693–697

Bastianetto S, Ramassamy C, Doré S, Christen Y, Poirier J, Quirion R (2000a) The *Ginkgo biloba* extract (EGb 761) protects hippocampal neurons against cell death induced by beta-amyloid. Eur J Neurosci 12(6): 1882–1890

Bastianetto S, Zheng WH, Quirion R (2000b) The *Ginkgo biloba* extract (EGb 761) protects and rescues hippocampal cells against nitric oxide-induced toxicity: involvement of its flavonoid constituents and protein kinase C. J Neurochem 74(6):2268–2277

Bäurle P, Suter A, Wormstall H (2009) Safety and effectiveness of a traditional ginkgo fresh plant extract – results from a clinical trial. Forsch Komplementmed 16(3):156–161

Bernatoniene J, Kucinskaite A, Masteikova R, Kalveniene Z, Kasparaviciene G, Savickas A (2009) The comparison of anti-oxidative kinetics in vitro of the fluid extract from maidenhair tree, motherwort and hawthorn. Acta Pol Pharm 66(4):415–421

Bidon C, Lachuer J, Molgó J, Wierinckx A, de la Porte S, Pignol B, Christen Y, Meloni R, Koenig H, Biguet NF, Mallet J (2009) The extract of *Ginkgo biloba* EGb 761 reactivates a juvenile profile in the skeletal muscle of sarcopenic rats by transcriptional reprogramming. PLoS One 4(11):e7998

Biggs ML, Sorkin BC, Nahin RL, Kuller LH, Fitzpatrick AL (2010) *Ginkgo biloba* and risk of cancer: secondary analysis of the Ginkgo Evaluation of Memory (GEM) Study. Pharmacoepidemiol Drug Saf 19(7): 694–698

Birks J, Grimley Evans J (2009) *Ginkgo biloba* for cognitive impairment and dementia. Cochrane Database Syst Rev (1):CD003120

Blecharz-Klin K, Piechal A, Joniec I, Pyrzanowska J, Widy-Tyszkiewicz E (2009) Pharmacological and biochemical effects of *Ginkgo biloba* extract on learning, memory consolidation and motor activity in old rats. Acta Neurobiol Exp (Wars) 69(2):217–231

Bown D (1995) Encyclopaedia of herbs and their uses. Dorling Kindersley, London, 424 pp

Brinkley TE, Lovato JF, Arnold AM, Furberg CD, Kuller LH, Burke GL, Nahin RL, Lopez OL, Yasar S, Williamson JD (2010) Effect of *Ginkgo biloba* on blood pressure and incidence of hypertension in elderly men and women. Am J Hypertens 23(5): 528–533

Brunetti L, Orlando G, Menghini L, Ferrante C, Chiavaroli A, Vacca M (2006) *Ginkgo biloba* leaf extract reverses amyloid beta-peptide-induced isoprostane production in rat brain in vitro. Planta Med 72(14): 1296–1299

Burschka MA, Hassan HA, Reineke T, van Bebber L, Caird DM, Mösges R (2001) Effect of treatment with *Ginkgo biloba* extract EGb 761 (oral) on unilateral idiopathic sudden hearing loss in a prospective randomized double-blind study of 106 outpatients. Eur Arch Otorhinolaryngol 258(5):213–219

Bustanji Y, Al-Masri IM, Mohammad M, Hudaib M, Tawaha K, Tarazi H, Alkhatib HS (2011) Pancreatic lipase inhibition activity of trilactone terpenes of *Ginkgo biloba*. J Enzyme Inhib Med Chem 26(4): 453–459

Canis M, Olzowy B, Welz C, Suckfüll M, Stelter K (2011) Simvastatin and *Ginkgo biloba* in the treatment of subacute tinnitus: a retrospective study of 94 patients. Am J Otolaryngol 32(1):19–23

Cesarani A, Meloni F, Alpini D, Barozzi S, Verderio L, Boscani PF (1998) *Ginkgo biloba* (EGb 761) in the treatment of equilibrium disorders. Adv Ther 15(5): 291–304

Chan WH (2005) Ginkgolides induce apoptosis and decrease cell numbers in mouse blastocysts. Biochem Biophys Res Commun 338(2):1263–1267

Chan WH (2006) Ginkgolide B induces apoptosis and developmental injury in mouse embryonic stem cells and blastocysts. Hum Reprod 21(11):2985–2995

Chan WH, Hsuuw YD (2007) Dosage effects of ginkgolide b on ethanol-induced cell death in human hepatoma G2 cells. Ann N Y Acad Sci 1095:388–398

Chávez-Morales R, Jaramillo-Juárez F, Posadas Del Río F, Reyes-Romero M, Rodríguez-Vázquez M, Martínez-Saldaña M (2010) Protective effect of *Ginkgo biloba* extract on liver damage by a single dose of CCl4 in male rats. Hum Exp Toxicol [Epub ahead of print]

Chen J, Wang X, Zhu J, Shang Y, Guo X, Sun J (2004) Effects of *Ginkgo biloba* extract on number and activity of endothelial progenitor cells from peripheral blood. J Cardiovasc Pharmacol 43(3):347–352

Chen JS, Huang PH, Wang CH, Lin FY, Tsai HY, Wu TC, Lin SJ, Chen JW (2011) Nrf-2 mediated heme oxygenase-1 expression, an antioxidant-independent mechanism, contributes to anti-atherogenesis and vascular protective effects of *Ginkgo biloba* extract. Atherosclerosis 214(2):301–309

Chi JD, Xu LX (1998) Chemical constituents of flavonoids in the leaf of *Ginkgo biloba* L. Zhongguo Zhong Yao Za Zhi 23(1):40–41 (In Chinese)

Chi JD, He XF, Liu AR, Xu LX (1997) HPLC determination of six flavonoid constituents in *Ginkgo biloba* leaves. Yao Xue Xue Bao 32(8):625–628

Cho JH, Sung JH, Cho EH, Won CK, Lee HJ, Kim MO, Koh PO (2009) *Ginkgo biloba extract* (EGb 761) prevents ischemic brain injury by activation of the Akt signaling pathway. Am J Chin Med 37(3):547–555

Chuarienthong P, Lourith N, Leelapornpisid P (2010) Clinical efficacy comparison of anti-wrinkle cosmetics containing herbal flavonoids. Int J Cosmet Sci 32(2):99–106

Chung SY, Cheng FC, Lee MS, Lin JY, Lin MC, Wang MF (2006) *Ginkgo biloba* leaf extract (EGb761) combined with neuroprotective agents reduces the infarct volumes of gerbil ischemic brain. Am J Chin Med 34(5):803–817

Cieza A, Maier P, Pöppel E (2003) The effect of *Ginkgo biloba* on healthy elderly subjects. Fortschr Med Orig 121(1):5–10 (In German)

dal Belo SE, Gaspar LR, Maia Campos PM, Marty JP (2009) Skin penetration of epigallocatechin-3-gallate and quercetin from green tea and *Ginkgo biloba* extracts vehiculated in cosmetic formulations. Skin Pharmacol Physiol 22(6):299–304

D'Andrea G, Bussone G, Allais G, Aguggia M, D'Onofrio F, Maggio M, Moschiano F, Saracco MG, Terzi MG, Petretta V, Benedetto C (2009) Efficacy of Ginkgolide B in the prophylaxis of migraine with aura. Neurol Sci 30(Suppl 1):S121–S124

Dartigues JF, Carcaillon L, Helmer C, Lechevallier N, Lafuma A, Khoshnood B (2007) Vasodilators and nootropics as predictors of dementia and mortality in the PAQUID cohort. J Am Geriatr Soc 55(3): 395–399

DeFeudis FV, Drieu K (2000) *Ginkgo biloba* extract (EGb 761) and CNS functions: basic studies and clinical applications. Curr Drug Targets 1(1):25–58

DeKosky ST, Williamson JD, Fitzpatrick AL, Kronmal RA, Ives DG, Saxton JA, Lopez OL, Burke G, Carlson MC, Fried LP, Kuller LH, Robbins JA, Tracy RP, Woolard NF, Dunn L, Snitz BE, Nahin RL, Furberg CD (2008) *Ginkgo biloba* for prevention of dementia, a randomized controlled trial. JAMA 300(19):2253–2262

Deng YK, Wei F, An BQ (2009) Effect of *Ginkgo biloba* extract on plasma vascular endothelial growth factor during peri-operative period of cardiac surgery. Zhongguo Zhong Xi Yi Jie He Za Zhi 29(1):40–42 (In Chinese)

Ding F, Wang Y, Liu Y, Yao W, Yu S, Xu G (2009) Effects of *Ginkgo biloba* extract (EGB) on acute myocardial ischemia induced by isoproterenol in rats. Zhongguo Zhong Yao Za Zhi 34(7):900–903 (In Chinese)

Dodge HH, Zitzelberger T, Oken BS, Howieson D, Kaye J (2008) A randomized placebo-controlled trial of *Ginkgo biloba* for the prevention of cognitive decline. Neurology 70(19 Pt 2):1809–1817

Dong XX, Hui ZJ, Xiang WX, Rong ZF, Jian S, Zhu CJ (2007) *Ginkgo biloba* extract reduces endothelial progenitor-cell senescence through augmentation of telomerase activity. J Cardiovasc Pharmacol 49(2): 111–115

Drabaek H, Petersen JR, Wïnberg N, Hansen KF, Mehlsen J (1996) The effect of *Ginkgo biloba* extract in patients with intermittent claudication. Ugeskr Laeger 158(27):3928–3931 (In Danish)

Drew S (2001) Effectiveness of *Ginkgo biloba* in treating tinnitus: double blind, placebo controlled trial. Br Med J 322:73

Duan FR, Yuan BQ (2008) Effect of extracts of *Ginkgo biloba* leaf on learning-memory ability and NMDA receptor 1 expression in the hippocampus in rats with kindling-induced epilepsy. Zhongguo Dang Dai Er Ke Za Zhi 10(3):367–370 (In Chinese)

Duke JA, Ayensu ES (1985) Medicinal plants of china. Vols 1 &2. Reference Publications, Inc, Algonac, 705 pp

Eckert A, Keil U, Kressmann S, Schindowski K, Leutner S, Leutz S, Müller WE (2003) Effects of EGb 761 *Ginkgo biloba* extract on mitochondrial function and oxidative stress. Pharmacopsychiatry 36(Suppl 1): S15–S23

Ekman L, Fransson D, Claeson P, Johansson M (2009) Development of an alternative method for determination of terpene lactones in ginkgo dry extract. Pharmeur Bio Sci Notes 2009(1):67–71

Elsabagh S, Hartley DE, Ali O, Williamson EM, File SE (2005) Differential cognitive effects of *Ginkgo biloba* after acute and chronic treatment in healthy young volunteers. Differential cognitive effects of *Ginkgo biloba* after acute and chronic treatment in healthy young volunteers. Psychopharmacology (Berl) 179(2):437–446

Emir H, Kaptan ZK, Samim E, Sungu N, Ceylan K, Ustun H (2009) The preventive effect of *Ginkgo biloba* extract in myringosclerosis: study in rats. Otolaryngol Head Neck Surg 140(2):171–176

Erbil G, Ozbal S, Sonmez U, Pekcetin C, Tugyan K, Bagriyanik A, Ozogul C (2008) Neuroprotective effects of selenium and *Ginkgo biloba* extract (EGb761) against ischemia and reperfusion injury in rat brain. Neurosciences (Riyadh) 13(3):233–238

Erdogan H, Fadillioğlu E, Kotuk M, Iraz M, Tasdemir S, Oztas Y, Yildirim Z (2006) Effects of *Ginkgo biloba* on plasma oxidant injury induced by bleomycin in rats. Toxicol Ind Health 22(1):47–52

Esposito M, Carotenuto M (2011) Ginkgolide B complex efficacy for brief prophylaxis of migraine in school-aged children: an open-label study. Neurol Sci 32(1): 79–81

Fan L, Tao GY, Wang G, Chen Y, Zhang W, He YJ, Li Q, Lei HP, Jiang F, Hu DL, Huang YF, Zhou HH (2009) Effects of *Ginkgo biloba* extract ingestion on the pharmacokinetics of talinolol in healthy Chinese volunteers. Ann Pharmacother 43(5):944–949

Fehske CJ, Leuner K, Müller WE (2009) *Ginkgo biloba* extract (EGb761) influences monoaminergic neurotransmission via inhibition of NE uptake, but not MAO activity after chronic treatment. Pharmacol Res 60(1):68–73

Feng X, Zhang L, Zhu H (2009) Comparative anticancer and antioxidant activities of different ingredients of *Ginkgo biloba* extract (EGb 761). Planta Med 75(8):792–796

Fernandes ES, Pinto RM, de Paula Reis JE, de Oliveira Guerra M, Peters VM (2010) Effects of *Ginkgo biloba* extract on the embryo-fetal development in Wistar rats. Birth Defects Res B Dev Reprod Toxicol 89(2):133–138

Fontana L, Souza AS, Del Bel EA, Oliveira RM (2005) *Ginkgo biloba* leaf extract (EGb 761) enhances catalepsy induced by haloperidol and L-nitroarginine in mice. Braz J Med Biol Res 38(11):1649–1654

Fransen HP, Pelgrom SM, Stewart-Knox B, de Kaste D, Verhagen H (2010) Assessment of health claims, content, and safety of herbal supplements containing *Ginkgo biloba*. Food Nutr Res 54. doi: 10.3402/fnr.v54i0.5221

Fu LM, Li JT (2011) A systematic review of single Chinese herbs for Alzheimer's disease treatment. Evid Based Complement Alternat Med 2011:Article ID 640284

Fu L, Li N, Mill RR (1999) Ginkgoaceae engler. In: Wu ZY, Raven PH (eds) Flora of China. Vol. 4 (Cycadaceae through Fagaceae). Science Press, Beijing

Gardner CD, Taylor-Piliae RE, Kiazand A, Nicholus J, Rigby AJ, Farquhar JW (2008) Effect of *Ginkgo biloba* (EGb 761) on treadmill walking time among adults with peripheral artery disease: a randomized clinical trial. J Cardiopulm Rehabil Prev 28(4): 258–265

Grässel E (1992) Effect of *Ginkgo-biloba* extract on mental performance. Double-blind study using computerized measurement conditions in patients with cerebral insufficiency. Fortschr Med 110(5):73–76 (In German)

Gu X, Xie Z, Wang Q, Liu G, Qu Y, Zhang L, Pan J, Zhao G, Zhang Q (2009) Transcriptome profiling analysis reveals multiple modulatory effects of *Ginkgo biloba* extract. FEBS J 276(5):1450–1458

Gulec M, Iraz M, Yilmaz HR, Ozyurt H, Temel I (2006) The effects of *Ginkgo biloba* extract on tissue adenosine deaminase, xanthine oxidase, myeloperoxidase, malondialdehyde, and nitric oxide in cisplatin-induced nephrotoxicity. Toxicol Ind Health 22(3):125–130

Guo L, Mei N, Liao W, Chan PC, Fu PP (2010) *Ginkgo biloba* extract induces gene expression changes in xenobiotics metabolism and the Myc-centered network. OMICS 14(1):75–90

Haguenauer JP, Cantenot F, Koskas H, Pierart H (1986) Treatment of equilibrium disorders with *Ginkgo biloba* extract. A multicenter double-blind drug vs. placebo study. Presse Med 15(31):1569–1572 (In French)

Haines DD, Varga B, Bak I, Juhasz B, Mahmoud FF, Kalantari H, Gesztelyi R, Lekli I, Czompa A, Tosaki A (2011) Summative interaction between astaxanthin, *Ginkgo biloba* extract (EGb761) and vitamin C in suppression of respiratory inflammation: a comparison with ibuprofen. Phytother Res 5(1):128–136

Hamann KF (2007) Special ginkgo extract in cases of vertigo: a systematic review of randomised, double-blind, placebo controlled clinical examinations. HNO 55(4):258–263 (In German)

Hao Y, Sun Y, Xu C, Jiang X, Sun H, Wu Q, Yan C, Gu S (2009a) Improvement of contractile function in isolated cardiomyocytes from ischemia-reperfusion rats by ginkgolide B pretreatment. J Cardiovasc Pharmacol 54(1):3–9

Hao YR, Yang F, Cao J, Ou C, Zhang JJ, Yang C, Duan XX, Li Y, Su JJ (2009b) *Ginkgo biloba* extracts (EGb761)

inhibits aflatoxin B1-induced hepatocarcinogenesis in Wistar rats. Zhong Yao Cai 32(1):92–96 (In Chinese)

Hao CY, Wang FZ, Duan HB (2010) Myocardial damage and change of mitochondrial Mn-superoxide dismutase activity in craniocerebral injured rats and the interventing effect of *Ginkgo biloba* extract. Zhongguo Zhong Xi Yi Jie He Za Zhi 30(3):299–302 (In Chinese)

He J, Lin J, Li J, Zhang JH, Sun XM, Zeng CM (2009) Dual effects of *Ginkgo biloba* leaf extract on human red blood cells. Basic Clin Pharmacol Toxicol 104(2):138–144

Hedrick UP (ed) (1972) Sturtevant's edible plants of the world. Dover Publications, New York, 686 pp

Hilton M, Stuart E (2004) *Ginkgo biloba* for tinnitus. Cochrane Database Syst Rev (2):CD003852

Holstein N (2001) Ginkgo special extract EGb 761 in tinnitus therapy. An overview of results of completed clinical trials. Fortschr Med Orig 118(4):157–164 (In German)

Hori TR, Ridge RW, Tulecke W, Del Tredi P (1997) *Ginkgo biloba*, a global treasure: from biology to medicine. Springer, Tokyo, 427 pp

Hsu CL, Wu YL, Tang GJ, Lee TS, Kou YR (2009) *Ginkgo biloba* extract confers protection from cigarette smoke extract-induced apoptosis in human lung endothelial cells: role of heme oxygenase-1. Pulm Pharmacol Ther 22(4):286–296

Hsuuw YD, Kuo TF, Lee KH, Liu YC, Huang YT, Lai CY, Chan WH (2009) Ginkgolide B induces apoptosis via activation of JNK and p21-activated protein kinase 2 in mouse embryonic stem cells. Ann N Y Acad Sci 1171:501–508

Hu SY (2005) Food plants of China. The Chinese University Press, Hong Kong, 844 pp

Huang XS, Liu X, Chen DZ (2006) Effect of *Ginkgo biloba* extract preconditioning on discordant cardiac xenografts. Zhongguo Zhong Xi Yi Jie He Za Zhi 26(Suppl):108–111

Huang X, Whitworth CA, Rybak LP (2007) *Ginkgo biloba* extract (EGb 761) protects against cisplatin-induced ototoxicity in rats. Otol Neurotol 28(6):828–833

Ihl R, Tribanek M, Bachinskaya N (2010) Baseline neuropsychiatric symptoms are effect modifiers in *Ginkgo biloba* extract (EGb 761®) treatment of dementia with neuropsychiatric features. Retrospective data analyses of a randomized controlled trial. J Neurol Sci 299(1–2):184–187

Ihl R, Bachinskaya N, Korczyn AD, Vakhapova V, Tribanek M, Hoerr R, Napryeyenko O and GOTADAY Study Group (2011) Efficacy and safety of a once-daily formulation of *Ginkgo biloba* extract EGb 761 in dementia with neuropsychiatric features: a randomized controlled trial. Int J Geriatr Psychiatry (Epub ahead of print)

Ivetic V, Popovic M, Naumovic N, Radenkovic M, Vasic V (2008) The effect of *Ginkgo biloba* (EGb 761) on epileptic activity in rabbits. Molecules 13(10): 2509–2520

Janssen IM, Sturtz S, Skipka G, Zentner A, Garrido MV, Busse R (2010) *Ginkgo biloba* in Alzheimer's disease: a systematic review. Wien Med Wochenschr 160(21–22):539–546

Jiang XY, Qian LP, Zheng XJ, Xia YY, Jiang YB, Sun DY (2009) Interventional effect of *Ginkgo biloba* extract on the progression of gastric precancerous lesions in rats. J Dig Dis 10(4):293–299

Jin T, Chen YW, Howard A, Zhang YZ (2008) Purification, crystallization and initial crystallographic characterization of the *Ginkgo biloba* 11S seed globulin ginnacin. Acta Crystallogr Sect F Struct Biol Cryst Commun 64(Pt 7):641–644

Kang JH, Park KH, Kim YJ, Kim JH (2003) The neuroprotective effect of ginexin (*Ginko biloba* extract) on rat retinal ganglion cell in optic nerve crush injury model. Invest Ophthalmol Vis Sci 44:E-Abstract 130

Kang JW, Kim JH, Song K, Kim SH, Yoon JH, Kim KS (2010) Kaempferol and quercetin, components of *Ginkgo biloba* extract (EGb 761), induce caspase-3-dependent apoptosis in oral cavity cancer cells. Phytother Res 24(Suppl 1):S77–S82

Kaptan ZK, Emir H, Gocmen H, Uzunkulaoglu H, Karakas A, Senes M, Samim E (2008) *Ginkgo biloba*, a free oxygen radical scavenger, affects inflammatory mediators to diminish the occurrence of experimental myringosclerosis. Acta Otolaryngol 17:1–6

Kaschel R (2009) *Ginkgo biloba*: specificity of neuropsychological improvement–a selective review in search of differential effects. Hum Psychopharmacol 24(5): 345–370

Kasper S, Schubert H (2009) *Ginkgo biloba* extract EGb 761 in the treatment of dementia: evidence of efficacy and tolerability. Fortschr Neurol Psychiatr 77(9):494–506

Keles MS, Demirci N, Yildirim A, Atamanalp SS, Altinkaynak K (2008) Protective effects of N-acetylcysteine and *Ginkgo biloba* extract on ischaemia-reperfusion-induced hepatic DNA damage in rats. Clin Exp Med 8(4):193–198

Kennedy DO, Scholey AB, Wesnes KA (2000) Dose-dependent cognitive effects of acute administration of *Ginkgo biloba* to healthy young volunteers. Psychopharmacology 151(4):416–423

Kim SJ (2001) Effect of biflavones of *Ginkgo biloba* against UVB-induced cytotoxicity in vitro. J Dermatol 28(4):193–199

Kim SJ, Lim MH, Chun IK, Won YH (1997) Effects of flavonoids of *Ginkgo biloba* on proliferation of human skin fibroblast. Skin Pharmacol 10(4):200–205

Kim HK, Son KH, Chang HW, Kang SS, Kim HP (1999) Inhibition of rat adjuvant-induced arthritis by ginkgetin, a biflavone from *Ginkgo biloba* leaves. Planta Med 65(5):465–467

Kim MS, Lee JI, Lee WY, Kim SE (2004) Neuroprotective effect of *Ginkgo biloba* L. extract in a rat model of Parkinson's disease. Phytother Res 18(8):663–666

Kim TE, Kim BH, Kim J, Kim KP, Yi S, Shin HS, Lee YO, Lee KH, Shin SG, Jang IJ, Yu KS (2009) Comparison of the pharmacokinetics of ticlopidine between administration of a combined fixed-dose tablet formulation of ticlopidine 250 mg/ginkgo extract 80 mg, and concomitant administration of ticlopidine 250-mg and ginkgo extract

80-mg tablets: an open-label, two-treatment, single-dose, randomized-sequence crossover study in healthy Korean male volunteers. Clin Ther 31(10):2249–2257

Kim BH, Kim KP, Lim KS, Kim JR, Yoon SH, Cho JY, Lee YO, Lee KH, Jang IJ, Shin SG, Yu KS (2010) Influence of *Ginkgo biloba* extract on the pharmacodynamic effects and pharmacokinetic properties of ticlopidine: an open-label, randomized, two-period, two-treatment, two-sequence, single-dose crossover study in healthy Korean male volunteers. Clin Ther 32(2):380–390

Koh PO (2009) *Ginkgo biloba* extract (EGb 761) prevents increase of Bad-Bcl-XL interaction following cerebral ischemia. Am J Chin Med 37(5):867–876

Koh PO (2010) *Gingko biloba* extract (EGb 761) prevents cerebral ischemia-induced p70S6 kinase and S6 phosphorylation. Am J Chin Med 38(4):727–734

Koltermann A, Hartkorn A, Koch E, Fürst R, Vollmar AM, Zahler S (2007) *Ginkgo biloba* extract EGb 761 increases endothelial nitric oxide production in vitro and in vivo. Cell Mol Life Sci 64(13):1715–1722

Koltermann A, Liebl J, Fürst R, Ammer H, Vollmar AM, Zahler S (2009) *Ginkgo biloba* extract EGb(R)761 exerts anti-angiogenic effects via activation of tyrosine phosphatases. J Cell Mol Med 13(8B):2122–2130

Kotil K, Uyar R, Bilge T, Ton T, Kucukhuseyin C, Koldas M, Atay F (2008) Investigation of the dose-dependent antivasospasmic effect of *Ginkgo biloba* extract (EGb 761) in experimental subarachnoid hemorrhage. J Clin Neurosci 15(12):1382–1386

Koutsilieri E, Riederer P (2007) Excitotoxicity and new antiglutamatergic strategies in Parkinson's disease and Alzheimer's disease. Parkinsonism Relat Disord 13: S329–S331

Kuller LH, Ives DG, Fitzpatrick AL, Carlson MC, Mercado C, Lopez OL, Burke GL, Furberg CD, DeKosky ST, Ginkgo Evaluation of Memory Study Investigators (2010) Does *Ginkgo biloba* reduce the risk of cardiovascular events? Circ Cardiovasc Qual Outcomes 3(1):41–47

Kwak WJ, Han CK, Son KH, Chang HW, Kang SS, Park BK, Kim HP (2002) Effects of ginkgetin from *Ginkgo biloba* leaves on cyclooxygenases and in vivo skin inflammation. Planta Med 68(4):316–321

Kwon SH, Nam JI, Kim SH, Kim JH, Yoon JH, Kim KS (2009) Kaempferol and quercetin, essential ingredients in *Ginkgo biloba* extract, inhibit interleukin-1beta-induced MUC5AC gene expression in human airway epithelial cells. Phytother Res 223(12): 1708–1712

Lau AJ, Chang TK (2009) Inhibition of human CYP2B6-catalyzed bupropion hydroxylation by *Ginkgo biloba* extract: effect of terpene trilactones and flavonols. Drug Metab Dispos 37(9):1931–1937

Lee TF, Chen CF, Wang LC (2004) Effect of ginkgolides on beta-amyloid-suppressed acetylocholine release from rat hippocampal slices. Phytother Res 18(7): 556–560

Leistner E, Drewke C (2010) *Ginkgo biloba* and ginkgotoxin. J Nat Prod 73(1):86–92

Li K, Yao P, Zhou SL, Song FF (2005) Protective effects of extract of *Ginkgo biloba* against ethanol-induced oxidative injury in rat testes. Wei Sheng Yan Jiu 34(5):559–562 (In Chinese)

Li L, Stanton JD, Tolson AH, Luo Y, Wang H (2009) Bioactive terpenoids and flavonoids from *Ginkgo biloba* extract induce the expression of hepatic drug-metabolizing enzymes through pregnane X receptor, constitutive androstane receptor, and aryl hydrocarbon receptor-mediated pathways. Pharm Res 26(4): 872–882

Li X, Hu Y, Fu Y, Ying Y, Chen G (2010a) Effect of *Ginkgo biloba* extract on glucose uptake of diaphragm in diabetic rats. Zhongguo Zhong Yao Za Zhi 35(3):356–359 (In Chinese)

Li XS, Fu YQ, Zhou B, Hu Y, Chen GR (2010b) Effects of *Ginkgo biloba* extraction on contraction capacity of diaphragm from type 2 diabetic rats. Zhongguo Ying Yong Sheng Li Xue Za Zhi 26(2):249–251 (In Chinese)

Li S, Luo J, Wang X, Guan BC, Sun CK (2011) Effects of *Ginkgo biloba* extracts on NMDA-activated currents in acutely isolated hippocampal neurons of the rat. Phytother Res 25(1):137–141

Lim H, Son KH, Chang HW, Kang SS, Kim HP (2006) Effects of anti-inflammatory biflavonoid, ginkgetin, on chronic skin inflammation. Biol Pharm Bull 29(5):1046–1049

Lin H, Wang H, Chen D, Li J, Gu Y (2008) Effect of extract of *Ginkgo biloba* leaves on proliferation of SCs cultured in vitro. Zhongguo Xiu Fu Chong Jian Wai Ke Za Zhi 22(9):1047–1050 (In Chinese)

Liu JX, Cong WH, Xu L, Wang JN (2004) Effect of combination of extracts of ginseng and *Ginkgo biloba* on acetylcholine in amyloid beta-protein-treated rats determined by an improved HPLC. Acta Pharmacol Sin 25(9):1118–1123

Liu F, Zhang J, Yu S, Wang R, Wang B, Lai L, Yin H, Liu G (2008) Inhibitory effect of *Ginkgo biloba* extract on hyperhomocysteinemia-induced intimal thickening in rabbit abdominal aorta after balloon injury. Phytother Res 22(4):506–510

Liu HJ, Wang XL, Zhang L, Qiu Y, Li TJ, Li R, Wu MC, Wei LX, Rui YC (2009a) Inhibitions of vascular endothelial growth factor expression and foam cell formation by EGb 761, a special extract of *Ginkgo biloba*, in oxidatively modified low-density lipoprotein-induced human THP-1 monocytes cells. Phytomedicine 16(2–3):138–145

Liu R, Li F, Liang R, Wang L, Wang Y (2009b) Influence of Ginkgo extract on metabolism of simvastatin. Zhongguo Zhong Yao Za Zhi 34(12):1578–1581 (In Chinese)

Liu XP, Luan JJ, Goldring CE (2009c) Comparison of the antioxidant activity amongst *Ginkgo biloba* extract and its main components. Zhong Yao Cai 32(5):736–740

Longpré F, Garneau P, Christen Y, Ramassamy C (2006) Protection by EGb 761 against beta-amyloid-induced neurotoxicity: involvement of NF-kappaB, SIRT1, and

MAPKs pathways and inhibition of amyloid fibril formation. Free Radic Biol Med 41(12):1781–1794

Lucinda LM, de Oliveira TT, Salvador PA, Peters VM, Reis JE, Guerra Mde O (2010a) Radiographic evidence of mandibular osteoporosis improvement in Wistar rats treated with *Ginkgo biloba*. Phytother Res 24(2):264–267

Lucinda LM, Vieira BJ, Oliveira TT, Sá RC, Peters VM, Reis JE, Guerra MO (2010b) Evidences of osteoporosis improvement in Wistar rats treated with *Ginkgo biloba* extract: a histomorphometric study of mandible and femur. Fitoterapia 81(8):982–987

Luo Y (2006) Alzheimer's disease, the nematode *Caenorhabditis elegans*, and *Ginkgo biloba* leaf extract. Life Sci 78(18):2066–2072

Luo Y, Smith JV, Paramasivam V, Burdick A, Curry KJ, Buford JP, Khan I, Netzer WJ, Xu H, Butko P (2002) Inhibition of amyloid-beta aggregation and caspase-3 activation by the *Ginkgo biloba* extract EGb761. Proc Natl Acad Sci USA 99(19):12197–12202

Lyon MR, Cline JC, Totosy de Zepetnek J, Shan JJ, Pang P, Benishin C (2001) Effect of the herbal extract combination *Panax quinquefolium* and *Ginkgo biloba* on attention-deficit hyperactivity disorder: a pilot study. J Psychiatry Neurosci 26(3):221–228

Ma K, Xu L, Zhang H, Zhang S, Pu M, Jonas JB (2009) Dosage dependence of the effect of *Ginkgo biloba* on the rat retinal ganglion cell survival after optic nerve crush. Eye 23:1598–1604

Ma K, Xu L, Zhang H, Zhang S, Pu M, Jonas JB (2010) The effect of *Ginkgo biloba* on the rat retinal ganglion cell survival in the optic nerve crush model. Acta Ophthalmol 88(5):553–557

Mahadevan S, Park Y (2008) Multifaceted therapeutic benefits of *Ginkgo biloba* L.: chemistry, efficacy, safety, and uses. J Food Sci 73(1):R14–R19

Major RT (1967) The *Ginkgo*, the most ancient living tree. Science 157:1270–1273

Martin R, Mozet C, Martin H, Welt K, Engel C, Fitzl G (2011) Effect of *Ginkgo biloba* extract (EGb 761) on parameters of oxidative stress in different regions of aging rat brain after acute hypoxia. Aging Clin Exp Res. (Epub ahead of print)

Meston CM, Rellini AH, Telch MJ (2008) Short- and long-term effects of *Ginkgo biloba* extract on sexual dysfunction in women. Arch Sex Behav 37(4): 530–547

Miyakawa T, Sawano Y, Miyazono K, Hatano K, Tanokura M (2007) Crystallization and preliminary X-ray analysis of ginkbilobin-2 from *Ginkgo biloba* seeds: a novel antifungal protein with homology to the extracellular domain of plant cysteine-rich receptor-like kinases. Acta Crystallogr Sect F Struct Biol Cryst Commun 63(Pt 9):737–739

Mohamed ME, Frye RF (2010) Inhibition of intestinal and hepatic glucuronidation of mycophenolic acid by *Ginkgo biloba* extract and flavonoids. Drug Metab Dispos 38(2):270–275

Morgenstern C, Biermann E (2002) The efficacy of Ginkgo special extract EGb 761 in patients with tinnitus. Int J Clin Pharmacol Ther 40(5):188–197

Mukai R, Satsu H, Shimizu M, Ashida H (2009) Inhibition of p-glycoprotein enhances the suppressive effect of kaempferol on transformation of the aryl hydrocarbon receptor. Biosci Biotechnol Biochem 73(7):1635–1639

Napryeyenko O, Sonnik G, Tartakovsky I (2009) Efficacy and tolerability of *Ginkgo biloba* extract EGb 761 by type of dementia: analyses of a randomised controlled trial. J Neurol Sci 283(1–2):224–229

Nathan PJ, Ricketts E, Wesnes K, Mrazek L, Greville W, Stough C (2002) The acute nootropic effects of *Ginkgo biloba* in healthy older human subjects: a preliminary investigation. Hum Psychopharmacol 17(1):45–49

Nevado J, Sanz R, Sánchez-Rodríguez C, García-Berrocal JR, Martín-Sanz E, González-García JA, Esteban-Sánchez J, Ramírez-Camacho R (2010) *Ginkgo biloba* extract (EGb761) protects against aging-related caspase-mediated apoptosis in rat cochlea. Acta Otolaryngol 130(10):1101–1112

Nicolaï SP, Gerardu VC, Kruidenier LM, Prins MH, Teijink JA (2010) From the Cochrane library: *Ginkgo biloba* for intermittent claudication. Vasa 39(2): 153–158

Niederhofer H (2009) First preliminary results of an observation of *Ginkgo biloba* treating patients with autistic disorder. Phytother Res 23(11):1645–1646

Niederhofer H (2010) *Ginkgo biloba* treating patients with attention-deficit disorder. Phytother Res 24(1): 26–27

O'Brien JT, Burns A (2011) Clinical practice with anti-dementia drugs: a revised (second) consensus statement from the British Association for Psychopharmacology. J Psychopharmacol 25(8):997–1019

Oliveira DR, Sanada PF, Saragossa Filho AC, Innocenti LR, Oler G, Cerutti JM, Cerutti SM (2009) Neuromodulatory property of standardized extract *Ginkgo biloba* L. (EGb 761) on memory: behavioral and molecular evidence. Brain Res 1269:68–89

Ozgoli G, Selselei EA, Mojab F, Majd HA (2009) A randomized, placebo-controlled trial of *Ginkgo biloba* L. in treatment of premenstrual syndrome. J Altern Complement Med 15(8):845–851

Parsad D, Pandhi R, Juneja A (2003) Effectiveness of oral *Ginkgo bilob*a in treating limited, slowly spreading vitiligo. Clin Exp Dermatol 28(3):285–287

Patel BP, Hamadeh MJ (2009) Nutritional and exercise-based interventions in the treatment of amyotrophic lateral sclerosis. Clin Nutr 28(6):604–617

Pérez CM (2009) Can *Ginkgo biloba* combat diseases? P R Health Sci J 28(1):66–74

Peters H, Kieser M, Hölscher U (1998) Demonstration of the efficacy of *Ginkgo biloba* special extract EGb 761 on intermittent claudication–a placebo-controlled, double-blind multicenter trial. Vasa 27(2): 106–110

Pinto Mda S, Kwon YI, Apostolidis E, Lajolo FM, Genovese MI, Shetty K (2009) Potential of *Ginkgo biloba* L. leaves in the management of hyperglycemia and hypertension using in vitro models. Bioresour Technol 100(24):6599–6609

Pittler MH, Ernst E (2000) *Ginkgo biloba* extract for the treatment of intermittent claudication: a meta-analysis of randomized trials. Am J Med 108:276–281

Qaâdan F, Nahrstedt A, Schmidt M, Mansoor K (2010) Polyphenols from *Ginkgo biloba*. Sci Pharm 78(4): 897–907

Qa'dan F, Mansoor K, Al-Adham I, Schmidt M, Nahrstedt A (2011) Proanthocyanidins from *Ginkgo biloba* leaf extract and their radical scavenging activity. Pharm Biol 49(5):471–476

Quaranta L, Bettelli S, Uva MG, Semeraro F, Turano R, Gandolfo E (2003) Effect of *Ginkgo biloba* extract on preexisting visual field damage in normal tension glaucoma. Ophthalmology 110(2):359–362

Rajaraman G, Yang G, Chen J, Chang TK (2009) Modulation of CYP1B1 and CYP1A1 gene expression and activation of aryl hydrocarbon receptor by *Ginkgo biloba* extract in MCF-10A human mammary epithelial cells. Can J Physiol Pharmacol 87(9):674–683

Rangel-Ordóñez L, Nöldner M, Schubert-Zsilavecz M, Wurglics M (2010) Plasma levels and distribution of flavonoids in rat brain after single and repeated doses of standardized *Ginkgo biloba* extract EGb 761®. Planta Med 76(15):1683–1690

Reisser CH, Weidauer H (2001) *Ginkgo biloba* extract EGb 761 or pentoxifylline for the treatment of sudden deafness: a randomized, reference-controlled, double-blind study. Acta Otolaryngol 121(5):579–584

Rhein V, Giese M, Baysang G, Meier F, Rao S, Schulz KL, Hamburger M, Eckert A (2010) *Ginkgo biloba* extract ameliorates oxidative phosphorylation performance and rescues abeta-induced failure. PLoS One 5(8):e12359

Rojas P, Montes S, Serrano-García N, Rojas-Castañeda J (2009) Effect of EGb761 supplementation on the content of copper in mouse brain in an animal model of Parkinson's disease. Nutrition 25(4):482–485

Russo V, Stella A, Appezzati L, Barone A, Stagni E, Roszkowska A, Delle Noci N (2009) Clinical efficacy of a *Ginkgo biloba* extract in the topical treatment of allergic conjunctivitis. Eur J Ophthalmol 19(3):331–336

Ryu KH, Han HY, Lee SY, Jeon SD, Im GJ, Lee BY, Kim K, Lim KM, Chung JH (2009) *Ginkgo biloba* extract enhances antiplatelet and antithrombotic effects of cilostazol without prolongation of bleeding time. Thromb Res 124(3):328–334

Sakakibara H, Ishida K, Grundmann O, Nakajima J, Seo S, Butterweck V, Minami Y, Saito S, Kawai Y, Nakaya Y, Terao J (2006) Antidepressant effect of extracts from *Ginkgo biloba* leaves in behavioral models. Biol Pharm Bull 29(8):1767–1770

Saleem S, Zhuang H, Biswal S, Christen Y, Doré S (2008) *Ginkgo biloba* extract neuroprotective action is dependent on heme oxygenase 1 in ischemic reperfusion brain injury. Stroke 39:3389–3396

Salehi B, Imani R, Mohammadi MR, Fallah J, Mohammadi M, Ghanizadeh A, Tasviechi AA, Vossoughi A, Rezazadeh SA, Akhondzadeh S (2010) *Ginkgo biloba* for attention-deficit/hyperactivity disorder in children and adolescents: a double blind, randomized controlled trial. Prog Neuropsychopharmacol Biol Psychiatry 34(1):76–80

Satvat E, Mallet PE (2009) Chronic administration of a *Ginkgo biloba* leaf extract facilitates acquisition but not performance of a working memory task. Psychopharmacology (Berl) 202(1–3):173–185

Schneider R, Welt K, Aust W, Löster H, Fitzl G (2009) Cardiac ischemia and reperfusion in spontaneously diabetic rats with and without application of EGb 761: II. Interstitium and microvasculature. Histol Histopathol 24(5):587–598

Seo BS, Lee HJ, Mo JH, Lee CH, Rhee CS, Kim JW (2009) Treatment of postviral olfactory loss with glucocorticoids, *Ginkgo biloba*, and mometasone nasal spray. Arch Otolaryngol Head Neck Surg 135(10):1000–1004

Shi C, Zhao L, Zhu B, Li Q, Yew DT, Yao Z, Xu J (2009) Protective effects of *Ginkgo biloba* extract (EGb761) and its constituents quercetin and ginkgolide B against beta-amyloid peptide-induced toxicity in SH-SY5Y cells. Chem Biol Interact 181(1):115–123

Shi C, Fang L, Yew DT, Yao Z, Xu J (2010a) *Ginkgo biloba* extract EGb761 protects against mitochondrial dysfunction in platelets and hippocampi in ovariectomized rats. Platelets 21(1):53–59

Shi C, Wu F, Yew DT, Xu J, Zhu Y (2010b) Bilobalide prevents apoptosis through activation of the PI3K/Akt pathway in SH-SY5Y cells. Apoptosis 15(6):715–727

Shi C, Xiao S, Liu J, Guo K, Wu F, Yew DT, Xu J (2010c) *Ginkgo biloba* extract EGb761 protects against aging-associated mitochondrial dysfunction in platelets and hippocampi of SAMP8 mice. Platelets 21(5):373–379

Shiao NH, Chan WH (2009) Injury effects of ginkgolide B on maturation of mouse oocytes, fertilization, and fetal development in vitro and in vivo. Toxicol Lett 88(1):63–69

Siegel G, Schäfer P, Winkler K, Malmsten M (2007) *Ginkgo biloba* (EGb 761) in arteriosclerosis prophylaxis. Wien Med Wochenschr 157(13–14):288–294

Singh B, Kaur P, Gopichand Singh RD, Ahuja PS (2008) Biology and chemistry of *Ginkgo biloba*. Fitoterapia 79(6):401–418

Singh V, Singh SP, Chan K (2010) Review and meta-analysis of usage of ginkgo as an adjunct therapy in chronic schizophrenia. Int J Neuropsychopharmacol 13(2): 257–271

Smith JV, Luo Y (2003) Elevation of oxidative free radicals in Alzheimer's disease models can be attenuated by *Ginkgo biloba* extract EGb 761. J Alzheimers Dis 5(4):287–300

Smith PF, Maclennan K, Darlington CL (1996) The neuroprotective properties of the *Ginkgo biloba* leaf: a review of the possible relationship to platelet-activating factor (PAF). J Ethnopharmacol 50(3):131–139

Smith PF, Zheng Y, Darlington CL (2005) *Ginkgo biloba* extracts for tinnitus: more hype than hope? J Ethnopharmacol 100(1–2):95–99

Snitz BE, O'Meara ES, Carlson MC, Arnold AM, Ives DG, Rapp SR, Saxton J, Lopez OL, Dunn LO, Sink KM, DeKosky ST, Ginkgo Evaluation of Memory

(GEM) Study Investigators (2009) *Ginkgo biloba* for preventing cognitive decline in older adults: a randomized trial. JAMA 302(24):2663–2670

Solomon PR, Adams F, Silver A, Zimmer J, DeVeaux R (2002) Ginkgo for memory enhancement. A randomized controlled trial. JAMA 288:835–840

Son JK, Son MJ, Lee E, Moon TC, Son KH, Kim CH, Kim HP, Kang SS, Chang HW (2005) Ginkgetin, a biflavone *from Ginkgo biloba* leaves, inhibits cyclooxygenases-2 and 5-lipoxygenase in mouse bone marrow-derived mast cells. Biol Pharm Bull 28(12): 2181–2184

Stackman RW, Eckenstein F, Frei B, Kulhanek D, Nowlin J, Quinn JF (2003) Prevention of age-related spatial memory deficits in a transgenic mouse model of Alzheimer's disease by chronic *Ginkgo biloba* treatment. Exp Neurol 184(1):510–520

Stoll S, Scheuer K, Pohl O, Müller WE (1996) *Ginkgo biloba* extract (EGb 761) independently improves changes in passive avoidance learning and brain membrane fluidity in the aging mouse. Pharmacopsychiatry 29(4):144–149

Szczurko O, Shear N, Taddio A, Boon H (2011) *Ginkgo biloba* for the treatment of Vitiligo vulgaris: an open label pilot clinical trial. BMC Complement Altern Med 11(1):21

Taki Y, Yamazaki Y, Shimura F, Yamada S, Umegaki K (2009) Time-dependent induction of hepatic cytochrome P450 enzyme activity and mRNA expression by bilobalide in rats. J Pharmacol Sci 109(3): 459–462

Tang F, Nag S, Shiu SY, Pang SF (2002) The effects of melatonin and *Ginkgo biloba* extract on memory loss and choline acetyltransferase activities in the brain of rats infused intracerebroventricularly with beta-amyloid 1–40. Life Sci 71(22):2625–2631

Tang D, Zhang Z, Gao Y, Wei Y, Han L (2009) Protective effects of serum containing *Ginkgo biloba* extract on glomerulosclerosis in rat mesangial cells. J Ethnopharmacol 124(1):26–33

Tang D, Yang D, Tang A, Gao Y, Jiang X, Mou J, Yin X (2010) Simultaneous chemical fingerprint and quantitative analysis of *Ginkgo biloba* extract by HPLC-DAD. Anal Bioanal Chem 396(8):3087–3095

Tchantchou F, Xu Y, Wu Y, Christen Y, Luo Y (2007) EGb 761 enhances adult hippocampal neurogenesis and phosphorylation of CREB in transgenic mouse model of Alzheimer's disease. FASEB J 21(10):2400–2408

Tchantchou F, Lacor PN, Cao Z, Lao L, Hou Y, Cui C, Klein WL, Luo Y (2009) Stimulation of neurogenesis and synaptogenesis by bilobalide and quercetin via common final pathway in hippocampal neurons. J Alzheimers Dis 18(4):787–798

Trompezinski S, Bonneville M, Pernet I, Denis A, Schmitt D, Viac J (2010) *Ginkgo biloba* extract reduces VEGF and CXCL-8/IL-8 levels in keratinocytes with cumulative effect with epigallocatechin-3-gallate. Arch Dermatol Res 302(3):183–189

U.S. Department of Agriculture, Agricultural Research Service (2010) USDA national nutrient database for standard reference, release 23. Nutrient Data Laboratory Home Page. http://www.ars.usda.gov/ba/bhnrc/ndl

Usai S, Grazzi L, Andrasik F, Bussone G (2010) An innovative approach for migraine prevention in young age: a preliminary study. Neurol Sci 31(Suppl 1): S181–S183

van Beek TA (2002) Chemical analysis of *Ginkgo biloba* leaves and extracts. J Chromatogr A 967(1):21–55

van Beek TA, Montoro P (2009) Chemical analysis and quality control of *Ginkgo biloba* leaves, extracts, and phytopharmaceuticals. J Chromatogr A 1216(11): 2002–2032

Vilar JB, Leite KR, Chen CL (2009) Antimutagenicity protection of *Ginkgo biloba* extract (Egb 761) against mitomycin C and cyclophosphamide in mouse bone marrow. Genet Mol Res 8(1):328–333

Vitolo O, Gong B, Cao Z, Ishii H, Jaracz S, Nakanishi K, Arancio O, Dzyuba SV, Lefort R, Shelanski M (2009) Protection against beta-amyloid induced abnormal synaptic function and cell death by Ginkgolide. J Neurobiol Aging 30(2):257–265

Wadsworth TL, McDonald TL, Koop DR (2001) Effects of *Ginkgo biloba* extract (EGb 761) and quercetin on lipopolysaccharide-induced signaling pathways involved in the release of tumor necrosis factor-alpha. Biochem Pharmacol 62(7):963–974

Walesiuk A, Braszko JJ (2010) Gingkoselect alleviates chronic corticosterone-induced spatial memory deficits in rats. Fitoterapia 81(1):25–29

Wang XX, Shang YP, Chen JZ, Zhu JH, Guo XG, Sun J (2004) Effects of *Ginkgo biloba* extract on number and activity of endothelial progenitor cells from peripheral blood. Yao Xue Xue Bao 39(8):656–660 (In Chinese)

Wang F, Jiang K, Li Z (2007a) Purification and identification of genistein in *Ginkgo biloba* leaf extract. Se Pu 25(4):509–513 (In Chinese)

Wang J, Zhou S, Bronks R, Graham J, Myers S (2007b) Supervised exercise training combined with *Ginkgo biloba* treatment for patients with peripheral arterial disease. Clin Rehabil 21(7):579–586

Wang D, Li X, Yu X, Shi Y, Yin L (2009a) Expression of tissue transglutaminase on renal interstitial fibrosis rats and intervention of GBE. Zhongguo Zhong Yao Za Zhi 34(9):1133–1136 (In Chinese)

Wang L, Shen T, Zhou CF, Yu JF, Zhu QX (2009b) Changes of nitric oxide after trichloroethylene irritation in hairless mice skin and protection of *Ginkgo biloba* extract and vitamin E. Zhonghua Lao Dong Wei Sheng Zhi Ye Bing Za Zhi 27(4):207–210 (In Chinese)

Wang BS, Wang H, Song YY, Qi H, Rong ZX, Wang BS, Zhang L, Chen HZ (2010a) Effectiveness of standardized *Ginkgo biloba* extract on cognitive symptoms of dementia with a six-month treatment: a bivariate random effect meta-analysis. Pharmacopsyhiatry 43(3): 86–91

Wang CG, Dai Y, Li DL, Ma KY (2010b) *Ginkgo biloba* leaf extract action in scavenging free radicals and reducing mutagenicity and toxicity of cigarette smoke in vivo. J Environ Sci Health A Tox Hazard Subst Environ Eng 45(4):498–505

Wang X, Zhang Z, Liu A (2010c) Effect of GBE50 on experimental arrhythmias. Zhongguo Zhong Yao Za Zhi 35(2):199–203 (In Chinese)

Weinmann S, Roll S, Schwarzbach C, Vauth C, Willich SN (2010) Effects of *Ginkgo biloba* in dementia: systematic review and meta-analysis. BMC Geriatr 10:14

Weng XJ, Chen LL, Zhang HQ (2008) Effect of total flavonoid in leaves of *Ginkgo biloba* on the apoptosis of eosinophil in broncho alveloar lavage fluid. Yao Xue Xue Bao 43(5):480–483 (In Chinese)

Winter JC (1998) The effects of an extract of *Ginkgo biloba*, EGb 761, on cognitive behavior and longevity in the rat. Physiol Behav 63(3):425–433

Witte S, Anadere I, Walitza E (1992) Improvement of hemorheology with *Ginkgo biloba* extract. Decreasing a cardiovascular risk factor. Fortschr Med 110(13):247–250 (In German)

Wu Y, Gu YM (2007) Effect of EGb and quercetin on culture neonatal rat cardiomyocytes hypertrophy and mechanism. Zhongguo Ying Yong Sheng Li Xue Za Zhi 23(2):138–142 (In Chinese)

Wu XY, Yang LQ, Chen J (2003) Determination in *Ginkgo biloba* extract and its preparations by high performance liquid chromatography. Yao Xue Xue Bao 38(11):846–849 (In Chinese)

Wu Y, Wu Z, Butko P, Christen Y, Lambert MP, Klein WL, Link CD, Luo Y (2006) Amyloid-beta-induced pathological behaviors are suppressed by *Ginkgo biloba* extract EGb 761 and ginkgolides in transgenic *Caenorhabditis elegans*. J Neurosci 26(50): 13102–13113

Wu ZM, Yin XX, Ji L, Gao YY, Pan YM, Lu Q, Wang JY (2008) *Ginkgo biloba* extract prevents against apoptosis induced by high glucose in human lens epithelial cells. Acta Pharmacol Sin 29(9):1042–1050

Xie ZG, Wu XW, Zhuang CR, Chen F, Wang Z, Wang YK, Hua X (2009a) Protective effects of *Ginkgo biloba* extract on morphology and function of retinal ganglion cells after optic nerve transection in guinea pigs. Zhong Xi Yi Jie He Xue Bao 7(10):940–946 (In Chinese)

Xie ZQ, Liang G, Zhang L, Wang Q, Qu Y, Gao Y, Lin LB, Ye S, Zhang J, Wang H, Zhao GP, Zhang QH (2009b) Molecular mechanisms underlying the cholesterol-lowering effect of *Ginkgo biloba* extract in hepatocytes: a comparative study with lovastatin. Acta Pharmacol Sin 30(9):1262–1275

Xiong GZ, Ye KH, Huang JL, Ye CL (2009) Effects of extracts of *Ginkgo biloba* leaves on apelin concentration and expressions in plasma and myocardium of rats with myocardial ischemia injury. Zhong Yao Cai 32(8):1238–1241 (In Chinese)

Xu Z, Wu W, Lan T, Zhang X (2009) Protective effects of extract of *Ginkgo biloba* on adriamycin-induced heart failure and its mechanism: role of ghrelin peptide. Zhongguo Zhong Yao Za Zhi 34(21):2786–2789 (In Chinese)

Xu J, Sun C, Ma H, Wang L, Zhang J, Zhang Y, Wu L (2010) Neuroprotective effect of extract of *Ginkgo biloba* against excitotoxicty compared with ginkgolide B in neuron cell of rat. Zhongguo Zhong Yao Za Zhi 35(1):114–117

Yan FL, Zheng Y, Zhao FD (2008) Effects of *Ginkgo biloba* extract EGb761 on expression of RAGE and LRP-1 in cerebral microvascular endothelial cells under chronic hypoxia and hypoglycemia. Acta Neuropathol 116(5):529–535

Yancheva S, Ihl R, Nikolova G, Panayotov P, Schlaefke S, Hoerr R, GINDON Study Group (2009) *Ginkgo biloba* extract EGb 761(R), donepezil or both combined in the treatment of Alzheimer's disease with neuropsychiatric features: a randomised, double-blind, exploratory trial. Aging Ment Health 13(2):183–190

Yang LQ, Wu XY, Chen J (2002) Determination of ginkgolic acids by high performance liquid chromatography. Yao Xue Xue Bao 37(7):555–558 (In Chinese)

Yang YL, Su YW, Ng MC, Chao PK, Tung LC, Lu KT (2007) Extract of *Ginkgo biloba* EGb761 facilitates extinction of conditioned fear measured by fear-potentiated startle. Neuropsychopharmacology 32(2): 332–342

Yang TH, Young YH, Liu SH (2011) EGb 761 (*Ginkgo biloba*) protects cochlear hair cells against ototoxicity induced by gentamicin via reducing reactive oxygen species and nitric oxide-related apoptosis. J Nutr Biochem 22(9):886–894

Yao Z, Drieu K, Papadopoulos V (2001) The *Ginkgo biloba* extract EGb 761 rescues the PC12 neuronal cells from beta-amyloid-induced cell death by inhibiting the formation of beta-amyloid-derived diffusible neurotoxic ligands. Brain Res 889(1–2):181–190

Yao ZX, Han Z, Drieu K, Papadopoulos V (2004) *Ginkgo biloba* extract (Egb 761) inhibits beta-amyloid production by lowering free cholesterol levels. J Nutr Biochem 15(12):749–756

Yao P, Liu LG, Jia WB, Song FF, Zhou S, Zhang X, Sun X (2005) Effect of flavonoids of *Ginkgo biloba* on anti-oxidizing system of mice after acute alcohol administration. Wei Sheng Yan Jiu 34(3):303–306 (In Chinese)

Yapar K, Cavuşoğlu K, Oruç E, Yalçin E (2010) Protective role of *Ginkgo biloba* against hepatotoxicity and nephrotoxicity in uranium-treated mice. J Med Food 13(1):179–188

Yeh KY, Pu HF, Kaphle K, Lin SF, Wu LS, Lin JH, Tsai YF (2008) *Ginkgo biloba* extract enhances male copulatory behavior and reduces serum prolactin levels in rats. Horm Behav 53(1):225–231

Yeh YC, Liu TJ, Wang LC, Lee HW, Ting CT, Lee WL, Hung CJ, Wang KY, Lai HC, Lai HC (2009) A standardized extract of *Ginkgo biloba* suppresses doxorubicin-induced oxidative stress and p53-mediated mitochondrial apoptosis in rat testes. Br J Pharmacol 156(1):48–61

Yeh KY, Liu YZ, Tai MY, Tsai YF (2010) *Ginkgo biloba* extract treatment increases noncontact erections and central dopamine levels in rats: role of the bed nucleus of the stria terminalis and the medial preoptic area. Psychopharmacology (Berl) 210(4):585–590

Yeung EY, Sueyoshi T, Negishi M, Chang TK (2008) Identification of *Ginkgo biloba* as a novel activator of pregnane X receptor. Drug Metab Dispos 36(11): 2270–2276

Yoshitake T, Yoshitake S, Kehr J (2010) The *Ginkgo biloba* extract EGb 761(R) and its main constituent flavonoids and ginkgolides increase extracellular dopamine levels in the rat prefrontal cortex. Br J Pharmacol 159(3):659–668

Yuan F, Yu R, Yin Y, Shen J, Dong Q, Zhong L, Song L (2010) Structure characterization and antioxidant activity of a novel polysaccharide isolated from *Ginkgo biloba*. Int J Biol Macromol 46(4):436–439

Zehra U, Tahir M, Lone KP (2010) *Ginkgo biloba* induced malformations in mice. J Coll Physicians Surg Pak 20(2):117–121

Zhang LY, Wang YL (2008) Effects of EGb761 on hippocamal synaptic plasticity of vascular dementia rats. Zhongguo Ying Yong Sheng Li Xue Za Zhi 24(1):36–40 (In Chinese)

Zhang CQ, Zhu YH, Wang J, Liang B, Xu HW, Qin CY (2007) The effect of *Ginkgo biloba* extract on portal hypertension and hepatic microcirculation in rats. Zhonghua Gan Zang Bing Za Zhi 15(4):245–248 (In Chinese)

Zhang Y, Chen AY, Li M, Chen C, Yao Q (2008) *Ginkgo biloba* extract kaempferol inhibits cell proliferation and induces apoptosis in pancreatic cancer cells. J Surg Res 148(1):17–23

Zhang J, Sun HM, Bai LM, Xu H, Wu HX, Cui L (2009a) Effect of *Ginkgo biloba* pingchan recipe on neuronal nitric oxide synthase mRNA expression in the brain of mouse models of Parkinson disease. Nan Fang Yi Ke Da Xue Xue Bao 29(8):1735–1740 (In Chinese)

Zhang Q, Wang GJ, A JY, Wu D, Ma B, Du Y (2009b) Application of GC/MS-based metabonomic profiling in studying the lipid-regulating effects of *Ginkgo biloba* extract on diet-induced hyperlipidemia in rats. Acta Pharmacol Sin 30(12):1674–1687

Zhang WF, Tan YL, Zhang XY, Chan RC, Wu HR, Zhou DF (2011) Extract of *Ginkgo biloba* treatment for tardive dyskinesia in schizophrenia: a randomized, double-blind, placebo-controlled trial. J Clin Psychiatry 72(5):615–621

Zhao RY, Li YH, Gao W, Wang YX (2005) Experimental study about the protective effect of *Ginkgo biloba* extract on renal acute ischemia-reperfusion injury in rats. Zhongguo Zhong Yao Za Zhi 30(23):1859–1862 (In Chinese)

Zhao Z, Liu N, Huang J, Lu PH, Xu XM (2011) Inhibition of cPLA2 activation by *Ginkgo biloba* extract protects spinal cord neurons from glutamate excitotoxicity and oxidative stress-induced cell death. J Neurochem 116(6):1057–1065

Zhong ZM, Yu L, Weng ZY, Hao ZH, Zhang L, Zhang YX, Dong WQ (2007) Therapeutic effect of *Ginkgo biloba* leaf extract on hypercholestrolemia in children with nephrotic syndrome. Nan Fang Yi Ke Da Xue Xue Bao 27(5):682–684 (In Chinese)

Zhou LJ, Song W, Zhu XZ, Chen ZL, Yin ML, Cheng XF (2000) Protective effects of bilobalide on amyloid beta-peptide 25–35-induced PC12 cell cytotoxicity. Acta Pharmacol Sin 21(1):75–79

Zhou W, Chai H, Lin PH, Lumsden AB, Yao Q, Chen C (2004) Clinical use and molecular mechanisms of action of extract of *Ginkgo biloba* leaves in cardiovascular diseases. Cardiovasc Drug Rev 22(4): 309–319

Zhou C, Li X, Du W, Feng Y, Kong X, Li Y, Xiao L, Zhang P (2010a) Antitumor effects of ginkgolic acid in human cancer cell occur via cell cycle arrest and decrease the Bcl-2/Bax ratio to induce apoptosis. Chemotherapy 56(5):393–402

Zhou ZY, Tang SQ, Zhou YM, Luo HS, Liu X (2010b) Antioxidant and hepatoprotective effects of extract of *Ginkgo biloba* in rats of non-alcoholic steatohepatitis. Saudi Med J 31(10):1114–1118

Zhou L, Meng Q, Qian T, Yang Z (2011) *Ginkgo biloba* extract enhances glucose tolerance in hyperinsulinism-induced hepatic cells. J Nat Med 65(1):50–56

Zuo XC, Zhang BK, Jia SJ, Liu SK, Zhou LY, Li J, Zhang J, Dai LL, Chen BM, Yang GP, Yuan H (2010) Effects of *Ginkgo biloba* extracts on diazepam metabolism: a pharmacokinetic study in healthy Chinese male subjects. Eur J Clin Pharmacol 66(5):503–509

Gnetum gnemon

Scientific Name

Gnetum gnemon **L.**

Synonyms

Gnetum acutatum Miq., *Gnetum brunonianum* Griff., *Gnemon domestica* Rumph., *Gnetum gnemon* var. *domesticum* MGF, *Gnetum gnemon* var. *laurinum* Blume, *Gnetum gnemon* var. *lucidum* Blume, *Gnetum gnemon* var. *majusculum* Blume, *Gnetum gnemon* var. *ovalifolium* (Poir.) Blume, *Gnetum gnemon* var. *stipitatum* MGF, *Gnetum gnemon* var. *sylvestris* Parl., *Gnetum gnemon* var. *volubile* MGF, *Gnetum griffithii* Parl., *Gnetum ovalifolium* Poir., *Gnemon silvestris* Rumph., *Gnetum silvestris* Brogn., *Gnetum vinosum* Elmer.

Family

Gnetaceae

Common/English Names

Daeking Tree, Gnetum Nut, Gnemon Tree, Jointfir, Joint-Fir Spinach, Melinjo Nut, Paddy Oats, Spanish Jointfir, Tulip Tree, Two Leaf

Vernacular Names

Borneo: Sabong;
Burmese: Hyinbyin, Tanyin-Ywe;
Chinese: Guan Zhuang Mai Ma Teng, Xian Zhou Mai Ma Teng;
Fiji: Bele Sukau, Bui No Vodre, Mosokau, Sikau, Sukau, Sukau Buli, Sukau Mata, Sukau Motu;
French: Gnetum À Feuilles Comestibles;
India: Ganemoe, Genemo (Assamese);
Indonesia: Melinjo, Emping Melinjo, Belinjo, Meninjo, Bagu, Bagoe, Blinjo, Eso, So, Trangkil, (Javanese), Ki-Trangkil, Maninjo, Tangkil Sake (Sundanese);
Khmer: Voë Khlaèt;
Malaysia: Belinjau, Belingar, Maninjau, Meninjau, Melinjau, Sejunteh, Sokat;
Papua New Guinea: Tulip (General), Ambiam, Ambiamtupee (Maring);
Philippines: Bago, Bago Banago, Banago (Bataan, Tayabas, and Camarines);
Solomon Islands: Dae, Daefasia, Daemalefo (Kwara'Ae), Zua, Dae Fasia, King Tree;
Thailand: Peesae;
Vietnamese: Bét, Rau Bép, Rau Danh, Gắm Cay.

Origin/Distribution

The tree is native to Assam, Cambodia, Vietnam, Thailand, Peninsular Malaysia, Philippines, Papua New Guinea, Fiji, Solomon Islands (Santa

T.K. Lim, *Edible Medicinal And Non-Medicinal Plants: Volume 3, Fruits*,
DOI 10.1007/978-94-007-2534-8_2, © Springer Science+Business Media B.V. 2012

Anna), and Vanuatu (Pentecost, Ambae, Maewo, Torres Islands). Cultivated elsewhere including southeast Xizang, W Yunnan (Yingjiang Xian) in China.

Agroecology

It occurs in moderately dry to humid tropical forest to lower montane forests from near sea-level to 1,700 m elevation in the Malesian and Melanesian regions. It is usually found near river and stream banks in both it s natural habitat and cultivated situations and lowland rainforests in all soil types. It thrives in areas with mean annual rainfall of 3,000–5,000 mm distributed uniformly throughout the year or seasonally and mean annual temperatures 22–30°C. It prefers well drained and moist soil generally slightly acid to neutral soils but grows in light to heavy soils (sands, sandy loams, loams, sandy clay loams, sandy clays, clay loams, and clays).

Plate 1 Leaves, inflorescences and fruit of Melinjo

Edible Plant Parts and Uses

Gnetum gnemon is widely used in Malay-Indonesian cuisine. Young leaves, inflorescences, young and ripe fruits are cooked in vegetable dishes (Plates 1, 2, 3, 4, 5, 6, 7 and 8). In Malaysia, the young leaves and shoots are popularly used with sea food like fish and prawns in coconut milk or used in spicy soups. Seeds are consumed raw, boiled, roasted or processed by pounding the heated kernels into flat cakes or crackers. In Indonesia and the east coast states of Peninsular Malaysia, the cakes are sun-dried and used to prepare the crispy snack called *emping* which is prepared by cooking in oil. *Emping* is eaten with rice or gado-gado and also eaten as a snack with tea or coffee. *Emping* is being exported by Indonesia. Melinjau is also an important ingredient for *Sayur Asem* which comprises peanuts, jackfruit, melinjau, young carambola fruit, chayote, long bean and tamarind. The fleshy coating of the seed is also fried which produces a chewy snack. In Vanuatu, the leaves and male inflorescence cones (Plate 6) are boiled and flavoured with coconut cream for consumption.

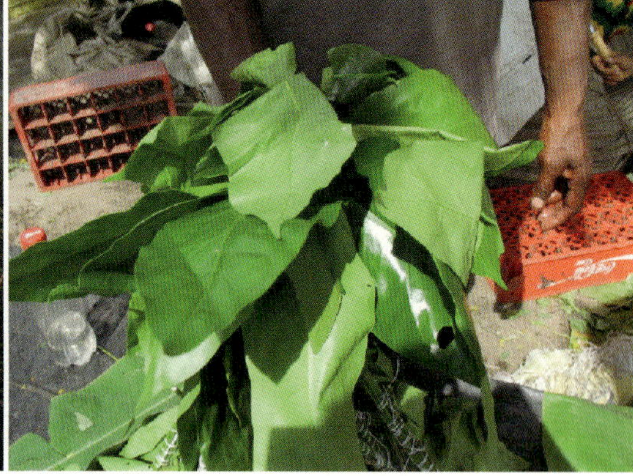

Plate 2 Bundles of Melinjo leaves on sale in local markets (*left*: Java, *right*: Papua New Guinea)

Plate 3 Melinjo leaves on sale in Kelantan market

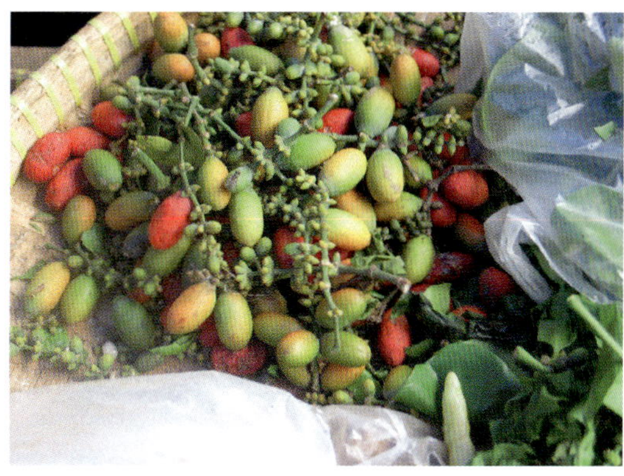

Plate 6 Ripe melinjo fruit on sale in local market in Java

Plate 4 Male and female inflorescences

Plate 7 Mature, unripe melinjo fruits on sale in Sarawak market

Plate 5 Ripe, red Melinjo fruit

Plate 8 Melinjo male inflorescences on sale in local market in Papua New Guinea

In Fiji, young leaves are eaten cooked with coconut milk and the fruits are also eaten. In the Philippines, the fruits are used as a substitute for coffee. In Papua New Guinea, the leaves and young cones are cooked with meat (game, pork) and often with a sauce made from red pulp of *Pandanus conoideus*. The leaves are also used to wrap food.

Botany

It is a small to medium-size, much-branched, dioecious evergreen shrub or tree, 5–10 m high with an erect trunk grey trunk clad with numerous whorls of branches down to the base. The leaves are opposite, simple, elliptical to oblong 8–20 cm long and 3–10 cm wide, entire; reddish-brown when young maturing shining dark green (Plates 1, 2 and 3). Inflorescence solitary and auxiliary with flowers in whorls at the nodes (Plates 1, 4 and 8). Female flowers, globose, tipped, in clusters of 5–8 at each inflorescence node. Male flowers in cylindrical, auxiliary catkin-like clusters, 3–6 cm × 2.5–3 mm, involucral collars clearly separated, to 1 cm apart, each collar with 50–80 flowers. The nut-like, ellipsoid fruit is 1–3.5 cm long, shortly apiculate first light green (Plates 4 and 7) turning to yellow (Plate 1) then to orange red, red to purple when ripe (Plates 5 and 6), containing one large and horny seed surrounded by a thin, orange-red, fleshy outer covering.

Nutritive/Medicinal Properties

The nutrient composition of the fruit of *melinjau (Gnetum gnemon)* per 100 g edible portion (pulp) based on analyses made in Sarawak (Voon and Kueh 1999) was reported as: water 72.6%, energy 92 kcal, protein 5.2%, fat 2.1%, carbohydrates 13.3%, crude fibre 5.2%, ash 1.8%, P 82 mg, K 624 mg, Ca 68 mg, Mg 52 mg, Fe 15.6 mg, Mn 34 ppm, Cu 1.9 ppm, Zn 11.8 ppm, vitamin C 2.9 mg.

The nutrient composition of the leaves of *melinjau (Gnetum gnemon)* per 100 g edible portion based on analyses made in Sarawak (Voon and Kueh 1999) was reported as : water 81.7%, energy 57 kcal, protein 4.2%, fat 1.5%, carbohydrates 6.6%, crude fibre 4.7%, ash 1.3%, P 68 mg, K 419 mg, Ca 94 mg, Mg 37 mg, Fe 3.8 mg, Mn 41 ppm, Cu 1.5 ppm, Zn 12.1 ppm, vitamin C 1.5 mg. the leaves were reported to have the following amino acid composition (g% total amino acid): 12.1 g asparagine, 5.0 threonine, 6.0 g serine, 13.9 g glutamic acid, 8.2 g proline, glycine 5.5 g, alanine 6.3 g, cysteine 0.2 g, valine 4.5 g, methionine 1.8 g, isoleucine 3.3 g, leucine 8.5 g, tyrosine 5.0, phenyalanine 5.4 g, histidine 2.1 g, lysine 6.3 g, tryptophan 0.1 g, argineine 5.8 g (Yeoh and Wee 1998).

Raw seed kernels of *Gnetum gnemon* were found to contain moisture 11.4, crude oil 4.0 and protein 6.9% (Berry 1980). The corresponding levels for the leaves were 4.5, 4.3 and 16.8%. Cyclopropene fatty acids were found in the seeds and leaves. The concentration for total these acids were 51.62, 37.87 and 46.91% of the total fatty acids in the oil of seeds, leaves and keropok (a product made from the seeds), respectively (Berry 1980).

Other Phytochemicals

Iliya et al. (2003c) isolated 6 flavonostilbenes (gnetoflavanols A, B, C, D, E and F) from *G. gnemon* roots. They also isolated 4 stilbene derivatives, gnemonols K and L (resveratrol trimers), M (isorhapontigenin dimer), and gnemonoside K (glucoside of resveratrol trimer) together with 11 known stilbenoids and a lignan from the acetone, methanol and 70% methanol soluble parts of the root of *Gnetum gnemon* which exhibited varying degree of antioxidant activity (Iliya et al. 2003a). Three stilbene trimers (gnemonols D, E, F) were isolated from the roots of *Gnetum gnemon* (Iliya et al. 2003b). Four new stilbene oligomers, gnemonols G, H, I, and J (1–4), were isolated from acetone extract of the roots along with five known stilbenoids, ampelopsin E, cis-ampelopsin E, gnetins C, D, and E (Iliya et al. 2002a, 2002b).

A 50% ethanol extract of the dried endosperms of melinjo yielded a new stilbenoid, named gnetin L, along with five stilbenoids gnetin C, gnemonosides A, C, and D, and resveratrol (Kato et al. 2009). All these stilbenoids exhibited 1,1-diphenyl-2-picrylhydrazyl (DPPH) radical scavenging activity similar to that of ascorbic acid and dl-α-tocopherol. These stilbenoids showed moderate antimicrobial activity except for gnemonoside A. Inhibition of lipase from porcine pancreas was manifested by four stilbenoids excluding gnemonoside A and resveratrol. Gnetin C, gnemonoside C and gnemonoside D inhibited the hydrolysis of starch by α-amylase from porcine pancreas. An ethanolic extract containing these stilbenoids also showed DPPH radical scavenging effect, lipase and α-amylase inhibition activity, and antimicrobial activity against food microorganisms and enterobacteria. The present study indicated that melinjo and extracts containing these constituents could be useful as health supplements. Resveratrol found in grapes and red wine and was reported to have chemopreventive, anti-cancer, anti-inflammatory, blood-sugar-lowering and other beneficial cardiovascular effects (Baur and Sinclair 2006; Bhat and Pezzuto 2002).

Gnemonol B isolated from *Gnetum gnemon* was found to exhibit strong antibacterial activities against vancomycin-resistant Enterococci (VRE) and methicillin-resistant *Staphylococcus aureus* (MRSA) (Sakagami et al. 2007). The MIC values of gnemonol B against five strains of VRE and nine strains of MRSA were 12.5 and 6.25 μg/ml, respectively. These compounds also showed synergistic effects when used in combination with commercially available antibiotics. The findings suggested that the application of the test compounds alone or in combination with antibiotics might be useful in controlling and treating VRE and MRSA infections.

Traditional Medicinal Uses

Melinjo leaf sap has been used medicinally to cure an eye complication in traditional medicine.

Other Uses

The wood of Melinjo is used as timber for the beams of houses and for the handles of tools. In Malaysia and Hong Kong, the wood is used for making paper, boxes, and house construction. The wood is used as fuel wood and pulped in Indonesia to make paper.

In Papua New Guinea, the bast fibres found underneath the outer bark of branches are used to make cordage for string bags (*bilum*), bowstring on musical instruments in the islands of Sumba, Indonesia, and for construction of fishing lines and fishnets in Malaysia.

Comments

Cultivated trees, belonging to *G. gnemon* var. *gnemon*, are larger in tree size and fruits. Other varieties, including *G. gnemon* var. *tenerum* Markgraf, are shrub-like with much smaller fruits and represent an important leafy vegetable in southern Thailand. Local Thai appellations for this variety are: Phak miang, liang, phak kariang.

Selected References

Athar M, Back JH, Tang XW, Kim KH, Kopelovich L, Bickers DR, Kim AL (2007) Resveratrol: a review of preclinical studies for human cancer prevention. Toxicol Appl Pharmacol 224(3):274–283

Backer CA, Van den Bakhuizen Brink RC Jr (1963) Flora of Java, vol 1. Noordhoff, Groningen, 648 pp

Baur JA, Sinclair DA (2006) Therapeutic potential of resveratrol: the in vivo evidence. Nat Rev Drug Discov 5:493–506

Berry SK (1980) Cyclopropene fatty acids in *Gnetum gnemon* (L.) seeds and leaves. J Sci Food Agric 31(7):657–662

Bhat KPL, Pezzuto JM (2002) Cancer chemopreventive activity of resveratrol. Ann N Y Acad Sci 957:210–229

Burkill IH (1966) A dictionary of the economic products of the Malay Peninsula. Revised reprint. 2 volumes. Ministry of Agriculture and Co-operatives, Kuala Lumpur, vol 1 (A–H) pp 1–1240, vol 2 (I–Z). pp 1241–2444

French BR (1986) Food plants of Papua New Guinea – a compendium. Australia and Pacific Science Foundation, Ashgrove, 408 pp

Fu L, Yu YF, Gilbert MG (1999) Gnetaceae Lindley. In: Wu ZY, Raven PH (eds) Flora of China, vol 4, Cycadaceae through Fagaceae. Science Press/Missouri Botanical Garden Press, Beijing/St. Louis

Iliya I, Ali Z, Tanaka T, Iinuma M, Furusawa M, Nakaya K, Murata J, Darnaedi D (2002a) Four new stilbene oligomers in the root of *Gnetum gnemon*. Helv Chim Acta 85(8):2538–2546

Iliya I, Tanaka T, Iinuma M, Ali Z, Furusawa M, Nakaya K, Shirataki Y, Murata J, Darnaedi D (2002b) Stilbene derivatives from two species of Gnetaceae. Chem Pharm Bull (Tokyo) 50(6):796–801

Iliya I, Ali Z, Tanaka T, Iinuma M, Furusawa M, Nakaya K, Murata J, Darnaedi D, Matsuura N, Ubukata M (2003a) Stilbene derivatives from *Gnetum gnemon* Linn. Phytochemistry 62(4):601–606

Iliya I, Ali Z, Tanaka T, Iinuma M, Furusawa M, Nakaya K, Shirataki Y, Murata J, Darnaedi D, Matsuura N, Ubukata M (2003b) Three new trimeric stilbenes from *Gnetum gnemon*. Chem Pharm Bull (Tokyo) 51(1):85–88

Iliya I, Tanaka T, Iinuma M, Nakaya K, Shirataki Y, Murata J, Darnaedi D, Matsuura N, Ubukata M (2003c) Six flavonostilbenes from *Gnetum africanum* and *Gnetum gnemon*. Heterocycles 60(1): 159–166

Kato E, Tokunaga Y, Sakan F (2009) Stilbenoids isolated from the seeds of melinjo (*Gnetum gnemon* L.) and their biological activity. Japan. J Agric Food Chem 57(6):2544–2549

Manner HI, Elevitch CR (2006) *Gnetum gnemon* (gnemon), ver. 1.1. In: Elevitch CR (ed) Species profiles for Pacific Island Agroforestry. Permanent Agriculture Resources (PAR), Holualoa, Hawai'i. http://www.traditionaltree.org

Ochse JJ, Bakhuizen van den Brink RC (1980) Vegetables of the Dutch Indies, 3rd edn. Ascher & Co, Amsterdam the Netherlands, 1016 pp

Peekal PG (1984) Gnetaceae. In: The flora of the Bismarck Archipelago for naturalists. Office of Forests/Division of Botany, Lae

Ridley HN (1922) The flora of the Malay Peninsula monocotyledons (concluded) gymnospermae general indices, vol V. L. Reeve and Co., Ltd., London

Saidin I (2000) Sayuran Tradisional Ulam dan Penyedap Rasa. Penerbit Universiti Kebangsaan Malaysia, Bangi, p 228 (In Malay)

Sakagami Y, Sawabe A, Komemushi S, Ali Z, Tanaka T, Iliya I, Iinuma M (2007) Antibacterial activity of stilbene oligomers against vancomycin-resistant enterococci (VRE) and methicillin-resistant *Staphylococcus aureus* (MRSA) and their synergism with antibiotics. Biocontrol Sci 12(1):7–14

Slik JWF (2006) Trees of Sungai Wain. National Herbarium Nederland. http://www.nationaalherbarium.nl/sungaiwain/

Verheij EWM, Sukendar (1992) *Gnetum gnemon* L. In: Verheij EWM, Coronel RE (eds) Plant resources of South-East Asia No 2. Edible fruits and nuts. Prosea Foundation, Bogor, pp 182–184

Von Reis S, Lipp FJ (1982) New plant sources for drugs and foods from the New York Botanical Garden. New York Botanical Garden. Herbarium. Harvard University Press, Cambridge

Voon BH, Kueh HS (1999) The nutritional value of indigenous fruits and vegetables in Sarawak. Asia Pac J Clin Nutr 8(1):24–31

Voon BH, Chin TH, Sim CY, Sabariah P (1988) Wild fruits and vegetables in Sarawak. Sarawak Department of Agriculture, Sarawak, 114 pp

Walter A, Sam C (2002) Fruits of oceania. ACIAR monograph no. 85. Australian Centre for International Agricultural Research, Canberra, 329 pp

Yeoh HH, Wee YC (1998) Total amino acid composition of tropical ferns and gymnosperms. Asian J Trop Biol 3(1):13–18

Carya illinoensis

Scientific Name

Carya illinoinensis **(Wangenheim) K. Koch.**

Synonyms

Carya angustifolia Sweet, *Carya diguetii* Dode, *Carya illinoensis* (Wangenheim) K. Koch, orth. var., *Carya olivaeformis* Nutall., *Carya oliviformis* (Michaux) Nuttall, *Carya pecan* (Marshall) Engl. & Graebn., *Carya tetraptera* Liebm., *Hicoria oliva* Raf., *Hicoria olivaeformis* (Michaux) Nuttall, *Hicoria pecan* (Marshall) Britton, *Hicorius pecan* Sargent, *Juglans angustifolia* Aiton, *Juglans cylindrica* Lam., *Juglans illinoinensis* Wangenheim, *Juglans olivaeformis* Hort. Paris ex Lam., *Juglans olivaeformis* Michaux, *Juglans oliviformis* Michaux, *Juglans pecan* Marshall.

Family

Juglandaceae

Common/English Names

Illinois Nut, Pecan, Pecan Nut, Soft-Shelled Hickory

Vernacular Names

Brazil: Nogueira-Japonêsa, Nogueira-Pecã, Noz-Japonêsa, Noz-Pecã, Pecã;
Chinese: Bao Ke Shan He Tao, Mei Guo Shan He Tao;
Czech: Ořechovec Pekanový;
Danish: Amerikansk Valnød, Pecannød, Pecantræ, Pekan;
Dutch: Pecannoot, Pekannoot;
Eastonian: Pekani-Hikkoripuu;
Finnish: Pekaanipähkinäpuu;
French: Noix De Pacanier, Noix De Pécan, Noix Pacane, Pacanier;
German: Hickorynußbaum, Pekannußbaum, Pekan-Nuß;
Italian: Noce Di Pecan;
Japanese: Peekannattsu;
Mexico: La Nuez Duquita, Nogal Liso, Nogalito (Spanish);
Polish: Orzesznik Pekan
Portuguese: Cária, Nogueira-Peca;
Russian: Černoj Pecan;
Slovaščina: Oreh Pecan;
Spanish: Nogal Americano, Pecán, Pecana, Pecanero;
Swedish: Pekannöt , Pekanträd;
Vietnamese: Quả Hồ Đào.

DOI 10.1007/978-94-007-2534-8_3,

Origin/Distribution

Pecan is native to south-central North America, in Mexico from Coahuila south to Jalisco and Veracruz, in north America from southern Iowa, Illinois and Indiana east to western Kentucky, North Carolina, South Carolina and western Tennessee, south through Georgia, Alabama, Mississippi, Louisiana, Texas, Oklahoma, Arkansas and Florida, and west into New Mexico. Pecan is widely cultivated in other areas including Australia, China, Brazil, Israel, Mexico, Peru and South Africa.

Agroecology

Pecan is a cool climate species. It can be grown from sea level up to altitudes of 800 m. It thrives in full sun, in areas with mean summer temperatures range as high as 27°C and mean winter temperatures from 10° to −1°C and average annual precipitation of 760–2,010 mm. It requires frost in the winter to overcome bud dormancy and permit proper vegetative growth in the spring, and the tree also require a period of two months of winter dormancy. It tolerates considerable winter cold without damage (−20°C), but require a long frost-free growing season. Late spring frost is injurious to young trees. Pecan nut performs best in areas of low to moderate humidity. Optimum temperature for photosynthesis is 27°C. It prefers well-drained, acidic to alkaline loam soils which are not subject to prolonged flooding but will grow on heavy textured soils, where it is limited to alluvial soils of recent origin. It is rarely grown on low and poorly drained clay flats and is not salt tolerant.

Edible Plant Parts and Uses

Pecan nuts are edible, with a rich, buttery flavour and have a high percentage of lipids (especially unsaturated fatty acids). They can be eaten fresh or used in cooking and in confectionery, particularly in sweet desserts, cakes, bread, candies and cookies but also in some savoury dishes. Pecan kernel pieces are also popularly eaten with ice-cream. One of the most common desserts with pecan as a main ingredient is the pecan pie, a traditional southern U.S. recipe. Pecans are also a major ingredient in praline candy, most often associated with New Orleans. Pecan oil was formerly used by native Americans to season food.

Botany

Pecan is a medium to large, deciduous, monoecious, perennial tree growing to 30–45 m high with a spread of 1.8–2.1 m. It has grey trunk which is shallowly furrowed, often scaly and flat-ridged with ascending branches forming an irregular, rounded crown. The twigs are grey brown and hairy when young but become rough and furrowed on older trees. Terminal buds are yellowish brown with 4 or more false-valved scales. Leaves are alternate, on 40–80 mm long petioles, imparipinnately compound with 7–15 finely serrate, glabrous or glabrescent often curved leaflets, lateral ones shortly petiolulate or sessile, blade ovate-lanceolate to elliptic-lanceolate or long elliptic, 7–18×2.5–4 cm, with scattered, peltate scales, abaxially pubescent or glabrescent, base oblique, broadly cuneate or subrounded, apex acuminate; terminal petiolule 5–25 mm. Flowers are borne in staminate and pistillate catkins. Staminate catkins are pendulous, 8–14 cm long and borne in clusters of threes and bear small yellowish-green flowers with sparsely pilose anthers; the female catkins are small, with three to six yellowish-green flowers clustered together at the end of new growth. The fruit is a nut-like drupe called tryma, brown, ovoid-ellipsoid, 3.75–5 cm by 2–3 cm, thin-shelled (Plates 1 and 2), and enveloped in a four-winged, 3–4 mm thick husk that starts out green and turns brown at maturity, at which time it splits off in four sections to release the thin-shelled nut. In a tryma, the epicarp and mesocarp separate as a somewhat fleshy or leathery rind from the hard, 2-valved ribbed endocarp (kernel) (Plate 3). The husks are produced from the exocarp tissue of the flower while the part known as the nut develops from the endocarp.

Plate 1 Pecan nut

Plate 2 Pecan nut with shell removed

Plate 3 Kernels of Pecan nut

Nutritive/Medicinal Properties

Analyses carried out in the United States (U.S. Department of Agriculture and Agricultural Research Service 2010) reported unroasted pecan nut kernels (minus the shells 47%) to have the following proximate composition (per 100 g edible portion): water 3.52 g, energy 691 kcal (2,889 kJ), protein 9.17 g, total lipid 71.97 g, ash 1.49 g, carbohydrates 13.86 g, total dietary fibre 9.6 g, total sugars 3.97 g, sucrose, 3.90 g, glucose 0.04 g, fructose 0.04 g, starch 0.46 g, Ca 70 mg, Fe 2.53 mg, Mg 121 mg, P 277 mg, K 410 mg, Na 0 mg, Zn 4.53 mg, Cu 1.2 mg, Mn 4.5 mg, F 10 μg, Se 3.8 μg, vitamin C 1.1 mg, thiamine 0.660 mg, riboflavin 0.130 mg, niacin 1.167 mg, pantothenic acid 0.863 mg, vitamin B-6 0.210 mg, total folate 22 μg, choline 40.5 mg, betaine 0.7 mg, vitamin A 56 IU, vitamin E (α-tocopherol) 1.40 mg, β-tocopherol 0.39 mg, γ-tocopherol 24.44 mg, δ-tocopherol 0.047 mg, vitamin K (phylloquinone) 3.5 μg, total saturated fatty acids 6.180 g, 16:0 (palmitic acid) 4.366 g, 18:0 (stearic acid) 1.745 g, 20:0 0.069 g; total monounsaturated fatty acids 40.801 g, 18:1 undifferentiated (oleic acid) 40.59 g; total polyunsaturated fatty acids 21.614 g, 18:2 undifferentiated (linoleic acid) 20.628 g, 18:3 undifferentiated (linolenic acid) 0.986 g, phytosterols 102 mg, stigmasterol 3 mg, campestrol 5 mg, β-sitosterol 89 mg, tryptophan 0.093 g, threonine 0.306 g, isoleucine 0.336 g, leucine 0.598 g, lysine 0.287 g, methionine 0.183 g, cystine 0.152 g, phenylalanine 0.426 g, tyrosine 0.215 g, valine 0.411 g, arginine 1.177 g, histidine 0.262 g, alanine 0.397 g, aspartic acid 0.929 g, glutamic acid 1.829 g, glycine 0.453 g, proline 0.363 g, serine 0.474 g, β-carotene 29 μg, β-cryptoxanthin 9 μg, lutein + zeaxanthin 17 μg.

From the above data, it is evident that pecan nuts are rich in calorific value, minerals like Mg, Zn, Cu, Mn, vitamin A, β-carotene, β-cryptoxanthin, vitamin Es (α-, β- and δ-tocopherol). It also contains selenium, lutein and zeaxanthin and the other minerals. Pecan is also a good source of protein, phytosterols (stigmasterol, campestrol β-sitosterol) and unsaturated fats like omega-6 fatty acids (linoleic acid). In a separate study, the order of the concentrations of the elements in all the nut (almond, Brazil nut pecan, macadamia and walnut) samples was found generally to be Mg > Ca > Fe > Cu > Cr > As > Se (Moodley et al. 2007). With maximum and minimum limits being set by the almond and pecan nut samples, Cr ranging from 0.94–2.02 μg/g was detected in the nut samples.

Nuts (tree nuts including pecan nut and peanuts) are an excellent source of vitamin E and magnesium (King et al. 2008). Individuals consuming nuts also have higher intakes of folate, β-carotene, vitamin K, lutein + zeaxanthin, phosphorus, copper, selenium, potassium, and zinc per 1,000 kcal. Regular nut consumption increases total energy intake, but the body weight of nut consumers is not greater than that of non-consumers. Nuts are an excellent source of phytochemicals (phyotsterols, phenolic acids, flavonoids, stilbenes, and carotenoids). The total phenolic constituents probably contribute to the total antioxidant capacity of nuts, which is comparable to broccoli and tomatoes.

Nuts including pecan contain bioactive constituents that elicit cardio-protective effects including phytosterols, tocopherols and squalene. The total oil contents from freshly ground Brazil nut, pecan, pine, pistachio and cashew nuts were found to range from 40.4–60.8% (w/w) while the peroxide values varied from 0.14–0.22 mEq O^2/kg oil (Ryan et al. 2006) . The most abundant monounsaturated fatty acid was oleic acid (C18:1), while linoleic acid (C18:2) was the most prevalent polyunsaturated fatty acid. The levels of total tocopherols ranged from 60.8–291.0 mg/g. Squalene ranged from 39.5 mg/g oil in the pine nut to 1377.8 mg/g oil in the Brazil nut. Beta-sitosterol was the most prevalent phytosterol, ranging in concentration from 1325.4–4685.9 mg/g oil. The study presented data that indicated nuts to be a good dietary source of unsaturated fatty acids, tocopherols, squalene and phytosterols.

Analyses conducted in Florida reported that on an edible portion basis, pecan was reported to have 2.1–6.4% moisture, 6.0–11.3% protein, 65.9–78.0% lipid, 3.3–5.3% total soluble sugars and 1.2–1.8% ash (Venkatachalam et al. 2007). With the exception of a high tannin (2.7%) in Texas seedling, pecan tannin content was found in a narrow range (0.6–1.85%). Unsaturated fatty acids (>90%) dominated pecan lipid composition with oleic (52.52–74.09%) and linoleic (17.69–37.52%) acids as the predominant unsaturated fatty acids. Location significantly influenced pecan biochemical composition. Pecan lipid content was negatively correlated with protein (r = −0.663) and total sugar (r = −0.625). Minor differences in subunit polypeptide profiles were found among samples. Rabbit polyclonal antibody-based immunoblotting experiments (Western blot) also illustrated similarity in polypeptide profiles with respect to immunoreactivity (Venkatachalam et al. 2007).

The main constituent of pecan nut kernels was found to be a high lipid content (70–79% wt/wt d. b.) with a large proportion of oleic acid (55–75% wt/wt) (Toro-Vazquez and Pérez-Briceño 1998). The concentration of α-, γ-, and δ-tocopherols in the oils extracted from Mexican native pecans, was substantially higher than those in the varieties commonly reported in the literature. However, the fatty acid and tocopherol composition of the pecan oil did not fully explain the high oxidative stability (OSI values from 8.5–10.8 h at 110°C) observed in some oils, which might indicate the presence of some other natural antioxidants with activity at 110°C. The oxidative stability, the melting and crystallization properties, and the viscosity of pecan oils were similar or superior to those of extra-virgin olive oil and unrefined sesame oil.

Other Phytochemicals

Volatile constituents of pecan leaves and nuts were found to comprise 38 compounds including 7 monoterpenes, 7 sesquiterpene hydrocarbonsm 11 terpene alcohols, 1 terpene aldehyde, 1 terpene ketone and 11 other aldehydes, alcohols, ketones and esters (Mody et al. 1976).

Examination of the petroleum ether, ether, and chloroform extracts of pecan leaves and petioles revealed the presence of unidentified phytosterols and a squalene-like substance in the unsaponified portions, and the presence of capric, lauric, myristic, palmitic, stearic, arachidic, oleic, linoleic, and linolenic acids in the saponified portions (Wilken and Cosgrove 1964). Nonhydrolyzable tannins containing a phloroglucinol and a catechol nucleus were found in the ethanol and methanol extracts. Investigation of the plant extracts revealed the presence of carbohydrates and the absence of discernible amounts of glycosides and alkaloids. A crystalline neutral substance obtained from a neutral lead acetate treated aqueous extract was identified as the m-inositol. A crystalline acidic substance isolated from an aqueous extract of the crude drug was identified as 3,4-dihydroxybenzoic acid.

Some pharmacological properties of pecan reported include:

Antioxidant Activity

Strong correlations were found in kernels between antioxidant capacity and total phenolics for both DPPH (r2 = 0.98) and ORAC (r2 = 0.75) antioxidant assays for six pecan cultivars (Villarreal-Lozoya et al. 2007). Antioxidant capacity ORAC values ranged from 372 to 817 μmol Trolox equivalents/g defatted kernel, corresponding to Desirable and Kanza cultivars, respectively. Condensed tannins ranged from 23 to 47 mg catechin equivalents/g defatted kernel and total phenolics from 62 to 106 mg of chlorogenic acid equivalents/g defatted kernel. After a consecutive basic-acid hydrolysis, gallic acid, ellagic acid, catechin and epicatechin were identified by HPLC. The total phenolics, antioxidant capacity and condensed tannins were 6, 4.5 and 18 times higher, respectively, for shells compared to kernels. The presence of phenolic compounds with high antioxidant capacity in kernels and shells indicated pecans to be an important dietary source of antioxidants.

In another study, significantly higher levels of total phenolics, condensed tannins and antioxidant activity measured through the ABTS and DPPH (30 minutes and 24 hour) systems were observed for the acetone extract (16.4 mg GAE/g; 31.2 mg CE/g; 235.3 μmol TEAC/g and 68.6 and 100.3 mg TEAC/g, respectively) of pecan kernel cake (do Prado et al. 2009b). The oxidation inhibition percentage in the β-carotene/linoleic acid system varied from 37.9% to 93.1%, with the acetone extract at 300 ppm showing significantly superior results. The samples with the greatest tendency to show red tones presented the highest levels of condensed tannins. Studies also indicated the high phenolic content and antioxidant activity of pecan nut shell infusion (do Prado et al. 2009a). The shell exhibited high fibre content (48%), the total phenolic content ranged from 116 to 167 mg GAE/g and the condensed tannin content was between 35 and 48 mg CE/g. The antioxidant activity varied from 1,112 and 1,763 μmol TEAC/g in the ABTS system. In the DPPH method, the antioxidant activity was from 305 to 488 mg TEAC/g (30 min reaction) and from 482 to 683 mg TEAC/g (24 h reaction). The oxidation inhibition percentage obtained in the β-carotene/linoleic acid system varied from 70% to 96%.

Antihypercholesterolemic Activity

Pecan is rich in natural plant sterols. Researchers (Ye et al. 2001) reported that pecans contain as much as 95 mg/g of plant sterols – 90% of which is in the form of β-sitosterol. Sterols are currently receiving attention clinically as dietary components with the ability to positively affect blood lipid profiles, benign prostatic hyperplasma and various disease states. Beta-sitosterol has been cited in multitudes of animal and human research studies as a food component that competes with the absorption of cholesterol in the body, and thus has the ability to lower blood cholesterol levels. Studies by Kornsteiner et al. (2006) and Kris-Etherton et al. (1999) indicated the presence of bioactive molecules, such as sterols and tocopherols, and a high content of total phenolic compounds, with possible natural antioxidant activity in pecan nut. Data suggested that there were non-fatty acid constituents in nuts including pecan that had additional cholesterol-lowering effects.

One study demonstrated that low-density lipoprotein cholesterol (LDL-C), was lowered in the pecan treatment group (7 women, 3 men, mean age = 45 ± 10 years) that consumed 68 g pecans per day for 8 weeks plus self-selected diets from 2.61 mmol/L at baseline to 2.35 at week 4 and to

2.46 at week 8 (Morgan and Clayshulte 2000). At week 8, total cholesterol and high-density lipoprotein cholesterol (HDL-C) in the pecan treatment group were significantly lower than in the control group (total cholesterol: 4.22 vs 5.02 mmol/L; HDL-C: 1.37 vs 1.47 mmol/L). Dietary fat, monounsaturated fat, polyunsaturated fat, insoluble fibre, magnesium, and energy were significantly higher in the pecan treatment group than in the control group. Body mass indexes and body weights were unchanged in both groups. Results indicated that pecans can be included in a healthful diet when energy intake and potential weight gain are addressed.

In a randomized, controlled, crossover feeding study, 24 subjects were assigned to 2 diets, each for 4 weeks: a control diet and a pecan-enriched (20% of energy) diet. Consumption of nuts was found to be associated with a reduced risk of coronary heart disease, and dietary intervention studies incorporating pecans show improved lipid profiles (Haddad et al. 2006). In a randomized, controlled, crossover feeding study cholesterol-adjusted plasma γ-tocopherol increased by 10.1%, α-tocopherol decreased by 4.6%, and malondialdehyde concentrations measured as thiobarbituric acid reactive substances decreased by 7.4% after participants were on the pecan diet. No changes were observed for ferric-reducing ability of plasma or Trolox equivalent antioxidant capacity values. The unsaturated fats in pecans were protected against oxidation by the high concentrations of γ-tocopherol and polymeric flavanols. These data provide some evidence for potential protective effects of pecan consumption in healthy individuals. In an earlier study involving 23 subjects were fed in random order with a diet rich in pecans (20% energy from pecans) and a diet free of nuts as reference, each for 4 weeks. The researchers found that the pecan-enriched diet significantly raised blood levels of γ-tocopherol compared to the AHA Step I diet (Haddad et al. 2001). Plasma α- and β-tocopherol levels decreased significantly on the pecan diet. While absolute plasma levels of γ-tocopherol did not change, the ratio of γ - tocopherol to cholesterol increased significantly on the pecan diet. Gamma tocopherol is an important antioxidant nutrient.

Morgan and Clayshulte (2000) compared the effects of a diet without nuts with another that included 68 g of pecans each day. Total cholesterol and HDL-C in the pecan treatment group were significantly lower than in the control group (total cholesterol: 4.22 vs 5.02 mmol/L; HDL-C: 1.37 vs 1.47 mmol/l). Dietary fat, monounsaturated fat, polyunsaturated fat, insoluble fibre, magnesium, and energy were significantly higher in the pecan treatment group than in the control group. Body mass indexes and body weights were unchanged in both groups. Another study by Rajaram et al. (2001) evaluated the effects of a Step 1 type diet (30% of energy in the form of fat) and compared them with those of another diet where 20% of the energy was replaced by pecans in 23 subjects (mean age: 38 y; 9 women, 14 men). Both diets improved the lipid profile, but the diet rich in pecans more significantly reduced the total cholesterol (6.7%), the LDL cholesterol (10.4%) and the triglycerides (11.1%), while the HDL cholesterol increased (2.5 mg/daily). Serum apolipoprotein B and lipoprotein (a) decreased by 11.6 and 11.1%, respectively, and apolipoprotein A1 increased by 2.2% when subjects consumed the pecan compared with the Step I diet. These differences were all significant. A 20% isoenergetic replacement of a Step I diet with pecans favourably altered the serum lipid profile beyond the Step I diet, without increasing body weight. Nuts such as pecans being rich in monounsaturated fat may therefore be recommended as part of prescribed cholesterol-lowering diet of patients or habitual diet of healthy individuals

Pecan was found to increase fibre and nutrient intake (Barloon et al. 2001). In an 8-week, randomized, controlled feeding trial, involving 40 hypercholesterolemic men and women between the ages of 22 and 71, all of whom had already been eating a relatively low-fat diet, were placed on either the AHA Step I diet or an isocaloric but higher-fat, pecan-based diet. There were no significant differences in negative risk factors (e.g., C - reactive protein levels or lipoprotein size/density) between diets. Researchers found that the pecan-rich diet significantly increased participants' levels of dietary fibre, thiamin, magnesium, copper and manganese and actually changed dietary copper and magnesium intakes from inadequate (on the AHA Step I diet) to adequate (on the pecan diet).

Chemoprotective Activity

Studies showed that pecan nut shell extract exhibited protective effects against toxicity induced by cyclophosphamide in the heart, kidney, liver, bladder, plasma and erythrocytes of rats (Benvegnu et al. 2010). Rats treated with cyclophosphamide exhibited an increase in lipid peroxidation and decrease in reduced glutathione (GSH) levels in all structures. Catalase activity was increased in the heart and decreased in liver and kidney. Also, cyclophosphamide treatment lowered plasmatic vitamin C levels and caused bladder macroscopical and microscopical damages. In contrast, co-treatment with pecan shell aqueous extract prevented the lipid peroxidation development and the GSH depletion in all structures, except in the heart and plasma, respectively. Catalase activity in the heart and liver as well as the plasmatic vitamin C levels remained unchanged. Additionally, the pecan shell aqueous extract prevented cyclophosphamide-induced bladder injury.

Allergy Problem

A case of a patient who experienced an acute vesicular cutaneous reaction after prolonged contact with pecans was reported (Joyce et al. 2006). This case illustrates the salient features of contact dermatitis and serves as a reminder that contact with allergenic foods can lead to hypersensitivity reactions.

Traditional Medicinal Uses

The leaves and bark are sometimes used as an astringent. *Carya illinoensis* is also known as Kiowa drug used as a tuberculosis remedy in traditional Mexican medicine. Studies confirmed that bark and leaf crude extracts of *Juglans regia*, *Juglans mollis*, *Carya illinoensis* and *Bocconia frutescens* showed in-vitro anti- *Mycobacterium tuberculosis* activity (Cruz-vega et al. 2008). Hexane bark extracts from *C. illinoensis*, *J. mollis* and *J. regia* were the most active with a minimal inhibitory concentration (MIC) of 31, 50 and 100 μg/ml, respectively. Ethanol bark extracts from *C. illinoensis* and *J. mollis* showed activity at 100 and 125 μg/ml, respectively. Leaf extracts had the lowest activity.

Other Uses

An edible oil is extracted from the kernels and used in the manufacture of cosmetics and certain drugs Pecan wood has been used for furniture, cabinetry, panelling, pallets, veneer, flooring, agricultural implements, tool handles, and for fuel.

The nut shells (husk) are used as paving for walks and driveways, fuel, mulches, soil conditioners, soft abrasives in hand soap, non-skid paints and metal polishes, and for tannin. They can also be ground into flour or used as fillers in plastic wood, adhesives, and dynamite. Recent studies showed that pecan nutshell has potential as an alternative biosorbent to remove Cu(II), Mn(II) and Pb(II) metallic ions from aqueous solutions (Vaghetti et al. 2009).

The polyphenolic extracts from pecan shell were found to markedly inhibit the mycelial growth of phytopathogenic fungi: *Pythium* sp., *Colletotrichum truncatum*, *Colletotrichum coccodes*, *Alternaria alternata*, *Fusarium verticillioides*, *Fusarium solani*, *Fusarium sambucinum*, and *Rhizoctonia solani* (Osorio et al. 2010). The waste pecan shell material may have potential as a fungicide for plant pathogens.

Comments

Today, the U.S. produces between 80% and 95% of the world's pecans.

Selected References

Barloon JL, Walzem RL, Storey JB, Macfarlane RD, Piziak VK (2001) High fat pecan-based diet as effective as step i diet to maintain plasma lipid and lipoprotein responses. Research presented at the American Heart Association Conference on Arteriosclerosis, Thrombosis and Vascular Biology, Arlington, VA, 11–13 May 2001

Benvegnu D, Barcelos RC, Boufleur N, Reckziegel P, Pase CS, Muller LG, Martins NM, Vareli C, Burger ME (2010) Protective effects of a by-product of the pecan nut industry (*Carya illinoensis*) on the toxicity induced by cyclophosphamide in rats *Carya illinoensis* protects against cyclophosphamide-induced toxicity. J Environ Pathol Toxicol Oncol 29(3):185–197

Cruz-Vega DE, Verde-Star MJ, Salinas-Gonzalez N, Rosales-Hernandez B, Estrada-Garcia I, Mendez-Aragon P, Carranza-Rosales P, Gonzalez-Garza MT, Castro-Garza J (2008) Antimycobacterial activity of *Juglans regia, Juglans mollis, Carya illinoensis and Bocconia frutescens*. Phytother Res 22(4):557–559

Do Prado ACP, Aragao AM, Fett R, Block JM (2009a) Antioxidant properties of pecan nut [*Carya illinoinensis* (Wangenh.) C. Koch] shell infusion. Grasas y aceites 60(4):330–335

Do Prado ACP, Aragao AM, Fett R, Block JM (2009b) Phenolic compounds and antioxidant activity of pecan [*Carya illinoinensis* (Wangenh.) C. Koch] kernel cake extracts obtained by sequential extraction. Grasas y aceites 60(5):458–467

Duncan WH, Duncan MB (1988) Trees of the Southeastern United States. The University of Georgia Press, Athens, p 322

Grauke LJ, Thompson TE (1996) Pecans and hickories. In: Janick J, Moore JN (eds) Fruit breeding, vol 3, Nuts. Wiley, New York, pp 278–239, 278 pp

Haddad E, Jambazian P, Tanzman J, Sabaté J (2001) Effect of a pecan rich diet on plasma tocopherol status. Abstract published in the March 2001 FASEB Journal. Research presented at the April Experimental Biology 2001 Meeting, Loma Linda University, School of public health, Loma Linda, California

Haddad E, Jambazian P, Karunia M, Tanzman J, Sabaté J (2006) A pecan enriched diet increases y-tocopherol/cholesterol and decreases thiobarbituric acid reactive substances in plasma of adults. Nutr Res 26(8): 397–402

Joyce KM, Boyd J, Viernes JL (2006) Contact dermatitis following sustained exposure to pecans (*Carya illinoensis*): a case report. Cutis 77(4):209–212

King JC, Blumberg J, Ingwersen L, Jenab M, Tucker KL (2008) Tree nuts and peanuts as components of a healthy diet. J Nutr 138:1736S–1740S

Kornsteiner M, Wagner KH, Elmadfa I (2006) Tocopherols and total phenolics in 10 different nut types. Food Chem 98:381–387

Kris-Etherton PM, Yu-Poth S, Sabaté J, Ratcliffe HE, Zhao G, Etherton TD (1999) Nuts and their bioactive constituents: effects on serum lipids and other factors that affect disease risk. Am J Clin Nutr 70:504–511

Lu A, Stone DE, Grauke LJ (1999) Juglandaceae A. Richard ex Kunth. In: Wu ZY, Raven PH (eds) Flora of China, vol 4, Cycadaceae through Fagaceae. Science Press/Missouri Botanical Garden Press, Beijing/St. Louis

Mody NV, Hedin PA, Neel WW (1976) Volatile components of pecan leaves and nuts, *Carya illinoensis* Koch. J Agric Food Chem 24(1):175–177

Moodley R, Kindness A, Jonnalagadda SB (2007) Elemental composition and chemical characteristics of five edible nuts (almond, Brazil, pecan, macadamia and walnut) consumed in Southern Africa. J Environ Sci Health B 42(5):585–591

Morgan WA, Clayshulte BJ (2000) Pecans lower low-density lipoprotein cholesterol in people with normal lipid levels. J Am Diet Assoc 100(3):312–318

Osorio E, Flores M, Hernández D, Ventura J, Rodríguez R, Aguilar C (2010) Biological efficiency of polyphenolic extracts from pecan nuts shell (*Carya illinoensis*), pomegranate husk (*Punica granatum*) and creosote bush leaves (*Larrea tridentata* Cov.) against plant pathogenic fungi. Ind Crops Prod 31(1): 153–157

Peterson JK (1990) *Carya illinoensis* (Wangenh.) K. Koch pecan. In: Burns RM, Honkala BH (Technical coordinators). Silvics of North America. Vol. 2. Hardwoods. Agric. Handbook 654. Washington, DC: U.S. Department of Agriculture, Forest Service. pp 205–210

Porcher MH et al (1995–2020) Searchable world wide web multilingual multiscript plant name database. The University of Melbourne, Australia. http://www.plantnames.unimelb.edu.au/Sorting/Frontpage.html

Rajaram S, Burke K, Connell B, Myint T, Sabaté J (2001) A monounsaturated fatty acid-rich pecan-enriched diet favourably alters the serum lipid profile of healthy men and women. J Nutr 131:2275–2279

Rehm S (1994) Multilingual dictionary of agronomic plants. Kluwer Academic Publishers, Dordrecht, 286 pp

Ryan E, Galvin K, O'Connor TP, Maguire AR, O'Brien NM (2006) Fatty acid profile, tocopherol, squalene and phytosterol content of Brazil, pecan, pine, pistachio and cashew nuts. Int J Food Sci Nutr 57(3–4): 219–228

Toro-Vazquez JF, Pérez-Briceño F (1998) Chemical and physicochemical characteristics of pecan (*Carya illinoensis*) oil native of the central region of Mexico. J Food Lipids 5:211–231

U.S. Department of Agriculture, Agricultural Research Service (2010) USDA National Nutrient Database for Standard Reference, Release 23. Nutrient Data Laboratory Home Page, http://www.ars.usda.gov/ba/bhnrc/ndl

Vaghetti JC, Lima EC, Royer B, da Cunha BM, Cardoso NF, Brasil JL, Dias SL (2009) Pecan nutshell as biosorbent to remove Cu(II), Mn(II) and Pb(II) from aqueous solutions. J Hazard Mater 162(1):270–280

Venkatachalam M, Kshirsagar HH, Seeram NP, Heber D, Thompson TE, Roux KH, Sathe SK (2007)

Biochemical composition and immunological comparison of select pecan [*Carya illinoinensis* (Wangenh.) K. Koch] cultivars. J Agric Food Chem 55(24):9899–9907

Villarreal-Lozoya JE, Lombardini L, Cisneros-Zevallos L (2007) Phytochemical constituents and antioxidant capacity of different pecan [*Carya illinoinensis* (Wangenh.) K. Koch] cultivars. Food Chem 102:1241–1249

Wickens GE (1995) Edible nuts, Non-wood forest products 5. FAO, Rome, 198 pp

Wilken LO Jr, Cosgrove FP (1964) Phytochemical investigation of *Carya illinoensis*. J Pharm Sci 53(4): 364–368

Ye L, Koehler PE, Eitenmiller RR (2001) Sterol content of peanuts, pecans and peanut products. Research paper presented at the 2000 Institute of Food Technologists Annual Meeting, Dallas, Texas, 10–14 June 2001

Juglans regia

Scientific Name

Juglans regia L.

Synonyms

Juglans duclouxiana Dode, *Juglans fallax* Dode, *Juglans kamaonia* (C. de Candolle) Dode, *Juglans orientis* Dode, *Juglans regia* var. *orientis* (Dode) Kitam., *Juglans regia* subsp. *kamaonica* (C. DC.) Mansf., *Juglans regia* var. *sinensis* C. de Candolle, *Juglans sinensis* (C. de Candolle) Dode.

Family

Juglandaceae

Common/English Names

Carparthian Walnut, Common Walnut, English Walnut, European Walnut, Hirsute Walnut, Madeira Nogal, Madeira Walnut, Persian Walnut, Walnut

Vernacular Names

Afghanistan: Charmaz;
Brazil: Nogueira;
Chinese: Hu Tao, Hu Tao Ren;
Czech: Ořešák Královský;
Danish: Almindelig Valnød, Valnød, Valnøddetræ, Valnødtræ;
Dutch: Noteboom, Okkernoot, Okkernoteboom, Walnoot, Walnooteboom;
Eastonian: Kreeka Pähklipuu;
Finnish: Jalopähkinät, Saksanpähkinä, Saksanjalopähkinä;
French: Noix, Noix Royale, Noix Commune, Noyer Commun, Noyer D'europe, Noyer De France, Noyer Royal;
German: Echter Walnußbaum, Europäischer Walnußbaum, Gemeine Walnuß, Nußbaum, Nussbaum, Christnuss, Echte Walnuss, Steinnuss, Wallnuss, Walnuss, Walnussbaum, Welschnuss-Baum, Wälsche Nuss;
Greek: Karithis;
Hungarian: (Közönséges) Dió(Fa), Királydió, Közömséges Dió, Nemes Dió, Pompás Dió;
India: Akhor, Akhrot (Hindi), Akrodu (Kannada), Akrottu (Malayalam) Aksotah (Sanskrit), Akrottu (Tamil), Akrotu (Telugu);
Italian: Noce, Noce Comune, Noceto;
Japanese: Perusha Gurumi, Kashi Gurumi, Kurumi;
Norway: Valnøtt, Ekte Valnøtt;
Pakistan: Akhrot;
Persian: Gerdoo;
Polish: Orzech Wloski, Orzech Włoski;
Portuguese: Nogueira-Comum, Noz;
Russian: Gretskii Orekh;
Slovakian: Orech Vlašský;

T.K. Lim, *Edible Medicinal And Non-Medicinal Plants: Volume 3, Fruits*,
DOI 10.1007/978-94-007-2534-8_4,

Slovenian: Navadni Oreh;
Spanish: Nuez, Nogal, Nogal Común, Nogal Europeo, Nogal Inglés;
Swedish: Valnöt, Valnöt-Arter, Valnötsträd, Äkta Valnöt;
Turkish: Ceviz.

Origin/Distribution

Walnut is indigenous to the region stretching from the Balkans eastward across Persia to the Himalayas and southwest China. The largest forests are found in Kyrgyzstan, where trees occur in extensive, nearly pure walnut forests at 1,000–2,000 m altitude. Walnut is widely cultivated throughout this region and elsewhere in temperate zone of the Old and New World. Important nut-growing regions include France, Serbia, Greece, Bulgaria, and Romania in Europe, China in Asia, California in North America, and Chile in South America. Walnut is also cultivated in Australia and New Zealand.

Agroecology

Walnut is a temperate to sub-temperate species. In its native range, it is found on Mountain slopes from 500 to 3,000 m altitude. It thrives in temperate Himalayas from 1,000 to 3,000 m elevation. It grows in climatic zones ranging from Cool Temperate Steppe to Wet through Subtropical Thorn to Moist Forest. Walnut is reported to tolerate annual precipitation of 300–1,500 mm, evenly distributed and annual temperature of 7.0–21.1°C. Rains in late spring and summer increase walnut blight infections. Although cold hardy, walnut is sensitive to high and low temperatures. Excessive temperature above 38°C causes sunburn on hulls and poor quality kernels; temperature of 27–32°C near harvest will result in well-filled kernels with high oil content; cool summer will cause kernel to shrivel excessively. Minimum temperature should not go below −29°C. When fully dormant, trees can withstand temperatures from −24°C to −27°C without serious damage, shoot tips are killed at low temperatures of −7°C to −9°C, and frost damages new growth and young fruits. Very high summer temperatures cause sunburn damage to kernels, slightly at 38°C, severe at 40.5–43.5°C. Walnut does best in deep, well-drained and moderately fertile soil and pH of 6.5–7 is ideal. It is sensitive to alkaline and saline soils; and is intolerant of waterlogged conditions. Good windbreak is needed in areas with strong winds.

Edible Plant Parts and Uses

Walnut seed kernels are eaten raw, roasted, or salted as snacks or used in cooking, confectionary, pastries, cakes, ice cream etc. The kernels can be ground into a meal and used as a flavouring in sweet and savoury dishes. Walnut when cold pressed yields a light yellow edible oil which is used in foods as flavouring, as salad dressings or cooking. Young green, unripe fruit are pickled in vinegar and eaten; in England these are called "pickled walnuts". Walnuts are preserved in sugar syrup and eaten whole in Armenian cuisine. In Italy, liqueurs called *Nocino* and *Nocello* are flavoured with walnuts. Walnuts are ground along with other ingredients to make walnut sauce in Georgia. Walnut sap is tapped from the tree in spring to make a form of sugar. The finely ground shells are used in the stuffing of 'agnolotti' pasta and also used as adulterant of spices. The dried green husks contain 2.5–5% ascorbic acid (vitamin C) – this can be extracted and used as a vitamin supplement. The leaves when dried and crushed can be used as tea.

Botany

A deciduous, monoecious tree up to 25 m tall with tomentose juvenile shoots. Leaves are imparipinnate, 17–40 cm long with 5–7 cm long, glabrescent petiole and rachis bearing . 5–9, softly tomentose, opposite to sub-opposite leaflets on 2–4 mm long petiolules. Leaflets are ovate to elliptic-ovate, 7–20 cm long, 3–8 cm

Plate 1 Walnut fruits and leaves

Plate 3 Whole walnut

Plate 2 Mature fruit dehiscing irregularly

Plate 4 Walnut seed (kernel) and shell

broad, with acute to acuminate apex, oblique to sub-rounded bases, glabrescent to pubescent on nerves below (Plate 1). Male spike is 6–15 cm long with bracts and bracteoles, 4 pubescent, ovate tepals with and 10–20 subsessile stamens and basifixed anthers. Female spikes with pistillate flowers in clusters of 3–9. Female flowers has a tomentose involucral tube of fused bract and bracteoles, obscurely 4-toothed and irregular at the margin, 4 linear tepals alternating with the teeth, inferior ovary with short 2 mm long style bearing 2 recurved, exserted stigma. Drupe is up to 5 cm long, ovoid to subglobose; epicarp green, glandular (Plate 1); irregularly dehiscent (Plate 2) containing a wrinkled-surface, hard, 2-valved endocarp (nut) (Plates 3 and 4) with seed (Kernel) 2 to 4-lobed at the base (Plates 4 and 5).

Nutritive/Medicinal Properties

The nutrient composition of raw walnut per 100 g edible portion excluding 55% shells refuse was reported as: water 4.07 g, energy 654 kcal (2738 kJ), protein 15.23 g, fat 65.21 g, ash 1.78 g, carbohydrates 13.71 g, dietary fibre 6.7 g, total

Plate 5 Walnut kernels

sugars 2.61 g, sucrose 2.43 g, glucose 0.08 g, fructose 0.09 g, starch 0.06 g, Ca 98 mg, Fe 2,91 mg, Mg 158 mg, P 346 mg, K 441 mg, Na 2 mg, Zn 3.09 mg, Cu 1.586 mg, Mn 3.414 mg, Se 4.9 mg, vitamin C 1.3 mg, thiamine 0.341 mg, riboflavin 0.150 mg, niacin 1.125 mg, pantothenic acid 0.570 mg, vitamin B-6 0.537 mg, total folate 98 μg, total choline 39.2 mg, betaine 0.3 mg, vitamin A 20 IU, vitamin E (α-tocopherol) 0.70 mg, β-tocopherol 0.15 mg, γ-tocopherol 20.83 mg, δ-tocopherol 1.89 mg, vitamin K (phylloquinone) 2.7 μg, total saturated fatty acids 6.126 g, 16:0 4.404 g, 18:0 1.659 g, 20:0 0.063 g; total monounsaturated fatty acids 8.933 g, 18:1 undifferentiated 8.799 g, 20:1 0.134 g; total polyunsaturated fatty acids 47.174 g, 18:2 undifferentiated 38.093 g, 18:3 undifferentiated 9.080 g; phytosterols 72 mg, stigmasterol 1 mg, campesterol 7 mg, β-sitosterol 64 mg, tryptophan 0.170 g, threonine 0.596 g, isoleucine 0.625 g, leucine 1.170 g, lysine 0.424 g, methionine 0.236 g, cystine 0.208 g, phenylalanine 0.711 g, tyrosine 0.406 g, valine 0.753 g, arginine 2.278 g, histidine 0.391 g, alanine 0.696 g, aspartic acid 1.829 g, glutamic acid 2.816 g, glycine 0.816 g, proline 0.706 g, serine 0.934 g, β-carotene 12 g, lutein + zeaxanthin 9 μg (U.S. Department of Agriculture, Agricultural Research Service 2010).

An analysis conducted in Portugal (Amaral et al. 2005) reported that fat was the predominant component in walnut ranging from 62.3 to 66.5%. Eighteen fatty acids were quantified. Polyunsaturated fatty acids and, in particular, linoleic acid were predominant. Another analysis in Portugal reported that the main constituent of fruits of cultivars c (cv. Franquette, Lara, Marbot, Mayette, Mellanaise and Parisienne) was fat ranging from 78.83 to 82.14%, being the nutritional value around 720 kcal per 100 g of fruits (Pereira et al. 2008). Linoleic acid was the major fatty acid reaching the maximum value of 60.30% (cv. Lara) followed by oleic, linolenic and palmitic acids. Beta-sitosterol, δ(5)-avenasterol, and campesterol were the major sterols found. The proximate nutrient composition of 18 walnut genotypes which were newly selected from Hizan (Bitlis) located in Eastern Anatolia, Turkey was found to be as follows (mean values): protein 18.1%, total fat 58.2%, linoleic acid (50.58–66.60%), oleic acid (14.88–28.71%), linolenic acid (9.16–16.42%), and other fatty acids (traces) (Muradoglu et al. 2010). The minimum and maximum macronutrient contents of walnut were determined as mg/100 g for K (911.0–684.3), P (434.7–356.2), Ca (756.7–388.2), Mg (444.0–330.8) and Na (48.9–26.1) while minimum and maximum micronutrient contents of walnut were determined for Fe (6.6–4.3), Cu (2.8–1.8), Mn (5.7–2.7) and Zn (4.3–2.7). The potassium contents were found to be higher than those of the other minerals in all kernels of the walnuts.

The analyses of tocopherol and tocotrienol composition of nine walnut cultivars (cvs. Arco, Franquette, Hartley, Lara, Marbot, Mayette, Mellanaise, Parisienne, and Rego). showed that all samples presented a similar qualitative profile composed of five compounds, α-tocopherol, β-tocopherol, γ-tocopherol, δ-tocopherol, and γ-tocotrienol (Amaral et al. 2005). Gamma-Tocopherol was the major compound in all samples, ranging from 172.6 to 262.0 mg/kg, followed by α- and δ-tocopherols, ranging from 8.7 to 16.6 mg/kg and from 8.2 to 16.9 mg/kg, respectively. Multivariate analysis of the data obtained showed the existence of significant differences in composition among cultivars

Eight cyclopropyl fatty acids were detected in walnut oil of Mediterranean walnuts (Hanus et al. 2008). Monocyclopropane acids: methyl 9-cyclopropyl-nonanoate, 6,7-methylene-, 8,9-methylene-, 9,10-methylene-, 11,12-methylene octadecanoates, and dicyclic acid - methyl 9,10,12,13-dimethylene octadecanoate, tricyclic acid - methyl 9,10,12, 13,15,16-trimethylene octadecanoate, and unsaturated - methyl 2-octyl-cyclopropene-1-octanoate were detected in walnut oil.

The total oil content of the walnuts from ten different cultivars ranged from 62.4 to 68.7% (Zwarts et al. 1999). The oleic acid content of the oils ranged from 14.3 to 26.1% of the total fatty acids, while the linoleic acid content ranged from 49.3 to 62.3% and the linolenic contents from 8.0 to 13.8%. In another study, the total oil content from 12 different cultivars in New Zealand ranged from 62.6 to 70.3% while the crude protein ranged from 13.6 to 18.1% (Savage 2001). Dietary fibre ranged from 4.2 to 5.2% while the starch content made up no more than 2.8% of the remaining portion of the kernel. The amino acid content of the walnuts was similar between cultivars and the patterns of essential amino acids were characteristic of a high quality protein. The oil yields from Turkish walnut kernels (Büyük Oba, Kaman-2, Kaman-5 varieties) varied between 61.4% and 72.8% (Özcan et al. 2010). The crude fibre contents of kernels ranged between 3.77% and 3.80% and crude protein contents of kernels ranged between 7.05% and 8.10%. The peroxide values of kernel oils varied between 3.18 meq/Kg and 3.53 meq/Kg and acidity values ranged between 0.35% and 0.56%. The main fatty acids of walnut kernel oils were oleic, linoleic, linolenic and palmitic acids. Linoleic acid contents of kernel oils varied between 49.7% and 55.5% and oleic acid contents ranged between 20.5% and 26.4%.

The Greek walnut oil was found to be rich in neutral lipids (96.9% of total lipids) and low in polar lipids (3.1% of total lipids). The neutral lipid fraction consisted mainly of triacylglycerides whereas the polar lipids mainly consisted of sphingolipids. GC-MS data showed that the main fatty acid was linoleic acid. Unsaturated fatty acids were found as high as 85%, while the percentage of the saturated fatty acids was found 15%. Two types of liposomes were prepared from the isolated walnut oil phospholipids and characterized as new formulations. These formulations may have future applications for encapsulation and delivery of drugs and cosmetic active ingredients.

Compared to most other nuts, which contain monounsaturated fatty acids, walnuts are unique because they are rich in n-6 (linoleate) and n-3 (linolenate) polyunsaturated fatty acids (Feldman 2002). Walnuts contain multiple health-beneficial components, such as having a low lysine:arginine ratio and high levels of arginine, folate, fibre, tannins, and polyphenols.

Twenty one volatile constituents were identified in Pakistani walnut (Abbasi et al. 2010): α-thujene, sabinene, *p*-cymene, 1,8-cineol, linalool, myretenal, pinocarveol, verbenol, myrtenol, *p*-cymen-8-ol, isobutyl cyanide, benzyl alcohol, α-cadinol, α-bisabolol, isopulegol, carvacrol, estragol, globulol, viridiflorol, nerolidol and neo-iso-3-thujanol.

Pharmacological activities of walnut and its plant parts reported are elaborated below.

Antioxidant Activity

Seven phenolic compounds, pyrogallol (1), p-hydroxybenzoic acid (2), vanillic acid (3), ethyl gallate (4), protocatechuic acid (5), gallic acid (6) and 3,4,8,9,10-pentahydroxydibenzo b,d] pyran-6-one (7), containing significant antioxidant activities were isolated and identified in walnut kernels (Zhang et al. 2009). The relative order of 1,1-diphenyl-2-picrylhydrazyl radical (DPPH) scavenging capacity for these compounds was $7 > 6 \geq 4 \geq 1 > \text{Trolox} \geq 5 > 3 > 2$. The results of this study suggested that the antioxidant activities of these phenolic compounds may be influenced by the number of hydroxyls in their aromatic rings. Walnut oil extracts exhibited in-vitro antioxidant capacity in a concentration-dependent manner using 1,1-Diphenyl-2-picrylhydrazyl radical (DPPH) scavenging and phosphomolybdenum complex assays (Abbasi et al. 2010). The extract exhibited adequate in vitro 'total antioxidant activity' and 'DPPH scavenging activity' relative to butylated hydroxytoluene (standard). The IC_{50} of the oil was calculated as 51.25 μl/ ml, relative to butylated hydroxytoluene (BHT), with *IC50* of 12.1 μl/ml.

Using the reducing power assay, the scavenging effect on DPPH (2,2-diphenyl-1-picrylhydrazyl) radicals and β-carotene linoleate model system, all the walnut fruit extracts of cultivars (cv. Franquette, Lara, Marbot, Mayette, Mellanaise and Parisienne) exhibited antioxidant capacity in a concentration-dependent manner being the lowest EC_{50} values obtained with extracts of cv. Parisienne (Pereira et al. 2008). Antioxidant activity of the leaf extracts was also similarly assessed (Pereira et al. 2007). In general, all of the studied walnut leaves cultivars presented high antioxidant activity (EC_{50} values lower than 1 mg/ml), being Cv. Lara the most effective one. Ten phenolic compounds were identified and quantified: 3- and 5-caffeoylquinic acids, 3- and 4-p-coumaroylquinic acids, p-coumaric acid, quercetin 3-galactoside, quercetin 3-pentoside derivative, quercetin 3-arabinoside, quercetin 3-xyloside and quercetin 3-rhamnoside.

Total phenols content of the green husk was determined by colorimetric assay and their amount ranged from 32.61 mg/g of GAE (cv. Mellanaise) to 74.08 mg/g of GAE t (cv. Franquette) (Oliveira et al. 2008). The antioxidant capacity of aqueous extracts of the green husks was assessed similarly. A concentration-dependent antioxidative capacity was verified in reducing power and DPPH assays, with EC_{50} values lower than 1 mg/ml for all the tested extracts. The antioxidant capacity of methanolic extract of Persian walnut green husk was assessed through reducing power assay, DPPH-scavenging effect, FRAP assay and oven test in sunflower oil (Rahimipanah et al. 2010). EC_{50} values of the extract in reducing power and DPPH assays were 0.19 and 0.18 mg/ml respectively. The FRAP values of extract and Trolox at concentration of 100 μg/ml didn't show any significant difference. The 400 ppm extract was as effective as 200 ppm BHA in retarding sunflower oil deterioration at 600°C. Total flavonoids and phenolics also were determined to be 144.65 mg quercetin and 3428.11 mg gallic acid equivalent per 100 g of dry sample respectively. The results indicated Persian walnut green husk to be a good source of antioxidant compounds to add to food.

Studies showed that 2,2′-Azobis'(2-amidino propane) hydrochloride (AAPH)-induced LDL oxidation was significantly inhibited by 87 and 38% with the highest concentration (1.0 μmol/L) of ellagic acid and walnut extract, respectively (Anderson et al. 2001). In addition, copper-mediated LDL oxidation was inhibited by 14 and 84% in the presence of ellagic acid and walnut extract, respectively, with a modest, significant LDL α-tocopherol sparing effect observed. Plasma thiobarbituric acid reacting substance (TBARS) formation was significantly inhibited by walnut extracts and ellagic acid in a dose-dependent manner, and the extracts exhibited a Trolox equivalent antioxidant activity (TEAC) value greater than that of α-tocopherol. LC-ELSD/MS analysis of the walnut extracts identified ellagic acid monomers, polymeric ellagitannins and other phenolics, principally nonflavonoid compounds. These results demonstrated that walnut polyphenolics were effective inhibitors of in-vitro plasma and LDL oxidation. The polyphenolic content of walnuts should be considered when evaluating their anti-atherogenic potential.

Studies on the antioxidant potential of aqueous extract of walnut bark showed that it had modulatory effect on cyclophosphamide-induced urotoxicity in Swiss albino male mice (Haque et al. 2003; Bhatia et al. 2006). Plant extract + cyclophosphamide group animals showed restoration in the level of cytochrome P450 content and in the activities of glutathione S-transferase (GST), glutathione peroxidase (GP) and catalase (CAT) in both liver and kidneys. But plant extract restored the activity of superoxide dismutase (SOD) and the level of reduced glutathione (GSH) in the kidneys only when compared with cyclophosphamide-treated animals. Plant extract treatment alone caused significant reduction in the content of cytochrome P450 in the kidneys mainly. The extract showed a significant increase in the level of GSH and in the activities of GP in both the tissues and CAT in liver only, whereas no significant change was observed in the activities of GST and SOD. cyclophosphamide treatment resulted in a significant increase in the lipid peroxidation in the liver and kidneys compared with controls, while the extract + cyclophosphamide treated group showed a significant decrease in the lipid peroxidation in liver and kidneys when compared with the cyclophosphamide-treated group. The study showed that the use of *J. regia* extract might be helpful in abrogation of cyclophosphamide toxicity during the chemotherapy. In a subsequent study, walnut bark extract treatment (150 mg/kg p.o. × 10 days) resulted in protective restoration of decreased antioxidants in (cyclophosphamide)-treated (18 mg/kg i.p. × 10 days) animals. Cyclophosphamide treatment decreased activities of CAT, GP, glutathione reductase and (GST) and in GSH content in urinary bladder and concomitantly significantly increased lipid peroxidation. Administration of extract restored all the antioxidants significantly and lowered the elevated lipid peroxidation in the bladder. A correlation between radical scavenging capacities of the extract with phenolic content was observed thus justifying its antioxidant potential against oxidative stress-mediated urotoxicity in mice. Its protective effect on cyclophosphamide -induced toxicity in bladder warranted further clinical investigations.

Three hydrolyzable tannins, glansrins A-C, together with adenosine, adenine, and 13 known tannins were isolated from the n-BuOH extract of walnuts (the seeds of *Juglans regia*) (Fukuda et al. 2003). Glansrins A-C were characterized as ellagitannins with a tergalloyl group, or related polyphenolic acyl group, based on spectral and chemical evidence. The 14 walnut polyphenols had superoxide dismutase (SOD)-like activity with EC_{50} 21.4–190 μM and a remarkable radical scavenging effect against 1,1-diphenyl-2-picrylhydrazyl (DPPH) (EC_{50} 0.34–4.72 μM).

Studies in rats showed that melatonin was found in walnuts and, when eaten, increased blood melatonin levels (Reiter et al. 2005). The increase in blood melatonin concentrations correlated with an increased antioxidative capacity of this fluid as reflected by augmentation of Trolox equivalent antioxidant capacity and ferric-reducing ability of serum values. Melatonin is a hormone that participates in the protection of nuclear and mitochondrial DNA.

Antihypercholesterolemic/ Cardiovascular Disease Activities

Though walnuts are energy rich, clinical dietary intervention studies show that walnut consumption does not cause a net gain in body weight when eaten as a replacement food (Feldman 2002). Five controlled, peer-reviewed, human clinical walnut intervention trials, involving approximately 200 subjects representative of the 51% of the adult population in the United States at risk of coronary heart disease were reviewed. The intervention trials consistently demonstrated walnuts as part of a heart healthy diet, lower blood cholesterol concentrations. These results were supported by several large prospective observational studies in humans, all demonstrating a dose response related inverse association of the relative risk of coronary heart disease with the frequent daily consumption of small amounts of nuts, including walnuts.

Studies showed that walnuts could reverse postprandial endothelial dysfunction associated with consumption of a fatty meal (Cortés et al. 2006). Compared with a Mediterranean diet, a walnut diet was shown to improve endothelial function in hypercholesterolemic patients. Supplemental walnuts, but not olive oil, counteracted the detrimental changes in FMD (flow-mediated dilation) associated with eating a fatty meal. Adding walnuts to a high-fat meal acutely improved FMD independently of changes in oxidation, inflammation, or plasma asymmetric dimethylarginine (ADMA). Nevertheless, unsaturated fatty acids and antioxidants in both olive oil and walnuts appeared to preserve the protective phenotype of endothelial cells. Endothelial dysfunction was found to be associated with coronary artery disease CAD and its risk factors and could be reversed by antioxidants and marine n-3 fatty acids as are found abundantly in walnuts. Substituting walnuts for monounsaturated fat in a Mediterranean diet was shown to improve endothelium-dependent vasodilation in hypercholesterolemic subjects. Further the scientists reported that the walnut diet significantly reduced total cholesterol (−4.4%) and LDL cholesterol

(−6.4%) (Ros et al. 2004). Cholesterol reductions correlated with increases of both dietary α-linolenic acid and LDL γ-tocopherol content, and changes of endothelium-dependent vasodilation correlated with those of cholesterol-to-HDL ratios. γ-tocopherol another strong antioxidant occurred abundant in walnuts. The walnut diet provided 90 to 135 mg/d of this vitamin E component and increased its serum level about two-fold. The scientists postulated LDL γ-tocopherol to be a marker of the bioavailability of the walnut constituents responsible for lowering cholesterol. Thus elucidating the cardioprotective effect of walnut intake beyond cholesterol lowering.

Epidemiological studies suggested that nut intake decreased coronary artery disease (CAD) risk. A more recent study found that the mechanisms responsible were associated with reduced coronary vascular disease (CVD) risk from walnut consumption that involved nonplasma lipid–related effects on endothelial function (Davis et al. 2006). Aortic cholesterol ester concentration, a measure of atherosclerotic plaque, was highest in the lowest α-tocopherol only group and declined significantly with increasing α-tocopherol. The aortic cholesterol ester of all walnut groups was decreased significantly relative to the lowest α-tocopherol only group but showed no dose response. The diets did not produce changes in the other vascular stress markers, whereas aortic endothelin-1 mRNA levels declined dramatically with increasing dietary walnuts but were unaltered in the α- tocopherol groups or γ-tocopherol group. The mechanisms underlying those results were postulated to be mediated in part by aortic endothelin 1–dependent mechanisms. The contrasting results between the α-tocopherol or γ-tocopherol diets and the walnut diets also make it unlikely that the non-plasma lipid–related CVD effects of walnuts were due to their α-tocopherol or γ-tocopherol content.

In another study, walnut meat tested appeared to be a functional food as it improved the antioxidant status of increased CHD-risk volunteers (Canales et al. 2007). Erythrocyte catalase (CAT), superoxide dismutase (SOD), total glutathione, reduced glutathione (GSH), oxidized glutathione (GSSG), and lipid peroxidation (LPO), and serum uric acid and paraoxonase-1 (PON1) were modified at increased CHD-risk individuals consuming walnut-enriched mea. Despite its high energy content, it also appeared adequate for overweight and obese people because it did not exert negative effect upon body weight. A significant interaction time X treatment was observed in all enzymes and substrates tested except HDL-C, uric acid and LPO. The treatment significantly increased CAT activity, total glutathione and GSSG. Significant gender X time X treatment interaction for total glutathione was found increasing at the end of the walnut meal period in male but not changing in female. Total glutathione and reduced glutathione/oxidized glutathione ratio (GSH/GSSG) ratio were lower in smokers. Hypercholesterolemics presented higher uric acid but no enzyme activities or substrate concentrations were different from those of normocholesterolemics.

In a randomized case–control study, 26 hyperlipidemic volunteers in Iran, (Zibaeenezhad et al. 2005), it was shown that frequent consumption of nuts in the daily diet was associated with a potentially decreased risk of coronary artery disease by decreasing the level of triglyceride (TG) and increasing the level of HDL. In the group consuming walnuts, 20 g per day for 8 weeks, the mean plasma TG level dropped by 17.1% from the baseline and HDL cholesterol also increased significantly by 9%. In another randomized, double blind case–control study (Zibaeenezhad et al. 2003), the scientists found that in hyperlipidemic patients (n = 29) receiving walnut oil encapsulated in 500 mg capsules, 3 g/day, for 45 days, plasma TG concentrations decreased by 19–33% of baseline in group A patients. No statistically significant change was observed in other measured parameters. Walnut oil was concluded to be a good antihypertriglyceridemic natural remedy.

In a health screening survey in France, (Lavedrine et al. 1999) it was found that a high level of HDL cholesterol and apolipoprotein apo A1 was associated with a high amount of walnut consumption (oil and kernel) in the regular diet, with a positive trend with increasing degree of walnut consumption. This association did not appear to be confounded by dietary animal fat and alcohol as measured in this study. Other blood lipids did not show significant associations with walnut consumption. The positive effect of walnut consumption on blood HDL cholesterol and apo A1 is of special interest since these lipid parameters had been shown to be negatively correlated with cardiovascular morbidity.

In a randomized, cross-over feeding trial involving 10 men with polygenic hypercholesterolemia in Spain (Munoz et al. 2001), the walnut diet reduced serum total and LDL cholesterol by 4.2%, and 6.0% respectively compared with the control diet. No changes were observed in HDL cholesterol, triglycerides, and apolipoprotein A-I levels or in the relative proportion of protein, triglycerides, phospholipids, and cholesteryl esters in LDL particles. The apolipoprotein B level declined in parallel with LDL cholesterol (6.0% reduction). Whole LDL, particularly the triglyceride fraction, was enriched in polyunsaturated fatty acids from walnuts (linoleic and alpha-linolenic acids). In comparison with LDL obtained during the control diet, LDL obtained during the walnut diet showed a 50% increase in association rates to the LDL receptor in human hepatoma HepG2 cells. LDL uptake by HepG2 cells was correlated with α linolenic acid content of the triglyceride plus cholesteryl ester fractions of LDL particles. Changes in the quantity and quality of LDL lipid fatty acids after a walnut-enriched diet facilitated receptor-mediated LDL clearance and may contribute to the cholesterol-lowering effect of walnut consumption.

In another randomized, crossover feeding trial in Spain (Zambone et al. 2000) the acceptability of walnuts and their effects on serum lipid levels and low-density lipoprotein (LDL) oxidizability in 49 free-living hypercholesterolemic persons was evaluated. Compared with the Mediterranean diet, the walnut diet produced mean changes of −4.1% in total cholesterol level, −5.9% in LDL cholesterol level, and −6.2% in lipoprotein(a) level. The mean differences in the changes in serum lipid levels were −0.28 mmol/L for total cholesterol level, −0.29 mmol/L for LDL cholesterol level, and −0.021 g/L for lipoprotein(a) level. Lipid changes were similar in men and women except for lipoprotein(a) levels, which decreased only in men. Low-density lipoprotein particles were enriched with polyunsaturated fatty acids from walnuts, but their resistance to oxidation was preserved. The studies concluded that substituting walnuts for part of the mono-unsaturated fat in a cholesterol-lowering Mediterranean diet further reduced total and LDL cholesterol levels in men and women with hypercholesterolemia.

In a recent meta-analysis exercise, high-walnut-enriched diets was found to significantly decrease total and LDL cholesterol for the duration of the short-term trials (Banel and Hu 2009). Thirteen studies representing 365 participants were included in the analysis. Diets lasted 4–24 week with walnuts providing 10–24% of total calories. When compared with control diets, diets supplemented with walnuts resulted in a significantly greater decrease in total cholesterol and in LDL-cholesterol concentrations. HDL cholesterol and triglycerides were not significantly affected by walnut diets more than with control diets. Other results reported in the trials indicated that walnuts provided significant benefits for certain antioxidant capacity and inflammatory markers and had no adverse effects on body weight.

Antidiabetic Activity

Fasting blood sugar decreased meaningfully in diabetic rats treated with *J. regia* and diabetic rats treated with glibenclamide (Asgary et al. 2008). Insulin level increased and glycosylated hemoglobin decreased significantly in diabetic groups receiving either glibenclamide or *J. regia* compared with the diabetic group with no treatment. The histological study revealed that the size of islets of Langerhans enlarged consequentially as compared with diabetic rats with no treatment. Effects of administering glibenclamide or extract of *J. regia* on all parameters discussed above showed no difference, and both tended to bring the values to near normal. The data showed that the ethanolic extract from leaves of *J. regia* had a dramatic antidiabetic effect on diabetes-induced rats. In a separate study, the methanolic extract of *J. regia* leaves was found to have a significant hypoglycemic action in alloxan-induced diabetic rats both in the short and long term (Teimori et al. 2010). Decrease in postprandial blood glucose level was observed at 8 hours after treatment and more in acarbose treated diabetic rats (53%), followed by the diabetic rats treated with 250 mg/kg of the extract (40%) and diabetic rats treated with 500 mg/kg of the extract as compared to the diabetic untreated control rats. There was also permanent postprandial blood glucose reduction in treated groups in comparison with the diabetic untreated control one during long term period. No change in the insulin and glut-4 genes expression was observed. However, in-vitro assay of α-glucosidase activity displayed inhibitory action of the extract, like acarbose the reference drug, but less effectively.

Antiinflammatory/ Antiatherogenic Activities

Walnut extract and its component ellagic acid were found to exhibit anti-inflammatory activity in human aorta endothelial cells and osteoblastic activity in the cell line KS483 (Papoutsi et al. 2008). Walnut extract and ellagic acid decreased significantly the TNF-alpha-induced endothelial expression of both VCAM-1 and ICAM-1. Both walnut extract (at 10–25 μg/ml) and ellagic acid induced nodule formation in KS483 osteoblasts. The results suggested that the walnut extract had a high anti-atherogenic potential and a remarkable osteoblastic activity, an effect mediated, at least in part, by its major component ellagic acid. Such findings supported the beneficial effect of a walnut-enriched diet on cardioprotection and bone loss.

Antiamyloidogenic Activity/ Alzheimer's Disease

Studies showed that that walnuts may reduce the risk or delay the onset of Alzheimer's disease by maintaining fibrillar amyloid beta-protein Aß in the soluble form (Chauhan et al. 2004). Fibrillar amyloid beta-protein (Aβ) is the principal component of amyloid plaques in the brains of patients with Alzheimer's disease. The walnut extract not only inhibited Aß fibril formation in a concentration and time-dependent manner but it was also able to defibrillize (breakdown) Aβ preformed fibrils. Further studies showed the anti-amyloidogenic compound in walnut to be an organic compound of molecular weight less than 10 kDa, and not a lipid nor a protein. It was proposed that polyphenolic compounds (such as flavonoids) present in walnuts may be responsible for its anti-amyloidogenic activity.

Anticancer/Antimutagenic Activity

Two new diarylheptanoids, juglanin A (1) and B (2), together with 15 known compounds, were isolated from the extract of the seed husks of walnuts (Liu et al. 2008a, b). Compounds 1, 2, and 10 displayed cytotoxic activity against human hepatoma (Hep G2) cells.

Juglone (5-hydroxy-1,4-naphthoquinone) and plumbagin (5-hydroxy-3-methyl-1,4-naphthoquinone), yellow pigments found in walnut, exhibited cytotoxicity against HaCaT keratinocytes (Inbaraj and Chignell 2004). The cytotoxicity of these quinones was due to two different mechanisms, namely, redox cycling and reaction with glutathione (GSH). The findings indicated that topical preparations containing juglone and plumbagin should be used with care as their use may damage the skin. However, it was probable that the antifungal, antiviral, and antibacterial properties of these quinones resulted from redox cycling. Herbal preparations derived from walnut have been used as hair dyes and skin colorants in addition to being applied topically for the treatment of acne, inflammatory diseases, ringworm, and fungal, bacterial, or viral infections.

The antimutagenic study using TA98 and TA100 tester strains of *Salmonella* revealed the water and acetone extracts of walnut to be more effective than the benzene and chloroform extracts in inhibiting the revertants induced by

2-aminoflourene (2AF) in TA100 tester strains (Kaur et al. 2003). The acetone extract of walnut exhibited a correlation of antimutagenic activities in the Ames assay with its antiproliferative effect in different cell lines, while the water extract exerted its effect distinctly in each cell line.

An 8-week walnut supplement randomised study was conducted on 21 subjects to examine effects of walnuts on serum tocopherols and prostate specific antigen (PSA) (Spaccarotella et al. 2008). Following the 8-week supplement an increase in serum gamma-tocopherol was noted concomitant with significant decrease in the α-topherol:gammatocopherol ratio and a trend towards an increase in the ratio of free PSA:total PSA. The results suggested that walnuts may improve biomarkers of prostate and vascular status. Tocopherols may protect against prostate cancer and cardiovascular disease (CVD).

Antiviral Activity

Studies in China reported that *J. regia* had anti-HIV activity Two extracts from *J. regia* were found to possess anti-HIV activity (Liu et al. 2008a, b). Extract B affected HIV-1 gp-41 fusing protein and extract E affected HIV-1 integrase respectively.

Antimicrobial Activity

Walnut fruit aqueous extract exhibited differential antibacterial activity against Gram positive (*Bacillus cereus, Bacillus subtilis, Staphylococcus aureus*) and Gram negative bacteria (*Pseudomonas aeruginosa, Escherichia coli, Klebsiella pneumoniae*) and fungi (*Candida albicans, Cryptococcus neoformans*) (Pereira et al. 2008). In another study, the antimicrobial capacity of the green husk extracts was investigated against Gram positive and Gram negative bacteria, and fungi (Oliveira et al. 2008). All the extracts inhibited the growth of Gram positive bacteria, with *Staphylococcus aureus* the most susceptible with MIC of 0.1 mg/ml. The results obtained indicated that walnut green husks may become an important source of compounds with health protective potential and antimicrobial activity. Walnut leaf extract was also found to selectively inhibit the growth of Gram positive bacteria, with *Bacillus cereus* being the most susceptible (MIC 0.1 mg/ml). Gram negative bacteria and fungi were resistant to the extracts at 100 mg/ml (Pereira et al. 2007).

Bark and leaf crude extracts of *Juglans regia, Juglans mollis, Carya illinoensis* and *Bocconia frutescens* showed in-vitro anti-*Mycobacterium tuberculosis* activity (Cruz-Vega et al. 2008). Hexane bark extracts from *C. illinoensis, J. mollis* and *J. regia* were the most active with a minimal inhibitory concentration (MIC) of 31, 50 and 100 μg/ml, respectively. Leaf extracts had the lowest activity. None of the aqueous extracts exhibited antimycobacterial activity.

Psidium guajava and *Juglans regia* leaf extracts were found to have inhibitory effect on bacteria isolated from acne lesions (Qadan et al. 2005) The zones of inhibition due to the *Psidium guajava* and *Juglans regia* leaf extracts ranged from 15.8 to 17.6 mm against *Propionibacterium acnes*, 11.3–15.7 mm against *Staphylococcus aureus* and 12.9–15.5 mm against *Staphylococcus epidermidis,* respectively. These zones of inhibition were significantly higher than those of tea tree oil and equivalent in case of Staphylococci spp., but less in case of *Propionibacterium acnes*, to those obtained from doxycycline or clindamycin. The study concluded that *Psidium guajava* and *Juglans regia* leaf extracts may be beneficial in treating acne especially when they are known to have anti-inflammatory activities.

Walnut bark extract showed a broad spectrum antimicrobial activity in a dose dependent manner (Alkhawajah 1997). It inhibited the growth of several species of pathogenic micro-organisms representing Gram-positive bacteria (*Staphylococcus aureus* and *Streptococcus mutans*), Gram-negative bacteria (*Esherichia coli* and *Pseudomonas aeruginosa*) and a pathogenic yeast (*Candida albicans*). The extract had either synergistic or additive action when tested with a wide range of antibacterial drugs. It also increased the pH of saliva. The author asserted that brushing teeth with this bark extract may improve oral hygiene, prevent plaque and caries formation, and reduce the incidence of gingival and periodontal infections.

Studies found that walnut extract inhibited the growth of *Escherichia coli* in-vitro and showed a potential action to increase the SnCl2 effect on plasmid DNA (Santos-Filho et al. 2008). The extract failed to induce modifications in the DNA mobility in agarose gel but was capable in decreasing the distribution of 99mTc on the blood cell compartment probably due to redoxi properties.

Lung-Protective Activity

Prophylactic treatment of methanolic extract of walnut kernel by gavage to Wistar rats for 1 week prior to cigarette smoke extract (CSE) exposure through intratracheal instillation protected against CSE-induced acute lung toxicity (Qamar and Sultana 2011). The extract significantly decreased the levels of lactate dehydrogenas, total cell count, total protein and increased the GSH (reduced glutathione) level in bronchoalveolar lavage fluid. It also significantly restored the levels of glutathione reductase, catalase and reduced the xanthine oxidase activity in lung tissue. Total polyphenolic content of walnut kernel extract was found to be 96 mg gallic acid equivalent (GAE)/g dry weight of extract. In DPPH (2,2-diphenyl-1-picrylhydrazyl) assay, the extract exhibited high free radical scavenging potential.

Allergy Problem

English walnuts had been implicated in severe, IgE-mediated food allergy in humans. Co-administration of walnut polyphenolic-enriched extract with antigen and alum increased serum concentrations of antigen-specific IgE and IgG1 in BALB/c mice (Comstock et al. 2010). Serum IgG2a/2b levels were similar between mice receiving ovalbumin/alum and ovalbumin/alum with polyphenolics. Thus, walnut polyphenolic extract enhanced the Th2-skewing effect of an aluminum hydroxide adjuvant. This indicated that walnut polyphenolic compounds may play a role in allergic sensitization of genetically predisposed individuals.

Traditional Medicinal Uses

The walnut tree has a long history of traditional medicinal uses. Various parts of walnut have been reported in folkloric medicine to treat a wide range of ailments and complaints. The fruit, when young and unripe, makes a wholesome, anti-scorbutic pickle, the vinegar in which the green fruit has been pickled proving a capital gargle for sore and slightly ulcerated throats. Walnut catsup embodies the medicinal virtues of the unripe nuts. The shell is anodyne and astringent. It is used to treat diarrhoea and anaemia. The husk, shell and peel are sudorific, especially if used when the walnuts are green and unripe nut has anthelmintic property. The seeds are antilithic, diuretic and stimulant. They are used internally to treat lower back pain, frequent urination, weakness of legs, chronic cough, asthma, constipation due to dryness or anaemia and stones in the urinary tract. Externally, the seeds are pulverised into a paste and applied as a poultice to areas of dermatitis and eczema. The seed oil is anthelmintic and the oils also used in the treatment of menstrual problems and dry skin conditions. In traditional Chinese medicine, walnut seeds are primarily considered a kidney tonic. They are also considered beneficial to the brain, back, and skin, and to relieve constipation caused by dehydration. The cotyledons are used in the treatment of cancer. Male inflorescences are made into a broth and used in the treatment of coughs and vertigo. The leaves are described as alterative, anthelmintic, anti-inflammatory, astringent and depurative. They are used internally to treat constipation, chronic coughs, asthma, diarrhoea, dyspepsia. The leaves are also used to treat skin ailments and detoxify the blood. The bark and leaves have alterative, laxative, astringent and detergent properties, and are used in the treatment of skin disorders: as remedy for scrofulous diseases, herpes, eczema, etc., and for healing indolent ulcers. The bark and root bark are anthelmintic, astringent and detergent. The bark, dried and powdered, and made into a strong infusion, is a useful purgative.

Other Uses

Walnut trees are often grown as ornamental and are cultivated in parks and large gardens.

Walnut is valued for its attractive high quality wood besides its edible fruit. The wood is hard, dense, tight-grained and polishes to a very smooth finish and thee colour ranges from creamy white in the sapwood to a dark chocolate colour in the heartwood. The wood is excellent for furniture, carving, and rifle stocks. Walnut burrs are commonly used to create bowls and other turned pieces. Veneer sliced from walnut burl is one of the most valuable and highly prized by cabinet makers and prestige car manufacturers. Walnut is also used in lutherie, i.e. making guitar bodies.

Walnut fruit, when dry pressed, yields valuable oil used in paints, ink and in soap-making. Walnut ink is dark brown in colour and is made by boiling the whole fruit or letting it oxidize. It can also be used to stain wood. The nuts can be used as a wood polish by rubbing the kernel into the wood to release the oils after cracking the shell. Young fruits are also used as fish poison. The rind of unripe fruits provides a good source of tannin.

Walnut shells provide a rich yellow-brown to dark brown dye that is used for dyeing fabric, as a colouring and tonic for dark hair and for other purposes The shells may be used as antiskid agents for tires, blasting grit, and in the preparation of activated carbon. Flour made from walnut shells is widely used in the plastics industry. Walnut shells can be used to clean soft metals, fiber-glass, plastics, wood and stone. Other uses include cleaning automobile and jet engines, electronic circuit boards, and paint and graffiti removal. This soft grit abrasive is well suited for air blasting, de-burring, de-scaling, and polishing operations because of its elasticity and resilience. The shell is used widely in oil well drilling for lost circulation material in making and maintaining seals in fracture zones and unconsolidated formations. Walnut shells are added to paint to give it a thicker consistency for "plaster effect". It is used as filler in dynamite. Occasionally, it is used as a cosmetic cleaner in soap and exfoliating cleansers.

A decoction of leaves, bark, and husks are used with alum for staining wool brown. Crushed leaves, or a decoction is used as insect repellant. The leaves contain juglone, which has been shown to have insecticidal and herbicidal properties. Juglone is also secreted from the roots of the tree; it exhibits allelopathic effects i.e. inhibitory effect on the growth of many other plants. Twigs and leaves are lopped for fodder in India. Bark of the tree and the fruit rind are dried and used as a tooth cleaner. The bark provides gums and in the local market it is sold under the name of 'dandasa'.

Walnuts also have religious significance. In Jammu, India it is used widely as a *prasad* (offering) to Mother Goddess Vaisnav Devi and, generally, as a dry food in the season of festivals such as Diwali.

Comments

Globally, USA is the leading walnut producing country followed by China, Iran, Turkey, Ukraine, Romania, India, France, Yugoslavia and Greece.

Selected References

Abbasi MA, Ali Raza A, Riaz T, Shahzadi T, Aziz-ur-Rehman, Jahangir M, Shahwar D, Siddiqui SZ, Chaudhary AR, Ahmad N (2010) Investigation on the volatile constituents of *Juglans regia* and their in vitro antioxidant potential. Proc Pak Acad Sci 47(3):137–141

Alkhawajah AM (1997) Studies on the antimicrobial activity of *Juglans regia*. Am J Chin Med 25(2): 175–180

Amaral JS, Alves MR, Seabra RM, Oliveira BP (2005) Vitamin E composition of walnuts (*Juglans regia* L.): a 3-year comparative study of different cultivars. J Agric Food Chem 53(13):5467–5472

Anderson KJ, Teuber SS, Gobeille A, Cremin P, Waterhouse AL, Steinberg FM (2001) Walnut polyphenolics inhibit in vitro human plasma and LDL oxidation. J Nutr 131(11):2837–2842

Asgary S, Parkhideh S, Solhpour A, Madani H, Mahzouni P, Rahimi P (2008) Effect of ethanolic extract of *Juglans regia* L. on blood sugar in diabetes-induced rats. J Med Food 11(3):533–538

Banel DK, Hu FB (2009) Effects of walnut consumption on blood lipids and other cardiovascular risk factors: a meta-analysis and systematic review. Am J Clin Nutr 90(1):56–63

Bhatia K, Rahman S, Ali M, Raisuddin S (2006) In vitro antioxidant activity of *Juglans regia* L. bark extract and its protective effect on cyclophosphamide-induced urotoxicity in mice. Redox Rep 11(6):273–279

Bown D (1995) Encyclopaedia of herbs and their uses. Dorling Kindersley, London, 424 pp

Bull P, Jackson D, Bedford T (1985) Edible tree nuts in New Zealand. V.R. Ward, Government Printer, Wellington, 79 pp

Canales J, Benedi M, Nus J, Librelotto J, Sanchez-Montero M, Sanchez-Muniz FJ (2007) Effect of walnut-enriched restructured meat in the antioxidant status of overweight/obese senior subjects with at least one extra CHD-risk factor. J Am Coll Nutr 26(3): 225–232

Chauhan N, Wang KC, Wegiel J, Malik MN (2004) Walnut extract inhibits the fibrillization of amyloid beta-protein, and also defibrillizes its preformed fibrils. Curr Alzheimer Res 1(3):183–188

Chopra RN, Nayar SL, Chopra IC (1986) Glossary of Indian medicinal plants. (Including the supplement). Council Scientific Industrial Research, New Delhi, 330 pp

Comstock SS, Gershwin LJ, Teuber SS (2010) Effect of walnut (*Juglans regia*) polyphenolic compounds on ovalbumin-specific IgE induction in female BALB/c mice. Ann NY Acad Sci 1190(1):58–69

Cortés B, Núñez I, Cofán M, Gilabert R, Pérez-Heras A, Casals E, Deulofeu R, Ros E (2006) Acute effects of high-fat meals enriched with walnuts or olive oil on postprandial endothelial function. J Am Coll Cardiol 48:1666–1671

Cruz-Vega DE, Verde-Star MJ, Salinas-González N, Rosales-Hernández B, Estrada-García I, Mendez-Aragón P, Carranza-Rosales P, González-Garza MT, Castro-Garza J (2008) Antimycobacterial activity of *Juglans regia, Juglans mollis, Carya illinoensis* and *Bocconia frutescens*. Phytother Res 22(4):557–559

Davis P, Valacchi G, Pagnin E, Shao Q, Gross HB, Calo L, Yokoyama W (2006) Walnuts reduce aortic ET-1 mRNA levels in hamsters fed a high-fat, atherogenic diet. J Nutr 136(2):428–432

Duke JA (1983) *Juglans regia* L. In: Handbook of energy crops. Purdue University, Center for New Crops & Plants Products, Lafayette, http://newcrop.hort.purdue.edu/newcrop/duke_energy/Juglans_regia.html

Duke JA, Ayensu ES (1985) Medicinal plants of China, vol 1 & 2. Reference Publications, Inc, Algonac, 705 pp

Facciola S (1990) Cornucopia. A source book of edible plants. Kampong Publications, Vista, 677 pp

Feldman EB (2002) The scientific evidence for a beneficial health relationship between walnuts and coronary heart disease. J Nutr 132(5):1062S–1101S

Fukuda T, Ito H, Yoshida T (2003) Antioxidative polyphenols from walnuts (*Juglans regia* L.). Phytochemistry 63(7):795–801

Grieve M (1971) A modern herbal, vol 2. Penguin. Dover Publications, New York, 919 pp

Hanus LO, Goldshlag P, Dembitsky VM (2008) Identification of cyclopropyl fatty acids in walnut (*Juglans regia* L.) oil. Biomed Pap Med Fac Univ Palacky Olomouc Czech Repub 152(1):41–45

Haque R, Bin-Hafeez B, Parvez S, Pandey S, Sayeed I, Ali M, Raisuddin S (2003) Aqueous extract of walnut (*Juglans regia* L.) protects mice against cyclophosphamide-induced biochemical toxicity. Hum Exp Toxicol 22(9):473–480

Inbaraj JJ, Chignell CF (2004) Cytotoxic action of juglone and plumbagin: a mechanistic study using HaCaT keratinocytes. Chem Res Toxicol 17(1):55–62

Kaur K, Michael H, Arora S, Härkönen PL, Kumar S (2003) Studies on correlation of antimutagenic and antiproliferative activities of *Juglans regia* L. J Environ Pathol Toxicol Oncol 22(1):59–67

Lavedrine F, Zmirou D, Ravel A, Balducci F, Alary J (1999) Blood cholesterol and walnut consumption: a cross-sectional survey in France. Prev Med 28(4): 333–339

Liu JX, Di DL, Wei XN, Han Y (2008a) Cytotoxic diarylheptanoids from the pericarps of walnuts (*Juglans regia*). Planta Med 74(7):754–759

Liu ZM, Wen RX, Ma HT, Yang YS, Wang XL, Lu XH, Li ZL (2008b) Investigation on inhibition of HIV III B virus with extractions of *Juglans regia*. Zhongguo Zhong Yao Za Zhi 33(21):2535–2538, Article in Chinese

Muñoz S, Merlos M, Zambón D, Rodríguez C, Sabaté J, Ros E, Laguna JC (2001) Walnut-enriched diet increases the association of LDL from hypercholesterolemic men with human HepG2 cells. J Lipid Res 42(12):2069–2076

Muradoglu F, Oguz HI, Yildiz K, Yilmaz H (2010) Some chemical composition of walnut (*Juglans regia* L.) selections from Eastern Turkey. Afr J Agric Res 5(17): 2379–2385

Oliveira I, Sousa A, Ferreira IC, Bento A, Estevinho L, Pereira JA (2008) Total phenols, antioxidant potential and antimicrobial activity of walnut (*Juglans regia* L.) green husks. Food Chem Toxicol 46(7):2326–2331

Ozcan MM, Iman C, Arslan D (2010) Physico-chemical properties, fatty acid and mineral content of some walnuts (*Juglans regia* L.) types. Agric Sci 1(2):62–67

Papoutsi Z, Kassi E, Chinou I, Halabalaki M, Skaltsounis LA, Moutsatsou P (2008) Walnut extract (*Juglans regia* L.) and its component ellagic acid exhibit anti-inflammatory activity in human aorta endothelial cells and osteoblastic activity in the cell line KS483. Br J Nutr 99(4):715–722

Pereira JA, Oliveira I, Sousa A, Valentão P, Andrade PB, Ferreira IC, Ferreres F, Bento A, Seabra R, Estevinho L (2007) Walnut (*Juglans regia* L.) leaves: phenolic compounds, antibacterial activity and antioxidant potential of different cultivars. Food Chem Toxicol 45(11):2287–2295

Pereira JA, Oliveira I, Sousa A, Ferreira IC, Bento A, Estevinho L (2008) Bioactive properties and chemical composition of six walnut (*Juglans regia* L.) cultivars. Food Chem Toxicol 46(6):2103–2111

Qadan F, Thewaini AJ, Ali DA, Afifi R, Elkhawad A, Matalka KZ (2005) The antimicrobial activities of *Psidium guajava* and *Juglans regia* leaf extracts to acne-developing organisms. Am J Chin Med 33(2): 197–204

Qamar W, Sultana S (2011) Polyphenols from *Juglans regia* L. (Walnut) kernel modulate cigarette smoke extract induced acute inflammation, oxidative stress and lung injury in Wistar rats. Hum Exp Toxicol 30(6): 499–506

Rahimipanah M, Hamedi M, Mirzapour M (2010) Antioxidant activity and phenolic contents of Persian walnut (*Juglans regia* L.) green husk extract. Afr J Food Sci Technol 1(4):105–111

Reiter RJ, Manchester LC, Tan DX (2005) Melatonin in walnuts: influence on levels of melatonin and total antioxidant capacity of blood. Nutrition 21(9):920–924

Ros E, Isabel Núñez I, Pérez-Heras A, Serra M, Gilabert R, Casals E, Deulofeu R (2004) A walnut diet improves endothelial function in hypercholesterolemic subjects. A randomized crossover trial. Circulation 109: 1609–1614

Santos-Filho SD, Diniz CL, Do Carmo FS, Da Fonseca ADS, Bernardo-Filho M (2008) Influence of an extract of *Juglans regia* on the growth of *Eschericia coli*, on the electrophoretic profile of plasmid DNA and on the radiolabeling of blood constituents. Braz Arch Biol Technol 51:63–168

Savage GP (2001) Chemical composition of walnuts (*Juglans regia* L.) grown in New Zealand. Plant Foods Hum Nutr 56(1):75–82

Spaccarotella KJ, Kris-Etherton PM, Stone WL, Bagshaw DM, Fishell VK, West SG, Lawrence FR, Hartman TJ (2008) The effect of walnut intake on factors related to prostate and vascular health in older men. Nutr J 7:13

Teimori M, Kouhsari SM, Ghafarzadegan R, Hajiaghaee R (2010) Study of hypoglycemic effect of *Juglans regia* leaves and its mechanism. J Med Plant 9(6):57–65

Tsamouris G, Hatziantoniou S, Demetzos C (2002) Lipid analysis of Greek walnut oil (*Juglans regia* L.). Z Naturforsch C 57(1–2):51–56

U.S. Department of Agriculture, Agricultural Research Service (2010) USDA National Nutrient Database for Standard Reference, Release 23. Nutrient Data Laboratory Home Page, http://www.ars.usda.gov/ba/bhnrc/ndl

Uphof JCTh (1968) Dictionary of economic plants, 2nd edn (1st edn 1959). Cramer, Lehre, 591 pp

Wu ZY, Raven PH (eds) (1999) Flora of China, vol 4 (Cycadaceae through Fagaceae). Science Press, Beijing, and Missouri Botanical Garden Press, St. Louis

Zambón D, Sabaté J, Muñoz S, Campero B, Casals E, Merlos M, Laguna JC, Ros E (2000) Substituting walnuts for monounsaturated fat improves the serum lipid profile of hypercholesterolemic men and women. A randomized crossover trial. Ann Intern Med 132(7):538–546

Zhang Z, Liao L, Moore J, Wu T, Wang Z (2009) Antioxidant phenolic compounds from walnut kernels (*Juglans regia* L.). Food Chem 113(1):160–165

Zibaeenezhad MJ, Rezaiezadeh M, Mowla A, Ayatollahi SM, Panjehshahin MR (2003) Antihypertriglyceridemic effect of walnut oil. Angiology 54(4):411–414

Zibaeenezhad MJ, Shamsnia SJ, Khorasani M (2005) Walnut consumption in hyperlipidemic patients. Angiology 56(5):581–583

Zwarts L, Savage GP, McNeil DL (1999) Fatty acid content of New Zealand-grown walnuts (*Juglans regia* L.). Int J Food Sci Nutr 50(3):189–194

Litsea garciae

Scientific Name

Litsea garciae **Vidal.**

Synonyms

Cylicodaphne garciae (Vidal) Nakai, *Lepidadenia kawakamii* (Hayata) Masam., *Litsea aurea* Kosterm., *Litsea kawakamii* Hayata, *Tetradenia kawakamii* (Hayata) Nemoto ex Makono & Nemoto.

Family

Lauraceae

Common/English Name

Engkala (common name)

Vernacular Names

Borneo: Kangkala, Engkala, Medang, Pangalaban, Ta'ang;
Indonesia: Malai (Bangka), Wuru Lilin (Javanese), Kelimah (Punan, Malinau, Kalimantan), Bua Talal (Lundaye, Kalimantan), Kelime, Kelimie (Merap, Kalimantan), Mail (Kenyah Uma' Lung, Kalimantan), Bua' Vengolobon (Abai, Malinau, Kalimantan), Telal, Bua' Telau (Lengilu', Malinau, Kalimantan), Wi Lahal (Berau, Malinau, Kalimantan), Kelima (Pua' Malinau, Kalimantan), Mali (Leppo'ke' Malinau, Kalimantan), Mali (Leppo' Ma'ut, Pujungan, Kalimantan), Mali (Uma' Long, Pujungan, Kalimantan), Beva' Mali, Mali (Uma' Lasan, Pujungan, Kalimantan), Mali (Uma' Badeng, Pujungan, Kalimantan), Kayu Mali (Uma' Bakung, Pujungan, Kalimantan), Malei, Mali (Penan Benalui, Pujungan, Kalimantan), (Tangkalak (Sundanese));
Malaysia: Engkala, Pengalaban Pengolaban (Sarawak), Ta'ang (Bidayuh, Sarawak), Pengolaban (Sabah);
Philippines: Kupa (Bagobo), Pipi (Bikol), Bagnolo (Tagalog);
Taiwan: Lan Yu Mu Jiang Zi (Chinese).

Origin/Distribution

The species is native to the Sarawak and southwest Sabah regions in Malaysia, Kalimantan in Indonesia and the Philippines. In Borneo, it is found throughout the island. It has also been described from Taiwan. It is cultivated in Indonesia (Java and Bangka).

Agroecology

It forms the mid-storey or sub-canopy trees. In Borneo, it occurs on disturbed sites in mixed dipterocarp forests up to 200 m elevation. In the Philippines, it occurs from low to medium altitude forest, also in secondary forest. It is found

T.K. Lim, *Edible Medicinal And Non-Medicinal Plants: Volume 3, Fruits*,
DOI 10.1007/978-94-007-2534-8_5, © Springer Science+Business Media B.V. 2012

often along rivers or on hillsides with sandy to clay soils and scattered near villages. It is also commonly cultivated. It is cultivated in Java especially in the mountain regions. In Bangka, it is one of the most common fruit trees.

Edible Plant Parts and Uses

The whole fruit is edible – the pulp and the thin skin. The pulp of the fruit is eaten either raw or steamed with rice. Fully ripe fruits are soft and eaten fresh after soaking in warm water for 5–10 min or by beating with a spoon. A pinch of salt enhances its flavour. The flesh is creamy white, soft textured and has a delicate flavour reminiscent of the taste of avocado. Unripe fruits (greenish white colour) are also sold in markets and are prepared by preserving in vinegar or salt solution.

Botany

A sub-canopy, evergreen, medium-sized, perennial tree growing to a height of 10–25 m high with a trunk of 40–50 cm diameter and stout, subglabrous branchlets. Leaves are alternate, simple, on a stout petiole, ovate-lanceolate or obovate-lanceolate to lanceolate, 25–40 cm by 6–15 cm, glabrous on both surfaces, apex obtuse, base cuneate to narrowly cuneate, penninerved, lateral veins of 16–20 pairs, prominently reticulate-veined on both surfaces (Plate 1), slivery shiny when young. Flowers are small, 15 mm diameter, yellowish-white, in axillary umbels, pedicel stout. Fruit is oblate (compressed globose), 22–30 mm across, pale greenish white turning to pink or red at maturity, seated on the cup-shaped perianth tube (cupule); the pedicel 1.5 cm long and stout (Plates 1, 2, 3 and 4). Flesh is creamy-white surrounding a single seed, 15–20 mm.

Nutritive/Medicinal Properties

The proximate nutrient composition of the fruit of *Litsea garciae* per 100 g edible portion based on analyses made in Sarawak (Voon et al. 1988; Voon and Kueh 1999) was reported as: water

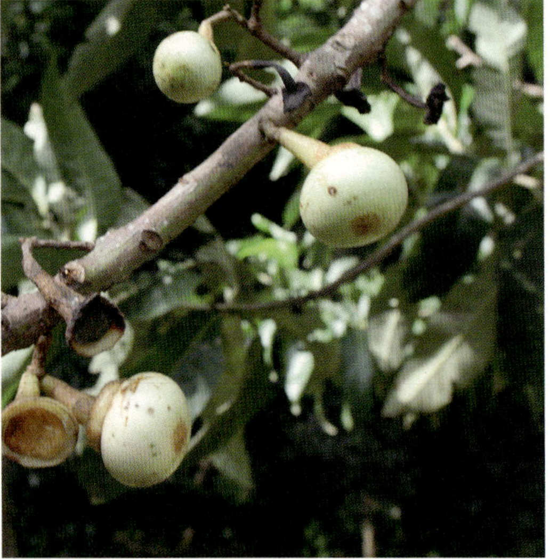

Plate 1 Unripe Engkala fruits

Plate 2 Ripe Engkala fruits and leaves

78.3%, energy 104 kcal, protein 1.4%, fat 6.8%, carbohydrates 10%, crude fibre 1.0%, ash 2.5%, P 26 mg, K 355 mg, Ca 7 mg, Mg 17 mg, Fe 0.5 mg, Mn 5 ppm, Cu 2.6 ppm, Zn 10.2 ppm and vitamin C 3.4 mg.

Traditional Medicinal Uses

Leaves, bark and wood chips have been used in traditional medicine especially in Sarawak. The Iban used the lightly burned bark to cure

Plate 3 Unripe whitish Engkala fruit with cupule at the base

Plate 4 Ripe red Engkala fruit seated on the green cupule

caterpillar stings. The Selako used a pounded poultice of the leaves or young shoots together with fennel seed and shallot to treat skin diseases and infections of the palms and skin. and to treat for skin burns. The leaves are warmed and poulticed against beri-beri by the Kayan. The Kelabit used warmed scrapings of the root bark as a poultice for sprains. The Iban use a poultice of pounded bark for boils and a decoction of the bark is taken for blood in stools. Young leaves or bark is pounded and poultice to extract pus from boils. Equal portion of the bark and durian bark is pounded and used as antidote on snakebite wounds. The Penan pounds the bark, warm and poulticed it for muscular pains and sprained ankles and knees.

Other Uses

Engkala provides useful timber for indoor construction. Oil can be extracted from the seeds and have been used to make soap and candles.

Comments

Engkala is propagated by seeds.

Selected References

Backer CA, van den Bakhuizen Brink RC Jr (1963) Flora of Java, vol 1. Noordhoff, Groningen, 648 pp

Chai PPK (2006) Medicinal plants of Sarawak. Lee Ming Press, Kuching, p 212

Huang P-H, Li J, Li X-W, van der Werff H (2008) *Litsea* Lamarck. In: Wu ZY, Raven PH, Hong DY (eds) Flora of China, vol 7, Menispermaceae through capparaceae. Science Press\Missouri Botanical Garden Press, Beijing\St. Louis

Jansen PCM, Jukema J, Oyen LPA, van Lingen RG (1992) Minor edible fruits and nuts. In: Verheij EWM, Coronel RE (eds) Plant resources of South-East Asia No 2. Edible fruits and nuts. Prosea Foundation, Bogor, pp 313–370

Kessler PJA, Sidiyasa K (1994) Trees of the Balikpapan-Samarinda area, East Kalimantan, Indonesia: a manual to 280 selected species, (Tropenbos series 7). Tropenbos Foundation, Wageningen, 446 pp

Merrill ED (1923–1925) An enumeration of Philippine flowering plants, vol 4. Government of the Philippine Islands, Department of Agriculture and Natural Resources, Bureau of Printing, Manila

Munawaroh E, Purwanto Y (2009) Studi hasil hutan non kayu di Kabupaten Malinau, Kalimantan Timur. (In Indonesian). Paper presented at the 6th Basic Science National Seminar, Universitas Brawijaya, Indonesia, 21 Feb 2009. http://fisika.brawijaya.ac.id/bss-ub/proceeding/PDF%20FILES/BSS_146_2.pdf

Puri RK (2001) Bulungan ethnobiology handbook. CIFOR, Bogor, 310 pp

Slik JWF (2006) Trees of Sungai Wain. Nationaal Herbarium Nederland. http://www.nationaalherbarium.nl/sungaiwain/

Voon BH, Kueh HS (1999) The nutritional value of indigenous fruits and vegetables in Sarawak. Asia Pac J Clin Nutr 8(1):24–31

Voon BH, Chin TH, Sim CY, Sabariah P (1988) Wild fruits and vegetables in Sarawak. Department of Agriculture, Sarawak, 114 pp

Persea americana

Scientific Name

Persea americana Miller.

Synonyms

Laurus persea L., *Persea americana* var. *angustifolia* Miranda, *Persea americana* var. *drymifolia* (Schltdl. & Cham.) S.F. Blake, *Persea americana* var. *nubigena* (L.O. Williams) L.E. Kopp, *Persea drymifolia* Schltdl. & Cham., *Persea edulis* Raf., *Persea floccosa* Mez, *Persea gigantea* L.O. Williams, *Persea gratissima* C.F. Gaertn., *Persea gratissima* var. *drimyfolia* (Schltdl. & Cham.) Mez, *Persea gratissima* var. *macrophylla* Meisn., *Persea gratissima* var. *oblonga* Meisn., *Persea gratissima* var. *praecox* Nees, *Persea gratissima* var. *vulgaris* Meisn., *Persea pleiogyna* Blake, *Persea nubigena* L.O. Williams, *Persea paucitriplinervia* Lundell, *Persea persea* (L.) Cockerell, *Persea steyermarkii* C.K. Allen.

Family

Lauraceae

Common/English Names

Alligator Pear, Alligator Pear, Avocado, Avocado Pear, Butter Pear, Guatemalan Avocado, Lowland Avocado, Mexican Avocado, Midshipman's Butter, Trapp Avocado, Vegetable Marrow, West Indian Avocado.

Vernacular Names

Amharic: Avocado;
Aztec: Ahuacaquahuitl, Ahuacatl;
Brazil: Abacate, Abacateiro;
Burmese: Htaw Bat, Kyese;
Chamorro: Alageta;
Chinese: E Li, Huang You Li, You Li, Xi Yin Du Lao Li, Zhang Li;
Cook Islands: 'Āpoka, Āpuka, 'Āvōta (Maori);
Creole: Zaboka;
Croatian: Americhki Avocado;
Czech: Hruškovec Přelahodný;
Danish: Avocado, Avogatpære;
Dutch: Advocaat;
Finnish: Avokado;
French: Avocat, Avocatier, Avocatier Du Guatemala, Avogado, Zaboka,Zabelbok;
German: Alligatorbirne, Avocado, Avocadobaum, Avocadobirne, Avocadopalme, Avocato-Birne;
Honduras: Hayi , Narimu, Zial;
Hungarian: Avokádó, Mexikói Avocado;
Icelandic: Avókadó;
Indonesia: Adpukat, Avocad, Avokad, Buah Apokat;
Italian: Avocado;
Japanese: Abokado, Perusea;
Khmer: 'Avôkaa;
Korean: Ah Bo K'a Do;

T.K. Lim, *Edible Medicinal And Non-Medicinal Plants: Volume 3, Fruits*,
DOI 10.1007/978-94-007-2534-8_6, © Springer Science+Business Media B.V. 2012

Malaysia: Avocado, Buah Apukado, Buah Mantega;
Marquesan: Avoka, Avoka;
Maya: On;
Nicaragua: Aguacate, Cura, Cupandra, Devora, Kulup, Sikia;
Niuean: Avoka;
Norwegian: Avokado;
Palauan: Bata;
Papua New Guinea: Bata;
Peru: Okh, Palta, Palto, Pultas;
Philippines: Abokado (Cebu), Aguacate (Spanish);
Polish: Awokado;
Portuguese: Abacate, Abacateiro;
Russian: Avokado;
Samoan: 'Avoka;
Spanish: Aguacate, Aguacate Olorosa, Avocado, Cura, Cupandra, Palta, Palto;
Swahili: Mparachichi, Mpea, Mwembe, Mafuta;
Swedish: Avokado, Advokatpäron, Aguakate, Aligatorpäron, Avocato;
Tahitian: Avoka, Avoka, Avoka, Avota;
Taiwan: Lao Li;
Thai: Awokhado;
Tongan: 'Avoka;
Vietnamese: Bó, Lê Daù;
West Africa: Avocado (Mandinka).

Origin/Distribution

The area of origin of avocado is in Central America – probably Mexico, Guatemala and Honduras. It was cultivated from the Rio Grande to central Peru before the arrival of Europeans. It is now cultivated elsewhere in the tropics and subtropics world-wide.

Agroecology

Avocados grow from sea level up to 2,250 m altitude. Avocados do best some distance from ocean influence but are not adapted to the desert interior. Avocados can be grown on a wide range of soils types such as red clay, sand, volcanic loam, lateritic soils, or limestone. It requires a well-drained aerated soil because the roots are intolerant of anaerobic conditions; water-logging for more than 24 hours can kill trees. Sites with underlying hardpan must be avoided. Avocados are intolerant to saline conditions but certain cultivars have shown considerable salt-tolerance in Israel. Optimum range of pH is from 5 to 7 but, in southern Florida, avocados are grown on limestone soils ranging from 7.2 to 8.3. Mexican and Guatemalan cultivars have shown chlorosis on calcareous soils in Israel. In areas of strong winds, wind-breaks are necessary as high winds may break the branches. Strong wind reduces humidity, dehydrates the flowers and interferes with pollination, and also causes many fruits to fall prematurely.

Depending on the race and varieties, avocados can thrive and perform well in climatic conditions ranging from true tropical to warmer parts of the temperate zone. avocados has a mean annual temperature of –4 to 40°C, Mean annual rainfall: 300–2,500 mm. The West Indian race and variety require a lowland tropical or near tropical (southern Florida) climate and relatively frost-free areas of the subtropics with high atmospheric humidity especially during flowering and fruit setting. Guatemalan types are native to cool, high-altitude tropics and are cold hardy (−1.1 to −3°C). Mexican types are native to dry subtropical plateaus and thrive in a Mediterranean climate. They are also cold hardy (−4.4 to −7.2°C). The Mexican race is the hardiest and the source of most of California avocados can be grown in the cooler parts of northern and inland California and along the Gulf Coast. Mexican hybrids are generally more cold tolerant than West Indian × Guatemalan hybrid varieties.

Edible Plant Parts and Uses

Ripe avocado pulp is eaten fresh. The flesh is typically greenish yellow to golden yellow when ripe. The flesh is prone to enzymatic browning after exposure to air. To avoid this, lime or lemon juice can be squeezed onto the pulp after peeling. The high oil content of the ripe avocado imparts a smooth, buttery texture to the flesh. The easily digestible flesh is rich in energy, minerals

Plate 1 (**a, b**) Different varieties of avocado

especially iron and vitamins A and B, monounsaturated fatty acids which have been documented as anti-cholesterol agents and has unusually high protein content for fruits (see below); providing a highly nutritious solid food, even for infants. The high oil content makes the avocado popular in vegetarian cuisine substituting for meat in sandwiches and salads. In North America, avocados are primarily served as salad vegetables. Avocado flesh may be sliced or diced and combined with tomatoes, cucumbers or other vegetables and served as a salad. Avocados are halved and garnished with seasonings, lime juice, lemon juice, vinegar, mayonnaise or other dressings and often the halves are stuffed with shrimp, crab or other seafood. In Chile, Caesar salad contains large slices of ripe avocado. In Kenya, the avocado is often eaten as a fruit, and is eaten alone, or mixed with other fruits in a fruit salad, or as part of a vegetable salad. The seasoned or unseasoned flesh is often used as a sandwich filling. A common breakfast in areas where avocados are grown is avocado on toast. This is made by mashing the avocado with some lemon juice, salt, and pepper and spreading it on hot, freshly toasted bread. In Australia and New Zealand, it is commonly served in sandwiches, on toast, or often with chicken. In Ghana, avocado is eaten alone in sliced bread as a sandwich. Avocado, cream cheese and pineapple juice may be blended as a creamy dressing for fruit salads. Avocado is a major ingredient in California rolls and other *Makizushi* ("Maki", or rolled sushi). In Latin American, it is common for halved avocado sprinkled with salt and consumed with tortillas

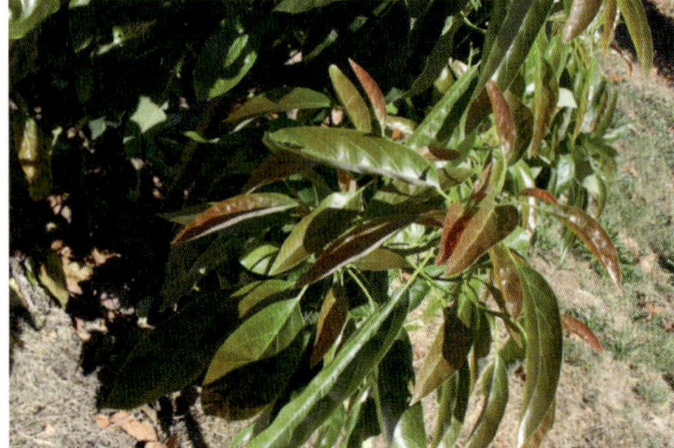

Plate 2 Young avocado leaves

Plate 3 (**a**, **b**) Avocados on sale in the local markets

Plate 4 Different avocado varieties on sale in local markets

and a cup of coffee as a complete meal. In Hawaii, avocado is sweetened with sugar and eaten combined with fruits such as pineapple, orange, grapefruit, dates, or banana.

In Peru, avocados are consumed with *tequeños* (rolled bread dough with white cheese) as mayonnaise, served as a side dish with *parrillas* (grilled dish), used in salads and sandwiches, or as a whole dish when filled with tuna, shrimps, or chicken. In Chile avocado is pureed and used in chicken, hamburgers, and hot dogs; and in slices for celery or lettuce salads. In Mexico, avocado slices are frequently added to hamburgers, *tortas* (Mexican sandwich), hot dogs, and *carne asada* (marinated steak) or combined with eggs (in scrambled eggs, tortillas or omelettes). In Mexico and Central America, avocados are served mixed with white rice, in soups, salads, or on the side of chicken and meat. Avocado is also eaten as an uncooked savoury dish mixed with herbs and/or spices, or as a sweetened dessert. In southern Africa, Avocado *Ritz* is a common dish. Ripe avocado pulp is the key ingredient in the very popular dip, *guacamole,* widely used for crackers, potato chips, bread or other snacks. The recipe for *guacamole* comprises ripe avocados, tomatoes, onion, lime or lemon juice and salt. Other common ingredients include coriander leaf, black pepper, cumin, jalapenos chillies and garlic.

Avocado flesh is usually not cooked as it becomes too bitter when cooked due to its tannin content and the flavour is impaired. However, the Hass cultivar can be cooked for a very short time without becoming overly bitter. Chunks of avocado may be added to hot foods such as soup, stew, chilli or omelettes just before serving. In Guatemalan restaurants, a ripe avocado is placed on the table when a hot dish is served and the diner scoops out the flesh and adds it in just before eating. For a gourmet breakfast, avocado halves are warmed in an oven at low heat, then topped with scrambled eggs and anchovies. Diced avocado can be added to lemon-flavored gelatine after cooling and before it is set or heated ice cream mixes and boiled custards after cooling.

In Brazil, the avocado is regarded more as a true fruit than as a vegetable and is used mostly mashed in sherbet, sorbets, ice cream, or milk shakes. In Vietnam, Philippines and Indonesia, a popular dessert drink is made with pureed avocado, sugar, milk or water and chocolate syrup is sometimes added. In Sri Lanka, treacle (palm sugar) is used instead. In Java, avocado flesh is thoroughly mixed with strong black coffee, sweetened and eaten as a dessert. A New Zealand recipe for avocado ice cream consist of avocado, lemon juice, orange juice, grated orange rind, milk, cream, sugar and salt. The mixture is frozen, beaten until creamy, and frozen again. In Suriname and Recife, avocado puree is used to thicken and flavour the liqueur *Advocaat.*

Avocado oil expressed from the flesh is rich in vitamins A, B, G and E. It is edible and has a digestibility coefficient of 93.8%. The tasty oil, characterized by a high content of unsaturated fatty acids, is highly esteemed as cooking-oil and baking-oil but has remained too costly to be utilized extensively as such. A decoction of the leaves has been used as tea but more for medicinal purposes.

Botany

An erect, medium tree, usually 10–12 m tall and may reach 20 m. Trunk is 30–60 cm across with gray-green, longitudinally fissured bark. Tree is almost evergreen, leaves being shed briefly in dry seasons at blooming time. Leaves are alternate; petiole 2–5 cm, sparsely pubescent. Leaf blade is simple, variable in shape, narrowly elliptic, elliptic, ovate, or obovate, 8–20×5–12 cm, usually somewhat glaucous lower surface, deep green above, leathery, sparsely yellowish brown pubescent above but very densely so below when mature, midrib conspicuously raised below, lateral veins 5–7 pairs, distinctly raised below, slightly elevated above, base cuneate or acute to subrounded, apex acute, margin entire (Plate 2). Flowers in cymose panicles 8–14 cm, most of them inserted on lower part of branchlet, pedunculate; peduncle 4.5–7 cm, peduncle and rachis densely yellowish-brown pubescent; bracts and bracteoles filiform, 2 mm, densely yellowish brown pubescent. Flowers yellow-greenish, 5–6 mm on 6 mm densely yellowish brown pubescent pedicels. Perianth densely yellowish brown pubescent outside and inside; perianth tube obconical, 1 mm; perianth lobes 6 in 2 whorls, oblong, 4–5 mm with obtuse apex, outer 3 smaller, all caducous. Fertile stamens 9 in 3 whorls; filaments 4 mm filiform, complanate, densely pilose, those of 3rd whorl each with 2 complanate ovate and orange glands at base, others glandless; anthers 4-celled; cells introrse in 1st and 2nd whorls, extrorse in 3rd whorl. Staminodes 3, of innermost whorl, sagittate-cordate, 0.6 mm, glabrous with pilose, 1.4 mm long stalk. Ovary ovoid, 1.5 mm, densely pubescent; style 2.5 mm, densely pubescent; stigma slightly dilated, discoid. Fruit yellow-green, deep-green or very dark-green, reddish-purple, or so dark a purple as to appear almost black, and is sometimes speckled with tiny yellow dots (Plates 1, 3 and 4), it may be smooth or pebbled, glossy or dull, thin or leathery and up to 6 mm thick, pliable or granular and brittle, large, usually pear-shaped, sometimes ovoid or globose, 8–18 cm; exocarp corky; mesocarp, bright-green fleshy, but generally entirely pale to rich-yellow, buttery and bland or nutlike in flavour. The single seed is oblate, round, conical or ovoid, large, 5–6.4 cm long, hard and heavy, ivory in colour but enclosed in two brown, thin, papery seedcoats often adhering to the flesh cavity, while the seed slips out readily.

Nutritive/Medicinal Properties

Food value of raw, avocado fruit (refuse 26% seed and skin) per 100 g edible portion was reported as follows (U.S. Department of Agriculture and Agricultural Research Service (2010)), water 73.23 g, energy 160 kcal (670 kJ), protein 2.00 g, total lipid (fat) 14.66 g, ash 1.58 g, carbohydrate 8.53 g; fibre (total dietary) 6.7 g, total sugars 0.66 g, sucrose 0.07 g, glucose 0.39 g, fructose 0.13 g, galactose 0.11 g, starch 0.11 g, minerals – calcium 12 mg, iron 0.55 mg, magnesium 29 mg, phosphorus 52 mg, potassium 485 mg, sodium 7 mg, zinc 0.64 mg, copper 0.190 mg, manganese 0.142 mg, flourine 7 mg, selenium 0.4 μg; vitamins – vitamin C (total ascorbic acid) 10 mg, thiamin 0.067 mg, riboflavin 0.130 mg, niacin 1.738 mg, pantothenic acid 1.389 mg, vitamin B-6 0.257 mg, folate (total) 81 μg, vitamin E (α-tocopherol) 2.07 mg, β-tocopherol 0.05 mg, γ-tocopherol 0.33 mg, δ-tocopherol 0.02 mg, vitamin K (phylloquinone) 21 μg, β-carotene 62 μg, α-carotene 24 μg, β-cryptoxanthin 28 μg, lutein + zeaxanthin 271 μg; lipids – fatty acids (total saturated) 2.126 g, 8:0 (caprylic acid) 0.001 g, 16:0 (palmitic acid) 2.075 g, 18:0 (stearic acid) 0.049 g; fatty acids (total monounsaturated) 9.799 g, 16:1 undifferentiated (palmitoleic acid) 0.698 g, 17:1 (heptadecenoic acid) 0.010 g, 18:1 undifferentiated (oleic acid) 9.066 g, 20:1 (gadoleic acid) 0.025 g; fatty acids (total polyunsaturated) 1.816 g, 18:2 undifferentiated (linoleic acid) 1.647 g, 18:3 undifferentiated (linolenic acid) 0.125 g, 18:3 n-3 c,c,c (α-linolenic acid), 0.111 g, 18:3 n-6 c,c,c (γ-linolenic acid), 0.015 g, 20:3 undifferentiated (eicosatrienoic acid) 0.016 g; stigmasterol 2 mg, campesterol 5 mg, β-sitosterol 76 mg; amino acids – tryptophan 0.025 g, threonine 0.073 g, isoleucine 0.084 g, leucine 0.143 g, lysine 0.132 g, methionine 0.038 g, cystine 0.027 g, phenylalanine 0.232 g, tyrosine 0.049 g, valine 0.107 g, arginine 0.088 g, histidine 0.049 g, alanine 0.109 g, aspartic acid 0.236 g, glutamic acid 0.287 g, glycine 0.104 g, proline 0.098 g and serine 0.114 g.

In avocado pulp, the predominant species of monoglycosyldiglycerides (MGD) were m/z 796.6 (oleic/linolenic and linoleic/linoleic acids) and m/z 800.4 (stearic/linoleic and oleic/oleic acids) (Pacetti et al. 2001). One of the main diglycosyldiglycerides (DGD) both in the pulp and seed was m/z 958.5 (oleic/linolenic); however, the pulp was also rich in m/z 962.4 (oleic/oleic), whereas in the seed, m/z 934.5 (palmitic/linoleic and palmitoleic/oleic) and m/z 960.5 (oleic/linoleic and stearic/linolenic) were more abundant. In the seed, the main phospholipid classes (phosphatidic acid (PA), phosphatidylcholine (PC), phosphatidylethanolamine (PE) and phosphatidylinositol (PI)) contained always palmitic/linoleic acids. α-linolenic acid was contained as MGD (linolenic/linolenic) and DGD (linolenic/linolenic), more present in the pulp than in the seed. The major molecular species of glycocerebrosides (GCer) in the pulp and seed carried hydroxy-palmitic acid (C16:0)/4,8-sphyngadienine (d18:2).

Carotenoids and chlorophylls identified in Hass avocado skin, flesh, and oil were lutein, α-carotene, β-carotene, neoxanthin, violaxanthin, zeaxanthin, antheraxanthin, chlorophylls a and b, and pheophytins a and b with the highest concentrations of all pigments in the skin (Ashton et al. 2006). Chlorophyllides a and b were identified in the skin and flesh tissues only. As the fruit ripened and softened, the skin changed from green to purple/black, corresponding to changes in skin hue angle, and a concomitant increase in cyanidin 3-O-glucoside and the loss of chlorophyllide a. In flesh tissue, chroma and lightness values decreased with ripening, with no changes in hue angle. The levels of carotenoids and chlorophylls did not change significantly during ripening. As fruit ripened, the total chlorophyll level in the oil from the flesh sections (outer dark green, middle pale green, and inner yellow flesh-nearest the seed) remained constant but declined in the oil extracted from the skin.

Hass avocado the most commonly consumed variety in the southwest United States was found to contain the highest content of lutein among commonly eaten fruits as well as measurable amounts of related carotenoids (zeaxanthin, α-carotene, and β-carotene) (Lu et al. 2005). Lutein accounted for 70% of the measured carotenoids, and the avocado also contained significant quantities of vitamin E. A significant increase in total carotenoid and fat content of harvested Hass avocados from all regions in California (San Luis Obispo, Ventura, Riverside, and San Diego) was noted as the season progressed from January to September (Lu et al. 2009). Concentrations of violaxanthin, 9′-cis-neoxanthin, lutein, zeaxanthin, β-cryptoxanthin, α- and β-carotene, α-tocopherol were determined. The following carotenoids were identified: all-trans neoxanthin, all-trans violaxanthin, all-trans neochrome, 9′-cis-neoxanthin, lutein-5, 6 epoxide, and chrysanthemaxanthin. Four carotenoids not previously described in the avocado were quantified: neoxanthin, neochrome, lutein-5, 6-epoxide and chrysanthemaxanthin. The total content of carotenoids was highly correlated with the total fat content (r=0.99) demonstrating a remarkable degree of constancy of carotenoid intake per gram of fat content in the California Hass avocado.

Total lipids (TL) accounted for approximately 20% of the mesocarp of avocado (Takenaga et al. 2008). Further analysis showed that the neutral lipid (NL) fraction accounted for at least 95% of the TL, and almost 90% of NL was triacylglycerol. Monoenoic acids accounted for at least 65% of the total fatty acids, and oleic acid, which is regarded as an especially important functional component of avocado accounted for approximately 50% of the monounsaturated fatty acids.

The yield of total volatiles of *P. americana* var Moro was 9.6 mg/kg fruit pulp (Pino et al. 2004). Terpenoids were the predominant class of constituents. Major terpenoids compounds were (Z)-nerolidol, (E,E)-α-farnesne, β-caryophyllene, caryophyllene oxide and α-copaene. Other non-terpenoid major compounds were (E,E)-2,4-decadienal and (Z)-3-hexenol.

A complex pattern of chloroplast pigments was found to be common to pulp, peel and leaves of avocado (Gross et al. 1973). This was due to the presence of chlorophyll in the ripe fruit and its peel. Two chromoplast-specific pigments were found in ripe pulp and were tentatively identified as α-citraurin and mimulaxanthin. The identity of the latter and of neochrome, was established inter alia by mass spectrometry (MS). The xanthophylls occurred esterified to a major extent in pulp but entirely free in peel. In additional studies, the identity of chrysanthemaxanthin as a major pigment in avocado pulp was confirmed by MS and the identity of neoxanthin similarly established (Gross et al. 1974). A carbonyl pigment was identified as 3-hydroxy-sintaxanthin. Two new UV fluorescent apocarotenoids were isolated, one of these was assigned the structure 5,8-epoxy-5,8-dihydro- 10′-apo-β-caroten-3, 10′-diol. The other had an acid labile pentaene chromophore, and structure had been tentatively assigned on similar evidence. These represented the first natural allyic apocarotenols with structures established.

Avocado seed was found to contain 13.6% tannin, 13.25% starch. Amino acids in the seed oil were reported as: capric acid, 0.6; myristic, 1.7; X, 13.5; palmitic, 23.4; X, 10.4; stearic, 8.7; oleic, 15.1; linoleic, 24.1; linolenic, 2.5%. The dried seed contained 1.33% of a yellow wax containing sterol and organic acid (Morton 1987). The seed and the roots were found to contain an antibiotic which prevented bacterial spoilage of food.

Peptone, β-galactoside, glycosylated abscisic acid, alkaloids, cellulose, polygalacto urease, polyuronoids, cytochrome P-450, and volatile oils were reported to be present in this plant (Yasir et al. 2010). The flowers of avocado yielded an acylated flavonol 3-O-trans-p-coumaroylkaempferol, in addition to quercetin 3-O-rhamnoside and isorhamnetin 3-O-glucoside (Kruthiventi and Krishnaswamy 2000).

The bark was reported to contain 3.5% of an essential oil with an anise odor and made up largely of methyl chavicol with a little anethole (Morton 1987).

Avocado is not only nutritionally rich, the fruit and other plant parts have many pharmacological attributes that impart many health and medicinal benefits.

Antioxidant Activity

Nutritional and bioactive values of avocado were found comparable to indices in durian and mango (Gorinstein et al. 2010). These fruits were found to contain high, comparable quantities of basic nutritional and antioxidant compounds, and to possess high antioxidant potentials. All fruits showed a high level of correlation between the contents of phenolic compounds (flavonoids and phenolic acids) and the antioxidant potential as evaluated by three-dimensional fluorescence, FTIR (Fourier transform infrared) spectroscopy and radical scavenging assays. The contents of total fibre, proteins and fats were significantly higher in avocado, and carbohydrates were significantly lower in avocado than in the two other fruits. The wavelength numbers of FTIR spectra for three investigated fruits were in the same range (1,700–600/cm) as for catechin and gallic acid, used as standards. Similarities were found between durian, mango and avocado in polyphenols (9.88, 12.06 and 10.69, mg gallic acid equivalent/g dry weight) and in antioxidant assays such as CUPRAC (27.46, 40.45 and 36.29, μM Trolox equivalent (TE)/g d.w.) and FRAP (23.22, 34.62 and 18.47, μoM TE/g d.w.), respectively.

Potent antioxidant activity was observed in methanol extracts of avocado seeds (Yuko et al. 2003). The extracts contained (+)-catechin and (−)-epicatechin as major active components. Results of AMVN (2,29-Azobis 2,4-dimethylvaleronitrile)-induced methyl linolate peroxidation assays showed that lipid peroxidation of the isolates at concentration of 0.1 mM (21% and 17%, respectively) were lower than that of caffeic acid at the same concentration.

In general, after refrigerated storage, a significant decrease in fatty acid content was observed in avocado slices/halves (Plaza et al. 2009). The main fatty acid identified and quantified in avocado was oleic acid (about 57% of total content), whereas β-sitosterol was found to be the major sterol (about 89% of total content). Vacuum/halves and air /slices were the samples that maintained better this content. With regard to phytosterols, there were no significant changes during storage. Antioxidant activity showed a slight positive correlation against stearic acid content. At the end of refrigerated storage, a significant increase in antiradical efficiency (AE) was found for vacuum samples. AE values were quite similar among treatments. The results indicated that minimal processing could be a useful tool to preserve health-related properties of avocado fruit.

The leaves of *Persea americana* yielded bioactive phytoconstituents isorhamnetin, luteolin, rutin, quercetin and apigenin (Owolabi et al. 2010). Their antioxidant activity was compared with BHT (butylated hydroxytoluene) by evaluating their ability to scavenge free radical using DPPH (1,1-diphenyl-2-picryl-hydrazyl) and H_2O_2 systems. The IC_{50} (mg/ml) of quercetin was 4.82×10^{-5}; rutin, 1.37×10^{-4}; luteolin, 3.34×10^{-4} and isorhamnetin, 4.41×10^{-4}. H_2O_2 scavenging activity of the compounds was in the following order: quercetin > rutin > isorhamnetin > luteolin > apigenin > BHA. The results revealed that the leaves of *P. americana* contained antioxidant activity which may be helpful in preventing the progress of various oxidative stress related diseases. Similary, Asaolu et al. (2010a) reported that the methanol extracts of the leaves of *Persea americana* possessed significant antioxidant activities as determined by DPPH, nitric oxide and reducing power radical-scavenging activity assays. *Persea americana* was found to have higher radical scavenging activity than *Cnidosculous aconitifolius* but the phenol content of *Cnidosculous aconitifolius* was higher. *Persea americana* was observed to possess more flavonoids than *Cnidosculous aconitifolius.* Phytochemical constituents of both aqueous and methanol extract of avocado leaves included sterol, tannin, saponin, flavonoids, alkaloids, phenol, phlobatannin, anthraquinones, triterpenes and cardiac glycosides in varying amounts (Asaolu et al. 2010c). Antioxidants in the methanolic extract of *Persea americana* included: carotenes (mg/100 g) 85.23, ascorbic acid (mg/100 g) 19.20, α-tocopherol (μmol/l) 20.34, glutathione peroxidase (μg/l) 18.50, superoxide dismutase (μg/l) 22.50 and catalase (μg/mg) 21.50. Antioxidants in the aqueous extract of *Persea americana* included carotenes (mg/100 g) 21.11, ascorbic acid (mg/100 g) 6.28, α-tocopherol (μmol/l) 5.24, glutathione peroxidase (μg/l) 3.10, superoxide dismutase (μg/l) 11.50 , and catalase (μg/mg) 10.34.

Antihypercholesterolemic, Antihyperlipidemic Activity

Avocado pulp and leaves exhibited antihypercholesterolemic and antihyperlipidemic activity.

A two-diet, cross-over design trial involving 8 phenotype IV and 8 phenotype II dyslipidemia patients was carried out to examine the effects of avocado on plasma lipid concentrations (Carranza et al. 1995). A diet rich in monounsaturated fatty acids (DRCA) using avocado as their major source (30% of the total calories were consumed as fat, 75% of the total fat from the avocado), with restriction of saturated fat and less of 300 mg of cholesterol per day for 4 weeks was evaluated. Patients were also given a low-saturated fat diet without avocado (DRSA). In phenotype II, both DRCA and DRSA significantly reduced total cholesterol and LDL-cholesterol levels. In phenotype IV, DRCA produced a mild reduction in triglyceride levels while DRSA increased them. On HDL-cholesterol concentrations, DRCA produced a significant increase in both phenotypes while DRSA did it only in phenotype IV. The results showed avocado to be an excellent source of monounsaturated fatty acids in diets designed to treat hypercholesterolemia with some advantages over low-fat diets with a greater amount of carbohydrates. In another study, 15 healthy and 30 hypercholesterolemic subjects (15 of them with associated type 2 diabetes mellitus) were administered an avocado enriched diet (2,000 Kcal, lipids 53% MFA 49 g, saturated/unsaturated ratio 0.54), and 7 non-diabetic hypercholesterolemic individuals received an isocaloric control diet (MFA 34 g, saturated/unsaturated ratio 0.7) (Lopez Ledesma et al. 1996). The results showed that in healthy individuals a 16% decrease of serum total cholesterol level after the high monounsaturated fatty acids (MUFA) diet, while it increased after the control diet. In hypercholesterolemic subjects a significant decrease of serum total cholesterol (17%), LDL-cholesterol (22%) and triglycerides (22%), and increase of HDL-cholesterol (11%) levels occurred with the avocado diet, while no significant changes were found with the control diet. The data showed that high lipid, high MUFA-avocado enriched diet could improve lipid profile in healthy and especially in mild hypercholesterolemic patients, even if hypertriglyceridemia (combined hyperlipidemia) was present.

Food consumption and body weight gain were lower in rats fed avocado pulp as the dietary fiber source with or without 10 g/kg cholesterol compared with those fed cellulose (control) (Naveh et al. 2002). The total dietary fibre content of fresh avocado fruit of the Ettinger variety was 5.2 g/100 g. Approximately 75% was insoluble, and 25% soluble. Relative cecum weight was higher in avocado-fed rats. Plasma and hepatic cholesterol levels did not differ in rats fed diets without cholesterol, but plasma cholesterol was greater in avocado-fed than in cellulose-fed rats that consumed cholesterol. Regardless of dietary cholesterol, hepatic total fat levels, as evaluated histologically, but not directly, were lower in avocado-fed rats. The data suggested the presence of an appetite depressant in avocado and that avocado pulp interfered with hepatic fat metabolism.

Studies showed there were no significant differences in the overall body weight gain of the hypercholesterolemic rats compared to normal control (Brai et al. 2007b). Liver to body weight ratio, plasma glucose, total cholesterol (T-CHOL), and LDL-CHOL levels were significantly elevated in rats fed hypercholesterolemic diet compared to normal controls. The administration of aqueous and methanolic leaf extracts of *P. americana* daily for 8 weeks induced reductions in plasma glucose (16% and 11%, respectively), T-CHOL (8% and 5%, respectively), and LDL-CHOL (19% and 20%, respectively) in the treated rats compared to the hypercholesterolemic controls. Also, plasma HDL-CHOL concentrations increased by 85% and 68%, respectively, in the aqueous and methanolic extract-treated rats compared to the hypercholesterolemic controls. These results suggested that aqueous and methanolic leaf extracts of *P. americana* lowered plasma glucose and influenced lipid metabolism in hypercholesterolemic rats with consequent lowering of T-CHOL and LDL-CHOL and a restoration of HDL-CHOL levels. This could represent a protective mechanism against the development of atherosclerosis.

The administration of the aqueous and methanolic avocado leaf extracts daily for 8 weeks resulted in 14 and 25% reduction, respectively, in the body weight gain of the treated rats compared to the hyperlipidaemic control (Brai et al. 2007a).

Mean liver weights were markedly increased in rats fed hyperlipidaemic diet (groups B, C and D: 70, 69 and 57%, respectively) compared to normal control rats. The methanolic extract provoked a minimal (8%) decrease in mean liver weight compared to the hyperlipidaemic control rats. It was hypothesized that *P. americana* leaf extracts increased catabolism of lipids accumulated in adipose tissue causing a decrease in body weight but did not influence liver lipid levels in rats.

Healthy subjects (n = 11/study) were recruited for two crossover, postprandial studies to assess the effect of avocado addition (150 g) as a lipid source to salsa on lycopene and β-carotene absorption in Study 1, and the absorption of lutein, α-carotene, and β-carotene from salad in Study 2 (Unlu et al. 2005). The addition of avocado to salsa enhanced lycopene and β-carotene absorption, resulting in 4.4 and 2.6 times the mean area under the concentration-versus-time curve (AUC) after intake of avocado-free salsa, respectively. In Study 2, supplementing 150 g avocado or 24 g avocado oil to salad similarly enhanced α-carotene, β-carotene, and lutein absorption, resulting in 7.2, 15.3, and 5.1 times the mean AUC after intake of avocado-free salad, respectively (150 g avocado). Neither the avocado dose nor the lipid source affected carotenoid absorption. The authors concluded that adding avocado fruit could significantly enhance carotenoid absorption from salad and salsa, which was attributed primarily to the lipids present in avocado.

In another study, rats that received avocado had about 27% lower triglycerides plasma levels whereas their HDL-cholesterol was 17% higher as compared to control group (Méndez and Hernández 2007). The HDL-cholesterol decrease was attributed to a lower content of protein, particularly of apo AI, with a concomitant higher proportion of phospholipids in HDL isolated from avocado group. HDL-cholesterol structural modifications induced by avocado were not related to modifications of lecithin cholesterol acyltransferase (LCAT) and phospholipid transfer protein (PLTP) activity activities, but occurred in parallel with higher serum levels of paraoxonase type-1 (PON1) activity when compared to the controls (57.4 vs. 43.0 μmol/min/ml serum). LCAT plays a key role in the reverse cholesterol transport (RCT) process by converting cholesterol to cholesteryl ester to form mature HDL-cholesterol particles. The findings indicated that the inclusion of avocado in the diet decreased plasma triglycerides, increased HDL-cholesterol plasma levels and modified HDL-cholesterol structure. The latter effect may have enhanced the antiatherogenic properties of HDL-cholesterol since PON1 (a high density lipoprotein) activity also increased as a consequence of avocado.

Studies in New Zealand reported that white rabbits fed a semipurified diet containing 0.2% cholesterol and 14% fat for 90 days showed that coconut oil (saturated fatty acids) was the most atherogenic fat, while corn oil (mainly PUFA) was only slightly less atherogenic than either olive or avocado oils. Avocado oil comprising mainly monounsaturated fatty acids (MUFA) was of the same order of atherogenicity as olive oil (MUFA) (Kritchevsky et al. 2003). Percentage of serum HDL cholesterol was highest in the rabbits fed the two monounsaturated fats.

A methanol extract of avocado fruits showed potent inhibitory activity against acetyl-CoA carboxylase, a key enzyme in fatty acid biosynthesis (Hashimura et al. 2001). The active principles were isolated and identified as (5E,12Z,15Z)-2-hydroxy-4-oxoheneicosa-5,12,15-trienyl (1), (2R,12Z,15Z)-2-hydroxy-4-oxoheneicosa-12,15-dienyl (2), ($2R^*,4R^*$)-2,4-dihydroxyheptadec-16-enyl (3) and ($2R^*,4R^*$)-2,4-dihydroxyheptadec-16-ynyl (4) acetates by instrumental analyses. The IC_{50} of the compounds were 4.0×10^{-6}, 4.9×10^{-6}, 9.4×10^{-6}, and 5.1×10^{-6}M, respectively.

Persea americana seed was found to show hypolipidemic effect (Asaolu et al. 2010b). Acute administration of cholesterol resulted in the elevation of total cholesterol (TC), triglyceride (TG), low density lipoprotein cholesterol (LDLC), very low density lipoprotein cholesterol (VLDLC) and reduction in high density lipoprotein cholesterol (HDLC). However, treatment with various doses of the methanolic extract of the seeds of *Persea americana* caused a significant reduction in the levels of TC, TG, LDLC and VLDLC while the levels of HDLC increased significantly. These effects were dose dependent as marked changes were observed at the highest concentration (300 mg/kg) of the methanolic extract of avocado seeds.The results suggest that *Persea americana* seeds may serve as possible alternative treatment for hyperlipemia and hypertension. Studies by Imafidon and Amaechina, (2010) showed that the different dose of *P. americana* aqueous seed extract, significantly reduced blood pressures of the hypertensive rats. Reduction in total cholesterol, LDL and triacylglycerol levels were observed at the 500 mg/kg body weight of seed extract in the plasma, kidney, liver and heart. The results suggested that the use of aqueous avocado seed extract in the treatment of hypertension may produce a favourable lipid profile at the 500 mg/kg dose level.

Cardiovascular Activity

Studies showed that avocado oil-rich diet induced a slightly higher angiotensin II-induced blood pressure response in the male Wistar rats as compared to the control rats (Salazar et al. 2005). In cardiac microsomes, avocado oil induced an increase in oleic acid content (13.18% versus 15.46%), while in renal microsomes, the oil decreased α-linolenic acid content (0.34% versus 0.16%), but increased the arachidonic acid proportion (24.02% versus 26.25%), compared to control. In conclusion, avocado oil-rich diet modified the fatty acid content in cardiac and renal membranes in a tissue-specific manner. The rise in renal arachidonic acid suggested that diet content could be a key factor in vascular responses.

Dietary supplementation of ASU (avocado-soyabean unsaponifiables) may be beneficial to prevent or ameliorate ischemic cerebral vascular disease (Yaman et al. 2007). Studies showed that malondialdehyde (MDA) and nitric oxide (NO) levels increased in group II rats (fed with standard diet) compared with group I rats (controls). In group III (rats fed with standard diet plus ASU pills for 10 days), MDA and NO levels decreased as compared to group II. Superoxide dismutase (SOD) and catalase (CAT) activities increased in group III as compared to group II rats. The number of apoptotic neurons was lower in group III as compared to group II rats. The findings suggested that avocado-soyabean unsaponifiables could decrease oxidative stress and apoptotic changes in ischemic rat hippocampus.

P. americana aqueous leaf extract (PAE) (25–800 mg/ml) produced concentration-dependent, significant, negative inotropic and negative chronotropic effects on guinea pig isolated electrically driven left and spontaneously beating right atrial muscle preparations, respectively (Ojewole et al. 2007). Further, PAE reduced or abolished, in a concentration-dependent manner, the positive inotropic and chronotropic responses of guinea pig isolated atrial muscle strips induced by noradrenaline and calcium (Ca^{2+}). PAE (50–800 mg/ml) also significantly reduced or abrogated, in a concentration-dependent manner, the rhythmic, spontaneous, myogenic contractions of portal veins isolated from healthy normal Wistar rats. Like acetylcholine, the plant extract (25–800 mg/ml) produced concentration-related relaxations of isolated endothelium-containing thoracic aortic rings pre-contracted with noradrenaline. The vasorelaxant effects of PAE in the isolated, endothelium-intact aortic rings were markedly inhibited or abolished by N(G)-nitro-L-arginine methyl ester (a nitric oxide synthase inhibitor). Further, PAE (25–400 mg/kg i.v.) caused dose-related, transient but significant reductions in the systemic arterial blood pressure and heart rates of the anaesthetised normotensive and hypertensive rats used. The results of this laboratory animal study indicated that PAE caused bradycardia, vasorelaxation and hypotension in the mammalian experimental models used. The vasorelaxant action of PAE was endothelium dependent, and was, therefore, possibly dependent on the synthesis and release of nitric oxide (NO). The vasorelaxant effects of PAE appeared to contribute significantly to the hypotensive (antihypertensive) effects of the plant extract. In another study, the intravenous administration of avocado leaf extract to anaesthetized normotensive rats induced a marked fall in mean arterial blood pressure which lasted 2–3 min (Adeboye et al. 1999). The short duration of action may be due to rapid metabolism.

Studies showed that the aqueous leaves extract of *Persea americana* produced significant vasorelaxation on isolated rat aorta and that the effect was dependent on the synthesis or release of endothelium-derived relaxing factors (EDRFs) as well as the release of prostanoid (Owolabi et al. 2005). The extract also reduced vasoconstriction probably by inhibiting Ca^{2+} influx through calcium channels.

Anticancer Activity

Brine shrimp lethality-directed fractionation of the 95% EtOH extract of the powdered dried bark of *Persea americana* var. *americana* afforded one new C 20 alkyl-alkene acetonyl methyl ester designated persealide (Ye et al. 1996). Persealide showed moderate cytotoxicity against three solid tumour cell lines: human lung carcinoma (A-549), human breast carcinoma (MCF-7) and human colon adenocarcinoma (HT-29). Three major bioactive, cytotoxic constituents (1–3) were isolated from unripe avocado fruit (Oberlies et al. 1998). Compounds 1–3 exhibited activity against six human tumour cell lines in culture and showed selectivity for human prostate adenocarcinoma (PC-3) cells with compound 3 being nearly as potent as adriamycin. Also, when tested against yellow fever mosquito larva, compound 3 was more effective than rotenone, a natural botanical insecticide and positive control.

Phytochemicals extracted from avocado flesh into a chloroform partition (D003) selectively induced cell cycle arrest, inhibit growth, and induce apoptosis in precancerous and cancer cancer cell lines but not normal, human oral epithelial cell lines by modulation of reactive oxygen species (ROS) (Ding et al. 2007). Results indicated that phytochemicals extracted with chloroform from avocado fruits targeted multiple signaling pathways and increase intracellular reactive oxygen leading to apoptosis. In a more recent study, Ding et al. (2009), observed that treatment of human oral cancer cell lines containing high levels of ROS with D003 increased ROS levels two to threefold and induced apoptosis. In contrast, ROS levels increased only 1.3-fold, and apoptosis was not induced in the normal cell lines containing much lower levels of basal ROS. When cellular ROS levels in the malignant cell lines were reduced by N-acetyl-l-cysteine (NAC), cells were resistant to D003 induced apoptosis. NAC also delayed the induction of apoptosis in dominant negative FADD (Fas-associated death domain)-expressing malignant cell lines. D003 increased ROS levels via mitochondrial complex I in the electron transport chain to induce apoptosis. Normal human oral epithelial cell lines transformed with HPV16 E6 or E7 expressed higher basal levels of ROS and became sensitive to D003.

An acetone extract of avocado containing these carotenoids (lutein, zeaxanthin, α-carotene, and β-carotene) and tocopherols was shown to inhibit the growth of both androgen-dependent (LNCaP) and androgen-independent (PC-3) prostate cancer cell lines in vitro (Lu et al. 2005). Incubation of PC-3 cells with the avocado extract led to G(2)/M cell cycle arrest accompanied by an increase in p27 protein expression. Lutein alone did not reproduce the effects of the avocado extract on cancer cell proliferation. In common with other colorful fruits and vegetables, the avocado contains numerous bioactive carotenoids. Because the avocado also contained a significant amount of monounsaturated fats, these bioactive carotenoids were likely to be absorbed into the bloodstream, where in combination with other diet-derived phytochemicals they may contribute to the significant cancer risk reduction associated with a diet of fruits and vegetables (Lu et al. 2005).

The novel plant toxin, persin was characterised, with in-vivo activity in the mammary gland and a p53-, estrogen receptor-, and Bcl-2-independent mode of action (Butt et al. 2006). Using a lactating mouse model persin was confirmed to have a similar cytotoxicity for the lactating mammary epithelium. Persin was previously identified from avocado leaves as the toxic principle responsible for mammary gland-specific necrosis and apoptosis in lactating livestocks.

It was well known that when lactating livestock ate avocado leaves they developed non-infectious mastitis and agalactia (Oelrichs et al. 1995). This was associated with extensive coagulation necrosis of the secretory acinar epithelium and interstitial oedema, congestion, and haemorrhage. Similar lesions had been produced in mammary glands of lactating mice fed a diet containing a small percentage of freeze-dried avocado leaf. Tests had shown that persin at the dose rate of 60–100 mg/kg had the same effect on mammary glands in lactating mice as leaves from avocado. The Guatemalan variety in doses exceeding 20 g fresh leaf per kg bodyweight, produced damage to the mammary gland with decreased milk production in goats (Craigmill et al. 1989). The lesions were characterised by oedema and reddening, with clots in the large ducts. Microscopically, there was widespread degeneration and necrosis of the secretory epithelium, the necrotic cells sloughing into the lumen. Further in -vitro studies using a panel of human breast cancer cell lines showed that persin, a novel plant toxin from avocado leaves, selectively induced a G2-M cell cycle arrest and caspase-dependent apoptosis in sensitive cells (Butt et al. 2006). The latter was dependent on expression of the BH3-only protein Bim. Bim is a sensor of cytoskeletal integrity, and there was evidence that persin acts as a microtubule-stabilizing agent. Due to the unique structure of the compound, persin could represent a novel class of microtubule-targeting agent with potential specificity for breast cancers.

One known, (2R)-(12Z,15Z)-2-hydroxy-4-oxoheneicosa-12,15-dien+ ++-1-yl acetate (1), and two novel compounds, persenone A (2) and B (3), were isolated from avocado fruit, as inhibitors of superoxide (O^{2-}) and nitric oxide (NO) generation in cell culture systems (Kim et al. 2000b). They showed marked inhibitory activities toward NO generation induced by lipopolysaccharide in combination with interferon-gamma in mouse macrophage RAW 264.7 cells. Their inhibitory potencies of NO generation (1, $IC_{50}=3.6$; 2, $IC_{50}=1.2$; and 3, $IC_{50}=3.5$ μM) were comparable to or higher than that of a natural NO generation inhibitor, docosahexaenoic acid (DHA; $IC_{50}=4.3$ μM). Furthermore, compounds 1–3 and DHA (docosahexaenoic acid) markedly suppressed tumour promoter 12-O-tetradecanoylphorbol-13-acetate-induced O^{2-}generation in differentiated human promyelocytic HL-60 cells (1, $IC_{50}=33.7$; 2, $IC_{50}=1.4$; 3, $IC_{50}=1.8$; and DHA, $IC_{50}=10.3$ μM). The compounds were found to be suppressors of both $NO^{-}$and O^{2-}-generating biochemical pathways but not to be radical scavengers. The results indicated these compounds to be unique antioxidants, preferentially suppressing radical generation, and thus may be promising as effective chemopreventive agent candidates in inflammation-associated carcinogenesis.

Antiosteoarthritic Activity

Reviews of scientific studies (Ernst 2003; Christensen et al. 2008) had shown that avocado/soybean unsaponifiables (ASU) to be some of the more promising remedies for osteoarthritis. The product (ASU) had been approved as a prescription drug in France for several years and has now been introduced in Denmark as a food supplement. Three of the four randomised, placebo-controlled, double-blind trials suggested efficacy of ASU for improving the symptoms of osteoarthritis (OA) (Ernst 2003). The majority of rigorous trial data available to date suggested that ASU was effective for the symptomatic treatment of OA but more research was warranted. In a 3-month, prospective, randomized, double-blind, placebo-controlled, parallel-group trial involving the 164 included patients, 163 were evaluable, 80 in the avocado/soybean unsaponifiables group and 83 in the placebo group, The number of patients who took back nonsteroidal antiinflammatory drugs (NSAIDs) therapy was significantly smaller in the group treated by ASU (33; 43.4%) than in the placebo group (53; 69.7%) (Blotman et al. 1997). The functional index showed a significantly greater improvement in the active drug group (−2.3) than in the placebo group (−1.0). Pain scores over time were similar in the two groups. Overall patient ratings were significantly better in the active drug group. After 6 weeks, ASU reduced the need for NSAID in patients with lower limb OA.

In another prospective, randomized, double-blind, placebo-controlled, parallel-group, multi-center trial with a 6-month treatment period, 85 patients who received avocado/soybean unsaponifiables (ASU) treatment showed significant symptomatic efficacy over the 79 who received placebo in the treatment of osteoarthritis (OA) (Maheu et al. 1998). The effect acted from month 2 and showed a persistent effect after the end of treatment. Nonsteroidal antiinflammatory drug (NSAID) consumption was slightly lower in the ASU group. Fewer patients in the ASU group required NSAIDs (48%, versus 63% in the placebo group). The success rate was 39% in the ASU group and 18% in the placebo group. Overall functional disability was significantly reduced in the ASU group. Improvement appeared more marked in patients with hip OA than knee OA. In a subsequent pilot randomized, double-blind, placebo-controlled trial involving 163 patients were included: 102 men and 61 women (mean age 63.2±8.7 years) failed to demonstrate a structural effect of ASU in hip OA (Lequesne et al. 2002). However, in a post-hoc analysis, ASU significantly reduced the progression of joint space loss as compared with placebo in the subgroup of patients with advanced joint space narrowing. These results suggested that ASU could have a structural effect but require confirmation in a larger placebo-controlled study in hip OA. Based on the available evidence, patients may be recommended to give ASU a chance for e.g., 3 months (Christensen et al. 2008). Meta-analysis data from 4 randomized controlled trials involving 664 OA patients with either hip (41.4%) or knee (58.6%) OA allocated to either 300 mg ASU (336) or placebo (328) supported better chances of success in patients with knee OA than in those with hip OA.

Studies showed that ASU pretreatment fully prevented the osteoarthritic osteoblast-induced inhibition of matrix molecule production and significantly increased type II collagen mRNA level, suggesting that this compound may promote OA cartilage repair by acting on subchondral bone osteoblasts (Henrotin et al. 2006). This finding constituted a new mechanism of action for this compound, known for its beneficial effects on cartilage. Studies demonstrated that the anti-inflammatory activity of ASU was not restricted to chondrocytes, but also affected monocyte/macrophage-like cells that served as a prototype for macrophages in the synovial membrane (Au et al. 2007). ASU reduced TNF-alpha, IL-1beta, COX-2, and iNOS expression in LPS-activated chondrocytes to levels similar to nonactivated control levels. The suppression of COX-2 and iNOS expression was paralleled by a significant reduction in PGE(2) and nitrite, respectively, in the cellular supernatant. ASU also reduced TNF-alpha and IL-1beta expression in LPS-activated monocyte/macrophage-like cells. These observations provided a scientific rationale for the pain-reducing and anti-inflammatory effects of ASU observed in osteoarthritis patients. Studies also demonstrated that ASU expressed a unique range of activities, which could counteract deleterious processes involved in OA, such as inflammation (Gabay et al. 2008). ASU decreased matrix metalloproteinases-3 and −13 expressions and prostaglandin E(2) (PGE(2)) release and prevented the degradation of I-kappa B alpha. There was an inhibition of the IL1beta-induced binding of p50/p65 complexes to NF-kappaB responsive elements in response to ASU. Finally, among the different mitogen-activated protein kinases known to be induced by IL1beta, ERK1/2 was the sole kinase inhibited by ASU.

Antimicrobial and Larvicidal Activities

All four antibacterial stereoisomers of 16-heptadecene-1, 2, 4-triol were synthesized from (S)-2, 2-dimethyl-l, 3-dioxolane-4-ethanal (Sugiyama et al. 1982). In comparison with synthetic products, the naturally occurring antibacterial triol in the avocado fruit was determined as (2R, 4R)-16-heptadecene-1, 2, 4-triol.

Methanol extracts from *Persea americana* and *Gymnosperma glutinosum* plants were observed to possess antimycobacterial activity; however, *P. americana* extracts possessed higher antimicrobial activity against *Mycobacterium tuberculosis* strains than those of *G. glutinosum*, as determined by their respective MICs of 125 μg/ml versus 250 μg/ml against H37Ra strain, respectively, and 62.5 μg/ml versus 250 μg/ml against H37Rv strain respectively (Gomez-Flores et al. 2008). It was also observed that hexane fraction of *P. americana* extract caused MICs of 31.2 μg/ml against H37Ra and H37Rv strains, whereas *G. glutinosum* GGF5 hexane extract fraction induced MICs of 125 μg/ml against H37Ra strain, fraction GGF6 caused MICs of 250 μg/ml against H37Ra and H37Rv strains, and fraction GGF7 induced MICs of 125 μg/ml against H37Ra and H37Rv strains. MICs of control clofazimine were 0.31 μg/ml against H37Ra and H37Rv strains respectively.

The hexane and methanol extracts from avocado seeds showed LC_{50} values of 2.37 and 24.13 mg/ml respectively In toxicity tests on *Artemia salina* (Leite et al. 2009). The extracts tested were also active against all the yeast strains tested in vitro, with differing results such that the minimum inhibitory concentration of the hexane extract ranged from 0.625 to 1.25 mg/ml, from 0.312 to 0.625 mg/ml and from 0.031 to 0.625 mg/ml, for the strains of *Candida* spp, *Cryptococcus neoformans* and *Malassezia pachydermatis*, respectively. The minimal inhibitory concentration for the methanol extract ranged from 0.125 to 0.625 mg/ml, from 0.08 to 0.156 mg/ml and from 0.312 to 0.625 mg/ml, for the strains of *Candida* spp., *Cryptococcus neoformans* and *Malassezia pachydermatis*, respectively. Against *Aedes aegypti* larvae, the *LC50* results obtained were 16.7 mg/ml for hexane extract and 8.87 mg/ml for methanol extract from avocado seeds.

Microbiological assay demonstrated that AV119, a patented blend of two sugars from avocado induced the aggregation of yeast cells and inhibited the invasiveness of *Malassezia furfur,* without affecting its growth (Donnarumma et al. 2007). *Malassezia furfur*, is a dimorphic, lipid-dependent yeast that is part of the normal human cutaneous commensal flora. Real-time PCR analysis demonstrated that AV119 was able to modulate the HBD-2 (Human β-defensin 2) response in treated keratinocytes, reaching a maximum after 48-hours treatment, and induced the recovery of a satisfactory proinflammatory response in human keratinocytes. As AV119 could induce aggregation of yeast cells, thus inhibiting their penetration into the keratinocytes, the sugar could be used in the preparation of cosmetics or pharmacological drugs to inhibit colonization of the skin by pathogenic strains of *M. furfur*.

Methanolic extracts of *Persea americana* and other Mexican plants (*Annona cherimola*, *Guaiacum coulteri*, and *Moussonia deppeana*) showed highest inhibitory effect MIC <7.5 to 15.6 μg/ml against *Helicobacter pylori*, the major etiological agent of chronic active gastritis and peptic ulcer disease linked to gastric carcinoma (Castillo-Juárez et al. 2009).

Antiviral Activity

An infusion of *Persea americana* leaves strongly inhibited herpes simplex virus type 1 (HSV-1), Aujeszky's disease virus (ADV) and adenovirus type 3 (AD3) in cell cultures (De almeida et al. 1998). The active principles identified were two new flavonol monoglycosides, kaempferol and quercetin 3-O-α-D-arabinopyranosides, along with the known kaempferol 3-O-α-L-rhamnopyranoside (afzelin), quercetin 3-O-α-L-rhamnopyranoside (quercitrin), quercetin 3-O-β-gluco-pyranoside and quercetin. The known quercetin 3-O-β-galactopyranoside was identified in a mixture. Afzelin and quercetin 3-O-α-D-arabinopyranoside showed higher activity against acyclovir-resistant HSV-1. Chlorogenic acid significantly inhibited the HSV-1 replication without any cytotoxicity. However, all the substances tested were less active than the infusion or fractions. The same substances did not affect ADV replication.

Antiinflammatory Activity

Persenone A, isolated from avocado fruit, was found to be an effective inhibitor of both nitric oxide (NO) and superoxide (O^{2-}) generation in cell culture systems (Kim et al. 2000a, c). Persenone A at concentration of 20 μM almost completely suppressed both iNOS and COX-2 protein expression in a mouse macrophage cell line RAW 264.7. In mouse skin, double treatments with persenone A (810 nmol) significantly suppressed double 12-O-tetradecanoylphorbol-13-acetate (TPA, 8.1 nmol) application-induced hydrogen peroxide (H_2O_2) generation. Treatment with persenone A before the second TPA treatment was sufficient to inhibit H_2O_2 generation, while the first treatment was not. Results thus suggested persenone A as a possible agent to prevent inflammation-associated diseases including cancer.

The aqueous extract of *Persea americana* leaves produced a dose-dependent inhibition of both phases of formalin pain test in mice, a reduction in mouse writhing induced by acetic acid and an elevation of pain threshold in the hot plate test in mice (Adeyemi et al. 2002). The extract also produced a dose-dependent inhibition of carrageenan-induced rat paw oedema. The results obtained confirmed that the extract possessed analgesic and anti-inflammatory effects.

Hepatoprotective Activity

Avocado exerted a protective effect against liver injury when fed to rats with liver damage caused by d-galactosamine, a powerful liver toxin (Kawagishi et al. 2001). As measured by changes in the levels of plasma alanine aminotransferase (ALT) and aspartate aminotransferase (AST), avocado showed extraordinarily potent liver injury suppressing activity. Five active compounds were isolated and their structures determined. These were all fatty acid derivatives, of which three, namely, (2E,5E,12Z,15Z)-1-hydroxyheneicosa-2,5,12,15-tetraen-4-one; (2E,12Z,15Z)-1-hydroxyheneicosa-2,12,15-trien-4-one; and (5E,12Z)-2-hydroxy-4-oxoheneicosa-5,12-dien-1-yl acetate, were novel.

Studies showed that paracetamol at a dose of 2 g/kg body weight induced acute hepatotoxicity in rats 7 hours after oral administration as evident by the increase in serum alanine aminotransferase and aspartate aminotransferase activities (Epoyun et al. 2006). This was also associated with depletion of hepatic glutathione GSH, decrease in glutathione-S-transferase activity and decrease in the activities of antioxidant enzymes (superoxide dismutase and catalase). The methanol leaf extract of PA dose-dependently protected against acute hepatotoxicity induced by paracetamol by increasing the activity of the antioxidant enzymes and preventing GSH depletion. The results indicated that the extract protected against paracetamol-induced hepatotoxicity presumably via antioxidant action.

Studies showed that the lyophilisate of avocado seeds had an anti-icteric activity in rats intoxicated by carbon tetrachloride, and force-fed on a daily basis with 14 mg/100 g PV of lyophilisat of avocado almonds diluted with 5 ml of distilled water (Assane et al. 2001). This was reflected in the stimulation of the liver to conjugate and eliminate bilirubin, as well as hepatoprotective activity characterised by the normalisation of the aminotransferases enzymes and the healing of hepatic lesions within 10 days.

Hypoglycaemic, Antidiabetic Activity

A decoction of *P. americana* seeds was found to have a hypoglycemic effect (Koffi et al. 2009). At 20 g/l the decotion reduced hyperglycaemia in alloxan diabetic rats. There is some short-term stabilization and a slight return of hyperglycemia after the treatment is stopped. In another study, the n-hexane, chloroform and methanol extracts of *Persea americana* seeds caused significant blood glucose lowering effect more than the glibenclamide in the single dose alloxanized rats but in the double dose alloxanized rats, the glibenclamide showed more significant blood glucose lowering effect than the extracts (Okonta et al. 2007). The results indicated that *P. americana* seed extract could lower blood glucose levels in mild hyperglycemia but not in severe hyperglycemia.

Studies in Nigeria reported that the administration of aqueous extract of *P. americana* leaf (100–200 mg/kg) to alloxan-diabetic rats produced a significant reduction in blood glucose level in a dose-dependent fashion after a single dose of the extract, as well as following prolonged treatment (7 days) compared to the control group (Antia et al. 2005). Maximum antidiabetic activity was reached at 6 h after a single dose of the extract, producing 60.02% reduction in blood glucose levels. However, the hypoglycemic effect of the extract was incomparable to that of the reference drug chlorpropamide. The extract (100–200 mg/kg) produced a sustained significant antidiabetic activity during prolonged treatment (7 days). A sustained significant reduction in the blood glucose levels of the treated rats was observed throughout the period of treatment. The observation confirmed the use of this plant in ethnomedical practice for diabetes management and demonstrated that long-term treatment for 7 days was more effective than single dose acute treatment. In other studies, Edem (2009), Edem et al. (2009) found that in alloxan induced diabetic rats, blood glucose levels were significantly reduced by 73.26–78.24% on consumption of avocado seed extracts, with greater effect exhibited by the 600 mg/kg extract. In normal rats, blood glucose levels were significantly reduced by 34.68–38.9% on consumption of the seed extract. Histological studies showed a degenerative effect on the pancreatic islet cells of diabetic rats. The results suggested restorative (protective) effect of the extract on pancreatic islet cells and that administration of aqueous extract of avocado seed may contribute significantly to the reduction of blood glucose levels and can be useful in the treatment of diabetes.

Avocado ethanolic leaf extract (PAE) induced dose-dependent hypoglycaemic responses in streptozotocin-induced diabetic rats while subchronic PAE treatment additionally increased hepatic glycogen concentrations (Gondwe et al. 2008). Acute PAE infusion decreased urine flow and electrolyte excretion rates, whilst subchronic treatment reduced plasma creatinine and urea concentrations. These results supported the ethnomedicinal use of *P. americana* leaf extract in diabetes management.

Anticonvulsant Activity

Persea americana leaf aqueous extract (100–800 mg/kg i.p.) significantly delayed the onset of, and antagonized, pentylenetetrazole (PTZ)-induced seizures in mice like the reference anticonvulsant agents, phenobarbitone and diazepam used (Ojewole and Amabeoku 2006). The plant's leaf extract (100–800 mg/kg i.p.) also significantly antagonized picrotoxin (PCT)-induced seizures, but only weakly antagonized bicuculline (BCL)-induced seizures. Although the data obtained in the present study did not provide conclusive evidence, the researchers asserted that avocado leaf aqueous extract appeared to produce its anticonvulsant effect by enhancing GABAergic neurotransmission and/or action in the brain. The findings of this study indicated that *Persea americana* leaf aqueous extract possessed an anticonvulsant property, and thus lend pharmacological credence to the suggested ethnomedical uses of the plant in the management of childhood convulsions and epilepsy.

Wound Healing Activity

In the excision wound model, complete healing (full epithelialisation) was observed on average on day 14 in the rats who receive oral or topical treatment of *Persea americana* fruit extract (300 mg/kg/day) (Nayak et al. 2008) . In contrast, the controls took approximately 17 days to heal completely. The extract-treated wounds were found to epithelialise faster than the controls. Wet and dry granulation tissue weight and the hydroxyproline content of the tissue obtained from extract-treated animals used in the dead space wound model were significantly higher compared with the controls. Rate of wound contraction, epithelialisation time together with the hydroxyproline content and histological observations supported the traditional use of *Persea americana* in the management of wound healing.

Antipsoriasis Activity

In a randomized, prospective clinical trial, the effects of the vitamin D(3) analog calcipotriol were evaluated against those of a recently developed vitamin B(12) cream containing avocado oil in an intra-individual right/left-side comparison involving 13 patients, 10 men and 3 women, with chronic plaque psoriasis (Stücker et al. 2001). There was a more rapid development of beneficial effects with the use of calcipotriol in the initial 8 weeks, although differences in effects were significant only at the time point of therapy week 8. After 12 weeks, neither the PASI (Psoriasis Area and Severity Index) score nor 20-MHz sonography showed significant differences between the two treatments. While the efficacy of the calcipotriol preparation reached a maximum in the first 4 weeks and then began to subside, the effects of the vitamin B(12) cream containing avocado oil remained at a constant level over the whole observation period. This would indicate that the vitamin B(12) preparation containing avocado oil may be suitable for use in long-term therapy, a hypothesis further supported by the fact that the investigator and the patients assessed the tolerability of the vitamin B(12) cream containing avocado oil as significantly better in comparison with that of calcipotriol. The results of this clinical trial provided evidence that the recently developed vitamin B(12) cream containing avocado oil had considerable potential as a well-tolerated, long-term topical therapy of psoriasis.

Antivenom Activity

Total inhibition of hemorrhage was observed with the ethanolic, ethyl acetate and aqueous extracts of *Persea americana*, and other plants in mice injected intradermally with *Bothrops asper* venom or venom-extract mixtures (Castro et al. 1999). Chemical analysis of these extracts identified catechines, flavones, anthocyanins and condensated tannins, which were postulated to be responsible for the inhibitory effect observed, probably owing to the chelation of the zinc required for the catalytic activity of venom's hemorrhagic metalloproteinases.

Preventive Periodontal Actvity

Avocado/soyabean unsaponifiables (ASU) were found to modulate the effect of interleukin-1beta (IL-1beta) on transforming growth factor-beta(TGF-beta1, TGF-beta2) and bone morphogenetic protein-2 (BMP-2) expression in human periodontal ligament (HPL) and human alveolar bone (HAB) cells in culture (Andriamanalijaona et al. 2006). In periodontal disease, interleukin-1beta (IL-1beta) is responsible for the matrix breakdown through excessive production of degrading enzymes by periodontal ligament fibroblasts and osteoblasts. Transforming growth factor-beta plays an important role in tissue regeneration as one of the factors capable of counteracting IL-1beta effects. The data indicated that IL-1beta strongly decreased the expression of TGF-beta1 and TGF-beta2 by HPL cells. ASU was capable of opposing the cytokine effect. In HAB cells, TGF-beta1 and BMP-2 mRNA levels were downregulated by the cytokine. ASU was found to reverse the IL-1beta-inhibiting effect. In contrast, the cytokine stimulated the production of TGF-beta2 in alveolar bone cells, with no significant effect of ASU. The results indicated that the IL-1beta-driven erosive effect in periodontitis could be enhanced by a decreased expression of members of the TGF-beta family. The ASU stimulation of TGF-beta1, TGF-beta2, and BMP-2 expression may explain their promoting effects in the treatment of periodontal disorders, at least partly. These findings supported the hypothesis that ASU could exert a preventive action on the deleterious effects exerted by IL-1beta in periodontal diseases.

Agglutinating Activity

An extract from the seeds of *Persea americana* possessed an erythro-agglutinating activity (Meade et al. 1980). The agglutinin was devoid of specificity for carbohydrates, but interacted readily with basic proteins or basic polyamino acids. The interaction between the agglutinin and egg-white lysozyme was not inhibited by chaotropic salts, but was sensitive to relatively low concentrations of urea. Results suggested that, as opposed to other plant agglutinins, the active component from avocado was not a protein. Similarly, in contrast to many lectins, the agglutinin from avocado was not mitogenic for mouse lymphocytes. The agglutinin partially inhibited the mitogenesis of lymphocytes when the cells were treated with concanavalin A, or with bacterial lipopolysaccharide.

Genotoxic Activity

Crude extracts from avocado fruits and leaves were found to have potential genotoxic activity (Kulkarni et al. 2010). Chromosomal aberrations were observed in cultured human peripheral lymphocytes exposed to separately increasing concentrations of 50% methanolic extracts of *Persea americana* fruit and leaves. The groups exposed to leaf and fruit extracts, respectively, showed a concentration-dependent increase in chromosomal aberrations as compared to that in a control group. The mean percentage total aberrant metaphases at 100 mg/kg, 200 mg/kg, and 300 mg/kg concentrations of leaf extract were found respectively to be 58, 72, and 78, which were significantly higher than that in the control group (6). The mean percentage total aberrant metaphases at 100 mg/kg, 200 mg/kg, and 300 mg/kg concentrations of fruit extract were found to be 18, 40, and 52, respectively, which were significantly higher than that for control (6). Acrocentric associations and premature centromeric separation were the two most common abnormalities observed in both the exposed groups. The group exposed to leaf extracts also showed a significant number of a variety of other structural aberrations, including breaks, fragments, dicentrics, terminal deletion, minutes, and Robertsonian translocations. The group exposed to leaf extract showed higher frequency of all types of aberrations at equal concentrations as compared to the group exposed to fruit extract.

Traditional Medicinal Uses

The plant is used in traditional medicine for the treatment of various ailments, such as monorrhagia, hypertension, stomach ache, bronchitis, diarrhea, and diabetes (Yasir et al. 2010). *Persea americana* has been used in ethnomedicine in Trinidad and Tobago for the treatment of hypertension, jaundice and diabetes (Lans 2006). The fruit skin is antibiotic and has been employed as a vermifuge and remedy for dysentery. The pulp is believed to promote menstruation. Avocado is used in traditional medicine in Aboudé-Mandéké in the region of Agboville in Côte-d'Ivoire (West Africa) to treat diabetes (Koffi et al. 2009). Various morphological parts of *Persea americana* are widely used in African traditional medicines for the treatment, management and/or control of a variety of human ailments, including childhood convulsions and epilepsy (Ojewole and Amabeoku 2006). The avocado is also said to have spasmolitic and abortive properties. Oil extracted from the seeds has astringent properties, and an oral infusion of the leaves is used to treat dysentery. The seed is ground and made into an ointment used to treat various skin afflictions, such as scabies, purulent wounds, lesions of the scalp and dandruff. Pulverized seeds or bark, mix with oil are applied on affected area as counterirritant for rheumatism and neuralgia. A seed decoction is put into a tooth cavity to relieve toothache.

The leaves have been reported as an effective antitussive, antidiabetic, and relief for arthritis pain by traditional medicine practitioners of Ibibio tribe in South Nigeria (Antia et al. 2005). It is recommended for anemia, exhaustion, hypercholesterolemia, hypertension, gastritis, and gastroduodenal ulcers. The leaves are chewed as a remedy for pyorrhea. Leaf poultices are applied on wounds. Heated leaves are applied on the forehead to relieve neuralgia. The leaf juice has antibiotic activity. The aqueous extract of the leaves has a prolonged hypertensive effect. The leaf decoction is taken as a remedy for diarrhea, sore throat and hemorrhage; it allegedly stimulates and regulates menstruation. It is also drunk as a stomachic. In Cuba, a decoction of the new shoots is a cough remedy. A boiled decotion of leaves, or shoots of the purple-skinned type serves as an abortifacient. Sometimes a piece of the seed is boiled with the leaves to make the decoction.

Other Uses

Surplus avocado fruit is an important food source for pigs and other livestock. Recent studies reported that avocado peel carbon could be a lucrative technique for treatment of domestic wastewater as the adsorption capacity of avocado peel carbon (APC) was comparable with that of commercial activated carbon (CAC) for reduction of chemical oxygen demand (COD) and biological oxygen demand (BOD) concentration (Devi et al. 2008). The maximum percentage reduction of chemical oxygen demand (COD) and biological oxygen demand (BOD) concentration under optimum operating conditions in the coffee processing plant using APC was 98.20% and 99.18% respectively and with commercial activated carbon (CAC) this reduction was 99.02% and 99.35% respectively. The unripe fruit is poisonous and the ground-up seed mixed with cheese is used as a rat and mouse poison.

Watery extracts of the avocado leaves contain a yellowish-green essential oil. Honeybees gather a moderate amount of pollen from avocado flowers and provide a dark, thick honey favoured by those who like buckwheat honey or sugarcane syrup.

The seeds furnish a black sap, used as indelible ink. During the Spanish conquest era, the ink was used to write many important documents. The ink has also been used to mark cotton and linen textiles. Avocado oil expressed from the pulp and seeds is used as hair-dressing and is employed in making facial creams, hand lotions and fine soap and pharmaceutical products. It is said to filter out the tanning rays of the sun, is non-allergenic and is similar to lanolin in its penetrating and skins softening action. In Brazil, 30% of the avocado crop is processed for oil, two-thirds of which is utilized in soap, on-third in cosmetics. The pulp residue after oil extraction is usable as stockfeed.

Avocado wood has been used for house building (especially for house posts), light construction, furniture, cabinet making, agricultural implements, carving, sculptures, musical instruments, paddles, small articles like pen and brush holders, jewellery boxes and novelties. It also yields a good-quality veneer and plywood. However the brittle and susceptible to termite attack.

Methanol extract of the stem bark of *P. americana* and the n-hexane layer obtained from the methanol extract by solvent partitioning caused 100% mortality of pine wood nematode (PWN), *Bursaphelenchus xylophilus* at low concentrations of 125 and 63 μg/ml, respectively (Dang et al. 2010). One potent nematicidal compound was isolated from the stem bark and its chemical structure was determined to be isoobtusilactone A. The chemical showed a very strong nematicidal activity against PWN; it caused mortalities of over 97% at concentrations higher than 50 μg/ml. The results suggested that isoobtusilactone A may have the potential to be explored as a natural nematicide or be useful as a lead molecule for development of new nematicidal agents for controlling the pine wilt disease caused by *B. xylophilus*.

Comments

Persea americana has been botanically classified into three groups: (a) *Persea americana* Mill. var. *americana* (*Persea gratissima* Gaertn.), – the West Indian avocado; (b) *Persea americana* Mill. var. *drymifolia* Blake (*Persea drymifolia* Schlecht. & Cham.), – the Mexican avocado; (c) *Persea nubigena* var. *guatemalensis* L. Wms. (*Persea americana* Mill. var. *nubigena* (L.O. Willimas) L. E. Kopp) – the Guatemalan avocado.

Selected References

Adeboye JO, Fajonyomi MO, Makinde JM, Taiwo OB (1999) A preliminary study on the hypotensive activity of *Persea americana* leaf extracts in anaesthetized normotensive rats. Fitoterapia 70(1):15–20

Adeyemi OO, Okpo SO, Ogunti OO (2002) Analgesic and anti-inflammatory effects of the aqueous extract of leaves of *Persea americana* Mill. (Lauraceae). Fitoterapia 73(5):375–380

Andriamanalijaona R, Benateau H, Barre PE, Boumediene K, Labbe D, Compere JF, Pujol JP (2006) Effect of interleukin-1beta on transforming growth factor-beta and bone morphogenetic protein-2 expression in human periodontal ligament and alveolar bone cells in culture: modulation by avocado and soybean unsaponifiables. J Periodontol 77(7):1156–1166

Antia BS, Okokon JE, Okon PA (2005) Hypoglycemic activity of aqueous leaf extract of *Persea americana* Mill. Indian J Pharmacol 37:325–326

Asaolu MF, Asaolu SS, Adanlawo IG (2010a) Evaluation of phytochemicals and antioxidants of four botanicals with antihypertensive properties. Int J Pharm Biol Sci 1(2):1–7

Asaolu MF, Asaolu SS, Fakunle JB, Emman-Okon BO, Ajayi EO, Togun RA (2010b) Evaluation of in-vitro antioxidant activities of methanol extracts of *Persea americana* and *Cnidosculus aconitifolius*. Pak J Nutr 9(11):1074–1077

Asaolu MF, Asaolu SS, Oyeyemi AO, Aluko BT (2010c) Hypolipemic effects of methanolic extract of *Persea americana* seeds in hypercholestrolemic rats. J Med Med Sci 1(4):126–128

Ashton OB, Wong M, McGhie TK, Vather R, Wang Y, Requejo-Jackman C, Ramankutty P, Woolf AB (2006) Pigments in avocado tissue and oil. J Agric Food Chem 54(26):10151–10158

Assane M, Diop PA, Niang-Sylla M, Lopez-Sall P, Gueye PM, Charlevna A (2001) Study of the anti-icteric and hepatoprotective activity of *Persea gratissima* Gaertner (Lauraceae) seeds. Dakar Med 46(2):89–93, In French

Au RY, Al-Talib TK, Au AY, Phan PV, Frondoza CG (2007) Avocado soybean unsaponifiables (ASU) suppress TNF-alpha, IL-1beta, COX-2, iNOS gene expression, and prostaglandin E2 and nitric oxide production in articular chondrocytes and monocyte/macrophages. Osteoarthritis Cartilage 15(11):1249–1255

Blotman F, Maheu E, Wulwik A, Caspard H, Lopez A (1997) Efficacy and safety of avocado/soybean unsaponifiables in the treatment of symptomatic osteoarthritis of the knee and hip. A prospective, multicenter, three-month, randomized, double-blind, placebo-controlled trial. Rev Rhum Engl Ed 64(12):825–834

Brai BIC, Odetola AA, Agomo PU (2007a) Effects of *Persea americana* leaf extracts on body weight and liver lipids in rats fed hyperlipidaemic diet. Afr J Biotechnol 6(8):1007–1011

Brai BIC, Odetola AA, Agomo PU (2007b) Hypoglycemic and hypocholesterolemic potential of *Persea americana* leaf extracts. J Med Food 10(2):356–360

Butt AJ, Roberts CG, Seawright AA, Oelrichs PB, Macleod JK, Liaw TY, Kavallaris M, Somers-Edgar TJ, Lehrbach GM, Watts CK, Sutherland RL (2006) A novel plant toxin, persin, with in vivo activity in the mammary gland, induces Bim-dependent apoptosis in human breast cancer cells. Mol Cancer Ther 5(9):2300–2309

Carranza J, Alvizouri M, Alvarado MR, Chávez F, Gómez M, Herrera JE (1995) Effects of avocado on the level of blood lipids in patients with phenotype II and IV dyslipidemias. Arch Inst Cardiol Mex 65(4):342–348, In Spanish

Castillo-Juárez I, González V, Jaime-Aguilar H, Martínez G, Linares E, Bye R, Romero I (2009) Anti-*Helicobacter pylori* activity of plants used in Mexican traditional medicine for gastrointestinal disorders. J Ethnopharmacol 122(2):402–405

Castro O, Gutiérrez JM, Barrios M, Castro I, Romero M, Umaña E (1999) Neutralization of the hemorrhagic effect induced by *Bothrops asper* (Serpentes: Viperidae) venom with tropical plant extracts. Rev Biol Trop 47(3):605–616, In Spanish

Christensen R, Bartels EM, Astrup A, Bliddal H (2008) Symptomatic efficacy of avocado-soybean unsaponifiables (ASU) in osteoarthritis (OA) patients: a meta-analysis of randomized controlled trials. Osteoarthritis Cartilage 16(4):399–408

Council of Scientific and Industrial Research (CSIR) (1966) The wealth of India. A dictionary of Indian raw materials and industrial products. (Raw materials), vol 7. Publications and Information Directorate, New Delhi

Craigmill AL, Seawright AA, Mattila T, Frost AJ (1989) Pathological changes in the mammary gland and biochemical changes in milk of the goat following oral dosing with leaf of the avocado (*Persea americana*). Aust Vet J 66(7):206–211

Dang QL, Kwon HR, Choi YH, Choi GJ, Jang KS, Park MS, Lim CH, Ngoc LH, Kim JC (2010) Nematicidal activity against *Bursaphelenchus xylophilus* of isoobtusilactone A isolated from *Persea americana*. Nematology 12(2):247–253

De Almeida AP, Miranda MMFS, Simoni IC, Wigg MD, Lagrota MHC, Costa SS (1998) Flavonol monoglycosides isolated from the antiviral fractions of *Persea americana* (Lauraceae) leaf infusion. Phytother Res 12(8):562–567

Devi R, Singh V, Kumar A (2008) COD and BOD reduction from coffee processing wastewater using avocado peel carbon. Bioresour Technol 99(6):1853–1860

Ding H, Chin YW, Kinghorn AD, D'Ambrosio SM (2007) Chemopreventive characteristics of avocado fruit. Semin Cancer Biol 17(5):386–394

Ding H, Han C, Guo D, Chin YW, Ding Y, Kinghorn AD, D'Ambrosio SM (2009) Selective induction of apoptosis of human oral cancer cell lines by avocado extracts via a ROS-mediated mechanism. Nutr Cancer 61(3):348–356

Donnarumma G, Buommino E, Baroni A, Auricchio L, De Filippis A, Cozza V, Msika P, Piccardi N, Tufano MA (2007) Effects of AV119, a natural sugar from avocado, on *Malassezia furfur* invasiveness and on the expression of HBD-2 and cytokines in human keratinocytes. Exp Dermatol 16(11):912–919

Edem DO (2009) Hypoglycemic effects of ethanolic extracts of alligator pear seed (*Persea americana* Mill) in rats. Eur J Sci Res 33(4):669–678

Edem D, Ekanem I, Ebong P (2009) Effect of aqueous extracts of alligator pear seed (*Persea americana* Mill.) on blood glucose and histopathology of pancreas in alloxan-induced diabetic rats. Pak J Pharm Sci 22(3):272–276

Epoyun AA, Adepoju GKA, Ekor M (2006) Protective effect of the methanolic leaf extract of *Persea americana* (avocado) against paracetamol-induced acute hepatotoxicity in rats. Int J Pharmacol 2(4):416–420

Ernst E (2003) Avocado-soybean unsaponifiables (ASU) for osteoarthritis – a systematic review. Clin Rheumatol 22(4–5):285–288

Gabay O, Gosset M, Levy A, Salvat C, Sanchez C, Pigenet A, Sautet A, Jacques C, Berenbaum F (2008) Stress-induced signaling pathways in hyalin chondrocytes: inhibition by avocado-soybean unsaponifiables (ASU). Osteoarthritis Cartilage 16(3):373–384

Gomez-Flores R, Arzate-Quintana C, Quintanilla-Licea R, Tamez-Guerra P, Tamez-Guerra R, Monreal-Cuevas E, Rodríguez-Padilla C (2008) Antimicrobial activity of *Persea americana* Mill (Lauraceae) (avocado) and *Gymnosperma glutinosum* (Spreng.) Less (Asteraceae) leaf extracts and active fractions against *Mycobacterium tuberculosis*. Am Eurasian J Sci Res 3(2):188–194

Gondwe M, Kamadyaapa DR, Tufts MA, Chuturgoon AA, Ojewole JA, Musabayane CT (2008) Effects of *Persea americana* Mill (Lauraceae) [Avocado] ethanolic leaf extract on blood glucose and kidney function in streptozotocin-induced diabetic rats and on kidney cell lines of the proximal (LLCPK1) and distal tubules (MDBK). Methods Find Exp Clin Pharmacol 30(1):25–35

Gorinstein S, Haruenkit R, Poovarodom S, Vearasilp S, Ruamsuke P, Namiesnik J, Leontowicz M, Leontowicz H, Suhaj M, Sheng GP (2010) Some analytical assays for the determination of bioactivity of exotic fruits. Phytochem Anal 21(4):355–362

Gross J, Gabai M, Lifshitz A, Sklarz B (1973) Carotenoids in pulp, peel and leaves of *Persea americana*. Phytochemistry 12(9):2259–2263

Gross J, Gabai M, Lifshitz A, Sklarz B (1974) Structures of some carotenoids from the pulp of *Persea americana*. Phytochemistry 13(9):1917–1921

Hashimura H, Ueda C, Kawabata J, Kasai T (2001) Acetyl-CoA carboxylase inhibitors from avocado (*Persea americana* Mill) fruits. Biosci Biotechnol Biochem 65:1656–1658

Henrotin YE, Deberg MA, Crielaard JM, Piccardi N, Msika P, Sanchez C (2006) Avocado/soybean unsaponifiables prevent the inhibitory effect of osteoarthritic subchondral osteoblasts on aggrecan and type II collagen synthesis by chondrocytes. J Rheumatol 33(8):1668–1678

Hodgson RW (1950) The avocado – a gift from the Middle Americas. Econ Bot 4:253–293

Imafidon KE, Amaechina FC (2010) Effects of aqueous seed extract of *Persea americana* Mill. (avocado) on blood pressure and lipid profile in hypertensive rats. Adv Biol Res 4(2):116–121

Kawagishi H, Fukumoto Y, Hatakeyama M, He P, Arimoto H, Matsuzawa T, Arimoto Y, Suganuma H, Inakuma T, Sugiyama K (2001) Liver injury suppressing compounds from avocado (*Persea americana*). J Agric Food Chem 49(5):2215–2221

Kim OK, Murakami A, Nakamura Y, Kim HW, Ohigash H (2000a) Inhibition by (−)-persenone A-related compounds of nitric oxide and superoxide generation from inflammatory leukocytes. Biosci Biotechnol Biochem 64(11):2500–2503

Kim OK, Murakami A, Nakamura Y, Takeda N, Yoshizumi H, Ohigashi H (2000b) Novel nitric oxide and superoxide generation inhibitors, persenone A and B, from avocado fruit. J Agric Food Chem 48(5):1557–1563

Kim OK, Murakami A, Takahashi D, Nakamura Y, Torikai K, Kim HW, Ohigashi H (2000c) An avocado constituent, persenone A, suppresses expression of inducible forms of nitric oxide synthase and cyclooxygenase in macrophages, and hydrogen peroxide generation in mouse skin. Biosci Biotechnol Biochem 64(11): 2504–2507

Koffi N, Ernest AK, Dodiomon S (2009) Effect of aqueous extract of *Persea americana* seeds on the glycemia of diabetic rabbits. Eur J Sci Res 26(3):376–385

Kopp LE (1966) A taxonomic revision of the genus *Persea* in the western hemisphere. Mem N Y Bot Gard 14:1–117

Kritchevsky D, Tepper SA, Wright S, Czarnecki SK, Wilson TA, Nicolosi RJ (2003) Cholesterol vehicle in experimental atherosclerosis 24: avocado oil. J Am Coll Nutr 22(1):52–55

Kruthiventi AK, Krishnaswamy NR (2000) Constituents of the flowers of *Persea gratissima*. Fitoterapia 71(1):94–96

Kulkarni P, Paul R, Ganesh N (2010) In vitro evaluation of genotoxicity of avocado (*Persea americana*) fruit and leaf extracts in human peripheral lymphocytes. J Environ Sci Health 28(3):172–187

Lans CA (2006) Ethnomedicines used in Trinidad and Tobago for urinary problems and diabetes mellitus. J Ethnobiol Ethnomed 2:45

Leite JJG, Brito EHS, Cordeiro RA, Brilhante RSN, Sidrim JJC, Bertini LM, de Morais SM, Rocha MFG (2009) Chemical composition, toxicity and larvicidal and antifungal activities of *Persea americana* (avocado) seed extracts. Rev Soc Bras Med Trop 42(2):110–113

Lequesne M, Maheu E, Cadet C, Dreiser RL (2002) Structural effect of avocado/soybean unsaponifiables on joint space loss in osteoarthritis of the hip. Arthritis Rheum 47(1):50–58

Li XW, Li J, van der Werff H (2008) *Persea* Miller. In: Wu ZY, Raven PH, Hong DY (eds) Flora of China, vol 7, Menispermaceae through Capparaceae. Science Press/ Missouri Botanical Garden Press, Beijing/St. Louis

Lopez Ledesma R, Frati Munari AC, Hernandez Dominguez BC, Cervantes Montalvo S, Hernandez Luna MH, Juarez C, Moran Lira S (1996) Monounsaturated fatty acid (avocado) rich diet for mild hypercholesterolemia. Arch Med Res 27(4):519–523

Lu QY, Arteaga JR, Zhang Q, Huerta S, Go VL, Heber D (2005) Inhibition of prostate cancer cell growth by an avocado extract: role of lipid-soluble bioactive substances. J Nutr Biochem 16(1):23–30

Lu QY, Zhang Y, Wang Y, Wang D, Lee RP, Gao K, Byrns R, Heber D (2009) California Hass avocado: profiling of carotenoids, tocopherol, fatty acid, and fat content during maturation and from different growing areas. J Agric Food Chem 57(21):10408–10413

Maheu E, Mazières B, Valat JP, Loyau G, Le Loët X, Bourgeois P, Grouin JM, Rozenberg S (1998) Symptomatic efficacy of avocado/soybean unsaponifiables in the treatment of osteoarthritis of the knee and hip: a prospective, randomized, double-blind, placebo-controlled, multicenter clinical trial with a six-month treatment period and a two-month followup demonstrating a persistent effect. Arthritis Rheum 41(1):81–91

Meade NA, Staat RH, Langley SD, Doyle RJ (1980) Lectin-like activity from *Persea americana*. Carbohydr Res 78(2):349–363

Méndez OP, Hernández LG (2007) High-density lipoproteins (HDL) size and composition are modified in the rat by a diet supplemented with "Hass" avocado (*Persea americana* Miller). Arch Cardiol Mex 77:17–24, In Spanish

Morton J (1987) Avocado. In: Fruits of warm climates. Julia F. Morton, Miami, pp 91–102

Naveh E, Werman MJ, Sabo E, Neeman I (2002) Defatted avocado pulp reduces body weight and total hepatic fat but increases plasma cholesterol in male rats fed diets with cholesterol. J Nutr 132(7):2015–2018

Nayak BS, Raju SS, Chalapathi Rao AV (2008) Wound healing activity of *Persea americana* (avocado) fruit: a preclinical study on rats. J Wound Care 17(3):123–126

Oberlies NH, Rogers LL, Martin JM, McLaughlin JL (1998) Cytotoxic and insecticidal constituents of the unripe fruit of *Persea americana*. J Nat Prod 61(6): 781–785

Oelrichs PB, Ng JC, Seawright AA, Ward A, Schäffeler L, MacLeod JK (1995) Isolation and identification of a compound from avocado (*Persea americana*) leaves which causes necrosis of the acinar epithelium of the lactating mammary gland and the myocardium. Nat Toxins 3(5):344–349

Ojewole JAO, Amabeoku GJ (2006) Anticonvulsant effect of *Persea americana* Mill. (Lauraceae) (avocado) leaf aqueous extract in mice. Phytother Res 20(8):696–700

Ojewole JA, Kamadyaapa DR, Gondwe MM, Moodley K, Musabayane CT (2007) Cardiovascular effects of *Persea americana* Mill (Lauraceae) (avocado) aqueous leaf extract in experimental animals. Cardiovasc J Afr 18(2):69–76

Okonta M, Okonta L, Aguwa CN (2007) Blood glucose lowering activities of seed of *Persea americana* on alloxan induced diabetic rats. Nig J Nat Prod Med 11:26–28

Orwa C, Mutua A, Kindt R, Jamnadass R, Anthony S (2009) Agroforestree database: a tree reference and selection guide version 4.0. http://www.worldagroforestry.org/sites/treedbs/treedatabases.asp

Owolabi MA, Jaja SI, Coker HAB (2005) Vasorelaxant action of aqueous extract of the leaves of *Persea americana* on isolated thoracic rat aorta. Fitoterapia 76(6):567–573

Owolabi MA, Coker HAB, Jaja SI (2010) Bioactivity of the phytoconstituents of the leaves of *Persea americana*. J Med Plant Res 4(12):1130–1135

Pacetti D, Boselli E, Lucci P, Frega NG (2001) Simultaneous analysis of glycolipids and phospholids molecular species in avocado (*Persea americana* Mill) fruit. J Chromatogr A 115(1–2):241–245

Pacific Island Ecosystems at Risk (PIER) (1999) *Persea americana* Mill., Lauraceae http://www.hear.org/pier/species/persea_americana.htm

Pino JA, Mabot R, Rosado A, Fuentes V (2004) Volatile components of avocado (*Persea americana* Mill.) cv. Moro grown in Cuba. J Essent Oil Res 16(2): 139–140

Plaza L, Sánchez-Moreno C, de Pascual-Teresa S, de Ancos B, Cano MP (2009) Fatty acids, sterols, and antioxidant activity in minimally processed avocados during refrigerated storage. J Agric Food Chem 57(8):3204–3209

Porcher MH et al (1995–2020) Searchable world wide web multilingual multiscript plant name database. The University of Melbourne, Australia. http://www.plantnames.unimelb.edu.au/Sorting/Frontpage.html

Salazar MJ, El Hafidi M, Pastelin G, Ramírez-Ortega MC, Sánchez-Mendoza MA (2005) Effect of an avocado oil-rich diet over an angiotensin II-induced blood pressure response. J Ethnopharmacol 98(3):335–338

Stücker M, Memmel U, Hoffmann M, Hartung J, Altmeyer P (2001) Vitamin B(12) cream containing avocado oil in the therapy of plaque psoriasis. Dermatology 203(2):141–147

Sugiyama T, Sato A, Yamashita K (1982) Synthesis of all four stereoisomers of antibacterial component of avocado. Agric Biol Chem 46(2):481–485

Takenaga F, Matsuyama K, Abe S, Torii Y, Itoh S (2008) Lipid and fatty acid composition of mesocarp and seed of avocado fruits harvested at northern range in Japan. J Oleo Sci 57(11):591–597

U.S. Department of Agriculture, Agricultural Research Service (2010) USDA National Nutrient Database for Standard Reference, Release 23. Nutrient Data Laboratory Home Page, http://www.ars.usda.gov/ba/bhnrc/ndl

Unlu NZ, Bohn T, Clinton SK, Schwartz SJ (2005) Carotenoid absorption from salad and salsa by humans is enhanced by the addition of avocado or avocado oil. J Nutr 135(3):431–436

van der Werff H (2002) A synopsis of *Persea* (Lauraceae) in Central America. Novon 12(4):575–586

Whiley AW (1992) *Persea americana* Miller. In: Verheij EWM, Coronel RE (eds) Plant resources of South-East Asia. No. 2: edible fruits and nuts. Prosea Foundation, Bogor, pp 240–254

Williams LO (1977) The avocados, a synopsis of the genus *Persea*, subg *Persea*. Econ Bot 31:315–320

Yaman M, Eser O, Cosar M, Bas O, Sahin O, Mollaoglu H, Fidan H, Songur A (2007) Oral administration of avocado soybean unsaponifiables (ASU) reduces ischemic damage in the rat hippocampus. Arch Med Res 38(5):489–494

Yasir M, Das S, Kharya MD (2010) The phytochemical and pharmacological profile of *Persea americana* Mill. Pharmacogn Rev 4(7):77–84

Ye Q, Gu ZM, Zeng L, Zhao GX, Chang CJ, McLaughlin JL, Sastrodihardjo S (1996) Persealide: a novel, biologically active component from the bark of *Persea americana* (Lauraceae). Pharm Biol 34(1):70–72

Yuko M, Jun K, Takanori K (2003) Antioxidative constituents in avocado (*Persea americana* Mill.) seeds. J Jpn Soc Food Sci Technol 50(11):550–552

Barringtonia asiatica

Scientific Name

Barringtonia asiatica **(L.) Kurz.**

Synonyms

Agasta asiatica Miers, *Agasta indica Miers*, *Agasta splendida* Miers, *Barringtonia butonica* J. R. Forst. & G. Forst., *Barringtonia levequii* Jard., nom. nud., *Barringtonia littorea* Oken, *Barringtonia senequei* Jard., *Barringtonia speciosa* J. R. Forst. & G. Forst., *Barringtonia speciosa* J. R. Forster & G. Forster., *Butonica speciosa* Lam., *Huttum speciosum* Britten, *Mammea asiatica* Linnaeus, *Michelia asiatica* Kuntze, *Mitraria commersonia* J. F. Gmel.

Family

Lecythidaceae also placed in Barringtoniaceae

Common/English Names

Balubiton, Barringtonia, Box Fruit, Butong, Butun, Fish Poison Tree, Fish-Killer Tree, Fish-Poison Tree, Fish-Poison-Tree, Langasat, Lugo, Motong-Botong, Pertun, Putat Laut, Sea Poison Tree, Sea Putat, Vuton

Vernacular Names

Burmese: Kyi-Git;
Chinese: Bin Yu Rui , Mo Pan Jiao Shu;
Indonesia: Bitung (Northern Sulawesi), Butun (Javanese, Sundanese), Keben-Keben (Balinese);
Malaysia: Butong Butun, Pertun, Putat Air, Putat Laut, Putat Gajah;
Papua New Guinea: Maliou;
Philippines: Biton, Botong, Motong-Botong, Botong-Botong (Bikol), Bitung, Bituing (Bisaya), Biton (Cebu Bisaya), Biton (Chabacano), Lugo (Ibanag), Vuton (Ivatan), Balubiton (Panay Bisaya), Boton, Botong (Tagalog);
Russian: Barringtonia Aziatskaia, Barringtonia Prekrasnaia;
Spanish: Arbol De Los Muertos;
Samoan: Futu;
Taiwan: Yin Du Yu Rui;
Thai: Chik Le, Chik Ta Lae, Don Ta Lae;
Vietnamese: Bang Qua Vuong.

Origin/Distribution

Barringtonia asiatica is indigenous to the mangrove habitats in the tropics from Madagascar, to Malesia, Taiwan, Philippines, northern Australia and Polynesia.

T.K. Lim, *Edible Medicinal And Non-Medicinal Plants: Volume 3, Fruits*,
DOI 10.1007/978-94-007-2534-8_7, © Springer Science+Business Media B.V. 2012

Agroecology

The tree is tropical in requirement and grows in littoral sandy beaches, coral sand flats or river banks, in mangrove swamps at sea level and also inland near rivers on limestone hillsides. The tree is salt tolerant. The fruit is dispersed by floating at sea, where it can survive afloat for a few years.

Edible Plant Parts and Uses

In Indo China, fruits are eaten as a vegetable after prolonged cooking to remove the saponins.

Botany

A small to medium-sized, evergreen, perennial tree growing to 7–20 m high with a cylindrical bole of 30 cm diameter, fissured bark and stout branches. Leaves are opposite to sub-opposite, sessile, simple, dark green above, paler dull green below, obovate to obovate-oblong (Plate 1), 20–40 × 10–20 cm, leathery, shiny, base cuneate, margin entire, apex obtuse or broadly rounded with pinnate venation. Inflorescence are terminal racemes, erect, 5–15 cm, 5–10(–20)-flowered; bracts are ovate, 8–20 mm; bracteoles are triangular, 1.5–5 mm. Flower buds are 2–4 cm across and borne on 5–9 cm pedicels. Calyx is undivided, rupturing at anthesis into 2 or 3 unequal, rounded or acuminate, persistent lobes and a tube 3–5 mm. Petals 4, white, ovate or elliptic, 5–6 cm. Stamens about 100, in 6 whorls; tube 1.5–6 mm; filaments and style white, red-tipped. Ovary is 4-loculed with 4–5 ovules per locule. Fruit is broadly pyramidal, 4–5 angled, indehiscent, smooth, 9–11 cm, green or brown, apex tapering and crowned by calyx; pericarp spongy, fibrous (Plate 1). Seed is oblong, 4–5 cm.

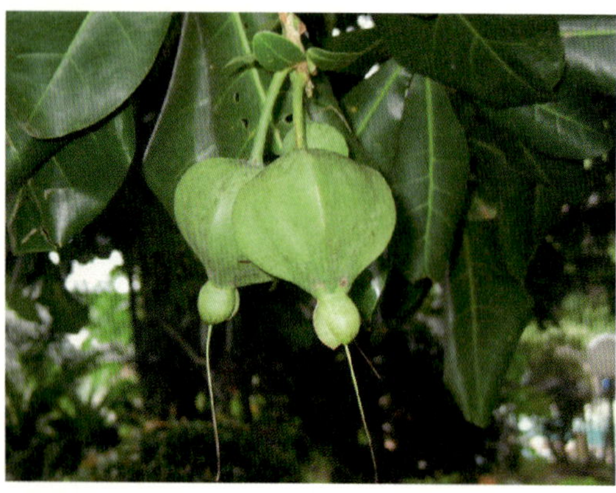

Plate 1 Fruits and leaves of box fruit

Nutritive/Medicinal Properties

The crude methanolic extract of *Barringtonia asiatica* (leaves, fruits, seeds, stem and root barks) and the fractions (petrol, dichloromethane, ethyl acetate and butanol) exhibited a very good level of broad spectrum antibacterial activity (Khan and Omoloso 2002). A number of fractions demonstrated antifungal activity against a number of fungi.

A crude aqueous extract of the seeds showed high biological activity using the brine shrimp hatchability and lethality assay (Mojica and Micor 2002). The LC_{50} obtained was better than the positive control used. Phytochemical tests showed the presence of terpenoids and saponins in the crude extract. It is possible that the seeds contain compounds with potential biological activity that can be used to treat cancer or tumour.

Various parts of the tree are used in folkloric medicine in its native area of habitation.

In the Philippines, the leaves are heated and applied externally for stomach-ache. In the Bismarck archipelago, the fresh fruit is scraped and applied topically to sores. The dried fruit is ground and mixed with water and taken for coughs, influenza, sore throats and bronchitis. Externally it is applied to wounds and a swollen spleen. In Australia, the aborigines used the plant to alleviate headache. In Fiji, a decoction of the leaves is used to treat hernia and a bark decoction used for constipation and epilepsy.

Other Uses

The tree is planted as shade and avenue and boulevard trees along the beach. The wood is of little economic value but in some places e.g. in Kediri, Java, in the Nicobars and the Philippines it is used for native huts. Also if impregnated it is suggested that it would make good tiles, paving blocks and cabinets.

All parts of the tree are poisonous, the active poisons including saponins. The fruit or seeds are used as fish poison. Ground seeds are used to stun fish for easy capture.

Two major saponins isolated from a methanol extract of the seeds of *Barringtonia asiatica* namely 3-*O*-{[β-D-galactopyranosyl(1→3)-β-D-glucopyranosyl(1→2)]-β-D-glucuronopyranosyloxy}-22-*O*-(2-methylbutyroyloxy)-15,16,28-trihydroxy-(3β, 15α,16α,22α)-olean-12-ene and 3-*O*-{[β-D-galactopyranosyl(1→3)-β-D-glucopyranosyl (1→2)]-β-D-glucuronopyranosyloxy}-22-*O*-[2 (*E*)-methyl-2-butenyloyloxy]-15,16,28-trihydroxy-(3β,15α,16α,22α)-olean-12-ene, exhibited antifeedant activity toward *Epilachna* larvae and has insecticidal potential (Herlt et al. 2002).

Comments

Its flowers emit a sickly, sweet smell that attracts bats and moths pollinators at night.

Selected References

Burkill IH (1966) A dictionary of the economic products of the Malay Peninsula. Revised reprint. 2 volumes. Ministry of Agriculture and Co-operatives, Kuala Lumpur, vol 1 (A–H) pp 1–1240, vol 2 (I–Z) pp 1241–2444

Cambie RC, Ash JE (1994) Fijian medicinal plants. CSIRO Australia, Collingwood

Chantaranothai P (1995) *Barringtonia* (Lecythidaceae) in Thailand. Kew Bull 50(4):677–694

Gutierrez HG (1980–1982) An illustrated manual of Philippine materia medica. 2 volumes. National Research Council of the Philippines, Tagig, Metro Manila, the Philippines, vol 1 (1980) pp 1–234, vol 2 (1982) pp 235–485

Herlt AJ, Mander LN, Pongoh E, Rumampuk RJ, Tarigan P (2002) Two major saponins from seeds *of Barringtonia asiatica*: putative antifeedants toward *Epilachna* sp Larvae. J Nat Prod 65(2):115–120

Khan MR, Omoloso AD (2002) Antibacterial, antifungal activities of *Barringtonia asiatica*. Fitoterapia 73(3):255–260

Mojica E-RE, Micor JRL (2002) Study of *Barringtonia asiatica* (Linnaeus) Kurz. Seed aqueous extract in *Artemia salina*. Int J Bot 3(3):325–328

Perry LM (1980) Medicinal plants of East and Southeast Asia. Attributed properties and uses. MIT Press, Cambridge/London, 620 pp

Qin H, Prance GIT (2007) *Barringtonia*. In: Wu ZY, Raven PH, Hong DY (eds) Flora of China, vol 13, Clusiaceae through Araliaceae. Science Press/ Missouri Botanical Garden Press, Beijing/ St. Louis

Yaplito MA (2002) *Barringtonia* Foster JR, Forster JG. In: van Valkenburg JLCH, Bunyapraphatsara N (eds) Plant resources of South-East Asia No. 12(2): Medicinal and poisonous plants 2, Prosea Foundation, Bogor, Indonesia, pp 101–107

Barringtonia edulis

Scientific Name

Barringtonia edulis **(Seemann) Miers.**

Synonyms

Barringtonia excelsa sensu Seemann non Blume, *Barringtonia seaturae* sensu Payens, *Butonica edulis* (Seemann) Miers, *Butonica samoensis* Miers, *Huttum edule* (Miers) Brittons.

Family

Lecythidaceae also in Barringtoniaceae

Common/English Names

Barringtonia, Cutnut, Pau Nut, Pao Nut

Vernacular Names

Fiji: Vutu, Vutu Valu, Vutu Kana, Vala, Vutu Kata; ***Papua New Guinea***: Vutu Ni Veikau, Pau, Pau nut; ***Rotuman***: Navele, Hufa'a

Origin/Distribution

B. edulis is indigenous to Melanesia – Papua New Guinea, Solomon Islands and Vanuatu. It is cultivated in villages and home-gardens in its native areas and in Fiji.

Agroecology

B. edulis is a lowland tropical species occurring naturally in dense and open forest, or on the edge of forests in mature fallow forest from sea level to 400 m. It is commonly planted around coastal and inland villages and in home gardens. It shares similar agro-ecological requirements as *B. procera* and *B. novae-hiberniae*.

Edible Plant Parts and Uses

Seeds kernels are eaten raw or cooked or roasted, flavour liken to that of raw peanuts. It provides a good source of energy and snacks for children. In Vanuatu, the kernels are pounded into *lap-lap*. The kernels can be stored for several months after being dried by binding in the bark of bourao (*Hibiscus tiliaceus*) and smoking over a fire. The mesocarp is insipid, fibrous, bland and is not eaten.

T.K. Lim, *Edible Medicinal And Non-Medicinal Plants: Volume 3, Fruits*,
DOI 10.1007/978-94-007-2534-8_8, © Springer Science+Business Media B.V. 2012

Botany

An evergreen monoecious tree, growing to 6–20 m high. Leaves occur in rosettes clustered towards the end of branches (Plates 1 and 2). Leaves are glossy, bright green, obovate-oblong to oblanceolate, 25–75 cm long by 8–30 cm wide, simple with slightly undulating margin, obtuse apex and tapering base on short, 0.5–1.5 cm petiole. Flowers are bisexual, pink or cream, showy, spirally arranged on the long 50–100 cm long rachis of a pendulous racemose inflorescence formed at the end of branches. Calyx consist of 3–4 persistent, imbricate green or crimson lobes inserted on rim of receptacle, or breaking at anthesis into 3–4 persistent pseudo-lobes, corolla consist of 4 cream coloured or rosy pink petals. Stamens, yellow or cream, many in 3–8 whorls. Style central and longer than the stamens. Fruit is ovoid, oblong or elongate, green, purplish or red, indehiscent, sub-sessile or shortly pedicellate (Plates 2, 3 and 4) containing one white kernel 2.5 cm by 1.5–3 cm (Plate 5).

Plate 1 Fruits formed from terminal inflorescence and leaves in whorls towards the shoot apex

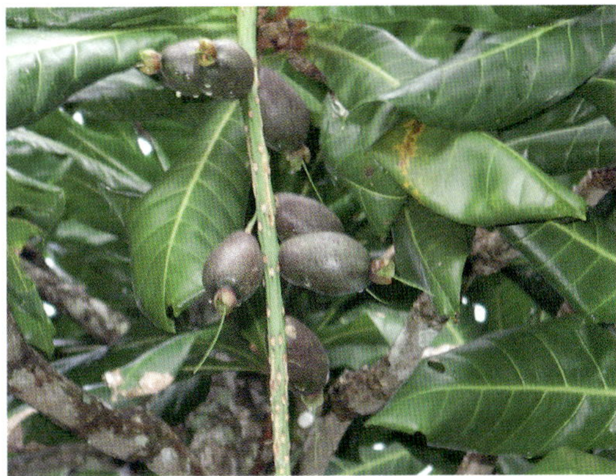

Plate 2 Shortly pedicellate to sub-sessile fruits

Plate 3 Ovoid, reddish-purple fruits with persistent calyx lobes and style

Plate 4 Close-up of fruit

Plate 5 Nut kernel with flesh removed

Nutritive/Medicinal Properties

The nutrient composition of raw cut-nut (*Barringtonia edulis*) per 100 g edible portion was reported as: water 38.9 g, protein 9.7 g, fat 11.8 g, sugars, 1.8 g, starch 23.3 g, ash 1.4 g, dietary fibre 10.2 g, β-carotene equivalent 36 μg, thiamin 0.15 mg, riboflavin 0.02 mg, niacin 2.6 mg, vitamin C 7 mg, Na 10 mg, K 410 mg, Ca 12 mg, Fe 2.4 mg, Mg 121 mg, Zn 2.3 mg, Cu 0.8 mg, Mn 0.8 mg, energy 243 kcal (1017 kJ) (English et al. 1996).

Fatty acid profile of *B. edulis* seed oil (mol% total fatty acids) was reported by Sotheeswaran et al. (1994) as follows: 16:0 (palmitic) 43.7%, 16:1 (palmitoleic) 1.8%, 18:0 (stearic) 24.8%, 18:1–9c (oleic), 20.4%, 18:1–11c (vaccenic) 0.6%, 18:2–9c,12 C (linoleic) 0.6%, 20:0 3.3%. Lipids of *B. edulis* seeds were found to contain no phytosterols, and contained mainly diacylglycerol.

The bark is used in traditional medicine for stomach ailments and gonorrhoea in the Solomon Islands. In Vanuatu, the leaves are used to treat ear inflammation, the sap from the trunk is used for ciguatera poisoning, for treating coughs and urinary complaints. The variety with red leaves is employed as abortifacient or as contraceptive.

Other Uses

Its non-edible uses are similar to that described for *B. procera*.

Comments

Red foliage varieties are also found in Vanuatu.

Selected References

Anonymous (1992) *Barringtonia edulis*. Curtis's Bot Mag 9(4):175–180

English RM, Aalbersberg W and Scheelings P (1996) Pacific Island Foods – Description and Nutrient Composition of 78 local foods. IAS Technical Report 96/02. ACIAR Report 9306. Institute of Applied Science, University of the South Pacific, Suva, Fiji. 94 pp

Evans B (1991) A variety collection of edible nut tree crops in Solomon Islands, vol 8, Research Bulletin. Dodo Creek Research Station, Honiara, 98 pp

French BR (1986) Food plants of Papua New Guinea – a compendium. Australia and Pacific Science Foundation, Ashgrove, 408 pp

Jebb M (1992) Edible Barringtonias. Curtis's Bot Mag 9(4):164–172

Pauku RL (2006) *Barringtonia procera* (cutnut), ver. 2.1. In: Elevitch CR (Ed) Species Profiles for Pacific Island Agroforestry. Permanent Agriculture Resources (PAR), Holualoa, Hawai'i. http://www.traditionaltree.org

Payens JP (1967) A monograph of the genus *Barringtonia* (Lecythidaceae). Blumea 15(2):157–263

Sotheeswaran S, Sharif SS, Moreau RA, Piazza GJ (1994) Lipids from the seeds of seven Fijian plant species. Food Chem 49:11–13

Walter A, Sam C (2002) Fruits of oceania, vol 85, ACIAR monograph. Australian Centre for International Agricultural Research, Canberra, 329 pp

Barringtonia novae-hiberniae

Scientific Name

Barringtonia novae-hiberniae **Lauterbach.**

Synonyms

Barringtonia brosimos Merr. & Perry, *Barringtonia excelsa* Guill., *Barringtonia oblongifolia* Kunth.

Family

Lecythidaceae also in Barringtoniaceae

Common/English Names

Cut Nut, Bush Cutnut, Wild Cutnut, Pau Nut

Vernacular Names

French: La Velle;
Papua New Guinea: Cut Nut, Pau Nut, Pao Nut (General), Tegeli (Tok Plesin, Madang), Laluan Pao (Tok Plesin, Morobe), Pulei, Purei (Tok Plesin Manus), Pala Paua, Hutun Pao (Tok Plesin, New Ireland), Kuanua Pao, Vutug Pao (Tok Plesin, New Britain), Siwai Hari (Tok Plesin, North Solomons);
Solomon Islands: Kat Nut (General), Borolong Sioko (Shortland Island), Tinga Hala, Fala, Kenu (New Georgia), Hara Hara (San Cristobal), Nua Nuado (Santa Cruz);
Vanuatu: Navele (Bislama), Va Rodh Vevingen (Tok Plesin).

Origin/Distribution

The species is native to Melanesia – Papua New Guinea, the Solomon Islands and Vanuatu.

Agroecology

Barringtonia novae-hiberniae is largely undomesticated. It is commonly encountered in secondary forests, fallow forests, and under coconut plantations and is less common around homes and within village surroundings. Its agro-ecological requirements would be similar to that described for *B. procera* and *B. edulis*.

Edible Plant Parts and Uses

The kernel of ripe fruit is eaten raw or cooked. The mesocarp of ripe fruit is not eaten.

T.K. Lim, *Edible Medicinal And Non-Medicinal Plants: Volume 3, Fruits*,
DOI 10.1007/978-94-007-2534-8_9,

Botany

A small to medium-sized tree, 7–1 m high with crown denser than that of *B. edulis*. Leaves on slender 2–7 cm long petiole, clustered in whorls towards the end of branches (Plate 1). Lamina is glabrous, shining green, lanceolate to narrowly obovate, 20–35 cm by 7–20 cm, simple, entire, with acuminate to obtuse apex and cuneate base, veins distinct, glossy green (Plates 1 and 2). Inflorescences terminal or lateral, pendulous racemes, 25–80 cm long, Flower pedicellate, calyx, green or red, 4 imbricate lobes persistent at anthesis circumscissile leaving a cup-shaped annular rim; corolla 4 pale green or yellowish petals; stamens many in 3–8 whorls, yellow tinged pink, style central, long and persistent. Fruit ellipsoid-ovoid, ovoid to broad-ovoid, 4–7 cm by 5–9 cm, green or reddish, indehiscent with one white kernel, 2.5 by 1.5–3.5 cm (Plates 2, 3 and 4).

Nutritive/Medicinal Properties

The nutrient composition of the kernel would be similar to that described for *B. edulis*.

Other Uses

Similar to that described for *B. edulis* and *B. procera*.

Plate 1 Leaves and young fruits

Plate 3 Developing pedicellate fruits

Plate 2 Detached leaf and fruits

Plate 4 Close-up fruits with persistent cup-shaped calyx rim, persistent and short pedicels

Comments

The fruit is usually propagated from seeds.

Selected References

Anonymous (1992) *Barringtonia novae-hiberniae*. Curtis's Bot Mag 9(4):172–173

Evans B (1991) A variety collection of edible nut tree crops in Solomon Islands, vol 8, Research bulletin. Dodo Creek Research Station, Honiara, 98 pp

French BR (1986) Food plants of Papua New Guinea – a compendium. Ashgrove, Australia and Pacific Science Foundation, 408 pp

Jebb M (1992) Edible Barringtonias. Curtis's Bot Mag 9(4):164–172

Pauku RL (2006) *Barringtonia procera* (cutnut), ver. 2.1. In: Elevitch CR (ed) Species Profiles for Pacific Island Agroforestry. Permanent Agriculture Resources (PAR), Holualoa, Hawai'i. http://www.traditionaltree.org

Payens JP (1967) A monograph of the genus *Barringtonia* (Lecythidaceae). Blumea 15(2):157–263

Walter A, Sam C (2002) Fruits of oceania, vol 85, ACIAR monograph. Australian Centre for International Agricultural Research, Canberra, 329 pp

Barringtonia procera

Scientific Name

Barringtonia procera **(Miers) Kunth.**

Synonyms

Barringtonia edulis (non Seemann) Bailey, *Barringtonia excelsa* auct. non Bl., *Barringtonia magnifica* Laut., *Barringtonia guppyana* Knuth, *Barringtonia schuchardtiana* K. Schum., *Barringtonia speciosa* K. Schum., *Butonica procera* Miers (basionym).

Family

Lecythidaceae, also placed in Barringtoniaceae

Common/English Names

Cutnut, Pau Nut

Vernacular Names

Fiji: Vutu, Vutukana;
French: La Velle;
Niue: 'Ai;
Papua New Guinea: Cut Nut, Pau Nut, Pao Nut, Pau, Pao (General), Tegeli (Tok Plesin, Madang), Laluan Pao (Tok Plesin, Morobe), Pulei, Purei (Tok Plesin Manus), Pala Paua, Hutun Pao (Tok Plesin, New Ireland), Kuanua Pao, Vutug Pao (Tok Plesin, New Britain), Siwai Hari (Tok Plesin, North Solomons);
Solomon Islands: Kat Nut (Pidgin), Fala, Aikenu (Kwara'ae, Malaita Island), Kenu (To'oabaita, Malaita Island), Vele (Varisi, Choiseul Island), Fara (Santa Ana, Santa Ana Island), Hara Hara (San Cristobal), Nua Nuado (Santa Cruz), Kino (Nduke, Kolombangara Island), Oneve (Marovo, New Georgia Island) Tinga Hala, Fala, Kenu (New Georgia Island), Fala (Maringe, Isabel Island), Nofe (Zabana, Isabel Island), Borolong Sioko (Shortland Island);
Tonga: 'Ai;
Vanuatu: Navele (Bislama), Va Rodh Vevingen (Tok Plesin);
Wallis & Futuna: Vutu Kai.

Origin/Distribution

Cutnut is indigenous to the Solomon Islands, Vanuatu, and Papua New Guinea. It is cultivated in Micronesia and Polynesia.

Agroecology

Its native habitat is characterized as wet humid tropical lowland rainforest with warm to hot temperatures throughout the year from 20–30°C. Prolonged low temperature below 20°C will affect

T.K. Lim, *Edible Medicinal And Non-Medicinal Plants: Volume 3, Fruits*,
DOI 10.1007/978-94-007-2534-8_10, © Springer Science+Business Media B.V. 2012

the growth and performance of the tree. It thrives in areas with fairly uniform annual rainfall which varies from 1500–4300 mm per annum. Cut nut is widespread and common at lower altitudes from sea level to 600 m and is found mainly in villages, food gardens and home gardens. Cut nut grows well in well-drained coastal coral soils with light to medium texture with high pH of 8; it abhors water-logged soils. It is also adaptable to a wide soil range from rocky soils, saline soils, poor infertile soils, light to heavy soils (sands, sandy loams, loams, sandy clay loams, sandy clays, clay loams, and clays) and tolerates a pH range from 5.1 to 8.5. It grows well in full sun or partial shade. It also has moderate tolerance to strong cyclonic winds.

Edible Plant Parts and Uses

The kernel is a highly nutritious seasonal food in Melanesia and is eaten raw, cooked or roasted as snack food; young leaves are sometimes eaten as a vegetable. In the western Solomon Islands, kernels are roasted and baked into puddings together with edible hibiscus (*Abelmoschus manihot*) and coconut cream.

Botany

A medium-sized, evergreen, erect, monoecious tree reaching heights of 8–24 m and with a crown diameter of 0.8–6 m. Leaves occur in whorls towards the end of branches. The leaf lamina is glossy green, glabrous, coriaceous, large, lanceolate, 21–67 cm by 5–20 cm, obtuse apex and tapering wedge-shaped base, simple, entire, slightly undulating margin sub-sessile to sessile, with prominent mid rib and lateral nerves (Plates 1 and 2). Inflorescence racemose and terminal, with a 30–110 cm long pendulous spike containing up to 150 densely packed flower buds, arranged spirally on the rachis, and varying in colours, typically from green to white or red (Plate 1). Flowers are bisexual, showy, sessile. Calyx, green or red, 4 lobes, connate at the basal portion, persistent at

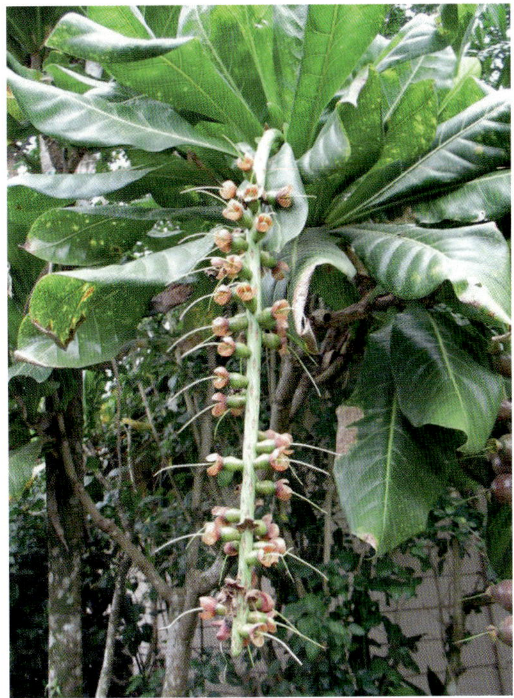

Plate 1 Large glossy green foliage and a string of young green fruits

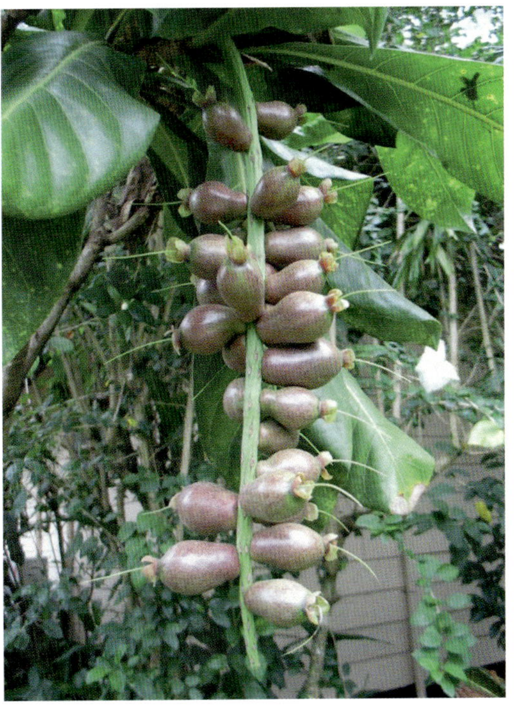

Plate 2 Well developed, sessile, reddish fruits

Plate 3 Sessile, obovoid or pyriform fruits

Plate 4 Close-up of the stylar end with the persistent style and calyx lobes

anthesis breaking into 2–3 persistent pseudo-lobes, corolla 4 yellowish-cream petals; stamens many in 3–8 whorls, yellow, style one, central, longer than stamens and persistent. Fruit elongate, obovoid or pyriform, 5–13 by 3–6 cm, sessile, glabrous, fibrous, green, red or purplish with one white kernel, 2–6 cm by 2–4 cm (Plates 3, 4 and 5).

Nutritive/Medicinal Properties

The nutrient composition of the kernel would be similar to that described for *B. edulis*.

In traditional medicine, the leaves were used to treat inflammation of the ear and headaches. Sap from the bark has been used for treating ciguatera poisoning, coughs, and urinary infections, and the red-leaved form is used as a contraceptive and abortifacient.

Plate 5 Broadly oblong fruit of a green variety

Other Uses

Fallen branches and felled trunks are used as fuel wood. Despite its poor quality, the wood is sometimes used for crafts and temporary light construction. The wood is sometimes used for making paddles in the Reef Islands, Temotu Province, Solomon Islands. The leaves are traditionally used for wrapping and parcelling nuts. Cut nut is useful for soil amelioration, as shade, and shelter and bee foraging plants. The kernel and mesocarp is used as feed for free-range chickens.

Comments

The species is propagated from seeds.

Selected References

Anonymous (1992) *Barringtonia procera*. Curtis's Bot Mag 9(4):174–175

Evans B (1991) A variety collection of edible nut tree crops in Solomon Islands, vol 8, Research bulletin. Dodo Creek Research Station, Honiara, 98 pp

French BR (1986) Food plants of Papua New Guinea – a compendium. Australia and Pacific Science Foundation, Ashgrove, 408 pp

Jebb M (1992) Edible Barringtonias. Curtis's Bot Mag 9(4):164–172

Pauku RL (2006) *Barringtonia procera* (cutnut), ver. 2.1. In: Elevitch CR (ed) Species profiles for pacific island agroforestry. Permanent Agriculture Resources (PAR), Holualoa, Hawai'i. http://www.traditionaltree.org

Payens JP (1967) A monograph of the genus *Barringtonia* (Lecythidaceae). Blumea 15(2):157–263

Walter A, Sam C (2002) Fruits of oceania, vol 85, ACIAR monograph. Australian Centre for International Agricultural Research, Canberra, 329 pp

Barringtonia racemosa

Scientific Name

Barringtonia racemosa **(L.) Spreng.**

Synonyms

Barringtonia apiculata R. Knuth, *Barringtonia caffra* E. Mey. ex R. Knuth, *Barringtonia celebesensis* R. Knuth, *Barringtonia ceylanica* Gardner ex C. B. Clarke, *Barringtonia ceylanica* (Miers) Gardner ex C. B. Clarke, *Barringtonia elongata* Korth., *Barringtonia excelsa* A. Gray, *Barringtonia inclyta* Miers ex B. D. Jacks, *Barringtonia insignis* Miq., *Barringtonia lageniformis* Merr. & L. M. Perry, *Barringtonia longiracemosa* C. T. White, *Barringtonia obtusangula* R. Knuth, *Barringtonia pallida* Koord. & Valeton, *Barringtonia racemosa* (L.) Blume ex DC. nom. illeg., *Barringtonia racemosa* Roxb. nom illeg., *Barringtonia racemosa* var. *elongata* Blume, *Barringtonia racemosa* var. *minor* Blume, *Barringtonia racemosa* var. *procera* Blume, *Barringtonia racemosa* var. *subcuneata* Miq., *Barringtonia rosaria* Oken, *Barringtonia rosata* R. Knuth, *Barringtonia rumphiana* R. Knuth, *Barringtonia salomonensis* Rech., *Barringtonia stravadium* Blanco, *Barringtonia terrestris* R. Knuth, *Barringtonia timorensis* Blume, *Butonica alba* Miers, *Butonica apiculata* Miers, *Butonica caffra* Miers, *Butonica ceylanica* Miers, *Butonica inclyta* Miers, *Butonica racemosa* Juss., *Butonica rosata* Miers, *Butonica rumphiana* Miers, *Butonica terrestris* Miers, *Eugenia racemosa* Linnaeus, *Huttum racemosum* Britten, *Megadendron ambiguum* Miers, *Menichea rosata* Sonn., *Michelia apiculata* Kuntze, *Michelia ceylanica* Kuntze, *Michelia racemosa* Kuntze, *Michelia rosata* Kuntze, *Michelia timorensis* Kuntze, *Stravadium album* Pers. nom. illeg., *Stravadium obtusangulum* Blume, *Stravadium racemosum* Sweet, *Stravadium rubrum* DC.

Family

Lecythidaceae also place in Barringtoniaceae

Common/English Names

Barringtonia, Brack-Water Mangrove, Common Putat, Fish-Killer Tree, Fish-Poison Tree, Fish-Poison Wood, Freshwater Mangrove, Hippo Apple, Powder-Puff Tree, Putat, Small-Leaved Barringtonia, Wild Guava

Vernacular Names

Afrikaans: Poeierkwasboom;
Burmese: Kye-Bin, Kyi;
Chinese: Yu Rui;
French: Bonnet D'évêque, Manondro;
India: Samudraphal, Kunda (Bengali), Ijjul, Norvisnee (Hindi), Ganigala Thora, Ganagilatora, Kanaginathora, Kanaginatora, Kempuganigilu, Kempukanagina, Neeruganagile, Neevaara, Nivar,

T.K. Lim, *Edible Medicinal And Non-Medicinal Plants: Volume 3, Fruits*,
DOI 10.1007/978-94-007-2534-8_11, © Springer Science+Business Media B.V. 2012

Samudraphala, (Kannada), Katampu, Samstaravati, Samstravadi, Samudracham, Samudrapad, Samudrappu, Samuthrachcham, Samudrakka, Samaskaravadi, Samudraccam, Samudrapoo, Samudrapu, Samuthraccham (Malayalam), Nivar, Sadphali (Marathi), Samudrapoo (Oriya), Hijjala, Nichula, Nipa, Samstravadi, Samudrapad, Samudraphala, Vishaya (Sanskrit), Aracakkini, Arattam, Arippiriyam, Arittiram, Calacakam, Calam, Calaparam, Camuttira, Camuttirakkatampu, Camuttirappalam, Camuttirappalavi, Camuttirappalavimaram, Cantakentam, Carucapam, Carucapikam, Carupam, Carusam, Cilesmanacani, Cilettumavisapakan, Citakantanam, Citakantanamaram, Cukataru, Cumpal, Cumpul, Cupuram, Curapi, Curapiyankam, Cutaru, Cuvetacamaram, Icataru, Icitaru, Icitarumaram, Icutaru, Icuvarataru, Intirapacitam, Intirapacitamaram, Isudaru, Isuvaradaru, Karaci, Karacimaram, Kaciram, Kadambam, Kadambu, Kalampakam, Kalampam, Karam, Karamam, Katampaccuvetam, Katampaccuvetamaram, Katampam, Katampuri, Katampurimaram, Katappaimaram, Katappamaram, Katotam, Kelivirutcam, Kogali, Kondalai, Kokali, Kontalai, Kontalai, Kontalam, Kontalankay, Kontavakkay, Koputtam, Kotalankay, Kucciram, Kuchidam, Mara, Maraam, Marakika, Marakikamaram, Maram, Maravam, Nipam, Palapatirappiriyam, Periyacali, Periyacalimaram, Perunceripan, Perunceripanmaram, Piriyakam, Piriyatati, Pitriyagam, Sametrapalam, Samuthram, Tarakatampam, Tenakam, Tevavirutcam, Tunikkatampamaram, Tunikkatampu, Varaki, Venkatampu, Venkatampumaram, Vicalam, Vicuvalopakarakam, Vilattaru (Tamil), Aare, Kadapa, Kanapa, Samudrapandu (Telugu);
Indonesia: Butan Darat, Butun Darat, Penggung, Putat Sungai;
Laotian: Som Pawng;
Malaysia: Putat Ayam, Putat Ayer, Putat Aying, Putat Kampong, Putat Padi, Putat Kerdil, Putat Darat;
Philippines: Putat (Bikol), Nuling, Tuba-Tuba (Cebu Bisaya), Paling (Ibanag), Putat (Maguindanao), Kasouai (Manobo), Putat (Samar-Leyte Bisaya), Kutkut-Timbalon, Putat (Sulu), Potat, Putad, Putat (Tagalog);
Portuguese: Massinhana;
Russian: Barringtonia Kistevidnaia;
Swahili: Mtomondo;
Thai: Chik Ban, Chik Suan;
Zulu: Ibhoqo.

Origin/Distribution

Barringtonia racemosa is found in the coastal areas of eastern Africa from Somalia to South Africa, through Madagascar and other Indian Ocean islands to tropical south Asia, southeast Asia, southern China, the Ryukyu Islands of Japan, northern Australia, Micronesia and Polynesia.

Agroecology

Barringtonia racemosa thrives in humid, moist and warm areas. It occurs primarily in primary and secondary forests in inundated flood plains on tidal river banks, or in mangrove swamp and estuarine localities. It is also found in undisturbed to slightly open (disturbed) mixed dipterocarp to sub-montane forests up to 1200 m as in Peninsular Malaysia and Borneo. It tolerates saline condition but does best in heavy clays, sandy loams or rich volcanic soils. It tolerates dry conditions but is intolerant of any frost.

Edible Plant Parts and Uses

Fruits and leaves are commonly sold in the local day and weekend markets (*pasar minggu*) in Peninsular Malaysia. Young leaves and fruit are edible. The young leaves are often eaten raw as *ulam* dipped in sambal, or cooked as vegetable in Malaysia. The fruit is also eaten as ulam or pickled. The fruit is also pounded to extract the starchy content, which is made into cakes.

Botany

Barringtonia racemosa is a small tree, 4–8 m high but can attain a height of 15 m (Plate 1). It has a rounded crown, greyish-brown, smooth or

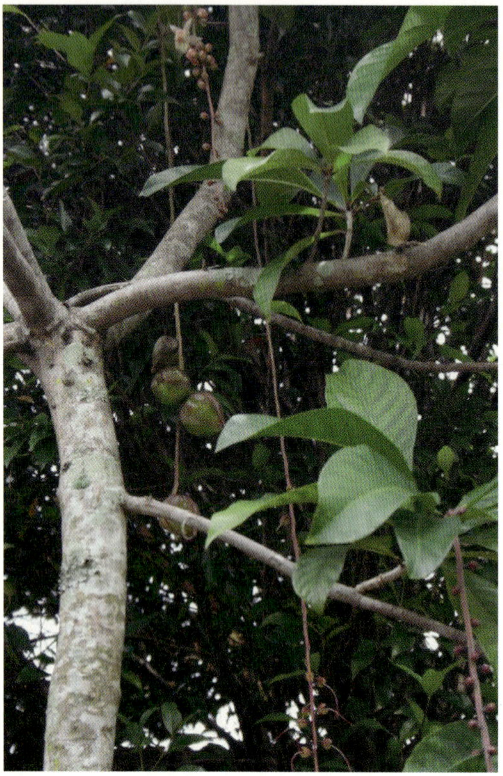

Plate 1 Fruits, inflorescence and leaves of *B. racemosa*

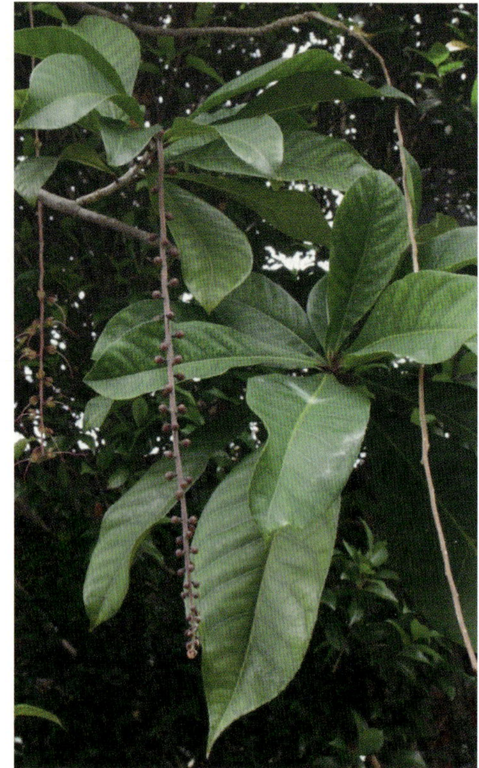

Plate 2 Close-up of leaves

fissured bark and spreading surface roots. Leaves are alternate, borne on glabrous, 2–15 mm long petioles, often clustered toward the end of branches (Plates 1 and 2). Leaf lamina deep green, large, obovate oblong, 8–35 cm by 4–14 cm, with cuneate base acute or acuminate apex and serrate-crenulate margins (Plates 1 and 2). Flowers produced in pendant, terminal or axillary many-flowered racemes up to 60–100 cm long (Plate 2). The flowers exude a pungent, putrid yet faintly sweet odour. Flowers white to pale pink, bisexual, 4-merous, sepals fused at base separate into 4 concave lobes, green tinged pink; petals elliptic to oblong-ovate, up to 3 x 1 cm, joined to staminal tube, stamens numerous with 3–3.5 cm long filaments, white or pinkish, forming a central mass; ovary 2–4 loculed with a red style. Fruit ovoid-cylindric, somewhat quadrangular with 4 ridges, 5–9 by 3–4 cm, with green, green tinged purplish ripening to purplish-red, pericarp with persistent calyx remnants and style (Plates 3, 4, 5 and 6). Each fruit has a large, ovoid, 2–3 cm seed surrounded by the fibrous, spongy flesh (Plate 5).

Nutritive/Medicinal Properties

Phytochemicals

Aerial plant parts of *B. racemosa* were found to contain high levels of phenolics and also carotenoids (lycopene, β-carotene) (Nurul Mariam et al. 2008; Hussin et al. 2009). Their studies showed that gallic acid and naringin were the major phenolic acids and flavonoids present in all different aerial parts of *B. racemosa*. Ferulic acid, rutin, kaempferol and luteolin were also detected in several aerial parts of *B. racemosa*. Overall results obtained from the biological assays suggested that *B. racemosa* is a source of bioactive compounds endowed with interesting biological activities, such as strongly antioxidants agents

Plate 3 Pendant cluster of fruits

Plate 4 Unripe fruits of kampong putat

and can be rated as good dietary sources of natural phenolic antioxidants.

Two new isomeric acylated oleanane-type triterpenoids along with three known compounds were isolated from the MeOH extract of the dried fruits of *Barringtonia racemosa*

Plate 5 Whole and halved fruit showing the large seed

Plate 6 Ripe fruits of kampong putat

(Gowri et al. 2009). The structures were characterized as racemosol A (1) [22α-acetoxy-3β,15α,16α,21β-tetrahydroxy-28-(2-methylbutyryl) olean-12-ene] and isoracemosol A (2) [21β-acetoxy-3β,15α,16α,28-tetrahydroxy-22α-(2-methylbutyryl)olean-12-ene]. The isolated compounds (1–5) were not active against HeLa and P388 D1 carcinoma cell lines.

Sun et al. (2006) isolated and identified five compounds from the stem bark: 3,3'-dimethoxy ellagic acid, dihydromyticetin, gallic acid, bartogenic acid and stigmasterol. In separate studies, two triterpenoids, olean-18-en-3β-O-E-coumaroyl ester and olean-18-en-3β-O-Z-coumaroyl ester, were isolated from the stem bark along with five known compounds, germanicol, germanicone, betulinic acid, lupeol, and taraxerol (Yang et al. 2006). An ethanolic extract of the roots of *Barringtonia racemosa* afforded two neo-clerodane-type diterpenoids, methyl-15, 16-epoxy-12-oxo-3,13(16),14-neo-clerodatrien-18,19-olide-17-carboxylate (nasimalun A, 1) and dimethyl-15,16-epoxy-3,13(16),14-neo-clerodatrien-17,18-dicarboxylate (17-carboxymethyl-hardwickiic acid methyl ester, nasimalun B, 2) (Hasan et al. 2000).

Antioxidant and Antiinflammatory Activities

The shoots and fruits of *Barringtonia racemosa* were found to have high polyphenolic contents (> 150 µ gallic acid equivalents/mg dried plant) and antioxidant activities when measured using the ferric reducing antioxidant power (FRAP) (>1.2 mM) and Trolox equivalent antioxidant capacity (TEAC) assays (>2.4 mM) (Razab and Abdul-Aziz 2010). A strong correlation was observed between the two antioxidant assays (FRAP vs TEAC) implying that the plant could both scavenge free radicals and reduce oxidants. There was also a strong correlation between the antioxidant activities and polyphenolic content suggesting the observed antioxidant activities were contributed mainly by the polyphenolics in the plant.

Scientific studies had shown that various aerial parts of *B. racemosa* tree possessed antioxidant and anti-inflammatory activities. Methanolic extracts of aerial parts of the plant contained relatively higher levels of total phenolics than ethanolic and boiling water extracts (leaf: 16.2 mg gallic acid equivalent/g freeze dried-weight (FDW) tissue, stick: 29.9 mg gallic acid equivalent/g FDW tissue, bark: 21.78 mg gallic acid equivalent/g FDW tissue) (Nurul Mariam et al. 2008). The ethanolic extracts in aerial parts gave higher levels of total flavonoid (leaf: 38.55 mg rutin/g FDW tissue, stick: 40.72 mg rutin/g FDW tissue, bark: 68.29 mg rutin/g FDW tissue). The amounts of β-carotene and lycopene were found higher in methanolic and ethanolic extracts of the leaf (342.2 µg β-carotene/g FDW tissue, 77.38 µg lycopene/g freeze dried-weight tissue; 356.9 µg β-carotene/g FDW tissue, 99.3 µg lycopene/g FDW tissue, respectively). The methanolic and ethanolic extracts in all aerial parts tested exhibited very strong antioxidant properties when compared to butylated hydroxytoluene (BHT), ascorbic acid and α-tocopherol in the free radical scavenging and reducing power assays. Another study found that different crude extracts of fully expanded leaf extracts of *B. racemosa* exhibited a very good level of nitric oxide (NO) inhibitory and antioxidant activities (Behbahani et al. 2007). Non polar extracts such as chloroform and hexane extracts were found to be strong inhibitors of NO at different concentrations (25, 50, 100 and 200 µg/ml) in comparison with polar extract (ethanol extract). Chloroform extract did not show cytotoxicity against RAW 264.7 cells at different concentrations in contrast to hexane and ethanol extracts. The chloroform and hexane extracts exhibited very strong antioxidant properties when compared to vitamin E (α-tocopherol), both extracts exhibited radical scavenging activity with an IC_{50} value of 54.29 and 63 µg/ml respectively. Good correlation between the antioxidant activity and the anti-inflammatory effect was observed. Results demonstrated that chloroform extract of *B. racemosa* leaf may have the potential to be used as anti-inflammatory and anti-oxidant agents and that the active compound in *B. racemosa* was lycopene.

Among the various extracts and fractions of *B. racemosa* fruit investigated preliminarily for carrageenan-induced acute inflammation in rats, the ethyl acetate fraction displayed potent anti-inflammatory activity (Patil et al. 2009). The bioactive constituent responsible for the observed pharmacological effects was bartogenic acid (BA). When evaluated for effectiveness against Complete Freund's Adjuvant-induced arthritis in rats, the results indicated that at doses of 2, 5 and 10 mg/kg/day p.o., BA protected rats against the primary and secondary arthritic lesions, body weight changes and haematological perturbations induced by Complete Freund's Adjuvant. The serum markers of inflammation and arthritis, such as C-reactive protein and rheumatoid factor, were also reduced in the BA-treated arthritic rats. The study validated the ethnomedicinal use of fruits of *B. racemosa* in the treatment of pain and inflammatory conditions. It further established the potent anti-arthritic effects of bartogenic acid.

The ethanolic extract of *Barringtonia racemosa* Roxb. (Lecythidaceae) fruits was found to have anti-inflammatory (Shika et al. 2010). The extract showed significant inhibition of carrageenan/formalin induced paw oedema at the three doses used. The activity of the extract was comparable to that of Indomethacin, the standard antiinflammatory drug.

Analgesic Activity

B. racemosa ethanolic extract showed significant inhibition of acetic acid induced writhing in mice at 125, 250 and 500 mg/kg doses almost comparable to the standard analgesic drug, acetyl salicylic acid (Shika et al. 2010).

Anticancer Activity

Seed extract of *B. racemosa* was shown to have anti-tumorous activity against Dalton's Lymphoma Ascitic cells when administered intraperitoneally daily in mice (Thomas et al. 2002). The optimum dose was found to be 6 mg/kg. This dose protected all the animals challenged with the tumour cells. The efficacy of the extract was found to be better than that of a standard drug, vincristine. The extract was found to be devoid of conspicuous acute and short-term toxicity to mice, when administered daily, (i.p.) for 14 days up to a dose of 12 mg/kg (which was double the concentration of optimum therapeutic dose). An ethanol extract of the plant leaves displayed cytotoxicity against the HelA (human cervila carcinoma) cell line with an IC_{50} value of 10 μg/ml (Mackeen et al. 1997). Quercetin 3-*O*-rutinoside (QOR) isolated from the fruits of *B. racemosa* showed dose- and time-dependent anti-proliferative activity in several leukemic cell lines with negligible effect on normal human peripheral blood mononuclear cell (PBMC) (Samanta et al. 2010). Results showed that QOR-induced apoptosis occurred preferentially on accumulation of cells in the sub-G_0 phase and genomic DNA fragmentation through the activation of mitochondria-dependent caspase cascade for the first time in T-lineage ALL cell line.

Antimicrobial Activity

A study reported that an ethanol extract of the roots of *B. racemosa* provided two novel clerodane diterpenoid nasimalun A and B that exhibited antibacterial activity (Khan et al. 2001). Various extracts of the different plant parts of *B. racemosa* exhibited antifungal activity (Hussin et al. 2009). Better antifungal activity was observed with the methanolic extracts in all aerial parts of *B. racemosa* that showed excellent inhibitory activity against all the fungi tested. The strongest inhibitory activity effect was observed with the methanolic extract of leaf against *Fusarium* sp. (53.45%), *Ganoderma lucidum* (34.57%), *Aspergillus* sp. (32.27%) and *Trichoderma koningii* (20.99%). Remarkable are also the specific effects of the boiling water extract of leaf against *Fusarium* sp. (51.72%) and with the ethanolic extract of bark against *Rhizopus* sp. (37.50%). None of the boiling water extracts of leaf, stick and bark showed inhibitory activity effect against *Ganoderma tropicum* and *T. koningii*. HPLC analysis of the extract of *B. racemosa* (leaves, sticks and barks) showed two different phenolic acids (gallic acid and ferrulic acid) and four different flavonoids (naringin, rutin, luteolin and kaempferol).

Antidiabetic Activity

Hexane, ethanol and methanol extracts of *B. racemosa* seeds displayed potent yeast and intestinal α-glucosidase inhibitory activities (Gowri et al. 2007). The methanol extract was found to be superior among them. However, none of the extracts exhibited pancreatic α-amylase inhibitory activity, rather the ethanol and methanol extracts accelerated the α-amylase enzyme activity. Also, bartogenic acid isolated from the methanol extract inhibited α-amylase.

Antinociceptive Activity

Aqueous bark extracts (500, 750, 1000 or 1500 mg/kg) of *B. racemosa* exhibited antinociceptive potential in male rats (Deraniyagala et al. 2003). The extract did not alter fertility, gestational length, peri- and neonatal development and appears to be non-teratogenic. The antinociceptive effect was mediated mainly via opioid mechanisms. Such inhibition of pain could arise from phenolic and steroidal constituents that were present in the extract.

Molluscicidal and Larvicidal Activity

The pericarp extracts of the fruit contained more potent molluscicidal components (LC_{50} =367.3–625.0 ppm) than the seed extracts (LC_{50}=530.53–704.27 ppm) (Adewunmi et al. 2001). The ranking order of toxicity for the pericarp extracts against the snail (*Biomphalaria glabrata*) was: CHC13 extract 367.3 ppm > ethyl acetate extract 390.3 ppm. > methanol extract 530.4 ppm > petroleum ether extract 704.27 ppm. The larvicidal activity was higher in the seed extracts (LC_{50}=588.44–1604.2 ppm) than in the pericarp extracts (LC_{50}=1507.0–4000.0 ppm). The ranking order of toxicity of the seed extract for the larvae of *Aedes aegypti* was: CHC13 extract 588.44 ppm > methanol extract 762.5 ppm > petroleum ether extract 1244.2 ppm > ethyl acetate extract 1604.2 ppm. The data indicated the potential role for *B. racemosa* in the control of snail and mosquito as intermediate hosts of schistosomiasis and dengue fever respectively.

Traditional Medicinal Uses

In South Africa, the Zulus used the fruit as a therapy for malaria. Extracts or preparations of the fruit have been used against malaria, cough, asthma, jaundice, headache, eye inflammation, diarrhoea and sores, and a bark decoction is applied externally to treat rheumatism. Seeds were used to treat eye inflammation and by midwives for parturition. In eastern Africa a root decoction was used as a febrifuge. In Malaysia, *Barringtonia racemosa* has been used widely in traditional medicine for anti-inflammation and anticancer. The leaves traditionally were used to treat high blood pressure and as a depurative. Pounded leaves alone or together with the root or bark have been used to treat chicken pox. The leaves have been used to treat high blood pressure and as a depurative. Warmed juice from the fruit has been used for treating ulcerated nose. In India, the fruit has been used for poulticing for sore throat and cutaneous eruptions and was found efficacious in coughs, asthma, and diarrhea. The roots have cooling properties and were used as deobstruent. The seeds are aromatic and found useful in treating colics and ophthalmic problems. Ethnomedical survey has shown that the seeds of *Barringtonia racemosa* are traditionally used in certain remote villages of Kerala (India) to treat cancer like diseases. In the Philippines, the bark has been used externally in decoction as an anti-rheumatic.

Other Uses

Barringtonia racemosa provides a medium-weight, semi-durable hardwood which is used for house posts, beams, general planking, flooring, boats, cabinet work, boxes, crates, agricultural implements, carts, wooden pallets, household utensils, mouldings, carving, turnery, wooden tiles and pavings. It is also suitable for plywood and veneer manufacture. The branches and wood is also suitable for fuel-wood. The fibres has been utilised in various kinds of hardboard, particle-board, blackboard and for pulp. The bark fibres have used for cordage. The bark also yields tannin. The seeds, bark, wood and roots contain the poison saponin and is used to stun fish. Extracts from the plant are reported effective against *Citrus* aphids. In Bengal, India, the seeds were used to poison people and coconut was said to be the antidote.

Comments

The tree is propagated from seeds.

Selected References

Adewunmi CO, Aladesanmi AJ, Adewoyin FB, Ojewole JAO, Naido N (2001) Molluscidial, insecticidal and piscicidal activities of *Barringtonia racemosa*. Nig J Nat Prod Med 5:56–58

Behbahani M, Ali AM, Muse R, Mohd NB (2007) Antioxidant and anti-inflammatory activities of leaves of *Barringtonia racemosa*. J Med Plants Res 1(5):95–102

Burkill IH (1966) A dictionary of the economic products of the Malay Peninsula. Revised reprint. 2 volumes. Ministry of Agriculture and Co-operatives, Kuala Lumpur, Malaysia. Vol 1. (A–H) pp. 1–1240, Vol 2. (I–Z). pp. 1241–2444

Chantaranothai P (1995) *Barringtonia* (Lecythidaceae) in Thailand. Kew Bull 50:677–694

Cheek M (2008) *Barringtonia racemosa*. (L.) Roxb. KwaZulu-Natal Herbarium. www.plantzafrica.com

Deraniyagala SA, Ratnasooriya WD, Goonasekara CL (2003) Antinociceptive effect and toxocological study of the aqueous bark extract of *Barringtonia racemosa* on rats. J Ethnopharmacol 86(1):21–26

Foundation for Revitalisation of Local Health Traditions. 2008. *FRLHT Database*. http://envis.frlht.org

Gowri PM, Tiwari AK, Ali AZ, Rao JM (2007) Inhibition of alpha-glucosidase and amylase by bartogenic acid isolated from *Barringtonia racemosa* Roxb. seed. Phytother Res 21(8):796–799

Gowri PM, Radhakrishnan SV, Basha SJ, Sarma AV, Rao JM (2009) Oleanane-type isomeric triterpenoids from *Barringtonia racemosa*. J Nat Prod 72(4):791–795

Hasan CM, Khan S, Jabbar A, Mohammad A, Rashid MA (2000) Nasimaluns A and B: neo-clerodane diterpenoids from *Barringtonia racemosa*. J Nat Prod 63(3):410–411

Hussin NM, Muse R, Ahmad S, Ramli J, Mahmood M, Sulaiman MR, Shukor MYA, Rahman MFA, Aziz KNK (2009) Antifungal activity of extracts and phenolic compounds from *Barringtonia racemosa* L. (Lecythidaceae). Afr J Biotechnol 8(12): 2835–2842

Kaume RN (2005) *Barringtonia racemosa* (L.) Spreng. [Internet] Record from Protabase. Jansen PCM, Cardon D (Eds) PROTA (Plant Resources of Tropical Africa / Ressources végétales de l'Afrique tropicale), Wageningen, Netherlands, http://database.prota.org/search.htm

Khan S, Jabbar A, Hasan CM, Rashid MA (2001) Antibacterial activity of *Barringtonia racemosa*. Fitoterapia 72(2):162–164

Mackeen MM, Ali AM, El-Sharkawy SH, Manap MY, Salle KM, Lajis NH, Kawazu K (1997) Antimicrobial and cytotoxic properties of some Malaysian traditional vegetables. Int J Pharmacog 35:174–178

Nurul Mariam H, Radzali M, Johari R, Syahida A, Maziah M (2008) Antioxidant activities of different aerial parts of putat (*Barringtonia racemosa* L.). Malaysian J Biochem Mol Biol 16(2):15–19

Patil KR, Patil CR, Jadhav RB, Mahajan VK, Patil PR, Gaikwad PS (2009) Anti-arthritic activity of bartogenic acid isolated from fruits of *Barringtonia racemosa* Roxb. (Lecythidaceae). Evid Based Complement Alternat Med, Advance Access published on September 21, 2009. doi: 10.1093/ecam/nep148

Payens JPDW (1967) A monograph of the genus *Barringtonia* (Lecythidaceae). Blumea 15(2):157–263

Perry LM (1980) Medicinal plants of East and Southeast Asia: attributed properties and uses. MIT Press, UK, 620 pp

Qin H, Prance GIT (2007) *Barringtonia*. In: Wu ZY, Raven PH, Hong DY (Eds). Flora of China. Vol 13. (Clusiaceae through Araliaceae). Science Press, Beijing, and Missouri Botanical Garden Press, St. Louis

Razab R, Abdul-Aziz A (2010) Antioxidants from tropical herbs. Nat Prod Commun 5(3):441–445

Saidin I (2000) Sayuran Tradisional Ulam dan Penyedap Rasa. Penerbit Universiti Kebangsaan Malaysia, Bangi, pp 228 (In Malay)

Samanta SK, Bhattacharya K, Mandal C, Pal BC (2010) Identification and quantification of the active component quercetin 3-*O*-rutinoside from *Barringtonia racemosa*, targets mitochondrial apoptotic pathway in acute lymphoblastic leukemia. J Asian Nat Prod Res 12(8):639–648

Shika P, Latha PG, Suja SR, Anuja GI, Shyamal S, Shine VJ, Sini S, Krishnakumar NM, Rajasekharan S (2010) Anti-inflammatory and analgesic activity of *Barringtonia racemosa* Roxb. fruits. Indian J Nat Prod 1(3):356–361

Sun HY, Long LJ, Wu J (2006) Chemical constituents of mangrove plant *Barringtonia racemosa*. Zhong Yao Cai 29(7):671–672 (In Chinese)

Thomas TJ, Panikkar B, Subramoniam A, Nair MK, Panikkar KR (2002) Antitumour property and toxicity of *Barringtonia racemosa* Roxb. seed extract in mice. J Ethnopharmacol 82(2–3):223–227

Whitmore TC (1972) Lecythidaceae. In: Whitmore TC (ed) Tree flora of Malaya volume 2. Longman, Kuala Lumpur, pp 257–266

World Agroforestry Centre 2008. Agroforestry Database. www.worldagroforestry.org/sites/TreeDBS/Aft.asp

Yang Y, Deng Z, Proksch P, Lin W (2006) Two new 18-en-oleane derivatives from marine mangrove plant, *Barringtonia racemosa*. Pharmazie 61(4):365–366

Yaplito MA (2001) *Barringtonia* Forster JR, Forster JG In: van Valkenburg JLCH, Bunyapraphatsara N (Eds). Plant resources of South-East Asia No 12(2): medicinal and poisonous plants 2. Prosea Foundation, Bogor, Indonesia. pp 101–107

Barringtonia scortechinii

Scientific Name

Barringtonia scortechinii King.

Synonyms

Barringtonia scortechinii var. *globosa* Craib.

Family

Lecythidaceae, also placed in Barringtoniaceae

Common/English Name

Putat

Vernacular Names

Borneo: Langsat Burung, Putat, Tempalang, Tempalong, Terakot, Kayuh Hat (Punan);
Malaysia: Putah Gajah, Putat Hutan, Putat Tuba (Peninsular), Tampalang (Sabah).

Origin/Distribution

The species is native to Thailand, Peninsular Malaysia, Sumatra, Borneo (Sarawak, Brunei, Sabah, West- and East-Kalimantan) and the Philippines.

Agroecology

A tropical species. In Borneo, it occurs naturally in undisturbed to slightly open mixed dipterocarp to sub-montane forests from near seal level up to 1200 m altitude. It is usually found on hillsides and ridges with sandy to clay soils. In secondary forests it is usually present as a pre-disturbance remnant tree. In Peninsular Malaysia and Peninsular Thailand, it is found in evergreen, deciduous and bamboo forests along streams, rivers and seashores at all elevations up to 1000 altitude.

Edible Plant Parts and Uses

The fruits are edible and used as a spice to flavour food. The young leaves are also eaten raw as *ulam* dipped in sambal, or cooked as vegetable in Malaysia.

Botany

A small to medium tree, reaching 21 m high, with 120 cm girth and a dense large crown and many branches. The bole is straight, sometimes fluted at base with smooth, reddish-brown bark and branches. Young twigs are maroon-red and glabrous (Plate 1). Leaves are crowded at twig tips, alternate, simple, elliptic, 9–14 by 4–5 cm, penni-veined with 7–10 pairs of secondary veins, glabrous, thinly coriaceous, dull-green, margin

T.K. Lim, *Edible Medicinal And Non-Medicinal Plants: Volume 3, Fruits*,
DOI 10.1007/978-94-007-2534-8_12, © Springer Science+Business Media B.V. 2012

Plate 1 Maroon twigs and emerging leaves

Plate 2 Mature elliptic, dull green, glabrous leaves with weakly toothed margins

mostly entire wavy very weakly toothed (Plates 2 and 3). Young leaves are maroon to bronze coloured (Plates 1 and 3). Flowers pendulous in racemose spikes 20–55 cm long formed behind leaves or terminally at the shoot tip. Flowers hermaphrodite, pink-yellowish petals, calyx and ovary wall four-angled, petals 4, ovary inferior and 4-loculed, numerous long, protruding stamens. Fruit green–purplish tinted drupe, winged and angled becoming ovoid or globose, 5–9 cm long.

Nutritive/Medicinal Properties

No information has been published on the nutritive value of the edible fruit or leaves.

In traditional medicine, a decoction of the leaves is drunk after childbirth to cleanse the

Plate 3 Junvenile, bronze-coloured leaves with weakly toothed margin

blood and stimulate uterus contraction (Ong and Nordiana 1999).

Other Uses

The wood is used for particle board and roofing.

Comments

The plant is propagated from seeds.

Selected References

Burkill IH (1966) A dictionary of the economic products of the Malay Peninsula. Revised reprint. 2 volumes. Ministry of agriculture and co-operatives, Kuala Lumpur, Malaysia. Vol 1. (A–H) pp 1–1240, Vol 2. (I–Z) pp 1241–2444

Chantaranothai P (1995) *Barringtoni*a (Lecythidaceae) in Thailand. Kew Bull 50(4):677–694

Ong HC, Nordiana M (1999) Malay ethno-medico botany in Machang, Kelantan, Malaysia. Fitoterapia 70:502–530

Payens JPDW (1967) A monograph of the genus *Barringtonia* (Lecythidaceae). Blumea 15:157–260

Slik JWF (2006) Trees of Sungai Wain. Nationaal HerbariumNederland.http://www.nationaalherbarium.nl/sungaiwain/

Whitmore TC (1972) Lecythidaceae. In: Whitmore TC (ed) Tree flora of Malaya volume 2. Longman, Kuala Lumpur, pp 257–266

Bertholletia excelsa

Scientific Name

Bertholletia excelsa Humb. & Bonpl.

Synonyms

Bartholessia excelsa Silva Manso, *Bertholletia nobilis* Miers.

Family

Lecythidaceae

Common/English Names

Brazil Nut, Brazilnut Tree, Butternut, Creamnut, Paranut, Para Nut

Vernacular Names

Alemão: Paranuss;
Bolivia: Tapa;
Brazil: Castanheira, Castanha Verdadeira, Castanheiro, Castanha-Do-Brazil, Castaña Del Brasil, Castanha-Do-Pará, Tocari (Portuguese);
Colombia: Conduiro, Jiturede (Huitoto), Castana Del Maranon (Spanish), Tt-Wa (Ticuna);
Cuba: Coquito Del Brazil;
Czech: Juvie Ztepilá;
Danish: Paranød;
Eastonian: Kõrge Parapähklipuu;
French: Noyer De Para, Noyer Du Brésil, Noix Du Brésil, Noix De Pará, Chatãigne Du Bresil, Amande D'amérique;
German: Paranußbaum, Brasilianishe Kastanien, Paranuss;
Greek: Karidia Brazilias;
Italian: Noce Del Brasilie;
Peru: Castaña;
Polish: Orzesznica Wyniosła;
Portuguese: Castanha-Do-Brasil, Castanha-Do-Pará;
Slovašcina: Brazilski Orešček;
Spanish: Castaña Del Brazil, Castaño De Pará, Nuez Del Brazil;
Surinam: Braziliaansche Noot, Brazilnoot, Ingie Noto, Kokelekoo, Para Noot, Tetoka, Toeka, Totoka;
Swedish: Paranöt;
Venezuela: Almendra, Iuvia, Jibia, Juvia, Matamatá De Altura, Yubia.

Origin/Distribution

Bertholletia excelsa is indigenous to tropical Amazonia – French Guiana, Surinam, Guyana, Bolivia, Venezuela, Brazil, Peru, and Amazonian Colombia. It forms large forests on the banks of the Amazons and Rio Negro, and likewise about Esmeraldas, on the Orinoco. It is most prevalent in the Brazilian states of Marahao, Mato Grosso,

T.K. Lim, *Edible Medicinal And Non-Medicinal Plants: Volume 3, Fruits*,
DOI 10.1007/978-94-007-2534-8_13, © Springer Science+Business Media B.V. 2012

Acre, Para, Rondonia, and the Amazonas. Brazil nuts are harvested almost entirely from wild trees. Brazil nuts have been harvested from plantations but production is low and it is currently not economically viable.

Agroecology

Brazil nut grows mostly wild in the hot wet, humid equatorial rainforests in South America with annual rainfall from 2000 to 3000 mm and mean daily temperatures of 27–32°C. Brazil nut forms extensive forests on the banks of the Amazons and Rio Negro, and likewise about Esmeraldas, on the Orinoco. It is grows on well-drained clay or sandy clay soils.

Edible Plant Parts and Uses

Brazil nut kernel is consumed fresh or roasted as a snack, dessert nut and is used in confectionery (cakes, biscuits, chocolates, etc.) and as food by the locals in its native range and as a edible cooking oil. The oil can also be used for salad dressing. This food is a valuable source of calories, fat, and protein for much of the Amazon's rural and tribal peoples. Indigenous tribes eat the nuts raw or grate them and mix them into gruels. In the Brazilian Amazon, the nuts are grated with the thorny stilt roots of *Socratea* palms into a white mush known as *leite de castanha* and then stirred into manioc flour.

Brazil nuts is an important economic commodity which are exported to North America, Europe and elsewhere. The improved functional properties of succinylated Brazil nut kernel globulin could be explored in a variety of food formulations such as high protein drinks, soups, bakery and meat products as well as in salad dressings and mayonnaise as an emulsifier (Ramos and Bora 2005). The residue of oil extraction gives a high protein flour that can be mixed with common bread flour or used in animal feeding. The high saturated fat content of Brazil nuts is among the highest of all nuts, surpassing even macadamia nuts. Because of the resulting rich taste, Brazil nuts can often substitute for macadamia nuts or even coconuts in food recipes. The high fat content of the nuts adversely impact on its keeping quality as shelled nuts turn rancid rapidly.

Botany

A large, deciduous tree reaching 30–60 m high with a straight, greyish, cylindrical, smooth trunk with a diameter of 1–2 m (Plate 1). Branching occurs way up the tree with a large emergent crown of long, robust branches. Leaves are alternate, simple, glabrous, coriaceous, large, entire or crenate, oblong, 20–35 cm by 10–15 cm, venation pinnate, with 29–45 pairs of parallel lateral veins (Plate 2). Flowers are small, hermaphrodite, creamy-white, zygomorphic, epigynous, in axillary or terminal panicles 20–40 cm long with one or two orders of branching; each flower has a

Plate 1 Tall, straight tree habit

Plate 2 Leaves with 29–45 pairs of parallel lateral veins

Plate 4 Brazil nut fruits

Plate 3 Brazil nut flowers

Plate 5 Shell of old Brazil nut fruit and seeds

two-parted, deciduous calyx, lobes 0.8–1.4 cm long, six unequal cream-colored petals to 3 cm long, and numerous stamens united into a broad, hood-shaped mass (androecial hood) arching over covering the style and stigma and tightly appressed to the top of the inferior ovary (Plate 3). Fruit a large spherical to subglobose, circumcissile, woody pyxidium, 10–16 cm diameter, dark brown, lenticellate, verrucose, calycine ring obscure, supracalycine zone 3–4 cm long, 4-locular, the operculum 0.5–0.7 cm in diameter, apiculate, falling away at maturity leaving a small opening narrower than the seeds (Plates 4 and 5). Fruits weigh 2–2.2 kg and are borne at the end of thick branches. Fruit has a hard, woody shell 8–12 mm thick, and inside contains 8–24 3-angled seeds (nut) 3.5–5 cm by 2.5–1.8 cm, with a woody, thick, indurate and rugose seed coat that is finely costate enclosing a pale brownish-white kernel inside (Plates 5, 6 and 7).

Plate 6 Brazil nut (seeds)

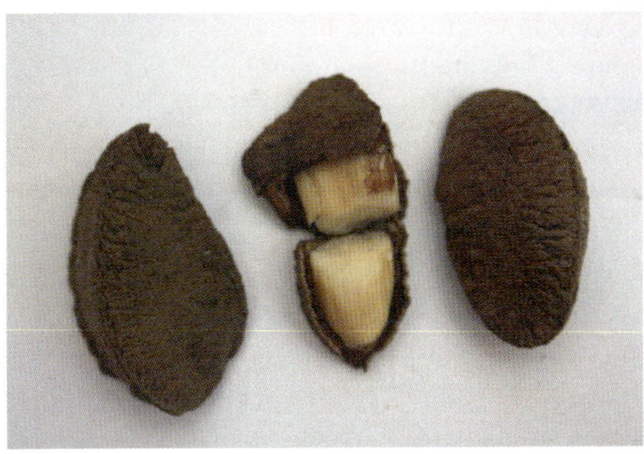

Plate 7 Tough, woody seed coat removed to expose the edible kernel

Nutritive/Medicinal Properties

Proximate food value of dried, unblanched Brazil nut (*Bertholletia excelsa*) (seeds) per 100 g edible portion (49% shells as refuse) (U.S. Department of Agriculture and Agricultural Research 2010) is as follows: water 3.48 g, energy 656 kcal (2743 kj), protein 14.32 g, total lipid (fat) 66.43 g, ash 3.51 g, carbohydrate 12.27 g, fibre (total dietary) 7.5 g, sugars (total) 2.33 g, sucrose 2.33 g, starch 0.25 g; calcium 160 mg, iron 2.43 mg, magnesium 376 mg, phosphorus 725 mg, potassium 659 mg, sodium 3 mg, zinc 4.06 mg, copper 1.743 mg, manganese 1.223 mg, selenium 1917 μg; vitamin C (total ascorbic acid) 0.7 mg, thiamin 0.617 mg, riboflavin 0.035 mg, niacin 0.295 mg, pantothenic acid 0.184 mg, vitamin B-6 0.101 mg, total folate 22 μg, choline (total) 28.8 mg, betaine 0.4 mg, vitamin E (α-tocopherol) 5.73 mg, γ- tocopherol 7.87 mg, δ-tocopherol 0.77 mg; lipids: fatty acids (total saturated) 15.137 g, 14:0 (myristic acid) 0.052 g, 16:0 (palmitic acid) 9.085 g, 17:0 (margaric acid) 0.047 g, 18:0 (stearic acid) 5.794 g, 20:0 (arachidic acid) 0.160 g; fatty acids (total monounsaturated) 24.548 g, 16:1 undifferentiated (palmitoleic acid) 0.229 g, 17:1 (heptadecenoic acid) 0.044 g, 18:1 undifferentiated (oleic acid) 24.223 g, 20:1 (gadoleic acid) 0.052 g; fatty acids (total polyunsaturated) 20.577 g, 18:2 undifferentiated (linoleic acid) 20.543 g, 18:3 undifferentiated (linolenic acid) 0.035 g, 18:3 n-3 c,c,c (α-linolenic acid) 0.017 g, 18:3 n-6 c,c,c 0.017 g; amino acids: tryptophan 0.141 g, threonine 0.362 g, isoleucine 0.516 g, leucine 1.155 g, lysine 0.492 g, methionine 1.008 g, cystine 0.367 g, phenylalanine 0.630 g, tyrosine 0.420 g, valine 0.756 g, arginine 2.148 g, histidine 0.386 g, alanine 0.577 g, aspartic acid 0.346 g, glutamic acid 3.147 g, glycine 0.718 g, proline 0.657 g and serine 0.683 g.

Brazil nuts are rich in fats – total saturated 15.7 g/100 g, total monounsaturated 24.548 g/100 g and total polyunsaturated 20.577 g/100 g, proteins and amino acids. The proteins found in Brazil nuts are very high in sulfur-containing amino acids like cysteine (8%) and methionine (18%) and are also extremely rich in glutamine, glutamic acid, and arginine. The presence of these amino acids (chiefly methionine) enhances the absorption of selenium and other minerals in the nut. Brazil nuts are also rich in saponins. The extracts of the pericarp of Brazil nuts contain tannins, particularly ellagotannin and gallotannins.

Proximate analysis of Brazil nut kernel showed 68.6 and 16.5% lipid and protein contents, respectively (Ramos and Bora 2003). The protein content in defatted kernel flour increased to about 46.7%. Extraction resulted in solubilising about 81% proteins, out of which nearly 72% proteins (globulin) were recovered. Unextracted and soluble proteins accounted for 18.7 and 9.2% of total proteins, respectively. In relation to functional properties, the globulin showed pH of minimum solubility at 5.0. The globulin presented lower solubility (0.9–41.3%) in the pH range of 5.0–10.0 but above and below this pH range, there was a considerable improvement. The water and oil absorption capacities were 2.0 and 1.4 g/g globulin.. Emulsifying capacity varied with the pH, being maximum at pH 3.0 (94.5 ml/100 mg of globulin) and minimum (38.5 ml/100 mg globulin) at its isoelectric pH (5.0). Emulsion activity and stability also followed similar behaviour as that of emulsifying capacity. Brazil nut storage proteins were found to consist of 2 S albumin and the globulins, 7 S vicilin, and an 11 S legumin (Sharma et al. 2010). When probed with anti-Brazil nut seed protein rabbit polyclonal antibodies, 7 S globulin exhibited higher immunoreactivity than 2 S albumin and 11 S globulin.

Antioxidant Activity

Brazil nuts are an excellent source of selenium and a good source of magnesium, potassium, phosphorus, zinc and thiamine. The high selenium content of the Brazil nut, makes it a healthy food qualified as an antiradical protector (Chunhieng et al. 2004). Brazil nut contained 126 ppm of selenium. Selenium was found to be distributed in the nut protein fractions. The water-extracted fraction, which represented 17.7% of the cake protein, was the richest in selenium with 153 ppm. Further Analysis by HPLC-MS showed that selenium was linked by a covalent bond to two amino acids to form selenomethionine and selenocystine (Vonderheide et al. 2002). Selenomethionine was demonstrated to be the most abundant of these seleno-amino acids. Studies in New Zealand reported consumption of 2 Brazil nuts daily to be as effective for increasing selenium status and enhancing blood glutathione peroxidase (GPx) activity as 100 μg Se as selenomethionine (Thomson et al. 2008). Inclusion of this high-selenium food in the diet could avoid the need for fortification or supplements to improve the selenium status.

Soluble phenolics content in brown skin of Brazil nuts was found to be 1236.07 as compared to 406.83 in kernel and 519.11 mg/100 g in whole nut (John and Shahidi 2010). Bound phenolics content of brown skin was also 86- and 19-folds higher than kernel and whole nut, respectively. Similarly extracts from the brown skin exhibited the highest antioxidant activity as evaluated by Trolox equivalent antioxidant capacity (TEAC), 1,1-diphenyl-2-picrylhydrazyl (DPPH) radical and hydroxyl radical scavenging activities using electron paramagnetic resonance (EPR), reducing power, and oxygen radical scavenging capacity (ORAC). Free- and bound phenolics identified and quantified included nine phenolic acids and flavonoids and their derivatives (gallic acid, gallocatechin. protocatechuic acid, catechin, vanillic acid, taxifolin, myricetin, ellagic acid, and quercetin). However, some phenolics were present only in the bound form. Furthermore, the phenolics were dominant in the brown skin.

Brazil nut oil was found to contain a high proportion of unsaturated fatty acids 75.6% that could be compared to that of walnut or olive oils (83%) (Chunhieng et al. 2008). The high content in linoleic acid (39.3%) and linolenic acid (36.1%) provides this oil with some interesting dietary characteristics. Compared to other edible oils it contained very high levels of β-tocopherols (88.3%) but lower levels of α-tocopherol 11.3% and γ-tocopherols (0.4%). Studies had established a hierarchy of the antioxidant properties of tocopherols and concluded that α-tocopherol, β-tocopherol, γ-tocopherol and δ-tocopherol exerted descending antioxidant properties (Kaiser et al. 1990). Nevertheless, it had been shown that β-tocopherol was as effective as α-tocopherol in physical quenching of O_2 but had a very low chemical reactivity. This tocopherol homologue might be particularly suitable for biological conditions in which an accumulation of oxidation products might weaken the antioxidant defence. The phytosterol composition of Brazil nut oil was found similar to olive and almond nut oils. Its high content in sitosterol could be useful as anti-cholesterol medicine. Brazil nut oil was also found to contain an interesting composition of phospholipid fatty acids comparable to sunflower lecithin. Based on these findings, Brazil nut oil has the potential to have high commercial value in the health food industry with application in some areas of medical science.

Antiinflammatory and Thyroid Activities

Selenium, acting as a strong antioxidant, was found to help strengthen the immune system and its response to infections (Arthur 2000). The antioxidant effects of selenium were suggested to be mediated through the glutathione peroxidases (GPx)4 that removed potentially damaging lipid hydroperoxides and hydrogen peroxide. At least five of these peroxidases had now been identified as operating in different cell and tissue compartments. Thus, selenium was found to act as an antioxidant in the extracellular space, the cell cytosol, in association with cell membranes and specifically in the gastrointestinal tract, all with potential to influence immune processes. Selenium appeared to reduce inflammation by stimulating the activity of glutathione peroxidase (an enzyme that neutralizes free radicals), improving phagocytic activity of white blood cells and decreasing the production of pro inflammatory prostaglandins. Thus, selenium inhibited the action of prostaglandins responsible for inflammatory reactions in the body. Studies also reported that high selenium (about 5 times the Recommended Daily Allowance) decreased the amount of active thyroid hormone in their blood and led to a steady slow weight gain (Hawkes and Keim 2003). The converse was true also in that low selenium increased their active thyroid hormone levels and led to a steady, slow weight loss.

Anticancer/Antiviral Activity

Brazil nuts naturally contain very high concentrations of selenium. Dietary selenium, including Brazil nuts, had been associated with protection against tumour development in laboratory animal studies (Chang et al. 1995). At high amounts, selenium could decrease the growth rate of the tumourous cells. Studies reported that selenium reduced the risk of cancer in two ways (Combs et al. 2001). It helped prevent free radical damage in the body with it's antioxidant abilities. Secondly, it was believed to prevent tumour growth by enhancing immune cell activity and suppressing development of blood vessels to the tumour. Selenium has many important functions in the human body. Selenium is found in the active site of many enzymes such as thioredoxin reductase, which catalyzes oxidation/reduction reactions. Glutathion peroxidase, an enzyme that helps to prevent the oxidation process, requires selenium for its formation. Ip and Lisk (1994) reported on the results of two mammary cancer prevention experiments in the rat dimethylbenz[a] anthracene model by continuous feeding of selenium-rich Brazil nut diet (processed to a smooth-textured nut material for mixing in the diet). A dose-dependent inhibitory response was observed at dietary selenium concentrations of 1–3 μg/g. Interestingly, Brazil nut was found to be just as powerful as sodium selenite, if not more so, at similar levels of dietary selenium intake in mammary cancer protection gland and plasma. The magnitude of tissue selenium accumulation was proportional to the amount of Brazil nut added to the diet. Supplementation with Brazil nut as the sole source of selenium produced an efficient gradient of two selenoenzymes (glutathione peroxidase and type I 5'-deiodinase) restoration at 0.05–0.2 μg/g of dietary selenium. A parallel comparison with sodium selenite indicated that the selenium in Brazil nut and selenite selenium were equally bioactive.

Ethanolic extract of *B. excelsa* was found to exhibit selective inhibitory activity against vesicular stomatitis viruses (VSV) (Ali et al. 1996).

Antiatherogenic Activity

The consumption of Brazil nut could affect the plasma lipids and apolipoproteins and some functional properties of the antiatherogenic high-density lipoprotein (HDL) as it is rich in both monounsaturated fatty acids and polyunsaturated fatty acids and selenium. Studies showed that Brazil nut ingestion did not alter HDL, low-density lipoprotein cholesterol, triacylglycerols, apolipoprotein A-I, or apolipoprotein B concentrations (Strunz et al. 2008). As expected, plasma selenium was significantly increased. However, the consumption of Brazil nuts for short duration by normolipidemic subjects in comparable amounts to those tested for other nuts did not alter serum lipid profile. The only alteration in HDL function was the increase in cholesteryl ester transfer. This latter finding may be beneficial because it would improve the nonatherogenic reverse cholesterol transport pathway.

The results of studies by Stockler-Pinto et al. (2010) revealed that the investigated patients (81) on hemodialysis presented selenium (Se) deficiency and that the consumption of only one Brazil nut a day (5 g) for 3 months was effective to increase the Se concentration in the plasma and erythrocyte and glutathione peroxidase (GSH-Px) activity in these patients, thus improving their antioxidant status. The data confirmed the hypothesis that selenium may exert an anti-atherogenic influence by reducing oxidative stress. Also in patients who had undergone hemodialysis, large amounts of reactive oxygen species (ROS) were produced and, at higher concentrations, ROS were thought to be involved in the pathogenesis of cardiovascular disease. Brazil nut is the richest known natural food source of selenium.

Skin Healing Activity

The antioxidant action of extracts of Brazil nut in stimulating collagen synthesis or on free radicals therefore makes them particularly useful for combating the effects of skin ageing, such as wrinkles or the relaxation of the basal tissues of the skin, or for improving skin healing. Brazil nut oil is commonly used as a cosmetic agent, for instance, in skin creams, lotions and soaps (UNCTAD 2005). It helps to lubricate and moisturize the skin, provides antioxidant benefits with its high selenium content, helps prevents dryness, and leaves skin soft, smooth, and hydrated.

Trypanocidal Activity

Acetone and methanol extracts of Brazil nut stem bark showed significant in-vitro trypanocidal activity against trypomastigote form of *Trypanosoma cruzi* (Campos et al. 2005). At the concentration of 500 μg/ml, the parasites were reduced by 100% and 90.3% respectively by both extracts, whereas the triterpene betulinic acid isolated from hexane extract produced 75.4%.

Radioactivity

The Brazil nut has been reported to be one of the world's most radioactive foods, due to the tree's accumulation of radium from the soil into the nut. Various parts of the Brazil nut tree had been determined to contain high amounts of selenium, barium and radium (Ra-226 and Ra-228) (Turner et al. 1958; Penna-Franca et al. 1968; Smith 1971; Parekh et al. 2008). The reported concentration of Ra-226 and Ra-228 vary, but overall, the radium concentrations in Brazil nuts were reported to be 1000 times higher than those in other foods. Turner et al. (1958) reported approximately 1.8 pCi/g Ra-226 in Brazil nut. Penna-Franca et al. (1968) reported 0.075–3.6 pCi/g Ra-226 in the nut (3.1–113.5 pCi/g in ash of nut) and 0.16–3.6 pCi/g Ra-228 in the nut (5.3–114.5 pCi/g in ash of nut) and also high levels of barium. They found that in general, samples richer in barium had the highest concentration of radium. In samples of different parts of the same tree a certain parallelism was noticed between barium and radium contents. The high radioactivity of Brazil nuts was explained, by its tree capacity of absorbing and concentrating barium, which is the natural carrier of radium. Smith (1971) reported that the concentrations of radium found in the leaves and fruit were higher than had been previously reported, the highest values (up to 6.6 pCi/g Ra-226/kg) occurring in the seed endosperm; the soils, however, did not contain abnormally high levels of radioactivity. Strontium and barium also accumulated in the endosperm, and the ratios of all the heavy alkaline earth ions to calcium in this tissue were 20 times higher than in some stem tissues. He suggested that specific chelation of these ions as a possible explanation of the abnormally high concentrations occurring in the brazil nut fruit. More recently, Parekh et al. (2008) found regardless of geographic origin, all brazil nuts contained measurable amounts of selenium, barium, and radium. The concentration range of Se (2–20 μg/g) and that of Ba (96–1990 μg/g) each varied by more than an order of magnitude in the nuts, while Ra-226 (17–27 mBq/g) and Ra-228 (18–31 mBq/g) activities were comparable and within a factor of two of one another. The greatest concentrations of the elements were measured in nuts from Bolivia, for Barium; Brazil, for Radium; and northern South America, for Selenium. Only the northern South American nuts contained Caesium (Cs)-C-137. Gabay and Sax (1969) determined that most of the radium of ingested Brazil nuts was not retained by the body. Bull et al. (2006) found 500 mBq thorium-228 in a routine faecal sample of a worker for plutonium and americium analysis. Investigations revealed that the worker consumed ~25 g/day of nuts, including Brazil nuts. A sample of these nuts was analysed and found to contain activities of 228Th in sufficient quantity to account for the faecal activity. They postulated that the intake of 228Th was accompanied by similar activity of the parent 228Ra, and biokinetic calculations showed that decay of 228Ra in-vivo would produce sufficient 228Th to account for the observed urine activity.

Traditional Medicinal Uses

The woody husks of Brazil seed pods have been used in Brazilian folk medicine to brew into tea to treat stomach aches, and the tree bark is brewed into tea to treat liver ailments. Brazil nut oil is used as an emollient.

Other Uses

Indigenous people in the Amazon basin used the empty fruit shell, to carry around small smoky fires to discourage attacks of black flies, as cups to collect rubber latex from tapped trees, and as drinking cups. The fruit shell is also used for the production of handicrafts. Brazil nut oil is used as cooking oil by the rural people and lamp oil. The oil is used in the cosmetic industry as a cosmetic ingredient in soaps, shampoos and hair conditioning/repair products and skin care products such creams, lotion, ointments. Brazil nut oil is also used as a lubricant in clocks, for making artists' paints, and in the cosmetics industry. Products containing Brazil nut oil are found, for instance, on the shelves of the Body Shop in United Kingdom and Europe. Brazil nut is also used as insect repellent and livestock feed. The tree also furnishes an excellent high quality timber.

Studies showed that Brazil nut shells may be useful as adsorbent either for basic or acid dyes such as methylene blue and indigo carmine respectively (de Oliveira Brito et al. 2010).

Comments

Bolivia is the world's largest producer of shelled Brazil nuts.

Selected References

Ali AM, Mackeen MM, El-Sharkawy SH, Abdul Hamid J, Ismail NH, Ahmad F, Lajis MN (1996) Antiviral and cytotoxic activities of some plants used in Malaysian indigenous medicine. Pertanika J Trop Agric Sci 19(2/3):129–136

Arthur JR (2000) The glutathione peroxidases. Cell Mol Life Sci 57:1825–1835

Bull RK, Smith TJ, Phipps AW (2006) Unexpectedly high activity of 228Th in excretion samples following consumption of Brazil nuts Radiat. Prot Dosim 121(4):425–428

Campos FR, Januário AH, Rosas LV, Nascimento SKR, Pereira PS, França SC, Cordeiro MSC, Toldo MPA, Albuquerque S (2005) Trypanocidal activity of extracts and fractions of *Bertholletia excelsa*. Fitoterapia 76(1):26–29

Chang JC, Gutenmann WH, Reid CM, Lisk DJ (1995) Selenium content of Brazil nuts from two geographic locations in Brazil. Chemosphere 30(4):801–802

Chunhieng T, Pétritis K, Elfakir C, Brochier J, Goli T, Didier Montet D (2004) Study of selenium distribution in the protein fractions of the Brazil nut, *Bertholletia excelsa*. J Agric Food Chem 52(13):4318–4322

Chunhieng T, Hafidi A, Pioch D, Brochier J, Montet D (2008) Detailed study of Brazil nut (*Bertholletia excelsa*) oil micro-compounds: phospholipids, tocopherols and sterols. J Braz Chem Soc 19(7):1374–1380

Combs GF, Clark LC, Turnbull BW (2001) An analysis of cancer prevention by selenium. Biofactors 14:153–159

de Oliveira Brito SM, Andrade HM, Soares LF, de Azevedo RP (2010) Brazil nut shells as a new biosorbent to remove methylene blue and indigo carmine from aqueous solutions. J Hazard Mater 174(1–3):84–92

Gabay JJ, Sax NI (1969) Retention of radium due to ingestion of Brazil nuts. Health Phys 16(6):812–813

Hawkes WC, Keim NL (2003) Dietary selenium intake modulates thyroid hormone and energy metabolism in men. J Nutr 133:3443–3448

Instituto Nacional de Pesquisas da Amazonia (INPA) (1986) Food and Fruit Bearing Forest Species 3: examples from Latin America. Forestry Paper 44-3, Food and Agriculture Organization of the United Nations, Rome. 332 pp

Ip C, Lisk DJ (1994) Bioactivity of selenium from Brazil nut for cancer prevention and selenoenzyme maintenance. Nutr Cancer 21(3):203–212

John JA, Shahidi F (2010) Phenolic compounds and antioxidant activity of Brazil nut (*Bertholletia excelsa*). J Funct Foods 2(3):196–209

Kaiser S, Di Mascio P, Murphy P, Sies HME (1990) Physical and chemical scavenging of singlet oxygen by tocopherols. Arch Biochem Biophys 277:101–108

Mori SA, Prance GT (1990a) Lecythidaceae—Part II. The zygomorphic-flowered New World genera (*Couroupita, Corythophora, Bertholletia, Couratari, Eschweilera, & Lecythis*). Fl Neotrop 21(2):1–376

Mori SA, Prance GT (1990b) Taxonomy, ecology, and economic botany of the Brazil nut (*Bertholletia excelsa* Humb. and Bonpl.: Lecythidaceae). Adv Econ Bot 8:130–150

Mortensen A, Skibsted LH (1997) Relative stability of carotenoid radical cations and homologue tocopheroxyl radicals. A real time kinetic study of antioxidant hierarchy. FEBS Lett 417:261–266

Parekh PP, Khan AR, Torres MA, Kitto ME (2008) Concentrations of selenium, barium, and radium in Brazil nuts. J Food Compos Anal 21(4):332–335

Penna-Franca E, Fiszman M, Lobao N, Costa-Ribeiro C, Trindade H, Dos Santos PL, Batista D (1968) Radioactivity of Brazil nuts. Health Phys 14(2):95–99

Ramos CMP, Bora PS (2003) Extraction and functional characteristics of Brazil nut (*Bertholletia excelsa* HBK) globulin. Food Sci Technol Int 9(4):265–269

Ramos CMP, Bora PS (2005) Functionality of succinylated Brazil nut (*Bertholletia excelsa* HBK) kernel globulin. Plant Foods Hum Nutr 60(1):1–6

Sharma GM, Mundoma C, Seavy M, Roux KH, Sathe SK (2010) Purification and biochemical characterization of Brazil nut (*Bertholletia excelsa* L.) seed storage proteins. J Agric Food Chem 58(9):5714–5723

Smith KA (1971) The Comparative uptake and translocation by plants of calcium, strontium, barium, and radium. I. *Berthollectias excelsa* (Brazil nut tree). Plant Soil 34(1):369–379

Stockler-Pinto MB, Mafra D, Farage NE, Boaventura GT, Cozzolino SM (2010) Effect of Brazil nut supplementation on the blood levels of selenium and glutathione peroxidase in hemodialysis patients. Nutrition 26(11–12):1065–1069

Strunz CC, Oliveira TV, Vinagre JC, Lima A, Cozzolino S, Maranhão RC (2008) Brazil nut ingestion increased plasma selenium but had minimal effects on lipids, apolipoproteins, and high-density lipoprotein function in human subjects. Nutr Res 28(3):151–155

Thomson CD, Chisholm A, McLachlan SK, Campbell JM (2008) Brazil nuts: an effective way to improve selenium status. Am J Clin Nutr 87(2):379–384

Turner RC, Radley JM, Mayneord VW (1958) The naturally occurring alpha ray activity of foods. Health Phys 1:268–275

U.S. Department of Agriculture, Agricultural Research Service (2010) USDA National Nutrient Database for Standard Reference, Release 23. Nutrient Data Laboratory Home Page, http://www.ars.usda.gov/ba/bhnrc/ndl

UNCTAD (2005) Market Brief in the European Union for Selected Natural Ingredients Derived from Native Species *Bertholletia excelsa* (Brazil Nut, Castaña) Vegetable oil. http://www.underutilized-species.org/Documents/Publications/

Vonderheide AP, Wrobel K, Kannamkumarath SS, B'hymer C, Montes-Bayon M, Ponce De Leon C, Caruso JA (2002) Characterization of selenium species in Brazil nuts by HPLC-ICP-MS and ES-MS. J Agric Food Chem 50(2):5722–5728

Couroupita guianensis

Scientific Name

Couroupita guianensis **Aubl.**

Synonyms

Couratari pedicellaris Rizzini, *Couroupita acreensis* R. Knuth, *Couroupita antillana* Miers, *Couroupita froesii* R. Knuth, *Couroupita guianensis* var. *surinamensis* (Mart. ex O. Berg) Eyma, *Couroupita idolica* Dwyer, *Couroupita membranacea* Miers, *Couroupita peruviana* O. Berg, *Couroupita saintcroixiana* R. Knuth, *Couroupita surinamensis* Mart. ex O. Berg, *Couroupita venezuelensis* R. Knuth, *Lecythis bracteata* Willd., *Pekea couroupita* Juss. ex DC. nom. inval.

Family

Lecythidaceae

Common/English Name

Cannonball tree

Vernacular Names

Brazil: Abricó-De-Macaco, Amêndoa-Dos-Andes, Bola-De-Canhão, Castanha-De-Macaco, Cuia-De-Macaco, Macacarecuia, (Portuguese);
Columbia: Coco Sachapura, Maraca;
Costa Rica: Bala De Canon,
Dutch: Kanonskogelboom;
French: Arbre À Boulet De Canon;
French Guiana: Kouroupitoumou;
German: Kanonenkugelbaum;
India: Kaman Gola, Nagkeshar (Bengali), Nagalinga, Shiv Kamal, Shivalinga (Flowers), Tope Gola (Hindu), Lingada Mara, Nagalingam, Nagalinga Pushpa (Flowers), (Kannada), Shivalingam (Marathi), Naagalingam (Tamil), Nagamalli (Flowers), Mallikarjuna (Flowers) (Telugu);
Indonesia: Sala;
Panama: Coco Sachapura, Granadillo De Las Huacas;
Peru: Ayahuma, Ayahúman;
Portuguese: Abricó De Macaco, Castanha De Macaco, Cuia De Macaco, Macacarecuia;
Spanish: Bola De Canon, Coco De Mono;
Surinam: Boskalebas, Iwadaballi, Kaupe;
Thai: Sala Lankaa;

T.K. Lim, *Edible Medicinal And Non-Medicinal Plants: Volume 3, Fruits*,
DOI 10.1007/978-94-007-2534-8_14, © Springer Science+Business Media B.V. 2012

Venezuela: Coco De Mono, Mamey Hediono, Muco, Taparo De Monte;
Vietnamese: Đầu Lân, Hàm Rồng, Ngọc Kỳ Lân, Sala .Vô Ưu.

Origin/Distribution

Cannonball tree is indigenous to the tropical rainforest of north-eastern South America, especially in the Amazon Basin. The tree is planted in gardens elsewhere in the tropics such as in India and Thailand Bolivia, Colombia, Costa Rica, Ecuador, Honduras, Panama, Peru, United States, Venezuela. The species is frequently planted as a botanical curiosity in other tropical and subtropical botanical gardens in many parts of the world.

Agroecology

C. guianensis is a hygrophyte and heliophyte. It is most frequently found in wet areas of lowland forests and river banks subjected to periodic flooding. Although a plant of moist soils, it thrives under dry conditions.

Edible Plant Parts and Uses

Fruits are edible and are occasionally eaten, but the smell of the white flesh discourages most people from trying them (Anon 2005). The pulp of the fruit is vinous, white, acid and not disagreeable. Fresh pulp is used by natives to prepare a cooling medicinal drink (Nelson and Wheeler 1937).

Botany

Cannonball tree is a large deciduous, unbuttressed tropical tree, growing to 35 m high. Leaves occur in clusters at the ends of the branches on 0.5–3 cm long petioles; lamina simple, narrowly obovate to elliptic, 8–30(–57) cm by 3–10 cm, glabrous on upper surface, pubescent on veins on lower surface, base cuneate and apex acute to acuminate, margin entire with 15–25 pairs of secondary veins. Inflorescences cauline, unbranched racemes, sometimes branched and paniculate (Plate 2), may cover entire trunk; pedicel/hypanthium 1–6 cm long. Flowers zygomorphic, 5–6 cm across; calyx –6 broadly ovate lobes with ciliate margins; petals 6 yellow with pink or red bases basally on the lower surface; androecium extended on one side into a flat, pale yellow pink tinted hood, stamens in a staminal ring, the ring stamens with white filaments and pale yellow anthers, the hood staminodes often white at the base, pink for most of length, and yellow at apex; ovary 6-locular (Plates 1, 2 and 3). Fruits indehiscent, globose, 12–25 cm diameter, woody, dark –brown, capsule falling from tree at maturity (Plates 1 and 4). Seeds up to 300, embedded in pulp, the pulp oxidizing bluish-green when exposed to air, the testa with trichomes.

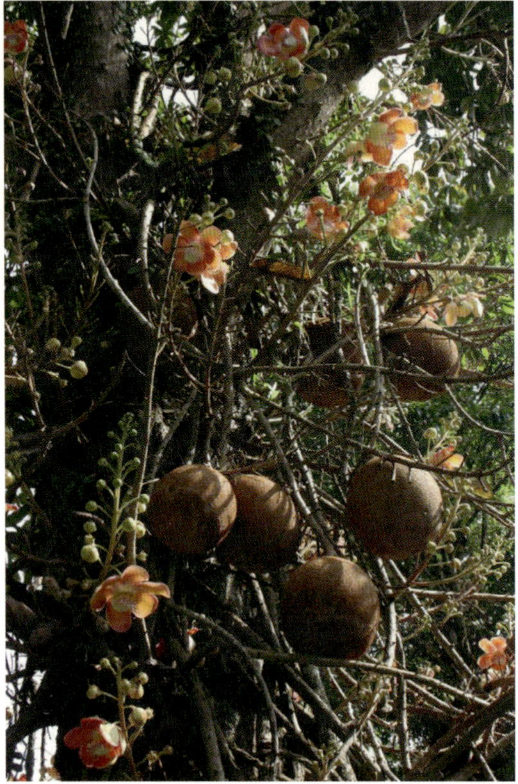

Plate 1 Flowers and fruits of cannonball tree

Plate 2 Cannonball inflorescence

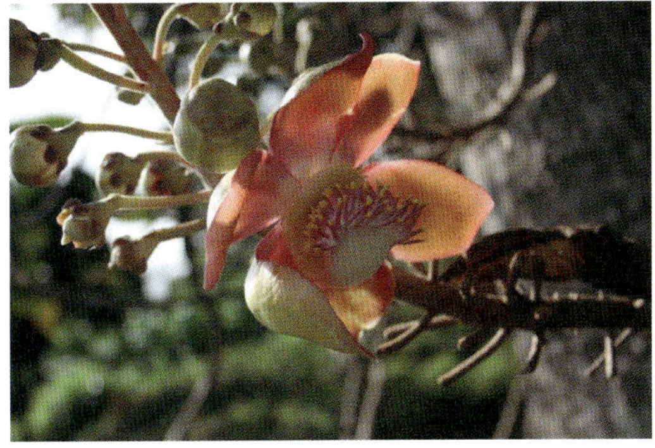

Plate 3 Close-up of cannonball flower

Plate 4 Cannonball tree laden with fruits

Nutritive/Medicinal Properties

The dried fruit of *Couroupita guianensis* yielded the following indolic constituents: tryptanthrin (6,12-dihydro-6, 12-dioxoindolo[2,1-b]quinazoline), indigo, indirubin, and isatin (Bergman et al. 1977; 1995). The fruit pulp was reported to contain sugar, gum, malic, citric and tartaric acid (Lans et al. 2001). The fruit pulp was also reported to contain citric hydrazide, γmalic hydrazide, lycopin and capric acid (Nelson and Wheeler 1937).

The seeds of *Couroupita guianensis* were found to contain 32% oil and 19.0% protein (Dave et al. 1985). The seed oil was found to have an iodine value of 126.1, saponification value of 185.7, peroxide value of 0.8 and acid value of 2.4. The fatty acid composition (wt%) comprised: C8:0 (caprylic), 0.3%; C10:0 (capric), 1.5%; C12:0 (lauric), 0.6%;C14:0 (Myristic), 1.2%; C14:1 (myristoleic), 0.9% ;C16:0 (palmitic), 6.3%; C16:1 (palmitoleic), 1.0%; C18:0 (stearic), 3.4%; C18:1 (oleic), 3.3%; and C18:2 (linoleic), 81.5%.

Forty-one volatile compounds were identified in *Couroupita guianensis* flowers, among which eugenol (41.6%), linalool (14.9%), (E,E)-farnesol (10.3%) and nerol (9.8%) were dominant

(Wong and Tie 1995). From the flowers of *Couroupita guianensis*, an aliphatic hydrocarbon and stigmasterol were isolated (Rane et al. 1994).

Some pharmacological properties of the aerial plant parts have been reported and are discussed below.

Antinociceptive Activity

Crude ethanol extract (CEE) and hexane, dichloromethane, ethyl acetate, and butanol fractions of *Couroupita guianensis* leaves exhibited significantly inhibited the number of contortions induced by acetic acid (Pinheiro et al. 2010). All fractions showed antinociceptive activity in the tail flick model, with the hexane and ethyl acetate the most potent and long acting fractions. In the hot plate method the highest effect observed was at the dose of 100 mg/kg from all fractions. Administration of naloxone (opioid receptor antagonist) inhibited the antinociceptive effect of fractions. Pre-treatment of mice with atropine (muscarinic receptor antagonist) reduced the antinociceptive activity of CEE and its fractions, the exception being the dichloromethane fraction. Mecamylamine (nicotinic receptor antagonist) did not inhibit the effect of dichloromethane fraction. L-NAME (L-nitro arginine methyl ester), a nitric oxide synthase inhibitor, reduced the anti-hyperalgesic effect of all fractions, but the most prominent effect was observed in the antinociceptive activity caused by CEE and butanol fraction. Results obtained demonstrated that *Couroupita guianensis* CEE and its fractions had antinociceptive activity that was mediated, at least in part, by opioid and cholinergic systems and nitric oxide pathway.

Wound Healing Activity

Sanjay et al. 2007 reported that ethanolic extract of whole plant of *Couroupita guianensis* (CGEE) (barks, leaves, flowers and fruits) accelerated the wound healing process in rats by decreasing the surface area of the wound and increasing the tensile strength. Nitrofurazone ointment was used as a positive control. Complete epithelization was observed within 15 days with CGEE. Measurements of the healed area and the hydroxyproline level were in agreement.

Antimicrobial Activity

Methanol extracts of leaves, flowers, fruit, stem and root barks, and stem and root heartwood of the plant were found to inhibit growth of 12 gram positive, 12 gram negative, and one protozoan (Khan et al. 2003). Most activity was in the petrol fractions of the flowers, fruit, and stem bark; the ethyl acetate fraction of the flowers, and stem and root bark; and the dichloromethane fractions of the stem and root barks. Some fractions of the stem bark and flowers exhibited antifungal activity. The ethanolic extract of whole plant of *Couroupita guianensis* (CGEE) (barks, leaves, flowers and fruits) was found to have moderate antibacterial activity against Gram positive (*Staphylococcus aureus*) and Gram Negative Bacteria (*Escherichia coli, Pseudomonas aeruginosa* and *Klebsiella pneumoniae* (Sanjay et al. 2007).

Anthelmintic Activity

The ethanolic flower extracts of *Couroupita guianensis* was found to be more effective than the chloroform and acetone extract in in-vitro anthelmintic activity on adult earthworm, *Pheritima phosthuma* (Rajamanickam et al. 2009). The anthelmintic activity was comparable with the standard drug piperazine citrate.

Traditional Medicinal Uses

Extract from the plant have been used to treat colds and stomach aches. Juice made from the leaves is used to cure skin diseases. The Shamans of South America have even used tree parts for treating malaria. The inside of the fruit can disinfect wounds and young leaves used to relieve toothache.

Other Uses

The flowers of Cannonball tree have a wonderful smell and can be used to scent perfumes and cosmetics. The hard shells of the fruit are sometimes used as containers.

The pulp of the fruit of *C. guianensis* is used to feed animals such as chickens, muscovy ducks, turkeys, and pigs.

Its wood is used to manufacture toys, boxes, parquet blocks, rackets, casting moulds and light artefacts.

Couroupita guianensis is considered a sacred tree in India by Hindus. Buddhist scriptures tell us that the Lord Buddha was born under the shade of a Sala Tree and died between two of these trees.

Comments

The fruit falls from the tree and cracks open when it hits the ground causing a sound of a small explosion.

Selected References

Anon (2005) *Couroupita guianensis* (Lecythidaceae). National Tropical Botanical Garden. http://ntbg.org/plants/plant_details.php?plantid=3487#

Bergman J, Egestad B, Jan-Olof Lindström J-O (1977) The structure of some indolic constituents in *Couroupita guaianensis* Aubl. Tetrahedron Lett 18(30):2625–2626

Bergman J, Lindström J-O, Tilstam U (1995) The structure and properties of some indolic constituents in *Couroupita guianensis* aubl. Tetrahedron 41(14): 2879–2881

Dave GR, Patel RM, Patel RJ (1985) Characteristics and composition of seeds and oil *of Couroupita guianensis* Aubl. from Gujarat, India. Fett Seifen Anstr 87(3):111–112

Hedrick UP (1972) Sturtevant's edible plants of the world. Dover Publications, New York, 686 pp

Khan MR, Kihara MA, Omoloso D (2003) Antibiotic activity of *Couroupita uianensis*. J Herbs Spices Med Plant 10(3):95–108

Lans C, Harper T, Georges K, Bridgewater E (2001) Medicinal and ethnoveterinary remedies of hunters in Trinidad. BMC Complement Altern Med 1:10

Lorenzi H (2002) Brazilian trees a guide to the identification and cultivation of Brazilian native trees, vol 1, 4th edn. Instituto Plantarum De Estudos Da lora Ltda, Brazil, 384 pp

Mori SA, Prance GT (1990) Lecythidaceae—Part II. The zygomorphic-flowered new world genera (*Couroupita, Corythophora, Bertholletia, Couratari, Eschweilera, & Lecythis*). Fl Neotrop 21(2):1–376

Nelson EK, Wheeler DH (1937) Some constituents of the cannonball fruit (*Couroupita guianensis* Aubl.). J Am Chem Soc 59(12):2499–2500

Pinheiro MM, Bessa SO, Fingolo CE, Kuster RM, Matheus ME, Menezes FS, Fernandes PD (2010) Antinociceptive activity of fractions from *Couroupita guianensis* aubl. leaves. J Ethnopharmacol 127(2): 407–413

Rajamanickam V, Rajasekaran A, Darlin quine S, Jesupillai M, Sabitha R (2009) Anthelmintic activity of the flower extract of *Couroupita guianensis*. Internet J Altern Med 8:1

Rane JB, Vahanwala SJ, Golatkar SG, Ambaye RY, Khadse BG (1994) Chemical examination of the flowers of *Couroupita guianensis* Aubl. Indian J Phar Sci 56(1):72–73

Sanjay PU, Jayaveera KN, Ashok Kumar CK, Kumar GS (2007) Antimicrobial, wound healing and antioxidant potential of *Couroupita guianensis* in rats. Pharmacologyonline 3:269–281

Wong KC, Tie DY (1995) Volatile constituents of *Couroupita guianensis* aubl. flowers. J Essent Oil Res 7(2):225–227

Lecythis ollaria

Scientific Name

Lecythis ollaria **Loefling.**

Synonyms

Eschweilera cordata (O. Berg) Miers, *Lecythis cordata* O. Berg, *Lecythis ollaria* L. nom. illeg., *Lecythis ollaria* Saldanha nom. illeg., *Lecythis ollaria* Spruce nom. illeg., *Lecythis ollaria* Vell. nom. illeg.

Family

Lecythidaceae

Common/English Names

Monkey Pot, Paradise Nut, Sabucaia Nut

Vernacular Name

Venezeula: Coco De Mono.

Origin/Distribution

The species is native to north-central Venezuela. It is cultivated in northern South America and tropical Central America and in trial plantings in SE Asia.

Agroecology

A tree of dry tropical forests and savannah.

Edible Plant Parts and Uses

The seeds are edible and delicious but may be toxic when they come from trees that grow on soils with high selenium accumulation.

Botany

Small to medium tree with fissured bark, growing to 20 m tall; twigs pubescent when young. Leaf, alternate, simple, glabrous, coriaceous, ovate, 6–15 × 4–8 cm, subsessile, apex acute to obtuse, base rounded to cordate, margin minutely crenulated, abaxial side pubescent (Plates 1 and 2). Inflorescence racemose, unbranched or branched 2–3 orders with pubescent rachises, with 15–25 flowers subtended by ovate bracteoles. Flowers, 5–8 cm wide, calyx 6 broadly ovate, lobes, petals 6 broadly obovate, white to yellowish white, androecium hood dorsi-ventrally expanded with antherless appendages, staminal ring with 242–372 stamens, hypanthium ferruginous-tomentose, ovary 3–5 locules with 5–15 ovules per locule, style short 2 mm long. Fruit cup-shaped or pot-shaped, rugose, dirty brown, woody capsule, 3.5–6 by 5.5–9 cm, pericarp 12 mm thick, calyx lobes persisting as a woody rim (Plates 1, 2 and 3). Seeds brown, with 4 longitudinal impressed veins.

T.K. Lim, *Edible Medicinal And Non-Medicinal Plants: Volume 3, Fruits*,
DOI 10.1007/978-94-007-2534-8_15,

Plate 1 Monkey-pot fruits and foliage

Plate 3 Empty woody, cup-shaped capsules

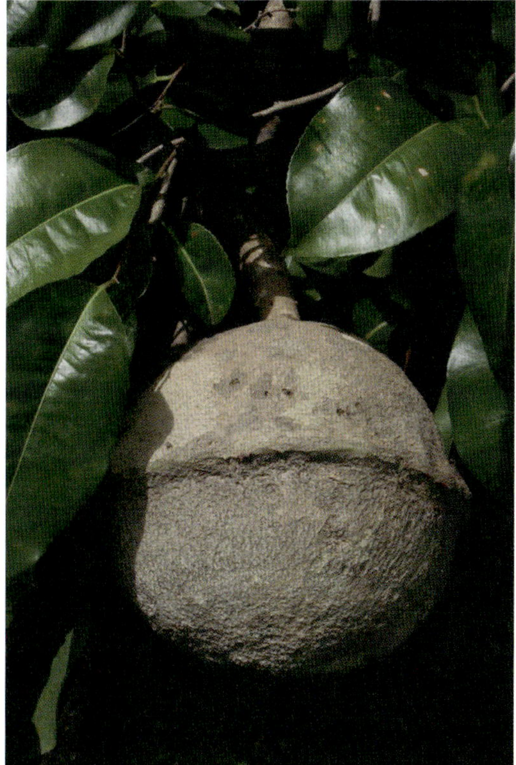

Plate 2 Close-up of the globose, monkey pot fruit

Nutritive/Medicinal Properties

High selenium concentrations (7–12 g/kg dry mass) were found in the seeds of the selenium-accumulator plant *Lecythis ollaria* (Hammel et al. 1996). Of the extracted selenium only 9% was shown to be firmly bound to proteins at pH 4.5 and 29% at pH 7.5. For the protein-bound selenium, concentrations of 0.7 g and 2.4 g per kg of seeds and 40 and 25 g per kg of extractable protein were determined at pH 4.5 and 7.5, respectively. The element was found to be present in extremely selenium-rich proteins with molecular masses below 20 kDa. Selenium was found in different parts (leaves, bark, capsules and seeds) of the *Lecythis ollaria* as well as the soil where the tree was growing (Ferri et al. 2004). Seeds show the highest content in selenium (about 5 g/kg) which was dependent on the maturation extent of the fruit. In seeds, about half of the total selenium content was found to be soluble in water while the remaining was involved in protein structure. In the aqueous fraction, the prevailing form of selenium appeared to be seleno-cystathionine with much lower amounts of Se(VI) and Se(IV).

Several cases of acute selenium intoxication from a natural source has been noted in the literature. A 54-year-old Venezuelan man suffered anxiety, chills, diarrhea, fever, anorexia, and weakness after consuming 70–80 "Coco de Mono" (*Lecythis ollaria*) nuts (Kerdel-Vegas 1964). Eight days after consuming the nuts, he suffered extensive loss of scalp and body hair. Following the observations of clinical cases of temporary alopecia (defluvium capillorum) after the ingestion of "coco de mono" nuts and associated symptoms of acute intoxication, fever and diarrhea and various neurological manifestations, the marked cytotoxic effect of "coco de mono" was determined in-vitro in mice fibroblasts.

Subsequent studies identified the pharmacologically active agent as selenocystathionine (Aronow and Kerdel-Vegas 1965; Kerdel-Vegas 1965; Kerdel-Vegas et al. 1965). More recently, two previously healthy women developed nausea, vomiting, headache and dizziness for several days, a massive hair loss about 2 weeks later and a discoloration of the fingernails following ingestion of *L. ollaria* nuts (Müller and Desel 2010). Detailed diagnostic procedures did not reveal any pathological results. Delayed quantitative determination of selenium in blood, however revealed toxic values (in case I: 479 μg/l of serum, 8 weeks after ingestion, and in case II 300 μg/L of serum, 9 weeks after ingestion).

Ethanolic extract of *Lecythis ollaria* was found to exhibit cytotoxicity in Hela cells (Ali et al. 1996).

In Caracas, the fruits are sold by vendors of herbal medicines, as water placed in them produces a depilatory effect (Mori and Prance 1990).

Other Uses

The timber is hard and valuable, used for house-frames, wharves and sluices. The urn-shaped, woody capsules are used as pots, vases and other containers. Its bark separates into thin sheets, like paper, used by the natives for cigarette wrappers.

Comments

The species is widespread both ecologically and geographically and is not listed as threatened.

Selected References

Ali AM, Mackeen MM, El-Sharkawy SH, Abdul Hamid J, Ismail NH, Ahmad F, Lajis MN (1996) Antiviral and cytotoxic activities of some plants used in Malaysian indigenous medicine. Pertanika J Trop Agric Sci 19(2/3):129–136

Aronow L, Kerdel-Vegas F (1965) Seleno-cystathionine, a pharmacologically active factor in the seeds of *Lecythis ollaria*. Nature 205:1185–1186

Burkill IH (1966) A dictionary of the economic products of the Malay Peninsula. Revised reprint. 2 volumes. Ministry of Agriculture and Co-operatives, Kuala Lumpur, Malaysia, vol 1 (A–H) pp 1–1240, vol 2 (I–Z) pp 1241–2444

Ferri T, Coccioli F, De Luca C, Callegari CV, Morabito R (2004) Distribution and speciation of selenium in *Lecythis ollaria* plant. Microchem J 78(2):195–203

Hammel C, Kyriakopoulos A, Behne D, Gawlik D, Brätter P (1996) Protein-bound selenium in the seeds of coco de mono (*Lecythis ollaria*). J Trace Elem Med Biol 10(2):96–102

Kerdel-Vegas F (1964) Generalized hair loss due to the ingestion of "coco de mono" (*Lecythis ollaria*). J Invest Dermatol 42:91–94

Kerdel-Vegas F (1965) The depilatory and cytotoxic action of "coco de mono" (*Lecythis ollaria*) and its relationship to chronic seleniosis. Econ Bot 20(2):187–195

Kerdel-Vegas F, Wagner T, Russell PG, Grant NH, Alburn HE, Clark DE, Miller JA (1965) Structure of the pharmacologically active factor in the seeds of *Lecythis ollaria*. Nature 205:1186–1187

Marsh TD (1937) Sapucaia nut *Lecythis* sp. Malayan Agric J 25:18–22

Mori SA, Prance GT (1990) Lecythidaceae—part II. The zygomorphic-flowered New World genera (*Couroupita, Corythophora, Bertholletia, Couratari, Eschweilera,* & *Lecythis*). Flora Neotrop 21(2):1–376

Müller D, Desel H (2010) Acute selenium poisoning by paradise nuts (*Lecythis ollaria*). Hum Exp Toxicol 29(5):431–434

Pires O'Brien J (1998) *Lecythis ollaria*. In: IUCN 2011. IUCN Red list of threatened species. Version 2011.1.www.iucnredlist.org

Py C, Fouqué A (1963) Les cultures fruitières de Porto Rico. Fruits d'Outre Mer 18:325–336

Lecythis pisonis

Scientific Name

***Lecythis pisonis* Cambess.**

Synonyms

Couroupita crenulata Miers, *Couroupita lentula* Miers, *Jacapucaya brasiliensis* Macgrave, *Lecythis amapaensis* Ledoux, *Lecythis amazonum* Mart. ex O. Berg, *Lecythis densa* Miers, *Lecythis hoppiana* R. Knuth, *Lecythis marcgraaviana* Miers, *Lecythis ollaria* Spruce, *Lecythis ollaria* Vellozo, *Lecythis paraensis* Huber, *Lecythis paraensis* Huber ex Ducke, *Lecythis pilaris* Miers, *Lecythis pisonis* Cambessedes subsp. *usitata* (Miers). Mori & Prance, *Lecythis setifera* Miers, *Lecythis sphaeroides* Miers, *Lecythis urnigera* Mart. ex O. Berg, *Lecythis usitata* Miers, *Lecythis usitata* var. *paraensis* (Huber ex Ducke) R. Knuth, *Lecythis usitata* Miers var. *tenuifolia* R. Knuth, *Lecythis velloziana* Miers, *Pachylecythis egleri* Ledoux.

Family

Lecythidaceae

Common/English Names

Cream Nut, Monkey Pot, Paradise Nut, Sapucaia Nut

Vernacular Names

Brazil: Caçamba-Do-Mato, Castanha-De-Sapucaia, Castanha Sapucaia, Cumbuca-De-Macaco, Marmita De Macaco, Quatete, Sapucaia, Sapucaia-Vermelha (Portuguese);
French: Canari Macaque, Marmite De Singe, Noix De Sapucaia.
Spanish: Olla De Mono, Castaña De Monte, Nuez Del Paraiso, Sapucaya.

Origin/Distribution

The species is native to Amazonia and Brazil.

Agroecology

A canopy to emergent tree, found isolated in periodically inundated and non-inundated forest in coastal Brazil from Pernambuco to São Paulo and in Amazonia.

T.K. Lim, *Edible Medicinal And Non-Medicinal Plants: Volume 3, Fruits*,
DOI 10.1007/978-94-007-2534-8_16, © Springer Science+Business Media B.V. 2012

Edible Plant Parts and Uses

The aromatic, sweet-tasting oleaginous seeds (nuts) are eaten fresh, boiled or roasted.

Botany

Large, deciduous trees 40–50 high with greyish to dark-brown bark and deep vertical fissures. Juvenile twigs glabrous or puberulous. Leaves pinkish-red, cream-coloured or pale green when young tuning dark green when nature. Lamina simple, narrowly to broadly ovate to elliptic, chartaceous, apex acuminate, base obtuse, margin crenate, petiole 5–12 mm long (Plates 1 and 2). Inflorescence 6–20 flowered racemes arising on twigs below the leaves, rachis 3–15 cm long, pedicel 5–12 mm subtended by caducous bract with two ovate bracteoles. Flower, 3–7 cm across, calyx with 6 ovate, purplish lobes, petals of 6 obovate, sub-equal rose-purple petals, sometimes white, androecium hood flat, white or purple, staminal ring with 115–350 stamens with 1–2 mm long, white or yellow filaments with white or yellow anthers; hypanthium glabrous or puberulent; ovary 4-locular with 6–15 ovules per locule, style short 1–2 mm long. Fruits woody, globose, oblong or turbinate, 6–15 cm by 8–30 cm, dehiscent capsule, pericarp 3 cm thick, calycine zone often prominent or indistinct (Plates 1, 2 and 3). Seeds 10–30 per fruit, fusiform, 4–6 by 2.5–3 cm, with 8–12 sulci, the basal cord-like funicle is enveloped by fleshy white aril.

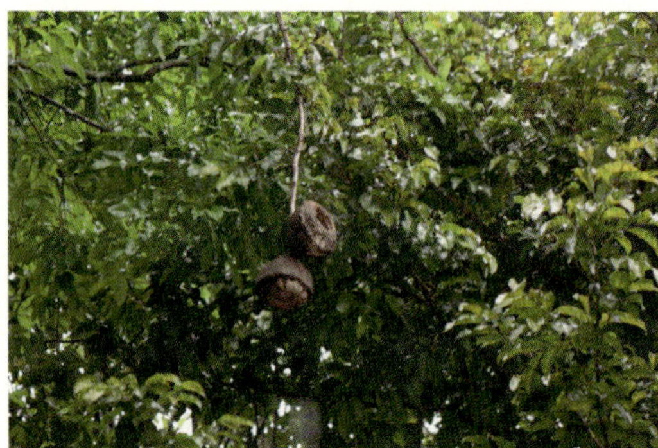

Plate 1 Sapucaia nuts and dense foliage

Plate 3 Close-up of empty woody capsule

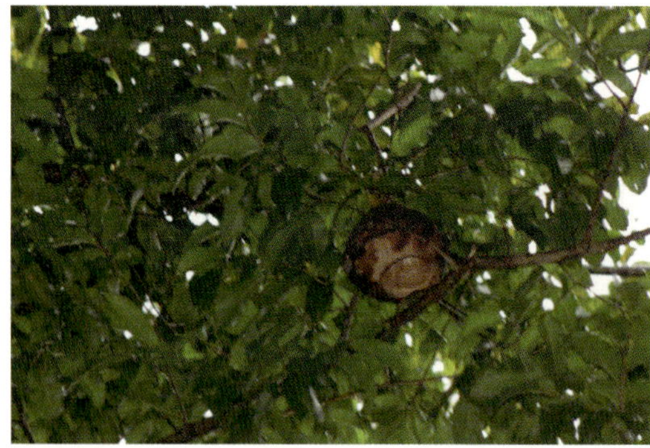

Plate 2 A dehisced empty wood capsules devoid of seeds

Nutritive/Medicinal Properties

Proximate nutrient composition of sapucaia (*Lecythis pisonis*) nuts, expressed as grams per 100 g raw matter was reported by Denadai et al. (2007) as follows: moisture 5.4 g, ash 3.80 g, crude lipid 60.61 g, total sugars 4.42 g, protein (Nx6.25) 20.47 g, total dietary fibre 5,67 g, energy 645.05 kcal, saturated fatty acids lauric (C12:0) 0.10 g, myristic (C14:0) 0.10 g, palmitic (C16:0) 12.14 g, stearic (C18:0) 6.13 g, monounsaturated fatty acids 34.22 g, palmitoleic (C16:1 ω7) 0.19 g, oleic (C18:1 ω9) 33.94 g, cis 11-eicosanoic (C20:1 ω 11) 0.10 g, polyunsaturated fatty acids 42.73 g, linoleic (C18:2 ω6) 42.5 g, α linmoleic (C18:3 ωα) 0.19 g; minerals – Na 5.28ug/g, Fe 32.65 ug/g, Mn 80.69ug/g, Zn

40.37ug/g, Cu 32.76ug/g, Ca 1.72 mg/g, Mg 2.179 mg/g, P 8.75 mg/g, K 8.90 mg/g. The protein content of sapucaia defatted nut meal was 66% with globulins (58.7%) as the major component while glutelins, albumins and prolamins accounted for 20.2, 20.1, and 1.0%, respectively. The proteins from sapucaia nuts contained adequate levels of phenylalanine, lysine, leucine, methionine, valine, and arginine, and the other amino acids were found in high or moderate amounts, based on the FAO/WHO standards for children. The proteins also contained adequate amounts of essential amino acids for pre-school children and all the essential amino acids for adults. The amino acid composition in mg/g protein found was: alanine 87.9, arginine 169.1, aspartate 110.1, cystine 15.4, glutamate 275.3, glycine 157.1, histidine 19.3, isoleucine 29.5, leucine 113.8, lysine 82.1, methionine 89.6, phenylalanine 106.3, proline 229, serine 686.5, threonine 34.4, tryptophan, tryrosine 32.5 and valine 62.0. In the study, no hemagglutination or inhibition were observed indicating low or non-detectable levels of lectin and proteinase inhibitors and demonstrating that the nuts analysed were free of these major anti-nutritional factors. Further the in-vitro digestibility of globulins by a multi-enzymatic complex was significant. These findings suggested that sapucaia nuts may be a new source of proteins for human consumption, a potential functional and nutritional agent, and an economically important oil source.

In another study, *Lecythis pisonis* nut samples from four different regions in São Paulo State, Brazil were found to have high lipid content (34.2–61.3%), an iodine value and an index of refraction equivalent to the corn oils (Vallilo et al. 1999). Among the macronutrients, high levels of P (5.2–6.2 mg/g) and Sn (69.1–77.0 μg/g) were observed in all the nuts. High levels of Pb (3.3–3.8 μg/g), Cu (2.9–3.3 μg/g), Zn (2.6–3.8 μg/g) and Mn (4.0–11.6 μg/g) were determined.

Other Uses

The tree is cultivated as an ornamental in tropical botanical gardens throughout the world. The wood is utilised for construction.

Comments

Flowers are pollinated by carpenter bees and the fruit is propagated from seeds.

Selected References

Anonymous (1938) Sapucaia nut, *Lecythis* spp. M. A. H. A. Magazine, Kuala Lumpur 8(3):107–111

Burkill IH (1966) A dictionary of the economic products of the Malay Peninsula. Revised reprint. 2 volumes. Ministry of Agriculture and Co-operatives, Kuala Lumpur, Malaysia, vol 1 (A–H) pp 1–1240, vol 2 (I–Z) pp 1241–2444

Denadai SMS, Hiane PA, Marangoni S, Baldasso PA, Miguel AMRO, Macedo MLR (2007) In vitro digestibility of globulins from sapucaia (*Lecythis pisonis* Camb.) nuts by mammalian digestive proteinases. Ciênc Tecnol Aliment Campinas 27(3):535–543

Lorenzi H (2002) Brazilian trees a guide to the identification and cultivation of Brazilian native trees, vol 3, 4th edn. Instituto Plantarum De Estudos Da lora Ltda, Brazil, 384 pp

Mori SA, Prance GT (1990) Lecythidaceae-Part II. The zygomorphic-flowered New World genera (*Couroupita, Corythophora, Bertholletia, Couratari, Eschweilera, & Lecythis*). Flora Neotrop 21(2):1–376

Vallilo MI, Tavares M, Aued-Pimentel S, Campos NC, Moita Neto JM (1999) *Lecythis pisonis* Camb. Nuts: oil characterization, fatty acids and minerals. Food Chem 66(2):197–200

Lecythis zabucaja

Scientific Name

Lecythis zabucaja Aubl.

Synonyms

Lecythis crassinoda Miers, *Lecythis davisii* Sandwith, *Lecythis davisii* var. *gracilipes* Eyma, *Lecythis hians* A.C. Smith, *Lecythis lecomtei* Pampanini, *Lecythis tumefacta* Miers, *Lecythis validissima* Miers, *Lecythis venusta* Miers.

Family

Lecythidaceae

Common/English Names

Monkey Pot Nut, Monkeynut, Paradise Nut, Potnut, Sapucaia Nut, Sapucaya Nut, Zabucaya Nut

Vernacular Names

Brazil: Castanha Sapucala, Sapucaia;
Cuba: Coquitos Del Brazil;
French: Canari Macaque, Marmite De Singe, Noix De Paradis, Noix De Sapucai;
French Guiana: Canari Macque, Koutapatou (Paramka), Zabucaio;
German: Paradiesnuß;
Guyana: Kume, Monkeypot, Wadaduri;
Spanish: Coco De Mono, Nuez Del Paraíso, Nuez Sapucaia, Olla De Mono;
Suriname: Zabuca, Kawatapatoe;
Venezuela: Coco De Mono, Tinajito.

Origin/Distribution

L. zabucaja is indigenous to the equatorial rainforests of the Guianas, E Venezuela, central and W Amazonia.

Agroecology

Paradise nut thrives in equatorial climate with mean annual maximum of 25°C and a minimum of 23.5°C and mean annual rainfall of 1,900–4,000 mm with 1–2 months of dry season. It grows on a fairly wide rang of soil types but grows best in deep, moist, organic rich alluvial soils up to 500 m altitude. It is usually found near water courses but abhors areas with extensive inundated flooding and is intolerant of waterlogged soils.

T.K. Lim, *Edible Medicinal And Non-Medicinal Plants: Volume 3, Fruits*,
DOI 10.1007/978-94-007-2534-8_17,

Edible Plant Parts and Uses

The kernel of seeds (nuts) of ripe fruits are edible after removal of the shell. The kernels are rated as the best in the world and is much sweeter than Brazil nuts. The kernels are also used in chocolates and other confections.

Botany

L. zabucajo is a tall, large, erect, deciduous tree, growing 35–55 m high (Plate 1), with a bole of 20–30 m and a diameter of 0.6–1.8 m, cylindrical with the base somewhat buttressed or swollen and brown to greyish-brown, deeply fissured bark. Twigs glabrous or puberulous when young. The leaves are simple, entire, elliptical, elliptical-obovate to oblong-lanceolate, 4–15 cm long by 2.5–6.7 cm wide, acuminate tips, obtuse bases and creanate margins, glabrous and deciduous. Flowers are white, bisexual with six sepals and petals, numerous stamens, inferior ovary with 4 locules and numerous ovules and formed in terminal racemes. Fruit variable in shape fruit size and form, subglobose, globose, turbinate to urn-shaped, woody capsule, 15–20 cm by 22–26 cm, with prounced calycine thickenings, with a thick, woody lid (operculum) which is shed when mature releasing the 12–40 seeds (nuts) which fall to the ground (Plates 2 and 3). The seed (nut) is fusiform, 5–8 cm by 2.5 cm, with a thin brown shell with longitudinal ridges enclosing an ivory-white, creamy, edible and sweet kernel.

Plate 1 Tree habit of paradise nut

Plate 2 Top view of empty, woody paradise nut capsule

Plate 3 Side-view of woody Paradise nut capsule

Nutritive/Medicinal Properties

Paradise nuts are as nutritious as Brazil nuts and contain 61% edible fat and 20% protein (García-Ramis 1991). Kernels of paradise nut (*Lecythis zabucajo*) have 2S seed proteins which are rich in the sulphur amino acids, especially methionine 14.0% and 4.3% cysteine (Zuo and Samuel 1996). The 2S proteins are synthesized as a precursor polypeptide of 18 kDa and processed stepwise to form the 9- and 3-kDa mature subunit polypeptides and have properties similar to those of the Brazil nut (*Bertholletia excelsa*) 2S proteins. Paradise nut kernel also has crystalline globulin protein (Vennesland et al. 1937).

Other Uses

Bees pollinate the flowers and provide a much-relished honey with the nectar. The seed oil is used in soap in Brazil. The tree also provides useful timber in Guiana and Suriname and tannin, besides being valued for its vegetable oil and fats.

Comments

The tree is propagated from seeds.

Selected References

García-Ramis G (1991) Sapucaia nuts - the genus *Lecythis*. WANATCA Yearbook 16:72–77

Hammond DS, Gourlet-Fleury S, van der Hout P, ter Steege H, Brown VK (1996) A compilation of known Guianan timber trees and the significance of their dispersal mode, seed size and taxonomic affinity to tropical rain forest management. For Ecol Manage 83(1–2):99–116

Kennard WC, Winter HF (1960) Some fruits and nuts for the tropics. USDA Agric Res Serv Misc Pub 801:78–80

Marsh TD (1937) Sapucaia nut *Lecythis* sp. Malayan Agric J 25:18–22

Martin FW, Campbell CW, Ruberte R (1987) Perennial edible fruits of the tropics: an inventory. U.S. Department of Agriculture, Agriculture Handbook No. 642

Mori SA (1990) *Lecythis*. In: Mori SA, Prance GT (eds) Flora neotropica: monograph no. 21. Lecythidaceae part II. New York Bot. Garden, New York, pp 267–333, 376 pp

Mori SA, Prance GT (1990) Lecythidaceae—part II. The zygomorphic-flowered New World genera (*Couroupita*, *Corythophora*, *Bertholletia*, *Couratari*, *Eschweilera*, & *Lecythis*). Flora Neotrop 21(2):1–376

Py C, Fouqué A (1963) Les cultures fruitières de Porto Rico. Fruits d'Outre Mer 18:325–336

Vennesland B, Blaugh MB, Saunders F (1937) Studies in proteins. V. A crystalline globulin from the paradise nut, *Lecythis zabucayo*. J Am Chem Soc 59(1):174

Zuo WN, Samuel SM (1996) Purification and characterization of the methionine-rich 2S seed proteins from the Brazil nut family (Lecythidaceae). J Agric Food Chem 44(5):1206–1210

Michelia mediocris

Scientific Name

Michelia mediocris **Dandy.**

Synonyms

Magnolia macclurei (Dandy) Figlar, *Magnolia mediocris* (Dandy) Figlar, *Michelia macclurei* Dandy, *Michelia macclurei* var. *sublanea* Dandy, *Michelia mediocris* var. *angustifolia* G. A. Fu, *Michelia rubriflora* Y. W. Law & R. Z. Zhou, *Michelia subulifera* Dandy, *Michelia tonkinensis* A. Chev.

Family

Magnoliaceae

Common/English Name

Michelia

Vernacular Names

Chinese: Bai Hua Han Xiao;
Vietnamese: Giới Xanh.

Origin/Distribution

The species is native to Cambodia, Vietnam, Southern China (Hainan, Guangdong, Guangxi, Guizhou).

Agroecology

Michelia mediocris occurs in wet evergreen forest slopes and margins in mixture of other hardwoods on low to high mountains from 400–1,000 m elevation mainly in areas with an annual rainfall of 1,000–2,000 mm and an average annual temperature of 20–25°C, latitude: 11°–22° N. The tree thrives on well drained, moist and fertile soils developed on acid magma as in Central Vietnam and on the Western Highland, on grey soils developed from old alluvia as in the South of Vietnam, and on many others soils developed from argillaceous schists or mica-schists as in the northern provinces of Central Vietnam or on soils on metamorphic rocks as in the mountains of North Vietnam. In Vietnam, it is endemic to dense primary or secondary tropical and sub-tropical evergreen forests of Northern and Central provinces, such as: Lao Cai, Thanh Hoa, Nghe An, Ha Tinh, Kon Tum, Gia Lai. A fast growing species when grown under full sun and coppice well.

T.K. Lim, *Edible Medicinal And Non-Medicinal Plants: Volume 3, Fruits*,
DOI 10.1007/978-94-007-2534-8_18, © Springer Science+Business Media B.V. 2012

Edible Plant Parts and Uses

In Vietnam, the fruit pulp and seeds are dried, crushed or ground into powder and mixed with salt and used as dip for cooked meat - dog, chicken or pork. The spice is aromatic, peppery and strongly fragrant and can be used to flavour coffee or tea like cardamon.

Botany

An evergreen, erect, much-branched, perennial tree, up to 35–37 m tall, with 70–90 cm trunk diameter and greyish brown, glabrous bark and buttressed base. The trunk is straight, cylindrical (Plate 1) and branches and branchlets are tomentose. Petiole is 1.5–3 cm, without a stipular scar. Leaves are simple, alternate, irregularly fixed on branches (Plate 2), leaf-blade coriaceous, glabrous, 8–20 cm long and 5.5 cm wide, rhomboid-elliptic, tip shortly acuminate, base cuneate, abaxially grayish white and puberulous, adaxially glabrous. Flower-bud is ellipsoid, enclosed by 3 caducous, spathaceous bracts and puberulous. Flower is solitary, terminal or opposite leaf, white with 9 spatulate tepals, 1.8–2.2 × 0.5–0.8 cm, stamens 1–1.5 cm long with connective exserted into a 3–4 mm tip and anthers 0.8–1.4 cm, gynophore 3–5 mm, gynoecium cylindric and 1 cm; carpels 7–14 with 4 or 5 ovules per carpel. Fruit 3–5 separated carpels, blackish brown when matured, 2–3.5 cm; mature carpels obovoid, ellipsoid, or globose, slightly compressed, 1–2 cm, white lenticellate, apex with an obtuse beak with 3–5 seeds, descent when ripe. Seeds pale brown with longitudinal furrows, 5–8 mm × 5 mm (Plate 3) with bright red testa.

Plate 1 Young tree with a tall, straight, cylindrical trunk

Plate 2 Alternate, simple, rhombic-elliptic leaves

Plate 3 Seeds removed from the fruit

Nutritive/Medicinal Properties

Nutritive value of the spice has not been reported. The oily, fragrant seeds are used locally to treat belly-ache.

Other Uses

It is a highly valued and productive indigenous Vietnamese timber species and resistant to termite. The wood is used for making furniture, house construction and carvings. It can be used for structures requiring medium strength, mainly in construction, communication, transport and for structures requiring resistant to collision and vibration.

Comments

The fruit and seeds have good potential as a spice.

Selected References

Hoang XT, Nguyen DM (2000) The physiological and ecological characters of *Tarrietia javanica* Blume (Huynh) and *Michelia mediocris* Dandy (Gioi xanh) to develop forest plantation technology. Forest Science and Technology Research Results. Forest Science Institute of Vietnam.http://www.mekonginfo.org/mrc_en/contact.nsf/0/7E626C0ACCF45B56C72566010071 1149/$FILE/List_of_Report_EN.html

Le DK, Nguyen XL, Nguyen HN, Ha HT, Hoang SD, Nguyen HQ, Vu VM (2003) Forest tree species selection for planting programmes in Vietnam. Forest sector support programme and partners. Ministry of Agriculture and Rural Development, Hanoi, Nov 2003, p 118

Liu Y, Xia N, Nooteboom HP (2008) Magnoliaceae. In: Wu ZY, Raven PH, Hong DY (eds) Flora of China, vol 7, Menispermaceae through Capparaceae. Science Press/Missouri Botanical Garden Press, Beijing/St. Louis

Nguyen BC (2002) *Michelia mediocris*. In: Do DS, Nguyen HN (eds) Use of indigenous tree species in reforestation in Vietnam. Agricultural Publishing House, Hanoi, pp 84–90

Bunchosia armeniaca

Scientific Name

Bunchosia armeniaca **(Cav.) DC.**

Synonyms

Brysonima nitida Ruiz & Pav. ex G. Don, *Bunchosia angustifolia* A. Juss., *Bunchosia pilocarpa* Rusby, *Malpighia armenica* Cav. (basionym).

Family

Malpighiaceae

Common/English Names

Bunchosia, Green Plum, Monk's Plum, Peanut Butter Fruit, Peanut Butter Tree

Vernacular Names

Brazil: Ameixa-Do-Peru, Ameixa –Do-Para, Caferana, Cafezinho, Caramel, Ciruela (Portuguese);
Columbia: Chico Mamey, Ciruela, Ciruela De Fraile, Ciruela Verde;
French: Bunchoise Des Andes, Bunchosie Des Andes;
Japanese: Ameishia, Bunchosia, Bunchoshia, Bunchoshia arumeniaka, Bunkoshia, Piinattsu bataa furuutsu;
Lithuanian: Abrikosinė andenė;
Peru: Cansaboca, Huánuco;
Portuguese: Falso guaraná, Fruta-manteiga-de-amendoim, Guaraná-rana, Manteiga-de-amendoim, Tártago;
Quechuan: Usan;
Russian: Bunkhoziia abrikosovaia;
Spanish: Ciruela De Fraile, Ciruela de monte, Ciruela silvestre, Ciruela verde, Mamay de terra fria.

Origin/Distribution

Bunchosia armeniaca is native to the Andean countries of Columbia, Ecuador, Peru, Bolivia. It is exotic to Brazil but is commonly cultivated there and other south American countries (Lorenzi et al. 2006). It also has been introduced to elsewhere in the tropics.

Agroecology

In its native range, peanut butter fruit is found at altitudes from 0 to 2,400 m. It thrives in full sun in moist, fertile loamy soil rich in organic matter with pH range of 6–7.6. It tolerates partial shade and is quite cold tolerant but briefly tolerant of freeze down to –2°C.

T.K. Lim, *Edible Medicinal And Non-Medicinal Plants: Volume 3, Fruits*,
DOI 10.1007/978-94-007-2534-8_19, © Springer Science+Business Media B.V. 2012

Plate 1 Yellow Bunchosia flowers

Plate 2 Immature fruits and mature leaves

Plate 3 Large cluster of Bunchosia fruits

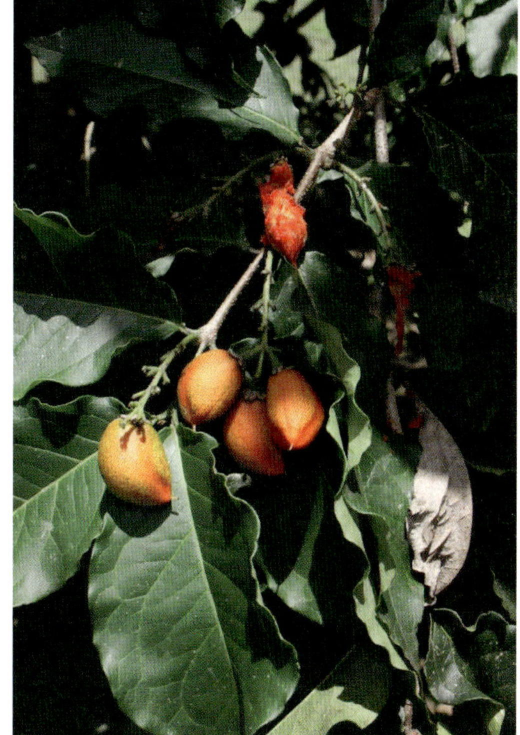

Plate 4 Ripe Bunchosia fruits and exposed fruit showing the orangey-red pulp

Edible Plant Parts and Uses

The fruit pulp is sweet and creamy and has a flavour similar to that of peanut butter, hence the name. Ripe fruit is mostly eaten fresh and is also used for jellies, jams, muffins, or preserves and to flavour drinks and milk shakes. The fruit can be refrigerated and the pulp can be frozen.

Botany

Bunchosia armeniaca, is a small, highly ornamental and hardy tropical, evergreen perennial tree or shrub, up to 5 m tall, with multiple trunks. Leaves are simple, opposite, ovate or oblong, chartaceous, up to 17 cm long and 10 cm wide with undulating margin (Plates 2, 3 and 4) and frequently have paired glands on the petiole or base of the blade and epipetiolar stipules. The flowers are mostly bisexual, actinomorphic,

Plate 5 Close-view of ripe Bunchosia fruits

12–15 mm across and borne in compact, axillary racemose inflorescences. Sepals, 5, persistent with one or more of sepals usually have one or more conspicuous glands. Petals, 5, lemon-yellow, 5–6 mm long, usually clawed and the limb is typically fringed or lacerate (Plate 1). Stamens usually 10, distinct or basally connate stamens in two whorls with half of them commonly reduced to staminodes. The gynoecium consists of a single compound pistil of almost always 3 carpels, 3 distinct styles, and a superior ovary with 3 locules, each containing a single pendulous, axile ovule. Fruits borne in clusters. Fruit is an indehiscent, 2.5 cm long, ovoid, ellipsoid to obovate berry with pale green exocarp (Plate 2) turning orange, or red at maturity (Plates 3, 4 and 5) and with 2 or 3 one-seeded pyrenes embedded in orangey-red or red-coloured pulp (Plate 4).

Nutritive/Medicinal Properties

No information has been published on its nutritive and medicinal value and uses.

Other Uses

The tree is mainly cultivated as a fruit tree and as an ornamental tree in gardens.

Comments

The fruit is usually propagated from seeds.

Selected References

Anderson WR (1981) Malpighiaceae. In: The Botany of the Guayana Highland—Part XI. Mem. New York Botanical Garden, 32: pp 21–305

Anderson WR (2001) Malpighiaceae. In: Berry PE, Yatskievych K, Holst BK (eds) Flora of the Venezuelan Guayana, vol 6. Missouri Botanical Garden Press, St. Louis, pp 82–185

Anderson WR (2007) Malpighiaceae. In: Hammel BE, Grayum MH, Herrera C, Zamora N (eds) Manual de Plantas de Costa Rica, vol 6. Monographs in Systematic Botany from the Missouri Botanical Garden, 111: pp 253–312

Anderson WR, Anderson C, Davis CC (2006) Malpighiaceae. http://herbarium.lsa.umich.edu/malpigh

Campbell RJ (1996) South American fruits deserving further attention. In: Janick J (ed) Progress in new crops. ASHS Press, Arlington, pp 431–439

Facciola S (1990) *Cornucopia.* A source book of edible plants. Kampong Publ., Vista, 677 pp

Fouqué A (1973) Espèces fruitières d'Amérique tropicale. XV. Fruits 28(7–8):548–558

Lorenzi H, Bacher L, Lacerda M, Sartori S (2006) Brazilian fruits & cultivated exotics (for consuming in natura). Instito Plantarum de Etodos da Flora Ltda, Brazil, 740 pp

Porcher MH et al (1995–2020) Searchable world wide web multilingual multiscript plant name database. The University of Melbourne, Australia.http://www.plantnames.unimelb.edu.au/Sorting/Frontpage.html

Malpighia emarginata

Scientific Name

Malpighia emarginata DC.

Synonyms

Malpighia berteroana Spreng., *Malpighia biflora* Poir., *Malpighia dicipiens* Sessé & Moc., *Malpighia emarginata* Moç. & Sessé ex DC., *Malpighia fallax* Salisb., *Malpighia glabra* var. *acuminata* A. Juss., *Malpighia glabra* var. *antillana* Urb. & Nied., *Malpighia glabra* L., *Malpighia glabra* var. *guatemalensis* Nied., *Malpighia glabra* var. *lancifolia* Nied., *Malpighia glabra* var. *typica* Nied., *Malpighia glabra* var. *undulata* (A. Juss.) Nied., *Malpighia lanceolata* Griseb., *Malpighia lucida* Pav. ex A. Juss., *Malpighia lucida* Pav. ex Moric., *Malpighia myrtoides* Moritz ex Nied., *Malpighia neumanniana* A. Juss., *Malpighia nitida* Mill., *Malpighia oxycocca* var. *biflora* (Poir.) Nied., *Malpighia peruviana* Moric., *Malpighia punicifolia* L., *Malpighia punicifolia* var. *lancifolia* Nied., *Malpighia punicifolia* var. *obovata* Nied., *Malpighia punicifolia* var. *vulgaris* Nied., *Malpighia retusa* Benth., *Malpighia semeruco* A. Juss., *Malpighia umbellata* Rose, *Malpighia undulata* A. Juss., *Malpighia uniflora* Tussac, *Malpighia urens* var. *lanceolata* (Griseb.) Griseb., *Malpighia virgata* Pav.

Family

Malpighiaceae

Common/English Names

Acerola, Barbados Cherry, Dwarf Barbados Cherry, Garden Cherry, Huesito, Indian Cherry, Manzanita, Native Cherry, Puerto Rico Cherry, Surinam Cherry, West Indian Cherry

Vernacular Names

Brazil: Acerola, Cereja-Das-Antilhas, Cereja-Do-Pará;
Barbados: Acerola, Buesito (Spanish);
Czech: Malipigie Lysá;
Danish: Barbadoskirsebær;
Dutch: Geribde Kers;
Eastonian: Sile Malpiigia;
French: Acerolier, Cerisier Carré, Cerise De Cayenne, Cerise De St. Domingue, Cerisier De Barbade, Cerisier Des Antilles, Cerisier De St. Domingue;
German: Acerola-Kirsche, Antillenkirsche, Barbados-Kirsche, Westindische Kirsche;
India: Vallaria Simeyaranelli;
Panama: Cereza, Cereza Colorada, Cereza De La Sabana, Grosella (Spanish);

T.K. Lim, *Edible Medicinal And Non-Medicinal Plants: Volume 3, Fruits*,
DOI 10.1007/978-94-007-2534-8_20, © Springer Science+Business Media B.V. 2012

Papiamento: Whimaruku Machu;
Philippines: Malpi (Tagalog);
Portuguese: Cerejeira, Cerejeira Das Atilhas, Cereja Do Pará , Gineira Da Jamaica;
Spanish: Acerola, Cereza De Barbados, Cereza Acerola De Las Antillas, Cereza De Jamaica, Camaroncito, Chercese, Heusito Semeruco, Escobillo;
Surinam: Shimarucu (Dutch);
Thai: Choeri (Bangkok);
Venezuela: Cemeruco, Semeruco (Spanish);
Vietnamese: So'ri.

Origin/Distribution

Acerola is indigenous to the Lesser Antilles from St. Croix to Trinidad, also Curacao and Margarita and neighbouring Central America including Mexico and southern Texas and northern South America as far south as Bahia in Brazil. It has become naturalized in Cuba, Jamaica and Puerto Rico after cultivation, and is commonly grown in house yards in the Bahamas and Bermuda, and to a lesser extent in Central and South America. It is now cultivated globally in the tropics and subtropics in Australia, Brazil, Canary Islands, Cuba, Ethiopia, French Guiana, Ghana, Hawaii, India, Indonesia, Jamaica, Madagascar, Pakistan, Peru, Philippines, Puerto Rico, Sri Lanka, Surinam, Taiwan, Thailand, Myanmar, Venezuela, Vietnam and Zanzibar.

Agroecology

Acerola is adaptable to climatic conditions in the tropics and subtropics. Mature tree can withstand short periods of frost down to –2°C but young plants are killed by frost at –1°C. Acerola is naturally adapted to both medium- and low-rainfall regions. It cannot tolerate water-logging and is drought tolerant but adequate water supply is essential for good yield. Acerola does best on deep, well-drained, friable and rich soils from pH 5.6–6.5. Acidic soils below pH 5.6 require liming to avoid calcium deficiency and to increase yield while calcareous soils above pH 6.5 need micronutrient supplementation.

Edible Plant Parts and Uses

The acid fruit can be eaten fresh out of hand but is usually processed into juice, puree, syrup, jelly, jams and other preserves. The puree may be dried or frozen for future use. The puree can be used as a topping on cake, pudding, ice cream or sliced bananas, or used in other culinary products. The fresh juice can be used in punch, sherbet and is used as a vitamin C supplement to other fruit juices and beverages. The fresh juice will prevent darkening of bananas sliced for fruit cups or salads. The juice is also used in gelatine desserts, sweets, ice and made into liquors and wine. Wine made from Acerola cherries in Hawaii was found to retain 60% of the ascorbic acid. Acerola was used as one of a combination of four flavours: Açai *(Euterpe oleracea)*, pomegranate and blueberry for vodka. Acerola flavour is also used in Tic Tac (mint sweet) dragées (sugar-coated confection).

A blended beverage based on green coconut (*Cocos nucifera*) water, pineapple (*Ananas comosus*) and acerola (*Malpighia emarginata*) pulps as a ready to drink product was recently developed (Pereira et al. 2009). The formulation composed of 65% green coconut water, 15% of pineapple and 20% acerola pulp was selected based on the best combination of nutritional components.

Botany

An evergreen small tree or bushy shrub reaching heights of 3–6 m with hairy branches and a short trunk of 10 cm diameter. The leaves are simple, bright glossy green, ovate-lanceolate (Plates 1 and 2), 5–10 cm long by 2.5–5.5 cm wide, emarginate, apiculate, obtuse or rounded at the apex, acute or cuneate at the base, undulating margin, tomentose when young becoming glabrous. The flowers are produced in sessile or short-peduncled,

Plate 1 Developing unripe and ripe fruits and emarginate leaves of acerola

Plate 2 Ripe fruit, flower and leaves of acerola

Plate 3 Close up of flower

axillary or terminal cymes. Flowers are 1–1.5 cm across, bisexual, with 5 glandular sepals, 5 pink or lavender, spoon-shaped, fringed and clawed petals (Plates 2 and 3), stamens with long

Plate 4 Ripe acerola fruit

Plate 5 Harvested acerola fruit for sale in the local market

filaments connate at the base and a trilocular ovary with 3 free styles. Fruit is a drupe, oblate to round, shallowly 3-lobed with a thin epicarp which turns from green to pale greenish yellow to pink and glossy red when ripe (Plates 1, 2, 4, and 5). The seeds are embedded in the orange-coloured, juicy and acid pulp. The 3 small, rounded seeds each have 2 large and 1 small fluted wings forming triangular, corrugated inedible "stones".

Nutritive/Medicinal Properties

Nutrient value of raw, acerola fruit per 100 g edible portion (exclude 20% refuse comprising 18% seed and 2% stem end) was reported as follows U.S. Department of Agriculture and Agricultural Research (2010) water 91.41 g, energy 32 kcal (134 kJ), protein 0.4 g, total lipid (fat) 0.30 g, ash

0.20 g, carbohydrate 7.69 g; total dietary fibre 1.1 g, minerals – calcium 12 mg, iron 0.20 mg, magnesium 18 mg, phosphorus 11 mg, potassium 146 mg, sodium 7 mg, zinc 0.10 mg, copper 0.086 mg, Se 0.6 μg; vitamins - vitamin C (total ascorbic acid) 1677.6 mg, thiamin 0.020 mg, riboflavin 0.060 mg, niacin 0.400 mg, pantothenic acid 0.309 mg, vitamin B-6 0.009 mg, folate (total) 14 μg, vitamin A 767 IU, lipids - fatty acids (total saturated) 0.068 g, 14:0 (myristic acid) 0.002 g, 16:0 (palmitic acid) 0.048 g, 18:0 (stearic acid) 0.016 g; fatty acids (total monounsaturated) 0.082 g, 16:1 undifferentiated (palmitoleic acid) 0.001 g, 18:1 undifferentiated (oleic acid) 0.081 g; fatty acids (total polyunsaturated) 0.090 g, 18:2 undifferentiated (linoleic acid) 0.046 g and 18:3 undifferentiated (linolenic acid) 0.044 g.

Analyses conducted on acerola fruit in Brazil reported that vitamin C content was very high ranging from 695–4,827 mg/100 g much higher than other fruits like pineapple, araçá (*Eugenia stipitata*), cashew, guava, kiwi, orange, lemon, and strawberry (Mezadri et al. 2006). The other nutritive constituents were determined as follows: proteins (0.21–0.80 g/100 g), fats (0.23–0.80 g/100 g), carbohydrates (3.6–7.80 g/100 g), mineral salts (iron 0.24, calcium 11.7, phosphorus 17.1 mg/100 g) and vitamins (thiamin 0.02, riboflavine 0.07, pyridoxine 8.7 mg/100 g). Acerola was also reported to have carotenoids and bioflavonoids which contributed to its high antioxidant capacity and provide important nutritive value. Another analysis reported that the average content of vitamin C was 1.79 g/100 g of pulp, it was higher than that in other fruits, like pineapple, araçá, cashew, guava, kiwi, orange, lemon, and strawberry but lower than the camu-camu sylvestral fruit of Amazônia. The contents of moisture, carbohydrate, fibre, lipids and minerals in the acerola were not significantly different when compared to other fruits (Visentainer et al. 1997).

In the ripe acerola fruit (*Malpighia emarginata*), four major carotenoids were identified (β-carotene, β-cryptoxanthin, lutein, and violaxanthin) together with other minor carotenoids (neoxanthin, antheraxanthin, neochrome, luteoxanthin, auroxanthin, β-cryptoxanthin-5,6-epoxide, β-cryptoxanthin-5,8-epoxide, cis-β-carotene, and cis-lutein) (Mezadri et al. 2005). An average composition for the ripe fruit per 100 g fresh weight was estimated as follows: β-carotene 536.55 g, β-cryptoxanthin 417.46 g, lutein 99.21 g, violaxanthin 395.33 g and total minor carotenoids 197.33 g. Vitamin A values are similar to those described in tomatoes and some tropical fruits such as guava and papaya. After juice-making, including a pasteurization stage as thermal processing, decreases in carotenoid content were observed as well as progress of cis-isomers and structural rearrangement of xanthophylls containing 5,6 epoxide groups. Two anthocyanins, cyanidin-3-α-O-rhamnoside (C3R) and pelargonidin-3-α-O-rhamnoside (P3R), and quercitrin (quercetin-3-α-O-rhamnoside), were isolated from fruit (Hanamura et al. 2005).

One hundred and fifty constituents were identified in the aroma concentrate of acerola fruit, from which furfural, hexadecanoic acid, 3-methyl-3-butenol, and limonene were found to be the major constituents (Pino and Marbot 2001). The amounts of esters, 3-methyl-3-butenol, and their various esters were thought to contribute to the unique flavour of the acerola fruit.

Numerous scientific studies conducted showed acerola to possess potent antioxidant, antihyperglycaemic, anticancer, skin-whitening and antimicrobial activities.

Antioxidant Activity

Acerola is known to be one of the best fruits rich in ascorbic acid and polyphenols. Studies reported that acerola polyphenols such as anthocyanin, cyanidin-3-α-O-rhamnoside (C3R) and quercitrin (quercetin-3-α-O-rhamnoside), isolated from acerola fruit, exhibited strong radical scavenging activity (Hanamura et al. 2005). A novel flavonoid, leucocyanidin-3-O-β-D-glucoside, trivial name "aceronidin" isolated from green mature acerola puree exhibited stronger 1,1-diphenyl-2-picrylhydrazyl (DPPH) radical quenching activity than that of α-tocopherol and comparable to that of flavonoids (Kawaguchi et al. 2007). In the yeast α-glucosidase inhibitory assay, aceronidin showed significantly greater inhibition than the other flavonoids tested. In the human salivary α-amylase inhibitory assay, aceronidin showed inhibition activity. Thus, these results indicated aceronidin to be a potent antioxidant that may be valuable as an inhibitor of sugar catabolic enzymes. Experimental evidence showed that acerola fruit extract had strong inhibitory effect on nitric oxide production in activated macrophages which could be partly attributable to the inhibition of iNOS (inducible nitric oxide synthase) expression, and scavenging of O^{2-} and NO radicals (Wakabayashi et al. 2003).

Among 11 frozen fruits' pulps tested for its antioxidant potency, acerola was the highest scoring domestic fruit, meaning it had the most antioxidant potency, with a TEAC (Trolox equivalent antioxidant activity) score of 53.2 mmol/g (Kuskoski et al. 2006). The descending order of antioxidant capacity was acerola > mango > strawberry > grapes > açaí (*Euterpe oleracea*) > guava > mulberry > graviola (*Annona muricata*) > passion fruit > cupuaçu (*Theobroma grandiflorum*) > pineapple. A detailed nutrition analysis showed acerola juice contained 32 times the amount of vitamin C in orange juice (over 3,000% as much).

Antihyperglycaemic Activity

Studies found that the inhibitory profiles of isolated polyphenols from acerola fruit such as cyanidin-3-α-O-rhamnoside (C3R) and pelargonidin-3-α-O-rhamnoside (P3R), except quercitrin (quercetin-3-α-O-rhamnoside) towards α-glucosidase activity were low; all polyphenols strongly inhibited advanced glycation end product (AGE) formation (Hanamura et al. 2005). In separate studies, the crude acerola polyphenol fraction (C-AP) was found to significantly suppress the plasma glucose level after administering both glucose and maltose, suggesting that C-AP had a preventive effect on hyperglycemia in the postprandial state (Hanamura et al. 2006). The mechanism for this effect was considered to have been both suppression of the intestinal glucose transport and the inhibition of α-glucosidase.

Anticancer/Antimicrobial Activities

Japanese scientists showed that various organic solvent fractions of acerola fruit exhibited varying degrees of cytotoxicity and antibacterial activities (Motohashi et al. 2004). Higher cytotoxic activity was concentrated in fractions A4 and A6 (acetone extract), and H3 and HE3 (hexane extract). These four fractions showed higher cytotoxic activity against tumour cell lines such as human oral squamous cell carcinoma (HSC-2) and human submandibular gland carcinoma (HSG), when compared with that against normal cells such as human periodontal ligament fibroblasts (HPLF) and human gingival fibroblasts (HGF). HE2 (hexane extract), AE2 (ethyl acetate extract), AE3, AE4, AE5, A8, A9 and A10 showed relatively higher anti-bacterial activity on the Gram-positive *Staphylococcus epidermidis* ATCC 1,228 but were ineffective on the representative Gram-negative species *Esherichia coli* and *Pseudomonas aeruginosa*. The fractions were inactive against *Helicobacter pylori*, two representative *Candida* species, and human immunodeficiency virus (HIV). H3, H4 and HE3, which displayed higher tumour-specific cytotoxicity, also showed higher multidrug resistance (MDR) reversal activity, than (±)-verapamil as positive control. The tumour specific cytotoxic activity and MDR reversal activity of acerola indicated its possible application for cancer therapy.

Skin Whitening Activity

Hanamura et al. (2008) also reported separately that crude acerola polyphenols (C-AP) had skin-lightening activity. They showed that C-AP significantly lightened the UVB-irradiated skin pigmentation. Furthermore, treatment with C-AP reduced the content of melanin in B16 melanoma cells, suggesting that the in-vivo skin-lightening effect of C-AP was due to the suppression of melanin biosynthesis in melanocytes. In addition, the scientists found that C-AP could effectively inhibit mushroom tyrosinase activity, the main constituents responsible for this effect were thought to be anthocyanins such as cyanidin-3-α-O-rhamnoside (C3R) and pelargonidin-3-α-O-rhamnoside (P3R). Their findings indicated that the skin-lightening effect of C-AP could be partly attributed to the suppression of melanogenesis through the inhibition of tyrosinase activity in melanocytes. They asserted that an oral ingestion of crude acerola polyphenol may therefore be efficacious for reducing UVB-induced hyper-pigmentation by inhibiting the tyrosinase in melanocytes. Based on its tyrosinase inhibitory activity, acerola showed potential application in skin whitening in the cosmetic industry. Acerola fruit is being used in various cosmetic and health products.

Toxicological Study

In a toxicological evaluation (Hanamura and Aoki 2008), the total polyphenol content of crude acerola polyphenols (C-AP) was found to be 57.7% with the main polyphenols being proanthocyanidin and cyanidin-3-α-O-rhamnoside In the acute oral toxicological test, no deaths or abnormalities at necropsy on day 14 were observed, confirming that the minimum fatal dose of C-AP is greater than 2,000 mg/kg body weight. In both subacute and subchronic toxicological tests, no death was recorded and the body weights and food intakes of the rats did not differ significantly from the control groups. Besides, there were no abnormal clinical signs related to administration of C-AP in any of the experimental animals. These results provided an important reference for the safety of acerola polyphenols as a food supplement for human consumption.

Traditional Medicinal Uses

In folk medicine, acerola fruits are considered beneficial to patients with liver ailments, diarrhoea and dysentery, as well as those with coughs or colds. Acerola juice is used as a gargle to relieve sore-throat.

Other Uses

Acerola makes a good hedge plant and has become a poplar bonsai plant especially in Taiwan. The bark contains 20–25% tannin and has been employed in the leather industry. It provides a hard and heavy timber.

Comments

Erstwhile, *M. emarginata* and *M. glabra* were considered as different species, the former has leaves with emarginate tips and the latter with glabrous foliage, but now *M. glabra* is considered by most authors to be a synonym of *M. emarginata*. Both species have emarginate leaves and rounded or apiculate leaves on the same tree.

Selected References

Backer CA, Van der Bakhuizen Brink RC Jr (1963) Flora of Java, vol 1. Noordhoff, Groningen, 648 pp

Council of Scientific and Industrial Research (1962) The wealth of India. A dictionary of Indian raw materials and industrial products. (Raw materials 6). Publications and Information Directorate, New Delhi, India

Hanamura T, Aoki H (2008) Toxicological evaluation of polyphenol extract from acerola (*Malpighia emarginata* DC.) fruit. J Food Sci 73(4):T55–T61

Hanamura T, Hagiwara T, Kawagishi H (2005) Structural and functional characterization of polyphenols isolated from acerola (*Malpighia emarginata* DC.) fruit. Biosci Biotechnol Biochem 69(2):280–286

Hanamura T, Mayama C, Aoki H, Hirayama Y, Shimizu M (2006) Antihyperglycemic effect of polyphenols from acerola (*Malpighia emarginata* DC.) fruit. Biosci Biotechnol Biochem 70(8):1813–1820

Hanamura T, Uchida E, Aoki H (2008) Skin-lightening effect of a polyphenol extract from acerola (*Malpighia emarginata* DC.) fruit on UV-induced pigmentation. Biosci Biotechnol Biochem 72(12):3211–3218

Harjadi SS (1992) *Malpighia glabra* L. In: Verheij EWM, Coronel RE (eds) Plant resources of South-East Asia, no 2. Edible fruits and nuts. Prosea Foundation, Bogor, pp 198–200

Kawaguchi M, Tanabe H, Nagamine K (2007) Isolation and characterization of a novel flavonoid possessing a 4, 2''-glycosidic linkage from green mature acerola (*Malpighia emarginata* DC.) fruit. Biosci Biotechnol Biochem 71(5):1130–1135

Kuskoski EM, Asuero AG, Morales MT, Fett R (2006) Roseane. Frutos tropicais silvestres e polpas de frutas congeladas: atividade antioxidante, polifenóis e antocianinas. Cienc Rural 36(4):1283–1287 (Wild fruits and pulps of frozen fruits: antioxidant activity, polyphenols and anthocyanins)

Ledin RB (1958) The Barbados or West Indian cherry. Fla Agric Exp Stn Bull 594:1–28

Lorenzi H, Bacher L, Lacerda M, Sartori S (2006) Brazilian fruits & cultivated exotics (for consuming in Natura). Instituto Plantarum de Etodos da Flora Ltda, Brazil, 740 pp

Mezadri T, Pérez-Gálvez A, Hornero-Méndez D (2005) Carotenoid pigments in acerola fruits (*Malpighia emarginata* DC.) and derived products. Eur Food Res Technol 220(1):63–69

Mezadri T, Fernández-Pachón MS, Villaño D, García-Parrilla MC, Troncoso AM (2006) The acerola fruit: composition, productive characteristics and economic importance. Arch Latinoam Nutr 56(2):101–109, In Spanish

Morton JF (1987) Barbados cherry. In: Fruits of warm climates. Julia F. Morton, Miami, pp 204–207

Moscoco CG (1956) West Indian cherry - richest known source of natural vitamin C. Econ Bot 10:280–294

Motohashi N, Wakabayashi H, Kurihara T, Fukushima H, Yamada T, Kawase M, Sohara Y, Tani S, Shirataki Y, Sakagami H, Satoh K, Nakashima H, Molnár A, Spengler G, Gyémánt N, Ugocsai K, Molnár J (2004) Biological activity of Barbados cherry (acerola fruits, fruit of *Malpighia emarginata* DC) extracts and fractions. Phytother Res 18(3):212–223

Ostendorf FW (1963) The West Indian cherry. Trop Abstr 18(3):145–150

Pereira AC, Siqueira AM, Farias JM, Maia GA, Figueiredo RW, Sousa PH (2009) Development of mixed drink of coconut water, pineapple and acerola pulp. Arch Latinoam Nutr 59(4):441–447, In Spanish

Pino JA, Marbot R (2001) Volatile flavour constituents of acerola (*Malpighia emarginata* DC.) fruit. J Agric Food Chem 49(12):5880–5882

Purseglove JW (1968) Tropical crops: Dicotyledons 1 & 2. Longman, London, 719 pp

Tropicos Org. Missouri Botanical Garden Jan 2009. http://www.tropicos.org

U.S. Department of Agriculture, Agricultural Research Service (2010) USDA National Nutrient Database for Standard Reference, Release 23. Nutrient Data Laboratory Home Page, http://www.ars.usda.gov/ba/bhnrc/ndl

Visentainer JV, Vieira OA, Matsushita M, de Souza NE (1997) Physico-chemical characterization of acerola (*Malpighia glabra* L.) produced in Maringá, Paraná state, Brazil. Arch Latinoam Nutr 47(1):70–72, In Portuguese

Wakabayashi H, Fukushima H, Yamada T, Kawase M, Shirataki Y, Satoh K, Tobe T, Hashimoto K, Kurihara T, Motohashi N, Sakagami H (2003) Inhibition of LPS-stimulated NO production in mouse macrophage-like cells by Barbados cherry, a fruit of *Malpighia emarginata* DC. Anticancer Res 23(4):3237–3241

Abelmoschus esculentus

Scientific Name

Abelmoschus esculentus (L.) **Moench.**

Synonyms

Abelmoschus bammia Webb., *Abelmoschus multiformis* Wall., *Hibiscus esculentus* L., *Hibiscus ficifolius* Mill., *Hibiscus hispidissimus* Cheval., *Hibiscus longifolius* Roxb., *Hibiscus praecox* Forssk.

Family

Malvaceae

Common/English Names

Bindi, Gobbo, Gombo, Gumbo, Lady's Finger, Ladies' Fingers, Okra, Okro

Vernacular Names

Angola: Ngumbo; Kingombo (Kimbundu), Quiabo (Portuguese);
Arabic: Bâmiyah;
Argentina: Gombo, Ají Turco, Quimbombo, Ocra, Ruibarbo;
Belgium: Ketmie Comestible;
Benin: Fétri (Adja), Yabonou, Kpéwoko (Bariba), Ila (Dassa), La (Dendi), Wannan (Dompargo), Févi (Fon), Gombo (French), Fétri (Mina), Gniéhoun (Popo), Gninhoun (Sahoué);
Brazil: Quiabo (Portuguese);
Burkina Faso: Gombo;
Burmese: You Padi;
Cameroon: Ankoul (Bamiléké), Kingombo (Bantu);
Chinese: Ka Fei Kui, Ka Fei Kuie, Ch'aan K'e Ts'au Kw'ai, Ka Fei Huang Kui, Huang Su Kui, Huang Qiu Kui, Qiu Kui, Ch'iu K'ui;
Croatian: Jedilna Oslez;
Czech: Ibišek Jedlý;
Danish: Abelmoskus, Almindelig Okra, Hibiscus Art;
Democratic Republic Of Congo: Umvumba (Kinyarwanda), Ngaingai (Swahili);
Dominican Republic: Molondrón;
Dutch: Okra;
Eastonian: Söödav Muskushibisk;
Egypt: Bamya;
Ethiopia: Bamia;
French: Bamie-Okra, Gombo, Gombeaud, Gumbo, Ocra, Oseille De Gombo, Ketmie Comestible, Ketmie Gombo;
German: Bisameibisch, Essbarer Eibisch Ocker, Okra;
Ghana: Nkuruma (Asante-Twi);
Greek: Bamia;
Hebrew: Bamiya, Hibiscus Ne'echal;
Hungarian: Gombó;

T.K. Lim, *Edible Medicinal And Non-Medicinal Plants: Volume 3, Fruits*,
DOI 10.1007/978-94-007-2534-8_21, © Springer Science+Business Media B.V. 2012

India: Bhendi (Assamese), Bendi, Bhindi, Bhindi-Tori, Ram-Turi, Ramturai, Ram-Turi, Ram-Turai, Bhendi (Hindu), Bende-Kayi, Bhendekayi, Bende Kaayi, Bende Kayi, Bende Kaayi Gida, Bende Naaru (Kannada), Venda, Venta, Ventak-Kaya (Malayalam), Bhelendri Bhelendri (Manipuri), Bhendi, Ram-Turai, Bhajichi-Bhendi, Benda, Bhendo (Marathi), Bawrhsaiabe (Mizoram), Bendi (Oriya), Bhenda, Darvika, Gandhamula, Pitali, Tindisha (Sanskrit), Venaikkay, Vendaik-Kay, Vendi, Vendaikkai, Ventai, Vendai (Tamil), Benda-Kaya, Bendakaya, Vendakaya, Benda, Penda, Venda (Telugu), Bhindi (Urdu);
Indonesia: Okra, Kopi Arab;
Italian: Gombo, Ocra, Bammia D'egitto, Corna Di Greci;
Ivory Coast: Eponoufa (Aboure), Zabré (Gouro), Zapoya (Shien), Gombo;
Japanese: Okura, Amerika Neri, Kiku Kimo;
Khmer: Poot Barang;
Korean: Oh K'u Ra;
Laotian: Khua Ngwang;
Madagascar: Gombo, Mana;
Malaysia: Kacang Bendi, Sayur Bendi, Kacang Lender, Bendir;
Mexico: Angú, Chimbombó, Okra, Algalia;
Morocco: Mlûhiya (Moroccan), Gombo, Corne-Grec, Lady-Finger (French);
Mozambique: Monhatando;
Nepal: Ram Toriya, Van Lasun;
Niger: Gombo;
Niger-Congo: Kingombo (Bantu);
Nigeria: Okuru (Igbo), Kingombo (Bantu) Illa;
Norwegian: Grønsakhibisk;
Panama: Ñajú;
Papiamento: Guiambo;
Persian: Bamiyah;
Philippines: Okra, Saluyot A Bunga, Haluyot;
Polish: Czyli Okra, Ketmia Jadalna, Ketmia Czerwona (Red-Podded), Ketmia Zielona (Green-Podded);
Portuguese: Quiabo, Quingombo, Quiabeiro;
Romanian: Bamă;
Russian: Bamija;
Sierra Leone: Okro (Krio), Bonde, Bondei (Mende), A Lontho (Tyemne);
Senegal: Ñaod (Badyara), Va-Tienega, Va-Tieneka (Basari), Gi-Nyúwuď (Bedik), Kãndia (Crioulo), Kunéga, Kunégo (Diola);
Spanish: Bendee, Gombo, Guigambó, Molondrones, Ouiabeiro, Quingombo, Quimbombó;
Swedish: Okra;
Thai: Krachiap-Khieo, Krachiap-Mon, Bakhus-Mun;
Togo: Gombo;
Turkish: Bamya;
Vietnamese: Dau Bap, Bup Bap Muw Owp Tay;
West Africa: Nkruman;
Zimbabwe: Idelele (Ndebele), Derere, Derere Rechipudzi (Shona).

Origin/Distribution

There is mixed views on okra's origin. One view is that it is native to Asia from North India. Another view is that it originated from Africa – Egypt, Ethiopia or West Africa. In Egypt okra has been cultivated since 2,000 BC. Okra is now much cultivated throughout the tropical and subtropical regions mainly for culinary purposes. Today, okra is mainly cultivated in India, Turkey and Greece and also in Southeast Asia.

Agroecology

Okra thrives in warm tropical or subtropical climate with temperatures above 20°C on well-drained, rich light and heavy soils that include sandy loam, loam or clay soil with pH of 5.8–8. The plant is a short day plant, grows well in full sun and is rather cold tender. It is tolerant to drought and water-logging.

Edible Plant Parts and Uses

The immature fruits, flower buds, flowers and calyces, leaves and young shoots are used as vegetable (greens) and are eaten in various ways.

The tender, immature green pods are eaten and have a unique texture and sweet flavour. The pods are eaten cooked on their own or eaten in soups and curries or sliced and fried with various meat. The fresh young pods can also be lightly sautéed in olive oil and eaten. Okra is also dipped in batter, deep fried and eaten. Okra is also great when pickled. Okra is also dried and powdered to use as a thickener. In India, the pods are dried, sliced into sections, and then fried for a crunchy, almost bread-like snack. The pods, when cut, exude mucilage that is used to thicken soups, stews (gumbo) and sauces. In Vietnam, okra is the important ingredient in the dish *canh chua*. In Malaysia and Singapore, one popular Nyonya recipe consist of slicing cooked okra served with pounded dried shrimps and *sambal belachan* (pounded chillies and fermented prawn paste). Young fruits are added to sweet and sour curry. In Indonesia, immature fruits and young, tender shoots are eaten as *lalab* with rice. They are also frequently used for *sayor* and *tumis*. In Thailand, young fruits are eaten raw or cooked by steaming, boiling or blanching and served with chili sauces. In Japan, okra is popularly served with soy sauce and *katsuobushi* or as tempura. In India and Pakistan, okra is cut into pieces and stir fried with red onions and spices, sautéed or added to gravy-based preparations like *Bhindi Ghosh* or *sambar* in south India. In the Middle East countries and Greece, okra is widely used in a thick stew made with vegetables and meat.

Okra is also relished in Nigeria where okra soup (Draw soup) is a special delicacy with *Garri(eba)* or akpu. *Frango com quiabo* (chicken with okra) is a Brazilian dish that is especially famous in the region of Minas Gerais. In the Caribbean islands, okra is cooked and eaten in soup, often with fish. In Haiti, okra is cooked with rice and maize; and also used as a sauce for meat. It is also an important ingredient in *callaloo*, a Caribbean dish and the national dish of Trinidad & Tobago. Breaded, deep fried okra is popularly served in the southern United States. Gumbo, a hearty stew whose key ingredient is okra, is found throughout the Gulf Coast of the United States and in the South Carolina low country.

The young leaves flower buds, flowers and calyces can be eaten cooked as greens (spinach). Young, tender leaves are also eaten raw in salads and used for flavouring. The leaves can be dried, crushed into a powder and stored for later use. Seeds are cooked or ground into a meal and used in making bread or made into 'tofu' or 'tempeh'. The mature seeds are toasted, ground and used as a non-caffeinated substitute for coffee. Okra seeds also provide good quality edible oil high in unsaturated fats such as oleic acid and linoleic acid and protein. The root is edible but very fibrous and mucilaginous, without very much flavour.

Botany

An annual, tropical, erect herb with a robust terete stem growing to 1–2 m high with tender parts covered with bristles. Leaves spirally arranged, long-petioled (50 cm), pubescent, 40–50 cm long and broad, palmately lobed with 3–7 oblong-lanceolate lobes of variable depth, coarsely serrate, 5–7 nerved, hispid and with subulate, caduceus stipules (Plate 1). Flowers are axillary, solitary, very large (4–8 cm diameter) and borne on long (5–7 cm) pedicel. Epicalyx with 7–12 bracteoles, linear-lanceolate, acute and pubescent. The calyx is spathaceous, 5-lobed and deciduous,

Plate 1 Okra fruit and palmately-lobed leaves

Plate 2 Heap of okra fruit on sale in a local market

Plate 3 Close-up of okra fruits

the corolla has 5 obovate petals, pale yellow with a red or purple spot at the base of each petal, staminal tube straight, erect, yellowish-white with anthers throughout the length, style with short branches and 5–9 stigmas. The fruit is a cylindrical capsule, 5–35 cm long, with longitudinal ribs down its length, light green, hispid covered with fine hairs (Plates 1, 2 and 3) and contain numerous globular-reniform, 3–6 mm diameter, dark brown or dark grey, tuberculate seeds.

Nutritive/Medicinal Properties

The proximate value per 100 g edible portion of okra (minus 14% refuse consisting of crowns and tips) was reported as: water 90.17 g, energy 31 kcal (129 kJ), protein 2.00 g, total lipid 0.10 g, ash 0.70 g, carbohydrate 7.03 g, total dietary fibre 3.2 g, total sugars 1.2 g, sucrose 0.40 g, glucose 0.13 g, fructose 0.21 g, starch 0.34 g; minerals – Ca 81 mg, Fe 0.8 mg, Mg 57 mg, P 63 mg, K 303 mg, Na 8 g, Zn 0.60 mg, Cu 0.094 mg, Mn 0.990 mg, Se 0.7 μg; vitamins – vitamin C 21.1 mg, thiamine 0.0.2 mg, riboflavin 0.060 mg, niacin 1.0 mg, pantothenic acid 0.245 mg, vitaminB-6 0.215 mg, total folate 88 μg, total choline 12.3 mg, β-carotene 225 μg, vitamin A 375 IU, vitamin A RAE 19 μg, vitamin E (α-tocopherol) 0.36 mg, vitamin K (phylloquinone) 53 μg; total saturated fatty acids 0.026 g, 16:0 (palmitic acid) 0.022 g, 18:0 (stearic acid) 0.003 g; total monounsaturated fatty acids 0.017 g, 18:1 (oleic acid) 0.016 g; total polyunsaturated fatty acids 0.027 g, 18:2 undifferentiated (linoleic acid) 0.026 g, 18:3 undifferentiated (linolenic acid) 0.001 g; phytosterols 24 mg; amino acids – tryptophan 0.017 g, threonine 0.065 g , isoleucine 0.069 g, leucine 0.105 g, lysine 0.081 g, methionine 0.021 g, cystine 0.019 g, phenylalanine 0.065 g, tyrosine 0.087 g, valine 0.091 g, arginine 0.084 g, histidine 0.031 g, alanine 0.073 g, aspartic acid 0.145 g, glutamic acid 0.271 g, glycine 0.044 g, proline 0.045 g, serine 0.044 g; lutein + zeaxanthin 516 μg U.S. Department of Agriculture, Agricultural Research Service (2010).

Okra is a rich source of vitamins A, C and niacin and minerals such as Ca, Mg, K, P and Fe. It is also rich in proteins, amino acids and pectin and is low in calories and total fatty acids. It also contains panthothenic acid, vitamin B-6, β-carotene, phytosterols, lutein and zeaxanthin.

Eleven flavonol glycosides and two anthocyanins were identified from the flower petals of okra (Hedin et al. 1968). The flavonol glycosides were: quercetin 4'-glucoside, quercetin 7-glucoside, quercetin 5-glucoside, quercetin 3-diglucoside, quercetin 4'-diglucoside, quercetin 3-triglucoside, quercetin 5-rhamnoglucoside, gossypetin 8-glucoside, gossypetin 8-rhamnoglucoside, gossypetin 3-glucosido-8-rhamnoglucoside, and the anthocyanins were cyanidin 4'-glucoside and cyanidin 3-glucosido-4' glucoside. The two anthocyanins comprised 28.5% of the flavonoid content of the red flower but only traces in the white flower.

Pre-treatment's of okra seed was found to have an effect on the mineral and the functional properties of the okra seed flour (Adelakun et al. 2010). Soaking reduced all mineral investigated and are time dependent. Blanching reduced all mineral content except magnesium. Malting reduced P, K, Mg and Fe, while increase in Ca, Na, Zn and Mn were observed. Roasting increased all the mineral content except phosphorus and magnesium. Functional properties showed that all pre-treatment's resulted in increase in water and oil absorption capacities, decrease in emulsion ability and stability and decrease in foam capacity and stability except malting, which showed an increase in foam capacity.

Several pharmacological properties of various parts of the plant have been reported and are discussed below.

Antioxidant Activity

The major antioxidant compounds in okra fruit were identified to be quercetin derivatives and (−)-epigallocatechin (Shui and Peng 2004). It was found that about 70% of total antioxidant activity was contributed by four quercetin derivatives: quercetin 3-O-xylosyl (1''' -->2'') glucoside, quercetin 3-O-glucosyl (1''' -->6'') glucoside, quercetin 3-O-glucoside and quercetin 3-O-(6''-O-malonyl)-glucoside.

In-vitro antioxidant assay of methanol extract of okra fruits showed potent antioxidant/radical scavenging activities with 50% inhibitory concentration values of 25 and 43 μL when analyzed by the xanthine oxidase and 2-deoxyguanosine methods, respectively (Atawodi et al. 2009). Total phenolic content of okra pulp and seed extracts were found to be 10.75 mg GAE/100 g extract and 142.48 mg GAE/100 g extract respectively and these corresponded with scavenging activities. The predominant phenolic compound was procycanidin B2, followed by procycanidin B1 and rutin in seeds but in okra pulp the predominant compounds were identified as catechin, procycanidin B2, epicatechin and rutin.

The chemical composition and antioxidant activity of okra seeds was increased by roasting at 1,600°C for 10–60 minutes (Adelakun et al. 2009a). The range means obtained for protein, fat, ash, fibre and sugar contents were 42.14–38.10, 31.04–17.22, 4.06–3.42, 3.45–3.60 and 8.82–8.65, respectively. The antioxidant activity was significantly increased by roasting, while in vitro digestibility showed that most antioxidative activities were available in the intestinal phase of gastrointestinal tracts. Soaking and blanching also increased the nutrient composition of the seeds. The range mean obtained for protein, fat, ash and fibre contents were 46.10–38.99, 28.08–25.08, 3.95–3.15 and 3.76–3.10, respectively (Adelakun et al. 2009b). However, blanching decreased the antioxidant (DPPH radical scavenging) activity. Okra extract was found to have in-vitro non-enzymatic inhibition of lipid per oxidation in liposome's (Ansari et al. 2005). The okra extract showed a dose dependent activity. The extract also showed a free radical scavenging activity (FRSA).

Hypolipidemic Activity

Abelmoschus esculentus was found to have hypolipidemic activity (Trinh et al. 2008). Cholesterol levels decreased 56.45%, 55.65%, 41.13%, 40.50% and 53.63% respectively in mice orally administered with AE1 (dichloromethane okra plant extract), AE2 (methanol okra plant extract), AE3 (dichloromethane okra fruit extract), AE4 (methanol okra fruit extract) and simvastatin as compared to the tyloxapol–induce hyperlipidemic mice. Triglycerids levels in treated groups had no significant difference as compared to simvastatine group except methanolic extract from fruit (AE4) administered group.

Antiulcer Activity

The anti-ulcer activity of fresh okra fruits also had recently been reported and mucilage from immature fruits demonstrated cytoprotection effect against ethanol and pyloric-ligation (PL) induced ulcers in rats. The protective effects were comparable with that of the reference compound, sucralfate (Jadhav et al. 2008). The potent cytoprotective effects of mucilage was ascribed to its mechanical barrier-forming property as well as anti-acid secreting effects.

Antimicrobial Activity

Pre-treatment of *Helicobacter pylori* with a fresh okra juice preparation inhibited the bacterial adhesion almost completely (Lengsfeld et al. 2004). Lyophilization and reconstitution of an extract solution led to a reduction of this effect. The anti-adhesive qualities of okra were assumed to be due to a combination of glycoproteins and highly acidic sugar compounds making up a complex three-dimensional structure that is fully developed only in the fresh juice of the fruit. The anti-adhesive activity is therefore due to the blocking capacity of specific *Helicobacter* surface receptors that coordinate the interaction between host and bacterium. Neither of the active fractions showed inhibitory effects on bacterial growth in-vitro. The results provide the rationale for the use of okra fruit as a mucilaginous food additive against gastric irritative and inflammatory diseases in Asian medicine.

Hypoglycaemic Activity

Okra mucilage comprised acidic polysaccharides with associated protein and minerals (Woolfe et al. 1977). Hydrolysis of okra mucilage revealed that the polysaccharide was composed of galacturonic acid, galactose, rhamnose and glucose (1.3:1.O:0.1:0.1). The mucilage, named Okra-mucilage F, isolated from the immature fruits of *A. esculentus* is composed of partially acetylated acidic polysaccharide and protein in a ratio of approximately 8.1:1.0 (Tomoda et al. 1980). The mucilage was found to possess a main chain of the repeating structure (1→4)-O-α-(D-galactopyranosyluronic acid)-(1→2)-O-α-L-rhamnopyranose, and half the L-rhamnose residues in the main chain with branches composed of 4-O-β-D-galactopyranosyl D-galactopyranose at position 4. The mucilage exhibited hypoglycaemic activity (Tomoda et al. 1987).

Anti-nutrient Activity

Four trypsin inhibitors with molecular weight about 20,000 were isolated from seeds of okra and named OTI-A, OTI-B, OTI-C and OTI-D (Ogata et al. 1986). All of these compounds inhibited bovine trypsin strongly with the formation of complexes at a 1:1 M ratio and OTI-A, OTI-B, and OTI-C also weakly inhibited bovine chymotrypsin but not simultaneously. Subtilisin Carlsberg, pronase, papain, and porcine pepsin were not inhibited.

Tablet Formulation

A gel forming polysaccharide gum obtained from the pods of *Abelmoschus esculentus* was employed as a mini matrix in a new sustained release tablet formulations of furosemide (a diretic) and diclofenac sodium (a non steroidal anti-inflammatory agent) (Ofoefule and Chukwu 2001). The researchers stated that these new formulations which utilized a plant hydrogel as mini matrix may offer the advantage of simplicity and economy.

Absence of Gossypol Toxicity Effect

Early report that okra contains gossypol was dismissed by recent studies. Studies conducted in Denmark found that only plants belonging to the cotton tribe (Gossypieae) in Malvaceae contained gossypol (detection limit better than 0.001%) (Jaroszewski et al. 1992). In particular, no gossypol could be detected in *Abelmoschus esculentus* (okra), *Hibiscus tiliaceus*, *Hibiscus sabdariffa*, and *Hevea brasiliensis*, which were earlier claimed to contain the compound. Gossypol-containing plants usually produced dextrorotatory gossypol of varying optical purity; an enantiomeric excess of (−)-gossypol was detected in only one plant, *Gossypium barbadense*.

Allergic Dermatitis

Handling of okra fruit or leaves can cause allergic dermatitis. By questionnaire survey, 32 out of 52 workers (61.5%) reported previous or current skin diseases from okra cultivation (Matsushita et al. 1989). The sites of skin lesions were mainly the arms, fingers and fingertips. Positive patch test reactions with preparations of okra leaves or immature pods in 111 workers, compared to 63 control subjects, were significantly higher in okra workers than in controls, ranging from 9.8% to 30.0%. 37 out of 111 workers (33.3%) were diagnosed as having allergic contact dermatitis (n = 17; 15.3%) and irritant contact dermatitis (n = 18; 16.2%) from okra cultivation. Subsequent animal studies showed that the proteolytic enzyme of okra may be responsible for development of skin lesions, and that allergic contact dermatitis may also play a part in addition to irritant contact dermatitis (Manda et al. 1992).

Traditional Medicinal Uses

In Indian ethnomedicine, an infusion of the fruit mucilage is used for treating dysentery and diarrhoea in acute inflammation and irritation of the stomach, bowels, and kidneys catarrhal infections, ardour urinae, dysuria and gonorrhoeaa. The mucilage can be used as a plasma replacement in India. A decoction of the immature fruit is considered demulcent, diuretic and emollient. The leaves and roots are also demulcent, though less so than the fruit. When gathered in their green state and pounded, the pods make a valuable emollient poultice and so also the leaves. In India the seeds are reported to be antispasmodic, cordial and stimulant. An infusion of the roasted seeds has sudorific properties. An infusion of the roots is used in the treatment of syphilis. The juice of the roots is used externally in Nepal to treat cuts, wounds and boils.

Other Uses

The mucilage is used medicinally and technically. Okra is sometimes cultivated as fibre plant; the fibre obtained from the stems is used as a substitute for jute. It is also used in making paper and textiles.

Comments

Gloves should be worn when harvesting the pods because of the hairs on the pods which can be irritating and can cause dermatitis. The hairs are removed by washing.

Selected References

Adelakun OE, Oyelade OJ, Ade-Omowaye BI, Adeyemi IA, Van de Venter M (2009a) Chemical composition and the antioxidative properties of Nigerian okra seed (*Abelmoschus esculentus* Moench) flour. Food Chem Toxicol 47(6):1123–1126

Adelakun OE, Oyelade OJ, Ade-Omowaye BI, Adeyemi IA, Van de Venter M, Koekemoer TC (2009b) Influence of pre-treatment on yield chemical and antioxidant properties of a Nigerian okra seed (*Abelmoschus esculentus* Moench) flour. Food Chem Toxicol 47(3):657–661

Adelakun OE, Ade-Omowaye BIO, Adeyemi IA, Van De Venter M (2010) Functional properties and mineral contents of a Nigerian okra seed (*Abelmoschus esculentus* Moench) flour as influenced by pre-treatment. J Food Technol 8(2):39–45

Ansari NM, Houlihan L, Hussain B, Pieroni A (2005) Antioxidant activity of five vegetables traditionally consumed by south-Asian migrants in Bradford, Yorkshire, UK. Phytother Res 19(10):907–911

Atawodi SE, Atawodi JC, Idakwo GA, Pfundstein B, Haubner R, Wurtele G, Spiegelhalder B, Bartsch H, Owen RW (2009) Polyphenol composition and antioxidant potential of *Hibiscus esculentus* L. Fruit cultivated in Nigeria. J Med Food 12(6):1316–1320

Bell LA (1988) Plant fibres for papermaking. Liliaceae Press, McMinnville, Oregon. 60 pp

Chopra RN, Nayar SL, Chopra IC (1986) Glossary of Indian medicinal plants. (Including the supplement). Council Scientific Industrial Research, New Delhi, 330 pp

Foundation for Revitalisation of Local Health Traditions (2008) FRLHT Database. htttp://envis.frlht.org

Grieve M (1971) A modern herbal, vol 2. Dover publications/Penguin, New York, 919 pp

Hedin PA, Lamar PL, Thompson AC, Minyard JP (1968) Isolation and structural determination of 13 flavonoid glycosides in *Hibiscus esculentus* (okra). Am J Bot 55(4):431–437

Hill AF (1952) Economic botany, 2nd edn. McGraw-Hill Book Co., New York, 560 pp

Jadhav RB, Sonawane DS, Surana SJ (2008) Cytoprotective effects of crude polysaccharide fraction of *Abelmoschus esculentus* fruits in rats. Pharmacognosy Mag 4(15 (Suppl)):S130–S132

Jaroszewski JW, Stroem-Hansen T, Hansen SH, Thastrup O, Kofod H (1992) On the botanical distribution of chiral forms of gossypol. Planta Med 58(5):454–458

Joshi AB, Hardas MW (1976) Okra *Abelmoschus esculenta* (Malvaceae). In: Simmonds NW (ed) Evolution of crop plants. Longman, London, pp 194–195

Khomsug P, Thongjaroenbuangam W, Pakdeenarong N, Suttajit M, Chantiratikul P (2010) Antioxidative activities and phenolic content of extracts from okra (*Abelmoschus esculentus* L.). Res J Biol Sci 5(4):310–313

Kunkel G (1984) Plants for human consumption. An annotated checklist of the edible phanerogams and ferns. Koeltz Scientific Books, Koenigstein

Lengsfeld C, Titgemeyer F, Faller G, Hensel A (2004) Glycosylated compounds from okra inhibit adhesion of *Helicobacter pylori* to human gastric mucosa. J Agric Food Chem 52(6):1495–1503

Manda F, Tadera K, Aoyama K (1992) Skin lesions due to okra (*Hibiscus esculentus* L.): proteolytic activity and allergenicity of okra. Contact Dermatitis 26(2):95–100

Martin FW (1982) Okra, potential multi-purpose crop for the temperate zones and tropics. Econ Bot 36:340–345

Matsushita T, Aoyama K, Manda F, Ueda A, Yoshida M, Okamura J (1989) Occupational dermatoses in farmers growing okra (*Hibiscus esculentus* L). Contact Dermatitis 21(5):321–325

Ochse JJ, van den Bakhuizen Brink RC (1980) Vegetables of the Dutch Indies, 3rd edn. Ascher & Co., Amsterdam, 1016 pp

Ofoefule SI, Chukwu A (2001) Application of *Abelmoschus esculentus* gum as a mini-matrix for furosemide and diclofenac sodium tablets. Ind J Pharm Sci 63(6):532–535

Ogata F, Imamura H, Hirayama K, Makisumi S (1986) Purification and characterization of four trypsin inhibitors from seeds of okra, *Abelmoschus esculentus* L. Agric Biol Chem 50(9):2325–2333

Porcher MH et al (1995–2020) Searchable World Wide Web multilingual multiscript plant name database. The University of Melbourne, Australia. http://www.plantnames.unimelb.edu.au/Sorting/Frontpage.html

Sawyerr ES (1983) Medicinal plants of West Africa, vol 1. Honiara, [no publishers] 88 pp

Shui G, Peng LL (2004) An improved method for the analysis of major antioxidants of *Hibiscus esculentus* Linn. J Chromatogr A 1048(1):17–24

Siemonsma JS (1994) *Abelmoschus esculentus* (L.) Moench. In: Siemonsma JS, Piluek K (eds) Plant resources of South-east Asia no 8. Vegetables. Prosea Foundation, Bogor, pp 57–60

Tomoda M, Shimada K, Saito Y, Sugi M (1980) Plant mucilages. 26. Isolation and structural features of a mucilage, 'Okra-mucilage F', from the immature fruits of *Abelmoschus esculentus*. Chem Pharm Bull 28:2933–2940

Tomoda M, Shiniza N, Oshima Y, Takahashi M, Murakami M, Hikino H (1987) Hypoglycemic activity of twenty plant mucilages and three modified products. Planta Med 53(1):8–12

Trinh HN, Nguyen NQ, Tran TVA, Nguyen VP (2008) Hypolipidemic effect of extracts from *Abelmoschus esculentus* L. – Malvaceae on tyloxapol- induced hyperlipidemia in mice. Mahidol J Pharm Sci 35(1–4):42–46

U.S. Department of Agriculture, Agricultural Research Service (2010) USDA National Nutrient Database for Standard Reference, Release 23. Nutrient Data Laboratory Home Page, http://www.ars.usda.gov/ba/bhnrc/ndl

Usher G (1974) A dictionary of plants used by Man. Constable, London, 619 pp

Woolfe ML, Chaplid MF, Otchere G (1977) Studies on the mucilages extracted from okra fruits (*Hibiscus esculentus* L.) and baobab leaves (*Adansonia digitata* L.). J Sci Food Agric 28:519–529

Cola acuminata

Scientific Name

Cola acuminata (P. Beauv.) Schott & Endl.

Synonyms

Bichea solitaria Stokes, *Cola macrocarpa* (G. Don) Schott & Endl., *Cola pseudoacuminata* Engl., *Cola supfiana* Busse, *Colaria acuminata* (P. Beauv.) Raf., *Edwardia acuminata* (P. Beauv.) Kuntze, *Edwardia lurida* Raf., *Sterculia acuminata* P. Beauv., *Sterculia macrocarpa* G. Don.

Family

Malvaceae, also placed in Sterculiaceae

Common/English Names

Abate Cola, Cola Nut, Goora Nut, Kola Nut, Kola Nut Tree

Vernacular Names

Angola: Dikezu, Mukezu (Kimbundu), Coleira (Portuguese);
Brazil: Colateira, Gorra, Noz-De-Cola (Portuguese);
Benin: Golo (Adja), Goro (Bariba), Goro (Dendi), Goro (Djerma), Gbahundja, Goro, Igolo, Vi (Fon), Kolatier (French), Evi, Goro (Gèn), Avi, Gbahundja, Gbanja, Igolo (Goun), Gbaoundja, Gora (Nagot), Goro (Peuhl), Goro (Popo), Koro (Waama), Goro (Watchi), Awedi, Gbanja, Gbawundja, Obi (Yoruba);
Burkina Faso: Na, Nahé, Nafo (Abé), Alou, Halou (Adioukrou), Ehoussé (Agni), Lou (Attié), Gurésu (Bété), Buessé, Mbuessé (Bonoua), Apo (Ebrié), Colatier (French),Yétou (Krou), Guéré (Néyau), Ouré, Huré (Plabo), Oué, Wé (Trépo), Ihié (Wobé);
Cameroon: Tse, Lepi (Bamileke), Abel (Boulou), Abeu (Ewondo), Kolatier (French), Gorohi (Fulfulde);
Central African Republic: Le-Eghil, Éguéle (Bakwele), Bobelo (Bunguli), Banga (Gbaya);
Danish: Kolanød, Kolatræ;
Democratic Republic Of Congo: Eme (Balese), Liko, Sombou (Bira), Eme (Efe), Buda (Mayumbe), Iko, Sombou, Sonbou, Moko (Mbuti), Muti-Mabey (Yanzi), Dikasu;
Dutch: Kolaboom;
Finnish: Koolapu;
French: Arbre À Kola, Cola, Colatier, Kola, Kolatier, Kolatier Sauvage, Noix De Cola, Noix De Kola;
Gabon: Muali;
German: Cola-Pflanze, Kolabaum, Kolabaum-Nüsse;
Ghana: Bese (Asante-Twi), Bese (Kola);

T.K. Lim, *Edible Medicinal And Non-Medicinal Plants: Volume 3, Fruits*,
DOI 10.1007/978-94-007-2534-8_22, © Springer Science+Business Media B.V. 2012

Guinea: Toole (Guerze), Goro (Pular);
Italian: Cola;
Ivory Coast: Na, Nahé, Nafo (Abé), Alou, Halou (Adioukrou), Ehoussé (Agni), Lou (Attié), Gurésu (Bété), Buessé, Mbuessé (Bonoua), Apo (Ebrié), Colatier (French),Yétou (Krou), Guéré (Néyau), Huré, Ouré (Plabo), Oué, Wé (Trépo), Ihié (Wobé);
Mali: N'tabanokò;
Nigeria: Evbe (Binis), Ibon (Ibibio), Oji (Ndokwa Delta State), Obi, Obi-Abata (Ondo), Obi-Ajopa;
Polish: Kola Zaostrzona, Zatwar;
Portuguese: Coleira;
Republic Of Congo: Le-Eghil, Éguéle (Bakwele), Bobelo (Bunguli), Faux-Colatier (French), Banga (Gbaya), Ombili (Koukouya), Obessi, Ombili (Téké);
Spanish: Arbol De La Cola, Colatero, Nuez De Cola;
Sierra Leone: Tolo (Mende), An Kola (Temne);
Swedish: Kolaträd;
Togo: Kolatier (French);
Uganda: Engongoli (Rukonjo), Engongoli (Runyaruguru), Ngongolia (Swahili).

Origin/Distribution

According to Opeke (1992), southern Nigeria is considered the center of occurrence of *C. acuminata*, with its original area of distribution stretching from Nigeria to Gabon. *C. acuminata* is also found spontaneously in the mountainous areas of Angola, Zaïre and Cameroon. It has been cultivated for a long time in the islands of Principe and São Tomé. It was introduced to central and south America during the slave trade. It has also spread eastwards to Mauritius and Malaysia (Purseglove 1968; Russell 1955a, b).

Agroecology

Its agroecological requirements is similar to that described for *Cola nitida*. Being a tropical lowland rainforest tree, it thrives in areas with a hot and humid climate from near sea level to 300 m, with mean annual temperature of 23–30°C and mean annual precipitation of 1,200–1,800 mm with distinct long wet and short, dry seasons. It will tolerate drier conditions on sites with high water table and is also found at altitudes over 300 m on deep rich soils under heavy and evenly distributed rainfall. Although an understorey tree that grows under partial shade in the forest, it performs better in open areas, developing a better spreading crown which yields more fruits. It is also found in marshy areas. It does well in both light sandy soils and heavy clay soils, but grows best in well drained, moderately fertile soils.

Edible Plant Parts and Uses

In Africa, the kolanut is chewed for its alkaloid properties (caffeine, kolanin, and theobromin), which dispel sleep, thirst, and hunger. There seems to be a slight preference for white kolanuts over red ones (Russell 1955a, b). They are also roasted, pounded and can also be added to drinks such as tea or milk or cereal such as porridge.

Large quantities of the nuts are exported to Europe and North America, where they are used chiefly for flavouring cola drinks such as Coca-Cola, which are refreshing or stimulating substitutes for tea or coffee (Irvine 1961). Other food uses of kola include beverages such as kola wine, kola cocoa and kola chocolate – a type of chocolate containing cacao and kola powder in cocoa butterfat (Opeke 1992). Studies found that replacement of coca by kolanut in different ratios in instant beverages was feasible and organoleptically acceptable (Jayeola and Akinwale 2002). Kola nut extract is used in the food industry as a flavoring ingredient (Burdock et al. 2009). Present day consumption of kola nut extract is 0.69 mg/kg/day. According to Burdock et al. (2009) although a NOEL/NOAEL (no observed adverse effect level/ no observed effect level) cannot be defined for repeated oral exposure to kola nut extract from available data, U.S. consumers have a history of safe consumption of cola-type beverages containing kola nut extract that dates back at least to the late nineteenth century, with a significant global history of exposure to the intact kola nuts that dates centuries longer.

Botany

Tree slender, 7–13 (–20)m high, with rough grey bark, often branching close to the ground (Plate 1) with terete, smooth branches. Petioles 2–4 cm long. Leaves crowded near the shoot tip, simple, entire, elliptic-oblong to obovate-lanceolate, 16–27 by 6–11 cm, leathery, green to dark green, tapering at base, tip obtusely acuminate (Plate 2). Flowers several to numerous, polygamous, some flowers male, female and hermaphrodite on the same plant and arranged in terminal and axillary cymose panicles; fetid, apetalous, calyx lobes united half their length, lobes yellow to yellowish-white with red blotches or streaks emanating from the base of the inner surface (Plate 3). Male flower, stamina column slender much shorter than calyx, bearing a ring of 10-bilobed anthers, In the hermaphrodite flowers, the anthers are subsessile in a ring surrounding the base of a 5-loculed, oblong ovary. Fruit consists of five pods or follicles sometimes fewer by abortion, each follicle sessile to subsessile, oblong, obtuse or rostrate, coriaceous or woody, rough, tuberculate, 10–20 cm long by 6–8 cm broad, green turning yellowish-brown (Plates 4, 5 and 6). Seeds

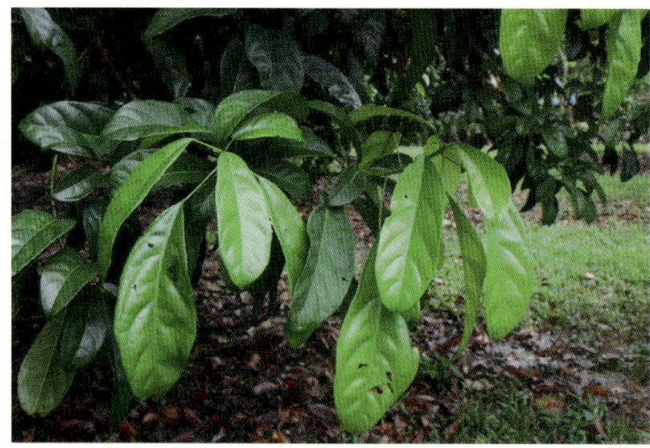

Plate 2 Large, elliptic-oblong to elliptic-obovate leaves with long petioles

Plate 1 Trunk branching close to the ground

Plate 3 Yellowish apetalous flowers with red blotches inside the calyx lobes

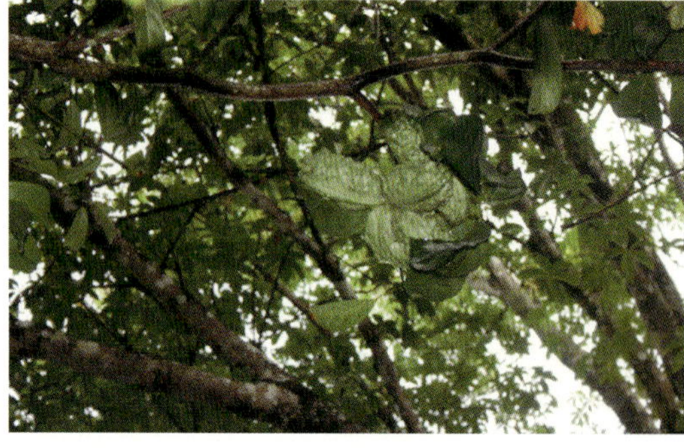

Plate 4 Immature star-shaped fruit

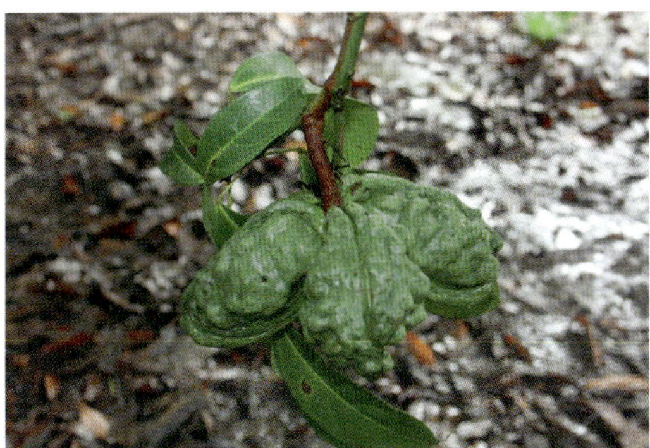

Plate 5 Side view of star-shape fruit

Plate 6 Close-view of a six-follicle, star-shaped fruit

5–12 per pod, oblong, obtuse or tetragonal, testa is red or purplish or white and cartilaginous, cotyledons 2–5, thick and horny.

Nutritive/Medicinal Properties

The seed of *C. acuminata* was found to comprise 13.5% water, 9.5% crude protein, 1.4% fat, 45% sugar and starch, 7.0% cellulose, 3.8% tannin and 3% ash and to be rich in caffeine (2.8%) and theobromine (0.05%) (Purseglove 1968). Oladokun (1989) reported the following nutrient composition for *Cola acuminata* nuts (based on nut weight 21–25 gm) as follows: lipid 4.55 mg, protein 59.57 mg, free sugars 395.76 mg, N 112.73 mg, P 9.97 mg, K 106.12 mg, Mg 22.52 mg, Ca 1.36 mg and Na 1.36 mg. Individual mineral nutrients were higher in heavy than light nuts. Cola nuts from *Cola acuminata* was reported contain up to 2.2% caffeine, and those from *Cola nitida* up to 3.5% caffeine (Seitz et al. 1992). Both contained less than 1% theobromine. Also reported were the polyphenols leucoanthocyanidin and cathecine and large amounts of starch (Seitz et al. 1992). Caffeine and cathecine were primarily present in the form of a caffeine-cathecine complex (especially in fresh nuts) that previously was wrongly thought to be a glycoside and named colanine (Seitz et al. 1992). Odebode (1996) found *C. nitida* and *C. acuminata* differed markedly in the amount of total phenol. The total phenol content was greater in *C. nitida* than *C. acuminata* confirming the general view that *Cola nitida* was more astringent than *Cola acuminata*, because astringency was related to the phenolic content of fruits.

Ibu et al. (1986) found that both *Cola nitida* and *C. acuminata* contained xanthine alkaloids found in tea and coffee. These xanthine derivatives included caffeine, theophylline and theobromine which were known to stimulate gastric acid secretion. They found that both cola species significantly induced gastric acid secretion and warned that peptic ulcer patients should not consume cola nuts. Adeyeye et al. (2007) found *Cola acuminata* to contain 356.24 mg/g protein total amino acids and 38.16% total acidic amino acids. Valine was most limiting and the percentage cystine level in the total sulphur amino acid was 44.27%. The high level of acidic amino acids in *C. acuminata* reinforced the recommendation that *C. acuminata* consumption should be avoided by peptic ulcer patients.

Some of the reported pharmacological properties are presented below.

Antioxidant Activity

Varieties of kola nuts (*Cola nitida* alba, *Cola nitida* rubra, and *Cola acuminata*) were found to contain considerable levels of (+)-catechin (27–37 g/kg), caffeine (18–24 g/kg), (−)-epicatechin (20–21 g/kg), procyanidin B 1 [epicatechin-(4β→8)-catechin] (15–19 g/kg), and procyanidin B2 [epicatechin-(4β→8)-epicatechin] (7–10 g/

kg) (Atawodi et al. 2007). Extracts of all varieties exhibited antioxidant capacity with IC_{50} values in the range 1.70–2.83 and 2.74–4.08 mg/ml in the hypoxanthine/xanthine oxidase and 2-deoxyguanosine HPLC-based assays, respectively. The authors asserted that caffeine was potentially the more effective cancer chemopreventive metabolite in terms of its antioxidant capacity.

Cardiovascular and CNS Activities

Cola acuminata had been reported to be used as a central nervous system stimulant (Reiling 1999). Reiling (1999) further reported that kolanut could curtail childhood asthma attack and lead to improvement. Preliminary observations on the cardiovascular system of the cat showed that aqueous extracts of the nut evoke a dose dependent differential response – the 8–9% hypertensive effect of 0.01 μg–1 mg/ml doses contrasting with the 18–71% hypotensive response of 10–1,000 mg/ml doses (Agatha et al. 1978). With higher doses a bradycardia developed and cardiac arrythmias were observed as a terminal event. Being diuretic, use of kola nut had been suggested for those with renal diseases, cardiac or renal edema and rheumatic and rheumatoid conditions (Agatha et al. 1978; Reiling 1999).

Studies showed that diastolic blood pressure was more responsive to changes in concentration of *Cola acuminata* extract with a significant concentration dependent increase in the arterial blood pressure of both the normotensive and hypertensive rats (Igbinovia et al. 2009). *C. acuminata* had been known to contain caffeine which had been found to be vasoactive and to augment the release of calcium from sarcoplasmic reticulum.

Antimicrobial Activity

Cola species were found to have antimicrobial activity. The leaf ethanol extracts of four *Cola* species were found to be more effective against the tested fungi than the bacteria at high concentrations (Sonibare et al. 2009). *C. acuminata* and *C. nitida* extracts exhibited activity against *Staphylococcus albus* at concentrations ranging from 10 to 150 mg/ml with comparable diameters of zone of inhibition of 7.3–16.0 for *C. acuminata* and 10.00–19.0 for *C. nitida*. Also, both *Cola* species demonstrated activities against *Candida albicans* and *Aspergillus niger* at concentrations ranging from 90 to 150 mg/ml. Only *C. acuminata* inhibited the growth of *Klebsiella pneumoniae* at the MIC of 90 mg/ml whereas, *Candida albicans* was inhibited by *C. acuminata*, *C. millenii* and *C. gigantea* at the MIC of 120 mg/ml. Phytochemical screening of the four species of *Cola* showed the presence of alkaloids, saponins, tannins and cardenolides that could have contributed to the antimicrobial activity.

Kola Nut Chewing and Cancer

Primary and secondary amines were determined in 3 kolanut species *Cola acuminata*, *C. nitida* and *Garcinia cola* (Atawodi et al. 1995). Dimethylamine, methylamine, ethylamine and isopentylamine were detected in all kola nut species, while pyrrolidine, piperidine and isobutylamine were detected in one or more species. Estimated average total daily intake of aliphatic amines by a typical kola nut chewer varied from 260 to 1,040 μg/day for secondary amines and from 2,430 to 9,710 μg/kg/day for primary amines. Methylating activity was found to be significantly higher in kola nuts (170–490 μg/kg) than ever been reported for a fresh plant product. Based on these data, the authors suggested that the possible role of kola nut chewing in human cancer aetiology should be explored in countries where kola nuts were widely consumed as stimulants.

Trypanostatic Activity

A proanthocyanidin isolated from *Cola acuminata* was found to potently induce growth arrest and lysis of bloodstream form trypanosomes of *Trypanosoma brucei* in-vitro in a dose – and time-dependent manner (Kubata et al. 2005). In a mouse model, it exhibited a trypanostatic effect that extended the life of infected, treated animals up to 8 days post-infection against the 4 days for infected, untreated animals. The proanthocyanidin showed a low cytotoxicity against mammalian cells whereas treated – blood stream forms showed massive enlargement of their flagellar pocket and lysosome-like structures caused by an intense formation of multivesicular bodies and vesicles within these organelles. The observed ultrastructural alterations caused rupture of plasma membranes and the release of cell contents, indicative of a necrotic process rather than a programmed cell death.

Traditional Medicinal Uses

Traditionally, the leaves, twigs, flowers, fruits follicles, and the bark of both *C. nitida* and *C. acuminata* were used to prepare a tonic as a remedy for dysentery, coughs, diarrhoea, vomiting (Ayensu 1978) and chest complaints (Irvine 1961). Kola is widely used as a treatment for whooping cough and asthma. *Citropsis articulata* and *Cola acuminata* were among 33 highly utilized traditional medicinal plants reported to be used for management of sexual impotence and erectile dysfunction (Kamatenesi-Mugisha and Oryem-Origa 2005). In Nigeria, a leaf decoction is used for diabetes (Abo et al. 2008), the fruit is eaten to clear the throat and as an antidote for sleep (Ogie-Odia and Oluowo 2009) In Sierra Leone, kola nuts are used for amenorrhoea and chewed as aphrodisiac (Lebbie and Gruies 1995). In Europe, cola nuts were once utilized to treat migraine headaches, neuralgia, vomiting and sea sickness.

Other Uses

The non-edible uses of *Cola acuminata* are similar to those described for *Cola nitida*.

As described for *Cola nitida*, chewing of kola nuts is a ubiquitous habit in the sub-Saharan countries of Africa, especially in northern Nigeria and Sudan. Kola chewing is akin to the social role of tea and coffee drinking or cigarette smoking in Western countries (Purseglove 1968; Rosengarten 1984; Russell 1955b). *C. acuminata* is widely used ceremonially and socially by the people of West and Central Africa. Among muslims in northern Nigeria and Sudan, kola is sacred. In some communities, kola is regarded as a sign of love and friendship. Kola is consumed in cultural, social and religious ceremonies like in dowry ceremonies, weddings, 'tontines', funerals, wake-keepings, etc. At a birth a kola tree may be planted for the new-born child. A kola tree is also often planted at the head of a grave as part of local death rites.

Comments

Many of the vernacular names listed for the African countries may also refer to *Cola nitida* or other *Cola* species.

Selected References

Abo KA, Fred-Jaiyesimi AA, Jaiyessimi AEA (2008) Ethnobotanical studies of medicinal plants used in the management of diabetes mellitus in South Western Nigeria. J Ethnopharmacol 115:67–71

Adeyeye EI, Asaolu SS, Aluko AO (2007) Amino acid composition of two masticatory nuts (*Cola acuminata* and *Garcinia kola*) and a snack nut (*Anacardium occidentale*). Int J Food Sci Nutr 58(4):241–249

Adjanohoun EJ, Adjakidjè V, Ahyi MRA, Aké Assi L, Akoègninou A, d'Almeida J, Apovo F, Boukef K, Chadare M, Cusset G, Dramane K, Eyme J, Gassita JN, Gbaguidi N, Goudote E, Guinko S, Houngnon P, Lo I, Keita A, Kiniffo HV, Kone-Bamba D, Musampa Nseyya A, Saadou M, Sodogandji T, De Souza S, Tchab A, Zinsou Dossa C, Zohoun T (1989) Contribution aux Études Ethnobotaniques et Floristiques en République Populaire du Bénin. Médecine Traditionelle et Pharmacopée. Agence de Coopération Culturelle et Technique, Paris, 895 pp

Agatha M, Brekenridge C, Soyemi EA (1978) Some preliminary observations on the efforts of kola nut on the cardiovascular system. Niger Med J 8(6):501–505

Agyare C, Asase A, Lechtenberg M, Niehues M, Deters A, Hensl A (2009) An ethnopharmacological survey and in vitro confirmation of ethnopharmacological use of medicinal plants used for wound healing in Gosomtwi-Atwima-Kwanwoma area. Ghana J Ethnopharmacol 125:393–403

Ajibesin KK, Ekpo BA, Bala DN, Essien EE, Adesanya SA (2008) Ethnobotanical survey of Akwa Ibom State of Nigeria. J Ethnopharmacol 115(3):387–408

Akendengue B, Louis AM (1994) Medicinal plants used by the Masango people in Gabon. J Ethnopharmacol 41:193–200

Atawodi SE, Mende P, Pfundstein B, Preussmann R, Spiegelhalder B (1995) Nitrosatable amines and nitrosamide formation in natural stimulants: *Cola acuminata*, *C. nitida* and *Garcinia cola*. Food Chem Toxicol 33:625–630

Atawodi SE, Pfundstein B, Haubner R, Spiegelhalder B, Bartsch H, Owen RW (2007) Content of polyphenolic compounds in the Nigerian stimulants *Cola nitida* ssp. *alba*, *Cola nitida* ssp. *rubra* A. Chev, and *Cola acuminata* Schott & Endl. and their antioxidant capacity. J Agric Food Chem 55(24):9824–9828

Ayensu ES (1978) Medicinal plants of West Africa. Reference Publication International, Algonac

Bossard E (1996) Quelques notes sur l'alimentation et les apports nutritionnels occultes en Angola. Garcia de Orta Sér Bot Lisboa 13(1):7–41

Bouquet A, Debray M (1974) Plantes Médicinales de la Côte d'Ivoire, 32. Mémoires/O.R.S.T.O.M, Paris, 231 pp

Burdock GA, Carabin IG, Crincoli CM (2009) Safety assessment of kola nut extract as a food ingredient. Food Chem Toxicol 47(8):1725–1732

Grønhaug TE, Glaeserud S, Skogsrud M, Ballo N, Bah S, Diallo D, Paulsen BS (2008) Ethnopharmacological survey of six medicinal plants from Mali, West-Africa. J Ethnobiol Ethnomed 4:26

Ibu JO, Iyama AC, Ijije CT, Ishmael D, Ibeshi M, Nwokediuko S (1986) The effect of *Cola acuminata* and *Cola nitida* on gastric acid secretion. Scand J Gastroenterol Suppl 124:39–45

Igbinovia ENS, Ugwu AC, Nwaopara AO, Otamere HO, Adisa WA (2009) The effects of *Cola acuminata* on arterial blood pressure. Pak J Nutr 8(2):148–150

Irvine FR (1961) Woody plants of Ghana: with special reference to their uses. Oxford University Press, London, 868 pp

Jayeola CO, Akinwale TO (2002) Utilization of kolanut and cocoa in beverage production. Nutr Food Sci 32(1):21–23

Kamatenesi-Mugisha M, Oryem-Origa H (2005) Traditional herbal remedies used in the management of sexual impotence and erectile dysfunction in western Uganda. Afr Health Sci 5(1):40–49

Kerharo J, Bouquet A (1950) Plantes Médicinales et Toxiques de la Côte d'Ivoire - Haute-Volta. Vigot Frères, Paris, 291 pp

Kubata BK, Nagamune K, Murakami N, Merkel P, Kabututu Z, Martin SK, Kalulu TM, Huq M, Yoshida M, Ohnishi-Kameyama M, Kinoshita T, Duszenko M, Urade Y (2005) *Kola acuminata* proanthocyanidins: a class of anti-trypanosomal compounds effective against *Trypanosoma brucei*. Int J Parasitol 35(1): 91–103

Lebbie AR, Guries RP (1995) Ethnobotanical value and conservation of sacred groves of the Kpaa Mende in Sierra Leone. Econ Bot 49(3):297–308

Odebode AC (1996) Phenolic compounds in the kola nut (*Cola nitida* and *Cola acuminata*) (Sterculiaceae) in Africa. Rev Biol Trop 44:513–515

Ogie-Odia EA, Oluowo EF (2009) Assessment of some therapeutic plants of the Abbi people in Ndokwa West L.G.A of Delta State, Nigeria. Ethnobot Leafl 13:989–1002

Oladokun MAO (1989) Nut weight and nutrient contents of *Cola acuminata* and *C. nitida* (Sterculiaceae). Econ Bot 43(1):17–22

Opeke LK (1992) Tropical tree crops. Spectrum Books Ltd, Ibadan

Porcher MH et al (1995–2020) Searchable world wide web Multilingual Multiscript Plant Name Database. Published by The University of Melbourne, Melbourne. http://www.plantnames.unimelb.edu.au/Sorting/Frontpage.html

Purseglove JW (1968) Tropical crops: dicotyledons, vols 1 & 2. Longman, London, 719 pp

Reiling J (1999) Therapeutics of kola. JAMA 282:1898–1899

Rosengarten F Jr (1984) The book of edible nuts. Walker and Company, New York, 384 pp

Russell TA (1955a) The kola nut of West Africa. World Crop 7:221–225

Russell TA (1955b) The kola of Nigeria and the Cameroons. Trop Agric Trinidad 32:210–240

Sawyerr ES (1983) Medicinal plants of West Africa, vol 1. [no publisher] Honiara, 88 pp

Seitz R, Gehrmann B, Kraus L (1992) Hagers Handbuch der Pharmazeutischen Praxis, vol 4, 5th edn. Springer, Berlin, pp 940–946

Sonibare MA, Soladoye MO, Esan OO, Sonibare OO (2009) Phytochemical and antimicrobial studies of four species of *Cola* Schott & Endl. (Sterculiaceae). Afr J Tradit Complement Altern Med 6(4):518–525

Sundstrom L (1966) The cola nut: functions in West African social life. Stud Ethnogr Ups 26:135–146

Tachie-Obeng E, Brown N (2004) *Cola nitida* and *Cola acuminata*: A state of knowledge report. In: Clark LE, Sunderland TCH (eds) The key non-timber forest products of Central Africa: state of the knowledge. USAID Technical Paper No. 122

Toyang NJ, Nuwanyakpa M, Ndi C, Django S, Kinyuy CW (1995) Ethnoveterinary medicine practices in the northwest province of Cameroon. IKDM 3(3):20–22

Uphof JCTh (1968) Dictionary of economic plants, 2nd edn (1st edn 1959). Cramer, Lehre, 591 pp

Van Eijnatten CLM, Roemantyo H (2000) Cola Schott & Endl. In: van der Vossen HAM, Wessel M (eds) Plant resources of South-East Asia No 16. Stimulants. Backhuys Publishers, Leiden, pp 78–83

Cola nitida

Scientific Name

Cola nitida **(Vent.) Schott & Endl.**

Synonyms

Cola vera K. Schum., *Sterculia nitida* Vent.

Family

Malvaceae, also placed in Sterculiaceae

Common/English Names

Bitter Cola, Bissy Nut, Cola, Cola Nut, Cola Seeds, Goora Nut, Kola Nut, Kola Nut Tree, Kola, Kola Seed, Sudan Cola Nut Tree

Vernacular Names

Africa: Bakuru (Banyun), Buguraba, Kaguru (Diola), Bese (Fanti), Goro (Fula), Goro (Hausa), Oji (Igbo), Bugurabu, Woro (Jola), Oro (Manding-Bambara), Goro, Kumkuo, Kuruo (Mandinka), Guro, Uro (Maninka), Gure (Moore), Chigban'bi (Nupe), Esele (Nzima), Goro (Tukulor), Bese, Bese-Pa (Twi), Goro (Wolof), Obi Gbanja (Yoruba);
Arabic: Guro, Woro;
Benin: Golo (Adja), Goro (Bariba), Goro (Dendi), Goro (Djerma), kolatier (French), Gbaoundja, Gora (Nagot), Goro (Peuhl), Goro (Popo), Koro (Waama);
Danish: Ægte Kola;
Finnish: Koolapuu;
French: Colatier, Kolatier;
German: Kolabaum;
Indonesia: Buah Kola;
Ivory Coast: Oussé (Ashanti), Goulé, Goulé Sou (Shien);
Polish: Kola Blyszczaca;
Sierra Leone: Toloi (Kpaa Mende);
Spanish: Colatero;
Swedish: Kolaträd.

Origin/Distribution

This species was originally distributed along the west coast of Africa from Sierra Leone to the Republic of Benin with the highest frequency and variability occurring in the forest areas of Côte d'Ivoire and Ghana (Opeke 1992). It is cultivated in Angola, Brazil, Chad, Congo, Democratic Republic of Congo, Equatorial Guinea, Ethiopia, Gabon, Guinea, India, Jamaica, Kenya, Mali, Mozambique, Senegal, Somalia, Sudan, Tanzania, Togo, Uganda, and Zimbabwe and introduced elsewhere into the tropics and subtritropics.

T.K. Lim, *Edible Medicinal And Non-Medicinal Plants: Volume 3, Fruits*,
DOI 10.1007/978-94-007-2534-8_23, © Springer Science+Business Media B.V. 2012

Agroecology

It is mainly grown between 6° and 7° north of the equator, but has also been found up to 10°N on the West Coast of Africa. It thrives in the hot tropical lowland forest with rainfall of 17,000 mm extending over a period of 8 months or more and a temperature of between 23°C and 28°C (Ekanade 1989). Kola can grow with annual rainfall of about 1,200 mm. The species have also been cultivated in the transitional zones in savannas (Opeke 1992). The species could be cultivated in drier areas where growth could only be made possible by the occurrence of wet land with a high water-table (Russell 1955). The species flourishes in both heavy and light soils as long as they have good depth. Better growth and fruiting is obtained in well-drained, fertile soils containing much humus.

Edible Plant Parts and Uses

Fresh kola seeds are chewed fresh as a masticatory stimulant and to prevent fatigue. Kolanut is used as a masticatory stimulant by Africans and has numerous uses in social, religious, ritual and ceremonial functions by the natives in the forest region of Africa (Nzekwu 1961; Opeke 1992; Asogwa et al. 2006). It is used during ceremonies related to marriage, child naming, installation of Chiefs, funeral and sacrifices made to the various gods of African mythology. It is commonly used individually, in groups or used ceremonially, presented to tribal chiefs or to guests. The juice is often consumed replacing coffee. Kola nuts are rich in xanthine alkaloids such as theobromine, caffeine and kolatin. Caffeine stimulates the body and kolatin stimulates the heart. It is preferred among East Asian Muslims, who are forbidden to drink alcohol especially during the fasting Ramadan month. The nuts are said to be restraining and to posses thirst-restraining properties. Chewing kola nut are also deemed to ease hunger pangs. It has been reported that small doses of kolanut increase mental activity, reduce the need to sleep, and also dispel hunger and thirst. It is for this reason that kolanut chewing has become very popular among students, drivers and many other consumers who need to remain active for unusually long period in Nigeria (Jayeola 2001).

Kolanut extracts are used in the manufacture of various non-alcoholic beverages, soft drinks, wines, chocolate and sweets (Beattie 1970; Ogutuga 1975; FAO 1982; Asogwa et al. 2006). The possible use of pulverized kola nuts for the preparation of hot non-alcoholic beverages has been reported (Eka 1971). It could also be used in jam and jelly production because of its high pectin content. Kola was used to make cola soft drinks, though today most of these mass-produced beverages use artificial flavourings. Exceptions include Barr's Red Kola, Red Bull's new Simply Cola, Harboe Original Taste Cola, Foxon Park Kola, Blue Sky Organic Cola, Whole Foods Market 365 Cola, Sprecher's Puma Kola, Virgil's Real Cola, and Cricket Cola (Kola plus green tea) (Wikipedia 2009). In 2007, United Kingdom supermarket Tesco launched an American Premium Cola that uses kola nuts, spices and vanilla. The Kolanuts are also processed into wine and forms the main ingredient in the manufacture of a heat tolerant (high melting point) chocolate. The red kola nuts are a potential source of food colourant. Cola nuts, in combination with coca-leaf are incorporated in brands of cocoa, tonic wines and other beverages.

Botany

Cola nitida is an understorey, medium-sized, evergreen tree, usually 9–12 m high but can attain a height of 20 m. The bole is cylindrical, often straight with greyish brown, longitudinal fissured bark and trunk diameter of 1.5 m across. Leaves are alternate, simple, glabrous and petiolate; petiole 1.2–10 cm long. Lamina is broadly oblong to broadly elliptic or elliptic-oblanceolate, 10–33 × 5–13 cm, with shortly acuminate apex, obtuse base, wavy margins, coriaceous, leathery, dark green and with 6–10 lateral nerves (Plates 1, 2). Inflorescence axillary, an irregularly branched panicles 5–10 cm long. Flowers polygamous with male, female or hermaphrodite on the same plant, 5-merous, apetalous, perianth

Plate 1 Cola leaves

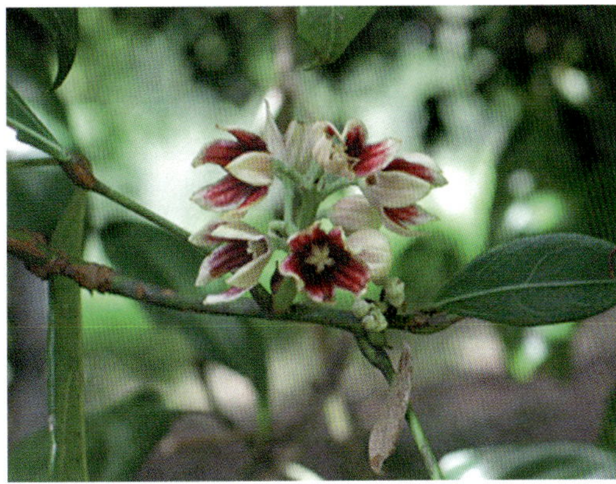

Plate 3 Close-up of Cola flowers

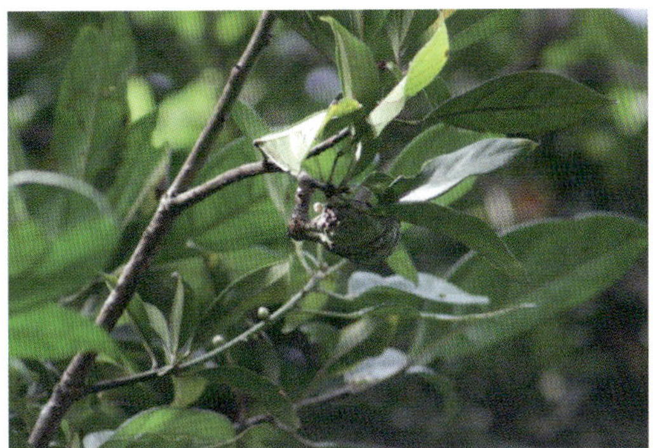

Plate 4 Young Cola fruit

Plate 2 Cola inflorescence and leaves

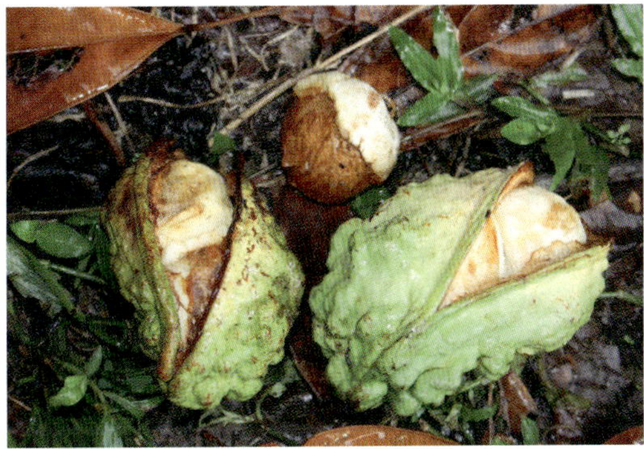

Plate 5 Mature Cola fruit and seeds

white or white on the outside and with blotches of purple on the inside (Plates 2, 3) or coloured. Staminate flowers with deeply lobed cup-shaped calyx, 2 cm across and numerous stamens in two whorls. Hermaphrodite flowers with calyx about 5 cm in diameter, 5 carpels and numerous rudimentary anthers at the base. Fruits oblong-ellipsoid, sessile follicles 13 × 7 cm, glossy green, smooth to the touch but knobbly with large tubercules (Plates 4, 5 and 6). Seeds 4–8 (10) per carpel, ovoid or subglobose 3–3.5 × 2–2.5 cm, either red or white with two cotyledons.

Plate 6 Close-view of Cola fruit

Nutritive/Medicinal Properties

Food value of raw *Cola nitida* nut per 100 g edible portion (Leung et al. 1968) was reported as: energy 148 cal, moisture 62.9%, protein 2.2, fat 0.4 g, total carbohydrate 33.7 g, fibre 1.4, ash 0.8 g, Ca 58 mg, P 86 mg, Fe 2 mg, β carotene 25 ug, thiamine 0.3 mg, riboflavin 0.3 mg, niacin 0.6 mg, and ascorbic acid 54 mg. Nutrient composition of kolanut (*C. nitida*) was reported by Odebunmi et al. (2009) as: moisture 66.4%, dry matter 33.60%, crude fat 5.71%, crude protein 2.63%, ash 1.50%, crude fibre 7.13% total carbohydrate 28.56%, and the minerals (mg/kg dry matter) – K 3,484 mg, Ca 124.4 mg, Mg 392 mg, Fe 16.43 mg, Zn 5.24 mg and P 411.43 mg. Oladokun (1989) reported the following nutrient composition for *Cola nitida* nuts (based on nut weight 21–25 g) as follows: lipid 37.05 mg, protein 611.63 mg, free sugars 2463.346 mg, N 115.40 mg, P 13.60 mg, K 103.59 mg, Mg 25.46 mg, Ca 3.55 mg and Na 1.54 mg. Individual mineral nutrients were higher in heavy than light nuts and nuts of *C. nitida* were heavier than those of *C. auminata*.

Fresh kola-nut contained 8.90% protein, 0.92% fat, 2.40% Ash and 1.50% caffeine. Also, the developed kola soft drink whose pH, specific gravity, total solids and caffeine were 5.40, 1.040, 10% and 0.01 respectively was acceptable to local tasters, making economic utilization of kolanut possible through the production of value-added products (Jayeola 2001).

Catechin, epicatechin, theobromine and caffeine were detected in *Cola* seeds (Niemenaka et al. 2008). Among the three species, *Cola acuminata*, *Cola nitida* and *Cola anomala*, *C. nitida* was found to be highest in flavonoid and caffeine content while *C. anomala* possessed high amount of theobromine. Catechin was the dominant flavonoid. Caffeine was the major alkaloid in *Cola* seeds and was considered as one of the signature compounds due to its concentration range. The average concentrations in *C. acuminata*, *C. nitida* and *C. anomala* accessions were 11,066, 13,761 and 7,013 mg/kg fresh weight respectively. Kola nut also contain a glucoside, kolanin, a heart stimulant. As the nut dries, kolanin is acted upon by an enzyme which gives rise to a red colour with the formation of kola-red – a compound very similar to phlobaphene (Burkill 1966).

Odebode (1996) found *C. nitida* and *C. acuminata* differed markedly in the amount of total phenol and that differences also existed between different coloured nut variants. The total phenol content was greater in *C. nitida* than *C. acuminata* . In *Cola nitida*, the quantity of total phenol in red nuts was up to three times that of white and pink nuts; but in *Cola acuminata* the difference was not significant. This investigation confirmed the general view that *Cola nitida* was more astringent than *Cola acuminata*, because astringency was related to the phenolic content of fruits (Odebode 1996).

Pharmacological properties of kola reported include:

Antioxidant Activity

Kolanut was found to contain appreciable levels of phenolic compounds such as (+)-catechin (27–37 g/kg), caffeine (18–24 g/kg), (−)-epicatechin (20–21 g/kg), procyanidin B1 [epicatechin-(4β→8)-catechin] (15–19 g/kg), and procyanidin B2 [epicatechin-(4β→8)-epicatechin] (7–10 g/kg) (Atawodi et al. 2007). Extracts of all kolanut varieties exhibited antioxidant capacity with IC_{50} values in the range 1.70–2.83 and 2.74–4.08 mg/ml in the hypoxanthine/xanthine oxidase and 2-deoxyguanosine HPLC-based assays, respectively. Of the major secondary plant metabolites present in kola nut extracts, caffeine was potentially the more effective cancer chemopreventive metabolite in terms of its antioxidant capacity.

Cardiovascular and CNS Activities

Kola nut was found to have stimulant effects on the central nervous system and heart. In humans it was reported to enhance alertness and physical energy, to elevate mood, to increase tactile sensitivity and to suppress the appetite (Umoren et al. 2009). Kola nut, a central nervous system stimulant , also contained caffeine. Caffeine also possessed toxic effects, extensions of their pharmacological effects (Carrillo and Bennitez 2000). The most serious caffeine-related CNS effects reported included seizures and delirium. Other symptoms affecting the cardiovascular system ranged from moderate increases in heart rate to more severe cardiac arrhythmia.

Autonomic changes included increased body temperature, increased blood pressure and increased respiratory rate. Consumption of a diet comprising kola nuts for 7 days elevated the mean arterial pressure (MAP) of normal rats (Osim and Udia 1993). Consumption of a diet comprising caffeine (equivalent to the amount confined in kola nuts given previously) for 7 days also elevated the MAP of rats. Further, the consumption of diet comprising only kola nuts for 7 days also elevated the MAP in rats. The results suggested that chronic consumption of kola nuts was capable of elevating mean arterial pressure in rats and that caffeine an active principle in kola nuts may be responsible. Kola nut also contained among other constituents, a heart stimulant, kolanin.

It was observed that a medium dose (5 mg/kg) of fresh kola-nut extract significantly increased the locomotor activity whereas a low dose (2.5 mg/kg) had no effect and a high dose (10 mg/kg) showed depressive effects on the locomotor activities (Ajarem 1990). The results suggested that the kola-nut extract induced biphasic changes in the locomotor activity of mice depending on the dose and the treatment duration. In a separate study, both the aqueous kola-nut extract (400 and 800 mg/kg) and caffeine (15 mg/kg) caused significant increases in the number of entries in Y maze test, but reduced the frequency of rearing (Ettarh et al. 2000). The extract did not significantly reduce the number of entries after 24 hours. It was suggested that kola-nut stimulated exploratory locomotor activity due to its caffeine content, but did not enhance habituation. In another study, kola mixtures showed no significant difference in influencing social behaviour, except in females treated with low dose (110 g food/10 g kola paste) of kola mixture. A significant increase in time spent for displacement was recorded (Al-Hazmi 2000). Animal studies showed that chronic consumption of kola nut and caffeine diets caused decrease in food intake and body weight (Umoren et al. 2009). Consumption of caffeine-diet also significantly decreased water intake and locomotor activity. The effect of kola nut-diets on water intake and locomotor activity was not significant. The authors concluded that the effect of kola nut on locomotor behaviour and water intake may not be due to caffeine only The results suggested that kola nut and caffeine could be used to combat hunger pangs and body weight in obese persons.

Ocular Activity

The ethnopharmacological effects of bolus ingestion of 30 g of *Cola nitida* was investigated on visually acute and healthy volunteers in order to determine its ocular implications or effects (Igwe et al. 2007). Results showed that *Cola nitida* had no effect on the pupil diameter, visual acuity and intraocular pressure but improved the near point of convergence by 43% and increased the amplitude of accommodation by 11% while existing heterophorias are ameliorated. The stimulating effect of *Cola nitida* extract appeared to allow near work to be done without stress. Somnolence and ocular muscle imbalance common features of the elderly could be ameliorated or relieved.

Elastase Reduction Activity

Studies showed that kola nut extract could reduce elastase liberation and activity from polymorphonuclear neutrophils in an infection by protecting α-1-proteinase inhibitor (alpha1PI) against inactivation from oxidants (Daels-Rakotoarison et al. 2003). Alpha1PI, is the main endogenous inhibitor of elastase helping to limit excessive elastase activity. By reducing elastase activity, the cola nut extract limits the injurious effects caused by this enzyme.

Antibacterial Activity

The methanol extract of root bark for both *Cola nitida* and *Cola milleni* were found to be potent against both *Mycobacterium bovis* and strains of *Mycobacterium vaccae* (Adeniyi et al. 2004). The minimum inhibitory concentration (MIC) of *C. nitida* against *M. bovis* was 125 μg/ml while the MIC of *C. milleni* against *M. bovis* is 62.5 μg/ml after at least 6 days of inhibition with growth index (GI) units lesser than or equal to the change in GI units inoculated with a 1/100 of the BACTEC inoculum for a control vial. The minimum inhibitory concentration of *C. milleni* against the six ATCC strain of *M. vaccae* ranged from 62.5 to 250 μg/ml while for *C. nitida* it ranged from 500 μg/ml to above 1,000 μg/ml.

The aqueous and alcoholic extracts of *C. nitida* bark were found to inhibit such organisms as *Staphylococcus aureus*, *Klebsiella pneumoniae*, *Proteus mirabilis*, *Pseudomonas aeruginosa*, β-haemolytic streptococci, *Escherichia coli* and *Neisseria gonorrhoeae* (Ebana et al. 1991).

Role in Malaria Etiology

Alaribe et al. (2003) found that kola nut ingested at a high concentration (about 35 g/day) mimicked malaria-like symptoms. It was found that 11 people without the malaria parasite in their blood before and after taking the kola nut complained of various malaria symptoms confirming that kola nut can mimic malaria-like symptoms. This quantity provided a high level of caffeine and cyanide in the circulation so that people with a low level of malaria parasite in them would notice active infection which otherwise may have been controlled by the host immune system. Lastly, the observed phenomenon could affect drug pressure and induce resistance to antimalarial drugs. Those that showed detectable parasite before ingesting kola had significant increase in parasite density. Statistical analysis showed a strong relationship between parasite increase and eating of the kola nut.

Cola Nut and Peptic Ulcer

Both cola species *Cola acuminata* and *Cola nitida* contain xanthines – the same type of alkaloids found in tea and coffee (Ibu et al. 1986). Common among these xanthine derivatives are caffeine, theophylline and theobromine that are known to stimulate gastric acid secretion. The study found that both cola species significantly induced gastric acid secretion. This corroborated clinical advice to peptic ulcer patients not to eat cola nuts.

Antigonadotropic Effects

A study reported that *C. nitida* bark was one of three plants with putative contraceptive and antigonadotropic effects (Benie et al. 2003; Benie and Thieulant 2004). Data showed that treatment of rats with the plant extract induced ovulation and oestrous cycle blockade at the dioestrous II stage. Analysis of the principal hormones involved in oestrous cycle regulation showed that the plant extracts decreased gonadotropin release (both luteinizing hormone and follicle stimulating hormone). In fact, *Afrormosia laxiflora*, *Pterocarpus erinaceus* and *C. nitida* extracts inhibited gonadotropin release as an anti-estrogen-like substance. In further studies, *C. nitida* extracts was found to diminish pituitary luteinizing hormone release in culture medium without acting on rat pituitary cell content (Benie and Thieulant 2004). The cola extract formed complexes with basic glycoproteins (but not with acid glycoproteins) and prevented them from entering the cells. Thus, kola-nut showed potential to be used as a natural fertility regulator.

Cancerous and Neurotoxic Effects

Kola nuts contain high amounts of N-nitroso compounds which are carcinogenic. In Nigeria, where the chewing of Kola nuts is a common practice, there is a high incidence of oral and gastrointestinal cancer which may be related to this habit. Dimethylamine, methylamine, ethylamine and isopentylamine were detected in all kola nut varieties, while pyrrolidine, piperidine and isobutylamine were detected in one or more varieties (Atawodi et al. 1995). Estimated average total daily intake of aliphatic amines by a typical kola nut chewer varied from 260 to 1,040 μg/day for secondary amines and from 2,430 to 9,710 μg/day for primary amines. Methylating activity of the nitrosated kola nuts, expressed as N-nitroso-N-methylurea equivalents, was significantly higher in kola nuts (170–490 μg/kg) than has ever been reported for a fresh plant product. These data suggested that the possible role of kola nut chewing in human cancer aetiology should be investigated in countries where kola nuts are widely consumed as stimulants.

In another study, kola-nut extract induced a number of overt neurotoxicological signs in male albino rats. A decrease in the total body weight and an increase in the absolute weights of the liver, kidney, brain and testis were observed after 18 weeks oral administration of kola-nut extract to the rats (Ikegwuonu et al. 1981). Total protein, RNA and DNA of these organs were significantly depressed. The activity of β-glucuronidase and β-galactosidase was induced only in the kidney, brain and testis of treated animals. While the levels of serum phosphomonoesterases and total cholesterol were significantly enhanced due to kola-nut intake, serum total and conjugated bilirubin levels were significantly decreased.

Traditional Medicinal Uses

In traditional folkloric medicine, the leaves, twigs, flowers, fruits follicles, and the bark of both *C. nitida* and *C. acuminata* were used to prepare a tonic as a remedy for coughs, diarrhœa, dysentery, vomiting (Ayensu 1978) and chest disorders (Irvine 1961). In the Ivory Coast, a macerated decoction of *C. nitida* is used to facilitate baby delivery and as aphrodisiac (Bouquet and Debray 1974). In Sierra Leone, an infusion of macerated leaves and salt is used for diarrhoea and asthma (Lebbie and Guries 1995). For headache and rheumatic pains, leaves of *Microdesmis puberula* are chewed with the nuts of *Cola nitida* and the paste used to massage joints and head; leaves are also ground with clay and rubbed on aching parts after the massage. In Benin, fresh nuts of *Colo nitida* are chewed as an aphrodisiac and to stimulate the muscles and nerves while the pounded fruit is used to treat ocular problems (Natabou Dégbé 1991). In Mali, macerated powdered bark or flowers of *Erythrina senegalensis* or powdered nuts or bark of *C. nitida* are used for amenorrhoea (Togola et al. 2008).

Other Uses

Besides edible products, kola nut has multifarious uses (Nzekwu 1961; Burkill 1966; Eka 1971; Ogutuga 1975; FAO 1982; Hamzat et al. 2000; Yahaya et al. 2001; Asogwa et al. 2006). Kolanut is used as an intercrop of cocoa to provide top shade. Kola nut is believed to act as a water purifier. Kola nut provides a source of colourant for cloth dyeing and for candles. It provides useful timber: pinkish-white sapwood and dull yellow heartwood which is suitable for making furniture, house and boat building, coach-work, plates, domestic utensils, musical instruments, gun stocks, wooden toys and games, joinery and carvings. The tree also provides good fuelwood. The kola nut testa, it has been suggested as a possible ingredient for making fertilizers The kola pod husk has also been utilized for the production of liquid soap. The most recent and remarkable advancement in kola by-product utilization is the use of kola pod husk in the replacement of up to 60% of the maize used in poultry feed formulations.

Cola nitida also has similar roles as *Cola acuminata* in social, religious and cultural customs and ceremonies.

Comments

Four subspecies have been recognised within *Cola nitida*: *alba*, *rubra*, *mixta* and *pallida by* Chevalier and Perrot (1911): one alba gives a white nut and is regarded as the best; the others gives pink, red or purple nuts (Burkill 1966).

Russell (1955) disagreed with Chevalier's criteria in differentiating the 4 sub-species within *Cola nitida* and reduced it to two distinctive kinds of *Cola nitida* described merely as 'white flowered' and 'red flowered'.

Selected References

Abbiw D (1990) Useful plants of Ghana. West African uses of wild and cultivated plants. Intermediate Technology and The Royal Botanic Gardens, Kew, 337 pp

Adeniyi BA, Groves MJ, Gangadharam PR (2004) In vitro anti-mycobacterial activities of three species of Cola plant extracts (Sterculiaceae). Phytother Res 18(5):414–418

Ajarem JS (1990) Effects of fresh kola-nut extract (*Cola nitida*) on the locomotor activities of male mice. Acta Physiol Pharmacol Bulg 16(4):10–15

Alaribe AAA, Ejezie GC, Ezedinachi ENU (2003) The role of kola nut (*Cola nitida*) in the etiology of malaria morbidity. Pharm Biol 41(6):458–462

Al-Hazmi MA (2000) Effects of the alkaloid fraction of kola nuts, *Cola nitida*, or its mixture, on the social behaviour of laboratory mice. Phyton (Buenos Aires) 67:93–101

Asogwa EU, Anikwe JC, Mokwunye FC (2006) Kola production and utilization for economic development. Afr Sci 7(4):217–222

Atawodi SE, Mende P, Pfundstein B, Preussmann R, Spiegelhalder B (1995) Nitrosatable amines and nitrosamide formation in natural stimulants: *Cola acuminata*, *C. nitida* and *Garcinia cola*. Food Chem Toxicol 33:625–630

Atawodi SE, Pfundstein B, Haubner R, Spiegelhalder B, Bartsch H, Wyn Owen R (2007) Content of polyphenolic compounds in the nigerian stimulants *Cola nitida* ssp. *alba*, *Cola nitida* ssp. *rubra* A. Chev, and *Cola acuminata* Schott & Endl. and their antioxidant capacity. J Agric Food Chem 55(24):9824–9828

Ayensu ES (1978) Medicinal plants of West Africa. Reference Publications International, Algona, 330 pp

Beattie GB (1970) Soft drink flavours: their history and characteristics. Cola or "kola" flavours. Flavour Ind 1(6):390–394

Benie T, Thieulant ML (2004) Mechanisms underlying antigonadotropic effects of some traditional plant extracts in pituitary cell culture. Phytomedicine 11(2–3): 157–164

Benie T, Duval J, Thieulant ML (2003) Effects of some traditional plant extracts on rat oestrous cycle compared with Clomid. Phytother Res 17(7):748–755

Bouquet A, Debray M (1974) Plantes Médicinales de la Côte d'Ivoire, 32. Mémoires/O.R.S.T.O.M, Paris, 231 pp

Burkill IH (1966) A dictionary of the economic products of the Malay Peninsula. Revised reprint. 2 volumes. Ministry of Agriculture and Co-operatives, Kuala Lumpur. Vol 1 (A–H), pp 1–1240; vol 2 (I–Z), pp 1241–2444

Carrillo JA, Bennitez J (2000) Clinically significant pharmacokinetic interactions between dietary caffeine and medications. Clin Pharmacokinet 39(12):127–153

Chevalier A, Perrot A (1911) Les Vegetaux utiles de l'Afrique tropicale Francaise. Challamell, Paris (Cited by Burkill, 1966)

Daels-Rakotoarison DA, Kouakou G, Gressier B, Dine T, Brunet C, Luyckx M, Bailleul F, Trotin F (2003) Effects of a caffeine-free *Cola nitida* nuts extract on elastase/alpha-1-proteinase inhibitor balance. J Ethnopharmacol 89:143–150

Ebana RUB, Madunagu BE, Ekpe ED, Otung IN (1991) Microbiological exploitation of cardiac glycosides and alkaloids from *Garcinia kola*, *Borreria ocymoides*, *Kola nitida* and *Citrus aurantifolia*. J Appl Bacteriol 71:398–401

Eka OU (1971) Chemical composition and use of kola nuts. J West Afr Sci Assoc 16:167–169

Ekanade O (1989) The effects of productive and non-productive kola, *Cola nitida* Vent. (Schott and Endlicher), on the status of major soil physical and chemical properties in S.W. Nigeria. Int Tree Crop J 5:279–294

Ettarh RR, Okoosi SA, Eteng MU (2000) The influence of kolanut (*Cola nitida*) on exploratory behaviour in rats. Pharm Biol 38:281–283

FAO (1982) Fruit-bearing forest trees. Forestry Paper No. 34. FAO, Rome. 177 pp

Hamzat RA, Olubamiwa O, Taiwo AA, Tiamiyu AK, Longe OG, Adeleye IOA (2000) Potentials of kola testa and pod husks in animal feeds. In: Book of Proceedings, 24th Annual NSAP Conference held at Umudike, Nigeria, 2000, p 112

Ibu JO, Iyama AC, Ijije CT, Ishmael D, Ibeshi M, Nwokediuko S (1986) The effect of *Cola acuminata* and *Cola nitida* on gastric acid secretion. Scand J Gastroenterol Suppl 124:39–45

Igwe SA, DAkunyili DN, Ikonne EU (2007) Ocular effects of acute ingestion of *Cola nitida* (Linn) on healthy adult volunteers. S Afr Optom 66(1):19–23

Ikegwuonu FI, Aire TA, Ogwuegbu SO (1981) Effects of kola-nut extract administration on the liver, kidney, brain, testis and some serum constituents of the rat. J Appl Toxicol 1(6):292–294

Irvine FR (1961) Woody plants of Ghana: with special reference to their uses. Oxford University Press, London, 868pp

Jayeola CO (2001) Preliminary studies on the use of kolanuts (*Cola nitida*) for soft drink production. J Food Technol Afr 6(1):25–26

Lebbie AR, Guries RP (1995) Ethnobotanical value and conservation of sacred groves of the Kpaa Mende in Sierra Leone. Econ Bot 49(3):297–308

Leung W-TW, Busson F, Jardin C (1968) Food composition table for use in Africa. FAO, Rome, 306 pp

Natabou Dégbé F (1991) Contribution à l'étude de la médecine et de la Pharmacopée traditionnelles au Bénin: Tentatives d'intégration dans le système de santé officiel. Thèse pour l'obtention du diplôme de Docteur en Pharmacie de l'Université Cheikh Anta Diop (Diplôme d'état)

Niemenaka N, Onomoa PE, Fotsoa Liebereib R, Ndoumou DO (2008) Purine alkaloids and phenolic compounds in three *Cola* species and *Garcinia kola* grown in Cameroon. S Afr J Bot 74(4):629–638

Nzekwu O (1961) Kola nut. Niger Mag 71:298–305

Odebode AC (1996) Phenolic compounds in the kola nut (*Cola nitida* and *Cola acuminata*) (Sterculiaceae) in Africa. Rev Biol Trop 44:513–515

Odebunmi EO, Oluwaniyi OO, Awolola GV, Adediji OD (2009) Proximate and nutritional composition of kola nut (*Cola nitida*), bitter cola (*Garcinia cola*) and alligator pepper (*Afromomum melegueta*). Afr J Biotechnol 8(2):308–310

Ogutuga DBA (1975) Chemical composition and potential commercial uses of kola nut *Cola nitida*. Ghana J Agric Sci 8:121–125

Oladokun MAO (1989) Nut weight and nutrient contents of *Cola acuminata* and *C. nitida* (Sterculiaceae). Econ Bot 43(1):17–22

Opeke LK (1992) Tropical tree crops. Spectrum Books Ltd, Ibadan

Osim EE, Udia PM (1993) Effects of consuming a kola nut (*Cola nitida*) diet on mean arterial pressure in rats. Int J Pharmacog 31:193–197

Russell TA (1955) The kola of Nigeria and the Cameroons. Trop Agric Trinidad 32:210–240

Tachie-Obeng E, Brown N (2004) *Cola nitida* and *Cola acuminata*: a state of knowledge report. In: Clark LE, Sunderland TCH (eds) The key non-timber forest products of Central Africa: state of the knowledge. USAID Technical Paper No. 122

Togola A, Austarheim I, Diallo A, Theiss D, Paulsen BS (2008) Ethnopharmacological uses of *Erythrina senegalensis*: a comparison of three areas in Mali, and a link between traditional knowledge and modern biological science. J Ethnobiol Ethnomed 4:6

Umoren EB, Osim EE, Udoh PB (2009) The comparative effects of chronic consumption of kola nut (*Cola nitida*) and caffeine diets on locomotor behaviour and body weights in mice. Niger J Physiol Sci 24(1):73–78

van Eijnatten CLM (1973) Kola. A review of the literature. Trop Abstr 28(8):541–550

Wikipedia (2009) Kola nut. http://en.wikipedia.org/wiki/Kola_nut. Accessed Aug 2010

World Agroforestry Centre (2008) Agroforestry database. http://www.worldagroforestry.org/sea/Products/AFDbases/af/index.asp. Accessed Aug 2010

Yahaya LE, Hamzat RA, Aroyeun SO (2001) Utilization of kola pod husk in liquid soap production. Moor J Agric Res 3(2):252–256

Grewia asiatica

Scientific Name

Grewia asiatica L.

Synonyms

Grewia hainesiana Hole, *Grewia obtecta* Wall., *Grewia rotundifolia* Thwaites, *Grewia subinaequalis* DC., *Grewia vestida* Wall., *Microcos lateriflora* L.

Family

Malvaceae, also placed in Sparmanniaceae, Tiliaceae

Common/English Names

Grewia, Falsa, Indian Phalsa, Parsa Phalsa

Vernacular Names

Bangladesh: Chandani Shewra;
French: Grain De Bébé;
India: Shukri (Gujarati), Dhamin, Falsa, Parusha, Phalsa, Shukri (Hindu), Dagala, Dadasala, Phulsha (Kannada), Phalsi (Konkani), Chadicha (Malayalam), Phalsa (Marathi), Pharosakoli (Oriya), Alpasthi, Dhanvana, Dharmana, Mriduphal, Parapara, Parushaka, (Sanskrit), Pharaho (Sindhi), Unnu (Tamil), Phutiki (Telugu), Phalsa (Urdu);
Khmer: Pophlië;
Laos: Nhap;
Nepali: Fussi, Phalsa;
Pakistan: Dhaman, Phalsa;
Philippines: Bariu'an (Iloko), Bariuan-Gulod;
Spanish: Falsa;
Thailand: Malai (Central), Ya-Khithut (Nakhon Ratchasima), Po Tao Hai (Chiang Mai);
Vietnam: Cò-Ke-Á.

Origin/Distribution

Phalsa is native to tropical southern Asia from Pakistan east to India and Nepal, Laos, Thailand and Cambodia. It has become naturalised and locally invasive in Australia and the Philippines.

Agroecology

Phalsa is a warm climate species. It grows on a wide range of soil types: sandy, clayey, calcareous soils but thrives best on loamy soils from near sea level to 1,000 m elevation. It tolerates light frost but suffers from leaf shedding and

T.K. Lim, *Edible Medicinal And Non-Medicinal Plants: Volume 3, Fruits*,
DOI 10.1007/978-94-007-2534-8_24, © Springer Science+Business Media B.V. 2012

grows in full sun. It is moderately drought tolerant but does better with irrigation during the dry periods and fruiting period.

Edible Plant Parts and Uses

Ripe fruits are eaten raw or used for making drinks, squash, sherbet or jams.

Botany

A spreading, woody shrub or small tree, 4–6 m high. Stem with greyish-white to brown bark and stellate tomentose young shoots. Leaves are alternate, spiral, simple, stipulate, petiolate, petiole 13–15 mm long. Stipules varying from linear to foliaceous and broadly falcate, caducous. Lamina is 10.5–2 cm long, 9–14 cm wide, undissected, varying from broad-cordate to obliquely ovate, base rounded or cordate or slightly oblique, apex acute or obtuse, margins serrate, 5(–6)-costate, greyish-tomentose beneath, scabrous above (Plates 2, 3). Flowers are arranged in inflorescences, in cymes or in umbels; predominantly yellow or red, regular, pedicellate, pedicel 11–14 mm long, perianth 2-whorled. Calyx to 12 mm long with 5 free sepals. Corolla to 10 mm long with 5 free petals. Stamens 50–100, free of the perianth, filaments free of each other, orange-yellow, turning purplish with oblong anthers. Ovary syncarpous, globose, strigose superior, 2–4-celled. Ovules 2 per cell, or 4 per cell. Style 5 mm long with obscurely 4-lobed stigma. Drupe more or less globose, 5–12 mm in diameter, mostly entire or obscurely 2-lobed, hairy, pale green to yellow turning to red to dark purple when ripe (Plates 1, 2 and 3), mesocarp fibrous and acidic.

Plate 1 Unripe and ripe Phalsa fruit

Plate 2 Ovate, 5-costate leaves

Plate 3 Close-up of Phalsa fruit

Nutritive/Medicinal Properties

The nutrient composition of fresh ripe phalsa fruit per100g edible portion was reported as: moisture 80.8 g, energy 72 kcal, protein 1.3 g, fat 0.9 g, carbohydrate 14.7 g, fibre 1.2 g, ash 1.1 g, vitamin A 20ug RAE (Retinol Activity Equivalent), total carotene 481 ug, vitamin C 22 mg, niacin 0.3 mg, Fe 3.1 mg, Ca 129 mg and P 39 mg (Gopalan et al. 2002).

The flowers were reported to contain grewinol, a long chain keto-alcohol, tetratricontane 22-ol 13-one (Lakshmi and Chauhan 1976). The phalsa seeds was found to produce approximately 5% yield of a bright yellow oil that contained 8%

palmitic acid, 11% stearic acid, 13.5% oleic acid, and 64.5% linoleic acid with 3% unsaponifiable (Morton 1987).

Some of the pharmacological properties of the plant reported are present below.

Antioxidant Activity

Extracts of *Grewia asiatica* leaves were also found to exhibit antioxidant activity (Gupta et al. 2007). The successive extracts such as petroleum ether, benzene, ethyl acetate, methanol, water and 50% crude methanol extracts exhibited IC_{50} values of 249.60, 16.19, 26.17, 27.38, 176.14 and 56.40 μg/ml, respectively in DPPH and 22.12, 27.00, 47.38, 56.85, 152.75 and 72.75 μg/ml, respectively in nitric oxide radical inhibition assays. These values were comparable with standards such as an ascorbic acid and quercetin.

G. asiatica fruit samples were found to have a potent in-vitro antioxidant activity which was positively correlated with its phenolic content (Asghar et al. 2008). The highest TEAC (Trolox equivalent antioxidant capacity) values of antioxidant activity were obtained for peel followed by pulp and seeds. When *G. asiatica* was stored at 0°C for duration of 1 month, the polyphenols were partly degraded and/or transformed into other products.

Antidiabetic Activity

Aqueous extracts of leaves of *Grewia asiatica* (200 mg/kg b.w.) showed more significant reduction in blood glucose level in alloxan induced diabetic Wister rats compared to control and glibenclamide (10 mg/kg b.w.) (Patil Priyanka et al. 2010).

Radioprotective Activity

G. asiatica fruit extract (GAE) was found to have radioprotective activities (Ahaskar et al. 2007; Sharma and Sisodia 2009; Sisodia and Singh 2009). After drug toxicity test, the oral administration of 700 mg/kg body weight/day of methanolic extract of *Grewia asiatica* (GAE) for 15 consecutive days before exposure to 10 Gy of γ-radiation was found to afford maximum protection as evidenced by the highest number of mice survivors after 30 days post irradiation (Ahaskar et al. 2007). At this dose level GAE was found to be effective against different levels of radiation doses. LD50/30 value of 6.21 for irradiation alone (control) and 9.53 for *Grewia asiatica* + irradiation group (experimental) was obtained; a dose reduction factor (DRF) 1.53 was calculated. The mice of experimental group exhibited significant modulation of radiation – induced decreases of reduced glutathione (GSH) and radiation – induced increase in lipid peroxidation (LPO) in the whole brain and liver at 24 hours after radiation exposure.

Studies showed that treatment of mice with GAE before and after irradiation caused a significant depletion in TBARS (thiobarbituric acid reactive substances) content followed by a significant elevation in glutathione and protein concentration in the intestine and testis of mice at all post-irradiation autopsy intervals in comparison to irradiated mice. Significant protection of DNA and RNA in testis was also noticed. GAE which contained contains anthocyanin-type cyanidin 3-glucoside, vitamins C and A, minerals, carotenes and dietary fibre was found to have strong radical scavenging activity in 2,2-diphenyl-1-picrylhydrazyl (DPPH) and O^{2-} assays and also showed in vitro radioprotective activity in protein carbonyl assay in a dose-dependent manner. Additional studies showed that prior/post-supplementation of GAE had radioprotective potential as well as neuroprotective properties against radiation (Sisodia and Singh 2009). Marked radiation induced changes in the amount of cerebellar lipid peroxidation (LPO), glutathione (GSH), protein, nucleic acids and histopathological changes could be significantly ameliorated specially at later intervals by supplementation of GAE prior to and post irradiation. Radiation induced deficits in learning and memory were also significantly ameliorated. The scientists (Sisodia et al. 2008a, b) further reported that radiation-induced augmentation in the levels of lipid peroxidation of mice cerebrum was significantly ameliorated by GAE pretreatment. Radiation-induced depletion in the level of glutathione and protein was prevented significantly by GAE administration. GAE post-treatment protected liver and blood against radiation-induced damage by inhibiting glutathione depletion and ameliorating lipid peroxidation levels that attended normal levels by day 30 post-treatment. Moreover, the magnitude of recovery from oxidative damage in terms of TBARS and GSH content was significantly higher in the irradiated+GAE-treated group. Studies showed that *Grewia asiatica* fruit pulp extract provided protection against radiation-induced alterations in blood of mice (Singh et al. 2008). Radiation induced deficit in different blood constituents glutathione, sugar, and protein levels in the serum could be significantly increased, whereas radiation induced elevation of LPO and cholesterol levels were markedly decreased in *Grewia asiatica* fruit pulp extract post-treated animals compared to the control group.

Hepatoprotective Activity

In a more recent study, *Grewia asiatica* fruit (GAE) was found to have hepatoprotective activity (Sharma and Sisodia 2010). Treatment of mice with GAE before and after irradiation caused a significant elevation in liver DNA and RNA level in comparison to irradiated mice which had a significant depletion in the DNA and RNA level at all intervals studied viz. 1–30 day. Photomicrograph of liver histology also showed that pre and post supplementation of GAE provided protection against radiation. Similarly counting of different type hepatocytes also showed that GAE protected the liver against radiation.

Antimicrobial Activity

The methanolic extract of leaves of *Grewia asiatica* was fodun to have antifungal and antiviral activity (Sangita et al. 2009). The extract exhibited activity against the following fungi in decreasing order of sensitivity: *Candida albicans*, *Aspergillus thiogenitalis*, *Penicillium notatum*, *Penicillium citrinum* and *Aspergillus niger*. *Candida albicans* was the most susceptible and *A. niger* the most resistant. The extract also showed maximum inhibitory activity at a concentration of 1,000 μg/ml against Urdbean leaf crinkle virus.

Uses in Ayurvedic Preparation

Gupta et al. (2010) confirmed that the aqueous extract of Jwarhar mahakashay Ayurvedic preparation (from the roots of *Hemidesmus indicus* R. Br., *Rubia cordifolia* L., *Cissampelos pareira* L.; fruits of *Terminalia chebula* Retz., *Emblica officinalis* Gaertn., *Terminalia bellirica* Roxb., *Vitis vinifera* L., *Grewia asiatica* L., *Salvadora persica* L. and granules of *Saccharum officinarum* L.) its antipyretic–analgesic effect with very low ulcerogenicity and toxicity the preparation had been used as a traditional antipyretic. Flavonoids, glycosides and tannins were later found to be present in the extract. The principal component was identified as 2-(1-oxopropyl)-benzoic acid, which is quite similar to the active compound found in the standard drug Aspirin (2-acetyl-oxy-benzoic acid).

Traditional Medicinal Uses

In traditional folkloric medicine, the fruit has been used as astringent, stomachic and cooling agent (Kirtikar and Basu 1989). When unripe, it has been reported to alleviate inflammation and was administered in respiratory, cardiac and blood disorders, as well as in fever. The fruit was also beneficial food throat ailments. Root bark has been prescribed for rheumatism and its infusion used as a demulcent. The leaves were applied on skin eruptions. Seeds of *G. asiatica* has been used as antifertility agent and was reported to have anti-implantation and abortifacient activities (Pokharkar et al. 2010). In traditional folkloric medicine in Bangladesh, *Grewia asiatica* plant is commonly used for gonorrhoea by the Garo tribe and local traditional healers in

Madhupur and Tangail district (Hossan et al. 2010). It is also used to treat lack of appetite, typhus, acidity, giddiness, diarrhoea, hypertension, stimulant, anorexia..In India, *G. asiatica* is also used for gonorrhoea and as astringent, demulcent, rheumatism, stomachic and tumour.

Other Uses

The leaves have been used as fodder. In Myanmar, the bark has been used as a soap substitute. A mucilaginous extract of the bark was found useful in clarifying sugar. Fibre extracted from the bark has been made into rope. The wood has been used for archers' bows, spear handles, shingles and poles for carrying loads on the shoulders Stems that have been pruned off served as garden poles and for basket-making.

Comments

The species has become locally invasive in Australia and the Philippines.

Selected References

Ahaskar M, Sharma KV, Singh S, Sisodia R (2007) Radioprotective effect of fruit extract of *Grewia asiatica* in Swiss albino mice against lethal dose of γ-irradiation. Asian J Exp Sci 21(2):297–310

Asghar MN, Khan IU, Sherin L, Ashfaq M (2008) Evaluation of antioxidant activity of *Grewia asiatica* berry using 2, 2,-azinobis-(3-ethylbenzothiazoline-6-sulphonicacid)andN,N-dimethyl-p-phenylenediamine radical cations decolourization assays. Asian J Chem (India) 20(7):5123–5132

Gopalan G, Rama Sastri BV, Balasubramanian SC (2002) Nutritive value of Indian foods. National Institute of Nutrition, Indian Council of Medical Research, Hyderabad

Gupta MK, Lagarkha RS, Sharma DK, Singh PK, Ansari RHS (2007) Antioxidant activity of the successive extracts of *Grewia asiatica* leaves. Asian J Chem (India) 19(5):3417–3420

Gupta M, Shaw BP, Mukherjee A (2010) A new glycosidic flavonoid from Jwarhar mahakashay (antipyretic) Ayurvedic preparation. Int J Ayurveda Res 1(2):106–111

Hossan MS, Hanif A, Agarwala B, Sarwar MS, Karim M, Taufiq-Ur-Rahman M, Jahan R, Rahmatullah M (2010) Traditional use of medicinal plants in Bangladesh to treat urinary tract infections and sexually transmitted diseases. Ethnobot Res Appl 8:61–74

Kirtikar KR, Basu BD (1989) Indian medicinal plants, vol I, 2nd edn. Published by Lalit Mohan Basu, Allahabad

Lakshmi V, Chauhan JS (1976) Grewinol, a keto-alcohol from the flowers of *Grewia asiatica*. Lloydia 39:372–374

Morton JF (1987) Phalsa. In: Fruits of warm climates. Julia F. Morton, Miami, pp 276–277

Parmar C (1992) *Grewia asiatica* L. In: Verheij EWM, Coronel RE (eds) Plant resources of South-East Asia No. 2: edible fruits and nuts. Prosea Foundation, Bogor, pp 184–186

Patil Priyanka S, Patel MM, Bhavsar CJ (2010) Comparative antidiabetic activity of some herbal plants extracts. Pharm Sci Monit 1(1):12–19

Pokharkar RD, Saraswat RK, Kotkar S (2010) Survey of plants having antifertility activity from Western Ghat area of Maharashtra state. J Herb Med Toxicol 4(2):71–75

Sangita K, Avijit M, Shilpa P, Shivkanya J (2009) Studies of the antifungal and antiviral activity of methanolic extract of leaves of *Grewia asiatica*. Pharm J 1(3):221–223

Sharma KV, Sisodia R (2009) Evaluation of the free radical scavenging activity and radioprotective efficacy of *Grewia asiatica* fruit. J Radiol Prot 29:429–443

Sharma KV, Sisodia R (2010) Hepatoprotective efficacy of *Grewia asiatica* fruit against oxidative stress in Swiss albino mice. Iran J Radiat Res 8(2):75–85

Singh S, Sharma KV, Ahaskar M, Sisodia R (2008) Protective role of *Grewia asiatica* on blood after radiation exposure in mice. J Complement Integr Med 5(1): Article 14

Sisodia R, Singh S (2009) Biochemical, behavioural and quantitative alterations in cerebellum of Swiss albino mice following irradiation and its modulation by *Grewia asiatica*. Int J Radiat Biol 85(9):787–795

Sisodia R, Ahaskar M, Sharma KV, Singh S (2008a) Modulation of radiation-induced biochemical changes in cerebrum of Swiss albino mice by *Grewia asiatica*. Acta Neurobiol Exp (Warsaw) 68(1):32–38

Sisodia R, Singh S, Sharma KV, Ahaskar M (2008b) Post treatment effect of *Grewia asiatica* against radiation-induced biochemical alterations in Swiss albino mice. J Environ Pathol Toxicol Oncol 27(1):113–121

Yadav AK (1999) Phalsa: a potential new small fruit for Georgia. In: Janick J (ed) Perspectives on new crops and new uses. ASHS Press, Alexandria, pp 348–352

Scaphium macropodum

Scientific Name

Scaphium macropodum (Miq.) Beumée ex K. Heyne.

Synonyms

Carpophyllum macropodum Miq. (basionym), *Firmiana affinis* Terr., *Scaphium affinis* (Mast.) Pierre, *Scaphium beccarianum* Pierre, *Scaphium lychnophorum* Pierre, *Sterculia beccariana* Pierre, *Sterculia lychnophora* Hance, *Sterculia macropoda* Hook.ex Kloppenburg-Versteegh.

Family

Malvaceae, also placed in Sterculiaceae

Common/English Name

Malva Nut, Makjong, Kembang Semangkok

Vernacular Names

Chinese: Ta Hai Tzeh, Tou Shai Hoi, Pandahai (Mandarin), Tai Hong Lam (Cantonese), Pong Tia Ha (Hokien);
Borneo: Berempayang, Kambang Sulih, Kembang Semangkok, Payang Karang;
Burmese: Thilaung, Samrung, Thibyu, Shaw, Thau-Thinbaw;
Indonesia: Tempang (Java), Kapas-Kapasan, Merpayang, Kepayang (Sumatra);
Khmer: Crap Chi Ling Leak, Som Rong, Samrang Si Phle, Som Vang, Som Rong Sva;
Malaysia: Kembang Semangkok Batu (Sabah), Boh Change (Semang),Makjong, Kembang Semangkok Batu, Kembang Semangkok Jantong Selayar;
Laotian: Mak Chong Ban, Crap Chi Ling Leak;
Singapore: Cheng T'ng Tree;
Thai: Makjong, Phunghtalai (Central), Samrong (South-eastern);
Vietnamese: Uoi, An Nam Tu, Dai Hai Tu, Huong Dao.

Origin/Distribution

Its natural distribution is found in Myanmar, Cambodia, Laos, Vietnam, Thailand, Peninsular Malaysia, Sumatra and Borneo (throughout the island).

Agroecology

Malva nut is a common shade-tolerant canopy tree in tropical rainforests such as undisturbed mixed dipterocarp forests, (peat)-swamp forests and sub-montane forests up to 1,200 m altitude,

T.K. Lim, *Edible Medicinal And Non-Medicinal Plants: Volume 3, Fruits*,
DOI 10.1007/978-94-007-2534-8_25, © Springer Science+Business Media B.V. 2012

usually on well-drained undulating hillsides and ridges where it often occurs in high densities. In secondary forests, it is usually present as a pre-disturbance remnant. It is found on rich clay and sandy soils and on rocky and shallow soils. Mast flowering and fruiting occurs every 3–4 years.

Edible Plant Parts and Uses

The gel made from malva nut (seed) is edible and locally malva nuts are used as ingredients in food dishes, beverages and as medicine. In Singapore, it is used in the dessert *cheng t'ng*. The malva seeds are sun dried and stored in ventilated conditions. Before consuming, the seeds are soaked in water where it forms a large quantity of mucilage. The mucilaginous gel from the outer seed coat is used to prepare a beverage, together with sugar or fruit juice. Malva nut juice in a can is widely sold in Thailand as a dietary health beverage.

Malva nut gum has been reported to have potential to improve yield and textural parameters of meat products. Studies in Thailand reported that Makjong gel can be used as a fat replacer in *Moo Yoo* (Thai emulsion-type pork sausage) product, one of the popular meat products in Ubon Ratchathani province Thailand (Juthong et al. 2007). Addition of 50% Makjong gel to *Moo Yoo* formulation resulted in decreased fat content, from 43.32% in control to 36.15% in the reduced fat product. Thai scientists showed that frankfurters with 0.2 g/100 g CMG (crude Malva gum) showed low cooking loss and had better textural properties than the frankfurters without CMG (Somboonpanyakul et al. 2007). Frankfurters lightness and redness were reduced due to CMG. Sensory analysis results indicated that the frankfurters with 0.2 g/100 g CMG were more firm and elastic. Overall, the study indicated the potential use of CMG to improve yield and textural parameters of meat products.

Botany

A large and straight upper canopy, perennial tree, over 45 m tall, with greyish-brown fissured bark and with a bole diameter up to 260 cm above the thick, spreading buttresses. Buttresses up to 2 m. twigs stout, glabrescent, with prominently raised large leaf scars; young shoots reddish-brown, pubescent. Stipules subulate, ferruginous, tomentose, caduceus. Leaves ovate to ovate-oblong, 15–25 cm by 7–12 cm, base shallowly cordate or cuneate to truncate, apex acute or acuminate, subcoriaceous, glabrous; nerves 4–7(–8) pairs, scalariform veins prominent on both surfaces; young leaves on saplings plamate, 3–5-lobed. Petioles 5–21 cm long, swollen at both ends. Inflorescence 3–21 cm long, pubescent, dense, axillary or terminal panicles. Perianth campanulate, 5-lobed, pubescent, greenish-white, faintly scented; stamens 10–13, in a whorl; anthers 2-celled, yellow. Pistil with 2 red carpels, androgynophore 2–4 mm, erect. Fruit a large follicle, 18–20 by 5–6 cm, soon dehiscing, boat-shaped and membranous, glabrous or hairy near base. Seeds ellipsoid, 3–3.5 cm by 1.4–2 cm, glabrous, becoming wrinkled and rugose when dry (Plate 1).

Plate 1 Malva nut

Nutritive/Medicinal Properties

Malva nut (fruit) has been reported to have medicinal properties. Soaked in water, malva seeds form a large quantity of gelatine. This gel is used as a medicine against intestinal infections, diarrhoea, throat aches, asthma, dysentery, fever, coughs, inflammation, but also for urinary ailments. Some people drink Malva nut juice as a dietary beverage because of its cooling properties and as a febrifuge. In Java, malva used with cinnamon and basil is used for treating sprue and cough and mixed with bitter aloes and seeds of *Vitex pteropoda* is used for dysentery.

It is also used, in China, as a traditional drug for the prevention of pharyngitis, treatment of tussis and constipation. Malva nut has cooling agent medicinal properties. It is used to treat dysentry, intestinal infections, coughing and sore throats (Lamxay 2001).

Chemical analysis of an alkaline extracted malva nut gum (Somboonpanyakul et al. 2006) revealed that it contained 62.0% carbohydrates, 8.3% ash and 8.4% protein. The major constituent monosaccharides of the gum were 31.9% arabinose, 29.2% galactose and 29.5% rhamnose. The gum also contained 6.4% uronic acid and small amounts of glucose, xylose and mannose. Malva seeds also contain alkaloids, such as sterculinine I and sterculinine II (Wang et al. 2003).

Makjong's extract exhibited high viscosity and gel strength which were concentration dependent. Gelation was observed when Makjong's extract concentration exceeded 0.75% (w/v) (Singthong et al. 2007). Makjong's extract (0.5%w/v) also formed gel under neutral conditions (pH 6–7). The addition of NaCl and CaCl2 significantly decreased the Makjong's extract viscosity. It was also observed that addition of sugar increased the viscosity and gel strength of Makjong's extract and excessive amount of sugar (>30%) resulted in decreasing viscosity and gel strength.

Other Uses

A valued and major source of kembang semangkok timber, moderately durable and very easy to treat and suitable for sawn timber, furniture manufacture, interior finishing, panelling, fancy veneer and flooring for residential dwelling. The fibrous inner bark is sometimes used as walling for houses.

Comments

In the past years, malva nuts surpassed cardamom as the second most important export product after coffee in Laos. Laos export malva nuts to China (leading importing country), Thailand, Vietnam, South-Korea and France.

Selected References

Burkill IH (1966) A Dictionary of the economic products of the Malay Peninsula. Revised reprint. 2 volumes. Ministry of Agriculture and Co-operatives, Kuala Lumpur. vol 1 (A–H), pp 1–1240, vol 2 (I–Z), pp 1241–2444

Choo KT, Gan S, Lim SC (1999) Timber notes – Light hardwoods VI (Dedali, Kedondong, Kelempayan, Kelumpang, Kembang Semangkok). *Timber Techonology Bulletin* No. 16. Timber Technology Centre (TTC), Forest Research Institute Malaysia, Kepong, Kuala Lumpur

Juthong T, Singthong J, Boonyaputthipong W (2007) Using mhakjong (*Scaphium macropodum*) gel as a fat replacer in Thai emulsion-type pork sausage (Moo Yo). Paper presented at Food Innovation Asia 2007 "Q" Food for Good Life, BITEC, Bangna, Bangkok, 14–15 June 2007

Kostermans AJGH (1953) The genera *Scaphium* Schott and Endl. & *Hildegardia* Schotto and Endl. (Sterculiaceae). J Sci Res Indones 2:13–23

Lamxay V (2001) Important non-timber forest products of Lao PDR. Lao PDR, Forest Research Center, Vientiane

Singthong J, Ounthuang M, Chommala K, Thongkaew C (2007) Extraction and functional properties of mhakjong extract. Paper presented at Food Innovation Asia 2007 "Q" Food for Good Life, BITEC, Bangna, Bangkok, 14–15 June 2007

Slik JWF (2006) Trees of Sungai Wain. Nationaal Herbarium Nederland. http://www.nationaalherbarium.nl/sungaiwain/. Accessed Feb 2010

Soerianegara I, Lemmens RHMJ (1994) Plant resources of South East Asia No. 5(1). Timber trees: major commercial timbers. Prosea Foundation, Bogor, p 383

Somboonpanyakul P, Wang Q, Cui W, Barbut S, Jantawat P (2006) Malva nut gum. (Part I): extraction and physicochemical characterization. Carbohydr Polym 64:247–253

Somboonpanyakul P, Barbut S, Jantawat P, Chinprahast N (2007) Textural and sensory quality of poultry meat batter containing malva nut gum, salt and phosphate. LWT Food Sci Technol 40(3):498–505

Vantomme P, Markkula A, Leslie RN (eds) (2002) Non-wood forest products in 15 countries of tropical Asia an overview. FAO, Bangkok

Wang RF, Yang XW, Ma CM, Shang MY, Liang JY, Wang X (2003) Alkaloids from the seeds of *Sterculia lychnophora* (Pangdahai). Phytochemistry 63:475–478

Wilkie P (2009) A revision of *Scaphium* (Sterculioideae, Malvaceae/Sterculiaceae). Edinb J Bot 66:283–328

Sterculia foetida

Scientific Name

Sterculia foetida L.

Synonyms

Clompanus foetida Kuntze, *Sterculia polyphylla* R.Br.

Family

Malvaceae, also placed in Sterculiaceae

Common/English Names

Bastard Poon Tree, Great Sterculia, Hazel Sterculia, Wild Almond, Poom Tree

Vernacular Names

China: Hsiang-P'ing-P'o Xiang Ping Po;
India: Jangli Badam (Bengali), Janglibadam (Hindu), Penari (Kannada), Gorapu-Badam, Gurapu-Vadam, Pee, Pottaikavalam (Tamil);
Indonesia: Kabu-Kabu, Kalupat (General), Jangkang, Kepuh, Kepoh, Poh (Java), Kepoh, Koleangka (Sundanese);
Kampuchea: Samrong;
Malaysia: Kelumpang, Kelumpang Jari, Kelapong, Kayu Lepong;
Myanmar: Letpan Shaw;
Philippines: Kalumpang (Bikol), Kalumpang (Bisaya), Bangad, Bangag, Bongog (Ibanag), Bangar, Bobor, Bubur (Iloko), Kurumpang (Maguindanao), Bangar (Negrito), Kalumpang (Pampangan), Bobo, Bobog, Bubog (Panay Bisaya), Bubog Kumpang (Sulu), Bubog, Kalumpang (Tagalog);
Spanish: Anacaguita;
Sri Lanka: Telembu (Sinhala);
Thailand: Chamahong, Homrong,Sam, Sam Rang;
Vietnamese: Trôm Hôi; Trôm Đất; (Cây) Quả Mõ.

Origin/Distribution

The species is found from East Africa to north Queensland, Australia – Australia, Bangladesh, Djibouti, Eritrea, Ethiopia, India, Indonesia, Laos, Kampuchea, Kenya, Malaysia, Myanmar, Oman, Pakistan, Philippines, Somalia, Sri Lanka, Tanzania, Thailand, Uganda, Yemen, Vietnam, Zanzibar. It is also cultivated elsewhere in the tropics including south China.

Agroecology

In its native range, it is found in primary and secondary forests, usually on river banks and sandstone rocks along the coasts, and in thickets and

T.K. Lim, *Edible Medicinal And Non-Medicinal Plants: Volume 3, Fruits*,
DOI 10.1007/978-94-007-2534-8_26, © Springer Science+Business Media B.V. 2012

open areas, from sea level to 1,000 m altitude. Although adaptable to most soils, it requires adequate moisture for optimum growth and development.

Edible Plant Parts and Uses

Seeds are usually cooked, roasted and eaten although they can be eaten raw. They have a pleasant, chestnut taste when roasted. The oil from the seed has also been used as culinary oil. The kernels have been used to adulterate cacao.

Botany

Medium tree to 30 m tall, 150 cm girth, bark greyish-brown, fissured or dippled (Plate 1). Buttress short to 90 cm. Twigs stout with prominent leaf scars. Leaves grouped together at apex, leaf petiole 10–20 cm, palmately compound, leaflets 7–9, elliptic-lanceolate 10 × 2.5 to 15 × 7.5 mm apex acuminate-caudate, acute base on very short petiolule (Plate 2); young leaves pinkish. Flowers 20–25 mm wide, foetid, woolly in loose, racemose panicles 10–15 cm long. Male and female flowers on separate trees. Flower apetalous, calyx deeply 5-partite divided, calyx tube red inside, glabrous, calyx lobes greenish-yellow turning to red, stamens joined into stamina column with anthers on top. Follicle 5–8 cm long and nearly as wide, woody-fibrous, stout, dehiscent with ventral suture, set in 2–5 clusters, green (Plate 3) to reddish when ripe with thick wall (Plate 4). Seeds 2.5–3 cm long, purple black, velvety, ellipsoid or oblongish, with small waxy yellow rudimentary aril at one end.

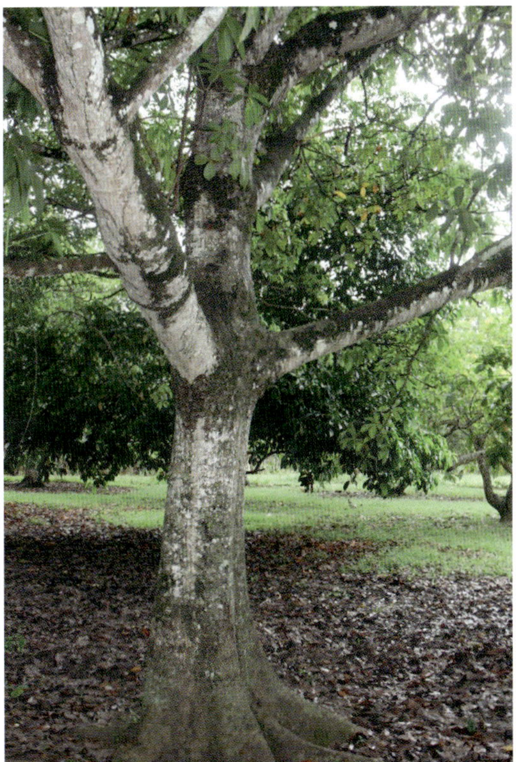

Plate 1 Sort-buttressed trunk and branches

Plate 2 Palmate compound leaves

Plate 3 Immature fruits

Plate 4 Hard, woody, thick rind of old fruits

Nutritive/Medicinal Properties

Seed kernels of *Sterculia foetida* were found to contain 53–55% of a pale yellow oil which polymerised rapidly at 240–250°C (Varma et al. 1956). *Sterculia foetida* oil was found to contain 71.8% of sterculic acid and minor proportions of oleic, linoleic, and saturated acids. The saturated component consisted mostly of myristic and palmitic acids. The oil had traces of tristearin (0.8%) and a major quantity of tristerculin (31.4%) together with different amounts of the glycerides of the type GS2U, GSU2, and GU3 of other fatty acids. The oil had a specific gravity at 40°C of 0.92399, refractive index 1.4662, acid value 5.7, saponification value 177.5, iodine value 74.

The seed oil of *S. foetida* was found to have 17.6% sterculic acid and 2.1% malvalic acid besides the conventional fatty acids (Badami et al. 1980). Cyclopropenoids fatty acids (CPFA) were found in *Sterculia striata* and *Sterculia foetida* seed oils (Aued-Pimentel et al. 2004). Fehling et al. (1998) developed a method to obtain pure malvalic (cis-8,9-methyleneheptadec-8-enoic) and sterculic (cis-9,10-methyleneoctadec-9-enoic) acid methyl esters from *Bombax munguba* and *Sterculia foetida* seed oils. Methyl malvalate and methyl sterculate were obtained with purities of 95–97% and 95–98%, respectively. A pale yellow oil was extracted from *S. foetida* seeds with a yield of 64.3% (Vipunngeun and Palanuvej 2009). Palmitic acid was found as dominant fatty acid (52%); oleic acid 14% and sterculic acid 10%. Oleic acid was proposed to be precursor of sterculic acid biosynthesis catalyzed by cyclopropane fatty acid synthase and desaturase.

Sterculia foetida seeds were found to have phosphatidylcholines and phosphatidylethanolamines which contained 58–83% of unsaturated fatty acids (Rao et al. 1989). Cyclopropane synthase from *Sterculia foetida* developing seeds was found to catalyze the addition of a methylene group from S-adenosylmethionine to the cisdouble bond of oleic acid (Bao et al. 2002). Further studies revealed that *S. foetida* cyclopropane synthase (SfCPA-FAS) was expressed in *S. foetida* seeds, but not in leaves, and was found to be a membrane protein localized to microsomal fractions (Bao et al. 2003).

The total ash value, acid insoluble ash value and water soluble ash value in powdered seeds of *Sterculia foetida* were found to be 3.9%, 0.76% and 0.84%, respectively (Shamsundar and Paramjyothi 1989). Crude fibre value was found to be 7.1%. Flavonoids, saponins and alkaloids were also present in the ethanol extract. Cyclopropene fatty acids were detected in the lipids of seeds and roots of *Ceiba pentandra* and *S. foetida* (Kaimal and Lakshminarayana 1970). Malvalic was the main acid in the root lipids of *S. foetida* and linoleic in the lipids of buds, flowers and trunk-wood of *C. pentandra*, trunk-wood and stems of *S. foetida*. Linolenic was the principal acid in the leaf lipids of *C. pentandra* and *S. foetida*.

Scutellarin, 6-O-β-D-glucuronyl scutellarein designated isoscutellarin, procyanidin-β-D-glucuronide, D-gluruconic acid, 6-O-β-D-glucuronyl luteolin and cyanidin-3-O-glucoside were isolated from leaves of *S. foetida* (Nair et al. 1977). Eight compounds were isolated from *S. foetida* leaves and identified as 5,7,8-tetrahydroxy-4′-methoxyflavone-8-O-β-D-glucoside (1), 5,7,8-tetrahydroxy-4′-methoxyflavone-7-O-β-D-glucoside (2), quercetin-3-O-β-D-glucoside (3), apigenin-6, 8-di-C-β-D-glucoside (4), puerarin (5), 5,7,8,3′-tetrahydroxy-4′-methoxyflavone (6), 5,7,8-tetrahydroxy-3′,4′-dimethoxyflavone(7),5,7,8-tetrahydroxy-4′-methoxyflavone (8) (Xia et al. 2009b). Two new flavonoid glycosides, hypolaetin 4′-methyl ether 8-O-β-d-glucuronide 2″-sulfate (1) and hypolaetin 4′-methyl ether 3′-O-β-d-glucoside (2), and a new phenylpropanoid glucose ester, 1,6-diferuloyl glucose (3), were isolated from the leaves of *Sterculia foetida* (Xia et al. 2009a).

Some pharmacological properties of the various plant parts are presented below.

Antiinflammatory and CNS (Central Nervous System) Activities

Taraxer-14-en-3β-ol was shown to be a bioactive ingredient in the leaves of *Sterculia foetida* (Naik et al. 2004). The alcohol its acetate and ketone showed anti-inflammatory activity against 12-O-tetradecanoylphorbol-3-acetate (TPA) induced mouse ear oedema with inhibition ratios of 60.0, 58.57 and 40.57 at 0.5 mg/ear, respectively. The percentage inhibition of inflammation increased with dose for each compound.

An alcoholic extract of *Sterculia foetida* leaves was found to reduced exploratory activity in mice (Mujumdar et al. 2000). Further it potentiated pentobarbitone sleeping time in normal and chronic pentobarbitone-treated mice. It also potentiated barbital sodium-induced hypnosis, indicating central nervous system depressant activity. The extract also exhibited significant anti-inflammatory activity in the acute carrageenan-induced rat paw edema and the chronic granuloma pouch models. However, it was devoid of analgesic activity in the tail flick model.

Luteolytic Activity

The results a study suggested that an extract of *Sterculia foetida* seeds, containing sterculic acid methyl ester could suppress luteal progesterone production or cause luteolysis of corpus luteum in ewes during early gestation (Tumbelaka et al.1994). The in-vivo suppressive effect of the extract on the corpus luteum may involve its ability to interfere with the conversion of pregnenolone to progesterone.

Mitogenic Activity

Sterculic acid triglycerides in the seed oil of *S. foetida* was found to significantly stimulate smooth muscle cell proliferation when tested on primary as well as subcultures of rabbit aorta smooth muscle cells (Leveille and Fischer-Dzoga 1982). Earlier, sterculic acid isolated as the methyl ester from *Sterculia foetida* oil at low concentration in the diet was shown to increase DNA synthesis and mitosis of hepatocytes in rainbow trout and rat (Scarpelli 1974).

Use in Controlled Release Formulation

Studies found that *Sterculia foetida* gum could be used a controlled release matrix polymer excipient (Chivate et al. 2008). The in-vitro release profiles indicated that tablets prepared from *Sterculia foetida* gum had higher retarding capacity than tablets prepared with hydroxymethylcellulose K15M prepared tablets. The differential scanning calorimetry results indicated that there were no interactions of *Sterculia foetida* gum with diltiazem hydrochloride.

Effects of Cyclopropene Fatty Acids on Reproduction

Studies by Nixon et al. (1977) found that unsaturated lipid in the diet enabled rats to cope with the effects of moderate levels of cyclopropene fatty acids, but the combination of cyclopropene in saturated lipid diets caused detrimental effects. *Sterculia foetida* oil (50% cyclopropene fatty acids) fed at 0.2% of the diet for three generations and 0.5% fed to first generation rats did not significantly affect breeding. *S. foetida* oil at 2% fed with 3% corn oil did not appreciably affect conception rate and litter size in the first litters, but reduced pup survival 36% in the first litters and 78% in the second litters. Replacement of the 3% corn oil with 3% animal fat in diet containing 2% *S. foetida* oil reduced litter rate from 87% to 7% in the first litter. The level of fetal cyclopropene doubled when animal fat replaced corn oil in the diet.

Traditional Medicinal Uses

In Indonesia, a bark decoction used as abortifacient. Roasted fruit skin or its ashes used as a decoction for gonorrhoea in Java and leaf infusion used as aperients for fever. In Java, leaves used for washing head, root infusion used for bathing sick child or patient with jaundice. Pounded leaves applied on broken limbs and dislocated joints. Heated oiled leaves are applied on abdomen of children to treat fever followed by placement of used leaves on the chest. In the Philippines, a leaf decoction is used to treat suppuratives cutaneous eruptions while a decoction of the bark is used in cases of dropsy and rheumatism as an aperient, diaphoretic, and diuretic. A decoction of the fruit is astringent. In India, the bark and leaves are regarded as aperient, diuretic and diaphoretic. a decoction of the fruit is mucilaginous and astringent. The seed oil is administered internally in itches and other skin diseases and is applied externally as a paste. In Ghana, seeds are employed as a purgative. Oil from the seed is extracted on a local scale to be used in medicine.

Other Uses

A timber tree used for plank production, boxes, doors of huts, furniture, canoes, boats, guitars and toys. It is also planted as avenue trees. The seed oil is used as illuminant in Indonesia. The bark fibres provides a weak rope/cord. A gum that resembles 'gum tragacanth', is obtained from the trunk and branches and is used for book binding and similar purposes. *S. foetida* leaves contain up to 2.66% calcium and are also a good source of protein and phosphorus, meeting nutritional requirements of ruminants.

Studies showed that the cyclopropene fatty acid (2n-octylcycloprop-1-enyl)-octanoic acid (I) exhibited insecticidal activity against three major stored product pests, namely, *Sitophilus oryzae*, *Callosobruchus chinensis*, and *Tribolium castaneum* (Rani and Rajasekharreddy 2010). The filter paper application of (2n-octylcycloprop-1-enyl)-octanoic acid (I) at 0.2 mg/cm^2 caused 100% mortality to all test insects 2 days after treatment (2 DAT). The insecticidal activities of the test compound were attributable to contact mode of action, although there was also significant fumigant toxicity. The results indicated that the bioactive compound isolated from *S. foetida* could act as a potent insecticide against *S. oryzae*, *C. chinensis*, and *T. castaneum* populations.

Cyclopropenoid fatty acids (CPE) isolated from *Sterculia foetida* oil was found to have antifungal activity (Schmid and Patterson 1988). CPE when incorporated into fungal cultures of plant pathogens, *Ustilago maydis*, caused inhibition of dry weight accumulation and sporidial number. Treated sporidia showed irregular wall deposition and a branched morphology. It also inhibited hyphal extension by *Rhizoctonia solani* and inhibited fatty acid desaturation by *Fusarium oxysporum*.

Comments

It is usually propagated by seeds which germinate readily and have no problem with long term storage.

Selected References

Aued-Pimentel S, Lago JHG, Chaves MH, Kumagai EE (2004) Evaluation of a methylation procedure to determine cyclopropenoids fatty acids from *Sterculia striata* St Hil. Et Nauds seed oil. J Chromatogr A 1054(1–2):235–239

Backer CA, Bakhuizen van den Brink Jr RC (1963) Flora of Java (spermatophytes only), vol. 1. Noordhoff, Groningen, the Netherlands, 648 pp

Badami RC, Patil KB, Subbarao YV, Sastri GSR, Vishvanathrao GK (1980) Cyclopropenoid fatty acids of *Sterculia* oils by gas liquid chromatography. Fette Seifen Anstrich 82:317–318

Bao X, Katz S, Pollard M, Ohlrogge J (2002) Carbocyclic fatty acids in plants: Biochemical and molecular genetic characterization of cyclopropane fatty acid synthesis of *Sterculia foetida*. Proc Natl Acad Sci USA 99(10):7172–7177

Bao X, Thelen JJ, Bonaventure G, Ohlrogge JB (2003) Characterization of cyclopropane fatty-acid synthase from *Sterculia foetida*. J Biol Chem 278(15): 12846–12853

Burkill IH (1966) A Dictionary of the Economic Products of the Malay Peninsula. Revised reprint. 2 volumes. Ministry of Agriculture and Co-operatives, Kuala Lumpur, Malaysia. Vol 1 (A–H), pp 1–1240; Vol 2 (I–Z), pp 1241–2444

Chivate AA, Poddar SS, Abdul S, Savant G (2008) Evaluation of *Sterculia foetida* gum as controlled release excipient. AAPS PharmSciTech 9(1):197–204

Chopra RN, Nayar SL, Chopra IC (1986) Glossary of indian medicinal plants. (Including the supplement). Council Scientific Industrial Research, New Delhi, 30 pp

de Padua LS, Lugod GC, Pancho JV (1977–1983) Handbook on Philippine medicinal plants, 4 vols. Documentation and Information Section, Office of the Director of Research, University of the Philippines at Los Banos, the Philippines

Fehling E, SchÖnwiese S, Klein E, Mukherjee KD, Weber N (1998) Preparation of malvalic and sterculic acid methyl esters from *Bombax munguba* and *Sterculia foetida* seed oils. J Am Oil Chemists Soc 75:1757–1760

Kaimal TNB, Lakshminarayana G (1970) Fatty acid compositions of lipids isolated from different parts of *Ceiba pentandra*, *Sterculia foetida* and *Hydnocarpus wightiana*. Phytochem 9:2225–2229

Kochummen KM (1972) Sterculiaceae. In: Whitmore TC (ed) Tree flora of Malaya, vol 2. Longman Malaysia Sdn, Bhd, pp 353–382

Lemmens RHMJ, Soerianegara I, Wong WC (1995) Plant resources of south-east asia No. 5(2) Timber trees: minor commercial timbers. Backhuys Publishers, Leiden, the Netherlands, pp 433–434

Leveille AS, Fischer-Dzoga K (1982) The mitogenic effects of methyl sterculate on aortic smooth muscle cells. Artery 11(3):207–224

Mujumdar AM, Naik DG, Waghole RJ, Kulkarni DK, Kumbhojkar MS (2000) Pharmacological studies on *Sterculia foetida* leaves. Pharm Biol 38(1):13–17

Naik DG, Mujumdar AM, Waghole RJ, Misar AV, Bligh SW, Bashall A, Crowder J (2004) Taraxer-14-en-3beta-ol, an anti-inflammatory compound from *Sterculia foetida* L. Planta Med 70(1):68–69

Nair RAG, Ramesh P, Subramanian SS (1977) Isoscutellarin and other polyphenols from the leaves of *Sterculia foetida*. Curr Sci 46:14–15

Nixon JE, Eisele TA, Hendricks JD, Sinnhuber RO (1977) Reproduction and lipid composition of rats fed cyclopropene fatty acids. J Nutr 107(4): 574–583

Perry LM (1980) Medicinal plants of east and southeast asia. Attributed properties and uses. MIT Press, Cambridge, Massachusetts, United States & London, United Kingdom, 620 pp

Rani PU, Rajasekharreddy P (2010) Insecticidal activity of (2n-octylcycloprop-1-enyl)-octanoic acid (I) against three coleopteran stored product insects from *Sterculia foetida* (L.). J Pest Sci 83(3):273–279

Rao YN, Prasad RBN, Rao SV (1989) Positional distribution of fatty acids on oilseed phosphatidylcholines and phosphatidylethanolamines. Eur J Lipid Sci Technol 94(12):482–484

Scarpelli DG (1974) Mitogenic activity of sterculic acid, a cyclopropenoid fatty acid. Science 185(4155): 958–960

Schmid KM, Patterson GW (1988) Effects of cyclopropenoid fatty acids on fungal growth and lipid composition. Lipids 23(3):248–252

Shamsundar SG, Paramjyothi S (1989) Preliminary pharmacognostical and phytochemical investigation on *Sterculia foetida* Linn. seeds. Afr J Biotechnol 9(13):1987–1989

Tumbelaka LI, Slayden OV, Stormshak F (1994) Action of a cyclopropenoid fatty acid on the corpus luteum of pregnant and nonpregnant ewes. Biol Reprod 50(2):253–257

Varma JP, Dasgupta S, Nath B, Aggarwal JS (1956) Composition of the seed oil *Sterculia foetida* L. J Am Oil Chemists Soc 34(9):452–454

Vipunngeun N, Palanuvej C (2009) Fatty acids of *Sterculia foetida* seed oil. J Health Res 23(3):157

Xia PF, Feng ZM, Yang YN, Zhang PC (2009a) Two flavonoid glycosides and a phenylpropanoid glucose ester from the leaves of *Sterculia foetida*. J Asian Nat Prod Res 11(8):766–771

Xia P, Song S, Feng Z, Zhang P (2009b) Chemical constituents from leaves of *Sterculia foetida*. Zhongguo Zhong Yao Za Zhi 34(20):2604–2606 (In Chinese)

Ya T, Gilbert MG, Dorr LJ (2007) Sterculiaceae. In: Wu ZY, Raven PH, Hong DY (eds) Flora of China, vol Hippocastanaceae through Theaceae, 12. Science Press/Missouri Botanical Garden Press, Beijing/ Louis

Sterculia monosperma

Scientific Name

Sterculia monosperma Ventenat.

Synonyms

Southwellia nobilis Salisb., *Sterculia nobilis* (Salisb.) Sm.

Family

Malvaceae, also placed in Sterculiaceae

Common/English Names

China Chestnut, Noble Bottle Tree Nut, Pheng Phok, Phoenix Eyes, Seven Sister's Fruit

Vernacular Names

Chinese: Ping Po, Feng Yan Guo, Jiu Ceng Pi

Origin/Distribution

The species is found from South China (southeast Fujian, south Guangdong, Guangxi and south Yunnan), Taiwan to Malaysia. The tree occurs mostly in the state of Perak in Malaysia.

Agroecology

The species thrives in the warm tropics and subtropics. In south China, it is found mostly in dense lowland and hilly forests. It is also cultivated.

Edible Plant Parts and Uses

Seeds (nuts) are boiled and eaten.

Botany

Tree medium to large, 10–30 m tall, with blackish-brown bark and thick, glabrous twigs. Leaves simple, entire, oblong, base rounded or obtuse, apex obtusely acuminate, coriaceous, 8.5–30 by 4–15 cm, petiole 2–6 cm (Plate 1). Panicles axillary or terminal, lax, pendant, nearly glabrous, 15–40 cm, axes thin and red, pedicels pubescent and red. Calyx creamy white or pale yellowish-red, calyx tube campanulate-turbinate, with incurved linear-lanceolate lobes (Plate 2). Male flowers many, androgynophore curved, glabrous, anthers yellow. Female flowers fewer, slightly larger with globose, 5-grooved ovary, densely hairy; style curved; stigma shallowly 5-lobed. Follicles 1–5, oblong-ovate, velvety, green turning red, 3.5–10 cm long, 1–4 seeded, thickly leathery, apex beaked (Plates 3–5). Seeds glossy, black-brown, ellipsoid or oblong, 2–2.5 cm.

T.K. Lim, *Edible Medicinal And Non-Medicinal Plants: Volume 3, Fruits*,
DOI 10.1007/978-94-007-2534-8_27, © Springer Science+Business Media B.V. 2012

Plate 1 Large, oblong leaves

Plate 2 Small creamy-white apetalous flowers with incurved calyx lobes

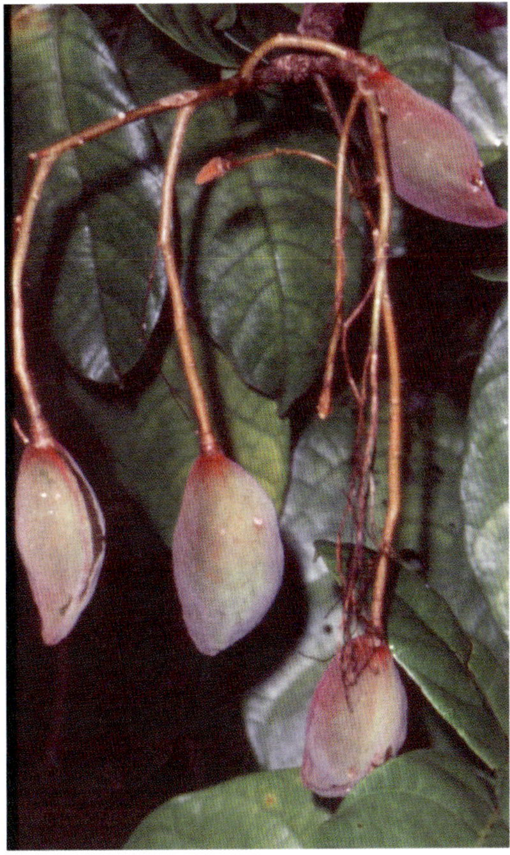

Plate 3 Developing velvety, beaked follicles

Nutritive/Medicinal Properties

China chestnuts (*Sterculia monosperma*) were found to contain cyclopropene fatty acids, CPFA (Berry 1982). The composition of the major fatty acids as methyl esters were 23.47% C16:0 (palmitic), 1.25% C16:1 (palmitoleic), 2.56% C18:0 (stearic), 24.89% C18:1 (oleic), 18.24% C18:2 (linoleic), 5.40% dihydrosterculic, 3.21% C18:3 (linolenic)+C20:0 (arachidic) and 19.15% sterculic. The proportion of CPFA in the oil did not decrease upon cooking the nuts.

China chestnut seed (*S. monosperma*) flour was found to have the following proximate composition in terms of percent dry matter (Noitang et al. 2009): fat (12.0%), protein (7.8%), carbohydrate (73.7%), fibre (5.5%), ash (1.0%), K 12.3 mg/g, P 2.30 mg/g, Mg 1.87 mg/g, S 0.88 mg/g, Ca 0.14 mg/g. The fatty acids profile was found to be composed of mainly palmitic (42%) and oleic acids (34%), with general long-chain fatty acids the other significant component by mass (13%). Glutamic acid (17.4%), aspartic acid (12.5%) and arginine (12.5%) were the three major amino acid constituents. The starch granules were quite round, about 10–15 μ diameter and composed of more than 35% (w/w) of amylose.

The fruit has been used in traditional Chinese medicine.

Other Uses

A minor timber tree.

Plate 4 Ripe dehiscent red follicles

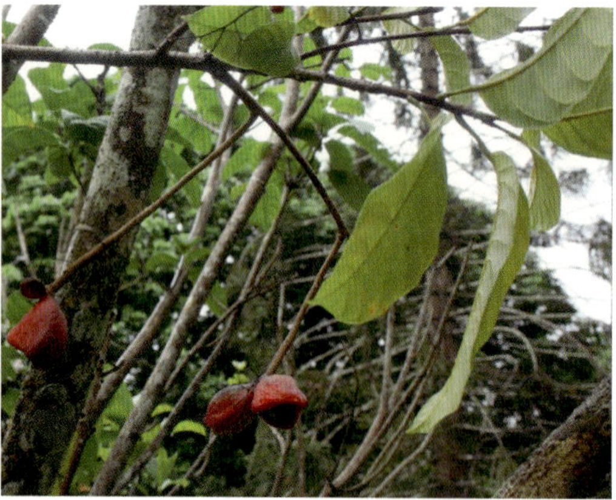

Plate 5 Close-view of ripe, red follicles

Comments

Two botanical varieties are recognised:

var. *monosperma* with rounded or obtuse leaf bases and brownish-black bark

var. *subspontanea* with cuneate or obtuse base and dark gray bark with sparse spots.

Selected References

Backer CA, Bakhuizen van den Brink Jr RC (1963) Flora of Java, (spermatophytes only), vol 1. Noordhoff, Groningen, the Netherlands, 648 pp

Berry SK (1982) Fatty acid composition and cyclopropene fatty acid content of China-chestnuts (*Sterculia monosperma*, Ventenat). J Am Oil Chemist Soc 59(1):57–58

Noitang S, Sooksai SA, Foophow T, Petsom A (2009) Proximate analysis and physico-chemical properties of flour from the seeds of the China chestnut, *Sterculia monosperma* Ventenat. Pak J Biol Sci 12(19):1314–9

Ya T, Gilbert MG, Dorr LJ (2007) Sterculiaceae. In: Wu ZY, Raven PH, Hong DY (eds) Flora of China, vol 12, Hippocastanaceae through Theaceae. Missouri Botanical Garden Press, St. Louis

Sterculia parviflora

Scientific Name

Sterculia parviflora **Roxb. ex G. Don.**

Synonyms

Sterculia holttumii Ridley, *Sterculia maingayi* Masters, *Sterculia obscura* K. Schumann.

Family

Malvaceae, also placed in Sterculiaceae

Common/English Names

Common Sterculia, Kelumpang

Vernacular Names

Malaysia: Kelumpang, Kelumpang Burung, Perupak, Rongga Jantan, Unting-Unting;
Thailand: Po-Khanum;
Vietnam: Trôm hoa thưa; Trôm hoa nho; Trôm maingay.

Origin/Distribution

The species is native to India, Indonesia, Malaysia, and Singapore.

Agroecology

A commonly scattered species of primary and secondary lowland forest and hill forest from near sea level to 500 m altitude. It is found on a variety of soils, including basalt and calcareous shale.

Edible Plant Parts and Uses

Seeds are cooked and eaten.

Botany

A medium to large, much-branched, briefly deciduous tree. It grows 7–20 m high, girth reaching 180 cm with small buttresses and greyish-brown or pinkish bark becoming flaky (Plate 1). Young twigs, slender, brown surfy. Leaves on 10–25 cm long petioles. Leaves simple, oblong to elliptic-oblong, 10–27 cm long by 5–15 cm wide, pointed tip and rounded or weakly cordate base, entire margin, glabrescent, with 7–10 pairs of secondary veins (Plates 2–3); stipules caducous. Flowers small, pale yellow with pink base (calyx cup) in pendulous axillary or terminal racemes. Flowers without petals, urceolate calyx cup with 5 lanceolate, pale yellow sepal lobes curved inwards and spreading outwards after anthesis, stamens (10) joined into a column with anthers at the top, ovary of partly free carpels (Plates 3–5). Fruit, oblong follicle,

T.K. Lim, *Edible Medicinal And Non-Medicinal Plants: Volume 3, Fruits*,
DOI 10.1007/978-94-007-2534-8_28, © Springer Science+Business Media B.V. 2012

Plate 1 Large to medium tree habit with flaky trunk.

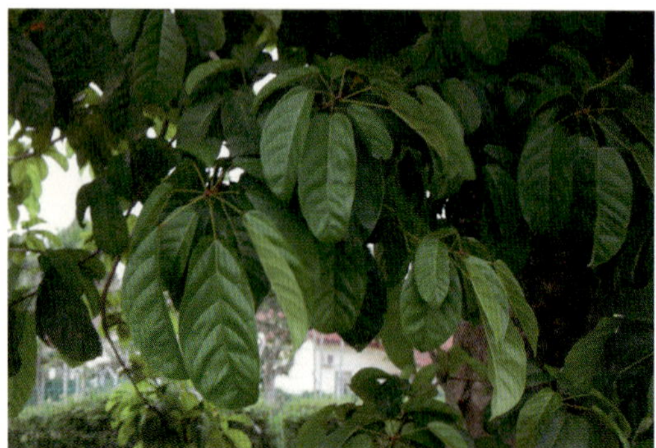

Plate 2 Large oblong leaves on long petioles

usually 5 clustered in star-shaped manner (Plates 6–8). Fruit 8–13 cm by 5–10 cm, velvety, pale greenish turning to yellowish flushed with pink and ripening to a brilliant orange-red, splitting exposing 3–5 velvety, black ellipsoid seeds 1–2 cm long (Plate 8).

Plate 3 A terminal inflorescence

Plate 4 Tiny apetalous flowers with pink calyx cup and yellowish, curved and spreading sepal lobes

Plate 5 Close up of racemose inflorescence and tiny flowers

Plate 6 Young greenish pink flushed follicles, 5 in star-shaped cluster

Plate 7 Maturing follicles yellowish flushed with pink

Plate 8 Ripe orange-red follicles split to expose the large black seeds

Nutritive/Medicinal Properties

No information has been published on its nutritive value and medicinal uses.

A new anthocyanin, pelargonidin-3-arabinoside, was found in the follicles of *Sterculia parviflora* (Lowry 1971). Galactosides and arabinosides were found to occur together in this species.

Other Uses

This species is utilized locally for ornamental purposes and the wood is used as minor timber for cart-wheels and rough work.

Comments

This species is not listed as a threatened species.

Selected References

Burkill IH (1966) A dictionary of the economic products of the Malay Peninsula. Revised reprint. 2 volumes. Ministry of Agriculture and Co-operatives, Kuala Lumpur, Malaysia. Vol. 1 (A–H), pp 1–1240; Vol. 2 (I–Z), pp 1241–2444

Chua LSL (1998) *Sterculia parviflora*. In: IUCN 2011. IUCN red list of threatened species. Version 2011.1. www.iucnredlist.org. Accessed March 2011

Kochummen KM (1972) Sterculiaceae. In: Whitmore TC (ed) Tree flora of Malaya, vol 2. Longman Malaysia Sdn, Kuala Lumpur, pp 353–382

Lemmens RHMJ, Soerianegara I, Wong WC (eds) (1995) Plant resources of South-East Asia No. 5(2) Timber trees: minor commercial timbers. Backhuys Publishers, Leiden, pp 433–434

Lowry JB (1971) Anthocyanins of some Malaysian *Sterculia* species. Phytochem 10(3):689–690

Theobroma bicolor

Scientific Name

Theobroma bicolor **Humb. & Bonpl.**

Synonyms

Cacao bicolor (Bonpl.) Poir., *Theobroma ovatifolia* Sessé & Moc. ex DC., *Tribroma bicolor* (Bonpl.) O.F. Cook.

Family

Malvaceae, also placed in Sterculiaceae

Common/English Names

Mocambo, Nicaraguan Cocoa, Patashte, Peru Cocoa, Peruvian Cacao, Tiger Cocoa, Wild Cacao

Vernacular Names

Brazil: Cacao Bravo, Cacau Bravo, Cacau-Do-Perú, Cubuassú, Cupua-I, Cupassú, Cupuassú, Cupuaçurana, Mocambo, Pataste (Portuguese);
Columbia: Bacao, Cacao Marraco, Cacao Pataste, Cacao Silvestre, Marraco (Spanish);
Czech: Kakaovník Dvoubarevný;
Ecuador: Balamit, Balanté, Cacao Blanco, Pataiste, Patas (Spanish);
German: Zweifarbiger Kakao;
Maya: Pataxte;
Mexico: Balamit, Balanté, Cacao Blanco, Cacao De Nicaragua, Cacao Malacayo, Patashte, Pataste Cimarrón, Pataste De Sapo, Patastle (Spanish);
Panama: Pataste (Spanish);
Peru: Cacau Bafú, Cacau Da Nova Granada, Macambo, Macao, Majambo, Najambu, Semilla De Macambo (Spanish), Papal, Wakam, Wakampe (Awajun);
Spanish: Petaste, Teta Negra.

Origin/Distribution

Its origin is still uncertain. One view believed that the species is native to Central America, another believed that it is indigenous to south America. The species is found in Belize, Brazil, Colombia, Costa Rica, Ecuador, El Salvador, Honduras, Mexico, Nicaragua, Guatemala, Panama, Peru and Venezuela. It also has been introduced to other tropical areas.

Agroecology

The tree thrives in a warm, humid, tropical environment with mean annual temperatures of 25–30°C and mean annual precipitation of 2,000–>3,000 mm from near sea level up to 1,000 m elevation. In its native range, it occurs in both dense and open forests. It grows best in partial or full shade in well-drained clays or loamy clays.

T.K. Lim, *Edible Medicinal And Non-Medicinal Plants: Volume 3, Fruits*,
DOI 10.1007/978-94-007-2534-8_29, © Springer Science+Business Media B.V. 2012

Edible Plant Parts and Uses

The sweetish, juicy, creamy pulp is eaten fresh, cooked or toasted. The pulp is also made into beverages like the maize-based drink, *chorote* and ice-cream. Traditionally, the pulp is mixed with achiote (*Bixa orellana*) and sugar to make a sweet dessert. The seeds are consumed roasted, boiled, used in pastry. The seed is also made into chocolate of inferior quality, although the cocoa butter fat extracted from the seed is of good quality.

The seeds can be used to replace cocoa seeds in the popular traditional Mexican beverage called *tejate* (Soleri and Cleveland 2007). The most common recipe of *tejate* comprises maize dough ground from grain processed with ashes, seeds of cacao (*T. cacao*), *pizle* (the seed of mamey, *Pouteria sapota*), and the aromatic flowers of a large, long-lived evergreen tree, *Quararibea funebris* known as *rosita* or *flor de cacao*, and often sweetened with cane sugar.

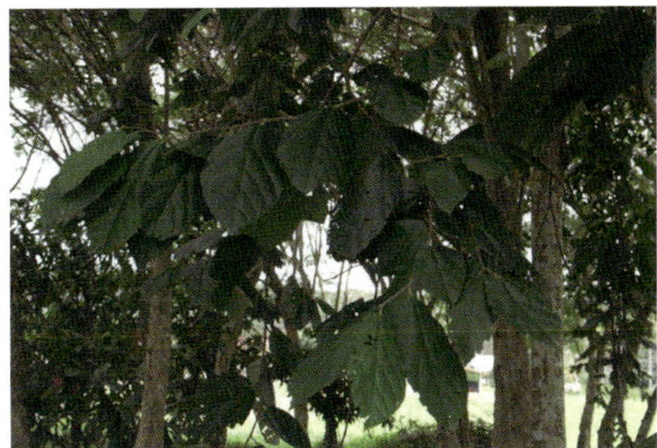

Plate 1 Trunk and foliage

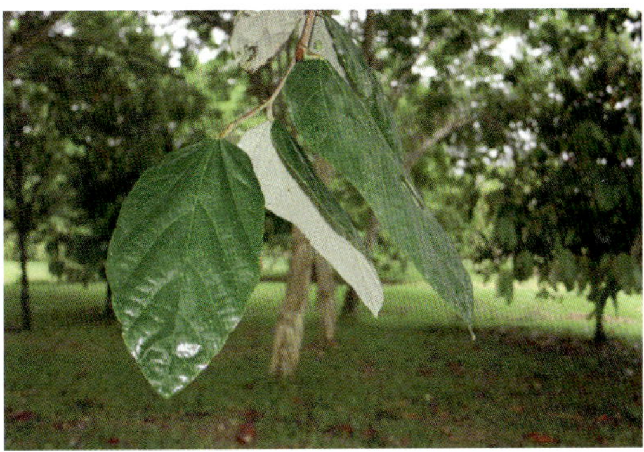

Plate 2 Close view of leaves

Botany

An evergreen, medium-sized tree, 25–30 m, trunk grey, becoming fissured, rough, lax crown composed of a few whorls of 3 branches. Leaves alternate, simple, stipules oblong-lanceolate, petioles 1.2–2.5 cm long, lamina dimorphic, those on the main stem broadly ovate-cordate, 12–35 cm by 12–16 cm, apex acuminate, base cordate palmate venation, those on lateral branches oblong to elliptic-ovate, apex acuminate, base cordate, lamina 30–50 cm by 20–40 cm wide, on 10–40 cm long petioles, upper surface pale green, lower silvery grey tomentose, venation pinnate (Plates 1–3). Inflorescence axillary on the leafy juvenile branches. Sepals 5, lanceolate to ovate-lanceolate, petals 5 smaller oblong-ovate, pinkish white to reddish, stamonides 5, stamens 5 compressed revurved filaments with bilocular anthers, ovary superior, oblong-ovate, 5-lobed, each locule with numerous ovules, styles 5 united, 1–2 mm long. Fruit oblong or ovoid-ellipsoid to subglobose, large, 10–25 cm long by 9–15 cm across, 500–3,000 g, green to greenish brown when immature ripening yellow or yellowish-brown, pericarp hard, longitudinally ribbed, grooves prominently nerved (Plates 4–6). Seeds arranged in five rows enclosed by fibrous, juicy cream to pale yellow pulp, 16–30 mm by 14–25 mm. Fruit fall to ground when ripe.

Nutritive/Medicinal Properties

The nutrient composition per 100 g fresh weight of mocambo seeds was reported as: energy 718 kcal (3,001 kJ), protein 3.37 g, carbohydrate 34.35 g, fat 53.7 g, ash 2.67 g, moisture 5.9% (Creed-Kanashiro et al. 2009).

The fat from the seeds of *T. bicolor* was found to consist of 96.5% triglyceride with only 2.5% diglyceride and 1.7% free fatty acid (Jee 1984). The major fatty acids found were 42.3% C18:0,

Plate 3 Young fruits

Plate 4 Mature fruits prominently ribbed, grooved and nerved

45.2% C18:1, and 6.0% C16:1. Most of the triglycerides were of carbon number 52 (18.0%) and 54 (77.6%). The major triglycerides comprised 38.6% 1-stearyl-2,3 diolein (SOO), 25.4% 2-oleyl-1,3 distearin (SOS) and 13.8% 1-palmito-2-oleyl-stearin (POS). Only 44.3% of the fat consisted of monounsaturated triglycerides.

Plate 5 Ripe fruit with yellowish creamy pulp

Plate 6 Pulp, seeds and the thick pericarp

A total of 65 and 123 volatile compounds were isolated from cacao maraco (*Theobroma bicolor*) fruit (Quijano and Pino 2009). Ethyl acetate, linalool and ethyl benzoate were the major components found.

While the seed of cacao blanco (*T. bicolor*) could be processed like that of *cacao de chocolate* to make chocolate, it was generally considered inferior, due to slight biochemical differences which were critical for formation of the aromas that define chocolate flavor (Reisdorff et al. 2004). The activities of the aspartic endopeptidase and the carboxypeptidase in *T. bicolor* and *T. grandiflorum* differed slightly from those in cocoa. The qualitative and quantitative differences between the globulins indicated a lower maximum yield of aroma precursors in *T. grandiflorum* and a higher maximum yield of aroma precursors in *T. bicolor*, compared to cocoa. The authors concluded that the quality of chocolate-like products made from the studied cocoa relatives could be improved by adapting fermentation procedures to particular biochemical features of these seeds.

Other Uses

The fruit pericarps may be cleaned, dried and used as bowls, containers and pots. The bad quality and discarded fruit are used as animal feed.

Comments

It has innumerable vernacular names in its native countries.

Selected References

Creed-Kanashiro H, Roche M, CERRÓN IT, Kuhnlein HV (2009) Traditional food system of an Awajun community in Peru. In: Kuhnlein HV, Erasmus B, Spigelski D (eds) Indigenous peoples'food systems: the many dimensions of culture diversity and environment for nutrition and health. Food and Agriculture Organization of the United Nations Centre for Indigenous Peoples' Nutrition and Environment, Rome, pp 59–91

Facciola S (1990) *Cornucopia*. A source book of edible plants. Kampong Publ, Vista, 677 pp

FAO (1986). Food and fruit bearing forest species 3 examples from Latin America forestry paper 44–3. Food and Agriculture Organization of the United Nations, pp 308

Jee MH (1984) Composition of the fat extracted from the seeds of *Theobroma bicolor*. J Am Oil Chemist Soc 61(4):751–753

Lorenzi H, Bacher L, Lacerda M, Sartori S (2006) Brazilian fruits & cultivated exotics (for consuming in natura). Instituto Plantarum de Etodos da Flora Ltda, Brazil, 740 pp

Macbride JF (1936–1971) Flora of Peru. Fieldiana: Botany Field museum of natural history, Chicago, 6 parts

Porcher MH et al. (1995–2020) Searchable world wide web multilingual multiscript plant name database. Published by the University of Melbourne, Melbourne. http://www.plantnames.unimelb.edu.au/Sorting/Frontpage.html. Accessed Feb 2009

Quijano CE, Pino JA (2009) Analysis of volatile compounds of Cacao Maraco (*Theobroma bicolor* Humb. et Bonpl.) fruit. J Essent Oil Res 21(3):211–215

Reisdorff C, Rohsius C, de Souza ADGC, Gasparotto L, Lieberei R (2004) Comparative study on the proteolytic activities and storage globulins in seeds of *Theobroma grandiflorum* (Willd ex Spreng) Schum and *Theobroma bicolor* Humb Bonpl, in relation to their potential to generate chocolate-like aroma. J Sci Food Agric 84:693–700

Soleri D, Cleveland DA (2007) Tejate: *Theobroma cacao* and *T. bicolor* in a traditional beverage from Oaxaca. Mexico Food Foodways 15(1–2):107–118

Theobroma cacao

Scientific Name

Theobroma cacao **L.**

Synonyms

Cacao guianensis Aubl., *Cacao minus* Gaertn., *Cacao sativa* Aubl., *Cacao theobroma* Tussac, *Theobroma cacao* fo. *leiocarpum* (Bernoulli) Ducke, *Theobroma cacao* subsp. *leiocarpum* (Bernoulli) Cuatrec., *Theobroma cacao* subsp. *pentagona* (Bernoulli) León, *Theobroma cacao* subsp. *sativa* (Aubl.) León, *Theobroma cacao* var. *leiocarpa* (Bernoulli) Cif., *Theobroma cacao* var. *typica* Cif., *Theobroma caribaea* Sweet, *Theobroma integerrima* Stokes, *Theobroma kalagua* De Wild., *Theobroma leiocarpum* Bernoulli, *Theobroma pentagonum* Bernoulli, *Theobroma saltzmanniana* Bernoulli, *Theobroma sapidum* Pittier, *Theobroma sativa* (Aubl.) Lign. & Le Bey, *Theobroma sativa* var. *leucosperma* A. Chev., *Theobroma sativa* var. *melanosperma* A. Chev., *Theobroma sphaerocarpum* A. Chev.

Family

Malvaceae, also placed in Sterculiaceae

Common/English Names

Cacao, Cacao Tree, Chocolate, Chocolate Tree, Cocoa, Cocoa Tree

Vernacular Names

Arabic: Kâkâû;
Aztec: Cacaoquahuatl, Cacaquatl;
Brazil: Árbore De Cacao, Árvore-Da-Vida, Cacaoeiro, Cacau, Cacau-Da-Mata, Cacueiro, Cacau Comum, Cacau Verdadeiro, (Portuguese);
Burmese: Kokoe;
Chinese: Ke Ke, Ke Ke Shu;
Czech: Kakaovník obecný;
Danish: Kakaotræ;
Dutch: Cacaoboom;
Eastonia: Harilik Kakaopuu;
Finnish: Kaakao Kaakaopuu;
French: Cacao, Cacaotier, Cacaoyer;
German: Kakao, Kakaobaum, Kakaopflanze;
Hungarian: Kakaó(Fa);
India: Kokkoo (Malayalam), Kakkavo, Kona Maram (Tamil);
Indonesia: Coklat;
Italian: Albero Del Cacao;
Japanese: Kakao, Kakao No Ki, Kokoa, Teoburaama Kakao;

T.K. Lim, *Edible Medicinal And Non-Medicinal Plants: Volume 3, Fruits*,
DOI 10.1007/978-94-007-2534-8_30, © Springer Science+Business Media B.V. 2012

Khmer: Kakaaw;
Malaysia: Koko, Poko Colat;
Mayan: Kakaw;
Norwegian: Kakaotre;
Philippines: Kakaw (Tagalog);
Polish: Kakaowiec;
Portuguese: Cacauí, Cupuaçu Da Mata, Cupuí;
Russian: Какао, Shokoladnoe Derevo;
Slovašcina: Kakav;
Spanish: Árbol Del Cacao, Caca, Cacao, Cacao Amarillo, Cacao Criollo, Cacao Del Monte, Cacao Dulce, Cacao Forastero, Cacao Morado, Cacao Ordinario, Cacaoeiro, Cacaotero, Cacaueira, Cacaueiro, Cacauzeiro, Calabacillo, Granos De Cacao;
Sri Lanka: Maikona Gaha (Sinhala);
Swedish: Kakao;
Thai: Kho Kho (Central Thailand), Ko Ko;
Turkish: Kakao;
Vietnamese: Ca Cao, Cây Ca Cao.

Origin/Distribution

Recent studies of *Theobroma cacao* genetics indicated that the plant originated in the upper basin of the Amazon and its headwaters in Peru, Ecuador, Columbia and Brazil. It was distributed by humans throughout Central America and Mesoamerica. Wild cocoa population are also present in the lower Amazon basin as well as along the Orinoco river basin in Venezuela and in the Guyanas. The Amazon cocoa population form the Forastero group and are distinct from the population in Central America that form the Criollo group. The latter also has its roots in the Amazon basin. The Amelonado Forastero cocoa in Bahia has its origin in the lower Amazon basin while the Trinatrio cocoa is a natural hybrid between Criollo and the Venezuelan Forastero. Besides Latin America, cocoa is grown extensively in West Africa (Ghana, Nigeria, Cameroon and Cote D'Ivoire) based on the Trinitario, Amelonado and Forastero cocoa materials; and in Indonesia, Malaysia and Papua New Guinea Trinitario and Forastero materials were introduced from the Caribbean and south America.

Agroecology

Cocoa flourishes in a hot, wet, humid tropical regime. It is grown within 20°N and 20°S of the Equator in the lowlands usually below 300 m but is also found but in sheltered valleys of Colombia at 900 m. Temperature varying between 30–32°C mean maximum and 18–21°C mean minimum but around 25°C is considered to be a favourable. It can't be grown commercially in areas where the minimum temperature fall below 100°C and annual average temperature is less than 21°C. It prefers areas with mean annual rainfall of 1,250–3,000 mm. and preferably between 1,500 and 2,000 mm. In its native habitat, cocoa is an understorey tree, growing best with partial overhead shade and uniformly high humidity. It tolerates brief periods of less than 3 months of drought (with <100 mm rain) as it is more sensitive to moisture stress than other tropical crops and is also intolerant of strong winds. In addition cocoa trees are sensitive to water logging. While they can with-stand brief flooding, they will not tolerate stagnant, water-logged conditions. The tree thrives in rich, organic, well-drained, moist, deep soils. An ideal cocoa soil should be at least 1.5 m deep, clay content 30–40%, with top soil having v 2% carbon, a cation exchange capacity of 120 mmol/kg and a base saturation of 35%. Such soils include well drained entisols (alluvial soils), deep and well-drained inceptisols (volcanic and other origins), red or yellowish ultisols and alfisols (mineral rich forest soils). Clay loams and sandy loams are also suitable. Shallow lateritic soils should be avoided. Cocoa is grown on soils with a wide range of pH from 6 to 7.5 where major nutrients and trace elements will be available. Cocoa does not do well in coastal sandy soils where coconut flourish.

Edible Plant Parts and Uses

Cacao beans (seeds) are the source of commercial cocoa, chocolate, and cocoa butter. The processing of cocoa beans involves opening the harvested pods, removal of mucilage, fermentation, drying, alkalization, roasting, milling, and pressing to express fat. From the defatted mass, the cocoa powder is obtained. In the United States, 'cocoa' often refers to cocoa powder, the dry powder made by grinding cocoa seeds and removing the cocoa butter from the dark, bitter cocoa solids. By itself it has an extremely bitter flavour. From cocoa powder containing 20–25% fat, a popular chocolate beverage is prepared and is derived from cocoa liquor treated with alkali before pressing to improve flavour and dispersibility. Low fat cocoa powders are used in the manufacture of confectionery, biscuits, ice-creams and other chocolate-flavoured products. Chocolate is made by mixing ingredients such as cocoa liquor, cocoa butter, sugar, milk powder, emulsifiers and other flavouring additives. There are three basic types of chocolates available in the market namely milk, dark and white chocolate. The cocoa-butter is a useful vegetable fat.

Most cocoa butter is used for the manufacture of chocolate and a small quantity is used for the manufacture of cosmetics, soap, tobacco and other pharmaceutical products. Cocoa butter has been described as the world's most expensive fat. The pulp surrounding the cocoa beans are edible but are not sought after liked the highly prized beans. The cocoa bean mucilage can be processed with the addition of sugar and cocoa placenta into marmalade which was found to be comparable to apricot marmalade (Anvoh et al. 2009). The liquefied pulp is also used by some cocoa producing countries to distill alcoholic spirits. The cocoa-pod husk can be hydrolysed under pressure for fermentation into alcoholic drinks. Pectins from cocoa hulls showed potential application in the food industry (Barazarte et al. 2008). Studies showed that strawberry jam could be made with the pectin extracted, but it was necessary to optimize the extraction parameters to increase its yield.

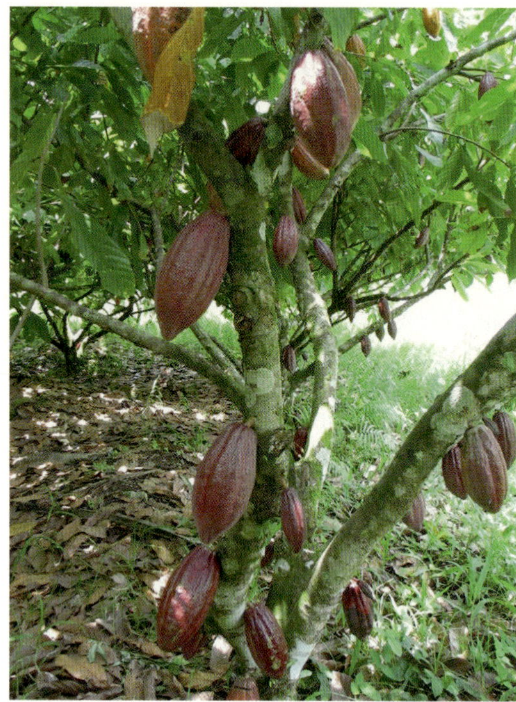

Plate 1 Slender banches, foliage and ellipsoid, red pods

Plate 2 Green pods rieneing to yellow

Botany

Evergreen, small tree, usually 4–8 m , rarely to 20 m high with a short, thick trunk, dark gray-brown bark , slender branches (Plates 1 and 2) and 2 m long tap root and a mass of surface-feeding roots. Branches occurs in whorls of 5, dimorphic; vertical chupons growing from the trunk have leaves arranged in five eighths phyllotaxy and the lateral branches (fans) have one half phyllotaxy. Branchlets are brown and puberulent. Stipules linear, caduceus. Leaves are large, alternate, distichous on normal branches, green, coriaceous or chartaceous on pubescent or tomentose petiole thickened pulvinate at ends (Plates 1 and 2). Leaf blade simple, entire, narrowly ovate to obovate-elliptic, 20–30 × 7–10 cm, both surfaces glabrous

Plate 3 Oblong red pods

Plate 4 Close-view of oblong pods

or sparsely stellate, base rounded to shallowly cordate, apex long acuminate. Inflorescence small and delicate, cymose occurring on trunk and branches. Flowers 18 mm in diameter on 10–14 mm pedicels. Sepals 5, triangular to narrowly lanceolate, whitish or reddish in colour. Petals 5, joined at the base into a cuplike structure, whitish-yellow with dark purple bands adaxially; ligules spathulate, yellowish. Stamens 5, fertile, alternating with 5 staminodes, the 2 whorls uniting to form a tube. Anthers bilobed. Ovary superior obovoid, slightly 5-angular, 5-celled; ovules 14–16 per locule, in 2 rows topped by a single, cylindrical style terminating in 5 sticky stigmatic surfaces. Fruit a drupe, variable in shape and colour, ovoid, oblong, ellipsoid sometimes pointed and constricted at the base or almost subglobose, with 5–10 furrows, green ripening to yellow, red or purplish ripening to same colour (Plates 1, 2, 3 and 4), with fruit wall (husk) up to 2 cm thick. Seeds (beans) embedded in white mucilaginous, acid sweet, edible pulp, 12–14 seeds per cell, ovoid, slightly flattened, 2.5 × 1.5 cm with leathery testa and white or purple cotyledons.

Nutritive/Medicinal Properties

Chemical component composition of cocoa seed was reported by Lopez and Dimick (1995) as: pulp:- water 82–87%, sugars 10–13%, pentosans 2–3%, citric acid 1–2%, salts 8–10%; cotyledons:- water 32–39%, cellulose 2–3%, starch 4–6%, pentosans 4–6%, fat 30–32%, proteins 8–10%, theobromine 2–3%, caffeine 1%, acids 1%, polyphenols 5–6%.

Cocoa seed had been reported to contain per 100 g 456 cal, 3.6 g H_2O, 12.0 g protein, 46.3 g fat, 34.7 g total carbohydrate, 8.6 g fibre, 3.4 g ash, 106 mg Ca, 537 mg P, 3.6 mg Fe, 30 µg β-carotene equivalent, 0.17 mg thiamine, 0.14 mg riboflavin, 1.7 mg niacin, and 3 mg ascorbic acid. According to CSIR (1976), the edible pulp of the fruit contained 79.7–88.5% water, 0.5–0.7% albuminoids, astringents, etc.; 8.3–13.1% glucose, 0.4–0.9% sucrose, a trace of starch, 0.2–0.4% non-volatile acids (as tartaric), 0.03% Fe_2O_3 and 0.4% mineral salts (K, Na, Ca, Mg). The shell contained 11.0% moisture, 3.0% fat, 13.5% protein, 16.5% crude fiber, 9.0% tannins, 6.0% pentosans, 6.5% ash, and 0.75 theobromine. Raw seeds contained 0.24 mg/100 g thiamine, 0.41 riboflavin, 0.09 pyridoxine, 2.1 nicotinamide, and 1.35 pantothenic acid. The component fatty acids of cocoa butter found were 26.2% palmitic and lower acids, 34.4 stearic and higher acids, 37.3% oleic acid, 2.1% linoleic and traces of isoleic. In g/100 g the individual amino acids in the water soluble fractions of unfermented and fermented beans are lysine 0.08, 0.56; histidine 0.08, 0.04; arginine 0.08, 0.03; threonine 0.14, 0.84; serine 0.88, 1.99; glutamic acid 1.02, 1.77; proline 0.72, 1.97; glycine 0.09, 0.35; alanine 1.04, 3.61; valine 0.57, 2.60; isoleucine 0.56, 1.68; leucine 0.45, 4.75; tyrosine 0.57, 1.27; and phenylalanine 0.56–3.36 g/100 g.

Cocoa seed extracts consistently revealed the presence of four pigmented anthocyanin compounds; in order of decreasing abundance, they were a cyanidin arabinoside, cyaniding galactoside, cyanidin rutinoside, and cyanidin pentoside (Cakirer et al. 2010).

The nutrient value of dry, unsweetened cocoa powder per 100 g edible portion was reported as water 3 g, energy 228 kcal (952 kJ), protein 19.60 g, lipids 13.70 g, ash 5.80 g, carbohydrate 57.90 g, total dietary fibre 33.2 g, total sugars 1.75 g, Ca 128 mg, Fe 13.86 mg, Mg 499 mg, P 734 mg, K 1,524 mg, Na 21 mg, Zn 6.81 mg, Cu 3.788 mg, Mn 3.837 mg, Se 14.3 μg, thiamine 0.078 mg, riboflavin 0.241 mg, niacin 2.185 mg, pantothenic acid 0.254 mg, vitamin B-6 0.118 mg, total folate 32 μg, total choline 12 mg, lutein+zeaxanthin 38 μg, vitamin E (α-tocopherol) 0.10 mg, vitamin K (phylloquinone) 2.5 μg, total saturated fatty acids, 8.070 g, 16:0 (palmitic) 3.690 g, 18:0 (stearic) 4.250 g, total monounsaturated fatty acids 4.570 g, 18:1 undifferentiated 4.570 g, total polyunsaturated fatty acids 0.440 g, 18:2 undifferentiated 0.440 g, tryptophan 0.293 g, threonine 0.776 g, isoleucine 0.760 g, leucine 1.189 g, lysine 0.983 g, methionine 0.202 g, cystine 0.239 g, phenylalanine 0.941 g, tyrosine 0.735 g, valine 1.177 g, arginine 1.111 g, histidine 0.339 g, alanine 0.904 g, aspartic acid 1.953 g, glutamic acid 2.948 g, glycine 0.879 g, proline 0.838 g, serine 0.846 g, caffeine 230 mg and theobromine 2,057 mg (U.S. Department of Agriculture and Agricultural Research 2010).

Anvoh et al. (2009) reported the cocoa bean mucilage to have the following nutrient content: 85.3% moisture, pH 3.14, glucose 214 g/L, total soluble solids 16.17° Brix, crude protein 7.2 g/L, ascorbic acid 18.3 mg/L, potassium 950 mg/L, calcium 171.5 mg/L, sodium, magnesium and phosphorus were lower, citric acid 9.1 mg/L, malic acid 3.6 mg/L, acetic acid 2.28 mg/L, and lower levels of fumaric, oxalic and lactic acid. Marmalade was produced with cocoa bean juice with additional sugar and cocoa placenta (11.5%) to the mucilage (44.72%). The output of the marmalade was 46.2%. Cellulose and fat contents were 5.36% and 5.23%, respectively. Total soluble solids were 67.14° Brix. Sensory analyses did not show any significant difference in taste, colour and consistency compared with a commercial apricot marmalade.

The major polyphenols found in cocoa beans include (−)-epicatechin, leucoanthocyanin L1, leucoanthocyanin L2, eucoanthocyanin L3, leucoanthocyanin L4, (+)-catechin, (−)-epigallocatechin, (+)-gallocatechin, 3-β-D-galactosidylcyanidin chloride, 3-α-L-arabinosidylcyanidin chloride and an unidentified compound showing blue fluorescence in UV radiation (Griffith 1960).

Among the plant parts, cocoa seeds (beans) had highest level of alkaloids with theobromine being the main alkaloid found: 1.39 and 2.03 g/100 g in seeds of Criollo and Costarrica varieties, respectively (Sotelo and Alvarez 1991). Caffeine was the second more important alkaloid in the cacao beans (180–920 mg/100 g of sample). Theophylline was also present in cocoa beans (357–367 mg/100 g of sample). Cocoa seeds also contained trypsin inhibitor (30–41 and 8–8.6 TUI/mg of sample) but no hemagglutinins. Fat was the predominant component of cocoa seeds. Unfermented and fermented cocoa beans contain p-hydroxybenzoic acid, vanillic acid, p-coumaric acid, ferulic acid, and syringic acid, while the fermented beans also contain protocatechuic, phenylacetic, phloretic acid and the lactone esculetin and o- and p-hydroxyphenyl acids (Duke 1983). Cocoa contains over 300 volatile compounds, including esters, hydrocarbonslactones, monocarbonyls, pyrazines, pyrroles, and others.

The amounts of methylxanthine derivatives caffeine, theobromine and theophylline also referred to as alkaloids, in cocoa had been reported to depend on various influencing factors, the most essential ones being processing procedures, genotype, geographical origin and cacao bean weight (Matissek 1997). Methylxanthines, natural plant stimulants, had been reported to contribute towards the typically bitter taste of cacao. According to the author, methylxanthines in cacao products have no negative effects on the health of humans because their amounts in chocolates and other processed foodstuffs containing cacao are so low that they account for only a very small proportion of the whole of the human diet.

Theobroma cacao seeds were found to contain an albumin fraction and a globulin fraction with proportions of 52% and 43%, respectively, of total seed proteins and no prolamin was detected as reported in earlier literature (Voigt et al. 1993). The 'glutelin fraction' described in the literature was found to consist of residual globulin. After fermentation, the proportion of the globulin fraction was considerably reduced. The seeds also contained a vicilin-like globulin, but apparently no legumin-class globulin. An aspartic endoprotease and a carboxypeptidase present in ungerminated cocoa seeds were found to be required for the proteolytic formation of the cocoa-specific aroma precursors (Voigt et al. 1994). Cocoa-specific aroma precursors were obtained by proteolytic digestion of the vicilinlike globulin but not by proteolysis of the albumin of cocoa seeds (Voigt et al. 1994; Kratzer et al. 2009). Recent studies found Albumin, globulin and glutelin protein fractions in unfermented and semi-fermented-dry cacao seeds were obtained by differential solubility extraction (Preza et al. 2010). Glutelins were the predominant fraction. In the albumin fraction, polypeptides of 42.3 and 8.5 kDa were found in native conditions. The globulin fraction presented polypeptides of 86 and 57 kDa in unfermented cacao seed that produced the specific-cacao aroma precursors, and after fermentation the polypeptides were of 45 and 39 kDa. The glutelin fraction presented proteins >200 kDa and globulins components <100 kDa in lesser proportion.

Food value of cocoa butter oil per 100 g edible portion was reported as follows (U.S. Department of Agriculture and Agricultural Research 2010): energy 884 kcal (3,699 kJ), protein 0 g, total lipid (fat) 100.00 g, total choline 0.3 mg, vitamin E (α tocopherol) 1.8 mg, vitamin K (phylloquinone) 24.7 μg; lipids – fatty acids (total saturated) 59.7 g, 16:0 (palmitic acid) 25.4 g, 18:0 (stearic acid) 33.20 g; fatty acids (total monounsaturated) 32.9 g, 16:1 undifferentiated (palmitoleic acid) 0.2 g, 18:1 undifferentiated (oleic acid) 32.6 g, fatty acids (total polyunsaturated) 3.00 g, 18:2 undifferentiated (linoleic acid) 2.80 g, 18:3 undifferentiated (linolenic acid) 0.100 g; phytosterols 201 mg. Cocoa butter also contained small amounts of sterols, β-sitosterol, stigmasterol and campesterol and tritepene alcohols (Fedeli et al. 1966). Staphylakis and Gegiou (1985) fractionated sterol lipids of cocoa butter into free sterols, steryl esters, steryl glucosides and acylated steryl glucosides 4-desmethyl, 4-methyl and 4,4′-dimethyl sterols or triterpene alcohols were isolated as free sterols. Esterified sterols amounted to 11.5% of the total sterols and glucosidic sterols to 16.3%.

Cocoa and its derived products: cocoa powder, cocoa liquor and chocolates (dark, milk and white) contained varying polyphenol contents with different levels of antioxidant capacity (Hii et al. 2009). Processing could influence the polyphenols contents of cocoa and its products. The polyphenols contributed about 12–18% of the dry weight of the whole cocoa bean (Hii et al. 2009). Main classes of polyphenolic compounds identified included simple phenols, benzoquinones, phenolic acids, acetophenones, phenylacetic acids, hydroxycinnamic acids, phenylpropenes, coumarines, chromones, naphtoquinones, xanthones, stilbenes, anthraquinones, flavonoids, lignans and lignins. Three main groups of cocoa polyphenols could be distinguished namely catechins (37%), anthocyanins (4%) and proanthocyanidins (58%). The main catechin is (−)-epicatechin comprising up to 35% of the total polyphenol content. Hii et al. (2009) reported that the range of total polyphenols recorded (range from 40.0 to 84.2 mgGAE/g) varied among geographical origins and also the planted varieties. The Criollo variety generally was found to have lower total polyphenol content since it lacked anthocyanins. Anthocyanins were hydrolyzed to anthocyanidins which polymerized with simple catechins to form complex tannins during fermentation. Kim and Keeney found concentration of (−)-epicatechin ranging from 2.66 to 16.52 mg/g for cocoa beans obtained from various countries.

Thirty-five odor-active volatile constituents in the flavor dilution (FD) factor range of 8–4,096 were isolated from cocoa powder (50 g; 20% fat content) (Frauendorfer and Schieberle 2006). Among them, 4-hydroxy-2,5-dimethyl-3(2H)-furanone (caramel-like), 2- and 3-methylbutanoic acid (sweaty, rancid), dimethyl trisulfide (cooked cabbage), 2-ethyl-3,5-dimethylpyrazine (potato-chip-like), and phenylacetaldehyde (honey-like) showed the highest FD factors. Quantitation of

31 key odorants revealed odor activity values (OAVs) (ratio of concentration to odor threshold) greater than 100 for the 5 odorants: acetic acid (sour), 3-methylbutanal (malty), 3-methylbutanoic acid, phenylacetaldehyde, and 2-methylbutanal (malty). In addition, another 19 aroma compounds showed OAVs >1. Unroasted and roasted Criollo cocoa beans were found to have 42 aroma compounds in the flavor dilution (FD) factor range of 1–4,096 for the unroasted and 4–8,192 for the roasted cocoa beans (Frauendorfer and Schieberle 2008). Compounds, 2- and 3-methylbutanoic acid (rancid) and acetic acid (sour) showed the highest FD factors in the unroasted beans, while 3-methylbutanal (malty), 4-hydroxy-2,5-dimethyl-3(2H)-furanone (caramel-like), and 2- and 3-methylbutanoic acid (sweaty) showed the highest FD factors in the roasted seeds. Quantitation of 30 odorants revealed concentrations above the odor threshold for 22 compounds in the unroasted and 27 compounds in the roasted cocoa beans, respectively. In particular, a strong increase in the concentrations of the Strecker aldehydes 3-methylbutanal and phenylacetaldehyde as well as 4-hydroxy-2,5-dimethyl-3(2H)-furanone was found, suggesting that these odorants should contribute most to the changes in the overall aroma after roasting. Various compounds contributing to the aroma of roasted cocoa beans, such as 3-methylbutanoic acid, ethyl 2-methylbutanoate, and 2-phenylethanol, were already present in unroasted, fermented cocoa beans and were not increased during roasting.

Cocoa and chocolate contain monomeric and oligomeric procyanidins including pentameric procyanidin (Hammerstone et al. 1999; Adamson et al. 1999). In addition to compounds described in the literature, such as epicatechin and catechin, quercetin, isoquercitrin (quercetin-3-O-glucoside) and quercetin-3-O-arabinose, other compounds were identified for the first time in cocoa samples, such as hyperoside (quercetin-3-O-galactoside), naringenin, luteolin, apigenin and some O-glucosides and C-glucosides of these compounds (Sánchez-Rabaneda et al. 2003). The following bioactive polyphenols were isolated from roasted cocoa nibs and found to be the contributors to cocoa astringent taste and many as potent antioxidants (Stark and Hofmann 2005; Stark et al. 2005; 2006a, b): theobromine and caffeine; the flavan-3-ols epicatechin, catechin, procyanidin B-2, procyanidin B-5, procyanidin C-1, [epicatechin-(4β→8)](3)-epicatechin, and [epicatechin-(4β→8)](4)-epicatechin; quercetin, naringenin, luteolin, and apigenin glycopyranosides; N-phenylpropenoyl amino acids (amino acids amides) namely (+)-N-[4′-hydroxy-(E)-cinnamoyl]-L-aspartic acid, (+)-N-[3′,4′-dihydroxy-(E)-cinnamoyl]-L-aspartic acid, (−)-N-[3′,4′-dihydroxy-(E)-cinnamoyl]-L-glutamic acid, (−)-N-[4′-hydroxy-(E)-cinnamoyl]-L-glutamic acid, (−)-N-[4′-hydroxy-(E)-cinnamoyl]-3-hydroxy-L-tyrosine, (+)-N-[4′-hydroxy-3′-methoxy-(E)-cinnamoyl]-L-aspartic acid, and (+)-N-(E)-cinnamoyl-L-aspartic acid, (−)-N-[3′,4′-dihydroxy-(E)-cinnamoyl]-3-hydroxy-l-tyrosine (clovamide), (−)-N-[4′-hydroxy-(E)-cinnamoyl]-L-tyrosine (deoxyclovamide), and (−)-N-[3′,4′-dihydroxy-(E)-cinnamoyl]-L-tyrosine (Stark et al. 2005), (+)-N-[3′,4′-dihydroxy-(E)-cinnamoyl]-L-aspartic acid, (+)-N-[4′-hydroxy-(E)-cinnamoyl]-L-aspartic acid, (−)-N-[3′,4′-dihydroxy-(E)-cinnamoyl]-L-glutamic acid, (−)-N-[4′-hydroxy-(E)-cinnamoyl]-L-glutamic acid, (−)-N-[4′-hydroxy-(E)-cinnamoyl]-3-hydroxy-L-tyrosine, (+)-N-[4′-hydroxy-3′-methoxy-(E)-cinnamoyl]-L-aspartic acid, and (+)-N-[(E)-cinnamoyl]-L-aspartic acid (Stark and Hofmann 2005) and N-[3′,4′-dihydroxy-(E)-cinnamoyl]-Ltryptophan, N-[4′-hydroxy-(E)-cinnamoyl]-L-tryptophan, and N-[4′-hydroxy-3′-methoxy-(E)-cinnamoyl]-L-tyrosine, from roasted cocoa powder (Stark et al. 2006a, b). Further studies by Stark et al. (2006a, b) demonstrated that the bitter-tasting alkaloids theobromine and caffeine, seven bitter-tasting diketopiperazines, seven bitter- and astringent-tasting flavan-3-ols, six puckering astringent N-phenylpropenoyl-l-amino acids, four velvety astringent flavonol glycosides, gamma-aminobutyric acid, β-aminoisobutyric acid, and six organic acids were the key organoleptics of the roasted cocoa nibs.

Composite samples of a broad range of chocolate- and cocoa-containing products marketed in the United States were characterized for percent fat (% fat), percent nonfat cocoa solids (% NFCS), antioxidant level by ORAC, total polyphenols, epicatechin, catechin, total monomers, and flavan-3-ol oligomers and polymers (procyanidins) (Miller et al. 2009) . On a gram weight basis epicatechin and catechin content of the products followed in decreasing order: cocoa powder > baking chocolate > dark chocolate = baking chips > milk chocolate > chocolate syrup. Analysis of the monomer and oligomer profiles within product categories presented two types of profiles: (1) products with high monomers and decreasing levels of oligomers and (2) products in which the level of dimers was equal to or greater than the monomers. Results showed a strong correlation ($R^2 = 0.834$) of epicatechin to the level of % NFCS and also very good correlations for N = 2–5 oligomers to % NFCS. A weaker correlation was observed for catechin to % NFCS ($R^2 = 0.680$). Other analyses showed a similar high degree of correlation with epicatechin and N = 2–5 oligomers to total polyphenols, with catechin being less well correlated to total polyphenols. PCA (Principal component analysis) also showed that most factors grouped closely together including the antioxidant activity, total polyphenols, and the flavan-3-ol measures with the exception of catechin and % fat in the product, which group separately. The cocoa-containing products tested ranged from cocoa powder with 227.34 mg of procyanidins per serving to 25.75 mg of procyanidins per serving for chocolate syrup. A study of commercial dark, milk and white chocolates on the Malaysian market found that dark chocolates exhibited the highest phenolics and flavonoids contents, followed by milk and white chocolates (Meng et al. 2009). Catechin and epicatechin were major flavonoids detected in dark chocolates. Theobromine was detected in dark and milk chocolates, but not in white chocolates. A high correlation ($R^2 = 0.93$) was found between total phenolic and flavonoid contents, indicating that the major phenolic compounds in dark chocolates belong to the flavonoid class.

(−)-Epicatechin was found in the range of 116.02–730.26 μg/g, whereas (+)-catechin was in the range of 81.40–447.62 μg/g in the Spanish commercial cocoa powder products studied (Andres-Lacueva et al. 2008). Among flavonols, quercetin-3-arabinoside and isoquercitrin were the major flavonols in the cocoa powder products studied, ranging from 2.10 to 40.33 μg/g and from 3.97 to 42.74 μg/g, respectively, followed by quercetin-3-glucuronide (0.13–9.88 μg/g) and quercetin aglycone (0.28–3.25 μg/g). Alkalinization treatment resulted in 60% loss of the mean total flavonoid content. Among flavanols, (−)-epicatechin presented a larger decline (67%) than (+)-catechin (38%), probably because of its epimerization into (−)-catechin, a less bioavailable form of catechin. A decline was also confirmed for di-, tri-, and tetrameric procyanidins. In the case of flavonols, quercetin presented the highest loss (86%), whereas quercetin-3-glucuronide, quercetin-3-arabinoside, and isoquercitrin showed a similar decrease (58%, 62%, and 61%, respectively). It was concluded that the large decrease found in the flavonoid content of natural cocoa powder, together with the observed change in the monomeric flavanol profile that resulted from the alkalinization treatment, could affect the antioxidant properties and the polyphenol bioavailability of cocoa powder products. The antioxidant polyphenols in cacao liquor, a major ingredient of chocolate and cocoa, had been characterized as flavan-3-ols and proanthocyanidin oligomers (Natsume et al. 2000). Individual cacao polyphenols, flavan-3-ols (catechin and epicatechin), and dimeric (procyanidin B2), trimeric (procyanidin C1), and tetrameric (cinnamtannin A2) proanthocyanidins, and galactopyranosyl-ent-(−)-epicatechin (2α→7, 4α→8)-(−)-epicatechin (Gal-EC-EC), were compared in various cocoa products. The profile of monomers (catechins) and proanthocyanidin in dark chocolate was similar to that of cacao liquor, while the ratio of flavan-3-ols to the total amount of monomeric and oligomeric polyphenols in the case of pure cocoa powder was higher than that in the case of cacao liquor or chocolate.

Huang and Barringer (2010) found in both alkalized-before-roasting and alkalized-after-roasting cocoa bean samples, there were significantly higher concentrations of alkylpyrazines for the samples with pH above 7 than pH below 7. At pH 8, the concentrations of 2,3-, 2,5-, and

2,6-dimethylpyrazine (DMP), 2,3,5-trimethylpyrazine (TrMP), 2,3,5,6-tetramethylpyrazine (TMP), and 2,3-diethyl-5-methylpyrazine (EMP) in the samples alkalized-before-roasting were higher than those in the samples alkalized-after-roasting. The concentrations of Strecker aldehydes and other volatiles followed a similar pattern as that of the alkylpyrazines. High pH favored the production of alkylpyrazines and Strecker aldehydes. Cocoa beans alkalized-before-roasting had higher levels of many important chocolate aroma volatiles than those alkalized-after-roasting. Thus, alkalizing before roasting should produce a stronger cocoa aroma. The higher the pH, the higher the concentrations of these important volatiles. The concentrations of total alcohols, acids, aldehydes, esters, ketones, and alkylpyrazines increased, peaked, and decreased within the timeframe used for typical roasting of alkalized and unalkalized Don Homero and Arriba cocoa beans Huang and Barringer (2011). The concentrations of alkylpyrazines and Strecker aldehydes increased as the roasting temperature increased from 120°C to 170°C. For most of the volatile compounds, there was no significant difference between Arriba and Don Homero beans, but Arriba beans showed higher concentrations of 2-heptanone, acetone, ethyl acetate, methylbutanal, phenylacetaldehyde, and trimethylpyrazine. For unalkalized Don Homero beans (pH 5.7), the time to peak concentration decreased from 13.5 to 7.4 min for pyrazines, and from 12.7 to 7.4 min for aldehydes as the roasting temperature increased from 120°C to 170°C.

Trans-resveratrol and trans-piceid were quantified in 22 cocoa liquors from 11 different countries. A very large range of concentrations was observed for trans-piceid (Jerkovic et al. 2010). The most concentrated sample (Arriba 06) reached 0.4 and 2.6 mg/kg of trans-resveratrol and trans-piceid, respectively, but in other cultivars stilbene levels were five times lower. Neither cis-resveratrol nor cis-piceid was found in cocoa liquors. A new resveratrol hexoside, trans-piceid-like hexoside was also detected. Among 19 top selling commercially available cocoa-containing and chocolate products in the US market, trans-resveratrol levels were highest in cocoa powders (1.85 μg/g), followed by unsweetened baking chocolates (1.24), semisweet chocolate baking chips (0.52), dark chocolates (0.35), milk chocolates (0.10), and chocolate syrups (0.09) (Hurst et al. 2008). These cocoa-containing and chocolate products have about 3–5 times more trans-piceid than trans-resveratrol. Levels of trans-piceid were highest in the cocoa powders (7.14 μg/g), followed by unsweetened baking chocolates (4.04), semisweet chocolate baking chips (2.01), dark chocolates (1.82), milk chocolates (0.44), and chocolate syrups (0.35). On an equal weight basis, cocoa powder had about half as much trans-resveratrol as the average California red wine. On a per serving basis, cocoa-containing and chocolate products had less trans-resveratrol than red wine and grape juice but more than roasted peanuts.

In addition to alkaloids (mainly theobromine), tannins, and other constituents, cocoa husk contains a pigment that is a polyflavone glucoside with a molecular weight of over 1,500, this pigment is claimed to be heat and light resistant, highly stable at pH 3–11, and useful as a food colorant; it was isolated at a 7.9% yield (Leung and Foster 1995).

Griffiths (1960, 1985) reported the presence of phenolic acids and flavonoids in various parts of the cocoa crop. Among the identified compounds cyanidin hydrochloride and gentisic acid (2,5-dihydroxybenzoic acid) were found in all tissues (viz. wood, bark, root, young and old leaf, flower, pod walls and bean) in varying level. cyanidin hydrochloride occurred in high levels in young leaf and bean cotyledon and high levels in green stem and young pod, whilst gentisic acid was found in high levels in mature leaf. Quercetin was found in sap wood, leaves, flower and pods with high levels in the flower and young pod. Caffeic acid was found in green stem, leaves, pods and bean with high levels in the leaf and pod. ρ-coumaric acid was found in green stem, leaves, flower, pods and bean with high levels in mature leaf and young pod. Kaempferol was detected in traces in young leaf; sinapic acid in mature leaf. Ferulic acid was detected in mature leaf and old pod wall. Of the distribution of methanol soluble polyphenols, leucocyanidin L_1, leucocynidin L_2 and (–) epicatechin were found in all the plant parts examined with high amounts of leucocyanidin L_1, leucocynidin L_2 in the young pod wall. Young pod walls and bean cotyledon had very high levels of (–) epicatechin whilst young leaf and flower had high levels. Rutin was found in moderate levels in young and old leaves. 3β-D-Galactosidyl cyaniding was found in very high level in bean cotyledon, moderate level in flower and traces in young leaf and young pod wall. Neochlorogenic acid was found in high level in mature leaf. Chlorogenic acid occurred in high level in young leaf and traces in sap wood, green stem and mature leaf. ρ-Coumaryl-quinic acid occurred in high amount in young leaf and traces in mature leaf and bean cotyledon.

The various pharmacological properties of cocoa and its products are presented below.

Cocoa, Chocolate and Cardiovascular Health

Cocoa and chocolate have recently been found to be rich plant-derived sources of antioxidant flavonoids particularly catechin and epicatechin with beneficial cardiovascular activities. Flavanols in cocoa and chocolate occur as monomers epicatechin and catechin can form links between C4 and C8, allowing them to assemble as dimers, oligomers, and polymers of catechins, the so-called procyanidins (Adamson et al. 1999). Procyanidins are also known as condensed tannins, which, through the formation of complexes with salivary proteins, are responsible for the bitterness of cacao.

The ways that cocoa and chocolate may exert its beneficial effects had been proposed to include antioxidant activity and nitric oxide activation, endothelial function improvement, vasodilation and blood pressure reduction, inhibition of platelet activity, reduction of insulin resistance, anti-hypercholesterolemic, antihyperlipidemic, and antiinflammatory effects (Ding et al. 2006; Engler and Engler 2006; Alspach 2007; Erdman et al. 2008; Corti et al. 2009). Wollgast and Anklam (2000) in their review listed the health beneficial effects of cocoa polyphenols to include anti-carcinogenic, anti-atherogenic, anti-ulcer, anti-thrombotic, antiinflammatory, anti-allergenic, immunomodulatory, antimicrobial, vasodilatory and analgesic effects.

The association of chocolate consumption with measured blood pressure (BP) and the incidence of cardiovascular disease was investigated from 1994 to 1998 in 19,357 participants (aged 35–65 years) free of myocardial infarction and stroke and not using antihypertensive medication (Buijsse et al. 2010). Mean systolic BP was 1.0 mmHg and mean diastolic BP 0.9 mmHg lower in the top quartile compared with the bottom quartile of chocolate consumption. The relative risk of the combined outcome of myocardial infarction and stroke for top vs. bottom quartiles was 0.61. Baseline BP explained 12% of this lower risk. The inverse association was stronger for stroke than for myocardial infarction. Cocoa flavanol, epicatechin, was found to confer cardioprotection in the setting of ischemia-reperfusion (I/R) myocardial injury (Yamazaki et al. 2008). After 10 days or pretreatment with epicatchin, a significant ~50% reduction in myocardial infarct size occurred. Epicatechin rats demonstrated no significant changes in hemodynamics. Tissue oxidative stress was reduced significantly in the epicatechin group vs. controls and matrix metalloproteinase-9 activity demonstrated limited increases in the infarct region with epicatechin. By 3 weeks, a significant 32% reduction in infarct size was observed with epicatchin pretreatment accompanied with sustained hemodynamics and preserved chamber morphometry. Studies in mice showed that epicatechin acted via opioid receptor stimulation and more specifically through the δ-opioid receptor to produce cardiac protection from ischemia-reperfusion injury (Panneerselvam et al. 2010).

Antioxidant Activity

Cocoa and chocolate were found to contain monomeric and oligomeric procyanidins including pentameric procyanidin (Hammerstone et al. 1999; Adamson et al. 1999) the procyanidin content of the samples was correlated to the antioxidant capacity measured using the ORAC assay as an indicator for potential biological activity (Adamson et al. 1999). Gu et al. (2006) determined total antioxidant capacity (AOC) (lipophilic and hydrophilic Oxygen Radical Antioxidant Capacity Fluorescein (ORAC-FL)), catechins, and procyanidins (monomer through polymers) of cocoa and chocolate products from major brands. Procyanidin (PC) content was found to be related to the non-fat cocoa solid (NFCS) content. The natural cocoa powders (average 87% of NFCS) contained the highest levels of AOC 826 μmol of trolox equivalent/g (TE/g) and PCs (40.8 mg/g). Alkalized cocoa (Dutched powders, average 80% NFCS) contained lower AOC (402 μmol of TE/g) and PCs (8.9 mg/g). Unsweetened chocolates or chocolate liquor (50% NFCS) contained 496 μmol of TE/g of AOC and 22.3 mg/g of PCs. Milk chocolates, which contain the least amount of NFCS (7.1%), had the lowest concentrations of AOC (80 μmol of TE/g) and PCs (2.7 mg/g). One serving of cocoa (5 g) or chocolate (15 or 40 g, depending upon the type of chocolate) was found to provide 2,000–9,100 μmol of TE of AOC and 45–517 mg of PCs, amounts that exceeded the amount in a serving of the majority of foods consumed in America. The monomers through trimers, contributed 30% of the total PCs in chocolates. Hydrophilic antioxidant capacity contributed >90% of AOC in all products. The correlation coefficient between AOC and PCs in chocolates was 0.92, suggesting that PCs were the dominant antioxidants in cocoa and chocolates. These results indicated that NFCS was correlated with AOC and PC in cocoa and chocolate products. Alkalizing dramatically decreased both the procyanidin content and antioxidant capacity, although not to the same extent.

The antioxidants in cocoa can prevent the oxidation of LDL-cholesterol, related to the mechanism of protection in heart disease (Weisberger 2001). Likewise, a few studies showed that reactive oxygen species associated with the carcinogenic processes was also inhibited. Wang et al. (2000) demonstrated that 2 hours after the intake of a procyanidin-rich chocolate containing 5.3 mg total procyanidin/g, of which 1.3 mg/g was (−)-epicatechin (epicatechin), plasma levels of epicatechin increased 133, 258 and 355 nmol/L in individuals who consumed 27, 53 and 80 g of chocolate, respectively. That the rise in plasma epicatechin levels was functionally significant was suggested by observations of trends for dose-response increases in the plasma antioxidant capacity and decreases in plasma lipid oxidation products. The above data supported the hypotheses that in healthy adults, (1) a positive relationship existed between procyanidin consumption and plasma procyanidin concentration and (*2*) the rise in plasma epicatechin contributed to the ability of plasma to scavenge free radicals and to inhibit lipid peroxidation.

Six categories of commercially available cocoa-containing and chocolate products: natural cocoa, unsweetened baking chocolate, dark chocolate, semisweet baking chips, milk chocolate, and chocolate syrup were analysed for antioxidant activity, polyphenol and procyanidin contents using four different methods: oxygen radical absorbance capacity (ORAC), vitamin C equivalence antioxidant capacity (VCEAC), total polyphenols, and procyanidins (Miller et al. 2006). All composite lots were further characterized for percent nonfat cocoa solids (NFCS) and percent fat. Natural cocoas had the highest levels of antioxidant activities, total polyphenols, and procyanidins followed by baking chocolates, dark chocolates and baking chips, and finally milk chocolate and syrups. The results showed a strong linear correlation between NFCS and ORAC (R^2=0.9849), total polyphenols (R^2=0.9793), and procyanidins (R^2=0.946), respectively. On the basis of principal component analysis, 81.4% of the sample set was associated with NFCS, antioxidant activity, total polyphenols, and procyanidins. The results indicated that, regardless of the product category, NFCS were the primary factor contributing to the level of cocoa antioxidants in the products tested.

In a comparative randomized double-blind crossover design or 20 volunteers, consumption of high-flavanol cocoa was found to lower concentration of plasma F(2)-isoprostanes, indicators of in-vivo lipid peroxidation especially when combined with physical excercise (Wiswedel et al. 2004).

Cocoa flavanol, (−)-epicatechin, was found to be a substrate of myeloperoxidase (MPO), as well as potent inhibitors of lipid peroxidation of low-density lipoprotein at micromolar concentrations (Schewe and Sies 2005). It strongly suppressed protein tyrosine nitration of low-density lipoprotein by MPO/nitrite or peroxynitrite. By mitigating undesirable MPO-mediated actions of nitrite, presumably via scavenging of the strong prooxidant and nitrating *NO2 radical, (−)-epicatechin may modulate NO metabolism in a favorable direction and thus counteract endothelial dysfunction.

Studies showed that a cocoa polyphenolic extract, ACTICOA powder (AP), exhibited preventive antioxidative effects on free radical production by leucocytes in rats and the cognitive performance (Rozan et al. 2007). The day after heat exposure, free radical production by leucocytes in rats treated with AP or vitamin E was significantly reduced as compared to control. The daily oral administration of AP or vitamin E protected rats from cognitive impairments after heat exposure by counteracting the overproduction of free radicals. Studies showed that the antioxidant activity, and flavan-3-ol levels of cocoa products investigated were stable over typical shelf lives of 1 year under controlled storage and over 2 years in ambient storage in the laboratory (Hurst et al. 2009). It was also found that 80 year old cocoa powder and 116 year old cocoa beans still possessed very high levels of antioxidant activity and flavan-3-ol content.

In a recent study of 20 healthy subjects fed a balanced diet for 4 weeks and supplemented with dark or white chocolate on the 14–27 day, detectable epicatechin levels and lower mononuclear blood cells DNA damage were observed only within 22 hours with dark chocolate (Spadafranca et al. 2010). Both effects were no longer evident after 22 hours. No effect was observed on plasma total antioxidant activity while white chocolate did not affect any variable. The results indicated that dark chocolate may transiently improve DNA resistance to oxidative stress, probably for flavonoid kinetics.

Antiatherosclerotic, Antihyperlipidemic Activities

(−)-Epicatechin, bioactive flavanol in coca was reported to exert protective action on two key processes in atherogenesis, oxidation of LDL and damage to endothelial cell by oxidized LDL (oxLDL) (Steffen et al. 2006). It attenuated or annulled LDL oxidation and counteracted deleterious actions of oxLDL on vascular endothelial cells. (−)-Epicatechin appeared to be a pleiotropic protectant for both LDL and endothelial cells. In addition to its major flavanol, (−)-epicatechin, chocolate also contained more (−)-catechin than (+)-catechin. The mean concentration of (−)-catechin in chocolate was 218 ± 126 mg/kg compared to 25 ± 15 mg/kg (+)-catechin (Donovan et al. 2006). At all concentrations, the intestinal absorption of (−)-catechin was lower than the intestinal absorption of (+)-catechin. Plasma concentrations of (−)-catechin were significantly reduced compared to (+)-catechin. In chocolate and cholate products, (+)-catechin was more bioavailable than (−)-catechin.

Kurosawa et al. (2005) found that after 6 months of dietary administration of cacao liquor polyphenols (CLP), at 1% (w/w) to the Kurosawa and Kusanagi-hypercholesterolemic (KHC) rabbits, a higher total cholesterol concentration was observed in the treatment group compared to the control group. However, no other effects were noted in lipid profiles in plasma or lipoproteins. The plasma concentration of thiobarbituric acid reactive substances (TBARS), a lipid-peroxidation index, was significantly decreased 1 month after the start of CLP administration compared to that of the control group. The antioxidative effect of CLP on LDL was observed from 2 to 4 months of administration. The area of atherosclerotic lesions in the aorta in the CLP group was significantly smaller than that in the control group and the tissue cholesterol and TBARS concentrations were lower in the CLP group than in the control group. The anti-atherosclerotic effect of CLP was confirmed both rheologically and histopathologically. CLP at a low concentration of 0.1 μg/ml significantly prolonged the lag time of LDL oxidation that was induced by a lipophilic azo-radical initiator, 2,2′-azobis(4-methoxy)-2,4-dimethyl-valeronitrile (V-70), or Cu(2+). The antioxidative effect of CLP was superior to those of the well-known antioxidative substances, vitamin C, vitamin E and probucol. The results indicated that CLP suppressed the generation of atherosclerosis, and its antioxidative effect appeared to have an important role in its anti-atherosclerotic activity.

An ex-vivo study showed that epicatechin, a major polyphenol in chocolate and chocolate extracts, to be a powerful inhibitor of plasma lipid oxidation due to polyphenols' ability to bind to lower density lipoproteins (Vinson et al. 2006). Conversely, the fat from chocolate alone was a pro-oxidant in this model. This was also demonstrated in an in-vivo human study. After consumption of dark chocolate and cocoa powder, the lower density lipoproteins isolated from plasma were protected from oxidation compared to the lipoproteins isolated after cocoa butter consumption, which were put under oxidative stress. In an animal model of atherosclerosis, cocoa powder at a human dose equivalent of two dark chocolate bars per day significantly inhibited atherosclerosis, lowered cholesterol, low-density lipoprotein, and triglycerides, raised high-density lipoprotein, and protected the lower density lipoproteins from oxidation. The authors concluded that chocolate had potential beneficial effects with respect to heart disease.

Hamed et al. (2008) found that 1 week consumption of dark chocolate improved lipid profiles and decreased platelet reactivity within the total group while reducing inflammation only in women indicating that regular dark chocolate ingestion may have cardioprotective properties. Following 7 days of regular dark chocolate intake, LDL was lowered by 6% and HDL rose by 9%. Adenosine diphosphate- and arachidonic acid-induced activated glycoprotein GPIIb/IIIa expression was reduced by dark chocolate. High-sensitivity C-reactive protein was also reduced by dark chocolate in women.

There is strong epidemiological evidence that the consumption of flavanol-rich cocoa products contributes to the prevention of cardiovascular diseases mediated partly by their strong antioxidant effects. However, in a recent review of 19 controlled intervention studies (two in patients with cardiovascular diseases), Scheid et al. (2010) found that in both, healthy participants and patients, total plasma antioxidant capacity as well as most markers of oxidative stress were not influenced by the intervention of cocoa consumption. Low-density lipoprotein oxidation ex-vivo and in-vivo decreased probably due to a direct beneficial effect of cocoa flavanols on lipid and protein oxidation. They asserted then that any recommendations for cocoa intake within preventive and therapeutic measures were not reasonable.

Mostofsky et al. (2010) conducted a prospective cohort study of 31,823 women aged 48–83 years without baseline diabetes or a history of heart failure or myocardial infarction, and found that moderate habitual chocolate intake was associated with a lower rate of heart failure hospitalization or death, but the protective association was not observed with intake of ≥1 servings per day. Over 9 years of follow-up, 419 women were hospitalized for incident heart failure (n=379) or died of heart failure (n=40). Compared with no regular chocolate intake, the multivariable-adjusted rate ratio of heart failure was 0.74 (95% CI, 0.58–0.95) for women consuming 1–3 servings of chocolate per month, 0.68 (95% CI, 0.50–0.93) for those consuming 1–2 servings per week, 1.09 (95% CI, 0.74–1.62) for those consuming 3–6 servings per week, and 1.23 (95% CI, 0.73–2.08) for those consuming ≥1 servings per day.

Studies showed that cocoa procyanidin had inhibitory effects on thrombin-induced expression and activation of matrix metalloproteinase (MMP)-2 in vascular smooth muscle cells (VSMC) (Lee et al. 2008). Expression and activation of matrix metalloproteinase (MMP)-2 play pivotal roles in the migration and invasion of human aortic vascular smooth muscle cells (VSMC) originating from normal human tissue, which is strongly linked to atherosclerosis. The results showed that cocoa procyanidins were potent inhibitors of mitogen-activated protein kinase (MEK) and membrane type-1 (MT1)-MMP, and subsequently inhibited the expression and activation of pro-MMP-2, and also the invasion and migration of VSMC, which may in part elucidate the molecular action of antiatherosclerotic effects of cocoa. Akita et al. (2008) found that after 6 months of dietary administration of cacao liquor polyphenols, heart rate and blood pressure of Kurosawa and Kusanagi-hypercholesterolaemic rabbits were lowered but plasma lipid concentrations were unchanged. The area of atherosclerotic lesions in the aorta in the cacao liquor polyphenol group was significantly smaller than that in the standard diet group. Additionally, cacao liquor polyphenols preserved parasympathetic nervous tone, although that in the standard diet group was significantly decreased with ageing. The authors concluded that cacao liquor polyphenols may play an important role to protect cardiovascular and autonomic nervous functions.

In a double-blind, placebo-controlled, randomized trial of the effects of dark chocolate and sweetened cocoa beverage of 101 participants for 6 weeks, no significant group (dark chocolate and cocoa or placebo)-by-trial (baseline, midpoint, and end-of-treatment assessments) interactions were found for the neuropsychological, hematological, or blood pressure variables examined (Crews et al. 2008). In contrast, the midpoint and end-of-treatment mean pulse rate assessments in the dark chocolate and cocoa group were significantly higher than those at baseline and significantly higher than the midpoint and end-of-treatment rates in the control group. This investigation failed to support the predicted beneficial effects of short-term dark chocolate and cocoa consumption on any of the neuropsychological or cardiovascular health-related variables included in this research. Consumption of dark chocolate and cocoa was, however, associated with significantly higher pulse rates at 3- and 6-week treatment assessments.

In a randomized crossover feeding trial 42 high-risk volunteers (19 men and 23 women; age: 69.7 ±11.5 years) were 40 g cocoa powder with 500 ml skim milk/day (C+M) or only 500 ml skim milk/day (M) for 4 weeks (Monagas et al. 2009). In monocytes, the expression of VLA-4 (very late antigen-4), CD40 (a co-stimulatory protein), and CD36 (integral membrane protein) was significantly lower after cocoa and milk intake than after milk intake. In addition, serum concentrations of the soluble endothelium-derived adhesion molecules P-selectin and intercellular adhesion molecule-1 were significantly lower (after cocoa and milk intake than after milk intake). The results suggested that the intake of cocoa polyphenols may modulate inflammatory mediators in patients at high risk of cardiovascular disease. These antiinflammatory effects may contribute to the overall benefits of cocoa consumption against atherosclerosis.

Djoussé et al. (2011) studied 2,217 participants of the NHLBI Family Heart Study using a cross-sectional design and found chocolate consumption might be inversely associated with prevalent calcified atherosclerotic plaque in the coronary arteries (CAC). There was an inverse association between frequency of chocolate consumption and prevalent CAC. Odds ratios (95% CI) for CAC were 1.0 (reference), 0.94 (0.66–1.35), 0.78 (0.53–1.13), and 0.68 (0.48–0.97) for chocolate consumption of 0, 1–3 times per month, once per week, and 2+ times per week, respectively, adjusting for age, sex, energy intake, waist-hip ratio, education, smoking, alcohol consumption, ratio of total-to-HDL-cholesterol, non-chocolate candy, and diabetes mellitus.

Wan et al. (2001) conducted a randomized, 2-period, crossover study in 23 healthy subjects fed 2 diets: an average American diet (AAD) controlled for fibre, caffeine, and theobromine and an AAD supplemented with 22 g cocoa powder and 16 g dark chocolate (CP-DC diet), providing approximately 466 mg procyanidins/day. LDL oxidation lag time was approximately 8% greater after the CP-DC diet than after the AAD. Serum total antioxidant capacity measured by oxygen radical absorbance capacity was approximately 4% greater after the CP-DC diet than after the AAD and was positively correlated with LDL oxidation lag time. HDL cholesterol was 4% greater after the CP-DC diet than after the AAD; however, LDL-HDL ratios were not significantly different. Twenty-four-hour urinary excretion of thromboxane B(2) and 6-keto-prostaglandin F(1)(alpha) and the ratio of the 2 compounds were not significantly different between the 2 diets. The results indicated that Cocoa powder and dark chocolate may favourably affect cardiovascular disease risk status by modestly reducing LDL oxidation susceptibility, increasing serum total antioxidant capacity and HDL-cholesterol concentrations, and not adversely affecting prostaglandins.

In a randomized, double-blind parallel arm study design, consumption of the phytosterol-enriched cocoa flavanol snack bars by subjects (n=32) with hypercholesterolemia but not control bars (n=35) for 6 weeks was associated with significant reductions in plasma total (4.7%) and LDL cholesterol (6%), and the ratio of total to high-density lipoprotein cholesterol (7.4%) (Polagruto et al. 2006). There were no changes in high-density lipoprotein cholesterol, triglycerides, or lipid-adjusted lycopene, β-cryptoxanthin, lutein/zeaxanthin, α-carotene levels, or levels of serum vitamins A or E. A significant reduction in lipid-adjusted serum β-carotene was observed in the phytosterol but not the no-phytosterol-added group. The data suggested that the incorporation of this snack food into a balanced diet represented a practical dietary strategy in the management of serum cholesterol levels.

Using a rat model of dietary-induced hypercholesterolemia, consumption of a new cocoa product rich in dietary fibre and naturally containing antioxidant polyphenols, with a hypercholesterolemic diet for 3 weeks improved the lipidemic profile and reduced lipid peroxidation (Lecumberri et al. 2007). The cocoa fibre diet showed an important hypolipidemic action, returning triacylglycerol levels in hypercholesterolemic animals to normal values. The hypocholesterolemic effect was also evident, reducing total and low-density lipoprotein cholesterol, yet basal values were not attained. Decreased lipid peroxidation in serum and liver as a consequence of cocoa intake was evident not only in hypercholesterolemic but also in normocholesterolemic animals. The findings suggested that the cocoa rich fibre with antioxidant polyphenols might contribute to a reduction of cardiovascular risk. Studies showed that Zucker fatty rats fed 5% soluble cocoa fibre -enriched diet had less weight gain and food intake than those fed the standard diet (Sánchez et al. 2010a). The group fed the fibre-enriched diet had lower values of the total cholesterol/high-density lipoprotein cholesterol ratio and triglyceride levels than the standard group. Fat apparent digestibility was also lower in the fibre group. Both systolic and diastolic blood pressure were decreased. Further, soluble cocoa fibre reduced plasma glucose and insulin, and as a consequence the insulin resistance was also decreased. The findings demonstrated that soluble cocoa fibre resulted in an improvement of the studied risk factors associated with cardiometabolic disorders.

In a study of 27 healthy subjects, 4 days of flavanol-rich cocoa induced consistent and striking peripheral vasodilation (Fisher et al. 2003). On day 5, pulse wave amplitude exhibited a large additional acute response to cocoa. In addition, intake of flavanol-rich cocoa augmented the vasodilator response to ischemia. Flavanol-rich cocoa also amplified the systemic pressor effects of NG-nitro-L-arginine methyl ester (L-NAME). The findings suggested that flavanol-rich cocoa induced vasodilation via activation of the nitric oxide system, provided a plausible mechanism for the protection that flavanol-rich foods induced against coronary events. In a 3 week clinical supplementation trial of 45 non-smoking, healthy volunteers, the group fed with dark chocolate or dark chocolate enriched with cocoa polyphenols, presented an increase in serum HDL cholesterol (11.4% and 13.7%, respectively), whereas in the white chocolate group there was a small decrease (−2.9%) (Mursu et al. 2004). The concentration of serum LDL diene conjugates, a marker of lipid peroxidation in vivo, decreased 11.9% in all three study groups. No changes were seen in the total antioxidant capacity of plasma, in the oxidation susceptibility of serum lipids or VLDL+LDL, or in the concentration of plasma F2-isoprostanes or hydroxy fatty acids. The results suggested that cocoa polyphenols may increase the concentration of HDL cholesterol, whereas chocolate fatty acids may modify the fatty acid composition of LDL and make it more resistant to oxidative damage.

In a double-blind, placebo-controlled, cross-over study of 49 normotensive adults (32 women, 17 men), regular consumption of the plant sterol-containing chocolate bar resulted in reductions of 2.0% and 5.3% in serum total and LDL cholesterol, respectively (Allen et al. 2008). Consumption of cocoa flavanol-containing dark chocolate bar also reduced systolic blood pressure at 8 week. Results indicated that regular consumption of chocolate bars containing plant sterol and cocoa flavanols as part of a low-fat diet may support cardiovascular health by lowering cholesterol and improving blood pressure. Cocoa powder being rich in polyphenols, such as catechins and oligomeric procyanidins, had been reported to have a hypocholesterolemic effect in humans (Yasuda et al. 2008a). Results of a 4 week study in rats by the authors suggested that oligomeric procyanidins from cocoa powder were the principal active components responsible for the hypocholesterolemic effect, and inhibited the intestinal absorption of cholesterol and bile acids through the decrease in micellar cholesterol solubility. Rats fed a high-cholesterol diet containing 1% polyphenol extract from cocoa powder had significantly lower plasma cholesterol concentrations, and had significantly greater fecal cholesterol and total bile acids excretion than the high cholesterol diet group. Rats fed with a high-cholesterol diet containing 0.024% catechin and 0.058% epicatechin did not influence plasma cholesterol concentrations, or fecal cholesterol or total bile acids excretion. Micellar solubility of cholesterol in vitro-was significantly lower for procyanidin B2 (dimer), B5 (dimer), C1 (trimer) and A2 (tetramer), the main components of polyphenol extract from cocoa powder, compared to catechin and epicatechin. Studies showed that rats fed a high cholesterol diet without cacao procyanidins had significantly higher plasma cholesterol level than the normal diet group (Osakabe and Yamagishi 2009). Supplementation of cacao procyanidins significantly decreased plasma cholesterol levels similar to those of the normal diet group. The liver cholesterol and triglyceride levels in all high cholesterol diet groups were significantly higher, but 1.0% cacao procyanidins supplementation significantly reduced this increase. The procyanidins dose-dependently reduced micellar solubility of cholesterol and this activity increased with increasing molecular weight. The results suggested that one of the mechanisms of cacao procyanidins lowering plasma cholesterol was inhibition of intestinal absorption of cholesterol.

In a comparative, double-blind study, 160 subjects who ingested either cocoa powder containing low-polyphenolic compounds were evaluated for plasma LDL cholesterol and oxidized LDL concentrations (Baba et al. 2007a). in 131 subjects with LDL cholesterol concentrations of > or =3.23 mmol/L at baseline, cocoa consumption decreased plasma LDL cholesterol, oxidized LDL, and apo B concentrations and increased plasma HDL cholesterol relative to baseline in the low-, middle-, and high-cocoa groups. The results suggested that polyphenolic substances derived from cocoa powder may contribute to a reduction in LDL cholesterol, an elevation in HDL cholesterol, and the suppression of oxidized LDL. Similar results were obtained in another study of 25 normocholesterolemic and mildly hypercholesterolemic human subjects (Baba et al. 2007b). A significantly greater increase in plasma HDL cholesterol (24%) was observed in the cocoa group than in the control group (5%). A negative correlation was observed between plasma concentrations of HDL cholesterol and oxidized LDL. At 12 week, there was a 24% reduction in dityrosine from baseline concentrations in the cocoa group.

In a randomised, in a double-blind, cross-over design of 21 volunteers (8 females and 13 males, 54.9 (se 2.2) years, BMI 31.6 (se 0.8) kg/m^2, systolic BP 134 (se 2) mmHg, diastolic BP (DBP) 87 (se 2) mmHg), consumption of cocoa flavanol was found to improve blood pressure responsiveness to exercise (Berry et al. 2010). Blood pressure response to exercise was attenuated by high cocoa flavanol consumption compared with low flavanol. Blood pressure increases were 68% lower for diastolic blood pressure and 14% lower for mean blood pressure. Endothelium-dependent flow-mediated dilatation measurements were higher after taking high cocoa flavanols than after taking low cocoa flavanols. The findings suggested that by facilitating vasodilation and attenuating exercise-induced increases in blood pressure, cocoa flavanols may decrease cardiovascular risk and enhance the cardiovascular benefits of moderate intensity exercise in at-risk individuals.

In an animal model of dietary-induced hypercholesterolemia, consumption of a cholesterol-rich diet containing the soluble cocoa fiber product (SCFP) as a source of dietary fibre and antioxidant polyphenols, resulted in lower food intake and body weight gain in comparison with control groups consuming cholesterol-free or cholesterol-rich diets with cellulose as dietary fibre (Ramos et al. 2008). SCFP diminished the negative impact of the cholesterol-rich diet, buffering the decrease of high density lipoprotein-cholesterol, and the increase of total and low density lipoprotein-cholesterol levels, and lipid peroxidation (malondialdehyde levels) induced by the fatty diet. SCFP also decreased triglyceride levels to values lower than those in the group fed the cholesterol-free diet. These results indicated the potential application of the SCFP as a dietary supplement or functional food ingredient.

In a randomized, placebo-controlled double-blind crossover study, 12 individuals with type 2 diabetes were randomized to 45 g chocolate with or without a high polyphenol content for 8 weeks and then crossed over after a 4-week washout period (Mellor et al. 2010). HDL cholesterol increased significantly with high polyphenol chocolate with a decrease in the total cholesterol: HDL ratio. No changes were seen with the low polyphenol chocolate in any parameters. Over the course of 16 weeks of daily chocolate consumption neither weight nor glycaemic control altered from baseline. The findings showed that high polyphenol chocolate was effective in improving the atherosclerotic cholesterol profile in patients with diabetes by increasing HDL cholesterol and improving the cholesterol:HDL ratio without affecting weight, inflammatory markers, insulin resistance or glycaemic control.

A meta-analysis of 8 randomized controlled trials (involving 215 participants) revealed that cocoa consumption significantly lowered LDL cholesterol by 5.87 mg/dL and marginally lowered total cholesterol by 5.82 mg/dL (Jia et al. 2010). However, no significant change was seen in LDL cholesterol in high-quality studies (3 studies included). Subgroup analyses suggested a cholesterol-lowering effect only in those subjects who consumed a low dose of cocoa and with cardiovascular disease risks. There was no evidence of a dose-effect relation, of any effect in healthy subjects, or of any change in HDL cholesterol. Overall, short-term cocoa consumption was found to significantly reduce blood cholesterol, but the changes were dependent on the dose of cocoa consumption and the healthy status of participants.

Antihypertensive and Amelioration of Endothelial Function Activity

In white rabbits, the polymeric procyanidins in cocoa extract caused an endothelium dependent relaxation that was mediated by the activation of nitric oxide (Karim et al. 2000). In a randomized, double-blind, placebo-controlled design conducted over a 2 week period in 21 healthy adult subjects, high-flavonoid chocolate flavonoid (213 mg procyanidins, 46 mg epicatechin) consumption was found to improve endothelium-dependent flow-mediated dilation of the brachial artery compared to low-flavonoid chocolate consumption (Engler et al. 2004). No significant differences were observed in the resistance to LDL oxidation, total antioxidant capacity, 8-isoprostanes, blood pressure, lipid parameters, body weight or body mass index between the two groups. Plasma epicatechin concentrations were markedly increased at 2 weeks in the high-flavonoid group but not in the low-flavonoid group. In a randomized, single-blind, sham procedure-controlled, cross-over study of 17 young healthy volunteers, consumption of dark chocolate acutely was found to decrease wave reflections, without affecting aortic stiffness, and that it may exert a beneficial effect on endothelial function in healthy adults (Vlachopoulos et al. 2005). Chocolate generated significant increase in resting and hyperemic brachial artery diameter throughout the study. Flow-mediated dilation

(FMD) of the brachial artery increased significantly at 60 min. The aortic augmentation index was significantly decreased with chocolate throughout the study indicating a decrease in wave reflections, whereas and carotid-femoral pulse wave velocity did not change to a significant extent.

Heiss et al. (2003) found that a single dose of a cocoa drink rich in flavan-3-ols transiently increased nitric oxide bioactivity in human plasma and significantly reversed endothelial dysfunction. The correlation between flow mediated dilation and levels of nitrosylated and nitrosated species suggested that flavan-3-ols induced arterial dilation via their effects on nitric oxide availability. This conclusion was supported by the negative results for the other vascular variables. Heiss et al. (2005) found that the circulating pool of bioactive nitric oxide and endothelium-dependent vasodilation was acutely increased in smokers following the oral intake of a flavanol-rich cocoa drink. The increase in circulating NO pool may contribute to beneficial vascular health effects of flavanol-rich food. The daily consumption of flavanol-rich cocoa drink for 7 days exhibited the potential to reverse endothelial dysfunction in a sustained and dose-dependent manner (Heiss et al. 2007). The daily consumption of a flavanol-rich cocoa drink (3×306 mg flavanols/day) over 7 days (n=6) resulted in continual flow-mediated dilation (FMD) increases at baseline (after overnight fast and before flavanol ingestion) and in sustained FMD augmentation at 2 h after ingestion. Increases observed in circulating nitrite, but not in circulating nitrate, paralleled the observed FMD augmentations. Generally applied biomarkers for oxidative stress (plasma, MDA, TEAC) and antioxidant status (plasma ascorbate, urate) remained unaffected by cocoa flavanol ingestion.

Results of randomized double-blind placebo-controlled study by Farouque et al. (2006) suggested that over a 6-week period, flavanol-rich cocoa did not modify vascular function in patients with established coronary artery disease. No acute or chronic brachial artery FMD (flow-mediated dilation) and SAC (systemic arterial compliance) were seen in groups fed flavanol-rich cocoa or placebo. No difference in soluble cellular adhesion molecules, forearm blood flow responses to ischaemia, exercise, sodium nitroprusside or acetylcholine chloride was seen in the group receiving flavanol-rich cocoa between baseline and 6 weeks.

The consumption of flavanol-containing milk chocolate by 38 free living, young (18–20 years old) male soccer players was significantly associated with a decrease in diastolic blood pressure (−5 mmHg), mean blood pressure (−5 mmHg), plasma cholesterol (−11%), LDL-cholesterol (−15%), malondialdehyde (−12%), urate (−11%) and lactate dehydrogenase (LDH) activity (−11%), and an increase in vitamin E/cholesterol (+12%) (Fraga et al. 2005). No relevant changes in these variables were associated with cocoa butter chocolate consumption. No changes in the plasma levels of (−)-epicatechin were observed following analysis of fasting blood samples. in healthy male adults the ingestion of flavanol-rich cocoa was found to be associated with acute elevations in levels of circulating NO species, an enhanced FMD response of conduit arteries, and an augmented microcirculation (Schroeter et al. 2006). (−)-Epicatechin and its metabolite, epicatechin-7-*O*-glucuronide, was identified as independent predictors of the vascular effects after flavanol-rich cocoa ingestion. Chronic consumption of a high-flavanol diet was associated with a high urinary excretion of NO metabolites, consistent with an augmented NO production or diminished degradation. Finally, in humans, epicatechins closely mimic the vascular effects of flavanol-rich cocoa, suggesting that they represent the primary mediator of the beneficial effect of cocoa flavanols on vascular function Similarly, in isolated aortic rings, concentrations of flavanols comparable to those occurring in plasma after cocoa intake induce endothelium-dependent relaxations. Endothelial dysfunction and platelet activation are cornerstones in the pathogenesis of atherothrombosis, leading to vasoconstriction, thrombus formation, and inflammation. Both active and passive cigarette smoking had consistently been shown to induce endothelial dysfunction (Hermann et al. 2006). The researchers found that in young healthy smokers, commercially available dark chocolate (74% cocoa), but not white chocolate, markedly improved flow-mediated vasodilation and improved plasma antioxidant status, suggesting that induction of eNOS and, in turn, elevated NO levels and a reduction in the production of reactive oxidant species contributed to the enhanced endothelial function under these conditions. In a cohort of 470 elderly men, cocoa intake was inversely associated with blood pressure, cardiovascular and all-cause mortality (Buijsse et al. 2006). Over the 15 years study period, men who consumed cocoa regularly had significantly lower blood pressure than those who did not. Over the course of the study, 314 men died, 152 due to cardiovascular diseases. Men in the group with the highest cocoa consumption were half as likely as the others to die from cardiovascular disease. Their risk remained lower even when considering other factors, such as weight, smoking habits, physical activity levels, calorie intake, and alcohol consumption. Chronic consumption of flavanol-rich cocoa was found to improve endothelial function and decrease vascular cell adhesion molecule in hypercholesterolemic postmenopausal women (Wang-Polagruto et al. 2006). Brachial artery hyperemic blood flow increased significantly by 76% after the 6-week cocoa intervention in the high cocoa flavanols group, compared with 32% in the low cocoa flavanols cocoa group. The 2.4-fold increase in hyperemic blood flow with high cocoa flavanols closely correlated ($r^2=0.8$) with a significant decrease (11%) in plasma levels of soluble vascular cell adhesion molecule-1. Similar responses were not observed after chronic use of low cocoa flavanols. The results suggested that reductions in plasma soluble vascular cell adhesion molecule-1 after chronic consumption of a flavanol-rich cocoa may be mechanistically linked to improved vascular reactivity. In a randomized, placebo-controlled, single-blind crossover trial of 45 healthy adults with mean age of 53 years, acute intake of both solid dark chocolate and liquid cocoa improved endothelial function and lowered blood pressure in overweight adults (Faridi et al. 2008). Endothelial function measured as flow-mediated dilatation improved significantly more with sugar-free than with regular cocoa. Sugar content may attenuate these effects, and sugar-free preparations may augment them. Data from a randomized, controlled, investigator-blinded, parallel-group, 18 week dietary trial involving 44 adults aged 56 through 73 years (24 women, 20 men) with untreated upper-range prehypertension or stage 1 hypertension indicated that inclusion of small amounts of polyphenol-rich dark chocolate as part of a usual diet efficiently reduced systolic and diastolic blood pressure and improved formation of vasodilative nitric oxide (Taubert et al. 2007a, b).

A natural flavonoid-enriched cocoa powder, commercially named CocoanOX was found to have potential to be used as a functional ingredient with antihypertensive effect (Cienfuegos-Jovellanos et al. 2009). The product was found to be rich in total procyanidins (128.9 mg/g), especially monomers, dimers, and trimers (54.1 mg/g), and mainly (−)-epicatechin (19.36 mg/g). This product produced a clear antihypertensive effect in spontaneously hypertensive rats, but these doses did not modify the arterial blood pressure in the normotensive Wistar-Kyoto rats. The initial values of diastolic blood pressure and systolic blood pressure were recovered in spontaneously hypertensive rats, respectively, 24 and 48 h post-administration of the different doses of CocoanOX or Captopril. A soluble cocoa fibre product (SCFP) was shown to exert the antihypertensive and antioxidant activities in spontaneously hypertensive rats (Sánchez et al. 2010b). Body weight gain was slower in the group treated with SCFP. SCFP increased liquid intake but decreased dry food intake in the rats. SCFP decreased plasma malondialdeyhde concentrations and slightly decreased plasma angiotensin converting enzyme activity. The results indicated that control of body weight and the control of increased angiotensin II may be involved in the antihypertensive effect of this product.

In a randomized, placebo-controlled, double-blind, crossover trial of a flavanol-rich cocoa drink in 20 individuals with essential hypertension, daily consumption of flavanol-rich cocoa for 2 weeks was not sufficient to reduce blood pressure or ameliorate insulin resistance in human subjects with essential hypertension (Muniyappa et al. 2008). Studies demonstrated that cocoa flavanols decreased vascular arginase-2 mRNA expression and activity in human umbilical vein endothelial cells (Schnorr et al. 2008). In a double-blind intervention study with cross-over design, dietary intervention with flavanol-rich cocoa caused diminished arginase activity in rat kidney and, erythrocyte arginase activity was lowered in healthy humans following consumption of a high flavanol beverage in-vivo. Increased arginase activity had been linked to low NO levels, and an inhibition of arginase activity had been reported to improve endothelium-dependent vasorelaxation.

In a randomized, placebo-controlled, single-blind crossover trial of 45 healthy adults with mane age 53 years, acute intake of both solid dark chocolate and liquid cocoa was found to improve endothelial function and lower blood pressure in overweight adults (Faridi et al. 2008). Sugar content may attenuate these effects, and sugar-free preparations may augment them. Solid dark chocolate and liquid cocoa (sugar-free and sugared) consumption improved endothelial function (measured as flow-mediated dilatation) compared with placebo. Blood pressure decreased after the intake of dark chocolate and sugar-free cocoa compared with placebo. Endothelial function improved significantly more with sugar-free than with regular cocoa.

In a randomised, single-blind, cross-over design 14 healthy overweight and obese subjects were randomised to either take 20 g dark chocolate with 500 mg polyphenols then 20 g dark chocolate with 1,000 mg polyphenols or vice-versa for 2 weeks separated by a 1-week washout period (Almoosawi et al. 2010). It was observed that the 500 mg polyphenol dose was equally effective in reducing fasting blood glucose levels, systolic BP (SBP) and diastolic BP (DBP) as the 1,000 mg polyphenol dose. There was also a trend towards a reduction in urinary free cortisone levels with both groups. No changes in anthropometrical measurements were noted. Desch et al. (2010b) conducted a meta-analysis of 10 randomized controlled trials comprising 297 individuals (healthy normotensive adults or patients with prehypertension/stage 1 hypertension) and confirmed the blood pressure-lowering capacity of flavanol-rich cocoa products in a larger set of trials than previously reported. However, significant statistical heterogeneity across studies could be found, and questions such as the most appropriate dose and the long-term side effect profile warrant further investigation before cocoa products could be recommended as a treatment option in hypertension.

Thirty-two men and 20 postmenopausal women with untreated mild hypertension (seated clinic blood pressure >130/85 and <160/100 mmHg) were randomized and instructed to consume daily a reconstituted cocoa beverage containing 33, 372, 712 or 1,052 mg per day of cocoa flavanols for 6 weeks in a double-blind, parallel comparison (Davison et al. 2010). Significant reductions in 24-h systolic (5.3 ± 5.1 mmHg), diastolic (3 ± 3.2 mmHg) and mean arterial BP (3.8 ± 3.2 mmHg) were observed only at the highest dose of 1,052 mg/day cocoa flavanols. No reduction in BP was seen at any other dose. One hundred and two patients with pre-hypertension/stage 1 hypertension and established cardiovascular end-organ damage or diabetes mellitus were randomly assigned to receive either 6 or 25 g/day of flavanol-rich dark chocolate for 3 months in a prospective randomized open-label blinded end-point design study (Desch et al. 2010a). Significant reductions in mean ambulatory 24-h blood pressure were observed between baseline and follow-up in both groups. There were no significant differences in blood pressure changes between groups. In the higher-dose group, a slight increase in body weight was noted.

In a randomized, controlled, double-masked, cross-over trial, 16 patients (64 ± 3 years of age) with coronary artery disease, endothelium-dependent vasomotor function, as measured by flow-mediated vasodilation of the brachial artery, was improved by 47% with high-flavanol (375 mg) dietary intervention compared with the low-flavanol (9 mg) dietary intervention (Heiss et al. 2010). After high flavanol intervention, the number of $CD34^{+}/KDR^{+}$ –CACs (circulating angiogenic cells), as measured by flow cytometry, increased 2.2-fold as compared with after the low flavanol intervention. The CAC functions, as measured by the capacity to survive, differentiate, proliferate, and to migrate were not different between the groups. High flavanols led to a decrease in systolic blood pressure and increase in plasma nitrite level. Overall, sustained improvements in endothelial dysfunction by regular dietary intake of flavanols are associated with mobilization of functional CACs. Van den Bogaard et al. (2010) conducted a double-blind, placebo-controlled 3-period crossover trial assigning 42 healthy individuals (age 62±4.5 years; 32 men) with office blood pressure of 130–159 mmHg/85–99 mmHg and low added cardiovascular risk to a random treatment sequence of dairy drinks containing placebo, flavanol-rich cocoa with natural dose consisting of 106 mg of theobromine, or theobromine-enriched flavanol-rich cocoa with 979 mg of theobromine. Treatment duration was 3 weeks with a 2-week washout. The primary outcome was that theobromine-enriched cocoa significantly increased 24-h ambulatory systolic blood pressure while lowering central systolic blood pressure.

Taubert et al. (2007b) conducted a meta-analysis of five randomized controlled studies of cocoa administration involving a total of 173 subjects with a median duration of 2 weeks. They found that consumption of foods rich in cocoa may reduce blood pressure, while tea intake appeared to have no effect. The meta-analysis of 15 studies with interventions of 2–18 weeks performed by Ried et al. (2010) suggested that dark chocolate (high flavanol) was superior to placebo in reducing systolic hypertension or diastolic prehypertension. Flavanol-rich chocolate did not significantly reduce mean blood pressure below 140 mmHg systolic or 80 mmHg diastolic. Egan et al. (2010) found that published studies on chocolate and blood pressure included a relatively small number of subjects, and the results were conflicting.

Antiplatelet Activity

Numerous studies had demonstrated platelet inhibitory properties of cocoa (Rein et al. 2000; Pearson et al. 2002; Innes et al. 2003; Murphy et al. 2003; Hermann et al. 2006; Heptinstall et al. 2006; Flammer et al. 2007; Bordeaux et al. 2007; Persson et al. 2011) In a study of 30 healthy volunteers, cocoa consumption of a polyphenol rich cocoa beverage suppressed ADP- or epinephrine-stimulated platelet activation and platelet microparticle formation (Rein et al. 2000). In a study of 30 volunteers, consumption of dark chocolate inhibited collagen-induced platelet aggregation in platelet rich plasma (Innes et al. 2003). White and milk chocolate had no significant effect on platelets. In a crossover design study of 16 healthy adults, flavanol-rich cocoa inhibited epinephrine-stimulated platelet activation and function (Pearson et al. 2002). These effects were qualitatively similar to aspirin, but less profound. Epinephrine-induced platelet function was inhibited by cocoa after 6 hours, inhibited 2 and 6 hours after aspirin, and after aspirin plus cocoa. Cocoa and aspirin, given separately, reduced epinephrine-stimulated GPIIb/IIIa-act expression at 2 and 6 hours, respectively, and at 2 and 6 hours when given together, suggesting an additive effective. Cocoa consumption had an aspirin-like effect on primary hemostasis. Holt et al. (2002) found significant increases in plasma flavanol concentrations, with a concurrent increase in the ratio of prostacyclin to leukotriene shortly after consumption of a small amount of chocolate. They also observed reductions in both ADP/collagen-stimulated and epinephrine/collagen-stimulated platelet-related primary hemostasis. Their results supported the hypothesis that small amounts of foods rich in flavonoids, such as chocolate, can transiently affect platelet function.

In a blinded parallel-designed study involving 32 healthy subjects, cocoa flavanol and procyanidin supplementation for 28 days significantly increased plasma epicatechin and catechin concentrations and significantly decreased platelet function (Murphy et al. 2003). Plasma concentrations of epicatechin and catechin in the active group increased by 81% and 28%, respectively, during the intervention period. The active group had significantly lower P selectin expression and significantly lower ADP-induced aggregation and collagen-induced aggregation than did the placebo group. Plasma ascorbic acid concentrations were significantly higher in the active than in the placebo, whereas plasma oxidation markers and antioxidant status did not change in either group. Flavonoids may be partly responsible for some health benefits, including antiinflammatory action and a decreased tendency for the blood to clot. The data supported the results of earlier acute studies that used higher doses of cocoa flavanols and poligomeric procyanidins from cocoa powder that inhibited platelet activation and function over 6 hours in humans.

In young healthy smokers, dark chocolate reduces platelet adhesion as assessed by a shear stress–dependent platelet test (Hermann et al. 2006). In an in-vitro study, several cocoa flavanols and their metabolites were shown to inhibit platelet aggregation, platelet-monocyte conjugate formation, platelet-neutrophil conjugate formation, and platelet activation (Heptinstall et al. 2006). Their effects were similar to those of aspirin and the effects of a cocoa flavanol and aspirin did not seem to be additive. The flavanols also inhibited monocyte and neutrophil activation, but this was not replicated by aspirin. 4′-O-methyl-epicatechin, one of the known metabolites of the cocoa flavanol (−)-epicatechin, was consistently effective as an inhibitor of platelet and leukocyte activation. In the second study, consumption of a flavanol-rich cocoa beverage also resulted in significant inhibition of platelet aggregation, platelet-monocyte conjugate formation, platelet-neutrophil conjugate formation, and platelet activation induced by collagen (Heptinstall et al. 2006). The inhibitory effects were related to their flavanol content. There was also inhibition of monocyte and neutrophil activation, but here it was concluded that cocoa constituents other than flavanols may contribute to the inhibition that was observed. The authors concluded that cocoa flavanols, their metabolites and possibly other cocoa constituents could modulate the activity of platelets and leukocytes in-vitro and ex-vivo. The research suggested that the consumption of certain cocoa products may provide a dietary approach to maintaining or improving cardiovascular health.

In a double-blind, randomized study of 22 heart transplant patients, dark chocolate was found to induce coronary vasodilation, improve coronary vascular function, and decrease platelet adhesion 2 hours after consumption (Flammer et al. 2007). Two hours after ingestion of flavonoid-rich dark chocolate, coronary artery diameter was increased significantly whereas it remained unchanged after control chocolate. Endothelium-dependent coronary vasomotion improved significantly after dark chocolate consumption. Platelet adhesion decreased from 4.9% to 3.8% in the dark chocolate group but remained unchanged in the control group. These immediate beneficial effects were paralleled by a significant reduction of serum oxidative stress and were positively correlated with changes in serum epicatechin concentration. Studies by Bordeaux et al. (2007) showed that chocolate consumers had longer Platelet Function Analyzer (PFA) closure times and decreased urinary 11-dehydro thromboxane B2 levels. Chocolate remained a significant independent predictor of both ex-vivo and in-vivo platelet function testing after adjusting for confounders. The authors concluded that even consuming modest amounts of commercial chocolate had important antiplatelet effects.

A randomized, single-blind design of 39 healthy men (age 23–40 years) showed that flavonoid-rich dark chocolate ingestion significantly improved coronary circulation in healthy adults, independent of changes in oxidative stress parameters, blood pressure and lipid profile, whereas non-flavonoid white chocolate had no such effects (Shiina et al. 2009). Flavonoid-rich dark chocolate consumption significantly improved coronary flow velocity reserve (CFVR) whereas non-flavonoid white chocolate consumption did not. Intake of dark (but not white) chocolate, malondialdehyde-modified low-density lipoprotein, triglyceride and heart rate significantly influenced the change of CFVR after 2 weeks of intake.

Recent studies by Persson et al. (2011) showed significant inhibition of angiotensin-converting enzyme activity and significant increase of NO were seen in human endothelial cells from umbilical veins (HUVEC) after 10 min of incubation with cocoa extract. In the study of 19 healthy volunteers, a significant inhibition of angiotensin-converting enzyme activity (mean 18%) was observed 3 hours after intake of 75 g dark chocolate containing 72% cocoa, but no significant change in NO was noted. According to angiotensin-converting enzyme genotype, significant inhibition of angiotensin-converting enzyme activity was seen after 3 h in individuals with genotype insertion/insertion and deletion/deletion (mean 21% and 28%, respectively). Data suggested that ingestion of dark chocolate containing high amount of cocoa inhibited angiotensin-converting enzyme activity in-vitro and in-vivo.

Antiinflammatory Activity

Predominantly experimental in-vitro evidence had demonstrated that some cocoa-derived flavanols could reduce the production and effect of pro-inflammatory mediators either directly or by acting on signalling pathways (Selmi et al. 2006). These flavanols had been found in model systems to possess potential anti-inflammatory activity relevant to cardiovascular health. The authors asserted that additional research in well-designed human clinical experiments, using cocoa properly characterized in terms of flavanol content are warranted.

Some cocoa and chocolates were found to be rich in (−)-epicatechin and its related oligomers the procyanidins from cocoa seeds were found to cause caused dose-dependent inhibition of isolated rabbit 15-lipoxygenase-1 with the larger oligomers being more active; the decamer fraction revealed an IC_{50} of 0.8 μM (Schewe et al. 2001). Among the monomeric flavanols, epigallocatechin gallate (IC_{50}=4 μM) and epicatechin gallate (5 μM) were more potent than (−)-epicatechin (IC_{50}=60 μM). (−)-Epicatechin and procyanidin nonamer also inhibited the formation of 15-hydroxy-eicosatetraenoic acid from arachidonic acid in rabbit smooth muscle cells transfected with human 15-lipoxygenase-1. In addition, epicatechin (IC_{50} approx. 15 μM) and the procyanidin decamer inhibited recombinant human platelet 12-lipoxygenase. These observations suggested general lipoxygenase-inhibitory potency of flavanols and procyanidins may decrease the leukotriene/prostacyclin ratio in humans and human aortic endothelial cells. Further studies showed that (−)-epicatechin and its low-molecular procyanidins inhibited both dioxygenase and LTA(4) synthase activities of human 5-lipoxygenase and that this action may contribute to a putative anti-inflammatory effect of cocoa products (Schewe et al. 2002). Mainly 5-hydroperoxy-6E,8Z, 11Z,14Z-eicosatetraenoic acid (5-HpETE) and hydrolysis products of 5,6-leukotriene A(4) (LTA(4)) was significantly inhibited by (−)-epicatechin in a dose-dependent manner with 50% inhibitory concentrations (IC_{50}) of 22 and 50 μmol/L, respectively. Among the procyanidin fractions isolated from cocoa seeds, only the dimer fraction and, to a lesser extent, the trimer through pentamer fractions exhibited comparable effects, whereas the larger procyanidins (hexamer through nonamer) were almost inactive.

Ramiro et al. (2005b) showed that cocoa flavonoids down-regulated inflammatory cytokines and chemokines and inhibited nitric oxide release from macrophages. Monocyte chemoattractant protein 1 and tumour necrosis factor alpha (TNF-α) were significantly and dose-dependently diminished by cocoa extract. The effect was higher than that produced by equivalent concentrations of epicatechin but was lower than that produced by isoquercitrin. a significantly greater inhibition of TNFalpha secretion was found when the cocoa extract was added prior to cell activation. Both cocoa extract and epicatechin decreased TNFalpha, interleukin (IL) 1alpha, and IL-6 mRNA expression, suggesting that their inhibitory effect on cytokine secretion was produced, in part, at the transcriptional level. The cocoa extract also significantly reduced nitric oxide secretion in a dose-dependent manner and with a greater effect than that produced by epicatechin. In a double-blind crossover study of 20 individuals at risk for cardiovascular diseases, ingestion of the high-flavanol cocoa drink but not the low-flavanol cocoa drink significantly increased plasma concentrations of nitroso compounds and flow-mediated dilation of the brachial artery (Sies et al. 2005). It was found that ingested flavonoids may reverse endothelial dysfunction through enhancement of NO bioactivity. Micromolar concentrations of

(−)-epicatechin or other cocoa flavan-3-ols were found to suppress lipid peroxidation in LDL induced by the proinflammatory proatherogenic enzyme myeloperoxidase in the presence of physiologically relevant concentrations of nitrite, an NO metabolite. Adverse effects of NO metabolites, such as nitrite and peroxynitrite, were thus attenuated. (−)-Epicatechin and other cocoa flavan-3-ols proved to be inhibitory to human 5-lipoxygenase, the key enzyme of proinflammatory cysteinyl leukotriene synthesis. Abbey et al. (2008) demonstrated that *Theobroma cacao* extract could repress stimulated calcitonin gene-related peptide (CGRP) release from trigeminal nerves by a mechanism that may involve blockage of calcium channel activity (Abbey et al. 2008). Release of calcitonin gene-related peptide (CGRP) from trigeminal nerves promotes inflammation in peripheral tissues and nociception. A cohort study of 824 subjects who ate chocolate regularly in the form of dark chocolate only was selected for the cohort study to ascertain the association of dark chocolate intake with serum C-reactive protein (CRP) (Di Giuseppe et al. 1983). After adjustment of confounding factors, dark chocolate consumption was inversely associated with CRP consumers of up to 1 serving (20 g) of dark chocolate every 3 days had serum CRP concentrations that were significantly lower than non-consumers or higher consumers. The findings suggested that regular consumption of small doses of dark chocolate may reduce inflammation. Serum C-reactive protein levels rose in response to inflammation.

In-vitro or ex-vivo studies had demonstrated that cocoa-based products were among the richest functional foods based upon flavanols and their influence on the inflammatory pathway (Selmi et al. 2008). Chronic and acute inflammation underlies the molecular basis of atherosclerosis. Several studies suggested that regular or occasional consumption of cocoa-rich compounds exerted beneficial effects on blood pressure, insulin resistance, vascular damage, and oxidative stress.

Studies showed that cacao extract suppressed pro-inflammatory pathways in activated T-cells (Jenny et al. 2009). Its impact on indoleamine 2,3-dioxygenase could relate to some the beneficial health effects ascribed to cacao. In mitogen-induced human peripheral blood mononuclear cells of healthy donors, enhanced degradation of tryptophan, formation of neopterin and interferon-gamma were almost completely suppressed by the cacao extracts at doses of > or =5 μg/ml. Cacao extracts had no effect on tryptophan degradation in lipopolysaccharide-stimulated myelomonocytic THP-1 cells. Studies by Kim et al. (2009) showed that cocoa a polyphenols suppressed tumour necrosis factor – alpha (TNF-α)-induced vascular endothelial growth factor expression by inhibiting phosphoinositide 3-kinase (PI3K) and mitogen-activated protein kinase kinase-1 (MEK1) activities in JB6 mouse epidermal cells. TNF-α is a pro-inflammatory cytokine that has a vital role in the pathogenesis of inflammatory diseases such as cancer and psoriasis. Vascular endothelial growth factor expression is associated with tumorigenesis, cardiovascular diseases, rheumatoid arthritis and psoriasis.

Studies by Cady and Durham (2010) suggested that a dietary supplement could upregulate mitogen-activated kinase (MAP) phosphatases MKP-1 and MKP-3, and that cocoa could prevent inflammatory responses in trigeminal ganglion neurons. In animals administered a cocoa-enriched diet, basal levels of the mitogen-activated kinase (MAP) phosphatases MKP-1 and MKP-3 were increased in neurons. Importantly, the stimulatory effects of acute or chronic peripheral inflammation on neuronal expression of the MAPK p38 and extracellular signal-regulated kinases (ERK) were significantly suppressed in response to cocoa. Likewise, dietary cocoa significantly repressed basal neuronal expression of calcitonin gene-related peptide (CGRP) as well as activated levels of the inducible form of nitric oxide synthase (iNOS), proteins implicated in the underlying pathology of migraine and migraine and temporomandibular joint (TMJ) disorders.

Reduction of Insulin Resistance

Grassi et al. (2005a) reported that consumption of flavanol-rich dark chocolate (DC) but not white chocolate for 15 days decreased blood pressure (BP) and insulin resistance in healthy subjects. Homeostasis model assessment of insulin resistance (HOMA-IR) was significantly reduced after dark than after white chocolate intake (0.94 versus 1.72 respectively). The quantitative insulin sensitivity check index (QUICKI) was significantly higher after dark than after white chocolate ingestion (0.398 compared with 0356). Systolic blood pressure was lower after dark than after white chocolate ingestion (107.5 compared with 113.9 Hg). Similarly Grassi et al. (2005b) found that consumption of flavanol-rich dark chocolate (DC) decreased blood pressure (BP) and reduced insulin resistance in patients with essential hypertension (EH). Dark chocolate but not white chocolate decreased ambulatory BP, decreased HOMA-IR but improved QUICKI, insulin sensitivity index (ISI), and flow-mediated dilation (FMD). Dark chocolate also decreased serum LDL cholesterol.

In another subsequent study, they (Grassi et al. 2008) reported that consumption of flavanol-rich dark chocolate (FRDC) or flavanol-free white chocolate (FFWC) at 100 g/daily for 15 days ameliorated insulin sensitivity and beta-cell function, decreased blood pressure, decreased total cholesterol and LDL cholesterol and increased flow-mediated dilation in hypertensive patients with impaired glucose tolerance. These findings suggested flavanol-rich, low-energy cocoa food products may have a positive impact on CVD risk factors.

Antidiabetic Activity

Dietary supplementation with cacao liquor proanthocyanidins (CLPr) was found to dose-dependently prevent the development of hyperglycemia in diabetic obese (db/db) mice (Tomaru et al. 2007). The levels of blood glucose and fructosamine were higher in the db/db mice than in the control db/+m mice fed a diet containing 0%, 0.5%, or 1.0% CLPr. In the db/db mice, a diet containing 0.5% or 1.0% CLPr decreased the levels of blood glucose and fructosamine compared with that containing 0% CLPr without significant effects on body weights or food consumption. The findings suggested that dietary intake of food or drinks produced from cacao beans might be beneficial in preventing the onset of type 2 diabetes. Results of studies suggested that proanthocyanidins derived from cacao (CLP) inhibited diabetes-induced cataract formation possibly by virtue of its antioxidative activity (Osakabe et al. 2004). Opacity was first detected in the lenses of the streptozotocin (STZ)-induced rats fed the normal dietary diet 5 weeks after STZ injection and cataracts had developed in the majority of these animals by 10 week. These changes were rarely seen in the STZ/CLP diet group. Histological examinations of the eyes of the STZ-treated normal diet group revealed focal hyperplasia of the lens epithelium and liquefaction of cortical fibers. There were similar but considerably less severe changes in the animals fed CLP. Hydroxynonenal (HNE), a marker of oxidative stress, was detected immunohistochemically in the lenses of the STZ-treated normal diet group, but not of those receiving CLP. Studies demonstrated that cocoa supplementation had an effect on postprandial glucose control but not for long term (4 weeks) in obese-diabetic (Ob-db) rats and could reduce circulating plasma free fatty acid and 8-isoprostane thereby enhancing the antioxidant defense system (Jalil et al. 2008). There were no significant differences in fasting plasma glucose and insulin level after 4 weeks of cocoa extract administration. Cocoa supplementation in Ob-db rats significantly reduced plasma glucose at 60 and 90 min compared to non-supplemented Ob-db rats. Plasma free fatty acid and oxidative stress biomarker (8-isoprostane) were significantly reduced after cocoa supplementation. Superoxide dismutase activity was enhanced in Ob-db compared to that in non-supplemented rats. However, no change was observed in catalase activity.

Diets rich in flavanols was found to reverse vascular dysfunction in diabetes, highlighting therapeutic potentials in cardiovascular disease (Balzer et al. 2008). In a feasibility study with ten diabetic patients a single ingestion of flavanol-containing cocoa was dose-dependently associated with significant acute increases in circulating flavanols and flow-mediated dilation (FMD) of the brachial artery. A 30-day, thrice-daily consumption of flavanol-containing cocoa increased baseline FMD by 30%, while acute increases of FMD upon ingestion of flavanol-containing cocoa continued to be manifest throughout the study. Treatment was well tolerated without evidence of tachyphylaxia. Endothelium-independent responses, blood pressure, heart rate, and glycemic control were unaffected.

In hyperglycaemic diabetic rats group, cocoa extract (1% and 3%) diets were found to significantly lower the serum glucose levels compared to the control (Ruzaid et al. 2005). Further, supplementation of 1% and 3% cocoa extract significantly reduced the level of total cholesterol in streptozotocin diabetic rats. In addition, 1%, 2%, and 3% cocoa extract diets also significantly lowered the total triglycerides. Serum HDL-cholesterol were augmented significantly in diabetic rats fed with 2% cocoa extract, while the LDL-cholesterol decreased significantly in the 1% treated group. The results indicated that cocoa extract may possess potential hypoglycaemic and hypocholestrolemic effects on serum glucose levels and lipid profiles, respectively. The results also found that the effect of cocoa extract was dose-dependent.

Antiobesity Activity

Animal studies showed that final body weights and mesenteric white adipose tissue weights were significantly lower in rats fed the real cocoa diet than in those fed the mimetic cocoa diet, and serum triacylglycerol concentrations tended to be lower in rats fed the real cocoa diet (Matsui et al. 2005). In white adipose tissue, cocoa ingestion also lowered the expression of genes for fatty acid transport-relating molecules, whereas it upregulated the expression of genes for uncoupling protein-2 as a thermogenesis factor. The authors concluded that ingested cocoa could prevent high-fat diet-induced obesity by modulating lipid metabolism, especially by decreasing fatty acid synthesis and transport systems, and enhancement of part of the thermogenesis mechanism in liver and white adipose tissue.

Overweight and obese adults were randomly assigned to high-flavanol cocoa (HF, 902 mg flavanols), HF and exercise, low-flavanol cocoa (LF, 36 mg flavanols), or LF and exercise for 12 weeks (Davison et al. 2008). A total of 49 subjects (M = 18; F = 31) completed the intervention. Compared to LF, HF increased brachial artery flow-mediated dilatation acutely (2 h post-dose) by 2.4% and chronically (over 12 weeks) by 1.6% and reduced insulin resistance by 0.31%, diastolic BP by 1.6 mmHg and mean arterial BP by 1.2 mmHg, independent of exercise. Regular exercise increased fat oxidation during exercise by 0.10 g/min and reduced abdominal fat by 0.92%. The results showed that although HF consumption was shown to improve endothelial function, it did not enhance the effects of exercise on body fat and fat metabolism in obese subjects. However, it may be useful for reducing cardiometabolic risk factors in this population.

The high caloric load of commercially available chocolate (about 500 kcal/100 g) may induce weight gain, a risk factor for hypertension, dyslipidemia, and diabetes. Surprisingly, in a 4-armed parallel design study, 49 healthy women showed no weight gain after daily consumption of 41 g chocolate, 60 g almonds, or almonds and chocolate together for 6 weeks (Kurlandsky and Stote 2006). Serum cholesterol concentrations showed no changes after 6 weeks; however, triacylglycerol levels were reduced by approximately 21%, 13%, 19%, and 11% in the chocolate, almond, chocolate and almond, and control groups, respectively. Circulating intercellular adhesion molecule levels decreased significantly by 10% in the treatment group consuming chocolate only. No significant changes were observed for vascular adhesion molecule and high-sensitivity C-reactive protein levels in any treatment group.

Immunomodulatory Activity

Cocoa liquid polyphenol (CLP) inhibited both hydrogen peroxide and superoxide anion, typical ROS, production by phorbol myristate acetate-activated granulocytes (Sanbongi et al. 1997). CLP also suppressed menadione-induced production of both hydrogen peroxide and superoxide anion in normal human peripheral blood lymphocytes (PBL). CLP treatment of normal peripheral blood lymphocytes in-vitro inhibited dose-dependently mitogen-induced proliferation of T cells and polyclonal Ig production by B cells. CLP treatment inhibited both IL-2 mRNA expression of and IL-2 secretion by T cells. The results suggested that antioxidant CLP had immunoregulatory effects. In-vitro-studies by Ramiro et al. (2005a) showed that IL-2 receptor alpha (CD25) expression on activated lymphoid cells was significantly reduced by epicatechin and cocoa extract in a dose-dependent manner, with the highest inhibition of about 50% when flavonoids were added 2 h before stimulation. IL-2 secretion was also inhibited by epicatechin and cocoa extract, exhibiting 60% and 75% of inhibition, respectively. Cocoa flavonoids were also able to enhance 3–4.5-fold IL-4 release. The results showed that cocoa extract down-regulated T lymphocyte activation and therefore the acquired immune response, a fact that could be important in some states of the immune system hyperactivity such as autoimmune or chronic inflammatory diseases.

Studies showed cocoa flavanols and procyanidins oligomers to be potent stimulators of both the innate immune system and early events in adaptive immunity (Kenny et al. 2007). Peripheral blood mono-nuclear cells (PBMCs), as well as purified monocytes and CD4 and CD8 T cells, isolated from healthy volunteers were cultured in the presence of cocoa flavanol fractions: short-chain flavanol fraction (SCFF), monomers to pentamers; and long-chain flavanol fraction (LCFF), hexamers to decamers. The isolated cells were then challenged with lipopolysaccharide. There was a marked increase of LPS-induced synthesis of interleukin IL-1β, IL-6, IL-10, and TNF-±in the presence of LCFF. LCFF and SCFF, in the absence of LPS, stimulated the production of granulocyte macrophage colony-stimulating factor. Further, LCFF and SCFF increased expression of the B cell markers CD69 and CD83. Kenny et al. (2009) demonstrated that certain flavanols and procyanidins isolated from cocoa could moderate a subset of signaling pathways derived from lipopolysaccharide (LPS) stimulation of polymorphonuclear cells (part of the innate arm of the immune system), mainly, polymorphonuclear cells oxidative bursts and activation markers, and they could influence select apoptosis mechanisms. They hypothesized that flavanols and procyanidins can decrease the impact of LPS on the N-formyl-Met-Leu-Phe-primed PMN ability to generate reactive oxygen species by partially interfering in activation of the mitogen-activated protein kinase pathway.

A 10% cocoa intake by weaned Wistar rats for 3 weeks induced significant changes in Peyer's patches and mesenteric lymph nodes lymphocyte composition and function, whereas a 4% cocoa diet did not cause significant modifications in either tissues (Ramiro-Puig et al. 2008). Cocoa diet strongly reduced secretory IgA (S-IgA) in the intestinal lumen, although IgA's secretory ability was only slightly decreased in Peyer's patches. In addition, the 10% cocoa diet increased T-cell-antigen receptor gammadelta cell proportion in both lymphoid tissues. The findings indicated that cocoa ingestion modulated intestinal immune responses in young rats, influencing gammadelta T-cells and secretory IgA production. Emerging data from in-vivo studies indicated that cocoa may have an immunomodulating effect (Ramiro-Puig and Castell 2009). Long-term cocoa intake in rats could affect both intestinal and systemic immune function. Studies in this line suggested that high-dose cocoa intake in young rats favoured the T helper 1 (Th1) response and increased intestinal gammadelta T lymphocyte count, whereas the antibody-secreting response decreased.

Animal studies showed that cocoa diets induced attenuation of antibody synthesis that may be attributable to specific down-regulation of the Th2 immune response (Pérez-Berezo et al. 2009). Four weeks after immunization, control rats produced anti-ovalbumin antibodies, arranged according to their amount and isotype as follows: IgG1 > IgG2a > IgM > IgG2b > IgG2c. Both cocoa diets studied (4% and 10%) down-modulated ovalbumin-specific antibody levels of IgG1 (main subclass associated with the Th2 immune response in rats), IgG2a, IgG2c and IgM isotypes. Conversely, cocoa-fed rats presented equal or higher levels of anti-OVA IgG2b antibodies (subclass linked to the Th1 response). Spleen cells from cocoa-fed animals showed decreased interleukin-4 secretion (main Th2 cytokine), and lymph node cells from the same rats displayed higher interferon-gamma secretion (main Th1 cytokine).

Hepatoprotective Activity

Studies showed that a diet rich in cocoa attenuated N-nitrosodiethylamine (DEN)-induced liver injury in rats (Granado-Serrano et al. 2009). The cocoa-rich diet prevented the fall in hepatic glutathione concentration, catalase and Glutathione peroxidase activities in DEN-injected rats, as well as diminished protein carbonyl content, caspase-3 activity, p-AKT and p-JNK levels, and increased glutathione S-transferase activity. However, cocoa administration did not annul the DEN-induced body weight loss and the increased levels of hepatic-specific enzymes and lactate dehydrogenase. Studies showed that treatment of human HepG2 in culture with cocoa polyphenolic extract significantly protected cells against oxidation stress (Martín et al. 2008). Pretreatment of cells with 0.05–50 μg/ml of cocoa polyphenolic extract (CPE) for 2 or 20 hours completely prevented cell damage and enhanced activity of antioxidant enzymes induced by a treatment with tert-butylhydroperoxide (t-BOOH). Increased reactive oxygen species (ROS) induced by t-BOOH was dose-dependently prevented when cells were pretreated for 2 or 20 hours with CPE. CPE treatment also activated survival signaling proteins, such as protein kinase B (AKT) and ERKs, and increased the activities of two antioxidant enzymes, glutathione peroxidase (GPx) and glutathione reductase (GR) (Martin et al. 2010). The findings indicated CPE to be an effective inductor of GPx and GR activities via ERK activation and that this up-regulation appeared to be required to attenuate t-BOOH-induced injury.

Studies by Arlorio et al. (2009) showed that cocoa inhibited drug-triggered liver cytotoxicity by inducing autophagy. Phenolic-rich extracts of both unroasted and roasted cocoa prevented Celecoxib-induced cell viability inhibition in MLP29 liver cells. The protective effect of cocoa against liver cytotoxicity by Celecoxib was probably accounted for by inducing the autophagic process, as shown by enhanced Beclin 1 expression and accumulation of monodansylcadaverine in autolysosomes. This fact suggested that apoptosis was prevented by inducing autophagy. The findings further suggested that cocoa could be added to the list of natural chemopreventive agents whose potential in hepatopathy prevention and therapy should be evaluated.

Anticancer/Antimutagenic Activities

Studies suggested that the polyphenol extracts and fractions of cacao bean husk were effective functional materials to be used in either preventing or inhibiting cancer (Lee et al. 2005). The extracts of cacao bean husk (especially, the 60% ethanol fraction after extraction with 50% acetone) containing 43 wt.% polyphenol exerted an excellent protective effect on hydrogen peroxide-induced inhibition of gap-junction intercellular communication (GJIC) in WB-F344 rat liver epithelial cells. The enhancement of GJIC by the extracts of was about ten-fold higher than that of a well-known dietary chemopreventive component, vitamin C. The extracts (especially, the 60% ethanol fraction) also suppressed DNA synthesis in all liver, stomach, and colon cancer cells.

Kenny et al. (2004b) reported that pentameric procyanidins isolated from cocoa inhibited the expression of the tyrosine kinase ErbB2 gene, thus slowing the growth of cultured human aortic endothelial cells (HAEC). Down-regulation of ErbB2 and inhibition of HAEC growth by cocoa procyanidins may have several downstream implications, including reduced vascular endothelial growth factor (VEGF) activity and angiogenic activity associated with tumour pathology and may offer important insight into the design of therapeutic agents that target tumours over expressing ErbB2. They further found that pentameric and octameric procyanidin fractions of cocoa inhibited the proliferation of human dermal microvascular endothelial cells (HDMECs), whereas the pentameric fraction modulated the activity of several crucial proteins in angiogenic signaling by altering their tyrosine phosphorylation (Kenny et al. 2004a). Similar to aortic endothelial cells, the pentameric procyanidin fraction down-regulated the expression of ErbB2 tyrosine kinase in HDMECs. Their findings suggested that polyphenols may influence endothelial growth signaling, thus affecting angiogenesis in-vitro and could have a beneficial effect for cells over-expressing ErbB2, such as in specific neoplasias.

A naturally occurring, cocoa-derived pentameric procyanidin (pentamer) was found to selectively inhibit the proliferation of human breast cancer cells (MDA MB-231, MDA MB-436, MDA MB-468, SKBR-3, and MCF-7) and benzo(a)pyrene-immortalized 184A1N4 and 184B5 cells (Ramljak et al. 2005). In contrast, normal human mammary epithelial cells in primary culture and spontaneously immortalized MCF-10A cells were significantly resistant. Pentamer caused significant depolarization of mitochondrial membrane in MDA MB231 cells but not the more normal MCF-10A cells, whereas other normal and tumour cell lines tested gave variable results. The results showed that breast cancer cells were selectively susceptible to the cytotoxic effects of pentameric procyanidin, and data suggested that inhibition of cellular proliferation by this compound was associated with the site-specific dephosphorylation or down-regulation of several cell cycle regulatory proteins. With the semi-fermented-dry cacao seed, it was observed that the albumin fraction showed antitumoral activity evaluated against murine lymphoma L5178Y in BALB/c mice. It caused significant decreases in the ascetic fluid volume and packed cell volume, inhibiting cell growth in 59.98% at 60% of the population (Preza et al. 2010). The findings suggested that the cacao seed protein fractions could be considered as source of potential antitumour peptides. Both the albumin and glutelin fraction of cacao seed showed the greatest antioxidant capacity due to free radical scavenging capacity evaluated by the ABTS + and ORAC-FL assays (Preza et al. 2010).

In-vitro data of studies by Lanoue et al. (2010) showed cocoa-derived flavanols to have limited effects on topoisomerise II activity and cellular proliferation in leukemia cancer cell lines. Of all the flavanols tested, the dimers (B2, B5 and a mix of both) exerted the greatest inhibition of topo II and inhibited cellular proliferation rates at concentrations similar to quercetin. However, in contrast to quercetin, the dimers did not function as topo II poisons. The authors predicted that these compounds were likely to have limited leukemogenic potential at physiological concentrations.

Studies showed that inhibition of cytochrome P450 (CYP) 1A activity by cacao products rich in polyphenols may prevent DNA damage by reducing metabolic activation of carcinogens (Ohno et al. 2009). While white chocolate did not modulate the numbers of revertant colonies produced by the mutagen, benzo[a]pyrene (B[a]P) in *Salmonella typhimurium* strain TA 98 treatment, milk chocolate and cacao powder extracts did. In contrast, none of the cacao products tested affected the number of revertant colonies when t-BuOOH was used as the mutagen in *S. typhimurium* strain TA 102.

Anti-Benign Prostatic Hyperplasia (BPH)/Antiprostate Activity

Studies found that Acticoa (Barry Callebaut France, Louviers, France) powder (AP), a cocoa polyphenolic extract, could prevent prostate hyperplasia induced by testosterone propionate (TP) in rats (Bisson et al. 2007a, b). TP significantly influenced the body weight gain of the rats and their food and water consumption, while AP at both doses tested reduced significantly these differences. TP significantly increased prostate size ratio, and this induced increase was significantly inhibited in AP-treated rats in comparison with positive controls in a dose-dependent manner. AP significantly reduced serum dihydrotestoterone level. The authors concluded that AP orally administered was effective for reducing established prostate hyperplasia, especially at the dose of 48 mg/kg/day and therefore may be beneficial in the management of BPH. Further studies showed that Acticoa powder at 24 mg/kg protected rats from prostate carcinogenesis when chronically given before the initiation and promotion phases of prostate tumour induction (Bisson et al. 2008). A reduction in the incidence of prostate tumours was observed for the chemo-induced+Acticoa powder 48 mg/kg-treated group in comparison with the chemo-induced+vehicle-treated group and no tumours were observed in the chemo-induced+Acticoa powder 24 mg/kg-treated group as in the not induced+vehicle-treated group after 9 months. The life span of the chemo-induced+Acticoa powder24-treated group was significantly increased in comparison with the chemo-induced+Acticoa powder 48 and the chemo-induced+vehicle-treated groups, close to the one of the not induced+vehicle-treated group. A significant reduction in the incidence of prostate tumours was also observed for the chemo-induced+Acticoa powder 24 and chemo-induced+Acticoa powder 48-treated groups in comparison with the chemo-induced+vehicle-treated group.

At the highest tested concentration (0.2%), cocoa polyphenols extracts induced a complete inhibition of growth of metastatic DU 145 and nonmetastatic 22Rv1 prostate cancer cell lines (Jourdain et al. 2006). Cocoa polyphenols extracts were more active against local cancer cells than against metastatic cells. Moreover, at the highest tested concentration, cocoa polyphenols extracts were not effective on a normal prostate cell lines. Beta-sitosterol induced low growth inhibition of both cancer cell lines. Cocoa polyphenols extracts, however, were significantly more active and showed a strong and fast inhibition of cell growth than β-sitosterol alone. No synergy or addition was observed when β-sitosterol was tested together with the cocoa polyphenols extract. The results showed that cocoa polyphenols extracts had an antiproliferative effect on prostate cancer cell growth but not on normal cells, at the highest tested concentration.

Photoprotective Activity

Dietary flavanols from cocoa were found to contribute to endogenous photoprotection, improve dermal blood circulation, and affect cosmetically relevant skin surface and hydration variables in women (Heinrich et al. 2006). Following exposure of selected skin areas to 1.25 × minimal erythemal dose (MED) of radiation from a solar simulator, UV-induced erythema was significantly decreased in the high flavanol group, by 15% and 25%, after 6 and 12 weeks of treatment, respectively, whereas no change occurred in the low flavanol group. The ingestion of high flavanol cocoa generated increases in blood flow of cutaneous and subcutaneous tissues, and increases in skin density and skin hydration. Neither of these variables was affected in the low flavanol cocoa group. Evaluation of the skin surface showed a significant decrease of skin roughness and scaling in the high flavanol cocoa group compared with those at week 12. In a crossover design study of ten healthy women, flavanol-rich cocoa consumption was found to significantly increase dermal blood flow and oxygen saturation (Neukam et al. 2007). Following ingestion of high flavanol cocoa, dermal blood flow was significantly enhanced by 1.7-fold at 2 hours and oxygen saturation was elevated 1.8-fold. No statistically significant changes were found upon intake of low flavanol cocoa. Maximum plasma levels of total epicatechin were observed 1 hour after intake of the high flavanol cocoa drink, 11.6 nmol/l at baseline, and 62.9 nmol/l at 1 hour. No change of total epicatechin was found in the low flavanol group. Williams et al. (2009) demonstrated using a double-blind in vivo study of 30 healthy subjects that regular consumption of a chocolate rich in flavanols conferred significant photoprotection and could thus be effective at protecting human skin from harmful UV effects. Conventional chocolate had no such effect. In the group that consumed high-flavanol chocolate, minimal erythema dose (MED) more than doubled after 12 weeks while in the LF chocolate group, the MED remained without significant change.

Studies using a model of ex-vivo human skin explants on which a cocoa polyphenol extract was applied showed that cocoa polyphenols exhibited a positive action on several indicators of skin elasticity and skin tonus, namely, glycosaminoglycans and collagen I, III and IV (Gasser et al. 2008). The efficacy was comparable to a commercially available product. Further, an enhancing effect of cocoa butter on activity of cocoa polyphenol was highlighted.

Antineurodegenerative and Neuroprotective Activity

Heo and Lee (2005) showed that the major flavonoids of cocoa, epicatechin and catechin, protected PC12 cells from amyloid beta protein (Abeta)-induced neurotoxicity and neuronal cell death. The results suggested that cocoa may have anti-neurodegenerative effect in addition to other known chemopreventive effects. A cocoa procyanidin fraction and procyanidin B2 (epicatechin-(4β-8)-epicatechin) – a major polyphenol in cocoa were found to have protective effects against apoptosis of PC12 rat pheochromocytoma (PC12) cells induced by hydrogen peroxide (Cho et al. 2008). The protective effect involved inhibiting the downregulation of Bcl-X(L) and Bcl-2 expression through blocking the activation of JNK and p38 MAPK. Cho et al. (2009) showed that cocoa procyanidin fraction and procyanidin B2 protected PC12 cells against 4-hydroxynonenal-induced apoptosis by blocking MKK4 activity as well as ROS accumulation. 4-hydroxynonenal (HNE), one of the aldehydic products of membrane lipid peroxidation, was reported to be elevated in the brains of Alzheimer's disease patients and mediated the induction of neuronal apoptosis in the presence of oxidative stress.

Neuronal cells incubated with cocoa extract or (−)-epicatechin, reduced ROS production in a dose-dependent manner, reaching 35% inhibition (Ramiro-Puig et al. 2009). pJNK and p38, involved in apoptosis, were down-regulated by cocoa extract and (−)-epicatechin with p38 inhibition reaching up to 70%. The results showed that cocoa extract and (−)-epicatechin may exert a neuroprotective action by reducing ROS production and modulating mitogen-activated protein kinases (MAPK) activation.

Anxiolytic and Antidepressant Activities

Animal studies showed that short term cacao mass administration had an anxiolytic effect but chronic consumption did not (Yamada et al. 2009). In a T-maze test, short-term cacao mass significantly abolished delayed avoidance latency compared with the control but did not change escape latency, suggesting that cacao mass administration reduced conditional fear-relating behaviour. Short-term cacao mass administration did not affect the concentration of brain monoamines, emotion-related neurotransmitters such as norepinephrine, serotonin and dopamine, in the rat brain. Contrariwise, chronic consumption of cacao mass tended to increase avoidance latency and did not change escape latency. Brain serotonin concentration and its turnover were enhanced by chronic consumption of cacao mass.

A clinical trial of 30 subjects with low to high anxiety traits, provided evidence that a daily consumption of 40 g of dark chocolate during a period of 2 weeks was sufficient to modify the metabolism of free living and healthy human subjects, as per variation of both host and gut microbial metabolism. Dark chocolate reduced the urinary excretion of the stress hormone cortisol and catecholamines and partially normalized stress-related differences in energy metabolism (glycine, citrate, trans-aconitate, proline, β-alanine) and gut microbial activities (hippurate and p-cresol sulfate).

A double blinded, randomised, clinical pilot crossover study comparing high cocoa liquor/polyphenol rich chocolate (HCL/PR) to simulated iso-calorific chocolate (cocoa liquor free/low polyphenols(CLF/LP)) on fatigue and residual function were conducted in 10 subjects with chronic fatigue syndrome (Sathyapalan et al. 2010). The data suggested that HCL/PR chocolate may improve symptoms in subjects with chronic fatigue syndrome. The residual function, as assessed by the London Handicap scale, also improved significantly after the HCL/PR arm and deteriorated after iso-calorific chocolate. The Hospital Anxiety and Depression score also improved after the HCL/PR arm, but deteriorated after CLF/CP.

Using the forced swimming test in rats, cocoa polyphenolic extract significantly reduced the duration of immobility at both doses of 24 mg/kg/14 days and 48 mg/kg/14 days (Messaoudi et al. 2008). No change of motor dysfunction was observed with the two doses tested in the open field. The results of the forced swimming test after a subchronic treatment and after an additional locomotor activity test confirmed the assumption that the antidepressant-like effect of cocoa polyphenolic extract in the forced swimming test model was specific and might be related to its content of active polyphenols.

Antimicrobial and Antiperiodontitic Activities

The cocoa polyphenols showed anti-periopathogenic bacteria (Ohshima et al. 2006). The results of the clinical trial showed that the ratio of each periopathogenic bacterium (*Porphyromonas gingivalis, Fusobacterium nucleatum*, and *Prevotella intermedia*) to total bacteria and volatile sulfur compounds levels significantly decreased after the cocoa-ingestion phase. The growth of periopathogenic bacteria was significantly suppressed by cocoa in a dose dependent manner within 1 h of incubation. On the contrary, the cocoa showed no antibacterial effect on the growth of indigenous streptococci. This pilot study indicated that a 2-week-cocoa-drinking could decrease periopathogenic bacteria from the oral cavity and improve oral malodor by reducing volatile sulfur compounds. Cocoa polyphenols were found to reduce biofilm formation by *Streptococcus mutans* and *Streptococcus sanguinis*, and inhibit acid production by *S. mutans* (Percival et al. 2006). Cocoa polyphenol dimer, tetramer, and pentamer inhibited the growth of *S. sanguinis*, whereas the growth of *S. mutans* was unaffected. However, pretreatment of surfaces with cocoa polyphenol pentamer (35 μM) reduced biofilm formation by *S. mutans* at 4 and 24 hours whereas the effects on *S. sanguinis* were less consistent. Cocoa polyphenol pentamer (500 μM) significantly reduced the terminal pH, and inhibited the rate of acid production by *S. mutans* at pH 7.0. Cocoa polyphenol pentamers were found to significantly reduce biofilm formation and acid production by *Streptococcus mutans* and *Streptococcus sanguinis* (Ferrazzano et al. 2009) indicating that polyphenols occurring in cocoa, coffee and tea could have a role in the prevention of cariogenic processes, due to their antibacterial action.

Studies showed that rats with experimental periodontitis that were fed a regular diet showed a time-dependent increase in the level of serum reactive oxygen metabolites (Tomofuji et al. 2009). They also exhibited an enhanced 8-hydroxydeoxyguanosine level and decreased reduced/oxidized glutathione ratio in the gingival tissue, inducing alveolar bone loss and polymorphonuclear leukocyte infiltration. Although experimental periodontitis was induced in the rats fed a cocoa-enriched diet, they did not display impairments in serum reactive oxygen metabolite level and gingival levels for 8-hydroxydeoxyguanosine and reduced/oxidized glutathione ratio. Alveolar bone loss and polymorphonuclear leukocyte infiltration after ligature placement were also inhibited by cocoa ingestion. The results suggest that consuming a cocoa-enriched diet could diminish periodontitis-induced oxidative stress, which, in turn, might suppress the progression of periodontitis.

Probiotic and Prebiotic Activities

In a randomized, double-blind, crossover, controlled intervention study, 22 healthy human volunteers were randomly assigned to either a high-cocoa flavanol (HCF) group (494 mg cocoa flavanols/day) or a low-cocoa flavanol (LCF) group (23 mg cocoa flavanols/day) for 4 weeks followed by a 4-week washout period before volunteers crossed to the alternant arm (Tzounis et al. 2011). Compared with the consumption of the LCF drink, the daily consumption of the HCF drink for 4 weeks significantly increased the bifidobacterial and lactobacilli prebiotic populations but significantly decreased clostridia counts. These microbial changes were paralleled by significant reductions in plasma triacylglycerol (and C-reactive protein concentrations. Furthermore, changes in C-reactive protein concentrations were linked to changes in lactobacilli counts ($R^2=-0.33$). The results showed that consumption of cocoa flavanols could significantly affect the growth of select gut microflora in humans, suggesting the potential prebiotic benefits associated with the dietary inclusion of flavanol-rich foods.

Possemiers et al. (2010) found chocolate to be a potential protective carrier for oral delivery of a microencapsulated mixture of probiotics, *Lactobacillus helveticus* and *Bifidobacterium longum*. Both dark and milk chocolates offered superior protection 91% and 80% survival in milk chocolate for *L. helveticus* and *B. longum*, respectively compared to 20% and 31% found in milk. Using a Simulator of the Human Intestinal Microbial Ecosystem (SHIME) model the two probiotics successfully reached the simulated colon compartments. This led to an increase in lactobacilli and bifidobacteria counts and the appearance of additional species in the fingerprints. The data indicated that the coating of the probiotics in chocolate provided an excellent solution to protect them from environmental stress conditions and for optimal delivery. The simulation with our gastrointestinal model showed that the formulation of a probiotic strain in a specific food matrix could offer superior protection for the delivery of the bacterium into the colon.

Sequential in vitro digestion of the water-insoluble cocoa fraction (WICF), a rich source of polyphenols, with gastrointestinal enzymes as well as its bacterial fermentation in a human colonic model system were carried out to investigate bioaccessibility and biotransformation of WICF polyphenols (Fogliano et al. 2011). In-vitro digestion solubilized 38.6% of WICF with pronase and Viscozyme L treatments releasing 51% of the total phenols from the insoluble material. This release of phenols did not cause a reduction in the total antioxidant capacity of the digestion-resistant material. In the colonic model, WICF significantly increased population of bifidobacteria and lactobacilli as well as butyrate production. Flavanols were converted into phenolic acids by the microbiota following a concentration gradient resulting in high concentrations of 3-hydroxyphenylpropionic acid (3-HPP) in the last gut compartment. The data showed that WICF may exert antioxidant action through the gastrointestinal tract despite its polyphenols being still bound to macromolecules and having prebiotic activity.

Antiviral Activity

Cacao husk lignin fractions, showed unexpectedly higher anti-human immunodeficiency virus (HIV) activity, as compared with the corresponding fractions from the cacao mass; the effect was comparable with that of popular anti-HIV compounds (Sakagami et al. 2008). The cacao husk lignin fractions also showed anti-influenza virus activity, but did not show antibacterial activity. The cacao husk lignin fractions synergistically enhanced the superoxide anion and hydroxyl radical scavenging activity of vitamin C. The fractions also stimulated nitric oxide generation by mouse macrophage-like cells, to a level higher than that attained by lipopolysaccharide (LPS). The results suggested the functionality of cacao husk lignin fractions as complementary alternative medicine.

Cocoa Consumption and Cognitive Performance

Studies showed an increase in the blood oxygenation level-dependent signal intensity in response to a cognitive task following ingestion of flavanol-rich cocoa (5 days of 150 mg of cocoa flavanols) by healthy young humans (Francis et al. 2006). No significant effects were evident in behavioral reaction times, switch cost, and heart rate after consumption of this moderate dose of cocoa flavanols. A single acute dose (450 mg flavanols) of flavanol-rich cocoa was found to increase the cerebral blood flow to gray matter, suggesting the potential of cocoa flavanols for treatment of vascular impairment, including dementia and strokes, and thus for maintaining cardiovascular health.

In a randomized, controlled, double-blinded, balanced, three period crossover trial involving 30 healthy adults, consumption of cocoa flavanols was found to result in acute improvements in mood and cognitive performance during sustained mental effort (Scholey et al. 2010). Assessments included the state anxiety inventory and repeated 10-minutes cycles of a Cognitive Demand Battery comprising of two serial subtraction tasks (Serial Threes and Serial Sevens), a Rapid Visual Information Processing (RVIP) task and a 'mental fatigue' scale, over the course of 1 hour. Consumption of both 520 and 994 mg cocoa flavanol significantly improved Serial Threes performance. The 994 mg cocoa flavanol beverage significantly speeded RVIP responses but also resulted in more errors during Serial Sevens. Increases in self-reported 'mental fatigue' were significantly attenuated by the consumption of the 520 mg cocoa flavanol beverage only.

Antihemolytic Activity

Studies showed that consumption of a flavonoid-rich cocoa beverage significantly reduced the susceptibility of erythrocytes to free radical-induced hemolysis (Zhu et al. 2005). The duration of the lag time, which reflected the capacity of cells to buffer free radicals, was increased. Consistent with the above, the purified cocoa flavonoids, epicatechin, catechin, Dimer B2 and the metabolite 3′-O-methyl epicatechin, exhibited dose-dependent protection against AAPH (2,2′-azobis-2-methyl-propanimidamide, dihydrochloride)-induced erythrocyte hemolysis at concentrations ranging from 2.5 to 20 μM.

Cocoa and Female/Male Reproductive Problems

Studies found that cocoa butter cream or cocoa butter lotion did not prevent striae gravidarum in pregnant women (Osman et al. 2008; Buchanan et al. 2010). In a multicentre, double-blind, randomised and placebo-controlled trial of 175 nulliparous women who completed the study, there were no significant difference in the development or severity of striae gravidarum in women who received the cocoa butter lotion or placebo (Osman et al. 2008). In another randomized, double-blind, placebo-controlled trial, there were no significant difference between women (150) using cocoa butter cream and women (150) receiving placebo, 44% of patients using cocoa butter cream compared with 55% of those using placebo (Buchanan et al. 2010).

Triche et al. (2008) studied the association of chocolate consumption with risk of preeclampsia in a prospective cohort study of 2,291 pregnant women who delivered a singleton livebirth between September 1996 and January 2000. Preeclampsia, a disorder with prominent cardiovascular manifestations, is a cause of maternal, fetal, and infant morbidity and mortality. They found that preeclampsia developed in 3.7% (n=63) of 1,681 women. Cord serum theobromine concentrations were negatively associated with preeclampsia (adjusted odd ratio (aOR=0.31)). Self-reported chocolate consumption estimates also were inversely associated with preeclampsia. Compared with women consuming under 1 serving of chocolate weekly, women consuming 5+ servings per week had decreased risk: aOR=0.81 with consumption in the first 3 months of pregnancy and 0.60 in the last 3 months. Their findings suggested that that chocolate consumption during pregnancy may lower risk of preeclampsia. Another study of 2,769 women pregnant between 1,959 and 1966, with liveborn infants of at least 28 weeks′ gestation did not support the previous finding that chocolate consumption was associated with a reduced occurrence of preeclampsia. Preeclampsia occurred in 68 (2.9%) of 2,105 eligible women. Adjusted odd ratio (Ors) for preeclampsia were near unity across most third-trimester theobromine concentrations. Adjusted ORs for preeclampsia according to theobromine concentration in serum at <20 weeks' gestation increased with increases in concentration, although estimates were imprecise. The authors asserted that unmeasured confounding or reverse causation may account for the positive association between early-pregnancy theobromine and preeclampsia. Saftlas et al. (2010) analyzed a prospective cohort study of subjects recruited from 13 prenatal care practices to ascertain the association of regular chocolate intake during pregnancy with reduced risks of preeclampsia and gestational hypertension (GH). They found that chocolate consumption was more frequent among normotensive (80.7%) than preeclamptic (62.5%) or Chocolate intake was more frequent among normotensive (80.7%) than preeclamptic (62.5%) or GH women (75.8%). Chocolate intake was associated with reduced odds of preeclampsia in the first and third trimester (aOR=0.55 and aOR=0.56 respectively). Only first trimester intake was associated with reduced odds of gestational hypertension (aOR=0.65).

Giannandrea (2009) reported that the consumption of cocoa in the period 1965–1980, was most closely correlated with the incidence of testicular cancer in young adults (r=0.859). An analogous significant correlation was also observed between early cocoa consumption and the prevalence rates of hypospadias in the period 1999–2003 (r=0.760). Based on the results of the ecological approach, the author suggested the need of further analytic studies to investigate the role of individual exposure to cocoa, particularly during the prenatal and in early life of the patients since various studies in animals reported that cocoa and theobromine, the main stimulant of cocoa, exerted toxic effects on the testis, inducing testicular atrophy and impaired sperm quality.

Gastroprotective and Lung-Protective Activities

In-vitro studies showed that cocoa procyanidin fraction can interact with cell membranes and protect Caco-2 cells from deoxycholic-induced cytotoxicity, oxidant generation, and loss of monolayer integrity (Erlejman et al. 2006). A hexameric procyanidin fraction (Hex), isolated from cocoa, at concentrations ranging from 2.5 to 20 μM partially inhibited deoxycholic-induced cell oxidants, alterations in the paracellular transport, and redistribution of the protein ZO-1 from cell-cell contacts into the cytoplasm. Similarly, Hex (5–10 μM) inhibited the increase in cell oxidants, and the loss of integrity of polarized Caco-2 cell monolayers induced by a lipophilic oxidant (2,2′-azobis 2,4-dimethylvaleronitrile). The results suggested that at the gastrointestinal tract, large procyanidins may exert beneficial effects in pathologies such us inflammatory diseases, alterations in intestinal barrier permeability, and cancer.

These results of studies in mice by Yasuda et al. (2008b) suggested that cacao liquor proanthocyanidins (CP) inhibited diesel exhaust particles (DEP)-induced lung injury by reducing oxidative stress, in association with a reduction in the expression of adhesion molecules. DEP-induced lung injury was characterized by neutrophil sequestration, edema and presence of numerous adducts of nitrotyrosine, N-(hexanonyl) lysine, 4-hydroxy-2-nonenal, and 8-OHdG. CPs prevented enhanced expression of vascular cell adhesion molecule-1 and intercellular adhesion molecule-1 caused by DEPs in the lung injury. Further, the level of thiobarbituric acid reactive substances in the lung was decreased by CP supplementation in the presence of DEPs.

Suppression of Dioxin Toxicity

In-vivo studies in mice showed that a cacao polyphenol extract (CPE) suppressed aryl hydrocarbon receptor transformation (Mukai et al. 2008). Dioxins enter the body through the diet and cause various toxicological effects through transformation of this aryl hydrocarbon receptor. CPE suppressed the 3-methylcholanthrene -induced transformation of aryl hydrocarbon receptor. CPE also suppressed 3-methylcholanthrene -induced cytochrome P4501A1 expression and NAD(P)H:quinone-oxidoreductase activity, but increased glutathione S-transferase activity. The results indicated that the ingestion of CPE suppressed the toxicological effects of dioxins in the body.

Bioavailability of Cocoa Butter, Cocoa Flavanol and N-Phenylpropenoyl-L-Amino Acids

Cocoa and chocolate were found also contain fats from cocoa butter; mainly stearic triglycerides (C18:0) that were less well absorbed than other fats, and were excreted in the faeces (Weisberger 2001). Thus, cocoa butter was less bioavailable and had minimal effect on serum cholesterol.

Studies in nine human volunteers showed that milk decreased urinary excretion but not plasma pharmacokinetics of cocoa flavan-3-ol metabolites (Mullen et al. 2009). Milk affected neither gastric emptying nor the transit time through the small intestine. Two flavan-3-ol metabolites were detected in plasma and four in urine. Milk had only minor effects on the plasma pharmacokinetics of an (epi)catechin-O-sulfate and had no effect on an O-methyl-(epi)catechin-O-sulfate. However, milk significantly lowered the excretion of four urinary flavan-3-ol metabolites from 18.3% to 10.5% of the ingested dose. The authors further elucidated that studies that showed protective effects of cocoa and those that showed no effect of milk on bioavailability used products that had a much higher flavan-3-ol content than the commercial cocoa used in their study.

Stark et al. (2008) found that 2 hours after consumption of a cocoa drink, 12 N-phenylpropenoyl-L-amino acids (NPA) were detected in the urine. The highest absolute amount of NPAs excreted with the urine was found for N-[4′-hydroxy-(E)-cinnamoyl]-L-aspartic acid, but the highest recovery rate (57.3% and 22.8%), that means the percentage amount of ingested amides excreted with the urine, were determined for N-[4′-hydroxy-(E)-cinnamoyl]-L-glutamic acid and N-[4′-hydroxy-3′-methoxy-(E)-cinnamoyl]-L-tyrosine. They found that these amides were metabolized neither via their O-glucuronides nor their O-sulfates.

Chocolate Consumption and Bone Density

Older women who consumed chocolate daily were found to have lower bone density and strength (Hodgson et al. 2008). Daily (> or =1 times/day) consumption of chocolate, in comparison to <1 time/week, was associated with a 3.1% lower whole-body bone density; with similarly lower bone density of the total hip, femoral neck, tibia, and heel; and with lower bone strength in the tibia and the heel. Confirmation of these findings could have important implications for prevention of osteoporotic fracture.

Traditional Medicinal Uses

Cacao is reported to be antiseptic, diuretic, ecbolic, emmenagogue, and parasiticidal and used as a folk remedy for alopecia, burns, cough, dry lips, eyes, fever, listlessness, malaria, nephrosis, parturition, pregnancy, rheumatism, snakebite, and wounds (Duke 1983). Cocoa butter is applied to wrinkles in the hope of correcting them (Leung and Foster 1995).

Other Uses

Cocoa tree can be intercroped with other crops such as coconut. The leaf litter can be used to improve soil. There is considerable nutrient cycling through the development of a deep leaf litter under the cocoa canopy. Cocoa pod husk can be burnt and the ash used as fertiliser. Cocoa pod husk can be used as fodder. The cocoa-pod husk has a low alkaloid content, while tannin is practically absent. The crude fibre content is low; it is completely unlignified and compares favourably with *Panicum maximum* and *Centrosema pubescens*. The cocoa bean testa can be used as fuel as it has a calorific value of 16,000–19,000 BTU/kg, a little higher than that for wood. The ash from pod husks contains potassium oxide, which can be extracted in the form of potassium hydroxide, a useful alkaline in the saponification process. Cocoa-bean fat from unfermented cocoa beans can be extracted and used in soap making.

Comments

There are three main cultivar groups of cocoa beans used to make cocoa and chocolate (Anonymous 2002). The most prized, rare, and expensive is the Criollo Group. Only 10% of chocolate is made from Criollo, which is less bitter and more aromatic than any other bean. 80% of chocolate is made using beans of the Forastero Group. Forastero trees are significantly hardier than Criollo trees, resulting in cheaper cacao beans. Trinitario, a hybrid of Criollo and Forastero, is used in about 10–15% of chocolate.

Selected References

Abbey MJ, Patil VV, Vause CV, Durham PL (2008) Repression of calcitonin gene-related peptide expression in trigeminal neurons by a *Theobroma cacao extract*. J Ethnopharmacol 115(2):238–248

Adamson GE, Lazarus SA, Mitchell AE, Prio RL, Cao G, Jacobs PH, Kremer BG, Hammerstone JF, Rucker RB, Ritter KA, Schmitz HH (1999) HPLC method for the quantification of procyanidins in cocoa and chocolate samples and correlation to total antioxidant capacity. J Agric Food Chem 47(10):4184–4188

Akita M, Kuwahara M, Itoh F, Nakano Y, Osakabe N, Kurosawa T, Tsubone H (2008) Effects of cacao liquor polyphenols on cardiovascular and autonomic nervous functions in hypercholesterolaemic rabbits. Basic Clin Pharmacol Toxicol 103(6):581–587

Allen RR, Carson L, Kwik-Uribe C, Evans EM, Erdman JW Jr (2008) Daily consumption of a dark chocolate containing flavanols and added sterol esters affects cardiovascular risk factors in a normotensive population with elevated cholesterol. J Nutr 138(4):725–731

Almoosawi S, Fyfe L, Ho C, Al-Dujaili E (2010) The effect of polyphenol-rich dark chocolate on fasting capillary whole blood glucose, total cholesterol, blood pressure and glucocorticoids in healthy overweight and obese subjects. Br J Nutr 103(6):842–850

Alspach G (2007) The truth is often bittersweet…: chocolate does a heart good. Crit Care Nurse 27(1):11–15

Andres-Lacueva C, Monagas M, Khan N, Izquierdo-Pulido M, Urpi-Sarda M, Permanyer J, Lamuela-Raventós RM (2008) Flavanol and flavonol contents of cocoa powder products: influence of the manufacturing process. J Agric Food Chem 56(9):3111–3117

Anonymous (2002) Xocoatl *Theobroma cacao*. www.xocoatl.org

Anvoh KYB, Bi AZ, Gnakri D (2009) Production and characterization of juice from mucilage of cocoa beans and its transformation into marmalade. Pak J Nutr 8(2):129–133

Arlorio M, Bottini C, Travaglia F, Locatelli M, Bordiga M, Coïsson JD, Martelli A, Tessitore L (2009) Protective activity of *Theobroma cacao* L. phenolic extract on AML12 and MLP29 liver cells by preventing apoptosis and inducing autophagy. J Agric Food Chem 57(22):10612–10618

Baba S, Natsume M, Yasuda A, Nakamura Y, Tamura T, Osakabe N, Kanegae M, Kondo K (2007a) Plasma LDL and HDL cholesterol and oxidized LDL concentrations are altered in normo- and hypercholesterolemic humans after intake of different levels of cocoa powder. J Nutr 137(6):1436–1441

Baba S, Osakabe N, Kato Y, Natsume M, Yasuda A, Kido T, Fukuda K, Muto Y, Kondo K (2007b) Continuous intake of polyphenolic compounds containing cocoa powder reduces LDL oxidative susceptibility and has beneficial effects on plasma HDL-cholesterol concentrations in humans. Am J Clin Nutr 85(3):709–717

Backer CA, Bakhuizen van den Brink RC Jr (1963) Flora of Java, (Spermatophytes only), vol 1. Noordhoff, Groningen, 648 pp

Balzer J, Rassaf T, Heiss C, Kleinbongard P, Lauer T, Merx M, Heussen N, Gross HB, Keen CL, Schroeter H, Kelm M (2008) Sustained benefits in vascular function through flavanol-containing cocoa in medicated diabetic patients a double-masked, randomized, controlled trial. J Am Coll Cardiol 51:2141–2149

Barazarte H, Sangronis E, Unai E (2008) Cocoa (*Theobroma cacao* L.) hulls: a possible commercial source of pectins. Arch Latinoam Nutr 58(1):64–70, In Spanish

Bartley BGD (2005) The genetic diversity of cacao and its utilization. CABI, Wallingford, 341 pp

Bergmann J (1969) The distribution of cacao cultivation in pre-Columbian America. Ann Assoc Am Geogr 59:85–96

Berry NM, Davison K, Coates AM, Buckley JD, Howe PR (2010) Impact of cocoa flavanol consumption on blood pressure responsiveness to exercise. Br J Nutr 103(10):1480–1484

Bisson JF, Hidalgo S, Rozan P, Messaoudi M (2007a) Preventive effects of ACTICOA powder, a cocoa polyphenolic extract, on experimentally induced prostate hyperplasia in Wistar-Unilever rats. J Med Food 10(4):622–627

Bisson JF, Hidalgo S, Rozan P, Messaoudi M (2007b) Therapeutic effect of ACTICOA powder, a cocoa polyphenolic extract, on experimentally induced prostate hyperplasia in Wistar-Unilever rats. J Med Food 10(4):628–635

Bisson JF, Guardia-Llorens MA, Hidalgo S, Rozan P, Messaoudi M (2008) Protective effect of Acticoa powder, a cocoa polyphenolic extract, on prostate carcinogenesis in Wistar-Unilever rats. Eur J Cancer Prev 17(1):54–61

Bordeaux B, Yanek LR, Moy TF, White LW, Becker LC, Faraday N, Becker DM (2007) Casual chocolate consumption and inhibition of platelet function. Prev Cardiol 10(4):175–180

Buchanan K, Fletcher HM, Reid M (2010) Prevention of striae gravidarum with cocoa butter cream. Int J Gynaecol Obstet 108(1):65–68

Buijsse B, Feskens EJM, Kok FJ, Kromhout D (2006) Cocoa intake, blood pressure, and cardiovascular mortality: the zutphen elderly study. Arch Intern Med 166(4):411–417

Buijsse B, Weikert C, Drogan D, Bergmann M, Boeing H (2010) Chocolate consumption in relation to blood pressure and risk of cardiovascular disease in German adults. Eur Heart J 31(13):1616–1623

Cady RJ, Durham PL (2010) Cocoa-enriched diets enhance expression of phosphatases and decrease expression of inflammatory molecules in trigeminal ganglion neurons. Brain Res 1323:18–32

Cakirer MS, Ziegler GR, Guiltinan MJ (2010) Seed color as an indicator of flavanol content in *Theobroma cacao* L. In: Bishop MR (ed) Chocolate, fast foods and sweeteners: consumption and health. Nova Science Pub. Inc, New York, pp 257–270, 344 pp

Cho ES, Lee KW, Lee HJ (2008) Cocoa procyanidins protect PC12 cells from hydrogen-peroxide-induced apoptosis by inhibiting activation of p38 MAPK and JNK. Mutat Res 640(1–2):123–130

Cho ES, Jang YJ, Kang NJ, Hwang MK, Kim YT, Lee KW, Lee HJ (2009) Cocoa procyanidins attenuate 4-hydroxynonenal-induced apoptosis of PC12 cells by directly inhibiting mitogen-activated protein kinase kinase 4 activity. Free Radic Biol Med 46(10): 1319–1327

Cienfuegos-Jovellanos E, Quiñones Mdel M, Muguerza B, Moulay L, Miguel M, Aleixandre A (2009) Antihypertensive effect of a polyphenol-rich cocoa powder industrially processed to preserve the original flavonoids of the cocoa beans. J Agric Food Chem 57(14):6156–6162

Corti R, Flammer AJ, Hollenberg NK, Lüscher TF (2009) Cocoa and cardiovascular health. Circulation 119(10):1433–1441

Council of Scientific and Industrial Research (CSIR) (1976) The wealth of India. A dictionary of Indian raw materials and industrial products, vol 10, Raw materials. Publications and Information Directorate, New Delhi

Crews WD Jr, Harrison DW, Wright JW (2008) A double-blind, placebo-controlled, randomized trial of the effects of dark chocolate and cocoa on variables associated with neuropsychological functioning and cardiovascular health: clinical findings from a sample of healthy, cognitively intact older adults. Am J Clin Nutr 87(4):872–880

Cuatrecasas J (1964) Cacao and its allies: a taxonomic revision of the genus *Theobroma*. Contrib US Natl Herb 35:495

Davison K, Coates AM, Buckley JD, Howe PR (2008) Effect of cocoa flavanols and exercise on cardiometabolic risk factors in overweight and obese subjects. Int J Obes (Lond) 32(8):1289–1296

Davison K, Berry NM, Misan G, Coates AM, Buckley JD, Howe PR (2010) Dose-related effects of flavanol-rich cocoa on blood pressure. J Hum Hypertens 24(9):568–576

Desch S, Kobler D, Schmidt J, Sonnabend M, Adams V, Sareban M, Eitel I, Blüher M, Schuler G, Thiele H (2010a) Low vs. higher-dose dark chocolate and blood pressure in cardiovascular high-risk patients. Am J Hypertens 23(6):694–700

Desch S, Schmidt J, Kobler D, Sonnabend M, Eitel I, Sareban M, Rahimi K, Schuler G, Thiele H (2010b) Effect of cocoa products on blood pressure: systematic review and meta-analysis. Am J Hypertens 23(1):97–103

Di Giuseppe R, Di Castelnuovo A, Centritto F, Zito F, De Curtis A, Costanzo S, Vohnout B, Duke JA (1983) *Theobromae cacao* L. Handbook of energy crops. Unpublished. http://www.hort.purdue.edu/newcrop/duke_energy/theobroma_cacao.html

Ding EL, Hutfless SM, Ding X, Girotra S (2006) Chocolate and prevention of cardiovascular disease: a systematic review. Nutr Metab (Lond) 3:2, 3

Djoussé L, Hopkins PN, Arnett DK, Pankow JS, Borecki I, North KE, Curtis Ellison R (2011) Chocolate consumption is inversely associated with calcified atherosclerotic plaque in the coronary arteries: the NHLBI Family Heart Study. Clin Nutr 30(1):38–43, Epub 2010 Jul 22

Donovan JL, Crespy V, Oliveira M, Cooper KA, Gibson BB, Williamson G (2006) (+)-Catechin is more bioavailable than (−)-catechin: relevance to the bioavailability of catechin from cocoa. Free Radic Res 40(10):1029–1034

Duke JA (1983) *Theobroma cacao* L. Handbook of Energy Crops. unpublished. http://www.hort.purdue.edu/newcrop/duke_energy/theobroma_cacao.html

Duke JA, DuCellier JL (1993) CRC handbook of alternative cash crops. CRC Press Inc, Boca Raton, 544 pp

Egan BM, Laken MA, Donovan JL, Woolson RF (2010) Does dark chocolate have a role in the prevention and management of hypertension?: commentary on the evidence. Hypertension 55(6):1289–1295

Engler MB, Engler MM (2006) The emerging role of flavonoid-rich cocoa and chocolate in cardiovascular health and disease. Nutr Rev 64(3):109–118

Engler MB, Engler MM, Chen CY, Malloy MJ, Browne A, Chiu EY, Kwak HK, Milbury P, Paul SM, Blumberg J, Mietus-Snyder ML (2004) Flavonoid-rich dark chocolate improves endothelial function and increases plasma epicatechin concentrations in healthy adults. J Am Coll Nutr 23(3):197–204

Erdman JW Jr, Carson L, Kwik-Uribe C, Evans EM, Allen RR (2008) Effects of cocoa flavanols on risk factors for cardiovascular disease. Asia Pac J Clin Nutr 17(Suppl 1):284–287

Erlejman AG, Fraga CG, Oteiza PI (2006) Procyanidins protect Caco-2 cells from bile acid- and oxidant-induced damage. Free Radic Biol Med 41(8):1247–1256

Faridi Z, Njike VY, Dutta S, Ali A, Katz DL (2008) Acute dark chocolate and cocoa ingestion and endothelial function: a randomized controlled crossover trial. Am J Clin Nutr 88(1):58–63

Farouque HM, Leung M, Hope SA, Baldi M, Schechter C, Cameron JD, Meredith IT (2006) Acute and chronic effects of flavanol-rich cocoa on vascular function in subjects with coronary artery disease: a randomized double-blind placebo-controlled study. Clin Sci (Lond) 111(1):71–80

Fedeli E, Lanzani A, Capella P, Jacini G (1966) Triterpene alcohols and sterols of vegetable oils. J Am Oil Chem Soc 43(4):254–256

Ferrazzano GF, Amato I, Ingenito A, De Natale A, Pollio A (2009) Anti-cariogenic effects of polyphenols from plant stimulant beverages (cocoa, coffee, tea). Fitoterapia 80(5):255–262

Fisher ND, Hughes M, Gerhard-Herman M, Hollenberg NK (2003) Flavanol-rich cocoa induces nitric-oxide-dependent vasodilation in healthy humans. J Hypertens 21(12):2281–2286

Flammer AJ, Hermann F, Sudano I, Spieker L, Hermann M, Cooper KA, Serafini M, Lüscher TF, Ruschitzka F, Noll G, Corti R (2007) Dark chocolate improves coronary vasomotion and reduces platelet reactivity. Circulation 116(21):2376–2382

Fogliano V, Corollaro ML, Vitaglione P, Napolitano A, Ferracane R, Travaglia F, Arlorio M, Costabile A, Klinder A, Gibson G (2011) In vitro bioaccessibility and gut biotransformation of polyphenols present in the water-insoluble cocoa fraction. Mol Nutr Food Res 55(Suppl 1):S44–S55

Fraga CG, Actis-Goretta L, Ottaviani JI, Carrasquedo F, Lotito SB, Lazarus S, Schmitz HH, Keen CL (2005) Regular consumption of a flavanol-rich chocolate can improve oxidant stress in young soccer players. Clin Dev Immunol 12(1):11–17

Francis ST, Head K, Morris PG, Macdonald IA (2006) The effect of flavanol-rich cocoa on the fMRI response to a cognitive task in healthy young people. J Cardiovasc Pharmacol 47(Suppl 2):S215–S220

Frauendorfer F, Schieberle P (2006) Identification of the key aroma compounds in cocoa powder based on molecular sensory correlations. J Agric Food Chem 54(15):5521–5529

Frauendorfer F, Schieberle P (2008) Changes in key aroma compounds of Criollo cocoa beans during roasting. J Agric Food Chem 56(21):10244–10251

Gasser P, Lati E, Peno-Mazzarino L, Bouzoud D, Allegaert L, Bernaert H (2008) Cocoa polyphenols and their influence on parameters involved in ex vivo skin restructuring. Int J Cosmet Sci 30(5):339–345

Genovese MI, Lannes SCDS (2009) Comparison of total phenolic content and antiradical capacity of powders and "chocolates" from cocoa and cupuassu. Ciênc Tecnol Aliment 29(4):810–814

Giannandrea F (2009) Correlation analysis of cocoa consumption data with worldwide incidence rates of testicular cancer and hypospadias. Int J Environ Res Public Health 6(2):568–578

Granado-Serrano AB, Martín MA, Bravo L, Goya L, Ramos S (2009) A diet rich in cocoa attenuates N-nitrosodiethylamine-induced liver injury in rats. Food Chem Toxicol 47(10):2499–2506

Grassi D, Lippi C, Necozione S, Desideri G, Ferri C (2005a) Short-term administration of dark chocolate is followed by a significant increase in insulin sensitivity and a decrease in blood pressure in healthy persons. Am J Clin Nutr 81(3):611–614

Grassi D, Necozione S, Lippi C, Croce G, Valeri L, Pasqualetti P, Desideri G, Blumberg JB, Ferri C (2005b) Cocoa reduces blood pressure and insulin resistance and improves endothelium-dependent vasodilation in hypertensives. Hypertension 46:398–405

Grassi D, Desideri G, Necozione S, Lippi C, Casale R, Properzi G, Blumberg JB, Ferri C (2008) Blood pressure is reduced and insulin sensitivity increased in glucose-intolerant, hypertensive subjects after 15 days of consuming high-polyphenol dark chocolate. J Nutr 138(9):1671–1676

Griffiths LA (1960) A comparative study of the seed polyphenols of the genus *Theobroma*. Biochem J 74(2):362–365

Griffiths LA (1985) Phenolic acids and flavonoids of *Theobroma cacao* L. Separation and identification by paper chromatography. Biochem J 70(1):120–125

Gu L, House SE, Wu X, Ou B, Prior RL (2006) Procyanidin and catechin contents and antioxidant capacity of cocoa and chocolate products. J Agric Food Chem 54(11):4057–4061

Hamed MS, Gambert S, Bliden KP, Bailon O, Singla A, Antonino MJ, Hamed F, Tantry US, Gurbel PA (2008) Dark chocolate effect on platelet activity, C-reactive protein and lipid profile: a pilot study. South Med J 101(12):1194

Hammerstone JF, Lazarus SA, Mitchell AE, Rucker R, Schmitz HH (1999) Identification of procyanidins in cocoa (*Theobroma cacao*) and chocolate using high-performance liquid chromatography/mass spectrometry. J Agric Food Chem 47(2):490–496

Heinrich U, Neukam K, Tronnier H, Sies H, Stahl W (2006) Long-term ingestion of high flavanol cocoa provides photoprotection against UV-induced erythema and improves skin condition in women. J Nutr 136(6):1565–1569

Heiss C, Dejam A, Kleinbongard P, Schewe T, Sies H (2003) Vascular effects of cocoa rich in flavan-3-ols. JAMA 290:1030–1031

Heiss C, Kleinbongard P, Dejam A, Perré S, Schroeter H, Sies H, Kelm M (2005) Acute consumption of flavanol-rich cocoa and the reversal of endothelial dysfunction in smokers. J Am Coll Cardiol 46(7):1276–1283

Heiss C, Finis D, Kleinbongard P, Hoffmann A, Rassaf T, Kelm M, Sies H (2007) Sustained increase in flow-mediated dilation after daily intake of high-flavanol cocoa drink over 1 week. J Cardiovasc Pharmacol 49(2):74–80

Heiss C, Jahn S, Taylor M, Real WM, Angeli FS, Wong ML, Amabile N, Prasad M, Rassaf T, Ottaviani JI, Mihardja S, Keen CL, Springer ML, Boyle A, Grossman W, Glantz SA, Schroeter H, Yeghiazarians Y (2010) Improvement of endothelial function with dietary flavanols is associated with mobilization of circulating angiogenic cells in patients with coronary artery disease. J Am Coll Cardiol 56(3):218–224, 2010 Jul 13

Heo HJ, Lee CY (2005) Epicatechin and catechin in cocoa inhibit amyloid beta protein induced apoptosis. J Agric Food Chem 53(5):1445–1448

Hermann F, Spieker LE, Ruschitzka F, Sudano I, Hermann M, Binggeli C, Lüscher TF, Riesen W, Noll G, Corti R (2006) Dark chocolate improves endothelial and platelet function. Heart 92(1):119–20

Heptinstall S, May J, Fox S, Kwik-Uribe C, Zhao L (2006) Cocoa flavanols and platelet and leukocyte function: recent in vitro and ex vivo studies in healthy adults. J Cardiovasc Pharmacol 47(Suppl 2):S197–S205

Hii CL, Law CL, Suzannah S, Misnawi, Cloke M (2009) Polyphenols in cocoa (*Theobroma cacao* L.). Asian J Food Ag-Ind 2(4):702–722

Hodgson JM, Devine A, Burke V, Dick IM, Prince RL (2008) Chocolate consumption and bone density in older women. Am J Clin Nutr 87(1):175–180

Holt RR, Schramm DD, Keen CL, Lazarus SA, Schmitz HH (2002) Chocolate consumption and platelet function. JAMA 287:2212–2213

Huang Y, Barringer SA (2010) Alkylpyrazines and other volatiles in cocoa liquors at pH 5 to 8, by selected ion flow tube-mass spectrometry (SIFT-MS). J Food Sci 75(1):C121–C127

Huang Y, Barringer SA (2011) Monitoring of cocoa volatiles produced during roasting by selected ion flow tube-mass spectrometry (SIFT-MS). J Food Sci 76(2):C279–C286

Hurst WJ, Glinski JA, Miller KB, Apgar J, Davey MH, Stuart DA (2008) Survey of the trans-resveratrol and trans-piceid content of cocoa-containing and chocolate products. J Agric Food Chem 56(18):8374–8378

Hurst WJ, Payne MJ, Miller KB, Stuart DA (2009) Stability of cocoa antioxidants and flavan-3-ols over time. J Agric Food Chem 57(20):9547–9550

Innes AJ, Kennedy G, McLaren M, Bancroft AJ, Belch JJ (2003) Dark chocolate inhibits platelet aggregation in healthy volunteers. Platelets 14:325–327

Jalil AM, Ismail A, Pei CP, Hamid M, Kamaruddin SH (2008) Effects of cocoa extract on glucometabolism, oxidative stress, and antioxidant enzymes in obese-diabetic (Ob-db) rats. J Agric Food Chem 56(17):7877–7884

Jenny M, Santer E, Klein A, Ledochowski M, Schennach H, Ueberall F, Fuchs D (2009) Cacao extracts suppress tryptophan degradation of mitogen-stimulated peripheral blood mononuclear cells. J Ethnopharmacol 122(2):261–267

Jerkovic V, Bröhan M, Monnart E, Nguyen F, Nizet S, Collin S (2010) Stilbenic profile of cocoa liquors from different origins determined by RP-HPLC-APCI(+)-MS/MS. Detection of a new resveratrol hexoside. J Agric Food Chem 58(11):7067–7074

Jia L, Liu X, Bai YY, Li SH, Sun K, He C, Hui R (2010) Short-term effect of cocoa product consumption on lipid profile: a meta-analysis of randomized controlled trials. Am J Clin Nutr 92(1):218–225

Jourdain C, Tenca G, Deguercy A, Troplin P, Poelman D (2006) In-vitro effects of polyphenols from cocoa and beta-sitosterol on the growth of human prostate cancer and normal cells. Eur J Cancer Prev 15(4):353–361

Karim M, McCormick K, Kappagoda CT (2000) Effects of cocoa extracts on endothelium-dependent relaxation. J Nutr 130:2105S–2108S

Kenny TP, Keen CL, Jones P, Kung HK, Schmitz HH, Gershwin ME (2004a) Cocoa procyanidins inhibit proliferation and angiogenic signals in human dermal microvascular endothelial cells following stimulation by low-level H2O2. Exp Biol Med 229(8):765–771

Kenny TP, Keen CL, Jones P, Kung HK, Schmitz HH, Gershwin ME (2004b) Pentameric procyanidins isolated from *Theobroma cacao* seeds selectively downregulate erbb2 in human aortic endothelial cells. Exp Biol Med 229(3):255–263

Kenny TP, Keen CL, Schmitz HH, Gershwin ME (2007) Immune effects of cocoa procyanidin oligomers on peripheral blood mononuclear cells. Exp Biol Med 232(2):293–300

Kenny TP, Shu SA, Moritoki Y, Keen CL, Gershwin ME (2009) Cocoa flavanols and procyanidins can modulate the lipopolysaccharide activation of polymorphonuclear cells in vitro. J Med Food 12(1):1–7

Kim H, Keeney PG (1984) (−)-Epicatechin content in fermented and unfermented cocoa beans. J Food Sci 49(4):1090–1092

Kim JE, Son JE, Jung SK, Kang NJ, Lee CY, Lee KW, Lee HJ (2009) Cocoa polyphenols suppress TNF-α-induced vascular endothelial growth factor expression by inhibiting phosphoinositide 3-kinase (PI3K) and mitogen-activated protein kinase kinase-1 (MEK1) activities in mouse epidermal cells. Br J Nutr 104(7):957–964, 2010 Oct

Klebanoff MA, Zhang J, Zhang C, Levine RJ (2009) Maternal serum theobromine and the development of preeclampsia. Epidemiology 20(5):727–732

Kratzer U, Frank R, Kalbacher H, Biehl B, Wöstemeyer J, Voigt J (2009) Subunit structure of the vicilin-like globular storage protein of cocoa seeds and the origin of cocoa- and chocolate-specific aroma precursors. Food Chem 113(4):903–913

Kurlandsky SB, Stote KS (2006) Cardioprotective effects of chocolate and almond consumption in healthy women. Nutr Res 26(10):509–516

Kurosawa T, Itoh F, Nozaki A, Nakano Y, Katsuda S, Osakabe N, Tsubone H, Kondo K, Itakura H (2005) Suppressive effects of cacao liquor polyphenols (CLP) on LDL oxidation and the development of atherosclerosis in Kurosawa and Kusanagi-hypercholesterolemic rabbits. Atherosclerosis 179(2):237–246

Lanoue L, Green KK, Kwik-Uribe C, Keen CL (2010) Dietary factors and the risk for acute infant leukemia: evaluating the effects of cocoa-derived flavanols on DNA topoisomerase activity. Exp Biol Med (Maywood) 235(1):77–89

Lecumberri E, Goya L, Mateos R, Alía M, Ramos S, Izquierdo-Pulido M, Bravo L (2007) A diet rich in dietary fiber from cocoa improves lipid profile and reduces malondialdehyde in hypercholesterolemic rats. Nutrition 23(4):332–341

Lee KW, Hwang ES, Kang NJ, Kim KH, Lee HJ (2005) Extraction and chromatographic separation of anticarcinogenic fractions from cacao bean husk. Biofactors 23(3):141–150

Lee KW, Kang NJ, Oak M-H, Hwang MK, Kim JH, Schini-Kerth VB, Lee HJ (2008) Cocoa procyanidins inhibit expression and activation of MMP-2 in vascular smooth muscle cells by direct inhibition of MEK and MT1-MMP activities. Cardiovasc Res 79(1):34–41

Leung AY, Foster S (1995) Encyclopedia of common natural ingredients used in food, drugs, and cosmetics. Wiley-Interscience, New York, 688 pp

Lopez AS, Dimick PS (1995) Cocoa fermentation. In: Rehm HJ, Reed G (eds) Biotechnology, 2nd edn. Wiley-VCH, Weinheim, pp 562–576

Martín MA, Ramos S, Mateos R, Granado Serrano AB, Izquierdo-Pulido M, Bravo L, Goya L (2008) Protection of human HepG2 cells against oxidative stress by cocoa phenolic extract. J Agric Food Chem 56(17):7765–7772

Martin FP, Rezzi S, Peré-Trepat E, Kamlage B, Collino S, Leibold E, Kastler J, Rein D, Fay LB, Kochhar S (2009) Metabolic effects of dark chocolate consumption on energy, gut microbiota, and stress-related metabolism in free-living subjects. J Proteome Res 8(12):5568–5579

Martín MA, Serrano AB, Ramos S, Pulido MI, Bravo L, Goya L (2010) Cocoa flavonoids up-regulate antioxidant enzyme activity via the ERK1/2 pathway to protect against oxidative stress-induced apoptosis in HepG2 cells. J Nutr Biochem 21(3):196–205

Marx F, Maia JGS (1991) Purine alkaloids in seeds of *Theobroma* species from the Amazon. Zeitschr Lebensmittel Forsch A 193(5):460–461

Matissek R (1997) Evaluation of xanthine derivatives in chocolate: nutritional and chemical aspects. Z Lebensm Unters Forsch A 205:175–184

Matsui N, Ito R, Nishimura E, Yoshikawa M, Kato M, Kamei M, Shibata H, Matsumoto I, Abe K, Hashizume S (2005) Ingested cocoa can prevent high-fat diet-induced obesity by regulating the expression of genes for fatty acid metabolism. Nutrition 21(5):594–601

Mellor DD, Sathyapalan T, Kilpatrick ES, Beckett S, Atkin SL (2010) High-cocoa polyphenol-rich chocolate improves HDL cholesterol in Type 2 diabetes patients. Diabet Med 27(11):1318–1321

Meng CC, Jalil AM, Ismail A (2009) Phenolic and theobromine contents of commercial dark, milk and white chocolates on the Malaysian market. Molecules 14(1):200–209

Messaoudi M, Bisson JF, Nejdi A, Rozan P, Javelot H (2008) Antidepressant-like effects of a cocoa polyphenolic extract in Wistar-Unilever rats. Nutr Neurosci 11(6):269–276

Miller KB, Stuart DA, Smith NL, Lee CY, McHale NL, Flanagan JA, Ou B, Hurst WJ (2006) Antioxidant activity and polyphenol and procyanidin contents of selected commercially available cocoa-containing and chocolate products in the United States. J Agric Food Chem 54(11):4062–4068

Miller KB, Hurst WJ, Flannigan N, Ou B, Lee CY, Smith N, Stuart DA (2009) Survey of commercially available chocolate- and cocoa-containing products in the United States. 2. Comparison of flavan-3-ol content with nonfat cocoa solids, total polyphenols, and percent cacao. J Agric Food Chem 57(19):9169–9180

Monagas M, Khan N, Andres-Lacueva C, Casas R, Urpí-Sardà M, Llorach R, Lamuela-Raventós RM, Estruch R (2009) Effect of cocoa powder on the modulation of inflammatory biomarkers in patients at high risk of cardiovascular disease. Am J Clin Nutr 90(5):1144–1150

Mostofsky E, Levitan EB, Wolk A, Mittleman MA (2010) Chocolate intake and incidence of heart failure: a population-based prospective study of middle-aged and elderly women. Circ Heart Fail 3(5):612–616

Motamayor JC, Risterucci AM, Lopez PA, Ortiz CF, Moreno A, Lanaud C (2002) Cacao domestication I: the origin of the cacao cultivated by the Mayas. Heredity 89:380–386

Mukai R, Fukuda I, Nishiumi S, Natsume M, Osakabe N, Yoshida K, Ashida H (2008) Cacao polyphenol extract suppresses transformation of an aryl hydrocarbon receptor in C57BL/6 mice. J Agric Food Chem 56(21):10399–10405

Mullen W, Borges G, Donovan JL, Edwards CA, Serafini M, Lean ME, Crozier A (2009) Milk decreases urinary excretion but not plasma pharmacokinetics of cocoa flavan-3-ol metabolites in humans. Am J Clin Nutr 89(6):1784–1791

Muniyappa R, Hall G, Kolodziej TL, Karne RJ, Crandon SK, Quon MJ (2008) Cocoa consumption for 2 wk enhances insulin-mediated vasodilatation without improving blood pressure or insulin resistance in essential hypertension. Am J Clin Nutr 88(6):1685–1696

Murphy KJ, Chronopoulos AK, Singh I, Francis MA, Moriarty H, Pike MJ, Turner AH, Mann NJ, Sinclair AJ (2003) Dietary flavanols and procyanidin oligomers from cocoa (*Theobroma cacao*) inhibit platelet function. Am J Clin Nutr 77(6):1466–1473

Mursu J, Voutilainen S, Nurmi T, Rissanen TH, Virtanen JK, Kaikkonen J, Nyyssönen K, Salonen JT (2004) Dark chocolate consumption increases HDL cholesterol concentration and chocolate fatty acids may inhibit lipid peroxidation in healthy humans. Free Radic Biol Med 37(9):1351–1359

Natsume M, Osakabe N, Yamagishi M, Takizawa T, Nakamura T, Miyatake H, Hatano T, Yoshida T (2000) Analyses of polyphenols in cacao liquor, cocoa, and chocolate by normal-phase and reversed-phase HPLC. Biosci Biotechnol Biochem 64(12):2581–2587

Neukam K, Stahl W, Tronnier H, Sies H, Heinrich U (2007) Consumption of flavanol-rich cocoa acutely increases microcirculation in human skin. Eur J Nutr 46(1):53–56

Ohno M, Sakamoto KQ, Ishizuka M, Fujita S (2009) Crude cacao *Theobroma cacao* extract reduces mutagenicity induced by benzo[a]pyrene through inhibition of CYP1A activity in vitro. Phytother Res 23(8):1134–1139

Ohshima T, Hirao C, Nishimura E, Kamei M, Maeda N (2006) Cocoa (*Theobroma cacao*) reduced periopathogenic bacteria and improved oral malodour. J Dent Res 85B, Abstract no 1363

Orwa C, Mutua A, Kindt R, Jamnadass R, Anthony S (2009) Agroforestree database: a tree reference and selection guide version 4.0 (http://www.worldagroforestry.org/sites/treedbs/treedatabases.asp)

Osakabe N, Yamagishi M (2009) Procyanidins in *Theobroma cacao* reduce plasma cholesterol levels in high cholesterol-fed rats. J Clin Biochem Nutr 45(2):131–136

Osakabe N, Yamagishi M, Natsume M, Yasuda A, Osawa T (2004) Ingestion of proanthocyanidins derived from cacao inhibits diabetes-induced cataract formation in rats. Exp Biol Med (Maywood) 229(1):33–39

Osman H, Usta IM, Rubeiz N, Abu-Rustum R, Charara I, Nassar AH (2008) Cocoa butter lotion for prevention of striae gravidarum: a double-blind, randomised and placebo-controlled trial. BJOG 115(9):1138–1142

Panneerselvam M, Tsutsumi YM, Bonds JA, Horikawa YT, Saldana M, Dalton ND, Head BP, Patel PM, Roth DM, Patel HH (2010) Dark chocolate receptors: epicatechin-induced cardiac protection is dependent on delta-opioid receptor stimulation. Am J Physiol Heart Circ Physiol 299(5):H1604–H1609

Pearson DA, Paglieroni TG, Rein D, Wun T, Schramm DD, Wang JF, Holt RR, Gosselin R, Schmitz HH, Keen CL (2002) The effects of flavanol-rich cocoa and aspirin on ex vivo platelet function. Thromb Res 106(4–5):191–197, 2002 May 15

Percival RS, Devine DA, Duggal MS, Chartron S, Marsh PD (2006) The effect of cocoa polyphenols on the growth, metabolism, and biofilm formation by *Streptococcus mutans* and *Streptococcus sanguinis*. Eur J Oral Sci 114(4):343–348

Pérez-Berezo T, Ramiro-Puig E, Pérez-Cano FJ, Castellote C, Permanyer J, Franch A, Castell M (2009) Influence of a cocoa-enriched diet on specific immune response in ovalbumin-sensitized rats. Mol Nutr Food Res 53(3):389–397

Persson IA, Persson K, Hägg S, Andersson RG (2011) Effects of cocoa extract and dark chocolate on angiotensin-converting enzyme and nitric oxide in human endothelial cells and healthy volunteers – a nutrigenomics perspective. J Cardiovasc Pharmacol 57(1):44–50

Polagruto JA, Wang-Polagruto JF, Braun MM, Lee L, Kwik-Uribe C, Keen CL (2006) Cocoa flavanol-enriched snack bars containing phytosterols effectively lower total and low-density lipoprotein cholesterol levels. J Am Diet Assoc 106(11):1804–1813

Porcher MH et al (1995–2020) Searchable World Wide Web Multilingual Multiscript Plant Name Database. The University of Melbourne, Australia. http://www.plantnames.unimelb.edu.au/Sorting/Frontpage.html

Possemiers S, Marzorati M, Verstraete W, Van de Wiele T (2010) Bacteria and chocolate: a successful combination for probiotic delivery. Int J Food Microbiol 141(1–2):97–103

Preza AM, Jaramillo ME, Puebla AM, Mateos JC, Hernández R, Lugo E (2010) Antitumor activity against murine lymphoma L5178Y model of proteins from cacao (*Theobroma cacao* L.) seeds in relation with in vitro antioxidant activity. BMC Complement Altern Med 10:61

Purseglove JW (1968) Tropical crops: dicotyledons, vol 1 & 2. Longman, London, 719 pp

Ramiro E, Franch A, Castellote C, Andrés-Lacueva C, Izquierdo-Pulido M, Castell M (2005a) Effect of *Theobroma cacao* flavonoids on immune activation of a lymphoid cell line. Br J Nutr 93(6):859–866

Ramiro E, Franch A, Castellote C, Pérez-Cano F, Permanyer J, Izquierdo-Pulido M, Castell M (2005b) Flavonoids from *Theobroma cacao* down-regulate inflammatory mediators. J Agric Food Chem 53(22):8506–8511

Ramiro-Puig E, Castell M (2009) Cocoa: antioxidant and immunomodulator. Br J Nutr 101(7):931–940

Ramiro-Puig E, Pérez-Cano FJ, Ramos-Romero S, Pérez-Berezo T, Castellote C, Permanyer J, Franch A, Izquierdo-Pulido M, Castell M (2008) Intestinal immune system of young rats influenced by cocoa-enriched diet. J Nutr Biochem 19(8):555–565

Ramiro-Puig E, Casadesús G, Lee HG, Zhu X, McShea A, Perry G, Pérez-Cano FJ, Smith MA, Castell M (2009) Neuroprotective effect of cocoa flavonoids on in vitro oxidative stress. Eur J Nutr 48(1):54–61

Ramljak D, Romanczyk LJ, Metheny-Barlow LJ, Thompson N, Knezevic V, Galperin M, Ramesh A, Dickson RB (2005) Pentameric procyanidin from *Theobroma cacao* selectively inhibits growth of human breast cancer cells. Mol Cancer Ther 4:537–546

Ramos S, Moulay L, Granado-Serrano AB, Vilanova O, Muguerza B, Goya L, Bravo L (2008) Hypolipidemic effect in cholesterol-fed rats of a soluble fiber-rich product obtained from cocoa husks. J Agric Food Chem 56(16):6985–6993

Rein D, Paglieroni TG, Wun T, Pearson DA, Schmitz HH, Gosselin R, Keen CL (2000) Cocoa inhibits platelet activation and function. Am J Clin Nutr 72:30–35

Ried K, Sullivan T, Fakler P, Frank OR, Stocks NP (2010) Does chocolate reduce blood pressure? A meta-analysis. BMC Med 8:39

Rozan P, Hidalgo S, Nejdi A, Bisson JF, Lalonde R, Messaoudi M (2007) Preventive antioxidant effects of cocoa polyphenolic extract on free radical production and cognitive performances after heat exposure in Wistar rats. J Food Sci 72(3):S203–S206

Ruzaid A, Amin I, Nawalyah AG, Hamid M, Faizul HA (2005) The effect of Malaysian cocoa extract on glucose levels and lipid profiles in diabetic rats. J Ethnopharmacol 98(1–2):55–60

Saftlas AF, Triche EW, Beydoun H, Bracken MB (2010) Does chocolate intake during pregnancy reduce the risks of preeclampsia and gestational hypertension? Ann Epidemiol 20(8):584–591

Sakagami H, Satoh K, Fukamachi H, Ikarashi T, Shimizu A, Yano K, Kanamoto T, Terakubo S, Nakashima H, Hasegawa H, Nomura A, Utsumi K, Yamamoto M, Maeda Y, Osawa K (2008) Anti-HIV and vitamin C-synergized radical scavenging activity of cacao husk lignin fractions. In Vivo 22(3):327–332

Sanbongi C, Suzuki N, Sakane T (1997) Polyphenols in chocolate, which have antioxidant activity, modulate immune functions in humans in vitro. Cell Immunol 177(2):129–136

Sánchez D, Moulay L, Muguerza B, Quiñones M, Miguel M, Aleixandre A (2010a) Effect of a soluble cocoa fiber-enriched diet in Zucker fatty rats. J Med Food 13(3):621–628

Sánchez D, Quiñones M, Moulay L, Muguerza B, Miguel M, Aleixandre A (2010b) Changes in arterial blood pressure of a soluble cocoa fiber product in spontaneously hypertensive rats. J Agric Food Chem 58(3):1493–1501

Sánchez-Rabaneda F, Jáuregui O, Casals I, Andrés-Lacueva C, Izquierdo-Pulido M, Lamuela-Raventós RM (2003) Liquid chromatographic/electrospray ionization tandem mass spectrometric study of the phenolic composition of cocoa (*Theobroma cacao*). J Mass Spectrom 38(1):35–42

Sathyapalan T, Beckett S, Rigby AS, Mellor DD, Atkin SL (2010) High cocoa polyphenol rich chocolate may reduce the burden of the symptoms in chronic fatigue syndrome. Nutr J 9:55

Scheid L, Reusch A, Stehle P, Ellinger S (2010) Antioxidant effects of cocoa and cocoa products ex vivo and in vivo: is there evidence from controlled intervention studies? Curr Opin Clin Nutr Metab Care 13(6):737–742

Schewe T, Sies H (2005) Myeloperoxidase-induced lipid peroxidation of LDL in the presence of nitrite. Protection by cocoa flavanols. Biofactors 24(1–4):49–58

Schewe T, Sadik C, Klotz LO, Yoshimoto T, Kühn H, Sies H (2001) Polyphenols of cocoa: inhibition of mammalian 15-lipoxygenase. Biol Chem 382(12): 1687–1696

Schewe T, Kühn H, Sies H (2002) Flavonoids of cocoa inhibit recombinant human 5-lipoxygenase. J Nutr 132(7):1825–1829

Schnorr O, Brossette T, Momma TY, Kleinbongard P, Keen CL, Schroeter H, Sies H (2008) Cocoa flavanols lower vascular arginase activity in human endothelial cells in vitro and in erythrocytes in vivo. Arch Biochem Biophys 476(2):211–215

Scholey AB, French SJ, Morris PJ, Kennedy DO, Milne AL, Haskell CF (2010) Consumption of cocoa flavanols results in acute improvements in mood and cognitive performance during sustained mental effort. J Psychopharmacol 24(10):1505–1514

Schroeter H, Heiss C, Balzer J, Kleinbongard P, Keen CL, Hollenberg NK, Sies H, Kwik-Uribe C, Schmitz HH, Kelm M (2006) (−)-Epicatechin mediates beneficial effects of flavanol-rich cocoa on vascular function in humans. Proc Natl Acad Sci U S A 103:1024–1029

Selmi C, Mao TK, Keen CL, Schmitz HH, Eric Gershwin M (2006) The anti-inflammatory properties of cocoa flavanols. J Cardiovasc Pharmacol 47(Suppl 2):S163–S171

Selmi C, Cocchi CA, Lanfredini M, Keen CL, Gershwin ME (2008) Chocolate at heart: the anti-inflammatory impact of cocoa flavanols. Mol Nutr Food Res 52(11):1340–1348

Shiina Y, Funabashi N, Lee K, Murayama T, Nakamura K, Wakatsuki Y, Daimon M, Komuro I (2009) Acute effect of oral flavonoid-rich dark chocolate intake on coronary circulation, as compared with non-flavonoid white chocolate, by transthoracic Doppler echocardiography in healthy adults. Int J Cardiol 131(3):424–429

Sieri S, Krogh V, Donati MB, de Gaetano G, Iacoviello L (2008) Regular consumption of dark chocolate is associated with low serum concentrations of C-reactive protein in a healthy Italian population. J Nutr 138:1934–1945

Sies H, Schewe T, Heiss C, Kelm M (2005) Cocoa polyphenols and inflammatory mediators. Am J Clin Nutr 81(1 Suppl):304S–312S

Sotelo A, Alvarez RG (1991) Chemical composition of wild *Theobrom*a species and their comparison to the cacao bean. J Agric Food Chem 39(11):1940–1943

Spadafranca A, Martinez Conesa C, Sirini S, Testolin G (2010) Effect of dark chocolate on plasma epicatechin levels, DNA resistance to oxidative stress and total antioxidant activity in healthy subjects. Br J Nutr 103(7):1008–1014

Staphylakis K, Gegiou D (1985) Free esterified and glucosidic sterols in cocoa butter. Lipids 20(11):723–728

Stark T, Hofmann T (2005) Isolation, structure determination, synthesis, and sensory activity of N-phenylpropenoyl-L-amino acids from cocoa (*Theobroma cacao*). J Agric Food Chem 53(13):5419–5428

Stark T, Bareuther S, Hofmann T (2005) Sensory-guided decomposition of roasted cocoa nibs (*Theobroma cocoa*) and structure determination of taste-active polyphenols. J Agric Food Chem 53(13):5407–5418

Stark T, Bareuther S, Hofmann T (2006a) Molecular definition of the taste of roasted cocoa nibs (*Theobroma cacao*) by means of quantitative studies and sensory experiments. J Agric Food Chem 54(15): 5530–5539

Stark T, Justus H, Hofmann T (2006b) Quantitative analysis of n-phenylpropenoyl-l-amino acids in roasted coffee and cocoa powder by means of a stable isotope dilution assay. J Agric Food Chem 54:2859–2867

Stark T, Lang R, Keller D, Hensel A, Hofmann T (2008) Absorption of N-phenylpropenoyl-L-amino acids in healthy humans by oral administration of cocoa (*Theobroma cacao*). Mol Nutr Food Res 52(10):1201–1214

Steffen Y, Schewe T, Sies H (2006) Myeloperoxidase-mediated LDL oxidation and endothelial cell toxicity of oxidized LDL: attenuation by (–)-epicatechin. Free Radic Res 40(10):1076–1085

Tang Y, Gilbert MG, Door LJ (2007) Sterculiaceae. In: Wu ZY, Raven PH, Hong DY (eds) Flora of China, vol 12, Hippocastanaceae through Theaceae. Science Press/Missouri Botanical Garden Press, St. Louis

Taubert D, Roesen R, Lehmann C, Jung N, Schömig E (2007a) Effects of low habitual cocoa intake on blood pressure and bioactive nitric oxide: a randomized controlled trial. JAMA 298:49–60

Taubert D, Roesen R, Schomig E (2007b) Effect of cocoa and tea intake on blood pressure: a meta-analysis. Arch Intern Med 167(7):626–634

Tomaru M, Takano H, Osakabe N, Yasuda A, Inoue K, Yanagisawa R, Ohwatari T, Uematsu H (2007) Dietary supplementation with cacao liquor proanthocyanidins prevents elevation of blood glucose levels in diabetic obese mice. Nutrition 23(4):351–355

Tomofuji T, Ekuni D, Irie K, Azuma T, Endo Y, Tamaki N, Sanbe T, Murakami J, Yamamoto T, Morita M (2009) Preventive effects of a cocoa-enriched diet on gingival oxidative stress in experimental periodontitis. J Periodontol 80(11):1799–1808

Triche EW, Grosso LM, Belanger K, Darefsky AS, Benowitz NL, Bracken MB (2008) Chocolate consumption in pregnancy and reduced likelihood of preeclampsia. Epidemiology 19(3):459–464

Tropicos Org. Nomenclatural and Specimen Database of the Missouri Botanical Garden. http://www.tropicos.org/Home.aspx

Tzounis X, Rodriguez-Mateos A, Vulevic J, Gibson GR, Kwik-Uribe C, Spencer JP (2011) Prebiotic evaluation of cocoa-derived flavanols in healthy humans by using a randomized, controlled, double-blind, crossover intervention study. Am J Clin Nutr 93(1):62–72

U.S. Department of Agriculture, Agricultural Research Service (USDA) (2010) USDA National Nutrient Database for standard reference, release 23. Nutrient Data Laboratory Home Page, http://www.ars.usda.gov/ba/bhnrc/ndl

van den Bogaard B, Draijer R, Westerhof BE, van den Meiracker AH, van Montfrans GA, van den Born BJ (2010) Effects on peripheral and central blood pressure of cocoa with natural or high-dose theobromine: a randomized, double-blind crossover trial. Hypertension 56(5):839–846

Vinson JA, Proch J, Bose P, Muchler S, Taffera P, Shuta D, Samman N, Agbor GA (2006) Chocolate is a powerful ex vivo and in vivo antioxidant, an antiatherosclerotic agent in an animal model, and a significant contributor to antioxidants in the European and American Diets. J Agric Food Chem 54(21):8071–8076

Vlachopoulos C, Aznaouridis K, Alexopoulos N, Economou E, Andreadou I, Stefanadis C (2005) Effect of dark chocolate on arterial function in healthy individuals. Am J Hypertens 18(6):785–791

Voigt J, Biehl B, Syed Wazir SK (1993) The major seed proteins of *Theobroma cacao* L. Food Chem 47(2):145–151

Voigt J, Heinrichs H, Voigt G, Biehl B (1994) Cocoa-specific aroma precursors are generated by proteolytic digestion of the vicilin-like globulin of cocoa seeds. Food Chem 50(2):177–184

Wan Y, Vinson JA, Etherton TD, Proch J, Lazarus SA, Kris-Etherton PM (2001) Effects of cocoa powder and dark chocolate on LDL oxidative susceptibility and prostaglandin concentrations in humans. Am J Clin Nutr 74(5):596–602

Wang JF, Schramm DD, Holt RR, Ensunsa JL, Fraga CG, Schmitz HH, Keen CL (2000) A dose-response effect from chocolate consumption on plasma epicatechin and oxidative damage. J Nutr 130:2115S–2119S

Wang-Polagruto JF, Villablanca AC, Polagruto JA, Lee L, Holt RR, Schrader HR, Ensunsa JL, Steinberg FM, Schmitz HH, Keen CL (2006) Chronic consumption of flavanol-rich cocoa improves endothelial function and decreases vascular cell adhesion molecule in hypercholesterolemic postmenopausal women. J Cardiovasc Pharmacol 47(Suppl 2):S177–S186

Weisberger JH (2001) Chemopreventive effects of cocoa polyphenols on chronic diseases. Exp Biol Med 226:891–897

Wessel M, Toxopeus H (2000) *Theobroma cacao* L. In: van der Vossen HA M, Wessel M (eds) Plant resources of South-East Asia No. 16 stimulants. Backhuys Publishers, Leiden, pp 113–121

Williams S, Tamburic S, Lally C (2009) Eating chocolate can significantly protect the skin from UV light. J Cosmet Dermatol 8(3):169–173

Wiswedel I, Hirsch D, Kropf S, Gruening M, Pfister E, Schewe T, Sies H (2004) Flavanol-rich cocoa drink lowers plasma F(2)-isoprostane concentrations in humans. Free Radic Biol Med 37:411–421

Wollgast J, Anklam E (2000) Polyphenols in chocolate: is there a contribution to human health? Food Res Int 33:449–459

Wood GAR, Lass RA (eds) (1985) Cocoa, 4th edn, Tropical agricultural series. Longman, London, 620 pp

Yamada T, Yamada Y, Okano Y, Terashima T, Yokogoshi H (2009) Anxiolytic effects of short- and long-term administration of cacao mass on rat elevated T-maze test. J Nutr Biochem 20(12):948–955

Yamazaki KG, Romero-Perez D, Barraza-Hidalgo M, Cruz M, Rivas M, Cortez-Gomez B, Ceballos G, Villarreal F (2008) Short- and long -term effects of (–)-epicatechin on myocardial ischemia-reperfusion injury. Am J Physiol Heart Circ Physiol 295(2):H761–H767

Yasuda A, Natsume M, Sasaki K, Baba S, Nakamura Y, Kanegae M, Nagaoka S (2008a) Cacao procyanidins reduce plasma cholesterol and increase fecal steroid excretion in rats fed a high-cholesterol diet. Biofactors 33(3):211–223

Yasuda A, Takano H, Osakabe N, Sanbongi C, Fukuda K, Natsume M, Yanagisawa R, Inoue K, Kato Y, Osawa T, Yoshikawa T (2008b) Cacao liquor proanthocyanidins inhibit lung injury induced by diesel exhaust particles. Int J Immunopathol Pharmacol 21(2):279–288

Zhu QY, Schramm DD, Gross HB, Holt RR, Kim SH, Yamaguchi T, Kwik-Uribe CL, Keen CL (2005) Influence of cocoa flavanols and procyanidins on free radical-induced human erythrocyte hemolysis. Clin Dev Immunol 12(1):27–34

Theobroma grandiflorum

Scientific Name

Theobroma grandiflorum (**Willd. ex Spreng.) K. Schum.**

Synonyms

Bubroma grandiflorum Willd. ex Spreng., *Guazuma grandiflora* (Willd. ex Spreng.) G. Don, *Theobroma grandiflora* Willd. ex Spreng., *Theobroma macrantha* Bernoulli, *Theobroma silvestre* Spruce ex K. Schum.

Family

Malvaceae, also placed in Sterculiaceae

Common/English Names

Brazilian Cocoa, Copoasu, Cupuassu, Cupuassu, Large-Flowered Cocoa

Vernacular Names

Brazil: Copoasu, Cupuaçú, Cupuaçuzeiro, Cupuarana, Cupu-Assu Cupuassu, Cupuassú, Cupu Do Mato, Copoaçú, Pupu, Pupuaçú (Portuguese);
Colômbia: Bacau, Kopoazu;
Costa Rica: Cacau Silvestre, Pataiste, Tete Negra;
Czech: Kakaovník Velkokvětý;
Portuguese: Cupuaçú, Cupuassú;
Suriname: Lupu;
Spanish: Copoasú, Copoasú, Copuazú, Cupassú, Cupuarana, Cupuasú.

Origin/Distribution

Cupuassu is indigenous to the Amazon basin in the southern and eastern Pará, covering the areas of the middle Tapajós, Xingu and Guamá, and reaching the northeast of Maranhão in Brazil (Cuatrecasas 1964). It is widely cultivated in the north of Brazil, with the largest production in Pará, followed by Amazonas, Rondônia and Acre. It is also cultivated in Colombia, Venezuela, Ecuador, and Costa Rica.

Agroecology

The cupuassu tree can be found throughout the Amazon region as part of spontaneous vegetations on non-inundated areas, particularly near existing or former settlements, in high-rainfall primary forests and along river banks. It can be found at elevations between sea level and 1,000 m. Cupuassu prefers a very humid environment of 77–88% with little temperature variation throughout the year, and mean annual temperatures of

T.K. Lim, *Edible Medicinal And Non-Medicinal Plants: Volume 3, Fruits*,
DOI 10.1007/978-94-007-2534-8_31, © Springer Science+Business Media B.V. 2012

between 21.6°C to 27.5°C, and rainfall of between 1,900 and 3,000 mm. Trees need lots of water for good growth.

Prolonged dry periods are harmful to the tree causing flower shedding and premature fruit fall. Rains following a period of drought will also cause fruit cracking (Diniz et al. 1984)

Edible Plant Parts and Uses

The delicious fruit pulp is eaten raw or prepared into a wide variety of beverages, juices, ice-cream, yoghurt, sorbet, preserves, sweets, jams, jellies, puddings and other desserts. The seeds have a high amount of fat and provide a good cocoa butter and can be used to produce chocolate and chocolate-like foodstuffs. Cupuaçu can replace cocoa in many day-to-day foods, especially for children, such as chocolate milk. Cupuaçu seeds can be made into cupulate, which looks and tastes just like chocolate but is cheaper and more resistant to heat.

Botany

An erect, evergreen, much-branched tree growing 5–15 m high with brown bark and tricomic branching (Plate 1). Leaves are simple, entire, alternate, coriaceous, narrowly ovate- to obovate-elliptic, 25–35 cm long and 6–10 cm wide, base rounded to shallowly cordate, apex long acuminate with a 9–10 pairs of lateral veins, bright-green, pubescent upper surface and grey underside (Plate 3). Young flushes are pink-bronze (Plate 2). Flowers bisexual, small, pentamerous, occurring in 3–5 flowered cymose inflorescence on trunk and main branches (Plates 3 and 4). Flower with a 5-lobed calyx divided nearly to base, 5 subtrapezoidal, purple petals, 5 stamens with bilocular anthers alternating with 5 staminodes and a pentagonal superior ovary with 5 locules containing numerous ovules. Fruit oblongish, obovate, subglobose to ellipsoidal, large, 20–25 cm long by 6–10 cm wide, weighing up to two kg, woody, hard, pubescent, rough, brown, fall to the ground when ripe (Plates 5–7). The endocarp is yellowish-white, pulpy, soft, aromatic, and acid-sweet enclosing 25–50 superposed seeds in five rows (Plate 8). Seed, subglobose, 2 cm across, covered in white testa, cotyledons fleshy, non-endospermous.

Plate 1 A young fruiting tree

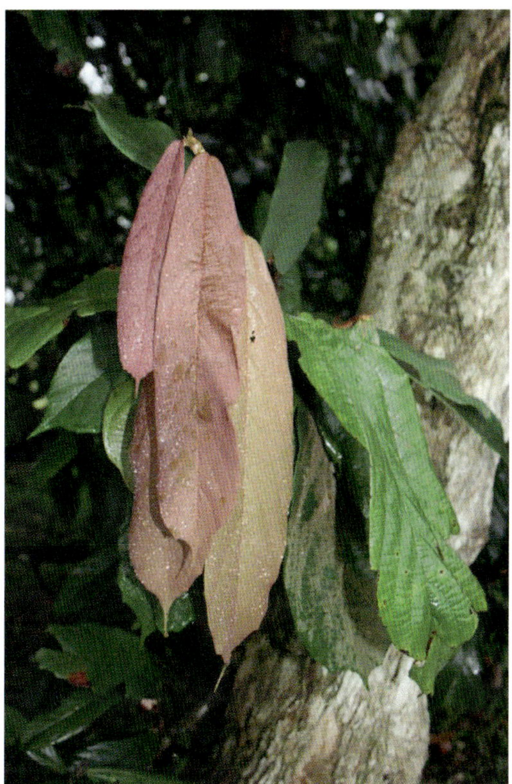

Plate 2 Juvenile pink leaves

Plate 3 Flower and mature leaves

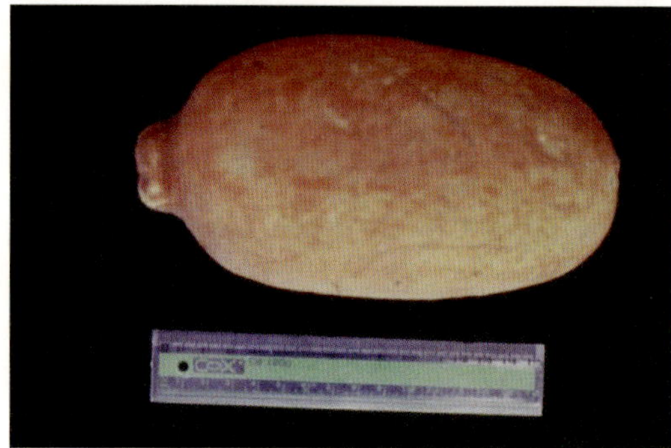

Plate 6 Large oblong-shaped fruit

Plate 4 Close-up of flower

Plate 7 Large, obovate and sub-globose fruit

Plate 5 A mature cupuassu fruit

Plate 8 Yellowish-white, creamy pulp covering the seeds

Nutritive/Medicinal Properties

Chemical composition of cupuassu fruit pulp was reported by Rogez et al. (2004) as follows: pH 3.4, dry matter (DM) 12.1%, protein 8.8% DM, lipids 12.7%DM, ash 5.3%DM, sugars 49%DM, glucose 6.9%, fructose 8.8% total dietary fibre 14.3%DM; amino acid content (g/100 g protein) – glycine 4.45 g, alanine 7.11 g, valine 6.06 g, leucine 6.82 g, isoleucine 4.42 g, proline 4.56 g, phenylalanine 4.64 g, tyrosine 3.9 g serine 4.73 g, threonine 4.09 g, cysteine

2.33 g, methionine 2.21 g, asparagines + aspartic acid 15.77 g, glutamine + glutamic acid 16.25 g, lysine 6.16 g,arginine 4.27 g, histidine 2.35 g; fatty acids (g/100 g total fatty acids analysed) - myristic acid 0.12 g, palmitic acid 55.22 g, palmitoleic acid 0.56 g, stearic acid 3.12 g, oleic acid 18.8 g, linoleic acid 3.08 g, α-linolenic acid 17.98 g, SFA/UFA ratio 1.4, MUFA/PUFA ratio 0.978,ω-6/ω-3 ratio 0.1713; minerals mg/100 g FW) - sodium 2.56 mg, potassium 34.27 mg, calcium 5.57 mg, magnesium 13.07 mg, phosphorus 15.73 mg, iron 0.432 mg, zinc 0.532 mg, copper 0.258 mg and manganese 0.21 mg.

Based on the above cupuassu pulp was reported to have a standard dry matter (12.1%) comparable with tropical fruits in general, its high protein content of 8.8% placed it among the rich type of tropical fruit (Rogez et al. 2004). Among the amino acids the limiting ones in cupuassu were methionine and arginine. Relatively high amount of lipids were found in the pulp. Regarding essential fatty acids, cupuassu was found to be poor in linoleic acid but relatively high in α-linolenic, an omega-3 fatty acid. Cupuassu has nutritional benefits since omega-3 fatty acids are generally low in diets based on meat, starchy sources, fruits and vegetable. The ratio ω-6/ω-3 highlight the relative importance of ω fatty acids, the recommended ratio being 5 or less. Fruit pulp was reported to have the following attributes, 82.20% moisture, 2.4% acidity, pH 3.3, reducing sugars 3%, brix 5.70%, pectin 0.39%, volatile 89%, vitamin C 23.12–26.20 mg, amino acids and 21.90 g (Diniz et al. 1984; Bermejo and León 1994; Nazare 1997).

The pectic fraction of the fruit pulp was found to compose mainly of a homogalacturonan highly esterified (DE 53%; DA 1.7%) with some rhamnogalacturonan insertions, carrying side chains containing galactose and arabinose (Vriesmann and Petkowicz 2009). The predominant component polysaccharide in the pulp, 91% of which was composed of high amylose starch (Vriesmann et al. 2009, 2010). The content of amylose found in the fraction was 71%, higher than those of common starches of cereal grains, tubers, roots, and other fruits. The presence of starch, as well as the presence of previously investigated pectin, conferred the high viscosity and gelling capability of the pulp. Further studies showed that starch could be cooperating in the formation of water-soluble pectin gels.

Among the volatile compounds, esters predominated in the cupuaçu fruit (Franco and Shibamoto 2000). Ethyl butyrate and hexanoate were the major compounds. Another research reported 56 volatile compounds were isolated from copoazú (*Theobroma grandiflorum*) fruit (Quijano and Pino 2007). Ethyl butanoate, ethyl hexanoate and linalool were the major constituents found. Boulanger and Crouzet (2000) reported that the main contributors to the floral falvour of cupuassu were linalool, a-terpinol, 2-phenyethanol, myrcene and limonene; diols and methoxy-2,5-dimethyl-3(2H)-furanone were believed to be the main contributors to the typical exotic odour.

The white pulp of the cupuaçu was found to be uniquely fragrant, and to contain theacrine (1,3,7,9-tetramethyluric acid) instead of the xanthines (caffeine, theobromine, and theophylline) found in cacao (Vasconcelos et al. 1975). Theacrine was also present in cupuassu seeds (Marx and Maia 1991). Studies showed that theacrine possessed potent sedative and hypnotic properties and its central nervous system effects were different from those of caffeine and theobromine (Xu et al. 2007). Theacrine could significantly prolong the sleeping time induced by pentobarbital, while caffeine and theobromine exhibited an inverted effect. Theobromine, theophylline, and caffeine, however were found in cupuassu seeds but in lower concentrations than in cocoa seeds (Lo Coco et al. 2007).

Seeds of *T. grandiflorum* contained mainly polysaccharides and lipid protein in the cotyledon mesophyll; it contained 129.5 mg/g protein, 542.2 mg/g lipids and 42.7 mg/g polyphenols (Martini et al. 2008). The cupuassu seed was found to be very rich in fats (± 60% dry weight), of which 91% were digestible by humans. The processing of the seed was similar to the processing of cocoa beans: fermentation, roasting, milling, and pressing to obtain fat. From the defatted mass, the cupuassu powder was obtained (Lannes 2003). During the fermentation of the cupuassu seed, the cotyledon's pH increased from 5.2 to

6.8, the phenolic compounds from 1.12 to 1.34 mg/100 g, the fats remained unchanged at 63.5% dry weight, and the colour changed from a beige cream to a dark reddish caramel. The flavour and aroma of the fermented seed was indistinguishable from that of fermented cocoa seed.

Cupuassu seed was found to contain 26.17% total protein, essential amino acids mg/g - threonine 59.20, valine 72.60, methionine 4.59, cysteine 23.31, isokeucine 45.8, leucine 84.1, tyrosine 37.06, phenyalanine 45.47, lysine 53.11, histidine 14.52, tryptophan 12.99; non-essentail amino acids mg/g – arginine 56.55, aspartic acid 123.81, serine 53.11, glutamic acid 148.26, proline 44.33, glycine 57.70 and alanine 47.76 (Carvalho et al. 2008). Fermentation and roasting promoted a slight reduction in the total protein and amino acid contents. The seeds and fermented beans presented four main protein bands with 15.5, 20.4, 27.1, and 33.6 kDa. The beans submitted to fermentation followed by roasting presented only one strong protein band with an apparent molecular weight of 21.0 kDa. The extractions for protein fractionation based on solubility did not result in pure protein fractions. Four main bands were observed in all isolated protein fractions corresponding to albumin, globulin, prolamin and glutelin bands (Carvalho et al. 2008).

New sulfated flavonoid glycosides, theograndins I (1) and II (2) were identified in cupuassu seeds (Yang et al. 2003). In addition, nine known flavonoid antioxidants, (+)-catechin, (−)-epicatechin, isoscutellarein 8-O-β-d-glucuronide, hypolaetin 8-O-β-d-glucuronide, quercetin 3-O-β-d-glucuronide, quercetin 3-O-β-d-glucuronide 6″-methyl ester, quercetin, kaempferol, and isoscutellarein 8-O-β-d-glucuronide 6″-methyl ester, were identified. Theograndin II (2) displayed antioxidant activity (IC_{50}= 120.2 μM) in the 1,1-diphenyl-2-picrylhydrazyl (DPPH) free-radical assay, as well as weak cytotoxicity in the HCT-116 and SW-480 human colon cancer cell lines with IC_{50} values of 143 and 125 μM, respectively. While compound 1 was less active as an antioxidant than 2, the known compounds were more potent in the DPPH assay (IC_{50} range 39.7-89.7 μM).

Polyphenolic compounds appeared to be related to the health benefits produced by the cocoa due to their antioxidant properties (Genovese and Lannes 2009). Cupuassu powder, prepared from *Theobroma grandiflorum* seeds, is a very promising cocoa powder substitute. In order to assess the potential health benefits of the cupuassu powder, a comparison was performed between cocoa, chocolate, and cupuassu powders in relation to the content of total phenolic compounds, flavonoids and DPPH scavenging capacity of methanolic extracts. Cupuassu "chocolates" (milk, dark, and white) were also analyzed. Results showed that the phenolic contents of cocoa and chocolate powders were more than three times higher than those of cupuassu powder; however, flavonoid contents were significantly lower (Genovese and Lannes 2009). The DPPH scavenging capacity varied hugely among the different samples, white cupuassu "chocolate" 0.5 μg of Trolox equivalent per 100 g (FW), milk cupuassu chocolate 4.5 μg, dark cupuassu chocolate 7.8 μg, cupuassu powder 13–120 μg (cocoa powder), and presented a significant correlation (r=0.977) with the total phenolic contents but not with the flavonoid contents (r=−0.035). The flavonoid content of cupuassu powder was more than three times higher than those of cocoa and chocolate powders. The cupuassu "chocolates" presented flavonoid contents ranging from 22 to 73 mg/100 FW showing that they could also represent a good source. Total flavonoids content in mg RE/100 g F.W. of 70% aqueous methanolic extracts from cupuassu powder was 590 mg, from coca powder 120 mg, chocolate powder 185 mg, white cupuassu chocolate 22 mg, milk cupuassu chocolate 61 mg, and dark cupuassu chocolate 73 mg.

Unreliable product quality had often thwarted attempts to commercialise chocolate-like wares from *T. grandiflorum* (cupuaçu) seeds (Reisdorff et al. 2004). This was found to be due to an insufficient aroma potential of cupuaçu seeds. The qualitative and quantitative differences between the globulins indicated a lower maximum yield of aroma precursors in *T. grandiflorum* and a higher maximum yield of aroma precursors in *T. bicolor*, compared to cocoa. There was also slight differences in activities of proteolytic enzymes (aspartic endopeptidase and the carboxypeptidase). The study indicated that the quality of chocolate-like products made from the cupuassu seeds could be improved by adapting fermentation procedures to particular biochemical features of the seeds.

Other Uses

The timber is commonly used for construction and joinery.

Comments

In Pará, three cultivars of cupuaçu are known: Redondo, with its rounded end, which is the most common; Mamorano, which has a pointed end and produces the biggest fruits; and Mamau, possibly a parthenocarpic mutant. Artificial hybrids between *T. grandiflorum* and *T. obovatum* produce fruits with the characteristics of cupuaçu, but which are smaller and less resistant to witches' broom disease.

Selected References

Bermejo JEH, León J (eds) (1994) Neglected crops: 1492 from a different perspective, vol 26, Plant Production and Protection. FAO, Rome, pp 205–209

Boulanger R, Crouzet J (2000) Free and bound flavour components of amazonian fruits: 2 cupuacu volatile compounds. Flav Fragr J 15:251–257

Carvalho AV, Garcia NHP, Farfan JA (2008) Proteins of cupuacu seeds (*Theobroma grandiflorum* Schum) and changes during fermentation and roasting. Ciênc Tecnol Aliment 28(4):986–993 (In Portuguese)

Cavalcante PB (1972) Frutas Comestíveis da Amazônia, vol 1, Publ. Avulsas No. 17 Museu paraense Emilia Goeldi Belém, Pará, Brasil, 84 pp (in Portuguese)

Cuatrecasas J (1964) Cacao and its allies: a taxonomic revision of the genus *Theobroma*. Contr US Natl Herb 35(6):379–614

Diniz TD de AS, Bastos TX, Rodrigues IA, Muller CH, Kato AK and Silva MMM de (1984) Condições climáticas em áreas de ocorrência natural e de cultivo de guaraná, cupuaçu, bacuri e castanha-do-Brasil. Belém: EMBRAPA-CPATU, 1984, 4pp (EMBRAPA-CPATU. Pesquisa em Andamento, 133). (in Portuguese)

Franco MRB, Shibamoto T (2000) Volatile composition of some Brazilian fruits: umbu-caja (*Spondias citherea*), camu-camu (*Myrciaria dubia*), araça-boi (*Eugenia stipitata*), and cupuaçu (*Theobroma grandiflorum*). J Agric Food Chem 48(4):1263–1265

Genovese MI, Lannes SCDS (2009) Comparison of total phenolic content and antiradical capacity of powders and "chocolates" from cocoa and cupuassu. Ciênc Tecnol Aliment 29(4):810–814

Lannes SCS (2003) Cupuassu – a new confectionery fat from Amazonia. Inform Am Oil Chemists Soc 14:40–41

Lo Coco F, Lanuzza F, Micali G, Cappellano G (2007) Determination of theobromine, theophylline, and caffeine in by-products of cupuacu and cacao seeds by high-performance liquid chromatography. Chromatogr Sci 45(5):273–275

Martini MH, Lenci CG, Figueira A, Tavares DDQ (2008) Localization of the cotyledon reserves of *Theobroma grandiflorum* (Willd. ex Spreng.) K. Schum., *T. subincanum* Mart., *T. bicolor* Bonpl. and their analogies with *T. cacao* L. Revista Brasil Bot 31(1):147–154

Marx F, Maia JGS (1991) Purine alkaloids in seeds of *Theobroma* species from the Amazon. Zeitschrift Lebensmittel Forsch A 193(5):460–461

Nazaré RFR de (1997) Processos aagroindustriais para o desenvolvimento de produtos de cupuaçu (*Theobroma grandiflorum*). In: Seminário Internacional Sobre Pimenta-Do-Reino E Cupuaçu. 1. 1996, Belém PA. Anais. Belém: Embrapa – Amazônia Oriental/JICA, 1997, pp 185–192. (Embrapa – Amazônia Oriental. Documentos, 89) (in Portuguese)

Quijano CE, Pino JA (2007) Volatile compounds of copoazú (*Theobroma grandiflorum* Schumann) fruit. Food Chem 104(3):1123–1126

Reisdorff C, Rohsius C, Claret de Souza AdG, Gasparotto L, Lieberei R (2004) Comparative study on the proteolytic activities and storage globulins in seeds of *Theobroma grandiflorum* (Willd ex Spreng) Schum and *Theobroma bicolor* Humb Bonpl, in relation to their potential to generate chocolate-like aroma. J Sci Food Agric 84(7):693–700

Rogez H, Buxant R, Mignolet E, Souza JNS, Silva EM, Larondelle Y (2004) Chemical composition of the pulp of three typical Amazonian fruits: araça-boi (*Eugenia stipitata*), bacuri (*Platonia insignis*) and cupuaçu (*Theobroma grandiflorum*). Eur Food Res Technol 218(4):380–384

Vasconcelos MNL, da Silva ML, Maia JGS, Gottlieb OR (1975) Estudo químico de sementes do cupuaçu. Acta Amazonica 5:293–295 (in Portuguese)

Vriesmann LC, de O Petkowicz CL (2009) Polysaccharides from the pulp of cupuassu (*Theobroma grandiflorum*): structural characterization of a pectic fraction. Carbohydr Polymer 77(1):72–79

Vriesmann LC, Silveira JLM, de O Petkowicz CL (2009) Chemical and rheological properties of a starch-rich fraction from the pulp of the fruit cupuassu (*Theobroma grandiflorum*). Mater Sci Eng C 29(2): 651–656

Vriesmann LC, Silveira JLM, de O Petkowicz CL (2010) Rheological behavior of a pectic fraction from the pulp of cupuassu (*Theobroma grandiflorum*). Carbohydr Polymer 79(2):312–331
Xu JK, Kurihara H, Liang Z, Yao XS (2007) Theacrine, a special purine alkaloid with sedative and hypnotic properties from *Cammelia assamica* var. *kucha* in mice. J Asian Nat Prod Res 9(6–8):665–672
Yang H, Protiva P, Cui B, Ma C, Baggett S, Hequet V, Mori S, Weinstein IB, Kennelly EJ (2003) New bioactive polyphenols from *Theobroma grandiflorum* ("cupuaçu"). J Nat Prod 66(11):1501–1504

Thaumatococcus daniellii

Scientific Name

Thaumatococcus daniellii **(Benn.) Benth.**

Synonyms

Donax daniellii (Horan.) Roberty, *Monostiche daniellii* (Benn.) Horan., *Phrynium daniellii* Benn.

Family

Marantaceae

Common/English Names

African Serendipity Berry, Katempfe, Miracle Berry, Miracle-Fruit, Miraculous Berry Sweet Prayer Plant, Yoruba Soft Cane Fruit

Vernacular Names

French: Fruit Miraculeux;
Ghana: Anworom;
Ivory Coast: Ndè-Tata (Akye), Urugua Méremné (Bakoué), Kohoun (Baba), Orofira (Leaf):
Nigeria: Ewe Eran (Yoruba), Ijowol, Toke;
Sierra Leone: Katamfe, Katembe; Ketenfe (Krio), Gbula – Leaf (Mende), Dane, Nɛni – Fruit (Mende).

Origin/Distribution

The species is a native of tropical west Africa – Sierra Leone, Ivory Coast, Gabon, Cameroon, Liberia, Guinea, Nigeria, Republic of the Congo, Equatorial Guinea, Ghana, Democratic Republic of the Congo Ghana. It has been introduced elsewhere into the tropics and naturalised in many countries including Australia.

Agroecology

The species thrives in large clumps as undergrowth in the hot, humid tropical rain forest in the lowlands and coastal zone of West Africa. In Australia, it has naturalized on the edges of tropical rainforest in Northern Australia.

Edible Plant Parts and Uses

The most popular use of *T. daniellii* aril is as a sweetener or taste modifier. From the aril of *T. daniellii*, an intensely sweet, non-toxic and heat stable protein – thaumatin – is extracted. In West Africa, the aril is traditionally used for

T.K. Lim, *Edible Medicinal And Non-Medicinal Plants: Volume 3, Fruits*,
DOI 10.1007/978-94-007-2534-8_32, © Springer Science+Business Media B.V. 2012

sweetening bread, over-fermented palm-wine, sour fruit desserts and sour food. The arils are also used to sweeten food such as *garri* (fermented cassava flour usually fried in palm oil), *pap* (maize-meal porridge) and tea.

Since the mid-1990s, it is used as sweetener and flavour enhancer by the food and confectionary industry in many countries, substituting synthetic sweeteners. It is used as sweetener or taste modifier in beverages, desserts, chewing gums and pet foods. Thaumatin has been approved as a food additive throughout Europe and in many other countries throughout the world.

Botany

A rhizomatous, perennial herb, up to 3–3.5 m high, with slender, creeping, single or forked rhizomes (Plate1). Leaf arise singly from each node of the rhizome. Petiole slender, subterete, 2–3 m high (Plate 3). Leaves ovate-elliptic, rounded-truncate at the base, shortly acuminate, 46 cm long by 30 cm wide, green, papery, with numerous parallel veins diverging from the midrib at 45° angle (Plates 1–2). Spikes with 10–12 flowers, arising from the lowest node, simple or branched, about 10 cm long; bracts imbricate, 4 cm long. Flowers pale violet, as long as the bracts; sepals broadly linear, 1 cm long; corolla-tube very short, lobes oblong, 2.5 cm long; ovary silky (Plate 4). Only 3–4 matured fruits are usually produced from the spike, formed at or below the ground. Fruit pyramidal or trigonal in shape, hard, 3 cm across, maturing from a dark-green through brown to crimson or bright-red colour when fully ripe (Plates 5–7), with 2–3 shining, black and hard seeds enveloped by the sticky, thin, transparent aril which contains the sweetening protein, thaumatin (Plate 7).

Plate 1 Clump of *Thaumatococcus dianellii*

Plate 2 Large ovate-elliptic leaves with numerous parallel veins

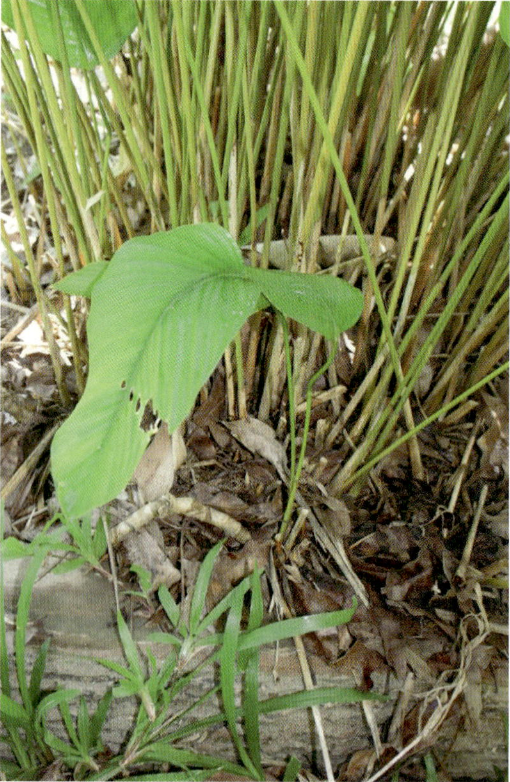

Plate 3 Long, slender leaf petioles

Plate 4 Pale violet small flowers

Plate 5 Crimson fruit formed at ground level

Plate 6 Detached crimson, hard fruit

Plate 7 Black hard seed covered by sticky thin, transparent aril

Nutritive/Medicinal Properties

Chemical Constituents

Thaumatin, a sweet-tasting protein was isolated from the arils of *Thaumatococcus daniellii* (Van der Wel and Loeve 1972). Thaumatin and monellin had been reported as the two sweetest compounds known to man - about 100,000 times sweeter than sugar on a molar basis and 3000 times on a weight basis (De Vos et al. 1985). Naturally occurring thaumatin was found to consist of six closely related proteins (I, II, III, a, b and c), all with a molecular mass of 22 kDa (207 amino acids) (Van der Wel and Loeve 1972; Ledeboer et al. 1984). The three-dimensional structure of thaumatin I had been determined at high resolution (De Vos et al. 1985; Ogata et al. 1992), revealing that the protein consisted of three domains: (a) an 11 strand, flattened β-sandwich (1–53, 85–127 and 178–207, domain I); (b) a small disulfide-rich region (54–84, domain III); and (c) a large disulfide-rich region (128–177, domain II).

Modification of lysine residues in the structure of thaumatin was found to reduce the sweetness of thamatin (Kaneko and Kitabatake 2000). Phosphopyridoxylation of lysine residues Lys78, Lys97, Lys106, Lys137 and Lys187 markedly reduced sweetness. The intensity of sweetness was returned to that of native thaumatin by dephosphorylation of these phosphopyridoxylated lysine residues except Lys106. These lysine residues occurred in thaumatin, but not in non-sweet thaumatin-like proteins, suggesting that these lysine residues were required for sweetness. The sweet-tasting proteins, thaumatin I and 11, isolated from the arils of *Thaumatococcus daniellii*, were found to lose their sweetness on heating (Korver et al. 1973).

Both proteins sustained reversible conformational changes as the temperature was increased. The native conformation was most stable at pH 5 and the conformational change was reversible up to 75°C even on prolonged heating.

Studies by Higginbotham et al. (1983) found that thaumatin when used as a flavour modifier and extender, and partial sweetener, was unlikely to be hazardous at the anticipated level of consumption. Thaumatin was readily digested prior to absorption in rats with no adverse effects resulting from its continuous administration to rats and dogs at dietary concentrations of 0%, 0.3%, 1.0% and 3.0% for 13 weeks. It was not teratogenic when administered orally to rats at 0, 200, 600 and 2,000 mg/kg body weight/day from day 6 to 15 of gestation and was without effect on the incidence of dominant lethal mutations when administered on five consecutive days to male mice at 200 and 2,000 mg/kg/day. The lack of mutagenic potential was confirmed in bacterial mutagenic assays with *Salmonella typhimurium* (strains TA1535, TA1537, TA1538, TA98 and TA100) and *Escherichia coli* WP2, at levels of addition of 0.05–50 mg/plate. In rats, thaumatin was found to be a weak sensitizer, comparable with egg albumen, when administered systemically but to be inactive when administered orally. Prick testing of laboratory personnel who had been intermittently exposed by inhalation to thaumatin for periods up to 7 years showed that 9.3% (13/140) responded positively to commercial thaumatin, while 30.7% were positive to *Dermatophagoides pteronyssinus* (house dust mite). Challenge tests in man did not demonstrate any oral sensitization.

The amounts of all three forms of sweet protein (To, TI, TII) increased during fruit maturation to reach a total of about 50 mg/g in mature aril tissue in *Thaumatococcus danielli* fruits from both the Ashanti and Kadjebe regions of Ghana (Mackenzie et al. 1985). Structural and immunological comparisons of the three thaumatin forms showed to be closely related to the other two forms which were known to differ at only five positions in their primary structure.

Aqueous extracts of the seed aril of *Thaumatococcus daniellii* were found to contain, in addition to the intensely sweet protein thaumatin, a cysteine protease designated thaumatopain (Cusack et al. 1991). Thaumatopain was found to be a monomeric protein of Mw 30,000. The protease strongly resembled papain in proteolytic activity, pH optima, susceptibility to inhibitors of cysteine proteases and in N-terminal sequence. Thaumatopain was found to be responsible for the cysteine protease activity previously attributed to thaumatin. Thaumatin was digested by thaumatopain at neutral to alkaline pH values. Thaumatin has no intrinsic proteolytic activity; the proteolytic activity in partially purified thaumatin preparations was attributable to a cysteine proteinase, thaumatopain, that couldbe separated from thaumatin by cation exchange chromatography (Stephen et al. 1991).

T. daniellii gel was found to contain residues of L-arabinose, D-xylose, D-glucuronic acid, and 4-O-methyl-D-glucuronic acid in the ratios 1.00:7.20:1.91:0.66, together with nitrogen (1%) and ash (3.1%) (Adesina and Higginbotham 1977). The ash-free gel contained 76% of pentose and 24% of uronic acid; 25% of the uronic acid occurred as the 4-O-methyl derivative.

The leaf, flower and fruit coat of *T. daniellii* were found to contain derivatives of flavonol and flavones (Adesina and Harborne 1978). These were absent in the rhizome and fruit parts – aril, gel and seed. No anthocyanins were found in the flower and a proanthocyanidin was detected in the fruit.

Antimicrobial Activity

The methanol extract of *Thaumatococcus danielli* leaves showed greater antimicrobial activity than the aqueous extract, against spoilage fungi in fresh white bread and 'eba' (a carbohydrate; >80% starch, indigenous staple food in southern Nigeria) (Adebayo and Kolawole 2010). The lowest minimum inhibitory concentration (MIC) of *T. danielli* was 25 mg/ml of the methanol extract against *Saccharomyces cerevisiae* and *Saccharomyces chevalieri* and its lowest minimum fungicidal concentration (MCC) was 50 mg/ml of the methanol extract against *S. cerevisiae* and *Penicillium* sp. In separate studies, Ojekale et al. (2007) reported that the leaf extracts contained alkaloids, tannins, saponins, anthraquinones, cardenolides and steroidal nucleus compounds. Antimicrobial screening of the leaf extract against selected microbes associated with food spoilage revealed a lack of activity against these microbes; viz; *Salmonella typhimurium, Shigella dysentarieae, Shigella sp., Escherichia coli, Staphylococcus aureus, Streptococcus lactis, Leuconostoc sp., Pediococcus cerevisae, Bacillus cereus, Candida krusei, Candida albicans, Aspergillus niger, Aspergillus flavus and Trichoderma konigii.*

Traditional Medicinal Uses

In the Ivory Coast and Congo, the fruit is used in folk medicine as a laxative and the seed as an emetic and for pulmonary problems, the leaf and root sap are used as sedative and for treating insanity. Leaf sap is also used as antidote against venoms, stings and bites.

Other Uses

In West Africa, The leaves are used for thatching roofs and in wrapping foods. The leaves are popularly used as a wrapping material for different categories of food, such as: unprocessed e.g. meat, kola nuts; semi-processed, e.g. fermented locust beans, and processed (cooked) foods like cooked rice, beans, *moin moin*, *pap*, maize-meal, pounded yam etc. The leaf stalk is used for weaving mats, fancy bags, slippers and as tools and building materials. In Gabon, it is cultivated in villages as fetish plant (for magic, religious and superstitious beliefs).

Comments

It is propagated from seeds or division of the rhizomes.

Selected References

Abbiw D (1990) Useful plants of Ghana. West African uses of wild and cultivated plants, Intermediate Technology and The Royal Botanic Gardens, Kew, 337 pp

Adebayo GJ, Kolawole LA (2010) In vitro activity of *Thaumatococcus daniellii* and *Megaphrynium macrostachyum* against spoilage fungi of white bread and 'Eba', an indigenous staple food in Southern Nigeria. Afr J Microbiol Res 4(11):1076–1081

Adebisi-Adelani O, Adeoye IB, Olajide-Taiwo FB, Usman JM, Agbarevoh P, Oyedele OO (2010) Gender analysis of production, potentials and constraints of *Thaumatococcus danielli* in Ekiti State. Continental J Agric Sci 4:54–59

Adegoke EA, Akinsaya A, Naqui SHZ (1968) Studies of Nigerian medicinal plants I A preliminary survey of plant alkaloids. J W Africa Sci Assoc 13:13–33

Adesina SK, Harborne JB (1978) The occurrence and identification of flavonoids in *Thaumatococcus daniellii*. Planta Med 34:323–327

Adesina SK, Higginbotham JD (1977) Studies on a novel polysaccharide gel from the fruit of *Thaumatococcus daniellii* (Benth.). Carbohydr Res 59(2):517–524

Arowosoge OGE, Popoola L (2006) Economic analysis of *Thaumatococcus danielli* (Benn.) Benth (miraculous berry) in Ekiti State, Nigeria. Int J Food Agric Environ 4(1):264–269

Bouquet A (1969) Féticheurs et Médecines Traditionnelles du Congo (Brazzaville). Mém. O.R.S.T.O.M., 36, 282 pp

Bouquet A, Debray M (1974) *Plantes Médicinales* de la Côte d'Ivoire. Mémoires O.R.S.T.O.M., 32, Paris, 231 pp

Burkill HM (1998) Useful plants of West Tropical Africa, vol 4. Families M-R. Royal Botanic Gardens, Kew, 969 pp

Csurhes S, Edwards R (1998) Potential environmental weeds in Australia: candidate species for preventative control. Biodiversity Group, Environment Australia, Canberra, 208 pp

Cusack M, Stephen AG, Roy P, Beynon RJ (1991) Purification and characterization of thaumatopain, a cysteine protease from the arils of *Thaumatococcus daniellii*. Biochem J 274:231–236

de Vos AM, Hatada M, van der Wel H, Krabbendam H, Peerdeman AF, Kim SH (1985) Three-dimensional structure of thaumatin I, an intensely sweet protein. Proc Natl Acad Sci USA 82(5):1406–1409

Edens L, Heslinga L, Klok R, Ledeboer MNJ, Toonen MY, Visser C, Verrips CT (1982) Cloning of cDNA encoding the sweet-tasting plant protein thaumatin and its expression in *Esherichia coli*. Gene 18(1):1–12

Green C (1999) Thaumatin: a natural flavour ingredient. World Rev Nutr Diet 85:129–132

Higginbotham JD, Snodin DJ, Eaton KK, Daniel JW (1983) Safety evaluation of thaumatin (Talin protein). Food Chem Toxicol 21(6):815–823

Hutchinson J, Dalziel MD, Keay RWJ (1954) Flora of west tropical Africa, 2nd edn. Crown Agents, London

Kaneko R, Kitabatake N (2000) Structure–sweetness relationship in thaumatin: importance of lysine residues. Chem Senses 26(2):167–177

Korver O, Van Gorkom M, Van Der Wel H (1973) Spectrometric investigation of thaumatin I and thaumatin II two sweet tasting proteins from *Thaumatococcus-daniellii*. Eur J Biochem 35: 554–558

Ledeboer AM, Verrips CT, Dekker BM (1984) Cloning of the natural gene for the sweet-tasting plant protein thaumatin. Gene 30:23–32

Mackenzie A, Pridham JB, Saunders NA (1985) Changes in the sweet proteins (Thaumatins) in *Thaumatococcus daniellii* fruits during development. Phytochem 24(11):2503–2506

Most BH, Summerfield RJ, Boxall M (1978) Tropical plants with sweetening properties. 2. *Thaumatococcus daniellii*. Econ Bot 32:321–335

Ogata CM, Gordon PF, De Vos AM, Kim SH (1992) Crystal structure of a sweet tasting protein thaumatin I, at 1.65 Å resolution. J Mol Biol 228:893–908

Ojekale AB, Makinde SCO, Osileye O (2007) Phytochemistry and anti-microbial evaluation of *Thaumatococcus danielli*, Benn. (Benth.) leaves. Nig Food J 25(2):176–182

Onwueme IC, Onochie BE, Sofowora EA (1979) Cultivation of *Thaumatococcus daniellii* – the sweetener. World Crops 3:106–111

Stephen AG, Powls R, Beynon RJ (1991) The relationship between thaumatin, a sweet protein and thaumatopain, a cysteine protease, from the arils of *Thaumatococcus daniellii*. Biochem Soc Trans 19(3):297S

Van Der Wel H, Loeve K (1972) Isolation and characterization of thaumatin I and thaumatin II the sweet tasting proteins from *Thaumatococcus daniellii*. Eur J Biochem 31:221–225

Lansium domesticum 'Duku Group'

Scientific Name

Lansium domesticum **Correa (Duku Group)** (Plate 6).

Synonyms

Aglaia dookkoo Griff.

Family

Meliaceae

Common/English Names

Duku, Malaysian Duku, Very Sweet Duku

Vernacular Names

Burmese: Duku;
Chinese: Da Guo Lan Sa;
Danish: Sød Duku;
Dutch: Doekoe, Kokosan;
French: Duku Doux À Large Fruit;
German: Süßer Duku, Dukubaum;
Indonesia: Ceroring (Bali), Dookkoo (Java, Sumatra), Duki, Duku;
Italian: Duku Dolce;
Japanese: Duku;
Korean: Long Kong;
Philippines: Duku (Tagalog);
Portuguese: Arbol-Do-Duku, Duku-Doce;
Spanish: Arbol De Duku, Duku Dulce;
Thai: Duku (Doogoo), Langsat Waan;
Vietnamese: Bòn Bon.

Origin/Distribution

Duku is indigenous to southeast Asia, throughout the Malay archipelago from Peninsular Thailand, Malaysia and in Indonesia. It is widely cultivated in southern Peninsular Malaysia.

Agroecology

As described for *Lansium domesticum* (Langsat/Dokong Group).

Edible Plant Parts and Uses

The thick, sweet, weakly aromatic aril of ripe fruit is eaten fresh and can be preserved in syrup.

Botany

Duku shares similar botanical characteristic as the langsat and dokong (duku langsat, lonkong) especially in tree habit, leaf and flower characters. Duku trees are also small, evergreen, slender with orthotropic branches bearing bright-green,

T.K. Lim, *Edible Medicinal And Non-Medicinal Plants: Volume 3, Fruits*,
DOI 10.1007/978-94-007-2534-8_33,

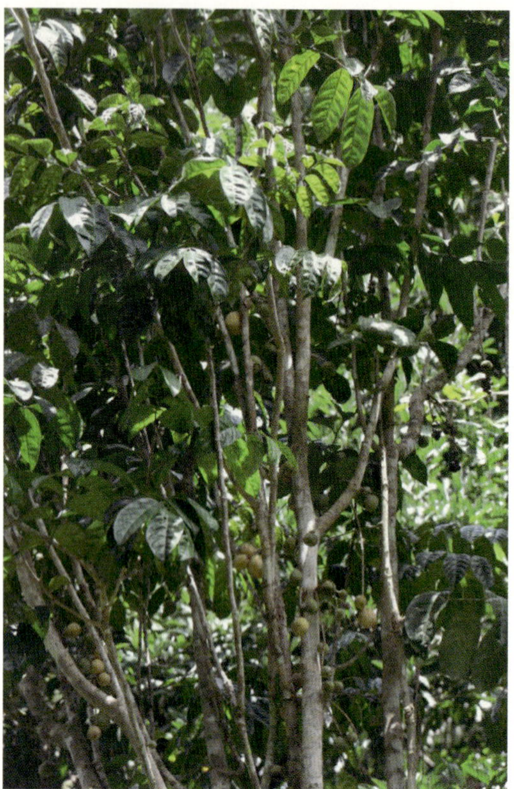

Plate 1 Duku tree, slender with orthotropic branches

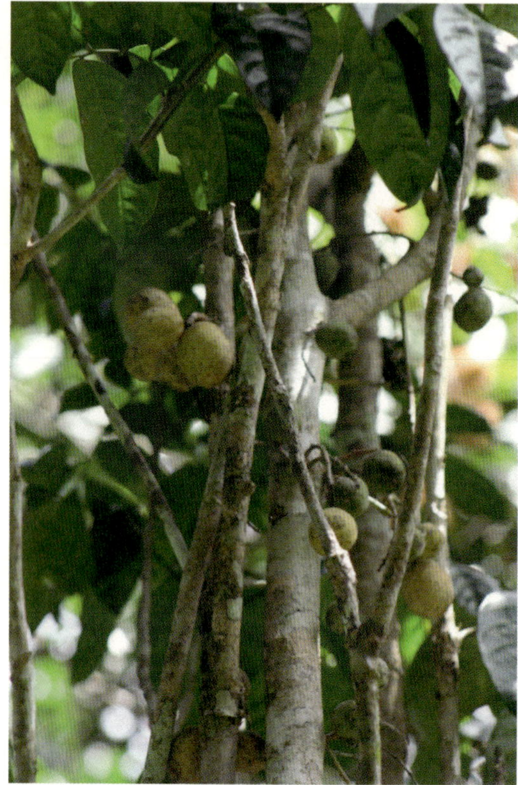

Plate 2 Large leaves and small clusters of 1–5 fruits

obovate or elliptic-oblong leaves (Plates 1 and 2). The inflorescence are usually solitary or in fascicles of 2–3 on the trunk or older branches, shorter with comparative less number of flowers which gives rise to fewer number of fruits 3–10 per cluster (Plates 2 and 3). Fruits are globose to subglobose, around 5–5.5 cm across, 5-loculed, with leathery, glabrous, non-lactiferous, 4–6 mm thick pericarp, dull green ripening to yellow, golden yellow or pale orangey yellow (Plates 2, 3, 4 and 5). The arils are also thick, juicy, translucent white (Plates 4 and 5), slightly aromatic with 0–3 green, bitter seeds. Fruit are also seedless as they are formed by apomixis (Prakash et al. 1977).

Nutritive/Medicinal Properties

Nutrient composition of *Lansium domesticum* (duku) fruit per 100 g edible portion was reported as energy 55 kcal, water 85.8 g, protein 0.9 g, fat 0.3 g, carbohydrate 12.1 g, fibre 0.4 g, ash 0.5 g, Ca 12 mg, P 30 mg, Fe 0.3 mg, Na 2 mg, K 142 mg, vitamin B1 0.10 mg, vitamin B2 0.14 mg, niacin 1.4 mg, vitamin C 1.7 mg (Tee et al. 1997).

The volatile constituents of both duku and langsat fruits were dominated by sesquiterpene hydrocarbons, the most abundant of which was germacrene-D (Wong et al. 1994).

For traditional medicinal uses see notes in *Lansium domesticum* 'Langsat/Dokong Group'.

Other Uses

As described for *Lansium domesticum* 'Langsat/Dokong Group'.

Comments

The local vernacular name, Duku, can be confusing, the 'Duku' of the Philippines and several types named duku in Indonesia in fact may appear

Plate 3 Dull green young fruit and bright golden yellow ripe duku fruit

Plate 4 Ripe, globose duku fruit

to belong to the duku-langsats. Similarly the langsep or pisitan of Java, which is usually equated with langsat, has in fact the thick skin of the duku.

Recently, Kiew et al. (2003) using amplified fragment length polymorphism (AFLP), confirmed that langsat, duku and duku-langsat belonged to a single species, *Lansium domesticum*. Also see notes under *Lansium domesticum* 'Langsat/ Dokong Group'.

Plate 5 Translucent white sweet aril and thick pericarp

Plate 6 Duku tree label

Selected References

Burkill IH (1966) A dictionary of the economic products of the Malay Peninsula. Revised reprint. 2 volumes. Ministry of Agriculture and Co-operatives, Kuala Lumpur. Vol 1 (A–H), pp 1–1240; Vol 2 (I–Z), pp 1241–2444

Corner EJH (1988) Wayside trees of Malaya, 3rd edn, 2 vols. The Malayan Nature Society, Kuala Lumpur, 774pp

Kiew R, Teo LL, Gan YY (2003) Assessment of the hybrid status of some Malesian plants using amplified fragment length polymorphism. Telopea 10(1): 225–233

Kostermans AJGH (1966) A monograph of Aglaia, sect. *Lansium* Kosterm (Meliaceae). Reinwardtia 7:221–282

Mabberley DJ (1985) Florae malesianae praecursores LXVII Meliaceae (Diverse Genera). Blumea 31:140–43

Mabberley DJ, Pannell CM (1989) Meliaceae. In: Ng FSP (ed) Tree flora of Malaysia, vol 4. Longman, Malaysia, pp 199–260

Molesworth Allen B (1967) Malayan fruits. An introduction to the cultivated species. Moore, Singapore, 245pp

Morton JF (1987) Fruits of warm climates. Florida Flair Books, Miami, pp 201–203

Porcher MH et al (1995–2020) Searchable world wide web multilingual multiscript plant name database. The University of Melbourne, Melbourne. http://www.plantnames.unimelb.edu.au/Sorting/Frontpage.html. Accessed December 2009

Prakash N, Lim AL, Manurung R (1977) Embryology of duku and langsat varieties of *Lansium domesticum*. Phytomorphology 27(1):50–59

Tee ES, Noor MI, Azudin MN, Idris K (1997) Nutrient composition of Malaysian foods, 4th edn. Institute for Medical Research, Kuala Lumpur, p 299

Wong KC, Wong SW, Siew SS, Tie DY (1994) Volatile constituents of the fruits of *Lansium domesticum* (duku and langsat) and *Baccaurea motleyana* (Muell. Arg.) Muell. Arg. (rambai). Flav Fragr J 9(6): 319–324

Yaacob O, Bamroongrugsa N (1992) *Lansium domesticum* Correa. In: Verheij EWM, Coronel RE (eds) Plant resources of South-East Asia No 2. Edible fruits and nuts. Prosea, Bogor, pp 186–190

Lansium domesticum 'Langsat-Lonkong Group'

Scientific Name

***Lansium domesticum* Correa 'Langsat-Longkong Group'**.

Synonyms

Synonyms listed applicable to all form or races of *Lansium domesticum*.

Aglaia aquea (Jack) Kosterm., *Aglaia domestica* (Correa) Pellegrin, *Aglaia dookoo* Griff., *Aglaia intricatoreticulata* Kosterm., *Aglaia merrillii* Elmer, *Aglaia sepalina* (Kosterm.) Kosterm., *Aglaia steenisii* Kosterm., *Lachanodendron domesticum* Nees, *Lansium aqueum* (Jack) M.Roem., *Lansium domesticum* var. *aqueum* Jack, *Lansium domesticum* var. *pubescens* Koord. & *Valet.*, *Lansium domesticum* var. *typicum* Backer, *Lansium javanicum* Koord. & Valet. ex Moll & Janss., *Lansium javanicum* M. Roem., *Lansium parasiticum* Sahni & Bennet, *Lansium parasiticum* var. *aqueum* (Jack) Sahni & Bennet, *Lansium pedicellatum* Kosterm., *Lansium sepalinum* Kosterm, *Taeniochlaena polyneura* Schellenberg.

Family

Meliaceae

Common/English Names

Dokong, Duku-Langsat, Lonkong (intermediate form), Langsat, Lansones, Lanzon (thin-skinned form), Duku (thick-skinned form, see preceding chapter)

Vernacular Names

Vernacular names listed may be applicable to all form or races of *Lansium domesticum*.

Burmese: Langsat;
Chinese: Lansa, Lan Sa Guo;
Costa Rica: Duki;
Cuba: Duku, Kokosan;
Danish: Langsat, Langsep;
Dutch: Doekoe, Langsep;
French: Lansium, Langsep;
German: Doko, Duku, Echter Lanzebaum, Langsta, Lansabaum, Lansibaum;
Honduras: Duki;
Indonesia: Dansot, Lansa, Lansat, Lansot, Lantat, Lasat, Lasot, Ranso (Alfurese, N. Sulawesi), Langsat (Acheh), Lansa (Bajo), Langsat (Bali), Babuna, Lonja (Baree), Langsat, Lansat, Lanset, Lantjat, Lassat (Batak), Dukem (Beak), Lasa (Bima), Langsa, Lese (Bugis), Londoto (Boeol, Manado, N. Sulawesi), Lalasat, Lasat, Lasate (Boeroe), Lasate, Lasete, Nasate (W. Ceram), Lakaolo, Lasato, Nasate (S. Ceram), Langeset,

T.K. Lim, *Edible Medicinal And Non-Medicinal Plants: Volume 3, Fruits*,
DOI 10.1007/978-94-007-2534-8_34, © Springer Science+Business Media B.V. 2012

Langsat, Lansat, Lanset, Lasat, Lawak, Lihat, Lehat, Losot, Ricat, Rihat, Rook (Dyak, Kalimantan), Langsat (Gajo), Bhulo (Gorontalo), Aha, Laha, Lasa, Lukama (N. Halmaheira), Lukem, (S. Halmaheira), Duku, Langsat, Langsep, Celoring (Java), Lansek (Kambang), Duku, Langsak, Langsat, Lasak, Pesen, Rarsak, Rasak, Ruku (Lampong), Doko, Duku, Langsep (Madura), Lasa (Makassar), Lasap Le (E. Makian), Duku (Malay-Jambi, Lingga, Singkep, Manado), Langsa, Lansa (Malay-Manado, Ambon), Langsat, Lansat (Malay-Manado), Bangkola, Lasa, Lase (Mandar), Langsek, Lansek (Minangkabau), Langsa (Mori), Lase (Nias), Seindefoer (Noefoer), Lasato, Lasatol (Oeilias), Lonsong (Sangir), Lasa (Sawoe), Lansat, Lantjat (Simaloer), Baunu, Lainsa, Lantja, Lasa (Soela), Dukuh, Kokosan, Pisitan (Sundanese), Lasa (Ternate), Lasa (Tidore);
Italian: Lansio, Lanzone;
Japanese: Ransa;
Korean: Lang Sat;
Malaysia: Dokong, Duku, Duku Hutan, Duku-Langsat, Langsat, Langsat Hutan, Longkong;
Philippines: Tubua (Bagobo), Lansones (Bikol), Boboa, Bulahan, Bukan (Bisaya), Buahan, Buan, Kalibongan (Manobo), Buahan (Sulu), Lansones (Tagalog);
Portuguese: Arbol-Do-Lanza;
Spanish: Arbol De Lanza, Lanzón;
Surinam: Duki;
Thai: Langsat, Lonkong;
Vietnamese: Bòn Bon.

See notes on botanical classification and vernacular names under Comments.

Origin/Distribution

The species is indigenous to southeast Asia, throughout the Malay archipelago from Peninsular Thailand, Malaysia, Indonesia to the island of Luzon in the Philippines. In Indonesia, langsat is found in Banyuwangi, Bangka, Palembang, West Kalimantan and in Sulawesi. Langsat is cultivated in a small scale in Vietnam, India, Sri Lanka, Hawaii, Australia, Surinam and Puerto Rico. It is common wild, naturalised and cultivated in these areas. Longkong is a major fruit crop in Thailand and is also widely cultivated in Peninsular Malaysia.

Agroecology

The species is strictly tropical and a lowland crop growing at altitudes from near sea level to 750 m. It thrives in a warm, humid and sheltered environment and is adverse to long dry spell. Areas with mean daily temperatures of 22–35°C and well-distributed annual rainfall of 2,000–3,500 mm is ideal for the crop. In its natural range it occurs as understorey of lowland primary forests. It is usually grown in mixed stands with companion tree crops and shading is beneficial especially during early establishment. The tree thrives on deep, rich, well-drained, sandy loam or other soils that are slightly acid to neutral and high in organic matter. Alkaline and calcareous soils are not suitable for the crop and it performs poorly on clayey soils and is intolerant of water-logging.

Edible Plant Parts and Uses

The succulent, delicious fruit aril is best eaten fresh, out of hand after peeling the skin. It can also be preserved in syrup or candied.

Botany

An evergreen, small, short-trunked, lactiferous sized tree, reaching heights of 10–20 m; cultivated trees are shorter 5–10 m high. The trunk is straight with a diameter of 30–40 cm, longitudinal furrows, greenish-brown and slightly scaly bark and finely pubescent, terete branchlets, crown slender, ovate to spreading. Branching are both orthotropic and plagiotropic (Plates 2 and 4). Leaves are alternate, imparipinnate, 30–50 cm long on a rachis thickened at the base and 15–30 cm long, with 5–7 alternate leaflets. Leaflets are alternate on 7–12 mm long petiolules, obovate or elliptic-oblong, apex shortly

Plate 1 Young and immature langsat fruit

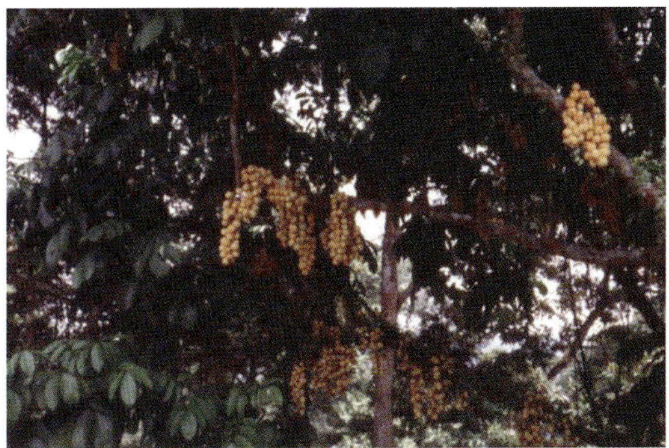

Plate 2 Leaves and plagiotrophic branches laden with langsat fruit clusters

acuminate, base narrowed, acute and slightly asymmetrical, margin entire, 12–25 cm long by 7–12.5 cm wide, pinnatinerved with midrib prominent below, coriaceous, shiny dark green above paler green below (Plates 2, 3 and 4). Inflorescence long, solitary or in fascicles of 2–10 on the trunk and older branches with numerous flowers. Flowers bisexual, sessile to shortly pediceled, small, yellowish in simple or branched racemes. Calyx is fleshy, cupular, 5-lobed, yellowish-green to light yellow; petals five erect, fleshy white or pale yellow, imbricate and glabrous, stamina tube subglobose, shorter than petals with one series of ten anthers; ovary globose, 4–5 loculed crowned by a short and thick, furrowed style and a broad orbicular, pubescent stigma. Fruit oblong-obovoid, ellipsoid to subglobose berry, 2–3.5 cm diameter, shortly pubescent, green turning to yellowish or yellowish-pale brown when ripe (Plates 1, 2, 3, 4, 5, 6 and 7), pericarp (skin) thin with latex (langsat) (Plates 1, 2 and 3) or without latex (dokong) (Plates 4, 5, 6 and 7), or thick (duku), five loculed containing translucent whitish aril

Plate 3 Ripe langsat fruit with drops of exuding white latex

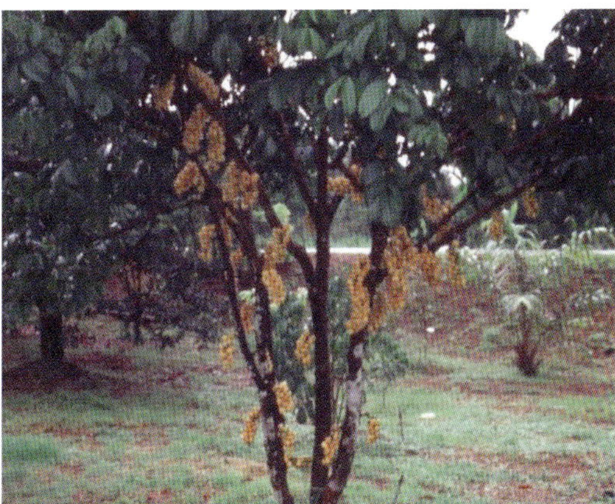

Plate 4 Trunk, orthotropic and plagiotropic branches laden with lonkong fruits

Plate 5 Closely packed cluster of ripe lonkong fruits

Plate 6 Thinned skinned and translucent aril of lonkong fruit

Plate 7 Non-lactiferous, thinned skinned lonkomg fruit with translucent, juicy aril

covering the seed. Fruits few to <10 (duku) to many 10–30 (langsat, longkong), borne in closely packed clusters. Usually, there are 1–3 seeds and the rest aborted or the fruit may have all seeds aborted or seedless as when formed apomitically (Prakash et al. 1977).

Langsat has smaller, usually 2–3 cm diameter, ellipsoid, yellow, thin-skinned (2–3 mm) fruits, which do not split but ooze a lot of latex when broken (Plates 1, 2 and 3).

Duku is larger 5–5.5 cm diameter, globose, golden yellow-brown, very thick-skinned 5–6 mm), but splits open and has no latex. It also has thicker and sweeter flesh.

Duku-langsat is intermediate in size, is round, brownish-yellow and has a skin that splits but which is much thinner than that of duku and sweet flesh. The tree is shorter and more prolific than the langsat or duku (Plates 4, 5, 6 and 7).

Nutritive/Medicinal Properties

Proximate nutrient composition of langsat/lonkong fruit per100g edible portion from analyses made in Thailand (Ministry of Public Health, Thailand 1987) was reported as moisture 82.9%, energy 66 cal, protein 0.9 g, fat 0.1 g, carbohydrate 15.3 g, fibre 0.3 g, Ca 5 mg, P 35 mg, Fe 0.7 mg, vitamin A 15 I.U., vitamin B1 0.08 mg, vitamin B2 0.02 mg, niacin 0.1 mg, vitamin C 46 mg. Nutrient composition of *Lansium domesticum* (duku) fruit per 100 g edible portion was also reported as: energy 34 kcal, water 90 g, protein 0.4 g fat 0.0 g, carbohydrate 8.2 g, fibre 0.9 g, ash 0.5 g, Ca 10 mg, P 20 mg, Fe 1.0 mg, Na 12 mg, K 230 mg, vitamin B1 0.05 mg, vitamin B2 0.02 mg, niacin 0.5 mg, vitamin C 13.4 mg (Tee et al. 1997).

Other Phytochemicals

More than 40 volatiles and static headspace of 36 volatiles of fresh longkong (original fresh pulp) were identified; 1,3,5-trioxane (33.8%), (E)-2-hexenal (23.8%), α-calacorene (6.9%), α-cubebene (5.4%), isoledene (4.7%), and copaene (4.0%) (Laohakunjit et al. 2007). In the heated fresh longkong, 1,3,5-trioxane was also found as the main compound (68.3%) among 23 volatile components followed by (E)-2-hexenal (5.1%), α-cubebene (2.3%), α-calacorene (2.3%), hexyl ester acetic acid (2.0%), δ-selinene (1.9%), and limonene (1.3%). The remaining volatiles (17) were found in a lower amount (0.24–1.2%).

Methylbutyrate, fruit, and sweet description odor was found in heated fresh pulp and in unheated pulp treated with salt. Although the relative peak of salted (NaCl) and salted and heated fruit comprised the total similar pattern and peak numbers of 35 and 23 volatiles, respectively, the main components were different (Laohakunjit et al. 2007). In fruits subjected to NaCl without heating, the major constituents were δ-selinene (17.7%), isoledene (13.9%), α-cubebene (12.5%), α-calacorene (10.8%), and epizoarene (9.9%). Epizoarene was not found in salted, heated samples. The total percent relative amount of volatiles found in NaCl-salted longkong was highest followed by normal fresh, fresh+heated, and NaCl+heated. Among the 45 total volatiles, acetic acid (sour aroma), (E)-2-hexenal (green, leaf aroma), limonene (citrus, lemon, orange aroma), 3-carene (green, lemon aroma), α-cubebene (citrus, fruit aroma), aromadendrene (wood aroma), isoledene (wood aroma), δ-selinene (herb aroma), β-panasinsene, α-calacorene (wood aroma), and α-cadinol (herb, wood aroma) were found in all four longkong fruit samples. The volatile constituents of both duku and langsat fruits were dominated by sesquiterpene hydrocarbons, the most abundant of which was germacrene-D (Wong et al. 1994).

Other phytochemical studies on *L. domesticum* reported the presence of tetranortriterpenoids from seeds (Nishizawa et al. 1985, 1988; Saewan et al. 2006; Fun et al. 2006; Mayanti et al. 2009, 2011), secoonocerane triterpenoid glycosides from fruit peel (Nishizawa et al. 1982, 1983), onoceranoid-type triterpenoids from fruit peels (Tanaka et al. 2002), onoceranoid-type triterpenoids from seeds (Saewan et al. 2006; Ragasa et al. 2006), onoceranoid-type triterpenoids from twigs (Dong et al. 2011), onoceranoid-type triterpenoids from bark (Tjokronegero et al. 2009; Supratman et al. 2010; Mayanti et al. 2011) onocerandiendione-type triterpenoids from seeds (Nishizawa et al. 1985) and cycloartanoid triterpene from leaves (Nishizawa et al. 1989). From the extract of the fruit peel, three secoonocerane triterpene glycosides named lansiosides A,B,C were isolated. (Nishizawa et al. 1982, 1983).

Three tetranortriterpenoids, dukulolides A, B and C were isolated as bitter principles from the seeds (Nishizawa et al. 1985). Three new tetranortriterpenoids, named dukunolides D, E, and F, were isolated from seeds of *Lansium domesticum* (Nishizawa et al. 1988). The seed extract of *L. domesticum* was found to be a rich source of limonoids (Saewan et al. 2006). Seco-Dukunolide F, methyl 2-[4-(3-furyl)-6b,10a-dihydroxy-3a,7,9,9-tetramethyl-6,10-dioxo-2,3,3a,6b,7,8,9,10,10a,11-decahydro-1aH,4H,6H-benzo[h][1]benzoxireno [3,2,1a-de]isochromen-8-yl]acetate, C27H32O9, a new tetranortriterpenoid was also isolated from the seeds (Fun et al. 2006). Recently, kokosanolide (8,14-secogammacera-7,14(27)-diene-3,21-dione) was isolated from the seed (Mayanti et al. 2009). In another study, five tetranorterpenoid, domesticulide A-E (1–5), were isolated from seed of *Lansium domesticum* together with 11 known triterpenoids (6–16) (Saewan et al. 2006). The seed extract of *L. domesticum* is a rich source of limonoids (Saewan et al. 2006). Six classes of the limonoids were isolated, including andirobin derivates (1–2), methyl angolensates (3, 4, 8, 9 and 10), mexicanolides (5–7), an azadiradione (11), onoceranoids (12–13) and dukunolides (14–16). Two tetranortriterpenoids, kokosanolide A (1) and C (2) were isolated from the seeds and three onoceranoid-type triterpenoids: kokosanolide B (3), 8,14-secogammacera-7,14-diene-3,21-dione (4) and a 1.5:0.5 mixture of 8,14-secogammacera-7,14(27)-diene-3,21-dione (5) and compound 4 were isolated from the bark of kokossan (*Lansium domesticum*) (Mayanti et al. 2011).

From the bark of *Lansium domesticum*, 8,14-secogammacera-7,14(27)-diene-3,21-dione-8,14-secogammacera-7,14-diene-3,21-dione was isolated (Tjokronegero et al. 2009) and more recently, kokosanolide B (14-Hydroxy-8,14-secogammacera-7-ene-3,21-dione) was isolated (Supratman et al. 2010).

Bioactive chemicals and associated pharmacological activities reported are presented below.

Antioxidant Activity

The fruit extract of *Lansium domesticum* was found to have antioxidant as evaluated by the DPPH free radical assay (Tilaar et al. 2007, 2008).

Antimicrobial Activity

The air-dried fruit peel of *Lansium domesticum* afforded five onoceroid triterpenes: 3 β-hydroxyonocera-8(26), 14-dien-21-one (1), α-y-onoceradienedione (2), lansiolic acid (3), lansionic acid (4) and lansioside C (5), while the air-dried seeds afforded 3 and germaorene D (6) (Ragasa et al. 2006). Antimicrobial tests on 1–6 gave the following results, compound 2 displayed high activity against *Pseudomonas aeruginosa*, while 1, 3, 4 and 5 had moderate activities against this microorganisms. 5 and 4 exerted moderate and low activities against *Bacillus subtilis*, respectively, while 3 and 5 had low activities against *Staphyloccus aureus*. All the compounds tested exhibited moderate activities against *Candida albicans* and *Aspergillus niger*, and low activities against *Trichophyton mentagrophytes*.

Skin Whitening Activity

The dried hydro-ethanol extract of *Lansium domesticum* fruit can be used as cosmetic (Tilaar et al. 2007, 2008). In-vitro studies showed that it had anti-tyrosinase activity. The dry extract re-dissolved in propylene glycol to obtain the final product, can be used as a skin care product for skin depigmentation and moisturizing. The recommended dose of the liquid extract is 2–5%. Trials on 30 female volunteers, aged 32–52 years old during 4 weeks showed that *Lansium* extract could significantly increase skin moisture content and decrease the skin melanin index. Dermatological safety evaluation using the Repeated Opened Patch Test (ROPT) showed that the extract did not cause any irritation or allergic skin reaction on all volunteers. The Singled Closed Patch Test (SCPT) showed that the extract at concentrations of 1% and 3% was safe while 5% concentration caused irritation in 1.9% of the volunteers.

Antileukotriene Activity

From the extract of the fruit peel, 3 secoonocerane triterpene glycosides named lansiosides A, B, C were isolated. (Nishizawa et al. 1982, 1983). Lansioside A was found to have a *N*-acetylglucosamine moiety as the sugar moiety. It effectively inhibited leukotriene D4-induced contraction of guinea pig ileum at 2.4 ppm concentration.

Cytotoxicity and Antitumour Activity

Three new onoceranoid triterpenes, lansionic acid, 3 β-hydroxyonocera-8(26),14-dien-21-one, and 21á-hydroxyonocera-8(26), 14-dien-3-one, together with lansic acid and methyl ester were isolated from the fruit peel of *Lansium domesticum*. Lansionic acid was concluded to be 3-dehydrolansionic acid. These triterpenoids exhibited moderate cytotoxicity against brine shrimp (*Artemia salina*) (Tanaka et al. 2002).

A new cycloartanoid triterpene, 3-oxo-24-cycloarten-21-oic acid, was isolated from leaves of *Lansium domesticum* (Nishizawa et al. 1989). Some of the natural product derivatives showed significant inhibitory activity of skin-tumour promotion on the basis of Epstein Barr virus activation.

Antibacterial Activity

Onoceranoid-type triterpenoids, represented by lamesticumin A (1), the ethanolysis product of lamesticumin A (2), lamesticumins B – F (3–7), lansic acid 3-ethyl ester (8), and ethyl lansiolate (9), along with four known analogues were isolated from the twigs *of Lansium domesticum* (Dong et al. 2011). Compounds 1–9 exhibited moderate antibacterial activity against Gram-positive bacteria.

Antiplasmodial Activity

The skin and aqueous leaf extracts of *Lansium domesticum* were found to reduce parasite populations of the drug sensitive strain (3D7) and the chloroquine-resistant strain (T9) of *Plasmodium falciparum* equally well (Yapp and Yap 2003). The skin extracts were also found to interrupt the life-cycle of the parasite. The data reported here indicated extracts of *L. domesticum* to be a potential source for compounds with activity towards chloroquine-resistant strains of *P. falciparum*. In Sabah, Malaysia the seed, leaf, and fruit skin extracts are used by indigenous tribes for treating malaria.

The seed extract of *L. domesticum* was found to be a rich source of limonoids (Saewan et al. 2006). Six classes of the limonoids were isolated, including andirobin derivates (1–2), methyl angolensates (3, 4, 8, 9 and 10), mexicanolides (5–7), an azadiradione (11), onoceranoids (12–13) and dukunolides (14–16). Compounds 2, 3, 4, 7, 8, 10, 11, and 15 exhibited antimalarial activity against *Plasmodium falciparum* with IC_{50}'s of 2.4–9.7 μg/ml (Saewan et al. 2006).

Traditional Medicinal Uses

In traditional medicine, the bark is deemed astringent and its decoction has been used for dysentery in Java, Borneo, and Malaya. A decoction of the bark and leaves has been employed in the treatment of dysentery, and the powdered bark is used as a poultice therapy for scorpion stings. Wood tar has been used for blackening teeth. Sap from the leaves has been used as eye drop for sore eyes. The juice of the bark and fruit skin is recorded as a Dyak arrow poison. The bitter seeds, ground and mixed with water, were given to children against worm. The seeds are also used as a febrifuge. In Peninsular Malaysia, the bitter seeds, crushed, were used to cure fevers among the Sakai. The resin from bark was prescribed for flatulence, for swellings and as an antispasmodic and was also deemed useful in the treatment of inflammation and colic of the gastro-intestinal tract. A tincture prepared from the dried rind, was also useful, as an antidiarrhoetic or anticolic.

Other Uses

The dried peel is burned in Java to give an aromatic smoke to repel mosquito and as incense in the rooms of sick people. The light brown wood is durable and used for house posts, tool handles and rafters.

Triterpenoid compounds from the seeds and bark were shown to have anti-feedant activity. Two tetranortriterpenoids, kokosanolide A (1) and C (2) were isolated from the seeds and three onoceranoid-type triterpenoids: kokosanolide B (3), 8,14-secogammacera-7,14-diene-3, 21-dione (4) and a 1.5:0.5 mixture of 8,14-secogammacera-7,14(27)-diene-3,21-dione (5) and compound 4 were isolated from the bark of kokossan (*Lansium domesticum*) (Mayanti et al. 2011). Compounds 1–5 exhibited moderate to strong antifeedant activity against the fourth instar larvae of *Epilachna vigintioctopunctata*.

Comments

A scan through past and recent literature revealed that *Lansium domesticum* Correa is a complex aggregate species of several varieties and plant forms and that its taxonomy may need to be revisited. Compounding this is the widespread confusion in the use of the numerous vernacular names. The same vernacular name being used for different varieties or races and plant types in different regions. Although many species have been ascribed to the genus *Lansium*, Mabberley (1985) recognized only three species, *Lansium membranaceum* Kosterm., *L. domesticum* Correa. agg. and *Lansium breviracemosum* Kosterm. In Peninsular Malaysia, the genus is represented by only one species, *Lansium domesticum* (Mabberley and Pannell 1989). In Java, two *Lansium* species have been recognised by Backer and van den Brink (1965): *L. domesticum* Correa, *L. humile* Hassk. and *L. domesticum* var. *pubescens* Koorders et Valeton was mentioned as a variety. Ridley (1922) mentioned *L. domesticum* var. *pubescens* which resemble more closely Kokosan in Java (Mabberley and Pannell 1989). Kostermans (1966) assigned duku as *Aglaia dokoo* and langsat as *Aglaia domesticum*. Two distinct plant types of *L. domesticum* were recognised by Corner (1988): langsat with small, sweet fleshed, and duku those with larger fruit, and seeds and thick pericarp. The duku-langsat (or duku Terengganu) is the intermediate type, generally regarded as superior to both duku and langsat (Mabberley 1985). Hassan (1994) recognised four different plant types: langsat, duku Trengganu (duku langsat), duku Johor, and dokong (longkong from Thailand). Ochse and Bakhuizen van den Brink (1931) divided *L. domesticum* into two varieties *L. domesticum* var. *typica* Backer and *L. domesticum* var. *pubescens* Kooders et Valeton and provided the following distinguishing features and vernacular names which compounded the complication and confusion.

a. var. *typica* – inflorescence rachises, young branchlets, under surface of leaves and calyx sparsely pubescent or sub-glabrous. Fruit – oblong-obvoid or ellipsoid, pericarp thin with little milky juice, seeds small, aril thick and smooth. Duku (Malay, Javanese, Madurese), Dukuh (Sundanese) doko (Madurese);
b. var. *pubescens* Koorders et Valeton- inflorescence rachises, young branchlets, under surface of leaves and calyx densely pubescent. Fruit subglobose, pericarp thick with copious milky juice, large seeds, sour and thin aril. Kokosan (Malay, Sundanese), Pijitan (Malay), Pisitan (Sundanese), Langsep (Javanese, Madurese), langsat, celoring (Javanese).

Using RAPD (random amplification of polymorphic DNA) analysis, 85 accessions were separated into three clusters one comprising 56 accessions which possessed thin-skinned fruit (mostly Dokong and Langsat), while the second had 28 accessions (mostly Duku-langsat, Duku Terengganu and Duku Johor) with thick fruit skin and the third comprising only one accession, namely Duku hutan (Song et al. 2000). They postulated that the four main types of *L. domesticum* investigated in the study could be different species of *Lansium* or sub-species of *L. domesticum*. However, botanists (Corner 1988; Mabberley 1985; Yaacob and Bamroongrugsa 1992; Bamroongrugsa 1994) held the view that there are two major plant forms and one intermediate form: duku form with thick pericarp, large, sweet, non-lactiferous fruit; dokong, duku-langsat intermediate form with thinner pericarp, medium-sized, sweet, non-lactiferous fruit; and langsat form with thin pericarp, small-to medium-sized, acid sweet to sweet, lactiferous fruit.

Recently, Kiew et al. (2003) using amplified fragment length polymorphism (AFLP), confirmed that langsat, duku and duku-langsat belonged to a single species. Five primer pairs were used resulting in 269 unambiguous bands. Ten bands were specific to duku and 20 bands were found in all the langsat and duku-langsat samples, while 78 bands were common to all three. Neither langsat nor duku-langsat had any unique bands. In addition, the genetic similarity between samples was very high: 91.5–100% for duku and 85.5–100% for langsat/duku-langsat. This can be ascribed to duku and langsat being apomictic. Their findings support the view held that all three are cultivars or races of a single species *L. domesticum*.

Selected References

Backer CA, van den Brink Bakhuizen RC Jr (1965) Flora of Java (spermatophytes only), vol 2. Wolters-Noordhoff, Groningen, 641 pp

Bamroongrugsa N (1994) Longkong: Another plant type of *Lansium domesticum* Corr. In: Prosiding Seminar 'Penaman Duku Terengganu, Dokong dan Salak'. MARDI-Jabatan Pertanian, pp 29–33

Bureau of Plant Industry (2005) Medicinal plants of the Philippines. Department of Agriculture Republic of Philippines. http://www.bpi.da.gov.ph/Publications/mp/mplants.html

Burkill IH (1966) A dictionary of the economic products of the Malay Peninsula. Revised reprint. 2 volumes. Ministry of Agriculture and Co-operatives, Kuala Lumpur. Vol 1 (A–H), pp 1–1240; Vol 2 (I–Z), pp 1241–2444

Burkill IH, Haniff M (1930) Malay village medicine, prescriptions collected. Gard Bull Straits Settlements 6:165–321

Corner EJH (1988) Wayside trees of Malaya, 3rd edn, 2 volumes. The Malayan Nature Society, Kuala Lumpur, 774pp

Dong SH, Zhang CR, Dong L, Wu Y, Yue JM (2011) Onoceranoid-type triterpenoids from *Lansium domesticum*. J Nat Prod 74(5):1042–1048

Fun H-K, Chantrapromma S, Boonnak N, Chaiyadej K, Chantrapromma K, Yu X-L (2006) *seco*-Dukunolide F: a new tetranortriterpenoid from *Lansium domesticum* Corr. Acta Cryst E62:o3725–o3727

Hassan MZS (1994) Pencaman langsat, duku Terengganu, duku Johor dan dokong (*Lansium domesticum* Corr.) melalui sifat-sifat tampang pokok. (Identification of duku Terengganu, duku Johor dan dokong (*Lansium domesticum* Corr.) through vegetative characters of the plants. In: Prosiding Seminar 'Penaman Duku Terengganu, Dokong dan Salak'. MARDI-Jabatan Pertanian, pp 103–111

Kiew R, Teo LL, Gan YY (2003) Assessment of the hybrid status of some Malesian plants using amplified fragment length polymorphism. Telopea 10(1):225–233

Kostermans AJGH (1966) A monograph of *Aglaia*, sect. *Lansium* Kosterm. (Meliaceae). Reinwardtia 7:221–282

Laohakunjit N, Kerdchoechuen O, Matta FB, Silva JL, Holmes WE (2007) Postharvest survey of volatile compounds in five tropical fruits using headspace-solid phase microextraction (HS-SPME). Hortscience 42(2):309–314

Mabberley DJ (1985) Florae malesianae praecursores LXVII Meliaceae (Diverse Genera). Blumea 31: 140–143

Mabberley DJ, Pannell CM (1989) Meliaceae. In: Ng FSP (ed) Tree flora of Malaysia, vol 4. Longman, Malaysia, pp 199–260

Mayanti T, Supratman U, Mukhtar MR, Awang K, Ng SW (2009) Kokosanolide from the seed of *Lansium domesticum* Corr. Acta Cryst E65, o750

Mayanti T, Tjokronegoro R, Supratman U, Mukhtar MR, Awang K, Hamid AA, Hadi A (2011) Antifeedant triterpenoids from the seeds and bark of *Lansium domesticum* cv Kokossan (Meliaceae). Molecules 16:2785–2795

Ministry of Public Health, Thailand (1987) Nutritional composition table of fruits in Thailand. Nutrition Division, Department of Health, Ministry of Public Heath, Thailand

Molesworth Allen B (1967) Malayan fruits. An Introduction to the Cultivated Species. Moore, Singapore, 245pp

Morton J (1987) Langsat. In: Morton JF (ed) Fruits of warm climates. Florida Flair Books, Miami, pp 201–203

Nishizawa M, Nishide H, Hayashi Y (1982) The structure of lansioside A: a novel triterpenes glycoside with amino-sugar from *Lansium domesticum*. Tetrahedron Lett 23:1349–1350

Nishizawa M, Nishide H, Kosela S, Hayashi Y (1983) Structure of lansiosides: biologically active new triterpene glycosides from *Lansium domesticum*. J Org Chem 48(24):4462–4466

Nishizawa M, Nademoto Y, Sastrapradja S, Shiro M, Hayashi Y (1985) Structure of dukunolides, bitter principles of *Lansium domesticum*. J Org Chem 50: 5487–5490

Nishizawa M, Nademoto Y, Sastrapradja S, Shiro M, Hayashi Y (1988) New tetranortriterpenoid from the seeds of *Lansium domesticum*. Phytochemistry 27:237–239

Nishizawa M, Emura M, Yamada H, Shiro M, Chairul, Hayashi Y, Tokuda HV (1989) Isolation of a new cycloartanoid triterpene from leaves of *Lansium domesticum* novel skin-tumor promotion inhibitors *Tetrah*. Letters 30(41):5615–5618

Ochse JJ, Bakhuizen van den Brink RC (1931) Fruits and fruitculture in the Dutch East Indies. G. Kolff & Co, Batavia-C, 180 pp

Porcher MH et al. (1995–2020) Searchable world wide web multilingual multiscript plant name database. Published by The University of Melbourne, Melbourne. http://www.plantnames.unimelb.edu.au/Sorting/Frontpage.html

Prakash N, Lim AL, Manurung R (1977) Embryology of duku and langsat varieties of *Lansium domesticum*. Phytomorphology 27(1):50–59

Ragasa CY, Labrador P, Rideout JA (2006) Antimicrobial terpenoids from *Lansium domesticum*. Philippine Agric Sci 89(1):101–105

Ridley HN (1922–1925) The flora of the Malay Peninsula. 5 volumes. Government of the Straits Settlements and Federated Malay States. L. Reeve & Co., London

Saewan N, Sutherland JD, Chantrapromma K (2006) Antimalarial tetranortriterpenoids from the seeds of *Lansium domesticum* Corr. Phytochemsitry 67(20): 2288–2293

Song BK, Clyde MM, Wickneswari R, Normah MN (2000) Genetic relatedness among *Lansium domesticum* accessions using RAPD markers. Ann Bot 86:299–307

Supratman U, Mayanti T, Awang K, Mukhtar MR, Ng SW (2010) 14-Hydroxy-8,14-secogammacera-7-ene-3,21-dione from the bark of *Lansium domesticum* Corr. Acta Cryst E66, o1621

Tanaka T, Ishibashi M, Fujimoto H, Okuyama E, Koyano T, Kowithayakorn T, Hoyashi M, Komiyama K (2002) New onoceranoid triterpene constituents from *Lansium domesticum*. J Nat Prod 65(11):1709–1711

Tee ES, Noor MI, Azudin MN, Idris K (1997) Nutrient composition of malaysian foods, 4th edn. Institute for Medical Research, Kuala Lumpur, 299pp

Tilaar M, Wong LW, Ranti AS, Wasitaatmadja, S, Suryaningsih M, Junardy FD, Maily (2007) In search of naturally derived whitening agent-pragmatic approach. In: Asian Societies of Cosmetic Scientists 8th conference ASCS 2007. Delivering Science to the Depths of Asian Skin, Singapore, 2007, pp 116–117

Tilaar M, Wong LW, Ranti AS, Wasitaatmadja SM, Suryaningsih, Junardy FD, Maily (2008) Review of *Lansium domesticum* Corrêa and its use in cosmetics. Bol Latinoam Caribe Plant Med Aromaticas 7(4): 183–189

Tjokronegero R, Mayanti T, Supratman U, Mukhtar MR, Ng SW (2009) 8,14-Secogammacera-7,14(27)-diene-3,21-dione-8,14-secogammacera-7,14-diene-3,21-dione (1.5/0.5) from the bark of *Lansium domesticum* Corr. Acta Cryst E65, o144

Wong KC, Wong SW, Siew SS, Tie DY (1994) Volatile constituents of the fruits of *Lansium domesticum* (duku and langsat) and *Baccaurea motleyana* (Muell. Arg.) Muell. Arg. (rambai). Flav Fragr J 9(6): 319–324

Yaacob O, Bamroongrugsa N (1992) *Lansium domesticum* Correa. In: Verheij EWM, Coronel RE (eds) Plant resources of south-east asia No 2. Edible fruits and nuts. Prosea, Bogor, pp 186–190

Yapp DTT, Yap SY (2003) *Lansium domesticum*: skin and leaf extracts of this fruit tree interrupt the lifecycle of *Plasmodium falciparum*, and are active towards a chloroquine-resistant strain of the parasite (T9) in vitro. J Ethnopharmacol 85(1):145–150

Sandoricum koetjape

Scientific Name

Sandoricum koetjape (Burm.F.) Merr.

Synonyms

Melia koetjape Burm.f., *Sandoricum indicum* Cav., *Sandoricum maingayi* Hiern, *Sandoricum nervosum* Blume, *Sandoricum nervosum* (Vahl) M.J. Roem., *Sandoricum vidalii* Merr., *Trichilia nervosa* Vahl.

Family

Meliaceae

Common/English Names

Kechapi, Lolly Fruit, Santol, Sentol, Wild Mangosteen

Vernacular Names

Brunei: Klampu;
Burmese: Thitto;
Chinese: Suan-Ming-Kou, Suan Ming Guo;
French: Faux Mangostan, Sandorique, Mangoustanier Sauvage;
German: Sandoribaum, Falsche Mangostane, Sandorie-Baum;
Guam: Santor, Wild Mangosteen;
Honduras: Santol;
India: Sayai, Sevai (Tamil), Sevamanu;
Indonesian: Kecapi, Ketuat, Sentul;
Khmer: Kôm Piing Riech;
Laotian: Tong, Toongz;
Malaysia: Klampu (Sarawak), Kecapi, Kelampu, Ranggu, Sentieh, Sentol, Setol, Sentul, Setul, Setui, Kechapi, Ketapi;
Philippines: Santor (Most Dialects), Katul (Sambali), Santul (Tagalog) Bagosantol, Biot, Magsantol, Malabobonau, Malarambo, Malasantol (Timber);
Sri Lanka: Donka (Sinhalese);
Thailand: Katon, Kra Thon, Sa Thon, Satawn;
Vietnamese: Sau Chua, Sau Tia, Sau Do.

Origin/Distribution

The santol is probably native to Indochina and Peninsular Malaysia, and has been introduced into India, Borneo, Indonesia, Mauritius, the Andaman Islands, and the Philippines where it has become naturalized. It is commonly cultivated throughout these regions and the fruits are seasonally abundant in the local markets. It has also been introduced into China, Taiwan, Australia and into a few locations in Central America and Southern Florida.

T.K. Lim, *Edible Medicinal And Non-Medicinal Plants: Volume 3, Fruits*,
DOI 10.1007/978-94-007-2534-8_35, © Springer Science+Business Media B.V. 2012

Agroecology

This species is found scattered in primary or sometimes secondary tropical rain forests below 1,000 m. It also occurs in dry as well as moist lowland dipterocarp forest. It can be grown in acid sandy soil and oolitic limestone. The tree is completely intolerant to frost.

Edible Plant Parts and Uses

The ripe fruit is usually consumed raw plain or with spices added. Both the rind, as well as the pulp which clings tightly to the seeds, is edible and can be eaten straight off the tree. With the seeds removed, the pulp is also cooked and candied or made into marmalade, jam or preserved in syrup. Santol marmalade in glass jars is exported from the Philippines to Oriental food dealers in the United States and probably elsewhere. Very ripe fruits are naturally vinous and are fermented with rice to make an alcoholic drink. Young fruits are candied in Malaysia by paring, removing the seeds, boiling in water, then boiling a second time with sugar. Santol seeds are inedible and may cause complications such as intestinal perforation if swallowed.

Botany

Santol is a deciduous, small to large tree, up to 50 m tall with a straight trunk, flaky or fissured, lenticillate, greyish to pale pinkish-brown bark which exude a milky latex when bruised. The tree bole is sometimes straight but often crooked or fluted, branchless for up to 18–21 m and with a trunk diameter up to 100 cm. Leaves are trifoliate, exstipulate and arranged spirally; leaflets are entire, elliptic to oblong-ovate, 20–25 cm long, blunt at the base and pointed at the apex, glossy, dark-green (Plates 5 and 6). Flowers occur in axillary stalked thyrse. Flowers are bisexual, 4–5 merous, greenish, yellowish, or pinkish-yellow; calyx truncate to shallowly lobed; petals free; staminal tube cylindrical, carrying 10 anthers; disk tubular; ovary superior, 4–5-locular with 2 ovules in each cell and style-head lobed. The fruit is a globose or oblate capsule, with prominent or shallow wrinkles extending a short distance from the base, 4.5–7.5 cm across, yellowish to brownish-golden, sometimes blushed with pink (Plates 1, 2, 3, 4, 5 and 6). The downy rind

Plate 1 Fruit of a greenish-yellow globose variety of Santol

Plate 2 Fruit of a brownish-golden globose variety of Santol

Plate 3 Fruit of pinkish golden oblate variety of Santol

Plate 4 Fruit still connected by the rachis of the stalked thryse

Plate 5 Santol fruits and leaves

may be thin or thick and contains a thin, milky juice. It is edible, as is the white, translucent, juicy, sweet, subacid or sour pulp. The pulp encloses 3–5 brown, inedible seeds 2 cm long, tightly clinging or sometimes free from the pulp.

Plate 6 Santol fruit and trifoliolate leaf

Nutritive/Medicinal Properties

Food nutrient composition of fruit of santol per 100 g edible portion is reported as: energy 57 kcal, moisture 84.5%, protein 0.4 g, fat 0.7 g, carbohydrates 13.9 g, dietary fibre 1.0 g, ash 0.5 g, Ca 9 mg, P 17 mg, Fe 1.2 mg, Na 3 mg, K 328 mg, B-carotene equivalent 5 mg, thiamin 0.05 mg, riboflavin 0.03 mg, niacin 0.09 mg, ascorbic acid 14 mg (Leung et al. 1972).

Nutrient composition of pickled santol fruit per 100 g edible portion is: energy 88 kcal, water 74.3 g, protein 0.6 g fat 0.1 g, carbohydrate 21.1 g, fibre 1.6 g, ash 2.3 g, Ca 7 mg, P 7 mg, Fe 0.8 mg, Na 340 mg, K 50 mg, carotenes 5 μg. retinol equivalent 1 μg, vitamin B1 0.01 mg, vitamin B2 0.04 mg, niacin 0.5 mg, vitamin C 9.3 mg (Tee et al. 1997).

Bryononic acid and two new ring-A secotriterpenoid acids, secobryononic acid and secoisobryononic acid were isolated from *Sandoricum koetjape* stem bark (Kosela et al. 1995). The seeds of *S. koetjape* contain limonoids (antifeedant compounds) (Powell et al. 1991). Two limonoids, sandoricin and 6-hydroxysandoricin were found to be effective limonoid antifeedants. Three trijugin-type limonoids were isolated from the leaves namely sandrapins A, B and C (Ismail et al. 2003b). Two new additional trijugin-type limonoids, sandrapins D and E, analogues of the previously reported sandrapins A-C, were isolated as minor components from the leaves of *Sandoricum koetjape* (Ito et al. 2004). Two new andirobin-type limonoids, named sandoripin A and sandoripin B were isolated from the leaves (Pancharoen et al. 2009).

Reported pharmacological properties of santol include the following:

Anticancer/Antiviral Activities

Studies found that some chemical extracts from santol stems and bark exhibited anti-cancer properties in-vitro. Japanese researchers found that the stem of *Sandoricum koetjape* contained triterpenes with anti-cancer activity (Kaneda et al. 1992). Triterpenes 3-oxo-olean-12-en-29-oic acid (a novel natural product) and katonic acid isolated from the stems demonstrated significant cytotoxic activity against cultured P-388 cells (ED_{50} values of 0.61 and 0.11 μg/ml, respectively). Significant, albeit less intense, cytotoxicity was also observed with a variety of cultured human cancer cells. Three bioactive compounds from *S. koetjape*, namely 3-oxo-olean-12-en-29-oic acid, katonic acid and, koetjapic acid were found to exhibit inhibitory activity against DNA polymerase β with IC_{50} values of 22.0, 36.0 and 20.0 μM respectively (Sun et al. 1999). Studies also reported two ichthyotoxic triterpenoids isolated from *Sandoricum koetjape* bark, koetjapic acid and 3-oxo-olean-12-en-29-oic acid together with non-toxic katonic acid exhibited remarkable inhibitory effect on Epstein-Barr virus early antigen (EBV-EA) activation induced by 12-O-tetradecanoylphorbol 13-acetate (TPA) (Ismail et al. 2003a). Of the triterpenoids active in-vitro, koetjapic acid appeared to be a promising cancer chemopreventive agent, since it significantly delayed tumour promotion in two-stage mouse skin carcinogenesis induced by 7,12-dimethylbenz(a)anthracene and promoted by TPA.

The n-hexane extract of *S. koetjape* exhibited cytotoxic and apoptotic properties on breast cancer cell lines (Aisha et al. 2009a). The n-hexane extract showed a dose dependent growth inhibition of all tested breast cell lines (MCF-7, MDA-MB-231, T47D and MCF-10A) with IC_{50} values between 44 and 48 μg/ml. At 100 μg/ml, the extract induced apoptotic cell death of MCF-7 by inducing activity of the effector caspases 3 and 7. The n-hexane extract of *S. koetjape* stem bark was also found to possess both anti-angiogenic and apoptotic properties on colon cancer cell lines (Aisha et al. 2009b). At 100 μg/ml, the extract showed 94% inhibition of the outgrowth of the blood vessels from the rat aorta rings. The extract also showed a dose dependent growth inhibition of all tested cell lines, IC_{50} values against HCT-116, human umbilical vein endothelial cell (HUVEC), CCD-18CO (human colonic fibroblasts) and HT-29 were 14, 23, 50 and 52 μg/ml, respectively. At 50 μg/ml, the extract potently induced apoptotic cell death of HCT-116 colon cancer cell line by inducing caspases 3 and 7 activity.

Antiinflammatory Activity

Several parts of the santol plant were found to have antiinflammatory effects (Rasadah et al. 2004). The compounds – 3-oxo-12-oleanen-29-oic acid and katonic acid from stems of *Sandoricum koetjape* were found to be the bioactive principles responsible for the antiinflammatory activity. The percentage of inhibition exhibited by 3-oxo-12-oleanen-29-oic acid was almost equivalent to indomethacin.

Traditional Medicinal Uses

The fruit, leaves, bark, roots, sap of the santol tree is variously used medicinally in traditional medicine. The preserved fruit pulp is employed medicinally as an astringent. The seed contains an amorphous bitter principle. The pounded leaves are sudorific when applied to the skin for skin infections or rashes. Leaves are used to make a decoction against diarrhea, fever and as a tonic after childbirth. In the Philippines, fresh leaves are placed on the body to cause sweating and a patient is bathed in a santol herbal tea to bring down fevers. A poultice of powdered bark is used to treat ringworm. Bark contains traces of a bitter principle, a toxic alkaloid and the bitter sandoricum acid. The aromatic roots are employed as an anti-diarrheic, anti-spasmodic, carminative, antiseptic, astringent, stomachic and are prescribed as a general tonic after childbirth. Bitter roots, bruised with vinegar and water, is a carminative; used for diarrhea and dysentery. Fresh or recently dried roots are swallowed for colic. In Malaysia, Traditionally, a decoction of the bark is used by Malays as a tonic after childbirth.

Other Uses

The santol tree makes an excellent shade tree. Besides being planted for its fruits, it is often planted for aesthetic purposes along avenues and in parks and also use as boundary or barrier or support tree. The tree is important in soil conservation, its roots form vesicular arbuscular mycorrhizae which improves the soil. The fruits are used as fish bait in Sarawak and the seeds has insecticidal effect. Sentol also provide valuable moderately hard and heavy timber which is useful for house-posts, interior construction, light-framing, barrels, cabinetmaking, carvings, ceiling, framing, furniture, hat-racks, posts, sculpture, boats, carts, sandals, clogs, butcher's blocks, household utensils, fences and carvings. The fragrant wood is used in perfumery. The dried heartwood yields 2 triterpenes – katonic acid and indicic acid–and an acidic resin. The bark yield tannin which is used in tanning fishing lines.

Extracts from santol seeds also have insecticidal properties. The seeds of *S. koetjape* contain limonoids (antifeedant compounds) (Powell et al. 1991). Two limonoids, sandoricin and 6-hydroxysandoricin were found to be effective limonoid antifeedants when fed to fall armyworm (*Spodoptera frugiperda*) or the European corn borer (*Ostrina nubilalis*) larvae. Reduced growth rates and increased times to pupation were evident at lower dose levels while significant mortality was noted at higher dose levels.

Comments

S. koetjape was formerly divided into 2 or 3 species based on the colour of the old leaves, however there appears to be no correlation with other characters and this distinction has been abandoned.

Selected References

Aisha AFA, Alrokayan SA, Abu-Salah KM, Darwis Y, Abdul Majid AMS (2009a) In vitro cytotoxic and apoptotic properties of the stem bark extract of *Sandoricum koetjape* on breast cancer cells. Int J Cancer Res 5:123–129

Aisha AFA, Sahib HB, Abu Salah KM, Darwis Y, Abdul Majid AMS (2009b) Cytotoxic and anti-angiogenic properties of the stem bark extract of *Sandoricum koetjape*. Int J Cancer Res 5:105–114

Backer CA, van den Bakhuizen Brink RC Jr (1965) Flora of Java (Spermatophytes only), vol 2. Wolters-Noordhoff, Groningen, 641 pp

Burkill IH (1966) A dictionary of the economic products of the Malay Peninsula. Revised reprint. 2 volumes. Ministry of Agriculture and Co-operatives, Kuala Lumpur, Malaysia. Vol. 1 (A–H), pp. 1–1240; Vol. 2 (I–Z), pp. 1241–2444

Council of Scientific and Industrial Research (1972) The wealth of India. A dictionary of Indian raw materials and industrial products (Raw materials 9). Publications and Information Directorate, New Delhi

Idris S (1998) *Sandoricum* Cav. In: Sosef MSM, Hong LT, Prawirohatmodjo S (eds) Plant resources of South-East Asia No 5(3), timber trees: lesser known timbers. Prosea Foundation, Bogor, pp 497–500

Ismail IS, Hideyuki I, Teruo M, Hiroshi H, Fumio E, Harukuni T, Hoyoku N, Takashi Y (2003a) Ichthyotoxic and anticarcinogenic effects of triterpenoids from *Sandoricum koetjape* bark. Biol Pharm Bull 26(9):1351–1353

Ismail IS, Ito H, Hantano T, Taniguchi S, Yoshida T (2003b) Modified limonoids from the leaves of *Sandoricum koetjape*. Phytochemistry 64(8): 1345–1349

Ito H, Hatano T, Taniguchi S, Yoshida T (2004) Two new analogues of trijugin-type limonoids from the leaves of *Sandoricum koetjape*. Chem Pharm Bull (Tokyo) 52(9):1145–1147

Kaneda N, Pezzuto JM, Kinghorn AD, Farnsworth NR, Santisuk T, Tuchinda P, Udchachon J, Reutrakul V (1992) Plant anticancer agents, L. cytotoxic triterpenes from *Sandoricum koetjape* stems. J Nat Prod 55: 654–659

Kosela S, Yulizar Y, Chairul TM, Askawa Y (1995) Secomultiflorane-type triterpenoid acids from stem bark of *Sandoricum koetjape*. Phytochemistry 38(3):691–694

Leung WTW, Butrum RR, Huang Chang F, Narayana Rao M, Polacchi W (1972) Food composition table for use in East Asia. FAO, Rome, 347 pp

Molesworth Allen B (1967) Malayan fruits. An introduction to the cultivated species. Moore, Singapore, 245 pp

Morton J (1987) Santol. In: Fruits of warm climates. Florida Flair Books, Miami, pp 199–201

Pancharoen O, Pipatanapatikarn A, Taylor WC, Bansiddhi J (2009) Two new limonoids from the leaves of *Sandoricum koetjape*. Nat Prod Res 23(1):10–16

Powell RG, Mikolajczak KL, Zilkowski BW, Mantus EK, Cherry D, Clardy J (1991) Limonoid antifeedants from seed of *Sandoricum koetjape*. J Nat Prod 54(1): 241–246

Rasadah MA, Khozirah S, Aznie AA, Nik MM (2004) Anti-inflammatory agents from *Sandoricum koetjape* Merr. Phytomedicine 11(2–3):261–263

Sotto RC (1992) *Sandoricum koetjape* (Burm.F.) Merr. In: Verheij EWM, Coronel RE (eds) Plant resources of South-East Asia, No. 2. Edible fruits and nuts. Prosea Foundation, Bogor, pp 284–287

Sun DA, Starck SR, Locke EP, Hecht SM (1999) DNA polymerase β inhibitors from *Sandoricum koetjape*. J Nat Prod 62(8):1110–1113

Tee ES, Noor MI, Azudin MN, Idris K (1997) Nutrient composition of Malaysian foods, 4th edn. Institute for Medical Research, Kuala Lumpur, 299 pp

Usher G (1974) A dictionary of plants used by man. Constable, London, 619 pp

Artocarpus hypargyreus

Scientific Name

***Artocarpus hypargyreus* Hance ex Benth** (Plate 3).

Synonyms

None

Family

Moraceae

Common/English Names

Kwai Muk, Silver-back Artocarpus, Sweet Artocarpus

Vernacular Names

China: Bai Gui Mu, Pai Kuei Mu.

Origin/Distribution

Kwai Muk is native to Sothern China, in Hong Kong, Fujian, Guangdong, Guangxi, Hainan, South Hunan, Jiangxi and southeast Yunnan.

Agroecology

In its native range, Kwai Muk occurs in broad-leaved, evergreen forests at elevations of 100–1,700 m. The tree is frost sensitive. Young trees are injured by brief periods of low temperature −2.2°C to 1.2°C, while mature tree will tolerate drop in temperatures down to −3°C to–4°C as experienced in Florida. It will grow on most soils provided they are well-drained, but thrives best in mildly acid sandy soils. It will grow in calcareous soils but do suffer from chlorosis associated with iron, manganese and zinc deficiencies. It grows in full sun and partial shade.

Edible Plant Parts and Uses

Ripe fruit has an excellent flavour and is eaten fresh or preserved with salt or sugar syrup. The fruit can be dried, dried fruit still retains the good texture and flavour.

Botany

An evergreen, perennial tree, 10–25 m tall with d.b.h of 40 cm and a dense rounded crown. Young twigs are greyish and puberulent and stipules caduceus and linear. Leaves , alternate, elliptic to elliptic-ovate, 8–15 × 4–7 cm, leathery, acuminate

T.K. Lim, *Edible Medicinal And Non-Medicinal Plants: Volume 3, Fruits*,
DOI 10.1007/978-94-007-2534-8_36, © Springer Science+Business Media B.V. 2012

Plate 1 Ripening Kwai Muk fruit and foliage

Plate 2 Red-fleshed Kwai Muk fruit and leaves with 6–7 secondary veins

tip base cuneate, glabrous and entire margin, veins conspicuous with 6–7 lateral veins, glossy dark green above, duller green below (Plates 1 and 2), stem and leaves exude white latex when bruised. Inflorescences axillary, solitary. Male inflorescences are ellipsoid to obovoid, 1.5–2 × 1–1.5 cm; bracts shield-shaped. Male flowers have 4 spatulate, densely pubescent lobes adnate to bracts and ellipsoid anther. Female inflorescence develops into a fruiting syncarp. Syncarp is pale green turning to pale yellow to golden yellow, globose to subglobose, 3–4 cm cross, brown pubescent and papillate; peduncle 3–5 cm, shortly pubescent (Plates 1 and 2). Seeds 1–7 embedded in orange-red to red, soft, pleasantly subacid pulp (Plate 2).

Nutritive/Medicinal Properties

No information has been published on its nutritive and medicinal values.

Plate 3 Tree label

Fourteen compounds were isolated from the ethanol extract of Kai Muk root bark: 3β-acetoxy-lupenol, docosanoic acid, pentacosanoic acid, tricosanoic acid, lupeol, methyl betulinate, palmitic acid, β-sitosterol, betulin, artonin A, daucosterol, oxyresveratrol [1 –(2,4-dihydroxyphenyl)-2–(3,5-dihydroxyphenyl) ethylene], (+)- catechin, and(+)- afzelechin-3-O-α-L-rhamnopyranoside (Chen et al. 2007). Compounds 1, 5, 6 and 9 belonged to a class of lupine alkyl sapogenins.

Artocarpus hypargyreas lectin (AHL) was purified from the seeds of *Artocarpus hypargyreus* (Zhou et al. 1995). The lectin was found to consist of two kinds of subunits with molecular weights of 15,000 and 19,000 respectively. The lectin agglutinated erythrocytes of the animals and human beings of A, B, O and AB blood groups. The lectin was sensitive to heat, but stable to changes of pH within the range of 4.5–9.5. AHL (*Artocarpus hypargyreus* lectin) was found to bind to two kinds of glycoproteins, pepsin and Con A. (Wu et al. 2000). The affinity of AHL for pepsin was very high. AHL could also bind to hyaluronic acid, β lactoglobulin, and bovine serum albumin. The combination of AHL with pepsin, hyaluronic acid, β lactoglobulin, or bovine serum albumin was strongly inhibited by Me Gal, and also inhibited by Me Man, D Gal, and Raf. The combination of AHL with Con A was strongly inhibited by Me Man, secondly by Me Glc, D Man, D Fru, and D Glc. The results suggested AHL to contain O glycoside linkage binding site.

Other Uses

The tree is small and attractive and lends itself well to landscaping on small urban properties. A minor timber species.

Comments

The species is vulnerable and is threatened by habitat loss.

Selected References

Campbell CW (1984) The Kwai Muk, a tropical fruit tree for southern Florida. Proc Fla State Hort Soc 97:318–319

Campbell CW, Popenoe J, Ozaki HY (1962) Adaptation trials of tropical and subtropical fruits on sandy soils in Broward County, Florida. Proc Fla State Hort Soc 75:361–363

Campbell CW, Knight RJ Jr, Zareski NL (1977) Freeze damage to tropical fruits in southern Florida in 1977. Proc Fla State Hort Soc 90:254–257

Chen LM, Xie P, Xiao QQ, Liu QH, Luo R, Ouyang S, Wu HL (2007) Chemical constituents of *Artocarpus hypargyreus*. Chin Tradit Herb Drugs 38(6):815–835 (in Chinese)

Hu SY (2005) Food plants of China. The Chinese University Press, Hong Kong, 844 pp

Jarrett FM (1960) Studies in *Artocarpus* and allied genera, IV. A revision of *Artocarpus* subgenus *Pseudojaca*. J Arnold Arbor 41:132–133

Ledin RB (1957) Tropical and subtropical fruits in Florida (other than *Citrus*). Econ Bot 11:349–376

Morton JF (1987) In: Morton JF (ed) Fruits of warm climates. Florida Flair Books, Miami, 505 pp

Mowry H, Toy LR, Wolfe HS (1958) Miscellaneous tropical and subtropical Florida fruits. Fla Agric Ext Serv Bull 156A:26–28

Sun W (1998) *Artocarpus hypargyreus*. In: IUCN 2011. IUCN red list of threatened species. Version 2011.1. <www.iucnredlist.org>

Wu YS, Zhang H, Zhou SF, Deng Y, Zhou DY, Lin WZ (2000) Characterization of *Artocarpus hypargyreus* lectin interacting with glycoproteins. Chin J Biochem Mol Biol 16(2):210–214 (in Chinese)

Wu Z, Zhou ZK, Gilbert MG (2003) Moraceae Link. In: Wu ZY, Raven PH, Hong DY (eds) Flora of China. Vol. 5 (Ulmaceae through Basellaceae). Science Press/Missouri Botanical Garden Press, Beijing/St. Louis

Zhou DY, Yang EB, Deng Y, Zhou SF, Lian XN (1995) Purification and characterization of lectin from the seeds of *Artocarpus hypargyreus* Hance. Acta Biochim Biophys Sin 27(1):61–66 (in Chinese)

Artocarpus altilis

Scientific Name

Artocarpus altilis **(Parkinson) Fosberg**

Synonyms

Artocarpus communis J. R. Forster & G. Forster, *Artocarpus incisa* L.f., *Artocarpus incisus* (Thunberg) Linnaeus f., *Artocarpus laevis* Hassk., *Artocarpus leeuwenii* Diels, *Artocarpus rima* Blanco, *Communis incisa, Radermachia incisa* Thunberg, *Saccus communis* Kuntze, *Sitodium altile* Parkinson, *Sitodium incisum* Thunb.

Family

Moraceae

Common/English Names

Breadfruit, Breadnut, Breadnut Tree

Vernacular Names

Banaban: Te Mai;
Chinese: Mian Bao Shu, Mian Bao Guo;
Cook Island: Kuru;
Danish: Brødfrugt, Brødfrugttrae;
Dutch: Broodvrucht (Fruit), Broodboom (Tree);
Eastonian: Hõlmine Leivapuu;
Fijian: Uto, Uto Sori, Kulu;
French: Arbre A Pain, A Pain, Chataignier, Châtaignier De Malabar;
German: Brotfrucht, Brotfruchtbaum, Echter Brotfruchtbaum;
Guatemala: Mazapan (Seedless), Castana (With Seeds);
Hawaii: 'Ulu;
Honduras: Mazapan (Seedless), Castana;
India: Bakri-Chajhar, Kathal, Khanun (Hindu), Gujjeki (Kannada), Jivi Kadgi (Konkani), Nirphanas Vilayatiphanas (Marathi), Nagadamani (Sanskrit);
Indonesia: Sukun, Kulur, Kelur, Timbul (Bali);
Italian: Artocarpo, Albero Del Pane;
Khmer: Sakee, Khnuaor-Samloo;
Kiribati: Mei, Mai;
Kosrae: Mos;
Malaysia: Sukun, Kuror, Kelor;
Marquesas: Mei, Mai
Mariana Islands: Lemae;
Marshall Islands: Mei, Mai;
Mexico: Castano De Malabar (Yucatan);
Micronesia: Mei, Mai;
Palau: Meduu;
Papiamento: Frut'e Pan;
Papua New Guinea: Kapiak;
Peru: Marure;
Philippines: Rimas (Tagalog);
Portuguese: Fruta-Pão, Pão De Massa;
Puerto Rico: Panapen (Seedless), Pana De Pepitas (With Seeds);
Rotoman:' Ulu;
Samoan: Ulu;

T.K. Lim, *Edible Medicinal And Non-Medicinal Plants: Volume 3, Fruits*,
DOI 10.1007/978-94-007-2534-8_37, © Springer Science+Business Media B.V. 2012

Solomon Islands: Bia, Bulo, Nimbalu;
Spanish: Árbol Del Pan, Arbor De Pan, Fruta De Pan (Fruit), Pan De Pobre, Pana, Pana De Pepita, Panapén;
Vanuatu: Beta;
Society Islands: Uru;
Swedish: Brödfruktträd;
Taiwan: Luo Mi Shu, Mian Bao Shu, Mian Bao Guo Shu;
Thailand: Sa Ke, Khanun-Samphor;
Tongan: Mei, Mai;
Tuvalu: 'Ulu;
Venezuela: Pan De Ano, Pan De Todo El Ano, Pan De Palo, Pan De Name, Topan, Tupan;
Vietnam: Sakê.

Origin/Distribution

The breadfruit is believed to be native to a vast area extending from the Indo-Malayan Archipelago the Philippines and the Moluccas through New Guinea to Western Micronesia. At least two species (*A. camansi* and *A. mariannensis*) and at least two different events (vegetative propagation coupled with human selection in Melanesia and Polynesia, and introgressive hybridization in Micronesia) were involved in the origins of breadfruit (Zerega et al. 2005). Most Melanesian and Polynesian cultivars appear to have arisen over generations of vegetative propagation and selection from *A. camansi*. In contrast, most Micronesian breadfruit cultivars appear to be the result of hybridization between *A. camansi*-derived breadfruit and *A. mariannensis*. Breadfruit widely planted in tropical Asia and has became a popular staple food crop and abundantly planted in garden areas, mature fallow forests, village tree groves, and in home gardens on both high islands and atolls throughout Melanesia, Polynesia, and Micronesia.

Agroecology

The breadfruit is tropical in its ecological requirement. It can be found growing from near sea-level to 1,550 m altitude in areas with mean annual temperatures of 15–40°C and mean annual rainfall from 1,000 to 3,500 mm and relative humidity of 70–80%. The latitudinal limits are approximately 17° N and S extending to both tropics of Cancer and Capricorn.

Breadfruit is adaptable to a wide range of soil types from deep, fertile, well-drained soil – sands, sandy-loams, loams, sandy clay loams to sandy coralline soils near the sea and on coralline limestone. In New Guinea, the breadfruit tree occurs wild along waterways and on the margins of forests in the flood plain, and often in freshwater swamps. Breadfruit tolerates brief periods of water-logging and sprout from roots after a small fire. It does best at near neutral 6–8 pH. It is sun-loving, though young trees tolerate 20–50% partial shading. The tree is salt tolerant, relatively drought tolerant for a few months but is cold sensitive (minimum temperature tolerated 10°C).

Edible Plant Parts and Uses

Breadfruit is a staple food in Polynesia and Micronesia, and as a supplementary staple in most of Melanesia. Immature and ripe fruit, seeds, young leaves and ripe blossoms eaten. Immature, half-ripe and ripe fruit and seeds from ripe fruits are eaten after boiling, roasting, baking or frying. Cooked or raw fruit can be preserved in pits or by sun-drying. Breadfruit can be commercially dehydrated by tunnel drying or freeze-drying and the waste from these processes constitutes a highly-digestible stock feed.

When still firm, breadfruit can be boiled, baked, fried, sautéed, sliced and stir-fried. It can be diced and added to a wide variety of main dishes as is used for potato, yam, taro, sweet potato. It can be made into chips, patties, salads, soups, curries, stews, casseroles, chowder, puddings, buns, breads, dessert and other savoury dishes. When half ripe it can be baked or lightly fried in oil. Whole fruits can be cooked in an open fire, then cored and filled with other foods such as coconut milk, sugar and butter, cooked meats, or other fruits. A soft ripe fruit is used in pies, cakes, biscuits, bread, puddings or other dessert. The pulp scraped from soft, ripe breadfruits is combined with coconut milk (not coconut water), salt and sugar and baked to make a pudding. Breadfruit can also be dried, made into flour by pounding or by grinding or frozen to preserve if for latter use. Breadfruit can be fermented by burial in layers of leaves in a pit, fermented fruit, mixed with coconut cream and baked into sour bread. Fermented breadfruit mash goes by many names such as *mahr*, *ma*, *masi*, *furo*, and *bwiru*, among others. Some popular items are breadfruit fritters consisting of mature fruit, egg, milk powder, chopped onion, bread fruit and fish salad, roast chicken and breadfruit stuffing, breadfruit in beef stew. Another common product is a mixture of cooked or fermented breadfruit mash mixed with coconut milk and baked in banana leaves. The steamed fruit is sometimes sliced, rolled in flour and fried in deep fat. Breadfruit is also candied, or sometimes prepared as a sweet pickle.

In Malaysia, firm-ripe fruits are peeled, sliced and fried in syrup or palm sugar until it is crisp and brown. Filipinos enjoy the cooked fruit with coconut and sugar. In Hawaii, under-ripe fruits are diced, boiled, and served with butter and sugar, or salt and pepper, or diced and cooked with other vegetables, bacon and milk as a chowder. In the Bahamas, breadfruit soup is made by boiling under-ripe chunks of breadfruit in water with cooked salt pork, chopped onion, white pepper and salt, to which is then added milk and butter, and a dash of sherry. The dried fruit has been made into flour and improved methods have been explored in Barbados and Brazil breadfruit flour is used in combination wheat flour in bread making. Breadfruit flour is much richer than wheat flour in lysine and other essential amino acids. In Jamaica, the flour is boiled, sweetened, and eaten as porridge for breakfast. Soft or overripe breadfruit is best for making chips and these are being manufactured commercially in Trinidad and Barbados. Some breadfruit is canned in Dominica and Trinidad for shipment to London and New York. In Hawaii, breadfruit can substitute for taro in poi, resulting in "breadfruit poi" called *poi 'ulu*. In Puerto Rico, it is called *pana* or *panen* which is served with a combination of sauteed *bacalao* (salted cod fish), olive oil and onions. In the Dominican Republic, breadfruit bread is called *buen pan*.

Breadfruit seeds are cooked in salted water, roasted with salt over a fire or in hot coals or baked before eating. In West Africa, they are sometimes made into a puree. In Costa Rica, the cooked seeds are sold by street vendors. In Jamaica, Puerto Rico and the South Pacific, fallen male flower spikes are boiled, peeled and eaten as vegetables or are candied by recooking, for 2–3 h, in syrup; then rolled in powdered sugar and sun-dried. Young leaves are cooked with coconut cream and salt.

Botany

A large, branch, evergreen canopy perennial tree often growing to 30 m high with a clear trunk to 6 m, a cylindrical bole diameter of 60 cm and rough greyish-brown bark. All parts of the plant yield a sticky latex. Leaves are clustered at end of branches and arranged spirally up the branchlet, on 8–12 cm petiole. Leaf blade 12–59 cm long by 10–47 cm wide, but juvenile leaves often larger, usually deeply pinnately lobed (Plates 1 and 2) with up to 13 lobes cut from 1/3 to 4/5 of the way to midrib, rarely nearly entire with a praemorse apex, varying in size and shape on the same tree, thickly coriaceous, below pale green, above dark green and glossy, glabrous, to moderately pubescent, juvenile leaves may be densely pubescent, with pale or colourless, rough-walled hairs on midrib, adaxial and abaxial blade. Inflorescence axillary and solitary, flowers on an unbranched axis (male inflorescences drooping, cylindrical to club-shaped) or flowers arising from a single point (female inflorescences stiffly upright, globose or cylindrical) (Plate 2). Male flowers with tubular perianth apically 2-lobed, pubescent, lobes lanceolate; anther elliptic. Female flowers with tubular perianth, ovoid ovary and along, apically 2-branched style. Fruiting syncarp indehiscent globose, ovoid to cylindrical, 9–29 cm long by 6–20 cm wide (Plates 1, 2, 3 and 4); surface colour typically green turning yellowish-brown when ripe, rarely pinkish; surface typically flat, especially in seedless cultivars with anthocarp

Plate 1 Leaves, inflorescence and very young bread fruit

Plate 2 Breadfruit and leaves

Plate 3 Close-up of globose breadfruit

apices rounded and barely protruding, but sometimes shortly echinate, especially in fertile cultivars with conical, flexuous anthocarp apices up to 5 mm at the base and 3–5 mm long; flesh creamy white to pale yellow, dense due to fusion between medial portions of adjacent perianths; frequently seedless with tiny aborted ovules but some cultivars with few to several developed, typically ellipsoid, oblong, or reniform achenes (seeds) with a hard, dull light brown or shiny dark brown wall (Plates 5 and 6). The fruit is borne singly or in clusters of two or three at the branch tips and are stalked (2.5–12.5 cm long).

Nutritive/Medicinal Properties

Analyses carried out in the United States (U.S. Department of Agriculture and Agricultural Research 2010) reported that raw, mature breadfruit (excluding the core 9% and skin 13%) had the following nutrient composition (per 100 g value): water 70.65 g, energy 103 kcal (431 kJ), protein 1.07 g, total lipid 0.23 g, ash 0.93 g, carbohydrates 27.12 g, total dietary fibre 4.9 g, total sugars 11.00 g, Ca 17 mg, Fe 0.54 mg, Mg 25 mg, P 30 mg, K 490 mg, Na 2 mg, Zn 0.12 mg, Cu 0.084 mg, Mn 0.060 mg, Se 0.6 μg, vitamin C 29 mg, thiamine 0.110 mg, riboflavin 0.030 mg, niacin 0.900 mg, pantothenic acid 0.457 mg, vitamin B-6 0.100 mg, total folate 14 μg, choline 9.8 mg, vitamin A 0 IU, vitamin E (α-tocopherol) 0.1 mg, vitamin K (phylloquinone) 0.5 μg, total saturated fatty acids 0.048 g, 16:0 (palmitic) 0.031 g, 18:0 (stearic) 0.017 g; total monounsaturated fatty acids 0.034 g, 16:1 undifferentiated (palmitoleic) 0.02 g, 18:1 undifferentiated (oleic) 0.032 g; total polyunsaturated fatty acids 0.066 g, 18:2 undifferentiated (linoleic) 0.048 g, 18:3 undifferentiated (linolenic) 0.018 g; threonine 0.052 g, isoleucine 0.064 g, leucine 0.065 g, lysine 0.037 g, methionine 0.010 g, cystine 0.009 g, phenylalanine 0.026 g, tyrosine 0.019 g, valine 0.047 g, lutein + zeaxanthin 22 μg.

The raw, mature breadfruit seed (excluding 32% shell) had the following nutrient composition (per 100 g value): water 56.27 g, energy 191 kcal (799 kj), protein 7.40 g, total lipid 5.59 g, ash 1.50 g, carbohydrates 29.24 g, total dietary fibre 5.29 g, Ca 36 mg, Fe 3.67 mg, Mg 54 mg, P 175 mg, K 941 mg, Na 25 mg, Zn 0.90 mg, Cu 1.148 mg, Mn 0.142 mg, vitamin C 6.6 mg, thiamine 0.482 mg, riboflavin 0.301 mg, niacin 0.438 mg, pantothenic acid 0.877 mg, vitamin

Plate 4 (**a**, **b**) Breadfruit on sale in a local market in PNG (*left*) and Fiji (*right*)

Plate 5 Breadfruit seeds

Plate 6 Close-view of breadfruit seeds

B-6 0.320 mg, total folate 53 μg, choline 9.8 mg, vitamin A 256 IU, total saturated fatty acids 1.509 g, 16:0 (palmitic) 0.999 g, 18:0 (stearic) 0.510 g; total monounsaturated fatty acids 0.712 g, 16:1 undifferentiated (palmitoleic) 0.030 g, 18:1 undifferentiated (oleic) 0.682 g; total polyunsaturated fatty acids 2.997 g, 18:2 undifferentiated (linoleic) 2.290 g, 18:3 undifferentiated (linolenic) 0.687 g; tryptophan 0.123 g, threonine 0.385 g, isoleucine 0.443 g, leucine 0.563 g, lysine 0.570 g, methionine 0.096 g, cystine 0.116 g, phenylalanine 0.797 g, tyrosine 0.577 g, valine 0.534 g, arginine 0.494 g, histidine 0.207 g, alanine 0.336 g, aspartic acid 0.817 g, glutamic acid 1.036 g, glycine 0.465 g, proline 0.369 g and serine 0.496 g (U.S. Department of Agriculture and Agricultural Research 2010).

Breadfruit is a high source of starch carbohydrate (65% by weight) sugars, dietary fibre, a fair source of B vitamins, thiamine and niacin and fair in riboflavin, vitamin C, calcium, potassium, phosphorus and copper. Yellow fleshed varieties also provide carotenoids.

From the above, it is evident that the seeds are rich in carbohydrate, fibre, proteins, amino acids, minerals like Ca, Fe, P, K, Mg and Cu and vitamin A and a fair source of thiamine, riboflavin and niacin and also have folate, vitamin B-6 and pantothenic acid. The leaves were also reported to be good sources of vitamin c, iron, phosphorus and calcium. Percent recoveries using high-performance liquid chromatography (HPLC) and gas chromatography (GC) of amino acids, fatty acids, and carbohydrates were 72.5%, 68.2%, and 81.4%, respectively from breadfruit (Golden and Williams 2001). The starch content of the breadfruit was 15.52 g/100 g fresh weight.

The proximate composition of the peel, pulp and core of breadfruit revealed that the highest moisture, ash, protein, fat and crude fibre contents can be found in the core while the pulp contained the highest levels of carbohydrate, starch, nitrogen free extract and organic matter (Adewusi et al. 1995). Total free and reducing sugars were highest in the core and lowest in the peel. Sucrose, glucose and fructose followed a similar pattern as the reducing sugars while the flatus-producing oligosaccharides raffinose (0.1%) and stachyose (0.05%) were present in the core. Only raffinose was present (0.05%) in both the peel and the pulp. Extracted starch from the breadfruit pulp was 58% of the total starch content on dry weight basis with minimal levels of ash, fat, protein and 98.6% starch. The extracted starch was 98% pure and contained only 2.3% damaged starch. The starch swelling and solubility properties showed a two-stage pattern while the Brabender amylograms showed patterns very typical of starches from most normal non-waxy cereals.

The starch yield of *Artocarpus altilis* fruit was 18.5 g/100 g (dw) with a purity of 98.86%, 27.68 and 72.32% of amylose and amylopectin, respectively (Rincón and Padilla 2004). Swelling power, water absorption and solubility values were higher than that of corn and amaranth starch. The amylographic study showed a gelatinization temperature at 73.3°C, with high stability during heating and cooling cycles. Breadfruit starch could also be categorized in the group of mixed short chain branched/long chain branched glucan starches, this agreed with digestibility results that showed a high degree of digestibility in vitro. These results might be advantageous in medical and food use. In a recent study, the starch contains 10.83% moisture, 0.53% crude protein, 0.39% fat, 22.52% amylose, 77.48% amylopectin and 1.77% ash contents (Akanbi et al. 2009). The average of the breadfruit starch particle size was 18 μm, pH 6.5, bulk density 0.673 g/ml, and dispersibility 40.67%. The swelling power of the breadfruit starch increased with increase in temperature, but there was a rapid increase in the swelling power from 70°C to 80°C. The pasting temperature of the starch paste was 84.05°C, setback and breakdown values were 40.08 and 7.92 RVU respectively. The peak viscosity value was 121.25 RVU while final viscosity value was 153.42 RVU. This study concluded that breadfruit starch has an array of functional, pasting and proximate properties that can facilitate its use in so many areas where the properties of other starches are acceptable.

It has also been processed into starches and into flour.

Proximate composition showed a significant difference in the raw and processed breadfruit seed flour samples (Okorie 2010). The moisture content was lowest (14.77%) in the roasted seed flour and highest (24.08%) in the boiled seed flour. Percentage ash, fat and protein were highest (3.66, 3.74 and 4.67%) in the raw flour while ash and fibre contents were least (2.75% and 1.81%), respectively in the boiled flour. Carbohydrate was highest (87.29%) in the boiled and least (85.60%) in the raw flour. Boiling and roasting indicated that the vitamin C content and the mineral contents were significantly higher in the raw seed flour. The effects of boiling and roasting with regards to loss and retention of the nutrients differed significantly with only the roasting retaining more of the nutrients than boiled seed flour. Sodium and potassium contents of boiled (0.27 and 0.75 mg/l), respectively and roasted (0.34 and 0.78 mg/l, respectively) seed flours compared well with the raw flour (0.37 and 0.83 mg/l, respectively).

Other Phytochemicals

Chemical investigation of the fruits of *A. altilis* furnished 12 compounds 1–12, stilbenes (1–3), arylbenzofuran (4), flavanone (5), flavones (6–8), triterpenes (9 and 10) and sterols (11 and 12) identified as (E)-2,3′,4,5′-stilbenetetrol (1), 4′-[3-methyl-1(E)-butenyl]-(E)-2,3′,4,5′-stilbenetetrol (2), (3-methyl-2-butenyl)-(E)-2,3′,4,5′-stilbenetetrol (3) 3′,5′,6-trihydroxy-2-phenylbenzofuran (4), 2′,4′,5,7-tetrahydroxyflavanone (5), 2′,4′,5,7-tetrahy droxyflavone (6), 2′,4′,5,7-tetrahydroxy-6-

[3-methyl-1(E)-butenyl]flavone (7), 2′,4′,5,7-tetrahydroxy-6-(3-methyl-2-butenyl)flavone (8), 3b-acetoxyolean-12-en-11-one (9), cycloartenyl acetate (10), sitosterol (11) and sitosterol b-D-glucopyranoside (12) (Amarasinghe et al. 2008). Earlier, cyclopropane-containing sterol: cycloartenol, triterpene: α-amyrin and cycloartane triterpenes: cycloart-23-ene-3β,25-diol and cycloart-25-ene-3β,24-diol were isolated from *Artocarpus altilis* fruit (Altman and Zito 1976). A disordered mixture of two compounds, friedelan-3β-ol and friedelin, were isolated from *Artocarpus altilis* (Fun et al. 2007).

A novel stilbene with dimethylchromene ring, 3,2′,4′-trihydroxy-6″,6″-dimethyl-pyrano(3″, 2″:4,5)-trans- stilbene, named artocarbene, was isolated from the heartwood (Shimizu et al. 1997). A new prenylated chalcone, 3′ ′,3′ ′-dimethylpyrano[3′,4′]2,4,2′-trihydroxychalcone; two flavonoid derivatives, (−)-cycloartocarpin and (−)-cudraflavone A, and eight known flavonoids, isobacachalcone, morachalcone A, gemichalcones B and C, artocarpin, cudraflavone C, licoflavone C and (2S)-euchrenone a(7) were isolated from the heartwood (Han et al. 2006). Six prenylated flavones, including one new compound, were isolated and identified from the stem bark extracts (Shamaun et al. 2010). The new prenylated flavone hydroxyartocarpin (1) was characterized as 3-(γ,γ-dimethylallyl)-6-isopentenyl-5,8,2′,4′-tetrahydroxy-7-methoxyflavone and the known compounds were artocarpin (2), morusin (3), cycloartobiloxanthone (4), cycloartocarpin A (5) and artoindonesianin V (6).

Two prenylflavones, artonins E and F and cycloartobiloxanthone were isolated from the bark (Hano et al. 1990). Three new prenylflavones, KB-1, KB-2 and KB-3 and a known flavones, morusin were isolated from the bark (Fujimoto et al. 1990). Five new isoprenylated phenols, artonols A, B, C, D and E were isolated from the bark of *Artocarpus communis* along with four known compounds artonin E, cycloartobiloxanthone, artonin K, and artobiloxanthone (Aida et al. 1997). Five prenylflavonoids similar to glabridin were isolated from breadfruit: artonin E, morusin, licochalcone A, licoricidin and licorisoflavan A (Fukai et al. 2003).

The roots and stem were found to contain prenylflavones: isocyclomorusin, isoccyclomulberrin and cycloatlisin and flavonoids: cyclomurusin, cyclomulberrin and engeletin (Chen et al. 1993). Nine prenylated flavones: artocarpin, cycloartocarpin, and chaplashin were isolated from the dichloromethane extract of the root and stems, whereas morusin, cudraflavone B, cycloartobiloxanthone, artonin E, cudraflavone C and artobiloxanthone were found in the root barks (Boonphong et al. 2007). Two known compounds, β-sitosterol and cudraflavone A, a new triterpenoid ester, lupeol acetate, a new pyranodihydrobenzoxanthone, named artomunoxanthone, and a novel quinonoid pyranobenzoxanthone, named artomunoxanthentrione were isolated from the root bark (Shieh and Lin 1992). From the root bark a novel pyranodihydrobenzoxanthone epoxide, named artomunoxanthotrione epoxide (Jong et al. 1992) and a 3-prenyl flavonoid, artonin V (Hano et al. 1994) were isolated. Five new prenylflavonoids, artocommunols CA, CB, CC, CD and CE, were isolated from the cortex of the roots of *Artocarpus communis*, along with the known compound cyclomorusin (Chan et al. 2003). Four flavonoids, dihydroartomunoxanthone, artomunoisoxanthone, cyclocomunomethonol and artomunoflavanone, together with three known compounds, artochamins B, D and artocommunol CC were isolated from the cortex of the roots of breadfruit (Weng et al. 2006).

Five geranyl dihydrochalcones, 1-(2,4-dihydroxyphenyl)-3-{4-hydroxy-6,6,9-trimethyl-6a,7,8,10a-tetrahydro-6H-dibenzo[b,d]pyran-5-yl}-1-propanone (2); 1-(2, 4-dihydroxyphenyl)-3-[3,4-dihydro-3,8-dihydroxy-2-methyl-2-(4-methyl-3-pentenyl)-2H-1-benzopyran-5-yl]-1-propanone (4); 1-(2, 4-dihydroxyphenyl)-3-[8-hydroxy-2-methyl-2-(3,4-epoxy-4-methyl-1-pentenyl)-2H-1-benzopyran-5-yl]-1-propanone (5); 1-(2,4-dihydroxyphenyl)-3-[8-hydroxy-2-methyl-2-(4-hydroxy-4-methyl-2-pentenyl)-2H-1-benzopyran-5-yl]-1-propanone (8);and 2-[6-hydroxy-3, 7-dimethylocta-2(E),7-dienyl]-2′,3,4,4′-tetrahydroxydihydrochalcone (9); along with four known geranyl flavonoids (1, 3, 6, 7), were isolated from the leaves of *Artocarpus altilis* (Wang et al. 2007).

Three new geranyl chalcone derivatives including isolespeol, 5′-geranyl-2′,4′,4-trihydroxychalcone and 3,4,2′,4′-tetrahydroxy-3′-geranyldihydrochalcone, together with two known compounds lespeol and xanthoangelol, were isolated from the leaves of breadfruit (Fang et al. 2008).

Studies isolated isoprenylated phenolic compounds (prenylflavones) from the plant which showed antioxidant, hypotensive, anti-inflammatory, antimicrobial, and antitumour activities (Chen et al. 1993).

Antioxidant Activity

Two new prenylflavonoids, cyclogeracommunin (1) and artoflavone A (2), were isolated from the cortex of roots of *Artocarpus communis* (Lin et al. 2009). Compounds 1, 2 and other isolated known compounds, artomunoisoxanthone (3), artocommunol CC (4), artochamin D (5), artochamin B (6), and dihydroartomunoxanthone (7), all showed inhibition of oxidative DNA damage. Compound 2 significantly showed 1,1-diphenyl-2-picrylhydrazyl (DPPH)-scavenging activity with IC_{50} values of 24.28, while compound 1 significantly displayed inhibitory effect on xanthine oxidase (XO) activity with IC_{50} value of 73.31 μM. These findings indicated compounds 1–7 to be promising antioxidants.

Anticancer Activity

Three prenylflavones, KB-1 (1), KB-2 (2) and KB-3(3) isolated from the bark exhibited strong cytotoxic activities against leukemia cells (L-1210) in-vitro (Fujimoto et al. 1990). Nine prenylated flavones isolated from the roots of *Artocarpus altilis* exhibited moderate cytotoxicity against KB (human oral epidermoid carcinoma) and BC (human breast cancer) cell lines (Boonphong et al. 2007). Cytotoxicities of compounds 1–9, cycloartocarpin (1), artocarpin (2), and chaplashin (3) were isolated from the dichloromethane extract of the root stems, whereas morusin (4), cudraflavone B (5), cycloartobiloxanthone (6), artonin E (7), cudraflavone C (8) and artobiloxanthone (9) against KB (human oral epidermoid carcinoma), BC (human breast cancer), and Vero cell lines were similar, with the IC_{50} values of 2.9–14.7 μg/ml. The prenylated flavonoid, 1-(2,4-dihydroxyphenyl)-3-[8-hydroxy-2-methyl-2-(4-methyl-3-pentenyl)-2 H-1-benzopyran-5-yl] 1-propanone from the dichloromethane extract of breadfruit leaves showed significant cytotoxicity against murine P-388 leukemia cells (Lotulung et al. 2008).

Of five geranyl dihydrochalcones isolated from the leaves of *Artocarpus altilis* compounds: 1-(2,4-dihydroxyphenyl)-3-{4-hydroxy-6,6,9-trimethyl-6a,7,8,10a-tetrahydro-6 H-dibenzo[b,d] pyran-5-yl}-1-propanone; 1-(2,4-dihydroxyphenyl)-3-[3,4-dihydro-3,8-dihydroxy-2-methyl-2-(4-methyl-3-pentenyl)-2 H-1-benzopyran-5-yl]-1-propanone; and 2-[6-hydroxy-3,7-dimethy locta-2(E),7-dienyl]-2′,3,4,4′-tetrahydroxydihydro chalcone exhibited moderate cytotoxicity against SPC-A-1 (non-small cell lung cancer), SW-480 (human colon cancer), and SMMC-7721 (hepatocellular carcinoma) human cancer cells (Wang et al. 2007). The geranyl chalcone, isolespeol, isolated from the leaves of breadfruit showed the highest inhibitory activity with an IC_{50} value of 3.8 μM in SW 872 human liposarcoma cells (Fang et al. 2008). Treatment of SW 872 human liposarcoma cells with isolespeol caused the loss of mitochondrial membrane potential. Results indicated that isolespeol induced apoptosis in SW 872 cells through FAa- and mitochondria-mediated pathways. The results of recent studies showed that *A. altilis* diethylether wood extract had a cytotoxic effect on breast cancer cells (T47D) in a concentration-dependent manner, with an extract IC_{50} 6.19 μg/ml (Arung et al. 2009). The wood extract, which primarily contained artocarpin, reduced cell viability by inducing apoptosis and sub-G1 phase formation in human breast cancer T47D cells in vitro.

Antitubercular Activity

Bioactivity study of the crude CH2Cl2 extract of the roots of *A. altilis* revealed that it possessed antitubercular activity against *Mycobacterium tuberculosis* with a minimum inhibitory concentration (MIC) of 25 μg/ml (Boonphong et al. 2007). All nine prenylated flavones: cycloartocarpin (1), artocarpin (2), and chaplashin (3) isolated from the dichloromethane extract of the root stems, and morusin (4), cudraflavone B (5), cycloartobiloxanthone (6), artonin E (7), cudraflavone C (8) and artobiloxanthone (9) isolated from the roots of *Artocarpus altilis* exhibited antitubercular activity with a minimum inhibitory concentration (MIC) ranging from 3.12 to 100 μg/ml. Among the isolated metabolites, compounds 2 and 3 were the most potent antitubercular agents possessing MIC value of 3.12 μg/ml comparable to that of the standard drug, kanamycin (MIC value of 2.5 μg/ml).

Alpha-Reductase Inhibitory Activity

The methanol extract of heartwood of breadfruit showed potent 5 α-reductase inhibitory activity (Shimizu et al. 2000a). Chlorophorin (IC_{50}=37 μM) and artocarpin (IC_{50}=85 μM) showed more potent inhibitory effects than did α-linolenic acid, which is known as a naturally occurring potent inhibitor. Structure-activity investigations suggested that the presence of an isoprene substituent (prenyl and geranyl) would enhance 5-α-reductase inhibitory effects. To enhance the penetration of artocarpin, extracted from the heartwood, into the deeper layers of the skin where androgen receptors abound, the compound was formulated in microparticles to help in transfollicular delivery (Pitaksuteepong et al. 2007). However, artocarpin formulated in microparticles did not show significant systemic action compared to the dermal application of an artocarpin solution and a flutamide preparation (1 mg) as positive control. A geranylated chalcone isolated from leaves showed potent 5-α-reductase inhibitory activity (Shimizu et al. 2000c). 5-α-reductase inhibitor drugs are used in benign prostatic hyperplasia, prostate cancer (Montironi et al. 1996), and baldness (Dallob et al. 1994).

Skin Whitening Activity

The methanol extracts of heartwood extract of breadfruit showed potent inhibitory activity on tyrosinase activity which was equivalent to kojic acid (Shimizu et al. 1998). The extract apparently inhibited melanin biosynthesis of both cultured B16 melanoma cells without any cytotoxicity and in the back of a brown guinea pig without skin irritation. Thus, the potentiality of the extracts of heartwood of breadfruit both as material of a useful skin whitening agent and as a remedy for disturbances in pigmentation is evident. Tyrosinase inhibitory activity-guided fractionation led to the isolation of seven active compounds including a new compound which has been characterized as 6-(3″-methyl-1″-butenyl)-5,7,2′,4′-tetrahydroxyflavone, named isoartocarpesin. Other active compounds were (+)-dihydromorin, chlorophorin, (+)-norartocarpanone, 4-prenyloxyresveratrol, artocarbene, and artocarpesin, These compounds were probably responsible for the melanin biosynthesis inhibitory effects. The structure-activity relationships of several flavonoids, stilbenes and related 4-substituted resorcinols, obtained from *Artocarpus altilis* suggested that specific natural or synthesized compounds having the 4-substituted resorcinol skeleton hadpotent tyrosinase inhibitory ability (Shimizu et al. 2000b). Kinetic studies indicated that specific compounds having the 4-substituted resorcinol skeleton exhibited competitive inhibition of the oxidation of DL-β-(3,4-dihydroxyphenyl) alanine (DL-DOPA) by mushroom tyrosinase. In another study, the prenylated flavonol, artocarpin, from the heartwood of breadfruit exhibited efficient skin lightening effect following topical application of artocarpin to UV-stimulated hyperpigmented dorsal skins of brownish guinea pigs (Shimizu et al. 2002)

Separate studies indicated that the ether extract of breadfruit heartwood had the potential of acting as a skin-lightening agent for application in cosmetics (Donsing et al. 2008). Artocarpin from heartwood of breadfruit was extracted by using diethyl ether or methanol; the artocarpin content found in ether extract was 45.19% w/w, whereas that in the methanol extract was 19.61% w/w. The ether extract showed dose-dependent tyrosinase-inhibitory activity. The obtained IC_{50} value was 10.26 μg/ml, while kojic acid, a well-known tyrosinase inhibitor, provided an IC_{50} of 7.89 μg/ml.

A. incisus extract at a concentration of 2–25 μg/ml was able to decrease the melanin production of the melanocyte B16F1 cells. The obtained micrograph also confirmed that the extract did not change the cell morphology but reduced the melanin content by inhibiting melanin synthesis, whereas the purified artocarpin at a concentration of 4.5 μg/ml caused changes in cell morphology. Additionally, the extract exhibited antioxidant activity in a dose-dependent manner at an EC_{50} of 169.53 μg/ml, as evaluated by the DPPH assay.

Antiinflammatory Activity

The antiinflammatory activities of the isolated flavonoids, cycloartomunin (1), cyclomorusin (2), dihydrocycloartomunin (3), dihydroisocycloartomunin (4), cudraflavone A (5), cyclocommunin (6), and artomunoxanthone (7), and cycloheterohyllin (8), artonins A (9) and B (10), artocarpanone (11), artocarpanone A (12), and heteroflavanones A (13), B (14), and C (15) isolated from *Artocarpus communis* and *A. heterophyllus* exhibited antiinflammatory activities (Wei et al. 2005), Compound 4 significantly inhibited the release of β-glucuronidase and histamine from rat peritoneal mast cells stimulated with ρ-methoxy-N-methylphenethylamine (compound 48/80). Compound 11 significantly inhibited the release of lysozyme from rat neutrophils stimulated with formyl-Met-Leu-Phe (fMLP). Compounds 8, 10, and 11 significantly inhibited superoxide anion formation in fMLP-stimulated rat neutrophils while compounds 2, 3, 5, and 6 evoked the stimulation of superoxide anion generation. Compound 11 exhibited significant inhibitory effect on NO production and iNOS protein expression in RAW 264.7 cells. The potent inhibitory effect of compound 11 on NO production in lipopolysaccharide (LPS)-activated macrophages, probably through the suppression of iNOS protein expression. A new prenylated chalcone, 3″,3″-dimethylpyrano[3′,4′]2,4,2′-trihydroxychalcone (1). Two flavonoid derivatives, (−)-cycloartocarpin (9) and (−)-cudraflavone A (10) and eight known flavonoids, isobacachalcone (2), morachalcone A (3), gemichalcones B (4) and C (5), artocarpin (6), cudraflavone C (7), licoflavone C (8), and (2S)-euchrenone a(7) (11) was isolated from the heartwood of breadfruit. (Han et al. 2006). Compounds 1–4, 6, and 11 exhibited potent inhibitory activity on nitric oxide production in RAW264.7 LPS-activated mouse macrophage cells with IC_{50} values of 18.8, 6.4, 16.4, 9.3, 18.7, and 12.3 μM, respectively.

Cathepsin K Inhibition Activity

The methanol/dichloromethane extract from bud covers of breadfruit was shown to have activity in a cathepsin K inhibition assay (Patil et al. 2002). In addition to the three known flavonoids isolated, two new compounds a dimeric dihydrochalcone, cycloaltilisin 6 (2), and a new prenylated flavone, cycloaltilisin 7 exhibited IC_{50} values of 98 and 840 nM, respectively in cathepsin K inhibition. Cathepsin K is a cysteine protease that plays an essential role in osteoclast function and in the degradation of protein components of the bone matrix by cleaving proteins such as collagen type I, collagen type II and osteonectin (Skoumal et al. 2005). Cathepsin K is reported to play a role in bone remodelling and resorption in diseases such as osteoporosis, osteolytic bone metastasis and rheumatoid arthritis.

Antiplatelet Activity

Four flavonoids, dihydroartomunoxanthone (1), artomunoisoxanthone (2), cyclocomunomethonol (3) and artomunoflavanone (4), together with three known compounds, artochamins B (5), D and artocommunol CC (6) were isolated from the cortex of the roots of breadfruit (Weng et al. 2006). Of the compounds tested in human platelet-rich plasma, compounds 1, 5 and 6 showed significant inhibition of secondary aggregation induced by adrenaline. It was concluded that the antiplatelet effect of 1, 5 and 6 was mainly due to an inhibitory effect on thromboxane formation.

Antiatherosclerotic Activity

Studies demonstrated that the ethyl acetate extract of *Artocarpus altilis* showed cytoprotective activities in human U937 cells incubated with oxidized LDL (OxLDL) using the 4-[3-(4-iodophenyl)-2-(4-nitrophenyl)-2 H-5-tetrazolio]-1, 3-benzene disulfonate (WST-1) assay (Wang et al. 2006) . The main cytoprotective components was identified as β-sitosterol and six flavonoids. The cytoprotective effect offers good prospects for the medicinal applications of *A. altilis*.

Antinephritic Activity

Five prenylflavonoids similar to glabridin (1–5), isolated from *Artocarpus communis* exhibited antinephritis activity in mice with glomerular disease (Masugi-nephritis) (Fukai et al. 2003). Oral administrations of artonin E (2) or licochalcone A (4) for 10 days (30 mg/kg/day) reduced the amount of urinary protein excretion compared to nephritic mice. ESR spectroscopy demonstrated that morusin (1) and licorisoflavan A (5) increased the radical intensity of sodium ascorbate by about two times. Morusin, licoricidin (3), licochalcone A and licorisoflavan A showed weak scavenging activity against superoxide anion radical.

Antimalarial Activity

Prenylated flavones cycloartocarpin (1), artocarpin (2), and chaplashin (3) isolated from the dichloromethane extract of the root stems, and morusin (4), cudraflavone B (5), cycloartobiloxanthone (6), artonin E (7), cudraflavone C (8) and artobiloxanthone (9) isolated from the roots of *Artocarpus altilis all* showed moderate antiplasmodial (IC_{50}=3.5 μg/ml) activity against the parasite *Plasmodium falciparum* (Boonphong et al. 2007). All nine compounds isolated, except 8, exhibited moderate antiplasmodial activity with the IC_{50} values ranging from 1.9 to 4.3 μg/ml.

Antiviral Activity

Breadfruit seed was found to contain lectin and jacalin (Pineau et al. 1990). The breadfruit lectin displayed the same IgAl and IgD precipitation specificity as jacalin. It also stimulated in vitro proliferation of human peripheral blood mononuclear cells. The results suggested that lectins from both species *Artocarpus altilis* (breadfruit) and *Artocarpus heterophyllus* (jackfruit) were very similar. The lectin jacalin was shown to specifically stimulate $CD4^+$ lymphocytes (Lafont et al. 1997). This lectin, presented a peptide highly similar to a sequence of the HIV external glycoprotein, interacted with the oligosaccharide side chains of CD4 and was able to inhibit in-vitro HIV infection, and triggered cell signalling directly via CD4 (Lafont et al. 1997)

Adverse Effects

Hyperglycaemic Activity

The findings of a laboratory animal study indicated that breadfruit root bark aqueous extract induced acute hyperglycaemia in Wistar rats, and that it disrupted the biochemical variables of the rat pancreas and liver (Adewole and Ojewole 2007b). Hepatic glycogen contents significantly increased, while hepatic hexokinase (HXK) and glucokinase (GCK) activities activities significantly decreased. The root-bark extract was also found to disrupt the ultrastructural characteristics and architecture of hepatocytes as well as oxidative energy metabolism (Adewole and Ojewole 2007a). Ultrastructural changes observed within the hepatocytes of rats included disrupted mitochondria with increase in lipid droplets, extensive hepatocellular vacuolation, scanty rough endoplasmic reticulum and ribosomes. Also there was overall inhibition of oxidative phosphorylation of the liver mitochondria

Negative Chronotropic Activity

Ethyl acetate soluble extracts from the leaves of the breadfruit, *Artocarpus altilis* exerted a weak, negative chronotropic effect and significantly reduced left ventricular pulse pressure in vivo in the rat (Young et al. 1993). On right ventricular myocardial strips, the same extracts produced a significant negative inotropic effect. This indicated that the in vivo effects might be due in part, to a direct inotropic effect on the myocardium. An in-vivo side effect was extensive intravascular haemolysis and consequent haemoglobinuria which could be caused by the vehicle, the extract, or a combined effect of the two. The mechanism of action of the inotropic agent was not cholinergic, and may involve a decoupling of excitation and contraction

Traditional Medicinal Uses

Breadfruit is used in many traditional folk medicines. In the Philippines, breadfruit is laxative and heated fruit slices are used for furuncles. A decoction of leaves is used for baths in rheumatism while a decoction of bark is used for dysentery. Tree latex is used for hernia in children.

In Trinidad and the Bahamas, a decoction of the breadfruit leaf is believed to lower blood pressure, and is also said to relieve asthma. Crushed leaves are applied on the tongue as a treatment for thrush. The leaf juice is employed as ear-drops. Ashes of burned leaves are used on skin infections. A powder of roasted leaves is employed as a remedy for enlarged spleen. The crushed fruit is poulticed on tumours to "ripen" them. Toasted flowers are rubbed on the gums around an aching tooth. The latex is used on skin diseases and is bandaged on the spine to relieve sciatica. Diluted latex is taken internally to overcome diarrheal. *Artocarpus altilis* was used for hypertension in ethnomedicines of Trinidad and Tobago (Lans 2006).

The *Artocarpus altilis* roots have been used as a component in Thai folk remedies for venereal diseases and cancer (Boonphong et al. 2007) whereas in Taiwan, the stems and roots have been used traditionally for the treatment of liver cirrhosis, hypertension, antiinflammatory, and detoxifying effects (Chen et al. 1993).

Other Uses

The breadfruit trees are planted as windbreaks, ornamentals and as shade trees in coffee plantations.

The wood has been valued a source of commercial timber. The wood is resistant to termites and marine worms and used for canoe hulls and occasionally in house construction. The wood is also used for carving statues, bowls and other wooden objects. Inner bark is used to make bark cloth called 'tapa'. In Samoa, Micronesia, and the Philippines The inner bast fibres are used to make strong cordage used for fishing and animal harnesses. The latex is used for chewing gum, as bird-lime, as seal-lime to prepare wooden surfaces for painting or to caulk boats to make them watertight. Leaves are used for wrapping food for cooking, for parcelling of fresh food, and as plates. Dried inflorescences are burnt as a mosquito repellent. Under-ripe fruits are cooked for feeding to pigs. Uncooked soft ripe fruits are used as animal feed.

Comments

There is a great diversity of seeded and seedless cultivars. More abundant are the seedless varieties.

Selected References

Adewole SO, Ojewole JA (2007a) *Artocarpus communis* Forst. Root-bark aqueous extract- and streptozotocin-induced ultrastructural and metabolic changes in hepatic tissues of wistar rats. Afr J Tradit Complement Altern Med 4(4):397–410

Adewole SO, Ojewole JO (2007b) Hyperglycaemic effect of *Artocarpus communis* Forst (Moraceae) root bark aqueous extract in Wistar rats. Cardiovasc J Afr 18(4):221–227

Adewusi SRA, Udio J, Osuntogun BA (1995) Studies on the carbohydrate content of breadfruit (*Artocarpus communis* Forst) from South-Western Nigeria. Starch-Starke 47(8):289–294

Aida M, Yamaguchi N, Hano Y, Nomura T (1997) Constituents of the Moraceae plants 30. Artonols A, B, C, D, and E, five new isoprenylated phenols from the bark of *Artocarpus communis* Forst. Heterocycles 45(1):163–175

Akanbi TO, Nazamid S, Adebowale AA (2009) Functional and pasting properties of a tropical breadfruit (*Artocarpus altilis*) starch from Ile-Ife, Osun State, Nigeria. Int Food Res J 16:151–157

Altman LJ, Zito WS (1976) Sterols and triterpenes from the fruit of *Artocarpus altilis*. Phytochemistry 15:829–830

Amarasinghe NR, Jayasinghe L, Hara N, Fujimoto Y (2008) Chemical constituents of the fruits of *Artocarpus altilis*. Biochem Syst Ecol 36(4):323–325

Arung ET, Shimizu K, Kondo R (2006) Inhibitory effect of isoprenoid-substituted flavonoids isolated from *Artocarpus heterophyllus* on melanin biosynthesis. Planta Med 72(9):847–850

Arung ET, Wicaksono BD, Handoko YA, Kusuma IW, Yulia D, Sandra F (2009) anti-cancer properties of diethylether extract of wood from sukun (*Artocarpus altilis*) in human breast cancer (T47D) cells. Trop J Pharm Res 8(4):317–324

Boonphong S, Baramee A, Kittakoop P, Puangsombat P (2007) Antitubercular and antiplasmodial prenylated flavones from the roots of *Artocarpus altilis*. Chiang Mai J Sci 34(3):339–344

Chan SC, Ko HH, Lin CN (2003) New prenylflavonoids from *Artocarpus communis*. J Nat Prod 66(3):427–430

Chen CC, Huang YL, Ou JC, Lin CF, Pan TM (1993) Three new prenylflavones from *Artocarpus altilis*. J Nat Prod 56:1594–1597

Commission SP (1983) Breadfruit – a food for all seasons. SPC South Pacific Foods Leaflet, No, 9

Dallob AL, Sadick NS, Unger W, Lipert S, Geissler LA, Gregoire SL, Nguyen HH, Moore EC, Tanaka WK (1994) The effect of finasteride, a 5 α-reductase inhibitor, on scalp skin testosterone and dihydrotestosterone concentrations in patients with male pattern baldness. J Clin Endocrinol Metab 79:703–706

Donsing P, Limpeanchob N, Viyoch J (2008) Evaluation of the effect of Thai breadfruit's heartwood extract on melanogenesis-inhibitory and antioxidation activities. J Cosmet Sci 59(1):41–58

Ersam T, Achmad SA, Ghisalberti EL, Hakim EH, Tamin R (2000) Some phenolic compounds from *Artocarpus altilis* (Park.) Fosb. In: Proceedings of international symposium on the role chemistry and environment, Padang, 2000, pp 113–119

Fang SC, Hsu CL, Yu YS, Yen GC (2008) Cytotoxic effects of new geranyl chalcone derivatives isolated from the leaves of *Artocarpus communis* in SW 872 human liposarcoma cells. J Agric Food Chem 56(19): 8859–8868

Fujimoto Y, Zhang XX, Kirisawa M, Uzawa J, Sumatra M (1990) New flavones from *Artocarpus communis* Forst. Chem Pharm Bull 38:1787–1789

Fukai T, Satoh K, Nomura T, Sakagami H (2003) Antinephritis and radical scavenging activity of prenylflavonoids. Fitoterapia 74(7–8):720–724

Fun HK, Boonnak N, Chantrapromma S (2007) Cocrystal of friedelan-3β-ol and friedelin. Acta Crystallogr 63(4):2014–2016

Golden KD, Williams OJ (2001) Amino acid, fatty acid, and carbohydrate content of *Artocarpus altilis* (breadfruit). J Chromatogr Sci 39(6):243–250

Han AR, Kang YJ, Windono T, Lee SK, Seo EK (2006) Prenylated flavonoids from the heartwood of *Artocarpus communis* with inhibitory activity on lipopolysaccharide-induced nitric oxide production. J Nat Prod 69(4):719–721

Hano Y, Yamagami Y, Kobayashi M, Isohata R, Nomura T (1990) Artonins E and F, two new prenylflavones from the bark of *Artocarpus communis* Forst. Heterocycles 31(5):877–882

Hano Y, Inami R, Nomura T (1994) Constituents of Moraceae plants 20. A novel flavone, artonin V, from the root bark of *Artocarpus altilis*. J Chem Res Syn 9:348–349

Jong TT, Lin CN, Shieh WL (1992) A pyranodihydrobenzoxanthone epoxide from *Artocarpus communis*. Phytochemistry 31(7):2563–2564

Lafont V, Hivroz C, Carayon P, Dornand J, Favero J (1997) The lectin jacalin specifically triggers cell signaling in CD4+T lymphocytes. Cell Immunol 181(1):23–29

Lans CA (2006) Ethnomedicines used in Trinidad and Tobago for urinary problems and diabetes mellitus. J Ethnobiol Ethnomed 2:45–55

Lin KW, Liu CH, Tu HY, Ko HH, Wei BL (2009) Antioxidant prenylflavonoids from *Artocarpus communis* and *Artocarpus elasticus*. Food Chem 115(2):558–562

Lotulung PD, Fajriah S, Hanafi M, Filaila E (2008) Identification of cytotoxic compound from *Artocarpus communis* leaves against P-388 cells. Pak J Biol Sci 11(21):2517–2520

Montironi R, Valli M, Fabris G (1996) Treatment of benign prostatic hyperplasia with 5-α-reductase inhibitor: morphological changes in patients who fail to respond. J Clin Pathol 49(4):324–328

Morton JF (1987) Breadfruit. In: Fruits of warm climates. Florida Flair Books, Miami, pp 58–63

Nomura T, Hano Y, Aida W (1998) Isoprenoid substituted flavonoids from *Artocarpus* plants (Moraceae). Heterocycles 47(2):1179–1205

Ochse JJ, Bakhuizen van den Brink RC (1980) Vegetables of the Dutch Indies, 3rd edn. Ascher & Co., Amsterdam the Netherlands, 1016 pp

Okorie SU (2010) Chemical composition of *Artocarpus communis* (breadfruit) seed flour as affected by processing (boiling and roasting). Pak J Nutr 9(5):419–421

Patil AD, Freyer AJ, Killmer L, Offen P, Taylor PB, Votta BJ, Johnson RK (2002) A new dimeric dihydrochalcone and a new prenylated flavones from the bud covers of *Artocarpus altilis*: potent inhibitors of cathepsin K. J Nat Prod 65(4):624–627

Pineau N, Pousset JL, Preud'homme JL, Aucouturier P (1990) Structural and functional similarities of breadfruit seed lectin and jacalin. Mol Immunol 27(3):237–240

Pitaksuteepong T, Somsiri A, Waranuch N (2007) Targeted transfollicular delivery of artocarpin extract from *Artocarpus incisus* by means of microparticles. Eur J Pharm Biopharm 67:639–645

Ragone D (1997) Breadfruit *Artocarpus altilis* (Parkinson) Fossberg. (Promoting the conservation and use of underutilized and neglected crops, 10). International Plant Genetic Resources Institute, Rome, 77 pp

Ragone D (2003) Breadfruit. In: Caballero B, Trugo L, Finglas P (eds) Encyclopedia of food sciences and nutrition. Academic, San Diego, California, pp 655–661

Ragone D (2006) *Artocarpus altilis* (breadfruit), ver. 2.1. In: Elevitch CR (ed) Species profiles for Pacific Island agroforestry. Permanent Agriculture Resources (PAR), Hōlualoa, http://www.traditionaltree.org

Rajendran R (1992) *Artocarpus altilis* (Parkinson) Fosberg. In: Verheij EWM, Coronel RE (eds) Plant resources of South-East Asia No 2. Edible fruits and nuts. Prosea Foundation, Bogor, pp 83–86

Rincón AM, Padilla FC (2004) Physicochemical properties of Venezuelan breadfruit (*Artocarpus altilis*) starch. Arch Latinoam Nutr 54(4):449–456

Shamaun SS, Rahmani M, Hashim NM, Ismail HB, Sukari MA, Lian GE, Go R (2010) Prenylated flavones from *Artocarpus altilis*. J Nat Med 64(4):478–481

Shieh WL, Lin CN (1992) A quinonoid pyranobenzoxanthone and pyranodihydrobenzoxanthone from *Artocarpus communis*. Phytochemistry 31(1):364–367

Shimizu K, Kondo R, Sakai K (1997) A stilbene derivative from *Artocarpus incisus*. Phytochemistry 45:1297–1298

Shimizu K, Kondo R, Sakai K, Lee SH, Sato H (1998) The inhibitory components from *Artocarpus incisus* on melanin biosynthesis. Planta Med 64(5):408–412

Shimizu K, Fukuda M, Kondo R, Sakai K (2000a) The 5-α-reductase inhibitory components from heartwood of *Artocarpus incisus*: structure-activity investigations. Planta Med 66(1):16–19

Shimizu K, Kondo R, Sakai K (2000b) Inhibition of tyrosinase by flavonoids, stilbenes and related 4-substituted resorcinols: structure-activity investigations. Planta Med 66(1):11–15

Shimizu K, Kondo R, Sakai K, Buabam S, Dilokkunanant U (2000c) A geranylated chalcone with 5 α-reductase inhibitory properties from *Artocarpus incisus*. Phytochemistry 54(8):737–739

Shimizu K, Kondo R, Sakai K, Takeda N, Nagahata T (2002) The skin-lightening effects of artocarpin on UVB-induced pigmentation. Planta Med 68(1):76–79

Skoumal M, Haberhauer G, Kolarz G, Hawa G, Woloszczuk W, Klingler A (2005) Serum cathepsin K levels of patients with longstanding rheumatoid arthritis: correlation with radiological destruction. Arthritis Res Ther 7:R65–R70

U.S. Department of Agriculture, Agricultural Research Service (2010) USDA national nutrient database for standard reference, release 23. Nutrient Data Laboratory home page. http://www.ars.usda.gov/ba/bhnrc/ndl. Accessed Feb 2010

Walter A, Sam C (2002) Fruits of oceania. ACIAR monograph No 85. Australian Centre for International Agricultural Research, Canberra, 329 pp

Wang Y, Deng T, Lin L, Pan Y, Zheng X (2006) Bioassay-guided isolation of antiatherosclerotic phytochemicals from *Artocarpus altilis*. Phytother Res 20(12):1052–1055

Wang Y, Xu K, Lin L, Pan Y, Zheng X (2007) Geranyl flavonoids from the leaves of *Artocarpus altilis*. Phytochemistry 68(9):1300–1306

Wei BL, Weng JR, Chiu PH, Hung CF, Wang JP, Lin CN (2005) Antiinflammatory flavonoids from *Artocarpus heterophyllus* and *Artocarpus communis*. J Agric Food Chem 53(10):3867–3871

Weng JR, Chan SC, Lu YH, Lin HC, Ko HH, Lin CN (2006) Antiplatelet prenylflavonoids from *Artocarpus communis*. Phytochemistry 67(8):824–829

Young RE, Williams LAD, Gardner MT, Fletcher CK (1993) An extract of the leaves of the breadfruit *Artocarpus altilis* (Parkinson) Fosberg exerts a negative inotropic effect on rat myocardium. Phytother Res 7(2):190–193

Zerega NJC, Ragone D, Motley TJ (2005) Systematics and species limits of breadfruit (*Artocarpus*, Moraceae). Syst Bot 30(3):603–615

Artocarpus anisophyllus

Scientific Name

***Artocarpus anisophyllus* Miq.**

Synonyms

Artocarpus anisophyllus var. *sessifolius* Kochummen, *Artocarpus klidang* Boerl., *Artocarpus superbus* Becc.

Family

Moraceae

Common/English Name

Entawak

Vernacular Names

Borneo: Bakil, Bintawak, Danging, Bintawa, Entawa, Entawak, Mantawa, Mentawa, Pepuan, Puan, Tarap ikal;
Brunei: Tarapikul;
Peninsular Malaysia: Keledang Babi;
Sabah: Terap Ikal;
Sarawak: Bintawak, Bintawa (Iban);

Origin/Distribution

The species is native to Malaysia, Sumatra, Borneo (throughout the island), and the Philippines.

Agroecology

It occurs in undisturbed to slightly disturbed mixed dipterocarp lowland forests up to 700 elevations. It is also found in secondary forests usually present as a pre-disturbance remnant. In its natural habitat, it occurs on alluvial sites near rivers and streams, but also on hillsides and ridges. It occurs on sandy to clay soils, also on calcareous soils.

Edible Plant Parts and Uses

Deep orange pulp is sweet and soft and is eaten fresh. Seeds can be eaten roasted or boiled.

Botany

Evergreen, medium, mid-canopy tree up (Plate 1) to 30 m high, with a 180 cm girth and old tree with a heavy, dark green, rounded crown. Bole with spreading buttress up to 2.5 m high. Bark is grey, smooth to dimple. Stem with white sap.

T.K. Lim, *Edible Medicinal And Non-Medicinal Plants: Volume 3, Fruits*,
DOI 10.1007/978-94-007-2534-8_38,

Plate 1 Foliage and sparse branching of young tree

Plate 3 Dried imparipinnate leaf under surface

Plate 2 Dried pinnate leaf (*top surface*)

Twigs are stout, buds are large, onion-shaped and covered with brown hairs. Stipules are approximately 100 mm long and hairy. Leaves are imparipinnate, 30–90 cm long with 8–12 pairs of stalked alternate leaflet and a terminal one, leaflet of two sizes small pair alternating with large pair, leathery, oblong glabrous, apex pointed, base equal, secondary nerves 7–20 pairs, raised distinctly below (Plates 1, 2 and 3). Flowers are yellowish, 2 mm diameter, placed within an elongate, compressed, fused inflorescence head on leafy twig. Male and female flower head in same leaf axil. Male head finger-like, 5–7.5 cm by 2 cm on a 7.5–10 cm long stalk. Fruit 10–12.5 cm by 7.5–10 cm, large, subglobose, yellowish-brown, syncarp with blunt thick 1.25 cm long spines (Plates 4 left and right) with numerous seed inside. Fruit stalk 4–9 cm long. Seed is ellipsoid, 17 mm by 10 mm and covered with deep orange pulp.

Nutritive/Medicinal Properties

Little is known about its nutritive value and medicinal attributes.

The Iban in Sarawak mix ash from the burnt leaves in a little cooking oil and apply to cuts, wound, boils and itching scalp in children.

Other Uses

The wood (keledang wood) is of fairly good quality and is used for gun stocks, hilts and parang sheaths. The bark is locally used as rope for backpacks. The Selako native community in Sarawak hang leaves on the door to prevent evil spirits from entering the premise.

Comments

It is propagated by seeds.

Plate 4 Mature (*left*) and ripe bitanwak fruit (*right*)

References

Browne FG (1955) Forest trees of Sarawak and Brunei and their products. Government Printing Office, Kuching

Chai PPK (2006) Medicinal plants of Sarawak. Lee Ming Press, Kuching, p 212

Jarrett FM (1959a) Studies in *Artocarpus* and allied genera. III. A revision of *Artocarpus* subgenus *Artocarpus*. J Arnold Arbor 40(2):113–155

Jarrett FM (1959b) Studies in *Artocarpus* and allied genera, III. A revision of *Artocarpus* subgenus *Artocarpus* (cont.). J Arnold Arbor 40(4):327–368

Kessler, PJA (ed) (2000) Secondary forest trees of Kalimantan, Indonesia. A manual to 300 selected species. Tropenbos-Kalimantan Series 3. Tropenbos Foundation, Leiden, 404pp

Kochummen KM (1978) Moraceae. In: Ng FSP (ed) Tree flora of Malaya, vol 3. Longman, Kuala Lumpur, pp 119–168

Slik, JWF (2006) Trees of Sungai Wain. Nationaal Herbarium Nederland, Leiden. http://www.nationaalherbarium.nl/sungaiwain/. Accessed Feb 2010

Voon BH, Chin TH, Sim CY, Sabariah P (1988) Wild fruits and vegetables in Sarawak. Sarawak Department of Agriculture, Kuching, p 114

Artocarpus camansi

Scientific Name

Artocarpus camansi Blanco.

Synonyms

Artocarpus incisa L. f. var. *muricata* Becc., *Artocarpus leeuwenii* Diels.

Family

Moraceae

Common/English Name

Breadnut

Vernacular Names

Caribbean: chataigne;
French: Chataignier;
Indonesia: Kaluweh, Kaluwih, (Java), Kolor, (Madurese), Timbul, (Malay), Kelewih, Kelur, Kulur, Sikun (Sundanese);
Malaysia: Kulur, Kelur, Kulor, Kuror;
Marquesas: Mei Kakano;
Papua New Guinea: Kapiak;
Philippines: Kamansi, Kamongsi, Pakok, Dulugian, Pakau, Kolo, Ugod;
Puerto Rico: Pana De Pepitas;
Spanish: castaña;
Sri Lanka: kos del.

Origin/Distribution

Breadnut is indigenous to New Guinea and Moluccas, and probably naturalized in the Philippines. Cultivated in Indonesia, Malaysia, the Caribbean Islands, Pacific Islands, tropical Central and South America, and coastal West Africa. It is now found only in cultivation in the Philippines, where it is typically grown as a backyard tree. It is infrequently grown in the Pacific islands outside of its native range. A few trees can be found in New Caledonia, Pohnpei, the Marquesas, Tahiti, Palau, and Hawai'i, mainly introduced by immigrants from the Philippines.

Agroecology

Breadnut, like breadfruit, has a wide range of adaptability to tropical ecological conditions and thrives well wherever breadfruit is grown. It will not grow where the temperatures go below 5°C with mean annual rainfall of 1,300–3,800 mm. It grows best in equatorial lowlands below 600–650 but also occur at elevations up to 1,550 m. The tree prefers light, well-drained soils (sands, sandy loams, loams, and sandy clay loams) with pH of 6–7.5 and in full sun.

T.K. Lim, *Edible Medicinal And Non-Medicinal Plants: Volume 3, Fruits*,
DOI 10.1007/978-94-007-2534-8_39, © Springer Science+Business Media B.V. 2012

Plate 3 (**a**, **b**) Piles of immature fruit being sold in local markets in Java

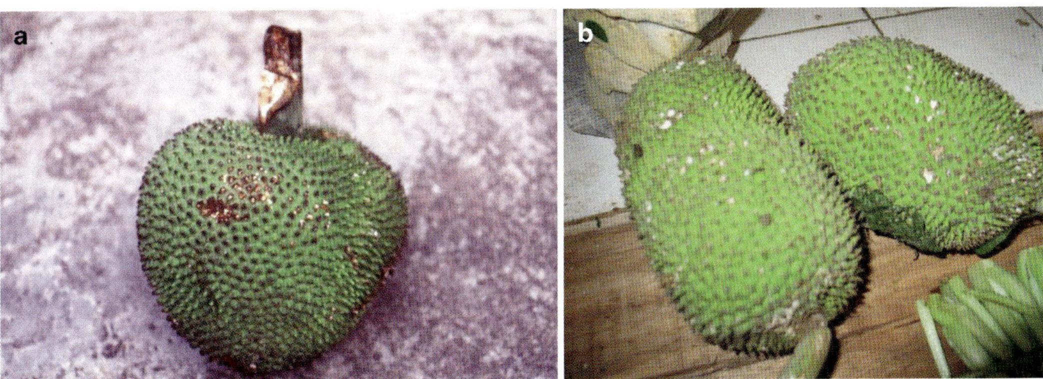

Plate 4 (**a**, **b**) Mature fruit of sub-globose and oval fruited varieties

Plate 5 Developing fruit of globose variety and leaf

Edible Plant Parts and Uses

The fruit and seeds are edible. Breadnut is grown primarily for its nutritious seed as its pulp is scanty. The seeds are rich in starch and are eaten either boiled or roasted. Boiled or roasted breadnut seeds are delicious and taste like chestnut. Boiled seeds make an excellent snack. The young immature and half-ripe fruits (Plates 3, 4 and 5) and seeds (Plate 5) are peeled, thinly sliced and cooked as a vegetable in soups, stews or curries. In Indonesia, the seeds are also made into *kripik* a fried delicacy. In the Philippines, cookies are made from camansi seed flour.

Botany

Evergreen perennial tree grows to height of 10–20 m with a bole diameter of 1 m and a trunk growing to 5 m high before branching (Plate 1) and buttresses at the base. It has a more open branching structure than breadfruit (*Artocarpus* altilis) or dugdug (*A. mariannensis*). A sticky, white, milky latex is present in all parts of the tree. Leaves are alternate, large, 40–60 cm long by 25–45 cm wide leathery, ovate to oblong-ovate typically pinnately lobed with 4–6 pairs of lobes and sinuses cut half way to the midrib (Plate 2). Juvenile leaves may be larger. Leaves are densely pubescent, with many white or reddish-white hairs on upper and lower veins, lower leaf surface, and petiole. Lamina is dull green with green veins. Two large green stipules enclose the bud, turning yellow before dehiscing. Staminate and pistillate flowers occur on the same tree at the ends of branches, with the male inflorescence appearing first. Male flowers are club-shaped, up to 3 cm in diameter and 25–35 cm long. Thousands of tiny flowers with two anthers are attached to a central spongy core. Female inflorescences consist of 1,500–2,000 reduced flowers attached to a spongy core. Unlike breadfruit, the individual flowers do not fuse together along their length. The fruit is a large fleshy syncarp, oval or ovoid to subglobose, 13–20 cm by 7–12 cm, surface colour green to yellowish-green when ripe, fruit surface echinate with narrowly conical, flexuous anthocarp apices up to 5 mm at the base and 5–15 mm long (Plates 3, 4 and 5). Fruit flesh is yellowish white, fragrantly aromatic when ripe, spongy and less dense due to partial fusion between medial portions of the perianths of adjacent flowers. The fruit contains numerous seeds (achenes), from 12 to as many as 150, comprising 30–50% or more of the total fruit weight (Plates 6 and 7). The seeds are rounded or flattened and about 2.5 cm long. They have a thin, light-brown, outer seed coat that is marked with darker veins.

Nutritive/Medicinal Properties

The spiky breadnut fruits have little pulp and are primarily grown for their large, nutritious seeds. The seed is a good source of protein (13–20%) and low in fat (6–29%) compared to nuts such as almond, brazil nut, and macadamia nut (Ragone 2006). The fat extracted from the seed is a light yellow, viscous liquid at room temperature with

Plate 6 Seedy fruit with little perianth flesh

Plate 7 Cooked seeds sold in a local market in Kuching

a characteristic odour similar to that of peanuts. It has a chemical number and physical properties similar to those of olive oil. Its seeds are a good source of minerals and contain more niacin than most other nuts. In 100 g edible portion, four amino acids, methionine (3.2 g), leucine (2.6 g), isoleucine (2.4 g), and serine (2.1 g) comprised 50% of 14 amino acids analyzed (Ragone 2006).

Nutritional composition of breadnut seeds per 100 g edible portion (dry weight basis): water 56.0–66.2%, protein 13.3–19.9 g, carbohydrate 76.2 g, fat 6.2–29.0 g, calcium 66–70 mg, potassium 380–1,620 mg, phosphorus 320–360 mg, iron 8.7 mg, magnesium 10.0 mg, sodium 1.6 mg, niacin 8.3 mg (McIntoch and Manchew 1993; Negron de Bravo et al. 1983; Quijano and Arango 1979). Proximate nutritive composition of fresh and boiled breadnut seeds per 100 g edible portion are boiled seeds in brackets after fresh nuts: moisture 60.15 mg (61.59 mg), crude protein 6.92 mg (6.89 mg) crude fat 3.65 mg (4.20 mg), ash 3.62 mg (3.42 mg), total dietary fibre 10.99 mg, (8.30 mg), total carbohydrate 25.67 mg (23.90 mg), Na, 119.18 mg (204.61 mg), K 599.97 mg (34.62 mg), Ca 10.70 mg (9.62 mg), Fe 1.24 mg (1.18 mg), P 4.30 mg (4.28 mg) Mg 44.78 mg (46.66 mg), Zn 0.74 mg (0.69 mg), Cu 0.34 mg (0.34 mg) and Mn 0.34 mg (0.36 mg) (Williams and Badrie 2005). Adeleke and Abiodun (2010) reported the following proximate composition for breadnut seeds on a wet weight basis: protein 4.87%, fat 3.48%, carbohydrate 26.11%, ash 3.43%, crude fiber 1.20%, moisture 60.96%, P 363 mg/100 g, K 325 mg/100 g, Na 248 mg/100 g, Ca 185 mg/100 g, Mg 1.48 mg/100 g, Fe 0.05 mg/100 g, Mn 1.2 mg/100 g and Cu 0.12 mg/100 g. The predominant essential amino acids (mg/g N) determined in the seeds were leucine 392 mg, phenylalanine 312 mg, arginine 293 mg, isoleucine 245 mg and lysine 275 mg followed by tyrosine 185 mg, histidine 167 mg, cystine 112 mg, methionine 95 mg and tryptophan 24 mg. The oil was rich in palmitic fatty acid 21.4%, oleic 12.4% and linolenic acid 14.8% and also had stearic 2.0%, arachidonic 1.9% and lauric acid 1.7%. Lactic and citric acids were the predominant organic acids while malic, acetic and butyric acids were present in trace amounts. The breadnut seeds could be used as composite flour and the oil could be a good source of edible oil for human consumption.

A lectin was isolated and purified from the seeds of *Artocarpus camansi* (Occena et al. 2007). The purified lectin was found to be a non-blood type specific since it agglutinated all human blood types. It agglutinated erythrocytes without trypsin treatment and addition of metal ions. The isolated lectin did not show any specificity towards the ten sugars used in this study. The molecular weight of the purified lectin was found to be around 2,000 kDa.

The breadnut tree probably has medicinal properties similar to breadfruit. In the Philippines, decoction of the bark is reported to be used as a vulnerary and for dysentery.

Other Uses

The wood is lightweight, flexible, and easy to work and is carve into statues, bowls, fishing floats, and other objects. The tree is also used as fuel-wood. Dried male flowers can be burnt to repel mosquitoes and other flying insects. Its latex and inner bark can be utilized in the same way as breadfruit.

Comments

The tree is propagated from seeds.

Selected References

Adeleke RO, Abiodun OA (2010) Nutritional composition of breadnut seeds (*Artocarpus camansi*). Afr J Agric Res 5(11):1273–1276

Coronel RE (1986) Promising fruits of the Philippines. University of the Philippines at Los Baños, College of Agriculture, Laguna

McIntoch C, Manchew P (1993) The breadfruit in nutrition and health. Trop Fruits Newslett 6:5–6

Negron de Bravo E, Graham HD, Padovani M (1983) Composition of the breadnut (seeded breadfruit). Caribb J Sci 19:27–32

Occena IV, Mojica ERE, Merca FE (2007) Isolation and partial characterization of a lectin from the seeds of *Artocarpus camansi* Blanco. Asian J Plant Sci 6(5):757–764

Ochse JJ, Bakhuizen van den Brink RC (1980) Vegetables of the Dutch Indies, 3rd edn. Ascher & Co, Amsterdam, 1016pp

Powell JM (1976) Ethnobotany. In: Paijmans K (ed) New Guinea vegetation. Part III. Elsevier, Amsterdam, pp 106–184

Quijano J, Arango GJ (1979) The breadfruit from Colombia – a detailed chemical analysis. Econ Bot 33(2):199–202

Ragone D (2003) Breadfruit. In: Caballero B, Trugo L, Finglas P (eds) Encyclopedia of food sciences and nutrition. Academic Press, San Diego, pp 655–661

Ragone D (2006) *Artocarpus camansi* (breadnut), version 2.1. In: Elevitch CR (ed) Species profiles for pacific island agroforestry. Permanent Agriculture Resources (PAR), Holualoa. http://www.traditionaltree.org. Accessed November 2009.

Ragone D (1997) Breadfruit *Artocarpus altilis* (Parkinson) Fossberg. (*Promoting the Conservation and Use of Underutilized and Neglected Crops* (10). International Plant Genetic Resources Institute, Rome, 77 pp

Tipan MC, Regatalio ED (2008) Acceptability of camansi (*Artocarpus camansi*) cookies stored in polypropylene cellophane of varying thickness. Philippine J Crop Sci 38th Annual CSSP Scientific Conference, , Iloilo City (Philippines), 12–16 May 2008 pp 126–127.

Williams K, Badrie N (2005) Nutritional composition and sensory acceptance of boiled breadnut *(Artocarpus camansi* Blanco) seeds. J Food Technol 3(4): 546–551

Zerega NJC, Ragone D, Motley TJ (2005) Systematics and species limits of breadfruit (*Artocarpus*, Moraceae). Syst Bot 30(3):603–615

Artocarpus dadah

Scientific Name

Artocarpus dadah Miq.

Synonyms

Antiaris palembanica Miq., *Artocarpus erythrocarpus* Korth. ex Miq., *Artocarpus inconstantissimus* (Miq.) Miq., *Artocarpus lakoocha* Roxburgh, *Artocarpus lakoocha* var. *malayanus* King, *Artocarpus mollis* Miq., *Artocarpus peltatus* Merr., *Artocarpus reniformis* Becc., *Artocarpus rufa* Miq., *Artocarpus rufescens* Miq., *Artocarpus tampang* Miq., *Ficus inconstantissima* Miq., *Ficus tampang* Miq., *Saccus dadah* Kuntze.

Family

Moraceae

Common/English Names

Dadak, Tampang Bulu

Vernacular Names

Malaysia: Tampang Bulu, Tampang Manis, Keledang Berok;
Indonesia: Dadak, Darak, Tampang, Tampang Dadak, Tampang Telor;
Thailand: Hat-Lukyai, Hat-Rum, Thangkhan.

Origin/Distribution

The species is indigenous to Malaysia and Indonesia. The species was also found in four large islands (Tarutao, Adang, Rawi and Lipe) part of Satun Province off the southwest coast of Thailand (Congdon et al. 1981).

Agroecology

The species is common in lowland evergreen and deciduous forests and open country, up to 1,000 m elevation. It is frequently found in villages throughout Malaya, Sumatra, Simalur, Banka and Borneo.

Edible Plant Parts and Uses

The ripe fruit is edible.

Botany

A small to medium-sized tree up to 24 m high, girth 120 cm, bark grey with distant fissures, twigs brown pubescent. Leaves on stalk

T.K. Lim, *Edible Medicinal And Non-Medicinal Plants: Volume 3, Fruits*,
DOI 10.1007/978-94-007-2534-8_40, © Springer Science+Business Media B.V. 2012

Plate 1 Immature fruit and leaves

Plate 3 Overripe fruit

Plate 2 Ripe subglobose fruit

Plate 4 Deep pink flesh

1.25–2.5 cm long, pubescent petioles; lamina oblong, leathery, 10–30 by 5–15 cm, tip acuminate, base rounded unequal side, broadly wedge shaped, margin entire, juvenile leaves deeply lobed, secondary veins 10–20 pairs distinct above, prominently raised below, green above (Plate 1), glaucous and hairy below. Flower heads on leafy twigs. Male heads yellowish, round, 0.8–2 cm on 0.8–2 cm stalk. Female heads 1.25–2 cm wide on 2.5–3.25 cm long stalk. Fruit roundish or lumpy, finally velvety or appearing smooth not spiny or chequered, 2.8–8.5 cm (Plates 1, 2 and 3) wide with deep pink flesh (Plate 4), on 1.25–3.75 cm long stalk. Seeds ellipsoid, 12 by 8 mm.

Nutritive/Medicinal Properties

No information has been published on its nutritive values.

Fractionation of an ethyl acetate-soluble extract of the bark of *Artocarpus dadah* afforded three new prenylated stilbenoid derivatives, 3-(γ,γ-dimethylallyl) resveratrol (1), 5-(γ, γ-dimethylallyl) oxyresveratrol (2), 3-(2,3-dihydroxy-3-methylbutyl) resveratrol (3), and a new benzofuran derivative, 3-(γ, γ-dimethylpropenyl) moracin M (4), along with six known compounds, oxyresveratrol, (+)-catechin, afzelechin-3-O-α-L-rhamnopyranoside, (−)-epiafzelechin, dihydromorin, and epiafzelechin-(4β→8)-epicatechin (Su et al. 2002). From an ethyl acetate-soluble extract of the twigs, compound 4 and two new neolignan derivatives, dadahols A (5) and B (6), as well as 10 known compounds, oxyresveratrol, (+)-catechin, afzelechin-3-O-α-L-rhamnopyranoside, resveratrol, steppogenin, moracin M, isogemichalcone B, gemichalcone B, norartocarpetin, and engeletin were isolated. The inhibitory effects against both cyclooxygenase-1 (COX-1) and –2 (COX-2) and in a mouse mammary organ culture assay of the isolates were determined.

Oxyresveratrol, a derivative compound of stilbene, was isolated from the root wood of *Artocarpus dadah* (Suhartati et al. 2009a). The compound exhibited high cytotoxicity against leukemia P-388 cancer cell and displayed high bioactivity against *Bacillus subtillis*, *Escherichia coli*, *Rhizopus oligosporus*, and *Fusarium oxysporum*. Morusin, a flavonoid prenylated at C-3 compound, isolated from the root bark of *Artocarpus dadah* exhibited high cytotoxycity against murine leukemia cell P-388 with an IC_{50} value of 3.1 μg/ml (Suhartati et al. 2009b).

Other Uses

A minor timber tree.

Comments

The tree is uncommon and has been listed as threatened.

Selected References

Congdon G, Sirirugsa P, Lojanapiwatna V, Wiriyachitra P (1981) A contribution to the Thai phytochemical survey II. J Sci Soc Thai 7:87–90

Jarrett FM (1959a) Studies in *Artocarpus* and allied genera. III. A revision of *Artocarpus* subgenus *Artocarpus*. J Arnold Arbor 40(2):113–155

Jarrett FM (1959b) Studies in *Artocarpus* and allied genera, III. A revision of *Artocarpus* subgenus *Artocarpus* (cont.). J Arnold Arbor 40(4):327–368

Kessler PJA, Sidiyasa K, Amriansyah, Zainal A (1995) Checklist of secondary forest trees in East and South Kalimantan, Indonesia. The Tropenbos Foundation, Wageningen, 84pp

Kochummen KM (1978) Moraceae. In: Ng FS (ed) Tree flora of Malaya, vol 3. Longman, London, pp 119–168

Slik JWF (2006) Trees of Sungai Wain. Nationaal Herbarium Nederland, Leiden. http://www.nationaalherbarium.nl/sungaiwain/

Lemmens RHMJ, Soerianegara I, Wong WC (1995) Plant resources of South-East Asia No 5(2). Timber trees: minor commercial timbers. Backhuys Publishers, Leiden, 655pp

Ng PKL, Wee YC (1994) The Singapore Red data book. Threatened plants and animals of Singapore. The Nature Society, Singapore, 343pp

Ridley HN (1922–1925) The flora of the Malay Peninsula, vol 5, Government of the straits settlements and federated Malay States. L. Reeve & Co, London

Su BN, Cuendet M, Hawthorne ME, Kardono LB, Riswan S, Fong HH, Mehta RG, Pezzuto JM, Kinghorn AD (2002) Constituents of the bark and twigs of *Artocarpus dadah* with cyclooxygenase inhibitory activity. J Nat Prod 65(2):163–169

Suhartati T, Yandri AS, Hadi S (2009a) The bioactivity test of oxyresveratrol, a bioactive compound, isolated from the root wood of *Artocarpus dadah*. Int J Pure App Chem 4(3):223–229

Suhartati T, Yandri AS, Hdi S, Fatriyadi FS (2009b) Morusin, a bioactive compound from the root bark of *Artocarpus dadah*. Eur J Sci Res 38(4):643–648

Verheij EWM, Coronel RE (eds) (1992) Plant resources of south-east Asia, No. 2. Edible fruits and nuts. Prosea Foundation, Bogor, pp 209–211

Artocarpus elasticus

Scientific Name

Artocarpus elasticus **Reinw. ex Blume.**

Synonyms

Artocarpus blumei Trecul, *Artocarpus blumei* var. *kunstleri* (King) Boerl., *Artocarpus corneri* Kuchummen, *Artocarpus jarrettiae* Kochummen, *Artocarpus kunstleri* King, *Artocarpus pubescens* Blume, *Artocarpus scortechinii* King.

Family

Moraceae

Common/English Names

Bendo, Wild Breadfruit-Tree

Vernacular Names

Borneo: Bakil, Danging, Pekalong, Pilang, Talun, Tap, Tekalong, Tarap, Terap, Terap Hutan, Terap Munyit;
Dutch: Wilde Broodboom;
Indonesia: Bendo, Boendah, Benda, Benda Ketan, Benda Kebo (Java), Terap (Sumatra), Terap (Kalimantan), Benda, Tereup (Sundanese);
Malaysia: Jerami, Ahbat (Sakai), Mendi, Ho (Semang), Talun, Kilid (Kelabit, Sarawak), Takalong, Terap Nasi, Mengkoterap Nasi, Tuka, Terap Nasi, Teureup, Terap;
Thailand: Aw (Peninsular).

Origin/Distribution

The species is indigenous to Myanmar, Thailand, Peninsular Malaysia, Sumatra, Java, Lesser Sunda Islands, Borneo (throughout the island) and the Philippines.

Agroecology

Strictly a tropical species, it is common in undisturbed to slightly disturbed mixed dipterocarp, keranga and sub-montane forests up to 1,300 m altitude. It occurs on alluvial sites, hillsides and ridges with sandy to clay soils. In secondary forests, it is usually present as a pre-disturbance remnant. It is cultivated in Java, Kalimantan and Malaysia in regions with short dry periods.

T.K. Lim, *Edible Medicinal And Non-Medicinal Plants: Volume 3, Fruits*,
DOI 10.1007/978-94-007-2534-8_41, © Springer Science+Business Media B.V. 2012

Edible Plant Parts and Uses

Fruits and seeds are eaten. Fruits have an unpleasant rancid odour, but the white pulp is sweet and somewhat savoury. The seeds are fried or roasted like peanuts and eaten.

Botany

Mid-canopy evergreen tree with a buttressed trunk, reaching 40 m high and with a bole diameter of 83 cm. The bark is grey-brown, smooth to scaly with ring scars. Stem with white sap. Twigs are stout and have adpressed hairs and prominent stipular scars. Stipule with silky brown hairs and covers the10–18 cm long buds. Leaves with 5–10 cm long, hairy petioles, spirally arranged, lamina – stiff and coriaceous, entire, broadly elliptic, 22–55 by 12–30 cm wide, densely pubescent on the underside and pubescent on the upper surface. Leaves of sapling and young growth larger and deeply and variously incised into pointed lobes, 60–120 cm long. Flower heads, unisexual on leafy twigs. Male heads, 15–20 cm by 2.5–4 cm, finger-like, furrowed, yellow turning to brown with sessile, small, 1 mm diameter flowers with two obconical perianth and 1 stamen. Female heads upright, barrel-shaped, 10 cm by 7.5 cm, green turning cream-colored, with woolly, shaggy stout, soft, recurved spines. The fruit is a syncarp, formed by the enlarged connate perianths which are adnate to the axis of the inflorescence. Fruit is cylindric, 11.5 cm by 5.5 cm, densely covered with shaggy, woolly spines, yellow brown (Plate 1) with a rancid smell when ripe. Seeds ellipsoid, 10 mm × 6 mm and pulp (perianth) is white.

Nutritive/Medicinal Properties

No nutritive information has been published on the edible fruit.

Recent studies carried out in Portugal, Taiwan and Indonesia reported that the bark and wood of *A. elasticus* yielded many prenylated flavonoid that exhibited antioxidant, cytotoxic, anti-cancer and anti-malarial activities.

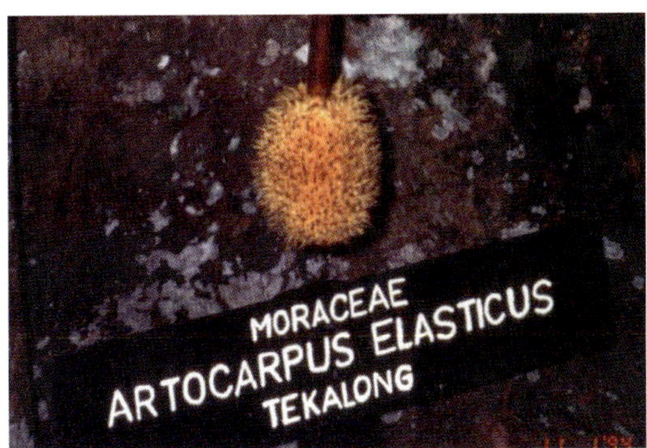

Plate 1 Tekalong fruit

Antioxidant Activity

Studies with the artelastin, a prenylated flavonoid from the wood of *Artocarpus elasticus* revealed that it inhibited production of reactive oxygen species (ROS) by human polymorphonuclear neutrophils (PMNs) and nitric oxide (NO) by J774 murine macrophage cell line (Cerqueira et al. 2008). Artelastin was found to be an inhibitor of ROS production due to a strong O_2-scavenging activity Artelastin showed also to be an inhibitor of NO production without NO scavenging activity. Prenylflavonoids isolated from *Artocarpus elasticus* were found to be promising antioxidants (Lin et al. 2009). Cycloartelastoxanthone, artelastoheterol, cycloartobiloxanthone and artonol A, all showed inhibition of oxidative DNA damage. Compounds, cycloartelastoxanthone, artelastoheterol and cycloartobiloxanthone significantly showed 1,1-diphenyl-2-picrylhydrazyl (DPPH)-scavenging activity with IC_{50} values of 18.7, 42.2and 26.8 μM, respectively, while compound 11 significantly displayed inhibitory effects on xanthine oxidase (XO) activity with IC_{50} values of 43.3 μM.

Anticancer Activity

Researchers in Portugal isolated a number of prenylated flavonoids from the wood of *Artocarpus elasticus*. The prenylated flavones were isolated artelastin, artelastochromene , artelasticin, artelastocarpin and artocarpesin from the wood (Kijjoa et al. 1996), followed subsequently by artelastinin, artelastofuran, artelasticin and cyclocommunin (Kijjoa et al. 1998). The two flavonoids artelastocarpin and carpelastofuran and the prenylated flavonoids artelastin, artelastochromene, artelasticin, artocarpesin, and cyclocommunin isolated earlier from this species were active against 3 human cancer cell lines, the cytotoxic effect varying from strong to moderate and with artelastin showing the most potent activity (Cidade et al. 2001). Artelastin, was further evaluated for its capacity to inhibit the growth of 52 human tumour cell lines, representing nine different tumour types (Pedro et al. 2005). A pronounced dose-dependent growth inhibitory effect was detected in all the cell lines, with GI_{50} values ranging from 0.8 to 20.8 μM. Artelastin cytotoxic inhibition of tumour growth was related to its action in disrupting microtubules and interfering with DNA replication in MCF-7 human breast cancer cells (GI_{50}=6.0 μM). GI_{50} is concentration that causes 50% growth inhibition. Artelastin exerted a biphasic effect in the DNA synthesis of MCF-7 cells, a stimulatory effect at low concentrations (below GI_{50}) for short times of exposition (6 and 24 h), and an inhibitory effect at high concentrations (above GI50). Furthermore, artelastin acted as a cytotoxic rather than a cytostatic compound. Massive cytoplasmatic vacuoles were detected in cells after artelastin treatment. Together with these morphological alterations, cells showed the presence of abnormal nuclear morphologies, and occasionally nuclear condensation, which were identified as apoptotic by TUNEL assay. Further studies with the prenylated flavones revealed that the flavones exhibited an antiproliferative effect on two human breast cancer cell lines: estrogen-dependent ER (+) (estrogen receptor positive) MCF-7 and the estrogen-independent ER (−) (estrogen receptor negative) MDA-MB-231 (Pedro et al. 2006). The findings indicated that these flavones could induce autophagy in both ER (+) and ER (−) human breast cancer cell lines.

The prenylated flavones also displayed an effect on the mitogenic response of human lymphocytes to PHA (phorbol 12-myristate 1 3-acetate). Researchers found that the most potent flavones possessed a 3,3-dymethylallyl group (prenyl) at C-8, such as artelastin, which exhibited the highest antiproliferative activity (Cerqueira et al. 2003). Studies revealed that artelastin had an irreversible inhibitory effect on the PHA-induced lymphocyte proliferation. Artelastin was also shown to be a potent inhibitor of both T- and B-lymphocyte mitogen induced proliferation although B-mitogenic response was the more sensitive one. Artelastin decreased the production of IFN-γ (interferon-γ), IL-2 (interleukin-2), IL-4 (interleukin-4) and IL-10 (interleukin-10) in ConA-stimulated splenocytes, but artelastin had no effect on apoptosis of splenocytes.

A new oxepinoflavone, artoindonesianin E1 was isolated from the wood of *Artocarpus elasticus*, along with four known prenylated flavones: artocarpin (2), cycloartocarpin, cudraflavones A and C (Musthapa et al. 2009). Artoindonesianin E1 exhibited cytotoxicity against murine leukemia P-388 cells with an IC_{50} 5.0 μg/ml. In Taiwan, scientists isolated more new prenylated flavonoids, artelastoheterol, artelasticinol, cycloartelastoxanthone, artelastoxanthone, and cycloartelastoxanthendiol, along with five known compounds, from the root bark of *Artocarpus elasticus* (Ko et al. 2005). The previously known compound, Artonol A exhibited cytotoxic activity against the A549 human cancer cell line, with an ED_{50} value of 1.1 μg/ml.

Antimalarial Activity

Three new prenylated flavones, trivially named artoindonesianin E-1 (1), artoindonesianin Z-4 (2) and artoindonesianin Z-5 (3), beside 17 known prenylated flavones were isolated from *A. heterophyllus, A. elasticus* and *A. lanceifolius* (Musthapa et al. 2010). Antimalarial effect of some selected isolated flavone derivatives showed that artonin E exhibited very strong inhibition (IC_{50}=0.1 mg/ml) against K1 strain in comparison against 3D7 strain of *Plasmodium falciparum* (IC_{50} 0.3 mg/ml). A related prenylated flavone, namely 12-hydroxyartonin E, exhibited strong inhibition (IC_{50}=0.9 mg/ml) against K1 strain, but weak inhibition (IC_{50}=14.3 mg/ml) against 3D7 strain. In addition, the other isolated flavone derivatives showed moderate inhibition with IC_{50} of r 2.1, 1.6, 3.6, 1.3, 6.7 and 2.1 mg/ml respectively against both strains of *P. falciparum*, except for flavanone-3-ol derivative (dihydromorin), which had no inhibition.

Traditional Medicinal Uses

Various parts of the tree have been used in traditional folkloric medicine in Southeast Asia. Leaves were given to consumptive patients and to nursing mothers in Malaysia. In Sarawak, young shoots are prescribed as a remedy for vomiting blood in the Iban community. Shoots are pounded together with shoots of *Alstonia* sp. (pelai) and boiled to make a herbal tea. Latex was used for dysentery in Java, roots used as an aperient decoction. Inner bark used for poulticing ulcers.

Other Uses

The light hardwood timber is suitable for light construction, posts, beams, joists, rafters, flooring, plywood, core veneer, packing boxes and crates, wooden pallets (expendable type), non-striking tool handles, pattern making, panelling, mouldings, furniture, joinery, and fishing boats.

Fibres from the bark used for making, strings, basket straps, shirts, coarse blanket and loin cloths by the jungle tribes in Malaysia who also make fishing nets and lines and for binding their bamboo blow-pipes. The fibres have also been reported as a material for paper and used for straps on baskets and blankets. The bark is also use to line rice-bins and use to make house walls. Latex is used for bird-lime. The oil of the seeds is used to prepare hair-oil.

Comments

The tree is propagated from seeds.

Selected References

Backer CA, van den Brink B, Jr RC (1965) Flora of java (spermatophytes only), vol 2. Wolters-Noordhoff, Groningen, p 641

Burkill IH (1966) A dictionary of the economic products of the Malay Peninsula. Revised reprint 2 volumes. Ministry of agriculture and co-operatives, Kuala Lumpur. Vol 1 (A–H), pp. 1–1240; Vol 2 (I–Z), pp 1241–2444

Cerqueira F, Cordeiro-da-Silva A, Araújo N, Cidade H, Kijjoa A, Nascimento MS (2003) Inhibition of lymphocyte proliferation by prenylated flavones: artelastin as a potent inhibitor. Life Sci 73(18):2321–2334

Cerqueira F, Cidade H, van Ufford L, Beukelman C, Kijjoa A, Nascimento MS (2008) The natural prenylated flavone artelastin is an inhibitor of ROS and NO production. Int Immunopharmacol 8(4):597–602

Chai PPK (2006) Medicinal plants of Sarawak. Lee Ming Press, Kuching, p 212

Cidade HM, Nacimento MS, Pinto MM, Kijjoa A, Silva AM, Herz W (2001) Artelastocarpin and carpelastofuran, two new flavones, and cytotoxicities of prenyl flavonoids from *Artocarpus elasticus* against three cancer cell lines. Planta Med 67(9):867–870

Jarrett FM (1959a) Studies in *Artocarpus* and allied genera. III. A revision of *Artocarpus* subgenus *Artocarpus*. J Arnold Arbor 40(2):113–155

Jarrett FM (1959b) Studies in *Artocarpus* and allied genera, III. A revision of *Artocarpus* subgenus *Artocarpus* (cont.). J Arnold Arbor 40(4):327–368

Kijjoa A, Cidade HM, Pinto MMM, Gonzalez MJTG, Anantachoke C, Gedris TE, Herz W (1996) Prenylflavonoids from *Artocarpus elasticus*. Phytochemistry 43(3):691–694

Kijjoa A, Cidade HM, Gonzalez MJTG, Afonso CM, Silva AMS, Herz W (1998) Further prenylflavonoids from *Artocarpus elasticus*. Phytochemistry 47(5): 875–878

Ko HH, Lu YH, Yang SZ, Won SJ, Lin CN (2005) Cytotoxic prenylflavonoids from *Artocarpus elasticus*. J Nat Prod 68(11):1692–1695

Kochummen KM (1978) Moraceae. In: Ng FSP (ed) Tree flora of Malaya, vol 3. Longman, London, pp 119–168

Lin KW, Liu CH, Tu HY, Ko HH, Wei BL (2009) Antioxidant prenylflavonoids from *Artocarpus communis* and *Artocarpus elasticus*. Food Chem 115(2):558–562

Musthapa I, Juliawaty LD, Syah YM, Hakim EH, Latip J, Ghisalberti EL (2009) An oxepinoflavone from *Artocarpus elasticus* with cytotoxic activity against P-388 cells. Arch Pharm Res 32(2):191–194

Musthapa I, Hakim EH, Juliawaty LD, Syah YM, Achmad SA (2010) Prenylated flavones from some Indonesian *Artocarpus* and their antimalarial properties. Int J Phytomed Relat Ind 2:2

Pedro M, Ferreira MM, Cidade H, Kijjoa A, Bronze-da-Rocha E, Nascimento MS (2005) Artelastin is a cytotoxic prenylated flavone that disturbs microtubules and interferes with DNA replication in MCF-7 human breast cancer cells. Life Sci 77(3):293–311

Pedro M, Lourenco CF, Cidade HM, Kijjoa A, Pinto M, Nascimento MSJ (2006) Effects of natural prenylated flavones in the phenotypical ER (+) MCF-7 and ER (–) MDA-MB-231 human breast cancer cells. Toxicol Lett 164(1):24–36

Slik JWF (2006) Trees of Sungai Wain. Nationaal Herbarium Nederland, Leiden. http://www.nationaalherbarium.nl/sungaiwain/

Voon BH, Chin TH, Sim CY, Sabariah P (1988) Wild fruits and vegetables in Sarawak. Sarawak Department of Agriculture, Kuching, p 114

Artocarpus glaucus

Scientific Name

Artocarpus glaucus **Blume.**

Synonyms

Artocarpus biformis Miq., *Artocarpus denisoniana* King, *Artocarpus glaucescens* Trecul, *Artocarpus glaucescens* var. *tephrophylla* (Miq.) Miq., *Artocarpus tephrophylla* Miq., *Artocarpus zollingeriana* Miq.

Family

Moraceae

Common/English Name

Buruni

Vernacular Names

Borneo: Buruni, Galing, Tampang Wangi, Telangking, Nangka Pipit, Tampang Buwah, Sembir, Tiwu Landu, Padau Paya.

Origin/Distribution

The species occurs in Malesia - Peninsular Malaysia, Sumatra, Java, Lesser Sunda Islands, Borneo (Sarawak, Sabah, West-, Central- and East-Kalimantan), Moluccas, and the Northern Territory of Australia.

Agroecology

In Malaysia and Indonesia, it occurs in undisturbed, humid mixed dipterocarp forests up to 100 m altitude on alluvial sites near rivers and streams. In secondary forests, it is usually present as a pre-disturbance remnant. In the north-western Northern Territory, Australia, it is found in monsoon forests along water courses.

Edible Plant Parts and Uses

The fruit has been reported as edible.

Botany

A small tree to 20 m and girth to 100 cm, with dark brown, smooth, lenticellate bark. Stem

T.K. Lim, *Edible Medicinal And Non-Medicinal Plants: Volume 3, Fruits*,
DOI 10.1007/978-94-007-2534-8_42, © Springer Science+Business Media B.V. 2012

Plate 1 Lumpy fruit and leaves

Plate 2 Close-up of fruit and flesh

with white sap. Leaves alternate, on 1–2.5 cm long, puberulous petioles with pubescent stipules; lamina simple, ovate, oblong elliptic to elliptic, 10–18 cm long, 5–8 cm wide, attenuate to acuminate at apex, cuneate to rounded at base, often slightly asymmetric, entire, glabrous above, appressed puberulous to glabrescent dark-green above, and grey-glaucous below (mature leaves), penni-veined with 6–13 pairs arching, lateral veins, prominent below (Plate 1). Inflorescence solitary or paired in leaf axils, individual flowers sessile. Male flower head 1.2–4 cm long, narrowly oblong or clavate on short 1–2 mm long stalk. Male flowers 0.6 mm long; tepals 2 or 3; stamen 0.8 mm long, filament cylindrical. Female head globose. Fruit a syncarp subglobose, often lobed and lumpy, to 3 cm diameter, surface papillate or nearly smooth, shortly pubescent, green maturing to orange, orange-red (Plates 1 and 2); peduncle 3–4 mm long.

Nutritive/Medicinal Properties

No information has been published on its nutritive value and medicinal uses.

The plant is not rich in flavanoids. A flavan-3-ol, cathecin and a 3-prenylflavone, cudraflavone C were isolated from *A. glaucus* (Hakim et al. 2006).

Other Uses

The wood is used for house building.

Comments

The species is found in the wild and not cultivated.

Selected References

Chew WL (1989) Moraceae. In: Flora of Australia, vol. 3. Hamamelidales to Casuarinales. Australian Government Publishing Service, Canberra, 219 pp

Dixon DJ (2011) Moraceae. In: Short DS and Cowie ID (eds) Flora of the Darwin region. Northern territory herbarium, Department of Natural Resources, Environment, Arts and Sports, Darwin Vol 1pp. 1–26

Hakim EH, Achmad SA, Juliawaty LD, Makmur L, Syah YM, Aimi N, Kitajima M, Takayama H, Ghisalberti EL (2006) Prenylated flavonoids and related compounds of the Indonesian *Artocarpus* (Moraceae). J Nat Med 60(3):161–184

Jarrett FM (1959a) Studies in *Artocarpus* and allied genera. III. A revision of *Artocarpus* subgenus *Artocarpus*. J Arnold Arbor 40(2):113–155

Jarrett FM (1959b) Studies in *Artocarpus* and allied genera, III. A revision of *Artocarpus* subgenus *Artocarpus* (cont.). J Arnold Arbor 40(4):327–368

Kochummen KM (1978) Moraceae. In: Ng FSP (ed) Tree flora of Malaya, vol 3. Longman, London, pp 119–168

Ridley HN (1922–1925) The flora of the Malay Peninsula, vol 5, Government of the straits settlements and federated Malay States. L. Reeve & Co, London

Slik JWF (2006) Trees of Sungai Wain. Nationaal Herbarium Nederland, Leiden. http://www.nationaalherbarium.nl/sungaiwain/

Verheij EWM, Coronel RE (eds) (1992) Plant resources of South-East Asia, No. 2. Edible fruits and nuts. Prosea Foundation, Bogor, pp 209–211

Artocarpus heterophyllus

Scientific Name

Artocarpus heterophyllus Lamarck

Synonyms

Artocarpus integrifolia var. *heterophylla* (Lam.) Pers., *Artocarpus brasiliensis* Gomez, *Artocarpus heterophylla* Lam., *Artocarpus integrifolia* auct., *Artocarpus integrifolia* sensu Trimen non. L. f., *Artocarpus integrifolia* var. *glabra* Stokes, *Artocarpus jaca* Lam., *Artocarpus maxima* Blanco, *Artocarpus philippinensis* Lam., *Polyphema jaca* Lour., *Saccus arboreus major* Rumph., *Sitodium cauliflorum* Gaertn., *Tsjaka-maram* Rheede.

Family

Moraceae

Common/English Names

Jack, Jackfruit, Jack Tree, Jak, Jakfruit, Jak Nut

Vernacular Names

Banaban: Te Mai N-Inria;
Bangladesh: Cāṭtal, Enchor (Unripe Fruit) (Bengali);
Borneo: Bedug, Nangka, Nangka Batu;
Brazil: Jaqueira (Portuguese);
Burmese: Peignai;
Chinese: Bo Luo Mi, Mu Bo Luo, Shu Bo Luo;
Danish: Jackfrugttrae, Jackfrugt, Jacktræ;
Dutch: Siri Broodboom;
Dominican Republic: Guenpan, Guenpan De Masa (Fleshy), Guenpan De Semilla (Seeded);
Eastonian: Erilehine Leivapuu, Vili: Jaka;
Fijian: Uto Ni Idia;
French: Jacquier, Arbre À Pain, Arbre À Pain À Graines;
German: Indischer Brotfruchtbaum, Jackfrucht, Jackfrutchbaum;
Hungarian: Jákafa, Kenyérfa;
Icelandic: Saðningaraldin;
India: Kathal (Assamese), Kathal (Bengali), Katahar (Bhojpuri), Cakki, Katahal, Kathal, Kathai, Kanthal, Katoi, Phannas (Hindu); Phannasa (Gujerati), Alasa, Bokke Gida, Gujja, Halasu, Halasina Hannu, Halasina Mara, Halasu Jaaka, Kantaka Phala, Kojje, Koovi, Panasero, (Kannada), Ponos (Konkani), Lamphong (Kuki), Chakka, Pilavu, Plavu, Tsjaka-Maram (Malayalam), Theibong (Manipuri), Phanas, Fannas, Phanas, Phunnus, Panas (Marathi), Lamkhuang (Mizoram), Lamkhuang, Panas, Ponoso (Oriya), Apuspaphala, Kantakiphala, Panasa, Panasah, Panasam, Pansa, Skandhaphala (Sanskrit), Acani, Aiyinipila, Anjili, Atcaravirutcatti, Ayirankanni, Cakkai, Cantakaputitam, Cantakaputitamaram, Cenkarippala, Cenkarippalamaram, Cikavaram, Cuvaturacayanam, Cuvatuvaki, Paala, Palaasu,

T.K. Lim, *Edible Medicinal And Non-Medicinal Plants: Volume 3, Fruits*,
DOI 10.1007/978-94-007-2534-8_43, © Springer Science+Business Media B.V. 2012

Pala, Palamaram, Pila, Ekaram, Ekaramaram, Ekaravalli, Iracalam, Kantakapalam, Kantapalam, Kurtekam, Kurtekamaram, Kuttippala, Kuttippila, Marican, Matika, Matukamaram, Mirutankapalam, Muppurakkani, Mutpala, Mutpurakkani, Narpala, Narpalamaram, Nattuppala, Palaa, Palacam, Palampala, Palampalamaram, Palaviruccakam, Palavirutcam, Palavumaram, Panacamaram, Paricatti, Pilavu, Piramataru, Pukam, Tagar, Umaipporikam, Umaipporikamaram, Varkkai, Varukkai, Varukkarai, Varukkaippala, Varukkaraimaram, Vatakapavirutti (Tamil), Panas, Panasa, Vaerupanasa, Veru Panasa (Telugu), Gujje (Unripe), Pilakkai (Ripe) (Tulu);
Indonesia: Nangke (Alas, Sumatra), Anaane (Ambon, Maluku), Angga, Mangka, Nangga, Nangka (N. Sulawesi), Pana, Panah, Panaih, Panas (Acheh, Sumatra), Nangka (Bali), Nnka (Bare, Kalimantan), Naka, Nangka, Pinasa, Sibodak (Batak), Nanga, Nangga (Bima, West Nusa Tenggara), Panasa (Boeginisch, Sulawesi), Nango (Boeol, Gorontalo, Sulawesi), Naang, Nakan, Nakane, Nakang (Boeroe, Maluku), Tehele Kaloeen (East Ceram, Maluku), Anaa Ane, Ain Nad Wakane, Inaale, Naka Kota, Nongga, Tafela (West Ceram, Maluku), Amnaalo, Anaato, Tajena (South Ceram, Maluku), Batuk, Baduk, Enaduk, Maauk, Naka, Nangka (Dyak, Kalimantan), Nangka (Flores), Langge (Gorontalo, Sulawesi), Naka (Halmahera, Maluku), Nangka, Nongka (Javanese), Ua Malai (Kisar, Maluku), Belaso, Benaso, Lamasa, Malasa, Menaso (Lampong, Sumatra), Uruwane (Leti, Maluku), Nangka (Madurese), Cidu (Makassar, Sulawesi), Nakale (Makian, North Sulawesi), Nangka (Malay), Nanakang, Nangka (Mandar, Sulawesi), Nangka (Mori, Sulawesi), Nad (Nias, Sumatra), Anad, Anad Wakan, Annal, Anaalo, Nangka (Oelias, Maluku), Nangka (Salajar, Sulawesi), Nangka (Sangir, Sulawesi), Nangke (Sasak, Lombok), Nangga (Saoesoe, Sulawesi Tengah), Hoka, Naga (Sawoe or Sawu, Nusa Tenggara Timor), Anaha, Anasah (Simaloer, North West Sumatra), Naka, Naki, Nangga, Ndeile (Soela, Maluku), Nanga, Nangga Sumba, Nangka (Sumba), Naka, Nakak, Nakat (Solor, Nusa Tenggara Barat), Naka (Ternate, Maluku) Naka (Tidore, Maluku), Kuloh, Naka, Taijonis, Sosak, Nangka Bubor, Keledang (Timor), Kroor, Naka, Naknak (North Papua Barat), Lamasa, Malasa, Menaso, Benaso (Sumatra), Nangka (Sundanese);
Italian: Falso Albero Del Pane;
Japanese: Paramitsu;
Khmer: Khnor, Knol;
Korean: Baramil;
Laotian: Miiz, Miiz Hnang, Mai Mi, Mak Mi, Mi;
Madagascar: Ampalibe (Malagasy);
Malaysia: Nangka;
Maldivian: Sakkeyo;
Mexico: Yaka;
Nepali: Rukh Kutaherr;
Papua New Guinea: Kapiak;
Persian: Derakhte Nan;
Philippines: Langka, Nanka, Nangka (Bisaya), Nangka (Ibanag), Nangka (Iloko), Yangka (Pampangan), Nanka (Sulu), Langka, Nangka (Tagalog);
Portuguese: Jaca, Jaqueira;
Samoan: Ulu Initia;
Sri Lanka: Kos Varaka, Vela (Sinhalese);
Spanish: Árbol Del Pan, Fruta Del Pobre, Jaca, Jaka, Jaqueiro Jaca Buen Pan, Pan De Fruta, Rima;
Swahili: Fenesi;
Swedish: Jackträd, Jackfrukt;
Thailand: Banun, Khanun, Makmi, Makmee, Maak Mee, Maak Laang;
Uganda: Fene (Luganda);
Vietnamese: Mit;
Yapese: Dapanapan.

Origin/Distribution

Jackfruit is native to south Asia – India, Bangladesh, Nepal and Sri Lanka. It was introduced to and has become naturalized in Malaysia and Indonesia. It is commercially grown and sold in South Asia, Southeast Asia and northern Australia. It is also grown in parts of Hawaii, Brazil, Suriname, Madagascar, and in the islands of West Indies such as Jamaica and Trinidad. It is the national fruit of Bangladesh and Indonesia.

Agroecology

The tree grows well in an equatorial to subtropical region at elevations up to 1,600 m with average annual rainfall of 1,000–2,400 mm and temperatures of 24–35°C. It thrives in full sun in well-drained, acid to neutral soils of moderate fertility with pH 5.0–7.5. The tree does not tolerate water logging or poor drainage. It is quite drought and frost tolerant.

Edible Plant Parts and Uses

Immature and ripe fruit perianths and seeds are eaten. The nutritious seeds (jak nuts) are eaten raw, grilled, roasted or boiled, and eaten like chestnuts or ground and blended with wheat flour for baking, or cooked in dishes. Unripe fruits are eaten as vegetables; the pulp and immature seeds are sliced, boiled and cooked in soups and curries, pickled or canned. In India and Sri Lanka, young green jack fruit is sliced after peeling and pickled in brine to which is added vinegar and spices. Cooked, young jackfruit is relished in the cuisines of India, Sri Lanka, Bangladesh, Malaysia, Indonesia, Kampuchea and Vietnam.

Notable examples of food and dessert dishes made with jackfruit include:

From India: *Enchorer Torkari* – curry with unripe jackfruit (West Bengal, also Bangladesh); *Kathal Subji* – spicy vegetable with raw jackfruit (Assam, Punjab and Uttar Pradesh, India); *Kathal Aachar* – pickle made of jackfruit (Uttar Pradesh and Assam); *Ghariyo, Pilakkai Kandbu and Dosa* – sweet jackfruit dishes (Mangalore,); *Chakka Pradaman* – jackfruit pudding (Kerala); *Idiyan chakka* – whole tender jackfruit, pounded and cooked with spices (Kerala); *Chakka aviyal, Chakka erisseri* – sliced unripe jackfruit and other vegetables (Kerala); *Jackfuit Pappad* – Jackfruit Pappad as a snack (Mangalore); *Chakka Varatti* – jackfruit jam (Kerala); *Chakka Vatta l* – jackfruit chips (Kerala); *Panasa Koora* – traditional jackfruit curry (coastal Andhra Pradesh); *Gujjeda Kajipu* – dry spicy curry of raw jackfruit (Mangalore); *Saath* – jackfruit pappadam, served as a snack (Goa and Mangalore); *Jackfruit Halwa* – a sweet made from ripe jackfruit (Udupi); *Raithey* – sour curry with boiled unripe jackfruit, raw mango and raw papaya, part of Navayath cuisine from Bhatkal, Karnataka; *Kathal Subzee* – spicy vegetable with raw jackfruit fried (Uttar Pradesh or Punjab); *Fanas Poli* – sun dried jackfruit pulp with sugar (Konkan); *Fansa Nevaryo* – sweet Navayath dish of rice ground with ripe jackfruit, stuffed with coconut and jaggery, wrapped in banyan leaves, and steamed (Bhatkal, Karnataka); *Chakkapuzhukku* (mashed jackfruit); *Panasapattu* (raw jackfruit curry); *Kadubus* (steamed dumplings made from a puree of ripe fruit, semolina and jaggery); *Sole saaru* (a curry made with semi-ripe jackfruit); raw jackfruit *pulao* (rice or wheat); *pilaf inida* (rice dish); HalasinaKai Palya (raw jackfruit gravy); *Dahi Kathal* (jackfruit yoghurt); jackfruit-potato *bhaji* (vegetable fritters), Kathal Masala curry, jackfruit kebabs.

From Sri Lanka: *Kiri Kos*: a creamy jackfruit curry cooked with coconut milk; *Polos Ambul*: unripe jackfuits cooked with spices.

From Malaysia: Sayur Nangka; Malaysian-style green Jackfruit; jackfruit egg curry; fried long beans with jackfruit.

From Thailand: *Kaeng Khiew Wan Gai* (Thai Sweet Green Chicken Stew), Jack Fruit Soup (Soup *Kanoon)*, Thai jackfruit curry with bell peppers, cashews, and lime leaf; *Ruam Mit* a light dessert with young coconut meat, an assortment of cooked strips of rice flour, syrup and a dash of coconut milk topped with shaved ice, traditional jackfruit ice-cream.

From Philippines: *Humba Nangka*: curry made from unripe jackfruit and coconut milk (Bohol); *Ginataang Langka sa Alimango* (jackfruit with crab in coconut milk (Bicol); *Sinh To Mit* (Filipino jackfruit sherbet).

From Indonesia: Sayur Gori – young jackfruit in coconut milk; *Gudeg* – a traditional dish from Yogyakarta; *Lodeh* – a traditional Indonesian vegetable dish with coconut milk; *Gule Nangka* – a traditional Indonesia spicy curry; *Sayur asam* – an optional ingredient using jackfruit with tamarind.

From Vietnam: *Goi mít* (or *Mít tron*) – jackfruit salad dish in Central Vietnam.

Nhút mít – salted jackfruit popular in Central Vietnam.

Some of the Jackfruit seed dishes include: *Panasapandu Ullikaram* [jackfruit seed sabji (spicy vegetable dish)]; Jackfruit Seeds *Poriyal* (dry curry without sauce), Jackfruit Seeds and Drumstick Leaves Thoran, Jackfruit Seeds-Drumstick Curry, *Chakkakkuru Mulakittathu* (Jackfruit seed curry), *Phanas Kadbole* – jackfruit seeds with flour, turmeric, ajwain, chilli powder, *Kalupol Maluwa* – jackfruit seeds cooked with spices and mixed with scraped coconut (Sri Lanka).

The fleshy perianths of ripe fruit are eaten fresh (Plate 6) or frozen or made into various delicacies *dodol* and *kolak* in Java, chutney, jam, jelly, paste or preserved as candies or crystallised dried fruit slices, jackfruit pancakes, jackfruit cakes, jackfruit *burfi*, jackfruit *tamale*. The ripe pulp is eaten with ice cream or to flavour ice cream, smoothies, milk-shakes and beverages or made into jackfruit honey or processed into powder and used for preparing drinks and sweets. The jackfruit puree can be used to produce a good quality drum-dried powder by incorporating 2.65% of soy lecithin and 10.28% of gum Arabic into the jackfruit puree (40% v/w water) (Pua et al. 2007). Flour from jackfruit seed can be prepared by dry milling. The flour had good capacities for water absorption (205%) and oil absorption (93%). Substitution of wheat flour with the seed flour, at the level of 5%, 10% and 20% markedly reduced the gluten strength of the mixed dough (Tulyathana et al. 2002). It is high in protein and carbohydrate contents and can be used with wheat flour for making white bread, chips, pappadums and various food formulations.

Botany

An evergreen, perennial, medium-sized, much branched, under-storey tree growing to 10–25 m tall, with a trunk diameter of 30–60 cm. Mature trees has tubular roots and thick blackish brown bark and a spreading conical to dome-shaped canopy when mature. Branchlets are furrowed to smooth, 2–6 mm thick and, glabrous. All parts of the tree exude milky latex when bruised. Stipules are amplexicaul, ovate, with or without bent, reflexed pubescence, caducous, with conspicuous annular scar. Leaves are spirally arranged and borne on 1–3 cm long petiole. Leaf blade is elliptic to obovate, large, 7–16 × 3–7 cm, entire, simple, glossy, glabrous, and leathery, base cuneate, apex blunt to acuminate; deeply lobed on new growth of young trees. Jackfruit is monoecious, having male and female inflorescences (or "spikes") on the same tree. Male and female spikes are borne separately on short, stout stems that sprout from older branches and the trunk. Male spikes are found on younger branches above female spikes. Male spikes are dense, fleshy, cylindrical to clavate, and up to 10 cm in length. Flowers are tiny, pale green when young, turning darker with age. Female flowers are larger, elliptic or rounded, with a tubular calyx, one-celled ovary and a globose fleshy rachis. Jackfruit has a compound or multiple fruit (syncarp) that is ellipsoid to oblong-cylindric to globose, 30–100 cm by 25–50 cm wide with a green to yellow brown exterior rind that is composed of hexagonal, bluntly conical tubercles that cover a thick, rubbery, whitish to yellowish wall (Plates 1, 2 , 3, 4 and 5). The acid to sweetish, yellow to orangey-yellow (when ripe) flesh (perianth) surrounds each seed (Plates 2, 3 and 6). The heavy fruit is held together by a central fibrous core. Fruit weighs 4.5–50 kg. Seeds are light brown to brown, rounded, 2–3 cm by 1–1.5 cm enclosed in a thin, whitish membrane.

Nutritive/Medicinal Properties

Analyses carried out in the United States report that raw, mature jackfruit (minus the skin and seeds) has the following proximate composition (per 100 g value): water 73.23 g, energy 94 kcal (393 kJ), protein 1.47 g, total lipid 0.30 g, ash 1.00 g, carbohydrates 24.01 g, total dietary fibre 1.60 g, Ca 34 mg, Fe 0.60 mg, Mg 37 mg, P 36 mg, K 303 mg, Na 3 mg, Zn 0.42 mg, Cu 0.187 mg, Mn 0.197 mg, Se 0.6 mcg, vitamin C

Plate 1 (**a**, **b**) Immature jackfruit higher up the tree (*left*) and close to the trunk base (*right*)

Plate 2 Fleshy perianths removed from ripe jackfruit (*left*) and ripe jackfruit sliced opened

Plate 4 Ripe jackfruit on sale in Papua New Guinea market

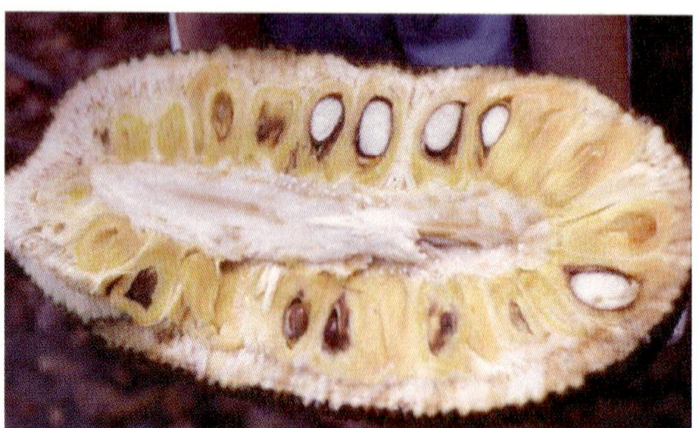

Plate 3 Ripe jackfruit halved to show the perianths and seeds

607 mg, thiamin 0.030 mg, riboflavin 0.110 mg, niacin 0.400 mg, vitamin B-6 0.108 mg, total folate 14 μg, vitamin A 297 IU, total saturated fatty acids 0.063 g, total monounsaturated fatty acids 0.044 g and total polyunsaturated fatty acids 0.086 g (U.S. Department of Agriculture and Agricultural Research 2010).

Jackfruit fleshy aril is rich in vitamin A and K, low in total lipids and saturated fatty acids and contain fair amounts of vitamin C, carbohydrates and fibre. It also contains all the essential minerals, folate, thiamine, riboflavin, niacin and vitamin B-6.

Recent studies showed that jackfruit is a good source of provitamin A carotenoids, though not as good as papaya. The carotenes β-carotene, α-carotene, β-zeacarotene, α-zeacarotene and β-carotene-5,6-epoxide and a dicarboxylic carotenoid, crocetin, were identified, corresponding theoretically to 141.6 retinol equivalents (RE) per 100 g (Chandrika et al. 2004). The findings showed that the biological conversion of provitamin A in jackfruit kernel appeared satisfactory and thus increased consumption of ripe jackfruit

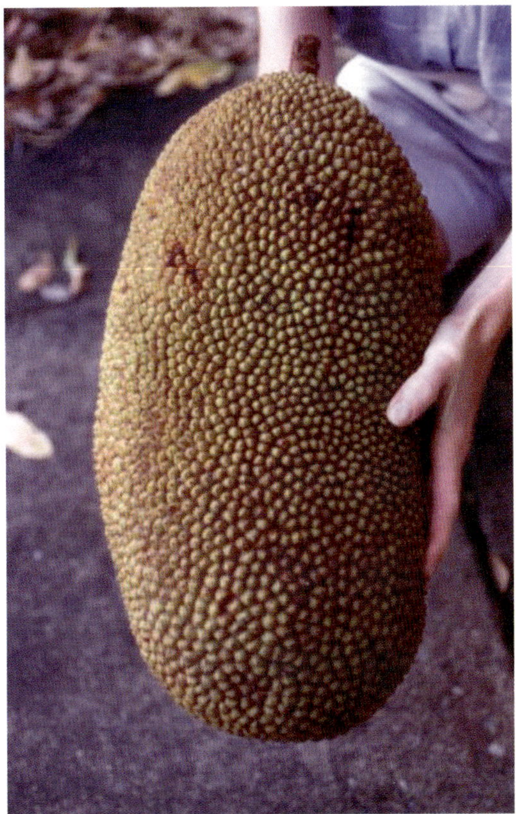

Plate 5 Close-up of external rind of ripe jackfruit

Plate 6 Package of excised golden yellow perianths

was advocated as part of a strategy to prevent and control vitamin A deficiency in Sri Lanka. Another study reported that the main carotenoids in the fruit were all-trans-lutein (24–44%), all-*trans*-β-carotene (24–30%), all-*trans*-neoxanthin (4–19%), 9-*cis*-neoxanthin (4–9%) and 9-*cis*-violaxanthin (4–10%) (de Faria et al. 2009). Total carotenoid in some batches ranged from 129.0 to 150.3 μg/100 g. The provitamin A values ranged from 3.3 to 4.3 μg RAE/100 g.

The major components identified in the aroma concentrate of "hard jackfruit" variety were isopentyl isovalerate (28.4%) and butyl isovalerate (25.6%) (Maia et al. 2004). The aroma concentrate of "soft jackfruit" was dominated by isopentyl isovalerate (18.3%), butyl acetate (16.5%), ethyl isovalerate (14.4%), butyl isovalerate (12.9%) and 2-methylbutyl acetate (12.0%). Wong et al. (1992) found that Jackfruit contained 45 volatile components of which 32 have not been reported previously. Esters represented a high proportion (31.9%) of the jackfruit volatiles and are important contributors to the flavour of the fruit.

The seeds of *A. heterophyllus* are rich in protein; the protein content was found to be higher than those from high protein animal sources such as beef and marine fishes (Ajayi 2008). The seeds have high carbohydrate content and could act as source of energy for animals if included in their diets. The oil content of the seeds was 11.39%. The seeds were found to be good sources of mineral elements. Potassium 2470.00 ppm was the prevalent mineral followed by sodium, magnesium and then calcium. The seeds also contain reasonable quantity of iron, 148.50 ppm.

Starch isolated and purified from jackfruit seeds had a starch yield 25–40% of total solids (Bobbio et al. 1978). The starch was characterized by rounded or bell-shaped granules ranging in size from 7 to 11 μm, with an amylose content of 28.1% and a D-glucose composition of more than 99%. The starch formed a highly rigid gel. Jackfruit seeds at two stages of ripeness were found to have a high protein content (ca. 22%), starch yield of 14% with a purity of 81% and amylase content of 12.27% (Madrigal-Aldana et al. 2011). The size of starch granules for seeds of physiological mature fruit ranged between 3 and 9.5 μm and from consumption ripeness jackfruit between 3 and 12 μm. The starch granules showed birefringence with diverse shapes such as semi-oval or bell shapes.

The effects of a jackfruit meal comprising of flesh (80% available carbohydrate) and seeds (20% available carbohydrate) was conducted in a random cross over design of 10 healthy subjects (Hettiaratchi et al. 2011). Jackfruit seed was found to a good source of starch (22%) and dietary fibre (11.1) and contained 4.7% protein (FW) and 8% resistant starch (FW). Jackfruit meal elicited a GI of 75 and was categorized as a low GI meal. The low GI could be due to the collective contributions from dietary fibre, slowly available glucose and un-gelatinised (intact) starch granules in the seeds.

Other Phytochemicals

Several flavones colouring matters were isolated from the wood of *A. heterophyllus*, morin, dihydromorin, cynomacurin, artocarpin, isotocarpin, cycloartocarpin, artocarpesin, oxydihydroartocarpesin, artocarpetin, norartocarpetin, cycloartinone, artocarpenone (Dave and Venkataraman 1956; Dave et al. 1960, 1961; Rao et al. 1973).

Arung et al. (2006a, b, 2010a, b, c) isolated the following compounds from *A. hetrophyllus* wood: artocarpanone, artocarpin, norartocarpin, artocarpesin, albanin A, cudraflavone B, cudraflavone C, 6-prenylapigenin, kuwanon C, brosimone I, 6-prenylapigenin and 3-prenyl luteolin. Zheng et al. (2008) isolated from Jakfruit wood, a furanoflavone, 7-(2,4-dihydroxyphenyl)-4-hydroxy-2-(2-hydroxy propan-2-yl)-2, 3-dihydrofuro(3, 2-g)chromen-5-one (artocarpfuranol) (1), together with 14 known compounds, dihydromorin (2), steppogenin (3), norartocarpetin (4), artocarpanone (5), artocarpesin (6), artocarpin (7), cycloartocarpin (8), cycloartocarpesin (9), artocarpetin (10), brosimone I (11), cudraflavone B (12), carpachromene (13), isoartocarpesin (14), and cyanomaclurin (15). Four new phenolic compounds, including one isoprenylated 2-arylbenzofuran derivative, artoheterophyllin A (1), and three isoprenylated flavonoids, artoheterophyllin B (2), artoheterophyllin C (3), and artoheterophyllin D (4), together with 16 known compounds, were isolated from the ethanol extract of the twigs (Zheng et al. 2009). From the root bark, several flavonoids were isolated, two novel 2′,4′,6′-trioxygenated flavanones, heteroflavanones A and B, characterized as 5,2′-dihydroxy-7,4′-dimethoxyflavanone and 8-(γ,γ-dimethylallyl)-5,2′,4′-trihydroxy-7-methoxyflavone, respectively (Lu and Lin 1993); two novel 2′,4′,6′-trioxygenated flavanones, heteroflavanones A and B, elucidated as 5-hydroxy-7,2′,4′,6′-tetramethoxyflavanone and 8-(γ,γ-dimethylallyl)5-hydroxy-7,2′,4′,6′-tetramethoxyflavanone (Lu and Lin 1993); a novel 2′,4′,6′-trioxygenated flavanone, named heteroflavanone C, 8-(γ,γ-dimethylallyl)-5,7-dihydroxy-2′,4′,6′-trimethoxyflavanone, a new prenylflavonoid, named cycloartocarpin A, a new long chain fatty ester, 9-hydroxytridecyl docosanoate, and four known compounds, β-sitosterol, betulin, ursolic acid and betulinic acid (Lu and Lin 1994); two new flavonoids characterized as 5,2′-dihydroxy-7,4′-dimethyoxyflavanone and 8-(γ,γ-dimethylallyl)-5,2′,4′-trihydroxy-7-methoxyflavone and nine known flavonoids (Lin et al. 1995); six prenylflavonoids, including two new prenylflavones have been characterized as 8-(γ,γ-dimethylallyl)-5,4′-dihydroxy-7,2′-dimethoxyflavone and 3,3′-di-(γ,γ-dimethylallyl)-5,7,2′,5′-tetrahydroxy-4′-methoxyflavone (Chung et al. 1995) and a novel phenolic compound heterophylol (Lin and Lu 1993). Novel natural Diels-Alder-type adducts artonin C, artonin D ((Hano et al. 1990), artonin X (Shinomiya et al. 1995) and cycloartenone (Dayal and Seshadri 1974) were isolated from the root bark. Nine flavonoids, artocarpin (1), cudraflavone C (2), 6-prenylapigenin (3), kuwanon C (4), norartocarpin (5), albanin A (6), cudraflavone B (7), brosimone I (8) and artocarpanone (9) (Arung et al. 2010c) and a new prenylated flavonoid, 3-prenyl luteolin were isolated from the wood of *Artocarpus heterophyllus*.

From the ether extract of dried latex of *Artocarpus heterophyllus*, two tetracyclic triterpenoids, 9, 19-cyclolanost-3-one-24,25-diol (24 R) and 9,19-cyclolanost-3-one-24,25-diol (24S) together with two known compounds, cycloartenone and cycloartenol were isolated (Barik et al. 1994). Cyclomorusin, cycloartocarpin, brosimone, stearic acid, palmitic acid, linoleic acid, squalene, β-sitosterol, and stigmasterol were isolated from the trunk of *A. heterophyllus*

(Chen et al. 2010). Artocarpesin, norartocarpetin and oxyresveratrol were isolated from the fruit (Fang et al. 2008).

Other phytochemicals isolated from *A. heterophyllus* included: two isoprenylflavones compounds were identified 6-(3-methyl-1-butenyl)-5,2′,4′-trihydroxy-3-isoprenyl-7-methoxy flavone and 5,7,2′,4′-tetrahydroxy-6-isoprenylflavone (Sato et al. 1996); artocarpine, artocarpetin, artocarpetin A, c artonin A, artonin B, ycloheterophyllin diacetate and cycloheterophyllin peracetate (Ko et al. 1998); cycloartomunin, cyclomorusin, dihydrocycloartomunin, dihydroisocycloartomunin, cudraflavone A, cyclocommunin, artomunoxanthone and cycloheterohyllin, artonins A and B, artocarpanone, artocarpanone A, and heteroflavanones A, B and C (Wei et al. 2005); artocarpesin, norartocarpetin and oxyresveratrol from the fruit (Fang et al. 2008), Three new prenylated flavones, trivially named artoindonesianin E-1 (1), artoindonesianin Z-4 (2) and artoindonesianin Z-5 (3), beside 17 known prenylated flavones (Musthapa et al. 2010); and lectins from jackfruit seeds, jacalin (Roque-Barreira and Campos-Neto 1985; Suresh et al. 1997) and artocarpin (Miranda Santos et al. 1991; Suresh et al. 1997; Pratap et al. 2002). However, artocarpin is also the trivial name of an isoprenoid-substituted flavonoid compound from jackfruit wood. To surmount the existing confusion in the literature concerning the trivial names used until now to designate the lectin and by the increasing interest in its biomedical applications, specially those concerning the immunomodulation activity exerted by the lectin, triggered by the recognition of glycoconjugates on the surface of cells of the innate immunity, Pereira-da-Silva et al. (2008) proposed a new nomenclature Artin M as a rational substitution for the name artocarpin, the D-mannose-binding lectin from jackfruit seeds. The new nomenclature proposed for the lectin referred to both its origin and its specificity on sugar recognition.

The plant has been reported to possess antibacterial, anti-inflammatory, antidiabetic, antioxidant and immunomodulatory properties (Prakash et al. 2009). *Artocarpus heterophyllus* is an important source of compounds like morin, dihydromorin, cynomacurin, artocarpin, isoartocarpin, cyloartocarpin, artocarpesin, oxydihydroartocarpesin, artocarpetin, norartocarpetin, cycloartinone, betulinic acid, artocarpanone and heterophylol which are useful in fever, boils, wounds, skin diseases, convulsions, diuretic, constipation, ophthalmic disorders and snake bite etc (Prakash et al. 2009).

Antioxidant Activity

Ethanol extracts of *A. heterophyllus*, showed antioxidant ability in scavenging 1, 1-diphenyl-2-picrylhydrazyl (DPPH) free radical. The IC_{50} of the ethanolic extracts of *A. heterophyllus*, was 410 μg/ml (Soubir 2007). Studies demonstrated that prenylated flavonoids have more antioxidative effect than non-prenylated flavonoids (Toda and Shirataki 2006). The prenylated flavonoid, 5,7,4′-trihydroxy-6,8-diprenylisoflavone, isolated from *Artocarpus heterophyllus* exhibited stronger inhibitory effect of on lipid peroxidation by interaction of haemoglobin and hydrogen peroxide than that of genistein, a non-prenylated isoflavone. Another study, found that *Artocarpus heterophyllus* has prenylflavones, which are potent antioxidants. Among them, artocarpine, artocarpetin, artocarpetin A, and cycloheterophyllin diacetate and peracetate had no effect on iron-induced lipid peroxidation in rat brain homogenate and did not scavenge the stable free radical 1,1-diphenyl-2-picrylhydrazyl (DPPH) (Ko et al. 1998). In contrast, cycloheterophyllin and artonins A and B inhibited iron-induced lipid peroxidation in rat brain homogenate and scavenged DPPH. They also scavenged peroxyl radicals and hydroxyl radicals that were generated by 2, 2'-azobis (2-amidinopropane) dihydrochloride and the Fe^{3+}-ascorbate-EDTA-H_2O_2 system, respectively. However, they did not inhibit xanthine oxidase activity or scavenge superoxide anion, hydrogen peroxide, carbon radical, or peroxyl radicals derived from 2,2′-azobis(2,4-dimethylvaleronitrile) in hexane.

Moreover, cycloheterophyllin and artonins A and B inhibited copper-catalysed oxidation of human low-density lipoprotein, as measured by fluorescence intensity, thiobarbituric acid-reactive substance and conjugated-diene formations and electrophoretic mobility. The research concluded that cycloheterophyllin and artonins A and B served as powerful antioxidants against lipid peroxidation when biomembranes were exposed to oxygen radicals.

Jackfruit pulp contained lower amount of total phenolics (0.46 mg GAE/g) compared to *A. odoratissimus* flesh (4.39 mg GAE/g) (Abu Bakar et al. 2009). Jackfruit seeds contained higher amounts of total phenolics than the edible portion (Soong and Barlow 2004). Ethanol extract of the jackfruit pulp afforded e 0.46 mg GAE/g total phenolics compared with water 0.25 mg GAE/g, methanol 0.21 mg GAE/g and acetone 0.18 mg GAE/g (Jagtap et al. 2010). Water was found to be the best solvent for extracting flavonoid compound 1.20 mg RE/g compared with acetone 0.19 mg RE/g, ethanol, 0.23 mg RE/g, methanol 0.24 mg RE/g. For the antioxidant activity using DPPH, minimum IC_{50} and maximum IC_{50} for the methanolic extract was 0.4 mg/ml and 0.7 mg/ml respectively. The water extract gave comparable results. The jackfruit pulp extract (5 mg/ml) showed higher ability to reduce Fe^{3+} to Fe^{2+}, 1.7 mM TEAC/g for the methanol extract and 1.4 mM TEAC/g for the water extract. Reducing ability of extracts correlated with phenolic and flavonoid contents. All extracts exhibited antioxidant activity and it was dose dependent in the order: methanol >ethanol >water >acetone. The concentrations of jackfruit pulp that caused 50% inhibition (IC_{50}) were as follows 3.43 mg/ml for methanolic extract, 3.6 mg/ml for ethanolic extract and 3.9 g/ml for the water extract. However, all jackfruit pulp extracts exhibited lower free radical activity than the standard ascorbic acid

Total water extract, ethyl acetate, and aqueous fractions from the leaves of *Artocarpus heterophyllus* showed significant antioxidant activity tested in different in vitro systems (DPPH, ABTS, FRAP, and Fe^{2+} chelating activity assay) (Loizzo et al. 2010). In particular, in DPPH assay *A. heterophyllus* total water extract exhibited a strong antiradical activity with an IC_{50} value of 73.5 μg/ml while aqueous fraction exerted the highest activity in FRAP assay (IC_{50} value of 72.0 μg/ml). Results showed jackfruit leaves to be a new potential source of natural antioxidants. In another study, the n-butanol and ethyl acetate fractions of jackfruit leaf extract exhibited antioxidant activity (Omar et al. 2011). Both fractions markedly scavenged diphenylpicrylhydrazyl radical and chelate Fe^{2+} in vitro. Isoquercitrin flavonoid was isolated from n-butanol fraction.

Antiinflammatory Activity

Studies reported that flavonoids including cycloartomunin, cyclomorusin, dihydrocycloartomunin, dihydroisocycloartomunin, cudraflavone A, cyclocommunin, artomunoxanthone and cycloheterohyllin, artonins A and B, artocarpanone, artocarpanone A, and heteroflavanones A, B and C isolated from *Artocarpus communis* and *Artocarpus heterophyllus* showed various degree of antiinflammatory activities by their inhibitory effects on the chemical mediators released from mast cells, neutrophils, and macrophages (Wei et al. 2005). Three phenolic compounds, artocarpesin [5,7,2′,4′-tetrahydroxy-6-(3-methylbut-3-enyl) flavone], norartocarpetin (5,7,2′,4′-tetrahydroxyflavone) and oxyresveratrol [trans-2,4,3′,5′-tetrahydroxystilbene] isolated from the fruit exhibited potent antiinflammatory activity in lipopolysaccharide (LPS)-activated RAW 264.7 murine macrophage cells (Fang et al. 2008). The studies indicated that artocarpesin suppressed the LPS-induced production of nitric oxide (NO) and prostaglandin E 2 (PGE 2) through the down-regulation of inducible nitric oxide synthase (iNOS) and cyclooxygenase 2 (COX-2) protein expressions. The studies indicated that artocarpesin may provide a potential therapeutic approach for inflammation-associated disorders. The methanolic extract of jackfruit bark exhibited significant dose–dependent antiinflammatory activity on carrageenan induced models, in albino rats (Lakheda et al. 2011).

Skin Whitening Activity

The flavonoid compound artocarpanone isolated from Jackfruit wood extract strongly inhibited both mushroom tyrosinase activity and melanin production in B16 melanoma cells (Arung et al. 2006a, b). This compound could provide a strong candidate as a remedy for hyperpigmentation in human skin. Other isoprenoid-substituted flavonoids Artocarpin, cudraflavone C, 6-prenylapigenin, kuwanon C, norartocarpin and albanin A isolated from the wood inhibited melanin biosynthesis in B16 melanoma cells with little or no cytotoxicity. In another study, among the compounds isolated from Jackfruit wood, a furanoflavone, 7-(2,4-dihydroxyphenyl)-4-hydroxy-2-(2-hydroxy propan-2-yl)-2, 3-dihydrofuro(3, 2-g)chromen-5-one (artocarpfuranol) (1), together with 14 known compounds, dihydromorin (2), steppogenin (3), norartocarpetin (4), artocarpanone (5), artocarpesin (6), artocarpin (7), cycloartocarpin (8), cycloartocarpesin (9), artocarpetin (10), brosimone I (11), cudraflavone B (12), carpachromene (13), isoartocarpesin (14), and cyanomaclurin (15), compounds 1–6 and 14 showed strong mushroom tyrosinase inhibitory activity with IC_{50} values lower than 50 μM (Zheng et al. 2008). The inhibition was more potent than kojic acid (IC_{50} = 71.6 μM), a well-known tyrosinase inhibitor. In addition, extract of *A. heterophyllus* exhibited anti-browning effect on fresh-cut apple slices. The results provided preliminary evidence supporting the potential of this natural extract as anti-browning agent in food systems. Zheng et al. (2009) found that four new phenolic compounds, including one isoprenylated 2-arylbenzofuran derivative, artoheterophyllin A (1), and three isoprenylated flavonoids, artoheterophyllin B (2), artoheterophyllin C (3), and artoheterophyllin D (4), isolated from the ethanol extract of the twigs did not show significant inhibitory activities against mushroom tyrosinase compared to kojic acid. It was found that similar compounds, such as norartocarpetin and artocarpesin in the twigs and woods of *A. heterophyllus*, contributed to their tyrosinase inhibitory activity.

By activity-guided fractionation of *A. heterophyllus* wood extract, norartocarpetin (1) and artocarpesin (2) were isolated (Arung et al. 2010a). These compounds were found to have 4-substituted resorcinol moiety in B ring, an important substructure for revealing the tyrosinase inhibitory activity. Also, the effect of albanin A (3) with a 4-substituted resorcinol moiety at B ring with prenyl substituent at C-3 position, was examined for comparison. The IC_{50} values of mushroom tyrosinase inhibitory activity of norartocarpetin, artocarpesin and albanin A were 1.7, 8.5 and 463 μM, respectively. In melanin formation inhibition on B16 melanoma cells, the IC_{50} of these compounds (1–3) were 209.1, 45.1 and 40.1 μM, respectively. As a result of cytotoxicity-guided fractionation, nine flavonoids, artocarpin (1), cudraflavone C (2), 6-prenylapigenin (3), kuwanon C (4), norartocarpin (5), albanin A (6), cudraflavone B (7), brosimone I (8) and artocarpanone (9) were identified from the methanol extract of the wood of *Artocarpus heterophyllus* (Arung et al. 2010c). Structure -cytotoxic activity investigation on B16 melanoma cells using the isolated compounds (1–9) and structurally related compounds indicated that the isoprenoid-substituted moiety of flavonoids enhanced their cytotoxicity and, its attached position and the number of the isoprenoid-substituted moiety per molecule influence their cytotoxicity. A new prenylated flavonoid, 3-prenyl luteolin isolated from wood of *Artocarpus heterophyllus* showed anti-melanogenesis activity (Arung et al. 2010b). The prenylated moiety at C-3 position was found to play an important role for tyrosinase inhibition.

Hypoglycaemic and Hypolipidemic Activities

Kotowaroo et al. (2006) found that the aqueous jackfruit leaf extract significantly inhibited α-amylase activity in rat plasma. The highest inhibitory activity (27.20%) was observed at a concentration of 1,000 μg/ml. Results from the study indicated that *Artocarpus heterophyllus* could act as a 'starch blocker' thereby reducing post-prandial glucose peaks. Another earlier study reported that the extracts of both *Artocarpus heterophyllus* leaves and *Asteracanthus longifolia* significantly improved glucose tolerance in the normal subjects and the diabetic patients when investigated at oral doses equivalent to 20 g/kg of starting material (Fernando et al. 1991).

The leaf extract of *Artocarpus heterophyllus* was found to cause a hypoglycaemic effect at a dose of 50 mg/kg, both in normal and alloxan-diabetic rats (Chandrika et al. 2006). The hypoglycaemic effect was at its maximum 2 h after flavonoid fraction administration, and multiple dosing maintained the activity for a week. The hypoglycaemic effect of the flavonoid fraction of leaf (49%) was higher than that of tolbutamide (27.0%), a sulphonyl urea drug commonly used for treatment of hyperglycaemia. Administering the flavonoid fraction for 3 months had no significant effects on liver function while the histology of liver, kidney and heart revealed no damage. These results indicated that the total flavonoid content of *A. heterophyllus* leaf exhibited a non-toxic and significant hypoglycaemic activity in male Wistar rats and may therefore be responsible for the previously reported antidiabetic activity.

In normoglycemic rats, administration of a single dose (20 mg/kg) of the ethylacetate (EA) fraction of the mature leaves of *A. heterophyllus* resulted in a significant reduction in the fasting blood glucose concentration and a significant improvement in glucose tolerance, compared to the controls (Chackrewarthy et al. 2010). In streptozotocin -induced diabetic rats, chronic administration of the EA fraction daily for 5 weeks resulted in a significant lowering of serum glucose, cholesterol and triglyceride levels. Compared to control diabetic rats, the extract-treated rats had 39% less serum glucose, 23% lower serum total cholesterol and 40% lower serum triglyceride levels and 11% higher body weight at the end of the fifth week. The percentage reductions in the serum parameters mediated by the test fraction were comparable with those produced by glibenclamide (0.6 mg/kg), the reference drug used in this study. It was concluded that *A. heterophyllus* leaves contained one or more hypoglycemic and hypolipidemic principles which may have potential to be developed further for the treatment of diabetes specifically associated with a hyperlipidemic state. In a recent study, the n-butanol and ethyl acetate fractions of jackfruit leaf extract elicited hypoglycemic, and hypolipidemic effects in streptozotocin (STZ)-diabetic rats (Omar et al. 2011). Administration of both fractions to streptozotocin-diabetic rats appreciably decreased fasting blood glucose, lipid peroxides, percent of glycosylated hemoglobin A1C, triglycerides, total cholesterol, low-density lipoprotein cholesterol and elevated insulin, high density lipoprotein cholesterol, and protein content. The hypoglycemic and hypolipidemic effects were suggested to be mediated in an antioxidative pathway by flavonoids. Isoquercitrin flavonoid was isolated from n-butanol fraction.

Antimicrobial Activity

The crude methanolic extracts of the stem and root barks, stem and root heart-wood, leaves, fruits and seeds of *Artocarpus heterophyllus* and their fractions exhibited a broad spectrum of antibacterial activity (Khan et al. 2003). The butanol fractions of the root bark and fruits were found to be the most active. None of the fractions were active against the fungi tested. Studies in Japan showed that methanol extract from *Artocarpus heterophyllus* showed the most intensive antibacterial activity (Sato et al. 1996). Two active isoprenylflavones compounds were identified 6-(3-methyl-1-butenyl)-5,2′,4′-trihydroxy-3-isoprenyl-7-methoxy flavone and 5,7,2′,4′-tetrahydroxy-6-isoprenylflavone. Both isolates completely inhibited the growth of primary cariogenic bacteria at 3.13–12.5 μg/ml. They also exhibited the growth inhibitory effects on plaque-forming streptococci. These phytochemical isoprenylflavones would be potent compounds for the prevention of dental caries. Another research found that the wood extract had sufficient antibacterial activity against certain Gram positive bacteria *Bacillus subtilis*, *Bacillus cereus*, *Staphylococcus aureus* and gram negative *Escherichia coli*, and *Klebsiella pneumonia*

(Indrayan et al. 2004). Total water extract, ethyl acetate, and aqueous fractions from the leaves of *Artocarpus heterophyllus* were found to exhibit antibacterial activities against some foodborne pathogens such as *Escherichia coli*, *Listeria monocytogenes*, *Salmonella typhimurium*, *Salmonella enterica*, *Bacillus cereus*, *Enterococcus faecalis*, and *Staphylococcus aureus* (Loizzo et al. 2010). The minimum inhibitory concentration (MICs) of extract and fractions determined by the agar dilution method ranged from 221.9 μg/ml for ethyl acetate fraction to 488.1 μg/ml for total extract. In the agar diffusion method the diameters of inhibition were 12.2 for the total extract, 10.7 and 11.5 for ethyl acetate and aqueous fractions, respectively.

Anticancer Activity

Artocarpin [6-(3-methyl-1-butenyl)-5,2′,4′-trihydroxy-3-isoprenyl-7-methoxyflavone] isolated from wood of jack fruit exhibited potent cytotoxic activity on human T47D breast cancer cells (Arung et al. 2010d). Artocarpin caused a reduction of cell viability in a concentration-dependent manner and an alteration of cell and nuclear morphology. Further, the percentage of the sub-G1 phase formation was elevated dose-dependently. Artocarpin induced activation of caspase 8 and 10 and activated capase 3. The data indicated that artocarpin induced apoptosis in T47D cells possibly via an extrinsic pathway.

Antiplasmodial Activity

Three new prenylated flavones, trivially named artoindonesianin E-1 (1), artoindonesianin Z-4 (2) and artoindonesianin Z-5 (3), beside 17 known prenylated flavones were isolated from *A. heterophyllus*, *A. elasticus* and *A. lanceifolius* (Musthapa et al. 2010). Antimalarial effect of some selected isolated flavone derivatives showed that artonin E exhibited very strong inhibition (IC_{50} 0.1 mg/ml) against K1 strain in comparison against 3D7 strain of *Plasmodium falciparum* (IC_{50} 0.3 mg/ml). A related prenylated flavone, namely 12-hydroxyartonin E, exhibited strong inhibition (IC_{50} 0.9 mg/ml) against K1 strain, but weak inhibition (IC_{50} 14.3 mg/ml) against 3D7 strain. In addition, the other isolated flavone derivatives showed moderate inhibition with IC_{50} of r 2.1, 1.6, 3.6, 1.3, 6.7 and 2.1 mg/ml respectively against both the strain of *P. falciparum*, except for flavanone-3-ol derivative (dihydromorin), which had no inhibition.

Sexual Behaviour Activity

Studies conducted suggested that *A. heterophyllous* seeds did not have aphrodisiac action, at least, in rats (Ratnasooriya and Jayakody 2002). In a sexual behaviour study, using receptive female rats, an oral administration of 500 mg/kg dose of seed suspension in 1% methylcellulose (SS) markedly inhibited libido, sexual arousal, sexual vigour and sexual performance within 2 h. Further, the treatment induced a mild erectile dysfunction. These anti-masculine effects on sexual function were not evident 6 hours post treatment indicating rapid onset and offset of action. Further, these actions on the sexual behaviour was not due to general toxicity, liver toxicity, stress or reduction in blood testosterone level but due to marked sedative activity. In a mating study, SS failed to alter ejaculating competence and fertility.

Lectin-Related Properties

Recent laboratory studies showed that lectins found in jackfruit and its seeds may have antibacterial, antifungal, antiviral, anticancer, agglutinating, mitogenic and immunostimulative properties. The lectin, jacalin, possessed agglutinating activities for human and rat sperm as well as human red blood cells (Namjuntra et al. 1985). The abundance of source material for the production of jacalin, its ease of purification, yield and stability had made it an attractive cost-effective lectin (Aucouturier et al. 1989). It has found applications in diverse areas such as the isolation of human plasma glycoproteins (IgA1, C1-inhibitor, hemopexin, α2-HSG), the investigation of IgA-nephropathy, the analysis of O-linked glycoproteins and the detection of tumours. In addition, the lectin's mitogenic activity which is specific for human CD4 T-lymphocytes coupled with the proliferative response induced by jacalin appears to represent a new and interesting assay for a functional study of CD4 cells, with obvious applications in primary and acquired, especially AIDS immune deficiency states (Aucouturier et al. 1989).

Lectin: Immunomodulatory and Agglutination Activities

Miranda-Santos et al. (1991) purified two fractions derived from a crude extract of jackfruit seeds, one fraction was the D-galactose binding lectin, jacalin, and another D-mannose-binding protein which they designated 'Artocarpin'. They showed that the proliferative response of mouse spleen cells and human peripheral blood mononuclear cells and polyclonal activation of human and mouse B cells for the secretion of immunoglobulin were mediated by artocarpin. Artocarpin was found to be unique in its capacity to induce polyclonal activation of B cells in the absence of proliferation. BALB/c nu/nu spleen cells failed to proliferate which indicated this lectin to be a T cell-dependent B cell polyclonal activator. Jacalin and artocarpin, were found to have different physicochemical properties and carbohydrate-binding specificities; however, comparison of the partial amino-acid sequence of artocarpin with the known sequence of jacalin indicated close to 50% sequence identity, confirming the homology between the two lectins (Prakash et al. 2009). Pratap et al. (2002) found artocarpin to be a single chain protein with considerable sequence similarity with jacalin, however it exhibited different properties. Jacalin was the first lectin found to exhibit the β-prism I fold and composed of two polypeptide chains produced by a post-translational proteolysis which had been shown to be crucial for generating its specificity for galactose. Artocarpin was found to behave as a polyspecific lectin; it readily interacted with a wide range of monosaccharides covering galactose, N-acetylgalactosamine, mannose, glucose, sialic acid and N-acetylmuramic acid (Barre et al. 2004). Molecular docking confirmed this unexpected ability of artocarpin to interact with structurally different sugars.

The lectin, jacalin, was found to be both a potent T cell mitogen and an apparently T cell-independent activator of human B cells for the secretion of immunoglobulins (Roque-Barreira and Campos-Neto 1985). They demonstrated that IgA was probably the major serum constituent precipitated by jacalin and that no IgG or IgM can be detected in the precipitates. Jacalin is a D-Gal binding lectin and should be a useful tool for studying of serum, and secretory IgA immunoglobulin. The lectin jacalin was found to react by precipitation and western blotting with human IgA1 and IgD but not with IgA2 (nor IgG and IgM) (Aucouturier et al. 1987). Predominantly reactive carbohydrates are D-galactose and N-acetyl D-galactosamine. Jacalin with an apparent Mr of about 54,000 was suggested to made up of three non-glycosylated and one glycosylated non-covalently linked subunits. Hagiwara et al. (1988) found that Jacalin precipitated only with IgA1-containing samples, including monomers, polymers, monoclonal, polyclonal and secretory IgA1, but not IgA2 of both A2m(1) and A2m(2) allotypes, nor with IgG1, 2, 3 and 4, IgM, IgD, and IgE; after neuraminidase treatment, only IgA1 and IgD were precipitated. Jacalin had a relatively broad pH range of activity in both precipitation and agglutination of IgA1-latex. Among 39 types of sugar tested, 10 displayed inhibitory activity, decreasing in the following order: p-nitrophenyl-α-D-galactopyranoside, 1-O-methyl-α-D-galactopyranoside, D-melibiose, p-nitrophenyl-β-D-galactopyranoside, GalNAc, stachyose, 1-O-methyl-α-D-mannopyranoside, D-galactose, D-galactosamine and 1-O-methyl-α-D-glucopyranoside. IgA1, treated with neuraminidase or not, but not the other human Igs, was also an excellent inhibitor of agglutination, being more powerful than the best sugars studied. Only neuraminidase-treated IgD was inhibitory, but less so than IgA1. Mahanta et al. (1992) found that jacalin preferentially bound to α-linked non-reducing D-galactose. Jacalin was found to have a Thomsen-Friedenreich-antigen-specific lectin structure and to be made up of two types of chains, heavy and light, with M(r) values of 16,200±1,200 and 2,090±300 respectively. Jacalin was found to be highly specific for the α-O-glycoside of the disaccharide Thomsen-Friedenreich antigen (Gal β 1-3GalNAc), even in its sialylated form (Kabir 1995). This property had made jacalin suitable for studying various O-linked glycoproteins, particularly human IgA1.

Jacalin had attracted considerable attention for its diverse biological activities and had been recognized as a Galβ1→3GalNAc (T) specific lectin (Wu et al. 2003a), however information of its binding was limited to the inhibition results of monosaccharides and several T related disaccharides, but its interaction with other carbohydrate structural units occurring in natural glycans had not been characterized. Wu et al. (2003a) studied the binding profile of this lectin with glycan/ligand units. The most potent ligands for jacalin were found to be mTn, mT, and possibly Pα glycotopes, while GalNAcβ1→4Galβ1→, GalNAcα1→3Gal, GalNAcα1→3GalNAc, and Galα1→3Gal determinants were poor inhibitors. Thus, the overall binding profile of Jacalin could be defined in decreasing order as high density of mTn, and mTα >>>simple Tn cluster > monomericTα > monomericPα > monomeric Tn > monomeric T > GalNAc > Gal > Methylα1→Man >> Man and Glc (inactive). Their finding should aid in the selection of this lectin for biological applications.

Lectin and Antiviral Activity

Jacalin from *Artocarpus heterophyllus* was found to interact with the lymphocyte cell-surface molecule CD4, a known receptor for the human immunodeficiency virus type 1 (HIV-1) (Corbeau et al. 1994). Moreover, jacalin was able to block HIV-1 infection of CD4+ lymphoblastoid cells. Here we studied whether jacalin prevents HIV-1 gp120-CD4 interactions. They found (i) that jacalin did not inhibit HIV-1 Lai-induced syncytium formation that required gp120-CD4 interactions; (ii) that jacalin prevented neither rgp120 binding to cell-surface CD4 nor sCD4 binding to viral envelope proteins expressed at the surface of HI V-1-infected lymphoblastoid cells; (iii) that jacalin did not compete for binding to CD4 with anti-CD4 mAb specific for the CDR2- or CDR3-like regions of the D1 domain of CD4; (iv) that jacalin did not bind a recombinant soluble molecule containing the D1/D2 domains of CD4; and, (iv) that jacalin binding to CD4 was inhibited by sugars known to interact with the lectinic-site of jacalin. The data provided implications for the mechanism by which jacalin was able to block HIV-1 infection of CD4+ cells.

Jackfruit lectin (JFL) from *Artocarpus heterophyllus* was found to exhibit inhibitory activity in vitro with a cytopathic effect towards herpes simplex virus type 2 (HSV-2), varicella-zoster virus (VZV), and cytomegalovirus (CMV) (Wetprasit et al. 2000). The 50% inhibitory dose values from plaque reduction assay (inactivation) were 2.5, 5, and 10 μg/ml of JFL for HSV-2, VZV, and CMV, respectively. Lymphocyte proliferation was significantly increased in the presence of the JFL in the concentration range of 2.5–50 μg/ml, but was reduced at 500 μg/ml. It was found that CD16(+)/CD56(+) cells (natural killer cells) were induced among the primary lymphocyte subpopulations. These data suggested that JFL was mitogenic for NK lymphocyte (CD16(+)/CD56(+)) and also active against HSV-2, varicella-zoster virus, and cytomegalovirus.

Jackfruit crude extracts were found to stimulate human lymphocytes. Study of the proliferative response of cell populations from normal human peripheral blood to purified jacalin showed it to be mitogenic through an interaction with lymphocytes by its lectin-binding site, as shown by inhibition by IgA (Pineau et al. 1990). Jacalin failed to stimulate B cells to proliferate and to undergo plasma cell maturation. It induced a proliferation of CD4 (and not CD8) lymphocytes. Jacalin's uniqueness in being strongly mitogenic for human CD4+ T lymphocytes had made it a useful tool for the evaluation of the immune status of patients infected with human immunodeficiency virus (HIV)-1 (Kabir 1998).

Lectin and Antibacterial Activity

Jacalin and human immunoglobulin A was found to exert an opsonic effect on type II group B streptococci (Payne et al. 1990). Jacalin and IgA mediated phagocytosis of II/c GBS (type II group B) streptococci strain possessing the trypsin-sensitive and trypsin-resistant components of the c protein via receptors that were not dependent on divalent cations and that were not modulated by plating monocytes on antigen-antibody complexes.

Lectin and Anticancer Activity

Jacalin had already been known to exert anti-proliferative effects in human colon cancer cell line. Bhatia et al. (2005) revealed that Jacalin could induce apoptosis in non-small cell lung cancer cell line NCI-H520 in a dose–dependent manner.

Lectin and Antiparasitic Activity

Jacalin was found to have a potent adjuvant effect the mouse humoral immune response to trinitrophenyl and *Trypanosoma cruzi* and that the protective action of the *T. cruzi*-specific antibodies depended on the number of parasites used in the immunization protocol (Albuquerque et al. 1999). Immunization of mice with trinitrophenylated jacalin (TNP-JAC) in saline resulted in an antibody response to the TNP hapten that was 8 and 16 times higher than that found in mice immunized with TNP-human γ globulin (TNP-HGG) or TNP-bovine serum albumin (TNP-BSA), respectively. In addition, immunization with either a lysate or viable epimastigote forms of *T. cruzi* in the presence of jacalin resulted in a marked increase in the levels of anti-*T. cruzi* antibodies. The protective action of antibodies against acute infection by *T. cruzi* was evident when mice were immunized with 1.0×10^5 epimastigotes plus jacalin. These animals had a significantly lower parasitemia than those immunized with epimastigotes alone.

Traditional Medicinal Uses

Many parts of the plant including the bark, roots, leaves, fruit and latex are endowed with medicinal properties and used in traditional medicine.

In the Philippines, the ash of the leaves, after burning with corn and coconut shells, is applied on wounds and ulcers as cicatrizant. In India, the leaves are used in treating skin disease. The root is anti-asthmatic. The decoction of the root is used in diarrhoea and for fever. The root is also useful in skin diseases. The latex of the tree is used to promote healing in glandular swellings and in snake bites. Mixed with vinegar and applied to these swellings and to abscesses, it promotes absorption or suppuration. The ash of jackfruit leaves, burned, is used alone or mixed with coconut oil to heal ulcers. The wood has a sedative property; its pith is said to produce abortion. The unripe fruit is astringent and if eaten in large quantities, it produces diarrhoea. The ripe fruit is demulcent, nutritive, and laxative. The pulp envelopes or arils of the seeds are considered by the Chinese to be cooling, tonic, and nutritious. The starch of the seeds is given in bilious colic. The roasted seeds are believed to have aphrodisiac properties. However, some reproductively active young men in rural areas of Sri Lanka claimed that consumption of these seeds few hours prior to coitus disrupts sexual function. In Sarawak, ash from the leaves mixed in a little coconut oil and prescribed for scabies or kuris by the Iban; the Melanau used the same for applying on cuts and wounds. Hot water extract of mature jak leaves (*Artocarpus heterophyllus*) is recommended by Ayurvedic and traditional medical practitioners as a treatment for diabetes mellitus (Chandrika et al. 2006).

Other Uses

Jackfruit is often used as wind-break, shade and backyard tree. It is also known for its remarkable, durable, termite resistant timber which is widely used in the manufacture of furniture, doors and windows, in roof construction, for turnery, masts, oars, implements and brush backs. The wood is also used in the production of musical instruments such as gamelan in Indonesia, guitars in Cebu and the hull of kutiyapi, a boat lute in the Philippines. It is also used to make the body of the Indian drums *mridangam* and *kanjira*. The leaves and fruit waste provide valuable fodder for cattle, pigs, and goats. Wood chips of the heartwood after boiling with alum, yield a rich yellow dye commonly used for dyeing, wool, silk and the cotton robes of Buddhist priests. In Indonesia, splinters of the wood are put into the bamboo tubes collecting coconut toddy in order to impart a yellow tone to the sugar. Material isolated from the heartwood of *Artocarpus heterophyllus* was found to be of multiple diversified uses. It could be used as a direct dye for wool and silk (Indrayan et al. 2004). The materials derived from the heartwood could be used as a neutralization indicator in wider range of concentrations (10–2 to 4 N) than the conventional indicators. The bark is used for tanning and is occasionally used for making cordage and cloth. Tree latex is used as bird lime; and heated makes a good cement for chinaware and earthenware, and to caulk boats and holes in buckets. Although it could be used as a substitute for rubber, the latex contains 82.6–86.4% resins, which may have value in varnishes. Roots of old trees are greatly prized for carving and picture framing.

Comments

Jackfruit is propagated by seeds which germinate readily.

Selected References

Abu Bakar MF, Mohamed M, Rahmat A, Fry J (2009) Phytochemicals and antioxidant activity of different parts of bambangan (*Mangifera pajang*) and tarap (*Artocarpus odoratissimus*). Food Chem 113(2):479–483

Ajayi IA (2008) Comparative study of the chemical composition and mineral element content of *Artocarpus heterophyllus* and *Treculia africana* seeds and seed oils. Bioresour Technol 99(11):5125–5129

Albuquerque DA, Martins GA, Campos-Neto A, João S (1999) The adjuvant effect of jacalin on the mouse humoral immune response to trinitrophenyl and *Trypanosoma cruzi*. Immunol Lett 68(2–3):375–381

Arung ET, Shimizu K, Kondo R (2006a) Inhibitory effect of artocarpanone from *Artocarpus heterophyllus* on melanin biosynthesis. Biol Pharm Bull 29(9): 1966–1969

Arung ET, Shimizu K, Kondo R (2006b) Inhibitory effect of isoprenoid-substituted flavonoids isolated from *Artocarpus heterophyllus* on melanin biosynthesis. Planta Med 72(9):847–850

Arung ET, Shimizu K, Kondo R (2007) Structure-activity relationship of prenyl-substituted polyphenols from *Artocarpus heterophyllus* as inhibitors of melanin biosynthesis in cultured melanoma cells. Chem Biodivers 4(9):2166–2171

Arung ET, Shimizu K, Tanaka H, Kondo R (2010a) Melanin biosynthesis inhibitors from wood of *Artocarpus heterophyllus*: the effect of isoprenoid substituent of flavone with 4-substituted resorcinol moiety at B ring. Lett Drug DesDiscov 7(8):602–605

Arung ET, Shimizu K, Tanaka H, Kondo R (2010b) 3-Prenyl luteolin, a new prenylated flavone with melanin biosynthesis inhibitory activity from wood of *Artocarpus heterophyllus*. Fitoterapia 81(6):640–643

Arung ET, Wicaksono BD, Handoko YA, Kusuma IW, Shimizu K, Yulia D, Sandra F (2010c) Cytotoxic effect of artocarpin on T47D cells. J Nat Med 64(4): 423–429

Arung ET, Yoshikawa K, Shimizu K, Kondo R (2010d) Isoprenoid-substituted flavonoids from wood of *Artocarpus heterophyllus* on B16 melanoma cells: cytotoxicity and structural criteria. Fitoterapia 81(2):120–123

Aucouturier P, Mlhaesco E, Mihaesco C, Preud'homme JL (1987) Characterization of jacalin, the human IgA and IgD binding lectin from jackfruit. Mol Immunol 24(5):503–511

Aucouturier P, Pineau N, Brugier JC, Mihaesco E, Duarte F, Skvaril F, Preud'homme JL (1989) Jacalin: a new laboratory tool in immunochemistry and cellular immunology. J Clin Lab Anal 3(4):244–251

Backer CA, van den Bakhuizen Brink RC Jr (1965) Flora of Java (spermatophytes only), vol 2. Wolters-Noordhoff, Groningen, 641pp

Barik BR, Bhaumik T, Dey AK, Kundu AB (1994) Triterpenoids of *Artocarpus heterophyllus*. Phytochemistry 35(4):1001–1004

Barre A, Peumans WJ, Rossignol M, Borderies G, Culerrier R, Van Damme EJM, Rougé P (2004) Artocarpin is a polyspecific jacalin-related lectin with a monosaccharide preference for mannose. Biochimie 86(9–10):685–691

Jai and Bee (2007) Jiva for jackfruit – roundup. Jugalbandee. http://jugalbandi.info/2007/06/jihva-for-jackfruit-roundup/

Bhatia S, Mehta S, Majumdar S, Ghosh S (2005) Study on the effect of Jacalin on non-small cell lung cancer cell line NCI-H520. Indian J Med Res 121:92–93

Bobbio FO, El-Dash AA, Bobbio PA, Rodrigues LR (1978) Isolation and characterization of the physicochemical properties of the starch of jackfruit seeds (*Artocarpus heterorphyllus*). Cereal Chem 55:505–511

Burkill IH (1966) A dictionary of the economic products of the Malay Peninsula. Revised reprint. 2 volumes. Ministry of Agriculture and Co-operatives, Kuala Lumpur. Vol. 1 (A–H), pp 1–1240; Vol. 2 (I–Z), pp 1241–2444

Chackrewarthy S, Thabrew MI, Weerasuriya M, Jayasekera S (2010) Evaluation of the hypoglycemic and hypolipidemic effects of an ethylacetate fraction of *Artocarpus heterophyllus* (jak) leaves in streptozotocin-induced diabetic rats. Pharmacogn Mag 6:186–190

Chai PPK (2006) Medicinal plants of Sarawak. Lee Ming Press, Kuching, 212pp

Chandrika UG, Jansz ER, Warnasuriya ND (2004) Analysis of carotenoids in ripe jackfruit (*Artocarpus heterophyllus*) kernel and study of their bioconversion in rats. J Sci Food Agric 85(2):186–190

Chandrika UG, Wedage WS, Wickramasinghe SMDN, Fernando WS (2006) Hypoglycaemic action of the flavonoid fraction of *Artocarpus heterophyllus*. Afr J Tradit Complement Altern Med 3(2):42–50

Chen CY, Cheng MJ, Kuo SH, Kuo SY, Lo WL (2010) Secondary metabolites from the stems of *Artocarpus heterophyllus*. Chem Nat Comput 46(4):638–640

Chopra RN, Nayar SL, Chopra IC (1986) Glossary of Indian medicinal plants (including the supplement). Council Scientific Industrial Research, New Delhi, 330pp

Chung MI, Lu CM, Huang PL, Lin CN (1995) Prenylflavonoids of *Artocarpus heterophyllus*. Phytochemistry 40(4):1279–1282

Corbeau P, Haran M, Binz H, Devaux C (1994) Jacalin, a lectin with anti-HIV-1 properties, and HIV-1 gp120 envelope protein interact with distinct regions of the CD4 molecule. Mol Immunol 31(8):569–575

Dave KG, Venkataraman K (1956) The colouring matters of the wood of *Artocarpus integrifolia*. Pt. I. Artocarpin. J Sci Ind Res 15B(4):183–190

Dave KG, Telang SA, Venkataraman K (1960) The colouring matter of the wood of *Artocarpus integrifolia:* Pt. II. Artocarpetin, a new flavone, and artocarpanone, a new flavanone. J Sci Ind Res 19B(12):470–476

Dave KG, Mani R, Venkataraman K (1961) The colouring matters of the wood of *Artocarpus integrifolia*. Pt. 111. Constitution of artocarpin and synthesis of tetrahydroartocarpin dimethyl ether. J Sci Indus Res 20B(3):112–121

Dayal R, Seshadri TR (1974) Colourless compounds of the roots of *Artocarpus heterophyllus*. Isolation of new compound artoflavone. Indian J Chem 12:895–896

de Faria AF, de Rosso VV, Mercadante AZ (2009) Carotenoid composition of jackfruit (*Artocarpus heterophyllus*), determined by HPLC-PDA-MS/MS. Plant Foods Hum Nutr 64(2):108–115

Elevitch CR, Manner HI (2006) *Artocarpus heterophyllus* (jackfruit), ver. 1.1v. In: Elevitch CR (ed) Species profiles for Pacific Island agroforestry. Permanent Agriculture Resources (PAR), Holualoa, http://www.traditionaltree.org

Fang SC, Hsu CL, Yen GC (2008) Anti-inflammatory effects of phenolic compounds isolated from the fruits of *Artocarpus heterophyllus*. J Agric Food Chem 56(12):4463–4468

Fernando MR, Wickramasinghe N, Thabrew MI, Ariyananda PL, Karunanayake EH (1991) Effect of *Artocarpus heterophyllus* and *Asteracanthus longifolia* on glucose tolerance in normal human subjects and in maturity-onset diabetic patients. J Ethnopharmacol 31(3):277–282

Foundation for Revitalisation of Local Health Traditions (2008) FRLHT database. http://envis.frlht.org

Hagiwara K, Collet-Cassart D, Kunihiko K, Vaerman JP (1988) Jacalin: isolation, characterization, and influence of various factors on its interaction with human IgA1, as assessed by precipitation and latex agglutination. Mol Immunol 25(1):69–83

Hano Y, Aida M, Nomura T (1990) Two new natural Diels-Alder-type adducts from the root bark of *Artocarpus heterophyllus*. J Nat Prod 53(2):391–395

Hettiaratchi UPK, Ekanayake S, Welihinda J (2011) Nutritional assessment of a jackfruit (*Artocarpus heterophyllus*) meal. Ceylon Med J 56:54–58

Indrayan AK, Kumar R, Rathi AK (2004) Multibeneficial natural material: dye from heartwood of *Artocarpus heterophyllus* Lamk. J Indian Chem Soc 81(12):1097–1101

Jagtap UB, Panaskar SN, Bapat VA (2010) Evaluation of antioxidant capacity and phenol content in jackfruit (*Artocarpus heterophyllus* Lam.) fruit pulp. Plant Foods Hum Nutr 65(2):99–104

Kabir S (1995) The isolation and characterisation of jacalin [*Artocarpus heterophyllus* (jackfruit) lectin] based on its charge properties. Int J Biochem Cell Biol 27(2):147–156

Kabir S (1998) Jacalin: a jackfruit (*Artocarpus heterophyllus*) seed-derived lectin of versatile applications in immunobiological research. J Immunol Methods 212(2):193–211

Khan MR, Omoloso AD, Kihara M (2003) Antibacterial activity of *Artocarpus heterophyllus*. Fitoterapia 74(5):501–505

Ko FN, Cheng ZJ, Lin CN, Teng CM (1998) Scavenger and antioxidant properties of prenylflavones isolated from *Artocarpus heterophyllus*. Free Radic Biol Med 25(2):160–168

Kotowaroo MI, Mahomoodally MF, Gurib-Fakim A, Subratty AH (2006) Screening of traditional antidiabetic medicinal plants of Mauritius for possible α-amylase inhibitory effects in vitro. Phytother Res 20(3):228–231

Lakheda S, Devalia R, Jain UK, Gupta N, Raghuwansi AS, Patidar N (2011) Anti-inflammatory activity of *Artocarpus heterophyllus* bark. Der Pharmacia Sin 2(2):127–130

Lin CN, Lu CM (1993) Heterophylol, a phenolic compound with novel skeleton from *Artocarpus heterophyllus*. Tetrahedron Lett 34(51):8249–8250

Lin CN, Lu CM, Huang PL (1995) Flavonoids from *Artocarpus heterophyllus*. Phytochemistry 39(6):1447–1451

Loizzo MR, Tundis R, Chandrika UG, Abeysekera AM, Menichini F, Frega NG (2010) Antioxidant and antibacterial activities on foodborne pathogens of *Artocarpus heterophyllus* Lam. (Moraceae) leaves extracts. J Food Sci 75(5):291–295

Lu CM, Lin CN (1993) Two 2′,4′,6′-trioxygenated flavanones from *Artocarpus heterophyllus*. Phytochemistry 33(4):909–911

Lu CM, Lin CN (1994) Flavonoids and 9-hydroxytridecyl docosanoate from *Artocarpus heterophyllus*. Phytochemistry 35(3):781–783

Madrigal-Aldana DL, Tovar-Gómez B, Mata-Montes de Oca M, Sáyago-Ayerdi SG, Gutierrez-Meraz F, Bello-Pérez LA (2011) Isolation and characterization of Mexican jackfruit (*Artocarpus heterophyllus* L) seeds starch in two mature stages. Starch 63(6):364–372

Mahanta SK, Sanker S, Rao NV, Swamy MJ, Surolia A (1992) Primary structure of a Thomsen-Friedenreich-antigen-specific lectin, jacalin [*Artocarpus integrifolia* (jack fruit) agglutinin]. Evidence for the presence of an internal repeat. Biochem J 284(Pt 1):95–101

Maia JGS, Andrade EHA, Zoghbi MDGB (2004) Aroma volatiles from two fruit varieties of jackfruit (*Artocarpus heterophyllus* Lam). Food Chem 85(2):195–197

Miranda Santos IKFD, Mengel JO Jr, Bunn-Moreno MM, Campos-Neto A (1991) Activation of T and B cells by a crude extract of *Artocarpus integrifolia* is mediated by a lectin distinct from jacalin. J Immunol Methods 140(2):197–203

Molesworth AB (1967) Malayan fruits. An introduction to the cultivated species. Moore, Singapore, 245pp

Morton JF (1965) The jackfruit (*Artocarpus heterophyllus* Lam): its culture, varieties and utilization. Proc Fla State Hort Soc 78:336–343

Morton J (1987) Santol. In: Morton JF (ed) Fruits of warm climates. Florida Flair Books, Miami, pp 199–201

Musthapa I, Hakim EH, Juliawaty LD, Syah YM, Achmad SA (2010) Prenylated flavones from some Indonesian *Artocarpus* and their antimalarial properties. Int J Phytomed Related Ind 2:2

Namjuntra P, Muanwongyathi P, Chulavatnatol M (1985) A sperm-agglutinating lectin from seeds of Jack fruit (*Artocarpus heterophyllus*). Biochem Biophys Res Commun 128(2):833–839

Ochse JJ, van den Bakhuizen Brink RC (1980) Vegetables of the Dutch Indies, 3rd edn. Ascher & Co., Amsterdam, 1016pp

Omar HS, El-Beshbishy HA, Moussa Z, Taha KF, Singab AN (2011) Antioxidant activity of *Artocarpus heterophyllus* Lam. (jack fruit) leaf extracts: remarkable attenuations of hyperglycemia and hyperlipidemia in streptozotocin-diabetic rats. Sci World J 11:788–800

Payne NR, Concepcion NF, Anthony BF (1990) Opsonic effect of jacalin and human immunoglobulin A on type II group B streptococci. Infect Immun 58(11):3663–3670

Pereira-da-Silva G, Roque-Barreira MC, Van Damme EJM (2008) Artin M: a rational substitution for the names artocarpin and KM+. Immunol Lett 11(1–2): 114–115

Pineau N, Aucouturier P, Brugier JC, Preud'homme JL (1990) Jacalin: a lectin mitogenic for human CD4 T lymphocytes. Clin Exp Immunol 80(3):420–425

Prakash O, Kumar R, Mishra A, Gupta R (2009) *Artocarpus heterophyllus* (jackfruit): an overview. Pharmacogn Rev 3:353–358

Pratap JV, Jeyaprakash AA, Rani PG, Sekar K, Surolia A, Vijayan M (2002) Crystal structures of artocarpin, a Moraceae lectin with mannose specificity, and its complex with methyl-α-d-mannose: implications to the generation of carbohydrate specificity. J Mol Biol 317(2):237–247

Pua CK, Hamid NSA, Rusul G, Rahman RA (2007) Production of drum-dried jackfruit (*Artocarpus heterophyllus*) powder with different concentration of soy lecithin and gum arabic. J Food Eng 78(2):630–636

Rao AVR, Varadan M, Venkataraman K (1973) Colouring matter of the *A. heterophyllus*. Indian J Chem 11:219–299

Ratnasooriya WD, Jayakody JR (2002) *Artocarpus heterophyllus* seeds inhibits sexual competence but not fertility of male rats. Indian J Exp Biol 40(3): 304–308

Roque-Barreira MC, Campos-Neto A (1985) Jacalin: an IgA-binding lectin. J Immunol 134(3):1740–1743

Sato M, Fujiwara S, Tsuchiya H, Fujii T, Iinuma M, Tosa H, Ohkawa Y (1996) Flavones with antibacterial activity against cariogenic bacteria. J Ethnopharmacol 54(2–3):171–176

Shinomiya K, Aida M, Hano Y, Nomura T (1995) A Diels-Alder-type adduct from *Artocarpus heterophyllus*. Phytochemistry 40(4):1317–1319

Slik JWF (2006) Trees of Sungai Wain. Nationaal Herbarium Nederland, Leiden. http://www.nationaal herbarium.nl/sungaiwain/

Soepadmo E (1992) *Artocarpus heterophyllus* Lamk. In: Verheij EWM, Coronel RE (eds) Plant resources of South East Asia No 2. Edible fruits and nuts. Prosea Foundation, Bogor, pp 86–91

Soong YY, Barlow PJ (2004) Antioxidant activity and phenolic content of selected fruit seeds. Food Chem 88(3):411–417

Soubir T (2007) Antioxidant activities of some local Bangladeshi fruits (*Artocarpus heterophyllus*, *Annona squamosa*, *Terminalia bellirica*, *Syzygium samarangense*, *Averrhoa carambola* and *Olea europa*). Sheng Wu Gong Cheng Xue Bao 23(2):257–261

Suresh S, Rani PG, Pratap JV, Sankaranarayanan R, Surolia A, Vijayan M (1997) Homology between jacalin and artocarpin from jackfruit (*Artocarpus integrifolia*) seeds. Partial sequence and preliminary crystallographic studies of artocarpin. Acta Crystallogr D Biol Crystallogr D53(4):469–471

Thomas CA (1980) Jackfruit, *Artocarpus heterophyllus* (Moraceae), as source of food and income. Econ Bot 34:154–159

Toda S, Shirataki Y (2006) Inhibitory effect of prenylated flavonoid in *Euchresta japonica* and *Artocarpus heterophyllus* on lipid peroxidation by interaction of hemoglobin and hydrogen peroxide. Pharm Biol 44(4):271–273

Tulyathana V, Tananuwonga K, Songjinda P, Jaiboonb N (2002) Some physicochemical properties of jackfruit (*Artocarpus heterophyllus* Lam) seed flour and starch. Sci Asia 28:37–41

U.S. Department of Agriculture, Agricultural Research Service (2010) USDA National nutrient database for standard reference, release 23. Nutrient Data Laboratory home page, http://www.ars.usda.gov/ba/bhnrc/ndl

Wei BL, Weng JR, Chiu PH, Hung CF, Wang JP, Lin CN (2005) Antiinflammatory flavonoids from *Artocarpus heterophyllus* and *Artocarpus communis*. J Agric Food Chem 53(10):3867–3871

Wetprasit N, Threesangsri W, Klamklai N, Chulavatnatol M (2000) Jackfruit lectin: properties of mitogenicity and the inhibition of herpesvirus infection. Jpn J Infect Dis 53(4):156–161

Wikipedia (2009) Jackfruit. http://en.wikipedia.org/wiki/Jackfruit

Wong KC, Lim CL, Wong LL (1992) Volatile flavour constituents of chempedak (*Artocarpus polyphema* Pers.) fruit and jackfruit (*Artocarpus heterophyllus* Lam.) from Malaysia. Flav Fragr J 7(6):307–311

Wu AM, Wu JH, Lin LH, Lin SH, Liu JH (2003a) Binding profile of *Artocarpus integrifolia* agglutinin (Jacalin). Life Sci 72(2):2285–2302

Wu Z, Zhou ZK, Gilbert MG (2003b) Moraceae link. In: Wu ZY, Raven PH, Hong DY (eds) Flora of China, vol 5, Ulmaceae through Basellaceae. Science Press/Missouri Botanical Garden Press, Beijing/St. Louis

Zheng ZP, Cheng KW, To JT, Li H, Wang M (2008) Isolation of tyrosinase inhibitors from *Artocarpus heterophyllus* and use of its extract as antibrowning agent. Mol Nutr Food Res 52(12):1530–1538

Zheng ZP, Chen S, Wang S, Wang XC, Cheng KW, Wu JJ, Yang D, Wang M (2009) Chemical components and tyrosinase inhibitors from the twigs of *Artocarpus heterophyllus*. J Agric Food Chem 57:6649–6655

Artocarpus integer

Scientific Name

Artocarpus integer (Thunb.) Merr.

Synonyms

Artocarpus champeden (Lour.) Stokes, *Artocarpus hirsutissima* Kurz, *Artocarpus integer* var. *silvestris* Corner, *Artocarpus integrifolia* L.f., *Artocarpus integrifolia* var. *hirsuta* Stokes, *Artocarpus jaca* Lam., *Artocarpus macrocarpon* (Thunb.) Dancer, *Artocarpus polyphema* Pers., *Polyphema champeden* Lour., *Radermachia integra* Thunb., *Saccus arboreus minor* Rumph., *Sitodium macrocarpon* Thunb.

Family

Moraceae

Common/English Names

Champedan, Chempedak, Small Jackfruit

Vernacular Names

Borneo: Banturung Manuk, Bukoh, Chempedak, Mengkahai, Nakan, Pulutan, Temedak, Temedak Man;
Burmese: Sonekadat;
Fijian: Uto Ni Idia;
India: Kathal, Kathar (Hindi), Chakka (Malayalam), Atibrhatphala, Kantakiphala, Panasa (Sanskrit), Chakka, Pilual (Tamil) Panasapandu (Telugu);
Indonesia: Campedak, Cempedak, Comedak (Java), Tundak, Tuadak (Kalimantan), Tawerak, Anaane, Tuada (Sulawesi), Cempeudak, Cimpedak, Sibodak, Bikawau, Cubadak, Temedak, Kakan, Akam-Akam (Sumatra), Tamberak (West Papua), Campedak (Sundanese), Baroh (Lingga);
Japanese: Koparamitsu;
Malaysia: Chempedak, Campedak, Baroh, Bankong, Bongkong; Takah(Pangan), Deko (Sakai), Menelang (Semang);
Sri Lanka: Pani Varaka (Sinhalese);
Thailand: Champada;
Vietnamese: Mít Tố Nữ.

Origin/Distribution

Chempedak is indigenous to Thailand, Peninsular Malaysia, Sumatra, Java, Borneo (throughout the island), Sulawesi, Maluku, New Guinea. It is widely cultivated in these areas and introduced elsewhere in the tropics.

Agroecology

A. integer is strictly a tropical species. In its natural ecological range, it commonly grows as an understorey tree in undisturbed mixed

T.K. Lim, *Edible Medicinal And Non-Medicinal Plants: Volume 3, Fruits*,
DOI 10.1007/978-94-007-2534-8_44, © Springer Science+Business Media B.V. 2012

dipterocarp forests rainforest areas up to 500 m altitude or sometimes higher, where there is no distinct dry season. It occurs mostly on ridges with clay to sandy soils. In secondary forests it is usually present as a pre-disturbance remnant, or cultivated. The tree thrives on fertile well-drained soils, but prefers a fairly high water table. It can survive periodic water-logging.

Edible Plant Parts and Uses

The sweet, fragrant and flavoursome, fleshy perianth of the fruit is eaten fresh or preserved in syrup. In Malaysia, 1–2 perianth balls are coated with flour and deep fried to make into fritters for consumption as dessert or snacks. The ripe and unripe flesh is salted in Malaysia and used as a pickle called *jerami*. The perianths are also candied or made into chips by sun-drying. Young fruits are cooked in coconut milk and eaten as a curried vegetable or in soup or made into pickles. The seeds are rich in starch, eaten roasted or after boiling in salty water for 30 min and peeling off the seed coat. The seeds have a pleasant, nutty flavour. A starchy flour may be obtained from the seeds. The young, tender leaves are also cooked as vegetables in Sarawak.

Botany

Chempedak is an evergreen, medium–sized, mid-canopy, branched, monoecious tree growing to 10–30 m tall. The bark is greyish brown with bumps on the trunk and main limbs where leafy twigs are produced, which bears the fruits (Plate 1). Stem has white sap. Brown stiff, reflexed hairs 3 mm long cover the twigs, stipules and leaves. Twigs are 2.5–4.0 mm thick, terete with annulate stipular scars. Stipules 2, ovate-triangular-oblong, dark green and up to 9 cm long. Leaves are alternate and borne on 1–3 cm long petiole. The lamina is obovate to elliptic, 5–25 × 2.5–12 cm in size, and the base is cuneate to rounded, with entire margin, and acuminate apex, glossy green above and pale green and pubescent below. The lateral veins are in 6–10 pairs and curve forward. The inflorescences are solitary, unisexual, cauliflorous or ramiflorous and borne on the axillary position of short leafy shoots. Male heads are cylindrical, 3–5.5 × 1 cm, with small, 1 mm diameter, yellowish flowers with 1 straight stamen and 2 oblong, concave, whitish-yellow perianth segments. The female heads have small, tubular flowers, with perianths cohering at base and simple filiform styles with a 1.5 mm long exserted stigma. The fruit is a syncarp, formed by the enlarged connate perianths which are adnate to the axis of the inflorescence. The fruit is cylindrical to oblong cylindrical, densely beset with short pyramidal, faintly prominent tubercles, 20–35 cm × 10–15 cm in size and borne on 7–12 cm long peduncle. The fruit is yellowish, brownish, or orange-yellow, and smells strongly at maturity (Plates 2, 3, 4 and 5). Pericarps, including the seeds, are ellipsoid to oblong about 3 × 2 cm (Plate 6), seeds are encased by a membranous testa (Plate 7). Cotyledons are unequal, thick and fleshy.

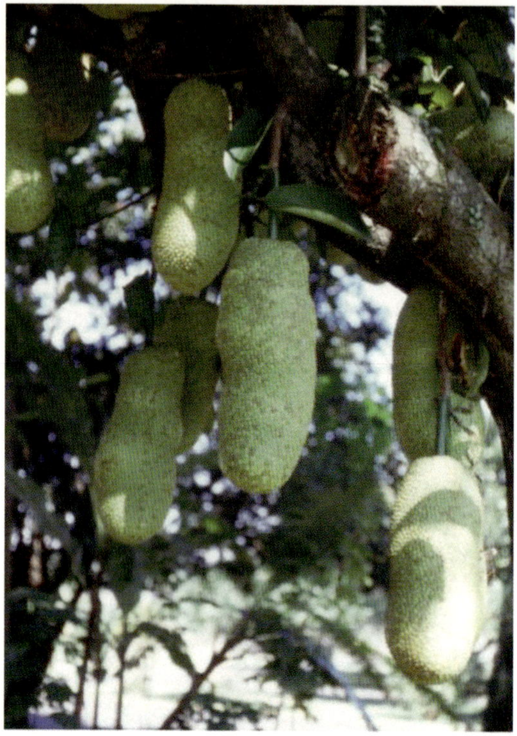

Plate 1 Immature chempedak fruit on a main branch

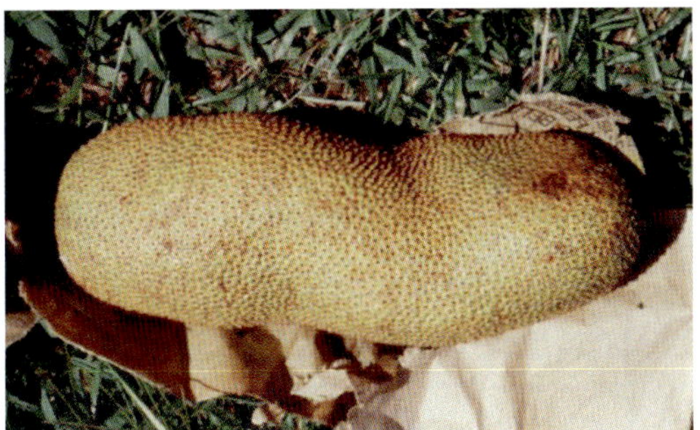

Plate 2 A ripe chempedak fruit

Plate 5 Ripe chempedak fruit in a Kuala Lumpur market

Plate 3 Ripe Mít Tố Nữ fruits in a Vietnamese market

Plate 6 Edible, yellow-orange, fleshy, aromatic pericarp (perianth) of chempedak

Plate 4 Ripe Chempadak fruit in a Sumatran market

Plate 7 Edible chempedak seeds with papery testa membranes

Nutritive/Medicinal Properties

The total fruit weight was found to vary from 0.6 to 3.5 kg and is generally smaller than the jackfruit. The total edible portion (perianths + seeds) is 25–50% of fresh fruit weight. The total weight of all perianths of a fresh fruit varied from 100 to 1,200 g. The proximate composition of the fleshy perianths on dry weight basis per 100 g edible portion was reported as: protein 3.5–7.0 g, fat 0.5–2 g, carbohydrates 84–87 g, fibre 5–6 g, and ash 2–4 g moisture (fresh weight basis) was 58–85% (Jansen 1992).

The proximate composition of seeds, per dry weight basis was reported as: protein 10–13%, fat 0.5–1.5%, carbohydrates 77–81%, fibre 4–6%, ash 3–4% and moisture (fresh weight basis) is 46–78% (Jansen 1992). Chempedak fruit volatiles contained 54 components of which 37.4%, were alcohols and 32.2% carboxylic acids (Wong et al. 1992). The main constituents were 3-methylbutanoic acid (28.2%) and 3-methylbutan-1-ol (24.3%). Other important flavour compounds include 2-acetyl-1-pyrroline and the tentatively identified 2,5-dimethyl-4-hydroxy-3 (2 H)-furanone.

The stem bark and roots of *Artocarpus integer* contain prenylated flavones which possess cytotoxic and antimalarial activities.

Anticancer Activity

From the roots the following prenylated flavones were isolated: artoindonesianin A, artoindonesianin B and artonin A. Artoindonesianins A and B exhibited cytotoxic activity against murine leukemia (P-388) cells (Hakim et al. 1999). Two isoprenylated flavones, artoindonesianins U and V were isolated from the heartwood and the latter showed strong cytotoxic activity against P-388 cell lines (Syah et al. 2004). Also isolated was a new prenylated flavone, named cyclochampedol, together with four known triterpenes-cycloeucalenol, glutinol, cycloartenone, and 24-methylenecycloartanone-as well as β-sitosterol (Achmad et al. 1996). This new flavone, cyclochampedol was found to be bioactive in the brine shrimp lethality assay. This assay is useful tool for preliminary assessment of cytotoxicity.

Antimalarial Activity

From the stem bark, the following prenylated flavones were isolated: artocarpones A and B, and seven known isoprenylated flavonoids, artonin A, cycloheterophyllin, artoindonesianin E, artoindonesianin R, heterophyllin, heteroflavanone C, and artoindonesianin A-2 (Widyawaruyanti et al. 2007). Among the flavones isolated, heteroflavanone had the most potent inhibitory activity against the growth of *Plasmodium falciparum* 3D7 clone, with an IC_{50} value of 1 nmol/L. From the heartwood, 4 prenylated flavones artoindonesianins Q-T were isolated (Syah et al. 2002).

In Thailand, researchers isolated an antimalarial prenylated stilbene, trans-4-(3-methyl-E-but-1-enyl)-3, 5, 2′, 4′-tetrahydroxystilbene with an EC50 of 1.7 μg/ml against *Plasmodium falciparum* in culture (Boonlaksiri et al. 2000).

Lectin and Cell Adhesion Activity

Purified lectins from seeds of six distinct clones of *Artocarpus integer* (lectin C) were shown to be structurally and functionally similar (Hashim et al. 1993). The lectins appeared to interact with several human serum proteins, with the predominance of the IgA1 and C1 inhibitor molecules. Interaction was not detected with IgA2, IgD, IgG and IgM. The lectin Cs were also shown to precipitate monkey, sheep, rabbit, cat, hamster, rat and guinea-pig serum. Due to their uniform properties, lectin C may provide better alternative to the *Artocarpus heterophyllus* lectin, jacalin, for use in future investigations. In another study, purified and crude extract of lectin C from six cultivars of *Artocarpus integer* seeds were found to consume complement and thus decreased the complement-induced haemolytic activity of sensitized sheep erythrocytes (Hashim et al. 1994). The change in the complement-mediated haemolytic activity was significantly decreased when incubation of the lectins was performed in the presence of melibiose. The reversal effect of the carbohydrate, a potent inhibitor of the lectin's binding to O-linked oligosaccharides of glycoprotein, demonstrated involvement of the lectins interaction with O-glycans of glycoproteins in the consumption of guinea-pig complement.

A mannose-binding lectin, termed chempedak lectin-M, was isolated from an extract of the crude seeds of chempedak (*Artocarpus integer*) (Lim et al. 1997). The mannose-binding lectin was composed of 16.8 kDa polypeptides with some of the polypeptides being disulphide-linked to give dimers. When tested with all isotypes of immunoglobulins, chempedak lectin-M demonstrated a selective strong interaction with human IgE and IgM, and a weak interaction with IgA2. The binding interactions of lectin-M were metal ion independent. The lectin was also shown to interact with horseradish peroxidase, ovalbumin, porcine thyroglobulin, human α1-acid glycoprotein, transferrin and α1-antitrypsin. It demonstrated a binding preference to Manα1–3Man ligands in comparison to Manα1–6Man or Manα1–2Man. Further stduides (Lim et al. 1998) showed that he lectin was the main mitogenic component in the crude extract of the chempedak seeds. It stimulated the proliferation of murine T cells at an optimal concentration of 2.5 μg/ml in a 3 day culture. Lectin-M appeared to be a T-cell mitogen as it does not induce significant DNA synthesis when cultured with spleen cells from the nude mouse. In the absence of T cells, the lectin was incapable of inducing resting B cells to differentiate into immunoglobulin secreting plasma cells. Further studies found chempedak lectin-M to be a lectin with high specificity and affinity for the core-mannosyl residues of the N-linked oligosaccharides of glycoproteins (Hashim et al. 2001). The lectin demonstrated strong interaction with haptoglobin β chain, orosomucoid, α1-antitrypsin, α2-HS glycoprotein, transferrin, hemopexin, α1B-glycoprotein, and the heavy chains of IgA, IgM and IgG of the human serum. The use of chempedak lectin-M to probe for serum glycoproteins that were separated in a 2-D gel electrophoresis and Western blotting technique may be conveniently applied to analyse the acute-phase and humoral immune responses simultaneously. Gabrielsen et al. (2010) found that the mannose-binding lectin from chempedak was a homotetramer with a single-monomer molecular weight of 16,800 Da. Chempedak mannose-binding lectin had successfully been used to detect altered glycosylation states of serum proteins.

A Galβ1–3GalNAc- and IgA1-reactive lectin was isolated and purified from the seeds of chempedak (*Artocarpus integer*) (Abdul-Rahman et al. 2002). The lectin demonstrated at least 95% homology to the N-terminal sequence of the α chains of a few other galactose-binding *Artocarpus* lectins. The two smaller subunits of the lectin, each comprised of 21 amino acid residues, demonstrated minor sequence variability. Their sequences were generally comparable to the β chains of the other galactose-binding *Artocarpus* lectins. When used to probe human serum glycopeptides that were separated by two-dimensional gel electrophoresis, the lectin demonstrated strong apparent interactions with glycopeptides of IgA1, hemopexin, α2-HS glycoprotein, α1-antichymotrypsin, and a few unknown glycoproteins. Immobilisation of the lectin to Sepharose generated an affinity column that may be used to isolate the O-glycosylated serum glycoproteins. The galactose-binding lectin from chempedak (*Artocarpus integer*) was found to consist of two chains: α and β (133 and 21 amino acids, respectively) (Gabrielsen et al. 2009). It was shown to recognize and bind to carbohydrates involved in IgA and C1 inhibitor molecules.

Traditional Medicinal Uses

In traditional folkloric medicine, the Iban in Sarawak apply a paste of the inner bark to heal wounds and prevent infection. In Peninsular Malaysia, juice of the roots has been used for fever. The ash from burnt leaves, maize and coconut shell has been used to treat ulcers. An infusion of the root ash is mixed with *Selaginella* ash and prescribed as a protective medicine after childbirth. The bark is used in poultices for painful feet, hands and for ulcers. In the Philippines, the leaves are heated and applied to wounds. The pith has been reported to cause abortion and the wood is sedative. Unripe fruit is astringent and ripe fruit laxative.

Other Uses

In Kerala and Bengal in India, the leaves are used as fodder. The ripe fruit is fed to cattle and the fruit, leaves and bark are used to feed elephants. The bark can be used to make rope and the resinous latex for the preparation of bird- lime and vanishing material. The bark is also rich in tannin.

The dark yellow to brown wood is strong and durable and is used for building construction, furniture, boats cabinet-work, brush-handles. In the Philippines, the wood is used for musical instruments, furniture, cabinet-work and tool-handles. The wood also makes a good fuel-wood. In Indo-China, the Buddhist used the timber for construction of sacred buildings on account of its yellow colour. With alum, the extract of heartwood provides a yellow dye that is moderately fast on silk. This dye is used in colouring the saffron-coloured robes of Buddhist monks in Indo-China.

The tree has been planted in conjunction with cash crops such as *Carica papaya*, in reforestation systems. The tree has been used as shade tree for coffee but was found to be not ideal.

Comments

Malaysia and Indonesia have developed many improved varieties of chempedak. Chempedak is usually propagated by seeds which exhibit a recalcitrant seed storage behaviour.

Selected References

Abdul-Rahman M, Karsani SA, Othman I, Abdul Rahman PS, Hashim OH (2002) Galactose-binding lectin from the seeds of chempedak (*Artocarpus integer*): sequences of its subunits and interactions with human serum O-glycosylated glycoproteins. Biochem Biophys Res Commun 295(4):1007–1013

Achmad SA, Hakim EH, Juliawaty LD, Makmur L, Suyatno Aimi N, Ghisalberti EL (1996) A new prenylated flavone from *Artocarpus champeden*. J Nat Prod 59(9):878–879

Backer CA, Bakhuizen van den Brink RC Jr (1965) Flora of java (Spermatophytes only), vol 2. Wolters-Noordhoff, Groningen, 641pp

Boonlaksiri C, Oonanant W, Kongsaeree P, Kittakoop P, Tanticharoen M, Thebtaranonth Y (2000) An antimalarial stilbene from *Artocarpus integer*. Phytochemistry 54(4):415–417

Burkill IH (1966) A dictionary of the economic products of the Malay Peninsula. Revised reprint, 2 volumes. Ministry of agriculture and co-operatives, Kuala Lumpur, Malaysia. Vol 1 (A–H), pp 1–1240; Vol 2 (I–Z), pp 1241–2444

Chai PPK (2006) Medicinal plants of Sarawak. Lee Ming Press, Kuching, p 212

Gabrielsen M, Riboldi-Tunnicliffe A, Abdul-Rahman PS, Mohamed E, Ibrahim WIW, Hashim OH, Isaacs NW, Cogdell RJ (2009) Crystallization and preliminary structural studies of chempedak galactose-binding lectin. Acta Crystallogr Sect F 65:895–897

Gabrielsen M, Abdul-Rahman PS, Isaacs NW, Hashim OH, Cogdell RJ (2010) Crystallization and initial X-ray diffraction analysis of a mannose-binding lectin from chempedak. Acta Crystallogr F66:592–594

Hakim EH, Fahriyati A, Kau MS, Achmad SA, Makmur L, Ghisalberti EL, Nomura T (1999) Artoindonesianins A and B, two new prenylated flavones from the root of *Artocarpus champeden*. J Nat Prod 62(4):613–615

Hakim EH, Achmad SA, Makmur L, Syah YM, Aimi N, Kitajima M, Takayama H, Ghisalberti EL (2006) Prenylated flavonoids and related compounds of the Indonesian *Artocarpus* (Moraceae). J Nat Med 60(3):161–184

Hashim OH, Gendeh GS, Jaafar MI (1993) Comparative analyses of IgA1 binding lectins from seeds of six distinct clones of *Artocarpus integer*. Biochem Mol Biol Int 29(1):69–76

Hashim OH, Gendeh GS, Cheong CN, Jaafar MI (1994) Effect of *Artocarpus integer* lectin on functional activity of guinea-pig complement. Immunol Invest 23(2):153–160

Hashim OH, Ahmad F, Shuib AS (2001) The application of *Artocarpus integer* seed lectin-M in the detection and isolation of selective human serum acute-phase proteins and immunoglobulins. Immunol Invest 30(2):131–141

Jansen PCM (1992) *Artocarpus integer* (Thunb.) Merr. In: Verheij EWM, Coronel RE (eds) Plant resources of South-East Asia No 2. Edible fruits and nuts. Prosea, Bogor, pp 91–94

Jarrett FM (1959a) Studies in *Artocarpus* and allied genera. III. A revision of *Artocarpus* subgenus *Artocarpus*. J Arnold Arboretum 40(2):113–155

Jarrett FM (1959b) Studies in *Artocarpus* and allied genera, III. A revision of *Artocarpus* subgenus *Artocarpus* (cont.). J Arnold Arbor 40(4):327–368

Lim SB, Chua CT, Hashim OH (1997) Isolation of a mannose-binding and IgE- and IgM-reactive lectin from the seeds of *Artocarpus integer*. J Immunol Methods 209(2):177–186

Lim SB, Kanthimathi MS, Hashim OH (1998) Effect of the mannose-binding *Artocarpus integer* lectin on the cellular proliferation of murine lymphocytes. Immunol Invest 27(6):395–404

Molesworth Allen B (1967) Malayan fruits. An introduction to the cultivated species. Moore, Singapore, p 245

Ochse JJ, Bakhuizen van den Brink RC (1931) Fruits and fruitculture in the Dutch East Indies. G. Kolff & Co, Batavia-C, 180pp

Slik JWF (2006) Trees of Sungai Wain. Nationaal Herbarium Nederland, Leiden. http://www.nationaalherbarium.nl/sungaiwain/

Syah YM, Achmad SA, Ghisalberti EL, Hakim EH, Makmur L, Mujahidin D (2002) Artoindonesianins Q-T, four isoprenylated flavones from *Artocarpus champeden* Spreng. (Moraceae). Phytochemistry 61(8):949–953

Syah YM, Achmad SA, Ghisalberti EL, Hakim EH, Mujahidin D (2004) Two new cytotoxic isoprenylated flavones, artoindonesianins U and V, from the heartwood of *Artocarpus champeden*. Fitoterapia 75(2):134–140

Widyawaruyanti A, Subehan SK, Kalauni SK, Awale S, Nindatu M, Zaini NC, Syafruddin D, Asih PBS, Tezuka Y, Kadota S (2007) New prenylated flavones from *Artocarpus champeden*, and their antimalarial activity in vitro. J Nat Med 61(4):410–413

Wong KC, Lim CL, Wong LL (1992) Volatile flavour constituents of chempedak (*Artocarpus polyphema* Pers.) fruit and jackfruit (*Artocarpus heterophyllus* Lam.) from Malaysia. Flavour Fragr J 7(6):307–311

World Agroforestry Centre (2008) Agroforestry database. http://www.worldagroforestry.org/sea/Products/AFDbases/af/index.asp

Artocarpus odoratissimus

Scientific Name

Artocarpus odoratissimus Blanco.

Synonyms

Artocarpus mutabilis Becc., *Artocarpus tarap* Becc.

Family

Moraceae

Common/English Names

Johey Oak, Marang, Terap

Vernacular Names

Borneo: Benturung, Jarap Hutan;
Brunei: Tarap;
Indonesia - Kalimantan: Pelah (Punam, Malinau), Kiran (Lun Daye, Mentarang), Kian, Kiang (Merap, Malinau,), Da'eng Kegheng (Kenya Uma' Lung), Monyet Hutan;
Philippines: Marang (Sulu), Madang (Lanao), Loloi (Tagalog);
Sabah: Tarap, Terap;
Sarawak: Pi-Ien (Bidayuh), Timadang (Dusun), Terap, Tarap, Marang, Lumuk, Lumuk Amat, Lumuk Ujut, Pingan (Iban), Keiran, Keran Beung, Keran Pukung, Keran Urung (Kelabit), Basut (Kenyah);
Thai: Khanun Sampalor.

Origin/Distribution

Marang is indigenous to Myanmar, Thailand, Peninsular Malaysia, Sumatra, Java, Lesser Sunda Islands and Borneo (throughout the island).

Agroecology

Marang occurs in undisturbed to slightly disturbed, mixed dipterocarp forest up to 800 m altitude between latitude 15° north and south. In secondary forests, it is usually present as a predisturbance remnant tree. It also occurs on alluvial sites near streams and rivers and hillsides and ridges, on sandy to clayey soils. It thrives best in regions with abundant and uniformly distributed rainfall on rich loamy, well-drained soils.

Edible Plant Parts and Uses

Ripe fruit perianth is sweet, juicy, fragrant and is eaten raw (fresh) or used as ingredient in cakes. The fruiting perianth also makes an excellent flavouring for ice-cream. The seeds are edible when

T.K. Lim, *Edible Medicinal And Non-Medicinal Plants: Volume 3, Fruits*,
DOI 10.1007/978-94-007-2534-8_45, © Springer Science+Business Media B.V. 2012

boiled or roasted and have a nutty flavor. Young fruits are cooked in coconut milk and eaten as a curried vegetable. Once opened, the perianth should be consumed quickly (in a few hours), as it loses flavour rapidly and the perianth oxidises.

Botany

Evergreen, branched, mid-canopy tree growing to 39 m high, with low buttresses a trunk diameter of 45 cm, bark grey to dark brown with sticky white latex and spreading canopy. Twigs have long, yellow to red, spreading hairs and stipule-scar rings. Stipules are ovate. 1–8 cm long, yellow to red and pubescent. Leaves are spirally arranged. Juvenile leaves pinnatifid. Mature leaves are broadly elliptic (Plate 1) to obovate, 16–50 cm by 11–28 cm, cuneate at base to slightly winged, margin entire or shallowly crenate, upper half often 3-lobed apex blunt or shortly acuminate, both surfaces sparsely pubescent, leathery with 13–15 pairs of lateral veins; petiole 2–3 cm long. Inflorescences axillary, solitary; Male and female flowers, small, 0.5 mm diameter, yellowish, on the same tree but in separate axillary inflorescence heads. Male heads solitary or in pairs, are ellipsoid to clavate and 4–11 × 2–6 cm. Female heads solitary with pubescent peltate bracts mostly shed and simple styles exserted to 1.5 mm. Fruit (syncarp) subglobose, 15–20 cm × 13 cm, green (Plate 2) turning yellow-brown when ripe, densely covered with stiff, hairy protuberances (Plates 3, 4, 5 and 6) of about 1 cm length; rind 8 mm thick; flesh (fruiting perianths) white, juicy, aromatic, sweet (Plates 5 and 6); peduncles 5–14 cm long. Pericarps (including the seeds) ellipsoid, about 15 mm × 8 mm.

Plate 1 Developing pendulous marang fruit and broadly elliptic, large leaves

Plate 2 Green, immature marang fruit

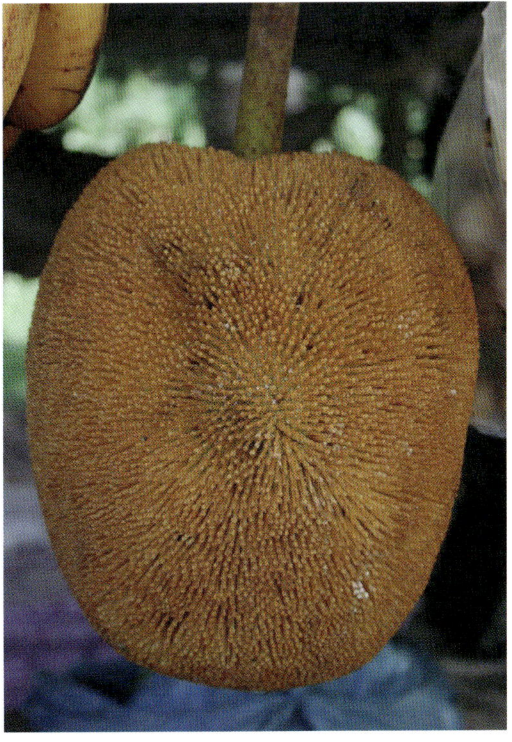

Plate 3 Ripe marang fruit

Plate 4 Marang fruits on sale in a local market

Plate 5 Ripe marang fruit sliced to show the white fleshy perianth

Plate 6 Close up of the white, fleshy, edible perianth

Nutritive/Medicinal Properties

The edible portion (i.e. the fleshy perianth) contains per 100 g edible portion contains: water 65.7–84.2 g, protein 0.8–1.47 g, fat 0.2–0.3 g, carbohydrates 32.4 g, ash 0.5–0.8 g, fibre 0.6–0.77 g, calcium 17 mg, phosphorus 35 mg, iron 2.1 mg and vitamin C 30 mg (Galang 1955).

A prenylated pyranoflavone derivative artosimmin (1) and traxateryl acetate was isolated from *Artocarpus odoratissimus* (Ee et al. 2010). Artosimmin was found to be significantly cytotoxic against cancer cell lines (HL-60 (Human promyelocytic leukemia cells) and MCF-7 (human breast adenocarcinoma cell line)) and also possessed antioxidant properties toward 1,1-diphenyl-2-picrylhydrazyl radical (DPPH). The seed extract of *A. odoratissimus* showed higher scavenging effect compared to the flesh (Abu Bakar et al. 2009). The same trend was also observed in FRAP assay. *A. odoratissimus* flesh contained higher amount of total phenolics (4.39 mg GAE/g) compared to Jackfruit pulp (0.46 mg GAE/g).

In traditional folkloric medicine, the Ibans in Sarawak have been reported to consume a decoction of the root for diarrhoea. Ash from the leaves is applied on scorpion stings and centipede bites. When added with a little coconut oil, the ash is used for scabies or kuris.

Other Uses

In the Iban community in Sarawak, leaves are used magically by hanging on the door to drive away evil female spirits from entering the premises to steal men's testicles.

Comments

Marang is usually propagated from seeds.

Selected References

Abu Bakar MF, Mohamed M, Rahmat A, Fry J (2009) Phytochemicals and antioxidant activity of different parts of bambangan (*Mangifera pajang*) and tarap (*Artocarpus odoratissimus*). Food Chem 113(2):479–483

Chai PPK (2006) Medicinal plants of Sarawak. Lee Ming Press, Kuching, 212pp

dela Cruz FS Jr (1992) *Artocarpus odoratissimus* Blanco. In: Coronel RE, Verheij EWM (eds) Plant resources of South-East Asia. No. 2: edible fruits and nuts. Prosea Foundation, Bogor, pp 94–96

Djarwaningsih T, Alonzo DS, Sudo S, Sosef MSM (1995) *Artocarpus* L. In: Lemmens RHMJ, Soerianegara I, Wong WC (eds) Plant resources of South-East Asia. No. 5(2): timber tree: minor commercial timber. Prosea Foundation, Bogor, pp 59–63

Ee GCL, Teo SH, Mawardi R, Lim CK, Lim YM, Bong CFJ (2010) Artosimmin – a potential anticancer lead compound from *Artocarpus odoratissimus*. Lett Org Chem 7(3):240–244

Galang FG (1955) Fruit and nut growing in the Philippines. AIA Printing Press, Malabon, pp 300–302

Jarrett FM (1959a) Studies in *Artocarpus* and allied genera, III. A revision of *Artocarpus* subgenus *Artocarpus*. J Arnold Arbor 40(2):113–155

Jarrett FM (1959b) Studies in *Artocarpus* and allied genera, III. A revision of *Artocarpus* subgenus *Artocarpus* (cont.). J Arnold Arbor 40(4):327–368

Munawaroh E, Purwanto Y (2009) Studi hasil hutan non kayu di Kabupaten Malinau, Kalimantan Timur (In Indonesian). In: Paper presented at the 6th basic science national seminar, Universitas Brawijaya, Indonesia, 21 Feb 2009. http://fisika.brawijaya.ac.id/bss-ub/proceeding/PDF%20FILES/BSS_146_2.pdf

Slik JWF (2006) Trees of Sungai Wain. Nationaal Herbarium Nederland, Leiden. http://www.nationaalherbarium.nl/sungaiwain/

Voon BH, Chin TH, Sim CY, Sabariah P (1988) Wild fruits and vegetables in Sarawak. Sarawak Department of Agriculture, Sarawak, 114pp

Artocarpus rigidus

Scientific Name

Artocarpus rigidus Blume.

Synonyms

Artocarpus cuspidatus Griff., *Artocarpus dimorphophylla* Miq., *Artocarpus echinata* Roxb., *Artocarpus kertau* Zoll. ex Miq., *Artocarpus muricata* Hunter, *Artocarpus rigidus* Wall., *Artocarpus varians* Miq.

Family

Moraceae

Common/English Names

Monkey Jack, Monkey Jackfruit (Plate 3)

Vernacular Names

Borneo: Buruni, Dadah, Keledang, Mayuh Dia, Pala Munsoh, Pala Musoh, Pujan;
Burmese: Taung Peing;
Chinese: Hou Mian Guo, Hou Mian Bao Shu;
Indonesia: Mandalika, Pujan (Kalimantan), Purian, Tawan, Tampuniek (Sumatra), Peussar (Sundanese), Purin (Bangka), Kosar, Mandalika (Java);
Khmer: Knor Prei;
Malaysia: Pala Munsoh (Iban), Gias, Jelatoh, Mendeleka, Pasal, Perian, Pontot, Purin, Perian, Peruput, Tampang, Tampuneh, Tampuret, Tapunet, Tawan, Temponek, Tempunai, Tempunik, Tunjong;
Thailand: Kanun Pan;
Vietnam: Ay Da Xóp, Ay Mit Nay;

Origin/Distribution

Monkey Jack is native to Myanmar, Thailand, Peninsular Malaysia, Sumatra, Java, Lesser Sunda Islands, Borneo (throughout the island). The tree is cultivated in parts of Southeast Asia.

Agroecology

The species thrives in a hot, wet and humid tropical climate. In it native range it occurs in undisturbed to disturbed mixed dipterocarp forests in the lowlands up to 500 m altitude. In secondary forests usually present as a pre-disturbance remnant tree. It occurs on sandy to clayey soils, but also on limestone in shade to partial shade.

T.K. Lim, *Edible Medicinal And Non-Medicinal Plants: Volume 3, Fruits*,
DOI 10.1007/978-94-007-2534-8_46, © Springer Science+Business Media B.V. 2012

Plate 1 Monkey jack tree habit and leaves

Edible Plant Parts and Uses

Fleshy perianth surrounding the seed is eaten fresh; the seed is roasted or boiled and eaten.

Botany

Upper canopy, evergreen tree up to 47 m tall and 79 cm bole and buttressed up to 3 m, bark greenish-grey, peeling off in flakes (Plates 1 and 2). Stem with white sap. Twigs stout, strongly rugose, glabrous to densely pubescent. Stipules ovate-lanceolate, 0.5–3 cm long. Leaves alternate, simple, penniveined, stalked,1–3 cm long, sparsely to densely pubescent, coriaceous, elliptic to oblong-elliptic, obovate to round, vary variable in size 6.5–26 by 3.5–15 cm, apex blunt to slightly pointed, base cuneate or rounded, margin entire, margin entire or shallowly crenate towards the apex (Plate 1); juvenile leaves pinnatifid; main veins prominent beneath. Inflorescences axillary, solitary; male head (sub)globose, 13–25 mm diameter, smooth, yellow on 1–5 mm stalk.; female head with pubescent peltate bracts being shed and simple styles exserted to ca. 5 mm. Syncarp globose, up to 7(−13)cm diameter, greenish –yellow ripening to dull orange, echinate from closely set, rigid, tapering, fluted, acute, hispidulous processes, 7–9 mm long; rind 1 cm thick; fruiting perianths orange and waxy (Plate 4), subacid, lacking odour; peduncle 8–25(−40)mm long. Pericarp (including the seed) ellipsoid, 12 mm × 7 mm.

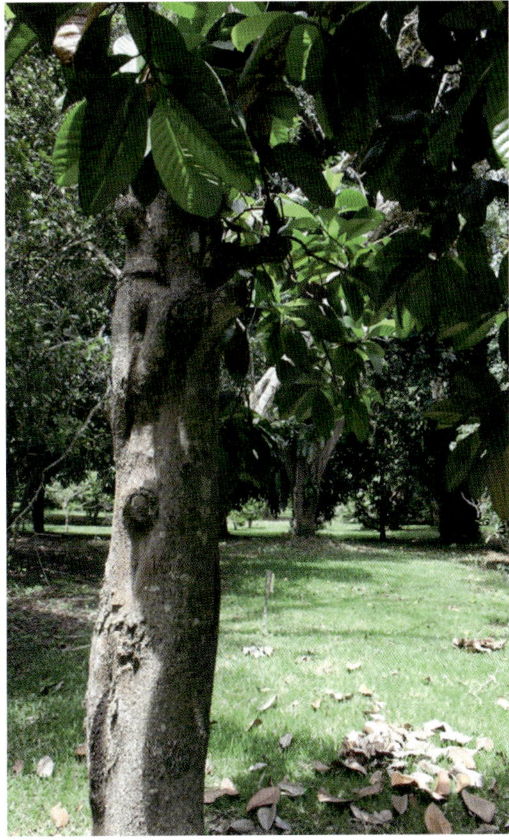

Plate 2 Monkey jack tree trunk

Plate 3 Tree label

Nutritive/Medicinal Properties

No information on its nutritive value has been published.

Four new isoprenylated flavone derivatives artonins M, N, O and P were isolated from the

Plate 4 Maokey jack fruit – echinate syncarp showing the edible orangey perianth inside

bark (Hano et al. 1993). Studies in Thailand reported that the roots contained phenolic compounds which exhibited antiplasmodial, antimicrobial and cytotoxic activities viz. flavonoid 7-demethylartonol E (1) and the chromone artorigidusin (2), together with four known phenolic compounds, the xanthone artonol B (3), the flavonoid artonin F (4), the flavonoid cycloartobiloxanthone (5), and the xanthone artoindonesianin C (6) (Namdaung et al. 2006). Compounds 1, 4, and 5 exhibited antiplasmodial activity against *Plasmodium falciparum* while all 6 compounds showed antimycobacterial activity against *Mycobacterium tuberculosis* with compound 4 being the most active compound (MIC 6.25 μg/ml). Compounds 1, 3, 5 and 6 also exhibited cytotoxic activity. Compounds 5 and 6 were active against KB tumour cells (sub-line of the keratin forming tumour cell line HeLa), whereas compounds 2, 5, and 6 showed varying toxicity to BC (breast cancer) cells. Compounds 1–3, 5, and 6 were active in the NCI-H187 (human small cell lung cancer) cytotoxicity assay, with compound 3 being the most active compound (IC_{50} 1.26 μg/ml).

The Phnong ethnic group in Kampuchea has been reported to add roots and bark of *A. rigidus* to water and to drink the decoction for stomach-ache.

Other Uses

The timber is useful and used in Peninsular Malaysia for furniture, house-beams and for boats in Sumatra. In Java, the latex is used for mixing with wax in calico-dyeing and as a veterinary medicine for wounds.

Comments

Monkey jack is vulnerable to habitat loss.

Selected References

Burkill IH (1966) A dictionary of the economic products of the Malay Peninsula. Revised reprint, 2 volumes. Ministry of agriculture and co-operatives, Kuala Lumpur. Vol 1 (A–H), pp 1–1240; Vol 2 (I–Z), pp 1241–2444

Hano Y, Inami R, Nomura T (1993) Components of the bark of *Artocarpus rigida* Bl. 2. Structures of four new isoprenylated flavone derivatives artonins M, N, O and P. Heterocycles 35:1341–1350

Jarrett FM (1959a) Studies in *Artocarpus* and allied genera, III. A revision of *Artocarpus* subgenus *Artocarpus*. J Arnold Arboretum 40(2):113–155

Jarrett FM (1959b) Studies in *Artocarpus* and allied genera, III. A revision of *Artocarpus* subgenus *Artocarpus* (cont.). J Arnold Arbor 40(4):327–368

Kochummen KM (1978) Moraceae. In: Ng FSP (ed) Tree flora of Malaya, vol 3. Longman Malaysia Sdn. Berhad, Kuala Lumpur, pp 119–168

Namdaung U, Arronrerk N, Suksamram S, Danwisetkanjana K, Saenboorueng J, Arichompu W, Suksamran A (2006) Bioactive constituents of the root bark of *Artocarpus rigidus* subsp. *rigidus*. Chem Pharm Bull (Tokyo) 54(10):1433–1436

Slik JWF (2006) Trees of Sungai Wain. Nationaal Herbarium Nederland, Leiden. http://www.nationaalherbarium.nl/sungaiwain/

Voon BH, Chin TH, Sim CY, Sabariah P (1988) Wild fruits and vegetables in Sarawak. Sarawak Department of Agriculture, Sarawak, 114pp

Artocarpus sericicarpus

Scientific Name

***Artocarpus sericicarpus* Jarrett.**

Synonyms

Artocarpus blumei auct. non Trecul ex Vidal, *Artocarpus elasticus* non Blume.

Family

Moraceae

Common/English Name

Pedalai

Vernacular Names

Philippines: Gumihan;
Sarawak: Terap Bulu, Pedalai (Iban).

Origin/Distribution

The pedalai is indigenous to Malaysia (Sarawak), the Philippines and Indonesia (Kalimantan, Sulawesi and the Moluccas (Maluku).

Agroecology

Pedalai like other *Artocarpus* species thrives in the humid tropics in areas with a mild monsoon climate. It occurs in tropical evergreen forests up to 500–1,000 m elevation. It is occasionally seen growing on the steep, clayey hillsides of the inland regions.

Edible Plant Parts and Uses

The fruit is eaten fresh. The sweet, creamy-white flesh (perianth) is easy to eat and like the marang (*Artocarpus odoratissimus*) the segments cling to the central core when the skin is removed. Fruit odour is not as strong as it is with the fruit of the marang. As with all the *Artocarpus* species, the seeds are also edible and those of the pedalai are considered to be some of the tastiest. They may be boiled, dried and roasted or fried.

Botany

A large, evergreen, rainforest tree up to 30–40 m high. Stipules are broadly lanceolate, 6–12 cm long. Leaves are elliptic to ovate, large, 20–70 cm by 10–50 cm, with entire to slightly crenate margin, dark green; juvenile leaves are more or less pinnatified, Inflorescences are axillary and solitary. Male heads are oblongoid, 3.5–10 cm by 1.5–2 cm wide; female heads are sub-globose,

T.K. Lim, *Edible Medicinal And Non-Medicinal Plants: Volume 3, Fruits*,
DOI 10.1007/978-94-007-2534-8_47, © Springer Science+Business Media B.V. 2012

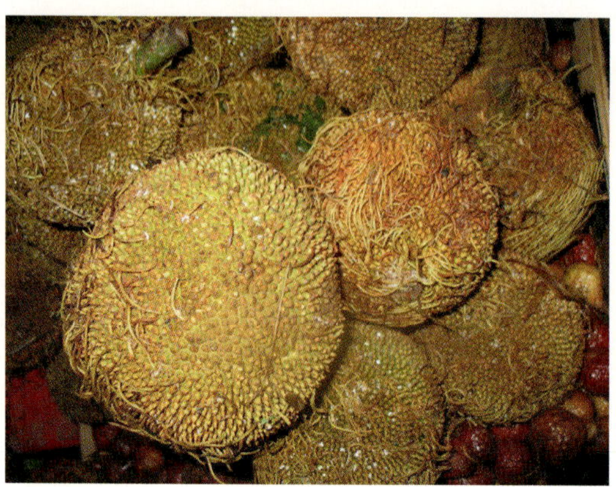

Plate 1 Pedalai fruit with thick, green peduncle, spiky skin and flexous-thread-like processes

4.5 cm by 4 cm with simple exserted style. Syncarp is sub-globose with a diameter of 8.5–12 cm, yellow brown, covered with long, flexous, soft, solid, thread-like processes; the peduncle is thick, green and 10–18 cm long (Plate 1). Inside the 2 mm thick spiky skin are the creamy-white, sweet, fragrant, fleshy perianth segments enclosing the seed. The seed (pericarp) are numerous, ellipsoid, brown, 1 cm by 0.6 cm.

Nutritive/Medicinal Properties

Nutrient composition per 100 g edible portion of fruit perianth segments was reported as follows: energy 119 kcal, moisture 69.3%, protein (fresh weight 1.7 g, dry weight 5.5 g), fat 0.3 g, carbohydrate 27.4 g, fibre 0.5 g, ash 1.7 g, P 33 mg, K 322 mg, Ca 22 mg, Mg 25 mg, Fe 8 μg, Mn 1 μg, Cu 2.5 μg, Zn 7.4 μg and vitamin C 1.8 mg (Voon et al. 1988; Voon and Kueh 1999).

Nutrient composition per 100 g edible portion of seed is reported as follows (Voon and Kueh 1999): energy 241 kcal, moisture 42.9%, protein (fresh weight 7.2 g, dry weight 12.6 g), fat 7.7 g, carbohydrate 35.7 g, fibre 5.5 g, ash 0.9 g, P 100 mg, K 159 mg, Ca 119 mg, Mg 57 mg, Fe 31 μg, Mn 6 μg, Cu 2.6 μg, Zn 24 μg.

Other Uses

The bark can be used to make bark cloth.

Comments

The tree is propagated from seeds.

Selected References

Chandlee DK (1988) A guide to *Artocarpus* fruits. Rare Fruit Council of Australia Newsletter 53(6): 12–18

Jarrett FM (1959a) Studies in *Artocarpus* and allied genera, III. A revision of *Artocarpus* subgenus *Artocarpus*. J Arnold Arbor 40(2):113–155

Jarrett FM (1959b) Studies in *Artocarpus* and allied genera, III. A revision of *Artocarpus* subgenus *Artocarpus* (cont.). J Arnold Arbor 40(4):327–368

Siebert B, Jansen PCM (1992) *Artocarpus* J.R. & G. Forster. In: Verheij EWM, Coronel RE (eds) Plant resources of South-East Asia, No. 2. Edible fruits and nuts. Prosea Foundation, Bogor, pp 79–83

Voon BH, Kueh HS (1999) The nutritional value of indigenous fruits and vegetables in Sarawak. Asia Pac J Clin Nutr 8(1):24–31

Voon BH, Chin TH, Sim CY, Sabariah P (1988) Wild fruits and vegetables in Sarawak. Sarawak Department of Agriculture, Sarawak, p 114

Artocarpus tamaran

Scientific Name

Artocarpus tamaran **Becc.** (Plate 6).

Synonyms

None.

Family

Moraceae

Common/English Name

Elephant Jack, Tamaran

Vernacular Names

Borneo: Tarap Tempunan, Tamarin, Tembaran; ***Sabah***: Talun, Tamaran, Timbangan; ***Sarawak***: Entawa (Iban).

Origin/Distribution

The species in native to Borneo.

Agroecology

It is a strictly tropical tree; found in evergreen tropical forests up to 650 m altitude.

Edible Plant Parts and Uses

The fleshy fruit perianth is edible sweet and pleasant. The seeds are edible boiled or roasted.

Botany

The evergreen tree reaches heights of 40 m with a tall erect trunk with fibrous brownish-grey bark (Plates 1 and 2). Twigs 5–10 mm thick , rugose with brown hairs, stipules 3–9 cm long lanceolate with brown hairs. Leaves large, 20–35 cm by 11–17 cm, ovate-elliptic to ovate, with acuminate tip and rounded base, margin entire or shallowly crenate, with 17–23 pairs of lateral veins prominent below on 3.5–4 cm long petioles (Plates 2 and 3); juvenile leaves deeply pinnatifid with sessile pinnae, emarginate at the base and a narrowly winged rachis. Inflorescence, unisexual, solitary in leaf axils. Male head, 7 cm by 1–1.5 cm, cylindric, obtuse completely covered with flowers and with solid processes with rufous, pilose apices, perianths tubular, 0.6 mm long and pubescent.

T.K. Lim, *Edible Medicinal And Non-Medicinal Plants: Volume 3, Fruits*,
DOI 10.1007/978-94-007-2534-8_48, © Springer Science+Business Media B.V. 2012

Plate 1 Tamaran trunk

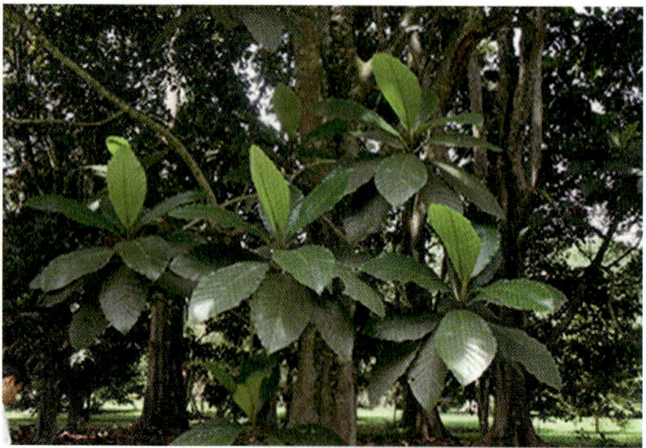

Plate 2 Large tamaran leaves

Female head with exserted, simple styles. Syncarp, cylindric, 10–14 cm by 5–8 cm (including processes) with 8 mm thick wall and on 5–10 cm long peduncle (Plate 4). Processes of two length, long flexous, filiform, solid 1 mm × 0.5 mm, and shorter conical and perforate 3 × 1 mm. Seeds (pericarps) numerous, ellipsoid, mauve-brown, 6 × 4 mm (Plate 5).

Plate 3 Close-view of tamaran leaf

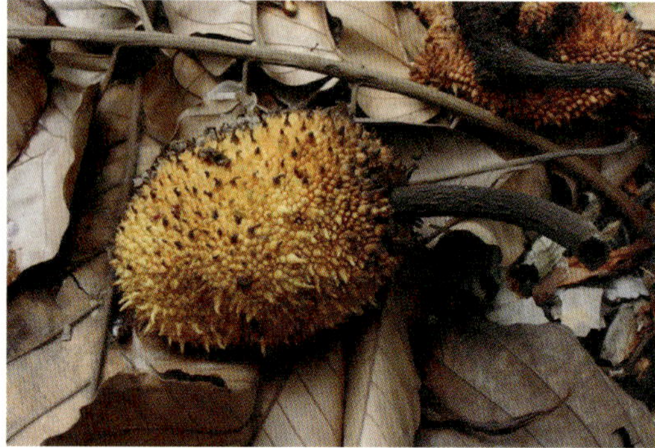

Plate 4 Syncarp–tuberculate with long and short processes

Plate 5 Ellipsoid seeds

Nutritive/Medicinal Properties

No information has been published on the nutritive value of the fruit or on the medicinal uses of the various plant parts although it has been listed

Plate 6 Tamaran tree label

Plate 7 Hat made from Tamaran bark cloth

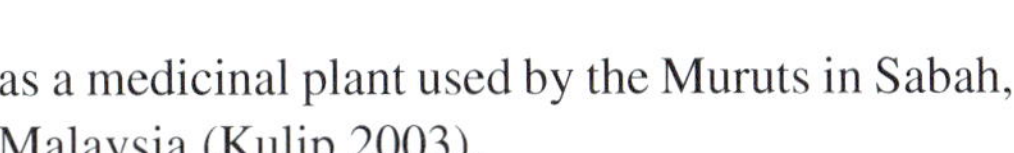

as a medicinal plant used by the Muruts in Sabah, Malaysia (Kulip 2003).

Other Uses

The bark is used to make cloth for traditional dresses (jackets) and hats (Plate 7).

Comments

The seeds germinate readily and is usually used for propagation.

Selected References

Jarrett FM (1959a) Studies in *Artocarpus* and allied genera I. General considerations. J Arnold Arbor 40(1):1–29

Jarrett FM (1959b) Studies in *Artocarpus* and allied genera, III. A revision of *Artocarpus* subgenus *Artocarpus*. J Arnold Arbor 40(2):113–155

Jarrett FM (1959c) Studies in *Artocarpus* and allied genera, III. A revision of *Artocarpus* subgenus *Artocarpus* (cont.). J Arnold Arbor 40(4):327–368

Kulip J (2003) An ethnobotanical survey of medicinal and other useful plants of Muruts in Sabah, Malaysia. Telopea 10(1):81–98

Ficus aspera

Scientific Name

Ficus aspera G. Forst. f.

Synonyms

Artocarpus cannonii W. Bull ex Van Houtte, *Ficus cannonii* (W. Bull ex Van Houtte) N. E. Br., *Ficus parcellii* hort.

Family

Moraceae

Common/English Names

Clown Fig, Mosaic Fig, Tongue Fig

Vernacular Name

Vanuatu: Nautschärob Abäng (Aneityum).

Origin/Distribution

The species is native to Vanuatu.

Agroecology

A tropical species but can be cultivated in sub-temperate areas and is intolerant of frost. It grows in full or partial sun on well-drained, aerated soils. It is quite drought tolerant and can withstand salt spray.

Edible Plant Parts and Uses

The ripe fruit is edible and is of local importance.

Botany

Deciduous or evergreen, branched tree to 20 m tall with villous, 3 cm diameter twigs. Leaves are marbled, speckled or dotted ivory white or cream on dark green, distichous, crenulate to crenate or entire, slightly rough to the touch above, hairy beneath, thin, strongly asymmetric, ovate to oblong-ovate to elliptic, to 32 cm long and 15.5 cm broad, obtusely subacuminate, base very oblique on 2 cm long petioles (Plate 1). Figs (syconium) axillary, usually ramiflorous or cauliflorous and are produced on warty tubercles. Male and female flowers are small and formed inside the syconium. Figs are globose, pubescent,

T.K. Lim, *Edible Medicinal And Non-Medicinal Plants: Volume 3, Fruits*,
DOI 10.1007/978-94-007-2534-8_49,

Plate 1 Variegated leaves and red globose fruit of *Ficus aspera*

to 25 mm in diameter, yellow to orange, ripening red (Plate 1).

Nutritive/Medicinal Properties

No information on its nutritive food value has been published as the species has no importance as an economic edible fruit.

Leaves of *F. aspera* have been used to treat ciguatera poisoning in traditional medicine in the Western Pacific islands (Bourdy et al. 1992) Ciguatera is a specific type of food poisoning associated with the ingestion of tropical fish, which, although normally safe for consumption, may at times contain high amounts of ciguatoxin, as well as other chemically related toxins. Because no symptomatic treatment has been totally satisfactory, folk remedies remain of great interest. *F. aspera* leaves are used to cook with fish to detoxify the fish poison, ciguatera in New Caledonia (Bourret 1981). *F. aspera* leaves are used to treat conjunctivitis in Aneityum, Vanuatu (Bradacs 2008).

Other Uses

F. aspera makes an interesting ornamental plant for both outdoors and indoors because of its unique variegated leaves and reddish fruits.

Comments

The species can be propagated from woody stem cuttings.

Selected References

Bourdy G, Cabalionb P, Amadea P, Laurenta D (1992) Traditional remedies used in the Western Pacific for the treatment of ciguatera poisoning. J Ethnopharmacol 36:163–174

Bourret D (1981) Bonnes Lantes de Nouvelle-Calédonie et des Loyautés. Les éditions du lagon, Noumea. 107 pp

Bradacs G (2008) Ethnobotanical survey and biological screening of medicinal plants from Vanuatu. Dissertation, [zur Erlangung des Doktorgrades der Naturwissenschaften (Dr. rer. nat.) der Naturwissenschaftlichen Fakultät IV. Chemie und Pharmazie der Universität Regensburg]

Clarke WC, Thaman RR (eds) (1993) Agroforestry in the Pacific Islands: systems for sustainability. United Nations University Press, Tokyo, 307 pp

Hedrick UP (1972) Sturtevant's edible plants of the world. Dover Publications, New York, 686 pp

Huxley AJ, Griffiths M, Levy M (eds) (1992) The new RHS dictionary of gardening, vol 4. Macmillan, London

Liberty Hyde Bailey Hortorium (1976) Hortus Third. A concise dictionary of plants cultivated in the United States and Canada. Liberty Hyde Bailey Hortorium/Cornell University/Wiley, New York, 1312 pp

Ficus auriculata

Scientific Name

Ficus auriculata **Loureiro.**

Synonyms

Covellia macrophylla Miq., *Ficus macrocarpa* H. Lév. & Vaniot, *Ficus macrophylla* Roxb., *Ficus regia* Miq., *Ficus roxburghii* Wall. ex Miq., *Ficus sclerocarpa* Griff.

Family

Moraceae

Common/English Names

Broadleaf Fig, Coconut-Strawberry Fig, Elephant Ear Fig Tree, Eve's Apron, Giant Indian Fig, Indian Big Leaf Fig, Roxburgh Fig

Vernacular Names

Brazil: Figueira da India (Portuguese);
Burmese: Demur, Lagum, Na Gum, Sin Thapan, Taba, Thu Hpak Lu Sang;
Chinese: Da Guo Rong, Ping Guo Rong;
German: Roxburgh-Feige;
India: Demur, Doomoor (Bengali), Thebol (Garo), Gular, Timbal, Timal, Timla, Tirmal, Tremal, Trimmal, phagoora, fagoora, tiamble, tunla (Hindi), Deshi Anjir (Marathi), theibal (Mizoram), Daduri, Timal, Trimbal, Trimal (Punjabi);
Indonesia: Terentang Langit (Sumatra);
Japanese: Oobai Chijiku;
Malaysia: Kalebok, Nangtan (Malacca), Pelir Musang;
Nepalese: Karrekan, Nimaaro, Timiilo;
Spanish: Higuera De Roxburgh, Higuera Del Himalaya;
Vietnamese: Cây Vả, Vả.

Origin/Distribution

The species is indigenous in India (Himalaya region), Bhutan, Nepal, Sikkim, Myanmar, Thailand, Vietnam, South China (South Guangdong, Guangxi, South West Guizhou, Hainan, South West Sichuan, Yunnan). Introduced and cultivated elsewhere.

Agroecology

In its native range, it is found in forests in moist valleys, or in midlands and mountainous areas, 100–1,700 (–2,100) m altitudes. It thrives in moist, friable soil, rich in organic matter.

T.K. Lim, *Edible Medicinal And Non-Medicinal Plants: Volume 3, Fruits*,
DOI 10.1007/978-94-007-2534-8_50, © Springer Science+Business Media B.V. 2012

Edible Plant Parts and Uses

Both ripe and near-ripe fruit are edible. Both peeled and unpeeled fruit are sold in the market in north and Central Vietnam (Plates 7 and 8). Near-ripe fruit is cut into slices and dried or sliced and diced into small bits in stir fried dishes in Vietnam particularly in Hue. Ripe fruit can be made into jams and curries. Young leaves are also eaten as vegetables in north Vietnam. In North-east India, the fruits are eaten raw.

Botany

The tree has an elongated and wide, open crown (Plate 1), growing to a height of 4–10 m with a stem diameter of 10–15 cm and greyish brown, rough bark. The stem is leafless in the middle of the stem and pubescent. Stipules are reddish purple, triangular-ovate, 1.5–2 cm, adaxially shortly pubescent. Leaves are alternate on thick, 8–15 cm long petiole and clustered towards the stem apex (Plate 1). Leaf blade is broadly ovate-cordate, large, 15–55 × 10–27 cm, thickly papery (Plates 2 and 6), abaxially pubescent, adaxially glabrous or puberulent on midvein or secondary veins, base

Plate 1 Leaves clustered towards the stem tip

Plate 2 Very large, broadly ovate-cordate leaves

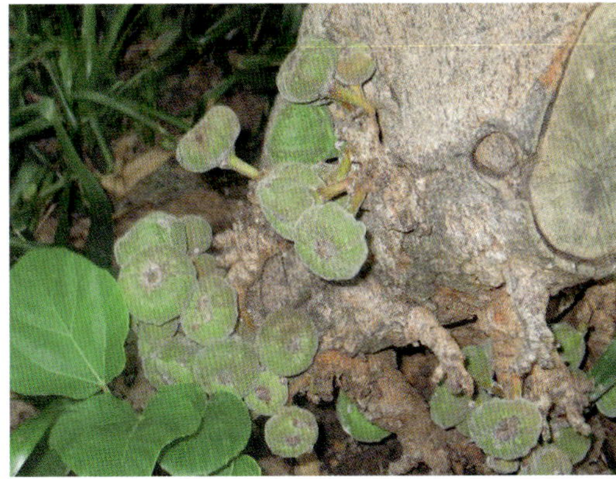

Plate 3 Cauliflorous fruit arising close to the trunk base

Plate 4 Ramiflorous fruit on the main branches

Plate 5 Elephant figs on sale in the market

Plate 6 Leaves on sale in the market

Plate 8 Peeled elephant figs on sale in the market

Plate 7 Pedicel end of unpeeled fruit is waxed to lengthen shelf-life

cordate or rounded, margin regularly shallowly dentate, apex obtuse and mucronate; basal lateral veins 4–6, secondary veins 3 or 4 on each side of midvein. Figs (syconium) are produced on specialized leafless branchlets near base of trunk (Plate 3) and main branches (Plate 4), depressed globose oblate, with 8–12 conspicuous longitudinal ridges, 3–6 cm in diameter, white, shortly pubescent when young, turning green, than brown to reddish brown (Plates 3, 4, 5, 7 and 8) and glabrescent when mature; peduncle 4–6 cm, thick, pubescent. Male and female flowers are small and formed inside the syconium.

Nutritive/Medicinal Properties

The methanol extracts of *F. auriculata* stem bark showed the presence of alkaloids, carbohydrates, saponins, glycosides, phytosterols, resins, phenols, tannins, diterpenes, flavonoids, proteins, and amino acids Antioxidant activity of methanol extract was found to be 84.088% at 0.1 mg/ml with IC_{50} value of 0.042 mg/ml (r2=0.9942) and that of chloroform extract was found to be 83.864% with IC_{50} value of 0.029 mg/ml (r2=0.955). However free radical scavenging activity of hexane extract was of 42% (r2=0.9036) in comparison of standard ascorbic acid used. Antibacterial activity of methanol extract was higher for *Escherichia coli* and hexane extract was highly effective against *Staphylococcus aureus.*

Mature fruit have digestive regulating properties and is diuretic and laxative in traditional medicine. In Vietnam, roots and leaves are used as antidote to poison and to reduce oedema. In India and Nepal, leaves are crushed and the paste is applied on the wounds. Leaves and roasted figs are also used in treating diarrhoea and dysentery. Stem bark juice is effective for diarrhoea, cuts and wounds. Root latex is used in treating mumps, cholera, diarrhoea and vomiting. Mixture of root powder of *F. auriculata* and bark of *Oroxylum indicum* is taken in jaundice.

Other Uses

Ficus auriculata is also an important tree fodder in the Himalayan region of Nepal and India. Its fodder quality is far superior to paddy straw, the main winter fodder in the rice growing regions of the Himalayas. In Nepal, the leaves are used for making plates for festive banquets. Also planted as an ornamental landscape plant in large gardens and parks in southeast Asia.

Comments

The species can be propagated from woody stem cuttings or air-layering.

Selected References

Burkill IH (1966) A dictionary of the economic products of the Malay Peninsula. Revised reprint, 2 volumes. Ministry of agriculture and co-operatives, Kuala Lumpur, Malaysia. vol 1(A–H):1–1240, vol 2(I–Z):1241–2444

Chen J, Su YS, Chen GQ, Wang WD (1999) Ethnobotanical studies on wild edible fruits in Southern Yunnan: folk names; nutritional value and uses. Econ Bot 53(1):2–14

Chhetri RB (2010) Ethnobotany of Moraceae in Meghalaya North-East India. Kathmandu Univ J Sci Eng Technol 6(1):5–10

Gaire BP, Lamichhane R, Sunar CB, Shilpakar A, Neupane S, Panta S (2011) Phytochemical screening and analysis of antibacterial and antioxidant activity of *Ficus auriculata* (Lour.) stem bark. Phcog J 3(21):49–55

Hu SY (2005) Food plants of China. The Chinese University Press, Hong Kong, 844 pp

Kunwar RM, Bussmann RW (2006) *Ficus* (Fig) species in Nepal: a review of diversity and indigenous uses. Lyonia 11(1):85–97

Manandhar NP (1991) Some additional note on wild food plants of Nepal. J Nat Hist Museum 12:19–32

Roder W, Rinzin, Gyeltshen T (2003) *Ficus auriculata*: its relative importance in Bhutan, farmers' preference and fodder quality. Agrofor Syst. 77(1):11–17

Tanaka Y, Nguyen VK (2007) Edible wild plants of Vietnam: the Bountiful Garden. Orchid Press, Bangkok, 175 pp

Wu Z, Zhou Z-K, Gilbert MG (2003) Moraceae link. In: Wu ZY, Raven PH, Hong DY (eds) Flora of China. Vol. 5 (Ulmaceae through Basellaceae). Science Press, Beijing, and Missouri Botanical Garden Press, St. Louis

Ficus carica

Scientific Name

Ficus carica L.

Synonyms

Caprificus insectifera Gasp., *Ficus kopetdagensis* Pachom., *Ficus latifolia* Salisb., *Ficus leucocarpa* Gasp., *Ficus macrocarpa* Gasp., *Ficus pachycarpa* Gasp., *Ficus praecox* Gasp. ex Guss., *Ficus sativa* Poit. & Torpin, *Ficus caprificus* Risso.

Family

Moraceae

Common/English Names

Fig, Common Fig, Cultivated Fig, Edible Fig, Wild Fig

Vernacular Names

Arabic: Anjir, Teen, Teen Barchomi, Ten;
Brazil: Figo, Figueira, Figueira-Comum, Figueira-Da-Europa, Figueira-Do-Reino (Portuguese);
Burmese: Thaphan, Thinbaw Thapan;
Chinese: Mo Fa Guo, Wu Hua Guo;
Cook Islands: Suke (Maori);
Croatian: Smokva, Smokvenica, Smokvina;
Czech: Fíkovník, Fíkovník Maloasijský, Fíkovník Smokvoň, Smokvoň;
Danish: Almindelig Figen, Figen;
Dutch: Echte Vijgeboom, Gewone Vijgeboom, Vijg, Vijgeboom;
Eastonian: Harilik Viigipuu;
Finnish: Viikuna;
French: Arbre À Carriques, Caprifiguier, Carique, Figue, Figue Commune, Figuier, Figuier Comestible, Figuier Commun;
German: Echte Feige, Echter Feigenbaum, Essfeige, Feige, Feigenbaum, Gemeiner Feigenbaum;
Hungarian: Füge(Fa), Fügefa;
India: Dumar (Bengali), Anjir (Gujarati), Anjeer, Anjir, Tin (Hindu), Anjoora, Anjeera, Anjura, Anjuri, Seeme Atthi, Shime-Atti, Shimeatti, Simeatti, Simeyam, Simeyatti, (Kannada), Shima-Atti, Shimayatti, Simayatti (Malayalam), Anjir, Anjeer (Marathi), Udumbara (Oriya), Anjeer, Anjir, Anjira, Manjula, Phalgu, Rajodumbara, Udumvara (Sanskrit), Anjura, Appira, Cevvatti, Cikappatti, Cimai Atti, Cimaiyatti, Seemai Aththi, Shimai-Atti, Shimeatti, Simaiyatti, Simie Attie Pullum, Tenatti, Ten-Atti, Ten Atti, Tacaiyatti, Utumparam (Tamil), Anjooramu, Anjooru, Anjuru, Manchi Medi, Mancimedi, Manjimedi, Medi, Modipatu, Seematti, Seemayatthi, Shima-Atti, Simaatti, Simayatti, Tene-Atti, Teneyatti, Theneyatthi (Telugu),Anjeer, Balas, Anjir, Anjeer (Khushk Dashti), Anjra (Khushk Dashti), Poast Darakht Anjir, Anjir Zard (Urdu);

T.K. Lim, *Edible Medicinal And Non-Medicinal Plants: Volume 3, Fruits*,
DOI 10.1007/978-94-007-2534-8_51, © Springer Science+Business Media B.V. 2012

I-Kiribati: Te Biku;
Italian: Fic, Fico, Fico Comune;
Japanese: Ichijiku;
Korean: Mu Hwa Gwa, Mu Hwa Gwa Na Mu;
Macedonian: Smoka;
Malaysia: Anjir;
Mangarevan: Pika;
Marshallese: Wojke Piik, Wõjke-Piik;
Nepalese: Anjiir;
Norwegian: Fiken;
Pakistan: Anjeer, Anjir, Injir;
Palauan: Uósech, Uosech;
Persian: Anjir, Anjeer;
Polish: Figowiec;
Portuguese: Behereira, Figueira, Figueira-Comum;
Rakahanga-Manihiki: Monamona, Monamona;
Russian: Inžir;
Samoan: Mati;
Serbian: Smoka, Smokva, Smokovnica;
Slovašcina: Figa, Figovec, Figovina, Smokva, Smokvovec;
Slovencina: Figovník Obyčajný;
Spanish: Breva, Higo, Higuera Común;
Swedish: Fikon, Fikonträd;
Tongarevan: Monamona;
Tuamotuan: Tute;
Vietnamese: Quả Vả, Vô Hoa Quả.

Origin/Distribution

Fig is indigenous in south-western Asia to northwest India. Figs have been cultivated by humans for over 5,000 years and was first known from Caria in south-western Asia It has naturalized all over the Mediterranean area and is now cultivated elsewhere in areas with a similar climate.

Agroecology

Fig thrives in the lowland in subtropical areas with light early spring rains and in tropical areas generally between 600 and 1,800 m altitudes. The semi arid tropical and subtropical regions of the world are suitable for fig cultivation with the assistance of irrigation. It tolerates winter temperatures as low as 10°C. Fig is quite tolerant light frosts. The top growth is susceptible to frost damage and can be killed back to the base in severe winters. Figs do not require winter chilling to break dormancy. It thrives in partial to full sun in the coastal and inland areas in the lowlands and highlands. Rains during fruit development and ripening are detrimental to the crop, causing the fruits to split. The fig can be grown on a wide range of soils: light sand, rich loams, clay or limestone, providing there is sufficient depth and good drainage. A heavy wet soil tends to encourage excessive plant growth at the expense of fruit production. Highly acid soils are unsuitable. The ideal pH should be between 6.0 and 6.5. Figs perform best on well-drained, reasonably fertile, organic matter rich soils. Figs are relatively salt-tolerant and can be grown along the coast near brackish water. Fig is intolerant of poorly drained, waterlogged conditions. Figs respond well to heavy mulching with organic materials to conserve moisture, improve soil structure and reduce root knot nematode levels. Fig also responds well to pruning and can be espaliered or pruned heavily in the dormant season for size control and to increase the main crop.

Edible Plant Parts and Uses

Fruits are relished as a popular and nutritious fruit both fresh and dried. When dried the fruits besides being consumed as snacks are also used for cooking for instance in speciality Chinese dishes and in herbal soups (Plate 6). Ripe fresh figs are eaten out of hand with or without peeling or they can be served with cream and sugar. Peeled or unpeeled, the fruits may be merely stewed or cooked in various ways, as in pies, in jams, puddings, cakes, bread or other bakery products, or added to ice cream mix. In Bengal, the fruit is called *Dumur*. It is cooked as a vegetable and is believed to be good for heart ailments. Ripe figs are also excellent for processing into fruit juice, fruit-tea, ratafia and canned beverages. Whole fruits can be preserved in sugar syrup, canned or processed into jam, marmalade, or paste. Fig paste (with added wheat and corn flour, whey, syrup, oils and other ingredients) forms the filling for the well known bakery product, "Fig Newton". The fruits are sometimes candied whole commercially. In Europe, western Asia, northern Africa and California, commercial canning and drying of figs are industries of great importance. Dried cull figs have been roasted and ground as a coffee substitute. In Mediterranean countries, low-grade figs are converted into alcohol. An alcoholic extract of dried figs has been used as a flavoring for liqueurs and tobacco. In China, leaves are gathered just before shedding, dried and used for herbal tea.

From fig latex, a protein-digesting enzyme ficin can be obtained, which is used for tenderizing meat, rendering fat, and clarifying beverages.

Botany

A robust, deciduous, dioecious shrub or small tree 5–10 m high with several spreading branches from a short, rough trunk. The bark is smooth, distinctly lenticellate, grey or dull white, young twigs glabrous or softly hairy and the root system is typically shallow and spreading. Stipules are red, ovate-lanceolate, about 1 cm long and caducous. Leaves are alternate and borne on 5–12 cm long, grooved petioles. Lamina is variable in shape and size, broadly ovate to nearly orbicular, (4-) 5–15 (–20) cm long by (3.5-) 5–15 (–18) cm wide, undivided or shallowly palmatifid to mostly palmatipartite, lobes spathulate with entire to apically few-dentate margin, 5-costate at the cordate base, apex acute to obtuse, margins undulate-dentate or dentate-crenate, sparsely to moderately hairy above, often densely hispid beneath especially on nerves, lateral nerves 6–8 (–9) pairs (Plate 1). Hypanthodia are axillary, solitary or paired, borne on up to 3 cm long peduncles, pyriform to globose, 1.5–2 cm in diameter, subsessile to sessile, subtended by 3, broadly deltoid basal bracts, apical orifice closed by 4–5, broadly deltoid, ciliate imbricate bracts. Male flowers are found near the apical orifice, with usually 4 united sepals, 4 stamens on long filaments with oval, exserted anthers. Female flowers are pedicellate, with 4 lanceolate-oblong lobed sepals, ovoid ovary with lateral style and bifd stigma. Achenes are small, 2 mm long and lenticular. Figs are usually pyriform-obovoid to subglobose, 2–5 (–8) cm in diameter, glabrous or shortly hispid, greenish-yellow, yellow, reddish to reddish-violet (Plates 2, 3, 4 and 5). The skin of the fig is thin,

Plate 1 Immature figs and palmitifid leaves

Plate 2 Ripening figs

Plate 3 Tray of ripe sub-globose figs

Plate 4 Fig sliced to show the inside

Plate 5 A pear shape fig variety

Plate 6 Dried figs much used in Chinese cuisine

the fleshy wall is whitish (Plate 4), pale-yellow, or amber, or more or less pink, rose, red or purple; juicy and sweet when ripe, gummy with latex when unripe.

Nutritive/Medicinal Properties

The nutrient composition of raw, ripe fig fruit per 100 g edible portion was reported as: water 79.1 g, energy 74 kcal (310 kJ), protein 0.75 g, total lipid (fat) 0.30 g, ash 0.66 g, carbohydrate19.18 g, total dietary fiber 2.9 g, total sugars 16.26 g, Ca 35 mg, Fe 0.37 mg, Mg 17 mg, P14mg, K 232 mg, Na 1 mg, Zn 0.15 mg, Cu 0.070 mg, Mn 0.128 mg, Se 0.2 μg, vitamin C 2 mg, thiamin 0.060 mg, riboflavin 0.050 mg, niacin 0.400 mg, pantothenic acid 0.300 mg, vitamin B-6 0.113 mg, total folate 6 μg, total choline 4.7 mg, vitamin A 142 IU (7 μg RAE), vitamin E (α-tocopherol) 0.11 mg, vitamin K (phylloquinone) 4.7 μg, total saturated fatty acids 0.060 g, 14:0 (myristic acid) 0.002 g, 16:0 (palmitic acid) 0.046 g, 18:0 (stearic acid) 0.012 g; total monounsaturated fatty acids 0.066 g, 18:1 undifferentiated (oleic acid) 0.066 g; total polyunsaturated fatty acids 0.144 g, 18:2 undifferentiated (linoleic acid) 0.144 g; phytosterols 31 mg, amino acids –tryptophan 0.006 g ,threonine 0.024 g, isoleucine 0.023 g, leucine 0.033 g, lysine 0.030 g, methionine 0.006 g, cystine 0.012 g, phenylalanine 0.018 g, tyrosine 0.032 g, valine 0.028 g, arginine 0.017 g, histidine 0.011 g, alanine 0.045 g, aspartic acid 0.176 g, glutamic acid 0.072 g, glycine 0.025 g, proline 0.049 g, serine 0.037 g; β- carotene 85 μg, lutein + zeaxanthin 9 μg (U.S. Department of Agriculture, Agricultural Research Service 2010).

The nutrient composition of dried fig of the variety Mission Valley per 100 g edible portion was reported as: water 30.05 g, energy 249 kcal (1,041 kJ), protein 3.30 g, total lipid (fat) 0.93 g, ash 1.86 g, carbohydrate 63.87 g, total dietary fibre 9.8 g, total sugars 47.92 g, sucrose 0.07 g, glucose 24.79 g, fructose 22.93 g, galactose 0.13 g, starch 5.07 g, Ca 162 mg, Fe 2.03 mg, Mg 68 mg, P 67 mg, K 680 mg, Na 10 mg, Zn 0.55 mg, Cu 0.287 mg, Mn 0.510 mg, Se 0.6 μg, vitamin C 1.2 mg, thiamin 0.085 mg, riboflavin 0.082 mg, niacin 0.619 mg, pantothenic acid 0.434 mg, vitamin B-6 0.106 mg, total folate 9 μg, total choline 15.8 mg, betaine 0.7 mg, vitamin A 10 IU, β-carotene 6 μg, vitamin E (α-tocopherol) 0.35 mg, β-tocopherol 0.01 mg, γ- tocopherol 0.37 mg, δ-tocopherol 0.01 mg, vitamin K (phylloquinone)

15.6 μg, total saturated fatty acids 0.144 g, 14:0 (myristic acid) 0.006 g, 16:0 (palmitic acid) 0.110 g, 18:0 (stearic acid) 0.01229 g; total mono-unsaturated fatty acids 0.066 g, 16:1 undifferentiated (palmitoleic acid) 0.001 g, 18:1 undifferentiated (oleic acid) 0.158 g; total polyunsaturated fatty acids 0.345 g, 18:2 undifferentiated (linoleic acid) 0.345 g; amino acids –tryptophan 0.020 g, threonine 0.085 g, isoleucine 0.089 g, leucine 0.128 g, lysine 0.088 g, methionine 0.034 g, cystine 0.036 g, phenylalanine 0.076 g, tyrosine 0.041 g, valine 0.122 g, arginine 0.077 g, histidine 0.037 g, alanine 0.134 g, aspartic acid 0.645 g, glutamic acid 0.295 g, glycine 0.108 g, proline 0.610 g, serine 0.128 g; lutein + zeaxanthin 32 μg (U.S. Department of Agriculture, Agricultural Research Service 2010).

From the nutrient data above figs are richly packed with nutrients – vitamins, fibre, protein, carbohydrates, minerals especially so for dried fig which possesses one of the highest plant sources of fibre, copper, manganese, magnesium, potassium, calcium, and vitamin K relative to human needs.

Thirty groups of various lipid compounds belonging to the classes of neutral lipids, glycolipids, and phospholipids were identified from fig fruit (Kolesnik et al. 1989). The main groups were triacylglycerols, free and esterified sterols, mono- and digalactosyldiglycerides, ceramide oligosides, cerebrosides, esterified sterol glycosides, and phosphatidylglycerols. In the fatty acid composition, linoleic, linolenic, oleic, and palmitic acids predominated (>90%). Ripe dried fruit of *F. carica* was found to contain alkaloids, flavonoids, coumarins, saponins, and terpenes (Vaya and Mahmood 2006; Teixeira et al. 2006).

Volatile compounds from the pentane extracts of figs were found to be different in receptive male figs, non-receptive male figs, receptive female figs and non-receptive female figs (Gibernau et al. 1997). Extracts from non-receptive figs were characterised by furanocoumarins (tentatively identified as angelicin and bergapten), sesquiterpene hydrocarbons (i.e. trans-caryophyllene, and a compound tentatively identified as germacrene D) and oxygenated sesquiterpenes (i.e. hydroxycaryophyllene), benzyl alcohol and benzylaldehyde. Extracts from receptive figs of both sexes were characterised by benzyl alcohol, linalool and linalool oxides (furanoid), cinnamic aldehyde, cinnamic alcohol and indole. Extracts from female receptive figs had in addition large amounts of pyranoid (linalool oxides), whereas an extract from male receptive figs contained eugenol and an unidentified sesquiterpene hydrocarbon. Differences between extracts from male and female figs appeared to be mainly qualitative due to pyranoid compounds, sesquiterpenes 1, 2 and 3 for female figs and eugenol and sesquiterpene 5 for male figs.

The phytosterols, campesterol, stigmasterol, and fucosterol in fig fruit were determined from the trimethysily ether (TMS) derivates of the unsaponifiable samples (Jeong and Lachance 2001). Fourteen compounds were separated from fig fruit; 13, 10, and 6 in bark, stem, and pith, respectively. Fatty acids in fig fruit, determined as their methyl esters, were myristic (14:0), palmitic (16:0), stearic (18:0), oleic (18:2), and linolenic (18:3) acids. Fig leaves were found to contain the following furocoumarins psoralen, bergapten and the coumarins umbelliferone, 4′,5′-dihydropsoralen and marmesin (Innocenti et al. 1982). The quantity of psoralen was greater than that of bergapten, umbelliferone, 4′,5′-dihydropsoralen and of marmesin. The leaves yielded two triterpenes 24-methylenecycloartanol and baurenol (Ahmed et al. 1988). The major flavonoids in *Ficus* leaf extract (70% ethanol) were quercetin and luteolin, with a total of 631 and 681 mg/kg extract, respectively (Vaya and Mahmood 2006). Leaves also contained calotropenyl acetate, methyl maslinate and lupeol acetate Saeed and Sabir (2002). Singab et al. (2010) found umbelliferone, caffeic acid, quercetin-3-O-β-d-glucopyranoside, quercetin-3-O-α-l-rhamnopyranoside, and kaempferol-3-O-α-l-rhamnopyranoside in fig leaves.

The two Portuguese white varieties ("Pingo de Mel" and "Branca Tradicional") of *F. carica* presented a similar profile composed of eight volatile compounds: acetaldehyde, ethyl acetate, methanol, ethanol, hexanal, limonene, (E)-2-hexenal and octanal in the leaves, pulp and peels (Oliveira et al. 2010b). The total volatile content was different among the vegetal materials, following the order leaves > peels > pulps. Methanol and ethanol were the major compounds in all samples.

Fig latex was found to contain caoutchouc (2.4%), resin, albumin, cerin, sugar and malic acid, rennin, proteolytic enzymes, diastase, esterase, lipase, catalase, and peroxidise (Morton 1987). Fig latex also contained 6-*O*-acyl-β-D-glucosyl-β-sitosterols, the acyl moeity being primarily palmitoyl and linoleyl with minor amounts of stearyl and oleyl (Rubnov et al. 2001), peptides (Maruyama et al. 1989), proteases , ficins A, B, C, D, and S (Sugiura and Sasaki. 1971; Sugiura et al. 1975(Azarkan et al. 2011). Seven phytosterols were determined in *F. carica* latex, with β-sitosterol and lupeol being the compounds present in higher concentrations (about 54 and 14%, respectively) (Oliveira et al. 2010a). A total of 18 fatty acids with saturated fatty acids being predominant (ca. 86.4% of total fatty acids). A total of 13 free amino acids were also identified with cysteine and tyrosine being the major ones (38.7 and 31.4%, respectively). Recoveries of phenolic compounds above 85% were obtained for chlorogenic acid, rutin, and psoralen from *Ficus carica* leaves using the sea sand extraction method (Teixeira et al. 2006).

The latex of *Ficus carica* constitutes an important source of many proteolytic components known under the general term of ficin and recent studies reinforced the view that they belonged to the cysteine proteases of the papain family (Azarkan et al. 2011). Five ficins forms were purified and characterised using a selective and reversible thiol-pegylation technique. The five ficin forms displayed different specific amidase activities against small synthetic substrates like dl-BAPNA and Boc-Ala-Ala-Gly-pNA, suggesting some differences in their active site organization. Enzymatic activity of the five ficin forms was completely inhibited by specific cysteine and cysteine/serine proteases inhibitors but was unaffected by specific serine, aspartic and metalloproteases inhibitors.

Fig leaves, fruit and latex have been reported in numerous scientific studies to possess antioxidant, anticancer, antiviral, hepatoprotective, hypoglycaemic, hypolipidaemic, hypocholesterolaemic, immunomodulatory, antipyretic, cholinesterase inhibition, anti-inflammatory, anti-wart, larvicidal, antibacterial and anthelmintic activities. Dried figs have been reported to have anti-platelet activities while fig latex has blood coagulation promoting properties.

Antioxidant Activity

Figs are also rich sources of flavonoids and polyphenols and contain many antioxidants. Figs were found to contain 1.090–1.110 mg/100 g fresh matter of total polyphenols (Vinson 1999). Studies conducted reported that fresh figs contained 360 mg catechin per 100 g fresh weight of free phenols, 486 mg catechin per 100 g fresh weight of total phenols while dried fig had 256 mg catechin per 100 g fresh weight free phenols and 320 mg catechin per 100 g fresh weight of total phenols (Vinson et al. 2005). They found in one study, a 40-gram portion of dried figs (two medium size figs) produced a significant increase in plasma antioxidant capacity as measured by TEAC for 4 hours after consumption, and overcame the oxidative stress of consuming high fructose corn syrup in a carbonated soft drink. Figs were found to be in-vivo antioxidants after human consumption. They reported that figs and dried plums had the best nutrient score among the dried fruits, and dates among the fresh fruits tested. Processing to produce the dried fruit significantly decreased the phenols in the fruits on a dry weight basis. Compared with vitamins C and E, dried fruits had superior quality antioxidants with figs and dried plums being the best. Fig antioxidants could enrich lipoproteins in the plasma and protect them from subsequent oxidation. Their findings suggested that dried fruits should be a greater part of the diet as they are dense in phenol antioxidants, minerals and nutrients, most notably fibre.

Studies showed that fruits of the dark coloured Mission Fig variety contained the highest levels of polyphenols, flavonoids, and anthocyanins and exhibited the highest antioxidant capacity (Solomon et al. 2006). Various concentrations of anthocyanins but a similar profile was found in all varieties (black, red, yellow, and green) studied. Cyanidin-3-O-rhamnoglucoside (cyanidin-3-O-rutinoside; C3R) was the main anthocyanin in all fruits. Color appearance of fig extract correlated well with total polyphenols, flavonoids, anthocyanins, and antioxidant capacity. Extracts of darker varieties showed higher contents of phytochemicals compared to lighter coloured varieties. Fruit skins contributed most of the above phytochemicals and antioxidant activity compared to the fruit pulp. Antioxidant capacity correlated well with the amounts of polyphenols and anthocyanins. In the dark-coloured Mission and the red Brown-Turkey varieties, the anthocyanin fraction contributed 36 and 28% of the total antioxidant capacity, respectively. C3R contributed 92% of the total antioxidant capacity of the anthocyanin fraction.

The major flavonoid content of leaf extracts (70% ethanol) from fig were quercetin and luteolin, with a total of 631 and 681 mg/kg extract, respectively. In another study The results clearly demonstrated that these leaf extracts had antioxidant capacity as evaluated by the phosphomolibdenum spectrophotometric method (Konyalioglu et al. 2005). Antioxidant capacity results were consistent with total flavonoid and phenol contents as evaluated by the aluminium chloride method and Folin-Ciocalteau reagent respectively. The α-tocopherol content of the n-hexane extract was found to be 3.2788%, whereas it was calculated as 0.0570% on a dry-weight basis of the leaves.

The leaves, pulps and peels of two Portuguese white varieties of *F. carica* (Pingo de Mel and Branca Tradicional) presented a similar phenolic profile that was composed of 3-O- and 5-O-caffeoylquinic acids, ferulic acid, quercetin-3-O-glucoside, quercetin-3-O-rutinoside, psoralen and bergapten. 3-O-caffeoylquinic acid and quercetin-3-O-glucoside (Oliveira et al. 2010c). Leaf organic acids profile were represented by oxalic, citric, malic, quinic, shikimic and fumaric acids, while in pulps and peels quinic acid was absent. The different plant parts exhibited activity against DPPH and nitric oxide radicals in a concentration-dependent way. However, only the leaves exhibited capacity to scavenge superoxide radical. Leaves were always the most effective part, which appeared to be related to phenolics compounds.

Anticancer Activity

A mixture of 6-*O*-acyl-β-D-glucosyl-β-sitosterols, the acyl moeity being primarily palmitoyl and linoleyl with minor amounts of stearyl and oleyl, were isolated as a potent cytotoxic agent from fig (*Ficus carica*) latex (Rubnov et al. 2001). Both the natural and the synthetic compounds showed in-vitro inhibitory effects on proliferation of various cancer cell lines.

Antiviral Activity

The water extract from the leaves of *Ficus carica* possessed distinct anti-HSV-1 (anti – Herpes simplex virus) effect (Wang et al. 2004). The MTC (maximum tolerated concentration) was 0.5 mg/ml, TDO (tryptophan 2,3-dioxygenase) was 15 mg/ml, and TI was 30 mg/ml. It possessed low toxicity and directly killing-virus effect on HSV-1. Latex of fig fruit was found to have antiviral activity (Lazreg Aref et al. 2011). The hexanic and hexane-ethyl acetate (v/v) extracts inhibited multiplication of herpes simplex type 1 (HSV-1), echovirus type 11 (ECV-11) and adenovirus (ADV) at concentrations of 78 μg/ml. All extracts had no cytotoxic effect on Vero cells at all tested concentrations.

Hepatoprotective Activity

The methanol extract of fig leaves at an oral dose of 500 mg/kg exhibited a significant hepatoprotective effect by lowering the serum levels of aspartate aminotransferase, alanine aminotransferase, total serum bilirubin, and malondialdehyde equivalent, an index of lipid peroxidation of the liver (Mohan et al. 2007). These biochemical observations were supplemented by histopathological examination of liver sections. The activity of extract was also comparable to that of silymarin, a known hepatoprotective. The activity of extract was also comparable to that of silymarin, a known hepatoprotective. In another study, there was significant reversal of biochemical (serum levels of glutamic oxaloacetate transaminase, glutamic pyruvic transaminase, bilirubin) changes, histological changes in liver and functional changes (sleeping times) induced by rifampicin treatment in rats by petroleum ether extract of fig leaves treatment, indicating promising hepatoprotective activity (Gond and Khadabadi 2008). Methanol extracts of *Ficus carica* (leaf and fruit) and *Morus alba* (root bark) showed potent antioxidant and hepatoprotective activities (Singab et al. 2010). The activity of the extracts was comparable to that of silymarin, a known hepatoprotective agent. In-depth chromatographic investigation of the most active extract (*Ficus carica* leaf extract) resulted in identification of umbelliferone, caffeic acid, quercetin-3-O-β-d-glucopyranoside, quercetin-3-O-α-l-rhamnopyranoside, and kaempferol-3-O-α-l-rhamnopyranoside.

Hypoglycaemic Activity

Ficus carica leaf extract showed a clear hypoglycaemic effect in diabetic rats. The extract decreased plasma glucose in diabetic (27.9 mmol/L to 19.6 mmol/L) but not in normal rats (Perez et al. 2000). Plasma insulin levels were decreased by treatment in non-diabetic rats from 4.9 μg/ml to 3.3 μg/ml. Glucose uptake (μmol/min) by rat hind quarters perfused was: 5.9 (untreated non-diabetic rats), 4.8 (treated non-diabetic rats, vs. untreated non-diabetic rats), 2.0 (untreated diabetic rats, vs. untreated non-diabetic rats) and 4.1 (treated diabetic rats) in absence of insulin; 7.0 (untreated non-diabetic rats), 8.3 (treated non-diabetic rats, vs. untreated non-diabetic rats), 5.0 (untreated diabetic rats, vs. untreated non-diabetic rats) and 6.4 (treated diabetic rats) in presence of insulin. Lactate released was lower in untreated diabetic vs. untreated non-diabetic rats. Pèrez et al. (2003) confirmed that in diabetic rats, the antioxidant status was compromised and that *Ficus carica* extracts tend to normalize it. Compared to normal animals, the diabetic rats presented significantly higher values for erythrocyte catalase normalized to haemoglobin levels for plasma vitamin E and fatty acids. Both fig fractions (basic and chloroform) tended to normalize the values of the diabetic animals' fatty acids and plasma vitamin E values. In another study, chloroform extract obtained from a decoction of *Ficus carica* leaves was found to improve the cholesterolaemic status of rats with streptozotocin-induced diabetes (Canal et al. 2000). Administration of the extract to rats with streptozotocin-induced diabetes led to a decline in the levels of total cholesterol and an decrease in the total cholesterol/HDL cholesterol ratio (with respect to the control group), together with a reduction of the hyperglycaemia

In a clinical study, the addition of FC (fig leaf decoction) to the breakfast diet in IDDM (insulin-dependent diabetes mellitus) patients could be of help to control postprandial glycemia (Serraclara et al. 1998). Post-prandial glycemia was significantly lower during supplementation with FC 156.6 mg/daily versus TC (non-sweet commercial tea) 293.7 mg/daily without preprandial differences 145.0 and 196.6 mg/daily, respectively. Medium average capillary profiles were also lower in the two sub-groups of patients during FC 166.7 mg/daily, and 157.1 mg/daily versus TC 245.8 mg/daily and 221.4 mg/daily. Average insulin dose was 12% lower for FC in the total group.

Antihyperlidemic/ Hypocholesterolemic Activity

Intraperitoneal (i./p.) administration of a *Ficus carica* leaf decoction in hypertriglyceridaemia induced rats reduced plasma triglyceride levels in the rats (Perez et al. 1999). Hypertriglyceridaemia was induced in rats following the protocol: a fasting period of 22 h, 2 h of oral (p.o.) administration of 20% emulsion of longchain triglycerides (LCT emulsion), both repeated once. The triglyceride plasma levels after LTC protocol decreased with time after treatment with fig leaf decoction. The plasma total cholesterol levels showed no significant differences in relation to baseline levels in the presence or absence of *Ficus carica* treatment either. The positive results suggested the presence in the fig leaf decoction of a compound or compounds that influence lipid catabolism. *Ficus carica* leaf extracts were found to decrease liver and serum cholesterol levels in hyperlipidemic rats (Rassouli et al. 2010). The lipid-lowering effect of total extract on liver cholesterol was more pronounced than that of serum. Phytochemical screening showed that total extract had a moderate level of flavonoids which could be suggested to account for observed hypocholestrolemic effects.

Immunomodulatory Activity

The leaf extract of *F. carica* was found to possess immunostimulant properties (Patil et al. 2010b). Administration of the ethanolic extract of the leaves of *Ficus carica* was found to ameliorate both cellular and humoral antibody response.

Antipyretic Activity

The ethanol extract of *Ficus carica*, at doses of 100, 200 and 300 mg/kg body wt. p.o., showed significant dose-dependent reduction in normal body temperature and yeast-provoked elevated temperature in albino rats (Patil et al. 2010a). The anti-pyretic effect of the ethanol extract of *Ficus carica* was comparable to that of paracetamol (150 mg/kg body wt., p.o.), a standard anti-pyretic agent.

Antimicrobial Activity

The acidic and phenolic fractions fractionated from the methanol extract of fig leaves showed strong antimicrobial activities, but not the basic and neutral fractions (Chung et al. 1995; Kang and Jung 1995). The degree of antimicrobial activities of phenolic fraction against selected bacteria were higher than those of acidic fraction, but that against yeasts and mould were almost equivalent to those of acidic fraction. The phenolic fraction was especially effective against *Staphylococcus aureus* and *Pseudomonas aeruginosa*. Four antimicrobial substances purified from the phenolic fraction which showed the strongest antimicrobial activities among the fractions from fig leaves, were identified as psoralen, bergapten, β-sitosterol and umbelliferone.

The methanol extract of unripened and ripe figs exhibited against food poisoning bacteria (Jeong et al. 2005). The methanol extracts of unripened figs showed stronger activity than that of the ripened figs especially against *Listeria monocytogenes, Salmonella enteriridis, Escherichia coli* 0157: H7, *Vibrio parahaemolyticus* and *Salmonella typhimurium* in 10 mg/ml. The systematic solvent fractions (n-hexane, chloroform, ethyl acetate, and butanol). showed stronger antibacterial activities than the methanol extract, even at the lower concentrations. The hexane fraction of ripened figs showed higher growth inhibition than those of unripened figs against *Listeria monocytogenes Escherichia coli* 0157: H7, *Yersinia enterocolitica* and *Vibrio parahaemolyticus*. The chloroform fraction showed strong antibacterial activity in all ripening stages against *E. coli* 0157: H7 and *V. parahaemolyticus*. The butanol fraction showed higher inhibition activity in unripened figs than in the ripened figs. The hexane and chloroform fractions showed inhibition activity of more than 75% against *E. coli* 0157:117 and *Vibrio parahaemolyticus* in 0.5 mg/ml. Jeong et al. (2009) found that the methanol extract of *F. carica* leaves showed strong antibacterial activity against the oral pathogens *Streptococcus gordonii, Streptococcus anginosus, Prevotella intermedia, Actinobacillus actinomycetemcomitans*, and *Porphyromonas gingivalis* with MIC of 0.156–0.625 mg/ml and MBC of 0.313–0.625 mg/ml. The combination effects of the methanol extract with ampicillin or gentamicin were synergistic against the oral bacteria. The synergistic effects of the methanolic extract with ampicillin or gentamicin combination against oral bacteria were presented as ≥4–8-fold reduction of MIC producing a synergistic effect as defined by FICI (fractional inhibitory concentration index) ≤ 0.375–0.5. The methanolic extract with ampicillin or gentamicin combination showed the strongest synergistic effect (FICI ≤ 0.375) against *Streptococcus sanguinis*, *Streptococcus sobrinus*, and *P. gingivalis*.

In another study, the methanolic extract of *Ficus carica* latex had no effect against bacteria except for *Proteus mirabilis* while the ethyl acetate extract had inhibition effect on the multiplication of five bacteria species (*Enterococcus fecalis, Citobacter freundei, Pseudomonas aeruginosa, Echerchia coli* and *Proteus mirabilis*) (Aref et al. 2010). For the opportunist pathogenic yeasts, ethyl acetate and chlorophormic fractions showed a very strong inhibition (100%); methanolic fraction had a total inhibition against *Candida albicans* (100%) at a concentration of 500 μg/ml and a negative effect against *Cryptococcus neoformans. Microsporum canis* was strongly inhibited with methanolic extract (75%) and totally with ethyl acetate extract at a concentration of 750 μg/ml.

Antiplatelet Activity

A study in rabbits showed the presence of spasmolytic activity in the ripe dried fruit of *Ficus carica* possibly mediated through the activation of K+ATP channels along with antiplatelet activity which provided sound pharmacological basis for its medicinal use in the gut motility and inflammatory disorders (Gilani et al. 2008).

Angiotension I-Converting Enzyme (ACE) Inhibition

The latex of fig tree was reported to contain three inhibitory peptides of angiotension I-converting enzyme (ACE) (Maruyama et al. 1989). These peptides were identified by the Ed man procedure and carboxypeptidase digestion to be: Ala-Val-Asn-Pro-Ile-Arg, Leu-Tyr-Pro-Val-Lys, and Leu-Val-Arg. The IC_{50} values of these peptides for ACE from rabbit lung were 13 μM, 4.5 μM, and 14 μM, respectively. Angiotensin is an oligopeptide hormone in the blood that causes blood vessels to constrict, and drives blood pressure up. It is part of the renin-angiotensin system.

Blood Coagulation Activity

Latex from the fig tree *Ficus carica* var. Horaishi was also found to contain proteases, Ficins A, B, C, D, and S (Sugiura and Sasaki 1971; Sugiura et al. 1975). Studies showed that Ficins A and S were immunochemically identical and also Ficins B and C resembled to each other. Ficins A and S, Ficin B, Ficin C, and Ficin D possessed common antigenic determinants at the rate of more than 38%, but the enzymes could not be distinguished. The immunochemical similarity of the enzymes to Ficin A was suggested to be in the order of Ficin A = Ficin S > Ficin B > Ficin C > Ficin D. Ficins are used as anthelminthic and for protein digestion. Studies showed that proteases in fig latex had an effect on human blood coagulation (Richter et al. 2002). Ficin derived from *Ficus carica* latex was found to shorten the activated partial thromboplastin time and the prothrombin time of normal plasmas and plasmas deficient in coagulation factors, except plasma deficient in factor X (FX) and generated activated FX (FXa) in defibrinated plasma. Separation of ficin from *Ficus carica* yielded six proteolytic fractions with a different specificity towards FX. Factor X was converted to activated FXβ by consecutive proteolytic cleavage. The cleavage pattern of FXa degradation products in the light chain was influenced by Ca2+ and Mn2+. The data suggested that the haemostatic potency of *Ficus* proteases was based on activation of human coagulation factor X.

Cholinesterase Inhibition Activity

Studies by Orhan et al. (2011) showed the n-hexane and acetone extracts of fig leaves exerted marked inhibition against both acetyl cholinesterase (AChE) and butyrylcholinesterase (BChE) However, they had low activity in the antioxidant tests of 2-diphenyl-1-picrylhydrazyl (DPPH) radical scavenging activity, metal-chelation capacity, and ferric-reducing antioxidant power (FRAP). The chloroform extract was found to be the richest in total flavonoid content (252.5 mg/g quercetin equivalent), while the n-butanol extract had the highest total phenol amount (85.9 mg/g extract gallic acid equivalent).

Antiwarts Activity

A prospective, open right/left comparative trial of fig tree latex therapy vs. local standard of cryotherapy was carried out on 25 patients with common warts (verruca vulgaris) (Bohlooli et al. 2007). It was found that fig tree latex therapy was marginally less effective than cryotherapy. Adverse effects were observed only in cryo-treated warts. Fig tree latex therapy of warts offers several beneficial effects including short-duration therapy, no reports of any side-effects, ease-of-use, patient compliance, and a low recurrence rate.

Nasal Gel Formulation

Basu and Bandyopadhyay (2010) prepared in-situ nasal gels of midazolam using three different concentrations (0.5%, 1.0% and 1.5% w/v) of *F. carica* mucilage (FCM) and synthetic polymers (hydroxypropylmethyl cellulose and Carbopol 934). Evaluation of FCM showed that it was as safe as the synthetic polymers for nasal administration. In-vivo experiments conducted in rabbits further confirmed that in situ nasal gels provided better bioavailability of midazolam than the gels prepared from synthetic mucoadhesive polymers. It was observed that the nasal gel containing 0.5% FCM and 0.5% sodium taurocholate exhibited appropriate rheological, mechanical and mucoadhesive properties and showed better drug release profiles. Further, this formulation produced no damage to the nasal mucosa that was used for the permeation study, and absolute bioavailability was also higher compared to gels prepared from synthetic polymers.

Antiinflammatory Activity

The study by Ali et al. (2011) validated the traditional claim with pharmacological data that fig leaves had anti-inflammatory and antioxidant activity The leaves are claimed to be effective in various inflammatory conditions like painful or swollen piles, insect sting and bites. Anti-inflammatory and antioxidant activity could be due to the presence of steroids and flavanoids in the leaves and the anti-inflammatory activity was attributed to its free radical scavenging activity. Ethanol extract of fig leaves at 600 mg/kg exhibited maximum anti-inflammatory effect equating to 75.9% in acute inflammation and in chronic study showed 71.66% reduction in granuloma weight (Patil and Patil 2011). The petroleum ether, chloroform and ethanol extracts all significantly reduced carrageenan-induced rat paw edema and cotton pellet granuloma methods, comparable to the stabdard drug indomethacin.

Anthelmintic Activity

The latex of *F. carica*, administered in doses of 3 ml/kg/day to mice, during three consecutive days, was effective in the removal of pinworms *Syphacia obvelata* (41.7%) but it did not significantly eliminate *Aspiculuris tetraptera* (2.6%) and *Vampirolepis nana* (8.3%) (de Amorin et al. 1999). The aqueous fruit extract of *Ficus benghalensis* was found to be more potent in killing all the test worms (*Pheretima posthuma*) than *F. religiosa* and *F. carica* fruit extracts (Sawarkar et al. 2011). *F. benghalensis* within an hour of post exposure was 100% effective while *F. carica* was effective at 2–3 hours of post exposure.

Larvicidal Activity

The milky sap of *F. carica* was found to have a significant toxic effect against early fourth-stage larvae of *Aedes aegypti* with an lethal concentration LC_{50} value of 10.2 μg/ml and an LC_{90} value of 42.3 μg/ml (Chung et al. 2011). Two natural furocoumarins, 5-methoxypsoralen and 8-methoxypsoralen were isolated from the milky sap of *F. carica*. The LC_{50} value of 5-methoxypsoralen and 8-methoxypsoralen were 9.4 and 56.3 μg/ml, respectively.

Allergic Reaction

The irritant potential of total methanolic extract and five triterpenoids newly isolated from the leaves of *Ficus carica* was investigated by open mouse ear assay (Saeed and Sabir 2002). Total methanolic extract, calotropenyl acetate, methyl maslinate and lupeol acetate showed potent and persistent irritant effects.

Studies showed that psoralen and bergapten were the only significant photoactive compounds, and were present in appreciable quantities in the leaf and shoot sap but were not detected in the fruit or its sap (Zaynoun et al. 1984). These compounds were found to be more concentrated in the leaf sap compared to the shoot sap. The psoralen levels were several times higher than those of bergapten. Lower concentrations of both compounds were present in autumn compared to spring and summer. The findings suggested that the allergic reaction was induced primarily by psoralen. The higher content of both photoactive compounds in spring and summer was partly responsible for the increased incidence of fig dermatitis during these seasons. Ingestion of the fruit did not cause photosensitiszation and the absence of photoactive furocoumarins in the fruit and its sap remained unexplained.

Two arborists developed acute blistering eruptions on their forearms, hands, and fingers after some contact with fig trees (Derraik and Rademaker 2007). The previous day, both men had pruned branches from a large fig tree, which had sustained damaged during a storm. The following morning, both complained of a burning discomfort which rapidly evolved into erythema and bullae on skin that had been in direct contact with the tree branches. These symptoms gradually resolved over 4–6 weeks.

Traditional Medicinal Uses

In Latin America, figs are commonly used as folk remedies. The fruit is considered mildly laxative, demulcent, diuretic, emollient, digestive, pectoral and to have aphrodisiac property. A decoction of the fruits is gargled to relieve sore throat; figs boiled in milk are repeatedly packed against swollen gums; the fruits are much used as poultices on tumours and other abnormal growths. unripe green fruits are cooked with other foods as a galactogogue and tonic. The roasted fruit is emollient and used as a poultice in the treatment of gumboils, dental abscesses. In India, the fruit is reported useful in inflammations and paralysis. *F. carica* is claimed to be useful in liver and spleen disorders, to cure piles and in treatment of gout. Locally the leaves are being used in the treatment of jaundice. The paste of figs is given in prescriptions, also as cooked vegetable for emaciation and debility, as a diuretic in urinary stones. The fruit in a medicated clarified butter is given for internal use in fever, consumption, asthma, epilepsy and insanity.

The leaf decoction is taken as a remedy for diabetes and calcifications in the kidneys and liver and as a stomachic. The leaves are also added to boiling water and used as a steam bath for painful or swollen piles. The latex from the stem is commonly applied on warts, corns, skin ulcers and sores, and taken as a purgative and vermifuge. It also has an analgesic effect against insect stings and bites. A decoction of the young branches is used as pectoral. The roots are used to treat leucoderma and ringworms.

Other Uses

In India, fig leaves are used for fodder and in the Mediterranean region also the fruits. The wood is pliable but porous and of little commercial value and has been used for hoops, garlands, ornaments etc. The fruits and other parts of the tree are important for folk medicine. In tropical America, the latex has been used as a detergent for washing dishes, pots and pans. It was an ingredient in some of the early commercial detergents for household use but was removed after many reports of irritated or inflamed hands in housewives. In southern France, fig leaves has been used as a source of woodland scented perfume material called "fig-leaf absolute".

Comments

Ficus carica has been divided into two lower taxa:

Ficus carica ssp. *carica* – Leaves are usually broader than long, deeply 3-lobed, lobes spathulate with entire to apically few dentate margin . Figs are mostly pyriform, subsessile shortly stipitate, glabrescent to minutely hairy.

Ficus carica ssp. *rupestris* (Hausskn. ex Boiss.) Browicz – Leaves are usually longer than broad, sometimes vary shallowly lobed with serrate margin. Figs are pyriform and stipitate to globose and stipeless, frequently densely short hairy.

Edible fig varieties fall under the following four categories:

(a) Common Fig (persistent). Figs develop parthenocarpically (without pollination) and are by far the most prevalent type of fig grown for fresh fig production. The fruit does not have true seeds and is primarily produced on current season wood.
(b) Smyrna figs (caducous). Smyrna fig varieties produce large edible fruit with true seeds. The *Blastophaga* wasp and Caprifigs are required for pollination and normal fruit development. If this fertilization process does not occur, fruit will not develop properly and will fall from the tree. Smyrna-type figs are commonly sold as dried figs.
(c) San Perdo (intermediate). This type of fig bears two crops of fruit in one season-one crop on the previous season's growth and a second crop on current growth. The first crop, called the Breba crop, is parthenocarpic and does not require pollination. Fruit of the second crop is the Smyrna type and requires pollination from the Caprifig. Usually grown as non-commercial fresh fig production.
(d) Caprifig. The Caprifig produces a small non-edible fruit; however, the flowers inside the Caprifig produce pollen. This pollen is essential for fertilizing fruit of the Smyrna and San Pedro types. The pollen is transported from the Caprifig to the pollen-sterile types by a *Blastophaga* wasp.

Selected References

Ahmed W, Khan AQ, Malik A (1988) Two triterpenes from the leaves of *Ficus carica*. Planta Med 54:481

Ali B, Mujeeb M, Aeri V, Mir SR, Faiyazuddin M, Shakeel F (2011) Anti-inflammatory and antioxidant activity of *Ficus carica* Linn. leaves. Nat Prod Res 1:1–6

Aref HL, Salah KBH, Chaumont JP, Fekih A, Aouni M, Said K (2010) In vitro antimicrobial activity of four *Ficus carica* latex fractions against resistant human pathogens. Pak J Pharm Sci 23(1):53–58

Azarkan M, Matagne A, Wattiez R, Bolle L, Vandenameele J, Baeyens-Volant D (2011). Selective and reversible thiol-pegylation, an effective approach for purification and characterization of five fully active ficin (iso) forms from *Ficus carica* latex. Phytochem 72(14–15): 1718–31

Backer CA, Bakhuizen van den Brink RC Jr (1965) Flora of Java (Spermatophytes only), vol 2. Wolters-Noordhoff, Groningen, 641 pp

Basu S, Bandyopadhyay AK (2010) Development and characterization of mucoadhesive in situ nasal gel of midazolam prepared with *Ficus carica* mucilage. AAPS PharmSciTech 11(3):1223–1231

Bohlooli S, Mohebipoor A, Mohammadi S, Kouhnavard M, Pashapoor S (2007) Comparative study of *Fig* tree efficacy in the treatment of common warts (verruca vulgaris) vs. cryotherapy. Int J Dermatol 46(5):524–526

Canal JR, Torres MD, Romero A, Pérez C (2000) A chloroform extract obtained from a decoction of *Ficus carica* leaves improves the cholesterolaemic status of rats with streptozotocin-induced diabetes. Acta Physiol Hung 87(1):71–76

Chiej R (1984) The Macdonald encyclopaedia ofmMedicinal plants. Macdonald & Co, London, 447 pp

Chevallier A (1996) The encyclopedia of medicinal plants. Dorling Kindersley, London, 336 pp

Chung DO, Kang SK, Chung HJ (1995) Purification and identification of antimicrobial substances in phenolic fraction of fig leaves. Han'guk Nonghwa Hakhoechi 38:293–296

Chung IM, Kim SJ, Yeo MA, Park SW, Moon HI (2011) Immunotoxicity activity of natural furocoumarins from milky sap of *Ficus carica* L. against *Aedes aegypti* L. Immunopharm Immunotoxicol 33(3):515–518

de Amorin A, Borba HR, Carauta JP, Lopes D, Kaplan MA (1999) Anthelmintic activity of the latex of *Ficus* species. J Ethnopharmacol 64(3):255–258

Derraik JGB, Rademaker M (2007) Phytophotodermatitis caused by with contact a fig tree (*Ficus carica*). J N Z Med Assoc 120, No 1259

Duke JA, Ayensu ES (1985) Medicinal plants of China. Vols 1&2. Reference Publications Inc, Algonac, 705 pp

Foundation for Revitalisation of Local Health Traditions (2008) FRLHT Database. htttp://envis.frlht.org

Gibernau M, Buser HR, Frey JE, Hossaert-Mckey M (1997) Volatile compounds from extracts of figs of *Ficus carica*. Phytochemistry 46(2):241–244

Gilani AH, Mehmood MH, Janbaz KH, Khan AU, Saeed SA (2008) Ethnopharmacological studies on antispasmodic and antiplatelet activities of *Ficus carica*. J Ethnopharmacol 119(1–2):1–5

Gond NY, Khadabadi SS (2008) Hepatoprotective activity of *Ficus carica* leaf extract on rifampicin-induced hepatic damage in rats. Indian J Pharm Sci 70:364–366

Grieve M (1971) A modern herbal, vol 2. Penguin/Dover publications, New York, 919 pp

Hu SY (2005) Food plants of China. The Chinese University Press, Hong Kong, 844 pp

Innocenti G, Bettero A, Caporale G (1982) Determination of the coumarinic constituents of *Ficus carica* leaves by HPLC. Farmaco Sci 37(7):475–485, In Italian

Jeong MR, Cha JD, Lee YE (2005) Antibacterial activity of Korean Fig (*Ficus carica* L.) against food poisoning bacteria. Korean J Food Cookery Sci 21:84–93

Jeong MR, Kim HY, Cha JD (2009) Antimicrobial activity of methanol extract from *Ficus carica* leaves against oral bacteria. J Bacteriol Virol 39(2):97–102

Jeong WS, Lachance PA (2001) Phytosterols and fatty acids in fig (*Ficus carica*, var. Mission) fruit and tree components. J Food Sci 66:278–281

Kang SK, Jung HJ (1995) Solvent fractionation of fig leaves and its antimicrobial activity. Agric Chem Biotechnol 38:289–292

Kolesnik A, Kakhniashvili TA, Zherebin YL, Golubev VN, Pilipenko LN (1989) Lipids of the fruit of *Ficus carica*. Chem Natl Comp 22(4):394–397

Konyalioglu S, Saglam H, KivcaK B (2005) α-Tocopherol, flavonoid, and phenol contents and antioxidant activity of *Ficus carica* leaves. Pharm Biol 43(8):683–686

Lazreg Aref H, Gaaliche B, Fekih A, Mars M, Aouni M, Chaumon PJ, Said K (2011) In vitro cytotoxic and antiviral activities of *Ficus carica* latex extracts. Nat Prod Res 25(3):310–319

Maruyama S, Miyoshi S, Tanaka H (1989) Angiotensin I-converting enzyme inhibitors derived from *Ficus carica*. Agric Biol Chem 53(10):2763–2767

Mohan GK, Pallavi E, Kumar BR, Ramesh M, Venkatesh S (2007) Hepatoprotective activity of *Ficus carica* Linn. leaf extract against carbon tetrachloride-induced hepatotoxicity in rats. DARU 15(3):162–166

Morton JF (1987) Fig. In: Morton Julia F (ed) Fruits of warm climates. Florida Flair Books, Miami, pp 47–50

Oliveira AP, Silva LR, Andrade PB, Valentão P, Silva BM, Gonçalves RF, Pereira JA, de Pinho PG (2010a) Further insight into the latex metabolite profile of *Ficus carica*. J Agric Food Chem 58(20): 10855–10863

Oliveira AP, Silva LR, Andrade PB, Valentão P, Silva BM, Pereira JA, de Pinho PG (2010b) Determination of low molecular weight volatiles in *Ficus carica* using HS-SPME and GC/FID. Food Chem 121(4): 1289–1295

Oliveira AP, Silva LR, de Pinho PG, Gil-Izquierdo A, Valentão P, Silva BM, Pereira JA, Andrade PB (2010c) Volatile profiling of *Ficus carica* varieties by HS-SPME and GC–IT-MS. Food Chem 123(2):548–557

Orhan IE, Ustün O, Sener B (2011) Estimation of cholinesterase inhibitory and antioxidant effects of the leaf extracts of Anatolian *Ficus carica* var. *domestica* and their total phenol and flavonoid contents. Nat Prod Commun 6(3):375–378

Pacific Island Ecosystems at Risk (PIER) (2004) *Ficus carica* L. Moraceae. www.hear.org/pier/species/ficus_carica.htm

Patil VV, Patil VR (2011) Evalution of anti-inflammatory activity of *Ficus carica* Linn leaves. Indian J Nat Prod Resour 2(2):151–155

Patil VV, Bhangale SC, Patil VR (2010a) Evaluation of anti-pyretic potential of *Ficus carica leaves*. Int J Pharm Sci Rev Res 2(2):48–50

Patil VV, Bhangale SC, Patil VR (2010b) Studies on immunomodulatory activity of *Ficus carica*. Int J Pharm Pharm Sci 2(1):97–99

Pérez C, Canal JR, Campillo JE, Romero A, Torres MD (1999) Hypotriglyceridaemic activity of *Ficus carica* leaves in experimental hypertriglyceridaemic rats. Phytother Res 13(3):188–191

Pèrez C, Canal JR, Torres MD (2003) Experimental diabetes treated with *Ficus carica* extract: effect on oxidative stress parameters. Acta Diabetol 40:3–8

Perez C, Dominguez E, Canal JR, Campillo JE, Torres MD (2000) Hypoglycaemic activity of an aqueous extract from *Ficus carica* (fig tree) leaves in streptozotocin diabetic rats. Pharm Biol 38(3):181–186

Porcher MH et al (1995–2020) Searchable world wide web multilingual multiscript plant name database. Published by The University of Melbourne, http://www.plantnames.unimelb.edu.au/Sorting/Frontpage.html

Purseglove JW (1968) Tropical crops: dicotyledons. 1&2. Longman, London, 719 pp

Rassouli A, Ardestani AF, Asadi F, Salehi MH (2010) Effects of fig tree (*Ficus carica*) leaf extracts on serum and liver cholesterol levels in hyperlipidemic rats. Int J Vet Res 2(1):77–80

Richter G, Schwarz HP, Dorner F, Turecek PL (2002) Activation and inactivation of human factor X by proteases derived from *Ficus carica*. Br J Haematol 119(4):1042–1051

Rubnov S, Kashman Y, Ruth Rabinowitz R, Schlesinger M, Mechoulam R (2001) Suppressors of cancer cell proliferation from fig (*Ficus carica*) resin: isolation and structure elucidation. J Nat Prod 64(7):993–996

Saeed MA, Sabir AW (2002) Irritant potential of triterpenoids from *Ficus carica* leaves. Fitoterapia 73:417–420

Sawarkar HA, Singh MK, Pandey AK, Biswas D (2011) In vitro anthelmintic activity of *Ficus benghalensis, Ficus carica* and *Ficus religiosa*: a comparative study. Int J Pharm Pharmaceut 3(Suppl 2):152–153

Serraclara A, Hawkins F, Pérez C, Domínguez E, Campillo JE, Torres MD (1998) Hypoglycemic action of an oral fig-leaf decoction in type-I diabetic patients. Diabetes Res Clin Pract 39(1):19–22

Singab AN, Ayoub NA, Ali EN, Mostafa NM (2010) Antioxidant and hepatoprotective activities of Egyptian moraceous plants against carbon tetrachloride-induced oxidative stress and liver damage in rats. Pharm Biol 48(11):1255–1264

Solomon A, Golubowicz S, Yablowicz Z, Grossman S, Bergman M, Gottlieb HE, Altman A, Kerem Z, Flaishman MA (2006) Antioxidant activities and anthocyanin content of fresh fruits of common fig (*Ficus carica* L.). J Agric Food Chem 54(20):7717–7723

Sugiura M, Sasaki M (1971) Studies on proteinases from *Ficus carica* var. Horaishi I: Purification and enzymatic properties of proteinases. Yakugaku Zasshi 91(4):457–466 (In Japanese)

Sugiura M, Sasaki M, Moriwaki C, Yamaguchi K (1975) Studies on proteinases from *Ficus carica* var. Horaishi. VI. Immunochemical comparison of ficins A, B, C, D and S. Chem Pharm Bull 23(9):1969–1975

Teixeira DM, Patão RF, Coelho AV, da Costa CT (2006) Comparison between sample disruption methods and solid-liquid extraction (SLE) to extract phenolic compounds from *Ficus carica* leaves. J Chromatogr A 1103:22–28

U.S. Department of Agriculture, Agricultural Research Service (2010) USDA National nutrient database for standard reference, release 23. Nutrient data laboratory home Page, http://www.ars.usda.gov/ba/bhnrc/ndl

USDA, ARS, National Genetic Resources Program (2004) Germplasm Resources Information Network - (GRIN) [Online Database]. National Germplasm Resources Laboratory, Beltsville, http://www.ars-grin.gov/cgibin/npgs/html/index.pl

Vaya J, Mahmood S (2006) Flavonoid content in leaf extracts of the fig (*Ficus carica* L.), carob (*Ceratonia siliqua* L.) and pistachio (*Pistacia lentiscus* L.). Biofactors 28(3–4):169–175

Vinson JA (1999) Functional food properties of figs. Cereal Foods World 44(2):82–87

Vinson JA, Zubik L, Bose P, Samman N, Proch J (2005) Dried fruits: excellent in vitro and in vivo antioxidants. J Am Coll Nutr 24(1):44–50

Wang G, Wang H, Song Y, Jia C, Wang Z, Xu H (2004) Studies on anti-HSV effect of *Ficus carica* leaves. Zhong Yao Cai 27(10):754–756 (In Chinese)

Wu Z, Zhou Z-K, Gilbert MG (2003) Moraceae link. In: Wu ZY, Raven PH, Hong DY (eds) Flora of China. Vol. 5 (Ulmaceae through Basellaceae). Science Press Beijing and Missouri Botanical Garden Press, St. Louis

Zaynoun ST, Aftimos BG, Abi AL, Tenekjian KK, Khalide U, Kurban AK (1984) *Ficus carica*; isolation and quantification of the photoactive components. Contact Dermatitis 11:21–25

Ficus dammaropsis

Scientific Name

Ficus dammaropsis Diels.

Synonyms

Dammaropsis kingiana Warb.

Family

Moraceae

Common/English Names

Dinnerplate Fig, Highland Breadfruit, Highlands Kapiak

Vernacular Names

Papua New Guinea: Hailands Kapiak (Tok Pisin), Suar, Suar Sur (Mendi), Mail (Kewa), Elu (Wiru), Minibi (Imbongu), Anugu (Duna).

Origin/Distribution

The species is indigenous in Papua New Guinea and West Papua.

Agroecology

It is found in temporary land clearings by streams and rivers and in secondary forests from 800 to 2,300 m altitude. It is cold tolerant. It is propagated from seeds.

Edible Plant Parts and Uses

Young leaves are commonly sold in the highland markets in Papua New Guinea. Young leaves are eaten cooked as vegetables with pig meat. Older leaves are used for wrapping food during cooking. The outside layer of the fruit is edible. The young fruit is boiled and eaten as a vegetable.

Botany

A small, perennial, much branched tree, 5–12 m high with strong branches containing milky latex. Young leaves are enclosed in two pale yellow sheath-like stipules. Leaves are coriaceous, simple, alternate, cordate, very large, 90 cm by 60 cm, with coarsely serrated margins and undulating wavy pleated surface (Plates 1, 2, 3, and 4) and a prominent midrib, variable in size and colour – reddish, yellowish, orangey to greenish and borne on stout petiole. Fruit is a fig syconium, arsing from the leaf axils, large up t o 15 cm diameter, globose, covered by overlapping lateral bracts

T.K. Lim, *Edible Medicinal And Non-Medicinal Plants: Volume 3, Fruits*,
DOI 10.1007/978-94-007-2534-8_52, © Springer Science+Business Media B.V. 2012

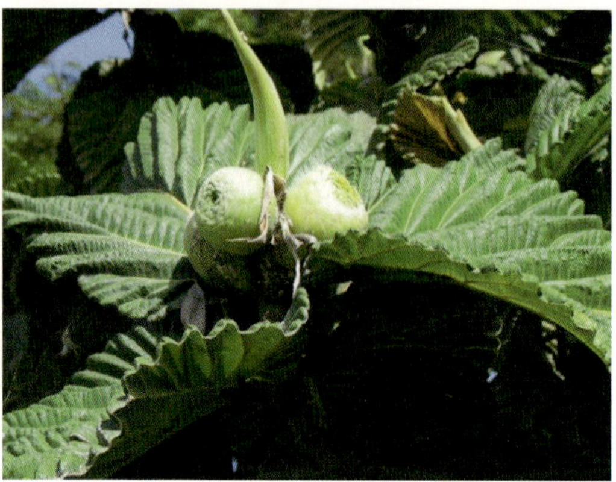

Plate 1 Cluster of immature globose fruit and very large leaves

Plate 2 Axillary fruit with overlapping bracts

Plate 3 Young leaves tied in bundle for sale as vegetables

Plate 4 Young leave with pleated, colourful lamina

(Plates 1 and 2), green when young turning rose-red to reddish brown or purple when ripe.

Nutritive/Medicinal Properties

No published information is available on its nutritive value and medicinal uses.

Other Uses

The bark is used for clothing.

Selected References

Corner EJH (1978) *Ficus dammaropsis* and the multibracteate species of *Ficus* sect *Sycocarpus*. Philos Trans R Soc B 281(982):373–406

French BR (1982) Growing food in the Southern Highlands province of Papua New Guinea. AFTSEMU (Agricultural Field Trials, Surveys, Evaluation and Monitoring Unit), World Bank Papua New Guinea Project

French BR (1986) Food plants of Papua New Guinea, a compendium. Australia Pacific Science Foundation, Sheffield, 407pp

Kambuou RN (1996) Papua New Guinea: country report. FAO international technical conference on plant genetic resources (Leipzig, 1996). Food and Agriculture Organization of the United Nations. http://www.fao.org/ag/AGP/agps/pgrfa/pdf/papuanew.pdf

Ficus pumila

Scientific Name

Ficus pumila L.

Synonyms

Ficus hanceana Maxim., *Ficus longipedicellata* H. Perrier, *Ficus minima* hort. ex Gard., *Ficus stipulata* Thunb., *Ficus scandens* Lamk, *Ficus repens* Hort., *Ficus repens* Hort. var. *lutchuensis* Koidz., *Plagiostigma pumila* Zucc., *Plagiostigma stipulata* Zucc., *Urostigma scandens* Liebm.

Family

Moraceae

Common/English Names

Climbing Ficus, Climbing Fig, Creeping Fig, Creeping Rubber Plant, Fig Ivy, Ticky Creeper, Tropical Ivy

Vernacular Names

Brazil: Falsa-Hera, Hera De China, Hera-Miúda, Mama De Pared (Portuguese);
Burmese: Kyauk Kat Nyaung New;
Chinese: Liang-Fen-Zi, Liang-Fen-Tzu (Cold Jelly Seed), Bing-Fen-Zi, Ping-Fen-Tsu (Ice Jelly Seed), Guasi-Man-Tou, Kuei-Man T'ou (Devil's Steamed Bread), Bi-Li, Pi-Li (Wall Litchi), Man Tu Luo, Yuan Bian Zhong;
Columbia: Uña, Uña De Gato, Uñita, Yedro;
Czech: Fíkovník Palistový, Fíkovník Šplhavý;
Danish: Hængefigen;
Dutch: Klimvijg;
Eastonian: Roomav Viigipuu;
Finnish: Kääpiököynnösviikuna;
French: Figuier Rampant, Figuier Nain Rampant;
German: Kletter-Feige, Kletterfeige, Kletter-Ficus;
Guatemala: Uña;
Indonesia: Daun Dolar, Karet Rambat;
Italian: Fico Rampicante, Fico Tappezzante;
Japanese: Aigyokushi, O-Itabi, Oo Itabi;
Lithuanian: Smulkusis Fikusas; Smulkieji Fikusai;
Philippines: Creeping Fig, Fig Ivy;
Polish: Figowiec Pnący;
Portuguese: Hera, Figueira-Trepadeira;
Puerto Rico: Paja De Colchón, Paz Y Justicia;
Serbian: Puzajući Fikus, Penjući Fikus;
Singapore: Wéntóu Xǔe;
Spanish: Higuera Trepadora, Ficus De China, Ficus Enano, Ficus Pumila, Ficus Rastrero, Ficus Tapizante, Ficus Trepador, Ficus Enano;
Sri Lanka: Tropical Ivy;
Swedish: Klätterfikus;
Taiwan: 'Àiyù Bīng', 'Àiyù Dòng', Aiyuzi (Mandarin), 'PeH-Ōe-Jī' (Min-Nan), "Ò-Giô (Hokien);
Turkish: Duvar Sarma, Duvar Sarmaşığı;

T.K. Lim, *Edible Medicinal And Non-Medicinal Plants: Volume 3, Fruits*,
DOI 10.1007/978-94-007-2534-8_53, © Springer Science+Business Media B.V. 2012

Thailand: Madueo Thao, Lin Suea;
Vietnam: Bị Lệ, Cây Trâu Cổ, Cây Xộp, Cơm Lênh, Mộc Liên,, Sung Thằn Lằn, Trâu Cổ, Vày Ốc, Mác Púp (Tày).

Origin/Distribution

The species is native to East Asia – south China (Anhui, Fujian, Guangdong, Guangxi, Guizhou, Henan, Hubei, Hunan, Jiangsu, Jiangxi, southern Shaanxi, Sichuan, Yunnan, Zhejiang), Taiwan, Japan (Honshu, Kyushu, Ryukyu Islands, Shikoku), and Vietnam. It is deemed as a garden escape in Nepal, New Zealand, Western Australia, Hawaii, Puerto Rico, Florida, Alabama, Georgia, Louisiana and Texas. In Sri Lanka and Japan, it is cultivated in orchards and plantations for the edible fruits, which are used for the production of jam and jellies.

Agroecology

The crop is a warm subtropical to sub-temperate species. It is found in altitudes below 1,000 m. It grows wild as a scandent vine or climber on rocks and tree trunks and is cultivated for covering walls, concrete and other surfaces. The plant prefers partial shade and humidity and is drought tolerant but poor salt tolerant.

Edible Plant Parts and Uses

Boiled fruits are made into grass jelly. In China, the syconia are picked ripe and placed in a porous bag to squeeze out the juice. The juice is cooked and then set into a gelatinous consistency called "*pai-liang-fen*." This jellylike material is cubed, mixed with water, syrup and flavourings and consumed as a refreshing drink in summer. It is canned and sold in Asian markets as "grass jelly" or "*ai-yu jelly*" or "ice-jelly" in Singapore. The jelly when combined with lime or lemon juice and sweeteners make a refreshing dessert in Taiwanese night markets and Singapore hawker centres. In Okinawa, Japan, the leaves are used as a herbal beverage. Grass Jelly is also extracted from boiled plants of *Mesona chinensis* in the mint family (Lamiaceae). In fact, cans of grass jelly often list this species on the label.

Botany

A vigorous evergreen, branched shrub, climber or scandent fig with milky latex, growing attached to walls, stucco, masonry, rocks, tree trunks by means of exudation secreted by the aerial roots. Branches are pubescent and stipules are lanceolate, with yellow brown silk-like hairs. Leaves are alternate, distichous, polymorphic; those on vegetative, young creeping stems are ovate-subcordate, 1–3 cm by1–2 cm wide, with obtuse apex, oblique cordate base, papery and glabrous (Plate 1). Leaves on older erect, woody branches are larger, oblong, ovate-oblong or obovate, 3–9 cm long by 1.5–4 cm wide, obtuse base, rounded or emarginate apex, faveolate and pilose beneath. Leaf veins are conspicuous, honey-comb-like, secondary veins three or four on each side of midvein, abaxially prominent, and adaxially impressed. Mature syncarp (synconium) are axillary on normal leafy branches, solitary, green turning to yellowish green to pale red or violet when mature, thick, fleshy, pear-shaped to ± globose or cylindric, 4–8 × 3–5 cm, shortly yellow pubescent when young, basally attenuate into a short stalk, apical pore truncate, navel-like, or acuminate (Plates 2 and 3). Male flowers are many, in several rows near apical

Plate 1 Juvenile and older leaves

Plate 2 Immature, pear-shaped fruit

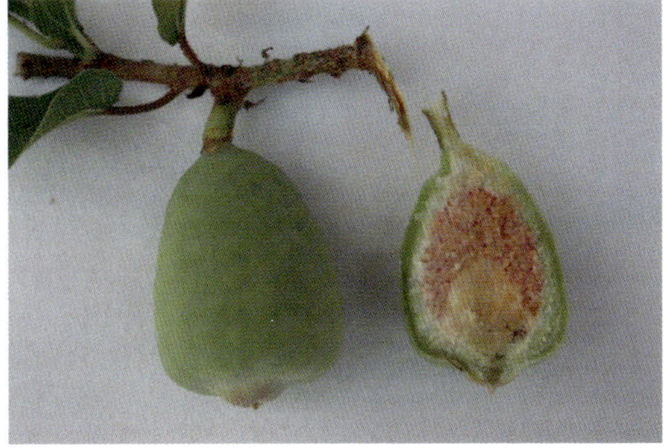

Plate 3 Fruit whole and halved to show the flowers inside

pore, pedicellate with two or three, linear calyx lobes and two stamens with short filaments. Gall flowers are pedicellate with three or four linear calyx lobes and short lateral style. Female flowers have long pedicel and four or five calyx lobes. Achenes are more or less globose, with adherent liquid.

Nutritive/Medicinal Properties

The fruit was found to contain sesquiterpenoid glucosides: pumilasides A, B and C; benzyl β-D-glucopyranoside, (E)-2-methyl-2-butenyl β-D-glucopyranoside and rutin (Kitajima et al. 2000). The fruit also contain acetylated dammarane-type triterpenoids: 3.β.-acetoxy-(20R,22E,24RS)-20,24-dimethoxydammaran-22-en-25-ol and 3,.β-acetoxy-(20S,22E, 24RS)-20,24-dimethoxydammaran-22-en-25-ol and related compounds, 3,.β-acetoxy-22,23,24, 25,26,27 -hexanordammaran-20-one and 3,.β-acetoxy-20,21,22,23,24,25,26,27- octanordammaran-17.β-ol (Kitajima et al. 1999). From the methanolic extract of *Ficus pumila* fruit, two sterols, (24S)-stigmast-5-ene-3β, 24-diol, (24S)-24-hydroxystigmast-4-en-3-one; two cycloartane-type triterpenoids, (24RS)-3β-acetoxycycloart-25-en-24-ol, (23Z)-3β-acetoxy-cycloart-23-en-25-ol and a euphane-type triterpenoid, (23Z)-3β-acetoxyeupha-7,23-dien-25-ol were obtained along with known sterols, triterpenoids and a furocoumarin (psoralene) (Kitajima et al. 1998a).

From a methanolic extract of fresh leaves of *Ficus pumila,* α-tocopherol-related compound was isolated together with α-tocopherol (Kitajima et al. 1998b). Additionally, two known sterols, 15 known triterpenoids and five known flavonoid glycosides were identified as constituents. The stems and leaves also contain coumarins, flavonoids and glycosides (Pistelli et al. 2000).

The gum obtained from the plant yields glucose, fructose and arabinose on hydrolysis (NIMM 1990). The fruit contains proteins and latex.

In studies conducted in Japan, four flavonoid glycosides were isolated from the leaves and identified as rutin (1 and 3), apigenin 6-neohesperidose (2), kaempferol 3-robinobioside (4) and kaempferol 3-rutinoside (5) (Leong et al. 2008). Among these compounds, rutin exhibited the strongest antioxidant activity in DPPH radical scavenging assay and superoxide radical inhibition assay. The preparation of Ooitabi leaves in water provide sufficient amount of flavonoid glycosides to the Okinawan although 50% of aqueous ethanol extracted these flavonoid glycosides more effectively. These results showed the potential of Ooitabi leaves as a natural source of antioxidant for health management.

A triterpene, neohopane isolated from the leaves of *Ficus pumila* exhibited antimicrobial activity at a concentration of 30 μg against *Escherichia coli, Pseudomonas aeruginosa,*

Bacillus subtilis and *Candida albicans* with an average antimicrobial index of 0.5, 0.3, 0.3 and 0.7, respectively (Ragasa et al. 1999). The chloroform extract of *Ficus pumila* afforded bergapten and oxypeucedanin hydrate (Juan et al. 1997). Bergapten was found to inhibit the growth of *Staphylococcus aureus, Eschericia coli* and *Salmonnella typhi*, but was inactive against *Trichophyton mentagrophytes, Mycobacterium pleie* and *Candida albicans*. Oxypeucedanin hydrate inhibited the growth of *Salmonella typhi*, but was found inactive against the other five microorganisms. When tested for antimutagenic activity by the use of the micronucleus test, bergapten was found to reduce the number of micronucleated polychromatic erythrocytes (MPCE) induced by mitomycin C by 44%, while oxypeucedanin hydrate reduced MPCE by 74%.

In folk medicine in Vietnam, the fruits and the leaves are considered to be tonic and are used in cases of impotence, lumbago and as a galactagogue. Furthermore, they are considered a treatment for rheumatism, anaemia, hematuria, chronic dysentery, haemorrhoids, menstrual disorders, osteodynia, dysuria, dyschezia, amenorrhoea, spermatorrhoea and hypogalactia. Externally the leaves are applied to carbuncles. The latex is reported to have anthelmintic properties. In cases of dropsy the plant ash is rubbed on the body. The branches are also effective in treating soreness of limbs, back pain, anaemia, hypogalactia, impetigo and boils. The young stem and leaves are used in the form of a decoction, extract or alcoholic maceration while the fruit in the form of a decoction, extract or candied fruit.

The leaves and fruit of *F. pumila* have been used in Chinese traditional herbal medicine since ancient times for treating bleeding, swelling, haemorrhoids, intestinal disorders and impotence and the bioactive compounds were found to be flavonoids. In Japan, the Okinawan folks use *Ficus pumila* leaves as a beverage or herbal medicine to treat diabetes and high blood pressure (Leong et al. 2008).

Other Uses

Creeping fig is widely cultivated as an ornamental and pot plant in sub-temperate, subtropical and tropical regions. It is also grown as a ground cover but needs regular pruning along the edges to keep it neat and within bounds. It is also well-suited for use in topiaries or hanging baskets. Various selections are commercially grown as a pot plant in temperate regions. In the Philippines, it is planted as nematode-resistant rootstock for *Ficus carica*.

Comments

This fig requires the fig wasp, *Blastophaga pumilae* for pollination.

Selected References

Backer CA, Bakhuizen van den Brink RC Jr (1965) Flora of Java (Spermatophytes only), vol 2. Wolters-Noordhoff, Groningen, 641 pp

Corner EJH (1965) Check-list of *Ficus* in Asia and Australia. Gardens' Bull Singapore 21:1–186

Doan TH, Do HB, Pham KM, Nguyen TT, Bui XC, Pham DM (eds) (1993) Les Plants Médicinales Au Vietnam [The medicinal plants of Vietnam], vol 2, Médicine Traditionelle et Pharmacopée [Traditional medicine and pharmacopoeia]. Agence de Coopération Culturelle et Technique, Paris, 189 pp

Hu S-Y (2005) Food plants of China. The Chinese University Press, Hong Kong, 844 pp

Juan EA, Rideout JA, Ragasa CY (1997) Bioactive furanocoumarin derivatives from *Ficus pumila* (Moraceae). Philipp J Sci 126(2):143–153

Kitajima J, Kimizuka K, Arai M, Tanaka Y (1998a) Constituents of *Ficus pumila* leaves. Chem Pharm Bull Tokyo 46(10):1647–1649

Kitajima J, Kimizuka K, Tanaka Y (1998b) New sterols and triterpenoids of *Ficus pumila* fruit. Chem Pharm Bull Tokyo 46(9):1408–1411

Kitajima J, Kimizuka K, Tanaka Y (1999) New dammarane-type acetylated triterpenoids and their related compounds of *Ficus pumila* fruit. Chem Pharm Bull Tokyo 47(8):1138–1140

Kitajima J, Kimizuka K, Tanaka Y (2000) Three new sesquiterpenoid glucosides of *Ficus pumila* fruit. Chem Pharm Bull Tokyo 48(1):77–80

Leong CNA, Masakuni T, Isao H, Hajime T (2008) Antioxidant flavonoid glycosides from the leaves of *Ficus pumila* L. Food Chem 109(2):415–420

National Institute of Materia Medica (1999) Selected medicinal plants in Vietnam, vol 1. Science and Technology Publishing House, Hanoi, 439 pp

National Institute of Materia Medica, Hanoi. (NIMM) (1990) Medicinal plants in Vietnam. World Health Organization, Regional Office for the Western Pacific, Manila, 410 pp

Pancho JV (1958) Notes on cultivated species of *Ficus* in the Philippines. Baileya 6:129–134

Perry LM (1980) Medicinal plants of East and Southeast Asia. Attributed properties and uses. MIT, Cambridge/London, 620 pp

Pistelli L, Chiellini EE, Morelli I (2000) Flavonoids from *Ficus pumila*. Biochem Syst Ecol 28(3):287–289

Ragasa CY, Juan E, Rideout JA (1999) A triterpene from *Ficus pumila*. J Asian Nat Prod Res 1(4):269–275

Rojo JP, Pitargue FC, Sosef MSM (1999) *Ficus pumila* L. In: de Padua LS, Bunyapraphatsara N, Lemmens RHMJ (eds) Plant resources of South-East Asia No. 12(1): medicinal and poisonous plants 1. Backhuys Publisher, Leiden, pp 285–286

Wu W-S, Ji X-P, Wang Y-F, Fang Y-L (2000) The extraction technology of total flavonoides from the leaves and receptacle of inflorescence of *Ficus pumila* L. J Plant Resour Environ 9(2):55–56, In Chinese

Wu Z, Zhou Z-K, Gilbert MG (2003) Moraceae link. In: Wu ZY, Raven PH, Hong DY (eds) Flora of China, vol 5, Ulmaceae through Basellaceae. Science Press/Missouri Botanical Garden Press, Beijing/St. Louis

Ficus racemosa

Scientific Name

Ficus racemosa **L.**

Synonyms

Covellia glomerata Miq., *Ficus glomerata Roxb., Ficus lucescens* Blume, *Ficus mollis* Miq.,*Ficus goolereana* Roxb., *Ficus semicostata* F.M.Bailey, *Ficus vesca* F. Muell. ex Miq.

Family

Moraceae

Common/English Names

Cluster Fig, Cluster Fig Tree, Cluster Tree, Country Fig, Goolar Fig, Redwood Fig

Vernacular Names

Brazil: Rumbodo (Portuguese);
Burmese: Umbar, Atti, Hpak-Lu, Jagyadumbar, Mayen, Taung Tha Phan, Thapan, Ye Thapan;
Chinese: Ju Guo Rong, Yu Dan Bo Luo;
Eastonian: Kobar-Viigipuu;
India: Dumur, Hpak-Lu, Jagyadumbar, Mayen, Taung Tha Phan, Thapan, Ye Thapan (Bengali), Ambar (Bombay), Gular, Umardo (Gujarati), Doomar, Domoor, Goolar, Gular, Udumbara, Umar, Umbar, Jagya Dumur (Hindu), Alhi, Atti, Atthimara, Rumadi (Kannada), Athi (Kerala), Athiathial, Atthi (Malayalam), Audumbar, Umbar, Umber (Marathi), Heibong (Manipuri), Dimri (Oriya), Gular, Hemadugdhaka, Jantuphala, Sadaphalah, Udumbar, Udumbara, Udumbarah, Yajnanga (Sanskrit), Anai, Athi, Atthi, Atti, Attee Marum, Malaiyin Munivan, Utumparam, Vellaiatthi (Tamil), Arri, Athi, Atti, Bodda, Maydi, Paidi, Udumbaramu (Telugu);
Indonesia: Crattock Loa, Elo (Java);
Laotian: Kok Dua Kieng;
Malaysia: Tangkol (Sarawak, Sabah);
Nepal: Dumri (Bankariya), Gular (Danuwar), Dumri, Gular (Nepali), Loa (Satar), Udumbara (Sanskrit), Gullar, Gullri (Tharu); Pakistan: Gular, Rumbal, Umber;
Singapore: Atteeka;
Sri Lanka: Attikka (Sinhalese);
Taiwan: Ju Guo Rong;
Thailand: Ma-Due-Au-Thum-Porn (Central), Duea Kliang, Duea Nam, Ma Duea, Ma Duea Chumphon;
Vietnamese: Cây Sung, Sung.

Origin/Distribution

The species is indigenous to the Indian subcontinent, southeast Asia, southern China, Papua New Guinea to Northern Australia Northern Territory, Queensland, north of Western Australia. It has been introduced into other tropical regions in south America.

T.K. Lim, *Edible Medicinal And Non-Medicinal Plants: Volume 3, Fruits*,
DOI 10.1007/978-94-007-2534-8_54, © Springer Science+Business Media B.V. 2012

Agroecology

A tropical species commonly found in moist lowland forests and moist open areas beside rivers and streams from 100 to 1,700 m altitude.

Edible Plant Parts and Uses

The figs are edible and used in various preserves and side-dishes. Leaves are eaten as vegetable.

The stem bark was found to have good potential to be used a as ingredient in nutra tea to derive beneficial effect attributed to its high amounts of phenolic compared to tea (Ahmed et al. 2010b). Sensory analysis of the nutra tea indicated no perceptible off-taste or off-aroma and the overall quality was similar to that of control (tea) and was acceptable in terms of all sensory attributes.

Botany

An erect, medium sized tree growing to 25–30 m, sometimes buttressed, with a trunk of 60–90 cm diameter and smooth, greyish-brownish bark. Stipules are ovate-lanceolate, 1.5–2 cm, membranous and pubescent. Leaves are alternate, elliptic-obovate, elliptic, or narrowly elliptic, 10–14 × 3–7 cm, leathery, abaxially pale green, pubescent when young, glabrescent, and scabrous, adaxially dark green and glabrous, base cuneate to obtuse, margin entire, apex acuminate to obtuse and borne on 2–3 cm petioles. Hypanthodia on 8–40 cm long peduncles, borne in large clusters from tubercles on the main trunk and main branches (cauliflorous) (Plate 1), turbinate to globose, to 2.5–4 cm diameter, immature receptacles velvety becoming glabrous with age and reddish orange (Plates 2 and 3); basal bracts three deltoid-ovate. Male, gall, and female flowers occur within the hypanthium. Male flowers are sessile in a row around the ostiole with three to four tepals tightly enclosing two stamens. Female flowers usually have three tepals, sessile with glabrous style, linear stigma

Plate 1 Fruit borne in clusters on main branches

Plate 2 Ripe (*orangey-red*) and unripe fruits of *Ficus racemosa*

Plate 3 *Ficus racemosa* fruits on sale in a local market in Vietnam

and substipitate ovary. Gall flowers have concaved stigma, pedicellate and are dispersed among female flowers. Seeds are lenticular, 1 mm long.

Nutritive/Medicinal Properties

The nutrient composition of *Ficus racemosa* fruit based on analyses made in Australia (Brand Miller et al. 1993) per 100 g edible portion is as follows: energy 44 kJ, moisture 81.9 g, , nitrogen 0.21 g, protein 1.3 g, fat 0.6 g, ash 0.6 g, available carbohydrate 0 g, Ca 72 mg, Cu 0.1 mg, Mg 35 mg, Fe 1.3 mg, P 47 mg, K 508 mg, Na 23 mg, Zn 0.3 mg, thiamin 0.05 mg, riboflavin 0.4 mg, niacin (derived form tryptophan or protein) 0.2 mg, vitamin C 1 mg.

The stem bark was found to have good potential to be used a as a beneficial ingredient in nutra tea (Ahmed et al. 2010b). On a dry matter basis, the total dietary fibre content was 20.5% of which major portion was contributed by insoluble dietary fibre (13.6%) and about 6.9% was soluble dietary fibre. The total starch content was low (7%) while, the bark contained exceptionally amounts of high total sugars (15%). The protein content was 3.9 g%. The lipid content was found to be 2.5% which also contained petroleum ether soluble phytochemical components including phytosterols such as β-sitosterol and triterpinoids such as β-amyrin and lupeol acetate. Vitamin C content was 61.7 mg. Potassium was the most abundant mineral (11,975 ppm) followed by chloride (7,475 ppm) and calcium (1,729 ppm). The bark was also a good source of other minerals and trace elements such as phosphorus (443 ppm) and iron (159.2 ppm), zinc (0.49 ppm), magnesium (196.2 ppm), sodium (255 ppm). Further, the bark powder was used as an ingredient in the preparation of tea and the bark incorporated tea (nutra tea) was found to contain significantly higher amounts of phenolic compounds compared to control tea indicating its usefulness in counteracting free radical induced oxidative damage within the body as phenolics are reported to scavenge free radicals, thus protecting the cells from oxidative damage. Sensory analysis of the nutra tea indicated no perceptible off-taste or off-aroma and the overall quality was similar to that of control (tea) and was acceptable in terms of all sensory attributes.

Scientific investigations have reported on many pharmacological properties of *Ficus racemosa.* The stem bark and leaves have been reported to show anti-diarrhoeal, antidiuretic, antitussive, anti-pyretic, antihyperglycemic, hepatoprotective, hypoglycaemic, antilipidperoxidative, antidureitic, antimicrobial, analgesic, anti-inflammatory, anthelmintic, antioxidant, radioprotective, wound healing, vasoprotective and anti-carcinogenic activities. Many of these activities need to be further investigated by proper clinical trials.

Antioxidant Activity

Root extracts of *F. racemosa* had antioxidant activities. Ethyl acetate extract of *Ficus racemosa* root exhibited antioxidant activity using various assays including 1,1-diphenyl, 2-picryl hydrazyl (DPPH) radical scavenging activity, hydroxyl radical scavenging activity, reducing capacity, and hydrogen peroxide scavenging activity (Sharma and Gupta 2008). The extract at 250 μg/ml concentration showed maximum scavenging activity of DPPH radical up to 73.11% and for hydrogen peroxide up to 65.42% at 1,000 μg/ml. Reducing power of the extract was also dose dependent. The measurement of total phenolic content indicated that 1 mg of the extract contained 8.8 μg equivalent of gallic acid. The antioxidant property of the extract may be due to presence of phenolic content. Another study reported that 70% acetone stem-bark extract of *F. racemosa* contained relatively higher levels of total phenolics than the other extracts (Manian et al. 2008) but methanol extracts had more hydrogen donating ability. Similar line of dose dependent activity has been maintained in all the samples in DPPH and OH scavenging systems. All the extracts exhibited antioxidant activity against the linoleic acid emulsion system (34–38%).

Antidiarrhoeal Activity

Extract of *Ficus racemosa* bark showed significant inhibitory activity against castor oil induced diarrhoea and PGE2 induced enteropooling in rats (Mukherjee et al. 1998). The extract also showed a significant reduction in gastrointestinal motility in charcoal meal tests in rats. The results supported the efficacy of *F. racemosa* bark extract used as anti-diarrhoeal agents. The bark decoction of *Ficus racemosa* caused a reduction in urinary Na+ level and Na+/K+ ratio, and an increase in urinary osmolarity indicating multiple mechanisms of action. The results provided scientific support for its claimed antidiuretic uses by some Sri Lankan traditional practitioners.

The latex of *F. racemosa* was found to exhibit significant inhibitory activity against castor oil-induced diarrhoea and enteropooling in latex treated rats (Bheemachari et al. 2007). It also exhibited significant reduction in gastrointestinal motility following charcoal meal in rats. The results obtained supported the traditional application of the latex as an antidiarrhoeal agent.

Antitussive Activity

The methanol extract of *Ficus racemosa* stem bark displayed antitussive potential against a cough induced model by sulphur dioxide gas in mice (Rao et al. 2003). The extract demonstrated significant antitussive activity at all tested dose levels when compared with the control. The antitussive activity of the extract was comparable to that of codeine phosphate (10 mg), a standard antitussive agent. The extract exhibited maximum inhibition of 56.9% at a dose of 200 mg/kg (p.o.) 90 minutes after administration.

Antipyretic Activity

Methanol extract of stem bark of *Ficus racemosa* at doses of 100, 200 and 300 mg/kg body weight showed significant dose-dependent reduction in normal body temperature and yeast-provoked elevated temperature in albino rats (Rao et al. 2002). The effect extended up to 5 hours after drug administration. The anti-pyretic effect of bark extract was comparable to that of paracetamol a standard anti-pyretic agent. Ethanolic extract of *F. racemosa* root was found top have antipyretic activity (Chomchuen et al. 2010). All doses of the ethanolic extract of *Ficus racemosa* root (EFR) significantly reduced the increased rectal temperature produced by lipopolysaccaride in rats and were found to be as potent as acetylsalicylic acid (ASA). In the Brewer's yeast-induced pyrexia model, all doses of EFR significantly reduced the pyrexia induced by yeast and EFR doses of 200 and 400 mg/kg were equally potent as ASA.

Hypolipidemic Activity

Oral administration of aqueous and ethanolic extracts of *Ficus racemosa* bark at a dose of 400 and 300 mg/kg body weight, respectively showed potent antihyperglycemic and antilipidperoxidative effects in alloxan induced diabetic (Sophia and Manoharan 2007; Vasudevan et al. 2007). The extracts also improved the antioxidants defense system in alloxan induced diabetic rats. The ethanol extract showed better effect than glibenclamide whereas the effect of whereas the aqueous extract was comparable to that of glibenclamide (reference drug). Findings indicated that both ethanol and aqueous extracts of the stem bark had pronounced antidiabetic and antilipidperoxidative effects in experimental diabetes and could therefore be used as an alternative remedy for the treatment of diabetes mellitus and its complications. Other studies also found that the methanol extract of the stem bark lowered glucose effect at doses of 200 and 400 mg/kg p.o.in normal and alloxan induced diabetic rats (Baslas and Agha 1985; Bhaskara et al. 2002; Rao et al. 2002b).

Hypoglycaemic/Antidiabetic Activity

Rahman et al. (1994) found that the petroleum ether extract of the stem bark completely inhibited the enzymes glucose-6-phosphatase and arginase, and stimulated the enzyme glucose-6-phosphate dehydrogenase from rat liver. Extracts from fruit and latex of the plant did not have any significant effect on blood sugar level of these diabetic rats. Of four bioactive components of the extract detected, lupeol acetate, β-sitosterol and β-amyrin were identified. β-sitosterol inhibited glucose-6-phosphate dehydrogenase at a higher rate than the other compounds.

The petroleum ether extract of *Ficus racemosa* leaves at doses of 200 and 400 mg/kg (orally) caused a reduction of blood glucose levels in streptozotocin-induced diabetic rats by 28.9% and 34.6% respectively at the end of 9 days (Mandal et al. 1997a). The results indicated that the petroleum ether extract of the leaves possesses significant hypoglycemic activity in hyperglycemic animals compared with glibenclamide as standard drug. In a recent study, the aqueous 80% ethanol extract and its water soluble fraction of *F. racemosa* fruit did not show any serum glucose lowering effect on non-diabetic and type 2 diabetic rats at the fasting condition, whereas the extract showed significant hypoglycaemic effect on the type 1 diabetic model rats (Jahan et al. 2009). Both the extract and fraction were consistently active in both non-diabetic and types 1 and 2 diabetic model rats when fed simultaneously with glucose load. On the contrary, they were ineffective in lowering blood glucose levels when fed 30 minutes prior to glucose load. The 1-butanol soluble part of the ethanol extract exhibited significant antioxidant activity in DPPH free radical scavenging assay. 3-O-(E)-caffeoyl quinate was isolated for the first time from this plant, which also showed significant antioxidant activity.

In a more recent study, untreated *F. racemosa* bark (FRB) significantly inhibited carbohydrate hydrolysing enzymes such as α-amylase, α-glucosidase, β-glucosidase, and sucrase in a dose-dependent manner (Ahmed and Urooj 2010b). Heat treatment of the sample comparably increased α-amylase, α-glucosidase, and sucrase inhibitory activities, while a marginal decrease in β-glucosidase inhibitory activity was observed; however, no statistical differences were noted. Untreated FRB showed IC_{50} values of 0.94% and 280, 212, and 367 μg/ml for α-amylase, α-glucosidase, β-glucosidase, and sucrase, respectively, while the IC_{50} values for heat treated FRB were 0.58% and 259, 223, and 239 μg/ml, respectively. Further, a significant correlation was observed between α-amylase, α-glucosidase, β-glucosidase, and sucrase inhibitory activities of both untreated and heat treated FRB. The results clearly demonstrated that inhibition of carbohydrate hydrolysing enzymes too be one mechanism through which *F. racemosa* stem bark exerted its hypoglycemic effect in-vivo. It also indicated the potential to explore the utilization of *F. racemosa* stem bark in the development of nutraceuticals and functional foods for the management of diabetes and related symptoms/disorders. In another recent study, Ahmed and Urooj (2010d) found that *Ficus racemosa* bark (FRB) exhibited significantly higher glucose-binding capacity than wheat bran (WB) and acarbose (ACB) and also exhibited significantly higher retardation of glucose diffusion compared to WB and ACB. In case of amylolysis kinetics the liberation of glucose was greatly inhibited by FRB, as reflected by a significantly lower glucose diffusion rate in the system containing FRB compared to the control and acarbose. Furthermore, FRB significantly increased the rate of glucose transport across the yeast cell membrane and also in isolated rat hemi-diaphragm. The findings indicated *F. racemosa* bark to possess strong hypoglycemic effect and its potential utilization as an adjunct in the management of diabetes mellitus.

Both the bark powder and aqueous bark extract of *F. racemosa* bark caused a significant decrease in blood glucose (54% and 66% respectively) and also significantly reduced serum cholesterol and triglyceride levels to the control levels in streptozotocin-induced diabetic rats (Ahmed and Urooj 2009). The aqueous extract was more effective and caused a significant reduction in TBARS (Thiobarbituric Acid Reactive Substances), AST (aspartate aminotransferase), ALT (alanine aminotransferase) levels compared to untreated diabetic rats. However, it did not reach control levels. A significant increase in glutathione concentrations over the control levels was also observed in rats treated with *F. racemosa* bark. It was concluded that *F. racemosa* bark had a significant hypolipidemic and hepatoprotective effect besides being a potent antihyperglycemic agent.

Antiinflammatory Activity

Ethanol extract of *Ficus racemosa* was found to contain a new compound (rel)-4,6-dihydroxy-5-[3-methyl-(E)-propenoic acid-3-yl]-7-β-glucopyranosyl-[2α,3β-dihydrobenzofuran]-(3,2: b)-[4α,5β-dihydroxy-6α-hydroxymethyltetrahydropyran] (racemosic acid) (Li et al. 2004). Racemosic acid showed potent inhibitory activity against COX-1 and 5-LOX in-vitro with IC_{50} values of 90 and 18 μM, respectively. Racemosic acid also demonstrated a strong antioxidant activity to scavenge ABTS free radical cations with an IC_{50} value of 19 μM. The extracts showed no cytotoxicity on the cell lines skin fibroblasts (1BR3), human Caucasian hepatocyte carcinoma (Hep G2) and human Caucasian promyelocytic leukaemia (HL-60).

In another study, ethanol leaf extract at a dose of 400 mg/kg produced maximum anti-inflammatory effects with 30.4%, 32.2%, 33.9% and 32% with carrageenin, serotonin, histamine and dextran-induced rat paw edema models respectively (Mandal et al. 2000a) In chronic model of cotton granuloma assay, it exhibited 41.5% reduction in granuloma weight. The results were comparable to that of Phenylbutazone, a prototype of a non-steroidal antiinflammatory agent.

Ethanol extract but not the petroleum ether extract of *F. racemosa* stem bark at the maximum dose (500 mg/kg) showed comparatively significant activity in tail flick method, significant inhibition of the writhes in writhing test and paw volume in carrageenan and egg albumin induced paw oedema test, and showed significant response in the hot plate method (Harer and Harer 2010). Similarly the hydro-alcoholic extract (100, 300, 500 mg/kg) showed significant response in analgesic and anti-inflammatory methods. The results obtained supported the use of stem bark in inflammatory and painful conditions.

Nephroprotective Activity

Studies demonstrated the chemopreventive effect of *Ficus racemosa* extract on KBrO3-mediated renal oxidative stress and cell promotion response in rats (Khan and Sultana 2005a, b). Treatment of rats orally with *Ficus racemosa* extract (200 mg/kg body weight and 400 mg/kg body weight) resulted in a significant decrease in xanthine oxidase, lipid peroxidation, γ-glutamyl transpeptidase and H_2O_2). There was significant recovery of renal glulathione content and antioxidant enzymes. There was also reversal in the enhancement of renal ornithine decarboxylase activity, DNA synthesis, blood urea nitrogen and serum creatinine. The results suggested that *Ficus racemosa* extract was a potent chemopreventive agent and suppressed potassium bromate-mediated nephrotoxicity in rats.

Radioprotective Activity

Studies reported that pre-treatment with different doses of ethanol extract of *Ficus racemosa* 1 hour prior to 2 Gy γ-radiation resulted in a significant decrease in the percentage of micronucleated binuclear V79 cells (Veerapur et al. 2009). Maximum radioprotection was observed at 20 μg/ml of FRE. The radioprotection was found to be significant when cells were treated with optimum dose of FRE (20 μg/ml) 1 hour prior to 0.5, 1, 2, 3 and 4 Gy γ-irradiation compared to the respective radiation controls. The ethanol extract exhibited significantly higher steady state antioxidant activity than aqueous extract and exhibited concentration dependent DPPH, ABTS, hydroxyl radical and superoxide radical scavenging and inhibition of lipid peroxidation with IC_{50} comparable with tested standard compounds The results indicated that the ethanol extract of *F. racemosa* acted as a potent antioxidant and a probable radioprotector.

Hepatoprotective Activity

The methanol extract of the bark of *F. glomerata* displayed potent in vitro antioxidant activity when compared to the root methanol extract (Channabasavaraj et al. 2008). In the in-vivo studies, the CCl(4) treated control rats showed a significant alteration in the levels of antioxidant and hepatoprotective parameters. The methanol extract of the bark when given orally along with CCl(4) at the doses of 250 and 500 mg/kg body weight showed a significant reversal of these biochemical changes towards the normal when compared to CCl(4)-treated control rats in serum, liver and kidney. The results were comparable to those observed for standard sylimarin. Histological studies also confirmed the same. The results indicated the potent hepatoprotective and antioxidant nature of the bark extract. Stem bark of *F. racemosa* was found to have hepatoprotective activity (Ahmed and Urooj 2010c). CCl4 administration to rats induced a significant decrease in serum total protein, albumin, urea and a significant increase in total bilirubin associated with a marked elevation in the activities of aspartate aminotransferase (AST), alanine aminotransferase (ALT) and alkaline phosphatase (ALP) as compared to control rats. Further, CCl4 intoxication caused significant increase in the TBARS and decrease in glutathione (GSH) levels in serum, liver and kidney. Pretreatment with petroleum ether (FRPE) and methanol (FRME) extract of *Ficus racemosa* stem bark restored total protein and albumin to near normal levels. Both the extracts resulted in significant decreases in the activities of AST, ALT and ALP, compared to CCl4-treated rats. However, a greater degree of reduction was observed in FRME pretreated group (FRPE 43%, 38%, and 33%; FRME 55%, 73%, and 38%). Total bilirubin content decreased from 2.1 mg/dL in CCl4-treated rats to 0.8 and 0.3 mg/dL in FRPE and FRME pretreated rats, respectively. The extracts improved the antioxidant status considerably as reflected by low TBARS and high GSH values. FRME exhibited higher hepatoprotective activity than a standard liver tonic (Liv52), while the protective effect of FRPE was similar to that of Liv52. The protective effect of *F. racemosa* was confirmed by histopathological profiles of the liver. The results indicated that *F. racemosa* possessed potent hepatoprotective effects against carbon tetrachloride-induced hepatic damage in rats.

An ethanol extract of the leaves of *Ficus racemosa* was evaluated for hepatoprotective activity in rats by inducing chronic liver damage by subcutaneous injection of 50% v/v carbon tetrachloride in liquid paraffin (Mandal et al. 1999). The biochemical parameters SGOT (serum glutamic oxaloacetic transaminase), SGPT (serum glutamic pyruvic transaminase), serum bilirubin and alkaline phosphatase were reverted back to normal by the injection. The activity of the leaf extract was also comparable to a standard liver tonic (Neutrosec).

Antiulcerogenic/Gastroprotective Activity

The fruit extract was found to have significant gastroprotective activity which might be due to gastric defence factors and phenolics might be the main constituents responsible for this activity (Rao et al. 2008). The extract showed dose dependent inhibition of ulcer index in pylorus ligation, ethanol and cold restraint stress – induced ulcers. It prevented the oxidative damage of gastric mucosa by blocking lipid peroxidation and by significant decrease in superoxide dismutase, H+K+ATPase and increase in catalase activity. HPTLC analysis showed the presence of 0.57% and 0.36% w/w of gallic acid and ellagic acid in the fruit extract. Kumar et al. (2010) found that the ethanol extract of *Ficus recemosa* fruits administered orally to rats at doses of 100, 200, 400 mg/kg body weight showed a dose dependent reduction in the incidence of ulcers in pyloric ligation, aspirin and ethanol induced gastric ulcer models in rats.

Analgesic Activity

Ethanol extracts of the leaf and bark exhibited analgesic activity at 100,300, 500 mg/kg in a dose dependent manner as evaluated by analgesiometer (Malairajan et al. 2006). The crude extracts of *Scoparia dulcis* and *Ficus racemosa* were found to have significant analgesic activity at the oral dose of 100 and 200 mg/kg b. wt. in mice (Zulfiker et al. 2010). In the hot plate test *S. dulcis* showed increased latency period compared to *F. racemosa* whereas in acetic acid induced writhing test *F. racemosa* showed reduced number of writhes than *S. dulcis* at two dose levels which were significant compared to control. The results obtained support the use of fruits of *F. racemosa* and whole herb of *S. dulcis* in painful conditions acting both centrally and peripherally. *F. racemosa* root is one of the five herbal roots in the 'Ben-Cha-Lo-Ka-Wi-Chian' herbal formula a famous antipyretic drug in Thai traditional medicine (Jongchanapong et al. 2010). Recent animal studies showed that the herbal mixture (400 mg/kg) produced a significant analgesic response in the hot-plate test, while all doses of BLW, except the lowest dose, produced significant analgesic responses in the tail-flick test. In acetic acid-induced writhing, the herbal mixture at doses of 200 and 400 mg/kg significantly decreased the mean writhing response in mice compared to vehicle controls. All the results demonstrated that Ben-Cha-Lo-Ka-Wi-Chian possessed both antipyretic and antinociceptive activities and also likely produced both central and peripheral analgesic responses.

Wound Healing Activity

Ethanol extract of the stem bark exhibited wound healing in excised and incised wound model in rats (Biswas and Mukherjee 2003). In a clinical trial of a proprietary composite ointment comprising *F. racemosa* as one of six herbal components, the ointment was found efficacious in controlling *Candida albicans* infections (Bhatt and Kora 1984). The ointment assisted in quicker epithelialisation, completely healing wound in 8–26 days of treatment.

Vasoprotective Activity

Two anthocyanin pigments were identified in the fruit of *F. racemosa*: peonidin −3-glucoside and pelargonidin −3-glucoside (Sarpate et al. 2009). The fruit anthocyanoside preparation (equivalent to 30% of anthocyanidins) demonstrated significant vasoprotective effect in rabbits, the skin capillary permeability increase due to chloroform, was reduced after i.p. (25–100 mg/kg) administration. The mixture of anthocyanosides in the methanol extract was more active that the proxerutin (25 mg/kg) a known protective microvascular drug.

Central Nervous System Activity

Both the cold aqueous extract (FRC) and the hot aqueous extract (FRH) of *Ficus racemosa* stem bark exhibited a dose dependent inhibition of rat brain acetylcholinesterase (AChE). FRH showed significantly higher cholinesterase inhibitory activity compared to FRC; however, both the extracts did not show 50% inhibition of AChE at the doses tested (200–1,000 μg/ml) (Ahmed and Urooj 2010a). The IC_{50} values of 1,813 and 1,331 μg/ml were found for FRC and FRH.

Antimicrobial Activity

Aqueous extract of stem bark was reported to have inhibitory activity against six species of fungi (*Trichophyton mentagrophytes, T. rubrum, T. soudanense, Candida albicans, Torulopsis glabrata*, and *Candida krusei*) (Vonshak et al. 2003). Leaf extract of *F. racemosa* exhibited significant antibacterial activity against *Bacillus subtilis, Pseudomonas aeruginosa* and *Staphylococcus aureus* (Mandal et al. 2000b). The petroleum ether extract was the most effective against the tested organisms. Various extracts of *F. racemosa* exhibited antibacterial activity (Ahmed and Urooj 2010c). In disk-diffusion assay chloroform, acetone and methanol extracts of *F. racemosa* stem bark showed moderate antibacterial against *Staphylococcus aureus, Bacillus cereus, Bacillus subtilis* compared to the positive control, while petroleum ether extract did not exhibit antibacterial activity against any of the organisms tested. Aqueous extract inhibited only *Bacillus subtilis*, while none of the extracts inhibited *Pseudomonas aeruginosa*. In agar-diffusion assay, both petroleum ether and aqueous extract did not show any inhibitory activity against any of the test organisms, while methanol extract showed moderate activity against *Staphylococcus aureus, Bacillus subtilis*, and *Escherichia coli*. Acetone extract showed moderate inhibition of *Staphylococcus aureus* and *Bacillus cereus*.

Anthelmintic/Antifilarial Activities

Bark extract of *Ficus racemosa* exhibited a dose-dependent inhibition of spontaneous motility (paralysis) of worms and evoked responses to pin-prick (Chandrashekhar et al. 2008). With higher doses (50 mg/ml of aqueous extract), the effects were comparable with that of 3% piperazine citrate. However, there was no final recovery in the case of worms treated with aqueous extract in contrast to piperazine citrate with which the paralysis was reversible and the worms recovered completely within 5 hours. The result showed that the aqueous bark extract possessed wormicidal activity and thus, may be useful as an anthelmintic.

Alcoholic and aqueous extracts of the fruits of *F. racemosa* were reported to exhibit antifilarial activity (Mishra et al. 2005). The concentrations required to inhibit the movement of the whole worm *Setaria cervi* and nerve muscle preparation for alcoholic extract of fruits of *F. racemosa* were 250 and 50 μg/ml, respectively, whereas aqueous extract caused inhibition of the whole worm and nerve muscle preparation at 350 and 150 μg/ml, respectively, suggesting a cuticular barrier. Both alcoholic and aqueous extracts caused death of microfilariae in-vitro. LC_{50} and LC_{90} were 21 and 35 μg/ml, respectively for alcoholic, which were 27 and 42 μg/ml for aqueous extracts.

Larvicidal Activity

The crude hexane, ethyl acetate, petroleum ether, acetone, and methanol extracts of the leaf and bark of *Ficus racemosa* exhibited larvicidal activity against the early fourth-instar larvae of *Culex quinquefasciatus* (Diptera: Culicidae) (Rahuman et al. 2008). All extracts showed moderate larvicidal effects; however, the highest larval mortality was found in bark acetone extract. Gluanol acetate, a tetracyclic triterpene, was isolated and identified as the active principle. Gluanol acetate was quite potent against fourth-instar larvae of *Aedes aegypti* L. (LC_{50} 14.55 and LC_{90} 64.99 ppm), *Anopheles stephensi* Liston (LC_{50} 28.50 and LC_{90} 106.50 ppm) and *C. quinquefasciatus* Say (LC_{50} 41.42 and LC_{90} 192.77 ppm).

Toxicity Studies

Recent toxicity studies (Jaykaran et al. 2009) indicated that aqueous extract of *Ficus racemosa* did not have lethal effect upon 100 times of the therapeutic dose in albino mice. Although not dose dependent, this extract produced significant abnormality in liver and kidney. Lipid changes occurred in the kidneys. Serum glutamic pyruvic transaminase level increased markedly. Fasting blood sugar continued to show inconsistent tendency toward hypoglycemia. Panwar et al. (2010) showed that administration of the aqueous extract of *F. racemosa* bark extract to Sprague dawley rats for 15–21 days (incremental as well as fixed 30 mg/100 g) in subacute toxicity study caused definitive liver damage. Serum creatinine and blood urea were increased significantly. Hepatotoxicity appeared to be reversible.

Traditional Medicinal Uses

All parts of *Ficus racemosa* (leaves, fruits, bark, latex, and sap of the root) are medicinally important in the traditional system of medicine like Ayurveda, Aiddha, Unani and homoeopathy in India (CSIR 1956; Kirtikar and Basu 1975; Nadkarni and Nadkarni 1982; Chopra et al. 1986; Vihari 1995; Joseph and Raj 2010). These parts of the plant are used in dysentry, diarrohoea, diabeties, stomachache, piles and as carminative and astringent and also as antioxidant and anticancer agent. The fruits are used as astringent, stomachic and carminative, to relieve dysentery, diarrhoea and for treatment of diabetes. Fruit paste is a good remedy for visceral obstruction and also useful in regulating diarrhea and constipation. Seed paste is taken in measles and smallpox and diarrhoea. The leaves powdered and mixed with honey is given in bilious infections, dysentery, diarrhoea, and as a mouth wash in spongy gum. Leaf latex and cow milk are mixed and used for boils, blisters and measles. Leaf juice is massaged into hair to check splitting. Infusion of leaves is used in menorrhoea. The milky sap of the plant is used by traditional healers as a popular, anti-inflammatory remedy for mumps and other inflammatory enlargements. The milky sap is also used to cure stomach-ache, diarrhoea, dysentery and piles, cholera and also claimed to be aphrodisiac. The astringent nature of the bark has been employed as a mouth wash in spongy gum and also internally in dysentery, menorrhagia and haemoptysis. The bark is antiseptic, antipyretic and vermicidal, and the decoction of bark is used in the treatment of various skin diseases, ulcers and diabetes. It is also used as a poultice in inflammatory swellings/boils, burns and regarded to be effective in the treatment of piles, dysentry, asthma, diarrhoea gonorrhea, gleet, menorrhagia, leucorrhea, hemoptysis and urinary diseases. Powdered stem bark is used to increase secretion of milk for lactating mother. In Sri Lankan indigenous system of medicine, paste of the stem bark is used in the treatment of skeletal fracture.

F. racemosa root is one of the five herbal roots in the 'Ben-Cha-Lo-Ka-Wi-Chian' herbal formula, an antipyretic drug for both children and adults in Thailand. (Chomchuen et al. 2010) The root has been used for the treatments of dysentery and pyresis in Thailand, and for diabetes in other countries. The Australian aborigines utilise the plant for treating hematuria, menorrhagia mumps, smallpox and inflammatory disorders (Lassak and McCarthy 1997). The roots are used in cases of dysentery and diabetes. Root sap is used as a remedy for heat stroke, chronic wounds and malaria in cattle.

Other Uses

In India, the tree is cultivated as host plant for lac insects, as shade tree for coffee, as rootstock for *Ficus carica,* the common fig and is regarded as a holy religious tree. The tree is planted for shade in gardens and parks elsewhere. The wood is often utilised in making cart frames, ploughs, box, fittings, match boxes and cheap furniture. The latex is used in production of water-resistant paper and as plasticizer for Hevea rubber. The leaves are commonly used as animal fodder.

Comments

The fruits are much sought after by the common Indian macaque.

Selected References

Acharya SK (1996) Folk uses of some medicinal plants of Pawannagar. Dang J Nat Hist Museum 15:25–36

Ahmed F, Urooj A (2009) Glucose-lowering, hepatoprotective and hypolipidemic activities of stem bark of *Ficus racemosa* in streptozotocin-induced diabetic rats. J Young Pharm 1:160–164

Ahmed F, Urooj A (2010a) Anticholinesterase activities of cold and hot aqueous extracts of *F. racemosa* stem bark. Phcog Mag 6(22):142–144

Ahmed F, Urooj A (2010b) Effect of *Ficus racemosa* stem bark on the activities of carbohydrate hydrolyzing enzymes: an in vitro study. Pharm Biol 48:518–523

Ahmed F, Urooj A (2010c) Hepatoprotective effects of *Ficus racemosa* stem bark against carbon tetrachloride-induced hepatic damage in albino rats. Pharm Biol 48(2):210–216

Ahmed F, Urooj A (2010d) In vitro studies on the hypoglycemic potential of *Ficus racemosa* stem bark. J Sci Food Agric 90(3):397–401

Ahmed F, Asha MR, Urooj A, Bhat KK (2010a) *Ficus racemosa* bark: nutrient composition, physicochemical properties and its utilization as nutra tea. Int J Nutr Metab 2(2):33–39

Ahmed F, Sharanappa P, Urooj A (2010b) Antibacterial activities of various sequential extracts of *Ficus racemosa* stem bark. Phcog J 2(8):203–206

Baslas RK, Agha R (1985) Isolation of a hypoglycaemic principle from the bark of *Ficus glomerata* Roxb. Himalayan Chem Pharm Bull 2(1):13–14

Bhaskara RR, Murugesan T, Sinha S, Saha BP, Pal M, Mandal SC (2002) Glucose lowering efficacy of *Ficus racemosa* bark extract in normal and alloxan diabetic rats. Phytother Res 16(6):590–592

Bhatt RM, Kora S (1984) Clinical and experiment study of Panchavalkal and Shatavari on burn wound sepsis-bacterial and fungal. J Nat Integ Med Assoc 26(5):131–133

Bheemachari J, Ashok K, Joshi NH, Suresh DK, Gupta VRM (2007) Antidiarrhoeal evaluation of *Ficus racemosa* Linn. latex. Acta Pharm Sci 49:133–138

Biswas TK, Mukherjee B (2003) Plant medicines of Indian origin for wound healing activity: a review. Int J Lower Extr Wounds 2(1):25–39

Brand Miller J, James KW, Maggiore P (1993) Tables of composition of Australian aboriginal foods. Aboriginal Studies Press, Canberra

Chandrashekhar CH, Latha KP, Vagdevi HM, Vaidya VP (2008) Anthelmintic activity of the crude extracts of *Ficus racemosa*. Int J Green Pharm 2:100–103

Channabasavaraj KP, Badami S, Bhojraj S (2008) Hepatoprotective and antioxidant activity of methanol extract of *Ficus glomerata*. Nat Med (Tokyo) 62(3):379–383

Chomchuen S, Singharachai C, Ruangrungsi N, Towiwat P (2010) Antipyretic effect of the ethanolic extract of *Ficus racemosa* root in rats. J Health Res 24(1):23–28

Chopra RN, Nayar SL, Chopra IC (1986) Glossary of Indian medicinal plants (Including the supplement). Council Scientific Industrial Research, New Delhi, 330 pp

Council of Scientific and Industrial Research (CSIR) (1956) The wealth of India. A dictionary of Indian raw materials and industrial products. (Raw Materials 4). Publications and Information Directorate, New Delhi

Ekanayake DT (1980) Indigenous system of medicine in Sri Lanka for the treatment of skeletal fracture. Sri Lanka For 14:145–152

Ghimire SK, Shrestha AK, Shrestha KK, Jha PK (2000) Plant resource use and human impact around RBNP. Nepal J Nat Hist Mus 19:3–26

Harer LS, Harer SP (2010) Evaluation of analgesic and anti-inflammatory activity of *Ficus racemosa* Linn. stem bark extract in rats and mice. Phcog J 2(6):65–70

Jahan IA, Nahar N, Mosihuzzaman M, Rokeya B, Ali L, Azad Khan AK, Makhmur T, Iqbal Choudhary M (2009) Hypoglycaemic and antioxidant activities of *Ficus racemosa* Linn. fruits. Nat Prod Res 23(4):399–408

Jansen PCM, Jukema J, Oyen LPA, van Lingen TG (1992) *Ficus racemosa* L. In: Verheij EWM, Coronel RE (eds) Plant resources of South-East Asia No. 2: edible fruits and nuts. Prosea Foundation, Bogor, p 335

Jaykaran, Bhardwaj P, Kantharia N, Yadav P, Panwar A (2009) Acute toxicity study of an aqueous extract of *Ficus racemosa* Linn. bark in albino mice. Internet J Toxicol 6:1

Jongchanapong A, Singharachai C, Palanuvej C, Ruangrungsi N, Towiwat P (2010) Antipyretic and antinociceptive effects of Ben-Cha-Lo-Ka-Wi-Chian remedy. J Health Res 24(1):15–22

Joseph B, Raj SJ (2010) Phytopharmacological properties of *Ficus racemosa* Linn – an overview. Int J Pharm Res 3(2):134–138

Khan N, Sultana S (2005a) Chemomodulatory effect of *Ficus racemosa* extract against chemically induced renal carcinogenesis and oxidative damage response in wistar rats. Life Sci 77:1194–1210

Khan N, Sultana S (2005b) Modulatory effect of *Ficus racemosa*: diminution of potassium bromate-induced renal oxidative injury and cell proliferation response. Basic Clin Pharm Toxicol 97(5):282–288

Kirtikar KR, Basu BD (1975) Indian medicinal plants. 4 vols, 2nd edn. Jayyed Press, New Delhi

Kumar A, Sharma US, Rao CV (2010) Experimental evaluation of *Ficus recemosa* Linn. Fruits extract on gastric ulceration. Int J Pharm Sci Rev Res 4(3):89–92

Kunwar RM, Bussmann RW (2006) *Ficus* (fig) species in Nepal: a review of diversity and indigenous uses. Lyonia 11(1):85–97

Lassak EV, McCarthy T (1997) Australian medicinal plants. Reed Books Australia, Victoria

Li RW, Leach DN, Myers SP, Lin GD, Leach GJ, Waterman PG (2004) A new anti-inflammatory glucoside from *Ficus racemosa* L. Planta Med 70:421–426

Malairajan P, Geetha Gopalakrishnan G, Narasimhan S, Jessi Kala Veni K (2006) Analgesic activity of some Indian medicinal plants. J Ethnopharmacol 106(3):425–428

Mandal SC, Mukherjee PK, Das J, Pal M, Saha BP (1997a) Hypoglycaemic activity of *Ficus racemosa* L. (Moraceae) leaves on strepozotocin-induced diabetic rats. Nat Prod Sci 3:38–41

Mandal SC, Mukherjee PK, Saha K, Pal M, Saha BP (1997b) Antidiarrhoeal evaluation of *Ficus racemosa* Linn. leaf extract. Nat Prod Sci 3(1):100–103

Mandal SC, Maity TK, Das J, Pal M, Saha BP (1999) Hepatoprotective activity of *Ficus racemosa* leaf extract on liver damage caused by carbon tetrachloride in rats. Phytother Res 13(5):430–432

Mandal SC, Maity TK, Das J, Saba BP, Pal M (2000a) Anti-inflammatory evaluation of *Ficus racemosa* Linn. leaf extract. J Ethnopharmacol 72(1–2):87–92

Mandal SA, Saha BP, Pal M (2000b) Studies on bacterial activity of *Ficus racemosa* leaf extract. Phytother Res 14(4):278–280

Manian R, Nagarajan Anusuya N, Siddhurajub P, Manian S (2008) The antioxidant activity and free radical scavenging potential of two different solvent extracts of *Camellia sinensis* (L.) O. Kuntz, *Ficus bengalensis* L. and *Ficus racemosa* L. Food Chem 107(3):1000–1007

Mishra V, Khan NU, Singhal KC (2005) Potential antifilarial activity of fruit extracts of *Ficus racemosa* Linn. against *Setaria cervi in vitro*. Indian J Exp Biol 43:346–350

Mukherjee PK, Saha K, Murugesan T, Mandal SC, Pal M, Saha BP (1998) Screening of anti-diarrhoeal profile of some plant extracts of a specific region of West Bengal, India. J Ethnopharmacol 60:85–89

Nadkarni KM, Nadkarni AK (1982) Indian materia medica with ayurvedic, unani-tibbi, siddha, allopathic, homeopathic, naturopathic & home remedies. 2 vol, 2nd edn. Sangam Books, Bombay

Panwar A, Jaykaran A, Chavda N, Saurabh M, Preeti Yadav P (2010) Subacute toxicity study of an aqueous extract of *Ficus racemosa* Linn. bark in rats. J Pharm Res 3(4):814–817

Rahman NN, Khan M, Hasan R (1994) Bioactive components from *Ficus glomerata*. Pure Appl Chem 66:2287–2290

Rahuman AA, Venkatesan P, Geetha K, Gopalakrishnan G, Bagavan A, Kamaraj C (2008) Mosquito larvicidal activity of gluanol acetate, a tetracyclic triterpenes derived from *Ficus racemosa* Linn. Parasitol Res 103(2):333–339

Rao RB, Anupama K, Swaroop KR, Murugesan T, Pal M, Mandal SC (2002a) Evaluation of anti-pyretic potential of *Ficus racemosa* bark. Phytomedicine 9:731–733

Rao RB, Murugesan T, Sinha S, Saha BP, Pal M, Mandal SC (2002b) Glucose lowering efficacy of *Ficus racemosa* bark extract in normal and alloxan diabetic rats. Phytother Res 16:590–592

Rao RB, Murugesan T, Pal M, Saha BP, Mandal SC (2003) Antitussive potential of methanol extract of stem bark of *Ficus racemosa* Linn. Phytother Res 17:1117–1118

Rao CV, Verma AR, Vijayakumar M, Rastogi S (2008) Gastroprotective effect of standardized extract of *Ficus glomerata* fruit on experimental gastric ulcers in rats. J Ethnopharmacol 115(2):323–326

Ratnasooriya WD, Jayakody JR, Nadarajah T (2003) Antidiuretic activity of aqueous bark extract of Sri Lankan *Ficus racemosa* in rats. Acta Biol Hung 54:357–363

Sarpate RV, Tupkari SV, Deore TK, Chandak BG, Nalle SC (2009) Isolation, characterization and microvascular activity of anthocyanins from *Ficus racemosa* fruits. Phcog Mag 19(5):78–82

Sharma SK, Gupta VK (2008) In vitro antioxidant studies of *Ficus racemosa* Linn. root. Phcog Mag 4(13(Suppl)):70–74

Siwakoti M, Siwakoti S (2000) Ethnobotanical uses of plants among the Satar tribes of Nepal. In: Maheswori JK (ed) Ethnobotany and medicinal plants of Indian subcontinent. Scientific publishers, Jodhpur, pp 79–108

Sophia D, Manoharan S (2007) Hypolipidemic activities of *Ficus racemosa* Linn. bark in alloxan induced diabetic rats. Afr J Trad Compl Altern Med 4(3):279–288

Vasudevan K, Sophia D, Balakrishnan S, Manoharan S (2007) Antihyperglycemic and antilipidperoxidative effects of *Ficus racemosa* (Linn.) bark extracts in alloxan induced diabetic rats. J Med Sci 7(3):330–338

Veerapur VP, Prabhakar KR, Parihar VP, Kandadi MR, Ramakrishana SB, Mishra B, Satish Rao BS, Srinivasan KK, Priyadarsini KI, Unnikrishnan MK (2009) *Ficus racemosa* stem bark extract: a potent antioxidant and a probable natural radioprotector. Evid Based Compl Alt 6(3):317–324

Vihari V (1995) Ethnobotany of cosmetics of Indo-Nepal border. Ethnobotany 7(1/2):89–94

Vonshak A, Barazani O, Sathiyamoorthy P, Shalev R, Vardy D, Golan-Goldhirsh A (2003) Screening South Indian medicinal plants for antifungal activity against cutaneous pathogens. Phytother Res 17(9):123–125

Yadav RKP (1999) Medicinal plants and traditional medicinal practice in the eastern part of Parsa district, Nepal. In: Proceeding of III national conference of science and technology, Royal Nepal Academy of Science and Technology, Kathmandu, pp 1421–1426

Zulfiker AHM, Rahman MM, Hossain MK, Hamid K, Mazumder MEH, Rana MS (2010) In vivo analgesic activity of ethanolic extracts of two medicinal plants – *Scoparia dulcis* L. and *Ficus racemosa* Linn. Biol Med 2(2):42–48

Ficus rubiginosa

Scientific Name

***Ficus rubiginosa* Desf. ex Vent.**

Synonyms

Ficus australis Willd. nom. illeg., *Ficus baileyana* Domin, *Ficus leichhardtii* (Miq.) Miq., *Ficus macrophylla* var. *pubescens* F.M.Bailey, *Ficus muelleri* (Miq.) Miq., *Ficus obliqua* var. *petiolaris* (Benth.) Corner, *Ficus platypoda* var. *angustata* (Miq.) Corner, *Ficus platypoda* var. *leichhardtii* (Miq.) R.J.F.Hend., *Ficus platypoda* var. *mollis* Benth., *Ficus platypoda* var. *petiolaris* Benth., *Ficus platypoda* var. *subacuminata* Benth., *Ficus rubiginosa* Desf. ex Vent. var. *rubiginosa*, *Ficus rubiginosa* var. *lucida* Maiden, *Ficus rubiginosa* var. *variegata* Guilf. nom. inval., *Ficus shirleyana* Domin, *Ficus* sp. A , *Mastosuke rubiginosa* (Vent.) Raf., *Urostigma rubiginosum* (Vent.) Gasp. *Urostigma muelleri* Miq.

Family

Moraceae

Common/English Names

Figwood, Illawarri Fig, Illawarra Fig, Little-Leaf Fig, Port Jackson Fig, Rusty Fig, Small-Leaved Rock Fig

Vernacular Name

German: Rost-Feige.

Origin/Distribution

Ficus rubiginosa is a native of eastern Australia. It occurs from north Queensland southwards along the eastern coastline of Australia to the vicinity of Bega on the South Coast of New South Wales.

Agroecology

F. rubiginosa is extremely hardy and robust. It is tolerant of many climates and extremes of soil fertility. It is found naturally on the edges of rainforest and gullies, rocky hillsides, cliffs, rock-faces, and brickwork on buildings. It is also found in harsh, arid, open sites in urban environments. As a hemi-epiphyte or lithophyte, *F. rubiginosa* can grow in low nutrient, harsh, arid sites. On fertile soils it can grow into a large, free standing tree. It can thrive on seasonally water-logged soils and once established, it can withstand periods of drought and low temperature of −1°C. for brief periods. It thrives in full sun or partial shade.

Edible Plant Parts and Uses

Ripe fruit is edible as a local bush food but is not very palatable and not tasty.

T.K. Lim, *Edible Medicinal And Non-Medicinal Plants: Volume 3, Fruits*,
DOI 10.1007/978-94-007-2534-8_55, © Springer Science+Business Media B.V. 2012

Botany

A broad spreading (10–12 m), evergreen tree reaching 10–15 m high with dark grey bark and scurfy-pubescent young twigs. Leaves are alternate, on 1–4 cm long petioles, equilateral, oval to obovate, 7.5–17.2 cm long and up to 6.2 cm broad, apex bluntly obtuse and the base broad and rounded, margin entire, coriaceous, green above, rusty-coloured below, surface prominently rubiginous above and below when young becoming glabrous with age (Plate 1). The stipules are up to 12.6 cm long, lanceolate, scarious on margins, glabrous within, but pubescent on the outer side. Syconia 7.4–17.3 mm long, 7.6–17.3 mm in diameter, glabrous, or minutely puberulous. Female florets embedded in wall of receptacle, sessile or pedicellate, tepals 3–5, stigma simple; male florets pedicellate, tepals 3–5, interspersed with the female and gall florets; gall florets sessile, pedicellate, tepals 3–5. Interfloral bracts are present. The figs are 1.5–2.5 cm in diameter, globose to oblate, green or rusty to yellowish, ripening to red or purplish-red, tipped with a small nipple and on a 2–5 mm stalk (Plate 1).

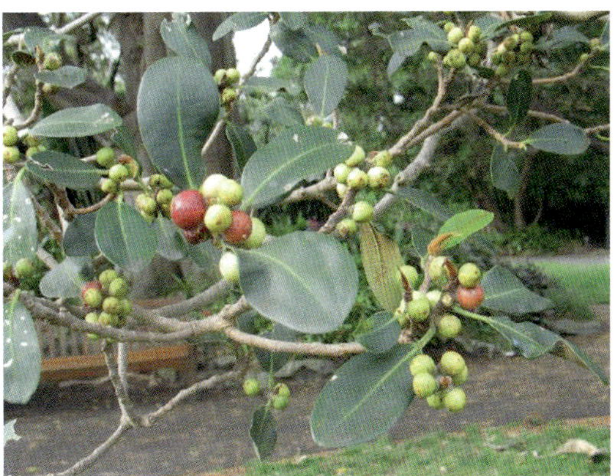

Plate 1 Ripe and immature fruits and leaves

Nutritive/Medicinal Properties

The fruit is important to desert Aboriginal tribes that anyone harming the plant could be killed (NDLG, undated).

The milky sap of figs was used as natural latex to cover wounds.

Other Uses

The tree is planted as avenue trees, landscape and shade trees in parks, golf courses and public gardens and is planted as a shade tree for livestock by pastoralists. It is popular for bonsai work despite its large leaves. *Ficus rubiginosa* is also suited for use as an indoor plant in low, medium or brightly-lit indoor spaces.

Comments

F. rubiginosa produces a powerful root system that can seriously damage urban infrastructure in the absence of adequate weed control measures.

Selected References

Dixon DJ (2003) A taxonomic revision of the Australian *Ficus* species in the section *Malvanthera* (*Ficus* subg. *Urostigma*: Moraceae). Telopea 10:141

Dixon DJ, Jackes BR, Bielig LM (2001) Figuring out the figs: the *Ficus obliqua-Ficus rubiginosa* complex (Moraceae: *Urostigma* sect. *Malvanthera*). Austral Syst Bot 14(1):133–154

Fairley A, Moore P (2000) Native plants of the Sydney district: an identification guide, 2nd edn. Kangaroo Press, Kenthurst, pp 62

Gilman EF, Watson DG (1993) *Ficus rubiginosa* rusty fig. fact sheet ST-257. Environmental Horticulture Department, Florida Cooperative Extension Service, Institute of Food and Agricultural Sciences, University of Florida, St. Petersburg, Florida. http://hort.ifas.ufl.edu/trees/FICRUBA.pdf

Halliday I (1989) A field guide to Australian trees. Hamlyn Australia, Melbourne, p 200

McCrone M (2006) Growing Port Jackson fig as bonsai in a warm temperate climate. ASGAP Aust Plant Bonsai Stud Group Newsl 11:3–4

Noosa & District Landcare Group (NDLG) (Undated) Native Species Recommended for Planting as Bush Tucker. Noosa & District Landcare group, Pomona. http://www.noosalandcare.org/support_documentation/BUSHTUCKER.pdf

Pacific Island Ecosystems at Risk (PIER) (1999). *Ficus rubiginosa* Desf. ex Vent., Moraceae. http://www.hear.org/pier/species/ficus_rubiginosa.htm

Ratcliffe D, Ratcliffe P (1987) Australian native plants for indoors. Little Hills Press, Crows Nest, p 141

Morus alba

Scientific Name

Morus alba **L.**

Synonyms

Morus alba f. *tatarica* Ser., *Morus alba* f. *macrophylla* (Loudon) C.K. Schneid., *Morus alba* f. *skeletoniana* (C.K. Schneid.) Rehder, *Morus alba* var. *constantinopolitana* Loudon, *Morus alba* var. *multicaulis* (Perr.) Loudon, *Morus alba* var. *purpurea* Bureau, *Morus alba* var. *tatarica* (L.) Ser., *Morus atropurpurea* Roxb., *Morus byzantina* Siebold, *Morus constantinopolitana* Poir., *Morus indica* L., *Morus intermedia* Perr., *Morus macrophylla* Moretti, *Morus morettiana* Jacq. ex Burr., *Morus multicaulis* Perr., *Morus nervosa* Del. ex Spach, *Morus pumila* Balb., *Morus rubra* Lour., *Morus tatarica* L.

Family

Moraceae

Common/English Names

Black-fruited mulberry, Chinese white mulberry, Mulberry, Mulberry tree, Mulberry bush, Russian mulberry, Silkworm mulberry, Silkworm tree, White-fruited mulberry, White mulberry

Vernacular Names

Arabic: El Ttuut, Tuth, Tuthtut;
Brazil: Amora, Amoreira-Branca (Portuguese);
Bulgarian: Chernitsia Biala;
Burmese: Posa;
Chinese: Bai Sang, Hong Sang, Sang, Sang Bai Pi, Sang Ye, Sang Shu, Sang Zhi;
Czech: Moruše Bílá, Morušovník Bílý;
Danish: Hvid Morbær, Morbær;
Dutch: Moerbei;
Eastonian: Valge Mooruspuu;
Dutch: Moerbei (Fruit), Moerbezie (Plant), Witte Moerbei;
Finnish: Valkomulperi;
French: Amomie, Mûre De Murier (Fruit), Mûrier (Plant), Mûrier Blanc;
German: Maulbeere (Fruit), Maulbeerbaum (Plant), Weiße Maulbeere;
Greek: Aspri Moria, Aspromuria;
Hungarian: Eperfa, Fehér Eperfa, Selyemeperfa, Szederfa;
India: Tut, Tunt (Bengali), Chimmu,Chinni, Chum, Chun, Karun, Sahtoot, Shahtoot, Shahtut, Shehtoot, Swa, Toot, Toota, Tooth,Tul, Tulklu, Tunt, Tut, Tutri, Tut (Cultivated) (Hindu), Shetun (Gujarati), Bilee Hippenerale, Bilee Uppu Nerale, Bili Uppu Naerale, Hippal Verali, Hippali Naerala, Tuti, Hippunerale, Hippu Naerale, Kambali Gida, Kambali Hannu, Korigida, Uppunute (Kannada), Malvari, Pattunulppulucceti (Malayalam), Kabrangchak Angouba (Manipuri), Tut, Tuti (Marathi), Thingtheihmu (Mizoram),

T.K. Lim, *Edible Medicinal And Non-Medicinal Plants: Volume 3, Fruits*,
DOI 10.1007/978-94-007-2534-8_56, © Springer Science+Business Media B.V. 2012

Thingtheihmu, Tootho (Oriya), Tooda, Toola, Tudah, Tula, Tutam,Tuta (Sanskrit), Kambali Poochi Chedi, Kambilipuch, Kamblichedi, Kampalicceti, Kampilippuccicceti, Mussuketi, Musukette, Musukotta, Pattuppuccimaram, Pattuppucci, Pattuppuchi (Tamil), Kambali Chettu, Malabary Aaku, Pippalipanducettu, Reshmicettu (Telugu);
Indonesia: Besaran, Lampung (Java), Bebesaran (Malay), Bebesaran (Sundanese);
Italian: Gelso (Plant), Gelso Bianco, Gelso Comune, Mora Di Gelso (Fruit), Moral Blanco, Morera Blanco, Morus, Moro (Fruit), Moro Bianco, Moro Da Carta;
Japanese: Guwa, Kuwa, Kara Guwa, Ma Guwa, Kara Yama Guwa;
Korean: Ppong, Pong Na µ;
Malaysia: Brbesaram, Besaram, Tut;
Nepali: Kimbu;
Norwegian: Hvitmorbær;
Pakistan: Tut, Tut Kishmishmi, Tutri;
Persian: Tuthtut;
Philippines: Mora, Moraya (Ibanag), Amingit (Igorot), Amoras (IloKo), Tanud, Tanyad (Igorot), Morera (Spanish), Moras (Tagalog);
Polish: Morwa Biała;
Portuguese: Amoreira-Branca, Amoreira (Plant), Amoreira Branca;
Russian: Shelkovitsa Belaia, Tut Belyi;
Slovašcina: Bela Murva, Murva Bela;
Slovencina: Moruša Biela;
Spanish: Mora, Mora Blanco, Mora De Árbol, Moral, Morera, Morera Blanca;
Swahili: Mforsadi, Mfurusadi;
Swedish: Vitt Mullbär;
Thai: Mon;
Vietnamese: Dâu, Dâu Tam, Dâu Cang (H`Mông), Tang, Mạy Mọn (Tày), Nan Phong (Dao).

Origin/Distribution

White mulberry is native to central, northern China and Korea. It is widely cultivated and naturalized in many warm-temperate and subtropical regions of Asia (India, Afghanistan), Europe and America.

Agroecology

White Mulberry thrives in a warm-temperate and subtropical climate with annual temperatures of 6–28°C. It is more cold-hardy than black mulberry and can survive sub-zero temperatures down to –24°C to –29°C. It is grown from sea level to 3,000 m elevation as occurs in northern India. It tolerates annual precipitation of 450–4,000 mm. Although it is somewhat drought tolerant, it needs to be watered in dry seasons for good fruit production. It grows well in full sun and is intolerant of strong winds.

White mulberry grows on wide variety of soils, but prefers a deep, well-drained soil, preferably a deep loam. Shallow soils such as those frequently found on chalk or gravel are to be avoided. Mulberries generally thrive with minimal fertilization.

Edible Plant Parts and Uses

The sweet, white, red or purplish-black fruits are eaten raw, frozen, dried or preserved. Mulberry fruits are used in the food industry to make different syrups, beverages, tonic wine, amaretto or vermouth wine, vinegar and different sweet products marmalade, chocolate, frosting, jelly and fondant. Mulberry fruit juice it is also used as natural alcoholic extract additive for food and pharmaceutical industries. Mulberry fruits after alcoholic fermentation and further distillation can be processed into an excellent hard alcoholic drink. The oil from the seeds is also edible. In Indonesia, young leafy shoots are eaten as a vegetable, particularly by nursing mothers. The leaves are eaten as *lalab* with rice or prepared as *sayor* (cooked with red chillies, salt, onions, dried shrimps, pepper and sugar) or with sweet potatoes.

Studies demonstrated that mulberry leaf powder (MLP) can be mixed with wheat flour (WF) in the optimum ratio of 1:4 to prepare *paratha*, the most common food item of breakfast and dinner in the Indian diet (Srivastava et al. 2003). The protein quality of the MLP-WF mix was estimated by measuring the Protein Efficiency Ratio, and was found to be 1.82 against a casein diet for which a value of 2.44 was observed. No adverse in-vivo effect on the growth of internal organs of rats (heart, liver, kidney and testes) was found. The storage stability of the mix was estimated for a period of 2 months in polyethylene bags at room temperature. A non-significant difference was observed between *paratha* prepared from fresh and stored mix. This indicated that mix can be stored for a period of 2 months at room temperature without loss of quality.

Plate 1 Morus alba with pink fruits and broadly ovate, serrated leaves

Botany

A monoecious, deciduous shrubs or trees, 3–10 m tall, sometime to 15 m high with a dense, compact, leafy crown and grey-brown, rough, shallowly vertical fissured bark. Branchlets orange-brown or dark green with reddish cast and pubescent. Stipules membranous, lanceolate, 5–9 mm and pubescent and petioles 2–5 cm and pubescent. Leaf lamina ovate to broadly ovate, to irregularly lobed, 5–15 cm long by 4–12 cm wide, lower surface sparsely pubescent along midvein or in axils of midvein and primary lateral veins, upper surface bright green and glabrous, base rounded to cordate, margin coarsely serrate to crenate, apex acute, shortly acuminate, or obtuse (Plates 1,2,4,5,6). Male catkins pendulous, 2–3.5 cm, densely white hairy. Male flowers: sepals pale green free, broadly ovate, cucullate, obtuse, glabrous to hairy; staminal filaments equal to sepals, with ovate, exserted globose to reniform anthers. Female catkins ovoid to ellipsoid, 1–2 cm, pubescent; peduncle 5–10 mm, pubescent. Female flowers sessile, sepals suborbicular, glabrous or ciliate on margins; ovary sessile, style absent, stigmas with mastoidlike protuberance with divergent, papillose branches. Sorosis ovoid, ellipsoid, or cylindric (10-) 15–25 mm long, 5–8 mm across, white

Plate 2 Red fruited Morus alba and leaves

Plate 3 Tree label of red fruited Morus alba

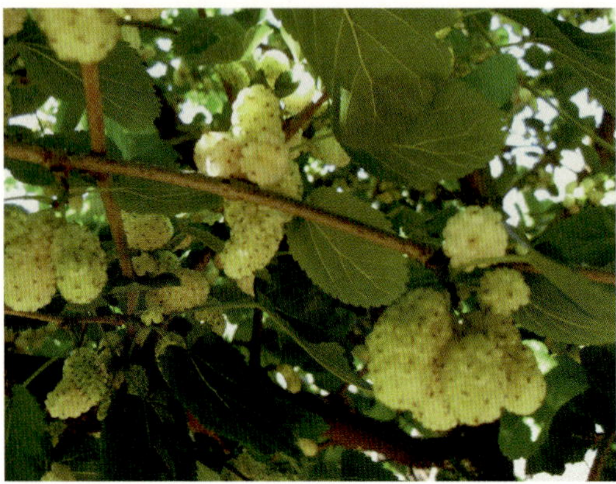

Plate 4 Whited fruited Morus alba cultivar

Plate 5 Close-up of white fruited Morus alba and leaves

Plate 6 Silkworms feeding on white mulberry leaves

(Plates 4–5) to pinkish-purple (Plates 1–2) or black, sweet, edible.

Nutritive/Medicinal Properties

Of 3 *Morus* species, *Morus alba* had the highest total fat content (1.10%), followed by *M. nigra* (0.95%) and *M. rubra* (0.85%), respectively (Ercisli and Orhan 2007). The major fatty acids in mulberry fruits were linoleic acid (54.2%), palmitic acid (19.8%) and oleic acid (8.41%), respectively. The total soluble solids content of *M. alba* was 20.4%, acidity 0.25%, pH 5.60, ascorbic acid 22.4 mg/100 g. Mineral compositions of the mulberry species were 0.83% nitrogen, 235 mg/100 g phosphorus, 1,141 mg/100 g potassium, 139 mg/100 g calcium, 109 mg/100 g magnesium, 60 mg/100 g sodium, 4.3 mg/100 g iron, 0.4 mg/100 g copper, 4.0 mg/100 g manganese and 3.1 mg/100 g zinc.

The ripe fruits of *Morus alba* (white mulberry), *Morus nigra* (black mulberry), *Morus laevigata* (large white fruit), and *Morus laevigata* (large black fruit), were found to have total dry weight 17.60 (21.97 mg/100 g, pH 3.2–4.78, and titratable acidity 0.84–2.0% citric acid) (Imran et al. 2010). Low riboflavin (vitamin B_2) and niacin (vitamin B_3) contents were recorded in all the fruits, while ascorbic acid (vitamin C) was in the range from 15.20 to 17.03 mg/100 g fresh weight (FW). The mulberry fruits were rich with regard to the total phenol and alkaloid contents, with values of 880–1,650 mg/100 g FW and 390–660 mg/100 g FW, respectively. Sufficient quantities of essential macro-(K, Ca, Mg, and Na) and micro-(Fe, Zn, and Ni) elements were found in all the fruits. K was the predominant element with concentration ranging from 1,270 to 1,731 mg/100 g, while Ca, Na, and Mg contents were 440–576, 260–280, and 240–360 mg/100 g, respectively. The decreasing order of micro-elements was Fe>Zn>Ni.

Other Phytochemicals

Plant unspecified part:- prenylated arylbenzofuran derivative 2-[3,5-Di-O-β-d-glucosyl-4-(3-

methylbut-2-enyl)phenyl]benzofuran-6-ol (Tian et al. 2010); sanggenol P, a new isoprenylated flavonoid, together with nine known ones, cyclomorusin, morusin, mulberrofuran G, sanggenol A, sanggenol L, sanggenol N, cyclomulberrin, cyclocommunol and ursolic acid (Geng et al. 2010); Diels-Alder type adducts, chalcomoracin and kuwanon J from *M. alba* callus tissues (Nomura et al. 2009).

Root: Kuwanon G, kuwanon H, kuwanon M, sanggenon C, sanggenon D, mulberrofuran G, mulberrofuran J, mulberrofuran Q, mulberroside A (Kimura et al. 1986; Kim et al. 2010b); morusin, kuwanon H, morusin-4′-glucoside, morusin-2′-glucoside, morusin-4′-glucoside, kuwanon H-6′-acetate, kuwanon H-7-acetate, triacetate mulberrofuran D (Luo et al. 1994); N-containing sugars: 1-deoxynojirimycin; N-methyl-1-deoxynojirimycin, fagomine, 3-epi-fagomine, 1,4-dideoxy-1,4-imino-D-arabinitol, 1,4-dideoxy-1,4-imino-D-ribitol, calystegin B2 (1 α,2 β,3 α,4 β-tetrahydroxy-nor-tropane), calystegin C1 (1 α,2 β,3 α,4 β,6 α-pentahydroxy-nor-tropane), 1,4-dideoxy-1,4-imino-(2-O-β-D-glucopyranosyl)-D-arabinitol, and nine glycosides of 1-deoxynojirimycin: 2-O-α-D-galactopyranosyl-1-deoxynojirimycin, 6-O-α-D-galactopyranosyl-1-deoxynojirimycin, 2-O-α-D-glucopyranosyl-1-deoxynojirimycin, 3-O-α-D-glucopyranosyl-1-deoxynojirimycin, 4-O-α-D-glucopyranosyl-1-deoxynojirimycin, 2-O-β-D-glucopyranosyl-1-deoxynojirimycin, 3-O-β-D-glucopyranosyl-1-deoxynojirimycin, 4-O-β-D-glucopyranosyl-1-deoxynojirimycin, 6-O-β-D-glucopyranosyl-1-deoxynojirimycin, a trihydroxy-nor-tropane alkaloid, calystegin A3; tetrahydroxy-nor-tropane alkaloids calystegins B1 and B2; Calystegin C1, a pentahydroxy-nor-tropane alkaloid (Asano et al. 1994); oxyresveratrol, oxyresveratrol 2,3′-O-β-D-diglucuronide;oxyresveratrol 2-O-β-D-glucuronide-3′-O-sulfate mulberroside A; cis-mulberroside A (Qiu et al. 1999); polyhydroxylated alkaloids, (2R,3R,4R)-2-hydroxymethyl-3,4-dihydroxypyrrolidine-N-propionamide (Asano et al. 2001); mulberroside A, oxyresveratrol (Chung et al. 2003); flavonoids: moralbanone, kuwanon S, mulberroside C, cyclomorusin, eudraflavone B hydroperoxide, oxydihydromorusin, leachianone G, α-acetyl-amyrin (Du et al. 2003); sanggenon C, sanggenon G, mulberrofuran C, kuwanon L, moracin O, moracin P (Cui et al. 2006); mulberroside A, 5,7,2V-trihydroxyflavanone-4V-O-h-dglucoside, albanol A, albanol B (El-Beshbishy et al. 2006); moracin M, steppogenin-4′-O-β-D-glucosiade, mullberroside A (Zhang et al. 2009a); flavanone glycoside, 5,2′-dihydroxyflavanone-7,4′-di-O-β-D-glucoside (steppogenin-7,4′-di-O-β-D-glucosiade) (Zhang et al. 2009b); albanol A (Kikuchi et al. 2010); stilbene glycosides, mulberroside A, cis-mulberroside A, resveratrol-4,3′-di-O-B-D-glucopyranoside, oxyresveratrol-2-O-b-D-glucopyranoside and oxyresveratrol-3′-O-b-D-glucopyranoside (Piao et al. 2011).

Twig: Mulberrin, cyclomulberrin, morusin, cyclomorusin, 2′, 4′,4, 2″-tetrahydroxy-3′-{3″-methylbut-3″-enyl-}-chalcone, mulberrofran G, scopoletin, moruchalcone A, kaempferol, ursolic acid, β-daucosterol (Xu et al. 2008); maclurin, rutin, isoquercitrin, resveratrol, and morin (Chang et al. 2011).

Leaf: Fagomine (Taniguchi et al. 1998); rutin, quercetin (Jia et al. 1999); quercetin, isoquercitrin (Doi et al. 2000); triterpenes (lupeol), sterols (β-Sitosterol), bioflavonoids (rutin, moracetin, quercetin-3-triglucoside, isoquercitrin), coumarins, volatile oil, alkaloids, amino acids, organic acids (Doi et al. 2001); prenylflavanes (prenylflavane 1, prenylflavane 2), glycoside of prenylflavane 1, isoquercitrin, astragalin, scopoline, skimming, roseoside II, benzyl D-glucopyranoside (Doi et al. 2001); mulberroside F (moracin M-6, 39-di-O-β-D-glucopyranoside) (Lee et al. 2002); 2-arylbenzofuran – chalcomoracin phytoalexin (Fukai et al. 2005); flavonol glycosides [quercetin 3-(6-malonylglucoside), rutin (quercetin 3-rutinoside) and isoquercitrin (quercetin 3-glucoside)] (Katsube et al. 2006); rutin, umbelliferone, chlorogenic acid, kaempferol, apigenin, luteolin (Chu et al. 2006); flavonol glycoside, quercetin 3-(6-malonylglucoside) (Enkhmaa et al. 2005); chlorogenic acid; caffeic acid; quercetin 3-0–β-D-glucopyranoside; kaempferol 3-0 –β-D-glucopyranoside (Kim et al. 2007); 1-deoxynojirimycin (DNJ) (Xie et al. 2008); flavonoids (rutin, moracetin, quercetin –3-triglucoside) (Kalantari et al. 2009); quercetin glycosides; chlorogenic acid (Katsube et al. 2009); pheophorbide compounds: pheophorbide a methyl ester, 132(S)-hydroxypheophorbide a methyl ester (Oh et al. 2010); four 2-arylbenzofuran derivatives, moracin V, moracin W, moracin X, moracin Y, moracin N, and moracin P (Yang et al. 2010a); chalcone derivatives morachalcone B and morachalcone C (Yang et al. 2010d); methyl esters pheophorbide and 132(S)-hydroxypheophorbide (Oh et al. 2010); two new flavanes, (2R,4S)-2′,4′-dihydroxy-2H-furan-(3″,4″:8,7)-flavan-4-ol and (2S)-2′,4′-dihydroxy-7-methoxyl-8-butyricflavane (Yang et al. 2010c); four Diels-Alder type adducts mulberrofuran F1, mulberrofuran F, chalcomoracin, kuwanon J, together with two chalcones morachalcone A, isobavachalcone and three flavones norartocarpetin, kuwanon C, 6-geranylapigenin (Yang et al. 2010b); chalcomoracin, a natural Diels-Alder type adduct with antibacterial activity and moracin N, a precursor of chalcomoracin (Gu et al. 2010); 11 flavonoids including new 3′-geranyl-3-prenyl-2′,4′,5,7-tetrahydroxyflavone (Dat et al. 2010).

Fruit: 4-O-α-D-galactopyranosyl-calystegine B, 3 β,6 β-dihydroxynortropane (Asano et al. 2001); cyanidin-3-O-β-D-glucopyranoside (C3G) (Kang et al. 2006a); anthocyanins: cyanidin 3-rutinoside, cyanidin 3-glucoside (Chen et al. 2006); five anthocyanins: cyanidin 3-O-(6″-O-α-rhamnopyranosyl-β-d-glucopyranoside) (C3RG),cyanidin3-O-(6″-O-α-rhamnopyranosyl-β-d-galactopyranoside) (C3RGa), cyanidin 3-O-β-d-glucopyranoside (C3G), cyanidin 3-O-β-d-galactopyranoside (C3Ga), cyanidin 7-O-β-d-glucopyranoside (C7G) (Du et al. 2008); flavonoids (quercetin 3-O-glucoside; quercetin 3-O-rutinoside, kaempferol 3-O-rutinoside, 5-O-caffeoylquinic acid (Pawlowska et al. 2008); anthocyanins, rutin, chlorogenic acids, lutein, δ-tocopherol, γ-tocopherol, α-tocopherol. (Mia et al. 2008); procatechuic acid, chlorogenic acid, 4-caffeolyquinic acid, taxifolin, rutin, quercetin) and three others (3,5-diCQA, taxifolin-hexoside, kamepferol-hexoside) (Zhang et al. 2008); resveratrol, oxyresveratrol, cyanidin-3-O-β-glucoside (Cy-3-glu), cyanidin-3-O-β-rutinoside (Cy-3-rut), rutin (Song et al. 2009).

The fruit and leaves of *M. alba* cultivar Hetianbaisang were found to be rich in 1-deoxynojirimycin (DNJ), resveratrol, oxyresveratrol, cyanidin-3-O-β-glucoside (Cy-3-glu), cyanidin-3-O-β-rutinoside (Cy-3-rut), and rutin (Song et al. 2009). The high contents of functional compounds in mulberry juice, fruits, and leaves implied that they might be potential sources for the development of functional drinks and food.

Mulberry leaf hydrocolloid extracted with water or sodium bicarbonate was found to compose of mainly carbohydrate with high levels of uronic acid indicating the ionic nature of the hydrocolloid (Lin and Lai 2009). Mulberry leaf hydrocolloid was found to possess unique rheology, particularly for the stiffer backbone as compared to other polysaccharide gums and would have great applications for food formulators. The rheological properties of mulberry leaf hydrocolloid was strongly affected by the ionic strength and ion types, it is expected that rheological behaviour of functional foods containing mulberry leaf hydrocolloids can be modulated by adjusting the ionic strength and ion types in food systems, and have merit for new applications in the future.

Antioxidant Activity

Mulberry fruit and leaves contain phytochemicals with antioxidant property. Five antioxidant anthocyanin compounds: cyanidin 3-O-(6″-O-α-rhamnopyranosyl-β-d-glucopyranoside) (C3RG), cyanidin 3-O-(6″-O-α-rhamnopyranosyl-β-d-galactopyranoside) (C3RGa), cyanidin 3-O-β-d-glucopyranoside (C3G), cyanidin 3-O-β-d-galactopyranoside (C3Ga) and cyanidin 7-O-β-d-glucopyranoside (C7G), were identified in the fruits of mulberry (*Morus alba.*) (Du et al. 2008). At the concentration of 0.10 mg/ml, the 1,1-diphenyl-2-picrylhydrazyl (DPPH) radical scavenging rates of C3G, C3Ga and C7G were about 88% of vitamin C, while C3RG and C3RGa were about 60%. Crude mulberry fruit extract had the same DPPH radical scavenging rate as vitamin C or the five anthocyanin monomers when the concentration reached 0.40 mg/ml, indicating that the crude mulberry extract to be an excellent antioxidant agent.

The phytochemical content of mulberry fruit varied among 27 cultivars tested, with total phenolic content, total anthocyanin content, total proanthocyanidin content, and peroxyl radical scavenging capacities ranging from 0.060–0.244, 0.001–0.056, 0.001–0.015, and 0.301–1.728, respectively (Mia et al. 2008). Good correlations were observed among the phenolic, anthocyanin, and proanthocyanidin contents and the radical scavenging capacities of mulberry fruits. Mulberry fruits were found to contain low amount of proanthocyanidins. The high total phenolic content of mulberry fruits were mainly contributed by anthocyanins, rutin, and chlorogenic acids. The lipid soluble antioxidants comprised significant amounts of lutein and δ- tocopheroland γ-tocopherol but low α-tocopherol.

In another study, the total polyphenol (TP) content in five major Korean mulberry varieties [M-1–M-5] was found to vary from 2,235 to 2,570 μg/g gallic acid equivalents, total anthocyanin (TA) content to vary from 1,229 to 2,057 μg/g, coloured anthocyanins (CA) from 126 to 190 μg/g, and total flavanol (TF) from 16.4 to 65.4 μg/g catechin equivalents except Mocksang (M-5). The ethanolic extract from mulberry fruit showed a rapid and concentration-dependent increase of antioxidant activity. The antioxidant activities of M-2 and M-4 varieties were higher than those of the others in a hemoglobin-induced linoleic acid system (Bae and Suh 2007). Six nonanthocyanin phenolics compounds (procatechuic acid, chlorogenic acid, 4-caffeolyquinic acid, taxifolin, rutin, quercetin) and three others (3,5-diCQA, taxifolin-hexoside, kamepferol-hexoside) were identified two Chinese *Morus alba* cultivars (Zhang et al. 2008). Rutin (111.38 and 90.79 μg/g FW) was the major nonanthocyanin phenolics in the two mulberry cultivars. Phenolic acids in cultivar da-10 (54.68 μg/g FW) were far more than in cultivar hongguo (14.93 μg/g FW). CQA and its isomers (40.02 μg/g FW) were the major phenolic acid in cultivar da-10. The nonanthocyanin phenolics in the two mulberry cultivars had higher antiradical activities in superoxide anion and DPPH radical and antioxidant activity in β-CLAMS assay than the Trolox. The authors suggested that results were relevant to not only the control of color stability and organoleptic characteristics of mulberry juice and wine but also the exploitation of the functional foods made from mulberry.

The radical scavenging activity of methanolic extract of ripe fruits of *Morus alba* (white mulberry), *Morus nigra* (black mulberry), *Morus laevigata* (large white fruit), and *Morus laevigata* (large black fruit) was concentration-dependent and showed a correlation with total phenolic constituents of the respective fruits(Imran et al. 2010).

Mulberry leaves (*Morus alba* var. *acidosa* G. cv. Taisang 2) exhibited antioxidant activity (Yen et al. 1996). The results of studies indicated that methanolic extract of mulberry leaves (MEML) showed stronger antioxidant activity and gave higher yields of extract than other organic solvents. The methanol extract exhibited 78.2% inhibition on peroxidation of linoleic acid, greater than that of α-tocopherol (72.1%) but equal to that of butylated hydroxyanisole. Of nine fractions of the methanol extract, two fractions possessed remarkable antioxidant activities. These two fractions showed 77.3% and 72.0% inhibition on peroxidation of linoleic acid, respectively. The antioxidant components of fractions I and II were identified as β-carotene and α-tocopherol, respectively.

White mulberry leaf was found to contain triterpenes (lupeol), sterols (β-Sitosterol), bioflavonoids(rutin,moracetin,quercetin-3-triglucoside and isoquercitrin), coumarins, volatile oil, alkaloids, amino acids and organic acids (Doi et al. 2001). Two novel prenylflavanes (1, 2) and a glycoside (3) of 1 were isolated along with six known compounds, isoquercitrin (4), astragalin (5), scopoline (6), skimming (7), roseoside II (8) and benzyl D-glucopyranoside (9), from the leaves of *Morus alba* (Doi et al. 2001). The inhibitory activities of compounds 1, 2 and 3 on the oxidation of human low density lipoprotein (LDL) were determined. Flavonoid content of mulberry leaves of 19 varieties varied from 11.7 to 26.6 mg/g in terms of rutin equivalent in spring leaves and 9.84–29.6 mg/g in autumn leaves (Jia et al. 1999). Fresh leaves gave more extract than air-dried or oven-dried ones. Mulberry leaves contained at least four flavonoids, two of which were identified as rutin and quercetin. The percentage superoxide ion scavenged by extracts of mulberry leaves, mulberry tender leaves, mulberry branches and mulberry bark were 46.5%, 55.5%, 67.5% and 85.5%, respectively, at a concentration of 5 μg/m. The scavenging effects of most mulberry extracts were greater than those of rutin (52.0%). Arabshahi-Delouee and Urooj (2007) reported that mulberry leaves could serve as a potential source of natural antioxidants due to their marked antioxidant activity. The methanol, acetone and water extracts of mulberry leaves showed varying degrees of efficacy in each assay in a dose-dependent manner. Methanolic extract with the highest amount of total phenolics, was the most potent antioxidant in all the assays used. The antioxidant activity of the extract remained unchanged at 50°C and was maximum at neutral pH. The extract stored at 5°C in the dark was stable for 30 days after which the antioxidant activity decreased gradually.

In another study, Kim et al. (2007) found that the ethyl acetate fraction of methanol extract of mulberry leaves showed the strongest antioxidant activity as assayed with the 2-deoxyribose oxidation and linoleic acid peroxidation. From the fraction were isolated chlorogenic acid, caffeic acid, quercetin 3-0–β-D-glucopyranoside and kaempferol 3-0–β-D-glucopyranoside. Overall, quercetin 3-0–β-D-glucopyranoside showed the strongest antioxidant activity by both the 2-deoxyribose oxidation and rat liver microsome peroxidation methods. In a recent study, air-drying temperature was found to impact on antioxidant activity of mulberry leaves (Katsube et al. 2009). DPPH radical scavenging activity and levels of polyphenolic compounds in mulberry leaves air-dried at 60°C or below were not significantly different from those of freeze-dried mulberry leaves, whereas both values in mulberry leaves air-dried at 70°C and over decreased significantly. Main compounds playing a central role in antioxidant activities in mulberry leaves were quercetin glycosides and chlorogenic acid. The results indicated that strict temperature control was important in the production of mulberry leaf products to maintain antioxidant activity and levels of polyphenolic compounds.

Studies showed that ethanolic extract of *Morus alba* twigs (EEMT) exhibited radical scavenging and reducing activity, as well as ferrous ion-chelating activity (Chang et al. 2011). In addition, EEMT also protected phospholipids against free radicals, indicating that EEMT could protect biomolecules from oxidative damage. The phenolic components present were found to be maclurin, rutin, isoquercitrin, resveratrol, and morin.

Anticancer Activity

Mulberry fruits, leaves and roots have bioactive chemicals with anti-cancerous property.

Anthocyanins, cyanidin 3-rutinoside and cyanidin 3-glucoside extracted from *Morus alba* fruit exerted a dose-dependent inhibitory effect on the migration and invasion, of highly metastatic A549 human lung carcinoma cells in absence of cytotoxicity (Chen et al. 2006). The results showed that cyanidin 3-glucoside and cyanidin 3-rutinoside treatments could decrease the expressions of matrix matalloprotinase-2 (MMP-2) and urokinase-plasminogen activator (u-PA) in a dose-dependent manner and enhance the expression of tissue inhibitor of matrix matalloprotinase-2 (TIMP-2) and plasminogen activator inhibitor (PAI). Further, a treatment of cyanidin 3-rutinoside and cyanidin 3-glucoside also resulted in an inhibition on the activation of c-Jun and NF-κB. Together, the result suggested that anthocyanins could decrease the in vitro invasiveness of cancer cells and therefore, may be of great value in developing a potential cancer therapy. In a separate study, Huang et al. (2008) reported that anthocyanins in mulberries fruit (MAC) exhibited inhibitory effects on the metastasis of B16-F1 melanoma cells under noncytotoxic concentrations. The antimetastatic effect of MACs was also evident in a C57BL/6 mice model. The findings suggested that MACs could mediate B16-F1 cell metastasis by reduction of MMP-2 and MMP-9 activities involving the suppression of the Ras/PI3K signaling pathway. MACs also inhibited the metastasis of B16-F1 cells in-vivo in the right groin of the C57BL/6 mice. Taken together, the findings proved the inhibitory effect of MACs on the growth and metastasis of B16-F1 cells. The results indicated that MACs might have potential for future application as an antimetastatic agent.

The cytotoxicity of 11 flavonoids isolated from *Morus alba* leaves including new 3′-geranyl-3-prenyl-2′,4′,5,7-tetrahydroxyflavone were evaluated against human cervical carcinoma HeLa, human breast carcinoma MCF-7, and human hepatocarcinoma Hep3B cells (Dat et al. 2010). Morusin (9) was the most potent with an IC_{50} value of 0.64 μM against human cervical carcinoma HeLa cells. In a recent study, two new chalcone derivatives named morachalcones B and C (1 and 2) isolated from the leaves of *Morus alba* displayed moderate cytotoxic activity against HCT-8 intestinal adenocarcinoma and BGC823 (human gastric gland carcinoma cells) cancer cell lines (Yang et al. 2010d). Yang et al. (2010a) reported that2-arylbenzofuran derivatives from the leaves were antitumourous. Of four new 2-arylbenzofuran derivatives, moracins V-Y (1–4), together with two known compounds, moracin N (5) and moracin P (6), compounds 2–5 exhibited varying cytotoxicity against several human cancer cell lines. Morachalcones B and C isolated from *M. alba* leaves displayed moderate cytotoxic activity against HCT-8 and BGC823 human cancer cell lines (Yang et al. 2010a). Three flavonoid compounds mulberrofuran F, mulberrofuran F and chalcomoracin isolated from *Morus alba* leaves showed cytotoxicity against several human cancer cell lines, A549 (Human lung adenocarcinoma epithelial cell line), Be17402 (hepatocarcinoma), BGC823, HCT-8 and A2780 (epithelial ovarian cancer) cell lines in-vitro (Yang et al. 2010b).

The water extract from root bark of Cortex Mori (CM, *Morus alba* L., Sangbaikpi), exhibited cytotoxic activity on K-562 (human erythromyeloblastoid leukemia), B380 human leukemia cells and B16 mouse melanoma cells at concentrations of >1 mg/ml (Nam et al. 2002). The tumour cells exposed to the extract underwent apoptosis. Overall, the water extract of CM induced apoptosis of tumour cells by inhibiting microtubule assembly but did not appear to exert G1 phase arrest. The extract had been known in Chinese traditional medicine to have antiphlogistic, diuretic, and expectorant properties. Skupień et al. (2008) found that *Morus alba* leaves were active against the sensitive leukaemic cell line HL60 and retained the in-vitro activity against multidrug resistant sublines (HL60/VINC and HL60/DOX). The 100% methanol, 50% aqueous methanol, 1-butanol (BuOH) and hot water extracts (W) of *M. alba* leaves containing phenolic compounds were found to inhibit the growth of HepG2 hepatoma cells (Naowaratwattana et al. 2010). Hundred percent MeOH, 50% MeOH, and BuOH extracts contained rutin, isoquercetin, and various derivatives of kaempferol and quercetin glycosides as their major constituents while the hot water extract contained primarily chlorogenic acid and caffeoylquinic acid derivatives. 2,2-Diphenyl-1-picrylhydrazyl radical scavenging activities were 70.0%, 45.8%, 41.0%, and 33.6%, and 50% inhibitory concentration values were 33.1, 79.4, 35.6, and 204.2 μg/ml for HepG2 cell proliferation inhibition for 100% MeOH, 50% MeOH, BuOH, and W extracts, respectively. The methanol extracts inhibited the growth of HepG2 hepatoma cells through coordinated actions of inducing cell cycle arrest in the G2/M phase (with p27(Kip1) protein expression), inhibiting topoisomerase IIα activity, and inducing cell apoptosis by activation of caspases.

A novel prenylated flavanone, 2′, 4′, 6′-tetrahydoroxy-6-geranylflavanone, isolated from ethyl acetate extracts of root of Kinuyutaka, a cultivar of *Morus alba* exhibited cytotoxic activity against rat hepatoma (dRLh84) cells with an IC_{50} of 52.8 μg/ml (Hisayoshi et al. 2004). A novel flavanone glycoside, 5,2′-dihydroxyflavanone-7,4′-di-O-β-D-glucoside (steppogenin-7,4′-di-O-β-D-glucosiade), isolated from the root bark of *Morus alba* exhibited significant inhibitory activity against human ovarian cancer HO-8910 cells' proliferation (Zhang et al. 2009b). IC_{50} of 48 hours and 72 hours were 3.68, 1.87 μmol/l, respectively.

Albanol A, isolated from the root bark extract of *Morus alba* showed potent cytotoxic activity (IC_{50} 1.7 μM) on human leukemia (HL60) cells and potent inhibitory activity on human DNA topoisomerase II (IC_{50} 22.8 μM) (Kikuchi et al. 2010). The results indicated that albanol A induced apoptotic cell death in HL60 via both the cell death receptor pathway by stimulation of death receptor, and the mitochondrial pathway by Topo II inhibition through caspase-2 activation. The results suggested that albanol A may be a promising lead compound for developing an effective drug for treatment of leukemia.

Antiviral Activity

A natural substance, 1-deoxynojirimycin, found in mulberry roots and seeds may be effective in the treatment of AIDS infection (Sergio 1989). The alcohol extract of San Baipi (*Morus alba* root), a traditional Chinese medicine for cough, asthma and other diseases, displayed activity against HIV in-vitro (Luo et al. 1994). Three anti-HIV flavonoids: morusin, kuwanon H and morusin-4′-glucoside were isolated. The anti-HIV activity and cytotoxicity of the derivatives morusin-2′-glucoside, morusin-4′-glucoside, kuwanon H-6′-acetate, kuwanon H-7-acetate and triacetate mulberrofuran D were also determined . Another research reported that a prenylated flavonoid, moralbanone, along with seven known flavonoid compounds kuwanon S, mulberroside C, cyclomorusin, eudraflavone B hydroperoxide, oxydihydromorusin, leachianone G and α-acetyl-amyrin isolated from the root bark of *Morus alba* differed in their antiviral activity (Du et al. 2003). Leachianone G showed potent antiviral activity (IC_{50} = 1.6 μg/ml), whereas mulberroside C showed weak activity (IC_{50} = 75.4 μg/ml) against Herpes Simplex type 1 virus (HSV-1).

Antibacterial Activity

Kuwanon G isolated from the ethyl acetate fraction of methanol extract of *Morus alba* exhibited antibacterial activity against oral pathogens (Park et al. 2003). MIC of kuwanon G against *Streptococcus mutans* causing dental caries was determined to be 8.0 μg/ml. Kuwanon G completely inactivated *S. mutans* at the concentration 20 μg/ml in 1 minutes. Kuwanon G also significantly inhibited the growth of other cariogenic bacteria such as *Streptococcus sobrinus*, *Streptococcus sanguis* and *Porpyromonas gingivalis* causing periodontitis. Transmission electron microscopy (TEM) of kuwanon G treated cells demonstrated remarkable morphological damage of the cell wall and condensation of the cytoplasm. The following prenylated flavonoids from *Morus alba*, kuwanon C, mulberrofuran G, albanol B showed strong antibacterial activity with 5–30 μg/ml of MICs, while morusin, sanggenon B and D, were effective to only Gram positive bacteria (Sohn et al. 2004).

Nine 2-arylbenzofurans isolated from *Morus* species including *M. alba* were tested for their antimicrobial activities against methicillin-sensitive *Staphylococcus aureus* (MSSA), methicillin-resistant *S. aureus* (MRSA), *Micrococcus luteus*, *Bacillus subtilis*, *Escherichia coli*, *Klebsiella pneumoniae* and *Pseudomonas aeruginosa*. Among these compounds, chalcomoracin (a leaf phytoalexin of mulberry tree) exhibited considerable antibacterial activity against MRSAs (MICs 0.78 μg/ml) (Fukai et al. 2005).

The purified compound, 1-deoxynojirimycin (DNJ) of *M. alba* exerted an eightfold greater reduction of MIC against *Streptococcus mutans* than the crude extract (MICs, 15.6 and 125 mg/l, respectively) (Islam et al. 2008). The extract strongly inhibited biofilm formation of *S. mutans* at its active accumulation and plateau phases. The purified compound led to a 22% greater reduction in alkali-soluble polysaccharide than in water-soluble polysaccharide. DNJ also distorted the biofilm architecture of *S. mutans*. The findings suugested a prospective role of DNJ as a therapeutic agent by controlling the overgrowth and biofilm formation of *S. mutans*.

Antidiabetic/Antihyperglycemic Activity

Numerous scientific research papers have been published on the antidiabetic/antihyperglycaemic activity of mulberry leaves and roots over the past two decades.

Hypoglycemic effects of hot water extracts of mulberry (*Morus alba*) leaves, or root cortex were observed in fasted and nonfasted streptozotocin (STZ)-induced diabetic mice at a single dose of 200 mg/kg (i.p.) (Chen et al. 1995). The leaf extract exhibited most potent hypoglycemic effects. The most potent fractions of Folium Mori and Cortex Mori Radicis were ethanol-insoluble extracts (A2). These A2 fractions demonstrated a fall in blood glucose levels of 24.6% and 60.5% at nonfasted STZ-mice, and 81.4% and 77.3% at fasted STZ-mice, respectively. Taniguchi et al. (1998) reported that fagomine, an N-containing pseudo-sugar derived from mulberry leaves was found to induce insulin secretion. Fagomine at more than 1 mmol/l significantly potentiated insulin secretion induced by 10 mmol/l glucose. The pseudo-sugar, however, did not affect the basal insulin secretion assessed at a glucose concentration of 3.5 mmol/l. The effects of fagomine on 10 mmol/l and 20 mmol/l glucose-induced insulin secretion were not significantly different. Fagomine (4 mmol/l) also potentiated glyceraldehyde-induced insulin secretion, but not the leucine-induced type. Glycolysis assessed by lactate production from glucose was significantly enhanced. The findings, suggested that fagomine potentiated glucose-induced insulin secretion through acceleration of some step(s) after the formation of glyceraldehyde 3-phosphate in the glycolytic pathway.

Mulberry therapy was found to exhibit potential hypoglycemic and hypolipidemic effects in diabetic patients (Andallu et al. 2001). Diabetic patients with mulberry therapy significantly improved their glycemic control vs. glibenclamide treatment. The results from pre- and post-treatment analysis of blood plasma and urine samples showed that the mulberry therapy significantly decreased the concentration of serum total cholesterol (12%), triglycerides (16%), plasma free fatty acids (12%), LDL-cholesterol (23%), VLDL-cholesterol (17%), plasma peroxides (25%), urinary peroxides (55%), while increasing HDL-cholesterol (18%). Although the patients with glibenclamide treatment showed marginal improvement in glycemic control, the changes in the lipid profile were not statistically significant except for triglycerides (10%), plasma peroxides (15%), and urinary peroxides (19%). Both treatments displayed no apparent effect on the concentrations of the glycosylated hemoglobin (Hb A(1)c) in diabetic patients. However, the fasting blood glucose concentrations of diabetic patients were significantly reduced by the mulberry therapy. Of three *Morus* species: *M. alba, M. alba* var. *nigra* and *M. nigra*; *M. alba* var. *nigra* (IC$_{5}$0 = 13.26 (12.86–13.66) mg/ml) and *M. alba* (IC_{50} = 17.60 (17.39–17.80) mg/ml) revealed appreciable α-amylase inhibitory activities in a concentration-dependent manner (Nickavar and Mosazadeh 2009). All the fractions (n-hexane, chloroform, ethyl acetate and aqueous fractions) of the most active extract (i.e. *M. alba* var. *nigra*) had potent inhibitory effects on the α-amylase activity. However, the lowest inhibitory potency was observed for the aqueous fraction.

Mulberry leaves possessed antihyperglycemic and antioxidant properties (Andallu and Varadacharyulu 2003). Mulberry-treated streptozotocin (STZ)-induced diabetic rats showed a significant decrease in fasting blood glucose concentrations indicating a good glycemic control. Increased lipid peroxidation and the activity of catalase (CAT) in erythrocytes observed in diabetic controls were significantly decreased by mulberry leaves (48% and 33%, respectively). Decreased GSH concentrations and the activity of glucose-6-phosphate dehydrogenase and antioxidant enzymes viz., glutathione peroxidase (GPx), glutathione reductase (GR), glutathione-S-transferase (GST) and superoxide dismutase (SOD) observed in uncontrolled diabetes were improved (52%, 69%, 151%, 95%, 24% and 106%) by mulberry treatment very efficiently.

Dried leaf powder of mulberry (*M. indica*) when given along with the diet at 25% level to streptozotocin-induced diabetic male Wistar albino rats for 8 weeks, controlled hyperglycemia, glycosuria, albuminuria and retarded onset of retinopathy (Andallu and Varadacharyulu 2002). Untreated diabetic rats showed hyperglycemia, glycosuria, albuminuria and developed lenticular opacity after 8 weeks of experimental period Treatment with dried mulberry leaf powder at 25% of the diet for a period of 8 weeks was found to be remarkably beneficial to STZ-diabetic rats as evidenced by controlled hyperglycemia and glycosuria (Andallu and Varadacharyulu 2007). In addition, mulberry leaves countered (reversed) the alterations in gluconeogenic substrates in STZ-diabetic rats as indicated by significant reduction in serum pyruvic and lactic acid levels, a significant increase in proteins and a significant decrease in free amino acid, urea, and creatinine levels in blood, and a decreased urinary excretion of urea and creatinine. Anomalies in the activities of hepatic gluconeogenic enzymes associated with impaired glucose homeostasis in STZ-diabetic rats were ameliorated by feeding the mulberry leaf-supplemented diet, indicating that control over hyperglycemia and associated complications in the diabetic state by mulberry leaves is by way of regulation of gluconeogenesis. With respect to all the parameters, mulberry leaves were more effective than the oral hypoglycemic drug glibenclamide.

Mulberry leaf extract, at a dose of 600 mg/kg body weight, exhibited therapeutic effects in diabetes-induced Wistar rats and restored the diminished β cell numbers (Mohammadi and Naik 2008). Blood glucose, glycosylated hemoglobin, triglyceride, LDL, VLDL, blood urea, cholesterol, were elevated in the diabetic group, but were brought to control group level in the diabetic group treated with 600 mg/kg body weight of mulberry leaf extract. The diameter of the islets and the number of β cells were reduced in the diabetic group; both parameters were brought to control group level after treatment with mulberry leaf extract. Studies showed that administration of *Morus alba* leaf aqueous extract exerted significant postprandial hypoglycemic effect in both non-obese diabetic and healthy rats, suggesting that the extract may be beneficial as food supplement to manage postprandial blood glucose (Park et al. 2009). Inhibitions of glucose transport as well as α-glucosidase in the small intestine were suggested as possible mechanisms related with the postprandial hypoglycemic effect of mulberry leaf extract.

Oku et al. (2006) reported that the extract from the leaves of *Morus alba* (ELM) containing 0.24% 1-deoxynojirimycin disaccharidase exhibited inhibitory effect on human and rat small intestinal disaccharidase activity. The sucrase activity of four human samples was inhibited by 96% and that of maltase and isomaltase by 95% and 99%, respectively. The activities of trehalase and lactase were inhibited by 44% and 38%, respectively. The ratio of the inhibitory effect for sucrase, maltase, isomaltase, trehalase and lactase was very similar among the four samples, and also that of resembled rat intestinal disaccharides. The inhibitory constant of the 1-deoxynojirimycin equivalent for sucrase, maltase and isomaltase was 2.1×10^{-4}, 2.5×10^{-4} and 4.5×10^{-4} mm, respectively. These results demonstrated that digestion was inhibited when an appropriate amount of ELM was orally ingested with sucrose or polysaccharide in man. When ELM was orally administered in a sucrose solution to fasted rats, the elevation in blood glucose was significantly suppressed, depending on the concentration of ELM given. These results suggested that mulberry leaf extract could be used as an ingredient in health foods and in foods that help to prevent diabetes. In another study, the aqueous ethanol extract from mulberry leaves (ME) inhibited postprandial hyperglycemia in normal Wistar rats (Miyahara et al. 2004). ME dose-dependently suppressed the postprandial rise of blood glucose in rats, when ME (0.02–0.5 g/kg) was given 0.5 h before the administration of carbohydrates such as sucrose, maltose and starch. The ME dose showing 50% inhibition of the increment of blood glucose (ED_{50}) was 0.11 g/kg for sucrose, 0.44 g/kg for maltose, and 0.38 g/kg for starch. ME and its basic fraction (MB) containing 1-deoxynojirimycin were assayed for their inhibitory effects (IC_{50}) on disaccharidase derived from the small intestine of rats. The IC_{50} value of ME was 3.2 μg/ml for sucrase, 10 μg/ml for isomaltase, and 51 microg/ml for maltase. The IC_{50} value of MB was 0.36 μg/ml for sucrase, 1.1 μg/ml for isomaltase, and 6.2 μg/ml for maltase. The IC_{50} value of 1-deoxynojirimycin as the principle component in ME was 0.015 μg/ml for sucrase and 0.21 μg/ml for maltase, and this value was comparable to the IC_{50} of voglibose. The inhibitory activity of ME in a-amylase was weak. The results suggested that ME strongly suppressed postprandial hyperglycemia after carbohydrate loading by inhibiting the activity of disaccharidases in the small intestine of rats. The effects of brewing time on dry weight content and μ-glucosidase inhibitory active component released mulberry tea were studied. Different tea products from mulberry leaves (*Morus alba*) showed significant differences in inhibitory activity against both sucrase and maltase (Hansawasdi and Kawabata 2006). The most effective enzyme inhibition was observed when 3–5 min brewing time was applied in tea preparation. In a Caco-2 cell culture experiment the tea reduced the liberated glucose contents in both apical and basal sides of the cell monolayers. It was concluded that hot water extract of mulberry leaves had inhibitory effect against α-glucosidases, sucrase and maltase enzymes, and had a potential to be consumed as antidiabetic herb tea.

Long-term administration of ethanolic extract of mulberry leaf (MA) was found to have antihyperglycemic, antioxidant and antiglycation effects in chronic streptozotocin-induced diabetic rats, which may be beneficial as food supplement for diabetics (Naowaboot et al. 2009a). Daily administration of 1 g/kg MA for 6 weeks decreased blood glucose by 22%, which was comparable to the effect of 4 U/kg insulin. Lipid peroxidation, measured as malondialdehyde and lipid hydroperoxide concentrations (3.50 and 3.76 μM, respectively) decreased significantly compared to non-treated control diabetic rats (8.19 and 7.50 μM, respectively). Hemoglobin A(1C), a biomarker for chronic exposure to high concentration of glucose, was also significantly decreased in the MA-treated group (6.78%) in comparison to untreated group (9.02%). The IC_{50} of in-vitro antiglycation and free radical scavenging activities of MA were 16.4 and 61.7 μg/ml, respectively. Vascular responses of the chronic diabetic rats to vasodilators, acetylcholine (3–30 nmol/kg) and sodium nitroprusside (1–10 nmol/kg) were significantly suppressed by 26–44% and 45–77% respectively, whereas those to vasoconstrictor, phenylephrine (0.01–0.1 μmol/kg) were significantly increased by 23–38% as compared to normal rats (Naowaboot et al. 2009b). Interestingly, the administration of 0.5 and 1 g/kg MA or 4 U/kg insulin significantly restored the vascular reactivities of diabetic rats. Moreover, 8 weeks of diabetes resulted in the elevation of malondialdehyde content in tissues (liver, kidney, heart, and aorta), and MA treatment significantly lessened this increase. The results provided evidence for the efficacy of MA in restoring the vascular reactivity of diabetic rats, the mechanism of which may be associated with the alleviation of oxidative stress. Dietary supplementation with 10 g of mulberry-leaf powder/kg or 1 g purified quercetin 3-(6-malonylglucoside)/kg in high-fat diet effectively suppressed blood glucose levels in obese mice (Katsube et al. 2010). An increased expression of glycolysis-related genes and suppression of thiobarbituric acid reactive substances concentrations in the liver of quercetin 3-(6-malonylglucoside) group compared to control mice were observed. The results showed that dietary consumption of quercetin 3-(6-malonylglucoside), the quantitatively major flavonol glycoside in mulberry leaves, improved hyperglycemia in obese mice and reduced oxidative stress in the liver.

Eighteen N-containing sugars were isolated from the roots of *M. alba* (Asano et al. 1994). These N-containing sugars were 1-deoxynojirimycin (1), N-methyl-1-deoxynojirimycin (2), fagomine (3), 3-epi-fagomine (4), 1,4-dideoxy-1,4-imino-D-arabinitol (5), 1,4-dideoxy-1,4-imino-D-ribitol (6), calystegin B2 (1 α,2 β,3 α,4 β-tetrahydroxy-nor-tropane, 7), calystegin C1 (1 α,2 β,3 α,4 β,6 α-pentahydroxy-nor-tropane, 8), 1,4-dideoxy-1,4-imino-(2-O-β-D-glucopyranosyl)-D-arabinitol (9), and nine glycosides of 1. These glycosides consisted of 2-O-α-D-galactopyranosyl-1-deoxynojirimycin (10) and

6-O-α-D-galactopyranosyl-1-deoxynojirimycins (11); 2-O-α-D-glucopyranosyl-1-deoxynojirimycin(12), 3-O-α-D-glucopyranosyl-1-deoxynojirimycin (13) and 4-O-α-D-glucopyranosyl-1-deoxynojirimycin (14); and 2-O-β-D-glucopyranosyl-1-deoxynojirimycin (15), 3-O-β-D-glucopyranosyl-1-deoxynojirimycin (16), 4-O-β-D-glucopyranosyl-1-deoxynojirimycin (17) and 6-O-β-D-glucopyranosyl-1-deoxynojirimycin (18). Compound 4 was found as a new member of polyhydroxylated piperidine alkaloids. It was found that the polyhydroxy-nor-tropane alkaloids possessed potent glycosidase inhibitory activities. Calystegin was a trihydroxy-nor-tropane, and calystegins B1 and B2 were tetrahydroxy-nor-tropane. Calystegin C1, a new member of calystegins, was reported as the first naturally occurring pentahydroxy-nor-tropane alkaloid. The inhibitory activities of these compounds were determined against rat digestive glycosidases and various commercially available glycosidases. Asano et al. (2001) reported new polyhydroxylated alkaloids, (2R,3R,4R)-2-hydroxymethyl-3,4-dihydroxypyrrolidine-N-propionamide from the root bark and 4-O-α-D-galactopyranosyl-calystegine B(2) and 3 β,6 βa-dihydroxynortropane from the fruits. Fifteen other polyhydroxylated alkaloids were also isolated. 1-Deoxynojirimycin, a potent α-glucosidase inhibitor, was concentrated 2.7-fold by silkworms feeding on mulberry leaves. Some alkaloids contained in mulberry leaves were potent inhibitors of mammalian digestive glycosidases but not inhibitors of silkworm midgut glycosidases, suggesting that the silkworm had enzymes specially adapted to enable it to feed on mulberry leaves. The possibility of preventing the onset of diabetes and obesity using natural dietary supplements containing 1-deoxynojirimycin and other α-glucosidase inhibitors in high concentration is of great potential interest. The comparison of mulberry leaves of different ages showed that the 1-deoxynojirimycin level was higher in mulberry shoots than young and mature leaves (Nuengchamnong et al. 2007). In another study, co-ingestion of mulberry leaf extract produced significant reductions in blood glucose increases for the initial 120 minutes of the study (Mudra et al. 2007). The mulberry-induced reduction in blood glucose presumably reflected the ability of mulberry to inhibit intestinal sucrase. The increased H2 observed with mulberry indicated that this supplement induced sucrose malabsorption. Reductions in blood glucose fluctuation with mulberry extract might reduce diabetes complications despite minor reduction of haemoglobin A1C. Some individuals might prefer an herbal over a pharmaceutical preparation, and such individuals might find mulberry extract more acceptable and better tolerated than acarbose or miglitol. In another study, consumption of 30 g of sucrose with 1.2 or 3.0 g of *M. alba* leaf extract by ten healthy female subjects effectively suppressed the elevations of postprandial blood glucose and insulin by inhibiting the intestinal sucrase, thus creating a prebiotic effect (Nakamura et al. 2009). The results suggested that the development of confections with *M. alba* leaf extract could contribute to the prevention and the quality of life for prediabetic and diabetic patients. Recent studies found that the in-vitro inhibitory activity of *M. alba* leaf extract on intestinal α-glucosidase was potent and that on intestinal α-amylase was very weak compared with acarbose (Kim et al. 2011). Sugar loading tests with starch, maltose, and sucrose showed that the extract may reduce postprandial increases in blood glucose by acting as an intestinal α-glucosidase inhibitor. Feeding tests suggested that MLE may exhibit fewer adverse side effects than other α-glucosidase inhibitors, such as abdominal flatulence and meteorism, which may be attributed to the impaired digestion of starch by strong inhibition of intestinal α-amylase.

Moracin M (1), steppogenin-4′-O-β-D-glucoside (2), mullberroside A (3) isolated from the root bark of *Morus alba* exhibited hypoglycemic effects in alloxan-diabetic mice (Zhang et al. 2009a). Compound 2 in a dose of 50 mg/kg exerted significant effect, 2 and 3 in a dose of 100 mg/kg exerted significant effect.

Antihyperlipidemic/Antiatherosclerotic Activity

Mulberry water extracts (MWEs) was found to have lipid-lowering effects (Liu et al. 2009). Plasma total cholesterol (TC) and triglyceride (TG) levels of hamsters fed high fat/cholesterol diets (HFCD) with MWEs were significantly reduced by about 30–37% and 16–35%, respectively, as compared to those without MWEs. Similar results were also measured in hepatic TC and TG of hamsters fed HFCD with MWEs. Low-density lipoprotein receptor (LDLR) gene expression and the uptake ability of low-density lipoprotein (LDL) in HepG2 cells were also upregulated by additions of MWEs. MWEs also decreased the gene expressions of enzymes involved in the TG and TC biosyntheses. Results suggested that hypolipidemic effects of MWEsweare via an enhancement of LDLR gene expression and the clearance ability of LDL and a decrease in the lipid biosynthesis. Therefore, MWEs could be used as a natural agent against hyperlipidemia.

Administration of freeze-dried powder of mulberry (*Morus alba*) fruit (MFP) to rats on a high-fat diet resulted in a significant decline in levels of serum and liver triglyceride, total cholesterol, serum low-density lipoprotein cholesterol, and a decrease in the atherogenic index, while the serum high-density lipoprotein cholesterol was significantly increased (Yang et al. 2010). In addition, the serum and liver content of thiobarbituric acid related substances, a lipid peroxidation product, significantly decreased, while the superoxide dismutase (SOD) of red blood cell and liver, as well blood glutathione peroxidase (GSH-Px) activities significantly increased. No significant changes in lipid profile in the serum and liver were observed in rats on a normal diet supplemented with MFP, but blood and liver antioxidant status improved, as measured by SOD and GSH-Px activity, and lipid peroxidation was reduced. These beneficial effects of MFP on hyperlipidaemia rats might be attributed to its dietary fibre, fatty acids, phenolics, flavonoids, anthocyanins, vitamins and trace elements content. Studies showed that *M. alba* leaves had antihyperlipidimc effects (Kobayashi et al. 2010). Plasma triglyceride and non-esterified fatty acid levels were significantly lower in the rats fed a high-fat diet and treated with *M. alba* leaves as compared with the untreated rats. DNA microarray analysis revealed that mulberry leaves upregulated expression of the genes involved in α-, β- and ω-oxidation of fatty acids, mainly related to the peroxisome proliferator-activated receptor signalling pathway, and suppressed the genes involved in lipogenesis. In addition, treatment with mulberry leaves augmented expression of the genes involved in the response to oxidative stress.

Consumption of mulberry leaves or their 1-butanol extract (MLBE) was found to reduce the concentration of serum lipids and atheromatous thickening of arterial intima in hypercholesterolemic rabbits (Doi et al. 2000). This was found to be attributed to the antioxidative activity of MLBE and isoquercitrin, the main component of MLBE. Quercetin, an aglycone of isoquercitrin, inhibited the formation of conjugated dienes and thiobarbituric acid reacting substances (TBARS) by copper-induced oxidative modification of rabbit and human LDLs. MLBE and isoquercitrin also inhibited the oxidation of LDL. The results indicated that mulberry leaves inhibited the oxidative modification of LDL and indicate that mulberry leaves may had prevented atherosclerosis. Enkhmaa et al. (2005) found that dietary consumption of mulberry (*Morus alba*) leaves and their major flavonol glycoside, quercetin 3-(6-malonylglucoside) (Q3MG), was found to attenuate the development of atherosclerotic lesions in LDL receptor-deficient (LDLR−/−) mice. The susceptibility of LDL to oxidative modification was significantly decreased in the Q3MG- and mulberry-treated mice, as evidenced by the 44.3% and 42.2% prolongation of the lag phase for conjugated diene formation compared with that of the control mice. The atherosclerotic lesion area in both the Q3MG- and mulberry-treated mice was significantly reduced by 52% compared with that of the controls. However, in the quercetin group, no protective effects were observed against LDL oxidation or atherosclerotic lesion formation. In conclusion, mulberry leaves attenuated the atherosclerotic lesion development in LDLR–/– mice through enhancement of LDL resistance to oxidative modification, and these antioxidative and antiatherogenic protective effects were attributed mainly to Q3MG, the quantitatively major flavonol glycoside in mulberry leaves.

Application of crude water extract of *Morus alba* in rat pups resulted in amelioration of the alterations of maternal serum glucose, LDL, HDL, total cholesterol and creatine phosphokinase activity as well as retinal neurotransmitters including acetylcholine (ACE), adrenaline (AD), nor-adrenaline (NAD), serotonin (5-HT), histamine (HS), dopamine (DA) and gamma amino butyric acid (GABA) (El-Sayyad et al. 2011). The retina of pups of either diabetic and/or hypercholesterolemia mothers exhibited massive alterations of retinal neurotransmitters and a striking incidence of cataract was detected in pups of either diabetic and/or hypercholesterolemic mothers. However, protection with *Morus alba* extract led to amelioration of the pathological alterations of retinal neurons and estimated neurotransmitters. The effects of the extract might be attributed to the hypoglycaemic, antihypercholesterolemic and anti-oxidative potential of flavonoids, the major components of the leaf extract.

Administration of flavonoids rich fraction of 70% alcohol extract of the Egyptian *Morus alba* root bark (MRBF-3) to rats for 10 days (600 mg/kg/day) significantly reduced the amount of the glucose from control level (379 mg/dl) to a lower level (155 mg/dl) and significantly increased the insulin level from control (10.8 μU/ml) to a high level (15.6 μU/ml) (Singab et al. 2005). Doses of 200 and 400 mg/kg/day were not significant. The measurement of produced lipid peroxides (expressed as the amount of thiobarbituric acid (TBA) reactive substance, nmol TBARS/ml serum) indicated antiperoxidative activity of MRBF-3. The oral administration of MRBF-3 to STZ-diabetic rats significantly decreased the lipid peroxides from 6.3 to 5.1 nmol TBARS/ml serum. The phytochemical investigation of MRBF-3 resulted in the isolation of four hydrophobic flavonoids with one or two isoprenoid groups: morusin, cyclomorusin, neocyclomorusin, and kuwanon E, a 2-arylbenzofuran, moracin M, and two triterpenes, betulinic acid and methyl ursolate. The data obtained from this study revealed that MRBF-3 may protect pancreatic β cells from degeneration and diminish lipid peroxidation. Consumption of MRBF-2 and (MRBF-3, in some extent) fractions of *M. alba* root bark 70% alcohol extract may act as a potent hypocholesterolemic nutrient and powerful antioxidant via the inhibition of LDL atherogenic modifications and lipid peroxides formation in hypercholesterolemic rats (El-Beshbishy et al. 2006). The results revealed that the administration of (MRBF-2 and/or MRBF-3) fractions resulted in alleviation of atherosclerotic state. Four compounds namely: mulberroside A, 5,7,2V-trihydroxyflavanone-4V-O-h-dglucoside and albanols A and B were isolated from MRBF-2. Administration of MRBF-3 significantly retained plasma and liver peroxides towards their normal levels, and also, produced significant increase in resistance towards major atherogenic modifications; namely LDL oxidation, LDL aggregation and LDL retention by 44%, 30%, and 33%, respectively.

Two *Morus alba* fruit extracts, MWEs (mulberry water extracts) and MACs (mulberry anthocyanin-rich extracts), were found to exhibit antioxidative and anti-atherosclerogensis abilities in-vitro (Liu et al. 2008). Data showed that MWEs and MACs were able to inhibit significantly the relative electrophoretic mobility (REM), ApoB fragmentation, and thiobarbituric acid reaction substances (TBARS) formation in Cu^{2+}-mediated oxidation LDL (low-density lipoprotein). MWEs and MACs also had the ability of 1,1-diphenyl-2-picrylhydrazyl (DPPH) radical scavenging for reducing the formation of free radicals mediated by copper ions. MWEs and MACs also could decrease macrophage death induced by oxLDL. Further, MWEs and MACs also could inhibit the formation of foam cells. Both MWEs and MACs showed a great ability of scavenging radicals, inhibition of LDL oxidation, and decrease in atherogenic stimuli in macrophages, while the efficacy of MACs was ten-fold greater than that of MWEs. The results demonstrated that anthocyanin components in mulberry extracts had potential in the prevention of atherosclerosis.

Atherosclerosis involves proliferation and migration of vascular smooth muscle cell (VSMC). Studies found that mulberry leaf extract (MLE) could effectively inhibit the migration of vascular smooth muscle cell by blocking small GTPase and Akt/NF-kappaB signals (Chan et al. 2009). Mulberry leaf extract (MLE) was found to be rich in polyphenols (44.82%), including gallic acid, protocatechuic acid, catechin, gallocatechin gallate, caffeic acid, epicatechin, rutin, and quercetin (Chan et al. 2009). MLE inhibited the migration of A7r5 cells in a dose- and time-dependent manner. MLE also inhibited the activities of matrix metalloproteinases (MMPs) MMP-2 and MMP-9, protein expressions, and phosphorylation of FAK and Akt, and protein expressions of small guanosine triphosphatases (GTPases: c-Raf, Ras, Rac1, Cdc42, and RhoA) in a dose-dependent manner. NF-kappaB expression was also inhibited by MLE.

Adipocytokine dysregulation is an important risk factor for atherosclerotic cardiovascular disease. Previous study showed that mulberry (*Morus alba*) leaf (ML) ameliorated atherosclerosis in apoE(−/−) mice. Treatment of db/db mice with ML was found to ameliorate adipocytokine dysregulation at least in part through inhibiting oxidative stress in white adipose tissue of db/db mice (Sugimoto et al. 2009). ML treatment increased the expression of adiponectin, and decreased the expression of TNF-α, MCP-1, and macrophage markers in white adipose tissue. ML decreased blood glucose, plasma triglyceride and lipid peroxides and the expression of NADPH oxidase subunits in white adipose tissue and liver. Co-treatment with ML and pioglitazone enhanced these effects, showing additive effects compared with pioglitazone. Further, their co-treatment attenuated the body weight increase observed under the pioglitazone treatment.The findings suggested that mulberry leaf may be a basis for a pharmaceutical for the treatment of the metabolic syndrome as well as reducing adverse effects of pioglitazone.

The results of studies by Lee et al. (2011) suggested that *Morus alba* could improve an atherogenic diet-induced hypertension, hyperlipidemia, and vascular dysfunction through inhibition of cell adhesion molecules expression and induction of vascular relaxation. Chronic treatment with low (100 mg/kg/day) or high (200 mg/day/kg) doses of water mulberry extract markedly attenuated hypertension and the impairments of acetylcholine-induced relaxation of aortic rings in rats fed an atherogenic diet. The extract reduced intima/media thickness in rats fed an atherogenic diet. The extract improved plasma levels of triglyceride (TG) and augmented plasma levels of high-density lipoprotein (HDL) and plasma low-density lipoprotein (LDL), but did not affect blood glucose levels. The extract suppressed increased cell adhesion molecules such as E-selectin, vascular cell adhesion molecule-1 (VCAM-1), and intracellular adhesion molecule-1 (ICAM-1) expression in the aorta.

Ethanol extract of mulberry leaves, inhibited human LDL oxidation induced by copper ion (Katsube et al. 2006). Antioxidants that prevent LDL from oxidation may reduce atherosclerosis. Three flavonol glycosides [quercetin 3-(6-malonylglucoside), rutin (quercetin 3-rutinoside) and isoquercitrin (quercetin 3-glucoside)] were identified as the major LDL antioxidant compounds. The results showed that quercetin 3-(6-malonylglucoside) and rutin were the predominant flavonol glycosides in the mulberry leaves.

Antiobesity Activity

In-vitro studies suggested that extracts of black, green, and mulberry teas could interfere with carbohydrate and triacylglycerol absorption via their ability to inhibit α-amylase, α-glucosidase, sodium-glucose transporters, and pancreatic lipase (Zhong et al. 2006). With the carbohydrate-containing meal, the tea extract resulted in a highly significant increase in breath-hydrogen concentrations, which indicated appreciable carbohydrate malabsorption. A comparison of hydrogen excretion after the carbohydrate-containing meal with that after the nonabsorbable disaccharide lactulose suggested that the tea extract induced malabsorption of 25% of the carbohydrate. The tea extract did not cause triacylglycerol malabsorption or any significant increase in symptoms This study provided the basis for additional experiments to determine whether the tea extract had clinical utility for the treatment of obesity or diabetes.

Separate studies reported that Ob-X, a mixture of three herbs, *Morus alba, Melissa officinalis* and *Artemisia capillaris* was found to regulate lipid metabolism, body weight gain and adiposity (Lee et al. 2008). Mice fed a high-fat diet for 12 weeks exhibited increases in body weight gain and adipose tissue mass compared with mice fed a low fat diet. However, feeding a high-fat diet supplemented with Ob-X significantly reduced these effects. Ob-X treatment also decreased the circulating levels of triglycerides and total cholesterol, and inhibited hepatic lipid accumulation. The data further demonstrate that Ob-X regulated body weight gain, adipose tissue mass, and lipid metabolism in part through changes in the expression of hepatic PPARalpha target genes. Ob-X, the anti-angiogenic herbal composition composed of *Melissa officinalis, Morus alba* and *Artemisia capillaries*, was found to inhibit differentiation of preadipocytes into adipocytes (Hong et al. 2011). These events were shown to be mediated by changes in the expression of genes involved in lipogenesis, angiogenesis, and the matrix metalloproteinases system. Studies ob/ob mice by Yoon and Kim (2011) showed that Ob-X, which had an anti-angiogenic activity, reduced body weight gain and visceral adipose tissue mass in genetically obese mice, providing evidence that obesity could be prevented by angiogenesis inhibitors. Ob-X treatment inhibited hepatic lipid accumulation and significantly decreased circulating glucose levels compared with controls.

Melanin-concentrating hormone receptor subtype 1 (MCH1) receptor binding studies showed that ethanol extract of *Morus alba* leaves (EMA) exhibited a potent inhibitory activity with IC_{50} value of 2.3 μg/ml (Oh et al. 2009a). EMA (10–100 μg/ml) also inhibited the intracellular calcium mobilization with the recombinant MCH1 receptors expressed in CHO cells. In an anti-obesity study with DIO mice, long term oral administrations of EMA for 32 consecutive days produced a dose-dependent decrease in body weight and hepatic lipid accumulation. The results suggested that chronic treatment with EMA exerted an anti-obesity effect in DIO mice, and its direct MCH1 receptor antagonism may contribute to decrease body weight.

Skin Whitening Activity

Morus alba leaves was found to inhibit melanin biosynthesis, which was closely related to hyperpigmentation (Lee et al. 2002). The methanol extract of dried *M. alba* leaves inhibited tyrosinase activity that converted dopa to dopachrome in the biosynthetic process of melanin. Mulberroside F (moracin M-6, 39-di-O-b-D-glucopyranoside), which was obtained after the bioactivity-guided fractionation of the extracts, showed inhibitory effects on tyrosinase activity and on the melanin formation of melan-a cells. This compound also exhibited superoxide scavenging activity that is involved in the protection against autooxidation. But its activity was low and was weaker than that of kojic acid. The results suggested that mulberroside F isolated from mulberry leaves could be used as a skin whitening agent. In another study, an organic layer prepared from the Chinese crude drug 'Sang-Bai-Pi' (*Morus alba* root bark) was found to contain inhibitory compounds for protein tyrosine phosphatase 1B (PTP1B) (Cui et al. 2006). The following bioactive compounds were isolated: sanggenon C (1), sanggenon G (2), mulberrofuran C (3) and kuwanon L (4) as PTP1B inhibitors, along with moracin O (5) and moracin P (6). Compounds 1–4 inhibited PTP1B with IC_{50} values ranging from 1.6 to 16.9 μM.

Extract from root cortices of mulberry (*Morus alba*) was separated into permeate and retentate fractions using a microfiltration/ultrafiltration membrane system (Yu et al. 2007). The clarification degree, antioxidant effects, and whitening capability of the permeates were increased as compared to those of feed. A higher content of active compounds, such as chlorogenoic acid and p-hydroxybenzoic acid was postulated for the higher antioxidant and whitening capabilities in permeate of extracts from root cortices of mulberry. Extracts from a hybrid Mulberry plant obtained from *Morus alba* L. and *Morus rotundiloba* Koidz, were found to have in-vitro anti-tyrosinase activity and to be a potential new source of Thai whitening agent (Nattapong and Omboon 2008). Betulinic acid, with anti-inflammatory and anti-tyrosinase activities, was also detected.

Mulberroside A, a glycosylated stilbene, was isolated and identified from the ethanol extract of the roots of *Morus alba* (Kim et al. 2010b). Mulberroside A and oxyresveratrol, the aglycone of mulberroside A, showed inhibitory activity against mushroom tyrosinase with an IC_{50} of 53.6 and 0.49 μM, respectively. The tyrosinase inhibitory activity of oxyresveratrol was thus approximately 110-fold higher than that of mulberroside A. Inhibition kinetics showed mulberroside A to be a competitive inhibitor of mushroom tyrosinase with L-tyrosine and L-DOPA as substrate. Oxyresveratrol showed mixed inhibition and non-competitive inhibition against L-tyrosine and L-DOPA, respectively, as substrate. The results indicated that the tyrosinase inhibitory activity of mulberroside A was greatly enhanced by the bioconversion process.

Two active pheophorbide compounds identified as pheophorbide a methyl ester and 132(S)-hydroxypheophorbide a methyl ester, isolated from fractionation of the methanol extract of *Morus alba* leaves, exhibited potent inhibitory activity in binding of europium-labelled melanin-concentrating hormone (MCH) to the human recombinant MCH-1 receptor (IC_{50} value; 4.03 and 0.33 μM, respectively) (Oh et al. 2010). Besides binding activity, the pheophorbides inhibited MCH-mediated extracellular signal-regulated kinase (ERK) phosphorylation in Chinese hamster ovary cells expressing human MCH-1 receptor. The results suggested that bith pheophorbides acted as modulators of MCH-1 receptor and MCH-mediated ERK signalling. Recent studies showed that purple-colored *Morus alba* fruit extract was found to contain the highest levels of anthocyanin and strongest antioxidant as well as anti-tyrosinase properties compared with other colored mulberry fruit extracts (Aramwit et al. 2010). Light or heat exposure by incubation of the mulberry fruit extract at 70°C for 10 h significantly decreased total anthocyanin and ascorbic acid content and led to a corresponding increase of the IC_{50} values. Studies showed that ethanolic extract of *Morus alba* twigs (EEMT) in the range of 0–60 μg/ml, exhibited tyrosinase inhibitory activity of EEMT that was superior to that of the ethanolic extract of mulberry root bark (EEMR) (Chang et al. 2011). The phenolic components present in the twigs were found to be maclurin, rutin, isoquercitrin, resveratrol, and morin.

Antiinflammatory Activity

Mulberroside A and oxyresveratrol obtained from Mori Cortex (*M. alba* root cortex) showed an inhibitory effect against FeSO4/H_2O_2-induced lipid peroxidation in rat microsomes and a scavenging effect on 1,1-diphenyl-2-picrylhydrazyl radical (Chung et al. 2003). Mulberroside A and oxyresveratrol exhibited potent anti-inflammatory effects significantly reducing paw edema. Exposure of lipopolysaccharide (LPS)-stimulated murine macrophage cell line RAW 264.7 to oxyresveratrol inhibited nitrite accumulation in the culture medium. Oxyresveratrol also inhibited the LPS-stimulated increase of inducible nitric oxide synthase (iNOS) expression in a concentration-dependent manner; however, it had little effect on iNOS enzyme activity, suggesting that the inhibitory activity of oxyresveratrol is mainly due to the inhibition of iNOS expression rather than iNOS enzyme activity. Oxyresveratrol significantly inhibited LPS-evoked nuclear translocation of NF-kappaB and cyclooxygenase-2 (COX-2) activity in RAW 264.7 cells. The results suggested that the antiinflammatory properties of oxyresveratrol might be correlated with inhibition of the iNOS expression through down-regulation of NF-kappaB binding activity and significant inhibition of COX-2 activity.

Morus alba leaf methanolic extract and its fractions (chloroform, butanol, and aqueous fractions) were found to inhibit NO production in LPS-activated RAW264.7 macrophages without an appreciable cytotoxic effect at concentration from 4 to 100 μg/ml (Choi and Hwang 2005). LPS-induced PGE2 production was significantly reduced only by butanol fraction. The butanol fraction exhibited inhibitory activities on COX-2 and iNOS. In addition, *M. alba* leaf extract and its fractions significantly decreased the production of TNF-α. The findings suggested that *M. alba* leaf extract seemed to be able in suppressing inflammatory mediators. Resveratrol, a polyphenol compound and prominent anti-inflammatory agent found in *Morus alba* fruits inhibited LPS-induced interleukin-8 production in a dose-dependent manner in human monocytic cell line, THP-1 (Oh et al. 2009b). The data showed that resveratrol inhibited IL-8 secretion by blocking MAPK phosphorylation and NF-kappaB activation. In inflammatory diseases, IL-8 plays a central role in the initiation and maintenance of inflammatory response. Sanggenon C and O, two Diels-Alder type adducts isolated from *Morus alba* were found to have anti-inflammatory activity (Dat et al. 2011). In LPS-stimulated RAW264.7 cells, both sanggenon C and O inhibited NO production and iNOS expression by suppressing NF-κB activity and IκBα activation. Sanggenon O showed stronger inhibition than the diastereomer sanggenon C.

Studies showed that ethyl acetate fractions of *Andrographis paniculata, Angelica sinensis* and *Morus alba* (three traditional Chinese medicine herbs) significantly decreased nuclear factor kappa B (NF-κB) luciferase activity and also the secretion of NO (nitric oxide) and prostaglandin E2 (PGE2) in lipopolysaccharide/interferon-gamma (LPS/IFN-γ) stimulated mouse peritoneal macrophages (Chao et al. 2009). In contrast, they did not affect IFN-γ luciferase activity or IFN-γ production in concanavalin A (Con A)-activated mouse splenocytes. The results indicated that the antiinflammatory properties of these plant extracts might have resulted from the inhibition of pro-inflammatory mediators (e.g., NO and PGE2), at least in part via suppression of a signaling pathway such as NF-κB and acted in a cell type dependent fashion.

Cudraflavone B, a prenylated flavonoid in *Morus alba* roots, was found to cause significant inhibition of inflammatory mediators in selected in vitro models and could be used for development as a nonsteroidal anti-inflammatory drug lead (Hošek et al. 2011). The compound was identified as a potent inhibitor of tumour necrosis factor α (TNFα) gene expression and secretion by blocking the translocation of nuclear factor κB (NF-κB) from the cytoplasm to the nucleus in macrophages derived from a THP-1 human monocyte cell line. The NF-κB activity reduction resulted in the inhibition of cyclooxygenase 2 (COX-2) gene expression. The compound acted as a COX-2 and COX-1 inhibitor with higher selectivity toward COX-2 than indomethacin.

Antiplatelet and Endothelial Function Activity

The effects of various phenolic compounds isolated from the root bark of the mulberry tree affected rat platelet cyclooxygenase and lipoxygenase products formed from (1–14C) arachiodonic acid (Kimura et al. 1986). Kuwanons G and H, sanggenon C, and mulberrofuran Q at concentrations of 10^{-3} to 10^{-4} M inhibited the formation of 12-hydroxy-5,8,10-heptadecatrienoic acid (HHT) and thromboxane B2 (cyclooxygenase products); however, they increased the formation of 12-hydroxy-5,8,10,14-eicosatetraenoic acid (12-HETE) (a lipoxygenase product). Sanggenon D and mulberrofuran J at concentration of 10^{-3} M inhibited the formation of HHT, thromboxane B2, and 12-HETE. Mulberrofuran G at a concentration of 10^{-3} M inhibited the formation of HHT, thromboxane B2, and 12-HETE, while at 10^{-5} M, it inhibited the formation of 12-HETE without affecting the formations of HHT and thromboxane B2. Kuwanon M and mulberroside A did not affect arachidonate metabolism in rat platelets.

Chloroform extract of mulberry twig dose-dependently inhibited platelet aggregation by arachidonic acid, and reduced vascular tension of phenylephrine preconstricted aorta rings with or without endothelium (Ling et al. 2010). The etyl acetate and n-butanol extracts dose-dependently inhibited NO production stimulated by LPS/IFN-gamma suggesting an anti-inflammatory effect.

Studies in isolated rat thoracic rings showed that ethyl acetate extract from leaves of *Morus alba* had dual vasoactive effects (Xia et al. 2008). The relaxation in endothelium-intact and -denuded aortas precontracted by high K(+) or phenylephrine was greater than the contraction in endothelium-denuded aortas precontracted by phenylephrine. The relaxation was mediated by inhibition of voltage- and receptor-dependent Ca(2+) channels in vascular smooth muscle cells, while the contraction occurred via activation of ryanodine receptors in the sarcoplasmic reticulum.

Studies showed that treatment of human endothelial cells (HEC) exposed to resistin (a cytokine) with *M. alba* extract and cucurmin inhibited significantly P-selectin and fractalkine expression beside inhibiting the increase in the intracellular ROS level (Pirvulescu et al. 2011). The extracts reduced NADPH activation and monocytes adhesion to HEC. The results indicate that *Morus alba* and curcumin targeted resistin-induced human endothelial activation partly via antioxidant mechanisms and suggested that they may represent therapeutic agents in vascular disease mediated by resistin.

Immunomodulatory Activity

A polysaccharide isolated from *Morus alba* root bark (PMA) was found to enhance proliferation of splenic lymphocytes in a synergistic manner in the presence of mitogens (Kim et al. 2000). However, PMA suppressed primary IgM antibody production from B cells, which was activated with lipopolysaccharide, a polyclonal activator, or immunized with a T-cell dependent antigen sheep red blood cells. The observations showed that the immunomodulating activity of PMA increased lymphocyte proliferation and that PMA decreased antibody production from B cells, which was distinct from those of other plant-originated polysaccharides.

Studies showed that *Morus alba* leaf extract at low dose and high dose of 100 mg/kg and 1 g/kg increased the levels of serum immunoglobulins and prevented the mortality induced by bovine *Pasteurella multocida* in mice (Bharani et al. 2010). It also increased the circulating antibody titre in indirect haemagglutination test. In contrast, it showed significant increase in the phagocytic index in carbon clearance assay, a significant protection against cyclophosphamide induced neutropenia and increased the adhesion of neutrophils in the neutrophil adhesion test. The authors concluded that *Morus alba* increased both humoral immunity and cell mediated immunity.

Antidyskinesia Activity

Studies showed that *Morus alba* leaf extract had a protective effect against haloperidol-induced orofacial dyskinesia and oxidative stress (Nade et al. 2010). Long-term treatment with haloperidol, a typical neuroleptic, induced neurodegeneration caused by excitotoxicity and oxidative stress, which played an important role in the development of orofacial dyskinesia. Concomitant treatment of methanol leaf extract (100–300 mg/kg, i.p.) and haloperidol (1 mg/kg, i.p.) attenuated the increase in vacuous chewing movements and tongue protrusions induced by haloperidol. The extract showed a marked effect on behavioral parameters altered by haloperidol treatment. Similar treatment with extract attenuated haloperidol-induced lipid peroxidation and nitrite and normalized superoxide dismutase, catalase, and protein in comparison to the control group.

Antiulcerogenic Activity

Studies indicated that *M. alba* extract exhibited significant anti-ulcerogenic activity in rats

(Abdulla et al. 2009). Animals pretreated with either plant extracts (250 or 500 mg/kg body weight) or omeprazole had statistically significant reduction of gastric mucosal damage reduction of edema and leucocytes infiltration of the submucosal layer compared to negative control animals.

Neuroprotective/Cognitive Activity

A water-soluble fraction from mulberry leaves (ML water fraction) exhibited cholinomimetic effect on the circulatory and autonomic nervous systems, which were compared with those of acetylcholine a reference drug (Lee et al. 2005), Intravenous administration of acetylcholine or a mulberry leaves (ML water fraction) produced temporary depressor and tachycardiac responses in a dose-dependent manner in unrestrained, conscious Sprague-Dawley rats. The systemic hemodynamic effects of acetylcholine and the ML water fraction were almost completely blocked by pretreatment with atropine, a muscarinic antagonist. The depressor responses to acetylcholine and ML water fraction were slightly enhanced and prolonged by pretreatment with neostigmine, an anticholinesterase, whereas the tachycardiac responses were remarkably blocked by pretreatment with pentolinium, a ganglionic blocking agent. In vitro experiments using the ileum isolated from rats showed that acetylcholine and a ML water fraction increased ileal contractility in a dose-dependent manner. The increases in ileal contractility were also completely abolished in the presence of atropine. Finally, the specific binding of [^{3}H] quinuclidinyl benzilate, a muscarinic antagonist, to rat cortical synaptic membranes was inhibited by ML water fraction in a concentration-dependent manner with an IC_{50} value of 9.5 mg/ml. The results suggested that the effects of a ML water fraction were mediated through direct stimulation of muscarinic cholinergic receptors by unknown cholinomimetic substance(s) contained in that fraction.

Mulberry leaves (ML) with enhanced accumulation of γ-aminobutyric acid (GABA) (GAML) exhibited neuroprotective actions against cerebral ischemia in vitro and in vivo (Kang et al. 2006b). Several neurological disorders such as Alzheimer's and Parkinson's diseases have been attributed to γ-aminobutyric acid (GABA) depletion in the brain. GABA enhanced the potential of neuroprotection in the PC12 cells damaged by H(2)O(2)-induced oxidation. GAML reduced the cytotoxicity in the PC12 cells against oxygen glucose deprivation-induced cerebral ischemic condition. The neuroprotective effect of GAML was further demonstrated in-vivo using middle cerebral artery occlusion brain injury model. Overall, the results suggested that the anaerobic treatment of ML makes GAML enhance the neuroprotection effect against in-vivo cerebral ischemia such as in vitro. cyanidin-3-O-β-d-glucopyranoside (C3G) from the mulberry fruits exhibited neuroprotective effects on neuronal cell damage (Kang et al. 2006a). A 1% HCl-MeOH mulberry fruit extract was shown to have a cytoprotective effect on PC12 cells that had been exposed to hydrogen peroxide. The extract inhibited the cerebral ischemic damage caused by oxygen glucose deprivation (OGD) in PC12 cells. The neuroprotective effect of the mulberry fruit extract was further demonstrated in vivo using a mouse-brain-injury model with a transient middle cerebral artery occlusion (MCAO). C3G was isolated as a neuroprotective constituent from the mulberry fruit extract. Compared with the control group, C3G had neuroprotective effects on the PC12 cells exposed to hydrogen peroxide in vitro and on cerebral ischemic damage in vivo.

Mulberry extract (ME), rich in phenolics and anthocyanins, was found to have a beneficial effect on the induction of antioxidant enzymes and on the promotion of cognition in senescence-accelerated mice (SAMP) (Shih et al. 2009). Studies showed that the senescence-accelerated mice fed the mulberry extract (ME) supplement demonstrated significantly less amyloid beta protein and showed improved learning and memory ability in avoidance response tests. ME-treated mice showed a higher antioxidant enzyme activity and less lipid oxidation in both the brain and liver, as compared to the control mice. Furthermore, treatment with ME decreased the levels of serum aspartate aminotransferase, alanine aminotransferase, triglyceride and total cholesterol that increase with ageing. The hepatoprotective effect of ME appeared to occur through a mechanism related to regulation of the mitogen-activated protein kinases and activation of the nuclear factor-erythroid 2 related factor 2, where the latter regulates the induction of phase 2 antioxidant enzymes and reduction of oxidative damage. Overall, supplementation of ME might be advantageous to the induction of an antioxidant defense system and for the improvement of memory deterioration in ageing animals. Chronic restraint stress produced cognitive dysfunction, altered behavioral parameters, increased leucocytes count, superoxide dismutase, lipid peroxidation, glucose and corticosterone levels, with concomitant decrease in catalase, and glutathione reductase activities in whole rat brain (Nade and Yadav 2010). Gastric ulceration, adrenal gland and spleen weights were also used as the stress indices. All these restraint stress induced perturbations were mitigated by ethyl acetate soluble fraction of *Morus alba*.

In recent studies, a 70% ethanol extract of *Morus alba* fruit (ME) exhibited protective effects against neurotoxicity in-vitro and in- vivo Parkinson's disease (PD) models (Kim et al. 2010a). Parkinson's disease, one of the most common neurodegenerative disorders, is characterised by the loss of dopaminergic neurons in the substantia nigra pars compacta (SNpc) to the striatum (ST), and involves oxidative stress. Mulberry fruit (*Morus alba*) is commonly eaten in Korea, and has long been used in traditional oriental medicine. In SH-SY5Y (neuroblastoma) cells stressed with 6-hydroxydopamine (6-OHDA), ME significantly protected the cells from neurotoxicity in a dose-dependent manner. Other assays demonstrated that the protective effect of ME was mediated by its antioxidant and anti-apoptotic effects, regulating reactive oxygen species and NO generation, Bcl-2 and Bax proteins, mitochondrial membrane depolarisation and caspase-3 activation. In mesencephalic primary cells stressed with 6-OHDA or 1-methyl-4-phenylpyridinium (MPP+), pre-treatment with ME also protected dopamine neurons, showing a wide range of effective concentrations in MPP+-induced toxicity. In the sub-acute mouse PD model induced by 1-methyl-4-phenyl-1,2,3,6-tetrahydropyridine (MPTP), ME showed a preventative effect against PD-like symptoms (bradykinesia) in the behavioural test and prevented MPTP-induced dopaminergic neuronal damage in an immunocytochemical analysis of the SNpc and ST. The results indicated that ME has neuroprotective effects in in-vitro and in-vivo PD models, and that it may be useful in preventing or treating Parkinson's disease.

Studies indicated that the methanolic extract of *Morus alba* leaves possessed antidopaminergic activity (Yadav and Nade 2008). The extract produced significant dose dependent potentiation of haloperidol (1 mg/kg, i.p.) and metoclopramide (20 mg/kg, i.p.) induced catalepsy in mice. The extract signifi cantly reduced number of fi ghts and increased latency to fi ghts in foot shockin-duced aggression; it also decreased amphetamine (1 mg/kg, i.p.) induced stereotyped behavior in a dose dependent manner. The sleeping time induced by phenobarbitone (50 mg/kg, i.p.) too was prolonged. The extract inhibited contractions produced by dopamine on isolated rat vas deferens.

Anxiolytic/Sedative Activity

Bioactive compounds in the "Mori Cortex": – root cortex of *M. alba* (traditional Chinese medicine) were identified as follows: mulberroside A and cis-mulberroside A were identified in plasma, oxyresveratrol, oxyresveratrol 2,3′-O-β-D-diglucuronide and oxyresveratrol and oxyresveratrol 2-O-β-D-glucuronide-3′-O-sulfate in urine and mulberroside A, oxyresveratrol, oxyresveratrol 2,3′-O-β-D-diglucuronide and oxyresveratrol and oxyresveratrol 2-0-β-D-glucuronide-3′-O-sulfate in bile (Qiu et al. 1999). The origins of these metabolites were considered to be mainly derived from mulberroside A in Mori Cortex. Oxyresveratrol showed a relaxant effect on contractions of Guinea pig tracheal smooth muscle induced by histamine and its IC_{50} was 25 μg/mg.

The results of a study by Yadav et al. (2008) suggested that a methanolic extract of *M. alba* leaves may possess an anxiolytic effect. Methanolic extract of *M. alba* leaves significantly increased the number and duration of head poking in the hole-board test. In the elevated plus-maze, the extract significantly increased the exploration of the open arm in similar way to that of diazepam. At a dose of 200 mg/kg i.p. the extract significantly increased both the time spent in and the entries into the open arm by mice. Further, in the open field test, the extract significantly increased rearing, assisted rearing, and number of squares traversed, all of which are demonstrations of exploratory behavior. In the light/dark paradigm, the extract produced significant increase in time spent in the lighted box as compared to vehicle. The spontaneous locomotor activity count, measured using an actophotometer, was significantly decreased in animals pretreated with *M. alba* extract, indicating a remarkable sedative effect of the plant.

Studies showed that aqueous extract of *Morus alba* leaves green tea (ME) possessed an antidepressant- without an anxiolytic-like effect, however, at high doses (500 or 1,000 mg/kg), the extract showed sedative effect and altered other functions such as muscle strength, animal activity in the maze and pain response (Sattayasai et al. 2008).

Nephroprotective/Uricosuric Activity

Five prenylflavonoids similar to glabridin (1–5), isolated from *Morus alba*, *Artocarpus communis*, *Glycyrrhiza uralensis* and *G. inflata*, exhibited antinephritis activity in mice with glomerular disease (Masugi-nephritis) (Fukai et al. 2003). Oral administrations of artonin E or licochalcone A for 10 days (30 mg/kg/day) reduced the amount of urinary protein excretion compared to nephritic mice. ESR spectroscopy demonstrated that morusin and licorisoflavan A increased the radical intensity of sodium ascorbate by about two times. Morusin, licoricidin, licochalcone A and licorisoflavan A showed weak scavenging activity against superoxide anion radical.

Mulberroside A, a major stilbene glycoside of *Morus alba* was found to have uricosuric and nephroprotective effects and have potential treatment of hyperuricemia with renal dysfunction (Wang et al. 2011). This was mediated in part by mitigation of the expression alterations of renal organic ion transporters in hyperuricemic mice. Mulberroside A at 10, 20, and 40 mg/kg decreased serum uric acid levels and increased urinary urate excretion and fractional excretion of uric acid in oxonate-induced hyperuricemic mice. it reduced serum levels of creatinine and urea nitrogen (10–40 mg/kg), urinary N-acetyl-β-D-glucosaminidase activity (10–40 mg/kg), β(2)-microglobulin (10–40 mg/kg) and albumin (20–40 mg/kg), and increased creatinine clearance (10–40 mg/kg) in hyperuricemic mice. In addition, mulberroside A downregulated mRNA and protein levels of renal glucose transporter 9 (mGLUT9) and urate transporter 1 (mURAT1), and upregulated mRNA and protein levels of renal organic anion transporter 1 (mOAT1) and organic cation and carnitine transporters (mOCT1, mOCT2, mOCTN1, and mOCTN2) in hyperuricemic mice.

Hepatoprotective Activity

Morus alba alcoholic leaf extract exerted a protective potential in carbon tetrachloride induced liver injury (Kalantari et al. 2009). The hydroalcoholic extract at dose of 800 mg/kg exhibited a significant liver protective effect by lowering the serum levels of aspartate aminotransferase (AST) and alanine aminotransferase (ALT) level, decreasing the sleeping time and resulting in less pronounced histopathological destruction of the liver architecture, there was no fibrosis and inflammation, as compared with CCl4 group. It was speculated that mulberry flavonoids (rutin, moracetin, quercetin-3-triglucoside) were responsible for the observed protective effects.

Adaptogenic Activity

Morus alba root extract possessed adaptogenic property (Nade et al. 2009). Pre-treatments with the ethyl acetate soluble fraction of methanol extract of *M. alba* roots (25, 50 and 100 mg/kg, p.o.) significantly attenuated the chronic stress (CS)-induced perturbations (Nade et al. 2009). Chronic stress significantly induced cognitive deficit, mental depression and hyperglycemia and increased blood corticosterone levels, gastric ulcerations and adrenal gland weight, but decreased the splenic weight. The results indicated that *M. alba* possessed significant adaptogenic activity, indicating its possible clinical utility as an anti-stress agent.

Anaphylactic Activity

Hot-water extract from the root bark of *Morus alba* (HEMA) significantly inhibited systemic anaphylaxis induced by the compound 48/80 in mice. HEMA also significantly inhibited the passive cutaneous anaphylaxis activated by anti-CGG IgE (Chai et al. 2005). HEMA had no cytotoxicity on rat peritoneal mast cells (RPMC). The results suggested that hot-water extract from the root bark of *Morus alba* (HEMA) inhibited the compound 48/80- or anti-chicken γ globulin (CGG) IgE-induced mast cell activation and its inhibitory effects on mast cell activations were favorably comparable to disodium cromoglycate. HEMA could be a candidate for effective therapeutic tools of allergic diseases.

Antivenom Activity

Morus alba leaf extract was found to possess potent antisnake venom property, especially against the local and systemic effects of *Daboia russelii* venom (Chandrashekara et al. 2009). The extract completely abolished the in-vitro proteolytic and hyaluronolytic activities of the venom. Edema, hemorrhage and myonecrotic activities were also neutralized efficiently. Additionally, the extract partially inhibited the pro-coagulant activity and completely abolished the degradation of alpha chain of human fibrinogen.

Traditional Medicinal Uses

In Traditional Chinese Medicine, all parts of the tree are used as medicine in one way or another. The fruit is used in the treatment of urinary incontinence, dizziness, tinnitus, insomnia due to anaemia, neurasthenia, hypertension, diabetes, premature greying of the hair, and constipation in the elderly. The fruit has a tonic effect on kidney energy and is used to tonify the blood. In India the syrup of the fruits is useful as a refrigerant in fevers and as an expectorant in coughs and sore throat. A drink made of its juice is cooling and refreshing and is a cure for dry throat and thirst.

The leaves are antibacterial, astringent, diaphoretic, hypoglycaemic, odontalgic and ophthalmic. Leaves are taken internally in the treatment of colds, influenza, eye infections and nosebleeds An injected extract of the leaves have been be used in the treatment of elephantiasis, purulent fistulae and in tetanus following oral doses of the leaf sap mixed with sugar. The bruised leaves are used in wounds and insect bites, and are thought to promote the growth of hair. In Korea, *Morus alba* leaves have been traditionally administered as natural therapeutic agent for the alleviating dropsy and diabetes (Oh et al. 2009a, b) and for hypertension and general weakness. The leaves are diaphoretic. Made into a strong decoction, they are used for sweating feet, dropsy, and intestinal disorders.

The stems and twigs are considered antirheumatic, antispasmodic, diuretic, hypotensive and pectoral. The twigs are considered prophylactic against all forms of cold, and are also diuretic and pectoral. They are used in the treatment of rheumatic pains and spasms, especially of the upper half of the body, high blood pressure. A tincture of the bark is used to relieve toothache. Native Americans used infusions made from the bark of *Morus alba* medicinally in various ways: as a laxative, as a treatment for dysentary, and as a purgative (Moerman 1998). In Korea, the bark is used for treating cough, anasarca with olignia, dysuria, constipation and elsewhere for fever, headache, red dry and sore eyes. The bark is used to treat cough, wheezing, edema, and to promote urination. The root bark is antiasthmatic, antitussive, diuretic, expectorant, hypotensive and sedative. It is used internally in the treatment of asthma, coughs, bronchitis, oedema, hypertension and diabetes. The root-bark is regarded as a restorative, tonic, and astringent remedy in nervous disorders and used as an emollient in Malaya. The bark is anthelmintic and purgative, it is used to expel tape worms. The juice of the fresh bark is use in epilepsy in children and in dribbling of saliva. Extracts of the plant have antibacterial and fungicidal activity. The milky sap of the tree is used in aphthous stomatitis in infants and incised wounds caused by snake, centipede, and spider bites. A lye made of the ashes of mulberry wood is used as a stimulant and escharotic in scaly skin diseases and unhealthy granulations.

Other Uses

White Mulberry leaves are the preferred natural food for silkworms (Plate 6). The leaves are also cut for food for livestock (cattle and goats) Mulberry is an excellent and handsome shade tree along streets or as living fence. Wood and bark are used in China and Turkey for tannery and paper fabrication, and also for fuel, natural colouring, alcohol ennobling and in cosmetic products elsewhere. Mulberry provides excellent timber which is used for sports good, furniture and is pollarded to furnish round wood. A kind of cellulose whiskers extracted from the branch-barks of mulberry (*Morus alba*) was reported to have potential applications in the fields of composites as a reinforcing phase, as well as in pharmaceutical and optical industries as additives (Li et al. 2009).

Comments

See alno notes on *Morus rubra* and *M. nigra*.

Selected References

Abdulla MA, Ali HM, Ahmed KAA, Noor SM, Ismail S (2009) Evaluation of the anti-ulcer activities of *Morus alba* extracts in experimentally-induced gastric ulcer in rats. Biomed Res 20(1):35–39

Andallu B, Varadacharyulu NC (2002) Control of hyperglycemia and retardation of cataract by mulberry (*Morus indica* L.) leaves in streptozotocin diabetic rats. Indian J Exp Biol 40(7):791–795

Andallu B, Varadacharyulu NC (2003) Antioxidant role of mulberry (*Morus indica* L. cv. Anantha) leaves in streptozotocin-diabetic rats. Clin Chim Acta 338(1–2): 3–10

Andallu B, Varadacharyulu NC (2007) Gluconeogenic substrates and hepatic gluconeogenic enzymes in streptozotocin-diabetic rats: effect of mulberry (*Morus indica* L) leaves. J Med Food 10(1):41–48

Andallu B, Suryakantham V, Lakshmi Srikanthi B, Reddy GK (2001) Effect of mulberry (*Morus indica* L.) therapy on plasma and erythrocyte membrane lipids in patients with type 2 diabetes. Clin Chim Acta 314(1–2):47–53

Arabshahi-Delouee S, Urooj A (2007) Antioxidant properties of various solvent extracts of mulberry (*Morus indica* L.) leaves. Food Chem 102(4):1233–1240

Aramwit P, Bang N, Srichana T (2010) The properties and stability of anthocyanins in mulberry fruits. Food Res Int 43(4):1093–1097

Asano N, Oseki K, Tomioka E, Kizu H, Matsui K (1994) N-containing sugars from *Morus alba* and their glycosidase inhibitory activities. Carbohydr Res 259(2):243–255

Asano N, Yamashita T, Yasuda K, Ikeda K, Kizu H, Kameda Y, Kato A, Nash RJ, Lee HS, Ryu KS (2001) Polyhydroxylated alkaloids isolated from mulberry trees (*Morus alba* L.) and silkworms (*Bombyx mori* L.). J Agric Food Chem 49(9):4208–4213

Bae SH, Suh HJ (2007) Antioxidant activities of five different mulberry cultivars in Korea. LWT – Food Sci Technol 40(6):955–962

Bharani SE, Asad M, Dhamanigi SS, Chandrakala GK (2010) Immunomodulatory activity of methanolic extract of *Morus alba* Linn. (mulberry) leaves. Pak J Pharm Sci 23(1):63–68

Bown D (1995) Encyclopaedia of herbs and their uses. Dorling Kindersley, London, 424 pp

Burkill IH (1966) A dictionary of the economic products of the Malay Peninsula, Revised reprint, 2 vols. Ministry of agriculture and co-operatives, Kuala Lumpur, vol 1 (A–H), pp 1–1240, vol 2 (I–Z), pp 1241–2444

Chai OH, Lee MS, Han EH, Kim HT, Song CH (2005) Inhibitory effects of *Morus alba* on compound 48/80-induced anaphylactic reactions and anti-chicken gamma globulin IgE- mediated mast cell activation. Biol Pharm Bull 28(10):1852–1858

Chan KC, Ho HH, Huang CN, Lin MC, Chen HM, Wang CJ (2009) Mulberry leaf extract inhibits vascular smooth muscle cell migration involving a block of small GTPase and Akt/NF-kappaB signals. J Agric Food Chem 57(19):9147–9153

Chandrashekara KT, Nagaraju S, Nandini SU, Basavaiah KK (2009) Neutralization of local and systemic toxicity of *Daboia russelii* venom by *Morus alba* plant leaf extract. Phytother Res 23(8):1082–1087

Chang LW, Juang LJ, Wang BS, Wang MY, Tai HM, Hung WJ, Chen YJ, Huang MH (2011) Antioxidant and antityrosinase activity of mulberry (*Morus alba* L.) twigs and root bark. Food Chem Toxicol 49(4): 785–790

Chao WW, Kuo YH, Li WC, Lin BF (2009) The production of nitric oxide and prostaglandin E2 in peritoneal macrophages is inhibited by *Andrographis paniculata, Angelica sinensis* and *Morus alba* ethyl acetate fractions. J Ethnopharmacol 122(1):68–75

Chen F, Nakashima N, Kimura I, Kimura M (1995) Hypoglycemic activity and mechanisms of extracts from mulberry leaves (folium mori) and cortex mori radicis in streptozotocin-induced diabetic mice. Yakugaku Zasshi 115:476–482

Chen P-N, Chu S-C, Chiou H-L, Kuo WH, Chiang CL, Hsieh Y-S (2006) Mulberry anthocyanins, cyanidin 3-rutinoside and cyanidin 3-glucoside, exhibited an inhibitory effect on the migration and invasion of a human lung cancer cell line. Cancer Lett 235(2): 248–259

Choi EM, Hwang JK (2005) Effects of *Morus alba* leaf extract on the production of nitric oxide, prostaglandin E2 and cytokines in RAW264.7 macrophages. Fitoterapia 76(7–8):608–613

Chu QC, Lin M, Tian XH, Ye JN (2006) Study on capillary electrophoresis–amperometric detection profiles of different parts of *Morus alba* L. J Chromatogr A 1116(1–2):286–290

Chung KO, Kim BY, Lee MH, Kim YR, Chung HY, Park JH, Moon JO (2003) In-vitro and in-vivo anti-inflammatory effect of oxyresveratrol from *Morus alba* L. J Pharm Pharmacol 55(12):1695–1700

Council of Scientific and Industrial Research (CSIR) (1962) The wealth of India. A dictionary of Indian raw materials and industrial products. (Raw materials 6). Publications and Information Directorate, New Delhi

Cui L, Na M, Oh H, Bae EY, Jeong DG, Ryu SE, Kim S, Kim BY, Oh WK, Ahn JS (2006) Protein tyrosine phosphatase 1B inhibitors from *Morus* root bark. Bioorg Med Chem Lett 16(5):1426–1429

Dat NT, Binh PT, le Quynh TP, Van Minh C, Huong HT, Lee JJ (2010) Cytotoxic prenylated flavonoids from *Morus alba*. Fitoterapia 81(8):1224–1227

Dat NT, Xuan Binh PT, Phuong Quynh LT, Huong HT, Minh CV (2011) Sanggenon C and O inhibit NO production, iNOS expression and NF-κB activation in LPS-induced RAW264.7 cells. Immunopharmacol Immunotoxicol (in press)

Doi K, Kojima T, Fujimoto Y (2000) Mulberry leaf extract inhibits oxidative modification of rabbit and human low-density lipoprotein. Biol Pharm Bull 23(9):1066–1071

Doi K, Kojima T, Makino M, Kimura Y, Fujimoto Y (2001) Studies on the constituents of the leaves of *Morus alba* L. Chem Pharm Bull (Tokyo) 49(2):151–153

Du J, He ZD, Jiang RW, Ye WC, Xu HX, But PP (2003) Antiviral flavonoids from the root bark of *Morus alba* L. Phytochemistry 62(8):135–138

Du Q, Zheng J, Xu Y (2008) Composition of anthocyanins in mulberry and their antioxidant activity. J Food Compos Anal 21(5):390–395

Duke JA, Ayensu ES (1985) Medicinal plants of China, vol 1&2. Reference Publications Inc., Algonac, 705 pp

El-Beshbishy HA, Singab ANB, Sinkkonen J, Pihlaja K (2006) Hypolipidemic and antioxidant effects of *Morus alba* L. (Egyptian mulberry) root bark fractions supplementation in cholesterol-fed rats. Life Sci 78:2724–2733

El-Sayyad HI, El-Sherbiny MA, Sobh MA, Abou-El-Naga AM, Ibrahim MA, Mousa SA (2011) Protective effects *of Morus alba* leaves extract on ocular functions of pups from diabetic and hypercholesterolemic mother rats. Int J Biol Sci 7(6):715–728

Enkhmaa B, Shiwaku K, Katsube T, Kitajima K, Anuurad E, Yamasaki M, Yamane Y (2005) Mulberry (*Morus alba* L) leaves and their major flavonol quercetin 3-(6-malonylglucoside) attenuate atherosclerotic lesion development in LDL receptor-deficient mice. J Nutr 135:729–734

Ercisli S, Orhan E (2007) Chemical composition of white (*Morus alba*), red (*Morus rubra*) and black (*Morus nigra*) mulberry fruits. Food Chem 103(4): 1380–1384

Facciola S (1990) Cornucopia. A source book of edible plants. Kampong Publ, Vista, 677 pp

Foundation for Revitalisation of Local Health Traditions (2008) FRLHT Database. htttp://envis.frlht.org

Fukai T, Satoh K, Nomura T, Sakagami H (2003) Antinephritis and radical scavenging activity of prenylflavonoids. Fitoterapia 74(7–8):720–724

Fukai T, Kaitou K, Terada S (2005) Antimicrobial activity of 2-arylbenzofurans from *Morus* species against methicillin-resistant *Staphylococcus aureus*. Fitoterapia 76:708–711

Geng C, Yao S, Xue D, Zuo A, Zhang X, Jiang Z, Ma Y, Chen J (2010) New isoprenylated flavonoid from *Morus alba*. Zhongguo Zhong Yao Za Zhi 35(12):1560–1565 (In Chinese)

Gu XD, Sun MY, Zhang L, Fu HW, Cui L, Chen RZ, Zhang DW, Tian JK (2010) UV-B induced changes in the secondary metabolites of *Morus alba* L. leaves. Molecules 15(5):2980–2993

Hansawasdi C, Kawabata J (2006) α-Glucosidase inhibitory effect of mulberry (*Morus alba*) leaves on Caco-2. Fitoterapia 77(7–8):568–573

Hisayoshi K, Masashi Y, Norio D, Koichi S (2004) A novel cytotoxic prenylated flavonoid from the root of *Morus alba*. J Insect Biotechnol Sericol 73(3):113–116

Hong Y, Kim MY, Yoon M (2011) The anti-angiogenic herbal extracts Ob-X from *Morus alba, Melissa offici-*

nalis, and *Artemisia capillaris* suppresses adipogenesis in 3T3-L1 adipocytes. Pharm Biol 49(8):775–783
Hošek J, Bartos M, Chudík S, Dall'Acqua S, Innocenti G, Kartal M, Kokoška L, Kollár P, Kutil Z, Landa P, Marek R, Závalová V, Žemli✱ka M, Šmejkal K (2011) Natural compound cudraflavone B shows promising anti-inflammatory properties in vitro. J Nat Prod 74(4):614–619
Huang HP, Shih YW, Chang YC, Hung CN, Wang CJ (2008) Chemoinhibitory effect of mulberry anthocyanins on melanoma metastasis involved in the Ras/PI3K pathway. J Agric Food Chem 56(19):9286–9293
Imran M, Khan H, Shah M, Khan R, Khan F (2010) Chemical composition and antioxidant activity of certain *Morus* species. J Zhejiang Univ Sci B11(12):973–980
Islam B, Khan SN, Haque I, Alam M, Mushfiq M, Khan AU (2008) Novel anti-adherence activity of mulberry leaves: inhibition of *Streptococcus mutans* biofilm by 1-deoxynojirimycin isolated from *Morus alba*. J Antimicrob Chemother 62(4):751–757
Jia ZS, Tang MC, Wu JM (1999) The determination of flavonoid contents in mulberry and their scavenging effects on superoxide radicals. Food Chem 64(4):555–559
Kalantari H, Aghel N, Bayati M (2009) Hepatoprotective effect of *Morus alba* L. in carbon tetrachloride-induced hepatotoxicity in mice. Saudi Pharm J 17(1):90–94
Kang TH, Hur JY, Kim HB, Ryu JH, Kim SY (2006a) Neuroprotective effects of the cyanidin-3-O-beta-d-glucopyranoside isolated from mulberry fruit against cerebral ischemia. Neurosci Lett 391(3):122–126
Kang TH, Oh HR, Jung SM, Ryu JH, Park MW, Park YK, Kim SY (2006b) Enhancement of neuroprotection of mulberry leaves (*Morus alba* L.) prepared by the anaerobic treatment against ischemic damage. Biol Pharm Bull 29(2):270–274
Katsube T, Imawaka N, Kawano Y, Yamazaki Y, Shiwaku K, Yamane Y (2006) Antioxidant flavonol glycosides in mulberry (*Morus alba* L.) leaves isolated based on LDL antioxidant activity. Food Chem 97:25–31
Katsube T, Tsurunaga Y, Sugiyama M, Furuno T, Yamasaki Y (2009) Effect of air-drying temperature on antioxidant capacity and stability of polyphenolic compounds in mulberry (*Morus alba* L.) leaves. Food Chem 113(4):964–969
Katsube T, Yamasaki M, Shiwaku K, Ishijima T, Matsumoto I, Abe K, Yamasaki Y (2010) Effect of flavonol glycoside in mulberry (*Morus alba* L.) leaf on glucose metabolism and oxidative stress in liver in diet-induced obese mice. J Sci Food Agric 90(14):2386–2392
Kikuchi T, Nihei M, Nagai H, Fukushi H, Tabata K, Suzuki T, Akihisa T (2010) Albanol A from the root bark of *Morus alba* L. induces apoptotic cell death in HL60 human leukemia cell line. Chem Pharm Bull (Tokyo) 58(4):568–571
Kim HM, Han SB, Lee KH, Lee CW, Kim CY, Lee EJ, Huch H (2000) Immunomodulating activity of a polysaccharide isolated form Mori cortex radicis. Arch Pharm Res 23:240–242
Kim YC, Kim MY, Takaya Y, Niwa M, Chung SK (2007) Phenolic antioxidants isolated from mulberry leaves. Food Sci Biotechnol 16(5):854–857
Kim HG, Ju MS, Shim JS, Kim MC, Lee SH, Huh Y, Kim SY, Oh MS (2010a) Mulberry fruit protects dopaminergic neurons in toxin-induced Parkinson's disease models. Br J Nutr 26:1–9
Kim JK, Kim M, Cho SG, Kim MK, Kim SW, Lim YH (2010b) Biotransformation of mulberroside A from *Morus alba* results in enhancement of tyrosinase inhibition. J Ind Microbiol Biotechnol 37(6):631–637
Kim GN, Kwon YI, Jang HD (2011) Mulberry leaf extract reduces postprandial hyperglycemia with few side effects by inhibiting α-glucosidase in normal rats. J Med Food 14(7–8):712–717
Kimura Y, Okuda H, Nomura T, Fukai T, Arichi S (1986) Effects of phenolic constituents from the mulberry tree on arachidonate metabolism in rat platelets. J Nat Prod 49:639–644
Kobayashi Y, Miyazawa M, Kamei A, Abe K, Kojima T (2010) Ameliorative effects of mulberry (*Morus alba* L.) leaves on hyperlipidemia in rats fed a high-fat diet: induction of fatty acid oxidation, inhibition of lipogenesis, and suppression of oxidative stress. Biosci Biotechnol Biochem 74(12):2385–2395
Lee SH, Choi SY, Kim HC, Hwang JS, Lee BG, Gao JJ, Kim SY (2002) Mulberroside F isolated from the leaves of *Morus alba* inhibits melanin biosynthesis. Biol Pharm Bull 25(8):1045–1048
Lee JS, Chung SH, Lee YS, Jin CB (2005) Bioanalysis and biotransformation cholinomimetic properties of a water-soluble fraction from mulberry leaves in rats. Biomol Ther 13(1):26–31
Lee J, Chae K, Ha J, Park BY, Lee HS, Jeong S, Kim MY, Yoon M (2008) Regulation of obesity and lipid disorders by herbal extracts from *Morus alba, Melissa officinalis,* and *Artemisia capillaris* in high-fat diet-induced obese mice. J Ethnopharmacol 115(2):263–270
Lee YJ, Choi DH, Kim EJ, Kim HY, Kwon TO, Kang DG, Lee HS (2011) Hypotensive, hypolipidemic, and vascular protective effects of *Morus alba* L in rats fed an atherogenic diet. Am J Chin Med 39(1):39–52
Li RJ, Fei JM, Cai YR, Li YF, Feng JQ, Yao JM (2009) Cellulose whiskers extracted from mulberry: a novel biomass production. Carbohydr Polym 76(1):94–99
Lin HY, Lai LS (2009) Isolation and viscometric characterization of hydrocolloids from mulberry (*Morus alba* L.) leaves. Food Hydrocolloids 23(3):840–848
Ling S, Zhang H, Zhang D, Zhang L, Bian K (2010) Characterizing effects of solvent specific *Morus alba* components on rat platelet aggregation, vascular tension and macrophage nitrite production. Zhongguo Zhong Yao Za Zhi 35(22):3024–3028 (In Chinese)

Liu LK, Lee HJ, Shih YW, Chyau CC, Wang CJ (2008) Mulberry anthocyanin extracts inhibit LDL oxidation and macrophage-derived foam cell formation induced by oxidative LDL. J Food Sci 73(6):H113–H121

Liu LK, Chou FP, Chen YC, Chyau CC, Ho HH, Wang CJ (2009) Effects of mulberry (*Morus alba* L.) extracts on lipid homeostasis in vitro and in vivo. J Agric Food Chem 57(16):7605–7611

Luo SD, Nemec J, Ning BM, Li QX (1994) Anti-HIV flavonoids from *Morus alba* L. Int Conf AIDS. 7–12 Aug 1994; 10: 203 abstract no. PB0240

Mia I, Lee BL, Ong CN, Liu X, Huang D (2008) Peroxyl radical scavenging capacity, polyphenolics, and lipophilic antioxidant profiles of mulberry fruits cultivated in southern China. J Agric Food Chem 56(20):9410–9416

Miyahara C, Miyazawa M, Satoh S, Sakai A, Mizusaki S (2004) Inhibitory effects of mulberry leaf extract on postprandial hyperglycemia in normal rats. J Nutr Sci Vitaminol 50:161–164

Moerman D (1998) Native American ethnobotany. Timber, Oregon, 927 pp

Mohammadi J, Naik PR (2008) Evaluation of hypoglycemic effect of *Morus alba* in an animal model. Indian J Pharmacol 40(1):15–18

Mudra M, Ercan-Fang N, Zhong L, Furne J, Levitt M (2007) Influence of mulberry leaf extract on the blood glucose and breath hydrogen response to ingestion of 75 g sucrose by type 2 diabetic and control subjects. Diabetes Care 30:1272–1274

Nade VS, Yadav AV (2010) Anti-stress effect of ethyl acetate soluble fraction of *Morus alba* in chronic restraint stress. Pharm Biol 48(9):1038–1046

Nade VS, Kawale LA, Naik RA, Yadav AV (2009) Adaptogenic effect of *Morus alba* on chronic footshock-induced stress in rats. Indian J Pharmacol 41(6):246–251

Nade VS, Kawale LA, Yadav AV (2010) Protective effect of *Morus alba* leaves on haloperidol-induced orofacial dyskinesia and oxidative stress. Pharm Biol 48(1):17–22

Nakamura M, Nakamura S, Oku T (2009) Suppressive response of confections containing the extractive from leaves of *Morus alba* on postprandial blood glucose and insulin in healthy human subjects. Nutr Metab (Lond) 6:29

Nam SY, Yi HK, Lee JC, Kim JC, Song CH, Park JW, Lee DY (2002) Cortex Mori extract induces cancer cell apoptosis through inhibition of microtubule assembly. Arch Pharm Res 25(2):191–196

Naowaboot J, Pannangpetch P, Kukongviriyapan V, Kongyingyoes B, Kukongviriyapan U (2009a) Antihyperglycemic, antioxidant and antiglycation activities of mulberry leaf extract in streptozotocin-induced chronic diabetic rats. Plant Foods Hum Nutr 64(2):116–121

Naowaboot J, Pannangpetch P, Kukongviriyapan V, Kukongviriyapan U, Nakmareong S, Itharat A (2009b) Mulberry leaf extract restores arterial pressure in streptozotocin-induced chronic diabetic rats. Nutr Res 29(8):602–608

Naowaratwattana W, De-Eknamkul W, De Mejia EG (2010) Phenolic-containing organic extracts of mulberry (*Morus alba* L.) leaves inhibit HepG2 hepatoma cells through G2/M phase arrest, induction of apoptosis, and inhibition of topoisomerase IIα activity. J Med Food 13(5):1045–1056

Nattapong S, Omboon L (2008) A new source of whitening agent from a Thai Mulberry plant and its betulinic acid quantitation. Nat Prod Res 22(9):727–734

Natural Products Research Institute (1998) Medicinal plants in the Republic of Korea. Seoul National University, WHO Regional Publications, Western Pacific series no 21. 316 pp

Nickavar B, Mosazadeh G (2009) Influence of three *Morus* species extracts on α-amylase activity. Iranian J Pharm Res 8(2):115–119

Nomura T, Hano Y, Fukai T (2009) Chemistry and biosynthesis of isoprenylated flavonoids from Japanese mulberry tree. Proc Jpn Acad Ser B Phys Biol Sci 85(9):391–408

Nuengchamnong N, Ingkaninan K, Kaewruang W, Wongareonwanakij S, Hongthongdaeng B (2007) Quantitative determination of 1-deoxynojirimycin in mulberry leaves using liquid chromatography–tandem mass spectrometry. J Pharm Biomed Anal 44(4): 853–858

Ochse JJ, Bakhuizen van den Brink RC (1980) Vegetables of the Dutch Indies, 3rd edn. Ascher & Co., Amsterdam, 1016 pp

Oh KS, Ryu SY, Lee S, Seo HW, Oh BK, Kim YS, Lee BH (2009a) Melanin-concentrating hormone-1 receptor antagonism and anti-obesity effects of ethanolic extract from *Morus alba* leaves in diet-induced obese mice. J Ethnopharmacol 122(2):216–220

Oh YC, Kang OH, Choi JG, Chae HS, Lee YS, Brice OO, Jung HJ, Hong SH, Lee YM, Kwon DY (2009b) Anti-inflammatory effect of resveratrol by inhibition of IL-8 production in LPS-induced THP-1 cells. Am J Chin Med 37(6):1203–1214

Oh BK, Oh KS, Kwon KI, Ryu SY, Kim YS, Lee BH (2010) Melanin-concentrating hormone-1 receptor binding activity of pheophorbides isolated from *Morus alba* leaves. Phytother Res 24(6):919–923

Oku T, Yamada M, Nakamura M, Sadamori N, Nakamura S (2006) Inhibitory effects of extractives from leaves of *Morus alba* on human and rat small intestinal disaccharidase activity. Br J Nutr 95:933–938

Park KM, You JS, Lee HY, Baek NI, Hwang JK (2003) Kuwanon G: an antibacterial agent from the root bark of *Morus alba* against oral pathogens. J Ethnopharmacol 84(2):181–185

Park JM, Bong HY, Jeong HI, Kim YK, Kim JY, Kwon O (2009) Postprandial hypoglycemic effect of mulberry leaf in Goto-Kakizaki rats and counterpart control Wistar rats. Nutr Res Pract 3(4):272–278

Pawlowska AM, Oleszek W, Braca A (2008) Quali-quantitative analyses of flavonoids of *Morus nigra* L. and *Morus alba* L. (Moraceae) fruits. J Agric Food Chem 56(9):3377–3380

Piao SJ, Chen LX, Kang N, Qiu F (2011) Simultaneous determination of five characteristic stilbene glycosides in root bark of *Morus albus* L. (Cortex Mori) using high-performance liquid chromatography. Phytochem Anal 22(3):230–235

Pirvulescu MM, Gan AM, Stan D, Simion V, Calin M, Butoi E, Tirgoviste CI, Manduteanu I (2011) Curcumin and a *Morus alba* extract reduce pro-inflammatory effects of resistin in human endothelial cells. Phytother Res (in press)

Qiu F, Kano Y, Yao XS (1999) Elucidation of the bioactive constituents in traditional Chinese medicine "Mori Cortex". Stud Plant Sci 6:281–289

Sattayasai J, Tiamkao S, Puapairoj P (2008) Biphasic effects of *Morus alba* leaves green tea extract on mice in chronic forced swimming model. Phytother Res 22(4):487–492

Sergio W (1989) Mulberry roots and seeds may be effective in the treatment of AIDS. Med Hypotheses 29(1):75–76

Shih PH, Chan YC, Liao JW, Wang MF, Yen GC (2009) Antioxidant and cognitive promotion effects of anthocyanin-rich mulberry (*Morus atropurpurea* L.) on senescence-accelerated mice and prevention of Alzheimer's disease. J Nutr Biochem 21(7):598–605

Singab ANB, El-Beshbishy HA, Yonekawa M, Nomura T, Fukai T (2005) Hypoglycemic effect of Egyptian *Morus alba* root bark extract: effect on diabetes and lipid peroxidation of streptozotocin-induced diabetic rats. J Ethnopharmacol 100(3):333–338

Skupień K, Kostrzewa-Nowak D, Oszmiański J, Tarasiuk J (2008) In vitro antileukaemic activity of extracts from chokeberry (*Aronia melanocarpa* [Michx] Elliott) and mulberry (*Morus alba* L.) leaves against sensitive and multidrug resistant HL60 cells. Phytother Res 22(5):689–694

Sohn HY, Son KH, Kwon CS, Kwon GS, Kang SS (2004) Antimicrobial and cytotoxic activity of 18 prenylated flavonoids isolated from medicinal plants: *Morus alba* L., *Morus mongolica* Schneider, *Broussnetia papyrifera* (L.) Vent, *Sophora flavescens* Ait and *Echinosophora koreensis* Nakai. Phytomedicine 11(7–8):666–672

Song W, Wang HJ, Bucheli P, Zhang PF, Wei DZ, Lu YH (2009) Phytochemical profiles of different mulberry (*Morus* sp.) species from China. J Agric Food Chem 57(19):9133–9140

Srivastava S, Kapoor R, Thathola A, Srivastava RP (2003) Mulberry (*Morus alba*) leaves as human food: a new dimension of sericulture. Int J Food Sci Nutr 54: 411–416

Sugimoto M, Arai H, Tamura Y, Murayama T, Khaengkhan P, Nishio T, Ono K, Ariyasu H, Akamizu T, Ueda Y, Kita T, Harada S, Kamei K, Yokode M (2009) Mulberry leaf ameliorates the expression profile of adipocytokines by inhibiting oxidative stress in white adipose tissue in db/db mice. Atherosclerosis 204(2):388–394

Taniguchi S, Asano N, Tomino F, Miwa I (1998) Potentiation of glucose-induced insulin secretion by fagomine, a pseudo-sugar isolated from mulberry leaves. Horm Metab Res 30:679–683

Tian HY, Guang XH, Zeng Y, Tan JB, Li FS, Liu GR, Tan GS, Zhou YJ (2010) A new prenylated arylbenzofuran derivative from *Morus alba* L. Chin Chem Lett 21(3): 329–331

Wang CP, Wang Y, Wang X, Zhang X, Ye JF, Hu LS, Kong LD (2011) Mulberroside a possesses potent uricosuric and nephroprotective effects in hyperuricemic mice. Planta Med 77(8):786–794

Wu Z, Zhou Z-K, Gilbert MG (2003) Moraceae link. In: Wu ZY, Raven PH, Hong DY (eds) Flora of China, vol 5, Ulmaceae through Basellaceae. Science Press/ Missouri Botanical Garden Press, Beijing/St. Louis

Xia M, Qian L, Zhou X, Gao Q, Bruce IC, Xia Q (2008) Endothelium-independent relaxation and contraction of rat aorta induced by ethyl acetate extract from leaves of *Morus alba* (L.). J Ethnopharmacol 120(3):442–446

Xie H, Wu F, Yang Y, Liu J (2008) Determination of 1-deoxynojirimycin in *Morus alba* L. leaves using reversed-phase high performance liquid chromatography fluorescence detection with pre-column derivatization. Se Pu 26(5):634–636 (In Chinese)

Xu YL, Li XE, Zou YX, Chen JJ (2008) Studies on chemical constituents from twigs of *Morus atropurpurea*. Zhongguo Zhong Yao Za Zhi 33(21):2499–2502 (In Chinese)

Yadav AV, Nade VS (2008) Anti-dopaminergic effect of the methanolic extract of *Morus alba* L. leaves. Indian J Pharmacol 40(5):221–226

Yadav AV, Kawale LA, Nade VS (2008) Effect of *Morus alba* L. (mulberry) leaves on anxiety in mice. Indian J Pharmacol 40:32–36

Yang X, Yang L, Zheng H (2010a) Hypolipidemic and antioxidant effects of mulberry (*Morus alba* L.) fruit in hyperlipidaemia rats. Food Chem Toxicol 48(8–9):2374–2379

Yang Y, Gong T, Liu C, Chen RY (2010b) Four new 2-arylbenzofuran derivatives from leaves of *Morus alba* L. Chem Pharm Bull 58:257–260

Yang Y, Wang HQ, Chen RY (2010c) Flavonoids from the leaves of *Morus alba* L. Yao Xue Xue Bao 45(1):77–81 (In Chinese)

Yang Y, Zhang T, Xiao L, Chen RY (2010d) Two novel flavanes from the leaves of *Morus alba* L. J Asian Nat Prod Res 12(3):194–198

Yang Y, Zhang T, Xiao L, Yang L, Ruoyun Chen R (2010e) Two new chalcones from leaves of *Morus alba* L. Fitoterapia 81(6):614–616

Yen GC, Wu SC, Duh PD (1996) Extraction and identification of antioxidant components from the leaves of mulberry (*Morus alba* L.). J Agric Food Chem 44(7):1687–1690

Yeung H-C (1985) Handbook of Chinese herbs and formulas. Institute of Chinese Medicine, Los Angeles

Yoon M, Kim MY (2011) The anti-angiogenic herbal composition Ob-X from *Morus alba, Melissa officinalis,* and *Artemisia capillaris* regulates obesity in genetically obese ob/ob mice. Pharm Biol 49(6):614–619

Yu ZR, Hung CC, Weng YM, Su CL, Wang BJ (2007) Physiochemical, antioxidant and whitening properties of extract from root cortices of mulberry as affected by membrane process. LWT – Food Sci Technol 40(5):900–907

Zhang W, Han F, He J, Duan C (2008) HPLC-DAD-ESI-MS/MS analysis and antioxidant activities of nonanthocyanin phenolics in mulberry (*Morus alba* L.). J Food Sci 73(6):C512–C518

Zhang HQ, Sun S, Xia B, Wu FH (2009a) In vivo hypoglycemic effects of phenolics from the root bark of *Morus alba*. Fitoterapia 80(8):475–477

Zhang M, Wang RR, Chen M, Zhang HQ, Sun S, Zhang LY (2009b) A new flavanone glycoside with anti-proliferation activity from the root bark of *Morus alba*. Chin J Nat Med 7(2):105–107

Zhong L, Furne JK, Levitt MD (2006) An extract of black, green and mulberry teas causes malabsorption of carbohydrate but not triacylglycerol in health controls. J Clin Nutr 84:551–555

Morus nigra

Scientific Name

Morus nigra L.

Synonyms

Morus japonica Siebold, *Morus laciniata* Mill., *Morus scabra* Moretti.

Family

Moraceae

Common/English Names

Black-Fruited Mulberry, Black Mulberry, Common Mulberry, Indian Mulberry, Persian Mulberry, Silkworm Mulberry, Sycamine

Vernacular Names

Afrikaans: Swartmoerbei;
Brazil: Amoreira Negra;
Bulgarian: Chernitsia Cherna;
Chinese: Hei Sang, Kui Guo;
Creole: Mi;
Dutch: Zwarte Moerbei;
Finnish: Mustamulperi;
French: Mûrier Noir, Mûres;
German: Schwarze Maulbeere, Schwarzer Maulbeerbaum;
Greek: Mavri Moria;
India: Shatut (Hindu);
Italian: Gelso Nero, Moro Nero;
Japanese: Kuro Guwa, Kuro Mi Guwa;
Portuguese: Amoreira Negra;
Russian: Šelkovica Černaja, Shelkovitsa Chernaia;
Spanish: Mora De Árbol, Mora Negra, Moral, Moral De Los Robles, Moral Negro, Morera Negra;
Swahili: Mforsadi;
Swedish: Svart Mullbär;
Turkish: Kara Aĝ;
Vietnamese: Dâu Bầu Đen.

Origin/Distribution

Black mulberry is indigenous to the Transcaucasian region in south-western Asia and has been domesticated long before the Roman times that its precise natural range is unknown. It has naturalized in the Mediterranean area and south-eastern USA. Black mulberry is cultivated in the warm temperate regions of Europe, Asia (mainly Iran, Iraq, India, Sri Lanka), America (Mexico, southern USA, Brazil, Peru, Colombia) and Hawaii for its edible fruits.

T.K. Lim, *Edible Medicinal And Non-Medicinal Plants: Volume 3, Fruits*,
DOI 10.1007/978-94-007-2534-8_57, © Springer Science+Business Media B.V. 2012

Agroecology

Black mulberry grows from sea level to 2,000 m elevation. It is more fastidious, faring less well in cold climates although it is tolerant to −10°C. It has a low chilling requirement of a few hundreds chill hours. It does not do well in the hot tropical zone with wet, humid summers. The tree is quite drought tolerant. A regular water supply is recommended if natural rainfall is not sufficient during the growing season for optimum production. The tree thrives in a well-drained, moist, light, loamy soil with pH of 6–6.5 in a sunny position. It does not perform well on chalky or gravelly soils.

Edible Plant Parts and Uses

The ripe fruits are eaten fresh, frozen, dried or made into sherbet. The ripe fruits are stewed and used for the production of beverages, syrup, sauce, wine, vinegar, conserves and jam. Mature slightly unripe fruits are preferred for tarts and pies. Black mulberries blend well with other fruits, especially pears and apples. Mulberry fruit juice it is also used as natural alcoholic extract additive for food and pharmaceutical industries. From the mulberry fruits after alcoholic fermentation and further distillation it is made a perfect hard alcoholic drink. In Greece, *Mouro* is the spirit beverage that is processed from the distillation of fermented fruits. In Devonshire, UK, the fruits are sometimes mixed with cider during fermentation, giving the drink a pleasant taste and deep red colour. The dried fruit can be pounded to a fine powder and mixed with the flour for bread and for sweet products (marmalade, chocolate, frosting, jelly and fondant). An edible oil can be obtained from mulberry seeds. Fruits are used as natural colouring in food, nutritive supplements and as source of flavones. The roots are also used as natural colouring in food. The twig, branches, stem wood are also similarly used and for alcohol ennobling in food industry.

Botany

A monoecious or dioecious, much-branched, deciduous tree growing 6–10 m tall with grey-brown bark and pale brown, pubescent branchlets. Stipules are lanceolate, membranous, brown pubescent and petioles pubescent and 1.5–2.5 cm long. Leaf lamina is leaf blade broadly ovate, unlobed, 6–12(−20) × 7–11 cm, thick, base cordate, margin regularly and coarsely serrate, apex acute to shortly acuminate, lower surface pale green and sparsely pubescent, upper surface dark green and coarse (Plates 1, 2). Male catkins are cylindric, 2–4 cm long and pubescent. Female catkins are ellipsoid, 2–2.5 cm long on a short peduncle. Female flowers with indistinct style and bifid, pubescent without stigma mastoid-like protuberance. Sorosis is ellipsoid, 2–2.5 × 1.5–2.5 cm, pink turning dark purplish red to purplish black when ripe (Plates 3, 4 and 5).

Nutritive/Medicinal Properties

Food value of raw, black mulberries (refuse 0%) per 100 g edible portion was reported as follows (U.S. Department of Agriculture, Agricultural Research Service 2010): water 87.86 g, energy 43 kcal (180 kJ), protein 1.44 g, total lipid (fat) 0.39 g, ash 0.69 g, carbohydrate 9.80 g; fibre (total dietary) 1.7 g, total sugars 8.10 g; minerals – calcium 39 mg, iron 1.85 mg, magnesium 18 mg,

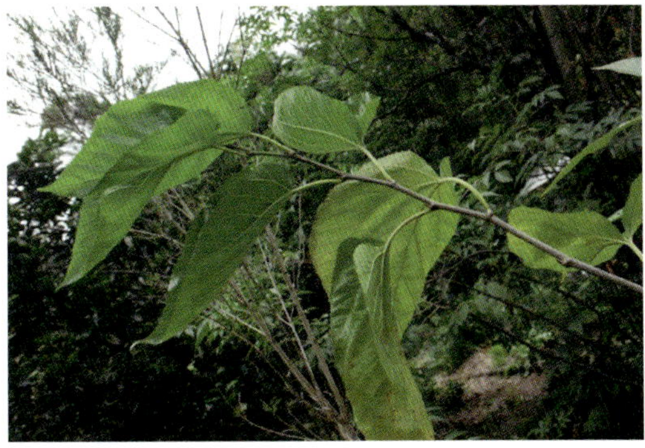

Plate 1 Leaves of black mulberry

Plate 2 Black mulberry seedling

Plate 3 Immature black mulberries

Plate 4 Ripe black mulberries

Plate 5 Close-up of ripe black mulberries

phosphorus 38 mg, potassium 194 mg, sodium 10 mg, zinc 0.12 mg, copper 0.060 mg, selenium 0.6 μg; vitamins – vitamin C (total ascorbic acid) 36.4 mg, thiamin 0.029 mg, riboflavin 0.101 mg, niacin 0.620 mg, vitamin B-6 0.050 mg, folate (total) 6 μg, total choline 12.3 mg, vitamin A 25 IU, β-carotene 9 μg, α-carotene 12 μg, lutein + zeaxanthin 136 μg, vitamin E (α-tocopherol) 0.87 mg, vitamin K (phylloquinone) 7.8 μg, lipids – fatty acids (total saturated) 0.027 g, 16:0 (palmitic acid) 0.006 g, 18:0 (stearic acid) 0.020 g; fatty acids (total monounsaturated) 0.041 g, 18:1 undifferentiated (oleic acid) 0.041 g; fatty acids (total polyunsaturated) 0.207 g, 18:2 undifferentiated (linoleic acid) 0.206 g and 18:3 undifferentiated (linolenic) 0.001 g.

Of 3 common *Morus* species, *Morus nigra* had the highest total phenolic and flavonoid contents 1,422 mg gallic acid equivalents/100 g fresh matter and 276 mg quercetin equivalents/100 g fresh matter respectively (Ercisli and Orhan 2007). *Morus alba* had the highest total fat content (1.10%), followed by *M. nigra* (0.95%) and *M. rubra* (0.85%). The major fatty acids in mulberry fruits were linoleic acid (54.2%), palmitic acid (19.8%) and oleic acid (8.41%), respectively. The total soluble solids content of the mulberry species varied between 15.9% (*M. rubra*) and 20.4% (*M. alba*), acidity between 0.25% (*M. alba*) and 1.40% (*M. nigra*), pH between 3.52 (*M. nigra*) and 5.60 (*M. alba*), ascorbic acid 19.4 mg/100 g (*M. rubra*) and 22.4 mg/100 g (*M. alba*), respectively. Mineral compositions of the mulberry species were 0.83% nitrogen, 235 mg/100 g phosphorus, 1,141 mg/100 g potassium, 139 mg/100 g calcium, 109 mg/100 g magnesium, 60 mg/100 g sodium, 4.3 mg/100 g iron, 0.4 mg/100 g copper, 4.0 mg/100 g manganese and 3.1 mg/100 g zinc.

The ripe fruits of *Morus alba* (white mulberry), *Morus nigra* (black mulberry), *Morus laevigata* (large white fruit), and *Morus laevigata* (large black fruit), were found to have total dry weight 17.60(21.97 mg/100 g, pH 3.2–4.78, and titratable acidity 0.84–2.0% citric acid) (Imran et al. 2010). Low riboflavin (vitamin B(2)) and niacin (vitamin B(3)) contents were recorded in all the fruits, while ascorbic acid (vitamin C) was in the range from 15.20 to 17.03 mg/100 g fresh weight (FW). The mulberry fruits were rich with regard to the total phenol and alkaloid contents, with values of 880–1,650 mg/100 g FW and 390–660 mg/100 g FW, respectively. Sufficient quantities of essential macro-(K, Ca, Mg, and Na) and μ-(Fe, Zn, and Ni) elements were found in all the fruits. K was the predominant element with concentration ranging from 1,270 to 1,731 mg/100 g, while Ca, Na, and Mg contents were 440–576, 260–280, and 240–360mg/100 g, respectively. The decreasing order of μ-minerals was Fe > Zn > Ni.

The fruit and leaves of *M. nigra* cultivar Shanxiguosang were found to be rich in 1-deoxynojirimycin (DNJ), resveratrol, oxyresveratrol, cyanidin-3-O-β-glucoside (Cy-3-glu), cyanidin-3-O-β-rutinoside (Cy-3-rut), and rutin (Song et al. 2009). The high contents of functional compounds in mulberry juice, fruits, and leaves implied that they might be potential sources for the development of functional drinks and food.

Four flavanoid compounds (quercetin 3-O-glucoside, quercetin 3- O-rutinoside, kaempferol 3- O-rutinoside, and 5- O-caffeoylquinic acid) were isolated from black mulberries fruit (Pawloska et al. 2008). The red pigment of *M. nigra* fruits contained four anthocyanins identified as cyanidin 3- O-glucoside, cyanidin 3-O-rutinoside, pelargonidin 3- O-glucoside, and pelargonidin 3- O-rutinoside.

A total of 18 flavour compounds were identified by GC/MS in the three black mulberry (*Morus nigra*) cultivars from the Aegean region of Turkey (Elmac and Altug 2002). Total sugar content varied between 11.3% and 16.2%, pH between 3.60 and 3.80 and total acidity between 1.51% and 1.79%. The sensory assessment indicated that black mulberry has fruity, sweet, sour, musky and woody flavour notes and fruity, acid, musky, leafy and woody-fresh aroma characteristics. Ethyl linolenate was the most effective compound contributing to the unique flavour of black mulberry.

Black mulberry juice and syrup could be considered as a rich source of anthocyanins and phenols, with a high potassium and citric acid content and low sodium level, and recommended for young children and elderly people (Darias-Martín et al. 2003). The characterisation of black mulberry juice showed that the quality of the jauice has the following quality parameters pH 3.28, titratable acidity (g/l citric acid) 21.07, TSS (°Brix) 15.63, malic acid (g/l) 1.44, lactic acid (g/l) 0.19, citric acid (g/l) 25.20, au 280 nm 35.30, anthocyanins (mg/l) 2120.00, potassium 3361.36 ppm, sodium 98.37 ppm and probable alcoholic grade (%) 9.18.

Nine phytochemicals were isolated and identified from the bark of *M. nigra*: olcancolic acid, apigenin, cyclocommunol, morusin, cyclomorusin, kuwanon C, daucosterol, ursolic acid, β-sitosterol (Wang et al. 2007). Three new compounds including two flavonoids and a new 2-phenylbenzofuran, named morunigrols A-C, together with three known compounds albafuran A, albafuran B, and mulberrofuran L, were isolated from the bark of *Morus nigra* (Wang et al. 2008).

Antioxidant Activity

Total phenolic content observed in black mulberry fruits was between 1,943 and 2,237 mg gallic acid equivalents/100 g fresh mass (Ercisli and Orhan 2008). The vitamin C content of genotypes varied between 14.9 and 18.7 mg/100 ml. The major fatty acids in all mulberry fruits were linoleic acid (53.57–64.41%) and palmitic acid (11.36–16.41%). Antioxidant activity of black mulberry genotypes was found between 63% and 76%, which lower than standard BHA and BHT. Regarding organic acid content, malic acid was the most predominant with a range of 123–218 mg/g followed by citric acid (21–41 mg/g).

Two major anthocyanins were isolated from the black mulberry (*Morus nigra*) growing in Brazil (Hassimotto et al. 2007). The two major pigments were cyanidin-3-glucoside (ca. 71% of the total) and cyanidin-3-glucosylrhamnoside (ca. 19% of the total), with no cinnamoyl groups. The crude methanolic extract (at 10 μM gallic acid equivalent), neutral flavonoids (NF) and acidic flavonoids (AF) eluate from polyamid solid-phase extraction (at 1 μM GAE), showed a higher antioxidant activity than quercetin and synthetic antioxidant (butyrate hydroxytoluene (BHT)) at 10 μM, by the ß-carotene bleaching method.

Black mulberry exhibited higher total phenolics (TP), total monomeric anthocyanin (TMA), total antioxidant capacity (TAC) and titratable acidity (TA), when compared to red mulberry (Özgen et al. 2009) . The average TP contents of *M. nigra* and *M. rubra* were 2,737 and 1,603 μg gallic acid equivalent in g fresh weight basis (GAE/g fw), respectively. *M. nigra* had the richest amount of anthocyanin with an average of 571 μg cy-3-glu/g fw. Overall, TAC averaged 10.5 and 12.0 mmol TE/L by the the trolox-equivalent antioxidant capacity (TEAC) and the ferric reducing antioxidant power (FRAP) methods, respectively. It was found that FRAP, TEAC, TP and TMA were significantly correlated ($r=0.64$–0.99) with each other. Fructose (5.27 g/100 ml) and glucose (5.81 g/100 ml) were determined to be the major sugars in both mulberries. *M. nigra* displayed a higher TA (2.05 g/100 ml) than *M. rubra* (0.78 g/100 ml), with citric acid as the major acid.

Studies showed that different extracts of *Morus nigra* fruit (fruit juice, hydroalcoholic and polyphenolic) had antioxidant effect of on haemoglobin glycosylation, peroxidative damage to human erythrocytes, liver hepatooytes of rats and human low-density lipoprotein (LDL) (Naderi et al. 2004). All three extracts inhibited haemoglobin glycosylation induced by glucose to differing degrees. The haemolysis of human erythrocytes induced by hydrogen peroxide was also inhibited. The production of malondialdehyde (MDA) during peroxidative damage to plasma membranes of isolated rat hepatocytes induced by tert-butyl hydroperoxide (tBH) was also inhibited. Inhibition of lipid peroxidation of LDL induced by copper (II) ion was achieved during the study. The results suggested that *Morus nigra* fruit had a protective action against peroxidative damage to biomembranes and biomolecules.

Recent studies showed that black mulberry (*Morus nigra*) genotypes had a higher bioactive content than purple mulberry *(Morus rubra*) genotypes (Ercisli et al. 2010). The average total phenolic content and total anthocyanins of black mulberry genotypes were 2,149 μg of gallic acid equivalent (GAE) per g and 719 μg of cyanidin 3-glucoside equivalent (Cy 3-glu) per g of fresh mass. In purple mulberry, these values were for GAE 1,690 μg/g and for Cy 3-glu 109 μg/g on fresh mass basis. The average antioxidant activity of black mulberry genotypes was also found to be higher than that of the purple ones according to FRAP assay (Trolox equivalent (TE) per fresh mass of black and purple mulberries was 13.35 and 6.87 μmol/g, respectively).

The radical scavenging activity of methanolic extract of ripe fruits of *Morus alba* (white mulberry), *Morus nigra* (black mulberry), *Morus laevigata* (large white fruit), and *Morus laevigata* (large black fruit)was concentration-dependent and showed a correlation with total phenolic constituents of the respective fruits(Imran et al. 2010).

Antinociceptive/Analgesic Activity

Morusin, the main prenylflavonoid present in the *Morus nigra* root barks exhibited promising antinociceptive or analgesic profile by the intraperitoneal route, being more potent than some standard drugs used as reference (de Souza et al. 2000). The mechanism by which the morusin exerted antinociceptive activity was postulated to involve the participation of the opioid system.

The dichloromethane extract from leaves of *M. nigra* at test doses of 100 g/kg and 300 g/kg, p.o. clearly demonstrated antinociceptive activity in the formalin, hot plate, and tail immersion tests as well as acetic acid-induced writhing test (Padilha et al. 2009). The extract administered at 300 mg/kg, p.o. had a stronger antinociceptive effect than indomethacin (5 mg/kg, p.o.) and morphine (10 mg/kg, p.o.), which supported previous claims for its traditional use.

Antiinflammatory Activity

In animal studies, *Morus nigra* leaf extract at doses of 100–300 mg/kg p.o. demonstrated antiinflammatory effects by reduced paw edema induced by carragenan and significantly inhibited the formation of granulomatous tissue (Padilha et al. 2010). The chemical compounds isolated including betulinic acid, β-sitosterol and germanicol, were postulated to be responsible for the antiinflammatory effect of the extract.

Skin Whitening Activity

2,4,2′,4′-tetrahydroxy-3-(3-methyl-2-butenyl)-chalcone (TMBC), a naturally occurring compound from *Morus nigra*, modulated melanogenesis by inhibiting tyrosinase (Zhang et al. 2009). TMBC inhibited the L-dopa oxidase activity of mushroom tyrosinase with an IC_{50} value of 0.95 μM, which was more potent than kojic acid (IC_{50} = 24.88 μM), a well-known tyrosinase inhibitor. The kinetic studies of tyrosinase inhibition revealed that TMBC acts as a competitive inhibitor of mushroom tyrosinase with L-dopa as the substrate. Furthermore, TMBC effectively inhibited both cellular tyrosinase activity and melanin biosynthesis in B16 melanoma cells without significant cytotoxicity. The inhibitory effect of TMBC on melanogenesis was attributed to the direct inhibition of tyrosinase activity, rather than the suppression of tyrosinase gene expression. Theresults indicated that TMBC may be a new promising pigmentation-altering agent for cosmetic or therapeutic applications.

It was found that *M. nigra* root extract contained some unknown natural products with potential tyrosinase inhibitory activity (Zheng et al. 2010). One new compound, 5′-geranyl-5,7,2′,4′-tetrahydroxyflavone, and 28 known phenolic compounds were isolatedfrom the (95% ethanol) extract of *M. nigra* roots. Nine compounds, 5′-geranyl-5,7,2′,4′-tetrahydroxyflavone, steppogenin-7-O-β-d-glucoside, 2,4,2′,4′-tetrahydroxychalcone, moracin N, kuwanon H, mulberrofuranG, morachalconeA, oxyresveratrol-3′-O-β-d-glucopyranoside and oxyresveratrol-2-O-β-d-glucopyranoside, showed better tyrosinase inhibitory activities than kojic acid. It was noteworthy that the IC_{50} values of 2,4,2′,4′-tetrahydroxychalcone and morachalcone A were 757-fold and 328-fold lower than that of kojic acid, respectively, suggesting a great potential for their development as effective natural tyrosinase inhibitors.

Hypoglycaemic/Antidiabetic Activity

Studies showed that alcoholic leaf extract of *Morus nigra* given orally was found to possess hypoglycaemic effect in streptozotocin-diabetic rats (Eidi et al. 2005). A significant reduction of serum glucose concentration in streptozotocin-diabetic rats was observed but the level of insulin in normal and diabetic rats was not altered. Mulberry hydro-alcoholic extract at the dose of 400 mg/kg (for a week) and 600 mg/kg (for 2 months) elicited a significant decrease in the blood glucose levels in the treated streptozotocin-induced diabetic rats compared to control (Hoseini et al. 2009). Of 3 *Morus* species *M. alba*, *M. alba* var. *nigra* and *M. nigra*, *M. alba* var.

nigra (IC_{50} = 13.26 (12.86–13.66) mg/ml) and *M. alba* (IC_{50} = 17.60 (17.39–17.80) mg/ml) revealed appreciable α-amylase inhibitory activities in a concentration-dependent manner (Nickavar and Mosazadeh 2009). All the fractions (n-hexane, chloroform, ethyl acetate and aqueous fractions) of the most active extract (i.e. *M. alba* var. *nigra*) had potent inhibitory effects on the α-amylase activity. However, the lowest inhibitory potency was observed for the aqueous fraction.

Adipogenesis Promoting Activity

Ten new isoprenylated flavonoids, nigrasins A-J (1–10), and three known compounds were isolated from the twigs of *Morus nigra* (Hu et al. 2011). Compounds 8 and 9 promoted adipogenesis, characterized by increased lipid droplet and triglyceride content in 3T3L1 cells, and induced up-regulation of the expression of adipocyte-specific genes, aP2 and GLUT4.

Lectin Activity

Morniga M is a jacalin-related and mannose-specific lectin isolated from the bark of the mulberry (*Morus nigra*) (Wu et al. 2004) The binding affinity of morniga M for ligands could be ranked in decreasing order as follows: groups carrying multiple N-glycans with oligomannosyl residues >>N-glycopeptide with a single trimannosyl core>Tri-Man oligomer [Man α 1→6(Man α 1→3) Man], Penta-Man oligomer [Man α 1→6(Manα1→3)Man α 1→6(Man α 1→3) Man] αMan α 1→2, 3 or 6 Man>Man>GlcNAc, Glc >> L-Fuc, Gal, GalNAc (inactive), demonstrating the unique specificity of this lectin that may not only assist in our understanding of cell surface carbohydrate ligand-lectin recognition, but also provide informative guidelines for the application of this structural probe in biotechnological and clinical regimens, especially in the detection and purification of N-linked glycans.

Studies found that the lectin, morniga M from *Morus nigra* triggered lymphocyte activation, and induced cell death in activated PBMC (peripheral blood mononuclear cell), activated T lymphocytes, and Jurkat T leukemia cells (Benoist et al. 2009). The morniga M-induced cell death resulted, at least in part, from caspase-dependent apoptosis and FADD(Fas-associated Death Domain Protein) -dependent receptor-mediated cell death. morniga M, also triggered AICD (Activation induced cell death) of T lymphocytes.

Traditional Medicinal Uses

Black mulberry juice is taken to control type II diabetes mellitus in tradtional medicine (Darias-Martín et al. 2003) Apart from that, this fruit has been used for the treatment of mouth, tongue and throat inflammations. sore throat and swollen vocal chords have been treated with fruit juice. *Morus nigra* has been used to relieve pain in Brazilian folk medicine (Padilha et al. 2009). *M. nigra* has laxative and antipyretic properties. *M. nigra* leaves are used in pharmacy for their astringent properties. A leaf, flower or root decoction can be gargled for diabetes. The bark is a reputed anthelmintic, used to expel tapeworms.

Other Uses

The tree are used as ornamentals in landscape planting, as living fences/boundaries and windbreak. When grown in plantation they can be used for biogas production and phytoremediation. Pot cultivation for small indoor plants and bonsai trees is also popular. Besides providing natural colouring for use in the food industry, the roots also provide anthelmintic, antibacterial and antiviral products for the pharmaceutical industry. The wood is yellowish-brown and hard and is used for furniture, in joinery for articles subject to wear, for lathe work, and in the manufacture of barrels, caskets, snuffboxes and cups. The twig, branches, stem wood are used in cosmetics –hair lotions; moisture products for skin, quality paper, dyestuffs, fuel wood, and also provide natural colouring for use in the food industry and for alcohol ennobling. A textile fibre called 'artificial cotton' is extracted from the bark in Japan.

The leaves are used for herbal teas, biogas production, compost, as sources of carotene, amino acids, medicine, leaf protein, phytoecdysteroids and livestock feeds for ruminants (goats, cows, sheep), rabbits, poultry. Although inferior to those of *M. alba*, the leaves of *M. nigra* are also used for raising silkworms. A yellow dye is obtained from the leaves used mainly for enhancing the sheen on silk.

Comments

See also notes on *Morus alba* and *M. rubra*.

Selected References

Benoist H, Culerrier R, Poiroux G, Ségui B, Jauneau A, Van Damme EJ, Peumans WJ, Barre A, Rougé P (2009) Two structurally identical mannose-specific jacalin-related lectins display different effects on human T lymphocyte activation and cell death. J Leukoc Biol 86(1):103–114

Chopra RN, Nayar SL, Chopra IC (1986) Glossary of Indian medicinal plants. Including the supplement. Council Scientific Industrial Research, New Delhi, 330 pp

Council of Scientific and Industrial Research (CSIR) (1962) The wealth of India. A dictionary of Indian raw materials and industrial products. Raw materials, vol 6. Publications and Information Directorate, New Delhi

Darias-Martín J, Lobo-Rodrigo G, Hernández-Cordero J, Díaz-Díaz E, Díaz-Romero C (2003) Alcoholic beverages from black mulberry. Food Technol Biotechnol 41(2):173–176

de Padilha M M, Vilela FC, da Silva MJD, dos Santos MH, Santos MH, Alves-da-Silva G, Giusti-Paiva A (2009) Antinociceptive effect of the extract of *Morus nigra* leaves in mice. J Med Food 12(6):1381–1385

de Padilha MM, Vilela FC, Rocha CQ, Dias MJ, Soncini R, dos Santos MH, Alves-da-Silva G, Giusti-Paiva A (2010) Antiinflammatory properties of *Morus nigra* leaves. Phytother Res 24(10):1496–1500

de Souza MM, Bittar M, Cechinel-Filho V, Yunes RA, Messana I, Monache FD, Ferrari F (2000) Antinociceptive properties of morusin, a prenylflavonoid isolated from *Morus nigra* root bark. Z Naturf 55(3–4):256–260

Eidi M, Eidi A, Fallahyan F (2005) Hypoglycaemic effects of *Morus nigra* L. leaves in normal and diabetic rats. In: Mrozikiewicz PM, Mrozikiewicz A, Orme M (eds) Proceedings of the 7th Congress of the European Association for Clinical Pharmacology and Therapeutics, Poznan, Poland, 25–29 June 2005

Elmac Y, Altug T (2002) Flavour evaluation of three black mulberry (*Morus nigra*) cultivars using GC/MS, chemical and sensory data. J Sci Food Agric 82(6):632–635

Ercisli S, Orhan E (2007) Chemical composition of white (*Morus alba*), red (*Morus rubra*) and black (*Morus nigra*) mulberry fruits. Food Chem 103(4): 1380–1384

Ercisli S, Orhan E (2008) Some physico-chemical characteristics of black mulberry (*Morus nigra* L.) genotypes from Northeast Anatolia region of Turkey. Sci Hortic 116(1):41–46

Ercisli S, Tosun M, Duralija B, Voća S, Sengul M, Turan M (2010) Phytochemical content of some black (*Morus nigra* L.) and purple (*Morus rubra* L.) mulberry genotypes. Food Technol Biotechnol 48(1): 102–106

Hassimotto NMA, Genovese MI, Lajolo FM (2007) Identification and characterisation of anthocyanins from wild mulberry (*Morus nigra* L.) growing in Brazil. Food Sci Technol Int 13(1):17–25

Hoseini HF, Saeidnia S, Gohari AR, Yazdanpanah M, Hadjiakhoondi A (2009) Investigation of antihyperglycemic effect of *Morus nigra* on blood glucose level in streptozotocin diabetic rats. Pharmacologyonline 3:732–736

Hu X, Wu JW, Zhang XD, Zhao QS, Huang JM, Wang HY, Hou AJ (2011) Isoprenylated flavonoids and adipogenesis-promoting constituents from *Morus nigra*. J Nat Prod 74(4):816–824

Huxley AJ, Griffiths M, Levy M (eds) (1992) The new RHS dictionary of gardening (4 Vols). MacMillan, London

Imran M, Khan H, Shah M, Khan R, Khan F (2010) Chemical composition and antioxidant activity of certain *Morus* species. J Zhejiang Univ Sci B11(12):973–980

Naderi GA, Asgary S, Sarraf-Zadegan N, Oroojy H, Afshin-Nia F (2004) Antioxidant activity of three extracts of *Morus nigra*. Phytother Res 18(5):365–369

Nickavar B, Mosazadeh G (2009) Influence of three *Morus* species extracts on α-amylase activity. Iran J Pharm Res 8(2):115–119

Özgen M, Serçe S, Kaya C (2009) Phytochemical and antioxidant properties of anthocyanin-rich *Morus nigra* and *Morus rubra* fruits. Sci Hortic 119(3): 275–279

Pawlowska AM, Oleszek W, Braca A (2008) Quali-quantitative analyses of flavonoids of *Morus nigra* L. and *Morus alba* L. (Moraceae) fruits. J Agric Food Chem 56(9):3377–3380

Porcher MH et al (1995–2020) Searchable world wide web multilingual multiscript plant name database. The University of Melbourne, Melbourne. http://www.plantnames.unimelb.edu.au/Sorting/Frontpage.html

Song W, Wang HJ, Bucheli P, Zhang PF, Wei DZ, Lu YH (2009) Phytochemical profiles of different mulberry (*Morus* sp.) species from China. J Agric Food Chem 57(19):9133–9140

U.S. Department of Agriculture, Agricultural Research Service (2010) USDA National Nutrient Database for Standard Reference, Release 23. Nutrient Data Laboratory Home Page. http://www.ars.usda.gov/ba/bhnrc/ndl

Wang L, Wang HQ, Chen RY (2007) Studies on chemical constituents from bark of *Morus nigra*. Zhong Zhong Yao Za Zhi 32(23):2497–2499 (In Chinese)

Wang L, Cui XQ, Gong T, Yan RY, Tan YX, Chen RY (2008) Three new compounds from the barks of *Morus nigra*. J Asian Nat Prod Res 10(9–10): 897–902

Wu Z, Zhou Z-K, Gilbert MG (2003) Moraceae link. In: Wu ZY, Raven PH, Hong DY (eds) Flora of China, vol 5, Ulmaceae through Basellaceae. Science Press/Missouri Botanical Garden Press, Beijing/St. Louis

Wu AM, Wu JH, Singh T, Chu KC, Peumans WJ, Rougé P, Van Damme EJM (2004) A novel lectin (Morniga M) from mulberry (*Morus nigra*) bark recognizes oligomannosyl residues in n-glycans. J Biomed Sci 11:874–885

Zhang X, Hu X, Hou A, Wang H (2009) Inhibitory effect of 2,4,2′,4′-tetrahydroxy-3-(3-methyl-2-butenyl)-chalcone on tyrosinase activity and melanin biosynthesis. Biol Pharm Bull 32(1):86–90

Zheng ZP, Cheng KW, Zhu Q, Wang XC, Lin ZX, Wang MF (2010) Tyrosinase inhibitory constituents from the roots of *Morus nigra*: a structure-activity relationship study. J Agric Food Chem 58(9):5368–5373

Morus rubra

Scientific Name

Morus rubra L.

Synonyms

Morus argutidens Koidz., *Morus canadensis* Poir., *Morus riparia* Raf., *Morus rubra* β *tomentosa* Bur. ex DC., *Morus rubra* var. *tomentosa* (Rafinesque) Bureau, *Morus scabra* Willd., *Morus tomentosa* Raf.

Family

Moraceae

Common/English Names

American Mulberry, Black-Fruited Mulberry, Red-Fruited Mulberry, Red Mulberry

Vernacular Names

Chinese: Hong Sang, Song She Shue;
French: Mûrier Rouge, Mûrier Sauvage;
German: Roter Maulbeerbaum Rote Maulbeere;
Greek: Kokkini Muria;
Italian: Gelso Rosso, Moro Rosso;
Japanese: Aka Mi Guwa, Ke Guwa;
Russian: Shelkovitsa Krasnaia;
Spanish: Mora De Árbol, Mora Roja, Moral Rojo;
Polish: Morwa Czerwona;
Swedish: Mullbär.

Origin/Distribution

The red mulberry native to eastern North America, from northernmost Ontario and Vermont south to southern Florida and west to southeast South Dakota and central Texas. Although red mulberry is common in the United States, it is considered rare in Massachusetts, Ontario and Vermont. It is listed as an endangered species in Canada.

Agroecology

Red mulberry is adapted to temperate climatic requirements with total annual rainfall of 1,000–2,000 mm. In its native range it is found in moist forests and thickets on the floodplains, river valleys, and moist hillsides at elevations below 600 m. It is rated as being moderately tolerant of brief periods of flooding. Red mulberry has a minimum requirement of 140 frost free days and can tolerate sub zero temperature down to −36°C. It grows in full sun but is tolerant to partial shading. It is quite drought tolerant but is intolerant of fire.

Red mulberry grows on a variety soils that include alfisols, inceptisols, spodosols, and

T.K. Lim, *Edible Medicinal And Non-Medicinal Plants: Volume 3, Fruits*,
DOI 10.1007/978-94-007-2534-8_58, © Springer Science+Business Media B.V. 2012

ultisols. It thrives best on well-drained, moist soils with pH of 5–7.

Edible Plant Parts and Uses

Red mulberry is relished as dessert for its large, sweet fruits. The fruits also are used in jellies, jams, pies, syrup and drinks. In Turkey all three common *Morus species*, *M. alba, M. rubra* and *M. nigra* are widely consumed fresh as well as dried, processed into jam, marmelade, pekmez (a traditional Turkish food), wine, juices, paste and ice cream.

Botany

A deciduous shrubs or trees,10–15 m high sometimes to 20 m with up to 50 cm trunk diameter, gray-brown bark and red-brown to light greenish brown, lenticellate branchlets usually glabrous. Leaves are alternate, green, stipules linear, 10–13 mm, thin, pubescent; petiole 2–2.5 cm, glabrous or pubescent. Leaf lamina highly variable in shape, broadly ovate, irregularly lobed, or no lobes, 10–18 cm long by 8–12 cm wide, base rounded to nearly cordate, sometimes oblique, margins serrate or crenate, apex abruptly acuminate; upper surface scabrous and lower surface pubescent (Plates 1, 2 and 3) . Male and female flowers are dioecious, small, pale green in catkins. Male catkins pendant, 3–5 cm long; male flowers with sepals connate at base, green tinged with red, 2–2.5 mm, pubescent outside, ciliate toward tip with four stamens. Female catkins 8–12 × 5–7 mm femal flowers with calyx tightly surrounding a green, broadly ellipsoid or obovoid, glabrous ovary; style with divergent, branches whitish, sessile, and topped by papillose stigma. Fruit a compound syncarp greenish-pink, pink, red ripening to black or deep purple, cylindric, 2–4 cm × 1 cm with fleshy calyx surrounding achenes (Plates 1, 2 and 3); achenes yellowish, oval, flattened, 2 mm and smooth

Plate 1 Immature red mulberries and irregularly lobed, toothed leaves

Nutritive/Medicinal Properties

Mean chemical composition, antioxidant activity and anthocyanin profile of purple mukberry (*M. rubra*) fruits reported by Koca et al. (2008) as

Plate 2 Broadly ovate, unlobed, toothed leaves

Plate 3 Immature and ripe red mulberries

173.10 g/kg dry matter, 13.50° Brix soluble solids, 4.92 pH, 4.0 g/kg total acidity, 127.18 g/kg total sugar, 120.95 g/kg reducing sugar, 5.92 g/kg non-reducing sugar, 12.60 g/kg crude protein, 2107.47 mg/kg potassium, 889.04 mg/kg calcium, 194.04 mg/kg magnesium, 118.94 mg/kg sodium, 28.50 mg/kg iron, 5.20 mg/kg zinc, 3.49 mg/kg manganese and 3.09 mg/kg copper. Average values for the natural antioxidants were found as 28.42 mg/kg ascorbic acid, 193.85 mg/kg total anthocyanins and 1308.07 mg/kg total phenolics. The ferric reducing/antioxidant power (FRAP) assay was used to measure the total antioxidant activity of purple mulberry and the average value was obtained as 33.90 μmol/g. Cyanidin 3-glucoside was the predominant anthocyanin. Another analysis conducted by Ercisli and Orhan 2007 reported the total fat content of red mulberry as 0.85% and the major fatty acids in mulberry fruits were linoleic acid (54.2%), palmitic acid (19.8%) and oleic acid (8.41%); the total soluble solids content as 15.9%, ascorbic acid 19.4 mg/100 g. Mineral compositions were 0.83% nitrogen, 235 mg/100 g phosphorus, 1,141 mg/100 g potassium, 139 mg/100 g calcium, 109 mg/100 g magnesium, 60 mg/100 g sodium , 4.3 mg/100 g iron, 0.4 mg/100 g copper, 4.0 mg/100 g manganese and 3.1 mg/100 g zinc. Özgen et al. (2009) reported that red mulberry exhibited lower total phenolics (TP), total monomeric anthocyanin (TMA), titratable acidity (TA) and Total antioxidant capacity (TAC) than black mulberry. The average TP content was 1,603 μg gallic acid equivalent in g fresh weight basis (GAE/g fw). Fructose (5.27 g/100 ml) and glucose (5.81 g/100 ml) were determined to be the major sugars in both mulberries. *M. rubra* displayed a lower TA (0.78 g/100 ml), with citric acid as the major acid.

Antiatherosclerotic Activity

Oral administration of aqueous leaves extract of *M. rubra* to streptozotocin-induced diabetic rats (100, 200 and 400 mg/kg body weight per day for a period of 30 days) produced significant fall in fasting blood glucose (FBG) in a dose-dependent manner (Sharma et al. 2010b). Treatment with the extract (400 mg/kg) showed significant improvement in body weight and serum lipid profile i.e., total cholesterol, triglyceride, HDL-cholesterol, LDL-cholesterol and VLDL-cholesterol, when compared with diabetic control. Endothelial dysfunction parameters (sVCAM-1, Fibrinogen, total NO levels and oxidized LDL), apolipoprotein A and apolipoprotein B were significantly reversed to near normal, following treatment with the extract. The study showed that aqueous leaf extract of *Morus rubra* (400 mg/kg) significantly improved the homeostasis of glucose and fat and possessed significant anti-atherosclerotic activity.

Antidiabetic Activity

M. rubra leaf extract orally administered to diabetic rats exerted a dose-dependent fall in fasting blood glucose (Sharma et al. 2010a). Treatment with 400 mg/kg extract produced a significant reduction in glycosylated haemoglobin with a concomitant elevation in plasma insulin and C-peptide levels. In erythrocytes, as well as liver, the activity of antioxidant enzymes and content of reduced glutathione were found to be significantly enhanced, while levels of serum and hepatic lipid peroxides were suppressed in extract-fed diabetic rats. Histopathological examination of pancreatic tissue revealed an increased number of islets and beta-cells in extract-treated diabetic rats. The results suggested the potential of *M. rubra* leaf extract to control over hyperglycaemia and dyslipidaemia. It also demonstrated antioxidant activity.

Oral Medicine

Juice of the fruit of *M. rubra* was found suitable as a transport/temporary storage medium for the maintenance of periodontal ligament cell viability of avulsed teeth (Özan et al. 2008). The efficacy of 4.0% and 2.5% *M. rubra* at 3, 6, and

12 hours was found to be significantly better than Hank's balanced salt solution (HBSS),. At 24 hours, 4% *M. rubra* was found to be similar to HBSS, but 2.5% *M. rubra* was found to be significantly worse than HBSS. The results showed that juice of the fruit sample of *M. rubra* studied at a concentration of 4% was a more effective storage medium than other groups like Hank's balanced salt solution (HBSS), phosphate-buffered saline (PBS), and tap water.

Traditional Medicinal Uses

Native American tribes have been reported to use infusions of red mulberry bark medicinally to stop dysentery, as a laxative, and as a purgative; infusions of the root have been used for weakness and urinary problems; and tree sap rubbed directly on the skin as treatment for ringworm (Moerman 1998).

Other Uses

Red mulberry fruits have been used as fodder for pigs and other domestic animals. The leaves are used as fodder for silkworms. The wood is used locally for fence-posts because the heartwood is relatively durable. Other uses of the wood include farm implements, cooperage, furniture, interior finish, and caskets.

Comments

Morus rubra is very similar to the white mulberry, *M. alba*, in morphological features but can be distinguished from the latter by leaf characters as follows (Wunderlin 1997)

Morus alba – Leaf blade abaxially glabrous or with pubescence only along major veins or in tufts in axils of principal lateral veins and midribs, adaxially glabrous to sparsely pubescent.

Morus rubra – Leaf blade abaxially pubescent or puberulent, adaxially with short, stiff, antrorsely appressed trichomes, usually scabrous.

Selected References

Core EL (1974) Red mulberry, *Morus rubra* L. In: Gill JD, Healy WM (Comp) Shrubs and vines for northeastern wildlife. USDA Forest Service, General Technical Report NE-9. Northeastern Forest Experiment Station, Upper Darby, pp 106–107

Ercisli S, Orhan E (2007) Chemical composition of white (*Morus alba*), red (*Morus rubra*) and black (*Morus nigra*) mulberry fruits. Food Chem 103(4):1380–1384

Huxley AJ, Griffiths M, Levy M (eds) (1992) The new RHS dictionary of gardening (4 Vols). MacMillan, London

Koca I, Ustun NS, Koca AF, Karadeniz B (2008) Chemical composition, antioxidant activity and anthocyanin profiles of purple mulberry (*Morus rubra*) fruits. J Food Agric Environ 6(2):39–42

Moerman D (1998) Native American ethnobotany. Timber, Oregon, 927 pp

Moore DM, Thomas WP (1977) Red mulberry/*Morus rubra* L. In: Halls LW (ed) Southern fruit-producing woody plants used by wildlife. USDA Forest Service, General Technical Report SO-16. Southern Forest Experiment Station, New Orleans, pp 55–56

Ozan F, Tepe B, Polat Z, Er K (2008) Evaluation of in vitro effect of *Morus rubra* (red mulberry) on survival of periodontal ligament cells. Oral Surg Oral Med Oral Pathol Oral Radiol Endodontol 105(2):e66–e69

Özgen M, Serçe S, Kaya C (2009) Phytochemical and antioxidant properties of anthocyanin-rich *Morus nigra* and *Morus rubra* fruits. Sci Hortic 119(3):275–279

Rehder A (1947) Manual of cultivated trees and shrubs hardy in Northern America exclusive of the subtropical and warmer temperate regions, 2nd edn. Macmillan, New York, 996 pp

Sharma SB, Gupta S, Ac R, Singh UR, Rajpoot R, Shukla SK (2010a) Antidiabetogenic action of *Morus rubra* L. leaf extract in streptozotocin-induced diabetic rats. J Pharm Pharmacol 62(2):247–255

Sharma SB, Tanwar RS, Rini AC, Singh UR, Gupta S, Shukla SK (2010b) Protective effect of *Morus rubra* L. leaf extract on diet-induced atherosclerosis in diabetic rats. Indian J Biochem Biophys 47:26–31

Wunderlin RP (1997) Moraceae link. In: Flora of North America Editorial Committee (ed) Flora of North America, North of Mexico, vol 3, Magnoliophyta: Magnoliidae and Hamamelidae. Oxford University Press, New York, 590 pp

Prainea limpato

Scientific Name

***Prainea limpato* (Miq.) Beumée ex K.Heyne**.

Synonyms

Artocarpus limpato Miq., *Artocarpus papuanus* (Becc.) Renner, *Ficus diepenhorstii* (Miq.) King, *Parartocarpus papuana* S.Moore, *Prainea cuspidata* Becc., *Prainea limpato* var. *longipedunculata* Kochummen, *Prainea microcephala* J. J.Sm., *Prainea multinervia* Merr., *Prainea papuana* Becc., *Urostigma diepenhorstii* Miq.

Family

Moraceae

Common/English Name

Limpato

Vernacular Names

Borneo: Buruni, Ematak, Karon;
East Kalimantan: Gimpango, Limpato;
Peninsular Malaysia: Buah Lampato, Tampang;
Sabah: Kesusu;
Sumatra: Limpatoe.

Origin/Distribution

The species is native to Peninsular Malaysia, Sumatra, Borneo (Sarawak, Sabah, East-Kalimantan), Moluccas, New Guinea.

Agroecology

The species thrives in the hot, humid tropics and is found in undisturbed to slightly disturbed (open) mixed dipterocarp lowland forest up to 300 m altitude. It usually occurs on hillsides with clay to sandy soils, and also common on limestone. In secondary forests it is usually present as a pre-disturbance remnant.

Edible Plant Parts and Uses

The fruits are edible, sweetish and tart. Seeds (nuts) are reported edible in Papua New Guinea.

Botany

Dioecious, evergreen, canopy tree up to 30 m high and 60 cm girth with grey-brown bole, cracking to scaly, and crown width from 3.0 to 5.8 m. Inner bark is orange with white latex and sapwood yellow-brown. Twigs 2–5 mm thick, finely rugose, adpressed puberulent. Leaves are alternate simple, glabrous to hairy below on

T.K. Lim, *Edible Medicinal And Non-Medicinal Plants: Volume 3, Fruits*,
DOI 10.1007/978-94-007-2534-8_59, © Springer Science+Business Media B.V. 2012

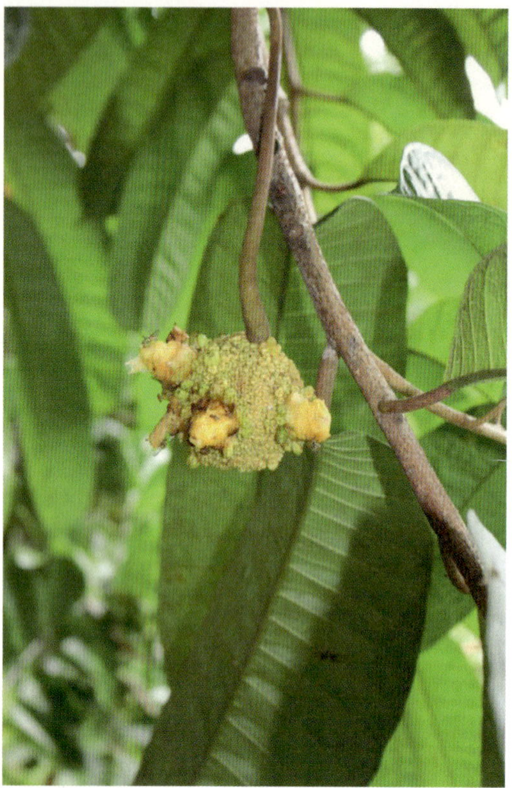

Plate 1 Female head globose with ripe fruiting perianths

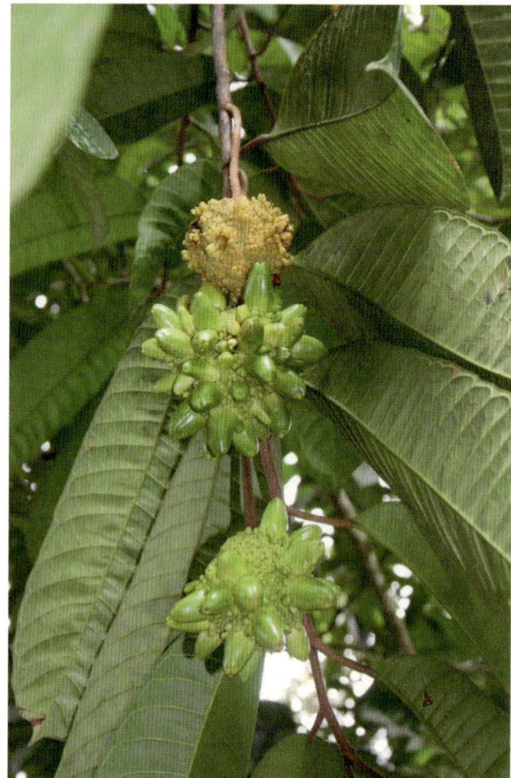

Plate 2 Globose male (*upper*) and female head with unripe protruding fruiting perianths

10–15 mm petioles; lamina oblong-elliptic, obovate elliptic or elliptic, 10–33 cm by 4–13 cm, base cuneate to rounded, apex mucronulate, penni-veined, 12–14 pairs of lateral nerves distinct below (Plate 3) on 8–22 mm long petioles. Inflorescence unisexual, capitate, heads globose to subglobose, solitary or paired in leaf axils, flowers mixed with numerous peltate, clavate or spathulate interfloral bracts, perianth free enclosing a single stamen or ovary. Male head 15 cm across, globose, perianth tubular, 1.4 mm long, perforate with 1.8 mm long stamen and bracts 0.3 mm peltate and ciliate on 3–4 cm long peduncle. Mature female head 3.5–4.5 cm across, globose, loosely covered by numerous flowers and yellow-green spathulate, ciliate bracts and fruiting perianths on long, puberulent peduncle 7.5–254 cm. Fruiting perianths, 8–20, ellipsoid 1.5 by 1 cm, protrude out of the globose head amongst the mass of unfertilised greenish-yellow flowers imparting an unusual peculiar and unique, irregularly radiating stellate mass (Plates 1, 2), remaining unfertilised perianths 4 mm long, with clavate apices and pubescent. Fruiting perianth is green turning to golden yellow to bright orangey-yellow when ripe (Plates 1, 2). Seed large attached laterally near base of the fruiting perianth.

Nutritive/Medicinal Propeties

No published information is available on the nutritive value of the fruit or the medicinal uses of the plant.

Plate 3 Large, coriaceous and penni-veined leaves

Other Uses

In Papua New Guinea, the bark is used as tapa cloth and for wrapping eggs and wood.

Comments

Prainea papuana is reduced to a subspecies, *Prainea limpato* (Miq.) K. Heyne subsp. *papuana* (Becc.) C.C. Berg.

Selected References

Berg CC (2005) Flora Malesiana precursor for the treatment of Moraceae 8: other genera than *Ficus*. Blumea 50(3):535–550

Jarrett FM (1959a) Studies in *Artocarpus* and allied genera I. General considerations. J Arnold Arbor 40(1):1–29

Jarrett FM (1959b) Studies in *Artocarpus* and allied genera, II. A revision of *Prainea*. J Arnold Arbor 40:30–37

Kochummen KM (1978) Moraceae. In: Ng FSP (ed) Tree flora of Malaya, vol 3. Longman, Kuala Lumpur, pp 119–168

Slik JWF (2006) Trees of Sungai Wain. Nationaal Herbarium Nederland, Leiden. http://www.nationaalherbarium.nl/sungaiwain/

Treculia africana

Scientific Name

Treculia africana Desc. var. *africana*.

Synonyms

Artocarpus africana Sim, *Ficus welsitschii* Miq., *Ficus whytei* Stapf. ex Johnston, *Treculia affona* N.E.Br., *Treculia africana var. nitida* Engl., *Treculia africana var. engleriana* (De Wild. et Th. Dur.) Engl., *Treculia centralis* A.Chev., *Treculia dewevrei* De Wild. et Th. Dur., *Treculia erinacea* A.Chev.

Family

Moraceae

Common/English Names

African-Boxwood, African Breadfruit, African Breadnut, Okwa Tree, Wild Jackfruit

Vernacular Names

French: Abre Á Pain D' Afrique;
German: Okwabaum;
Ghana: Bibiritun, Debeletun, Memrentun;
Nigeria: Afon, Ediang, Ukwa;
Portuguese: Isaquiente;
Senegal: Brebretim (Wolof);
Swahili: Mwaya;
Tanzania: Ezeya;
Uganda: Muzinda (Luganda).

Origin/Distribution

The species is native to tropical Africa, from Senegal to Mozambique, Madagascar, Angola, Benin, Cameroon, Central African Republic, Congo, Cote d'Ivoire, Democratic Republic of Congo, Equatorial Guinea, Gabon, Gambia, Guinea-Bissau, Liberia, Madagascar, Malawi, Mozambique, Nigeria, Sao Tome et Principe, Senegal, Sierra Leone, Sudan, Tanzania, Togo, Uganda and Zambia.

Agroecology

A native of tropical Africa. In its native range, *Treculia* is usually found near streams or in moist, swampy areas and in riverine forests from 0 to 1,500 m altitude. It is not very light demanding and will grow in partial shade in a wide variety of soils and climatic conditions including subtropical.

T.K. Lim, *Edible Medicinal And Non-Medicinal Plants: Volume 3, Fruits*,
DOI 10.1007/978-94-007-2534-8_60, © Springer Science+Business Media B.V. 2012

Edible Plant Parts and Uses

The fruit and seeds are eaten fully ripe. African breadfruit can be made into wine or fruit yoghurt where the fleshy pulp is blended and mixed with milk, pasteurized and fermented. The seeds are extracted after macerating the fruit in water. The seeds are dried, roasted, fried or boiled and eaten as dessert nut or ground to a meal known as breadfruit flour. Breadfruit flour can be used to produce a variety of sweetened baked goods and pastries (Anazonwu-Bello 1981) including cookies, buns, cakes, biscuits, bread rolls, doughnuts, chips, spring rolls, pasta, weaning (baby) foods, porridge and breakfast cereals. The flour is also suitable for the preparation of a product which tastes much like mushroom soup. A non-alcoholic beverage, almond milk, can be prepared from powdered seeds, which is recommended as a breakfast drink in Nigeria. The seed is also a rich source of vegetable oil, protein carbohydrate, as well as several minerals, proteins and vitamins, and is a potential commercial raw material for the production of edible vegetable oils and margarine.

Plate 1 Leaves and globose, immature fruit of *Treculia africana*

Plate 2 Close-view of the globose immature fruit

Botany

An evergreen tree, 5–40 m high with a dense spreading crown, fluted trunk of 2–3 m girth and rough, dark brown bark that exudes white latex when bruised. Branchlets are terete and lenticellate. Leaves are alternate, simple and borne on short, 1.5 cm petioles. Lamina is narrowly lanceolate to elliptic-lanceolate, 4–8 cm long by 1–2.5 cm wide, acutely cuspidate, sub-coriaceous, green, glabrous above and paler green below with some hairs along 10–18 pairs of veins (Plates 1, 2 and 3). Young leaves are reddish-bronze. Flower heads are globose, 2.5–10 cm across, yellowish-brown, axillary and terminal to the final branches, with numerous small, white flowers, male and female flowers usually separate. Infructescences compound, globose, very large reaching 40–50 cm in diameter, on the trunk or main branches and may weigh up to 15 kg, green turning to yellow to yellowish-brown (Plates 1, 2, 4 and 5). It contains many orange-coloured seeds, about 1 cm across, buried in the spongy pulp. The outer surface is covered with rough pointed outgrowths.

Nutritive/Medicinal Properties

The proximate nutrient composition of *Treculia africana* raw seeds per 100 g edible portion is reported as: energy 377 kcal, moisture 9.2%, protein 12.6%, fat 5.6%, carbohydrate 70.4%, fibre 1.6%, ash 2.2 g, Ca 127 mg, P 317 mg (Leung et al. 1968). Another report stated that the chemical composition and nutritive value of the seeds of African breadfruit, *Treculia africana* var. *africana*, contained, in terms of dry weight, 13.4% protein, 18.9% lipid, 58.1% carbohydrate, 1.4% crude fibre and 2.1% ash; minerals Na, 7.0 mg%; Mg, 184 mg%; Ca, 17.5 mg%; K, 585 mg%;

Plate 3 Close view of leaves

Plate 4 Ripe fruit

P, 382 mg%; Cu, 3.9 mg%; Fe, 1.6 mg%; Cr, 0.20 mg% and Zn, 7.5 mg% (Edet et al. 1985). The moisture content was found to be 7.8% in terms of wet weight. Vitamins B1, 0.5 mg%; B2, 0.3 mg% C, 45.1 mg% and β-carotene, 6.0 mg% were found to be present in the seeds.

Plate 5 Cross section of the fruit

Anti-nutritional compounds found were oxalate (soluble), 2.5 mg% (total), 3.0 mg%; phytate, 2.0 mg%; tannin, 15.0 mg% and HCN, 1.1 mg%. Ekpenyong (1985) reported that the seeds of African breadfruit (*Treculia africana*) had high protein and oil contents (23% and 11%, respectively). The oil could be used for industrial purposes and for human consumption due to its high food energy value. Phosphorus and potassium contents were high but the low calcium content could lead to a low Ca/P ratio. Glutamic acid, aspartic acid and glycine were the most abundant amino acids, followed by lysine, leucine, threonine and valine. The content of sulphur-containing amino acids in African breadfruit was low. Its high content of essential amino acids suggested that it could be a suitable replacement for soybean in areas where the latter was scarce, or too expensive

Treculia africana seeds were found to be rich in protein; its protein content was reported higher than those from high protein animal sources such as beef and marine fishes (Ajayi 2008). The seeds had high carbohydrate content and could act as source of energy for animals if included in their diets. The oil content of the seeds was 18.54%. The results of the physicochemical properties of the seeds were comparable to those of conventional oil seeds such as groundnut and palm kernel oils and could be useful for nutritional and industrial purposes. The seeds were found to be good sources of mineral elements (Ajayi 2008). Potassium at 1,680 ppm was the prevalent mineral followed by sodium, magnesium and then calcium. It also contained reasonable quantity of iron. The crude protein and fat contents of the unprocessed (raw) seeds of *T. africana* were 20.1% and 13.7%, respectively (Giami et al. 2001). The level of phytic acid in the raw seed (1.19 mg/g) was lower than the levels found in some commonly consumed pulses in Nigeria. Albumin and globulin protein fractions were found to be the major seed proteins of African breadfruit seed, constituting 67.8% of the total protein of the raw seed. There were no significant differences between crude protein, ash and fat contents of the raw and heat processed samples. Boiling proved more effective than roasting for improving protein digestibility and for reducing the levels of trypsin inhibitor, phytic acid and polyphenols of the samples. The complete removal of these antinutrients, however, would require a more severe heat treatment of the seed, which in turn would profoundly reduce the nutritional value and availability of proteins, as demonstrated by the low values obtained for in vitro protein digestibility, protein fractions and protein extractability.

Raw flour of *T. africana* contained 20.1% crude protein, 2.5% total ash and 13.7% fat (Giami et al. 2000). Heat processing significantly improved in-vitro protein digestibility, and water and fat absorption capacities but decreased bulk density, nitrogen solubility, emulsion and foaming properties, trypsin inhibitor, and phytic acid and polyphenol contents of the samples. Boiling proved more effective than roasting for improving protein digestibility, emulsion capacity and foam stability and reducing antinutritional factor levels. Partial proteolysis increased nitrogen solubility, bulk density and water and fat absorption capacities but decreased foam capacity at hydrolysis levels greater than 35%. Fatty acid and peroxide values of the samples increased during storage. Compared to raw samples, heat processed samples had significantly lower and more acceptable peroxide values and free fatty acid contents and higher and more stable water (3.0 g/g sample) and fat (2.4 g/g sample) absorption capacities.

Roasting of the raw unprocessed seeds of *T. africana* did not bring about any significant difference in ether extract, crude protein, dietary fibre, ash or carbohydrate contents of the raw and roasted seeds (Lawal 1986). However, moisture content showed a significant decrease (16.1%). Mineral analysis showed increases of potassium (39.2%) and calcium (28.2%). Significant differences were observed for vitamins B1, B2 and C contents with 30.6%, 25.6% and 34.7% destruction, respectively. Amino acid analysis showed losses for lysine (12.1%), isoleucine (13.0%) and leucine (12.7%).

Studies reported significant increases in the protein and ash contents of fermented full fat and defatted fermented *Treculia africana* seed (TAS) flours compared with the other flours (Fafasi et al. 2004). Fermented TAS flours had 18.6% and 20.8% of its total dry matter as protein while the autoclaved, toasted and untreated TAS flours had 18.0 and 17.0%, 16.3% and 16.8%, and 15.1% and 18.1% protein respectively. The energy value of unprocessed full fat TAS flour (17.12 kJ/g) was significantly higher than the processed full fat TAS flours (15.5, 16.3 and 16.3 kJ/g for fermented, autoclaved and toasted TAS flours respectively). The percent utilisable energy due to protein was highest in the fermented TAS flour (20.2% and 23.9% for full fat and defatted samples), and the values were for other treatments: autoclaved TAS flour 18.6% and 18.8%, toasted TAS flour 16.9% and 18.6%, unprocessed TAS flour 14.8% and 20.3% respectively for the full fat and defatted samples. The K/Na ratio was $\gg 1$ while the Ca/P ratio was $\ll 1$ in all the flours. Zinc was the predominant mineral in the flours and highest in fermented TAS flour. Results showed significant reduction in oxalates, phytates, tannin and hydrocyanic acid contents of processed TAS flours. Hydrocyanic acid was not detected in the processed TAS flours. Blending of African breadfruit flour with corn and defatted soybean improved the protein content of the mixtures but increased the antinutritional factors (Nwabueze 2007). Extrusion cooking drastically reduced antinutritional factors, phytic acid and tannin content of the extrudates by 91%, 44% and 92%, respectively. The model developed produced the highest percent variation in TIA (80%) in comparison with either phytic acid (60%) or tannin (50%). Feed composition and screw speed were the most significant process variables affecting TIA. Effect of feed composition was quadratic, while that of screw speed was linear. The quadratic effects of screw speed and feed composition on tannin and phytic acid were also significant.

The protein, fat, ash and crude fibre contents of the flour blends prepared from the toasted African breadfruit seeds and wheat flours increased while carbohydrate and moisture contents decreased with increased levels of toasted African breadfruit flour in the blends (Akubor et al. 2000). The TABF showed significantly higher water absorption capacity, oil absorption capacity, foaming capacity and emulsion activity than wheat flour. These properties increased with increased amounts of TABF in the blends. All the flour blends exhibited a least gelation concentration of 8% (W/V). The bulk density (g/cm3) and wettability values of flour blends varied from 0.74 to 0.84 and 19 to 31, respectively. Heating increased the swelling capacity of the flour blends.

The yield of the oil from crushed *T. africana* seeds was 20.83% (Ajiwe et al. 1995). The oil was found to be an unsaturated, semi-drying oil which is unsaturated, with a high saponification value, acidic and requires purification.

The following compounds were identified in the leaves of *T. africana*: lupeol, lupeol acetate, apigenin, epiphyllocourmarin, catechin, 4-hydroxybenzoic acid and 6,9-dihyroxymegastigmane-3-one (Metuno et al. 2008).

Pharmacological properties of African breadfruit reported include the following:

Antidiabetic Activity

Recent studies showed that the lipid fractions and blood glucose level in the breadfruit fed rats were significantly lower than in the group fed on normal rat chow, except for LDL (low density lipid) which was not significantly different in the breadfruit fed normal rats and was higher in the diabetics-breadfruit fed rats (Oyelola et al. 2007). The HDL (high density lipoproteins)/LDL ratio indicated that consumption of breadfruit favours LDL level in both normoglycaemic and alloxan (or diabetic) treated rats compared to controls. The results demonstrated breadfruit seeds probably contained agents that reduced blood sugar, total cholesterol, and very low density lipids. A separate study reported that test fractions of aqueous acetone root bark extract had only a slight effect on blood sugar level of normal rats (Okwari et al. 2006). On short term and chronic administration in diabetic rats however, diethyl ether-soluble (DEF) and the water-soluble (WSF) fractions significantly reduced the fasting blood sugar levels at differing rates when compared with the control group of animals. The diethyl ether soluble fraction (10 mg/kg dose level) was found to exhibit the highest activity giving 69.4% reduction in blood sugar level which was in comparable range with the reference standard glibenclamide (0.5 mg/kg) which reduced blood sugar levels by 65.8% below the initial baseline values. The results supported the use of the plant by herbalists for diabetes control.

Antimicrobial Activity

Aqueous ethanol extract of the bark of *T. africana* was also reported to have antibacterial activity. The extracts were effective on the following bacterial organisms: *Salmonella typhi, Shigella dysentriae, Escherichia coli, Pseudomonas aeruginosa* and *Staphylococcus aureus* (Ogbonnia et al. 2008). The mean minimum inhibition concentration (MIC) of the extract ranged from 3.125 to 25 mg/ml for the different organisms tested. The extract exhibited minimum bactericidal concentration (MBC) of 50 mg/ml on *S. dysentriae* and *P. aeruginosa* only, while other tested bacteria strains required higher concentrations. Phytochemical screening revealed the presence of steroidal saponin glycosides as the major component, anthraquinone glycoside and polyphenols. The minimal inhibition concentration (MIC) values obtained with the crude leaf extracts varied from 78 to 156 μg/ml against 20 (95.24%) of the 21 tested microorganisms (Kuete et al. 2008). Compound

(2) catechin suppressed the growth of all the tested microorganisms (6 species of Gram-positive bacteria, 12 species Gram-negative bacteria and three *Candida* species). Other bioactive compounds, phyllocoumarin and 6, 9-dihydroxy-megastigmane-3-one (3) showed selective activity.

Leaf crude extract of the three *Treculia* species namely *Treculia obovoidea, Treculia africana* and *Treculia acuminata* as well as that from the twigs of *T. africana* were found to prevent the growth of *Mycobacterium smegmatis* and *Mycobacterium tuberculosis* (Kuete et al. 2010). The lowest MIC value (19.53 μg/ml) was recorded with extract of the leaves of *T. africana* on *Mycobacterium smegmatis*, and those of *T. africana* and *T. acuminata* against *M. tuberculosis*. All the extracts inhibited to various extent the anti-reverse transcriptase activity at 200 μg/ml. The best IC_{50} values, 31.1, 29.5 and 21.1 μg/ml were recorded respectively with the extracts of the leaves of *T. obovoidea, T. acuminata* and *T. africana*. Results of the antioxidant activity indicate a dose-dependent ability of sample to scavenge the DPPH radical. The lowest IC_{50} values were obtained with extracts of the leaves of *T. acuminata* (56.3 μg/ml) and *T. obovoidea* (55.9 μg/ml). Pronounced tumour-reducing activity was observed with the extracts of the leaves of *T. africana* (89.67%), *T. acuminata* (92.16%), *T. obovoidea* (96.67%) and that of the twigs of *T. acuminata* (87.18%). The overall results provided evidence that plants of the genus *Treculia* might be potential sources of antitubercular, anti-HIV and antitumour compounds.

Analgesic Activity

The crude extract of *Treculia africana* caused a decrease in rearing, grooming and locomotor activity in mice (Aderibigbe et al. 2010). It also potentiated ketamine-induced sleeping time and produced hypothermic effect in mice. The crude extract possessed sedative effect, which may be through increase in the activity of GABA in the brain. *Treculia africana* was found to possess sedative and strong analgesic properties (Aderibigbe and Agboola 2010). The results showed that *Treculia africana* extract reduced head dips, produced anxiogenic effect, reduced locomotor activity without effect on learning and memory and produced analgesic effect peripherally in mice.

Traditional Medicinal Uses

Ethno-medicinally, crude extracts from different parts of the tree have been employed either singly or in combination with other herbs in traditional African folk medicine in the treatment of various ailments. In Ghana, a root decoction is used as an anthelmintic and febrifuge. Decoctions from different plant parts are used as an anti-inflammatory agent. A bark decoction is used for cough and whooping cough, and ground bark with oil and other plant parts for swellings. Crushed leaves juice are applied on the tongue as a treatment for thrush in children; the latex is applied as an antibacterial agent in eardrops, and as chewing stick. The caustic latex is applied on carious teeth. Leaf decoctions were reported used in Trinidad and Bahamas to lower blood pressure and is used also in some communities as an effective treatment in stomach upset and other gastro intestinal infections. In Nigeria, herbalists have used the plant extract in an ancient recipe for diabetes. The plant is also used as a laxative, vermifuge, febrifuge, galactogue and as a therapy for leprosy.

Other Uses

T. africana has been used for intercropping in forestry systems and is also a promising backyard tree for the home. It is also a source of fuel wood and charcoal. The tree provides good mulch that can be used to improve the soil. The wood fibre is suitable for pulp and paper making. The heavy timber it provides is suitable for furniture, carving, turnery and inlay wood. In Ghana, it is used for furniture and joinery. The fruit pulp and bran can be used in livestock feed. Seed oil is a possible commercial raw material for the production of pharmaceuticals, soaps, Shampoos, alkyd resins, perfumes and paints besides providing an edible vegetable oil. The oil extracted from crushed breadfruit (*Treculia africana)* seeds could be used for making soap, hair shampoo and alkyd resin (Ajiwe et al. 1995). Leaves are used for fodder in Tanzania.

Comments

The species is propagated by seeds.

Selected References

Aderibigbe AO, Agboola OI (2010) Studies of behavioural and analgesic properties of *Treculia africana* in mice. Int J Biol Sci 4:2

Aderibigbe AO, Adeyemi IO, Agboo OI (2010) Central nervous system depressant properties of *Treculia africana* Decne. Ethnobot Leaflets 14:108–119

Ajayi IA (2008) Comparative study of the chemical composition and mineral element content of *Artocarpus heterophyllus* and *Treculia africana* seeds and seed oils. Bioresour Technol 99(11):5125–5129

Ajiwe VIE, Okeke CA, Agbo HU (1995) Extraction and utilization of breadfruit seed oil (*Treculia africana*). Bioresource Technol 53(2):183–184

Akubor PI, Isolokwu PC, Ugbane O, Onimawo IA (2000) Proximate composition and functional properties of African breadfruit kernel and flour blends. Food Res Int 33(8):707–712

Anazonwu-Bello JN (1981) Indigenous foods and nutritional adequacy. Paper presented at symposium on development of indigenous technology. Ministry of Science and Technology, Enugu

Edet EE, Eka OU, Ifon ET (1985) Chemical evaluation of the nutritive value of seeds of African breadfruit (*Treculia africana*). Food Chem 17(1):41–47

Ejiofor MAN, Obiajulu OR, Okafor JC (1988) Diversifying utilities of African breadfruit *(Treculia africana* Decne. subsp. *africana)* as food and feed. Int Tree Crop J 5:125–134

Ekpenyong TE (1985) Chemical composition and amino acid content of African breadfruit (*Treculia africana* Decne). Food Chem 17(1):59–64

Fasasi OS, Eleyinmi AF, Fasasi AR, Karim OR (2004) Chemical properties of raw and processed breadfruit (*Treculia africana*) seed flour. Food Agric Environ 2(1):65–68

Giami SY, Adindu MN, Akusu MO, Emelike JN (2000) Compositional, functional and storage properties of flours from raw and heat processed African breadfruit (*Treculia africana* Decne) seeds. Plant Foods Hum Nutr 55(4):357–368

Giami SY, Adindu MN, Hart AD, Denenu EO (2001) Effect of heat processing on in vitro protein digestibility and some chemical properties of African breadfruit (*Treculia africana* Decne) seeds. Plant Foods Hum Nutr 56(2):117–126

Kuete V, Metuno R, Ngameni B, Mbaveng AT, Ngandeu F, Bezabih M, Etoa F-X, Ngadjui BT, Abegaz BM, Beng VP (2008) Antimicrobial activity of the methanolic extracts and compounds from *Treculia africana* and *Treculia acuminata* (Moraceae). S Afr J Bot 74(1):111–115

Kuete V, Metuno R, Keilah PL, Tshikalange ET, Ngadjui BT (2010) Evaluation of the genus *Treculia* for antimycobacterial, anti-reverse transcriptase, radical scavenging and antitumor activities. S Afr J Bot 76(3):530–535

Lawal RO (1986) Effect of roasting on the chemical composition of the seeds of *Treculia africana*. Food Chem 22(4):305–314

Leung W-TW, Busson F, Jardin C (1968) Food composition table for use in Africa. FAO, Rome, 306 pp

Makinde MA, Elemo BO, Arukwe U, Peter P (1985) Ukwa seeds (*Treculia africana*) protein 1: chemical evaluation of the protein quality. J Agric Food Chem 33(1):70–72

Metuno R, Ngandeu F, Tchinda AT, Ngameni B, Kapche GDWF, Djemgou PC, Ngadjui BT, Bezabih M, Abegaz BM (2008) Chemical constituents of *Treculia acuminata* and *Treculia africana* (Moraceae). Biochem Syst Ecol 36(2):148–152

Nwabueze TU (2007) Effect of process variables on trypsin inhibitor activity (TIA), phytic acid and tannin content of extruded African breadfruit–corn–soy mixtures: a response surface analysis. LWT Food Sci Technol 40(1):21–29

Ogbonnia SO, Enwuru NV, Onyemenem EU, Oyedele GA, Enwuru CA (2008) Phytochemical evaluation and antibacterial profile of *Treculia africana* Decne bark extract on gastrointestinal bacterial pathogens. Afr J Biotechnol 7(10):1385–1389

Okwari OO, Ofem OE, Ettarh RR, Eyong EU (2006) Blood glucose level and lipid profile in rats fed on *Treculia africana* (breadfruit) diet: a sub-chronic study. Niger J Health Biomed Sci 5(2):21–25

Oyelola OO, Moody JO, Odeniyi MA, Fakeye TO (2007) Hypoglycemic effect of *Treculia africana* decne root bark in normal and alloxan-induced diabetic rats. Afr J Tradit Complement Altern Med 4(4):387–391

Turril WB (1952) *Treculia africana* Desc. var. *africana* [family Moraceae]. Flora of tropical East Africa, vol 1. Crown Agents for the Colonies, London

World Agroforestry Centre (2006) *Treculia africana*. AgroForestryTree Database, World Agroforestry Centre. http://www.worldagroforestry.org/sea/Products/AFDbases/AF/asp/SpeciesInfo.asp?SpID=1651

Moringa oleifera

Scientific Name

Moringa oleifera Lamk.

Synonyms

Anoma moringa Lour., *Guilandina moringa* Linn., *Hyperanthera arborea* J.F. Gmel., *Hyperanthera decandra* Willd., *Hyperanthera moringa* Vahl, *Moringa erecta* Salisb., *Moringa moringa* Millsp., *Moringa octogona* Stokes, *Moringa parvifolia* Noronha, *Moringa polygona* DC., *Moringa pterygosperma* C.F. Gaertn, *Moringa nux-ben* Perr., *Moringa zeylanica* Pers.

Common/English Names

Ben Oil Tree, Ben Nut, Ben Tree, Behn Tree, Behen Tree, Benzolive Tree, Cabbage Tree, Drumstick, Drumstick Tree, Horseradish Tree, Moringa, West Indian Ben

Vernacular Names

Arabic: Rawag, Ruwag, Shagara Al Ruwag, Shagara Al Ruway (Sudan), Alim, Halim, Habbah Ghaliah;
Benin: Yuru Ara, Yorwata, Yoroguma (Bariba), Windibudu (Dendi), Kpano, Kpatima, Patima, Yovokpatin, Yovotin (Fon), Tekpinda (Natemba), Guildandeni, Latj Iri, Legi-Lakili (Peul), Yori Kununfa (Waama), Ewé Ilé (Yoruba), Agun, Ayere, Ewé Igbale, Ewé Ile, Ewé Oyibo, Manyieninu, Oyibo (Yoruba-Nago);
Brazil: Cedra, Moringa;
Burkina Faso: Guilgandani, Gigandjah (Fulfuldé), Argentiga, Alsam Tiga (Moré);
Burmese: Daintha, Dandalonbin, Dan Da Lun, Dan Da Lun Bin;
Cameroun: Paizlava (Daggai), Guiligandja (Foulfoudé), Zogalagandi (Huasa), Gagawandalahai (Mafa), Djihiré (Mandara), Naa-Toukoré (Moundang), Chabané (Pokoko), Naa-Nko (Toupouri);
Chad: Kag N'dongue (Sara);
Chamorro: Katdes, Malongay, Marunggai;
Chinese: La Mu, La Mok (Cantonese);
Columbia: Angela;
Costa Rica: Marango (Spanish);
Cote D'ivoire: Arjanayiiri;
Cuba: Palo Jeringa, Palo De Tambor;
Czech: Moringa Olejná;
Danish: Behen, Behennød, Behennødtræ;
Dominican Republic: Palo De Aceite, Palo De Abejas, Libertad (Spanish);
Dutch: Benboom, Peperwortel Boom;
Eastonian: Õli-Rõikapuu;
El Salvador: Tebebrinto;
Ethiopia: Aleco, Shalchada, Shelagda (Konsoigna), Shiferaw (Amharigna), Aleko, Haleko (Giddigna), Kalan'gi (Hamer-Bena);
Fiji: Sajina;
French: Ben Ailée, Ben Ailé, Ben Oléifère, Benzolive, Moringa Ailée, Neverdie, Pois Quénique;

T.K. Lim, *Edible Medicinal And Non-Medicinal Plants: Volume 3, Fruits*,
DOI 10.1007/978-94-007-2534-8_61,

German: Behenbaum, Behennussbaum, Flugelsaniger, Meerrettichbaum, Moringaölbaum, Pferderettichbaum;
Ghana: Buid (Ashanyi), Atiuwuse, Babatsi, Kpotowuzie, Yevu-Ti (Ewe), Ownwukuow, Zangala (Dagari), Zingeridende (Hausa);
Guadaloupe: Moloko, Ben-Ailé;
Guam: Malungkai, Marronggai, Marungai, Marunggai, Malungay, Katdes;
Guatemala: Paraíso, Paraíso Blanco, Perlas (Spanish);
Guyana: Saijhan, Sijan;
Haiti: Bambou-Bananier, Ben Oleifere, Benzolive, Benzolivier, Graines Benne;
Honduras: Maranga, Maranga Calalu (Spanish);
Hungarian: Lóretekfa;
India: Saijna, Sajina (Assamese), Munga Ara, Shajna, Sojna, Sojne Danta, Sujana (Bengali), Midho Saragavo, Saragvo, Seeng Ni Phali, (Gujarati), Hargua, Mungana, Mungna, Rasunna, Sahijan, Sahjan, Sahinjan, Sainjana, Sainjna, Sajna Sanan Suhanjua, Sanan Suhanjuna,, Sanjana, Sanjna, Sanjano, Saonjna, Sargua, Segu, Segva, Segve, Senjana, Shajna, Shajnah, Soanjana, Soanjna, Soanjna Shevga, Sondna, Soujna, Sunara (Hindu), Guggala, Mochaka, Murunga, Mochaka Mara, Nugge, Nuggi, Nugge Mara (Kannada), Mashinga, Mushinga Saang (Konkani), Mouringou, Murina, Muringa, Muringai, Murinna, Murunna, Sigru, Tishnagandha (Malayalam), Shajna (Manipuri), Achajhada, Badadishing, Moshimg, Munagacha-Jhad, Munagachajhada, Mungai, Murungamul, Sainga, Shegat, Shegta, Shegva, Shevga, Shevaga, Shevgi, (Marathi), Munigha, Muniya, Shajna, Saijna, Sajana, Sajina, Soandal, Sujuna (Oriya), Saajinaa, Sanjina, Soanjana, Surajana (Punjabi), Akshiba, Aksiva, Bahala-Pallavah, Bahalah, Bahumula, Chaksushya, Chalusha, Damsamula, Danshamula, Dravinaandhata, Dvishigru, Dvisigru, Gandhaka, Haritapatra, Haritashaka, Jalaproya, Janapriya, Kakshivaka, Kalibaka, Kaminisha, Katukanda, Komalpatraka, Krishnagandha, Krishnashigru, Kshamadansha, Madhugunjana, Madhushigruka, Madhusravah, Mechaka, Mocaka, Mocha, Mochaka, Mukhabhanga, Mukhamlda, Mulakaparni, Murangi, Murungi, Nashana, Rochana, Ruchiranjana, Saigravam, Sanamaka, Sanbhanjana, Shakapatra, Shigru, Shigruka, Shobhanjana, Shobhataka, Sigru, Sigruh, Sitavhaya, Sobhanjana, Sobhanjanah, Strichittahari, Subhanjana, Sumula, Supatraka, Sutikshna, Svetamaricha, Svetashigru, Swetamaricha, Tikshnamula, Tikshnanandhaka, Tiksnagandha, Tilashigru, Ugra, Upadansha, Vanapallava, Vidradhinashana, (Sanskrit), Achuram, Asasuram, Cikkuru, Copancanam, Karunjanam, Kaykkirai, Kilavi, Kirancanam, Kiranjanam, Murangai, Murungai, Murunkai, Sikkuru, Suligai, Tavuselam, (Tamil), Mochakamu, Mulaga, Mulaga Chettu, Mulage, Munaga, Munaga Chettu, Sajana, Sigrupa, Sitavrykshamu, Tellamunaga, (Telugu), Noorggaee (Tulu), Sahajna (Urdu);
Indonesia: Kachang Kelur, Kelor, Lemunggai, Meringgai, Remunggai, Sajor Kelor, Semunggai, Smunggai, Tjelor (Bali), Kelor (Leaves) Klentang, Limaran (Fruit) (Javanese), Kelor (Leaves), Klentang, Kolentang (Fruit) (Sundanese), Marongghi (Leaves), Klentang (Fruit) (Madurese);
Italian: Been, Bemen, Sàndalo Ceruleo;
Japanese: Marungai, Marunga Oreifera, Wasabi No Ki, Wasabi No Ki;
Kenya: Mkimbo, Mlonge, Mlongo, Mronge, Mrongo, Mlongo, Mzunze, Mzungu, Shingo (Swahili);
Khmer: Ben Ailé, Daem Mrom, Daem Mrum;
Laotian: B'loum;
Madagascar: Anamambo, Anamorongo, Feliimorongo, Felikambo, Felikamoranga, Landihazo, Moringa, Moringy (Malagasy);
Malaysia: Kelur, Buah Kelur, Remunggai (Fruit), Daun Kelor, Daun Remunggai (Leaf);
Malawi: Cham'mwanba, Kangaluni (Chichewa), Nsangoa (Senna), Chakate,Kalokola, Maula Tengo, Mpundi, Muula, Mbula, Mpempu, Mpenba (Yao);
Mali: Anamambo, Anamorongo, Feliimorongo, Felikambo, Felikamoranga, Landihazo, Moringa, Moringy, Névrédé (Bambara);
Mauritius: Brede Mouroum;
Mexico: Árbol Del Ben, Arbol Do Los Aspáragos, Arbol De Las Perlas, Paraíso De Espana, Perla De La India, Perlas Del Oriente (Spanish);

Nepal: Sajiwan, Sitachini, Swejan;
Nicaragua: Marango;
Niger: Windi Bundu (Zarma), Zôgala Gandi (Huasa);
Nigeria: Gawara, Habiwal Hausa, Konamarade, Rini Maka (Fulani), Bagaruwar Maka, Bagaruwar Masar, Barambo, Barembo, Danga, Koraukin Zaila, Rimin Nacara, Rimin Turawa, Samarin, Shipka Hali, Shuka Halinka, Zogale, Zogall, Zogalla-Gandi, (Huasa), Ikwe Oyibo, Odudu Oyibo, Okwe Oyibo, Okwe Olu, Uhe, Oku-Ghara-Ite, Okochi Egbu (Ibo), Gergedi (Igala), Adagba Malero, Ewé Igbalé, Ewé Ilé, Ewe Igbale, Idagbo Monoyé (Yoruba);
Palauan: Malungkai;
Panama: Arbol Do Los Aspáragos, Babano Del Arbo, Ben, Jacinto (Spanish);
Philippines: Kalungai (Bikol), Alungai, Balungai, Dool, Malungit, Kalungai, Kamunggay (Bisaya), Marungai (Ibanag), Komkompilan, Marungai (Iloko), Dool, Malungit, Kamalungai (Pampangan), Kalamungai, Kamalongan (Panay Bisaya), Arunggai (Pangasingan), Kalungai, Kamalungua, Kamalungai, Malongai, Malungai, Malunggay Talbos, Mulangai, Mulangay, Mulanggay. (Tagalog), Marongoi (Sambali);
Portuguese: Acácia Branca, Marungo, Muringa, Moringuiera;
Puerto Rico: Ben, Jasmin Francés, Jazmin Francés, Resada (Spanish);
Russian: Moringa Oleifera;
Senegal: Binêbeddai (Diola), Névrédayo, Nédèdayo (Mandingue), Nebôday, Sap-Sap (Pulaar), Nébéday, Sap-Sap (Serere), Nobodai, Névoidai, Nébédai, Sap-Sap (Wolof);
Somalia: Dangap, Mirongo;
Spanish: Árbol Del Ben, Babano Del Arbo, Ben, Maranga, Morango, Paraíso, Paraíso Blanco;
Sri Lanka: Moo Rin Guu, Murungà, Murunga (Bean/Pod), Murunga Gasa (Tree), Murunga Kolaya, Murunga Kolle (Leaves);
Sudan: Alim, Halim, Ruwag, Shagara Al Ruwag (Arabic);
Surinam: Kelor, Peperwortel Boom;
Taiwan: La Mu (Chinese);
Tanzania: Mboga Chungu, Mjungu Moto, Mlonge (Swahili);
Thai: Ka Naeng Doeng, Ma Khon Kom, Ma Rum (Bean/Pod), Phak I Huem, Phak I Hum, Phak Nuea Kai, Phak Ma Rum (Leaves), Se Cho Ya;
Tibetan: Si-Gru;
Togo: Baganlua, Bagaelean (Dagomba), Kpotima, Yevu-Ti (Ewe), Mágurua Maser (Huasa), Yovovoti (Mina), Gambaduk (Moba);
Trinidad: Saijan;
Vietnamese: Chùm Ngây;
Zimbabwe: Mupulanga, Zakalanda (Tonga).

Family

Moringaceae

Origin/Distribution

Drumstick originated from India, southern edge of W Himalaya, and is now widely distribute throughout the tropics in South and central America, Africa, Asia and the Pacific islands.

Agroecology

Moringa oleifera is cultivated throughout the semi-arid tropics, tropics and sub-tropics from 250 to 1,600 m elevation. It thrives in well-drained, sandy-loamy soils with a high water table, but is drought resistant and can grow in arid and coastal areas. It also tolerates light frost.

Edible Plant Parts and Uses

All parts of the tree-leaves, tender young capsules (pods), immature seeds, flowers, fruits and young roots are edible. In Indonesia, the young leaves and flowers are collected, cooked and eaten like other vegetables or in soup, lalab or *sayur* and used like spinach. In India and Bangladesh, the tender drumstick leaves, finely chopped, make an excellent garnish for any vegetable dishes, *dals, sambars*, salads, etc. they can also be used in place of or with coriander. Tender drumstick leaves, finely chopped, make an excellent garnish for any vegetable dishes, *dals, sambars*, salads, etc. Tender leaves are used to flavour ghee and to enhance the shelf-life of ghee. In Africa, the leaves are preferred over the pods. The leaves are eaten as a salad, cooked, and in soups and sauces. In the Mascarene Islands it is known as '*brède mouroungue*' or '*brède médaille*'. In the Philippines, the tender leaves are most often added to a broth to make a simple and highly nutritious soup. They are also sometimes used as a characteristic ingredient in *tinola* – a traditional chicken dish, composed of chicken in a broth, moringa leaves, and either green papaya or another secondary vegetable. Leaves can easily be dried in the shade to reduce loss of vitamins and rubbed over a wire screen to make a powder that can be stored and conveniently added to soups, sauces, and food without changing their taste. In West Africa, some health programs fight malnutrition by promoting a number of measures including the use of *Moringa oleifera* leaf powder in the diet of children and pregnant and lactating women. Dachana et al. (2010) showed the possibility of utilizing dried *Moringa oleifera* leaves (DMO) to improve the nutritional characteristics of cookies. Protein, iron, calcium, β-carotene and dietary fibber contents increased with increasing amount of DML from 0% to 15%. Sensory evaluation showed that cookies incorporated with 10% DML powder were acceptable.

The flowers are cooked and consumed either mixed with other foods or fried in batter, butter or oil. In West Bengal and Bangladesh, the flowers are usually cooked with green peas and potato. In Africa, flowers are sometimes eaten as a vegetable, added to sauces or used to make tea. In Sudan the flowers are made into a paste by crushing and then fried.

The immature green pods are probably the most valued, they are chopped into smaller pieces and use in curries and stews. They are prepared in various types of curries mixed with coconut milk, poppy seeds and mustard in India and Bangladesh. The immature pod pieces can just be boiled, until semi-soft and consumed directly without any extra processing or cooking. It is used in curries, *sambars, kormas*, and *dals*, although it is also used to add flavour to cutlets, etc. Drumstick dal, is prepared by adding drumstick pulp to boiled *dal*, and beaten along with the dal before seasoning. Another popular South Indian dish is *sambar* or *sambhar* which is a spiced lentil soup. Sambar is usually cooked with toor dal, drumsticks and other locally grown vegetables. The spices used typically in this stew are turmeric, chilli powder and cumin among others. Sambar is eaten with rice just like the *drumstick dal*. Scraped drumstick pulp can be made into a tasty drumstick *bhurtha* (drumstick curry) in the same way as eggplant curry (*baingan bhurtha*). In Malaysia, one popular recipe is to cook the chopped pieces in coconut curry to which is added milk, shrimps or chicken pieces. The immature seeds and surrounding white material can be removed from larger pods and cooked in various ways. Older pods are added to sauces.

Mature seeds are fried and eaten like peanuts in Nigeria. The seeds are added locally to sauces for their bitter taste.The mature seed contains about 40% edible oil – 'Ben oil'. *Moringa* seed has a fairly soft kernel, so the oil can be extracted by hand using a screw press. The Ben oil from the seeds is said to be used for salads and other culinary purposes. Bread with 10% debittered Moringa seed (DBMS) flour and cookies with 20% DBMS grits had more protein, iron and calcium than wheat flour, and were found to be acceptable (Ogunsina et al. 2011). Incorporating Moringa seeds in baked foods may be exploited as a means of boosting nutrition in Africa and Asia where malnutrition is prevalent.

The gum from the bark is sometimes added to sauces to make them thicker. The young roots are shredded after removal of the bark and used as a condiment in the same way as horseradish; however, it contains the alkaloid spirochin, a potentially fatal nerve-paralysing agent (Morton 1991), so such practices should be strongly discouraged.

Botany

A small deciduous, much-branched tree to 10 m, with furrowed, grey bark that comes off in corky flakes, soft, white wood, tuberous pungent root

Plate 1 Mature pods and flowers

Plate 2 Leaves used as vegetables

and a thin crown. Leaves are alternate, 2-3-pinnate up to 60 cm long with 4–6 pairs of pinnae somewhat clustered towards the twig end (Plates 1 and 2). Petiole 4–15 cm long, petiolules 1–6 mm; leaflets elliptical or obovate, 0.5–3 m by 0.3–2 cm, thin, glabrous or puberulent, greyish-green. Inflorescences are paniculate, axillary, with numerous white to creamy, fragrant, zygomorphic flowers. Flower has a deeply –5-partite copular-cyathiform tube with oblong, reflex segments, five creamy-white petals, five perfect stamens alternating with five subulate staminodes, stalked, one-celled ovary with three longitudinal furrows and three placentae bearing a double row of ovules and a thin, curved white style. Fruit is elongated, pendulous, linear, dagger shaped, 3-angled, 9-ribbed (Plates 1 and 3), splitting into three valves. Seeds numerous, subglobose, 1–1.4 cm across, trigonous with three thin wings and embedded in whitish fleshy placenta.

Plate 3 Drumstick pods

Nutritive/Medicinal Properties

Tee et al. (1997) reported the nutrient composition of fresh remunggai pod per 100 g edible portion as: energy 40 kcal, water 84.6 g, protein 5.1 g fat 0.3 g, carbohydrate 4.1 g, fibre 6.1 g, ash 0.8 g, Ca 22 mg, P 31 mg, Fe 0.3 mg, Na 3 mg, K 208 mg, carotenes 75 μg, vitamin A (RE) 13 μg, vitamin B-1 0.05 mg, vitamin B-2 0.12 mg, niacin 0.2 mg, vitamin C 258 mg. Nutrient composition of raw drumstick fruit per100 g edible portion is reported as: moisture 86.9 g, energy 26 kcal, protein 2.5 g, fat 0.1 g, carbohydrate 3.7 g, fibre 4.8 g, ash 2.0 g, vitamin A RE 9.2 μg, vitamin A RAE 4.6 μg, total carotene 110 μg, vitamin C 120 mg, thiamine 0.05 mg, riboflavin 0.07 mg, niacin 0.2 mg, Ca 30 mg, Fe 0.2 mg and P 110 mg (Gopalan et al. 2002).

The leaves were reported to have the following nutrient composition per 100 g edible portion: energy 80 kcal, water 75.5 g, protein 5.8 g fat 1.1 g, carbohydrate 14 g, fibre 1.2, ash 2.4 g, Ca 261 mg, P 73 mg, Fe 3 mg, Na 9 mg, K 444 mg, carotenes 2,931 μg, vitamin A (RE) 489 μg, vitamin B-1 0.27 mg, vitamin B-20.72 mg, niacin 4.3 mg, vitamin C 106 mg (Tee et al. 1997). *M. oleifera* leaves were analysed to have crude protein 27.51%, crude fibre 19.25%, crude fat 2.23%, ash content 7.13%, moisture content s 76.53%, carbohydrate content 43.88%, and the calorific value 1,296.00 kJ/g (305.62 cal/g) (Oduro et al. 2008). Elemental analysis indicated the leaves contained appreciable levels of calcium 2,009 mg/100 g DM and iron 28.29 mg/100 g DM.

The leafy tips of *Moringa oleifera* contained per 100 g edible portion: water 78.66 g, energy 268 kJ (64 kcal), protein 9.4 g, fat 1.4 g, carbohydrate 8.28 g, total dietary fibre 2.0 g, Ca 185 mg, Mg 147 mg, P 112 mg, Fe 4.0 mg, Zn 0.6 mg, Cu 0.105 mg, Mn 1.063 mg, Se 0.9 ug; vitamin C 51.7 mg, vitamin A 7564 IU, thiamine 0.257 mg, riboflavin 0.660 mg, niacin 2.22 mg, folate 40 μg, pantothenic acid 0.125 mg; tryptophan 0.144 g, threonine 0.411 g, isoleucine 0.451 g, leucine 0.791 g, lysine 0.537 g, methionine 0.123 g, cystine 0.140 g, phenylalanine 0.487 h, tyrosine 0.347 g, valine 0.611 g, arginine 0.532 g, histidine 0.196 g, alanine 0.705 g, aspartic acid 0.920 g, glutamic acid 1.035 g, glycine 0.517 g, proline 0.451 g, serine 0.414 g. The raw fruit contained per 100 g edible portion: water 88.2 g, energy 155 kJ (37 kcal), protein 2.1 g, total lipid 0.2 g, carbohydrate 8.53 g, total dietary fibre 3.2 g, ash 0.07 g, Ca 30 mg, Mg 45 mg, P 50 mg, K 461 mg, Na 42 mg, Fe 0.36 mg, Zn 0.45 mg, Cu 0.084 mg, Mn 0.259 mg, Se 0.7 ug; vitamin A 74 IU, thiamine 0.053 mg, riboflavin 0.074 mg, niacin 0.620 mg, pantothenic acid 0.794 mg, vitamin B-6 0.120 mg, folate 44 μg, ascorbic acid 141.0 mg, total saturated fatty acids 0.033 g, total monounsaturated fatty acids 0.102 g, and total polyunsaturated fatty acids 0.003 g (U.S. Department of Agriculture and Agricultural Research 2010).

Alpha- and γ-tocopherol were found in the leaves, flowers and fresh beans of *Moringa oleifera* (Sánchez-Machado et al. 2006). D-mannose, D-glucose, protein, ascorbic acid, polysaccharide were found in the flowers (Pramanik and Islam 1998)

The crude protein contents of the extracted and unextracted leaves of *M. oleifera* were 43.5% and 25.1% respectively (Makkar and Becker 1996). The true protein contents of these leaves were 93.8% and 81.3% of the total crude protein (non protein nitrogen contents of 2.7% and 4.7% were observed in the extracted and unextracted leaves). All essential amino acids including sulfur-containing amino acids were higher than adequate concentration when compared with recommended amino acid pattern of FAO/WHO/UNO reference protein for a 2–5-year-old child.

As evident from the above, leaves are very rich in minerals and vitamins. The results revealed that the leaves contain an appreciable amount of nutrients and can be included in diets to supplement our daily nutrient needs. The young fruit is also reported to be high in protein, a fair source of calcium and iron and high in phosphorus and contains all the essential amino acids.

Nutrient composition of raw drumstick flowers per100 g edible portion was reported as: moisture 85.9 g, energy 50 kcal, protein 3.6 g, fat 0.8 g, carbohydrate 7.1 g, fibre 1.3 g, ash 1.3 g, Ca 51 mg and P 90 mg (Gopalan et al. 2002).

The mature *M. oleifera* seeds contained 332.5 g crude protein, 412.0 g crude fat, 211.2 g carbohydrate and 44.3 g ash per kg dry matter (Oliveira et al. 1999). The essential amino acid profile compared with the FAO/WHO/UNU scoring pattern requirements for different age groups showed deficiency of lysine, threonine and valine. The content of methionine + cysteine (43.6 g/kg protein), however, was exceptionally higher and close to that of human milk, chicken egg and cow's milk. The seed extract agglutinated rabbit erythrocytes but did not show trypsin inhibitor and urease activities (Oliveira et al. 1999).

The dry seeds contain on average: protein 29%, fibre 7.5% and oil 36–42%; of the total fatty acid content oleic acid 65–75%, behenic acid 9%, palmitic acid 9%, stearic acid 7% and small amounts of lignoceric acid and myristic acid (Polprasid 1994).

The oil concentration from *Moringa oleifera* variety Mbololo seeds from Kenya was found to range from 25.8% (cold press) to 31.2% (chloroform/methanol) (Tsaknis et al. 1999). The oil was found to contain high levels of unsaturated fatty acids, especially oleic (up to 75.39%). The dominant saturated acids were behenic (up to 6. 73%)

and palmitic (up to 6.04%). The oil was also found to contain high levels of β-sitosterol (up to 50.07%), stigmasterol (up to 17.27%), and campesterol (up to 15.13%). Alpha-, γ-, and δ-tocopherols were detected up to levels of 105.0, 39.54, and 77.60 mg/kg of oil, respectively. The induction period (at 120°C) of *M. oleifera* seed oil was reduced from 44.6% to 64.3% after degumming. The *M. oleifera* seed oil showed high stability to oxidative rancidity. In another study, the hexane-extracted oil content of *Moringa oleifera* seeds was found to range from 38.00% to 42.00% (Anwar and Bhanger 2003). Protein, fibre, and ash contents were found to be 26.50–32.00, 5.80–9.29, and 5.60–7.50%, respectively. Results of physical and chemical parameters of the extracted oil were as follows: iodine value, 68.00–71.80; refractive index (40°C), 1.4590–1.4625; density (24°C), 0.9036–0.9080 mg/ml; saponification value, 180.60–190.50; unsaponifiable matter, 0.70–1.10%. Tocopherols (α, γ, and δ) in the oil were up to 123.50–161.30, 84.07–104.00, and 41.00–56.00 mg/kg, respectively. The oil was found to contain high levels of oleic acid (up to 78.59%) followed by palmitic, stearic, behenic, and arachidic acid up to levels of 7.00, 7.50, 5.99, and 4.21%, respectively. The induction period (Rancimat, 20 l/h, 120°C) of the crude oil was 9.99 hour and reduced to 8.63 hour after degumming.

In a more recent study, the petroleum ether extracted oil from *Moringa oleifera* seeds ranged from 27.83% to 45.07% on kernel basis and 15.1–28.4% on whole seed basis in 20 different Indian clones (Banerji et al. 2009). Leaves and pods showed a good source of vitamin C. Oleic acid (C18:1) was found to be the major fatty acid being 78.91–85.52% as compared to olive oil, which is considered to be richest source of oleic acid. All the clones from India did not show any presence of behenic acid (C 22:0). The oil was also found to contain high levels of β-sitosterol which ranged from 42.29% to 47.94%, stigmasterol from 13.66% to 16.61%, and campesterol from 12.53% to 16.63%. The γ- and δ-tocopherol were found to be in the range of 128.0–146.95, 51.88–63.5 and 55.23–63.84 mg/kg, respectively.

Other Phytochemicals

The presence of gallic acid, chlorogenic acid, ellagic acid, ferulic acid, kaempferol, quercetin and vanillin were found in the leaf, fruit and seed extracts (Singh et al. 2009).

Flower: pterygospermin (Das et al. 1957); 9-octadecen- 1-ol, (Z) – (CAS) cis-9–octadecen–1–ol, oleol, satol, ocenol, sipo, decanoic acid and dodecanal (Nepolean et al. 2009).

Fruits: O-[2′-hydroxy-3′-(2″-heptenyloxy)]-propyl undecanoate and O-ethyl-4-[(α-L-rhamnosyloxy)-benzyl] carbamate along with methyl p-hydroxybenzoate and β-sitosterol (Faizi et al. 1998); phenolic glycosides; 4-[(2′-O-acetyl-α-l-rhamnosyloxy) benzyl]isothiocyanate, 4-[(3′-O-acetyl-α-l-rhamnosyloxy) benzyl]isothiocyanate, and S-methyl-N-{4-[(α-l-rhamnosyloxy)benzyl]}thiocarbamate, together with five known phenolic glycosides (4–8) (Cheenpracha et al. 2010).

Seeds: 4(α-L-rhamnosyloxy)benzyl isothiocyanate (Eilert et al., 1981); phenylacetonitrile (Villasenor et al. 1989a); 4(α-L-rhamnosyloxy) phenylacetonitrile, 4-hydroxyphenylacetontrile, and 4-hydroxyphenyl-acetamide (Villasenor et al. 1989b); O-ethyl-4-(α-L-rhamnosyloxy) benzyl carbamate, 4(α-L-rhamnosyloxy)-benzyl isothiocyanate, niazimicin, niazirin, β-sitosterol, glycerol-1-(9-octadecanoate), 3-O-(6′-O-oleoyl-β-D-glucopyranosyl)-β-sitosterol, and β-sitosterol-3-O-β-D-glucopyranoside, (Guevara et al. 1999); 4-(α-l-rhamnopyranosyloxy)-benzylglucosinolate (Bennet et al. 2003); β-sitosterol (Mahajan and Mehta 2011); roridin E, veridiflorol, 9-octadecenoic acid (Nepolean et al. 2009); -(α-L-rhamnopyranosyloxy)benzyl isothiocyanate, methyl N-4-(α-L-rhamnopyranosyloxy) benzyl carbamate (both known compounds), and 4-(β-D-glucopyranosyl-1 → 4-α-L-rhamnopyranosyloxy)-benzyl thiocarboxamide (Oluduro et al. 2010).

Leaves: two nitrile glycosides, niazirin and niazirinin, and three mustard oil glycosides,

4-[(4′-O-acetyl-α-L-rhamnosyloxy) benzyl] isothiocyanate, thiocarbamate glycosides niaziminin A and niaziminin A, and niaziminin B (Faizi et al. 1994; Faizi et al. 1995); phytate (Makkar and Becker 1996); three thiocarbamate (TC)- and isothiocyanate (ITC)-related compounds, 4-[(4'-O-acetyl-α-L-rhamnosyloxy) benzyl] ITC (Murakami et al. 1998); flavonoids - quercetin and kaempferol (Siddhuraju and Becker 2003); quercetin-3-O-glucoside, quercetin-3-O-(6' '-malonyl-glucoside), kaempferol-3-O-glucoside, kaempferol-3-O-(6' '-malonyl-glucoside),3-caffeoylquinicacid,5-caffeoylquinic acid (Bennet et al. 2003); quercetin, kaempferol, isorhamnetin (Yang et al. 2008); hexadecanoic acid, ethyl ester (CAS) ethyl palmitate, palmitic acid ethyl ester, 2,6-dimethyl-1, 7-octadiene-3-ol, 4-hexadecen-6-yne, (z)-(CAS), 2-hexanone, 3-cyclohexyliden-4-ethyl-E2- dodecenylacetate, Hi-oleic safflower oil (CAS) and safflower oil (Nepolean et al. 2009); B chlorogenic acid, rutin, quercetin glucoside, and kaempferol rhamnoglucoside, (Atawodi et al. 2010); saponins, tannins (Bukar et al. 2010); glycosides of pyrrole alkaloid (pyrrolemarumine 4″-O-α-l-rhamnopyranoside) and 4'-hydroxyphenylethanamide (marumosides A and B), along with eight known compounds; niazirin, methyl 4-(α-l-rhamnopyranosyloxy) benzylcarbamate, benzyl β-d-glucopyranoside, benzyl β-d-xylopyranosyl-(1→6)-β-d-glucopyranoside, kaempferol 3-O-β-d-glucopyranoside, quercetin 3-O-β-d-glucopyranoside, adenosine and l-tryptophan (Sahakitpichan et al. 2011).

Stem: 4-hydroxymellein, vanillin, ß-sitosterone, octacosanic acid and ß-sitosterol (Saluja et al. 1978); 4-(α-L-rhamnopyranosyloxy)-benzylglucosinolate (Bennett et al. 2003); several procyanidin (Atawodi et al. 2010).

Roots: 4-(α-L-rhamnosyloxy)benzyl isothiocyanate (Eilert et al. 1981); alkaloids(Mazumder et al. 1999); 4-(α-L-rhamnopyranosyloxy)-benzylglucosinolate and benzylglucosinolate (Bennett et al. 2003); aurantiamide acetate and 1,3-dibenzyl urea (Sashidhara et al. 2009); several procyanidin (Atawodi et al. 2010).

Gum: The purified, whole-gum exudate from the drum-stick plant was found to contain L-arabinose, D-galactose, D-glucuronic acid, L-rhamnose, D-mannose and D-xylose in the molar ratios of 14.5:11.3:3:2:1:1 (Bhattacharya et al. 1982); and leucoanthocyanin (Khare et al. 1997).

Pharmacolgical properties of the plant parts reported include:

Bioavailability of Vitamin A/Carotenes

Nambiar and Seshadri (2001) found that β-carotene from drumstick leaves was effective in overcoming vitamin A deficiency although serum vitamin A levels remained somewhat lower compared to the group replete with vitamin A acetate. A marked reduction in food intake, body weight, accompanied by clinical signs of vitamin A deficiency and a decline in serum vitamin A (29.2–19.1 μg/dl) and liver vitamin A (3.7–2.0 μg/dl) were seen in rats at the end of 4 weeks of feeding a vitamin A deficient diet. On repletion significant improvements in clinical signs, food intake and body weights were noted in the three groups fed with vitamin A (4,000 IU/kg diet) in the form of vitamin A acetate (group A), fresh drumstick leaves (group B) or dehydrated drumstick leaves (group C) compared to the baseline and at the end of 4 weeks of depletion. The gain in body weight was highest for the group replete with dehydrated drumstick leaves. Among the replete groups, the serum vitamin A was highest for group A (34.7 μg/dl) given synthetic vitamin A, compared to group B (25.8 μg/dl) and group C (28.2 μg/dl) given drumstick leaves. All these were significantly higher than the serum vitamin A values seen at the end of 4 weeks of depletion (19.1 μg/dl). A significant improvement was also observed in the liver retinol levels on repletion for 4 weeks in the three groups, compared to the vitamin A depleted rats. In terms of growth parameters, the fresh and dehydrated drumstick leaves were better than the synthetic vitamin A. The researchers therefore concluded that in developing countries like India, sources of vitamin A such as drumstick leaves were valuable in overcoming the problem of vitamin A deficiency.

In a later study, the bioavailability of β-carotene and lutein from fresh and lyophilized *M. oleifera* leaves were anlayzed by Pullakhandam and Failla (2007), using the coupled in vitro digestion/Caco-2 cell model. Beta-carotene and lutein were stable during simulated gastric and small intestinal digestion. The efficiency of micellarization of lutein during the small intestinal phase of digestion exceeded that of β-carotene. Addition of peanut oil (5% vol/wt) to the test food increased micellarization of both carotenoids, and particularly β-carotene. Caco-2 cells accumulated β-carotene and lutein from micelles generated during digestion of drumstick leaves in a time-and concentration-dependent manner. The relatively high bioaccessibility of β-carotene and lutein from drumstick leaves ingested with oil supported the potential use of this plant food for improving vitamin A nutrition and perhaps delaying the onset of some degenerative diseases such as cataracts

Antioxidant Activity

The following flavonoids were identified in drumstick leaves: quercetin 89.8 mg/100 g fw, kaempferol 36.3 mg/100 g fw, isorhamnetin 2.9 mg/100 g f.w. giving a total flavonoid content of 129 mg/100 g and a dry matter of 25.5% (Yang et al. 2008). Flavonoids are recognised for their antioxidant and antiproliferative effects which may protect the body from various diseases and disorders.

Studies suggested that the extracts of both mature and tender leaves of *Moringa oleifera* exhibited potent antioxidant activity against free radicals, prevented oxidative damage to major biomolecules and afforded significant protection against oxidative damage (Sreelatha and Padma 2009). The successive aqueous extracts of *Moringa oleifera* exhibited strong scavenging effect on 2, 2-diphenyl-2-picryl hydrazyl (DPPH) free radical, superoxide, nitric oxide radical and inhibition of lipid per oxidation. The aqueous extracts of leaf (LE), fruit (FE) and seed (SE) of *Moringa oleifera* were found to significantly inhibit the OH-dependent damage of pUC18 plasmid DNA and also inhibited synergistically with trolox, with an activity sequence of LE>FE>SE (Singh et al. 2009). The presence of gallic acid, chlorogenic acid, ellagic acid, ferulic acid, kaempferol, quercetin and vanillin were found in the extracts. The leaf extract had comparatively higher total phenolics content (105.04 mg gallic acid equivalents (GAE)/g), total flavonoids content (31.28 mg quercetin equivalents (QE)/g), and ascorbic acid content (106.95 mg/100 g) and showed better antioxidant activity (85.77%), anti-radical power (74.3), reducing power (1.1 ascorbic acid equivalents (ASE)/ml), inhibition of lipid peroxidation, protein oxidation, OH-induced deoxyribose degradation, and scavenging power of superoxide anion and nitric oxide radicals than did the FE, SE and standard α-tocopherol. Additionally, the leaf and fruit extracts were found to inhibit violacein production, a quorum sensing-regulated behavior in *Chromobacterium violaceum.*

The flavonoid fraction *of Moringa oleifera* leaves was found to be effective in preventing cataractogenesis in selenite model by enhancing the activities of antioxidant enzyme and sulfhydryl content, reducing the intensity of lipid peroxidation, and inhibiting free radical generation in rat pups (Sasikala et al. 2010). The total phenolic content of the flavonoid fraction of *Moringa oleifera* leaves was found to be 4.4 mg of catechin equivalent/g dried plant material. The extract showed remarkable activity on 2,2-diphenyl-picrylhydrazyl (IC_{50} 36 μg/ml) and in superoxide radical (IC_{50} 33.81 μg/ml) scavenging assays. The flavonoid fraction effectively prevented the morphological changes and oxidative damage in the lens. In another recent study, the methanol extract of the leaves of *M. oleifera* was found to contain chlorogenic acid, rutin, quercetin glucoside, and kaempferol rhamnoglucoside, whereas in the root and stem barks, several procyanidin peaks were detected. (Atawodi et al. 2010). With the xanthine oxidase model system, all the extracts exhibited strong in vitro antioxidant activity, with 50% inhibitory concentration (IC_{50}) values of 16, 30, and 38 μl for the roots, leaves, and stem bark, respectively. Similarly, potent radical scavenging capacity was observed when extracts were evaluated with the 2-deoxyguanosine assay model system, with IC_{50} values of 40,

58, and 72 μl for methanol extracts of the leaves, stem, and root barks, respectively. The high antioxidant/radical scavenging effects observed for different parts of *M. oleifera* provided justification for their widespread therapeutic use in traditional medicine in different continents.

Season, agroclimate and production location were found to impact on the antioxidant activity of *Moringa oleifera* leaves (Siddhuraju and Becker 2003; Iqbal and Bhanger 2006). All extracts (water, aqueous methanol, and aqueous ethanol) of freeze-dried *M. oleifera* leaves from different agroclimatic regions were capable of scavenging peroxyl and superoxyl radicals (Siddhuraju and Becker 2003). Among the three different moringa samples, both methanol and ethanol extracts of Indian origins showed the highest antioxidant activities, 65.1% and 66.8%, respectively, in the β-carotene-linoleic acid system. Nonetheless, increasing concentration of all the extracts had significantly increased reducing power, which may in part be responsible for their antioxidant activity. The major bioactive compounds of phenolics were found to be flavonoids such as quercetin and kaempferol. On the basis of the results obtained, oringa leaves were found to be a potential source of natural antioxidants due to their marked antioxidant activity. Overall, both methanol (80%) and ethanol (70%) were found to be the best solvents for the extraction of antioxidant compounds from Moringa leaves. Significant differences were observed in the antioxidant activity of the extracts from different locations and seasons in Pakistan (Iqbal and Bhanger 2006). Generally, samples from Mardaan exhibited highest antioxidant activity followed by Balakot, Chakwal, Jamshoro, and Nawabshah. Overall antioxidant efficacy was greater in December or March depending upon location, and least in June.

Anticancer Activity

Of six bioactive compounds isolated from ethanol extract of *M. oleifera* seeds, niazimicin was found to have potent antitumour promoting activity in the two-stage carcinogenesis in mouse skin using 7,12-dimethylbenz(a)anthracene (DMBA) as initiator and TPA (12-O-tetradecanoylphorbol-13-acetate) as tumour promoter (Guevara et al. 1999). From the results, niazimicin was proposed to be a potent chemo-preventive agent in chemical carcinogenesis.

Hydro-alcoholic extract of drumsticks of *Moringa oliefera* at doses of 125 mg/kg bodyweight and 250 mg/kg body weight for 7 and 14 days, respectively, modulated changes in Phase I (Cytochrome b(5) and Cytochrome p(450)) and Phase II (Glutathione-S-transferase) enzymes, anti-oxidant enzymes, glutathione content and lipid peroxidation in the liver of 6–8 weeks old female Swiss albino mice (Bharali et al. 2003). Significant increase in the activities of hepatic cytochrome b(5), cytochrome p(450), catalase, glutathione peroxidase, glutathione reductase, acid soluble sulfhydryl content (–SH) and a significant decrease in the hepatic malondehyde level were observed at both dose levels of treatment when compared with the control values. Glutathione-S-transferase activity was found to be significantly increased only at the higher dose level. Butylated hydroxyanisol (BHA) fed at a dose of 0.75% in the diet for 7 and 14 days (positive control) caused a significant increase (in the levels of hepatic phase I and phase II enzymes, anti- oxidant enzymes, glutathione content and a decrease in lipid peroxidation. The skin papillomagenesis studies demonstrated a significant decrease in the percentage of mice with papillomas, average number of papillomas per mouse and papillomas per papilloma bearing mouse when the animals received a topical application of the extract at a dose of 5 mg/kg body weight in the peri-initiation phase 7 days before and 7 days after DMBA (7,12-dimethylbenz[α]anthracene) application, Group II), promotional phase (from the day of croton oil application and continued till the end of the experiment, Group III) and both peri and post initiation stages (from 7 days prior to DMBA application and continued till the end of the experiment, Group IV) compared to the control group (Group I). The percentage inhibition of tumour multiplicity was recorded to be 27, 72, and 81 in Groups II, III, and IV, respectively. These findings were suggestive of a possible chemopreventive potential of *Moringa oliefera* extract against chemical carcinogenesis though a hepatic pathway.

Studies showed that in mouse pre-treated with the ethanolic extract of *Moringa oleifera* leaves for seven consecutive days at doses of 250, 500, 1,000 and 2,000 mg/kg b.w., the percentage of cyclophosphamide-induced micronuclei in polychromatic erythrocytes decreased with increasing concentration of the extract (Sathya et al. 2010). Results of comet assay showed similar decrease in DNA damage in mice pre-dosed with the extract. These results indicated the presence of chemopreventive phytoconstituents in the crude extract of *M. oleifera* leaves offering protection against cyclophosphamide-induced genotoxicity in the mouse. Recent studies found that the leaf extracts from *M. oleifera* exerted strong antiproliferation and potent induction of apoptosis of human tumour KB cells as well as morphological changes and DNA fragmentation (Sreelatha et al. 2011). In addition, the leaf extract at various concentrations was found to induce ROS production suggesting modulation of redox-sensitive mechanism. HPTLC analysis indicated the presence of phenolics such as quercetin and kaempferol. *Moringa oleifera* leaf extracts also significantly inhibited LPO formation and enhanced the activity of antioxidative enzymes such as superoxide dismutase and catalase in human tumour KB cells (Sreelatha and Padma 2011). The leaf extracts significantly reduced the incidence of comets in the oxidant stressed cells. MTT assay showed that hydrogen peroxide caused a marked decrease in the viability of KB cells whereas the leaf extracts effectively increased the viability of assaulted KB cells. The observed cytoprotective activity was probably due to the antioxidant properties of its constituents, mainly phenolics which correlated highly with antioxidant activity.

Antiviral Activity

Of the six bioactive compounds isolated from ethanol extract of the seeds: O-ethyl-4-(α-L-rhamnosyloxy)benzyl carbamate (1), 4(α-L-rhamnosyloxy)-benzyl isothiocyanate (2), niazimicin (3), niazirin (4), β-sitosterol (5), glycerol-1-(9-octadecanoate) (6), 3-O-(6′-O-oleoyl-β-D-glucopyranosyl)-β-sitosterol (7), and β-sitosterol-3-O-β-D-glucopyranoside (8), four compounds (2, 3, 7, and 8), showed inhibitory activity against Epstein-Barr virus-early antigen (EBV-EA) activation (Guevara et al. 1999). Compounds 2, 3 and 8 exhibited very significant activities.

Three known thiocarbamate (TC)-and isothiocyanate (ITC)-related compounds were isolated from the leaves of *Moringa oleifera* as inhibitors of tumour promoter teleocidin B-4-induced Epstein-Barr virus (EBV) activation in Raji cells (Murakami et al. 1998). However, only niaziminin among ten TCs including eight synthetic ones showed considerable inhibition against EBV activation. The structure-activity relationships indicated that the presence of an acetoxy group at the 4′-position of niaziminin is important and indispensable for inhibition. On the other hand, among the ITC-related compounds, naturally occurring 4-[(4′-O-acetyl-α-L-rhamnosyloxy) benzyl]ITC and commercially available allyl- and benzyl-ITC significantly inhibited activation, suggesting that the isothiocyano group to be a critical structural factor for activity.

Moringa oleifera was 1 of 11 Thai medicinal plant that inhibited plaque formation of herpes simplex virus type 1 (HSV-) by more than 50% at 100 μg/ml in a plaque reduction assay (Lipipun et al. 2003). The 11 plants were also effective against thymidine kinase-deficient HSV-1 and phosphonoacetate-resistant HSV-1 strains. These therapeutic efficacies were characterized using a cutaneous HSV-1 infection in mice. The extract of *M. oleifera* at a dose of 750 mg/kg/day significantly delayed the development of skin lesions, prolonged the mean survival times and reduced the mortality of HSV-1 infected mice as compared with 2% DMSO in distilled water. There was no significant difference between acyclovir and *M. oleifera* in mean survival times. Toxicity of the plant extract was not observed in treated mice. The results suggested that *M. oleifera* extract may be a possible candidate as anti-HSV-1 agent.

Burger et al. (2002) reported that *Moringa oleifera* dry leaf powder may be a valuable nutrient for the poor communities of Africa by boosting the immune system to fight infections and thereby enhancing the well-being of HIV+persons. Moringa is exceptionally rich in vitamins A/β-carotene, C, E and key elements including selenium, but also contains almost a full RDA of other nutrients required for a healthy lifestyle. Literature reports supported a synergism between Nutrition Acquired Immunodeficiency and Acquired Immunodeficiency Syndrome. This suggested that enhanced nutrition (such as that which can be achieved via Moringa) could benefit a person with AIDS. However, more clinical trials are needed to substantiate this. Recently, Monera and Maponga (2010) found that *Moringa oleifera* supplementation was common among HIV positive people. Sixty-eight percent (68%) of the study participants consumed *Moringa oleifera*. Of these, 81% had already commenced antiretroviral drugs. Friends or relatives were the most common source of a recommendation for use of the plant (69%). Most (80%) consumed *Moringa oleifera* to boost the immune system. The leaf powder was mainly used, either alone (41%) or in combination with the root and/or bark (37%).

Cardioprotective Activity

Chronic treatment with *M. oleifera* mitigated effects of isoproterenol (ISP)-induced hemodynamic perturbations (Nandave et al. 2009). Chronic *M. oleifera* treatment resulted in significant favourable modulation of the biochemical enzymes (superoxide dismutase, catalase, glutathione peroxidase, lactate dehydrogenase, and creatine kinase-MB) but failed to demonstrate any significant effect on reduced glutathione compared to the ISP control group. *Moringa* treatment significantly prevented the rise in lipid peroxidation in myocardial tissue. Furthermore, *M. oleifera* also prevented the deleterious histopathological and ultrastructural perturbations caused by ISP. Based on the results it was concluded that *M. oleifera* extract possessed significant cardioprotective effect, which may be attributed to its antioxidant, antiperoxidative, and myocardial preservative properties.

CNS (Central Nervous System) Activity

The methanol extract of the leaves of *Moringa oleifera* was found to produce a significant alteration in general behavioural pattern by head dip test, Y-maze test, evasion test, and reduction in muscle relaxant activity by rotarod test, chimney test and traction test in animal studies (Pal et al.1996). Beside these, the extract also potentiated pentobarbitone induced sleeping time and lowered body temperature in experimental animals.

Studies showed that pre-treatment with aqueous extract of Moringa root inhibited penicillin-induced seizure and markedly reduced locomotor activity in Holtzman strain adult albino rats (Ray et al. 2003). Chronic treatment with the extract significantly increased brain serotonin and decreased the dopamine level in cerebral cortex (CC), midbrain (MB), caudate nucleus (CN) and cerebellum (CB). Norepinephrine level was significantly decreased in CC but no appreciable change was observed in MB, CB and CN. Thus the central inhibitory effect of *Moringa oleifera* was elucidated in the light of the disturbed balance between serotonin, dopamine and norepinephrine. In another study, 5-hydroxytryptamine in *Moringa oleifera* (MO) was found to induce potentiation of pentobarbitone hypnosis in albino rats (Ray et al. 2004). MO (350 mg/kg) caused inhibition of awareness, touch response, motor activity, righting reflex, and grip strength. It significantly increased the PB (pentobarbital) sleeping time, serum 5-HT (5-hydroxytryptamine) level and alpha-wave activity. These observations indicated that the aqueous extract of MO potentiated PB induced sleeping time and increased the alpha-wave activity through 5-HT.

Recent studies showed that brain monoamines were altered discreetly in different brain areas after colchicine infusion in brain (Ganguly and Guha 2008). There was a decrease in norepinephrine (NE) level in cerebral cortex (CC), hippocampus (HC) and caudate nucleus (CN). Dopamine (DA) and serotonin (5-HT) levels were decreased in CC, HC and CN. The EEG studies showed a decrease in beta- and alpha-waves and increase in biphasic spike wave pattern in experimental Alzheimer rat model. After treatment with *Moringa oleifera* (MO) leaf extract the monoamine levels of brain regions were restored to near control levels. EEG studies showed an increase in beta-waves and a decrease in spike wave discharges. The findings indicated that *M. oleifera* might have a role in providing protection against Alzheimer's disease in the rat model by altering brain monoamine levels and electrical activity.

Antiinflammatory, Antiasthmatic, Antiarthritic Activities

Of the hot water infusions of flowers, leaves, roots, seeds and stalks or bark of *Moringa oleifera* tested for antiinflammatory and diuretic activities, the seed infusion showed a significant inhibition of acetylcholine-induced contraction with an ED_{50} of 65.6 mg/ml bath concentration, inhibition of carrageenan-induced edema at 1,000 mg/kg and diuretic activity at 1,000 mg/kg (Cáceres et al. 1992). Some activity was also demonstrated in the roots.

The roots of *Moringa oleifera* was found to contain anti-inflammatory principle that may be useful in the treatment of the acute inflammatory conditions (Ndiaye et al. 2002). At a dose of 750 mg/kg, the *Moringa oleifera* treatment significantly inhibited the development of oedema at 1, 3 and 5 hours (reduction by 53.5%, 44.6% and 51.1% respectively). Increasing the dose of *Moringa oleifera* to 1,000 mg/kg did not increase the inhibitory effect on oedema development at 1 and 3 hours, whereas this dose potentiated the oedema at 5 hours. Treatment with indomethacin significantly inhibited the development of oedema 1, 3 and 5 hours (49.1%, 82.1% and 46.9% respectively). The findings indicated that the aqueous root extract of *Moringa oleifera* at 750 mg/kg reduced the carrageen in induced oedema to similar extent as the potent antiinflammatory drug indomethacin.

Studies showed that in ovalbumin-sensitized model control guinea pigs, tidal volume was decreased, respiration rate increased, and both the total and differential cell counts in blood and bronchoalveolar lavage fluid were increased significantly compared with non-sensitized controls (Mahajan and Mehta 2008; Mahajan et al. 2009). N-butanol seed extract of *M. oleifera* treatment gave improvement in all parameters except bronchoalveolar lavage tumour necrosis factor-alpha and interleukin-4. Moreover, *Moringa* butanol treatment demonstrated protection against acetylcholine-induced bronchoconstriction and airway inflammation. The results indicated that the extract had an inhibitory effect on airway inflammation and possessed an antiasthmatic property through modulation of the relationship between Th1/Th2 cytokine imbalances. The results of these studies confirmed the traditional claim for the usefulness of this herb in the treatment of allergic disorders like asthma. Another study showed that asthmatic symptoms were found in TDI (toluene diisocyanate) control rats only, while both the ethanol *Moringa* seed extract (MOEE)- and dexamethasone (DXM) treated rats did not manifest any airway abnormality (Mahajan et al. 2007a). In MOEE-and DXM-treated rats, neutrophil and eosinophil levels in the blood were decreased significantly; levels of total cells and each different cell in their BAL fluid were markedly decreased as compared to those in TDI controls. TNF alpha, IL-4, and IL-6 were predominant in serum as well as in BAL fluids in TDI controls, but these levels were reduced significantly by MOEE treatment. The antioxidant activity in relation to anti-inflammatory activity of the extract and histopathological observations also reflected a protective effect. Based on the above findings and observations, the researchers concluded that *Moringa oleifera* may possess some beneficial properties that act against chemically stimulated immune-mediated inflammatory responses that are characteristic of asthma in the rat.

β-Sitosterol isolated from an n-butanol extract of *M. oleifera* seeds was found to possess anti-asthmatic actions that might be mediated by inhibiting the cellular responses and subsequent release/synthesis of Th2 cytokines (Mahajan and Mehta 2011). β-Sitosterol significantly increased the tidal volume and decreased the respiration rate of sensitized and challenged guinea pigs to the level of non-sensitized control guinea pigs and lowered both the total and differential cell counts, particularly eosinophils and neutrophils, in blood and bronchoalveolar lavaged fluid. Further, β-sitosterol treatment suppressed the increase in cytokine levels (TNFα, IL-4 and IL-5), with the exception of IL-6, in serum and in bronchoalveolar lavaged fluid detected in model control animals. Additionally, treatment with β-sitosterol protected against airway inflammation in lung tissue histopathology.

In a clinical study of 20 patients of either sex with mild-to-moderate asthma, treatment with finely powdered dried seed kernels of *M. oleifera* for 3 weeks produced significant improvement in forced vital capacity, forced expiratory volume and peak expiratory flow rate values by 32.97%, 30.05%, and 32.09%, respectively, in asthmatic subjects (Agrawal and Mehta 2008). Significant improvement was also observed in symptom score and severity of asthmatic attacks. None of the patients showed any adverse effects with *M. oleifera*. The results of the study suggested the usefulness of *M. oleifera* seed kernel in patients of bronchial asthma.

Following oral administration, the crude methanol extract of the root inhibited carrageenan-induced rat paw edema in a dose-dependent manner, with 50% inhibitory concentration - IC_{50} of 660 mg/kg (Ezeamuzie et al. 1996). The extract was much more potent, on the 6-day air pouch acute inflammation induced with carrageenan, with IC_{50} values of 302.0 and 315.5 mg/kg for the inhibition of cellular accumulation and fluid exudation, respectively. Maximum inhibition obtained with 600 mg/kg was 83.8% and 80.0%, respectively. When delayed (chronic) inflammation was induced in the 6-day air pouch model using Freund's complete adjuvant, the extract was still effective though less than in acute inflammation. In contrast, a moderate dose of indomethacin (5 mg/kg) inhibited the acute, but not the delayed form of air pouch inflammation. Acute toxicity tests in mice suggested very low toxicity. These results suggested that the root of *Moringa oleifera* contained antiinflammatory principle(s) that may be useful in the treatment of both the acute and chronic inflammatory conditions. The methanolic extracts of the root or leaf of *M. oleifera* were found to be effective in the reduction of pain induced by complete Freund's adjuvant (CFA) in rats (Manaheji et al. 2011). The potency of the root or leaf extracts of *M. oleifera* (300 and 400 mg/kg) was similar to that of indomethacin, resulting in significant reductions in both thermal hyperalgesia and mechanical allodynia in rats with CFA-induced arthritis compared with the control group. A comparison of single and combination therapies of root and leaf extracts also showed a synergistic effect on pain reduction.

Ethanolic extract of *M. oleifera* exhibited anti-arthritic activity in adjuvant-induced arthritis in adult female Wistar rats. With extract treatment, the percentage reduction in body weight was less, paw edema volume and arthritic index score was decreased significantly as compared to diseased control animals (Mahajan et al. 2007b). Serum levels of Rheumatoid Factor (RF), TNF-alpha, interleukin-1, and interleukin-6 showed decreased levels as compared to those in the diseased control group. Treatment with the extract also altered oxidative stress in relation to its anti-inflammatory activity. Histopathological observations showed mild or less infiltration of lymphocytes, angiogenesis and synovial lining thickening. From the results and observations, the scientitists concluded that *Moringa oleifera* possessed promising antiarthritic property.

A rare aurantiamide acetate (4) and 1,3-dibenzyl urea (5) were isolated and characterized from the roots of *M. oleifera* (Sashidhara et al. 2009). The isolated compounds inhibited the production of TNF-alpha and IL-2; further, compound 5 showed significant analgesic activities in a dose dependant manner. The authors asserted that the findings may help in understanding the mechanism of action of this traditional plant leading to control of activated mast cells on inflammatory conditions like arthritis, for which the crude extract had been used.

The ethyl acetate extract of *Moringa oleifera* fruits yielded three new phenolic glycosides; 4-[(2′-O-acetyl-α-l-rhamnosyloxy) benzyl]isothiocyanate (1), 4-[(3′-O-acetyl-α-l-rhamnosyloxy) benzyl]isothiocyanate (2), and S-methyl-N-{4-[(α-l-rhamnosyloxy)benzyl]}thiocarbamate (3), together with five known phenolic glycosides (4–8) (Cheenpracha et al. 2010). The antiinflammatory activity of isolated compounds was determined using the lipopolysaccharide (LPS)-induced murine macrophage RAW 264.7 cell line. It was also found that 4-[(2′-O-acetyl-α-l-rhamnosyloxy) benzyl]isothiocyanate (1) possessed potent NO-inhibitory activity with an IC_{50} value of 1.67 μM, followed by 2 (IC_{50} = 2.66 μM), 4 (IC_{50} = 2.71 μM), and 5 (IC_{50} = 14.4 μM), respectively. Western blots demonstrated these compounds reduced LPS-mediated iNOS expression. In the concentration range of the IC_{50} values, no significant cytotoxicity was noted. Structure-activity relationships following NO-release indicated: (1) the isothiocyanate group was essential for activity, (2) acetylation of the isothiocyanate derivatives at C-2′ or at C-3′ of rhamnose led to higher activity, (3) un-acetylated isothiocyanate derivatives displayed eight times less activity than the acetylated derivatives, and (4) acetylation of the thiocarbamate derivatives enhanced activity. These data indicated compounds 1, 2, 4 and 5 were responsible for the reported NO-inhibitory effect of *Moringa oleifera* fruits.

Antihyperlipidemia, Antihypercholesterolemic Activities

Studies found that administration of the crude leaf extract of *Moringa oleifera* along with high-fat diet decreased the high-fat diet-induced increases in serum, liver, and kidney cholesterol levels by 14.35% (115–103.2 mg/100 ml of serum), 6.40% (9.4–8.8 mg/g wet weight) and 11.09% (1.09–0.97 mg/g wet weight) respectively (Ghasi et al. 2000). The effect on the serum cholesterol was statistically significant. No significant effect on serum total protein was observed. However, the crude extract increased serum albumin by 15.22% (46–53 g/l). This value was also found to be statistically significant. It was concluded that the leaves of *Moringa oleifera* had definite hypocholesterolemic activity and the results supported its use in herbal medicine as a hypocholesterolemic agent in obese patients in India. Studies of Thai dietary extracts found that the potency of extracts from *Hibiscus sabdariffa, Moringa oleifera* and *Cucurbita moschata* at 100 μg/ml were similar to 0.4 μg/ml pravastatin in inhibiting HMG-CoA reductase and possibly reducing cholesterol biosynthesis (Duangjai et al. 2010).

In hypercholesterol-fed rabbits, at 12 weeks of treatment, Moringa leaf extract significantly lowered the cholesterol levels and reduced the atherosclerotic plaque formation to about 50% and 86%, respectively (Chumark et al. 2008). The extract also exhibited good antioxidant activity with IC_{50} of 78.15 and 2.14 μg/ml, in scavenging DPPH radicals and Trolox had respectively. The extract significantly prolonged the lag-time of CD (conjugated diene) formation and inhibited TBARS formation in both in vitro and ex vivo experiments in a dose-dependent manner. The results indicated the plant to possess antioxidant, hypolipidemic and antiatherosclerotic activities and to have therapeutic potential for the prevention of cardiovascular diseases. The hypocholesterolemic potency of *M. oleifera* extract at 100 μg/l was found to be similar to 0.4 μg/ml in inhibiting HMG-CoA reductase and possibly reducing cholesterol biosynthesis (Duangjai et al. 2010).

Moringa oleifera (MO) reduced iron-deficient diet-induced increases in serum and hepatic lipids with dose-dependent increases of serum quercetin and kaempferol, but did not prevent anaemia (Ndong et al. 2007b). By electron microscopy, in iron deficient hepatocytes, slightly swollen mitochondria and few glycogen granules were observed, but glycogen granules increased and mitochondria were normalized by treatment with MO. Furthermore, lipoproteins were observed in the Golgi complex under treatment with MO. These results suggested a possible beneficial effect of MO in the prevention of hyperlipidemia and ultrastructural changes in hepatocytes due to iron-deficiency.

Moringa oleifera and lovastatin were found to lower the serum cholesterol, phospholipid, triglyceride, VLDL, LDL, cholesterol to phospholipid ratio and atherogenic index, but increased the HDL ratio (HDL/HDL-total cholesterol) as compared to the corresponding control groups (Mehta et al. 2003). Treatment with *M. oleifera* or lovastatin in normal rabbits decreased the HDL levels. However, HDL levels were significantly increased or decreased in *M. oleifera*-or lovastatin-treated hypercholesterolaemic rabbits, respectively. Lovastatin-or *M. oleifera*-treated hypercholesterolaemic rabbits showed decrease in lipid profile of liver, heart and aorta while similar treatment of normal animals did not produce significant reduction in heart. *Moringa oleifera* was found to increase the excretion of faecal cholesterol. The study demonstrated that *M. oleifera* possessed a hypolipidemic effect.

Wound Healing Activity

Significant increase in wound closure rate, skin-breaking strength, and granuloma breaking strength, hydroxyproline content, granuloma dry weight and decrease in scar area was observed with 300 m/kg b.w. aqueous *M. oleifera* leaf extract in rats (Rathi et al. 2006). The prohealing actions appeared to be due to increased collagen deposition as well as better alignment and maturation. From the results obtained, the scientists concluded that the aqueous extract of *M. oleifera* had significant wound healing property.

Antiulcerogenic Activity

Root and leaf preparations showed a significant dose-dependent anti-ulcer activity in experimental aspirin (200 mg/kg)-induced gastric ulcers in male albino rats (Ruckmani et al. 1998a). The effect of the alkali preparation of the root seemed to be more pronounced than that of the fresh leaf juice. Another study showed that the methanol fraction of *M. oleifera* leaf extract possessed significant protective actions in acetylsalicylic acid, serotonin and indomethacin induced gastric lesions in experimental rats (Pal et al. 1995). A significant enhancement of the healing process in acetic acid-induced chronic gastric lesions was also observed with the extract-treated animals.

Antianaphylactic Activity

When administered 1 hour before compound 48/80 injection, *M. oleifera* ethanol extract (MOEE) at doses of 0.001–1.000 g/kg completely inhibited the inducible induced anaphylactic shock (Mahajan and Mehta 2007). MOEE significantly inhibited passive cutaneous anaphylaxis activated by anti-IgE antibody at a dose of 1 g/kg. When MOEE extract was given as pretreatment at concentrations ranging 0.1–100 mg/ml, the histamine release from the mast cells that was induced by the 48/80 was reduced in a dose-dependent manner. The results suggested a potential role for MOEE as a source of anti-anaphylactic agents for use in allergic disorders.

Hepatoprotective Activity

Aqueous and alcoholic extracts of root and flower of *M. oleifera* exhibited antihepatotoxic activity in paracetamol treated albino rats based on liver function assessed, liver to body weight ratio, serum levels of transaminase (SGPT, SGOT), alkaline phosphatase (SALP) and bilirubin (Ruckmani et al. 1998b). In another study, oral administration of the *M. oleifera* extract showed a significant protective action in liver damage induced by antitubercular drugs such as isoniazid (INH), rifampicin (RMP), and pyrazinamide (PZA) in rats (Pari and Kumar 2002). This was made evident by its effect on the levels of glutamic oxaloacetic transaminase (aspartate aminotransferase), glutamic pyruvic transaminase (alanine aminotransferase), alkaline phosphatase, and bilirubin in the serum; lipids, and lipid peroxidation levels in liver. This observation was supplemented by histopathological examination of liver sections. The results of this study showed that treatment with *M. oleifera* extracts or silymarin (as a reference) appeared to enhance the recovery from hepatic damage induced by antitubercular drugs. In another study enhanced hepatic marker enzymes and lipid peroxidation of antitubercular drug treatment (isoniazid, rifampicin, and pyrazinamide) in rats was found to be accompanied by a significant decrease in the levels of vitamin C, reduced glutathione, superoxide dismutase, catalase, glutathione peroxidase, and glutathione S-transferase (Kumar and Pari 2003). Administration of *Moringa oleifera* extract and silymarin significantly decreased hepatic marker enzymes and lipid peroxidation with a simultaneous increase in the level of antioxidants. It was speculated that *Moringa oleifera* extract exerted its protective effects by decreasing liver lipid peroxides and enhancing antioxidants.

The hepatoprotective activity of *Moringa oleifera* (MO) extract was observed following significant histopathological analysis and reduction of the level of alanine aminotransferase (ALT), aspartate aminotransferase (AST) and alkaline phosphatase (ALP) in male Sprague–Dawley rats pre-treated with MO compared to those treated with acetaminophen alone (Fakurazi et al. 2008). Meanwhile, the level of glutathione (GSH) was found to be restored in MO-treated animals compared to the groups treated with acetaminophen alone. These observations were comparable to the group pre-treated with silymarin prior to acetaminophen administration. Group that was treated with acetaminophen alone exhibited high level of transaminases and ALP activities besides reduction in the GSH level. The histological hepatocellular deterioration was also observed. The results from the present study suggested that the leaves of MO could prevent hepatic injuries from acetaminophen induced by preventing the decline of glutathione level. Separate studies showed that *M. oleifera* leaf extracts effectively suppressed CCl(4)-induced oxidative stress in liver slices (Sreelatha and Padma 2010). Treatment with *Moringa oleifera* extract increased the activities of antioxidant enzymes and glutathione content and reduced the levels of TBARS (thiobarbituric acid-reacting substances) significantly. Observed reduction in the level of lipid peroxides showed a decreased tendency of peroxidative damage.

Biochemical and histological studies showed that *Moringa oleifera* seed extract when orally administered daily reduced liver damage as well as symptoms of liver fibrosis induced by carbon tetrachloride (Hamza 2010). The administration of *Moringa* seed extract decreased the CCl(4)-induced elevation of serum aminotransferase activities and globulin level. The elevations of hepatic hydroxyproline content and myeloperoxidase activity were also reduced by *Moringa* treatment. Furthermore, immunohistochemical study showed that *Moringa* markedly reduced the numbers of smooth muscle alpha-actin-positive cells and the accumulation of collagens I and III in liver. *Moringa* seed extract showed significant inhibitory effect on 1,1-diphenyl-2-picrylhydrazyl free radical, as well as strong reducing antioxidant power. The activity of superoxide dismutase as well as the content of both malondialdehyde and protein carbonyl, oxidative stress markers, were reversed after treatment with *Moringa*. The results suggested that *Moringa* seed extract could act against CCl(4)-induced liver injury and fibrosis in rats by a mechanism related to its antioxidant properties, antiinflammatory effect and its ability to attenuate the hepatic stellate cells activation.

Studies found that co-administration of aqueous seed extract of *M. oleifera* (500 mg/100 g body weight/day for a period of 24 days) significantly prevented the arsenic-induced alteration of hepatic function markers and lipid profile (Chattopadhyay et al. 2010). Subchronic exposure to sodium arsenite (0.4 ppm/100 g body weight/day via drinking water for a period of 24 days) significantly increased activities of hepatic and lipid function markers such as alanine transaminase, aspartate transaminase, cholesterol, triglycerides, LDL along with a decrease in total protein and HDL. A marked elevation of lipid peroxidation in hepatic tissue was also evident from the hepatic accumulation of malondialdehyde and conjugated dienes along with suppressed activities in the antioxidant enzymes such as superoxide dismutase and catalase. Moreover, the degeneration of histoarchitecture of liver found in arsenic-treated rats was protected along with partial but definite prevention against DNA fragmentation induction. Similarly, generation of reactive oxygen species and free radicals were found to be significantly less along with restored activities of antioxidant enzymes in *M. oleifera* co-administered group with comparison to arsenic alone treatment group. The present investigation offered strong evidence for the hepatoprotective and antioxidative efficiencies of *M. oleifera* seed extract against oxidative stress induced by arsenic.

Antidiabetic Activity

Studies indicated that *Moringa oleifera* (MO) exhibited an ameliorating effect for glucose intolerance, and the effect might be mediated by quercetin-3-glucoside and fibre contents in MO leaf powder (Ndong et al. 2007a). The action of MO was greater in GK rats (Goto-Kakizaki (GK) rats, modelled type 2 diabetes) than in Wistar rats. In the glucose tolerance test showed MO significantly decreased blood glucose at 20, 30, 45 and 60 minutes for GK rats and at 10, 30 and 45 minutes for Wistar rats compared to the both controls after glucose administration. Furthermore, MO significantly decreased stomach emptying in GK rats. The results indicated that MO had an ameliorating effect for glucose intolerance, and the effect might be mediated by quercetin-3-glucoside and fibre contents in MO leaf powder. The action of MO was greater in GK rats than in Wistar rats. The study by Jaiswal et al. (2009) validated scientifically the widely claimed use of *M. oleifera* as an ethnomedicine to treat diabetes mellitus. The dose of 200 mg/kg of *M. oleifera* leaf aqueous extract was found to decrease blood glucose level (BGL) of normal animals by 26.7% and 29.9% during n fasting blood glucose (FBG), oral glucose tolerance test (OGTT) studies respectively. In streptozotocin induced sub and mild diabetic animals the same dose produced a maximum fall of 31.1% and 32.8% respectively, during OGTT. In case of severely diabetic animals FBG and post prandial glucose levels were reduced by 69.2% and 51.2% whereas, total protein, body weight and haemoglobin were increased by 11.3%, 10.5% and 10.9% respectively after 21 days of treatment. Significant reduction was found in urine sugar and urine protein levels from +4 and +2 to nil and trace, respectively.

Antimicrobial Activity

Extract of *M. oleifera* was found to inhibit the growth of pathogens of subcutaneous phycomycosis in humans and animals, namely *Basidiobolus haptosporus* and *Basidiobolus ranarum, Trichophyton rubrum* and *Trichophyton mentagrophytes*, at a 1: 10 dilution (Nwosu and Okafor 1995).

4(α-L-rhamnosyloxy)benzyl isothiocyanate was identified as an active antimicrobial agent from seeds of *Moringa oleifera* and *M. stenopetala* (Eilert et al. 1981). Seeds of *Moringa oleifera* contain a glucosinolate that on hydrolysis yields 4-(α-Lrhamnosyloxy)-benzyl isothiocyanate, an active bactericide and fungicide (Bennett et al. 2003). The seeds of *Moringa oleifera* yield a lower amount (4–5% of dry weight) of glucosinolate than those of *Moringa stenopetala* (8–10% of dry weight) and should therefore be used at a higher dosage. Roots of *M. oleifera* only contained this compound and benzyl isothiocyanate, but not pterygospermin as previously suggested. Defatted and shell free seeds of both species contained about 8–10% of 4(α-L-rhamnosyloxy) benzyl isothiocyanate, but this amount was produced from *M. oleifera* only when ascorbic acid was added during water extraction. The compound acted on several bacteria and fungi. The minimal bactericidal concentration in-vitro was 40 μmol/l for *Mycobacterium phlei* and 56 μmol/l for *Bacillus subtilis.* In another study, the purified dichloromethane extract of *M. oleifera* showed antibacterial activity against *Staphylococcus aureus*, *Escherichia coli, Pseudomonas aeruginosa* and *Klebsiella pneumoniae* but not the methanolic extract (Nantachit 2006). Antibacterial activity began at 5–10% W/V concentration.

Ethanol extracts of the seeds and leaves of *Moringa oleifera* showed anti-fungal activities in-vitro against dermatophytes such as *Trichophyton rubrum, Trichophyton mentagrophytes, Epidermophyton floccosum*, and *Microsporum canis* (Chuang et al. 2007). GC-MS analysis of the chemical composition of the essential oil from leaves showed a total of 44 compounds. Isolated extracts could be of use for the future development of anti-skin disease agents.

In-vitro studies demonstrated that the fresh leaf juice of *M oleifera* and aqueous extracts from the seeds inhibited the growth of *Pseudomonas aeruginosa* and *Staphylococcus aureus* and that extraction temperatures above 56°C inhibited this activity (Cáceres et al. 1991). No activity was demonstrated against four other pathogenic Gram-positive and Gram-negative bacteria and *Candida albicans*. No activity was demonstrated against six pathogenic dermatophytes. In a study using an experimental model of *Staphylococcus aureus* pyodermia, mice were treated with an ointment of *M. oleifera* or neomycin. The healing time with the *M. oliefera* ointment was similar to the reference drug, but shorter than untreated animals, indicating that *M. oleifera* could be an alternative treatment for skin infections (Cáceres and Lopez 1991). Small protein/peptide fractions from *M. oleifera* leaves was tested for antibacterial activity against *Escherichia coli*, *Klebsiella aerogenes*, *Klebsiella pneumoniae*, *Staphylococcus aureus* and *Bacillus subtilis* (Dahot 1998). Fraction P1, P2 and P3 showed strong inhibitory activity against *Escherichia coli*, *Staphylococcus aureus* and *Bacillus subtilis* but clear zone of inhibition was also noted against *Klebsiella aerogenes* with peptide 1. Fraction P2 showed significant zone of inhibition against *Aspergillus niger*.

The antimicrobial activities of *M. oleifera* leaves, flower and seeds were investigated in-vitro against fungi Gram negative and Gram positive bacteria (Nepolean et al. 2009) The seed extract MOS showed antibacterial activity against *Klebsiella pneumoniae*, *Staphylococcus aureus*, and Streptococcus. The ethanol leaf extract MOL1 showed antibacterial activity against most of the organisms like *Escherichia coli*, *Klebsiella pneumoniae*, *Enterobacter*, *Pseudomonas aeruginosa* and *Staphylococcus aureus*, however the aqueous leaf extract MOL2 samples showed antibacterial activity only against *Pseudomonas aeruginosa* and *Staphylococcus aureus*. The ethanol extract of *M. oleifera* was found to have better inhibition of *Salmonella typhi* than any other organic extracts. The flower extract, MOF showed activity against *Escherichia coli, Klebsiella pneumoniae and Proteus mirabilis* but not against *Salmonella typhi* A. The activities of these extracts were comparable to those of antibiotics, ciprofloxacin, cotrimoxazole and chloramphenicol, commonly used for treating typhoid fever. All the extracts except the MOF extract showed antifungal activity against *Candida albicans*. Ethanolic leaf extract of *M. oleifera* (MOL1), when analysed by GC-MS was found to have 15 major components like hexadecanoic acid, ethyl ester (CAS) ethyl palmitate, palmitic acid ethyl ester, 2,6-dimethyl-1, 7-octadiene-3-ol, 4-hexadecen-6-yne, (z)-(CAS) (Nepolean et al. 2009). The aqueous MOL2 extract contained 2-hexanone, 3-cyclohexyliden-4-ethyl - E2-dodecenylacetate, and Hi-oleic safflower oil (CAS) and safflower oil. MOS extract contained the following major compounds roridin E, veridiflorol, 9-octadecenoic acid. The flower extract MOF extract had 9-octadecen– 1-ol, (Z) – (CAS) cis-9–octadecen–1–ol, oleol, satol, ocenol, sipo, decanoic acid and dodecanal. Similarly, ethanolic extracts of *M. oleifera* demonstrated the highest activity, while the aqueous extracts showed the least activity against *Salmonella typhi*, causative agent of typhoid fever (Doughari et al. 2007). In earlier studies, Das et al. (1957) isolated a compound name pterygospermin from *M. oleifera* flowers which was known to possess antifungal activity.

The crude seed extract of *Moringa oleifera* was found to exhibit antimicrobial activity against *Escherichia coli, Pseudomonas aeruginosa, Staphylococcus aureus, Cladosporium cladosporioides*, and *Penicillium sclerotigenum* (Oluduro et al. 2010). Characterization and identification of the extract revealed the occurrence of three bioactive compounds: 4-(α-L-rhamnopyranosyloxy)benzyl isothiocyanate, methyl N-4-(α-L-rhamnopyranosyloxy) benzyl carbamate (both known compounds), and 4-(β-D-glucopyranosyl-1 → 4-α-L-rhamnopyranosyloxy)-benzyl thiocarboxamide. All the compounds at 5 mg/l had very high bactericidal activity against some of test pathogens. 4-(β-D-Glucopyranosyl-1 → 4-α-L-rhamnopyranosyloxy)benzyl thiocarboxamide was the most potent, with 99.2% inhibition *toward Shigella dysenteriae* and 100% toward *Bacillus cereus, E. coli* and *Salmonella typhi* within 4 hours of contact. The steam distillate of *M. oleifera* was found to exhibit antimicrobial activity (Kekuda et al. 2010). Among bacteria tested, strong inhibition was observed against *Eshcerichia coli* followed by *Staphyloccous aureus, Klebsiella pneumoniae, Pseudomonas aeruginosa* and *Bacillus subtilis*. Strong inhibition was found against *Aspergillus niger* followed by *A. oryzae, A. terreus* and *A. nidulans*.

The chloroform and ethanol extracts of seeds and leaf of *Moringa oleifera* exhibited antimicrobial activity against some selected food – borne microorganisms (Bukar et al. 2010). The leaf ethanol extract exhibited broad spectrum activity against *Escherichia coli, Pseudomonas aeruginosa, Staphylococcus aureus* and *Enterobacter aerogenes*. The MIC values ranged between 2.0 and >4.0 mg/ml for all the organisms. *M. oleifera* seed chloroform extract was only active against *E. coli* and *Salmonella typhimurium*. The MIC values ranged between 1.0 and >4.0 mg/ml for the tested organisms respectively. Antifungal activity result revealed 100% inhibition in growth of *Mucor* and *Rhizopus* species by *M. oleifera* seed chloroform extract at concentration of 1 mg/ml. Saponins were detected in all the extracts while tannins were only detected in *Moringa oleifera* leaf chloroform extract. The water extract of *Moringa oleifera* leaf stalk at dilution of 1,000, 700, 400 and 200 mg/ml exhibited only mild in-vitro activity against *Escherichia coli* and *Enterobacter aerogenes* (Thilza et al. 2001). *Pseudomonas aerogienosa, Staphylococcus albus, Staphylococcus aureus* and *Staphylococcus pyogenus* was resistant at these concentrations. The highest activity was produced against *Escherichia coli* at 1,000 mg/l concentration which was comparably less than that of the standard drug tetracycline (250 mg/ml). Studies by Rahman et al. (2010) suggested that the extracts from *Moringa oleifera* leaf could be a source of natural antimicrobials with potential applications in pharmaceutical industry to control coliform bacteria such as *Escherichia coli, Shigella dysenteriae, Salmonella* sp., *Enterobacter* sp., *Klebsiella pneumoniae* and *Serratia marcescens*. At the concentration of 300 μg/disc, the organic extracts of hexane, chloroform, ethyl acetate and methanol extracts of *Moringa oleifera* leaf exhibited a remarkable antibacterial effect against all the tested bacterial pathogens. The zones of inhibition against all the tested bacterial pathogens were found in the range of 8.0–23.2 mm, along with their respective minimum inhibitory concentration (MIC) values ranging from 62.5 to 1,000 μg/ml.

Aqueous and ethanolic seed extracts of *Moringa oleifera* exhibited antibacterial activity (inhibition halo >13 mm) against *Staphylococcus aureus, Vibrio cholerae, Escherichia coli* isolated from the whiteleg shrimp, *Litopenaeus vannmaei* (Viera et al. 2010). *E. coli* isolated from tilapia fish, *Oreochromis niloticus*, was sensitive to the ethanolic extract.

Antiurolithiatic Activity

Oral administrations of aqueous and alcoholic extracts of *Moringa oleifera* root-wood were found to reduce ethylene glycol induced urolithiasis in male wistar rats (Karadi et al. 2006). Ethylene glycol feeding resulted in hyperoxaluria as well as increased renal excretion of calcium and phosphate. Supplementation with aqueous and alcoholic extract of *Moringa oleifera* root-wood significantly reduced the elevated urinary oxalate, showing a regulatory action on endogenous oxalate synthesis. The increased deposition of stone forming constituents in the kidneys of calculogenic rats was also significantly lowered by curative and preventive treatment using aqueous and alcoholic extracts. The aqueous extract *Moringa oleifera* bark was found to have antiurolithiatic activity (Fahad et al. 2010). Oral administration of the bark extract at two doses of extract for prophylactic and curative uses resulted in significant reduction in the weight of bladder stones in albino Wistar rats with stones produced by zinc disc foreign body insertion in the bladder compared to the control group.

Radioprotective Activity

Moringa oleifera leaf extract was found to possess radioprotective activity. Healthy adult Swiss albino mice were injected intra-peritoneally with 150 mg/kg body weight of 50% methanolic extract (ME) of *M. oleifera* leaves, as a single dose, or in 5 daily fractions of 30 mg/kg each, and exposed to whole body gamma irradiation (RT, 4 Gy) 1 hour later (Rao et al. 2001). Pre-treatment with a single dose of 150 mg/kg of the methanol extract significantly reduced the percent aberrant cells to 2/3 rd that of RT alone group on day 1 and brought the values to normal range by day 7 post-irradiation. A similar effect was also seen for the micronucleated cells. Fractionated administration of ME (30 mg/kg × 5) gave a higher protection than that given by the same dose administered as a single treatment. The methanol extract also inhibited the Fenton reaction-generated free radical activity in-vitro in a concentration dependent manner. These results demonstrated that pre-treatment with the methanolic leaf extract of *M. oleifera* conferred significant radiation protection to the bone marrow chromosomes in mice leading to higher survival after lethal whole body irradiation.

Anticlastogenic Activity

Freeze-dried boiled *M. oleifera* pod was found not to have clastogenic activity in the mouse while it possessed anticlastogenic activity against both direct-acting (mitomycin C, MMC) and indirect-acting (7, 12-dimethylbenz(a)anthracene, DMBA) clastogens (Promkum et al. 2010). Freeze-dried boiled *M. oleifera* pod at 1.5%, 3.0% and 6.0% in the diets decreased the number of micronucleated peripheral reticulocytes (MNRETs) induced by both MMC and DMBA. However, the effect was statistically significant in the dose dependent manner only in the MMC-treated group.

Antiulcerogenic Activity

The methanol fraction of *M. oleifera* leaf extract was found to possess significant protective actions in acetylsalicylic acid, serotonin and indomethacin induced gastric lesions in experimental rats (Pal et al. 1995). A significant enhancement of the healing process in acetic acid – induced chronic gastric lesions was also observed with the extract-treated animals. Pre-treatment of adult Holtzman strain albino rats with aqueous leaf extract of *M. oleifera* (MO) (300 mg/kg body weight) for 14 days before aspirin-induced ulceration decreased mean ulcer index, increased both enterochromaffin cell count and serotonin (5-hydroxytryptamine; 5-HT) content in all ulcerated group, but treatment with ondansetron, a 5-HT3 receptor antagonist, along with MO pre-treatment increased mean ulcer index, decreased 5-HT content without any alteration in EC cell count (Debnath and Guha 2007; Debnath et al. 2011). The results suggested that the protective effect of MO on ulceration was mediated by increased enterochromaffin cell count and 5-HT levels which may act via 5-HT3 receptors on gastric tissue.

Hypotensive Activity

Bioassay directed fractionation of an ethanolic extract of *Moringa oleifera* leaves resulted in the isolation of four pure compounds, niazinin A, niazinin B, niazimicin, niaziminin A and niaziminin B (Gilani et al. 1994). Intravenous administration of either one of the compounds (1–10 mg/kg) produced hypotensive and bradycardiac effects in anaesthetized rats. Pre-treatment of the animals with atropine (1 mg/kg) completely abolished the hypotensive and bradycardiac effects of acetylcholine (ACh), whereas cardiovascular responses to the test compounds remained unaltered, ruling out the possible involvement of muscarinic receptor activation. In isolated guinea-pig atria all the compounds (50–150 μg/ml) produced negative inotropic and chronotropic effects. Each compound inhibited K^+–induced contractions in rabbit aorta as well as ideal contractions induced by ACh or histamine at similar concentrations. Spontaneous contractions of rat uterus were also inhibited equally by all compounds. The data indicated that the direct depressant action of these compounds exhibited on all the isolated preparations tested was probably responsible for its hypotensive and bradycardiac effects observed in-vivo. Moreover, spasmolytic activity exhibited by the constituents of the plant provided a scientific basis for the traditional uses of the plant in gastrointestinal motility disorders.

In a separate study, the aqueous extract of stem bark at lower concentrations (1–10 μg) produced a dose dependent positive inotropic effect and at higher concentrations (0.1–1 μg) a dose dependent negative inotropic effect on the isolated frog heart (Limaye et al. 1995). It also produced a dose dependent hypotensive effect on dog blood pressure. It failed to elicit any effect on isolated guinea-pig ileum, rat stomach fundus or frog rectus abdominis muscle. In another study, bioassay-guided analysis of an ethanol extract of *Moringa oleifera* leaves showing hypotensive activity led to the isolation of two nitrile glycosides, niazirin and niazirinin, and three mustard oil glycosides, 4-[(4′-O-acetyl-α-L-rhamnosyloxy) benzyl] isothiocyanate, niaziminin A, and niaziminin B (Faizi et al. 1994; Faizi et al. 1995). Isothiocyanate 4 and the thiocarbamate glycosides niaziminin A and B showed hypotensive activity while nitrile glycosides 1 and 2 were found to be inactive in this regard. In subsequent studies, the ethanolic and aqueous extracts of *Moringa oleifera* whole pods and seed were found to exhibit hypotensive activity (Faizi et al. 1998). The activity of the ethanolic extract of both the pods and the seeds was equivalent at the dose of 30 mg/kg. The ethyl acetate phase of the ethanolic extract of pods was found to be the most potent fraction at the same dose. Its bioassay-directed fractionation led to the isolation of thiocarbamate and isothiocyanate glycosides which were also the hypotensive principles of the pods as observed in case of *Moringa* leaves. Two new compounds, O-[2′-hydroxy-3′-(2″-heptenyloxy)]-propyl undecanoate (1) and O-ethyl-4-[(α-L-rhamnosyloxy)-benzyl] carbamate (2) along with the known substances methyl p-hydroxybenzoate (3) and β-sitosterol were also isolated in the study. The latter two compounds and p-hydroxybenzaldehyde showed promising hypotensive activity.

Antifertility Activity

The bark of *M. oleifera* tree, powdered, produced an abortifacient which caused violent uterine contractions with fatal results (Bhattacharya et al. 1978). This abortifacient agent is used in the Bengali area of India. Oral administration of aqueous extract of *M. oleifera roots* progressively increased the uterine wet weight of bilaterally ovariectomized rats (Prakash 1988; Shukla et al. 1988a, Shukla et al. 1988b; Shukla et al. 1989). This estrogenic activity was supported by stimulation of uterine histo-architecture as revealed by increases in the height of luminal epithelium, well developed glands, loose stroma and rich vascularity. The cervix showed metaplastic changes in the epithelium with marked keratinization. In the vagina, cornification was very prominent, rugae increased and stroma was loose. When the extract was given conjointly with estradiol dipropionate (EDP), there was a successive reduction in the uterine wet weight when compared to the gain with EDP alone and uterine histological structures were also inhibited. In the deciduoma test, the highest dose of 600 mg/kg interfered with the formation of deciduoma in 50% of the rats, showing some antiprogestational activity. Doses up to 600 mg/kg of the extract orally failed to induce a decidual response in the traumatized uterus of ovariectomized rats. The antifertility effect of the extract appeared to be due to multiple attributes. Administration of the extract caused a significant increase in the glycogen contents, protein concentration, activity of acid and alkaline phosphatase and the level of total cholesterol in all the organs at initial days of treatment. Biochemical observations supplemented with the histological findings correlated with the anti-implantation action of the aqueous extract in the light of its hormonal properties.

Immunomodulatory Activity

An immuno-enhancing polysaccharide isolated from the hot aqueous extract of mature pods of *Moringa oleifera* was found to contain only D-glucose as a monosaccharide constituent and the structure of the repeating unit of the polysaccharide was composed of only α-(1→4) linked glucan (Mondal et al. 2004). This polysaccharide showed significant macrophage activity through the release of nitric oxide on mouse monocyte cell line (J744.1) and also enhanced the number of albumin/globulin percentage, the phagocyte cells and respiratory burst cells in a fish system.

Studies showed that ethanolic extract (50%) of *M. oleifera* leaves (MOE) significantly reduced cyclophosphamide induced immuno-suppression by stimulating both cellular and humoral immunity (Gupta et al. 2010). MOE showed significant dose dependent increase in WBC, percent neutrophils, weight of thymus and spleen along with phagocytic index in normal and immunosuppressed mice. Administration of methanolic extract of *Moringa oleifera* (MEMO) (250 and 750 mg/kg, po) and *Ocimum sanctum* (100 mg/kg, po) significantly increased the levels of serum immunoglobulins and also prevented the mortality induced by bovine *Pasteurella multocida* in mice (Sudha et al. 2010). They also increased significantly the circulating antibody titre in indirect haemagglunation test. Moreover, MEMO produced significant increase in adhesion of neutrophils, attenuation of cyclophosphamide-induced neutropenia and an increase in phagocytic index in carbon clearance assay. From the above results, it was concluded that MEMO stimulated both cellular and humoral immune response. The low dose of MEMO was found to be more effective than the high dose. In another study, the ethanolic extract of *M. oleifera* seeds dose-dependently (50, 100 and 200 mg/kg) inhibited spleen weight as well as circulatory leukocyte and splenocyte counts in mice (Mahajan and Mehta 2010). The delayed-type hypersensitivity reaction was significantly inhibited by decreasing the mean foot pad thickness at 48 hours. The production of the humoral antibody titre was significantly ameliorated at a dose of 100 and 200 mg/kg respectively. Furthermore, the extract caused a down-regulation of macrophage phagocytosis due to carbon particles. Taken together, the results suggested that the seeds of *Moringa oleifera* possessed immunosuppressive activity.

Regulation of Hyperthyroidism

Following the administration of *Moringa oleifera* aqueous leaf extract 175 mg/kg b.w./day for 10 days in Swiss rats, serum triiodothyronine (T(3)) concentration and hepatic lipid peroxidation decreased with a concomitant increase in the serum thyroxine (T(4)) concentration, in female rats, while in males no significant changes were observed, suggesting that *Moringa oleifera* leaf extract is more effective in females than in the males (Tahiliani and Kar 2000). Similar trend was observed at the higher dose of 350 mg/kg, suggesting the inhibiting nature of *Moringa oleifera* leaf extract in the peripheral conversion of T(4) to T(3), the principal source of the generation of latter hormone. As the antiperoxidative effects were exhibited only by the lower dose and percent decrease in T(3)concentration was nearly the same by both the doses, it was suggested that the lower concentration of this plant extract may be used for the regulation of hyperthyroidism.

Mitigation of Fluoride Toxicity

Preliminary studies revealed tamarind and drumstick extracts had potential to mitigate fluoride toxicity (Ranjan et al. 2009). Aqueous extracts of *Tamarindus indica* fruit pulp (100 mg/kg body weight) and *M. oleifera* seeds (50 mg/kg body weight) administered orally once daily for 90 days lowered plasma fluoride concentrations in rabbits receiving fluorinated drinking water (200 mg NaF/l water). Cortical indices and metaphysial width in animals receiving extracts also revealed beneficial effects of the plant extracts. Changes in plasma biochemistry suggested less hepatic and renal damages in animals receiving plant extracts along with fluorinated water in comparison to that receiving fluorinated water alone.

Drug Potentiation Activity

Pharmacokinetic studies revealed that *M. oleifera* pod fraction-treated Swiss albino mice had significantly increased rifampicin plasma concentration as well as inhibited rifampicin-induced cytochrome P-450 activity (Pal et al. 2010). The results suggested that the bioavailability-enhancing property of MoAF may help to lower the dosage level and shorten the treatment course of the antibiotic, rifampicin.

Safety and Toxicological Studies

Safety and toxicological studies showed that LI85008F, a novel synergistic composition of *Moringa oleifera, Murraya koenigi*, and *Curcuma longa*, had broad spectrum safety in animal models (Krishnaraju et al. 2010). These herbs are well recognized and widely used in ayurvedic system of medicine for treating a variety of diseases and are also have been used for culinary purposes for thousands of years. The acute oral LD50 of LI85008F was greater than 5,000 mg/kg in female SD rats and no changes in body weight or adverse effects were observed following necropsy. Acute dermal LD50 of LI85008F was greater than 2,000 mg/kg. LI85008F was classified as non-irritating to skin in a primary dermal irritation study conducted using New Zealand Albino rabbits. LI85008F caused minimal irritation to eyes in a primary eye irritation test conducted on New Zealand Albino rabbits. A dose-dependent 28-day sub-acute toxicity study demonstrated no significant changes in selected organ weights. Evaluations on hematology, clinical chemistry, and histopathology did not show any significant adverse changes. The NOAEL (No observed adverse effect level) of LI85008F was found to be greater than 2,500 mg/kg body weight.

Larvicidal Activity

The methanol seed extract of *M. oleifera* exhibited larvicidal activity in the first to fourth instar larva of *Anopheles stephensi* mosquito, malarial vector, with LC_{50} 57.79 ppm, LC_{90} 125.93 ppm for first instar, 63.90 and 133.07 ppm for second instar, 72.45 and 139.82 ppm for the third, 78.93 and 143.20 ppm for the fourth instar (Prabhu et al. 2011). For the pupal satge LC_{50} was 67.77 ppm, and LC_{90} 141 ppm.

Genotoxic/Mutagenic Activity

A mutagenic compound 4(α-L-rhamnosyloxy) phenylacetonitrile was isolated from roasted seeds of *Moringa oleifera* Lam (Villasenor et al. 1989a). The results of the in-vivo Micronucleus Test showed that the number of micronucleated polychromatic erythrocytes (PCE)/1,000 PCE for this compound was higher than that of the solvent control, dimethylsulfoxide, and approximated that of the positive control, tetracycline, indicating 4(α-L-rhamnosyloxy) phenylacetonitrile to be a genotoxic compound. The following biosynthetically and chemically related compounds 4(α-L-rhamnosyloxy)phenylacetonitrile, 4-hydroxyphenylacetontrile, and 4-hydroxyphenyl-acetamide isolated from the roasted seeds of *Moringa oleifera*, exhibited mutagenic activity in albino mice (Villasenor et al. 1989b).

The concentration 0.2 μg/μl of *M. oleifera* seed extract recommended to treat water for humans was found not to a risk to human health as evaluated Ames, Kado, and cell-free plasmid DNA assays (Rolim et al. 2011). The mutagenicity detected in *Salmonella typhimurium* TA100 and TA102 at concentrations higher than 0.4 μg/μl was not due to water-soluble *Moringa oleifera* lectin isolated from the extract. The results suggested that the purified lectin could be an alternative for water treatment.

Toxicity and Antinutritional Activities

Methanolic extract of *M. oleifera* root was found to contain some alkaloids (total alkaloid 0.2%). Studies showed that the weekly moderate and high dose (>46 mg/kg body wt.) and daily /therapeutic high dose (7 mg/kg) of crude extract of the roots affected liver and kidney functions and haematological parameters (Mazumder et al. 1999). The crude extract at moderate dose level in weekly treatment changed serum aminotransferase and plasma cholesterol levels significantly. High dose in addition to the above parameters changed total bilirubin, non-protein nitrogen, and blood urea and plasma protein. High dose daily treatment and moderate and high dose weekly treatment of the crude extract increased white blood cell count and decreased clotting time significantly. However, the weekly dose (3.5 mg/kg) and low and moderate daily/therapeutic dose (3.5 and 4.6 mg/kg) did not produce adverse effects on liver and kidney functions. The root extract was found to contain the alkaloid spirochin, a potentially fatal nerve-paralysing agent (Morton 1991).

M. oleifera leaves were found to have negligible tannins; the saponins content (5.0% as diosgenin equivalent) was similar to that present in soyabean meal, and trypsin inhibitors and lectins were not detected (Makkar and Becker 1996). The phytate content was 3.1%. The ethanol extracted leaves were virtually free of tannins, lectins, trypsin inhibitors and saponins, and phytate content was 2.5%.

M. oleifera seed was found to contain an anti-nutritional factor. The seed extract agglutinated rabbit erythrocytes but did not show trypsin inhibitor and urease activities (Oliveira et al. 1999). Feeding rats with a diet containing the seed meal showed loss of appetite, impaired growth, lower NPU (net protein utilization) and enlargement of stomach, small intestine, caecum+colon, liver, pancreas, kidneys, heart and lungs and atrophy of thymus and spleen in comparison with rats fed on an egg-white diet. The results indicated that consumption of *M. oleifera* raw mature seeds should be viewed with some caution until suitable processing methods are developed to abolish the yet unknown adverse anti-nutritional factors.

Traditional Medicinal Uses

Various parts of this plant such as the leaves, roots, seed, bark, fruit, flowers and immature pods act as cardiac and circulatory stimulants, possess antitumour, antipyretic, antiepileptic, anti-inflammatory, antiulcer, antispasmodic, diuretic, antihypertensive, cholesterol lowering, antioxidant, antidiabetic, hepatoprotective, antibacterial and antifungal activities, and are being employed for the treatment of different ailments in the indigenous traditional system of medicine, particularly in South Asia.

In the Philippines, young leaves have been used as a galactagogoue. In India, the leaves as a poultice is reported useful in reducing glandular swellings. The leaves are said to have purgative properties. Eating of the leaves is also recommended in gonorrhoea on account of their diuretic action.

The fresh root is regarded as an acrid, pungent remedy which is stimulant and diuretic. In India and Indo-China the roots are regarded as antiscorbutic and when pounded are considered an effective poultice for inflammatory swellings. The root is rubefacient being applied externally in the form of a plaster as a counterirritant. The juice of the root, with milk is also useful as a decoction in hiccoughs, asthma, gout, lumbago, rheumatism, enlarged spleen or liver, internal and deep-seated inflammations, and calculous affections. A decoction or infusion of the root is an effective gargle. A decoction of the root is used in Nicaragua for dropsy. The bark is used as a rubefacient and vesicant and is a popular abortifacient in India. A decoction of the root-bark is used as a fomentation to relieve spasm, and is considered useful in calculous affections. In the Philippines, a decoction of the roots is used to cleanse sores and ulcers and is considered antiscorbutic and is also given to delirious patients; chewing the roots and applying it to snake bites will prevent the poison from spreading and the bark is used as a rubefacient remedy.

The flowers are used in India for catarrh, with or in lieu of young leaves, or young pods. The pods have anthelmintic properties and are administered in affection of the lever and spleen, in articulator pains, etc. In Siddha medicine, the drumstick seeds are used as a sexual virility drug for treating erectile dysfunction in men and also in women for prolonging sexual activity. The oil is used as an external application for rheumatism and used externally as a rubefacient. The gum, mixed with sesame oil, is recommended to be poured into the ears for the relief of otalgia. It is also said to produce abortion. The gum is also given for intestinal complaints in Java.

Other Uses

M. oleifera is a typical multipurpose tree species with a high economic potential (Anwar et al. 2007). It has a multifarious uses besides as a medicinal and food plant.

The tree is used for bee foraging, soil conservation, shade, windbreak, live fence, hedge tree, ornamental boundary marker and for fibres. Leaves and twigs can be used as livestock fodder especially for goats, camels and donkeys. The bark exudes a white to reddish gum ('Ben gum' or 'Moringa gum') with the properties of tragacanth, which serves for tanning and in calico printing. The soft, white wood burns smoke-free and yields a blue dye. In India its pulp has been used to make paper suitable for newsprint, wrapping, printing and writing papers, and for viscose rayon grade pulp for textiles and cellophane (Guha and Negi 1965).

The 'Ben-oil' from the seeds keeps its quality and so can lubricate precision machinery like watches. It is also used for illuminant, soap and cosmetics-perfumes and hair dressing. Ben oil has been shown to be particularly effective in the manufacture of soap producing a stable lather with high washing efficiency suitable for some African countries. The seed cake is considered unsuitable as animal feed because of the high content of alkaloids and saponins and is mainly used as fertilizer. The seeds and seed cake, a residue from oil extraction, can also be used for water purification. Studies demonstrated that the gum of *Moringa oleifera* could be used as a binder and release retardant in tablet formulation (Panda et al. 2008).

Water Purification

Traditionally drumstick seeds have been used to purify domestic household water in rural areas in Sudan (Jahn 1981; Jahn et al. 1986). Women collecting water would tie the ground seeds and suspend them in the turbid water overnight. The seed was found to contain a protein (cationic polyelectrolyte) that acted as a flocculant in water purification (Gassenschmidt et al.1995). Studies reported that a turbidity reduction of 80.0–99.5% paralleled by a primary bacterial reduction of 1–4 log units (90.00–99.99%) was obtained within the first 1–2 hours of treatment with the bacteria being concentrated in the coagulated sediment of *Moringa oleifera* seed material as a (Madsen et al.1987). During the 24 hours observation period a secondary bacterial increase due to regrowth in the supernatant water was consistently observed for *Salmonella typhimurium and Shigella sonnei,* in some cases for *Escherichia coli*, but not for *Vibrio cholerae, Streptococcus faecalis and Clostridium perfringens.* The seed also contained a natural polyelectrolytes non-protein flocculant that was more effective in clarifying and purifying turbid waters (Babu and Chaudhuri 2005). In laboratory tests, direct filtration of a turbid surface water (turbidity 15–25 NTU, heterotrophic bacteria 280–500 cfu/ml, and faecal coliforms 280–500 MPN 100/ml), with seeds of *M. oleifera* as coagulant, produced a substantial improvement in its aesthetic and microbiological quality (lower turbidity 0.3–1.5 NTU, heterotrophic bacteria 5–20 cfu/ml and faecal coliforms 5–10 MPN 100/ml). The method appeared suitable for home water treatment in rural areas of developing countries. These natural coagulants produced a 'low risk' water; however, additional disinfection or boiling should be practised during localised outbreaks/epidemics of enteric infections. Seeds of *Moringa oleifera* were found to contain small storage proteins able to flocculate particles in suspension in water and be used to improve water purification processes (Broin et al. 2002). It was able to aggregate montmorillonite clay particles as well as Gram-positive and Gram-negative bacteria.

Removal of Environmental/Industrial Pollutants and Heavy Metals

Among other natural flocculant/coagulant agents, *Moringa oleifera* seed extract exhibited ability to remove an anionic surfactant (Beltrán-Heredia and Sánchez-Martín 2009). Sodium lauryl sulphate was removed from aqueous solutions up to 80% through coagulation/flocculation process using *Moringa oleifera* seed extract. Temperature and were found to be not very important factors in removal efficiency. Extracts obtained by water soaking of *M. oleifera* intact seeds was found to contain a water soluble *M. oleifera* lectin (WSMoL) (Santos et al. 2005). The antioxidant component (WSMoAC) reduced 1,1-diphenyl-2-picrylhydrazyl radical (DPPH), was slower than catechin and was thermostable. The extracts showed a primary glycopolypeptide band of Mw 20,000; the main native acidic protein showed hemagglutinating activity. WSMoL may be involved in the seed coagulant properties.

Moringa oleifera seeds, was found to be an environmental friendly and natural coagulant for the pre-treatment of palm oil mill effluent (POME) (Bhatia et al. 2007). In coagulation-flocculation process, the *M. oleifera* seeds after oil extraction (MOAE) acted as an effective coagulant with the removal of 95% suspended solids and 52.2% reduction in the chemical oxygen demand (COD). The combination of MOAE with flocculant (NALCO 7751), increased the suspended solids removal to 99.3% and COD reduction was 52.5%. The coagulation-flocculation process at the temperature of 30°C resulted in better suspended solids removal and COD reduction compared to the temperature of 40°C, 55°C and 70°C. The MOAE combined with flocculant (NALCO 7751) reduced the sludge volume index (SVI) to 210 ml/g with higher recovery of dry mass of sludge (87.25%) and water (50.3%). In another study, water extract of *M. oleifera* seed acted as effective coagulants for water and wastewater treatment (Bhuptawat et al. 2007). Overall COD removals of 50% were achieved at both 50 and 100 mg/l *M. oleifera* doses. When 50 and 100 mg/l seed doses were applied in combination with 10 mg/l of alum, COD removal increased to 58% and 64%, respectively. The majority of COD removal occurred during the filtration process. The simple water extract may be obtained at minimal cost from the press cake residue remaining after oil extraction from the seed.

Another study evaluated the sorption properties of bioactive constituents of *Moringa oleifera* seeds for decontamination of cadmium at laboratory scale (Jamal et al. 2008). The maximum removal of cadmium was 72% by using 0.2 g/l of bioactive dosage. Among other natural flocculant/coagulant agents, *Moringa oleifera* seed extract ability to remove an anionic surfactant had been evaluated with interesting results. Sodium lauryl sulphate was removed from aqueous solutions up to 80% through coagulation/flocculation process. *Moringa oleifera* pods were found to be an effective sorbent for removal of organics and had been used extensively to accrue and then to pre-concentrate benzene, toluene and ethylbenzene in waste water sample (Akhtar et al. 2007). Studies showed that *M. oleifera* seeds had potential application as biosorbents in cadmium, lead, cobalt, copper and silver decontamination from aqueous effluents (Araújo et al. 2010).

Filtrate from crushed *Moringa oleifera* seeds was found to have cyanobactericidal activity (Lürling and Beekman 2010). High-density populations of the common bloom-forming cyanobacterium *Microcystis aeruginosa* chlorophyll-a concentrations of approximately 270 μg/l were reduced to very low levels within 2 weeks of exposure to >/=80 mg crushed seeds per litre. At the highest dosage of 160 mg/l, the Photosystem II efficiency dropped to zero rapidly and remained nil during the course of the experiment (14 days). Hence, under laboratory conditions, a complete wipe-out of the bloom could be achieved.

Biodiesel Potential

Moringa oleifera oil was found to have potential as acceptable feedstock for biodiesel. *M. oleifera oil* was found to have a high content of oleic acid (>70%) with saturated fatty acids comprising most of the remaining fatty acid profile (Rashid et al. 2008). As a result, the methyl esters (biodiesel) obtained from this oil exhibited a high cetane number of approximately 67, one of the highest found for a biodiesel fuel. Other fuel properties of biodiesel derived from *M. oleifera* such as cloud point, kinematic viscosity and oxidative stability were also determined and compared biodiesel standards such as ASTM D6751 and EN 14214.

Moringa oil extracted from the seed of the malunggay plant is now being tapped as source of biodiesel in the Philippines (Flores 2008). It is gaining preferable status over *Jatropha* as a source of biofuel. All parts of the malunggay plant are used whereas *Jatropha* is left with poisonous waste after oil extraction. Also, malunggay needs only 1–2 years for seedling maturation compared to Jatropha's 3–5 years. The maths of malunggay's commercial potential was found to be attractive: Seeds can be bought at P10 per kilo, and a hectare of malunggay seedlings could harvest 20,000 k in 2 years with a potential profit of P200,000.

Seed Treatment

Studies demonstrated that aqueous *Moringa* seed extract (AMSE) had potential as a seed biofungicide for groundnuts (Donli and Dauda 2003). Studies showed that the seed extract, AMSE at all the concentrations used except 1 g/l brought about significant reduction in the incidence of fungi on the seeds, such reduction increased as the dosage of AMSE increased. There were no significant differences in control between the highest concentration of AMSE (20 g l/l) and Apron Plus (metalaxyl + carboxin + furathiocarb) fungicide at the manufacturer's recommended dose. The sensitivity to AMSE of the fungi tested varied, *Mucor* sp being the most sensitive and *Aspergillus niger* the least, with *Rhizopus stolonifer* and *Aspergillus flavus* intermediate.

Comments

The tree is propagated from seeds and stem cuttings.

Selected References

Agrawal B, Mehta A (2008) Antiasthmatic activity of *Moringa oleifera* Lam: A clinical study. Indian J Pharmacol 40(1):28–31

Akhtar M, Moosa Hasany S, Bhanger MI, Iqbal S (2007) Sorption potential of *Moringa oleifera* pods for the removal of organic pollutants from aqueous solutions. J Hazard Mater 141(3):546–556

Anwar F, Bhanger MI (2003) Analytical characterization of *Moringa oleifera* seed oil grown in temperate regions of Pakistan. J Agric Food Chem 51(22):6558–6563

Anwar F, Latif S, Ashraf M, Gilani AH (2007) *Moringa oleifera*: a food plant with multiple medicinal uses. Phytother Res 21(1):17–25

Araújo CS, Alves VN, Rezende HC, Almeida IL, de Assunção RM, Tarley CR, Segatelli MG, Coelho NM (2010) Characterization and use of *Moringa oleifera* seeds as biosorbent for removing metal ions from aqueous effluents. Water Sci Technol 62(9):2198–2203

Atawodi SE, Atawodi JC, Idakwo GA, Pfundstein B, Haubner R, Wurtele G, Bartsch H, Owen RW (2010) Evaluation of the polyphenol content and antioxidant properties of methanol extracts of the leaves, stem, and root barks of *Moringa oleifera* Lam. J Med Food 13(3):710–716

Babu R, Chaudhuri M (2005) Home water treatment by direct filtration with natural coagulant. J Water Health 3(1):27–30

Banerji R, Bajpai A, Verma SC (2009) Oil and fatty acid diversity in genetically variable clones of *Moringa oleifera* from India. J Oleo Sci 58:9–16

Beltrán-Heredia J, Sánchez-Martín J (2009) Removal of sodium lauryl sulphate by coagulation/flocculation with *Moringa oleifera* seed extract. J Hazard Mater 164(2–3):713–719

Bennett RN, Mellon FA, Foidl N, Pratt JH, Dupont MS, Perkins L, Kroon PA (2003) Profiling glucosinolates and phenolics in vegetative and reproductive tissues of the multi-purpose trees *Moringa oleifera* L. (Horseradish tree) and *Moringa stenopetala* L. J Agric Food Chem 51(12):3546–3553

Bharali R, Tabassum J, Azad MR (2003) Chemomodulatory effect of *Moringa oleifera*, Lam, on hepatic carcinogen metabolising enzymes, antioxidant parameters and skin papillomagenesis in mice. Asian Pac J Cancer Prev 4(2):131–139

Bhatia S, Othman Z, Ahmad AL (2007) Pretreatment of palm oil mill effluent (POME) using *Moringa oleifera* seeds as natural coagulant. J Hazard Mater 145(1–2):120–126

Bhattacharya J, Guha G, Bhattacharya B (1978) Powder microscopy of bark–poison used for abortion: *Moringa pterygosperma* Gaertn. J Indian Forensic Sci 17:47–50

Bhattacharya SB, Das AK, Banerji N (1982) Chemical investigations on the gum exudates from Sajna (*Moringa oleifera*). Carbohydr Res 102(1):253–262

Bhuptawat H, Folkard GK, Chaudhari S (2007) Innovative physico-chemical treatment of wastewater incorporating *Moringa oleifera* seed coagulant. J Hazard Mater 142(1–2):477–482

Bosch CH (2004) *Moringa oleifera* Lam. [Internet] Record from Protabase. Grubben GJH, Denton OA

(Eds). PROTA (Plant Resources of Tropical Africa/ Ressources végétales de l'Afrique tropicale), Wageningen, the Netherlands. http://database.prota.org/search.htm

Broin M, Santaella C, Cuine S, Kokou K, Peltier G, Joet T (2002) Flocculent activity of a recombinant protein from *Moringa oleifera* Lam. Seeds Appl Microbiol Biotechnol 60:114–119

Bukar A, Uba A, Oyeyi TI (2010) Antimicrobial profile of *Moringa oleifera* Lam. extracts against some food – borne microorganisms. Bayero J Pure Appl Sci 3(1):43–48

Burger DJ, Fuglie L, Herzig JW (2002) The possible role of *Moringa oleifera* in HIV/AIDS supportive treatment. International Conference on AIDS. 2002 Jul 7–12, 14: abstract no. F12423

Burkill IH (1966) A Dictionary of the Economic Products of the Malay Peninsula. Revised reprint, vol 2. Ministry of Agriculture and Co-operatives, Kuala Lumpur, Malaysia, vol 1 (A–H), pp 1–1240, vol. 2 (I–Z), pp 1241–2444

Burkill HM (1998) Useful plants of west tropical Africa, vol 4. Families M-R. Royal Bot, Gardens, Kew, 969 pp

Cáceres A, Lopez S (1991) Pharmacological properties of *Moringa oleifera*. 3. Effect of seed extracts in the treatment of experimental pydermia. Fitoterapia 52(5):449–450

Cáceres A, Cabrera O, Morales O, Mollinedo P, Mendia P (1991) Pharmacological properties of *Moringa oleifera*. 1: preliminary screening for antimicrobial activity. J Ethnopharmacol 33(3):213–216

Cáceres A, Saravia A, Rizzo S, Zabala L, De Leon E, Nave F (1992) Pharmacologic properties of *Moringa oleifera*. 2: Screening for antispasmodic, antiinflammatory and diuretic activity. J Ethnopharmacol 36(3):233–237

Chattopadhyay S, Maiti S, Maji G, Deb B, Pan B, Ghosh D (2010) Protective role of *Moringa oleifera* (sajina) seed on arsenic-induced hepatocellular degeneration in female albino rats. Biol Trace Elem Res (in Press)

Cheenpracha S, Park EJ, Yoshida WY, Barit C, Wall M, Pezzuto JM, Chang LC (2010) Potential anti-inflammatory phenolic glycosides from the medicinal plant *Moringa oleifera* fruits. Bioorg Med Chem 18(17):6598–6602

Chopra RN, Nayar SL, Chopra IC (1956) Glossary of indian medicinal plants. (Including the Supplement). Council Scientific Industrial Research, New Delhi, 30 pp

Chuang PH, Lee CW, Chou JY, Murugan M, Shieh BJ, Chen HM (2007) Anti-fungal activity of crude extracts and essential oil of *Moringa oleifera* Lam. Bioresour Technol 98(1):232–236

Chumark P, Khunawat P, Sanvarinda Y, Phornchirasilp S, Morales NP, Phivthong-Ngam L, Ratanachamnong P, Srisawat S, Pongrapeeporn KU (2008) The in vitro and ex vivo antioxidant properties, hypolipidaemic and antiatherosclerotic activities of water extract of *Moringa oleifera* Lam. leaves. J Ethnopharmacol 116(3):439–446

Council of Scientific and Industrial Research (CSIR) (1962) The Wealth of India. A Dictionary of Indian Raw Materials and Industrial Products. (Raw Materials 6). Publications and Information Directorate, New Delhi, India

Dachana KB, Rajiv J, Indrani D, Prakash J (2010) Effect of dried moringa (*Moringa oleifera* Lam.) leaves on rheological, microstructural, nutritional, textural and organoleptic characteristics of cookies. J Food Qual 33(5):660–677

Dahot MU (1998) Antimicrobial activity of small protein of *Moringa oleifera* leaves. J Islamic Acad Sci 11(1):27–32

Das BR, Kurup PA, Narasimha Rao PL (1957) Antibiotic principle from *Moringa pterygosperma* Part VII. Antibacterial activity and chemical structure of compounds related to pterygospermin. Indian J Med Res 45:191–196

Debnath S, Guha D (2007) Role of *Moringa oleifera* on enterochromaffin cell count and serotonin content of experimental ulcer model. Indian J Exp Biol 45(8):726–731

Debnath S, Biswas D, Ray K, Guha D (2011) *Moringa oleifera* induced potentiation of serotonin release by 5-HT(3) receptors in experimental ulcer model. Phytomed 18(2–3):91–95

Donli PO, Dauda H (2003) Evaluation of aqueous *Moringa* seed extract as a seed treatment biofungicide for groundnuts. Pest Manag Sci 59(9):1060–1062

Doughari JH, Pukuma MS, De N (2007) Antibacterial effects of *Balanites aegyptiaca* L Drel. and *Moringa oleifera* Lam. on Salmonella typhi. Afr J Biotechnol 6(19):2212–2215

Duangjai A, Ingkaninan K, Limpeanchob N (2010) Potential mechanisms of hypocholesterolaemic effect of Thai spices/dietary extracts. Nat Prod Res 8:1–12

Eilert U, Wolters B, Nahrstedt A (1981) The antibiotic principle of seeds of *Moringa oleifera* and *Moringa stenopetala*. Planta Med 42(5):55–61

Ezeamuzie IC, Ambakederemo AW, Shode FO, Ekwebelem SC (1996) Antiinflammatory effects of *Moringa oleifera* root extract. Int J Phcog 34(3):207–212

Fahad J, Vijayalakshmi, Kumar MCS, Sanjeeva, Kodancha GP, Adarsh B, Udupa AL, Rathnakar UP (2010) Antiurolithiatic activity of aqueous extract of bark of *Moringa oleifera* (Lam.) in rats. Sc Res 2(4):352–355

Faizi S, Siddiqui BS, Saleem R, Siddiqui S, Aftab K, Gilani AH (1994) Isolation and structure elucidation of new nitrile and mustard oil glycosides from *Moringa oleifera* and their effect on blood pressure. J Nat Prod 57(9):1256–1261

Faizi S, Siddiqui BS, Saleem R, Siddiqui S, Aftab K, Gilani AH (1995) Fully acetylated carbamate and hypotensive thiocarbamate glycosides from *Moringa oleifera*. Phytochem 38(4):957–963

Faizi S, Siddiqui BS, Saleem R, Aftab K, Shaheen F, Gilani AH (1998) Hypotensive constituents from the pods of *Moringa oleifera*. Planta Med 64(3):225–228

Fakurazi S, Hairuszah I, Nanthini U (2008) *Moringa oleifera* Lam prevents acetaminophen induced liver injury through restoration of glutathione level. Food Chem Toxicol 46(8):2611–2615

Flores H (2008) Malunggay oil as biofuel. Philippine Star, April 11, 2008

Foundation for Revitalisation of Local Health Traditions (2008) *FRLHT* Database. http://envis.frlht.org

Ganguly R, Guha D (2008) Alteration of brain monoamines & EEG wave pattern in rat model of Alzheimer's disease & protection by *Moringa oleifera*. Indian J Med Res 128(6):744–751

Gassenschmidt U, Jany KD, Tauscher B, Niebergall H (1995) Isolation and characterization of a flocculating protein from *Moringa oleifera* Lam. Biochim Biophys Acta 1243(3):477–481

Ghasi S, Nwobodo E, Ofili JO (2000) Hypocholesterolemic effects of crude extract of leaf of *Moringa oleifera* Lam in high-fat diet fed wistar rats. J Ethnopharmacol 69(1):21–25

Gilani AH, Aftab K, Suria A, Siddiqui S, Saleem R, Siddiqui BS, Faizi S (1994) Pharmacological studies on hypotensive and spasmolytic activities of pure compounds from *Moringa oleifera*. Phyto Res 8(2):87–91

Gopalan G, Rama Sastri BV, Balasubramanian SC (2002) Nutritive value of indian foods. National Institute of Nutrition. Indian Council of Medical Research, Hyderabad

Guevara AP, Vargas C, Sakurai H, Fujiwara Y, Hashimoto K, Maoka T, Kozuka M, Ito Y, Tokuda H, Nishino H (1999) An antitumor promoter from *Moringa oleifera* Lam. Mutat Res 440(2):181–188

Guha SRD, Negi JS (1965) Wrapping, printing, and writing paper from *Moringa pterygosperma*. Indian Pulp Paper 20(6):377–379

Gupta A, Gautam MK, Singh RK, Kumar MV, Rao CV, Goel RK, Anupurba S (2010) Immunomodulatory effect of *Moringa oleifera* Lam. extract on cyclophosphamide induced toxicity in mice. Indian J Exp Biol 48(11):1157–1160

Hamza AA (2010) Ameliorative effects of *Moringa oleifera* Lam seed extract on liver fibrosis in rats. Food Chem Toxicol 48(1):345–355

Hu S-Y (2005) Food plants of China. The Chinese University Press, Hong Kong, 844 pp

Iqbal S, Bhanger MI (2006) Effect of season and production location on antioxidant activity of *Moringa oleifera* leaves grown in Pakistan. J Food Comp Anal 19:544–551

Jahn SAA (1981) Traditional water purification in tropical developing countries: existing methods and potential applications, GTZ Manual No.117. Pub. GTZ, Eschborn

Jahn SAA, Musnad HA, Burgstaller H (1986) The tree that purifies water: Cultivating multipurpose Moringaceae in the Sudan. In Wazeka R (ed) Unasylva No. 152 Genetics and the forests of the future. FAO Rome

Jaiswal D, Rai PK, Kumar A, Mehta S, Watal G (2009) Effect of *Moringa oleifera* Lam. leaves aqueous extract therapy on hyperglycemic rats. J Ethnopharmacol 123(3):392–396

Jamal P, Muyibi SA, Syarif WM (2008) Optimization of process conditions for removal of cadmium using bioactive constituents of *Moringa oleifera* seeds. Med J Malaysia 63 Suppl A:105–106

Karadi RV, Gadge NB, Alagawadi KR, Savadi RV (2006) Effect of *Moringa oleifera* Lam. root-wood on ethylene glycol induced urolithiasis in rats. J Ethnopharmacol 105(1–2):306–311

Kekuda TR, Mallikarjun PN, Swathi D, Nayana KV, Aiyar MB, Rohini TR (2010) Antibacterial and antifungal efficacy of steam distillate of *Moringa oleifera* Lam. J Pharm Sci Res 2(1):34–37

Khare GC, Singh V, Gupta PC (1997) A new leucoanthocyanin from *Moringa oleifera* gum. J Indian Chem Soc 74:247–248

Krishnaraju AV, Sundararaju D, Srinivas P, Rao CV, Sengupta K, Trimurtulu G (2010) Safety and toxicological evaluation of a novel anti-obesity formulation LI85008F in animals. Toxicol Mech Methods 20(2):59–68

Kumar NA, Pari L (2003) Antioxidant action of *Moringa oleifera* Lam. (drumstick) against antitubercular drugs induced lipid peroxidation in rats. J Med Food 6(3):255–259

Limaye DA, Nimbkar AY, Jain R, Ahmad M (1995) Cardiovascular effects of the aqueous extract of *Moringa pterygosperma*. Phytother Res 9:37–40

Lipipun V, Kurokawa M, Suttisri R, Taweechotipatr P, Pramyothin P, Hattori M, Shiraki K (2003) Efficacy of Thai medicinal plant extracts against herpes simplex virus type 1 infection in vitro and in vivo. Antiviral Res 60(3):175–180

Lürling M, Beekman W (2010) Anti-cyanobacterial activity of *Moringa oleifera* seeds. J Appl Phycol 22(4):503–510

Madsen M, Schlundt J, El Fadil O (1987) Effect of water coagulation by seeds of *Moringa oleifera* on bacterial concentrations. J Trop Med Hyg 90(3):101–109

Mahajan SG, Mehta AA (2007) Inhibitory action of ethanolic extract of seeds of *Moringa oleifera* Lam. on systemic and local anaphylaxis. J Immunotoxicol 4(4):87–94

Mahajan SG, Mehta AA (2008) Effect of *Moringa oleifera* Lam. seed extract on ovalbumin-induced airway inflammation in guinea pigs. Inhal Toxicol 20(10):897–909

Mahajan SG, Mehta AA (2010) Immunosuppressive activity of ethanolic extract of seeds of *Moringa oleifera* Lam. in experimental immune inflammation. J Ethnopharmacol 130(1):183–186

Mahajan SG, Mehta AA (2011) Suppression of ovalbumin-induced Th2-driven airway inflammation by β-sitosterol in a guinea pig model of asthma. Eur J Pharmacol 650(1):458–464

Mahajan SG, Mali RG, Mehta AA (2007a) Effect of *Moringa oleifera* Lam. seed extract on toluene diisocyanate-induced immune-mediated inflammatory responses in rats. J Immunotoxicol 4(2):85–96

Mahajan SG, Mali RG, Mehta AA (2007b) Protective effect of ethanolic extract of seeds of *Moringa oleifera*

Lam. against inflammation associated with development of arthritis in rats. J Immunotoxicol 4(1):39–47

Mahajan SG, Banerjee A, Chauhan BF, Padh H, Nivsarkar M, Mehta AA (2009) Inhibitory effect of n-butanol fraction of *Moringa oleifera* Lam. seeds on ovalbumin-induced airway inflammation in a guinea pig model of asthma. Int J Toxicol 28(6):519–527

Makkar HPS, Becker K (1996) Nutritional value and antinutritional components of whole and ethanol extracted *Moringa oleifera* leaves. Animal Feed Sci Tech 63(1):211–228

Manaheji H, Jafari S, Zaringhalam J, Rezazadeh S, Taghizadfarid R (2011) Analgesic effects of methanolic extracts of the leaf or root of *Moringa oleifera* on complete Freund's adjuvant-induced arthritis in rats. Zhong Xi Yi Jie He Xue Bao 9(2):216–222

Mazumder UK, Gupta M, Chakrabarti S, Pal D (1999) Evaluation of hematological and hepatorenal functions of methanolic extract of *Moringa oleifera* Lam. root treated mice. Indian J Exp Biol 37(6):612–4

Mehta K, Balaraman R, Amin AH, Bafna PA, Gulati OD (2003) Effect of fruits of *Moringa oleifera* on the lipid profile of normal and hypercholesterolaemic rabbits. J Ethnopharmacol 86(2–3):191–195

Mondal S, Chakraborty I, Pramanik M, Rout D, Islamm SS (2004) Structural studies of an immunoenhancing polysaccharide isolated from mature pods (fruits) of *Moringa oleifera* (Sajina). Med Chem Res 13:390–400

Monera TG, Maponga CC (2010) *Moringa oleifera* supplementation by patients on antiretroviral therapy. J Int AIDS Soc 13(Suppl 4), p 188

Morton JF (1991) The Horseradish Tree, *Moringa pterygosperma* (Moringaceae) - a boon to arid lands? Econ Bot 45:318–333

Murakami A, Kitazono Y, Jiwajinda S, Koshimizu K, Ohigashi H (1998) Niaziminin, a thiocarbamate from the leaves of *Moringa oleifera*, holds a strict structural requirement for inhibition of tumor-promoter-induced Epstein-Barr virus activation. Planta Med 64(4):319–323

Nambiar VS, Seshadri S (2001) Bioavailability trials of beta-carotene from fresh and dehydrated drumstick leaves (*Moringa oleifera*) in a rat model. Plant Foods Hum Nutr 56(1):83–95

Nandave M, Ojha SK, Joshi S, Kumari S, Arya DS (2009) *Moringa oleifera* leaf extract prevents isoproterenol-induced myocardial damage in rats: evidence for an antioxidant, antiperoxidative, and cardioprotective intervention. J Med Food 12(1):47–55

Nantachit K (2006) Antibacterial activity of the capsules of *Moringa oleifera* Lamk. (Moringaceae). Chiang Mai Univ J 5(3):365–368

Ndiaye M, Dieye AM, Mariko F, Tall A, Sall Diallo A, Faye B (2002) Contribution to the study of the anti-inflammatory activity of *Moringa oleifera* (Moringaceae). Dakar Med 47(2):210–212 (In French)

Ndong M, Uehara M, Katsumata S, Suzuki K (2007a) Effects of oral administration of *Moringa oleifera* Lam on glucose tolerance in goto-kakizaki and wistar rats. J Clin Biochem Nutr 40(3):229–233

Ndong M, Uehara M, Katsumata S, Sato S, Suzuki K (2007b) Preventive effects of *Moringa oleifera* (Lam) on hyperlipidemia and hepatocyte ultrastructural changes in iron deficient rats. Biosci Biotechnol Biochem 71(8):1826–1833

Nepolean P, Anitha J, Renitta RE (2009) Isolation, analysis and identification of phytochemicals of antimicrobial activity of *Moringa oleifera* Lam. Curr Biotica 3(1):33–38

Nwosu MO, Okafor JI (1995) Preliminary studies of the antifungal activities of some medicinal plants against *Basidiobolus* and some other pathogenic fungi. Mycoses 38:191–195

Ochse JJ, Bakhuizen van den Brink RC (1980) Vegetables of the Dutch Indies, 3rd edn. Ascher & Co., Amsterdam, 1016 pp

Oduro I, Ellis WO, Owusu D (2008) Nutritional potential of two leafy vegetables: *Moringa oleifera* and *Ipomoea batatas* leaves. Sci Res Essay 3(2):57–60

Ogunsina BS, Radha C, Indrani D (2011) Quality characteristics of bread and cookies enriched with debittered *Moringa oleifera* seed flour. Int J Food Sci Nutr 62(2):185–194

Oliveira JTA, Silveira SB, Vasconcelos IM, Cavada BS, Moreira RA (1999) Compositional and nutritional attributes of seeds from the multiple purpose tree *Moringa oleifera* Lamarck. J Sci Food Agric 79(6):815–820

Oluduro OA, Aderiye BI, Connolly JD, Akintayo ET, Famurewa O (2010) Characterization and antimicrobial activity of 4-(β-D-glucopyranosyl-1→4-α-L-rhamnopyranosyloxy)-benzyl thiocarboxamide; a novel bioactive compound from *Moringa oleifera* seed extract. Folia Microbiol (Praha) 55(5):422–426

Pal SK, Mukherjee PK, Saha BP (1995) Studies on the antiulcer activity of *Moringa oleifera* leaf extract on gastric ulcer models in rats. Phytother Res 9:463–465

Pal SK, Mukherjee PK, Saha K, Pal M, Saha BP (1996) Studies on some psychopharmacological actions of *Moringa oleifera* Lam. (Moringaceae) leaf extract. Phytother Res 10:402–405

Pal A, Bawankule DU, Darokar MP, Gupta SC, Arya JS, Shanker K, Mohangupta M, Yadav NP, Singh Khanuja SP (2010) nfluence of *Moringa oleifera* on pharmacokinetic disposition of rifampicin using HPLC-PDA method: a pre-clinical study. Biomed Chromatogr, In press

Panda DS, Choudhury NS, Yedukondalu M, Si S, Gupta R (2008) Evaluation of gum of *Moringa oleifera* as a binder and release retardant in tablet formulation. Indian J Pharm Sci 70(5):614–618

Pari L, Kumar NA (2002) Hepatoprotective activity of *Moringa oleifera* on antitubercular drug-induced liver damage in rats. J Med Food 5(3):171–177

Polprasid P (1994) *Moringa oleifera* Lamk. In: Siemonsma JS, Piluek K (eds) Plant resources of south-east asia no 8. vegetables. Prosea, Bogor, Indonesia, pp 213–215

Porcher MH et al (1995–2020) Searchable World Wide Web Multilingual Multiscript Plant Name Database. Published by The University of Melbourne. Australia. http://www.plantnames.unimelb.edu.au/Sorting/Frontpage.html

Prabhu K, Murugan K, Nareshkumar A, Ramasubramanian N, Bragadeeswaran S (2011) Larvicidal and repellent potential of *Moringa oleifera* against malarial vector, *Anopheles stephensi* Liston (Insecta: Diptera: Culicidae). Asian Pac J Trop Biomed 1(2):124–129

Prakash AO (1988) Ovarian response to aqueous extract of *Moringa oleifera* during early pregnancy in rats. Fitoterapia 59(2):89–96

Pramanik A, Islam SS (1998) Chemical investigation of aqueous extract of the mature and premature flowers of *Moringa oleifera* and structural studies of a polysaccharide isolated from its premature flowers. Indian J Chem 37B:676–682

Promkum C, Kupradinun P, Tuntipopipat S, Butryee C (2010) Nutritive evaluation and effect of *Moringa oleifera* pod on clastogenic potential in the mouse. Asian Pac J Cancer Prev 11(3):627–632

Pullakhandam R, Failla ML (2007) Micellarization and intestinal cell uptake of beta-carotene and lutein from drumstick (*Moringa oleifera*) leaves. J Med Food 10(2):252–257

Rahman MM, Rahman MM, Akhter S, Jamal MA, Pandeya DR, Haque MA, Alam MF, Rahman A (2010) Control of coliform bacteria detected from diarrhea associated patients by extracts of *Moringa oleifera*. Nepal Med Coll J 12(1):12–19

Ramachandran C, Peter KV, Gopalakrishnan PK (1980) Drumstick (*Moringa oleifera*): a multipurpose Indian vegetable. Econ Bot 34:276–283

Ranjan R, Swarup D, Patra RC, Chandra V (2009) *Tamarindus indica* L. and *Moringa oleifera* M. extract administration ameliorates fluoride toxicity in rabbits. Indian J Exp Biol 47(11):900–905

Rao AV, Devi PU, Kamath R (2001) In vivo radioprotective effect of *Moringa oleifera* leaves. Indian J Exp Biol 39(9):858–863

Rashid U, Anwar F, Moser BR, Knothe G (2008) *Moringa oleifera* oil: a possible source of biodiesel. Bioresour Technol 99(17):8175–8179

Rathi BS, Bodhankar SL, Baheti AM (2006) Evaluation of aqueous leaves extract of *Moringa oleifera* Linn for wound healing in albino rats. Indian J Exp Biol 44(11):898–901

Ray K, Hazra R, Guha D (2003) Central inhibitory effect of *Moringa oleifera* root extract: possible role of neurotransmitters. Indian J Exp Biol 41(11):1279–1284

Ray K, Hazra R, Debnath PK, Guha D (2004) Role of 5-hydroxytryptamine in *Moringa oleifera* induced potentiation of pentobarbitone hypnosis in albino rats. Indian J Exp Biol 42(6):632–635

Rolim LADMM, Macêdo MFS, Sisenando HA, Napoleão TH, Felzenszwalb I, Aiub CAF, Coelho LCBB, Medeiros SRB, Paiva PMG (2011) Genotoxicity Evaluation of *Moringa oleifera* Seed Extract and Lectin. J Food Sci 76(2):T53–T58

Ruckmani K, Davimani S, Jayakar B, Anandan R (1998a) Anti-ulcer activity of the alkali preparation of the root and fresh leaf juice of *Moringa oleifera* Lam. Ancient Sci Life 17(3):220–223

Ruckmani K, Kavimani S, Anandan R, Jayakar B (1998b) Effect of *Moringa oleifera* Lam. on paracetamol-induced hepatotoxicity. Indian J Pharm Sci 60(1):33–35

Sahakitpichan P, Mahidol C, Disadee W, Ruchirawat S, Kanchanapoom T (2011) nusual glycosides of pyrrole alkaloid and 4'-hydroxyphenylethanamide from leaves of *Moringa oleifera*. Phytochem [Epub ahead of print]

Saluja MP, Kapil RS, Popli SP (1978) Studies in medicinal plants: part VI chemical constituents of *Moringa oleifera* Lam. and isolation of 4-hydroxymellein. Indian J Chem 16B:1044–1045

Sánchez-Machado DI, López-Cervantes J, Vázquez NJ (2006) High-performance liquid chromatography method to measure alpha- and gamma-tocopherol in leaves, flowers and fresh beans from *Moringa oleifera*. J Chromatogr A 1105(1–2):111–114

Santos AF, Argolo AC, Coelho LC, Paiva PM (2005) Detection of water soluble lectin and antioxidant component from *Moringa oleifera* seeds. Water Res 39(6):975–980

Sashidhara KV, Rosaiah JN, Tyagi E, Shukla R, Raghubir R, Rajendran SM (2009) Rare dipeptide and urea derivatives from roots of *Moringa oleifera* as potential anti-inflammatory and antinociceptive agents. Eur J Med Chem 44(1):432–436

Sasikala V, Rooban BN, Priya SG, Sahasranamam V, Abraham A (2010) *Moringa oleifera* prevents selenite-induced cataractogenesis in rat pups. J Ocul Pharmacol Ther 26(5):441–447

Sathya TN, Aadarsh P, Deepa V, Balakrishna Murthy P (2010) *Moringa oleifera* Lam. leaves prevent cyclophosphamide-induced micronucleus and DNA damage in mice. Int J Phytomed 2:147–154

Shukla S, Mathur R, Prakash AO (1988a) Antifertility profile of the aqueous extract of *Moringa oleifera* roots. J Ethnopharmacol 22(1):51–62

Shukla S, Mathur R, Prakash AO (1988b) Biochemical and physiological alterations in female reproductive organs of cyclic rats treated with aqueous extract of *Moringa oleifera* Lam. Acta Eur Fertil 19(4):225–232

Shukla S, Mathur R, Prakash AO (1989) Histoarchitecture of the genital tract of ovariectomized rats treated with an aqueous extract of *Moringa oleifera* roots. J Ethnopharmacol 25(3):249–261

Siddhuraju P, Becker K (2003) Antioxidant properties of various solvent extracts of total phenolic constituents from three different agro climatic origins of drumsticks tree (*Moringa oleifera* Lam.) leaves. J Agri Food Chem 51(8):44–55

Singh BN, Singh BR, Singh RL, Prakash D, Dhakarey R, Upadhyay G, Singh HB (2009) Oxidative DNA damage protective activity, antioxidant and anti-quorum sensing potentials of *Moringa oleifera*. Food Chem Toxicol 47(6):1109–1116

Sreelatha S, Padma PR (2009) Antioxidant activity and total phenolic content of *Moringa oleifera* leaves in two stages of maturity. Plant Foods Hum Nutr 64(4):303–311

Sreelatha S, Padma PR (2010) Protective mechanisms of *Moringa oleifera* against CCl(4)-induced oxidative stress in precision-cut liver slices. Forsch Komplementmed 17(4):189–194

Sreelatha S, Padma PR (2011) Modulatory effects of *Moringa oleifera* extracts against hydrogen peroxide-induced cytotoxicity and oxidative damage. Hum Exp Toxicol [Epub ahead of print]

Sreelatha S, Jeyachitra A, Pdama PR (2011) Antiproliferation and induction of apoptosis by *Moringa oleifera* leaf extract on human cancer cells. Food Chem Toxicol [Epub ahead of print]

Stuart GU (2010) Philippine alternative medicine. Manual of Some Philippine Medicinal Plants. http://www.stuartxchange.org/OtherHerbals.html

Sudha P, Asdaq SM, Dhamingi SS, Chandrakala GK (2010) Immunomodulatory activity of methanolic leaf extract of *Moringa oleifera* in animals. Indian J Physiol Pharmacol 54(2):133–140

Sutherland J (1999) The *Moringa oleifera* pages. [Internet] Department of Engineering, University of Leicester, Leicester, United Kingdom. http://www.le.ac.uk./engineering/staff/Sutherland/moringa/moringa.htm. Accessed Jan 2004

Tahiliani P, Kar A (2000) Role of *Moringa oleifera* leaf extract in the regulation of thyroid hormone status in adult male and female rats. Pharmacol Res 41(3):319–323

Tee ES, Noor MI, Azudin MN, Idris K (1997) Nutrient composition of Malaysian foods, 4th edn. Institute for Medical Research, Kuala Lumpur, 299 pp

Thilza IB, Sanni S, Isah ZA, Sanni FS, Talle M, Joseph MB (2001) In vitro antimicrobial activity of water extract of *Moringa oleifera* leaf stalk on bacteria normally implicated in eye diseases. Academia Arena 2(6):80–82

Tsaknis J, Lalas S, Gergis V, Dourtoglou V, Spiliotis V (1999) Characterization of *Moringa oleifera* variety Mbololo seed oil of Kenya. J Agric Food Chem 47(11):4495–4499

U.S. Department of Agriculture, Agricultural Research Service (2010) USDA National Nutrient Database for Standard Reference, Release 23. Nutrient Data Laboratory Home Page, http://www.ars.usda.gov/ba/bhnrc/ndl

Viera GH, Mourão JA, Angelo AM, Costa RA, Vieira RH (2010) Antibacterial effect (in vitro) of *Moringa oleifera* and *Annona muricata* against gram positive and gram negative bacteria. Rev Inst Med Trop Sao Paulo 52(3):129–132

Villasenor IM, Finch P, Lim-Sylianco CY, Dayrit F (1989a) Structure of a mutagen from roasted seeds of *Moringa oleifera*. Carcinogenesis 10:1085–1087

Villasenor IM, Lim-Sylianco CY, Dayrit F (1989b) Mutagens from roasted seeds of *Moringa oleifera*. Mutat Res 224(2):209–212

Yang RY, Lin S, Kuo G (2008) Content and distribution of flavonoids among 91 edible plant species. Asia Pac J Clin Nutr 17(S1):275–279

Muntingia calabura

Scientific Name

Muntingia calabura L.

Synonyms

None

Family

Muntingiaceae, also placed in Elaeocarpaceae, Tiliaceae

Common/English Names

Calabur-Tree, Calabura, Capulin, Cotton Candy Tree, Jamaican Cherry, Jamfruit,, Jamfruit Tree, Ornamental Cherry, Panama Berry, Silkwood Tree, Singapore Cherry, Strawberry Tree, West Indian Cherry, Yumansa

Vernacular Names

Brazil: Calabura, Curumi, Curuminzeira, Pau-De-Seda;
Chamorro: Mansanita, Manzanilla, Manzanita;
Colombia: Chitato, Pasito;
Cook Islands: Venevene (Maori);
French: Bois Ramier, Cerisier De Panama;
India: Gasagase Hannina Mara (Kannada), Paachara (Marathi), Ten Pazham (Tamil), Nakkaraegu (Telugu);
Indonesia: Cerri, Kersen, Talok;
Khmer: Krâkhôb Barang;
Laotian: Khoom Sômz, Takhôb;
Malaysia: Kerukup Siam, Buah Ceri;
Naruan: Bin;
Palauan: Budo;
Peru: Cerezo Caspi, Yumanaza;
Philippines: Datiles (Bikol), Zanitas (Ibanag), Seresa, Zanitas (Iloko), Aratiles, Datiles, Latires, Ratiles (Tagalog);
Portuguese: Calbura, Páo De Seda;
Spanish: Bolaina, Cacaniqua, Capulín Blanco, Cereza, Memiso, Memizo, Nigua, Niguito, Yamanaza;
Tahitian: Monomona;
Thai: Takhop Farang;
Tongarevan: Venevene;
Vietnamese: Mât Sâm, Trúng Cá;
Yapese: Budo.

Origin/Distribution

It is native to southern Mexico, Central America, tropical South America to Peru and Bolivia, the Greater Antilles, St. Vincent and Trinidad. The plant has been widely introduced to almost all tropical regions.

T.K. Lim, *Edible Medicinal And Non-Medicinal Plants: Volume 3, Fruits*,
DOI 10.1007/978-94-007-2534-8_62, © Springer Science+Business Media B.V. 2012

Agroecology

M. calabura is a tropical species. It colonizes open disturbed sites in tropical lowlands. It thrives in poor soil, is tolerant of acidic and alkaline conditions and drought. However, it is intolerant of saline conditions. Its seeds are dispersed by birds and fruit bats.

Edible Plant Parts and Uses

The sweet ripe fruit is eaten fresh. In Mexico, the fruits are eaten and sold in local markets. The fruits can be processed into jams or cooked in tarts. The leaf infusion is drunk as a tea-like beverage.

Botany

A fast growing small, evergreen tree reaching heights of 3–12 m, with spreading branches pendent towards the tip. Leaves are simple, alternate, pubescent, sticky, distichous, oblong-ovate to broadly oblong-lanceolate, 4–15 cm long by 1–6 cm wide, with serrulate margins, pointed apex and asymmetrical base, one side rounded and the other acute (Plates 1–3). Stipules are linear, 5 mm long and caduceus. Flowers are bisexual, 2 cm in diameter, white, extra-axillary, solitary or in pairs with 5, green, reflexed, lanceolate, 1 cm long sepals, white, obovate, 1 cm long spreading rotate petals with many stamens with slender filaments and yellow anthers; ovar is stipitate, glabrous and 5-celled, crowned by a persistent, capitate, 5-ridge stigma (Plates 2, 3). The berry fruit is subglobose, about 1.5 cm wide, baccate, smooth, pale green turning red on ripening (Plates 1 and 2), fleshy, sweet and musky-flavoured with many small, 1–2 mm, elliptic greyish yellow seeds.

Plate 2 Fruit and flower

Plate 3 Flowers and leaves

Plate 1 Ripe and unripe fruit

Nutritive/Medicinal Properties

The nutritional value per 100 g of edible portion of the fruit was reported to contain approximately: moisture 77.8 g, protein 0.32 g, fat 1.56 g, fibre 4.6 g, ash 1.14 g, calcium 124 mg, phosphorous 84 mg, iron 1.18 mg, carotene 0.019 mg, thiamine 0.065 mg, riboflavin 0.037 mg, niacin 0.554 mg, and ascorbic acid 80.5 mg. The energy value was 380 kJ/100 g (Morton 1987).

A total of 42 volatile compounds were identified in the vacuum distillation extract of ripe fruits of *Muntingia calabura* which was dominated by alcohols (44.7%), esters (26.5%) and carbonyl compounds (23.3%) (Wong et al. 1996). Steam distillation-extraction resulted in the identification of 56 compounds, among which esters (31.4%), alcohols (15.9%), phenolic compounds (11.3%), sesquiterpenoids (10.6%) and furan derivatives (8.3%) were quantitatively significant. A potent odorant detected was 2-acetyl-1-pyrroline (1.3%). In both isolates methyl salicylate was the most abundant component.

Other phytochemicals isolated from the plant parts are mentioned below together with various pharmacological properties.

Antioxidant Activity

Muntingia calabura methanolic leaf extract showed strong reducing power and significant antioxidant activity (Siddiqua et al. 2010). In the DPPH radical scavenging assay, the IC_{50} value of the extract was found to be 22 μg/ml. The total phenolic content was found to be 0.903 for gallic acid when compare to 2.900 for tannic acid as the calibration standard. Total phenolics of *M. calabura* fruits were found to range from 1,486 mg GAE/100 of fresh weight to 358 mg GAE/100 fresh weight (Preethi et al. 2010). High levels of antioxidant activity of the hexane, chloroform, ethyl acetate, butanol and methanol fruit extracts were observed using 1,1-diphenyl–2-picryl hydroxyl (DPPH) radical, reducing power, ferric ion chelating assay, superoxideanion, and nitric oxide scavenging activity assays. A good correlation between antioxidant activity and total phenolic/flavonoid contents of the fruits was observed.

Anticancer Activity

Studies have shown that the bark, roots and leaves contain anticancer agents – flavanones and flavones from leaves, flavones from the bark and flavones, flavans and biflavans from roots which exhibited cytotoxic effects. From a cytotoxic ether-soluble extract of *Muntigia calabura* roots, 12 new flavonoids were isolated, namely seven flavans 1–7, three flavones 8, 10, and 12, and two biflavans 9 and 11 (Kaneda et al. 1991). Most of the compounds exhibited cytotoxic activity when tested against cultured P-388 leukemia cells, with the flavans being more active than the flavones. In addition, certain of these structurally related flavonoids exerted somewhat selective activities when evaluated against a number of human cancer cell lines. The cytotoxic flavonoids chrysin, 2′, 4′-dihydroxychalcone, and galangin 3, 7-dimethyl ether were isolated from the leaves and stems of *M. calabura* (Nshimo et al. 1993). These compounds were active against one or more of a panel of human and murine cell lines. Also isolated were the inactive compounds, 5, 7-dihydroxy-8-methoxyflavonol, tiliroside and buddlenoid.

Studies in Peru reported on the isolation from ethyl acetate extract of *M. calabura* leaves a flavanone with an unsubstituted B-ring, (2R,3R)-7-methoxy-3,5,8-trihydroxyflavanone (5), as well as 24 known compounds, which were mainly flavanones and flavones (Su et al. 2003). Of the compounds obtained, in addition to compound 5, (2S)-5-hydroxy-7-methoxyflavanone, 2′,4′-dihydroxychalcone, 4,2′,4′-trihydroxychalcone, 7-hydroxyisoflavone and 7,3′,4′-trimethoxyisoflavone were found to induce quinone reductase activity in an assay with cultured Hepa 1c1c7 (mouse hepatoma) cells.

In another study, two new flavones, 8-hydroxy -7,3′,4′,5′-tetramethoxyflavone and 8,4′-dihydroxy-7,3′,5′-trimethoxyflavone, together with 13 known compounds were isolated from the stem bark of *Muntingia calabura* (Chen et al. 2004). Among the isolates, 8-hydroxy-7,3′,4′,5′-tetramethoxyflavone, 8,4′-dihydroxy-7,3′,5′-trimethoxyflavone, and 3-hydroxy-1-(3,5-dimethoxy-4-hydroxyphenyl)propan-1-one demonstrated effective cytotoxicities (ED_{50} values = 3.56, 3.71, and 3.27 μg/ml, respectively) against the P-388 leukemia cell line in-vitro. Two new dihydrochalcones, 2′,4′-dihydroxy-3′-methoxydihydrochalcone, (−)-3′-methoxy-2′,4′,beta-trihydroxydihydrochalcone, a new flavanone, (2S)-(−)-5′-hydroxy-7,3′,4′-trimethoxyflavanone, and a new flavonol derivative, muntingone, along with 16 known compounds, were isolated from the leaves of *Muntingia calabura* (Chen et al. 2005). Among the isolates, (2 S)-5′-hydroxy-7,3′,4′-trimethoxyflavanone, 4′-hydroxy-7-methoxyflavanone, 2′,4′-dihydroxychalcone, and 2′,4′-dihydroxy-3′-methoxychalcone exhibited cytotoxicity (IC_{50} values <4 μg/ml) against P-388 and/or HT-29 (human colon cancer) cell lines in-vitro.

M. calabura leaves were found to possess potential antiproliferative and antioxidant activities that could be attributed to its high content of phenolic compounds (Zakaria et al. 2011). the aqueous and methanol extracts of *M. calabura* inhibited the proliferation of MCF-7 (human breast cancer), HeLa, HT-29, HL-60 and K-562 cancer cells while the chloroform extract only inhibited the proliferation of MCF-7, HeLa, HL-60 (human leukemia) and K-562 (myelogenous leukemia) cancer cells. All extracts of *M. calabura*, which failed to inhibit the MDA-MB-231 breast cancer cells proliferation, did not inhibit the proliferation of 3T3 (normal) fibroblast cells, indicating its safety. All extracts (20, 100 and 500 μg/ml) were found to possess antioxidant activity when tested using the DPPH radical scavenging and superoxide scavenging assays with the methanol, followed by the aqueous and chloroform, extract exhibiting the highest antioxidant activity in both assays. The total phenolic content for the aqueous, methanol and chloroform extracts were 2970.4, 1279.9 and 2978.1 mg/100 g gallic acid, respectively.

Antinociceptive Activity

Studies using animal models also confirmed that the leaves possessed antinociceptive, antiinflammatory and antipyretic activities. The scientists also found that *M. calabura* leaves possessed antinociceptive activity against chemically and thermally induced noxious stimuli (Zakaria et al. 2007a). The bioactive compound(s) responsible for its antinociceptive activity was/were heat-stable and worked partly via the opioid receptor system. Pre-treatment with naloxone (2 and 10 mg/kg) blocked the extract activity in both tests, indicating the involvement of the opioid receptor system in aqueous leaf extract antinociceptive activity. In further studies, the scientists found that *M. calabura* leaves peripheral antinociception involved at least in part, the activation of μ-opioid, β-adrenergic and muscarinic receptors and resisted the effect of extreme acidic and alkaline conditions as well as various enzymes (Zakaria et al. 2008). The leaf extract, administered sub-cutaneously at the concentrations of 5%, 50% and 100%, were found to show significant antinociceptive activity in a concentration-dependent manner. The extract exhibited significant decline in activity when pre-treated sub-cutaneously against 10 mg/kg naloxonazine, 10 mg/kg pindolol and 5 mg/kg atropine, but not 10 mg/kg ß-funaltreaxamine, 10 mg/kg naltrindole, 10 mg/kg phenoxybenzamine, 10 mg/kg bicuculine or 5 mg/kg mecamylamine, respectively. The extract exhibited significant increase in activity after pre-treatment at alkaline pH (pH 9 and 11) while maintaining the activity at the extreme acidic and alkaline conditions (pH 2 and pH 13), respectively. The extract activity was not altered after pre-treatment against α-amylase, protease, lipase or their combination, when compared to the dH20-pre-treated group, respectively. Additional studies using abdominal constriction test in mice demonstrated the involvement of L-arginine/nitric oxide/cyclic guanosine monophosphate (L-arginine/NO/cGMP) pathway in the aqueous extract of *Muntingia calabura* leaves antinociception activity (Zakaria et al. 2006b).

Antiinflammatory/Antipyretic Activities

The same research group using the formalin-, carrageenan-induced paw edema- and brewer's yeast-induced pyrexia tests in rats found that aqueous extract of *Muntingia calabura* leaves at concentrations of 10%, 50% and 100%, exhibited significant antinociceptive, antiinflammatory and antipyretic activities in a concentration-independent manner (Zakaria et al. 2007b). The studies supported the Peruvian folklore claims of its medicinal values.

Hypotensive Activities

Studies conducted in Taiwan found that leaf extract of *M. calubra* had cardiovascular effects. Water soluble extract (WSE) from the leaf of *M. calabura* elicited both a transient and delayed hypotensive effect via the production of nitric oxide in the anesthetized rats (Shih et al. 2006). Intravenous administration of the WSE (10, 25, 50, 75 or 100 mg/kg) produced an initial followed by a delayed decrease in systemic arterial pressure (SAP) in a dose-dependent manner. In contrast, the same treatment, had no appreciable effect on heart rate or the blood gas/electrolytes concentrations. In addition, activation of NO/sGC/cGMP (nitric oxide/soluble guanylate cyclise/cyclic guanosine monophosphate) signaling pathway may mediate the *M. calabura*-induced hypotension. In subsequent studies, the intravenous bolus administration of the n-butanol soluble fraction (BSF) (10–100 mg/kg) from methanol leaf extract of *M. calabura* produced biphasic dose-related antihypertensive and bradycardiac effects in spontaneously hypertensive rats (SHR) (Shih 2009). The cardiovascular depressive effects of BSF treatments were greater in SHR than in normotensive Wistar-Kyoto (WKY) rats. Both the initial and delayed antihypertensive and bradycardiac effects of BSF (25 mg/kg, i.v.) in SHR rats, were significantly blocked by pretreatment with a non-selective nitric oxide (NO) synthase (NOS) inhibitor, a soluble guanylyl cyclase (sGC) inhibitor, or a protein kinase G (PKG) inhibitor. Moreover, the initial effects of BSF in SHR rats were suppressed by pre-treatment with a selective endothelial NOS (eNOS) inhibitor; whereas the delayed effects were attenuated by a selective inducible NOS (iNOS) inhibitor. These results indicated that the BSF from the leaf of *M. calabura* elicited both transient and delayed antihypertensive and bradycardiac actions in SHR, which might be mediated through NO generated respectively by eNOS and iNOS.

Cardioprotective Activity

A recent study in India confirmed the protective effects *of M. calabura* leaf extract against isoproterenol-induced biochemical alterations in myocardial infarction in rats (Nivethetha et al. 2009). Isoproterenol significantly increased the activities of creatine phosphokinase, lactate dehydrogenase and the transaminases (aspartate transaminase and alanine transaminase), in serum with a concomitant decrease in these enzymes in tissue. Pretreatment with the aqueous leaf extract of *M. calabura* at a dose of 300 mg/kg body weight for 30 days had a significant effect on the activities of marker enzymes compared to the other groups. Serum uric acid level, which increased on isoproterenol administration, registered near normal values on treatment with the leaf extract under study.

Antibacterial Activity

Studies also reported that *M. calabura* possessed potential antibacterial property was comparable to the standard antibiotics used (Zakaria et al. 2006a, c). At all concentrations (10,000, 40,000, 70,000 and 100,000 ppm) tested, the aqueous extract of *M. calabura*, was effective against *Staphylococcus aureus* and *Kosuria rhizophila* while the methanol extract was effective against *Shigella flexneri*, *Bacillus cereus*, *Staphylococcus aureus*, *Proteus vulgaris*, *Aeromonas hydrophila* and *Kosuria rhizophila*. This activity was not observed with the chloroform extract. At the concentration of 40,000 ppm and above, the aqueous extract exhibited significant antibacterial activity against *Corneybacterium diphtheria*, *Proteus vulgaris*, *Staphylococcus epidermidis* and *Aeromonas hydrophila*; the methanol extract was effective against *Corneybacterium diphtheria and Listeria monocytogenes*; and the chloroform extract was effective against *Staphylococcus aureus*. The methanol extract of *M. calabura* leaves inhibited MSSA (methicillin sensitive *Staphylococcus aureus)* with MIC = 1,250 μg/ml, MBC = 1,250 μg/ml, and MRSA (methicillin-resistant *Staphylococcus aureus)* with MIC = 1,250 μg/ml and MBC = 1,250 μg/ml (Zakaria et al. 2010). The ethyl acetate partition of the methanol extract exhibited effective antibacterial activities with the MIC/MBC value of 156 and 313 μg/ml against MSSA and MRSA,. The ethyl acetate partition after fractionation process yielded 15 fractions (A1-A15) of which only fractions A9 to A15 effectively inhibited the growth of both MSSA and MRSA with MIC/MBC values ranging from 78 to 2,500 μg/ml.

Traditional Medicinal Uses

In traditional Filipino medicine, the flowers are regarded to have antispasmodic and emollient property and the flower decoction is used as an antiseptic and to treat abdominal cramps. Flower infusion is also taken to relieve headache and the first symptoms of a cold.

Other Uses

The results of a recent study indicated that *Muntingia calabura* was an attractive candidate for removing cationic dyes from the dye wastewater (Santhi et al. 2009). A new, low cost, locally available granular biomaterial prepared from a mixture of leaves, fruits and twigs of *Muntingia calabura* was able to remove cationic dyes, methylene blue, methylene red and malachite green from aqueous solution. The release of dyes into wastewater by various industries poses serious environmental problems due the persistent and recalcitrant nature of the dyes.

The timber from Jamaican cherry is reddish-brown, firm, compact, durable and light-weight and can be used for general carpentry, interior sheathing, casks and boxes. In Brazil, it is being considered as pulp for paper making. It could also be used as firewood as it ignites quickly, burns with intense heat and gives off very little smoke. The bark is stripped and used to produce ropes. The tree could be used for reforestation projects due to its ability to grow in poor soil and its effective establishment and rapid growth. It is also planted as ornamental trees along avenues.

Comments

In some areas it is deemed as a noxious weed because of its rapid establishment and growth on all soil types.

Selected References

Burkill IH (1966) A dictionary of the economic products of the Malay Peninsula. Revised reprint. 2 volumes. Ministry of agriculture and co-operatives, Kuala Lumpur, vol 1 (A–H) pp 1–1240, vol 2 (I–Z) pp 1241–2444

Chen J-J, Lin R-W, Duh C-Y, Huang H-Y, Chen I-S (2004) Flavones and cytotoxic constituents from the stem bark of *Muntingia calabura*. J Chin Chem Soc 51(3):665–670

Chen JJ, Lee HH, Duh CY, Chen IS (2005) Cytotoxic chalcones and flavonoids from the leaves of *Muntingia calabura*. Planta Med 71(10):970–973

Kaneda N, Pezzuto JM, Soejarto DD, Kinghorn AD, Farnworth NR, Santisuk T, Tuchinda P, Udchachon J, Reutrakul V (1991) Plant anticancer agents, XLVIII. New cytotoxic flavonoids from *Muntingia calabura* roots. J Nat Prod 54(1):196–206

Morton JF (1987) Jamaica cherry. In: Morton JF (ed) Fruits of warm climates. Florida Flair Books, Miami, pp 65–69

National Academy of Sciences (1980) Firewood crops: shrub and tree species for energy production. National Academy of Sciences, Washington, DC, 236 pp

Nivethetha M, Jayasri J, Brindha P (2009) Effects of *Muntingia calabura* L. on isoproterenol-induced myocardial infarction. Singapore Med J 50(3):300–302

Nshimo CM, Pezzuto JM, Kinghorn AD, Farnsworth NR (1993) Cytotoxic constituents of *Muntingia calabura* leaves and stems collected in Thailand. Pharm Biol 31(1):77–81

Pacific Island Ecosystems at Risk (PIER) (1999) *Muntingia calabura* L. Muntinginaceae. http://www.hear.org/Pier/species/muntingia_calabura.htm

Pongpangan S, Poobrasert S (1991) Edible and poisonous plants in Thai forests. O.S. Printing House, Bangkok, 176 pp

Preethi K, Vijayalakshmi N, Shamna R, Sasikumar JM (2010) In vitro antioxidant activity of extracts from fruits of *Muntingia calabura* Linn. from India. Pharmacognosy J 2(14):11–18

Santhi T, Manonmani S, Ravi S (2009) Uptake of cationic dyes from aqueous solution by biosorption onto granular *Muntingia calabura*. E-J Chem 6(3):737–742

Shih CD (2009) Activation of nitric oxide/cGMP/PKG signaling cascade mediates antihypertensive effects of *Muntingia calabura* in anesthetized spontaneously hypertensive rats. Am J Chin Med 37(6):1045–1058

Shih CD, Chen JJ, Lee HH (2006) Activation of nitric oxide signaling pathway mediates hypotensive effect of *Muntingia calabura* L. (Tiliaceae) leaf extract. Am J Chin Med 34(5):857–872

Siddiqua A, Premakumari KB, Sultana R, Vithya S (2010) Antioxidant activity and estimation of total phenolic content of *Muntingia calabura* by colorimetry. Int J Chem Tech Res 2(1):205–208

Stone BC (1970) The flora of Guam. Micronesica 6:1–659

Stuart GU (2010) Philippine alternative medicine. Manual of some Philippine medicinal plants. http://www.stuartxchange.org/OtherHerbals.html

Su N, Jung Park E, Vigo JS, Graham JG, Cabieses F, Fong HH, Pezzuto JM, Kinghorn AD (2003) Activity-guided isolation of the chemical constituents of *Muntingia calabura* using a quinone reductase induction assay. Phytochemistry 63(30):335–341

Subhadrabanhdu S (2001) Under utilized tropical fruits of Thailand. FAO Rap publication 2001/26, Bangkok

Verheij EWM (1992) *Muntigia calabura*. In: Verheij EWM, Coronel RE (eds) Plant resources of South-East Asia. No. 2: edible fruits and nuts. Prosea Foundation, Bogor, pp 223–225

Wong KC, Chee SG, Er CC (1996) Volatile constituents of the fruits of *Muntingia calabura* L. J Essent Oil Res 8(4):423–426

Zakaria ZA, Fatimah CA, Mat Jais AM, Zaiton H, Henie EFP, Sulaiman MR, Somchit MN, Thenamutha M, Kasthuri D (2006a) The in vitro antibacterial activity of *Muntingia calabura* extracts. Int J Pharmacol 2(3):290–293

Zakaria ZA, Sulaiman MR, Mat Jais AM, Somchit MN, Jayaraman KV, Balakhrisnan G, Fatimah CA (2006b) The antinociceptive activity of *Muntingia calabura* aqueous extract and the involvement of L-arginine/nitric oxide/cyclic guanosine monophosphate pathway in its observed activity in mice. Fundam Clin Pharmacol 20(4):365–372

Zakaria ZA, Zaiton H, Henie EFP, Mat Jais AM, Kasthuri D, Thenamutha M, Othman FW, Nazaratulmawarina R, Fatimah CA (2006c) The in vitro antibacterial activity of *Corchorus olitorius* and *Muntingia calabura* extracts. J Pharmacol Toxicol 1(2):108–114

Zakaria ZA, Mustapha S, Sulaiman MR, Mat Jais AM, Somchit MN, Fatimah CA (2007a) The antinociceptive action of aqueous extract from *Muntingia calabura* leaves: the role of opioid receptors. Med Princ Pract 16:130–136

Zakaria ZA, Nor Hazalin NA, Zaid SN, Ghani MA, Hassan MH, Gopalan HK, Sulaiman MR (2007b) Antinociceptive, anti-inflammatory and antipyretic effects of *Muntingia calabura* aqueous extract in animal models. Nat Med 61(4):443–448

Zakaria ZA, Somchit MN, Sulaiman MR, Mat Jais AM, Fatimah CA (2008) Effects of various receptor antagonists, pH and enzymes on *Muntingia calabura* antinociception in mice. Res J Pharmacol 2(3):31–37

Zakaria ZA, Sufian AS, Ramasamy K, Ahmat N, Sulaiman MR, Arifah AK, Zuraini A, Somchit MN (2010) In vitro antimicrobial activity of *Muntingia calabura* extracts and fractions. Afr J Microbiol Res 4(4):304–308

Zakaria ZA, Mohamed AM, Jamil NS, Rofiee MS, Hussain MK, Sulaiman MR, Teh LK, Salleh MZ (2011) In vitro antiproliferative and antioxidant activities of the extracts of *Muntingia calabura* leaves. Am J Chin Med 39(1):183–200

Musa acuminata subsp. *zebrina*

Scientific Name

***Musa acuminata* L. A. Colla subsp. *zebrina* (Van Houtte) R.E. Nasution.**

Synonyms

Musa acuminata subsp. *rubrobracteata* M. Hotta, *Musa acuminata* var. *alasensis* Nasution, *Musa acuminata* var. *bantamensis* Nasution, *Musa acuminata* var. *breviformis* Nasution, *Musa acuminata* var. *cerifera* (Backer) Nasution, *Musa acuminata* var. *longipetiolata* Nasution, *Musa acuminata* var. *nakaii* Nasution, *Musa acuminata* var. *rutilipes* (Backer) Nasution, *Musa acuminata* var. *violacea* Kurz, *Musa acuminata* var. *zebrina* (Van Houtte ex Planch.) Nasution, *Musa brieyi* De Wild., *Musa cavendishii* Lamb., *Musa cavendishii* var. *hawaiiensis* N.G.Teodoro, *Musa cavendishii* var. *pumila* N.G.Teodoro, *Musa cerifera* (Backer) Nakai, *Musa chinensis* Sweet nom. nud., *Musa javanica* Nakai, *Musa minor* Nakai, *Musa rhinozerotis* Kurz, *Musa rumphiana* Kurz, *Musa* × *paradisiaca* var. *pumila* G.Forst., *Musa* × *sapientum* var. *pumila* (N.G.Teodoro) Merr., *Musa simiarum* Miq., *Musa simiarum* var. *violacea* Kurz, *Musa sinensis* Sagot ex Baker, *Musa sundaica* Nakai, *Musa* × *paradisiaca* var. *pumila* Blanco nom illeg., *Musa zebrina* Van Houtte ex Planch., *Musa zebrina* f. *cerifera* Backer, *Musa zebrina* f. *rutilipes* Backer.

Family

Musaceae

Common/English Names

Banana Rojo, Blood Banana, Blood Leaf Banana, Red Banana Tree

Vernacular Name

Indonesia: Pisang Darah.

Origin/Distribution

Distribution of this species is found in Tanzania, tropical and subtropical Asia.

Agroecology

It grows in full sun to partial shade in a moist, well-drained soil rich in organic matter. The species is not cold hardy. For more details on agroecology refer to the genus *Musa*.

T.K. Lim, *Edible Medicinal And Non-Medicinal Plants: Volume 3, Fruits*,
DOI 10.1007/978-94-007-2534-8_63, © Springer Science+Business Media B.V. 2012

Edible Plant Parts and Uses

Fruits are edible but not very palatable. The species is regarded with the best eating flower head, tender young leaf buds and pith which are consumed as tasty vegetables.

Botany

A slender, erect, evergreen herbaceous perennial with a shiny not waxy, yellow-green pseudostem growing up to 3 m high with suckers close to the parent plant. Petiole up to 30–45 cm, petiole canal is open with spreading margins. The leaves are oblong, dark green with irregular bronze-red to purple blotches on the top surface and a prominent midrib which is pale green on the upper side and brownish on the underside (Plate 1). Inflorescence is terminal, oblique (30°) to horizontal on a 30 cm long, hairy peduncle and with horizontal rachis. Bracts, overlap when young, purple brown on external surface and orange red internal surface. Bracts are revolute before falling.

Plate 1 Red-green variegated leaves

Male bud top shaped. Male flowers with cream tepals with orangey-yellow lobes, 5 stamens. Female flowers with similar tepals, staminodes, cream coloured ovary and bright yellow style. Fruit curved upwards obliquely 45°, <15 cm long, with pointed apex and pronounced ridges. Twelve fruits per bunch. Fruit is orange coloured when ripe with a cream-coloured pulp and containing >20 angular seeds.

Nutritive/Medicinal Properties

No information has been published on its nutritive or medicinal values.

Other Uses

The species is cultivated more for its ornamental value than as an edible banana.

Comments

The species is propagated by the removal of offsets or from seed.

Selected References

Daniells J, Jenny C, Tomekpe K (2001) Musalogue: a catalogue of *Musa* germplasm. Diversity in the genus *Musa*. In: Arnaud E, Sharrock S (comp) International Network for the Improvement of Banana and Plantain, Montpellier

Govaerts R, Häkkinen (2009) World checklist of musaceae. In: The board of trustees of the royal botanic gardens, Kew. http://www.kew.org/wcsp/. Accessed 11 Nov 2009

Jarret RL (1987) Biochemical/genetic markers and their uses in the genus *Musa*. In: Persley GJ, De Langhe EA (eds) Banana and plantain breeding strategies, proceedings of an international workshop held at Cairns, Australia October 1986. ACIAR proceedings no. 21

Nasution RE (1991) A taxonomic study of *Musa acuminata* Colla with its intraspecific taxa in Indonesia. Memoirs Tokyo Univ Agric 32:1–122

Nelson SC, Ploetz RC, Kepler AK (2006) *Musa* species bananas and plantains, version 2.2. In: Elevitch CR (ed) Species profiles for Pacific Island agroforestry. Permanent Agriculture Resources (PAR), Hōlualoa. http://www.traditionaltree.org

Musa acuminata (AA Group) 'Lakatan'

Scientific Name

Musa acuminata **Colla (AA Group) 'Lakatan'.**

Synonyms

Musa acuminata Colla (Cavendish Group) cv. 'Lacatan', *Musa* × *paradisiaca* L. ssp. *sapientum* (L.) Kuntze var. *lacatan* Blanco.

Family

Musaceae

Common/English Names

Lakatan Banana, Pisang Berangan, Pisang Barangan

Vernacular Names

Indonesia: Pisang Barangan, Pisang Barangan Merah/Kuning;
Malaysia: Pisang Berangan, Pisang Berangan Merah/Kuning;
Philippines: Lakatan, Mapang;
Thailand: Kluai Hom Maew, Kluai Ngang Phaya;
Vietnam: Pisang Berangan – Malaysia (top) Pisang Barangan – Indonesia (Bottom left).

Origin/Distribution

The Lakatan Banana cultivar is reported to have originated from the Philippines.

Agroecology

Bananas and plantains including lakatan banana are adapted to the warm and humid tropics and subtropics between latitudes 30°N and 30°S. It thrives best in areas with mean annual temperature of 26–30°C. It will tolerate mean maximum temperature of 35–37°C. Banana is frost sensitive. At low temperatures of 16°C growth is impeded and at temperature of 13°C or lower chilling injury occurs, growth is halted, plant tissue is damaged and at temperature below −2°C the plant is killed. Banana prefers areas with mean annual rainfall of over 2,000–4,000 mm evenly distributed throughout the year. However, it will grow in areas with 500 mm rainfall with supplemental irrigation and if the soil is fertile. Banana does best in full sun and in flat lowlands but will also thrive on north to north-west facing gentle slopes. It grows from sea level to 1,800 m altitude. Banana plants are susceptible to strong winds and need to be protected in a sheltered locality or by good wind breaks which also protect against cold winds. Areas prone to frequent typhoon or cyclonic winds should be avoided. Strong winds above 25–50 km/h will cause considerable leaf shredding, leaf drying, distortion of

T.K. Lim, *Edible Medicinal And Non-Medicinal Plants: Volume 3, Fruits*,
DOI 10.1007/978-94-007-2534-8_64, © Springer Science+Business Media B.V. 2012

the crown and blow-over. Banana tolerates a wide range of soils, but well drained, fairly-deep, fertile loams are preferable. Heavy, clayey soils are suboptimal, especially if they are low in organic matter and aeration. Bananas can tolerate water-logged soils but will perform and produce poorly. Where water-logging is prevalent, bananas and plantains are grown on raised beds. Shallow, sandy soils should be avoided unless compost, mulch or other rich organic matter can be placed over the root zone. Bananas can be grown in tidal flats and will withstand salt spray but will yield poorly. Banana tolerates a wide range of soil acidity, pH 5.0–7.5 is optimal.

Edible Plant Parts and Uses

Lakatan or Pisang berangan is the most popular dessert banana cultivar in Malaysia and the Philippines. It is also relish as a dessert fruit in Indonesia. For more information on edible uses refer to notes under *Musa acuminata* (AAA Group) 'Dwarf Cavendish'.

Botany

Lakatan banana is a succulent, herbaceous plant with robust, shiny green to purplish-red pseudostem 3–4 m high and a girth of 50–80 cm. Leaves are large, oblong, entire, green, waxy surfaced, arranged spirally with yellow to green dorsal mid rib, petiole about 30 long and petiolar canal open with spreading margins. Inflorescence a terminal spike shooting out from the heart in the tip of the stem, complete with Female flowers occupy the lower 5–15 rows; above them may be some rows of hermaphrodite or neuter flowers; male flowers are borne in the upper rows; peduncle hairy about 45 cm long, rachis falling vertically. Bracts ovate, overlapping when young, externally purple-red, internally pink-purple. Male flower – compound tepal yellow, style slightly curved, stigma and ovary yellow or cream. Bunch weight varies from 12 to 20 kg with 8–12 hands and 12–20 fingers per hand. Fruit medium to large, 13–18 cm by 3.3–3.6 cm, rounded in transverse section, attractive golden-yellow, smooth skin, pulp yellow, fine, firm, aromatic and very sweet (Plate 1a, b). Flavour is excellent.

Nutritive/Medicinal Properties

Proximate nutrient composition of the edible portion (per 100 g) of pisang berangan was reported by Tee et al. (1997) as follows: energy 0.3 kcal., water 73.1 g, protein 1.0 g, fat 0.3 g, carbohydrate 24.2 g, fibre 0.5 g, ash 0.9 g, Ca 0 mg, P 5 mg, Fe 0.1 mg, Na 18 mg, K 233 mg, carotenes 219 ug, vitamin A 37 ug RE, vitamin B-1 0.03 mg, vitamin B-2 0.07 mg, niacin 0.8 mg and vitamin C 5.6 mg. Pisang berangan, (AA Group 'Lakatan') cultivar from Malaysia

Plate 1 (**a**) and (**b**) Ripe pisang berangan fruits

was found to contain 230 μg/100 g of β-carotene (Englberger et al. 2003).

For more information on nutritive values and medicinal properties refer to notes under *Musa acuminata* (AAA Group) 'Dwarf Cavendish'.

Other Uses

For information on other uses of banana refer to notes under *Musa acuminata* (AAA Group) 'Dwarf Cavendish'.

Comments

The cultivar is propagated from suckers.

Selected References

Arnaud E, Horry JP (eds) (1997) Musalogue: a catalogue of *Musa* germplasm. Papua New Guinea collecting missions, 1988–1989. International Network for the Improvement of Banana and Plantain, Montpellier

Daniells J, Jenny C, Tomekpe K (2001) Musalogue: a catalogue of *Musa* germplasm. Diversity in the Genus *Musa*. In: Arnaud E, Sharrock S (comp) International Network for the Improvement of Banana and Plantain, Montpellier. www.inibap.org/publications/musalogue.pdf

Dela Cruz FS, Gueco LS, Damasco OP, Huelgas VC, Banasihan IG, Lladones RV, Van den Bergh I, Molina AB (2007) Catalogue of introduced and local banana cultivars in the Philippines. IPB-UPLB, Bioversity – Philippines, and DA-BAR, Los Baños, 59 pp

Edison HS, Sutanto A, Hermanto C, Lakuy H, Rumsarwir Y (2002) The exploration of Musaceae in Irian Jaya (Papua) 2002. Research Institute for Fruits, Central Research Institute for Horticulture – INIBAP

Englberger L, Darnton-Hill I, Coyne T, Fitzgerald MH, Marks GC (2003) Carotenoid-rich bananas: a potential food source for alleviating vitamin A deficiency. Food Nutr Bull 24(4):303–318

Espino RRC, Jamaluddin SH, Silayoi B, Nasution RE (1992) *Musa* L. (Edible cultivars). In: Verheij EWM, Coronel RE (eds) Plant resources of South-East Asia 2. Edible fruits and nuts. Prosea Foundation, Bogor, pp 225–233

Hautea DM, Molina GC, Balatero CH, Coronado NB, Perez EB, Alvarez MTH, Canama AO, Akuba RH, Quilloy RB, Frankie RB, Caspillo CS (2004) Analysis of induced mutants of Philippine bananas with molecular markers. In: Jain SM, Swennen R (eds) Banana improvement: cellular, molecular biology, and induced mutations. Science Publishers, Enfield

Nasution RE (1994) *Musa* L. In: Siemonsma JS, Kasem P (eds) Plant resources of South-East Asia 8. Vegetables. Prosea Foundation, Bogor, pp 215–217

Nelson SC, Ploetz RC, Kepler AK (2006) *Musa* species bananas and plantains, version 2.2. In: Elevitch CR (ed) Species profiles for Pacific Island agroforestry. Permanent Agriculture Resources (PAR), Hōlualoa. http://www.traditionaltree.org

Ploetz RC, Kepler AK, Daniells J, Nelson SC (2007) Banana and plantain – an overview with emphasis on Pacific Island cultivars, version 1. In: Elevitch CR (ed) Species profiles for Pacific Island agroforestry. Permanent Agriculture Resources (PAR), Hōlualoa. http://www.traditionaltree.org

Robinson JC (1996) Bananas and plantains. Crop production science in horticulture. CAB International, Wallingford, 238 pp

Simmonds NW, Shepherd K (1955) The taxonomy and origins of the cultivated bananas. J Linn Soc Bot 55:302–312

Stover RH, Simmonds NW (1987) Bananas, vol 3. Longmans, London

Tee ES, Noor MI, Azudin MN, Idris K (1997) Nutrient composition of Malaysian foods, 4th edn. Institute for Medical Research, Kuala Lumpur, 299 pp

Uma S, Kalpana S, Sathiamoorthy S, Kumar V (2005) Evaluation of commercial cultivars of banana (*Musa* spp.) for their suitability for the fibre industry. Plant Genet Resour Newsl 142:29–35

Valmayor RV, Wagih ME (1996) *Musa* L. (Plantain and cooking banana). In: Flach M, Rumawas F (eds) Plant resources of South-East Asia 9, Plants yielding non-seed carbohydrates. Prosea Foundation, Bogor, pp 126–131

Valmayor RV, Silayoi B, Jamaluddin SH, Kusumo S, Espino RRC, Pascua OC (1990) Commercial banana cultivars in ASEAN. In: Hassan A, Pantastico EB (eds) Banana: fruit development, postharvest physiology, handling and marketing in ASEAN. ASEAN Food Handling Bureau, Kuala Lumpur, 159 pp

Valmayor RV, Jamaluddin SH, Silayoi B, Kusumo S, Danh LD, Pascua OC, Espino RRC (2000) Banana cultivar names and synonyms in Southeast Asia. International Network for the Improvement of Banana and Plantain – Asia and the Pacific Office, Los Banos

Wu D, Kress WJ (2000) Musaceae. In: Wu ZY, Raven PH (eds) Flora of China: flagellariaceae through marantaceae, vol 24. Science Press/Missouri Botanical Garden Press, Beijing/St. Louis, pp 297–313

Musa acuminata (AA Group) 'Sucrier'

Scientific Name

Musa acuminata **Colla (AA Group) 'Sucrier'**.

Synonyms

None

Family

Musaceae

Common/English Names

Banana, Gold Banana, Mas Banana, Pisang Mas

Vernacular Names of Cultivars

Brazil: Banana Ouro;
Columbia: Bocadillo;
Ecuador: Orito;
Federated States of Micronesia: Kudu;
French Polynesia: Peru, Fig, Tinito;
Guyana: Parika;
Hawaii: Lady's Finger;
India: Surya Kadali;
Indonesia: Mow Mei, Sarmi (Irian Jaya), Pisang Lempung, Pisang Mas, Pisang Susu;
Latin America: Cambur Titiaro, Date, Dedo De Dama, Fig, Golden Early, Guineo Blanco, Manices, Nino, Rose;
Malaysia: Pisang Mas, Pisang Susu;
Myanmar: Sagale Nget-Pyaw;
Philippines: Amas, Caramelo, Kamoros;
Pohnpei: Kudud;
Thailand: Kluai khai;
United States: Nino (Florida);
Vietnam: Choi Trung;
West Indies: Sucrier, Sucrier Fig, Fig, Datil, Honey, de Rosa, Fig Sucre.

Origin/Distribution

This cultivar originated in Malaysia. Sucrier is the most widely cultivated AA cultivar and is one of the world's most popular local bananas.

Agroecology

See notes in *Musa acuminata* (AA Group) 'Lakatan'.

Edible Plant Parts and Uses

An excellent and popular, sweet dessert banana that is highly esteemed in Malaysia, Indonesia and Thailand. The flowers are also eaten.

T.K. Lim, *Edible Medicinal And Non-Medicinal Plants: Volume 3, Fruits*,
DOI 10.1007/978-94-007-2534-8_65, © Springer Science+Business Media B.V. 2012

Plate 1 Unripe bunch of Pisang lempung

Plate 2 Ripe bunch of Pisang lempung

Botany

A short, erect herbaceous banana plant with pseudostem of 2.0–3.6 m high, girth of 40–65 cm, shiny green or yellowish green, with underlying purple-red pigmentation and watery or white sap. Suckers 3–6. Leaves large, elliptic-oblong with pinnate venations and entire margin, with brown blotches at the petiole base, petiole canal on the third leaf spreading or straight with erect margins. Inflorescence arising from the centre of the pseudostem, pendent, peduncle hairy, complete, male flowers at the distal end, female flowers at proximal end and neuter flowers in the middle. Male bracts large, ovate, imbricate, intermediate or pointed, external surface reddish-purple or brownish-purple and inside surface cream or white. Male flowers creamy-white. Bunch weight is 8–12 kg with 5–9 hands and 14–18 fingers per hand (Plates 1 and 2). Fruit is small, 6–12 cm by 3–4 cm diameter, plump with tapering or beaked apices, rounded in cross-section. The skin is thin, smooth, golden yellow when ripe (Plates 3–6). The flesh is firm deep yellow to orangey, very fragrant and very sweet.

Plate 3 Ripe hand of Pisang lempung

Nutritive/Medicinal Properties

Proximate nutrient composition of the edible portion (per 100 g) of pisang mas was reported by Tee et al. (1997) as follows: energy 99 kcal.,

Plates 4 and 5 Ripe hands of pisang mas (top view and bottom view)

Plate 6 Pisang susu, muntul

water 73.0 g, protein 1.4 g, fat 0.3 g, carbohydrate 22.9 g, fibre 1.7 g, ash 0.7 g, Ca 0 mg, P 3 mg, Fe 0.2 mg, Na 10 mg, K 342 mg, carotenes 380 μg, vitamin A 63 μg RE, vitamin B-1 0.04 mg, vitamin B-2 0.08 mg, niacin 0.5 mg and vitamin C 8.3 mg.

Ten yellow or yellow/orange-fleshed cultivars (Asupina, Kirkirnan, Pisang Raja, Horn Plantain, Pacific Plantain, Kluai Khai Bonng, Wain, Red Dacca, Lakatan, and Sucrier) were found to have significant carotenoid levels, potentially meeting half or all of the estimated vitamin A requirements for a non-pregnant, non-lactating adult woman within normal consumption patterns (Englberger et al. 2006). All were acceptable for taste and other attributes.

Five methyl esters obtained after transesterification of extracts of three Thai banana varieties, namely "Kluai Khai"(KK) Sucrier type, "Kluai Namwa"(KN) and "Kluai Hom"(KH) were identified (Meechaona et al. 2007). The components found in KK, KN and KH respectively were hexadecanoic acid methyl ester (43.17, 29.18, 30.57%); 9, 12, 15-octadecatrienoic acid methyl ester(35.93,30.46,39.68%);9,12-octadecadienoic acid methyl ester (14.35, 36.10, 21.82%); 9-hexadecanoic acid methyl ester (3.76, 3.34, 3.32%) and octadecanoic acid methyl ester (2.79, 0.92, 4.60%). The banana oils of KK, KN and KH showed moderate antioxidant activities as evaluated by DPPH assay and the IC_{50} values were determined as KK 90 μg/ml, KN 73 μg/ml, KH 8 μg/ml in comparison to vitamin E.

For more health and medicinal attributes seed notes under in *Musa acuminata* (AA Group) 'Lakatan'.

Other Uses

See notes in *Musa acuminata* (AA Group) 'Lakatan'.

Comments

Like all bananas, Sucrier cultivars are propagated vegetatively from suckers.

Selected References

Daniells J, Jenny C, Tomekpe K (2001) Musalogue: a catalogue of *Musa* germplasm. Diversity in the genus *Musa*. In: Arnaud E, Sharrock S (eds) International Network for the Improvement of Banana and Plantain, Montpellier

Edison HS, Sutanto A, Hermanto C, Rumsarwir Y, Lakuy H (2002) The exploration of Musaceae in Irian Jaya (Papua). Central Research Institute for Horticulture –INIBAP, Solok Research Institute for Fruits Solok, Indonesia, 58 pp

Englberger L, Wills RB, Blades B, Dufficy L, Daniells JW, Coyne T (2006) Carotenoid content and flesh color of selected banana cultivars growing in Australia. Food Nutr Bull 27(4):281–291

Espino RRC, Jamalussin SH, Silayoi B, Nasution RE (1992) *Musa* L. (Edible cultivars). In: Verheij EWM, Coronel RE (eds) Plant resources of South-East Asia no 2. Edible fruits and nuts. PROSEA, Bogor, pp 225–233

Gowen S (ed) (1995) Bananas and plantains. Chapman and Hall, London, 612 pp

Meechaona R, Sengpracha W, Banditpuritat J, Kawaree R, Phutdhawong W (2007) Fatty acid content and antioxidant activity of Thai bananas. Maejo Int J Sci Tech 1(2):222–228

Ploetz RC, Kepler AK, Daniells J, Nelson SC (2007) Banana and plantain – an overview with emphasis on Pacific Island cultivars, version 1. In: Elevitch CR (ed) Species profiles for Pacific Island agroforestry. Permanent Agriculture Resources (PAR), Hōlualoa. http://www.traditionaltree.org

Robinson JC (1996) Bananas and plantains. Crop production science in horticulture. CAB International, Wallingford, 238 pp

Stover RH, Simmonds NW (1987) Bananas, 3rd edn, Tropical agricultural series. Longman, Essex, 468 pp

Tee ES, Noor MI, Azudin MN, Idris K (1997) Nutrient composition of Malaysian foods, 4th edn. Institute for Medical Research, Kuala Lumpur, 299 pp

Valmayor RV, Wagih ME (1996) *Musa* L. (Plantain and cooking banana). In: Flach M, Rumawas F (eds) Plant resources of South-East Asia no. 9. Plants yielding non-seed carbohydrates. Prosea Foundation, Bogor, pp 126–131

Valmayor RV, Jamaluddin SH, Silayoi B, Kusumo S, Danh LD, Pascua OC, Espino RRC (2000) Banana cultivar names and synonyms in Southeast Asia. International Network for the Improvement of Banana and Plantain – Asia and the Pacific Office, Los Banos

Musa acuminata (AAA Group) 'Dwarf Cavendish'

Scientific Name

Musa acuminata **Colla (AAA Group) 'Dwarf Cavendish'.**

Synonyms

Musa acuminata L. A. Colla, *Musa acuminata* Colla (Cavendish Group) cv. 'Dwarf Cavendish', *Musa cavendishii* Lambert & Paxton var. *nana, Musa chinensis* R. Sweet, *Musa nana* auct. non J. de Loureiro, *Musa nana* J. de Loureiro, *Musa sinensis* P. A. Sagot, *Musa sinensis* P. A. Sagot ex J. G. Baker, *Musa sinensis* R. Sweet ex P. A. Sagot, *Musa* × *paradisiaca* L. cultigroup Dwarf Cavendish.

Family

Musaceae

Common/English Names

Chinese Banana, Canary Banana, Banana, Giant Cavendish Banana (Williams), Dwarf Cavendish, Grande Naine (Chiquita Banana), Robusta

Vernacular Names of Local Cultivars

Giant Cavendish

Australia: Giant Chinese, Grande Nain, Mons Mari, Tall Mons Mari;
Brazil: Nanico;
Central America: Robusta, Valery;
China: Bijiaw;
Cook Islands: Amoa Kauare, Amoa Taunga;
Dominica: Porto Rique;
Egypt: Maghrabi, Williams;
Fiji: Vaimama Leka, Veimama;
French Polynesia: Hamoa;
Guadeloupe: Poyo;
Hawaii: Williams, Williams Hybrid, Robusta, Taiwan, Valery;
India: Harichal;
Indonesia: Pisang Ambon Hijau;
Jamaica: Robusta, Valery,
Malaysia: Pisang Cina, Pisang Masak Hijau, Pisang Serendah;
New Guinea: Saina;
Philippines: Tumok;
Pohnpei: Utin Wai;
Samoa: Palagi (Also General Name For Cavendish Group);
Surinam: Congo;
Thailand: Kluai Hom Kiau, Kluai Khlong Chang;

T.K. Lim, *Edible Medicinal And Non-Medicinal Plants: Volume 3, Fruits*,
DOI 10.1007/978-94-007-2534-8_66, © Springer Science+Business Media B.V. 2012

Tonga: Siaine (Also Name For General Cavendish Group), Siaine Ha'Amoa;
West Indies: Giant Governor, Nain Gant, Porto Rique, Robusta;
Vietnam: Chuoi Tieu Nho.

Grande Nain (Gran Nain)

Indonesia: Pisang Ambon Jepang;
Philippines: Umalog;
Vietnam: Chuoi Va Huong.

Pisang Masak Hijau

Other common names:
'Hamakua' (Hawai'i); 'Bungulan'(Philippines); 'Lacatan' (western tropics); 'Pisang Buai','Pisang Embun Lumut' (Malaysia); 'Pisang Ambon Loemoet'(Indonesia); 'Kluai Hom Kiau' (Thailand); 'Thihmwe'(Burma/Myanmar); 'Sapumal Anamalu' (Sri Lanka); 'Bout Rond' and 'Giant Fig', 'Congo' (West Indies); 'Mestica'(Brazil); 'Monte risto' (Puerto Rico), 'Chuoi Tieu Cao#1' (Vietnam); 'Siaine' (Tonga, also general name for Cavendish Group); 'Amoa Kauare' (Cook Is.); 'Veimama' (Fiji) Dwarf Cavendish'.

Other common names:
'Cavendish', 'Chinese', 'Dwarf Chinese', 'Pake' (Hawai'i); 'Poot', 'Tampohin', Tampihan, 'Sulay Baguio' (Philippines); 'Jainaleka' (Fiji); 'Fa'i Palagi' (also refers to 'Giant Cavendish' in Samoa) (Samoa); 'Chuoi Duu' (IndoChina); 'Canary Banana', 'Dwarf Cavendish' (general); 'Pisang Serendah' (Malaysia); 'Pisang badak' (Indonesia); 'Kaina Vavina' (Papua New Guinea); 'Dwarf Cavendish' (Australia); 'Ai Keuk Heung Ngar Tsiu'(Hong Kong); 'Kluai Hom Khieo Khom', 'Kluai Hom Kom' (Thailand), 'Wet-ma-lut' (Burma/Myanmar); 'Banane Gabou' (Seychelles); 'Pacha Vazhai', 'Mauritius', 'Vamanakeli', Pachawara', 'Basrai', 'Kabulee', 'Bhusawal', 'Jahaji' (India); 'Binkehel', 'Nanukehel', 'Pandi' (Sri Lanka); 'Kinguruwe', 'Malindi' (Tanzania and Zanzibar), 'Nyoro' (Kenya); 'Giuba' (Somalia); 'Mouz siny', 'Moz Hindi', 'Hindi', 'Indian', 'Basrai' (Egypt); 'Bazrai' (Pakistan); 'Johnson' (Canary Islands); 'Camyenne' (Guinea); 'Guineo Enano', 'Petite Naine', 'Governor' (West Indies); 'Camburi Pigmeo', 'Enano' (lit. "dwarf") (Latin America); 'Ana', Ananica', 'Caturra'.

Origin/Distribution

Dwarf Cavendish is a triploid banana cultivar originally from Vietnam and China. It became the primary replacement for the Gros Michel banana in the 1950s after crops of the latter were decimated by the Panama disease (Persley and George 1996).

Agroecology

The optimal temperature for Cavendish bananas is 27–30°C. Growth is impeded below13°C and chlorophyll is damaged below 6°C. Hence, a frost-free site should be selected for planting. Regular monthly precipitation between 100 and 200 mm is ideal. Is ideal. Banana plants are susceptible to topple by strong winds exceeding 30 km/h so wind breaks or shelter is important. Bananas should be planted in full sun to ensure photosynthetic activity in the plant. Flat and fertile terrain with good drainage, and soils fairly deep and fertile soil with a pH of between 6 and 6.5 is ideal.

Edible Plant Parts and Uses

The fruits are popular as dessert fruit, sweet and aromatic. The ripe fruit is peeled and eaten out-of hand, or sliced and served in fruit cups and salads, fruit compote, sandwiches, ice-cream, custards and gelatins. Sliced ripe bananas, canned in syrup, for commercial use in frozen tarts, pies, gelatins and other products. Ripe bananas sliced longitudinally, baked, grilled or broiled, are served as an accompaniment for ham or other meats sometimes with a garnish of brown sugar or chopped peanuts. Ripe bananas may be thinly sliced and cooked with lemon juice and sugar to make jam or sauce. Whole, peeled bananas can be spiced with a mixture of vinegar, sugar, cloves and cinnamon and boiled long enough to become thick. Ripe bananas can also be mashed and made into smoothies, milk shakes, sorbets ice-cream, bread, muffins, cheesecakes, doughnuts, cakes, banana jams and cream pies. Ripe bananas can be dipped in batter and deep fried to make goreng pisang or banana fritters or sliced and dried to make banana chips. Banana fritters is one of the most popular banana based products which is consumed by all age groups in Malaysia, Singapore and Indonesia. They are usually taken with tea or coffee during breakfast or tea time. In the South Pacific islands, unpeeled, ripe bananas are traditionally cooked in hot stone ovens or in the embers of a fire, while unripe bananas are peeled, grated or sliced, sometimes mixed with coconut cream, wrapped in leaves and cooked in an oven (Goode 1974; Lancaster and Coursey 1984). In Samoa, ripe bananas are mashed, mixed with coconut cream, scented with *Citrus* leaves, and served as a thick, fragrant beverage usually to the chiefs (Massal and Barrau 1956). In Cameroon, green banana is boiled and cooked in a sauce of palm oil with fish, cooked meat, haricot or green beans and seasonings (Oke et al. 1990). In Uganda, where it is the staple, green banana is boiled with beans, onions, pepper and salt to prepare a dish called *akatogo*. Another dish called *Omuwumbo* is prepared by wrapping banana pulp in banana leaves, steaming for an hour and pressing it to a firm mass and eaten. Also, in many African countries, cooking banana is boiled or steam, then mashed, baked, dried or pounded to make *fufu* (Oke et al. 1990).

Banana can be made into a wide array of processed food products (Aurore et al. 2008). Ripe banana or plantain fruit is used to make sweet meats known as banana figs which are very popularly eaten as snacks in Africa (Oke et al. 1990). Traditionally, the ripe fruit is sun-dried or sometimes dried in ovens or over fires; usually as slices, although banana figs are sometimes prepared from whole fruits (Goode 1974; Mukasa and Thomas 1970; Simmonds 1966). Figs can also be prepared from peeled firm-ripe bananas, split lengthwise, sulphured, and oven-dried to a moisture content of 18–20% and wrapped individually in plastic (Morton 1987). In Uganda, dried banana slices known as *mutere* are prepared primarily as a famine reserve, the slices being stored and used only in times of need when they are cooked directly or first ground into a flour (Goode 1974; Mukasa and Thomas 1970). Dried bananas can be minced and used together with candied lemon peel in fruit cake and other bakery products.

In Western Samoa, unripe bananas are preserved by fermentation using a technique normally applied to breadfruit in the Pacific Islands (Cox 1980). The product is a fermented paste called *masi* which is formed into loaves, wrapped in leaves and baked.

Banana puree made from ripe bananas. The puree can be frozen, canned by the addition of ascorbic acid to prevent discoloration or aseptically packed. The puree is used for beverage industry, baby foods, snack foods, banana powder, cake, pie, ice cream, cheesecake, doughnuts, milk shakes jam, sauces and many other products. Banana puree is used extensively in the processing of straight banana drink, sweetened with sugar or unsweetened. Banana nectar is prepared from banana puree in which a cellulose gum stabilizer is added. It is homogenized, pasteurized and canned, with or without enrichment with ascorbic acid. Studies showed that fructose syrup could be obtained from banana starch by employing an enzymatic process (Hernandez-Uribe et al. 2008). The syrup displayed comparable characteristics to those of commercial syrups. In Costa Rica, ripe bananas are peeled and boiled slowly for hours to make a thick syrup called "honey" (Morton 1987). Banana jam is also made from banana puree by adding sugar, pectin and citric acid.

Starch made from rejected cooking bananas and plantains has potential for food systems that require high temperature processing such as jellies, sausages, canned and bakery products but not frozen products. The results of freeze–thaw stability suggested that banana starches cannot be used in frozen products (Bello-Pérez et al. 1999).

Banana or plantain flour is made domestically by sun-drying slices of unripe fruits and pulverizing/grinding (Suntharalingam and Ravindran 1993). Banana flour has high starch content and is gluten free. Banana flour has no banana flavour but it can be used as flour in the processing of snack foods and bakery products with the addition of other flavourings. Green banana flour, a complex-carbohydrates source, mainly of resistant starch can be obtained by drying unripe peeled bananas (first stage of ripening) in a dryer tunnel (Tribess et al. 2009). The resistant starch content of the flour produced varied from (40.9 g)/100 g to (58.5 g)/100 g, on dry basis (d. b.), and was influenced by the combination of drying conditions. Banana powder is produced from ripe banana puree commercially by spray-drying, or drum-drying (Meadows 2007). Banana flour has a high sugar and low starch content and can be used as a substitute for fresh banana in making traditional cakes or their premixes as well as in the processing of banana snacks, crackers or crisps. Good quality spray dried banana powder could be obtained from ripe 'Robusta' variety banana pulp by spray drying with an inlet temperature of 150°C and outlet temperature 100°C with added additives (Evelin et al. 2007). Spray dried banana powder could be used as food ingredient and natural flavouring agent. Green plantain flour can be used as a thickening agent for mango sauce (Ramírez and Pacheco-Delahaye 2000).

Quality snack products can be developed from dehydrated banana flours at ripening stages 4, 5 and 6 (peel colour) mixed separately at 40% banana to 60% rice flour levels (Gamlath 2008). Protein and mineral (except for zinc and copper) content of the products were significantly different from 4 to 6 of the ripening stages. Most of the essential amino acids in the extruded products increased significantly at the ripening stage of 6. All the products were within the acceptable range in the 9-point Hedonic scale showing the best texture and flavour scores for stage 4 and 6, respectively. The extruded products show potential as snack products because of their nutritional quality and sensory acceptability.

Deep fat fried and salted banana chips of improved keeping quality can been prepared by soaking raw-sliced bananas in 500 ppm or 1,000 ppm sodium metabisulfite solution for an hour, then blanching in boiling water and drying in a forced draft iso-temperature over at 70°C till bone dry (Adeva et al. 1968). The dehydrated slices are then deep fried. Fried banana chips may be one of the important potential banana products in Bangladesh being easily saleable snack food in the markets (Molla et al. 2009). Preparation of banana chips is very simple and can easily be processed in rural areas where modern facilities of processing do not exist. Jackson et al. (1996) found that crispiest chips could be produced from whole green bananas by blanching in water at 69°C and 22 min, then peeled, sliced and fried in oil. Bananas can be utilised to made *keropok* or crackers, which are popular snack foods among Malaysians. Crackers have conventionally been produced from fish, shrimp, squid, nuts and vegetables like chick pea and green peas, etc. (Yusoff and Mohd Zain 2000).

The acid treatment of unripe banana flour (UBF) was carried out to obtain a fibre-rich product UBF showed a total dietary fibre content of 17% and a total starch content of 73% (Aguirre-Cruz et al. 2008). The preparation of a fiber-rich product with UBF may be important for the development of food and medical products. Banana and plantain can be made into bread. Studies found no significant differences between the whole wheat bread and the plantain substituted bread up to 10% plantain flour substitution in all the sensory attributes tested viz. crust, taste, aroma, shape, internal texture, appearance and general acceptability (Olaoye et al. 2006) the plantain substituted bread had comparable sensory and nutritional qualities to the whole wheat bread while the plantain substituted bread had higher proteins contents than the latter. Yeast leavened bread can be prepared by adding banana powder (10–30%) to hard-red spring wheat (Mohamed et al. 2009). Based on the added banana powder amounts only, the prepared bread could deliver 42.87–128.6 mg potassium/30 g of bread (one regular slice) and 0.33–1.00 g of fibre. Unripe banana flour can be used as a food ingredient to make pasta (spaghetti) of high quality, on the basis of low-carbohydrate digestibility, and increased resistant starch and antioxidant phenolics contents (Ovando-Martinez et al. 2009). Formulations consisting of 100% durum wheat semolina (control) and mixtures of semolina:banana flour of 85:15, 70:30 and 55:45 were prepared for spaghetti processing. The addition of banana flour increased the indigestible fraction, antioxidant capacity and the content of phenolic compounds in the spaghetti. As a consequence of the compositional changes, a slow, low rate for the enzymatic hydrolysis of carbohydrates was observed.

Studies showed that wheat flour, yeast raised doughnuts substituted with 20% banana flour showed the highest score in overall acceptability (Chong and Noor Aziah 2008). Chemical analyses result indicated a higher percentage of total dietary fibre and caloric content in doughnut substituted with banana flour than the control (no banana flour).

Cookies were produced by substituting 10%, 20% and 30% of the wheat flour for unripe banana meal (Fasolin et al. 2007). In view of the high nutritional value of the cookies containing banana meal, with no significant alteration of their physical and sensorial characteristics, the use of this meal as a partial substitute of wheat flour was found to be viable and could be recommended in the preparation of alternative nutritionally enriched foods. Banana starch and its products such as cookies had higher resistant starch levels than those made with corn starch (Bello-Pérez et al. 2004). The cookies had lower available starch than the starches while banana starch had lower susceptibility to the in-vitro alpha-amylolysis reaction. Slowly digestible cookies were prepared from resistant starch-rich powder (RSRP) prepared from autoclave-treated lintnerized banana starch (Aparicio-Saguilán et al. 2007). Results revealed RSRP from banana starch as a potential ingredient for bakery products containing slowly digestible carbohydrates.

Studies indicated that in spite of the increased starch digestion rate, composite plantain starch noodles were found to be a better source of indigestible carbohydrates than pure wheat starch pasta with potential dietary applications (Osorio-Diaz et al. 2008). Cavendish banana pulp and peel flour was found useful as functional ingredients in yellow alkaline noodles (Ramli et al. 2009). Predicted GI (glycaemic index) values of cooked noodles were in the order; banana peel noodles<banana pulp noodles<control noodles. Since the peel flour was higher in total dietary fibre but lower in resistant starch contents than the pulp flour, the low pGI of banana peel noodles was mainly due to its high dietary fibre content. Banana pulp noodles could be prepared by partial substitution of wheat flour with green Cavendish banana pulp flour (Saifullah et al. 2009b). The GI of the banana pulp noodles (50) was found to be lower than the normal yellow noodles (53) and the tensile strength and elasticity values were higher than the control yellow noodles. Partial substitution of green banana pulp into noodles may be useful for controlling starch hydrolysis of yellow noodles. Similarly, banana peel (BP) noodles prepared by partial substitution of wheat flour with green Cavendish banana peel flour (Saifullah et al. 2009a). The tensile strength of BP noodles was similar to control yellow noodles but their elasticity was higher. Following *in-vitro* starch hydrolysis studies, it was found that the estimated GI of BP noodles was lower than control noodles. Partial substitution of banana peel into noodles may be useful for controlling starch hydrolysis of yellow noodles. Studies showed that the addition of a source of indigestible carbohydrates (banana flour) to wheat semolina (banana flour: wheat semolina ratio of 85:15, 70:30, and 55:45) to make spaghetti was possible and did not affect consumer preference (Agama-Acevedo et al. 2009). The use of banana flour decreased the lightness and diameter of cooked spaghetti, and increased the water absorption of the product. Hardness and elasticity of spaghetti were not affected by banana flour, but adhesiveness and chewiness increased as the banana flour level in the blend rose.

A combination of malted maize and soybean, roasted groundnut and cooking banana in the ratio of 50:15:15:20 gave a very recommendable weaning food for infants between the ages of 6 months and 2 years (Onyeka and Dibia 2002). Njoki and Faller (2001) developed a weaning food from a plantain/soy/corn blend by extrusion cooking. Extrusion cooking and pre-gelatinisation increase the energy content of cooking bananas and significantly reduces its bulkiness for weaning food (Bukusuba et al. 2008). Soybean and simsim addition was found to improve protein quality and quantity of banana-based weaning products.

The use of ultrasound as a pre-treatment prior to air-drying to dehydrate banana was found to be cost-effective and faster to produce dried banana fruits, which can be directly consumed or become part of foodstuffs like cakes, pastries and many others (Fernandes and Rodrigues 2007).

Since the early 1960s, Brazil has produced dehydrated banana flakes in vacuum sealed cans for the local markets and export to the USA and elsewhere (Morton 1987). The flakes are used on cereal, in baked goods, canapes, meat loaf and curries, desserts, sauces, and other products. Chiquita Banana Flakes, a free flowing product made from fresh, ripe bananas is widely marketed in the Americas. This product is advertised as 100% natural without any preservatives or additives and is easily used in products to increase flavour intensity.

Over ripe bananas can also be made into wine and vinegar. Banana vinegar is made by a 2-stage fermentation of banana juice prepared from banana pulp, the first stage involves the alcoholic fermentation of sugar substrate by yeast, usually *Saccharomyces cerevisiae* to produce alcohol and carbon dioxide (Adams 1980; Suresh and Ethiraj 1991). This is followed by the aerobic production of acetic acid and water from the alcohol by *Acetobacter aceti* and *Acetobacter rancens*. The alcoholic fermentation normally takes about a week while the acetification stage takes 1–2 months. Fermentation of banana juice from the pulp (*Musa sapentium*) with Bakers yeast yielded wine (Akubor et al. 2003). The wine produced had 5% (V/V) alcohol, 0.04% protein, 48° Brix SS, 0.85% TA and 1.4 mg/100 ml vitamin C. Sensory evaluation results showed there were no significant differences in flavour, taste, clarity and overall acceptability between banana wine and a reference wine. The banana wine was generally accepted.

In Burundi, banana is mainly used for the production of beer, which is also common in East Africa (Oke et al. 1990). In Rwanda, a local beer called *urgawa* is made by adding roasted sorghum flour to banana juice and the mixture left ferment for 1–2 days (Adriaens and Lozet 1951; Champion 1970). The juice before fermentation is sometimes consumed as a non-alcoholic drink while a very potent beverage is obtained by mixing honey with the banana pulp before fermentation. Domestic banana beer production is mad from ripened peel banana which are pressed and the juice fermented (Davies 1993). The beer produced has an alcohol content of 2–5% while stronger beer with alcohol content of 11–15° is sometimes made from undiluted banana juice. Studies showed that banana could be used as an adjunct and aromatic compound in beer brewing and assist in the development of new products as well as in obtaining concentrated worts (Carvalho et al. 2008, 2009). Traditionally, the raw materials for beer production are barley, hops, water, and yeast, but most brewers use also different adjuncts banana was found to be a raw material favorable to alcoholic fermentation being rich in carbohydrates and minerals and providing low acidity.

The terminal male bud of the wild banana, *M. balbisiana*, is marketed in Southeast Asia. It is often boiled whole after soaking an hour in salt water, or with several changes of water to reduce astringency, and eaten as a vegetable. The male bud of cultivated bananas is considered too astringent but it is, nevertheless, sometimes similarly consumed. The flowers may be removed from the bud and prepared separately. They are used in curries in Malaya and eaten with palm oil in West Tropical Africa.

The new shoots of young plants may be cooked as greens. The inner soft core of the banana stem is edible Banana pseudostem core constitutes about 10–15% of the whole and contains 1% starch, 0.68% crude fiber and 1% total ash (Morton 1987). It is often cooked and eaten as a vegetable in India and is canned with potatoes and tomatoes in a curry sauce. Circular slices about 1/2 in (1.25 cm) thick are treated with citric acid and potassium metabisulphite and candied.

In India, a solution of the ash from burned leaves and pseudostems is used as salt in seasoning vegetable curries. The ash contains roughly (per 100 g): potassium, 255 mg; magnesium, 27 mg; phosphorus, 33 mg; calcium, 6.6 mg; sodium, 51 mg (Morton 1987).

Dried green plantains, ground fine and roasted, have been used as a substitute for coffee.

Inflorescence and flowers are also eaten.

Botany

A comparatively short herbaceous banana plant with pseudostem reaching heights of 1.8–2.5 m high. The pseudostem stem is robust, yellow-green to green with milky sap and <3 suckers. Leaves are broad, intermediate with short petioles and brown patches at the petiole base and midrib light green in dorsal surface. Inflorescence complete with female (proximal end), neuter and male flowers (top, distal end), peduncle hairy and 30–60 cm long, rachis falling vertically. Male bud intermediate shape, bract apex pointed, external surface red purple, internal surface red, flower compound tepal cream-yellow, style curved at base, stigma yellow, ovary white with red-purple pigmentation. Male bracts and flowers are not shed. Average bunch weight is 15–18 kg with 8–12 hands and 14–20 fingers per hand. Fruit length 15–25 cm, pedicel 20 cm long, fruits curve upwards, transverse section rounded, fruit apex lengthily pointed, light green to yellow thin skin, flesh is white to cream to pale yellow, soft, fine textured, sweet and aromatic (Plates 1, 2, 3, 4 and 5).

Plate 1 Bunch of Williams Cavendish

Plate 2 Kluai Hom Kiau bananas on sale in a Thai market

Plate 3 Hand of pisang masak hijau

Plate 4 Hand of Pisang lumut

Plate 5 Ripe 'fingers' of Williams banana

Nutritive/Medicinal Properties

The dominant and most common banana cultivar consumed in the United States is the Chiquita Banana – cv. 'Grande Nain' which belongs to *Musa acuminata* (AAA) 'Dwarf Cavendish" subgroup. The following nutritive value of banana (*Musa acuminata*) was reported by U.S. Department of Agriculture, Agricultural Research Service (2010): water 74.91 g, energy 89 kcal (371 kJ), protein 1.09 g, total lipid 0.33 g, ash 0.82 g, carbohydrate 22.84 g, total dietary fibre 2.6 g, total sugars 12.23 g, sucrose 2.39 g, glucose 4.98 g, fructose 4.85 g, maltose 0.01 g, starch 5.38 g, Ca 5 mg, Fe 0.26 mg, Mg 27 mg, P 22 mg, K 358 mg, Na 1 mg, Zn 0.15 mg, Cu 0.078 mg, Mn 0.27omg, F 2.2 mg, Se 1 μg, vitamin C 8.7 mg, thiamine 0.031 mg, riboflavin 0.073 mg, niacin 0.665 mg, pantothenic acid 0.334 mg, vitamin B-6 0.367 mg, total folate 20 μg, total choline 9.8 mg, betaine 0.1 mg, vitamin A 3 μg RAE, vitamin A 64 IU, β-carotene 26 μg, α-carotene 25 μg, lutein + zeaxanthin 22 μg, vitamin E (α-tocopherol) 0.10 mg, γ-tocopherol 0.02 mg, δ-tocopherol 0.01 mg, vitamin K (phylloquinone) 0.5 μg, total saturated fatty acids 0.112 g 10:0 (capric) 0.001 g, 12:0 (lauric) 0.002 g, 14:0 (myristic) 0.002 g, 16:0 (palmitic) 0.102 g, 18:0 (stearic) 0.005 g; total monounsaturated fatty acids 0.032 g, 16:1 undifferentiated (palmitoleic) 0.010 g, 18:1 undifferentiated (oleic) 0.022 g; total polyunsaturated fatty acids 0.073 g, 18:2 undifferentiated (linoleic) 0.046 g, 18:3 undifferentiated (linolenic) 0.027 g; phytosterols 16 mg, tryptophan 0.009 g, threonine 0.028 g, isoleucine 0.028 g, leucine 0.068 g, lysine 0.050 g, methionine 0.008 g, cystine 0.009 g, phenylalanine 0.049 g, tyrosine 0.009 g, valine 0.047 g, arginine 0.049 g, histidine 0.077 g, alanine 0.040 g, aspartic acid 0.124 g, glutamic acid 0.152 g, glycine 0.038 g, proline 0.028 g and serine 0.040 g.

The average contents of the total volatiles from cultivars "Dwarf Cavendish", "Giant Cavendish", "Robusta" and "Williams" were found to be 93.0, 116.5, 157.3 and 157.0 mg/kg, respectively (Nogueira et al. 2003). The ester and alcoholic fractions appeared to play a dominant role in the organoleptic characteristics of banana fruit, presenting a substantial content ranging from 57.2 to 89.8 mg/kg and 19.0 to 47.7 mg/kg, respectively, in all cultivars from Madeira Island studied. 3-Methyl butyl butanoate ester was the major constituent15.8–20.5 mg/kg fresh fruit. Octyl acetate was the minor component (1.6e3.2 mg/kg fresh fruit). The most important alcohol was ethanol (23.9 mg/kg fresh fruit). Other fractions, such as carbonyl compounds, carboxylic acids were also associated with the aroma of banana. The carboxylic acids were associated with the ripened aroma. The carbonyl fraction with the alcohols, contributed to the woody or musty flavour. Hex-2(E)-enal and pentan-2-one were the major components of this fraction contributing to the herbal note (Schiota 1993). A total of 43 components were quantified in Cavendish banana essence (Jordan et al. 2001). Among them 26 contributed to the aromatic essence but isoamyl acetate, 2-pentanol acetate, 2-methyl-1-propanol, 3-methyl-1-butanol, 3-methylbutanal, acetal, isobutyl acetate, hexanal, ethyl butyrate, 2-heptanol, and butyl butyrate contributed and define the aroma in banana. Phenylpropanoids such as eugenol, methyl eugenol and elimicin were also found in the banana fruit but at very low concentration (<1 mg/kg fresh fruit). They contributed to the floral note of the ripening aroma (Macku and Jennings 1987). Twelve aromatic compounds (2 alcohols, 9 esters and 1 phenol) were identified in fresh Cavendish banana (Boudhrioua et al. 2003) Seven volatile compounds, four among the previous 12 (isoamyl alcohol, isoamyl acetate, butyl acetate and elemicine) and three other nonidentified compounds were selected by olfactometric analysis as characteristics of banana smell. Aromatic changes of the banana Cavendish were found during the drying isoamyl acetate, isoamyl alcohol and butyl acetate were found to strongly decreased during drying while some compounds increased or appeared to be formed at the end of drying. Elimicin was the most thermal resistant compound. Glycosidically bound volatile compounds were found in fresh Cavendish banana (Pérez et al. 1997). Of the 35 aglycons identified, alcohols such as decan-1-ol and 2-phenylethanol; acids such as 3-methylbutanoıc, benzoic acid were quantitatively the most important aglycons in glycosides isolated from fresh banana. Free volatile compounds and aglycons were identified in cultivars of *Musa* sp. grown in the French West Indies (Aurore et al. 2011). The main volatile compounds found in Cavendish banana were (E)-2-hexenal and acetoin, in Plantain: (E)-2-hexenal and hexanal, and in Frayssinette: 2, 3 butanediol and two diastereomer solerols. The most abundant of aglycons were 3-methylbutanol, 3-methyl-butanoic acid, solerol (two disatereoisomers) and acetovanillone. This compound, rarely identified in fruits, was detected for the first time in glycoconjugated volatile compounds of banana fruits.

Cavendish, (AAA Group), a common cultivar was found to contain low level of 21 μg/100 g of β-carotene (Englberger et al. 2003). Total carotene and pro-vitamin A RAE levels ranged from 150 μg and 8 RAE/100 g, respectively, for Cavendish Williams (Blades et al. 2003). Of the lipid extracts isolated from unripe and ripe banana pulp and peel, unsaturated acids, particularly linoleic and palmitoleic, decreased about threefold in the pulp while more than a two-fold increase in stearic acid occurred (Goldstein and Wick 1969). Generally, unsaturated fatty acids decreased in both the pulp and peel during ripening. The peel contained almost four times more lipid than the pulp.

The following 13 alcohols were identified in the volatiles in ripe banana: ethanol, propan-l-01, p-methylpropan-l-01, butan-l-01, pentan-2–01, 3-methylbutan-l-01, hexan-l-01, heptan-2–01, cis and tram hex-3-en-l-01, cis and trans hex-4-en-l-01, and cis pent-2-en-l-01 (Murray et al. 1968). 2-methylbutan-l-01 was shown to be associated with 3-methylbutan-l-01 in a ratio of 1:200. The following phytochemicals were identified in banana pulp (*Musa sapientum*): Cycloartenol, cycloeucalenol, 24-methylene cycloartanol, campesterol, β-sitosterol and stigmasterol (Knapp and Nicholas 1969b). The major triterpene was 24-methylene cycloartanol while β-sitosterol accounted for greater than 72% of the sterol fraction.

The peel of six varieties of banana and plantain: dessert banana (*Musa* AAA), plantain (*Musa* AAB) cooking banana (*Musa* ABB) and hybrid (*Musa* AAAB) were found to be rich in total dietary fibre (40–50%), proteins (8–11% DW), essential amino acids, poly-unsaturated fatty acids and potassium (Emaga et al. 2007). Leucine, valine, phenylalanine and threonine were essential amino acids in significant quantities. Lysine was the limiting amino acid. content of lipid varied from 2.2–10.9% and was rich in polyunsaturated fatty acids, particularly linoleic acid and α-linolenic acid. Peel of plantain was richer in starch than were the banana peels. Maturation of fruits involved increase in soluble sugar content and, at the same time, decrease in starch. Banana peels were also found to be a potential source of dietary fibres and pectins (Emaga et al. 2008). In all the stages of maturation, the pectin concentration in banana peels was higher than in plantain peels. Further, the galacturonic acid and methoxy group contents in banana peels were higher than in plantain peels. The average molecular weights of the extracted pectins were in the range of 132.6–573.8 kDa and were not dependant on peel variety, while the stage of maturation did not affect the dietary fibre yields and the composition in pectic polysaccharides. Plantain peels contained a higher amount of lignin but had a lower hemicellulose content than banana peels.

Banana peel is rich in phytochemical compounds, such as anti-oxidants. The total amount of phenolic compounds in banana (*Musa acuminata* AAA) peel was found to range from 0.90–3.0 g/100 gDW (Nguyen et al. 2003; Someya et al. 2002). Gallocatechin was found in a concentration of 160 mg/100 g DW in the peel (Someya et al. 2002). Ripe banana peel also contained anthocyanins delphinidin and cyanidin (Seymour 1993), and catecholamines such as the antioxidant dopamine (Kanazawa and Sakakibara 2000). Banana peels also contained carotenoids and their fatty acid esters (Subagio et al. 1996). The carotenoid content of the banana peel was in the range of 3–4 μg/g as lutein equivalent. The carotenoids were identified as lutein, β-carotene, α-carotene, violaxanthin, auroxanthin, neoxanthin, isolutein, β-cryptoxanthin and α-cryptoxanthin. Most of the oxygenated carotenoids were found to occur in the esterified form, mainly with myristate, and to a lesser extent with laurate, palmitate or caprate. Further, the peel also contained sterols and triterpenes, such as β-sitosterol, stigmasterol, campesterol, cycloeucalenol, cycloartenol, and 24-methylene cycloartanol (Knapp and Nicholas 1969a). 24-Methylene cycloartanol palmitate and an unidentified triterpene ketone were the major constituents. The ester represented approximately 30% of the total extractable lipid. Banana peel flour was found to be an important source of fibre (NDF), corresponding about 32% of its dried weight (Ranzani et al. 1996). The addition of this flour to a basal casein diet lowered its protein digestibility and increased the faecal bulk of the rats. However, it did not alter the protein quality. In addition, the growth of the rats fed diets containing banana peel did not differ from those fed control diet. These results suggested the feasibility of technological studies aiming the development of food products with banana peel.

The water-soluble polysaccharides isolated from the vascular gel of *Musa paradisiaca*, were fractionated (Mondal et al. 2001). Fractionated polymers contained arabinose, xylose and galacturonic acid as major sugars, together with traces of galactose, rhamnose, mannose and glucose residues. Methylation analysis revealed the presence of a highly branched arabinoxylan with a significant amount of terminal arabinopyranosyl units and an arabinogalactan type I pectin.

Some pharmacological properties of banana reported include:

Antioxidant Activity

A potent water-soluble antioxidant dopamine, one of the catecholamines was identified in the popular commercial Cavendish banana (Kanazawa and Sakakibara 2000). It suppressed the oxygen uptake of linoleic acid in an emulsion and scavenged diphenylpicrylhydrazyl radical. Dopamine exhibited greater antioxidative potency than glutathione, food additives such as butylated hydroxyanisole and hydroxytoluene, flavone luteolin, flavonol quercetin, and catechin, and had similar potency to the strongest antioxidants gallocatechin gallate and ascorbic acid. Banana contained dopamine at high levels in both the peel and pulp. Dopamine levels ranged from 80–560 mg per 100 g in peel and 2.5–10 mg in pulp, even in ripened bananas ready to eat. Another antioxidant, gallocatechin, was identified in the Cavendish banana (Someya et al. 2002) and exhibited potent antioxidant activity. Gallocatechin was more abundant in peel (158 mg/100 g dry wt.) than in pulp (29.6 mg/100 g dry wt.). The antioxidant activity of the banana peel extract, against lipid autoxidation, was stronger than that of the banana pulp extract. This result indicated bananas to be a good source of natural antioxidants for foods and functional food source against cancer and heart disease. Banana (*Musa acuminata* AAA) peel extracts exhibited a high capacity to scavenge 2,2-diphenyl-1-picrylhydrazyl (DPPH.) and 2,2′-azino-bis(3-ethylbenzothiazoline)-6-sulfonic acid (ABTS.+) free radicals, and they were also good lipid peroxidation inhibitors (González-Montelongo et al. 2010). Compared with methanol, ethanol, acetone, water, methanol:water or ethanol:water, acetone:water extracts were considerably more potent at inhibiting the peroxidation of lipids in the β-carotene/linoleic acid system or scavenging free radicals. However, aqueous extracts had a high capacity to protect lipids from oxidation in the thiobarbituric acid reactive substances (TBARS) test, as well as in the β-carotene bleaching assay. Moreover, acetone:water most efficiently extracted all extractable components (54%), phenolic compounds (3.3%), and anthocyanin compounds (434 μg cyanidin 3-glucoside equivalents/100 g freeze-dried banana peel). Banana peel also contained large amounts of dopamine and L-dopa, catecholamines with a significant antioxidant activity. However, ascorbic acid, tocopherols or phytosterols were not detected in the different extracts. The antioxidant activity of banana peel extracts from different cultivars was similar.

Studies showed that flavonoid rich fraction (FRF) of the two varieties of *Musa paradisiaca* banana (Palayamkodan and Rasakadali) exhibited significant hypolipidaemic and antioxidant activities (Krishnan and Vijayalakshmi 2005). Albino rats administered flavonoid rich fraction (FRF) of the two banana varieties exhibited decreased lipids and lipid peroxides levels. MDA and hydroperoxides were significantly diminished.

Yellow alkaline noodle prepared from 30% matured green banana (*Musa acuminata* × *balbisiana* Colla cv. Awak) flour with addition of 10% oat β-glucan resulted in significantly higher total dietary fibre, and especially insoluble dietary fibre, resistant starch and total starch contents (Chong and Noor Aziah 2010). Thirty percent of banana flour significantly) improved the antioxidant properties of noodles in terms of the total phenolic content and inhibition of peroxidation. Noodle incorporated with 30% banana flour and added oat β-glucan showed the lowest GI and carbohydrate digestibility rate, and higher concentrations of essential minerals (magnesium, calcium, potassium and phosphorus) and proximate components, with the exception of crude fat, when compared to the control. Sensory evaluation indicated that the quality of the 30% banana flour-substituted noodle was comparable to the control.

Antibacterial Activity

Crude extract of green Cavendish banana peel recorded more significant antibacterial and antioxidant activities than that of yellow peel using various solvents (Mokbel and Hashinaga 2005). Ethyl acetate and water soluble fractions of the green peel displayed high antimicrobial and antioxidant activity, respectively. Green banana peel displayed high antioxidant activity as measured by ß-carotene bleaching method, DPPH free radical and linoleic acid emulsion. Antioxidant activity of the water extracts was comparable to those of synthetic antioxidants such as butylated hydroxyanisole and butylated hydroxytoluene. Among all isolated components, ß-sitosterol, malic acid, succinic acid, palmatic acid, 12-hydroxystrearic acid, glycoside, d-malic and 12-hydroxystrearic acid were the most active against all the Gram-negative (*Salmonella enteritidis* and *Escherichia coli*) and Gram positive (*Staphylcoccus aureus, Bacillus subtilis*, and *Bacillus cereus*) bacterial species tested. The MIC of ß-sitosterol, d-malic and succinic acid varied between 140 and 750 ppm. Palmitic acid was comparatively less significant against all tested the bacteria species, while 12-hydroxystrearic acid recorded antimicrobial activity as measured by paper disk methods but not for MIC.

Antihypertensive Activity

Studies showed ripe banana pulp, administered daily (50 g/rat/day) together with standard food pellets, prevented an increase in blood pressure induced by the intramuscular injection of deoxycorticosterone enantate (DOC, 25 mg/rat) in rats given access to both water and 2% NaCl solution (Perfumi et al. 1994). The antihypertensive effect of banana was not related to reduced salt intake; contrariwise rats receiving banana during DOC-treatment consumed significantly larger amounts of salt relative to controls. Ritan-serin, a 5-HTjc receptor antagonist, partially inhibited the effect of banana on DOC-induced salt intake, suggesting that the effect may be partially mediated by serotonergic mechanisms. The finding suggested that an increase in central serotonin levels triggered by the high tryptophan and carbohydrate content of banana was responsible for the serotonin-mediated component of the natriorexic effect of banana. However, both the effect of banana on salt intake and that on blood pressure could not be entirely accounted for by its influence on endogenous serotonin levels; additional mechanisms should be evaluated. Rao et al. (1999) found that ripened and unripened 'Nendran', 'Rasthali', 'Poovan','Robusta', 'Bontha' and 'Safed Velchi' bananas exhibited angiotensin converting enzyme (ACE) inhibitory activity. The inhibition of ACE by different ripened banana cultivars was much more than that of unripened banana cultivars. In a study of normotensive volunteers, they observed a 10% fall in blood pressure after consumption of two bananas per day for a week. The findings suggested consumption of bananas may be beneficial to hypertensive individuals. Banana significantly decreased the rise of systolic blood pressure diastolic blood pressure, mean arterial blood pressure and plasma ACE activity in healthy volunteers subjected to cold stress induced hypertension without much effect on heart rate and peak expiratory flow rate (Sarkar et al. 1999).

Studies suggested that banana could prevent a deficiency of potassium, which might increase the risk of stroke (Singhal 2001; Levine and Coull 2002). Studies showed that a lower serum potassium level in diuretic users, and low potassium intake in those not taking diuretics were associated with increased stroke incidence among older individuals (Green et al. 2002). Lower serum potassium was associated with a particularly high risk for stroke in the small number of diuretic users with atrial fibrillation. Dietary potassium appeared to be the most important marker among nonusers of diuretics suggesting that increased water excretion prevented the potassium from being absorbed by the body. Potassium is an essential mineral needed to regulate water balance, acidity level and blood pressure. A lack of potassium may cause muscle weakness, irregular heartbeat, nausea or vomiting. Nutrient data showed that potassium is very rich in potassium and very low in sodium that is ideal for maintenance of normal blood pressure.

Anticancer Activity

A new bicyclic diarylheptanoid, rel-(3S,4aR, 10bR)-8-hydroxy-3-(4-hydroxyphenyl)-9-methoxy-4a,5,6,10b-tetrahydro-3H-naphtho[2,1-b] pyran (1), as well as four known compounds, 1,2-dihydro-1,2,3-trihydroxy-9-(4-methoxyphenyl)phenalene (2), hydroxyanigorufone (3), 2-(4-hydroxyphenyl)naphthalic anhydride (4), and 1,7-bis(4-hydroxyphenyl)hepta-4(E), 6(E)-dien-3-one (5), were isolated from an ethyl acetate-soluble fraction of the methanol extract of the fruits of *Musa* × *paradisiaca* cultivar (Jang et al. 2002). Using an in-vitro assay to determine quinone reductase induction and a mouse mammary organ culture assay, hydroxyanigorufone was found to have potential cancer chemopreventive activity.

Along with other fruits and vegetables, consumption of bananas may be associated with a reduced risk of colorectal cancer (Deneo-Pellegrini et al. 1996), leukemia in children (Kwan et al. 2004) and in women, breast cancer (Zhang et al. 2009) and renal cell carcinoma (Rashidkhani et al. 2005). In a case-control study conducted in Uruguay, of 61 food items, a reduction in risk of colorectal cancer was observed for total vegetable intake, total fruit intake, and lettuce, apple, and banana consumption. The strongest protection was observed for banana intake for consumption in the third tertile compared with the first (Deneo-Pellegrini et al. 1996). In a study on the association between fruits and vegetables and risk of renal cell carcinoma in a population-based prospective cohort study of Swedish women, dietary information from 61,000 women age 40–76 years were collected by a food-frequency questionnaire (Rashidkhani et al. 2005). The results suggested that high consumption of fruits and vegetables might be associated with reduced risk of renal cell carcinoma. Within the group of fruits, the strongest inverse association was observed for banana. Studies in China on the consumption of vegetable and fruit consumption on breast cancer risk was conducted on 438 cases (Zhang et al. 2009). The study found that consumption of individual vegetable and fruit groups such as dark green leafy vegetables, cruciferous vegetables, carrots and tomatoes, banana, watermelon/papaya/cantaloupe were all inversely and significantly related with breast cancer risk. An inverse association was also observed for vitamin A, carotene, vitamin C, vitamin E, and fibre intake. These data indicated that greater intake of vegetables and fruits was associated with a decreased risk of breast cancer among Chinese women in Guangdong.

Results of studies suggested that regular consumption of oranges/bananas and orange juice during the first 2 years of life was associated with a reduction in risk of childhood leukemia diagnosed between the ages of 2 and 14 years (Kwan et al. 2004).

A methanol extract of banana peel significantly suppressed the regrowth of ventral prostates and seminal vesicles induced by testosterone in castrated mice (Akamine et al. 2009). Further studies in the androgen-responsive LNCaP human prostate cancer cell line showed that the extract inhibited dose-dependently testosterone-induced cell growth, while the inhibitory activities of the extract did not appear against dehydrotestosterone-induced cell growth. These results indicated that methanol extract of banana peel could inhibit 5-α-reductase and might be useful in the treatment of benign prostate hyperplasia.

Hypoglycemic/Antidiabetic Activity

Studies in diabetic subjects showed that consumption of under-ripe bananas elicited a low glycaemic response compared to consumption of over-ripe bananas or white bread (Hermansen et al. 1992). The mean postprandial blood glucose response area to white bread (181 mmol/l × 240 minutes) was significantly higher compared with under-ripe banana (62 mmol/l × 240 minutes) and over-ripe banana (106 mmol/l × 240 minutes). Glycaemic indices of the under-ripe and over-ripe bananas differed (43 and 74). The mean insulin response areas to the three meals were similar: 6,618 pmol/l × 240 minutes (white bread), 7,464 pmol/l × 240 minutes (under-ripe banana) and 8,292 pmol/l × 240 minutes (over-ripe banana). Studies with seven male subjects with untreated noninsulin-dependent diabetes mellitus ingesting banana of varying degree of ripeness found that the plasma glucose, serum insulin, C-peptide, and plasma glucagon area responses varied little with ripeness of the bananas (Ercan et al. 1993). The glucagon area response was negative after glucose ingestion but was positive following banana ingestion.

Animal studies showed that the methanolic extract of mature, green fruits of *Musa paradisiaca* possessed hypoglycemic activity (Ojewole and Adewunmi 2003). The extract induced significant, dose-related reductions in the blood glucose concentrations of both normal and diabetic mice. Chlorpropamide (250 mg/kg p.o.), a reference antidiabetic agent, also produced reductions in the blood glucose concentrations of normal and diabetic mice. The results supported the folkloric use of the plant in the management and/or control of adult-onset, type-2 diabetic mellitus among the Yoruba-speaking people of South-Western Nigeria. In a blind within-subject crossover design of obese type 2 diabetic subjects, patients supplemented with native banana starch for 4 weeks lost more body weight than when they were on control (soymilk) treatment (Ble-Castillo et al. 2010). Plasma insulin and HOMA-I (homeostasis model assessment index) were reduced after native banana starch consumption, compared with baseline levels, but not significantly when compared to the control treatment.

Green bananas are known to contain substantial concentrations of resistant starch and are a common part of the Micronesian diet (Thakorlal et al. 2010). Resistant starch is a type of starch that is resistant to starch hydrolyzing enzymes in the stomach and thus behaves more like dietary fibre. Resistant starch had been shown to have beneficial effects in disease prevention including modulation of glycaemic index diabetes, cholesterol lowering capability and weight management. Studies found the following Micronesian banana cultivars: Utin Kerenis, Inahsio and Utin Ruk to contain the highest amounts of resistant starch (Thakorlal et al. 2010). Resistant starch found in banana may be beneficial in disease prevention, including modulation of glycaemic index (GI), diabetes, cholesterol lowering capability and weight management (Zhang et al. 2005). In-vitro colonic fermentation and glycemic response in healthy volunteers of different kinds of unripe banana flour obtained from the cooked pulp of unripe bananas (*Musa acuminata*, Nanicão variety) (UBM), and the unripe banana starch (UBS), obtained from isolated starch of unripe banana, plantain type (*Musa paradisiaca*) were investigated (Menezes et al. 2010). The flours presented high concentration of unavailable carbohydrates, which varied in the content of resistant starch, dietary fibre and indigestible fraction (IF). The in-vitro colonic fermentation of the flours was high, 98% for the UBS and 75% for the UBM. The increase in the area under the glycemic curve after ingestion of the flours was 90% lower for the UBS and 40% lower for the UBM than the increase produced after bread intake. These characteristics highlighted the potential of UBM and UBS as functional food ingredients.

Panda et al. (2009) found that serum glucose level, lipid profile, glucose tolerance, hepatic and muscle glycogen contents as well as the activities of hepatic hexokinase and glucose-6-phosphatase recovered significantly in streptozotocin-induced diabetic rats after oral administration of separate ethyl acetate fractions of *Eugenia jambolana* seeds (200 mg/kg of body) or *Musa paradisiaca* roots (100 mg/kg of body weight) or combined form for 90 days. The loss in body weight of diabetic animals was reversed and serum levels of insulin as well as C-peptide, which were found to be reduced in diabetic rats, increased significantly after oral administration of the fractions. The size and volume of pancreatic islets in diabetic treated rats increased significantly compared with the diabetic control group. Treatment of diabetic rats with the combined dose of the plant fractions was found to be more effective than treatment with the individual fraction. The plant fractions were found to be free from metabolic toxicity. Rai et al. (2009) found that a dose of 500 mg/kg body weight (bw) of *Cynodon dactylon* produced maximum falls of 23.2% and 22.8% in blood glucose levels of normoglycemic rats during studies of fasting blood glucose and glucose tolerance, respectively, whereas the same dose of *Musa paradisiaca* produced a rise of 34.9% and 18.4%. In diabetic rats during glucose tolerance tests, a fall of 27.8% and a rise of 17.5% were observed with the same dose of *C. dactylon* and *M. paradisiaca,* respectively. Laser-induced breakdown spectroscopy indicated that *C. dactylon* was rich in magnesium (Mg), whereas *M. paradisiaca* was comparatively rich in potassium (K) suggesting thereby the defined roles of these elements in diabetes management. The chemical composition of banana flour (BF) obtained from unripe banana (*Musa paradisiaca*) showed that total starch (73.36%) and dietary fibre (14.52%) comprised the highest constituents (Juarez-Garcia et al. 2006). Of the total starch, available starch was 56.29% and resistant starch 17.50%. BF bread exhibited higher protein and total starch content than control bread, but had higher lipid content. Appreciable differences were found in available, resistant starch and indigestible fraction between the bread studied, with BF bread showing higher resistant starch and indigestible fraction content. HI (hydrolysis index)-based predicted glycemic index for the BF bread was 65.08%, which was significantly lower than control bread (81.88%), suggesting a "slow carbohydrate" feature for the BF-based bread. Results indicated banana flour as a potential ingredient for bakery products containing slowly digestible carbohydrates.

Ripe yellow banana fruit was found to have an average glycemic index (GI) of 51 deemed to be low according to the standards set by Foster-Powell et al., (2002) and a glycemic load (GL) (GI × dietary carbohydrate content) of 13 suggesting that banana does not elevate blood sugar level. Just ripe banana (yellow with some green section) was found to have a lower GI of 42 and GL of 11, under ripe bananas a GI of 30 and GL of 6 and over ripe banana (with brown flecks) a GI of 48 and GL of 12 (Denyer and Dickinson 2010). The results suggested that just-ripe and under-ripe banana would be a good fruit choice for diabetic management. The higher the GL, the greater the expected elevation in blood glucose and in the insulinogenic effect of the food (Foster-Powell et al. 2002). Several prospective observational studies had shown that the chronic consumption of a diet with a high glycemic load was independently associated with an increased risk of developing type 2 diabetes, obesity, cardiovascular disease, and certain cancers (Foster-Powell et al. 2002). A prospective study of dietary glycemic load, carbohydrate intake, and risk of coronary heart disease was conducted in a cohort of 75,521 women aged 38–63 years with no previous diagnosis of diabetes mellitus, myocardial infarction, angina, stroke, or other cardiovascular diseases in 1984 with a follow-up of 10 years

(Liu et al. 2000). Epidemiologic data obtained suggested that a high dietary glycemic load from refined carbohydrates increased the risk of CHD, independent of known coronary disease risk factors. Ludwig, (2003) in a review of glycemic index and the regulation of body weigh advocated the potential utility of low glycemic index diets in the prevention and treatment of obesity and related complications.

Antidiarrheal Activity

Studies found that banana flakes could be used as a safe, cost-effective treatment for diarrhea in critically ill tube-fed patients (Emery et al. 1997). Banana flakes could be given concurrently with a workup for *Clostridium difficile* colitis, thereby expediting treatment of diarrhea. Thirty-one patients with diarrhea and receiving enteral feedings were randomized to receive either banana flakes or medical treatment for diarrhea. Both banana flakes and medical treatments reduced the severity of diarrhea in critically ill tube-fed patients. Over the course of treatment, mean diarrhea scores were 21.64 for the banana flake group and 25.41 for the medical group. The banana flake group had less diarrhea clinically, with 57% of the subjects diarrhea free on their last study day as opposed to 24% of the medically treated subjects. Results of a recent study supported the benefits of green plantain (*Musa paradisiaca*) in the dietary management of persistent diarrhea in hospitalized children, in relation to diarrheal duration, weight gain and costs (Álvarez-Acosta et al. 2009). In a prospective, in-hospital controlled trial, two different treatments viz. green plantain-based diet and traditional yogurt-based diet were administered to a sample of 80 children of both sexes, with ages ranging from 1–28 months, who had experienced ≥14 days of persistent diarrhea associated with *Aeromonas hydrophilia* and *Shigela flexneri*. The experimental group fed on a green plantain diet exhibited a significantly better response in: decreasingg stool output and consistency, stool weight, diarrhea duration, and increasing daily body weight gain than the yogurt-based diet group. The average duration of diarrhea in the plantain-based diet group was 18 hours shorter and it also had lower cost.

Protease Inhibition Activity

The proteolysis of casein by trypsin, chymotrypsin and papain was inhibited by ripened and unripened bontha, poovan, nendran, cavendish and rasthali bananas (Rao 1991). The inhibition of trypsin, chymotrypsin and papain by different ripened banana cultivars was much more than that of unripened banana cultivars. The trypsin and chymotrypsin inhibitory activity of ripened poovan was heat stable, resistant to pronase and partly stable to trypsin but the trypsin and chymotrypsin inhibitory activity of unripened poovan was stable to heat and resistant to pronase only. The partial stability of trypsin inhibitory activity and instability of papain inhibitory activity of ripened poovan to alkaline pH suggested that the inhibitory factors of trypsin and papain were dissimilar. The probable role of unripened banana papain inhibitors in curing stomach ulcers and antinutritional role of ripened banana trypsin inhibitors was also elaborated.

Antilithiatic Activity

The fresh juice of *Musa paradisiaca* (cv. Puttubale) stem showed antilithiatic activity (Prasad et al. 1993). The stones formed were mainly of magnesium ammonium phosphate with traces of calcium oxalate. Banana stem juice (3 ml/rat/day orally) was found to be effective in reducing the formation and also in dissolving the pre-formed stones of magnesium ammonium phosphate with traces of calcium oxalate in the urinary bladder of albino rats.

Antidyspepsia Activity

A controlled (non-blinded) study of 46 people suggested that banana powder, a traditional Indian food, may help treat non-ulcer dyspepsia (Aurora and Sharma 1990). After 8 weeks of treatment, 75% of subjects taking banana powder reported complete or partial symptom relief compared to 20% of those who received no treatment.

Lectin and Associated Activities (Immunomodulatory, Anticancer and Antiviral)

A lectin (BanLec-I) from banana (*Musa paradisiaca*) with a binding specificity for oligomannosidic glycans of size classes higher than (Man)6GlcNAc was isolated and purified (Koshte et al. 1990). It did not agglutinate untreated human or sheep erythrocytes, but it did agglutinate rabbit erythrocytes. BanLec-I stimulated T-cell proliferation. BanLec-I was found to be very effective as a probe in detecting glycoproteins.

One of the predominant proteins in the pulp of ripe bananas (*Musa acuminata* L.) and plantains (*Musa* spp.) was identified as a lectin (Peumans et al. 2000). The banana and plantain agglutinins were designated BanLec and PlanLec, respectively. Both BanLec and PlanLec were found to be dimeric proteins composed of two identical subunits of 15 kDa. They readily agglutinated rabbit erythrocytes and exhibited specificity towards mannose. Molecular cloning revealed that BanLec has sequence similarity to previously described lectins of the family of jacalin-related lectins. The banana lectin was found to be a powerful murine T-cell mitogen. Further studies found that the banana (*Musa acuminata*) lectin was different from other mannose/glucose binding lectins, such as concanavalin A and the pea, lentil and *Calystegia sepium* lectins (Goldstein et al. 2001). Although a glucose/mannose binding protein which recognized α-linked gluco-and manno-pyranosyl groups of polysaccharide chain ends, the banana lectin was shown to bind to internal 3-O-α-D-glucopyranosyl units. The lectin was also shown to bind to the reducing glucosyl groups of β-1,3-linked glucosyl oligosaccharides (e.g. laminaribiose oligomers). Additionally, banana lectin also recognized β1,6-linked glucosyl end groups (gentiobiosyl groups) occurring in many fungal β1,3/1,6-linked polysaccharides.

BanLec, the mannose-specific banana lectin was found to be unique in its specificity for internal α1,3 linkages as well as β1,3 linkages at the reducing terminal (Gavrovic-Jankulovic et al. 2008). The immunomodulatory potential of natural BanLec was recognized by a strong immunoglobulin G4 antibody response and T cell mitogen activity in humans. To explore its applicability in glycoproteomics and its modulatory potential, the gene of banana lectin was cloned, sequenced and a recombinant protein was produced in *Escherichia coli*. The specificity of rBanLec for detection of glycan structures was the same as for natural BanLec. Besides, the immunomodulatory potential of rBanLec and nBanLec were comparable as assessed by an inhibition assay and a human T cell proliferation assay where they induced a strong proliferation response in CD3+, CD4+, and CD8+ populations of human PBMCs. This recombinant BanLec was found to be a useful reagent for glycoproteomics and lectin microarrays, with a potential for modulation of the immune response. Further studies using the murine model found that rBanLec exhibited immunostimulatory potential (Stojanovi et al. 2010). It was demonstrated that the responses of Balb/c- and C57 BL/6-originated splenocytes to rBanLec stimulation differed both qualitatively and in intensity. Induced responses included T lymphocyte proliferation and intensive interferon-gamma secretion. Both phenomena were more distinct in Balb/c-originated cultures. Balb/c-originated lymphocytes produced interleukin (IL)-4 and IL-10 following rBanLec stimulation.

A one-step purification of high purity lectins from banana pulp was developed using sugar-immobilized gold nano-particles (Nakamura-Tsuruta et al. 2008). For example, a protein isolated from banana using Glcalpha-GNP (alpha-glucose-immobilized gold nano-particle) was identified as banana lectin by trypsin-digested peptide-MS finger printing method.

A homodimeric, fructose-binding lectin was isolated from *Musa acuminata* (Del Monte banana) was found to have cytokine-inducing activity (Cheung et al. 2009). The N-terminal amino acid sequence of its identical 15-kDa subunits was similar to lectins from other *Musa* species except for the deletion of the N-terminal glycine residue in Del Monte banana lectin. The hemagglutinating activity was stable up to 80°C and in the pH range of 1–13. The lectin was capable of eliciting a mitogenic response in murine splenocytes and inducing the expression of the cytokines interferon-gamma, tumour necrosis factor-alpha, and interleukin-2 in splenocytes. The lectin also inhibited proliferation of leukemia (L1210) cells and hepatoma (HepG2) cells and the activity of HIV-1 reverse transcriptase. The results suggested that the banana lectin could be developed into a useful anti-HIV, immunopotentiating and antitumour agent in view of its trypsin stability and thermostability. BanLec was found to inhibit primary and laboratory-adapted human immunodeficiency virus HIV-1 isolates of different tropisms and subtypes (Swanson et al. 2010). BanLec was found to possess potent anti-HIV activity, with IC_{50} values in the low nanomolar to picomolar range. Findings indicated that BanLec inhibited HIV-1 infection by binding to the glycosylated viral envelope and blocking cellular entry. The relative anti-HIV activity of BanLec compared favorably to other anti-HIV lectins, such as snowdrop lectin and Griffithsin, and to T-20 and maraviroc, two anti-HIV drugs currently in clinical use. These results indicated BanLec to be a potential component for an antiviral microbicide that could be used to prevent the sexual transmission of HIV-1.

Plant lectins, that occur in foods like wheat, corn, tomato, peanut, kidney bean, banana, pea, lentil, soybean, mushroom, rice, and potato have great potential as cancer therapeutic agents (De Mejía and Prisecaru 2005). They maintain full biological activity by resisting digestion, surviving gut passage, and binding to gastrointestinal cells and/or entering the blood stream intact. In-vitro, in-vivo, and in human case studies have confirmed several lectins to have anticancer and therapeutic efficacy, preferentially binding to cancer cell membranes or their receptors, causing cytotoxicity, apoptosis, and inhibition of tumour growth. Ingestion of lectins also sequesters the available body pool of polyamines, thereby thwarting cancer cell growth. They also affect the immune system by altering the production of various interleukins, or by activating certain protein kinases. Lectins can bind to ribosomes and inhibit protein synthesis. They also alter the cell cycle by inducing non-apoptotic G1-phase accumulation mechanisms, G2/M phase cell cycle arrest and apoptosis, and can activate the caspase cascade. Lectins can also downregulate telomerase activity and inhibit angiogenesis.

Antivenom Activity

Phospholipase A2 (PLA2), myotoxic and hemorrhagic activities, including lethality in mice, induced by crotalidae venoms were significantly inhibited when different amounts of *Musa paradisiaca* extract were mixed with these venoms before assays (Borges et al. 2005). Contrariwise, mice that received banana extract and venoms without previous mixture or by separate routes were not protected against venom toxicity. Polyphenols and tannins were suggested to be responsible for the in-vitro inhibition of the toxic effects of snake venoms. The scientists concluded that the banana extract exerted protection against the toxic effects of snake venoms in-vitro but not in-vivo.

Mutagenic Activity

Studies indicated that fruit peel extract from *M. paradisiaca* showed mutagenic effect in the peripheral blood cells of Swiss albino mice (Andrade et al. 2008). Animals treated orally with three different concentrations of the extract (1,000, 1,500, and 2,000 mg/kg body weight) significantly increased micronucleated polychromatic erythrocytes. The two higher doses of the extract of *M. paradisiaca* induced statistically significant increases in the average numbers of DNA damage in peripheral blood leukocytes. The polychromatic/normochromatic erythrocyte ratio scored in the treated groups was not statistically different from the negative control.

See also Notes in *Musa acuminata* × *balbisiana* (AAB Group) 'Horn Plantain'

Traditional Medicinal Uses

Fruits, leaves, peels, root, and stalks from banana plants have been used orally or topically as a medicine for treating diarrhoea and dysentery, in the healing of intestinal lesions in colitis (Stover and Simmonds 1987). The banana plant is sued in folkloric medicine for treating inflammation, pains and snake-bites by the Sumu (Ulwa) people of south-eastern Nicaragua (Coe and Anderson 1999).

All parts of the banana plant have medicinal applications: the flowers in bronchitis and dysentery and on ulcers; cooked flowers are given to diabetics; the astringent plant sap in cases of hysteria, epilepsy, leprosy, fevers, hemorrhages, acute dysentery and diarrhea, and it is applied on hemorrhoids, insect and other stings and bites; young leaves are placed as poultices on burns and other skin afflictions; the astringent ashes of the unripe peel and of the leaves are taken in dysentery and diarrhea and used for treating malignant ulcers; the roots are administered in digestive disorders, dysentery and other ailments; banana seed mucilage is given in cases of catarrh and diarrhea in India.

Alleged hallucinogenic effects of the smoke of burning banana peel have been investigated scientifically and have not been confirmed.

Other Uses

Banana plants can be planted as a nurse and shade crop for young cocoa, pepper and coffee plants. Also used as intercrop in rubber and oil palm plantations.

The banana plant because of its continuous reproduction is regarded by Hindus as a symbol of fertility and prosperity, and the leaves and fruits are deposited on doorsteps of houses where marriages are taking place. A banana plant is often installed in the corner of a rice field as a protective charm. Malay women bathe with a decoction of banana leaves for 15 days after childbirth. Early Hawaiians used a young plant as a truce flag in wars.

In the Philippines, the Pinatubo Negritos cut off a banana plant close to the ground, make a hollow in the top of the stump, which then fills with watery sap and drunk as an emergency thirst quencher. Central Americans obtain the sap of the red banana in the same manner and take it as an aphrodisiac.

Banana leaves are used as plates, as food wrappers for wrapping curry pastes, meat, fish and cooked food. The leaves are also used as a cooking foil for steaming rice dumplings, rice cakes and food such as the spicy fish dish *Otak-otak* in Malaysia. The leaves are also used as packing material for food and flowers.

Banana leaves are also used for lining cooking pits and for wrapping food for cooking or storage. A section of leaf often serves as an eye-shade. In Latin America, it is a common practice during rains to hold a banana leaf by the petiole, upside-down, over one's back as an "umbrella" or "raincoat". The leaves of the 'Fehi' banana are used for thatching, packing, and cigarette wrappers. The pseudostems have been fastened together as rafts. Split lengthwise, they serve as padding on banana inspection turntables and as cushioning to protect the bunches ("stems") during transport in railway cars and trucks. Seat pads for benches are made of strips of dried banana pseudostems in Ecuador. In West Africa, fibre from the pseudostem is valued for fishing lines. In the Philippines, it is woven into a thin, transparent fabric called "agna" which is the principal material in some regions for women's blouses and men's shirts. It is also used for making handkerchiefs. In Ceylon, it is fashioned into soles for inexpensive shoes and used for floor coverings.

Plantain fibre is said to be superior to that from bananas. In the mid-nineteenth century, there was quite an active banana fibre industry in Jamaica. Improved processes have made it possible to utilize banana fibre for many purposes such as rope, table mats and handbags. In Kerala, India, a kraft type paper of good strength has been made from crushed, washed and dried banana pseudostems which yield 48–51% of unbleached pulp. A good quality paper is made by combining banana fibre with that of the betel nut husk (*Areca catechu* L.).

But Australian investigators maintained that the yield of banana fibre was too low for extraction to be economical. Only 28–113 g could be obtained from 18–36 kg of green pseudostems; 132 ton of green pseudostems would yield only 1 ton of paper. Their conclusion was that the pseudostem had much greater value as organic matter chopped and left in the field.

In southern India, leaf sheaths are stripped into shreds, dried and used for tying packages and making garlands. In regions where bananas are grown, the large leaves may be used as umbrellas when the pseudostems are tied together to form a floatation device (Morton 1987). Plantain peel ash can be used as a source of alkali for solid soap production (Onyegbado et al. 2002). In Bhutan, banana leaves are utilised in house construction, roofing, and for building temporary sheds (Rinchen 1996). The leaves and stems are used for preserving fish, meat, butter, "pani" (beetle leaves), and other foodstuffs. The leaves and stems are also utilised as fodder for elephants and other animals. Banana stems are also used as conduits and to make rafts. The fibres from banana leaves make good paper, but they are not commonly used for this purpose in Bhutan.

Banana fibres extracted from the leaf petiole, sheaths and pseudostem have multifarious uses (Uma et al. 2005) that include: base material for the pulp and paper and industry for manufacture of certain papers – kraft paper, bank notes, tea bags; a natural absorbent for oil spillage; base material for bioremediation and recycling; natural water purifier; textiles – cloth, shirt, blouses, manufacture of ropes, string, cordage, yarns, used for making handbags, purse, shoes, fishing lines, baskets, trays, table mats, lamp shades and handicrafts.

Processes for the production of paper pulp, biogas, alcoholic beverages, etc., from banana waste have also been described (Tewari et al. 1986; Hammond et al. 1996; Joshi et al. 2001; Dhabekar and Chandak 2010). Processes for the production of bioethanol from banana peels and wastes by fermentation with yeasts, *Saccharomyces uvarum* (Joshi et al. 2001) and *Saccharomyces cerevisiae* (Tewari et al. 1986; Dhabekar and Chandak 2010) were developed. Ethanol yields from normal ripe bananas were determined as follows: whole fruit – 0.091, pulp – 0.082, and peel – 0.006 l/kg of whole fruit (Hammond et al. 1996). Ripeness effects on ethanol yield were measured as green – 0.090, normal ripe – 0.082, and overripe – 0.069 l/kg of green whole bananas.

Pleurotus ostreatus and *P. sajor-caju* oyster mushrooms were employed in the biological fermentation of banana waste to produce animal feed (Reddy et al. 2000). The biological efficiency of the fruiting body-yield of *P. sajor-caju* was 47.32% and 49.32% on the steam sterilized pseudostem and leaf biomass of banana, respectively. Spent compost of both these substrates, when analysed for their nutritive composition, showed 5.01% and 4.73% crude protein, 15.11% and 13.13% crude fibre, 0.74% and 1.02% fat, 8.26% and 7.94% ash and 0.08% and 0.10% silicon, respectively. These values are close to the values required in balance dietary cattle feed. Thus, left over compost after the harvest of fruiting bodies may serve as rich source of protein enriched animal feed. Banana agricultural waste (leaf biomass and pseudostems) can be used as substrate for the production of various lignolytic and cellulolytic enzymes such as laccase, lignin peroxidase, xylanase, endo-1,4-β-d-glucanase (CMCase) and exo-1,4-β-d-glucanase using *Pleurotus ostreatus* and *P. sajor-caju* mushrooms (Reddy et al. 2003). Banana wastes can be utilised as substrates for the production of edible mushrooms like waste banana peel was utilised for the production of valuable microfungal biomass rich in protein and fatty acids (Essien et al. 2005).

Reject banana waste was digested in batch reactors to produce methane gas (Clarke et al. 2008).

Dried banana peel, because of its 30–40% tannin content, is used to blacken leather. The ash from the dried peel of bananas and plantains is rich in potash and used for making soap. That of the burned peel of unripe fruits of certain varieties is used for dyeing. Studies showed that minced bananas peels could remove heavy metals namely copper and lead from river water similar to other purification materials (Castro et al. 2011).

Banana peels were found to be promising materials for adsorption removal of dyes from aqueous solutions (Annadurai et al. 2002). The adsorption capacities decreased in the order methyl orange (MO) > methylene blue (MB) > Rhodamine B (RB) > Congo red (CR) > methyl violet (MV) > amido black 10B (AB). An alkaline pH was favorable for the adsorption of dyes.

Every part of the banana and plantain plant (except the roots and suckers) can be and have been used to feed livestock in various parts of the world. The following banana and plantain materials have been fed with varying degrees of success to various types of livestock (Babatunde 1992):

- Fresh, green, chopped or unchopped green banana fruits with peels.
- Ripe, raw whole banana or plantain fruits.
- Dehydrated, sliced, milled, whole, green bananas or plantains.
- Cooked, green, whole banana and plantain fruits.
- Dehydrated, milled, green and ripe plantain or banana peels.
- Chopped, fresh, green plantain and banana fruits ensiled with molasses, grass, legume, rice bran or any other products that will increase their feeding value.
- Whole, fresh, green leaves, fed directly to animals or after being ensiled with an easily fermentable carbohydrate such as molasses.
- Banana and Plantain stalk or pseudostem, chopped and fed raw, or ensiled with easily fermentable carbohydrates, e.g. molasses.

Reject ripe bananas, supplemented with protein, vitamins and minerals, are commonly fed to swine. Green bananas are also used for fattening hogs but, because of the dryness and astringency and bitter taste due to the tannin content, these animals do not care for them unless they are cooked, which makes the feeding costs too high for most growers. Therefore, dehydrated green banana meal has been developed and, though not equal to grain, can constitute up to 75% of the normal hog diet, 40% of the diet of gestating sows. It is not recommended for lactating sows, nor are ripe bananas, even with a 40% protein supplement.

Beef cattle are very fond of green bananas whether they are whole, chopped or sliced. Because of the fruit's deficiency in protein, urea is added at the rate of 8.8 lbs (4 kg) per ton, with a little molasses mixed in to mask the flavor. But transportation is expensive unless the cattle ranch is located near the banana fields. A minor disadvantage is that the bananas are somewhat laxative and the cattle need to be washed down daily. With dairy cattle, it is recommended that bananas constitute no more than 20% of the feed.

In the Philippines, it has been found that meal made from dehydrated reject bananas can form 14% of total broiler rations without adverse effects. Meal made from green and ripe plantain peels has been experimentally fed to chicks in Nigeria. A flour from unpeeled plantains, developed for human consumption, was fed to chicks in a mixture of 2/3 flour and 1/3 commercial chickfeed and the birds were maintained until they reached the size of fryers. They were found thinner and lighter than those on 100% chickfeed and the gizzard lining peeled in shreds. It was assumed that these effects were the result of protein deficiency in the plantains, but they were more likely the result of the tannin content of the flour which interferes with the utilization of protein.

Leaves, pseudostems, fruit stalks and peels, after chopping, fermentation, and drying, yield a meal somewhat more nutritious than alfalfa press-cake. This waste material has been considered for use as organic fertilizer in Somalia. In Malaya, pigs fed the pseudostems are less prone to liver and kidney parasites than those on other diets.

Ravindran and Blair (1991) found that up to 25% of green banana fruit reject waste can be processed and incorporated in poultry diets without adverse effect. Tewe (1983) found that plantain peel meal can successfully be included in broiler rations with maize up to 7.5% beyond which it was detrimental. Studies in the Philippines found that banana rejects dehydrated and ground to form a meal could be utilized in the formulation of broiler rations (Velasco et al. 1983). Results over an 8 weeks feeding period indicated that the optimum level of banana meal incorporation in the diet was 25% banana meal replacing ground corn (or 14% banana meal of the total ration). At this concentration, the broilers showed comparable body weight gain, feed intake, feed conversion efficiency and carcass quality to those fed the control diet (no banana meal).

According to Le Dividich et al. (1978), cattle and pigs relished chopped, fresh, green banana or plantain fruits when the slices had been sprinkled with salt to elevate their inherent low levels of inorganic nutrients. For ensiling purposes, the chopped green bananas or plantains are preferred to the ripe fruits which lose some of their dry matter and, in particular sugars during ensiling. Clavijo and Maner, (1975) reported in their review paper, that pigs could successfully utilize bananas in the fresh or dried meal forms and that the pig would consume large quantities of bananas and grow well if those bananas were sufficiently ripe (Hernandez and Maner 1967, cited by Le Dividich et al. (1978). It was further shown that when the banana was fed green, combined with 30% protein supplement, they voluntarily consumed only about 50% as much as when fed ripe bananas. Ffoulkes and Preston (1978) found that the dry matter of banana leaves and pseudostems was relatively digestible for ruminants; 65% digestibility for leaves and 75% for the pseudostems. They recommended that urea and a highly digestible forage or sweet potato foliage should be used as supplements to pseudostems or leaves being fed as banana the leaves and pseudostems alone could not meet the maintenance requirements of ruminants. Geoffroy and Chenost (1973), found that banana meal and banana silage could be employed to replace the cereal in a concentrate diet for goats. Dehydrated, green, milled banana (banana pulp flour) had been successfully used as a source of starch in the preparation of calf feeds and specifically in the manufacture of milk replacers (Le Dividich et al. 1978). Viswanathan et al. (1989), found that dried banana stalk replacing 0%, 20%, 40% and 50% Paragrass hay can be fed to sheep without any detrimental affect on the health of the animals.

Comments

The name 'Dwarf Cavendish' is in reference to the height of the pseudostem, not the fruit (which are medium sized). Cavendish bananas especially Williams and Grande Nain (Chiquita banana) are the most popular of the edible banana cultivars and the dominant commercial cultivars for the export market of Central America, South America, the Caribbean, West Africa Philippines, Thailand and Indonesia. Cavendish bananas comprise over 40% of these fruit that are produced worldwide. The various local cultivars are similar except for their height and characteristics of the bunch and fruit.

References

Adams MR (1980) The small-scale production of vinegar from bananas. Publication Tropical Products Institute, No. G 132, 15 pp

Adeva LV, Gopez MD, Payumo EM (1968) Studies on the preparation and storage qualities of banana chips. Philip J Sci 97(1):27–35

Adriaens EL, Lozet F (1951) Contributions a l' etude des boissons fermentées indigenes au Ruanda. Bull Agric Congo Belge 42:931–950

Agama-Acevedo E, Islas-Hernandez JJ, Osorio-Diaz P, Rendon-Villalobos R, Utrilla-Coello RG, Angulo O, Bello-Pérez LA (2009) Pasta with unripe banana flour: physical, texture and preference study. J Food Sci 74(6):263–267

Aguirre-Cruz A, Alvarez-Castillo A, Yee-Madeira H, Bello-Pe´rez LA (2008) Production of fiber-rich powder by the acid treatment of unripe banana flour. J Appl Polym Sci 109(1):382–387

Akamine K, Koyama T, Yazawa K (2009) Banana peel extract suppressed prostate gland enlargement in testosterone-treated mice. Biosci Biotechnol Biochem 73(9):1911–1914

Akubor PI, Obio SO, Nwadomere KA, Obiomah E (2003) Production and quality evaluation of banana wine. Plant Foods Hum Nutr 58:1–6

Álvarez-Acosta T, Cira León LN, Acosta-González S, Parra-Soto H, Cluet-Rodriguez I, Rossell MR, Colina-Chourio JA (2009) Beneficial role of green plantain [*Musa paradisiaca*] in the management of persistent diarrhea: a prospective randomized trial. J Am Coll Nutri 28(2):169–176

Andrade CUB, Perazzo FF, Maistro EL (2008) Mutagenicity of the *Musa paradisiaca* (Musaceae)

fruit peel extract in mouse peripheral blood cells in vivo. Genet Mol Res 7(3):725–732

Annadurai G, Juang RS, Lee DJ (2002) Use of cellulose-based wastes for adsorption of dyes from aqueous solutions. J Hazard Mater 92(3):263–274

Aparicio-Saguilán A, Sáyago -Ayerdi SG, Vargas-Torres A, Tovar J, Ascencio-Otero TE, Bello- Pérez LA (2007) Slowly digestible cookies prepared from resistant starch-rich lintnerized banana starch. J Food Comp Anal 20(3–4):175–181

Aurora A, Sharma MP (1990) Use of banana in non-ulcer dyspepsia. Lancet 335:612–613

Aurore G, Parfait B, Fahrasmane L (2008) Bananas, raw materials for making processed food products. Trends Food Sci Technol 20:1–13

Aurore G, Ginies C, Ganou-parfait B, Renard CMGC, Fahrasmane L (2011) Comparative study of free and glycoconjugated volatile compounds of three banana cultivars from French West Indies: Cavendish, Frayssinette and Plantain. Food Chem 129(1):28–34

Babatunde GM (1992) Availability of banana and plantain products for animal feeding. In: Machin D, Nyvold S (eds) Roots, tubers, plantains and bananas in animal feeding. FAO, Rome

Bello-Pérez LA, Agama-Acevedo E, Sánchez-Hernández L, Paredes-López O (1999) Isolation and partial characterization of banana starches. J Agric Food Chem 47(3):854–857

Bello-Pérez LA, Sáyago-Ayerdi SG, Méndez-Montealvo G, Tovar J (2004) In vitro digestibility of banana starch cookies. Plant Foods Hum Nutr 59(2):79–83

Blades BL, Duffi cy L, Englberger L, Daniells JW, Coyne T, Hamill S, Wills RB (2003) Bananas and plantains as a source of provitamin A. Asia Pac J Clin Nutr 12(Suppl):S36

Ble-Castillo JL, Aparicio-Trápala MA, Francisco-Luria MU, Córdova-Uscanga R, Rodríguez-Hernández A, Méndez JD, Díaz-Zagoya JC (2010) Effects of native banana starch supplementation on body weight and insulin sensitivity in obese type 2 diabetics. Int J Environ Res Public Health 7(5):1953–1962

Borges MH, Alves DL, Raslan DS, Pilo-Veloso D, Rodrigues VM, Homsi-Brandeburgo MI, de Lima ME (2005) Neutralizing properties of *Musa paradisiaca* L. (Musaceae) juice on phospholipase A2, myotoxic, hemorrhagic and lethal activities of Crotalidae venoms. J Ethnopharmacol 98:21–29

Boudhrioua N, Giampaoli P, Bonazzi C (2003) Changes in aromatic components of banana during ripening and air-drying. Lebensm Wiss Technol 36(6):633–642

Bukusuba J, Muranga FI, Nampala P (2008) Effect of processing technique on energy density and viscosity of cooking banana: implication for weaning foods in Uganda. Int J Food Sci Technol 43(8):1424–1429

Carreño S, Aristizábal LM (2003) Utilisation de bananes plantain pour produire du vin. Info Musa 12(1):2–4

Carvalho GBM, Silva DP, Vicente A, Felipe MGA, Teixeira JA, Almeida E, Silva JB (2008) Banana: an alternative as adjunct and natural aromatic compound for beers. In: Markos J (ed) Proceedings of the 35th international conference of Slovak Society of Chemical Engineering. Tatranské Matliare, Slovak University of Technology, Bratislava (Slovakia), Paper no 279–1

Carvalho GBM, Silva DP, Bento CV, Vicente AA, Teixeira JA, Felipe MGA, Almeida E, Silva JB (2009) Banana as adjunct in beer production: applicability and performance of fermentative parameters. Biotechnol Appl Biochem 155(1–3):356–365

Castro RSD, Caetano L, Ferreira G, Padilha PM, Saeki MJ, Zara LF, Martines MAU, Castro GR (2011) Banana peel applied to the solid phase extraction of copper and lead from river water: preconcentration of metal ions with a fruit waste. Ind Eng Chem Res 50(6):3446–3451

Champion J (1970) Culture du bananier in Rwanda. Fruits 25:161–168

Cheung AH, Wong JH, Ng TB (2009) *Musa acuminata* (Del Monte banana) lectin is a fructose-binding lectin with cytokine-inducing activity. Phytomed 16(6–7):594–600

Chong LC, Noor Aziah AA (2008) Influence of partial substitution of wheat flour with banana (*Musa paradisiaca* var. *Awak*) flour on the physico-chemical and sensory characteristics of doughnuts. Int Food Res J 15(2):119–124

Chong LC, Noor Aziah AA (2010) Effects of banana flour and beta-glucan on the nutritional and sensory evaluation of noodles. Food Chem 119(1):34–40

Clarke WP, Radnidge P, Lai E, Jensen PD, Hardin MT (2008) Digestion of waste bananas to generate energy in Australia. Waste Manag 28(3):527–533

Clavijo H, Maner JH, (1975) The use of waste bananas for swine feed. In: Proceedings of the conference on animal feeds of tropical and sub-tropical origin. Tropical Products Institute, London, pp 99–106

Coe F, Anderson GJ (1999) Ethnobotany of the Sumu (Ulwa) of southeastern Nicaragua and comparisons with Miskitu plant lore. Econ Bot 53:363–383

Cox PA (1980) Two Samoan technologies for breadfruit and banana preservation. Econ Bot 34:181–185

Daniells J, Jenny C, Tomekpe K (2001) Musalogue: a catalogue of *Musa* germplasm. Diversity in the genus *Musa* (Arnaud E. Sharrock S compli.) International Network for the Improvement of Banana and Plantain, Montpellier, France, <www.inibap.org/publications/musalogue.pdf>

Davies G (1993) Domestic banana beer production in Mpigi District, 2nd edn. Infomusa, Uganda, pp 12–15

De Mejía EG, Prisecaru VI (2005) Lectins as bioactive plant proteins: a potential in cancer treatment. Crit Rev Food Sci Nutr 45(6):425–445

Dei-Tutu J (1975) Studies on the development of fatale mix, a plantain product. Ghana J Agric Sci 8(5): 153–157

Deneo-Pellegrini H, De Stefani E, Ronco A (1996) Vegetables, fruits, and risk of colorectal cancer: a case-control study from Uruguay. Nutr Cancer 25(3):297–304

Denyer G, Dickinson S (2010). Home of the glycemic index. http://www.glycemicindex.com/

Dhabekar A, Chandak A (2010) Utilization of banana peels and beet waste for alcohol production. Asiatic J Biotech Res 1:8–13

Dignan CA, Burlingame BA, Arthur JM, Quigley RJ, Milligan GC (1994) The Pacific Islands food composition tables. South Pacific Commission, Palmerston North, p 147

Emaga TH, Andrianaivo RH, Wathelet B, Tchango JT, Paquot M (2007) Effects of the stage of maturation and varieties on the chemical composition of banana and plantain peels. Food Chem 103(2):590–600

Emaga TH, Robert C, Ronkart SN, Wathelet B, Paquot M (2008) Dietary fibre components and pectin chemical features of peels during ripening in banana and plantain varieties. Bioresour Technol 99(10):4346–4354

Emery EA, Ahmad S, Koethe JD, Skipper A, Perlmutter S, Paskin DL (1997) Banana flakes control diarrhea in enterally fed patients. Nutr Clin Pract 12(2):72–75

Englberger L, Darnton-Hill I, Coyne T, Fitzgerald MH, Marks GC (2003) Carotenoid-rich bananas: a potential food source for alleviating vitamin A deficiency. Food Nutr Bull 24(4):303–318

Ercan N, Nuttall FQ, Gannon MC, Lane JT, Burmeister LA, Westphal SA (1993) Plasma glucose and insulin responses to bananas of varying ripeness in persons with noninsulin-dependent diabetes mellitus. J Am Coll Nutr 12(6):703–709

Eshun S (1977) Popular Ghanaian dishes. Ghana Publishing Corp, Tema, 107 pp

Espino RRC, Jamalussin SH, Silayoi B, Nasution RE (1992) *Musa* L. (edible cultivars). In: Verheij EWM, Coronel RE (eds) Plant resources of South-East Asia No 2. Edible fruits and nuts. PROSEA, Bogor, pp 225–233

Essien JP, Akpan EJ, Essien EP (2005) Studies on mould growth and biomass production using waste banana peel. Bioresour Technol 96(13):1451–1456

Evelin MA, Jacob JP, Vijayanand D (2007) Packaging and storage studies on spray dried ripe banana powder under ambient conditions. J Food Sci Technol 44(1):16–21

Fasolin LH, de Almeida GC, Castanho PS, Netto-Oliveira ER (2007) Cookies produced with banana meal: chemical, physical and sensorial evaluation. Cienc Tecnol Aliment 27(3):524–529

Fernandes FAN, Rodrigues S (2007) Ultrasound as pre-treatment for drying of fruits: dehydration of banana. J Food Eng 82(2):261–267

Ffoulkes D, Preston TR (1978) The banana plant as cattle feed: digestibility and voluntary intake of mixtures of sugarcane and banana forage. Trop Anim Prod 3:125–129

Foster-Powell K, Holt SHA, Brand-Miller JC (2002) International tables of glycemic index and glycemic loads values: 2002. Am J Clin Nutr 62:5–56

Gamlath S (2008) Impact of ripening stages of banana flour on the quality of extruded products. Int J Food Sci Technol 43(9):1541–1548

Gavrovic-Jankulovic M, Poulsen K, Brckalo T, Bobic S, Lindner B, Petersen A (2008) A novel recombinantly produced banana lectin isoform is a valuable tool for glycoproteomics and a potent modulator of the proliferation response in CD3+, CD4+, and CD8+ populations of human PBMCs. Int J Biochem Cell Biol 40(5):929–941

Geoffroy F, Chenost M (1973) Utilisation des dechets de banane par les ruminants en zone tropicale humide. Bull Tech Product Anim (2–3):67–75

Goldstein JL, Wick EL (1969) Lipid in ripening banana fruit. J Food Sci 34:482–484

Goldstein IJ, Winter HC, Mo H, Misaki A, Van Damme EJ, Peumans WJ (2001) Carbohydrate binding properties of banana (*Musa acuminata*) lectin II. Binding of laminaribiose oligosaccharides and beta-glucans containing beta1,6-glucosyl end groups. Eur J Biochem 68(9):2616–2619

González-Montelongo R, Lobo MG, González M (2010) Antioxidant activity in banana peel extracts: testing extraction conditions and related bioactive compounds. Food Chem 119(3):1030–1039

Goode PM (1974) Some local vegetables and fruits of Uganda. Dept. Agric, Entebbe

Green DM, Ropper AH, Kronmal RA, Psaty BM, Burke GL (2002) Serum potassium level and dietary potassium intake as risk factors for stroke. Neurology 59(3):314–320

Guerrero S, Alzamora SM, Gerschenson N (1994) Development of shelf-stable banana puree by combined factors: microbial stability. J Food Prot 57(10):902–907

Hammond JB, Egg R, Diggins D, Coble GC (1996) Alcohol from bananas. Bioresour Technol 56(1): 125–130

Hermansen K, Rasmussen O, Gregersen S, Larsen S (1992) Influence of ripeness of banana on the blood glucose and insulin response in type 2 diabetic subjects. Diabet Med 9(8):739–743

Hernandez J, Maner MH (1967) Evaluaci6n del Banano Verde, Maduro y Verde Cocido en Crecimiento y Engorde de Cerdos. Instituto Nacional de Investigaciones Agropecuarias. Experimento SDGP 2.2.3, Quito, Ecuador

Hernandez-Uribe JP, Rodrigez-Ambriz LA, Bello-Pérez LA (2008) Obtention of fructose syrup from plantain (*Musa paradisiaca* L.) starch: partial characterization. Interciencia 33(5):372–376

Jackson JC, Bourne MC, Barnard J (1996) Optimization of blanching for crispness of banana chips using response surface methodology. J Food Sci 61(1):165–166

Jang DS, Park EJ, Hawthorne ME, Vigo JS, Graham JG, Cabieses F, Santarsiero BD, Mesecar AD, Fong HH, Mehta RG, Pezzuto JM, Kinghorn AD (2002) Constituents of *Musa* × *paradisiaca* cultivar with the potential to induce the phase II enzyme, quinone reductase. J Agric Food Chem 50(22):6330–6334

Jordan MJ, Goodner KL, Shaw PE (2001) Volatile components in banana (*Musa acuminata* Colla cv.

Cavendish) and yellow passion fruit (*Passiflora edulis* Sims f. *flavicarpa* Degner) as determined by GC-MS and GC-olfatorymetry. Proc Fla State Hort Soc 114:153–157

Joshi SS, Dhopeshwarkar R, Jadhav U, Jadhav R, D'souza L, Dixit J (2001) Continuous ethanol production by fermentation of waste banana peels using flocculating yeast. Indian J Chem Technol 8(3):153–156

Juarez-Garcia E, Agama-Acevedo E, Sáyago-Ayerdi SG, Rodríguez-Ambriz SL, Bello-Pérez LA (2006) Composition, digestibility and application in breadmaking of banana flour. Plant Foods Hum Nutr 61(3):131–137

Kanazawa K, Sakakibara H (2000) High content of dopamine, a strong antioxidant, in Cavendish banana. J Agri Food Chem 48(3):844–848

Knapp FF, Nicholas HJ (1969a) Sterols and triterpenes of banana peel. Phytochem 8(1):207–214

Knapp FF, Nicholas HJ (1969b) The sterols and triterpenes of banana pulp. J Food Sci 34(6):584–586

Koshte VL, van Dijk W, van der Stelt ME, Aalberse RC (1990) Isolation and characterization of BanLec-I, a mannoside-binding lectin from *Musa paradisiaca* (banana). Biochem J 272(3):721–726

Krishnan K, Vijayalakshmi NR (2005) Alterations in lipids & lipid peroxidation in rats fed with flavonoid rich fraction of banana (*Musa paradisiaca*) from high background radiation area. Indian J Med Res 122:540–546

Kwan ML, Block G, Selvin S, Month S, Buffler PA (2004) Food consumption by children and the risk of childhood acute leukemia. Am J Epidemiol 160(11): 1098–1107

Lancaster PA, Coursey DG (1984) Traditional post harvest technology of perishable staples, vol No. 59, FAO Agricultural Services Bull. Food and Agriculture Organization of the United Nations, Rome

Le Dividich J, Geoffroy F, Canope I, Chenost M (1978) Using waste banana as animal feed, 1978. In: Ruminant nutrition: selected articles from the world animal review. FAO Animal Production and Health Paper 12. FAO, Rome, pp 160

Levine SR, Coull BM (2002) Potassium depletion as a risk factor for stroke: will a banana a day keep your stroke away? Neurology 59(3):302–303

Liu S, Willett WC, Stampfer MJ, Hu FB, Franz M, Sampson L, Hennekens CH, Manson JE (2000) A prospective study of dietary glycemic load, carbohydrate intake, and risk of coronary heart disease in US women. Am J Clin Nutr 71(6):1455–1461

Ludwig DS (2003) Dietary glycemic index and the regulation of body weight. Lipids 38(2):117–121

Macku C, Jennings WG (1987) Production of volatiles by ripening bananas. J Agri Food Chem 35:845–848

Manimegalai G, Premalatha MR (1996) Studies on banana toffees. In: Nayar NK, George TE (eds) Symposium on technological advancement in banana/plantain production and processing. Kerala Agricultural University, Mannuthy, p 44

Massal E, Barrau J (1956) Musaceae, pp 15–18. In: Food plants of the South Sea Islands. Tech. Pap. No. 94. South Pacific Commission, Noumea, New Caledonia, pp 51

Meadows AB (2007) Drum drying of banana pulp on the sorption isotherm and flexible packaging requirement. Nigerian Food J 25(2):130–140

Menezes EW, Dan MC, Cardenette GH, Goñi I, Bello-Pérez LA, Lajolo FM (2010) In vitro colonic fermentation and glycemic response of different kinds of unripe banana flour. Plant Foods Hum Nutr 65(4):379–385

Mohamed A, Xu J, Singh M (2009) Yeast leavened banana-bread: formulation, processing, colour and texture analysis. Food Chem 118(3):620–626

Mohapatra D, Mishra S, Sutar N (2010) Banana and its by-product utilisation: an overview. J Sci Ind Res 69:323–329

Mokbel MS, Hashinaga F (2005) Antibacterial and antioxidant activities of banana (Musa, AAA cv. Cavendish) fruits peel. Am J Biochem Biotechnol 1(3):125–131

Molla MM, Nasrin TAA, Islam MN (2009) Study on the suitability of banana varieties in relation to preparation of chips. J Agric Rural Dev 7(1&2):81–86

Mondal SK, Ray B, Thakur S, Ghosal PK (2001) Isolation, purification and some structural features of the mucilaginous exudate from *Musa paradisiaca*. Fitoterapia 72:263–271

Morton JF (1987) Banana. In: Julia F (ed) Fruits of warm climates. Morton, Miami, pp 29–46

Mukasa SK, Thomas DG (1970) Staple food crops in Uganda. pp. 139–153. In: Jameson JD (ed) Agriculture in Uganda, 2nd edn. Oxford University Press, London, p 395

Murray KE, Palmer JK, Whitfield FB, Kennett BH, Stanley G (1968) The volatile alcohols of ripe bananas. J Food Sci 33(6):632–634

Nakamura-Tsuruta S, Kishimoto Y, Nishimura T, Suda Y (2008) One-step purification of lectins from banana pulp using sugar-immobilized gold nano-particles. J Biochem 143(6):833–839

Newley P, Akehurst A, Campbell B (2008) Banana growing guide: cavendish bananas. NSW Department of Primary Industries, Sydney

Nguyen TBT, Ketsa S, van Doorn WG (2003) Relationship between browning and the activities of polyphenol oxidase and phenylalanine ammonia lyase in banana peel during low temperature storage. Posth Biol Technol 30(2):187–193

Njoki P, Faller JF (2001) Development of an extruded plantain/corn/soy weaning food. Int J Food Sci Technol 36(4):415–23

Nogueira JMF, Fernandes PJP, Nascimento AMD (2003) Composition of volatiles of banana cultivars from Madeira Island. Phytochem Anal 14(2):87–90

Ojewole JA, Adewunmi CO (2003) Hypoglycemic effect of methanolic extract of *Musa paradisiaca* (Musaceae) green fruits in normal and diabetic mice. Methods Find Exp Clin Pharmacol 25(6):453–456

Oke OL, Redhead J, Hussain MA (1990) Roots, tubers, plantains and bananas in human nutrition, vol No. 24, FAO Food and Nutrition Series. Food and Agriculture Organization of the United Nations, Rome

Olaoye OA, Onilude AA, Idowu OA (2006) Quality characteristics of bread produced from composite flours of wheat, plantain and soybeans. Afr J Biotechnol 5(11):1102–1106

Onyegbado CO, Iyagba ET, Offor OJ (2002) Solid soap production using plantain peel ash as source of alkali. J Appl Sci Environ Manag 6(1):73–77

Onyeka U, Dibia I (2002) Malted weaning food made from maize, soybean, groundnut and cooking banana. J Sci Food Agric 82(5):513–516

Osorio-Diaz P, Aguilar-Sandoval A, Agama-Acevedo E, Rendon-Villalobos R, Tovar J, Bello-Pérez LA (2008) Composite durum wheat flour/plantain starch white salted noodles: proximal composition, starch digestibility, and indigestible fraction content. Cereal Chem 85(3):339–343

Ovando-Martinez M, Sayago-Ayerdi S, Agama-Acevedo E, Goni I, Bello-Pérez LA (2009) Unripe banana flour as an ingredient to increase the undigestible carbohydrates of pasta. Food Chem 113(1):121–126

Panda DK, Ghosh D, Bhat B, Talwar SK, Jaggi M, Mukherjee R (2009) Diabetic therapeutic effects of ethyl acetate fraction from the roots of *Musa paradisiaca* and seeds of *Eugenia jambolana* in streptozotocin-induced male diabetic rats. Methods Find Exp Clin Pharmacol 31(9):571–584

Pérez AG, Cert A, Rios JJ, Olias JM (1997) Free and glycosidically bound volatile compounds from two bananas cultivars: valery and Pequen˜a Enna. J Agri Food Chem 45:4393–4397

Perfumi M, Massi M, De Caro G (1994) Effects of banana feeding on deoxycorticosterone-induced hypertension and salt consumption in rats. Pharm Bio 32(2): 115–125

Persley GJ, George P (1996) Banana improvement: research challenge and opportunity. World Bank Publications, Washington, DC, p 47

Peumans WJ, Zhang W, Barre A, Houlès Astoul C, Balint-Kurti PJ, Rovira P, Rougé P, May GD, Van Leuven F, Truffa-Bachi P, Van Damme EJ (2000) Fruit-specific lectins from banana and plantain. Planta 211(4):546–554

Porcher MH et al (1995–2020) Searchable world wide web multilingual multiscript plant name database. Published by The University of Melbourne, Australia, http://www.plantnames.unimelb.edu.au/Sorting/Frontpage.html

Prasad KV, Bharathi K, Srinivasan KK (1993) Evaluation of *Musa paradisiaca* Linn. cultivar – "Puttubale" stem juice for antilithiatic activity in albino rats. Indian J Physiol Pharmacol 37:337–341

Rai PK, Jaiswal D, Rai NK, Pandhija S, Rai AK, Watal G (2009) Role of glycemic elements of *Cynodon dactylon* and *Musa paradisiaca* in diabetes management. Lasers Med Sci 24(5):761–768

Ramírez D, Pacheco-Delahaye E (2000) Evaluación físico-química de salsas de mango (*Mangifera indica* L.) utilizando harina de plátano verde (*Musa* sp., grupo AAB) como agente espesante. Rev Facult Agron 26(1):53–65

Ramli S, Alkarkhi AF, Yeoh SY, Leong MT, Easa AM (2009) Effect of banana pulp and peel flour on physicochemical properties and in vitro starch digestibility of yellow alkaline noodles. Int J Food Sci Nutr 60(Suppl 4):326–340

Ranzani MR, Sturion GL, Bicudo MH (1996) Chemical and biological evaluation of ripe banana peel. Arch Latinoam Nutr 46(4):320–324, In Portuguese

Rao NM (1991) Protease inhibitors from ripened and unripened bananas. Biochem Int 24(1):13–22

Rao NM, Prasad KVSRG, Pai KSR (1999) Angiotensin converting enzyme inhibitors from ripened and unripened bananas. Curr Sci 76(1):86–88

Rashidkhani B, Lindblad P, Wolk A (2005) Fruits, vegetables and risk of renal cell carcinoma: a prospective study of Swedish women. Int J Cancer 113(3):451–455

Ravindran V, Blair R (1991) Feed resources for poultry production in Asia and the Pacific region. I. Energy sources. World's Poultry Sci J 47:213–231

Reddy GV, Kothari IL, Akhilesh M, Patel CR (2000) Solid state fermentation of banana waste into protein enriched animal feed by *Pleurotus sajor-caju*. Appl Biol Res 2(1/2):177–178

Reddy GV, Babu PR, Komaraiah P, Roy KRRM, Kothari IL (2003) Utilization of banana waste for the production of lignolytic and cellulolytic enzymes by solid substrate fermentation using two *Pleurotus* species (*P. ostreatus* and *P. sajor-caju*). Process Biochem 38(10):1457–1462

Rinchen D (1996) Bamboo, cane, wild banana, fibre, floss and brooms. In: Non-wood forest products of Bhutan. RAP Publication: 1996/6. Food and Agriculture Organization of the United Nations, Bangkok, Thailand

Robinson JC (1996) Bananas and plantains. Crop Production Science in Horticulture, CAB International, Wallingford, p 238

Saifullah R, Abbas FMA, Yeoh SY, Azhar ME (2009a) Utilization of banana peel as a functional ingredient in yellow noodle. Asian J Food Ag-Ind 2(3):321–329

Saifullah R, Abbas FMA, Yeoh SY, Azhar ME (2009b) Utilization of green banana flour as a functional ingredient in yellow noodle. Int Food Res J 16(3):373–379

Sarkar C, Bairy KL, Rao NM, Udupa EG (1999) Effect of banana on cold stress test & peak expiratory flow rate in healthy volunteers. Indian J Med Res 110:27–29

Schiota H (1993) New esteric components in the volatiles of banana fruit (*Musa sapientum* L.). J Agri Food Chem 41:2056–2062

Seymour GB (1993) Banana. In: Seymour G, Taylor J, Tucker G (eds) Biochemistry of fruit ripening. Chapman and Hall, London, pp 95–98

Simmonds NW (1966) Bananas, 2nd edn. Longmans, London

Singh S, Jain S, Singh SP, Singh D (2009) Quality changes in fruit jams from combinations of different fruit pulps. J Food Process Preserv 33(1):41–57

Singhal MK (2001) Banana potassium and stroke. Indian J Exp Biol 40(11):1322

Someya S, Yoshiki Y, Okubo K (2002) Antioxidant compounds from bananas (Musa Cavendish). Food Chem 79(3):351–354

Stojanović MM, Zivković IP, Petrusić VZ, Kosec DJ, Dimitrijević RD, Jankov RM, Dimitrijević LA, Gavrović-Jankulović MD (2010) In vitro stimulation of Balb/c and C57 BL/6 splenocytes by a recombinantly produced banana lectin isoform results in both a proliferation of T cells and an increased secretion of interferon-gamma. Int Immunopharmacol 10(1):120–129

Stover RH, Simmonds NW (1987) Bananas, 3rd edn, Tropical agricultural series. Longman, Essex, p 468

Subagio A, Morita N, Sawada S (1996) Carotenoids and their fatty-acid esters in banana peel. J Nutr Sci Vitaminol 42(6):553–566

Suntharalingam S, Ravindran G (1993) Physical and biochemical properties of green banana flour. Plant Foods Hum Nutr 4:19–27

Suresh ER, Ethiraj S (1991) Utilization of overripe bananas for vinegar production. Trop Sci 31(3):317–320

Swanson MD, Winter HC, Goldstein IJ, Markovitz DM (2010) A lectin isolated from bananas is a potent inhibitor of HIV replication. J Biol Chem 285(12):8646–8655

Tewari HK, Marwaha SS, Rupal K (1986) Ethanol from banana peels. Agric Wastes 16:135–146

Tewe OO (1983) Replacing maize with plantain peels in diets for broilers. Nutr Rep Int 28(1):23–29

Thakorlal J, Perera CO, Smith B, Englberger L, Lorens A (2010) Resistant starch in Micronesian banana cultivars offers health benefits. Pac Health Dialog 16(1):49–59

Tribess TB, Hernandez-Uribe JP, Méndez-Montealvo MGC, Menezes EW, Bello-Perez LA, Tadini CC (2009) Thermal properties and resistant starch content of green banana flour (*Musa cavendishii*) produced at different drying conditions. Food Sci Technol 42(5):1022–1025

Uma S, Kalpana S, Sathiamoorthy S, Kumar V (2005) Evaluation of commercial cultivars of banana (*Musa* spp.) for their suitability for the fi bre industry. Plant Genet Resour Newsl 142:29–35

U.S. Department of Agriculture, Agricultural Research Service (2010) USDA National Nutrient Database for Standard Reference, Release 23. Nutrient Data Laboratory Home Page, http://www.ars.usda.gov/ba/bhnrc/ndl

Valmayor RV, Wagih ME (1996) *Musa* L. (plantain and cooking banana). In: Flach M, Rumawas F (eds) Plant resources of South-East Asia No. 9. Plants yielding non-seed carbohydrates. Prosea Foundation, Bogor, pp 126–131

Valmayor RV, Jamaluddin SH, Silayoi B, Kusumo S, Danh LD, Pascua OC, Espino RRC (2000) Banana cultivar names and synonyms in Southeast Asia. International Network for the Improvement of Banana and Plantain – Asia and the Pacific Office, Los Banos, Laguna, Philippines

Velasco MNI, Bautista JG, Cruz LN, Ganac EC, Silverio VG, Avante DC (1983) A study of the utilization of banana meal as replacement for ground corn in poultry rations. Philipp J Anim Ind 38(1–2):11–17

Viswanathan R, Kadirvel, Chandrasekaran D (1989) Nutritive value of banana stalk (*Musa cavendishi*) as a feed for sheep. Anim Feed Sci Technol 22:327–332

Yusoff MS, Mohd Zain M (2000). Using banana in keropok product development. In: Wahab Z, Mahmud TMM, Siti Khalijah D, Nor 'Aini Mohd F, Mahmood M (eds) Proceedings of the first national banana seminar, Awana Genting, Malaysia, 1998/11/23–25, pp 314–321

Zainun CA (1999) Quality of ready-to-fry banana fritters. Paper presented at the National Horticulture Conference '99, Kuala Lumpur, Malaysia

Zhang P, Whistler RL, Bemiller JN, Hamaker BR (2005) Banana starch: production, physicochemical properties, and digestibility – a review. Carbohydr Polym 59(4):443–458

Zhang CX, Ho SC, Chen YM, Fu JH, Cheng SZ, Lin FY (2009) Greater vegetable and fruit intake is associated with a lower risk of breast cancer among Chinese women. Int J Cancer 125(1):181–188

Musa acuminata (AAA Group) 'Gros Michel'

Scientific Name

Musa acuminata **Colla (AAA Group) 'Gros Michel'**.

Synonyms

Musa acuminata L. cv 'Gros Michel', *Musa* × *paradisiaca* L. cv. "Gros Michel'.

Family

Musaceae

Common/English Names

Big Mike, Gros Michel Banana, Pisang Ambon Banana

Vernacular Names of Local Cultivars

Columbia: Banano, Guineo, Habano;
Fiji: Jainabalavau;
French Antilles: Makanguia, Raimbaud;
Hawaii: Bluefields;
Indonesia: Pisang Ambon Putih;
Malaysia: Pisang Ambon, Pisang Embun;
Mexico: Plantano Roatan;
Myanmar: Thihmwe;
Philippines: Ambon;
Papua New Guinea: Avabakor, Disu;
Puerto Rico: Guaran, Guineo Gigante;
Samoa: Fa'i Fia Palagi;
Spanish: Banano, Guineo Giganet, Platano Roatan;
Sri Lanka: Anamala;
Thailand: Kluai hom Dok Mai, Kluai Hom Thong;
Tonga: Siaine Fisi;
Vietnam: Chuoi Tieu Cao #2;
West Indies: Gros Michel.

Origin/Distribution

Gros Michel bananas originated from Martinique, Jamaica.

Agroecology

See notes under *Musa acuminata* (AA Group) 'Lakatan.

Edible Plant Parts and Uses

Pisang Ambon putih or Pisang Embun is the most important dessert cultivar in Indonesia and is also highly esteemed in Malaysia. It has good flavour and excellent keeping quality.

T.K. Lim, *Edible Medicinal And Non-Medicinal Plants: Volume 3, Fruits*,
DOI 10.1007/978-94-007-2534-8_67, © Springer Science+Business Media B.V. 2012

Plate 1 Fruit bunch of ripe Pisang Ambon Putih bananas

Plate 2 Hand of unripe Pisang Ambon Putih bananas

Plate 3 Hand of semi-ripe Pisang embun

Botany

Gros Michel is a tall, vigorous herbaceous plant; the pseudostem, is tall, robust and vigorous, (3-) 5–7 m high, shiny yellowish green with watery sap. Leaves are large and intermediate, with pink-purple dorsal midrib. Inflorescence complete, peduncle vary hairy, 30–60 cm long, rachis with a curve. Male bud ovoid, bract shape intermediate with young bracts slightly overlapping, external surface waxy, brownish-purple-red, internal surface red. Bunch weight is about 15–25 kg with 10–14 hands and 12–14 fingers per hand (Plate 1). Fruit is medium to large, 15–20 cm by 3.5–4 cm diameter, sharply curved upwards, fruit apex pointed with persistent style, transverse section with prominent ridges, skin is green turning yellow when ripe, smooth and moderately thick, pedicel 10 cm long (Plates 2 and 3). The flesh is cream coloured, moderately firm, fairly aromatic and sweet.

Nutritive/Medicinal Properties

Proximate nutrient composition of the edible portion (per 100 g) of pisang embun was reported by Tee et al. (1997) as follows: energy 84 kcal., water 78.2 g, protein 1.4 g, fat 0.4 g, carbohydrate 18.7 g, fibre 0.6 g, ash 0.7 g, Ca 5 mg, P 4 mg, Fe 0.3 mg, Na 13 mg, K 332 mg, carotenes 120ug, vitamin A 20ug RE, vitamin B-1 0.03 mg, vitamin B-2 0.03 mg, niacin 0.4 mg, vitamin C 4.3 mg.

Other Uses

See notes under *Musa acuminata* (AA Group) 'Lakatan.

Comments

Pisang embun is propagated vegetatively by suckers.

Selected References

Daniells J, Jenny C, Tomekpe K (2001) Musalogue: a catalogue of *Musa* germplasm. Diversity in the Genus *Musa*. In: Arnaud E, Sharrock S (comp) International Network for the Improvement of Banana and Plantain, Montpellier. <www.inibap.org/publications/musalogue.pdf>

Espino RRC, Jamalussin SH, Silayoi B, Nasution RE (1992) *Musa* L. (edible cultivars). In: Verheij EWM, Coronel RE (eds) Plant resources of south-East Asia no 2. Edible fruits and nuts. PROSEA, Bogor, pp 225–233

Gowen S (ed) (1995) Bananas and plantains. Chapman & Hall, London, 612 pp

Ploetz RC, Kepler AK, Daniells J, Nelson SC (2007) Banana and plantain – an overview with emphasis on Pacific Island cultivars, version 1. In: Elevitch CR (ed) Species profiles for Pacific Island agroforestry. Permanent Agriculture Resources (PAR), Hōlualoa. <http://www.traditionaltree.org>

Robinson JC (1996) Bananas and plantains. Crop production science in horticulture. CAB International, Wallingford, 238 pp

Stover RH, Simmonds NW (1987) Bananas, 3rd edn, Tropical agricultural series. Longman, Essex, 468 pp

Tee ES, Noor MI, Azudin MN, Idris K (1997) Nutrient composition of Malaysian foods, 4th edn. Institute for Medical Research, Kuala Lumpur, 299 pp

Valmayor RV, Wagih ME (1996) *Musa* L. (plantain and cooking banana). In: Flach M, Rumawas F (eds) Plant resources of South-East Asia no. 9. Plants yielding non-seed carbohydrates. Prosea Foundation, Bogor, pp 126–131

Valmayor RV, Jamaluddin SH, Silayoi B, Kusumo S, Danh LD, Pascua OC, Espino RRC (2000) Banana cultivar names and synonyms in Southeast Asia. International Network for the Improvement of Banana and Plantain – Asia and the Pacific Office, Los Banos

Musa acuminata (AAA Group) 'Red'

Scientific Name

***Musa acuminata* Colla (AAA Group) 'Red', *Musa acuminata* Colla (AAA Group) 'Red Dacca'**.

Synonyms

Musa acuminata Colla (AAA Group) cv. 'Red', *Musa sapientum* L. f. *rubra* Bail., *Musa sapientum* L. var. *rubra* (Firm.) Baker, *Musa rubra* Wall. ex Kurz., *Musa* × *paradisiaca* L. ssp. *sapientum* (L.) Kuntze var. *rubra*, *Musa acuminata* Colla (AAA Group) cv. 'Cuban Red', *Musa acuminata* Colla (Cavendish Group) cv. 'Cuban Red', *Musa acuminata* Colla (AAA Group) cv. 'Red Jamaican', *Musa acuminata* Colla (AAA Group) cv. 'Jamaican Red', *Musa acuminata* Colla (AAA Group) cv. 'Spanish Red'.

Family

Musaceae

Common/English Names

Red Dacca (Australia), Red Banana, 'Red' Banana (USA), Claret Banana, Cavendish Banana "Cuban Red", Jamaican Red Banana, Morado, Red Cavendish Banana, Rojo

Vernacular Names

Brazil: Banana Roxa;
Chinese: Hong Guo Jiao;
Columbia: Banano Color Clarete, Tafetan, Tafetan Morado;
Cook Islands: Kinaki Tangata;
Cuba: Guineo Morado, Indio;
Danish: Aeblebanan, Kubabanan, Rød Banan;
Dutch: Cubabanaan, Rode Banana;
East Africa: Neuse, Nyekundu Ya Kisungu, Mzungu Mwekundu;
Fiji: Jainadamu;
French: Banane De Cuba, Banane Rouge, Banane Violette;
French Guiana: Bacove Violette;
German: Kuba Banane, Weinrote Banane;
Guadeloupe: Figue Rose, Figue Rouge-Vin, Figue Violette;
Haiti: Figue Rouge;
Hawaii: Cuban Red, Pink Banana, Red;
India: Chenkadali, Lal Kela, Lal Kera (Hindu);
Indonesia: Pisang Susu Merah, Udang;
Italian: Banana Di Cuba, Banana Rosa;
Japan: Baracoa;
Malaysia: Pisang Raja Udang, Pisang Tembaga (Sabah);
Martinique: Figue Rose;
Myanmar: Shwe Nget Pyaw;
Papua New Guinea: Mossi, Rong Rong, Tabunatar (PNG 132), Kirkirnan, (PNG139);
Philippines: Morado;
Pohnpei: Akadahn Weitahta;

T.K. Lim, *Edible Medicinal And Non-Medicinal Plants: Volume 3, Fruits*,
DOI 10.1007/978-94-007-2534-8_68, © Springer Science+Business Media B.V. 2012

Portuguese: Bananeira Roxa, Banana Vermelha;
Puerto Rico: Colorado;
Samoa: Fa'I Suka, Fa'I Niue;
Seychelles: Rouge;
Spanish: Platano Colorado;
Sri Lanka: Rathambala;
Thailand: Kluai Nak;
Vietnam: Chuoi Com Lua;
West Indies: Claret, Colorado, Figue Rouge, Morado, Red.

Plate 1 Red Dacca bananas

Origin/Distribution

This banana is reported to have originated in the West Indies and Central America.

Plate 2 Pisang Udang

Agroecology

See notes in *Musa acuminata* (AA Group) 'Lakatan'.

Edible Plant Parts and Uses

A plump red or purplish-red banana, the Red Dacca has creamy to delicate orangey-yellow soft, starchy flesh and is eaten raw after peeling, whole or chopped and added to desserts and fruit salads. It can also be baked, fried and toasted. Red Dacca makes excellent banana smoothies and shakes and ideal for fruit compotes. It can also be barbequed whole and eaten.

The flowers of Red Dacca are also eaten cooked in curries, stews and soups.

Plate 3 Pisang Tembaga in Sabah

Botany

A herbaceous banana with red –green shiny pseudostem, 3–4.5 m, sap milky and pinkish-red or purplish-red petioles and midribs. Petiole canal on the 3rd leaf is open with spreading margins. Inflorescence emerge at the top of the plant, pendant with 58–62 cm long, hairy peduncle, complete with female, neuter and male flowers. Male bud ovoid, bracts imbricate, obtuse and split, waxy, purple-brown externally, internal surface red. Male flower tepal cream with pinkish pigmentation. Female flower, yellow tepals, with straight style, yellow stigma and ovary cream with purplish-red pigmentation. Fruit plump, orange-red, red, pink or purplish-red when mature (Plates 1, 2 and 3), flesh cream to pale orangey yellow when ripe, rounded in cross-section and blunt tipped. Number of fruits per hand up to 17, size of fruit 16–20 cm.

Nutritive/Medicinal Properties

Proximate nutrient composition of the edible portion (per 100 g) of pisang rajah udang (Pisang merah) was reported by Tee et al. (1997) as follows: energy 83 kcal., water 77.9 g, protein 0.9 g, fat 0.3 g, carbohydrate 19.3 g, fibre 0.8 g, ash 0.8 g, Ca 7 mg, P 18 mg, Fe 0.3 mg, Na 2 mg, K 233 mg, carotenes 219 μg, vitamin A 37 μg RE, vitamin B-1 0.03 mg, vitamin B-2 0.07 mg, niacin 0.8 mg and vitamin C 5.6 mg.

Pisang raja udang (AAA Group 'Red') cultivar from Malaysia was found to be a good source of vitamin A, containing 290 μg/100 g of β-carotene (Englberger, et al. 2003).

Thirty-nine volatile components were identified in the red-skinned, aromatic and sweet "Indio" banana fruit (Pino et al. 2003). Aliphatic esters comprised the largest class of volatiles. The major components were 2-heptyl acetate (14.4%), isoamyl acetate (9.0%), 2-methylbutyl acetate (8.9%) and 2-heptyl hexanoate (9.5%). The esters of 2-heptanol were important components for banana aroma. Apart from these esters, the other principal constituent was elemicin (6.5%), which also contributed to banana aroma. Acetates were also found in banana volatiles, but not as major components.

Oral administration of the ethanol extract of mature, green fruits of *Musa* AAA (Chenkadali) significantly decreased the elevated levels of serum triacylglycerol, cholesterol and alanine amino transferase (ALT) activity induced by alloxan in rats (Kaimal et al. 2010). Significant decrease was also observed in the elevated level of lipid peroxides while lowered GSH (reduced glutathione) content increased substantially in liver and pancreas. The effect was dose independent and rats treated with 500 mg extract/kg body weight showed comparable levels of serum triacylglycerol, cholesterol, ALT activity and liver lipid peroxides to that of normal control and glibenclamide treated groups. Although, there was no significant difference, treatment with 500 mg/kg body weight of the extract showed a higher content of GSH and lower level of lipid peroxides in pancreas compared with glibenclamide. Histopathological examination of pancreas and liver revealed regeneration of islet cells and hepatocytes respectively, which correlated with the biochemical findings. The findings suggested that ethanol extract of mature green fruits of *Musa* AAA (Chenkadali) had antioxidant and hypolipidaemic properties and may be used for treating diabetes mellitus.

For more health and medicinal attributes see notes in *Musa* (AA Group) 'Lakatan'.

Other Uses

See notes in *Musa acuminata* (AA Group) 'Lakatan'.

Comments

Red dacca bananas are extremely popular in Central America and less so in southeast Asia.

Selected References

Daniells J, Jenny C, Tomekpe K (2001) Musalogue: a catalogue of *Musa* germplasm. Diversity in the genus *Musa*. In: Arnaud E, Sharrock S (eds) International Network for the Improvement of Banana and Plantain, Montpellier. <www.inibap.org/publications/musalogue.pdf>

Englberger L, Darnton-Hill I, Coyne T, Fitzgerald MH, Marks GC (2003) Carotenoid-rich bananas: a potential food source for alleviating vitamin A deficiency. Food Nutr Bull 24(4):303–318

Espino RRC, Jamalussin SH, Silayoi B, Nasution RE (1992) *Musa* L. (edible cultivars). In: Verheij EWM, Coronel RE (eds) Plant resources of South-East Asia no 2. Edible fruits and nuts. PROSEA, Bogor, pp 225–233

Gowen S (ed) (1995) Bananas and plantains. Chapman & Hall, London, 612 pp

Kaimal S, Sujatha KS, George S (2010) Hypolipidaemic and antioxidant effects of fruits of *Musa* AAA (Chenkadali) in alloxan induced diabetic rats. Indian J Exp Biol 48(2):165–173

Pino JA, Ortega A, Marbot R, Aguero J (2003) Volatile components of banana fruit (*Musa sapientum* L.) "Indio" for Cuba. J Essent Oil Res 15:70–71

Ploetz RC, Kepler AK, Daniells J, Nelson SC (2007) Banana and plantain – an overview with emphasis on Pacific Island cultivars, version 1. In: Elevitch CR (ed)

Species profiles for Pacific Island agroforestry. Permanent Agriculture Resources (PAR), Hōlualoa. <http://www.traditionaltree.org>

Robinson JC (1996) Bananas and plantains. Crop production science in horticulture. CAB International, Wallingford, 238 pp

Stover RH, Simmonds NW (1987) Bananas, 3rd edn, Tropical agricultural series. Longman, Essex, 468 pp

Tee ES, Noor MI, Azudin MN, Idris K (1997) Nutrient composition of Malaysian foods, 4th edn. Institute for Medical Research, Kuala Lumpur, 299 pp

Valmayor RV, Wagih ME (1996) *Musa* L. (plantain and cooking banana). In: Flach M, Rumawas F (eds) Plant resources of South-East Asia no 9. Plants yielding non-seed carbohydrates. Prosea Foundation, Bogor, pp 126–131

Valmayor RV, Jamaluddin SH, Silayoi B, Kusumo S, Danh LD, Pascua OC, Espino RRC (2000) Banana cultivar names and synonyms in Southeast Asia. International Network for the Improvement of Banana and Plantain – Asia and the Pacific Office, Los Banos

Musa acuminata × balbisiana (AAB Group) 'Horn Plantain'

Scientific Name

***Musa acuminata* × *balbisiana* (AAB Group) 'Horn Plantain'**.

Synonyms

Musa corniculata Rumphias, *Musa corniculata* Lour., *Musa emasculata* de Briey ex De Wild., *Musa protractorachis* De Wild.

Family

Musaceae

Common/English Names

Horn Plantain, Pisang Tanduk, True Horn Plantain

Vernacular Names for Local Cultivars

Brazil: Banan Pacova;
Colombia: Dominico Gigante, Harton, Liberal;
Hawaii: Kiwipipi;
India: Monngil (Southern);
Indonesia: Neij Kumbi Gouji (Irian Jaya), Pisang Byar, Pisang Tanduk;
Malaysia: Pisang Lang, Pisang Tanduk;
Mexico: Planto Macho;
Nigeria: Ishitim;
Philippines: Tindok;
Puerto Rico: Plantano Comun;
Thailand: Kluai Klai, Klaui Nga Chang;
Vietnamese: Chuoi Sung Bo;
West Indies: Banane Corne.

Origin/Distribution

The genus *Musa*'s centre of origin is Asia (primarily southern and south-eastern) and horn plantains in Malesia.

Agroecology

Horn plantains like other plantains require mean annual rainfall of 1,500–2,500 mm evenly distributed throughout the year. Regions where dry season is longer than 2–3 months supplementary irrigation can be provided. Optimum temperatures for growth and development is from 26°C to 32°C, temperatures lower than 18°C or higher than 36°C is detrimental and will adversely impede growth. Horn plantain plants have weak rooting systems relative to the size of their above-ground parts and are easily toppled by strong winds and needs to be planted in wind-sheltered areas. Sunny conditions are favourable for plantains. Plantains including horn plantains thrive best on light, deep friable soils. Heavy clayey soils with impede drainage should be avoided.

T.K. Lim, *Edible Medicinal And Non-Medicinal Plants: Volume 3, Fruits*,
DOI 10.1007/978-94-007-2534-8_69, © Springer Science+Business Media B.V. 2012

Sandy loam and silty clay loam soils rich in organic matter are ideal for good growth and development.

Edible Plant Parts and Uses

Plantain fruits are usually not eaten raw. The near-ripe, green or ripe yellow fruits are eaten after some form of processing, for instance by boiling, grilling, steaming, frying or baking. The fruits are also made into banana chips. Dried green plantains, ground fine and roasted, have been used as a substitute for coffee.

In Malaysia, Singapore, Brunei and Indonesia, plantain is excellent for making deep-fried fritters popularly eaten as snack or dessert known as *goreng pisang*. Pisang goreng is called *godoh gedang* in Bali, *limpang limpung* in Java, *pisang rakit* in Sibolga, *and pisang kipas* in Pontianak. Ripe plantains are sliced length-wise or angle, dipped in batter and deep fried in vegetable oil till crispy brown. Similarly in Kerala, India, fried plantain fritters dipped in wheat flour batter with or without sesame seeds are called *Ethakka appam*, *pazham boli* or *pazham pori* and are popular snacks. In Africa, fried plantain is a popular dish wherever plantain is grown; it is called, *dodo* in Nigeria and Cameroon, *kewele* in Ghana and *aloco* in Côte d'Ivoire. Fried plantain may be served as a snack, a starter or as a side dish to a main course. In Columbia, Ecuador and Peru, fried green plantain slice are called *chifles*, this term is also used for thinner plantain chips. In Latin America, fried ripe plantains are called *amarillos*. Sweet-savoury strips of plantain, deep fried till golden brown and caramelized are very popular in Latin America; they are called *tajadas* in Venezuela, *tajadas de plátano* in Columbia *and platano madruo en tajadas* in Ecuador. In Venezuela, yo-yo is a traditional cuisine consisting of two slices of fried ripened plantain *(tajadas*) with a local cheese in the middle. In Latin America, twice fried, green plantain patties called *tostones* (or *patacones* in Ecuador, Columbia, Costa, Rica, Panama and Venezuela) are popular as a side, appetizer, or snack. In Cuba, Haiti, Dominican Republic and Puerto Rico, the *tostones* are dipped in Creole sauce from chicken, pork, beef, or shrimp before eating. In Puerto Rico, crisp fried plantain sliced at an angle, called *tostones de plátano* are eaten seasoned with seasoning (garlic, salt, ground pepper, ground oregano). In Venezuela, *patacones* made of length-wise sliced, oil fried plantain are much relished; when made form unripe plantains they are called *patacon verde* and from ripe plantains, *patacon amarillo*. In Nigeria, roasted plantain is called *boli*. The plantain is normally grilled and served in a hot palm oils sauce with roasted fish and ground peanuts. This dish is popular as a snack in southern and south-western Nigeria. Steam or boiled ripe fruit plantain slices in coconut milk and palm sugar makes a delicious dessert. Other popular plantain recipes in Malaysia and Singapore include a dessert porridge called *bubur talas ketan* and *cokodok* , also called *kuih kodok* or *jemput-jemput,* a traditional Malaysian, round fritters made from banana pulp and flour.

Ripe banana or plantain fruit is used to make sweet meats known as banana figs which are very popularly eaten as snacks in Africa (Oke et al. 1990). Traditionally, the ripe fruit is sun-dried or sometimes dried in ovens or over fires; usually as slices, although banana figs are sometimes prepared from whole fruits (Goode 1974; Mukasa and Thomas 1970; Simmonds 1966). In Bafoussam and Yaoundé, Cameroon, roasted or fried plantain, plantain chips, boiled plantain or banana and pounded plantain are popular (Ngoh Newilah et al. 2005). They are eaten with various sauces, vegetables and other food complements. Other preparations found in these regions include stuffed plantain or banana, plantain or banana porridges and traditional meals called *kondre* and *malaxé*.

In West Africa, whole boiled plantains are eaten alone after peeling the skin or served with a sauce of palm oil, meat, fish, vegetables and seasonings (Johnston 1958; Tezenas du Montcel 1979). The plantains are also pounded in a mortar after boiling to form a paste or dough known variously as *fufu, foofoo, foufou* or *foutou* which is eaten with soup or a sauce of meat and vegetables (Johnston 1958; Lassoudière 1973). In Cuba, *Fufu de platano* is a traditional popular lunch dish. It is a *fufu* made by boiling the plantains in water and mashing with a fork. The *fufu* is then mixed with chicken stock and *sofrito*, a sauce made from lard, garlic, onions, pepper, tomato sauce, a touch of vinegar and cumin.

In Rwanda, Tanzania and Uganda, *matooke* is a common plantain cuisine, made from peeled boiled, mashed green plantain and wrapped in plantain leaf and set in a cooking pot on the plantain stalks. A popular Caribbean plantain cuisine is *relleno de maduro*, sweet plantain mashed with egg and flour, butter, cheese, chopped onions, stuffed with raisins, olives, capers, ground meat and spices, then molded into a ball and fried. Some popular plantain recipes of Puerto Rico and the Dominican Republic are:

Alcapurrias de plátano (beef-filled plantain fritters) – green plantains, anniato oil, beef, lobster, shrimp, chicken or turkey fillings, yautia (tannier) and salt. The plantain and yautia are mashed into a dough and use as casing for the marinated meat fillings, the dish is serve hot.

Amarillos a la Moda (flaming ripe plantains a la mode) – luscious dessert of ripe plantain, butter, cinnamon, brown sugar, brandy wine, served with vanilla ice cream.

Arañitas (plantain bird's nest) – fried thin rasped strips of green plantains with adobo seasoning (garlic salt, ground pepper, ground oregano), served as side dish or appetiser.

Asopao de Pollo (rice and chicken gumbo with plantain balls) – green plantain, chicken pieces, white rice, potatoes, onion, tomato, cilantro, garlic, bay leaf olive oil with adobo (a blend of salt, oregano, garlic, black pepper and turmeric).

Bollitas de plátano (plantain balls) Add to any soup! Made of green plantain and wheat flour, for adding to any soup.

Canoas de amarillo (beef-filled whole ripe plantains) – ripe plantain with meat filling and cheese serve as light meal or eaten with rice and beans.

Cereal de plátano (plantain cereal) – grated green plantain, cooked with sugar, water, salt and milk.

Mangú (Dominican style boiled green plantains) – boiled green plantains with diced onions, olive oil, salt, vinegar and Swiss cheese . *Mangú* is usually served as a side-dish for beef, pork and fish dishes.

Maduros (fried sliced ripe plantains) – a side dish.

Guanimes de plátano (tamale-style plantain guanimes) – grated green plantain, sugar, salt, coconut milk, formed each guanime in sausage shapes and wrapped in plantain leaf, simmer in chicken broth or water.

Mofongo (mashed fried plantains) – mashed green plantain to which is added bacon and ground garlic, olive oil and salt to taste and fried.

Pan de plátano (ripe plantain bread) – ripe plantain, flour, margarine, eggs, baking soda, sugar, sour cream, vanilla baking powder and the mixture baked.

Pastelon de amarillos (Latin Lagsagne, Puerot Rico lagsagne) – traditional Puerto Rican/Dominican cuisine similar to lasagne, it uses sweet plantains to replace the pasta. *Pastelón de plátano* (plantain pie) – ingredients include ripe plantains, garlic, pitted olives, stuffed pepper or onion, chopped bell pepper, chopped onion, olive oil bay leaves, lean ground meat, raisins, cooking wine, olive oil, butter and cheese.

Piononos (ripe plantain with various fillings) – *Piononos* consist of a round outer shell made with ripe plantain slices, a filling which may be beef, chicken, lobster, shrimp or simply vegetables (vegan diet), chopped bell pepper, chopped onion, olive oil, bay leaves and salt. *Planto hervido* – boiled plantain.

Sopa de plátano (plantain soup (puree)) – green plantain grated into a paste and boiled, salt added to taste.

Tortilla de plátanos (plantain omelette) – sliced ripe plantain, diced ham, beaten eggs, sausages, sweet peas, salt and frying oil.

In Ghana, a type of pancake called *fatale* is prepared from a mixture of pounded ripe plantains and fermented wholemeal maize flour (Dei-Tutu 1975). The pounded plantain pulp is mixed with the fermented maize flour into a paste which is seasoned with onion, pepper, ginger and salt and then fried in palm oil. The pancake may be served with beans as a main meal or used on its own as a snack or dessert. A similar product known as *krakro* is made by mixing pounded plantains with corn flour, onion, ginger and salt. This mixture is left to rise for half an hour and then moulded into balls and fried (Eshun 1977).

Plantain blossoms are also eaten raw or cooked. In the Philippine, the blossoms are called *Puso ng Saging*. In Laos, the flowers are eaten raw in vermicelli soups and in Vietnam, the blossoms of the male inflorescence are eaten in salads and soups. In India, in the states of Andra Pradesh and Tamil Nadu, plantain flowers are used in curries, while in Karnataka, the blossoms are used to make sweet-sour *gojju* (a gravy dish).

The plantain pseudostem is also eaten particularly in southern India (Tamil Nadu, Assam, Kerala) but also in Burma, Thailand and Bali. When the bunch of bananas is harvested, the plant is cut down and layers peeled off to get to the tender central stalk, which looks much like a palm heart. It is called Thor in Bengali. In Assam, the plantain psuedostem"heart" is cooked with salt, ginger, garlic, turmeric, baking soda and oil to prepare the popular dish *Kol Posola Khar*.

For more edible food uses, see notes in *Musa acuminata* (AAA Group) 'Dwarf Cavendish'.

Botany

Plantains are triploid, tall perennial herb. When mature, the pseudostem reaches a height of 4–5 m with a girth of 80–90 cm, usually dull medium green with purple blotches. The leaves are large, fleshy-stalked, tender, smooth, oblong or elliptic, waxy green or with some purplish blotches on the underside, up to 275 cmt long and 60 cm wide with entire margin and pinnate venation. Colour of the midrib in the dorsal side is pink-purple and the erect (cigar) leaf surface is green on the dorsal side. Leaves vary from 4 to 15 and are arranged spirally. They unfurl, as the plant grows, and extend upward and outward. The inflorescence, a transformed growing point, is a terminal spike shooting out from the heart in the top of the stem. In horn plantains the inflorescence is incomplete, female flowers are present but neuter and male flowers are absent. Fruits green turning to yellow when ripe, are formed in pendent bunches consisting of 2–3 (–5) hands with 8–10 fingers per hand (Plates 1, 2 and 3). Each finger (fruit) is large, 25–35 cm by 6–7 cm diameter, curved like a horn, and weighs 250–600 g.

Nutritive/Medicinal Properties

Proximate nutrient composition of the edible portion (per 100 g) of pisang tanduk was reported by Tee et al., (1997) as follows: energy 118 kcal., water 69.5 g, protein 1.2 g, fat 0.4 g, carbohydrate 27.6 g, fibre 0.5 g, ash 0.8 g, P 31 mg, Fe 0.2 mg, Na 18 mg, K 405 mg, carotenes 500ug, vitamin A 83 μg RE, vitamin B-1, 0.07 mg, vitamin B-2 0.01 mg and vitamin C 16.3 mg.

Plate 1 (**a**, **b**) Ripe horn plantain 'Pisang tanduk'

Plate 2 (**a**, **b**) Unripe horn plantains

Plate 3 (**a**, **b**) Near ripe horn plantains

Pisang tandok (AAB, Horn plantain), a cultivar from Malaysia was found to be a good source of vitamin A, containing 370 μg/100 g of β-carotene (Englberger et al. 2003). Blades et al. (2003) found that four *Musa* cultivars (Horn Plantain (Pisang Tanduk), Kirkirnan, Asupina and Pisang Raja) to be good sources of provitamin A (>75 RAE/100 g) and hence potentially useful for the prevention of vitaminA deficiency. α-, all-trans-β and cis-β-carotene were the only provitamin A carotenoids present in detectable quantities in all samples and levels of these carotenes ranged from 61 to 1,055, 50 to 1412 and 7 to 85 μg /100 g banana, respectively.

Ten yellow or yellow/orange-fleshed cultivars (Asupina, Kirkirnan, Pisang Raja, Horn Plantain, Pacific Plantain, Kluai Khai Bonng, Wain, Red Dacca, Lakatan, and Sucrier) were found to have significant carotenoid levels, potentially meeting half or all of the estimated vitamin A requirements for a non-pregnant, non-lactating adult woman within normal consumption patterns (Englberger et al. 2006). All were acceptable for taste and other attributes.

Antiulcerogenic Activity

Orally administered plantain banana pulp powder (*Musa sapientum* var. *paradisiaca*) was shown to have significant anti-ulcerogenic activity in rats subjected to aspirin, indomethacin, phenylbutazone, prednisolone and cysteamine and in guinea-pigs subjected to histamine (Goel et al. 1986.

The powder not only increased mucosal thickness but also significantly increased [3H]thymidine incorporation into mucosal DNA. The study suggested that banana powder treatment not only strengthen mucosal resistance against ulcerogens but also promoted healing by inducing cellular proliferation.

The results of animal studies by Mukhopadhyaya et al. (1987) confirmed that plantain banana powder strengthened mucosal resistance and promoted the healing of ulcers. Rats of either sex treated with plantain powder (*Musa sapientum* var. *paradisiaca*) exhibited: (i) a significant increase in the [3H]thymidine incorporation into mucosal cell DNA; (ii) a significant increase in the total carbohydrate (sum of total hexoses, hexosamine, fucose and sialic acid) content of gastric mucosa; (iii) a significant decrease in gastric juice DNA and protein; (iv) a significant increase in the total carbohydrates and carbohydrate/protein ratio of gastric juice. Aspirin treatment to rats caused similar effects as banana on the [3H]thymidine incorporation into mucosal cell DNA but showed opposite effects on the other parameters. These results suggested that plantain treatment increased and aspirin decreased the gastric mucosal resistance as evidenced by a respective decrease and increase in gastric juice DNA, the latter serving as an index of the rate of mucosal shedding. Increased cellular mucus may be the factor for increased mucosal resistance.

The active anti-ulcerogenic ingredient extracted from unripe plantain banana (*Musa sapientum* L. var. *paradisiaca*) was identified as the flavonoid, leucocyanidin (Lewis et al. 1999; Lewis and Shaw 2001). Dried unripe plantain banana powder, the extracted leucocyanidin and a purified synthetic leucocyanidin demonstrated a significant protective effect against aspirin-induced erosions. Leucocyanidin and its synthetic hydroxyethylated and tetrallyl derivatives were found to protect the gastric mucosa from aspirin-induced erosions. Leucocyanidin and its hydroxyethylated and tetraallyl derivatives significantly increased mucus thickness. Plantain (*Musa sapientum* var. *paradisiaca*) methanol extract (50 mg/kg, twice daily for 5 days) showed significant antiulcer effect and antioxidant activity in gastric mucosal homogenates, where it reversed the increase in ulcer index, lipid peroxidation and super oxide dismutase values induced by cold stress (Goel et al. 2001). However, it did not produce any change in catalase values, which was significantly decreased by stress.

Supplementation of diet with unripe pawpaw or plantain extract significantly reduced the weight gained by rabbits compared with the control (Eriyamremu et al. 2007). The total phospholipids content (phosphatidylethanolamine, phosphatidylcholine, and sphingomyelin in the stomach and duodenum were significantly increased in the test groups compared to the control. The results suggested that unripe pawpaw meal and unripe plantain extract altered the phospholipid profile of the stomach and duodenum in ways which may affect membrane fluidity of these tissues and would have profound effect on the gastro-duodenal mucosa and thus have implication(s) for gastric and duodenal ulcers in rabbits.

Hypoglycaemic/Antidiabetic Activity

Oral administration of 0.15, 0.20 and 0.25 g/kg of chloroform extract of the *Musa sapientum* flowers (MSFEt) to diabetic rats for 30 days resulted in a significant reduction in blood glucose, glycosylated haemoglobin and an increase in total haemoglobin, but in the case of 0.25 g/kg the effect was highly significant (Pari and Umamaheswari 1999, 2000). It also prevented decrease in body weight. There was a significant improvement in glucose tolerance in animals treated with the flower extract and the effect was compared with glibenclamide. The decrease in thiobarbituric acid reactive substances (TBARS) and the increase in reduced glutathione (GSH), glutathione peroxidase (GSH-Px), superoxide dismutase (SOD) and catalase (CAT) clearly showed the antioxidant property of the flower extract. Oral administration of the ethanolic extract of *Musa sapientum* flowers showed significant blood glucose lowering effect at 200 mg/kg in alloxan induced diabetic rats (120 mg/kg, i.p.) and the extract was also found to significantly scavenge oxygen free radicals, viz., superoxide dismutase (SOD), catalase (CAT) and also protein, malondialdehyde and ascorbic acid in-vivo (Dhanabal et al. 2005). The antidiabetic activity observed may be attributed to the presence of flavonoids, alkaloids, steroid and glycoside principles. Methanolic extract of *Musa sapientum* var. *paradisiaca* (MSE) showed better ulcer protective effect in non-insulin dependent diabetes mellitus (NIDDM) rats compared with sucralfate (ulcer protective drug) and glibenclamide (antidiabetic drug) in cold-restraint stress-induced gastric ulcers (Mohan Kumar et al. 2006). NIDDM caused a significant decrease in gastric mucosal glycoprotein level without having any effect on cell proliferation. However, all the test drugs reversed the decrease in glycoprotein level in NIDDM rats, but cell proliferation was enhanced in case of banana extract alone. Both cold-restraint stress or NIDDM as such enhanced gastric mucosal LPO (lipid peroxidation), NO and SOD, but decreased CAT levels while CRS plus NIDDM rats caused further increase in LPO and NO level without causing any further changes in SOD and CAT level. MSE pretreatment of the banana extract showed reversal in the levels of all the above parameters better than glibenclamide. The results indicated that the ulcer protective effect of methanol banana extract could be due to its predominant effect on mucosal glycoprotein, cell proliferation, free radicals and antioxidant systems.

Analgesic Activity

Using hot plate method and tail immersion method, the aqueous and ethanol extracts of *Musa sapientum* stem were found to significantly increase reaction time as compared to vehicle treated group (Suvarna et al. 2009). Maximum analgesic effect was observed at 30 minutes interval for 100 and 200 mg/kg, i.p.

Wound Healing Activity

Studies in rats showed that aqueous and methanolic extracts of plantain (*Musa sapientum* var. *paradisiaca*) at 100 mg/kg, increased wound breaking strength and levels of hydroxyproline, hexuronic acid, hexosamine, superoxide dismutase, reduced glutathione in the granulation tissue (Agarwal et al. 2009). Both extracts decreased percentage of wound area, scar area and lipid peroxidation when compared with the control group and showed good safety profile.

Antimicrobial Activity

The Gram-positive bacteria (*Staphylococcus aureus* ATCC 25921, *S. aureus* and *Bacillus subtilis*) and all Gram-negative bacteria (*Escherichia coli* ATCC 25922, *Escherichia coli*, *Pseudomonas aeruginosa*, *Salmonella paratyphi*, *Shigella flexneri* and *Klebsiella pneumonia*) were susceptible to ethanolic extracts of unripe banana (*Musa sapientum*), lemon grass and turmeric while some namely *E. coli* ATCC 25922, *E. coli*, *P. aeruginosa* and *S. flexneri* were not susceptible to aqueous extracts of the three medicinal plants (Fagbemi et al. 2009). The minimum inhibition concentration (MIC) ranged from 4 to 512 mg/ml while the minimum bactericidal concentration (MBC) ranged from 32 to 512 mg/ml depending on isolates and extracting solvent. Ethanolic extracts showed greater antimicrobial activity than aqueous extracts. Four common medicinal plants including *Curcuma longa, Kaempferia parviflora, Allium sativum* and *Musa sapientum* exhibited inhibitory effects on the invasion of *Helicobacter pylori* (an aetiological agent of active chronic gastritis and peptic ulcer disease) to HEp-2 cells (Chaichanawongsaroj et al. 2009). Although, *Allium sativum* and *Musa sapientum* demonstrated marked anti-internalisation activities, the high concentrations of the extracts may have cytotoxic effects. Ethanol extract of *Musa sapientum* blossoms showed an antibacterial activity against *Bacillus subtilis* and *B. cereus* (Jahan et al. 2010). The chloroform and water extracts had negligible bacterial effect. The results showed that the chitosan-polyethylene glycol (CS-PEG) blended films encapsulated the ethanolic extracts of *Musa sapientum* were effective against the test organisms. With increasing concentrations of the ethanolic extract, the zone of inhibition also increased.

For more health and medicinal attributes see notes in *Musa acuminata* (AAA Group) 'Dwarf Cavendish'.

Other Uses

The non-edible uses of the fruit and various plant parts are as described for *Musa acuminata* (AAA Group) 'Dwarf Cavendish'.

Comments

See notes in *Musa acuminata* (AAA Group) 'Dwarf Cavendish'.

Selected References

Agarwal PK, Singh A, Gaurav K, Goel S, Khanna HD, Goel RK (2009) Evaluation of wound healing activity of extracts of plantain banana (*Musa sapientum* var. *paradisiaca*) in rats. Indian J Exp Biol 47(1):32–40

Anonymous (2011) About Puerto Rico: plantains. http://www.dollarman.com/puertorico/plantains.html

Blades BL, Dufficy L, Englberger L, Daniells JW, Coyne T, Hamill S, Wills RB (2003) Bananas and plantains as a source of provitamin A. Asia Pac J Clin Nutr 12(Suppl):S36

Chaichanawongsaroj N, Vilaichone R, Amornyingjarean S, Poovorawan Y (2009) Anti-*Helicobacter pylori* and anti-internalisation activities of Thai folk remedies used to treat gastric ailments. In: 19th European congress of clinical microbiology and infectious diseases, Helsinki 16–19 May 2009, Abstract number: P1389

Dei-Tutu J (1975) Studies on the development of fatale mix, a plantain product. Ghana J Agric Sci 8(5):153–157

Dhanabal SP, Sureshkumar M, Ramanathan M, Suresh B (2005) Hypoglycemic effect of ethanolic extract of *Musa sapientum* on alloxan induced diabetes mellitus in rats and its relation with antioxidant potential. J Herb Pharmacother 5(2):7–19

Edison HS, Sutanto A, Hermanto C, Rumsarwir Y, Lakuy H (2002) The exploration of Musaceae in Irian Jaya (Papua). Central Research Institute for Horticulture –INIBAP, Solok Research Institute for Fruits Solok, Indonesia, 58 pp

Englberger L, Darnton-Hill I, Coyne T, Fitzgerald MH, Marks GC (2003) Carotenoid-rich bananas: a potential food source for alleviating vitamin A deficiency. Food Nutr Bull 24(4):303–318

Englberger L, Wills RB, Blades B, Dufficy L, Daniells JW, Coyne T (2006) Carotenoid content and flesh color of selected banana cultivars growing in Australia. Food Nutr Bull 27(4):281–291

Eriyamremu GE, Asagba SO, Osagie VE, Ojeaburu SI, Lolodi O (2007) Phospholipid profile of the stomach and duodenum of normal rabbits fed with supplement of unripe pawpaw and unripe plantain (*M. sapientum*) extract. J Appl Sci 7(22):3536–3541

Eshun S (1977) Popular Ghanaian dishes. Ghana Publishing, Tema, 107 pp

Espino RRC, Jamalussin SH, Silayoi B, Nasution RE (1992) *Musa* L. (edible cultivars). In: Verheij EWM, Coronel RE (eds) Plant resources of South-East Asia no 2. Edible fruits and nuts. PROSEA, Bogor, pp 225–233

Fagbemi JF, Ugoji E, Adenipekun T, Adelowotan O (2009) Evaluation of the antimicrobial properties of unripe banana (*Musa sapientum* L.), lemon grass (*Cymbopogon citratus* S) and turmeric (*Curcuma* longa L.). Afr J Biotechnol 8(7):1176–1182

Goel RK, Gupta S, Shankar R, Sanyal AK (1986) Anti-ulcerogenic effect of banana powder (*Musa sapientum* var. *paradisiaca*) and its effect on mucosal. J Ethnopharmacol 18(1):33–44

Goel RK, Sairam K, Rao CV (2001) Role of gastric antioxidant and anti-*Helicobactor pylori* activities in antiulcerogenic activity of plantain banana (*Musa sapientum* var. *paradisiaca*). Indian J Exp Biol 39(7): 719–722

Goode PM (1974) Some local vegetables and fruits of Uganda. Department Agriculture, Entebbe

Gowen S (ed) (1995) Bananas and plantains. Chapman & Hall, London, 612 pp

Jahan M, Warsi MK, Khatoon F (2010) Concentration influence on antimicrobial activity of banana blossom extract-incorporated chitosan-polyethylene glycol (CS-PEG) blended film. J Chem Pharm Res 2(5): 373–378

Johnston BF (1958) The staple food economies of Western tropical Africa. Stanford University Press, Stanford

Lancaster PA, Coursey DG (1984) Traditional post harvest technology of perishable staples. FAO agricultural services bull no. 59. Food and Agriculture Organization of the United Nations, Rome

Lassoudibre A (1973) Le bananier plantain en Cote d'Ivoire. Fruits 28:453–462

Lewis DA, Shaw GP (2001) A natural flavonoid and synthetic analogues protect the gastric mucosa from aspirin-induced erosions. J Nutr Biochem 12(2): 95–100

Lewis DA, Fields WN, Shaw GP (1999) A natural flavonoid present in unripe plantain banana pulp protects the gastric mucosa from aspirin-induced erosions. J Ethnopharmacol 65(3):283–288

Mohan Kumar M, Joshi MC, Prabha T, Dorababu M, Goel RK (2006) Effect of plantain banana on gastric ulceration in NIDDM rats: role of gastric mucosal glycoproteins, cell proliferation, antioxidants and free radicals. Indian J Exp Biol 44(4):292–299

Mukasa SK, Thomas DG (1970) Staple food crops in Uganda. In: Jameson JD (ed) Agriculture in Uganda, 2nd edn. Oxford University Press, London, 395 pp

Mukhopadhyaya K, Bhattacharya D, Chakraborty A, Goel RK, Sanyal AK (1987) Effect of banana powder (*Musa sapientum* var. *paradisiaca*) on gastric mucosal shedding. J Ethnopharmacol 21(1):11–19

Ngoh Newilah G, Tchango Tchango J, Fokou E, Etoa FX (2005) Processing and food uses of bananas and plantains in Cameroon. Fruits 60(4):245–253

Oke OL, Redhead J, Hussain MA (1990) Roots, tubers, plantains and bananas in human nutrition. FAO food and nutrition series, no. 24. Food and Agriculture Organization of the United Nations, Rome

Pari L, Umamaheswari J (1999) Hypoglycaemic effect of *Musa sapientum* L. in alloxan-induced diabetic rats. J Ethnopharmacol 68(1–3):321–325

Pari L, Umamaheswari J (2000) Antihyperglycaemic activity of *Musa sapientum* flowers: effect on lipid peroxidation in alloxan diabetic rats. Phytother Res 14(2):136–138

Ploetz RC, Kepler AK, Daniells J, Nelson SC (2007) Banana and plantain – an overview with emphasis on Pacific Island cultivars, version 1. In: Elevitch CR (ed) Species profiles for Pacific Island agroforestry. Permanent Agriculture Resources (PAR), Hōlualoa. <http://www.traditionaltree.org>

Robinson JC (1996) Bananas and plantains. Crop production science in horticulture. CAB International, Wallingford, 238 pp

Simmonds NW (1966) Bananas, 2nd edn. Longmans, London

Stover RH, Simmonds NW (1987) Bananas, 3rd edn. Tropical Agricultural Series, Longman, 468 pp

Suvarna IP, Pramod IL, Anagha JM (2009) To study analgesic activity of stem of *Musa sapientum* Linn. J Pharm Res 2(9):1381–1382

Tee ES, Noor MI, Azudin MN, Idris K (1997) Nutrient composition of Malaysian foods, 4th edn. Institute for Medical Research, Kuala Lumpur, 299 pp

Tezenas du Montcel H (1979) Le bananier plantain au Cameroun. Fruits 34:307–313

Valmayor RV, Wagih ME (1996) *Musa* L. (plantain and cooking banana). In: Flach M, Rumawas F (eds) Plant resources of South-East Asia no. 9. Plants yielding non-seed carbohydrates. Prosea Foundation, Bogor, pp 126–131

Valmayor RV, Jamaluddin SH, Silayoi B, Kusumo S, Danh LD, Pascua OC, Espino RRC (2000) Banana cultivar names and synonyms in Southeast Asia. International Network for the Improvement of Banana and Plantain – Asia and the Pacific Office, Los Banos

Musa acuminata × *balbisiana* (ABB Group) 'Saba'

Scientific Name

Musa acuminata **×** ***balbisiana*** **Colla (ABB Group) 'Saba'.**

Synonyms

Musa × *paradisiaca* L. cultigroup Plantain cv. 'Saba', *Musa sapientum* L. var. *compressa* (Blanco) N.G.Teodoro.

Family

Musaceae

Common/English Names

Cardaba, Papaya Banana, Saba Banana, San Pablo, Sweet Plantain

Vernacular Names of Local Cultivars

Hawaii: Dippig;
Indonesia: Biu Gedang Saba (Bali), Kumro, Neij Sepatu Mogerei, Sepatu (Irian Jaya), Kepok, Giant Pisang Kepok, Pisnag Kuning Turangkog;
Malaysia: Pisang Abu Nepah, Pisang Nepah;
Philippines: Cardaba, Ripping, Saba;
Thailand: Kluai Hin;
Pohnpei: Inabaniko, Uht Kapakap;
United States: Praying Hands (Florida);
Vietnam: Chuoi Mat, Chuoi Sap.

Origin/Distribution

Saba banana, a triploid hybrid banana cultivar originated from the Philippines.

Agroecology

Saba banana like other banana cultivars is best grown in warm but moist areas, with temperature ranging from 18°C to 35°C and mean annual rainfall of 2,500 mm evenly distributed throughout the year. It grows best in full sun in well-drained, deep, moist, fertile soils rich in organic matter and soil pH between 5.5 and 6.5. Nonetheless due to their *Musa balbisiana* characteristics they tolerate dry soils and colder conditions better than other cultivated bananas. They can survive long dry seasons as long as adequate irrigation is provided. However, their fruits may not ripen under such conditions. If planted in places frequently visited by strong winds, wind breakers should be provided.

Edible Plant Parts and Uses

Saba bananas can be eaten raw or cooked. In the Philippines, ripe Saba bananas are consumed as a dessert or sweet snack, often simply boiled, in

T.K. Lim, *Edible Medicinal And Non-Medicinal Plants: Volume 3, Fruits*,
DOI 10.1007/978-94-007-2534-8_70, © Springer Science+Business Media B.V. 2012

syrup, or sliced lengthwise and fried, then sprinkled with sugar. They are also quite popular in this fried form (without the sugar) in the local dish, *Arroz a la Cubana*, consisting of minced picadillo-style seasoned beef, white rice, and fried eggs, with fried plantains on the side. Saba bananas are also popularly used in various traditional Filipino desserts and dishes like *Maruya* (thinly sliced Saba banana dipped in batter and deep fried,) *Halo-halo* (shaved ice, evaporated milk and caramelized Saba banana) and *Ginanggang* (grilled skewered bananas brushed with margarine and sprinkled with sugar). In addition, there is the equally popular *merienda* (light meal) snack, *Turrón*, where ripe plantains are sliced and then wrapped in lumpia wrapper (a thin rice paper) deep-fried and finished off with a brown sugar glaze. Saba bananas are also relished in Indonesia, Malaysia and Singapore in dishes like *Pisang Aroma* (similar to the Filipino *Turrón*) and *Pisang goreng* (fried bananas). In Sumatra, Saba banana pulp is grilled, compressed into flat pancakes and sprinkle with caramelized coconut shavings and served in banana leaves as shown in Plates 5–8.

Saba banana is also processed into a Filipino condiment known as *banana ketchup* or banana sauce comprised of mashed banana, sugar, vinegar, spice and dyed red to look like tomato ketchup. Saba banana are also made into banana chips, wine, vinegar and *pastilles de Saging* sweets.

The dark red inflorescences of Saba (banana hearts, locally known in the Philippines as "*Puso ng Saba*") are eaten.

Botany

Saba banana has a large, robust, shiny green pseudostem with a girth of 60–90 cm and reaching heights of 3–7 m, with dark brown underlying pigmentation and watery sap, with 4–8 suckers. The leaves are large, drooping, dark blue-green with pale green dorsal midrib and inward curved petiole canal. Inflorescence complete, peduncle length 50–75 cm, rachis falling vertically. Male bud intermediate to ovoid shape., young bracts overlap, bract apex intermediate to obtuse and split, external surface red purple, internal surface red. Flower compound tepal yellow, style straight and stigma and ovary cream-coloured. Bunch weight 14–22 kg with 7–12 hands per bunch and 12–20 fingers per hand. Fruits 8–13 cm long and 2.5–5.5 cm in diameter (Plate 3). Fruits angular with pointed tip, transverse section prominently ridged, dull green ripening to dull yellow and sweet, fine-textured creamy white to pale yellow pulp (Plates 1 & 2 and 4).

Plates 1 and 2 Unripe and semi ripe hands of Pisang Kepok

Plate 3 Bunch of Giant Kepok banana

Plate 6 Grilled pisang kepok is flattened into pancake-shape

Plate 4 Ripe hand of Pisang kepok

Plate 7 Stack of grilled pisang kepok pancakes

Plate 5 Grilling of peeled pisang kepok

Plate 8 Grilled pisang kepok served with caramelised coconut shavings

Nutritive/Medicinal Properties

According to Temanel (2007) Saba bananas provide the same nutritional value as potatoes. Saba is a starchy cultivar and is widely used as a raw material for several products in the Philippine food industry (Lustre et al. 1976). Starch degradation, the formation of reducing and non-reducing sugars and an increase in the moisture content, pectins and acidity of the pulp, were found to occur during ripening. Acetylene-induced ripening accelerated the rate of chemical change, without significantly affecting the final concentrations at which chemical constituents were present in the ripe pulp.

For more information on nutritive value of banana and medicinal attributes refer to notes in *Musa acuminata* (AA Group) 'Lakatan' and *Musa acuminata* (AAA Group) Dwarf Cavendish'.

Other Uses

Saba bananas are also cultivated as ornamental plants and shade trees for their large size and showy coloration. The waxy, green leaves are also used as traditional wrappings of native food dishes in Southeast Asia. Fibers can also be taken from the trunk and leaves and processed into ropes, mats, and sacks. For more information on other uses, refer to notes in *Musa acuminata* (AA Group) 'Lakatan'.

Comments

Saba banana is the most important cultivar for local markets in the Philippines and Indonesia but less so elsewhere.

Saba bananas namely Kluai saba, Kluai thin, previously classified as having BBB genome was reclassified as having ABB genome using PCR (polemerase chain reation)-restriction fragment length polymorphism PCR-RFLP technology (Arjcharoen et al. 2010). These results were in agreement with the report that identified Kluai Saba, Kluai Phama Haek Khuk and Kluai Hin as ABB (Boonruangrod et al. 2008).

Selected References

Arjcharoen A, Silayoi B, Wanichkul K, Apisitwanich S (2010) Variation of B genome in *Musa* accessions and their new identifications. Katsetsart J (Nat Sci) 44:392–400

Boonruangrod R, Desai D, Fluch S, Berenyi M, Burg K (2008) Identification of cytoplasmic ancestor genepools of *Musa acuminata* Colla and *Musa balbisiana* Colla and their hybrids by chloroplast and mitochondrial haplotyping. Theor Appl Genet 118(1):43–55

Daniells J, Jenny C, Tomekpe K (2001) Musalogue: a catalogue of *Musa* germplasm. Diversity in the genus *Musa*. In: Arnaud E, Sharrock S (comp) International Network for the Improvement of Banana and Plantain, Montpellier. www.inibap.org/publications/musalogue.pdf

Dela Cruz FS, Gueco LS, Damasco OP, Huelgas VC, Banasihan IG, Lladones RV, Van den Bergh I, Molina AB (2007) Catalogue of introduced and local banana cultivars in the Philippines IPB-UPLB, Bioversity – Philippines, and DA-BAR. Los Baños, 59 pp

Edison HS, Sutanto A, Hermanto C, Lakuy H, Rumsarwir Y (2002) The exploration of Musaceae in Irian Jaya (Papua) 2002. Research Institute for Fruits, Central Research Institute for Horticulture –INIBAP

Espino RRC, Jamalussin SH, Silayoi B, Nasution RE (1992) *Musa* L. (edible cultivars). In: Verheij EWM, Coronel RE (eds) Plant resources of South-East Asia no. 2. Edible fruits and uts. PROSEA, Bogor, pp 225–233

Gowen S (ed) (1995) Bananas and plantains. Chapman and Hall, London, 612 pp

Lustre AO, Soriano MS, Morga NS, Balagot AH, unac MM (1976) Physico-chemical changes in 'Saba' bananas during normal and acetylene-induced ripening. Food Chem 1(2):125–137

Ploetz RC, Kepler AK, Daniells J, Nelson SC (2007) Banana and plantain – an overview with emphasis on Pacific Island cultivars, ver. 1. In: Elevitch CR (ed) Species profiles for Pacific Island Agroforestry. Permanent Agriculture Resources (PAR), Hōlualoa. http://www.traditionaltree.org

Robinson JC (1996) Bananas and plantains. Crop production science in horticulture. CAB International, Wallingford, 238 pp

Stover RH, Simmonds NW (1987) Bananas, 3rd edn, Tropical Agricultural Series. Longman, Essex, 468 pp

Temanel BE (2007) Techno-guide for saba banana production in Cagayan Valley. http://www.openacademy.ph

Valmayor RV, Wagih ME (1996) *Musa* L. (plantain and cooking banana). In: Plant resources of South-East Asia no. 9. Plants yielding non-seed Carbohydrates. Prosea Foundation, Bogor, pp 126–131

Valmayor RV, Jamaluddin SH, Silayoi B, Kusumo S, Danh LD, Pascua OC, Espino RRC (2000) Banana cultivar names and synonyms in Southeast Asia. International Network for the Improvement of Banana and Plantain – Asia and the Pacific Office, Los Banos

Musa acuminata × *balbisiana* (AAB Group) 'Maia-Maoli-Pōpō'ulu'

Scientific Name

***Musa acuminata* × *balbisiana* Colla (AAB Group) 'Maia-Maoli-Pōpō'ulu'.**

Synonyms

None

Family

Musaceae

Common/English Name

Pacific Plantains

Vernacular Names of Local Cultivars

Maoli type
Cook Islands: Mangaro Torotea, Mangaro Manii, Mangaro Aumarei, Mangaro Taanga;
Hawaii: Kaualau, Mānai 'ula, Puhi;
Marquesas Islands: Māo'i, Māohi Huamene, Māo'i Koka, Māo'i Ku'uhua, Māo'i Pukiki, Mei'a Ma'ohi Hai;
Tahiti: Mā'ohi, Māohi Huamene, Māo'i Koka, Māo'i Ku'uhua, Māo'i Pukiki, Mei'a Ma'ohi Hai;
Tonga: Feta'u, Feta'u Hina, Hopa;

Pōpō'ulu' type
Cook Islands: Mangaro Akamou;
Hawaii: Hua Moa, Ka'io, Lahi, Lahilahi;
Samoa: Fa'i Samoa Fua Moa;
Tonga: Hopa, Putalinga.

Maoli-Pōpō'ulu' intermediate type
Pohnpei: Karat en Iap, Peleu;
Tonga: Hopa.

Pacific plantain
Australia: Pacific plantain;
Columbia: Comino, Pompo;
Ecuador: Comino, Maquerio, Pompo;
French Polynesia: Meia'a Ma'ohi Hai, Meia'a Mao'i Maita.

Origin/Distribution

Isozyme studies suggested that the genes contributed by the *M. acuminata* genome to the triploid Pacific plantain AAB subgroup were similar to those of the *acuminata/banksii* complex of Papua New Guinea (Lebot et al. 1993). Thus it was likely that the Pacific plantain subgroup, including the Hawaiian Maoli, Pōpō'ulu and Iholena cultivars, originated in Papua New Guinea/Melanesia, rather than in Asia or the Malay Archipelago. De Langhe and De Maret (1999) also suggested that the basic stock of the Polynesian Maia Maoli/Pōpō'ulu, originated probably in the Bismarck archipelago of Papua New Guinea.

T.K. Lim, *Edible Medicinal And Non-Medicinal Plants: Volume 3, Fruits*,
DOI 10.1007/978-94-007-2534-8_71, © Springer Science+Business Media B.V. 2012

Agroecology

Pacific plantains like bananas thrive best in warm, humid tropics with temperatures of 27–30°C and mean annual rainfall of 2,500–3,000 mm evenly distributed throughout the year. However, they tolerate slightly higher temperatures and drier conditions than bananas. They grow best in deep friable, well-drained loams with pH of 4.5–7.5. Pacific plantains are susceptible to strong winds which cause blow-downs.

Plate 1 Maoli plantains – plum with rounded ends

Edible Plant Parts and Uses

Pacific Plantains are eaten cooked – steamed, boiled, grill, baked or fried. Plantains can also be dried and ground into flour.

Plate 2 Maoli plantains – long with rounded ends and persistent style

Botany

Pacific plantains have comparatively short, 2–3 m high, herbaceous, shiny dark green to red-green or purple pseudostem with purple underlying pigmentation and watery sap. Numbers of suckers vary from 2 to 5. Leaf habit intermediate, with brown to dark brown blotches at the petiole base, dorsal midrib green to pink-purple to red-purple and petiole canals straight with erect margins. Inflorescence complete, peduncle 30–60 cm, glabrous, rachis hanging vertically, bracts are dehiscent leaving the male axis naked. Male bud normal and lanceolate, waxy with intermediate to pointed apex, young bracts slightly overlapping, external surface red-purple to purple-brown, internal surface orange-red to red. Male flower compound tepal white to cream, style curved, stigma cream, bright yellow to orange and ovary cream or green. Fruits sausage-shaped, 15–20 cm long, curved or straight with blunt or rounded ends, green turning yellow on ripening. The Maoli types are long, plump with rounded ends (Plates 1 and 2), whereas Pōpō'ulu types are fat and squat, and bluntly square at their tips (Plate 3). Pulp yellow to orange.

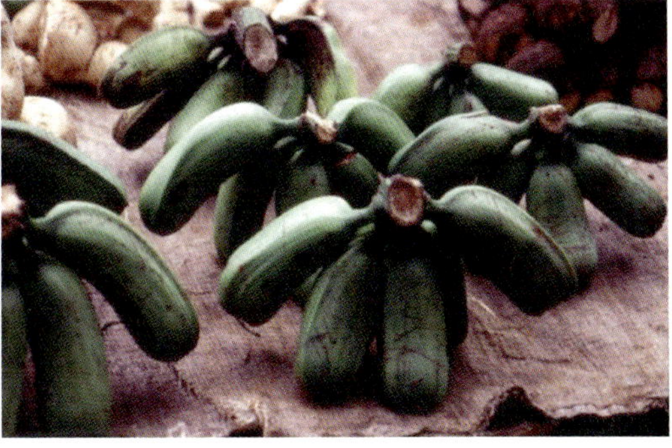

Plate 3 Pōpō'ulu plantains – plump, short and squat with blunt ends

Nutritive/Medicinal Properties

The proximate nutrient composition of plantains (*Musa* × *paradisiaca*) was reported as: water 65.28 g, energy 122Kcal (510 kJ), protein 1.30 g, fat 0.37 g, ash 1.17 g, carbohydrate 31.89 g, total dietary fibre 2.3 g, total sugars 15 g, Ca 3 mg, Fe 0.60 mg, P 34 mg, K 499 mg, Na 4 mg, Zn 0.14 mg, Cu 0.081 mg, Se 1.5 μg, vitamin C 18.4 mg, thiamine 0.052 mg, riboflavin 0.054 mg, niacin 0.686 mg, pantothenic acid 0.260 mg, vitamin B-6 0.299 mg, total folate 22 μg, total choline 13.5 mg, vitamin A 56 μg RAE, vitamin A 1,127 IU, β-caortene 457 μg, α-carotene 438 μg, lutein+zeaxanthin 30 μg, vitamin E (α-tocopherol) 0.14 mg, vitamin K (phylloquinone) 0.7 μg, total saturated fatty acids 0.143 g, 10:0 0.001 g, 12:0 0.002 g, 14:0 0.002 g, 16:0 0.096 g, 18:0 0.005 g, total monounsaturated fatty acids 0.032 g, 16:1 undifferentiated 0.009 g, 18:1 undifferentiated 0.021 g, total polyunsaturated fatty acids 0.069 g, 18:2 undifferentiated 0.043 g, 18:3 undifferentiated 0.025 g, tryptophan 0.015 g, threonine 0.034 g, isoleucine 0.036 g, leucine 0.059 g, lysine 0.060 g, methionine 0.017 g, cystine 0.020 g, phenylalanine 0.044 g, tyrosine 0.032 g, valine 0.046 g, arginine 0.108 g, histidine 0.064 g, alanine 0.051 g, aspartic acid 0.108 g, glutamic acid 0.116 g, glycine 0.045 g, proline 0.050 g and serine 0.041 g (US Department of Agriculture, Agricultural Research Service 2010).

Plantains from the Maoli and Pōpō'ulu subgroups were found to be tannin-free whilst the tannin content in green bananas and plantains belonging to the AA, AB, AAA, AAB, and ABB genotypes ranged between 0.3% and 2.1% of fresh matter (Santos et al. 2010). Epigallocatechin (EGC) was their major constituent (88–98%) followed by minor proportions of (−)-epicatechin (EC) (2–12%).

For medicinal attributes see notes under *Musa acuminata* (AA Group) "Lakatan".

Other Uses

See notes under *Musa acuminata* (AA Group) "Lakatan".

Comments

The history of the "Maia Maoli/Pōpō'ulu" group of bananas is closely tied to that of the Polynesian people.

Selected References

Arnaud E, Horry JP (eds) (1997) Musalogue: a catalogue of *Musa* germplasm. Papua New Guinea collecting missions, 1988–1989. International Network for the Improvement of Banana and Plantain, Montpellier

Daniells J, Jenny C, Tomekpe K (2001) Musalogue: a catalogue of *Musa* germplasm. Diversity in the genus *Musa*. In: Arnaud E, Sharrock S (comp) International Network for the Improvement of Banana and Plantain, Montpellier. www.inibap.org/publications/musalogue.pdf

De Langhe E (1996) Banana and plantain: the earliest fruit crops? INIBAP Annual report 1995. INIBAP, Montpellier, pp 6–8

De Langhe E, de Maret P (1999) Tracking the banana: significance to early agriculture. In: Gosden C, Hather J (eds) The prehistory of food. Routledge, London, pp 377–396

Lebot V, Aradhya KM, Manshardt R, Meilleur B (1993) Genetic relationships among cultivated bananas and plantains from Asia and the Pacific. Euphytica 67(3):163–175

Ploetz RC, Kepler AK, Daniells J, Nelson SC (2007) Banana and plantain – an overview with emphasis on Pacific Island cultivars, ver. 1. In: Elevitch CR (ed) Species profiles for Pacific Island agroforestry. Permanent Agriculture Resources (PAR), Hōlualoa. http://www.traditionaltree.org

Robinson JC (1996) Bananas and plantains. Crop production science in horticulture. CAB International, Wallingford, 238 pp

Santos J-RU, Bakry F, Brillouet J-M (2010) A preliminary chemotaxonomic study on the condensed tannins of green banana flesh in the *Musa* genus. Biochem Syst Ecol 38(5):1010–1017

Stover RH, Simmonds NW (1987) Bananas, 3rd edn. Tropical Agricultural Series, Longman, 468 pp

US Department of Agriculture, Agricultural Research Service (2010) USDA National Nutrient Database for Standard Reference, Release 23. Nutrient Data Laboratory Home Page. http://www.ars.usda.gov/ba/bhnrc/ndl

Valmayor RV, Wagih ME (1996) *Musa* L. (plantain and cooking banana). In: Flach M, Rumawas F (eds) Plant resources of South-East Asia 9, plants yielding non-seed carbohydrates. Prosea Foundation, Bogor, pp 126–131

Musa acuminata × *balbisiana* (AAB Group) 'Pisang Raja'

Scientific Name

Musa acuminata **×** ***balbisiana*** **Colla (AAB Group) 'Pisang Raja'**.

Synonyms

Musa regia Rumphias, *Musa* × *paradisiaca.*

Family

Musaceae

Common/English Name

Pisang Raja, Raja Banana

Vernacular Names

Indonesia: Biu Raja (Java), Ndondi (Irian Jaya), Pisang raja, Pisang raja bulu;
Malaysia: Pisang raja;
Papua New Guinea: Houdir, Kalamama wudu, Larip;
Philippines: Radja;
Pohnpei: Utin kerenis (kirou Rohi);
Thailand: Kluai Khai Boran;
Windward Islands: Grindy.

Origin/Distribution

The cultivar is indigenous to Malesia-Malaysia and Indonesia.

Agroecology

See notes in *Musa acuminata* (AA Group) 'Lakatan'.

Edible Plant Parts and Uses

Pisang Raja is highly prized in Indonesia and Malaysia. It can be eaten raw, but is more popularly used as a cooking banana in Malaysia and the Philippines. The fruit with skin removed, is immersed in batter and deep fried to make the famous *goreng pisang* (banana fritters).

Botany

Pseudostem short, 2.1–3.6 m high, with 65–77 cm girth, green to reddish-green, dull and robust with milky sap. Numer of suckers 4. Leaves green without blotches, drooping, mid-rib pale-green or yellow in dorsal surface. Inflorescence falling vertically, peduncle glabrous to hairy, rachis falling vertically, female proximal end, neuter and male flowers distal end. Male bud ovoid to

T.K. Lim, *Edible Medicinal And Non-Medicinal Plants: Volume 3, Fruits*,
DOI 10.1007/978-94-007-2534-8_72,

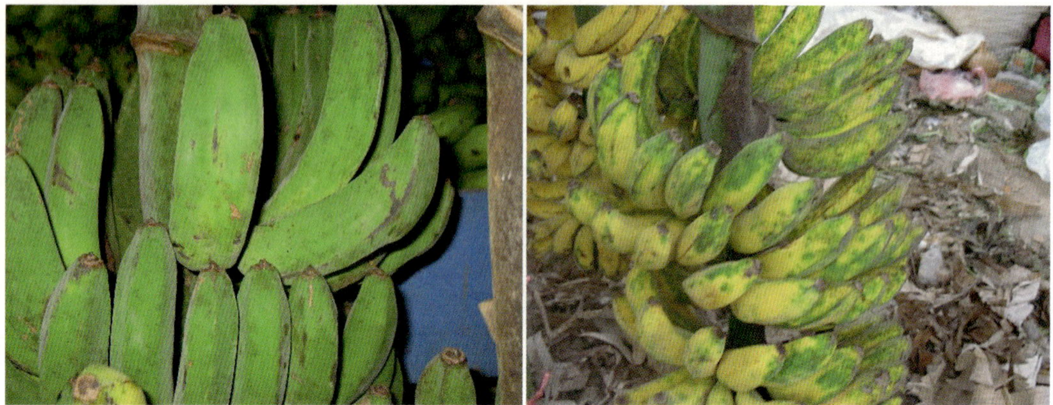

Plates 1 and 2 Pisang Raja Bulu, Indonesia (unripe ripe and half-ripe bananas)

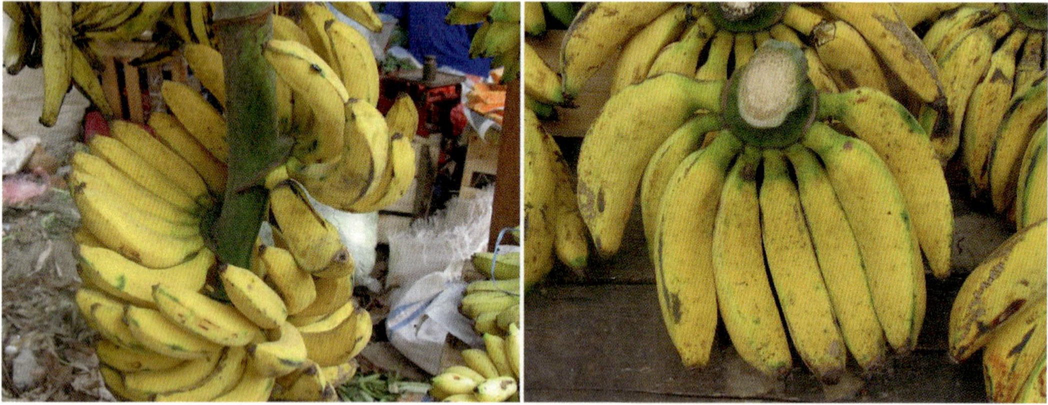

Plates 3 and 4 Bunch and hand of ripe Pisang Raja bananas

intermediate shape, young bract imbricate, bract apex obtuse and split, external surface pink-purple to red-purple, internal surface red to pink purple. Flower free tepal cream, compound tepal yellow, style straight, stigma cream to yellow, ovary white to cream. Fruit bunch has 6–9 hands with 14–17 fruits per hand, average bunch weight 12–16 kg (Plates 1–2). Fruit curved upwards, slightly angled, 15–20 cm long by 3–4.5 cm across, with thick coarse skin, green turning to yellow, traverse section slightly ridged, fruit apex lengthily pointed without floral remnants (Plates 3–4), pulp creamy-yellow, deep yellow to orangey yellow, texture coarse, very sweet.

Nutritive/Medicinal Properties

For nutritive values refer to that for banana under *Musa acuminata* (AAA Group) 'Dwarf Cavendish'.

Pisang raja buluh, a cultivar from Malaysia was found to be a good source of vitamin A, containing 420 μg/100 g of β-carotene (Englberger et al. 2003). Blades et al. (2003) found that four *Musa* cultivars (Horn Plantain, Kirkirnan, Asupina and Pisang Raja) to be good sources of provitamin A (>75 RAE/100 g) and hence potentially useful for the prevention of vitamin A deficiency.

Alpha-, all-trans-β and cis-β-carotene were the only provitamin A carotenoids present in detectable quantities in all samples and levels of these carotenes ranged from 61 to 1,055, 50 to 1,412 and 7 to 85 μg/100 g banana, respectively.

Ten yellow or yellow/orange-fleshed cultivars (Asupina, Kirkirnan, Pisang Raja, Horn Plantain, Pacific Plantain, Kluai Khai Bonng, Wain, Red Dacca, Lakatan, and Sucrier) were found to have significant carotenoid levels, potentially meeting half or all of the estimated vitamin A requirements for a non-pregnant, non-lactating adult woman within normal consumption patterns (Englberger et al. 2006). All were acceptable for taste and other attributes.

Other Uses

See notes in *Musa acuminata* (AA Group) 'Lakatan'.

Comments

The plant is propagated by off-sets (suckers).

References

Blades BL, Dufficy L, Englberger L, Daniells JW, Coyne T, Hamill S, Wills RB (2003) Bananas and plantains as a source of provitamin A. Asia Pac J Clin Nutr 12(Suppl):S36

Daniells J, Jenny C, Tomekpe K (2001) Musalogue: a catalogue of *Musa* germplasm. Diversity in the genus *Musa*. In: Arnaud E, Sharrock S (comp) International Network for the Improvement of Banana and Plantain, Montpellier. www.inibap.org/publications/musalogue.pdf

Dela Cruz FS, Gueco LS, Damasco OP, Huelgas VC, Banasihan IG, Lladones RV, Van den Bergh I, Molina AB (2007) Catalogue of introduced and local banana cultivars in the Philippines IPB-UPLB, Bioversity – Philippines, and DA-BAR. Los Baños, 59 pp

Edison HS, Sutanto A, Hermanto C, Lakuy H, Rumsarwir Y (2002) The exploration of Musaceae in Irian Jaya (Papua) 2002. Research Institute for Fruits, Central Research Institute for Horticulture – INIBAP

Englberger L, Darnton-Hill I, Coyne T, Fitzgerald MH, Marks GC (2003) Carotenoid-rich bananas: a potential food source for alleviating vitamin A deficiency. Food Nutr Bull 24(4):303–318

Englberger L, Wills RB, Blades B, Dufficy L, Daniells JW, Coyne T (2006) Carotenoid content and flesh color of selected banana cultivars growing in Australia. Food Nutr Bull 27(4):281–291

Espino RRC, Jamaluddin SH, Silayoi B, Nasution RE (1992) *Musa* L. (edible cultivars). In: Verheij EWM, Coronel RE (eds) Plant resources of South-East Asia 2. Edible fruits and nuts. Prosea Foundation, Bogor, pp 225–233

Nasution RE (1994) *Musa* L. In: Siemonsma JS, Kasem P (eds) Plant resources of South-East Asia 8. Vegetables. Prosea Foundation, Bogor, pp 215–217

Ploetz RC, Kepler AK, Daniells J, Nelson SC (2007) Banana and plantain – an overview with emphasis on Pacific Island cultivars, ver. 1. In: Elevitch CR (ed) Species profiles for Pacific Island Agroforestry. Permanent Agriculture Resources (PAR), Hōlualoa. http://www.traditionaltree.org

Robinson JC (1996) Bananas and plantains. Crop production science in horticulture. CAB International, Wallingford, 238 pp

Simmonds NW, Shepherd K (1955) The taxonomy and origins of the cultivated bananas. J Linn Soc Bot 55:302–312

Stover RH, Simmonds NW (1987) Bananas, 3rd edn, Tropical agricultural series. Longman, Essex, 468 pp

Valmayor RV, Wagih ME (1996) *Musa* L. (plantain and cooking banana). In: Flach M, Rumawas F (eds) Plant resources of South-East Asia 9, plants yielding non-seed carbohydrates. Prosea Foundation, Bogor, pp 126–131

Valmayor RV, Silayoi B, Jamaluddin SH, Kusumo S, Espino RRC, Pascua OC (1990) Commercial banana cultivars in ASEAN. In: Hassan A, Pantastico EB (eds) Banana: fruit development, postharvest physiology, handling and marketing in ASEAN. ASEAN Food Handling Bureau, Kuala Lumpur, 159 pp

Valmayor RV, Jamaluddin SH, Silayoi B, Kusumo S, Danh LD, Pascua OC, Espino RRC (2000) Banana cultivar names and synonyms in Southeast Asia. International Network for the Improvement of Banana and Plantain – Asia and the Pacific Office, Los Banos

Musa acuminata × *balbisiana* (AAB Group) 'Silk'

Scientific Name

***Musa acuminata* × *balbisaina* Colla (AAB Group) 'Silk'**.

Synonyms

Musa acuminata × *babisiana* Colla, *Musa berteri* Colla, *Musa berteroi* Colla, *Musa berteroniana* von Steudel, *Musa cliffortiana* L., *Musa dacca* P.F. Horaninow, *Musa* × *sapientum* L., *Musa paradisiaca* L. ssp. *sapientum* (L.) C.E.O Kuntze, *Musa* × *paradisiaca*, Kuntze var. *cubensis*, *Musa* × *paradisiaca* var. *dacca* (P. F. Horaninow) J. G. Baker ex K. M. Schumann, *Musa rosacea* N.J. von Jacquin.

Family

Musaceae

Common/English Names

Apple Banana, Latundan Banana, Manzana Banana, Silk Banana, Silk Fig Banana

Vernacular Names of Local Cultivars

Australia: Sugar (Queensland);
Brazil: Maca;
Cook Islands: Miti Ruki, Tiki;
East Africa: Kipukusu, Kipungusu;
Federated States of Micronesia: Utin Kuam, Utin Menihle, Uht Tikitik (Pohnpei);
Hawaii: Amorosa, Lady Finger, Manzana, Manzano;
India: Amrithapani, Digjowa, Dudhsagar, Honda, Kozhikodu, Madhuranga, Malbhog, Morthoman, Mutheli, Nanjankode, Rasabale, Rasthali, Saapkal, Sabari, Sabri, Sonkel, Suvandal, Thozhuvan;
Indonesia: Pisang Raja Sereh;
Latin America: Manzana, Manzano;
Malaysia: Pisang Rastali;
Myanmar: Htwa-bat;
Papua New Guinea: Avundumong, Maramba, Worodong;
Philippines: Amorosa, Cantong, Katungal, Latundan, Letondal, Tordan, Tundan, Turdan;
Sri Lanka: Kolikutt;
Thailand: Kluai Nam;
United States: Apple (Florida);
Vietnam: Chuoi Goong;

T.K. Lim, *Edible Medicinal And Non-Medicinal Plants: Volume 3, Fruits*,
DOI 10.1007/978-94-007-2534-8_73, © Springer Science+Business Media B.V. 2012

West Indies: Apple, Figue Pomme, Manzana, Manzano, Silk Fig;
Zanzibar: Pukusa.

Origin/Distribution

Philippines have been reported to be origin of this banana. Latundan is a common banana cultivar in the Philippines.

Agroecology

Silk fig banana shares similar growing requirements as the Lakatan and Cavendish bananas. It thrives best in lowland, humid tropical areas with rich, moist, well-drained soils well protected from strong winds. Nutrient poor and dry soils (as well as cooler weather) results in fruits that have dry, hard flesh and little flavour.

Edible Plant Parts and Uses

Silk Fig banana is a popular dessert banana in Malaysia, Indonesia and the Philippines. The pulp is sub acidic with an apple-like flavour. The skin of fully ripe fruit tends to split.

Botany

A comparatively short, herbaceous banana plant; pseudostem 3–3.6 m high, robust, shiny green-yellow with underlying pink purple pigmentation and watery sap. Leaves broad, intermediate, mid rib light green dorsal surface, large brown blotches at the petiole base. Inflorescence complete, peduncle very hairy and 30–60 cm long. Male bud intermediate shape, bract with obtuse split apex, red-purple externally and internal surface red. Flower compound tepal yellow or cream with pink pigmentation, style curved at a base, stigma orange and ovary cream. Average bunch weight is 10–14 kg with 5–9 hands and 12–16 fingers per hand. Fruit curved towards stalk, small to medium, 10–20 cm by 3–6 cm diameter, straight, skin yellow-green to yellow and thin, usually with tiny brownish dark speckles, skin tends to split once fully ripe, transverse section slightly ridged and apex lengthily pointed (Plates 1, 2, 3 and 4).

Plate 1 Fully ripe Pisang Raja Sereh (Java)

Plate 2 Semi-ripe Pisang Raja Sereh (Java)

Plate 3 Over-ripe Pisang Raja Sereh (Sumatra)

Plate 4 Pisang Rastali (Malaysia)

Flesh is white, soft textured, slightly subacid and distinctively fragrant in flavour. Fruits are smaller and pudgier than the Lakatan and the commercial Cavendish bananas.

Nutritive/Medicinal Properties

Proximate nutrient composition of the edible portion (per 100 g) of pisang rastali was reported by Tee et al. (1997) as follows: energy 95 kcal., water 74.7 g, f protein 1.6 g, fat 0.2 g, carbohydrate 21.8 g, fibre 1.0 g, ash 0.7 g, P 63 mg, Fe 0.2 mg, Na 30 mg, K 277 mg, carotenes 12ug, vitamin A 2ug RE, vitamin B-1, 0.04 mg, vitamin B-2 0.06 mg, niacin 3.2 mg and vitamin C 7.8 mg.

Other Uses

See notes in *Musa acuminata* (AA Group) "Lakatan".

Comments

Ripe pisang rastali fruits tend to split shortening its post-harvest shelf life.

Selected References

Daniells J, Jenny C, Tomekpe K (2001) Musalogue: a catalogue of *Musa* germplasm: diversity in the genus *Musa*. In: Arnaud E, Sharrock S (comp) International Network for the Improvement of Banana and Plantain, Montpellier. www.inibap.org/publications/musalogue.pdf

Espino RRC, Jamalussin SH, Silayoi B, Nasution RE (1992) *Musa* L. (edible cultivars). In: Verheij EWM, Coronel RE (eds) Plant resources of South-East Asia no 2. Edible fruits and nuts. PROSEA, Bogor, pp 225–233

Gowen S (ed) (1995) Bananas and plantains. Chapman & Hall, London, 612 pp

Ploetz RC, Kepler AK, Daniells J, Nelson SC (2007) Banana and plantain – an overview with emphasis on Pacific Island cultivars, ver. 1. In: Elevitch CR (ed) Species profiles for Pacific Island agroforestry. Permanent Agriculture Resources (PAR), Hōlualoa, Hawai'i. http://www.traditionaltree.org

Porcher MH et al (1995–2020) Searchable World Wide Web multilingual multiscript plant name database. The University of Melbourne, Melbourne. http://www.plantnames.unimelb.edu.au/Sorting/Frontpage.html

Robinson JC (1996) Bananas and plantains (crop production science in horticulture). CAB International, Wallingford, 238 pp

Stover RH, Simmonds NW (1987) Bananas, 3rd edn, Tropical agricultural series. Longman, Essex, 468 pp

Tee ES, Noor MI, Azudin MN, Idris K (1997) Nutrient composition of Malaysian foods, 4th edn. Institute for Medical Research, Kuala Lumpur, 299 pp

Valmayor RV, Wagih ME (1996) *Musa* L. (plantain and cooking banana). In: Flach M, Rumawas F (eds) Plant resources of South-East Asia no. 9. Plants yielding non-seed carbohydrates. Prosea Foundation, Bogor, pp 126–131

Valmayor RV, Silayoi B, Jamaluddin SH, Kusumo S, Espino RRC, Pascua OC (1990) Commercial banana cultivars in ASEAN. In: Hassan A, Pantastico EB (eds) Banana: fruit development, postharvest physiology, handling and marketing in ASEAN. ASEAN Food Handling Bureau, Kuala Lumpur, 159 pp

Valmayor RV, Jamaluddin SH, Silayoi B, Kusumo S, Danh LD, Pascua OC, Espino RRC (2000) Banana cultivar names and synonyms in Southeast Asia. International Network for the Improvement of Banana and Plantain – Asia and the Pacific Office, Los Banos

Musa acuminata × *balbisiana* (ABB Group) 'Bluggoe'

Scientific Name

Musa acuminata × ***balbisiana*** Colla (ABB Group) 'Bluggoe'.

Synonyms

None

Family

Musaceae

Common/English Names

Bluggoe, Bluggoe Banana, Bluggoe Plantain

Vernacular Names

Americas: Apple Plantain, Burro, Chato, Cachaco, Cuatrofilos, Hog Banana, Horse Banana, Largo, Majoncho;
Australia: Square Cooker, Mondolpin;
Cook Islands: Tarua Matie;
Cuba: Burro, Orinoco;
Dominican Republic: Horse Plantain;
East Africa: Mkojosi, Bokoboko, Kproboi, Muskat, Punda, Kidhozi, Kivivu;
Federated States of Micronesia: Awatwat, Pithothaw, Inaiso;
French Polynesia: Largo, Poro'ini, Poro'ini Pa'afa'afa'a, Poro'ini Hima'a umu;
Fiji: Jamani;
Hawaii: Largo;
India: Nalla Bontha;
Indonesia: Pisang Batu (Java);
Jamaica: Horse Plantain;
Malaysia: Pisang Abu Keling, Pisang Sabah (Sabah);
Myanmar: Hpi Gyan;
Philippines: Matavia;
Samoa: Fa'i Pata Samoa, Puataelo;
Sri Lanka: Mondan;
Thailand: Kluai Som;
Tonga: Pata Tonga;
Trinidad and Tobago: horse Plantain;
Vietnam: Chuoi Ngop Lun;
West Indies: Whitehouse Plantain, Chamaluco, Poteau, Cacambou,'Moko, Bluggoe, Buccament, Mafoubay.

Origin/Distribution

It is generally accepted that the Indo-Malaysia region is the main centre of diversity of *Musa* spp. as proposed by Simmons and Shepherd (1955) and that Malesia is the primary centre. Edible banana were suggested to be domesticated in south east Asia and spread throughout the tropics and subtropics across Americas, Africa, Asia and Australia.

T.K. Lim, *Edible Medicinal And Non-Medicinal Plants: Volume 3, Fruits*,
DOI 10.1007/978-94-007-2534-8_74, © Springer Science+Business Media B.V. 2012

Agroecology

As described for *Musa acuminata* Colla (AA Group) 'Lakatan'.

Edible Plant Parts and Uses

Bluggoe banana being a starchy banana is primarily used for cooking especially in desserts and sweet snacks. Uses are similar to the Saba and the horn plantains.

Botany

Pseudostem 3.55 high with a girth of 46.85 cm, shiny light green with watery sap. Leaf habit intermediate, Leaf petiole green, 56 cm long, with small brown-black blotches at the base and petiole canal with margins curved inwards. Leaf blade 220 cm by 57 cm wide, moderately waxy, midrib green, leaf base rounded, shiny green upper and dull green lower surface. Inflorescence (male bud) peduncle length 4.7 cm, green and glabrous. Male flower compound tepal pinkish-purple basic colour, compound tepal lobe yellow, free tepal oval, tinted pink, with yellow filament and anther. Bunch cylindrical and vertically pendant. Fruits large, 137 mm by 43 mm wide and 50 mm thick, angular, straight fruit that have long peduncles, 4–7 hands per bunch (Plates 1 and 2). Fruit dull green ripening to yellow, pulp soft, white, fruit skin peels off easily.

Nutritive/Medicinal Properties

Proximate nutrient composition of the edible portion (per 100 g) of pisang abu was reported by Tee et al. (1997) as follows: energy 124 kcal., water 67.3 g, protein 1.3 g, fat 0.1 g, carbohydrate 29.6, fibre 0.9 g, ash 0.8 g, Ca 0 mg, P 6 mg, Fe 0.1 mg, Na 6 mg, K 289 mg, carotenes 42 μg, vitamin A 7 μg RE, vitamin B-1 0.02 mg, vitamin B-2 0.02 mg, niacin 0.1 mg and vitamin C 12.1 mg.

Plate 1 Bunch of unripe Pisang Abu bananas

Plate 2 Ripe, plump, straight, angular pisang abu bananas with long peduncles

Other Uses

The waxy, green leaves are also used as traditional wrappings of native food dishes in Southeast Asia. Fibers can also be taken from the trunk and leaves and processed into ropes, mats, and sacks.

Studies by Dormound et al. (2001) found that the combination 40:60 (bluggoe banana (*Musa* ABB) pseudostem: corn silage), reduced the feeding cost per unit of gain of Jersey calves. Digestibility of dry matter and acid detergent fibre increased significantly when pseudostem was 40%.

Comments

Bluggoe banana is grown in many countries due to its excellent taste, high productivity, resistance to drought, Panama and sigatoka diseases, and good performance in poor soils.

Selected References

Dela Cruz FS, Gueco LS, Damasco OP, Huelgas VC, Banasihan IG, Lladones RV, Van den Bergh I, Molina AB (2007) Catalogue of introduced and local banana cultivars in the Philippines. IPB-UPLB, Bioversity – Philippines, and DA-BAR, Los Baños, 59 pp

Dormond H, Rojas A, Jiménez C, Quirós G (2001) Effect of increasing levels of bluggoe banana pseudostems added to corn silage as roughage, on Jersey calves growing in confinement, during the dry season. Agron Costarricense 24(2):31–40

Ploetz RC, Kepler AK, Daniells J, Nelson SC (2007) Banana and plantain – an overview with emphasis on Pacific Island cultivars, ver. 1. In: Elevitch CR (ed) Species profiles for Pacific Island agroforestry. Permanent Agriculture Resources (PAR), Hōlualoa, Hawai'i. http://www.traditionaltree.org

Robinson JC (1996) Bananas and plantains. Crop production science in horticulture. CAB International, Wallingford, 238 pp

Stover RH, Simmonds NW (1987) Bananas, 3rd edn, Tropical agricultural series. Longman, Essex, 468 pp

Simmonds NW, Shepherd K (1955) The taxonomy and origins of the cultivated bananas. J Linn Soc London Bot 55(359):302–12

Tee ES, Noor MI, Azudin MN, Idris K (1997) Nutrient composition of Malaysian foods, 4th edn. Institute for Medical Research, Kuala Lumpur, 299 pp

Valmayor RV, Wagih ME (1996) *Musa* L. (plantain and cooking banana). In: Flach M, Rumawas F (eds) Plant resources of South-East Asia no. 9. Plants yielding non-seed carbohydrates. Prosea Foundation, Bogor, pp 126–131

Valmayor RV, Jamaluddin SH, Silayoi B, Kusumo S, Danh LD, Pascua OC, Espino RRC (2000) Banana cultivar names and synonyms in Southeast Asia. International Network for the Improvement of Banana and Plantain – Asia and the Pacific Office, Los Banos

Musa coccinea

Scientific Name

Musa coccinea **H. C. Andrews**

Synonyms

Quesnelia lamarckii Baker.

Family

Musaceae

Common/English Names

Okinawa Torch, Red Flowering Banana, Red Flowering Thai Banana, Red Torch Banana, Scarlet Banana

Vernacular Names

Chinese: Ba Jiao Hong, Hong Jiao, Ye Jiao, Xiao Ba Jiao;
Spanish: Guineo De Fuego;
Taiwan: Mi-Jên Chiao;
Vietnamese: Chuoi Rung, Chuoi Tau.

Origin/Distribution

It is native to China (southeast Yunnan to Guangdong) to Vietnam.

Agroecology

In it native range, it is found in ravines and slopes, also cultivated in gardens; near sea level to 600 m. It prefers a well drained, acidic soil rich in organic matter and thrives in full sun to medium shade.

Edible Plant Parts and Uses

The plant and fruits have been deemed as edible in Vietnam (McElwee 2010). The red bracts are potential sources of natural food colorant and neutraceuticals (Pazmiño-Durán et al. 2001).

Botany

A slender, short perennial herb plant with a green, non-waxy pseudostem growing up to 1.5 m high and 5 cm in diameter at base, producing abundant suckers freely close to the parent plant. Petioles up to 35 cm., with narrow erect margins

T.K. Lim, *Edible Medicinal And Non-Medicinal Plants: Volume 3, Fruits*,
DOI 10.1007/978-94-007-2534-8_75, © Springer Science+Business Media B.V. 2012

Plate 1 *Musa coccinea* inflorescence

Plate 2 Close-up of *Musa coccinea* inflorescence

clasping the pseudostem. Leaf blade up to 100 cm long and 25 cm wide glossy dark green adaxially and paler green abaxially, apex rounded, bases asymmetrical and rounded mid-rib distinct, green. Inflorescence erect scarcely emerging from the sheath of the subtending leaf, the rachis glabrous, sterile bracts usually 2, bright scarlet with green leaflike tips (Plates 1 and 2). Flowers of the basal bracts female, upper flowers male. Female flowers 1–3 per bract; ovary dorsiventrally compressed, orange-yellow, glabrous; compound tepal 3.5 cm long, orange-yellow with green lobes, the lateral lobes oval-oblong with a spine-like dorsal appendage, centre lobe smaller; free tepal as long as the compound tepal, dorsally thickened, opaque, and orange in colour, staminodes short (about 1 cm.); style as long as the perianth with a bright orange clavate stigma. Male flowers 2 per bract, compound tepal bright orange with green tip and lobes, the lateral lobes with a spine-like dorsal appendage, the centre lobe shorter and without appendage, joined to the accessory teeth; free tepal as long as the compound, narrow oblong, dorsally thick, opaque, and orange in colour, laterally hyaline, obtuse at apex; stamens nearly as long as the perianth, not exserted. Fruit oblong, 4–5 cm long by 2–2.5 cm across, laterally compressed dorsiventrally, crowned by the persistent withered perianth, pericarp about 1.5 mm. thick, orange-yellow at full maturity with a waxy bloom; pulp white. Seeds cylindrical, about 6 mm. long, warty, black.

Nutritive/Medicinal Properties

The red bracts of *Musa coccinea* were found to contain the following anthocyanidins – pelargonidin and cyanidin. They are potential sources of natural food colorant and neutraceuticals (Pazmiño-Durán et al. 2001).

Other Uses

Cultivated as an ornamental and valued for its brilliant scarlet inflorescences which are used as cut flowers.

Comments

The plant is propagated by sucker off-setts.

Selected References

Champion J (1967) Notes et docuemts sur les bananiers et leur culture. Tome I: Botanique et Genetique. Institut Francais de Reserches Fruitieres Outre-Mer (IFAC), Editions SETCO, Paris

Cheesman EE (1950) Classification of the bananas. III. Critical notes on species. q. *Musa coccinea* Andrews. Kew Bull 5(1):29–31

Daniells J, Jenny C, Tomekpe K (2001) Musalogue: a catalogue of *Musa* germplasm. Diversity in the genus *Musa*. In: Arnaud E, Sharrock S (comp) International Network for the Improvement of Banana and Plantain, Montpellier

Govaerts R, Häkkinen (2009) World checklist of Musaceae. The Board of Trustees of the Royal Botanic Gardens, Kew. http://www.kew.org/wcsp/ . Accessed 11 Nov 2009

Liu A-Z, Li D-Z, Li X-W (2002) Taxonomic notes on wild bananas (*Musa*) from China. Bot Bull Acad Sin 43:77–81

McElwee PD (2010) Resource use among rural agricultural households near protected areas in Vietnam: the social costs of conservation and implications for enforcement. Environ Manage 45(1):113–131

Pazmiño-Durán EA, Giusti MM, Wrolstad RE, Glória MBA (2001) Anthocyanins from banana bracts (*Musa* × *paradisiaca*) as potential food colorants. Food Chem 73(2):327–332

Musa troglodytarum

Scientific Name

Musa troglodytarum L.

Synonyms

Musa fehi Bertero ex Vieill., *Musa × paradisiaca* subsp. *troglodytarum* (L.) K.Schum., *Musa × paradisiaca* var. *dorsata* G.Forst., *Musa rectispica* Nakai, *Musa × sapientum* subsp. *troglodytarum* (L.) Baker, *Musa × sapientum* var. *troglodytarum* (L.) Baker nom. illeg., *Musa seemannii* F.Muell., *Musa troglodytarum* var. *acutibracteata* MacDan., *Musa uranoscopos* Colla nom illeg., *Musa uranoscopos* Lour.

Family

Musaceae

Common/English Names

F'ei Banana, Fehi Banana, Hueta, Tahitian Red Cooking Banana

Vernacular Names

Chinese: Fei Shi Jiao;
Fiji: F'ei;
Hawaiian: Mai'A Hē'Ī, Mai'A Polapola;
Indonesia: Pisang Tongkat Langit, Pisang Tongkat Langit Papua;
New Caledonian: Dáak;
Tahiti: F'ei.

Origin/Distribution

Fe'i bananas are thought to have originated in the New Guinea area and from there were spread westward through the Pacific by human travellers (Stover and Simmonds 1987). The Fe'i banana cultivars range naturally from the Moluccas to French Polynesia. Fe'i bananas are particularly associated with the Marquesas and Society Islands (French Polynesia). Fe'i bananas were important staple and ceremonial foods since the Marquesas first settled from the Samoa-Tonga region about 250 BC and in Tahiti around 700–800 AD. Unfortunately, their prevalence has declined drastically in recent decades.

Agroecology

Like all bananas and plantains, Fei banana is well adapted to hot, wet, tropical region with well distributed mean annual rainfall of over 2,000 mm per annum. They are intolerant of low cold temperatures and grow best in full sunlight. In most areas, they require wind protection and are susceptible to being blown over. It tolerates a wide variety of soil conditions, but deep, well-drained alluvial soils are best suited for its cultivation.

T.K. Lim, *Edible Medicinal And Non-Medicinal Plants: Volume 3, Fruits*,
DOI 10.1007/978-94-007-2534-8_76, © Springer Science+Business Media B.V. 2012

In the Pacific islands, Fe'i banana is cultivated and persists sparingly in low elevation in mesic to wet valleys.

Edible Plant Parts and Uses

The deep yellow and orange pulp possesses high levels of β-carotene and are edible raw, but are better eaten cooked or roasted.

Botany

Pseudostem 3–10 m tall and 50–100 cm in circumference, predominantly brown below becoming bright non-waxy green above with a variable amount of brown streaking and blotches. Sap variable from watery cream to pink or bright red or purple. Rhizomes short, suckers arising more or less vertically from the ground and plants forming dense clumps with many pseudostems. Petiole pale green non-waxy, margins erect or reflexed more or less green. Leaf lamina bright green non-waxy above paler but also without wax below. Leaf base right-handed and cuneate to rounded. Inflorescence is terminal and upright with a glabrous, green velvety peduncle. Flowers of basal bracts female with 4 "hands" of 4–6 uniseriate flowers. Female flower with perianth as long as the ovary, free tepal very much shorter than compound tepal which is yellow, style is straight with orange coloured stigma and green ovary. Male bud in advanced blooming is ovate, the bracts slightly imbricate at the tip. Bracts are bright reddish-pink in colour in both surfaces or glossy green outside and paler green internally with yellow-golden tip. Male flowers about 6–10 per bract in two rows; compound tepal orange-yellow with smooth orange lateral lobes, free tepal boat shaped, stamens as along as long as the compound tepal, the filaments a little longer than the anthers. Fruit bunch held stiffly erect, variable from very dense to moderately lax, usually with 5 hands per bunch and 5–7 fruits per hand (Plates 1 and 2). Fruit 12–20 cm long, ridged, fruit apex rounded, coppery to orange colour, pedicel 1.5 cm long,

Plate 1 Harvested bunch of ripe Pisang Tongkat Langit

Plate 2 Terminal upright bunch of Pisang Tongkat Langit

no floral relict at fruit apex, pulp golden yellow to orange at maturity and no seed.

Nutritive/Medicinal Properties

Ripe Fe'i banana with yellow/orange flesh was reported to have the following nutrient composition: Ca 68.6 mg, K 253 mg, β-carotene equivalent 565–2,473 μg, riboflavin 0.47–14.30 mg, niacin 22.6 mg and vitamin E 1,55 mg (Englberger et al. 2003a, b). The yellow/orange-fleshed Asupina (a Fe'i banana cultigen) was found to contain the highest level (1,412 μg/100 g) of trans β-carotene, the most important provitamin A carotenoid, a level more than 20 times higher than that of cultivar Williams (Englberger et al. 2006b). Karat, another Fei banana cultigens was found to have high riboflavin level including high levels of uncharacterized flavonoids (Englberger et al. 2006a). Niacin and α-tocopherol were found at levels that may contribute importantly to dietary intake within normal patterns of consumption. Englberger et al. (2003a) also found that *Karat* and *Taiwang* bananas had low levels of iron and zinc similar to bananas elsewhere, 0.2 and 0.1 mg iron/100 g. Similarly, Shovic and Whistler (2001) found low mineral levels (0.2–0.5 mg iron /100 g and 0.1–0.3 mg zinc/100 g) in the bananas analysed.

Other Uses

Other non-edible uses of Fe'i bananas in the Pacific islands have been documented by MacDaniels (1947) and Pétard (1955). The reddish purple sap of the pseudostem is occasionally used as a dye and ink. Samoans decorate the margins of mats with thin strips of banana fibre died pink or red with the sap. The pseudostems are tied together to form a temporary raft for crossing lakes and streams. Fibrous material from the pseudostems and leaves are stripped off, dried and use for making ropes, plaiting mats, fans and other containers. The Fe'i harvesters used these ropes to bind bunches of fruit to the carrying poles. The dried leaves were used as bedding and for packing, and also made a good fuel for starting fires. Additionally, small thin pieces of the dried leaves could be used as cigarette papers.

Comments

Fe'i bananas are grouped in the section Australimusa, however their precise origins are still uncertain. *Musa maclayi* (based on morphology) and *Musa lolodensis* (based on DNA studies) were suggested as probable parents of the extant clones (Ploetz et al. 2007). Recent genetic studies indicated that they are closest genetically to these two species and to *Musa peekelii* suggesting that Fe'i bananas may be interspecific hybrids. Wong et al. (2002) based on results of AFLP suggested that species of section Rhodochlamys should be combined into a single section with species of section *Musa*, and likewise for species of section Australimnusa to be merged with those of section Callimusa.

Selected References

Edison HS, Sutanto A, Hermanto C, Lakuy H, Rumsarwir Y (2002) Tongkat Langit Papua. In: The exploration of Musaceae in Irian Jaya (Papua). Central Research Institute for Horticulture –INIBAP, Solok Research Institute for Fruits Solok, 58 pp

Englberger L, Aalbersberg W, Ravi P, Bonnin E, Marks GC, Fitzgerald MH, Elymore J (2003a) Further analyses on Micronesian banana, taro, breadfruit and other foods for provitamin A carotenoids and minerals. J Food Compos Anal 16:219–236

Englberger L, Schierle J, Marks GC, Fitzgerald MH (2003b) Micronesian banana, taro, and other foods: newly recognized sources of provitamin A and other carotenoids. J Food Compos Anal 16(1):3–19

Englberger L, Shierle J, Aalbersberg W, Hoffmnn P, Humphries J, Huang A, Lorens A, Levendusky A, Daniels J, Marks GC, Fitzgerald HM (2006a) Carotenoid and vitamin content of Karat and other and other Micronesian banana cultivars. Int J Food Sci Nutr 57(5–6):399–418

Englberger L, Wills RB, Blades B, Dufficy L, Daniells JW, Coyne T (2006b) Carotenoid content and flesh color of selected banana cultivars growing in Australia. Food Nutr Bull 27(4):281–291

Govaerts R, Häkkinen (2009) World checklist of *Musaceae*. The Board of Trustees of the Royal Botanic Gardens, Kew. http://www.kew.org/wcsp/. Accessed 11 Nov 2009

Häkkinen M, Väre H (2008) Typification and check-list of *Musa* L. names (Musaceae) with nomenclatural notes. Adansonia Sér 3 30:94

MacDaniels LH (1947) A study of the Fe'i banana and its distribution with reference to polynesian migrations (Bernice P. Bishop Museum Bulletin 190). Bishop Museum, Honolulu, 56 pp

Pétard PH (1955) Les plantes tinctoriales Polynésiennes. J Agric Trop Bot Appl 2:193–199

Ploetz RC, Kepler AK, Daniells J, Nelson SC (2007) Banana and plantain – an overview with emphasis on Pacific Island cultivars, ver. 1. In: Elevitch CR (ed) Species profiles for Pacific Island agroforestry. Permanent Agriculture Resources (PAR), Hōlualoa. <http://www.traditionaltree.org>

Sharrock S (2001) Diversity in the genus *Musa* – focus on *Australimusa*. INIBAP annual report 2000. Montpellier, pp 14–19

Shovic AC, Whistler WA (2001) Food sources of provitamin A and vitamin C in the American Pacific. Trop Sci 41:199–202

Stover RH, Simmonds NW (1987) Bananas. Longmans, London, 468 pp

Wagner WL, Herbst DR, Sohmer SH (1999) Manual of the flowering plants of Hawaii. Revised edition. Bernice P. Bishop Museum special publication. University of Hawai'i Press/Bishop Museum Press, Honolulu, 1919 pp (two volumes)

Wong S, Kiew R, Argent G, Set O, Lee SK, Gan YY (2002) Assessment of the validity of the sections in *Musa* (Musaceae) using ALFP. Ann Bot 90(2):231–238

Musa velutina

Scientific Name

***Musa velutina* H. (A.) Wendland & C. G. O. Drude**.

Synonyms

Musa dasycarpa W. S. Kurz.

Family

Musaceae

Common/English Names

Baby Pink Banana, Fuzzy Banana, Pink Banana, Pink Fruiting Banana, Pink Velvet Banana, Red Banana, Self-Peeling Banana

Vernacular/Local Names

Brazil: Bananeira-De-Jardim, Bananeira-Ornamental (Portguese);
Spanish: Guineo Rosado.

Origin/Distribution

Pink banana is native to India and Mynamar. Wild distribution is common at lower altitudes across Assam to Arun-achal Pradesh in northeast India and northern Myanmar.

Agroecology

It occurs in low altitude forest, in partial shade or full sun. It is cold tolerant down to 10°C. It thrives in free draining, moist soil, rich in organic matter.

Edible Plant Parts and Uses

It has edible pink fruit but is not very palatable. When the fruit is ripe the skins comes away hence the common name of self-peeling banana. In some areas of Northeast India, the male buds are collected and eaten as a vegetable.

Botany

A slender, short plant with a yellow-green pseudostem growing up to 1.5 m high and 7 cm

T.K. Lim, *Edible Medicinal And Non-Medicinal Plants: Volume 3, Fruits*,
DOI 10.1007/978-94-007-2534-8_77,

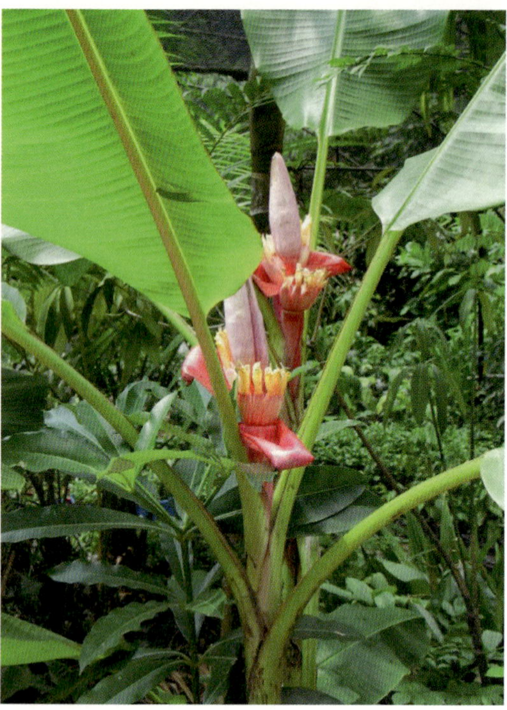

Plate 1 Velvet pink banana inflorescences

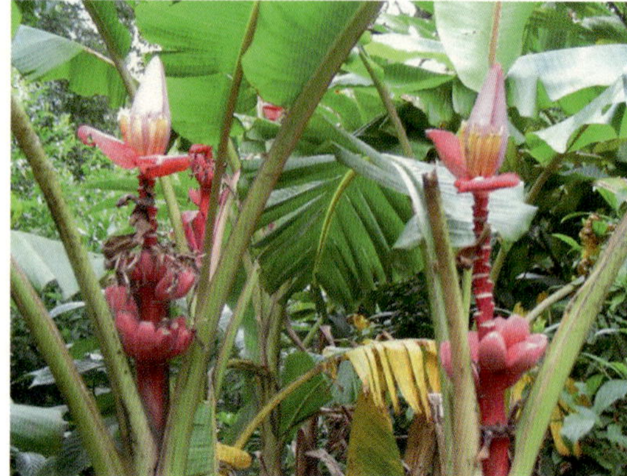

Plate 2 Pink banana fruits on a still flowering inflorescence

diameter at the base, producing suckers freely and rapidly. Petiole up to 30 cm, petiole canal margins straight with erect margins, petiole bases winged and not clasping the pseudostem. Leaf blade up to 1 m long and 35 cm wide glossy dark green adaxially and paler green abaxially, apex truncate, bases asymmetrical and pointed on both sides mid-rib distinct, green above, reddish below. Inflorescence erect with a red peduncle, 10 cm long and 2 cm diameter, densely covered with white pubescence, sterile red bract one, persistent during flowering (Plates 1 and 2). Female bud ovoid, up to 15 cm long and 6 cm wide, bracts pale pink in both surfaces, without imbrications and wax. Bracts are revolute. Basal flowers female, hermaphrodite, 3–5 per bract in a single row, ovary 3 cm long, pale pink, velvety, densely pubescent, compound tepal 3 cm long, orange pink in colour and the lobes orange, free tepal 3 cm long, boat shaped, yellowish, strongly grooved, stamens 5, whitish-green style with orange stigma. Male bud ovoid, 12 cm long and 5 cm wide, bracts pale pink in both surfaces. Male flowers about 5 per bract, falling with the bract, compound tepal 3.5 cm. long, orange-yellow, with a pink flush on the back, rudimentary ovary also pink, lobes of tepal orange, barely 2 mm. long the lateral ones with a minute dorsal appendage ; free tepal 3.5 cm long, 1 cm. wide, oblong, sometimes jagged-toothed towards the apex, with a short broad acumen; stamens 5, exserted, whitish, anthers yellow. Fruit bunch compact, with 4–5 hands and 4 fruits per hand on average, in 1 row (Plates 2 and 3). Fruit bright pink, rounded, 7 cm long, 3–4 cm across, pubescent, broadly truncate at apex, and sub-sessile at base (Plates 2 and 3), pericarp 3–4 mm, thick, splitting at maturity and separating in irregular strips from apex to base, exposing the central mass of white pulp and seeds. Seeds black, tuberculate, irregularly angulate-depressed, 4–6 mm across, 2–3 mm high, 80–90 per fruit.

Nutritive/Medicinal Properties

Kitdamrongsont et al. (2008) reported that the red-coloured bracts of male flowers of *M. balbisiana*, *M. velutina*, *M. laterita*, and *E. superbum* contained only nonmethylated anthocyanin, delphinidin-3-rutinoside, and cyanidin-3-rutinoside. Total anthocyanin content in the analysed bracts ranged from 0–119.70 mg/100 g bract fresh weight.

Plate 3 Pink banana fruits

They provide potential sources of natural food colorant and neutraceuticals (Pazmiño-Durán et al. 2001).

Other Uses

It is usually planted as an ornamental banana because of its pink inflorescence and fruit, and lush green foliage.

Comments

M. velutina can become a weed in agricultural fields because of its rapid suckering habits and its abundant seeds which are spread by birds and bats.

Selected References

Cheesman EE (1949) Classification of the bananas. III. Critical notes on species. J. *Musa velutina* Wendl. & Drude. Kew Bull 4(2):135–137

Govaerts R, Häkkinen (2009) World checklist of *Musaceae.* The Board of Trustees of the Royal Botanic Gardens, Kew. http://www.kew.org/wcsp/. Accessed 11 Nov 2009

Häkkinen M, Sharrock S (2002) Diversity in the genus *Musa* – focus on rhodochlamys. INIBAP Annual Report 2001. INIBAP, Montpelleir, pp 16–23

Hakkinen M, Vare H (2008) Taxonomic history and identity of *Musa dasycarpa, M. velutina* and *M. assamica* (Musaceae) in Southeast Asia. J Syst Evol 46(2):230–235

Kitdamrongsont K, Pothavorn P, Swangpol S, Wongniam S, Atawongsa K, Svasti J, Somana J (2008) Anthocyanin composition of wild bananas in Thailand. J Agric Food Chem 56(22):10853–10857

Liberty Hyde Bailey Hortorium (1976) Hortus third. A concise dictionary of plants cultivated in the United States and Canada. Wiley/Liberty Hyde Bailey Hortorium/Cornell University, New York, 1312 pp

Pazmiño-Durán EA, Giusti MM, Wrolstad RE, Glória MBA (2001) Anthocyanins from banana bracts (*Musa × paradisiaca*) as potential food colorants. Food Chem 73(2):327–332

Simmonds NW (1960) Notes on banana taxonomy. Kew Bull 14(2):198–212

Väre H, Häkkinen M (2009) Proposal to conserve the name *Musa velutina* against *M. dasycarpa* (Musaceae). Taxon 58(3):1009

Horsfieldia australiana

Scientific Name

Horsfieldia australiana S.T. Blake.

Synonyms

None

Family

Myristicaceae

Common/English Names

Bush Nutmeg, Cape Nutmeg, Horsfieldia

Vernacular Names

None

Origin/Distribution

The species is native to Papua New Guinea and Australia.

Agroecology

The species occurs in dense monsoon forest associated with freshwater streams, in sub-coastal sandy lowland or sandstone areas. It thrives in deep well-drained, sandy or alluvial soils.

Edible Plant Parts and Uses

Seed (nut) kernel has a coconut-like taste and is eaten raw or roasted. The kernel may have some potential in the restaurant industry or as snack food.

Botany

A dioecious, evergreen, medium-sized tree, 10–12–15 m high with a spreading crown, slight buttresses and brown, closely fissured rather thickly, scaly bark, yellowish when cut with watery sap turning red on exposure to air. Leaves are distichous, alternate, elliptic-lanceolate, large, simple, entire, with tapering base and apex, glossy green, dull pallid below. Flowers unisexual, yellow to orangey-yellow, borne in or above leaf axils in one to many flowered racemes or

T.K. Lim, *Edible Medicinal And Non-Medicinal Plants: Volume 3, Fruits*,
DOI 10.1007/978-94-007-2534-8_78, © Springer Science+Business Media B.V. 2012

Plate 1 Greenish, unripe fruit and foliage

Plate 2 Dehiscent capsule, aril covered seeds and seeds with red aril removed

panicles, with male and female flowers on different plants; male flowers spicily scented. Perianth tubular with 3–5 valvate lobes, stamens united into a column with hairy anthers in male flowers. Female flower with similar perianth, an ovoid ovary and no style. Fruit a dehiscent capsule, 2.8–3.5 cm by 1.5–1.8 cm, ovoid-oblong, glabrous, greenish-yellow turning to brown, splitting into two valves to reveal the brown mottled oblong-oval seed enveloped by a thin, red, entire aril coating (Plates 1 and 2).

Nutritive/Medicinal Properties

The proximate nutrient value of fresh fruit of *Horsfieldia australiana* per 100 g edible portion had been reported by Brand Miller et al. (1993) as: energy 1,838 kJ, moisture 3.6 g, nitrogen 1.7 g, protein 11 g, fat 43..8 g, ash 1.8 g, dietary fibre 3.8 g, Ca 144 mg, Cu 1.5 mg, Fe 4 mg, Mg 146 mg, K 325 mg, Na 30 mg, Zn 2.2 mg, niacin equivalent 1.8 mg.

Fruit has been deemed to be good for stomach complaints.

Other Uses

The tree provides useful timber.

Comments

The tree is propagated by seeds.

Selected References

Blake ST (1954) Botanical contributions of the Northern Australia regional survey II. Studies on miscellaneous Northern Australian plants. Aust J Bot 2(1):99–140

Brand Miller J, James KW, Maggiore P (1993) Tables of composition of Australian aboriginal foods. Aboriginal Studies Press, Canberra

Whitehead PJ, Gorman J, Griffiths AD, Wightman G, Massarella H, Altman J (2006) Feasibility of small scale commercial native plant harvests by indigenous communities. RIRDC Publication No 04/149, 186 pp

Myristica fatua

Scientific Name

Myristica fatua **Houtt.**

Synonyms

Myristica affinis Warb., *Myristica finschii* Warb., *Myristica incetilis* Rich. ex. A. Gray, *Myristica macrophylla* Roxb., *Myristica magnifica* Bedd., *Myristica masculas* Reinw. ex de Vrise, *Myristica mindanensis* Warb., *Myristica morindaefolia* Blume, *Myristica nivea* Merr., *Myristica platyphylla* A.C. Sm., *Myristica plumeriifolia* Elmer, *Myristica sericea* Warb., *Myristica spadicea* Blume, *Myristica spanogheana* Miq., *Myristica subcordata* Blume, *Myristica tomentosa* Thunb., *Myristica wallaceana Warb.*, *Myristica wenzellii* Merr.

Family

Myristicaceae

Common/English Names

Male Nutmeg, Mountain Nutmeg, Wild Nutmeg

Vernacular Names

Dutch: Mannetjes-Nooten;
India: Ramapatre (Kannada), Churapayin, Kothapayin, Kottapannu (Malayalam);
Indonesia: Pala Laki-Laki (Moluccas);
Malaysia: Pala Laki-Laki;
Samoan: Atone;
Vanuatu: Nandai.

Origin/Distribution

The plant is endemic to the Western Ghats – Gasthyamalai (west) and Central Malanad India; Kalimantan, Sulawesi, Moluccas in Indonesia; the Philippines; New Guinea; Solomon Island; and the south Pacific Islands – New Hebrides, Fiji, Samoa, Vanuatu and New Caledonia.

Agroecology

It is tropical in requirement. It is found in humid, swampy rainforest on poorly drained soils from 100 to 800 m elevation.

T.K. Lim, *Edible Medicinal And Non-Medicinal Plants: Volume 3, Fruits*,
DOI 10.1007/978-94-007-2534-8_79, © Springer Science+Business Media B.V. 2012

Edible Plant Parts and Uses

The dried nuts and arils are often used as spices. Seeds and mace are used as adulterated substitute of true nutmeg.

Botany

The tree is evergreen, dioecious, to 37 m high, with greyish-purplish rough, buttressed trunk often with aerial roots, pneumatophores or knee roots. Branchlets are terete, stout and rusty tomentose. Leaves are alternate, distichous, simple, 20–60 ×10–15 cm, elliptic-oblong, with acute or acuminate apex, rounded or subtruncate base, entire margin, coriaceous, with 12–18 pairs of distinct impressed nerves, tomentose when young becoming glabrous and borne on stout, 1.5–2.5 cm long petiole (Plate 1). Flowers are unisexual, urceolate, in dense umbellate racemes at the axils of persistent or fallen leaves. Inflorescence is sessile or on very short 5 mm long peduncle, flower bearing axis with 3- many crowded well-developed flowers on 4–8 mm pedicels at the apex. The perianth is widely campanulate, light yellow, 6–8 mm long with recurved, pubescent lobes shorter than the tube. Male flowers have 8- many stamens connate into a cylindrical column. Female inflorescence similar to the male one. Female flower has a densely hairy ovary and a sessile stigma. Fruit is an oblong-ovoid to ellipsoid capsule, 5.5–7 by 10.5 cm, densely rusty tomentose, with orangey-red, lacerated aril, weakly aromatic, enclosing one cylindrical seed. Fruits are formed in pendant clusters (Plate 1).

Plate 1 Fruit clusters and large leaves of wild nutmeg

Nutritive/Medicinal Properties

Nuts have been used for headaches and other sickness in folkloric medicine. The nuts were pounded with senna and used as a purgative in Peninsular Malaysia.

Mryistica fatua has antimycobacterial and anti-parasitic activities. In studies conducted in New Caledonia and Vanuatu, methanolic and dichloromethane extracts of *Myristica fatua* exhibited inhibitory activity against *Mycobacterium bovis* BCG strain at a concentration of 100 μg/ml (Billo et al. 2005). The almond extract from the nut was reported to display significant antiparasitic activity in-vitro against *Caenorhabditis elegans* with IC_{50} value of 6.6 μg/ml while the almond and aril extract had an IC_{50} value of 0.5–5 μg/ml against *Trypanosoma brucei brucei* (Desrivot et al. 2007).

Other Uses

Trees are used for wood carving especially totem poles in New Guinea.

Comments

The fruit is not as fragrant as the common nutmeg.

Selected References

Backer CA, Bakhuizen van den Brink RC Jr (1963) Flora of java, (spermatophytes only), vol 1. Noordhoff, Groningen, 648 pp

Billo M, Cabalion P, Waikedre J, Fourneau C, Bouttier S, Hocquemiller R, Fournet A (2005) Screening of some New Caledonian and Vanuatu medicinal plants for antimycobacterial activity. J Ethnopharmacol 96(1–2):195–200

BIOTIK (2006–2008) *Myristica fatua* Houtt. Myristicaceae. Biotik Organisation. http://www.biotik.org/india/species/m/myrifama/myrifama_en.html#debut

Burkill IH (1966) A dictionary of the economic products of the Malay Peninsula. Revised reprint, vol 2. Ministry of Agriculture and Co-operatives, Kuala Lumpur, Malaysia, vol 1. (A–H) pp 1–1240, vol 2. (I–Z), pp 1241–2444

Desrivot J, Waikedre J, Cabalion P, Herrenknecht C, Bories C, Hocquemiller R, Fournet A (2007) Antiparasitic activity of some New Caledonian medicinal plants. J Ethnopharmacol 112(1):7–12

Sinclair J (1968) The genus *Myristica* in Malesia and outside Malesia. Gard Bull Singapore 23:1–540

Wilde WJJO (1994) Taxonomic review of *Myristica* (Myristicaceae) in the Pacific. Blumea 38:349–406

Womersley JS (ed) (1978) Handbooks of the flora of Papua New Guinea, vol 1. Melbourne, Melbourne University Press, 278 pp

Myristica fragrans

Scientific Name

Myristica fragrans **Houtt.**

Synonyms

Myristica aromatica Sw., *Myristica laurifolia* var. *lanceolata* Hook.f., *Myristica moschata* Thunb., *Myristica officinalis* L. f.

Family

Myristicaceae

Common/English Names

Nutmeg, Mace, Nutmeg Tree

Vernacular Names

Amharic: Gabz, Gewz;
Arabic: Basbas, Basbasah, Fuljan, Jawz At-Tiyb, Josat Al Teeb, Josat Al-Tib, Jasat At-Tib, Jouza Al-Teeb, Jouza At-Teeb, Jousbuva, Jouzuttib, Jowz Buwwa;
Armenian: Meshgengouz, Mshkenkoyz;
Azerbaijani: Covuz, Covuz Qoz;
Basque: Intxaur Muskatu;
Brazil: Bicuiba, Flor De Noz Moscada, Noz Moscada (Portuguese);
Bulgarian: Indijsko Orekhche;
Burmese: Mutwinda, Zadeikpo;
Catalan: Nou Moscada;
Chinese: Dauh Kau Syuh, Dou Kou, Dhu, Rou Dou Kou, Rou Dou Kou Yi, Rou Guo, Rou Kou, Yu Guo, Yu Guo Hua, Yuhk Dauk Kau;
Croatian: Macis, Muškatni Cvjetić, Muškatni Oraščić;
Cyprus: Moschokarido;
Czech: Muškátovník Pravý, Muškátovník Vonný;
Danish: Muskat, Muskatblomme, Muskatnød, Muskatnødtræ;
Dhivehi: Thakoovah;
Dutch: Foelie (Mace), Muskaatboom, Nootmuskaat, Nootmuskaatboom;
Eastonian: Lõhnav Muskaadipuu, Maasis, Muskaatpähkel, Muskaatõis;
Esperanto: Miristiko Floro, Miristiko Nukso, Miristiki Semo, Muskato;
Finnish: Muskotti, Muskottikukka, Muskottipähkinä, Muskottipuu;
French: Fleur De Muscade, Macis, Muscade, Muscadier, Muscadier Commun, Muscadier Cultivé, Noix De Banda, Noix De Banda, Noix De Muscade, Noix Muscade, Pied De Muscade, Pied-Muscade Muscadier;
Gallegan: Noz Moscada;
Greek: Moschokarido;
Hebrew: Egoz Muskat, Egoz Musqat, Mays;
Hungarian: (Valódi) Muskátdió(Fa), Szerecsendió(Fa), Szerecsendió Virág;
Icelandic: Masi, Múskat, Múskathýði;
India: Jaiphal (Bengali), Jayfal, Jaypatri (Gujarati), Jaayphala, Jayphala, Japhal, Jaephal, Jaiphal,

T.K. Lim, *Edible Medicinal And Non-Medicinal Plants: Volume 3, Fruits*,
DOI 10.1007/978-94-007-2534-8_80, © Springer Science+Business Media B.V. 2012

Jaitri, Javitri, Kathal, Taifal (Hindi), Jajipatra, Jakayi, Jakayi Patri, Japatre, Jatiphala (Kannada), Jathikka (Malayalam), Jaypatri, Jayphal (Marathi), Jaiphal, Javatri (Punjabi), Jajipatri, Jajiphalam, Jatipatra, Jatiphala (Sanskrit), Atipalam, Jadikkai, Jatikka, Jati Pattiri, Jatippu,Pattiri, Sadhi-Kai (Tamil), Jajikaia, Japatri, Vicuiba (Telugu), Basbas, Jaiphal, Javitri (Urdu);
Indonesia: Fuli (Mace), Buah Pala, Bumbu (Mace in Javanese, Malay, Sundanese), Pala, Pala Banda, Sekar Pala (Mace in Javanese), Bunga Pala (Mace in Malay), Kambang Pala (Mace in Javanese, Malay, Sundanese), Kulit Pala, Kulumud Pala (Mace in Sundanese);
Italian: Mace, Noce Moscata;
Japanese: Mesu, Natumegu, Nikuzuku;
Khmer: Pôch Kak;
Korean: Meisu, Neotumek, Notumek, Yuktugu;
Laotian: Chan Th'e:D;
Latvian: Muskatrieksts;
Lithuanian: Kvapusis Muskatmedis, Macis, Muskatas Muskato Žiedai;
Malaysia: Buah Pala, Kembang Pala (Mace), Pala, Poko Pala;
Maltese: Nuċimuskáta;
Nepalese: Jaaiipatrii, Jayaphal;
Norwegian: Muskat, Muskatblomme, Muskatnøtt;
Papiamento: Netmuskat;
Persian: Basbaz, Djus Hendi, Jouz Hendi, Jouzboyah;
Philippinese: Duguan (Tagalog);
Polish: Gałka Muszkatołowa, Muszkat, Muszkatowiec;
Portuguese: Moscadeira, Nuz Moscada;
Romanian: Frunzișoară, Nucșoară;
Russian: Muskatnii Orekh, Muskatnyj Orekh, Muskatnyj Tsvet, Muskatnogo Orekha, Mushkatnoi Drechi, Sushonaya Shelukha;
Singapore: Buah Pala, Pokok Pala;
Slovak: Muškátovník Voňavý, Muškátový Kvet, Muškátový Orech;
Slovenian: Muškat, Muškatni Cvet, Muškatni Oreščeк;
Spanish: Corteza De La Nuez Moscada, Macia, Macis, Moscadero, Moscada, Neuz Moscada, Neuz Muscada, Nogal Moscado;
Sri Lanka: Sadhika, Wasa-Vasi (Sinhalese);
Swahili: Basibasi, Kungumanga;
Swedish: Muskott Muskotnöt, Musotblomma;
Thai: Chan-Thet (Central), Chan-Ban (Northern), Dok Chand, Dok Chand Nattes, Luk Chand;
Tibetan: Dza Ti, Zati;
Turkey: Besbase, Cevz Buva, Industan Djevisi;
Ukrainian: Muskatnyj Horikh;
Vietnamese: Dâu Khâu, Nhuc Dâu Khau;
Yiddish: Mushkat, Mushkatnoys, Muskat, Muskatnus.

Origin/Distribution

Nutmeg is indigenous to the Moluccas and Banda Islands in Indonesia.

Agroecology

Nutmeg grows wild on rich volcanic soils in lowland tropical rain forests in its indigenous range. Its cultivation as a crop is largely confined to islands in the hot, humid tropics at altitudes up to 4,000 m. Nutmeg prefers a warm and humid tropical climate, with average temperatures of 25–30°C and average annual rainfall of 2,000–3,500 mm without any real dry period. Flowering can be adversely affected by temperatures above 35°C and by hot dry winds. Frost always damages or kills the tree and makes commercial production impossible. Therefore, in the tropics the crop can only be grown below 1,000 m altitude. The superficial root system makes the tree very susceptible to wind damage. The crop can grow on any kind of soil provided there is sufficient water but without any risk of water-logging. Preferred soils are those of volcanic origin and soils with a high content of organic matter with pH 6.5–7.5.

Edible Plant Parts and Uses

The dry shelled seed (nutmeg) and dried red aril (mace) are sold and use as spices whole or ground. Both are aromatic, fragrant, sweet and warm. Mace is used in savoury dishes. Mace is used to flavour milk-based sauces e.g. *bechamel* and is widely used in processed food like sausages and *charcuterie*. It is also sparingly used in soups, ketchups, and sauces with fish, seafood such as potted shrimps and eggs. Mace is also complementary to puddings, cakes especially cheesecake and lemon curd tarts and drinks. It is also used to season pickles and chutneys. The mace powder is an important ingredient for the hot Indian curry spice mixture, *Garam Masala* (comprising coriander seeds, black peppercorns, cumin seeds, bay leaves, green cardamoms, ground mace, cloves and cinnamon stick).

In most countries grated nutmeg is used in small quantities to flavour confectionery but in western Europe it is also used in meat dishes and soups. Nutmeg is a flavoursome addition to cheese sauces and is best grated fresh. In Indian cuisine, nutmeg powder is used almost exclusively in sweet dishes. It is known as *Jaiphal* in most parts of India. It may also be used in small quantities in a curry spice mixture called *Kashirimi masala* which consists of green cardamoms, cinnamon stick, cloves, black peppercorns, caraway seeds, cumin seeds and ground nutmeg. Ground nutmeg is also an important ingredient in the Ethiopian spice mixture *Berbere* (ingredients include: red chillies, white cardamoms, cumin seeds, coriander seeds, fenugreek seeds, cloves, allspice berries, black peppercorns, ajowan seeds, ground ginger, ground nutmeg and salt). *Berbere* is used in many local Ethiopian cuisine from baked fish dishes to chicken stews. Ground nutmeg is also smoked in India. In Middle Eastern cuisine, nutmeg powder is often used as a spice for savoury dishes. In European cuisine, nutmeg and mace are used especially in potato dishes and in processed meat products; they are also used in soups, sauces and baked goods. Japanese varieties of curry powder include nutmeg as an ingredient. Nutmeg is a traditional ingredient in mulled cider, mulled wine, and eggnog. In Europe nutmeg is used in fillings for pasta especially those using spinach and cheese and it may be added to risotto, tomato sauce or sauces for fish or chicken pies. It is excellent in cheese sauce for a cauliflower cheese or onion sauce to serve with lamb chops and in creamy mashed potatoes. Nutmeg is a traditional flavouring for cakes, biscuits, gingerbreads, fruit or milk puddings. Grated nutmeg is excellent in cheery or apple pies. A sprinkle of freshly grated nutmeg and sugar adds a refreshing dimension to slices of well-chilled oranges. Ground nutmeg together with cinnamon and vanilla are important ingredient in Mocha on a Cloud – a coffee chocolate beverage.

Essential oils (mostly nutmeg oil from the seed and mace oil from the aril, but also from the bark, leaf and flower) and extracts (e.g. oleoresins) are often used in the canning industry, in soft drinks and in cosmetics. Nutmeg essential oil is extensively used as a flavour component in major food products that include bakery products, dehydrated soups, ice-cream, sauces and processed meats; the maximum permitted level in food is about 0.08%. It replaces ground nutmeg as it leaves no particles in the food. Nutmeg oil is also used to flavour tobacco. Nutmeg oil is mainly used in ketchups and soft-drinks, pharmaceutical products and male toiletries. Mace oil is a direct substitute for nutmeg oil.

The fleshy pericarps of the fruit are made into confectionery like jellies, marmalades, sweets, crystallized pieces and slices, pickles and preserves, which are very popular in West Java, Singapore and Malaysia. In Indonesia, the pericarp is cut into slices and eaten as a delicacy as *petis* (Sundanese) or rujak (Malay). It is also made into pickles and eaten as *pala asin* with *sambal* and rice or made into *pala manisan* (sweetmeat). The pericarp is used in Grenada to make a jam called "Morne Delice".

Botany

A dioecious, aromatic, evergreen tree, usually 5–13 m but can reach heights of 20 m. The bark exudes a sticky pink or red sap when injured. Twigs are slender, 1–2 mm in diameter towards the top. Leaves are alternate, simple, chartaceous; petiole about 1 cm long; blade elliptical to lanceolate, 5–15 cm × 3–7 cm, base acute, margin entire, apex acuminate, aromatic, upper leaf surface shiny dark green lower surface paler green (Plates 1 and 2). Inflorescences axillary, in umbellate cymes, male cymes 1-10-flowered, female

Plate 1 Nutmeg fruits and leaves

Plate 2 Glossy nutmeg leaves

cymes 1-3-flowered on the same tree. Flowers are fragrant, glabrescent with sparse, minute tomentum, pale yellow, with a 3-lobed perianth; male flowers, 5–7 mm long, on a slender pedicel less than 1 mm thick, with 8–12 stamens adnate to a column; female flowers up to 10 mm long, with a superior, sessile, 1-celled ovary with a single basal ovule. Fruit is subglobose with a long medial groove, berry or drupe-like, 5–8 cm long, fleshy, yellowish, splitting open into two halves when ripe, containing one seed (Plates 1, 2, 3, 4, and 5). Seed is broadly ovoid, 2–3 cm long, with a shiny dark brown, hard and stony, furrowed and longitudinally wrinkled shell, surrounded by a laciniate red aril (Plate 5) which become horny, brittle and a yellowish-brown colour when dried. Kernel has a small embryo and ruminate

Plate 3 Harvested nutmeg fruits

Plate 4 Close-up of whole nutmeg fruits

Plate 5 Nutmeg fruit halved to show the thick, firm flesh, red mace and brown seed

endosperm which contains many veins containing essential oil.

Nutritive/Medicinal Properties

Proximate nutrient composition of ground nutmeg spice per 100 g edible portion (U.S. Department of Agriculture, Agricultural Research Service 2010) is water 6.23 g, energy kcal 525 (2,196 kj), protein 5.84 g, total fat 36.31 g, ash 2.34 g, carbohydrate 49.29 g, total dietary fibre 20.8 g, total sugars 28.49 g, Ca 184 mg, Fe 3.04 mg, Mg 183 mg, P 213 mg, K 350 mg, Na 16 mg, Zn 2.15 mg, Cu 1.027 mg, Mn 2.9 mg, Se 1.6 μg, vitamin C 3 mg, thiamine 0.346 mg, riboflavin 0.057 mg, niacin 1.2990 mg, vitamin B-6 0.160 mg, folate 76 μg, choline 8.8 mg, vitamin A 102 IU, vitamin A 5 μg RAE, γ-tocopherol 0.53 mg, total saturated fatty acids 25.94 g, 12:0 (lauric) 0.370 g, 14:0 (myristic) 22.830 g, 16:0 (palmitic) 2.260 g, 18:0 (stearic) 0.170 g; total monounsaturated fatty acids 3.220 g, 16:1 undifferentiated (palmitoleic) 1.400 g, 18:1 (oleic) undifferentiated 1.590 g; total polyunsaturated fatty acids 0.350 g, 18:2 undifferentiated (linoleic) 0.350 g; phytosterols 62 mg, β-carotene 16 μg and β-cryptoxanthin 90 μg.

Nutmeg powder is rich in minerals like Ca, K, P, Fe, Mg, niacin and low in Na and also contain phytosterols, β-carotene and β-cryptoxanthin. FatB cDNAs were isolated form nutmeg seeds and found to accumulate predominantly myristate (14:0)-containing oils (Voelker et al. 1997). Nutmeg FatB hydrolysed C14-18 substrates in-vitro and expression in *Brassica napus* seeds led to an oil enriched in C14-18 saturates.

Fifteen compounds were isolated from nutmeg (seed of *Myristica fragrans*) and identified as myristicin (1), methyleugenol (2), safrole (3), 2, 3-dihydro-7-methoxy-2(3, 4-methylenedioxyphenyl)-3-methyl-5-(E) -propenyl-benzofuran (4), dehydrodiisoeugenol (5), 2, 3-dihydro-7-methoxy-2-(3-methoxy-4, 5-methylenedioxyphenyl) -3-methyl-5-(E)-propenyl-benzofuran (6), erythro-2-(4-allyl-2, 6-dimethoxyphenoxy)-1-(3, 4-dimethoxyphenyl) propane (7), erythro-2-(4-allyl-2, 6-dimethoxyphenoxy)-1-(3, 4, 5-trimethoxyphenyl) propane (8), erythro-2-(4-allyl-2, 6-dimethoxyphenoxy)-1-(3, 4-dimethoxyphenyl) propan-1-ol acetate (9), erythro-2-(4-allyl-2, 6-dimethoxyphenoxy)-1-(3, 4-dimethoxyphenyl) propan-1-ol (10), erythro-2-(4-allyl-2, 6-dimethoxyphenoxy)-1-(3, 4, 5-trimethoxyphenyl) propan-1-ol (11), 5-methoxy-dehydrodiisoeugenol (12), erythro-2-(4-allyl-2, 6-dimethoxyphenoxy)-1-(4-hydroxy-3-methoxyphenyl)-propan-1-ol (13), guaiacin (14) and a neolignan, threo-2-(4-allyl-2, 6-dimethoxyphenoxy)-1-(3-methoxy-5-hydroxyphenyl) propan-1-ol (15) named myrisisolignan (Yang et al. 2008). A new compound, a dehydrodiphenylpropanoid derivative of myristicin was fodun in nutmeg (Davis and Cooks 1982).

Daniel (1994) listed the following components of the essential oil (volatile) of nutmeg: aromatic ethers – eugenol, methyl eugenol, iso-eugenol, methyliso-eugenol, methoxy eugenol, safrole, myristicin, elemencin, iso-elemicin; terpenes-α-terpene, γ-terpine, α-pinene, β-pinene, α-phellandrene, β phellandrene, α-thujene, myrcene, terpinolene, camphene, uinonene(dipentene), sabinene, δ-carene; monoterpene alcohol – geraniol, α-terpineol, citronellol, 4-terpineol, β-terpineol, linalool; sesquiterpene – caryophyllene; terpinic esters – gernayl acetate, bornyl acetate, linalyl acetate; acids formic, butryric, octanoic, acetic and aromatic hydrocarbons – *p*-cymene, toluene. Additionally the following compounds were also found on the baisis of retention times during gas chromatography: cumene, camphor, menthyl isovalerate, cyclamen aldehyde and menthone.

The fixed oil or nutmeg butter was reported by Daniel (1994) to contain the following after removal of the essential oil component (12.5%): trimyristin 84%, unsaponifiable constituents 9.8%, oleic acid (as glyceride) 3.5%, resinous material 2.3%, linolenic acid (as glyceride) 0.64, formic, acetic and cerotic acids traces.

The volatile oil of nutmeg pericarp was found to contain 16 monoterpenes (60%), 9 monoterpene alcohols (29%), 8 aromatic ethers (7%), 3 sesquiterpenes (1%), 6 esters (1%) and 8 other minor components (Choo et al. 1999). The components were similar to those in nutmeg seed and mace oils but differ substantially in concentrations. The sabinene, myristicin and safrole concentrations were much lower while the terpinen-4-ol and α-terpineol contents were much higher than in nutmeg and mace oils.

Forty-eight compounds were identified from the essential oil of nutmeg extracted by supercritical carbon dioxide, and its main components have been found to be myristic acid, myristicin, terpinen-4-ol, α-pinene and safrole (Qiu et al. 2004). Thirty-eight compounds were identified for the essential oil obtained by steam distillation, and its main components have been found to be β-pinene, terpinen-4-ol, α-pinene, γ-terpinene and β-phellandrene.

The nutmeg seed was found to be rich in essential and fatty oils, with myristicin and myristic acid being the characteristic compounds in each group respectively (Spricigo et al. 1999). The steam distillation of ground nutmeg yielded 6.9 wt.% essential oil. Extraction of nutmeg with diethyl ether, yielded 46.5 wt.% of fatty material. Compounds identified in the essential oil of nutmeg included: α-pinene (11%), β-pinene 10.18%, sabinene 36.64%, myrcene 3.15%, limonene 3.77%, β-phellandrene 2.73%, γ-terpinene 1.88%, cymene 1.65%, 1-methyl-4(1-methylethyl)2-cyclohexen-1-ol (cis) 1.81%, 1-methyl-4(1-methylethyl)2-cyclohexen-1-ol (trans) 1.79%, terpinene-4-ol 3.68%, safrol 1.01%, methyleugenol 0.40%, elemicin, 3.29%, myristicin 6.98%. From these results the major components of the essential oil were found to be α-pinene, β-pinene, sabinene and myristicin, the characteristic aromatic component of nutmeg. The fatty acid analysis of the non-volatile fraction yielded high quantity of myristic acid C14:0 46.47%, as well as other low molecular weight saturated fatty acids C8:0 10.76%, C10:0 5.36%, C12:0 2.99%. Gas chromatographic-mass spectroscopy studies on essential oil resulted in the identification of 49 components representing 96.49% of the total amount, and the major component was sabinene (20.22%), followed by terpinen-4-ol (12.08%), safrole (10.32%), α-pinene (9.7%), β-phellandrene (6.56%), and γ-terpinene (5.93%) (Singh et al. 2005). The acetone extract showed the presence of 23 components representing 71.66% of the total amount. The major components were isocroweacin (18.92%), elemicin (17.68%), methoxyeugenol (8.13%), linoleic acid (4.12%), dehydrodiisoeugenol (4.06%), palmitic acid (2.8%), and trans-isoeugenol (2.76%). The following constituents were isolated from the aril of *Myristica fragrans*: 3-(3,4,5-trimethoxyphenyl)-2-(E)-propen-1-ol (1), 3-(3-methoxy-4,5-methylenedioxyphenyl)-2-(E)-propen-1-ol (2), 2,3-dihydro-7-methoxy-2-(3,4-dimethoxyphenyl)-3-methyl-5-(1-(E)-propenyl)benzofuran (3), fragransol-C (4), fragransol-D (5), 2,3-dimethyl-1,4-bis-(3,4-methylenedioxyphenyl)butan-1-ol (6), myristicanol-A (7) and myristicanol-B (8) (Hattori et al. 1988). Compounds 4, 5, 7 and 8 are new neolignans which may be formed by coupling of cinnamyl alcohol and propenylbenzene unis.

Steam distillation yielded 95–118 compounds from the essential oil of nutmeg (Wang et al. 2004). Seventy nine compounds were identified which comprised 95.18–98.70% of the total essential oil. Myristicin (39.63%) and terpene series were the major components.

Yuan et al. (2006) found that soaking nutmeg in water and roasting with bran before extraction of nutmeg essential oil by vapour distillation changed the quantity and quality of components in volatile oils. Thirteen new components occurred and four components disappeared in volatile oils after processing. The contents of methyleugenol and methylisoeugenol active ingredients were increased. The contents of myristicin and safrol toxic ingredients in volatile oils were decreased. Separate studies showed that monoterpenoids and their derivatives were the main components, and aromatic compounds the secondary components in the total essential oil of nutmeg processed by different traditional methods (steamed with water steam, roasted with flour, sauted with flour, roasted with talcum powder, roasted with loess, and roasted with bran) (Hang and Yang 2007). Fifty-eight to one hundred and four of the chromatographic peaks were detected, among them 76 compounds accounting for 98.32–99.99% of the total essential oil in nutmeg were identified, which were composed of 69.15–97.24% monoterpenoids and 2.06–25.51% aromatic compounds of the total essential oil, respectively. Roasting with bran before extraction was found to improve quality.

Other phytochemicals and associated pharmacological activities reported are presented below.

Neolignans and Biological Activity

Eight neolignans and five lignans were isolated from mace, the aril of *Myristica fragrans*, and their structures elucidated as erythro-2-(4″-allyl-2″,6″-dimethoxyphenoxy)-1-(3′,4′,5′-trimethoxyphenyl)propan-1,3-diol (1b),threo-2-(4″-allyl-2″-methoxyphenoxy)-1-(4′-hydroxy-3′-methoxyphenyl) propan-1-ol (2a),threo-1-(4′-hydroxy-3′-methoxyphenyl)-2-(2″-methoxy-4″-(1‴ (E)-propenyl) phenoxy) propan-1-ol (3a), itserythro form (3b), threo-1-(4′-hydroxy-3′-methoxyphenyl)-1-methoxy-2-(2″-methoxy-4″-(1‴-(E)-propenyl)phenoxy)-propane (4a), itserythro form (4b), fragransol-A (5), fragransol-B (6), fragransin D1 (7), fragransin D2 (8), fragransin D3 (9), fragransin E1 (10) and the known compound austrobailignan-7 (11). (Hada et al. 1988). The absolute configuration of (+)-erythro-(7S,8R)-Δ8′-4,7-dihydroxy-3,3′,5′-trimethoxy-8-O-4′-neolignan and (−)-erythro-(7R,8S)-Δ8′-4,7-dihydroxy-3,3′,5′-trimethoxy-8-O-4′-neolignan was determined by Kasahara et al. (1995). In addition, LiAIH4 reduction of the MTPA ester of erythro-(7S,8R)-delta 8′-4-acetoxy-7-hydroxy-3,3′, 5′-trimethoxy-8-O-4′-neolignan afforded (-)-(8R)-delta 8′-4-hydroxy-3,3′, 5′-trimethoxy)-8-O-4′-neolignan, and that of the MTPA ester of erythro-(7R,8S)-delta 8′-4-acetoxy-7-hydroxy-3,3′, 5′-trimethoxy-8-O-4′-neolignan afforded (+)-(8S)-delta 8′-4-hydroxy-3, 3′, 5′-trimethoxy)-8-O-4′-neolignan, respectively.

A neolignan, licarin-B, (C20H20O4), was isolated from nutmeg (seed of *M. fragrans*) (Kim et al. 1991). Juhász et al. (2000) reported the successful synthesis of four neolignans-fragnasols A, B, and C and dehydrodiisoeugenol, starting from the readily available phenol derivative isoeugenol from *Myristica fragrans*. Three new neolignans, named 1-deoxycarinatone (1), isodihydrocarinatidin (2), and isolicarin A (3), together with the known neolignan (+)-dehydrodiisoeugenol (4), were isolated from mace (the aril of *Myristica fragrans*) (Li and Yang 2007). Their structures were elucidated as 2-[(1S)-2-(4-hydroxy-3-methoxyphenyl)-1-methylethyl]-6-methoxy-4-(prop-2-enyl)phenol (1), 4-[(2R,3R)-2,3-dihydro-7-methoxy-3-methyl-5-(prop-2-enyl)benzofuran-2-yl]-2-methoxyphenol (2), and 4-{(2S,3R)-2,3-dihydro-7-methoxy-3-methyl-5-[(1E)-prop-1-enyl]benzofuran-2-yl}-2-methoxyphenol (3). Yang et al. (2008) isolated myrisisolignan fron nutmeg seed. Myrislignan, erythro-(1R,2S)-2-(4-allyl-2,6-dimethoxyphenoxyl)-1-(4-hydroxy-3-methoxyphenyl) propan-1-ol, a major acyclic neolignan in seeds *of Myristica fragrans* was reported to have anti-feeding activity (Li and Yang 2008a). Seven metabolites of myrislignan were produced by liver microsomes from rats pre-treated with sodium phenobarbital. These were identified as myrislignanometins A-G (2–8), respectively. Machilin A and structurally related lignans isolated from *Myristica fragrans* were found to stimulate osteoblast differentiation (Lee et al. 2009). Machilin A stimulated osteoblast differentiation via activation of p38 MAP kinase. Other lignans isolated from *Myristica fragrans* also stimulated osteoblast differentiation in MC3T3-E1 cells; the lignans included macelignan, machilin F, nectandrin B, safrole, licarin A, licarin B, myristargenol, and meso-dihydroguaiaretic acid. The data suggested that lignans isolated from *Myristica fragrans* had anabolic activity in bone metabolism. Yang et al. (2010) studied the intestinal permeability and transport of 10 neolignans isolated from *M. fragrans* using the Caco-2 cell monolayer model. Among the 10 neolignans, the 8- O-4′-type neolignans demonstrated high permeability while the benzofuran-type neolignans were of poor to moderate permeability. Among them, eight neolignans were transported mainly via passive diffusion. The findings indicated that the 8- O-4′-type neolignans were well-absorbed compounds and could be used as oral leading compounds in drug discovery. *M. fragrans* seeds were also found to have the diastereomers of (+)-licarin A and isolicarin A, neolignans which were detectectable in the rat plasma after intravenous administration (Li and Yang 2008b).

Antioxidant Activity

Of all the fractions of nutmeg methanol seed extract, the ethyl acetate fraction exhibited the highest antioxidant activity as assayed by DPPH radical scavenging activity (Maeda et al. 2006). The active compound in the fraction was found to be 4-allyl-2,6-dimethoxyphenol. Comparison of the chemical composition of free volatile compounds found in the nutmeg essential oil revealed only two common compounds (eugenol and terpinen-4-ol) (Jukic et al. 2006). The antioxidative activities of the essential oil and enzymatically released aglycones from nutmeg, were compared using two different assays: the 2,2′-diphenyl-1-picrylhydrazyl radical scavenging method (DPPH) and the ferric reducing/antioxidant power assay (FRAP). Both methods showed that the aglycone fraction possessed stronger antioxidant properties than free volatiles from the oil.

Studies showed that the spices namely cloves (*Syzygium aromaticum*), licorice (*Glycyrrhiza glabra*), mace (aril of *Myristica fragrans*) and greater cardamom (*Amomum subulatum*), exerted antioxidant activities at various concentrations (Yadav and Bhatnagar 2007a, b). None of the spices showed prooxidant properties. The effect of spices on the inhibition of LPO (lipid peroxidation) was concentration dependent. Cloves, mace and cardamom inhibited the initiation as well as propagation phases of $FeCl_3$ induced lipid peroxidation LPO, while licorice inhibited the initiation phase only. The reducing power of various spices increased with concentration. The percentage inhibition of superoxide radical generation by the spices was also observed to be concentration dependent. The results showed that spices used in the present study had significant ability to inhibit LPO due to their polyphenol content, strong reducing power and superoxide radical scavenging activity. Cloves showed the highest antioxidant activity probably due to the higher polyphenol content as compared to other spices. Metal chelating activity was significantly high with all the spice extracts except mace. Cloves showed the highest DPPH (1,1-Diphenyl-2-picrylhydrazyl) radical scavenging activity, followed by licorice, mace and cardamom. FRAP (ferric reducing/antioxidant power) values for cloves were also the highest, while other spices showed comparatively lesser FRAP values. The results showed the spices tested to be strong antioxidants and may have beneficial effects on human health. The antioxidative activity of phenylpropanoid compound extracts from nutmeg seed was evaluated using the 1,1-diphenyl-2-picrylhydrazyl radical-scavenging method, superoxide disumutase assay, ferric thiocyanate assay, and radical-scavenging effect assay with electron-spin resonance (Maeda et al. 2008). High antioxidant activity was found in monoterpenoid extracts including terpinene-4-ol (3), α-terpineol (4), and 4-allyl-2,6-dimethoxyphenol (12). Compound 12 expressed particularly high antioxidant activity.

In another study, six diarylbutane lignans 1–5 and one aryltetralin lignin isolated from the methanol (95%) extracts of *Myristica fragrans* seeds were found to have antioxidant activity (Kwon et al. 2008). 7-methyl ether diarylbutane lignan 4 was found to be new a compound. Due to its potency, compound 3 with an IC_{50} value of 2.6 μM in TBARS assay was tested for complementary in-vitro investigations, such as lag time (140 minutes at 1.0 μM), relative electrophoretic mobility (REM) of ox-LDL (inhibition of 80% at 20 μM and 72% at 10 μM), and fragmentation of apoB-100 (inhibition of 93% at 20 μM) on copper-mediated LDL oxidation. In macrophage-mediated LDL oxidation, the TBARS formation was also inhibited by compound three. Olaleye et al. (2006) also reported that nutmeg seed aqueous extract contained active principles with antioxidant properties. Studies showed that alkaloids, saponins, anthraquinones, cardiac glycosides, flavonoids and phlobatanins were present while tannins were absent in the aqueous extract. The phytate content was 564.11 mg/100 g while the antioxidant indices of 100 mg/100 g, 44% and 0.6 were obtained for the ascorbic acid value, free radical scavenging activity and reducing power, respectively. The results of the histopathological studies showed pathological features of various degrees in the rats' organs with severity corresponding to the concentration of extract. There was lymphoid depletion of the follicles in the spleen, degeneration of the germinal epithelial cells in the testes, bile duct proliferation and congestion of blood vessels in the liver, degeneration, necrosis with desquamation of tubular epithelial cells and congestion of renal blood vessels in the kidney and degeneration of myocardial fibres and myocardial necrosis in the heart in the treatment groups compared with the control. The results suggested that nutmeg popularly consumed as food and for various medicinal purposes may contain some active principles with antioxidant properties. However, prolonged use at high doses (400–500 mg/kg) could be very toxic to the studied organs.

Both the acetone extract and essential oil of nutmeg showed strong antioxidant activity in comparison with butylated hydroxyanisole (BHA) and butylated hydroxytoluene (BHT) (Singh et al. 2005). Their strong antioxidant activity was also confirmed by their inhibitory action in linoleic acid system, 2,2′-diphenyl-1-picrylhydracyl (DPPH) and reducing power. They showed strong scavenging activity in comparison with synthetic antioxidants. Gas chromatographic-mass spectroscopy studies on essential oil resulted in the identification of 49 components representing 96.49% of the total amount, and the major component was sabinene (20.22%), followed by terpinen-4-ol (12.08%), safrole (10.32%), α-pinene (9.7%), β-phellandrene (6.56%), and γ-terpinene (5.93%) (Singh et al. 2005). The acetone extract showed the presence of 23 components representing 71.66% of the total amount. The major components were isocroweacin (18.92%), elemicin (17.68%), methoxyeugenol (8.13%), linoleic acid (4.12%), dehydrodiisoeugenol (4.06%), palmitic acid (2.8%), and trans-isoeugenol (2.76%).

When maintained at room temperature, all the spice essential oils tested including nutmetmeg oil appeared endowed with good radical-scavenger properties in the DPPH assay (effectiveness order: clove ≫ cinnamon > nutmeg > basil ≫ oregano ≫ thyme) (Tomaino et al. 2005). When heated up to 180°C, nutmeg oil (but not the other essential oils under study) showed a significantly higher free radical-scavenger activity and evident changes in its chemical composition. All the essential oils tested appeared able to prevent α-tocopherol loss following oil heating at 180°C for 10 min (efficiency order: clove > thyme ≫ cinnamon > basil ≫ oregano > nutmeg). In conclusion, the essential oils under study exhibited good antioxidant properties and might be efficiently used to control lipid oxidation during food processing.

The DPPH radical scavenging capacity of the acetone extract of fresh mutmeg mace as well as its fractions was comparatively lower than that of green pepper (*Piper nigrum*) phenolics (Chatterjee et al. 2007). In contrast, these fractions had a greater ability to inhibit lipid oxidation than phenolics from pepper as revealed by β-carotene–linoleic acid assay. Acetone extract of nutmeg mace and its subsequent TLC isolated fractions were found to constitute mainly of lignans. A DNA protecting role of these compounds even at doses as high as 5 kGy further suggested the potential use of green pepper and fresh nutmeg mace and their extracts as a nutraceutical in preventing oxidative damage to cells.

Antioxidant capacity of nutmeg oil was confirmed by the 2,2-diphenyl-1-picrylhydrazyl (DPPH) free radical scavenging assay and the β-carotene-linoleic acid assay (Kim et al. 2010). The antioxidant EC_{50} values of the crude nutmeg oil dissolved in methanol were 2.4 μl/ml and 0.4 μl/ml, respectively. The former value was approximately equivalent to the free radical scavenging capacities of 462 μM BHT and 656 μM α-tocopherol, and the latter one was comparable to the inhibitive capacities of 43 μM BHT and 9 μM α-tocopherol against the oxidation of β-carotene and linoleic acid. Further investigations on three major antioxidant constituents of the nutmeg oil (i.e., eugenol, isoeugenol, and methoxyeugenol), produced the following trend. Their antioxidant activities in the DPPH assay decreased in the following order: eugenol > methoxyeugenol > BHT > isoeugenol > α–tocopherol, while in the β-carotene-linoleic acid assay, the antioxidant activities of the chemicals were in the following order: α–tocopherol > BHT > isoeugenol > methoxyeugenol > eugenol.

Antimicrobial Activity

The two antimicrobial resorcinols malabaricone B and malabaricone C isolated from mace, the dried seed covers of *Myristica fragrans*, exhibited strong antifungal and antibacterial activities (Orabi et al. 1991). Structure modifications by methylation or reduction resulted in diminished activity.

M. fragrans seeds were found to have antibacterial activity (Narasimhan and Dhake 2006). The seeds were powdered and extracted with chloroform to obtain trimyristin, which on saponification yielded myristic acid, and petroleum ether to obtain myristicin. All the constituents isolated from nutmeg exhibited good antibacterial activity against selected Gram-positive and Gram-negative organisms.

Nutmeg was also found to have antibacterial activity against cariogenic bacteria like *Streptococcus mutans* (Chung et al. 2006). The anticariogenic compound was successfully isolated from the methanol extract and identified as macelignan. The MIC of macelignan against *S. mutans* was 3.9 μg/ml, which was much lower than those of other natural anticariogenic agents such as sanguinarine (15.6 μg/ml), eucalyptol (250 μg/ml), mentol and thymol (500 μg.ml) and methyl slaicylate (1,000 μg/ml). Macelignan also possessed preferential activity against other oral microorganisms such as *Streptococcus sobrinus, Streptococcus salivarius, Streptococcus sanguis, Lactobacillus acidophilus* and *Lactobacillus casei* in the MIC range of 2–31.3 μg/ml. The specific activity and fast-effectiveness of mace lignan against oral bacteria strongly suggested that it could be employed as a natural antibacterial agent in functional foods or oral care products. Studies showed macelignan to be a potent natural antibiofilm agent against oral primary colonizers such as *Streptococcus mutans, Streptococcus sanguis* and *Actinomyces viscosus* (Yanti et al. 2008). In early dental plaque formation, these oral primary colonizers were found to initially attach to the pellicle-coated tooth surface to form a biofilm. The results showed that at 24 hours of biofilm growth, *S. mutans, A. viscosus* and *S. sanguis* biofilms were reduced by up to 30%, 30% and 38%, respectively, after treatment with 10 μg/ml macelignan for 5 minutes. Increasing the treatment time to 30 minutes resulted in a reduction of more than 50% of each of the single primary biofilms.

Ethanol extract of nutmeg strongly reduced the population of *Escherichia coli* O157 but not the non-pathogenic *E. coli* strains (Takikawa et al. 2002). Antibacterial activity by the nutmeg extract was also found against the enteropathogenic *E. coli* O111, but not against enterotoxigenic (O6 and O148) and enteroinvasive (O29 and O124) *E. coli*. Further, all *E. coli* O157 strains tested were found to be more sensitive to β-pinene than non-pathogenic *E. coli* strains. The essential oil of nutmeg showed complete zones of inhibition against *Fusarium graminearum* at all the tested doses (Singh et al. 2005). For other tested fungi and bacteria, the essential oil and acetone extract of nutmeg gave good to moderate zone inhibition.

Myristica fragrans was found to have potent antibacterial activity against *Helicobacter pylori* compared with other botanicals tested (Bhamarapravati et al. 2003; Mahady et al. 2005). *Helicobacter pylori* is now recognized as the primary etiological factor associated with the development of gastritis and peptic ulcer disease. In addition, *Helicobacter pylori* infections were also associated with chronic gastritis, gastric carcinoma and primary gastric B-cell lymphoma. Methanol extracts of *Myristica fragrans* (seed) and aril were found to have a MIC of 12.5 μg/ml while the leaf extract was less inhibitory with an MIC of 50 μg/ml against *Helicobacter pylori*.

Successful infection by *Listeria monocytogenes* was found to be dependent upon a range of bacterial extracellular proteins including a cytolysin termed listeriolysin O and phosphatidylcholine-specific phospholipase C. Five plant essential oils: bay, clove, cinnamon, nutmeg and thyme, significantly reduced the production of listeriolysin O by *Listeria monocytogenes* (Smith-Palmer et al. 2002). Firouzi et al. (2007) reported that in the broth culture system, nutmeg essential oil (EO) had a greater effect on *Listeria monocytogenes*

(MIC = 0.20 μl/ml) than did oregano EO (MIC = 0.26 μl/ml). However, oregano EO had a greater effect on *Yersinia enterocolitica* (MIC = 0.16 μl/ml) than did nutmeg EO (MIC = 0.25 μl/ml). In ready-to-cook Iranian barbecued chicken, the log CFU per gram of both bacteria after up to 72 hours of incubation was not decreased significantly by various combinations of oregano and nutmeg EOs (1, 2, and 3 μl/g) and storage temperatures (3, 8, and 20°C) when compared with control samples (without EOs).

Despite food irradiation being a known effective method to eliminate pathogens difficult to eradicate by conventional methods, consumers and industry at large had been reluctant to adopt it (Variyar et al. 2008). This was mainly attributed to some apprehensions regarding the safety of irradiated food. One such apprehension related to 2-alkylcylcyclobutanones, unique radiolytic products thought to be formed in minute quantities in food during radiation processing. Studies by Variyar et al. (2008) demonstrated the natural occurrence of 2-dodecylcyclobutanone and 2-tetradecylcyclobutanone in commercial nonirradiated as well as fresh cashew nut samples and 2-decylcyclobutanone as well as 2-dodecylcyclobutanone in nonirradiated nutmeg samples.

Anticancer Activity

Myristica fragrans was found to have anti-cancer activities. Mace of *M. fragrans* exhibited chemopreventive activity against DMBA-induced papillomagenesis in the skin of male Swiss albino mice (Jannu et al. 1991). Mice with skin papillomas induced by DBMA and croton oil had average tumours of 5.67 per tumour bearing mice. Their skin papilloma incidence was significantly reduced by 50% to 1.75 with 1% mace diet supplementation at the initial stage of tumorigenesis. Separate studies also reported that nutmeg mace reduced 3-methylcholanthrene (MCA)-induced carcinogenesis in the uterine cervix of virgin, young adult, Swiss albino mice (Hussain and Rao 1991). Oral administration of mace at the dose level of 10 mg/mouse per day for 7 days before and 90 days following carcinogen thread insertion, the cervical carcinoma incidence, as compared with that of the control (73.9%), was 21.4%. This decline in the incidence of carcinoma was highly significant. Dihydroguaiaretic acid Dihydroguaiaretic acid (DHGA), and Nordihydroguaiaretic acid (NDGA) isolated from the arils of *Myristica fragrans* suppressed leukemia, lung cancer and colon cancer in an in-vitro bioassay (Park et al. 1998). DHGA showed an inhibitory effect against the complex formation of the fos-jun dimer and the DNA consensus sequence with an IC_{50} value of 0.21 μmol. Nordihydroguaiaretic acid (NDGA) also inhibited fos-jun dimer action showing IC_{50} values of 7.9 μmol.

Methanol extract of *Myristica fragrans* induced apoptosis of Jurkat leukemia T cell line in a mechanisms involving SIRTI mRNA down-regulation (Chirathaworn et al. 2007). At the concentrations 50 and 100 μg/ml, the methanol extract significantly inhibited Jurkat cell proliferation and induced apoptosis. Macelignan was identified as a novel inhibitor of P-Glycoprotein (P-gp) activity and may be a promising lead compound for the rational design of more efficacious drugs to reverse multidrug resistance in cancer (Im et al. 2009). Macelignan (40 μM) increased the cellular accumulation of daunorubicin by approximately threefold in NCI/ADR-RES epithelial cancer cells, whereas it did not alter the cellular accumulation of daunorubicin in MCF-7 sensitive cells. Similarly, the presence of macelignan also enhanced significantly the cellular accumulation of rhodamine 123 in a concentration-dependent manner in NCI/ADR-RES cells. Furthermore, cancer cells were more susceptible to the cytotoxicity of vinblastine, a P-gp substrate, in the presence of macelignan. The results suggested that macelignan had inhibitory effects on P-gp mediated cellular efflux.

Myristicin, 1-allyl-3,4-methylenedioxy-5-methoxybenzene, a naturally occurring alkenylbenzene compound found in the nutmeg was found to have cytotoxic and apoptotic effects on the human neuroblastoma SK-N-SH cells

(Lee et al. 2005). A dose-dependent reduction in cell viability occurred at myristicin concentration ≥0.5 mM in SK-N-SH cells. The apoptosis triggered by myristicin was accompanied by an accumulation of cytochrome c and by the activation of caspase-3. The results obtained suggested that myristicin induced cytotoxicity in human neuroblastoma SK-N-SH cells by an apoptotic mechanism. Among two new phenolic compounds namely (-)-1-(2,6-dihydroxyphenyl)-9-[4-hydroxy-3-(p-menth-1-en-8-oxy)-phenyl]-1-nonanone (1) and (7 R,8 R)-7,8-dihydro-7-(3,4-dihydroxyphenyl)-3′-methoxy-8-methyl-1′-(E-propenyl)benzofuran (2) and other compounds namely(+)-δ-8′-7-acetoxy-3,4,3′,5′-tetramethoxy-8- O-4′-neolignan (3), and (7 S,8 S,7′ R,8′ S)-4,5′-dihydroxy-3,3′-dimethoxy-7,7′-epoxylignan (4) isolated from nutmeg fruit, compound 4 exhibited strong DPPH antioxidative activity and cytotoxicity against K-562 (myelogenous leukaemia) cells with IC50 values of 39.4 and 2.11 μM, respectively (Duan et al. 2009).

Antiulcerogenic Activity

Nutmeg extract was found to reduce peptic ulcer and its antiulcerogenic effect may be due to its calcium blocking activity (Jan et al. 2004, 2005). It was found that extract from *Myristica fragrans* reduced the volume, free and total acidity of gastric secretion when compared to carbachol. The extract had similar effects as Verampil. Over production of gastric acid in response to various stimuli was responsible for peptic ulceration in majority of patients. Calcium was found to play an important role in the release of various stimulant mediators. The seeds of *Myristica fragrans* were reported to have calcium channel blocking activity. Although its effect was less than cimetidine, the nutmeg extract could still be used as natural calcium channel antagonist. Calcium channel blockers are widely used in controlling contraction of cardiovascular smooth muscles, allergic reaction and prevention of premature labor.

Antiinflammatory Activity

Eugenol, a major constituent of nutmeg, showed anti-inflammatory activity in the rat paw carrageenin oedema test (Bennett et al. 1988). Nutmeg was found to have antiinflammatory, analgesic and antithrombotic activities. In rodents, the chloroform extract inhibited the carrageenan-induced rat paw oedema, produced a reduction in writhings induced by acetic acid in mice and offered protection against thrombosis induced by ADP/adrenaline mixture in mice (Olajide et al. 1999). In another study, the methanol extract (1.5 g/kg), ether fraction (0.9 g/kg), n-hexane fraction (0.5 g/kg) and its fractions Fr-II (0.19 g/kg) and Fr-VI (0.17 g/kg) of *Myristica fragrans* fruit mace showed a lasting antiinflammatory activity, and the potencies of these fractions were approximately the same as that of indomethacin (10 mg/kg) (Ozaki et al. 1989) Fr-VI was determined to be myristicin. These results suggested that the antiinflammatory action of mace was due to the myristicin that it contained.

Neuroprotective Activity

Of cyclooxygenase-2 and inducible nitric oxide synthase, that consequently resulted in the reduction of nitric oxide in lipopolysaccharide (LPS)-treated microglial cells. It also significantly suppressed the production of pro-inflammatory cytokine tumour necrosis factor-α and interleukin-6. These results suggested that macelignan possessed therapeutic potentials against neurodegenerative diseases with oxidative stress and neuroinflammation In subsequent studies, the reasearchers showed that daily administration of macelignan reduced the spatial memory impairments induced by the chronic lipopolysaccharide (LPS) infusions (Cui et al. 2008). Oral administrations of macelignan reduced the hippocampal microglial activation induced by chronic infusions of LPS into the fourth ventricle of Fisher-344 rat brains. The results indicated that macelignan may possess therapeutic potential for the prevention of Alzheimer's disease. Further studies by the researchers (Ma et al. 2009) showed that macelignan suppressed both the phosphorylations of mitogen-activated protein kinase (MAPK) and the degradation of inhibitory-kappa B (IkappaBalpha) and increases of nuclear NF-kappaB in LPS-stimulated BV-2 microglial cells. The results suggested that macelignan had an antiinflammatory effect on the affected brain through regulation of the inflammation through the MAPK signal pathway.

Hepatoprotective Activity

Studies showed that there was a significant increase in the cytosolic glutathione S-transferase (GST) activity in the liver of mice exposed to 1% 2,3-tert-butyl-4-hydroxyanisole (BHA) or nutmeg mace diet (Kumari and Rao 1989). In addition, there was a significant increase in the acid-soluble sulfhydryl (SH) content in the liver of mice fed on 1% BHA and 2% mace diets. GST protects the body by detoxifying potentially damaging xenobiotics. Aqueous suspension on nutmeg mace was found to modulate hepatic xenobiotic metabolizing enzymes in the F1 progeny of mice (Chhabra and Rao 1994). Dams receiving mace treatment and their F1 pups showed significantly elevated hepatic sulfhydryl content, glutathione S-transferase and glutathione reductase activities and cytochrome b5 content. Hepatic cytochrome P450 content decreased in dams ($P<0.05$) receiving the lower mace dose for 21 days and the F1 pups, but increased in dams receiving the higher dose for both time periods and the lower dose for 14 days. Only the 14-day-old pups of dams receiving either mace dose showed significantly levels of hepatic glutathione peroxidise.

Of 21 spices tested, nutmeg showed the most potent hepatoprotective activity when fed to rats with liver damage caused by lipopolysaccharide (LPS) plus d-galactosamine (D-GalN) (Morita et al. 2003). Myristicin, one of the major essential oils of nutmeg, was found to possess extraordinarily potent hepatoprotective. The hepatoprotective activity of myristicin was attributed at least in part, to the inhibition of TNF-a release from macrophages.

Sohn et al. (2007) reported that macelignan protected HepG2 cells against tert-butylhydroperoxide (t-BHP)-induced oxidative damage. The results showed that macelignan significantly reduced the cell growth inhibition and necrosis caused by t-BHP. Furthermore, macelignan ameliorated lipid peroxidation as demonstrated by a reduction in malondialdehyde formation in a dose-dependent manner. It was also found that macelignan reduced intracellular reactive oxygen species (ROS) formation and DNA damaging effect caused by t-BHP. These results strongly suggested that macelignan had significant protective ability against oxidative damage caused by reactive intermediates. In a subsequent paper, Sohn et al. (2008) reported that the protective effects of macelignan on cisplatin-induced hepatotoxicity may be associated with the mitogen activated protein kinase (MAPK) signaling pathway. Pretreatment with macelignan for 4 days significantly prevented the increased serum enzymatic activities of alanine and aspartate aminotransferase in a dose-dependent manner. Cisplatin-induced phosphorylation of c-Jun N-terminal kinase1/2 (JNK1/2) and extracellular signal-regulated kinase1/2 (ERK1/2) was abrogated by pretreatment with macelignan, however, that of p38 was not significantly affected. It was also found that macelignan attenuated the expression of phosphorylated c-Jun in cisplatin-treated mice. Accordingly, it was suggested that the hepatoprotective effects of macelignan could be related to activation of the MAPK signaling pathway, especially JNK and c-Jun, its substrate. The findings suggested that co-treatment of cisplatin with macelignan may provide more advantage than cisplatin treatment alone in cancer therapy.

Antidiarrhoel Activity

Barrowman et al. (1975) reported that nutmeg halted diarrhoea and steatorrhoea in a patient with medullary carcinoma of the thyroid with pulmonary metastases. The patient had diarrhoea and steatorrhoea with large amounts of prostaglandin-like material present in peripheral blood, and some was extracted from the tumour.

The diarrhoea which persisted after thyroidectomy responded to treatment with nutmeg. The results suggested that prostaglandins may be partly responsible for the diarrhoea associated with medullary carcinoma of the thyroid when lung metases were present and that nutmeg may have acted by interfering with the syntheisis or actions of prostaglandins. Eugenol, a major constituent of nutmeg, inhibited prostaglandin synthesis, and reduced the tone of isolated gut muscle and myometrium in man and laboratory animals (Bennett et al. 1988). With rats in-vivo, the compound reduced the rate of intestinal transit, the intestinal accumulation of fluid induced by prostaglandin E2, and the diarrhoea induced by castor oil.

Separate studies reported that nutmeg had good antidiarrheal, hypotensive, sedative effects and analgesic effect (Grover et al. 2002). Both nutmeg crude suspension (NMC) and petroleum ether (PE), but not aqueous extract (Aq), decreased the mean number of loose stools or increased the latency period. NMC increased intestinal tone while PE had no such effect. PE had no effect on guinea pig ileum, but inhibited the contraction produced by acetylcholine, histamine and prostaglandin. NMC but not PE extract showed a significant but weak analgesic effect. While PE effectively potentiated both phenobarbitone and pentobarbitone-induced sleeping time, NMC was considerably less effective. NMC administered intraduodenally did not produce much effect on blood pressure (BP), but potentiated the action of exogenously administered adrenaline and nor-adrenaline. On the other hand, PE in higher, but not lower, doses caused a precipitous fall in BP not blocked by atropine. Thus, overall extracts of nutmeg showed a good antidiarrheal effect, with a significant sedative property. The extracts possessed only a weak analgesic effect, with no harmful effects on blood pressure and ECG.

The extract from *Myristica fragrans* seeds (160 μg/ml) exhibited in-vitro inhibitory activity of 99.2% against human rotaviruses (Gonçalves et al. 2005). Rotaviruses have been recognized as the major agents of diarrhea in infants and young children in developed as well as developing countries. In Brazil, diarrhea is one of the principal causes of death, mainly in the infant population. The results indicated that *M. fragrans* extract could be useful in the treatment of human diarrhea if the etiologic agent was a rotavirus.

Radioprotective Activity

Nutmeg seed (MF) was found to have radioprotective effect (Sharma and Kumar 2007). Administration of MF to Swiss albino mice significantly enhanced hepatic glutathione (GSH) and decreased testicular lipid peroxidation (LPO) level whereas acid phosphatase (ACP) and alkaline phosphatase (ALP) activity did not show any significant alteration. Irradiation resulted in significant elevation in lipid peroxidation LPO level and acid phosphatase ACP activity, and decreased the hepatic glutathione GSH content and alkaline phosphatase ALP activity. MF pretreatment effectively protected against radiation induced biochemical alteration as reflected by a decrease in LPO level and ACP activity, and an increase in GSH and ALP activity.

Macelignan from *Myristica fragrans* was found to have a protective effect on immortalized human keratinocytes (HaCaT) against UVB damage (Anggakusuma et al. 2010). Macelignan at a concentration of 0.1–1 μM increased the viability of HaCaT cells following UVB irradiation and inhibited matrix metalloproteinase MMP-9 and inducible cyclogenase COX-2 expression in a concentration-dependent manner. An inhibitory effect was also seen in the signal transduction network, where macelignan treatment reduced the activation of UVB-induced mitogen-activated protein kinases (MAPKs), phosphatidylinositol 3-kinase/Akt (PI3K/Akt) and their downstream transcription factors.

Lignans present in the aqueous extract of fresh nutmeg mace exhibited immunomodulatory and radiomodifying activities in mammalian splenocytes (Checker et al. 2008). These macelignans inhibited the proliferation of splenocytes in response to polyclonal T cell mitogen concanavalin A (Con A). This inhibition of proliferation was due to cell cycle arrest in G1 phase and augmentation of apoptosis as shown by increase in pre-G1 cells. The increase in activation induced cell death by macelignans was dose dependent. It was found to inhibit the transcription of interleukin IL-2 and IL-4 genes in response to Con A. The production of IL-2, IL-4 and IFN-gamma cytokines was significantly inhibited by macelignans in Con A-stimulated lymphocytes in a dose dependent manner. Macelignans protected splenocytes against radiation-induced intracellular ROS production in a dose dependent manner. Macelignans was not cytotoxic towards lymphocytes. On the contrary, it significantly inhibited the radiation-induced DNA damage in splenocytes as indicated by decrease in DNA fragmentation.

Antidiabetic, Hypolipidemic/ Hypocholesteromic Activities

Myristica fragrans seed extract administration to hypercholesterolemic rabbits was found to reduce serum cholesterol and LDL Cholesterol by 69.1% and 76.3% respectively and also lowered cholesterol/phospholipid ratio by 31.2% and elevated the decreased HDL-ratio significantly (Sharma et al. 1995). Extract feeding also prevented the accumulation of cholesterol, phospholipids and triglycerides in liver, heart and aorta and dissolved atheromatous plaques of aorta by 70.9–76.5%. Fecal excretion of cholesterol and phospholipid were significantly increased in seed extract fed rabbits.

The ethanolic extract of nutmeg showed hypolipidaemic effect in albino rabbits experimentally induced with hyperlipidaemia (Ram et al. 1996). When compared with the control the levels of lipoprotein lipids were significantly lower in the experimental group administered with the extract (500 mg/kg daily) after 60 days; total cholesterol 573 vs. 209 mg/dl, low density lipoprotein (LDL) cholesterol 493 vs. 131 mg/dl, and triglycerides 108 vs. 67 mg/dl. High density lipoprotein (HDL) cholesterol levels were not significantly different. Total cholesterol:HDL ratio and LDL:HDL ratio were significantly lower in the experimental group. The ethanolic *Myristica fragrans* extract also showed platelet anti-aggregatory ability. There were significantly lower levels of total cholesterol in heart (3.7 vs. 2.2 mg/100 g) and liver (11.9 vs. 1.5 mg/100 g). The toxicity studies showed absence of any adverse effects on various haematological and biochemical parameters.

Myristica fragrans extract was found to ameliorate hyperglycemia and abnormal lipid metabolism in animal models (Arulmozhi et al. 2007). After 7 days of oral administration, the hydroalcoholic extract of fruits of *Myristica fragrans* at doses of 150 and 450 mg/kg, ameliorated the metabolic abnormalities caused by chlorpromazine as evidenced by significant reduction of glucose and triglyceride (TG) levels (maximal effect of 41% and 53% reduction of glucose and TG, respectively, at 450 mg dose) The standard antidiabetic rosiglitazone at 10 mg significantly reduced the TG (63%) and glucose (40%) levels in this model, while the standard antidiabetic glimepiride has exhibited 55% and 16% reduction in TG and glucose, respectively. In rats fed a high-cholesterol diet, *Myristica fragrans* extract significantly reduced the elevated TG (47% reduction at 450 mg,) and cholesterol (66.7% reduction at 450 mg), and also exhibited a reduction in hepatic TG secretion after tyloxapol administration.

Macelignan was found to have antidiabetic activity. Macelignan reduced serum glucose, insulin, triglycerides, free fatty acid levels, and triglycerides levels in the skeletal muscle and liver of db/db mice (Han et al. 2008). Further, macelignan significantly improved glucose and insulin tolerance in these mice, and without altering food intake, their body weights were slightly reduced while weights of troglitazone-treated mice increased. Macelignan increased adiponectin expression in adipose tissue and serum, whereas the expression and serum levels of tumour necrosis factor-alpha and interleukin-6 decreased. Macelignan downregulated inflammatory gene expression in the liver and increased AMP-activated protein kinase activation in the skeletal muscle of db/db mice. Strikingly, macelignan reduced endoplasmic reticulum (ER) stress and c-Jun NH(2)-terminal kinase activation in the liver and adipose tissue of db/db mice and subsequently increased insulin signalling. The results indicated that macelignan enhanced insulin sensitivity and improved lipid metabolic disorders by activating PPARalpha/gamma and attenuating ER stress, suggesting that it to have potential as an antidiabetes agent for the treatment of type 2 diabetes. A methanol extract of nutmeg seed yielded protein tyrosine phosphatase 1B (PTP1B) compounds meso-dihydroguaiaretic acid (1) and otobaphenol (2) (Yang et al. 2006). Compounds 1 and 2 inhibited PTP1B with IC_{50} values of 19.6 and 48.9 μM, respectively. Treatment with compound 1 on 32D cells overexpressing the insulin receptor (IR) resulted in a dose-dependent increase in the tyrosine phosphorylation of IR. The results indicated that compound 1 could act as an enhancing agent in intracellular insulin signaling, possibly through the inhibition of PTP1B activity. Protein tyrosine phosphatase 1B (PTP1B) was proposed as one of the drug targets for treating type 2 diabetes and obesity.

Antiobesity Activity

Nguyen et al. (2010) found that the total extract of *Myristica fragrans* (nutmeg) activated the AMP-activated protein kinase (AMPK) enzyme in differentiated C2C12 cells. AMPK is a potential therapeutic target for the treatment of metabolic syndrome including obesity and type-2 diabetes. As active constituents, seven 2,5-bis-aryl-3,4-dimethyltetrahydrofuran lignans, tetrahydrofuroguaiacin B (1), saucernetindiol (2), verrucosin (3), nectandrin B (4), nectandrin A (5), fragransin C(1) (6), and galbacin (7) were isolated from this extract. Among the isolates, compounds 1, 4, and 5 at 5 μM produced strong AMPK stimulation in differentiated C2C12 cells. A tetrahydrofuran mixture (THF) was also found to have a preventive effect on weight gain in a diet-induced animal model. The results suggested that nutmeg and its active constituents could be used not only for the development of agents to treat obesity and possibly type-2 diabetes but may also be beneficial for other metabolic disorders.

CNS and Antidepressant Activity

N-hexane extracts (5, 10, and 20 mg/kg) of *M. fragrans* elicited a significant antidepressant-like effect in mice, when assessed in both the forced swim test (FST) and the tail suspension test (TST) forced swim test (FST) and the tail suspension test (TST) (Dhingra and Sharma 2006) *M. fragrans* extract significantly decreased immobility periods of mice in both the FST and the TST. The 10 mg/kg dose was found to be most potent, as indicated by the greatest decrease in the immobility period compared with the control. Further, this dose of the extract was found to have comparable potency to imipramine (15 mg/kg i.p.) and fluoxetine (20 mg/kg i.p.). The extract did not have a significant effect on locomotor activity of mice. The antidepressant-like effect of the extract appeared to be mediated by interaction with the adrenergic, dopaminergic, and serotonergic systems.

The n-hexane extract of *M. fragrans* (MF) was found to have a memory enhancing effect (Parle et al. 2004). MF extract at the lowest dose of 5 mg/kg p.o. administered for three successive days significantly improved learning and memory of young and aged mice. This extract also reversed scopolamine- and diazepam-induced impairment in learning and memory of young mice. MF extract enhanced learning and retention capacities of both young and aged mice. The observed memory-enhancing effect may be attributed to a variety of properties (individually or in combination) the plant was reported to possess, such as antioxidant, antiinflammatory, or perhaps procholinergic activity. *M. fragrans* was found to have acetylcholinesterase- inhibiting activity. The n-hexane extract of *M. fragrans* seeds (5 mg/kg p.o.) administered for three successive days to young male Swiss albino mice was found to significantly decrease acetylcholinesterase activity as compared with a standard acetylcholinesterase-inhibiting drug, metrifonate (50 mg/kg i.p.) (Dhingra et al. 2006). Acetylcholinesterase is an enzyme that inactivates acetylcholine in the central cholinergic pathways that plays a prominent role in the learning and memory processes.

Skin Whitening Activity

Myristica fragrans was one of six active plant extract (final concentration 1 mg/ml in methanol) tested that exhibited more than 65% of inhibition of elastase activity (Lee et al. 1999). The extract had an IC_{50} of 284.1 μg/ml on the activity of human leukocyte elastase. *M. fragrans* was one of several plant extracts that showed inhibition of mushroom tyrosinase activity (Lee et al. 1997). Cho et al. (2008) found that macelignan isolated from *Myristica fragrans* exhibited in-vitro inhibition on melanogenesis and its related enzymes such as tyrosinase, tyrosinase-related protein-1 (TRP-1), and tyrosinase-related protein-2 (TRP-2) in melan-a murine melanocytes. The IC_{50} values of macelignan for melanogenesis and tyrosinase were 13 μM and 30 μM, respectively. Macelignan also significantly decreased tyrosinase, TRP-1, and TRP-2 protein expression. The results indicated that macelignan effectively inhibited melanin biosynthesis and thus could be employed as a new skin-whitening agent.

Antiplatelet Activity

Nutmeg oil was found to have anti-platelet aggregation activity (Janssen and Laeckman 1990). It was shown that eugenol and isoeugenol in the oil played the major role in the inhibitory activity of nutmeg. Medicinally, it appeared that nutmeg oil and nutmeg powder could be replaced by eugenol and/or isoeugenol.

Anxiogenic Activity

The n-hexane extract of *Myristica fragrans* (MF) seeds, acetone-insoluble part of the n-hexane extract (AIMF) and trimyristin (TM) were found to have anxiogenic activity(Sonavane et al. 2002b). The MF (10 and 30 mg/kg), AIMF (30, 100, and 300 mg/kg), and TM (10, 30, and 100 mg/kg) administered intraperitoneally exhibited anxiogenic activity in elevated plus-maze (EPM) paradigm. In the EPM test, MF, AIMF, and TM decreased the time spent by mice in the open arm and the entries in the open arm. Further, the effect of diazepam (1 mg/kg i.p.), serotonin 5-HT3 receptor antagonist, ondansetron (1 mg/kg i.p.), and 5-HT1A receptor agonist, buspirone (1 mg/kg i.p.), on the occupancy in open arm and entries in open arm was significantly reduced by TM. In the open-field test, AIMF as well as TM reduced the number of rearing and locomotion. Both TM and AIMF reduced the number of head pock in the hole-board test. Inhibition of anxiolytic activity of ondansetron (5-HT3 receptor antagonist), buspirone (5-HT1A receptor agonist), and diazepam [acting on gamma-aminobutyric acid (GABAA) receptor] suggesting a nonspecific anxiogenic activity of TM and also a link between 5-HT and GABA systems in the anxiogenic activity of trimyristin.

Anticonvulsant, Sedative and Cataleptic Activities

Myristica fragrans (MF) seed extract exhibited anticonvulsant, cataleptic and sedative activities in male albino mice (Sonavane et al. 2002a). MF inhibited seizures induced MF by maximum electroshock (MES), pentylenetetrazol (PTZ), picrotoxin, and lithium sulphate-pilocarpine nitrate (Li-Pilo). However, picrotoxin-induced seizures were not inhibited. The haloperidol-induced catalepsy was potentiated but motor coordination and pentobarbitone-induced sleep were not affected significantly. MF had complex actions on the central nervous system. Although it exhibited anticonvulsant activity against MES, PTZ and lithium-pilocarpine, it failed to inhibit picrotoxin-induced seizures. MF reduced central dopaminergic activity but was without any effect on pentobarbitone-induced sleep. A ligroin extract of nutmeg (*Myristica fragrans*) caused a significant increase in the duration of light and deep sleep in the young chicken (Sherry et al. 1982). The presence of trimyristin tended to increase the effect of the extract. The extract did not contain detectable amounts of myristicin, elemicin, safrole, or eugenol, which either individually or collectively were suggested to be the active agent of nutmeg.

Using established animal models, nutmeg oil showed a rapid onset of action and short duration of anticonvulsant effect (Wahab et al. 2009). It was found to possess significant anticonvulsant activity against electroshock-induced hind limb tonic extension. It exhibited dose dependent anticonvulsant activity against pentylenetetrazole-induced tonic seizures. It delayed the onset of hind limb tonic extensor jerks induced by strychnine. It was anticonvulsant at lower doses, whereas weak proconvulsant at a higher dose against pentylenetetrazole and bicuculline induced clonic seizures. Nutmeg oil was found to possess wide therapeutic margin, as it did not induce motor impairment when tested up to 600 μoL/kg in the inverted screen acute neurotoxicity test. Furthermore, the LD_{50} (2,150 μL/kg) value was much higher than its anticonvulsant doses (50–300 μL/kg). The results indicated that nutmeg oil may be effective against grand mal and partial seizures, as it prevented seizure spread in a set of established animal models. Slight potentiation of clonic seizure activity limited its use for the treatment of myoclonic and absence seizures.

Sexual Enhancement Activity

The extracts of the nutmeg and clove were found to stimulate the mounting behaviour of male mice, and also to significantly increase their mating performance (Tajuddin et al. 2003). The drugs were devoid of any conspicuous general short term toxicity. Oral administration of the ethanolic extract of *M. fragrans* at the dose of 500 mg/kg, produced significant augmentation of sexual activity in male rats (Tajuddin et al. 2005). It significantly increased the mounting frequency, intromission frequency, intromission latency and caused significant reduction in the mounting latency and post ejaculatory interval. It also significantly increased mounting frequency with penile anaesthetisation as well as erections, quick flips, long flips and the aggregate of penile reflexes with penile stimulation. The extract was also observed to be devoid of any adverse effects and acute toxicity. The resultant significant and sustained increase in the sexual activity of normal male rats without any conspicuous adverse effects indicated that the 50% ethanolic extract of nutmeg possessed aphrodisiac activity, increasing both libido and potency, which might be attributed to its nervous stimulating property. The present study thus provided a scientific rationale for the traditional use of nutmeg in the management of male sexual disorders.

Cognitive Activity

Oral administration of all the nutmeg extracts namely methanolic (ME), dichloromethane (DE), and hexane (HE) extracts at 500 mg/kg to mice caused a significant increase in locomotor activity, the i.p. administration of DE showed significant reduction in rectal temperature along with a significant increase in tail flick latency at 300 mg/kg (El-Alfy et al. 2009). A significant decrease in core body temperature was observed with HE at 100 mg/kg, while higher doses caused significant increases in hot plate latency. The results indicated that different behavioral effects were produced that varied by the type of nutmeg extract as well as by the route of administration.

Studies showed that inhalation of nutmeg seed essential oil at a dose of 0.5 ml/cage decreased locomotion of mice by 68.62%; and inhalation of 0.1 and 0.3 ml/cage inhibited locomotion by 62.81% and 65.33%, respectively (Muchtaridi et al. 2010). Generally, larger doses and longer administrations of nutmeg seed essential oil exhibited greater locomotor inhibition. The most concentrated compound in the plasma was myristicin. Half an hour after the addition of 1 ml/cage of nutmeg seed oil, the plasma concentration of myristicin was 3.7 μg/ml; 1 and 2 hours after the addition, the blood levels of myristicin were 5.2 μg/ml and 7.1 μg/ml, respectively. Other essential oil compounds identified in plasma were safrole (2-hour inhalation: 1.28 μg/ml), 4-terpineol (half-hour inhalation: 1.49 μg/ml, 1-hour inhalation: 2.95 μg/ml, 2-hour inhalation: 6.28 μg/ml) and fatty esters. The concentrations of the essential oil compounds in the blood plasma were relatively low (μg/ml or ppm). The findings suggested that the volatile compounds of nutmeg seed essential oil identified in the blood plasma may correlate with the locomotor-inhibiting properties of the oil when administered by inhalation.

Drug /Drug Interaction Activity

Mace and nutmeg were found to significantly inhibit cytochrome metabolic enzymes CYP3A4 or CYP2C9 activity (Kimura et al. 2010). Cytochrome P450 inhibition is a principal mechanism to pharmacokinetic drug-drug interactions. Furthermore, bioassay-guided fractionation of mace (*Myristica fragrans*) led to isolation and structural characterization of a new furan derivative (1) along with other 16 known compounds, including an acylphenol, neolignans, and phenylpropanoids. Among these isolates, (1 S,2R)-1-acetoxy-2-(4-allyl-2,6-dimethoxyphenoxy)-1-(3,4-dimethoxyphenyl)propane (9) exhibited the most potent CYP2C9 inhibitory activity with an IC_{50} value comparable to that of sulfaphenazole, a CYP2C9 inhibitor. Compound 9 competitively inhibited CYP2C9-mediated 4′-hydroxylation of diclofenac. The inhibitory constant (K_i) of compound 9 was determined to be 0.037 μM. Compound 9 was found to be 14-fold more potent than was sulfaphenazole.

Molluscicidal Activity

In-vivo and in-vitro treatments of trimyristin and myristicin (active molluscicidal components of *Myristica fragrans)* significantly inhibited the acetylcholinesterase (AChE), acid and alkaline phosphatase (ACP/ALP) activities in the nervous tissue of *Lymnaea acuminata* vector of liver flukes, *Fasciola gigantica* and *Fasciola hepatica* in cattle (Jaiswal et al. 2010). The results suggested that inhibition of AChE, ACP, and ALP by trimyristin and myristicin in the freshwater snail *Lymnaea acuminata* may be the cause of the molluscicidal activity of *Myristica fragrans*.

Hallucinogenic and Narcotic-Like Activities

Nutmeg had been known for its hallucinogenic properties for a long time (Weiss 1960) Considrable narcotic power had been attributed to myristicin. Weiss reported detailed psychic experiences of adult prison inmates following the ingestion of powdered nutmeg. Symptons included semi-stupor, periods of sleepiness, episodes of excitement and agitation and physical symptoms of vomining, dizziness and numbness. Fras and Friedman (1969) reported feelings of depersonalization and unreality, changes in perception, as well as illusions and hallucinations, especially visual, were the significant aspects of the subjective experience of an eighteen-year-old adolescent. The patient was also able to differentiate the effects of nutmeg from those of marihuana and morning-glory seeds, on the basis of a temporary break with reality which he experienced with nutmeg. Adults may abuse the hallucinogenic properties of nutmeg. Children may be at high risk at home, since nutmeg may be widely available as a cooking additive. In the course of its use in traditional medicine, overdose may occur.

Toxicity Studies

Several intoxications had been reported after an ingestion of approximately 5 g of nutmeg, corresponding to 1–2 mg myristicin/kg body weight (b.w.) (Hallström and Thuvander 1997). Although these intoxications may be ascribed to the actions of myristicin, it was likely that other components of nutmeg may also be involved. The acute toxicity of myristicin appeared to be low. No toxic effects were observed in rats administered myristicin perorally at a dose of 10 mg/kg b.w., while 6–7 mg/kg b.w. may be enough to cause psychopharmacological effects in man. A weak DNA-binding capacity was demonstrated, but there were no indications that myristicin exerted carcinogenic activity in short-term assays using mice. Intake estimations indicated that nonalcoholic drinks may be the most important single source of myristicin intake. Based on available data, it seemed unlikely that the intake of myristicin from essential oils and spices in food, estimated to a few mg per person and day in this report, would cause adverse effects in humans.

Swiss albino mice treated orally with 0.003 and 0.3 mg/day of mace during 7 days showed a significant increase in creatine phosphokinase level (Malti et al. 2008). The microscopic evaluation showed that mace induced morphological perturbation in mice's liver. The results also showed an inhibitory effect of glyceraldehyde 3-phosphate dehydrogenase and an important increase in the level of thiobarbituric acid reactive substances, succinate dehydrogenase activities and no change in catalase activities. All of these results showed that *M. fragrans* at 0.3 mg/g in mice affected energy metabolism and oxidative stress.

Seeds of nutmeg are used as spice, but they are also abused because of psychotropic effects described after ingestion of large doses. It was postulated that these effects could be attributable to metabolic formation of amphetamine derivatives from the main nutmeg ingredients elemicin (EL), myristicin (MY), and safrole (SA). Beyer et al. (2006) reported that in a case of a suspected nutmeg abuse, neither such amphetamine derivatives nor the main nutmeg ingredients could be detected in urine. However, the metabolites of EL, MY, and SA in urine were detected using gas chromatography–mass spectrometry. In the human urine sample, the following metabolites were identified: O-demethyl elemicin, O-demethyl dihydroxy elemicin, demethylenyl myristicin, dihydroxy myristicin, and demethylenyl safrole. As in the human urine sample, neither amphetamine derivatives nor the main nutmeg ingredients could be detected in the rat urine samples.

Macelignan which contained a small amount of the carcinogen safrole was found to have a toxic effect on the heart (Javaregowda et al. 2010). Macelignan was suggested to act as PPAR α/γ dual agonist for diabetic patients. The researchers confirmed that PPAR α/γ was expressed on H9c2 cells, neonatal cardiomyocytes and heart. Compared to control, macelignan significantly reduced the beating number and frequency in a concentration dependent manner. Cell viability was strikingly decreased at 16 hours after treatment. And the expression of autophagic markers as well as apoptotic molecules was maximized at 16 hours. A significant reduction of the systolic blood pressure was shown in both strains, and pulse was also strikingly decreased after its injection. However, the component of serum was not changed by macelignan. Therefore, the researchers stressed that the use of macelignan as an anti-cancer or anti-inflammatory agent should be carefully considered.

Traditional Medicinal Uses

Nutmeg (seed), mace and nutmeg essential oils have been used in traditional medicine since ancient times (Burkill 1966; Ozaki et al. 1989; Daniel 1994; van Gils and Cox 1994; Duke et al. 2002; Grover et al. 2002; Jaiswal et al. 2009).

In Western medicine nutmeg is sometimes used as a stomachic, stimulant, carminative as well as for intestinal catarrh and colic, to stimulate appetites, to control flatulence, and as an emmenagogue and abortifacient. Nutmeg and mace have been used in traditional Arabian medicine, against colds, fever, and general, respiratory complaints. In Arabia, utmeg has been used as stimulant digestive, tonic and aphrodisiac. It was regarded to be an anti-helminthic and was used for that purpose. Nutmeg is still used to treat flatulence, as a digestive aid, to improve appetite and to treat diarrhoea, vomiting and nausea. Since early times, the Hindus in India have deemed it as warmth-producing, stimulating, and good for digestion. It was used as a digestive aid, for freckles and skin blotches, as a carminative and an aphrodisiac. It was also used to treat tuberculosis. It was employed for headache, nerve fevers, cold fevers, foul breath, dysentery and intestinal weakness. Nutmeg was prescribed as an analgesic in neuritic pains, as a sedative in highly tense nervous states, and as a sedative and anti-spasmodic in asthma. In traditional Indian folkloric domestic medicine, nutmeg and mace have also been used in electuaries and tonics particularly for dysentery. Nnutmeg is used in small quantities to induce hypnotic effect in irritable children. It is also administered as an hypnotic and sedative in epileptic convulsions. In India, Peninsular Malaysia and Indonesia nutmeg has been used as tonic especially after child birth, also for indigestionand stomach complaints. In Malaysia and Singapore, nutmeg has been used as tonic for males, for malaria, rheumatism, sciatica and early stages of leprosy and various pains. The fruit wall has been prescribed in a draught for ulceration of the bones and in Moluccas for sprue.

Myristicin has been reported to be poisonous, taken into the intestine or subcutaneously it causes degeneration of the liver and inflammation of the intestines. Poisoning from excessive use of nutmegs has been reported to be due to myristicin.

Other Uses

Nutmeg pericarp waste can be used as substrate to grow the popular edible mushroom 'kulat pala' (*Volvariella volvacea* (Bull. ex Fr.) Sing), which possesses a light nutmeg flavour. Nutmeg essential oil is widely used in the food, perfumery, cosmetic and pharmaceutical industries. It is applied in medicinal drugs like cough syrups, tooth past, tobacco, perfume and shampoo formulations. In aromatherapy, it is used as a stimulator and energizer. Daniel (1994) listed the uses of nutemeg and its derivatives as follows:

- Camphene is used in the manufacture of camphor and pharmaceutical drugs.
- δ-pinene is used in the manufacture of camphor, solvents, plasticers, perfume bases and synthetic pine oil.
- dipentene is used as solvent, wetting and dispersing agent and in the manufacture of resins.
- δ-borneol which exits mainly as acetate ester rather than free alcohol is used in perfumery and incense making.
- i-terpineol is used as antiseptic, and in perfumes and soaps.
- Geraniol is used mainly in perfumery.
- Safrole is used in perfumery and the manufacture of heliotropin and soap and also as an antiseptic.
- Eugenol; is used in the manufacture of vanillin, perfumery and in dental analgesic.
- Isoeugenol is used in the manufacture of vanillin.

Nutmeg also contains a valuable, thick, yellow fat that is called nutmeg butter or fixed oil of nutmeg. Nutmeg butter is used to make candles and is important in certain salves, ointments, medicines and perfumery. Nutmeg butter, a fixed oil obtained by pressing the seeds, is used in ointments and perfumery. Approximately 74% (by weight) of nutmeg butter is trimyristin which can be turned into glycerol and myristic acid. Trymyristin can be used as replacement for cocoa butter, can be mixed with other fats like cottonseed oil or palm oil, and has applications as an industrial lubricant.

Nutmeg extract and essential oil have insecticidal and fungicidal properties and have potential to be developed into insecticides and fungicides. Nutmeg oil was one of several essential plant oils tested that caused death and larvae of the stored grain pulse beetle, *Callosobruchus chinensis* when fumigated (Chaubey 2008). These essential oils reduced the oviposition potential, egg hatching rate, pupal formation and emergence of adults of F(1) progeny of the insect when fumigated with sublethal concentrations. These essential oils also caused chronic toxicity as the fumigated insects caused less damage to the stored grains. In another study, *M. fragrans* extract was 1 of 11 plant extracts that were found effective against larvae of *Anopheles stephensi* as evidenced by low lethal concentration and lethal time (Senthilkumar et al. 2009). It was also effective against the adults. These plant extracts were easy to prepare, inexpensive, and safe for mosquito control which might be used directly as larvicidal and mosquitocidal agents in small volume aquatic habitats or breeding sites of around human dwellings.

Myristica fragrans seed compounds were also found to have potential insecticides or as leads for the control of cockroaches (Jung et al. 2007). Thirteen constituents of hexane-soluble fraction from a methanolic extract of the seeds from *Myristica fragrans* exhibited insectical activity against adult females of *Blattella germanica* (L.)

(Dictyoptera: Blattellidae). (1R)-(+) -camphor, (1 S)-(−)-camphor, dipentene, (1R)-(+)-3-pinene, and (+)-α-terpineol (0.10–0.14 mg/cm^2) were more toxic than propoxur (0.19 mg/cm^2). (E)-Sabinene hydrate and propoxur were almost equitoxic. Potent insecticidal activity also was observed with (R)-(+) -citronellal, (S)-(−) -citronellal, (R)-(−) -α-phellandrene, (1 S)-(−) -α-pinene, (1R)-(+) -α-pinene, and safrole (0.27–0.48 mg/cm^2).

A methanol extract of *Myristica fragrans* (nutmeg) seeds was found to reduce the development of various plant diseases (Cho et al. 2007). Three antifungal lignans were isolated from the methanol extract and identified as erythro-austrobailignan-6 (EA6), meso-dihydroguaiaretic acid (MDA) and nectandrin-B (NB). In-vitro antimicrobial activity of the three lignans varied according to compound and target species. *Alternaria alternata, Colletotrichum coccodes, C. gloeosporioides, Magnaporthe grisea, Agrobacterium tumefaciens, Acidovorax konjaci* and *Burkholderia glumae* were relatively sensitive to the three lignans. In vivo, all three compounds effectively suppressed the development of rice blast and wheat leaf rust. In addition, EA6 and NB were highly active against the development of barley powdery mildew and tomato late blight, respectively. Both MDA and NB also moderately inhibited the development of rice sheath blight. The crude essential oil of nutmeg at a concentration of 0.1% inhibited radial growth of *Colletotrichum gloeosporoides* (98%), *Colletotrichum musa* (97%), *Fusarium oxysporum* (75%), *Fusarium semitectum* (78%), *Aspergillus niger* (71%) and *Aspergillus glaucus* (60%) (Valente et al. 2011). Growth inhibition increased from 85 to 100% at a concentration of 0.3%.

Comments

The main producing countries today are Indonesia (East Indian Nutmeg) and Grenada (West Indian Nutmeg); while Indonesian nutmegs are mainly exported to Europe and Asia, Grenada nutmeg mostly finds its way into the USA.

Selected References

Anggakusuma, Yanti, Hwang JK (2010) Effects of macelignan isolated from *Myristica fragrans* Houtt. on UVB-induced matrix metalloproteinase-9 and cyclooxygenase-2 in HaCaT cells. J Dermatol Sci 57(2):114–122

Arulmozhi DK, Kurian R, Veeranjaneyulu A, Bodhankar SL (2007) Antidiabetic and antihyperlipidemic effects of *Myristica fragrans* in animal models. Pharm Biol 45(1):64–68

Backer CA, Bakhuizen van den Brink RC Jr (1963) Flora of Java, (spermatophytes only), vol 1. Noordhoff, Groningen, 648 pp

Barrowman JA, Bennett A, Hillebrand P, Rolles K, Pollock DJ, Wright T (1975) Diarrhoea in thyroid medullary carcinoma: role of prostaglandins and therapeutic effect of nutmeg. Br Med J 3(5974):11–12

Bennett A, Stamford IF, Tavares IA, Jacobs S, Capasso F, Mascolo N, Autore G, Romano V, Di Carlo G (1988) The biological activity of eugenol, a major constituent of nutmeg (*Myristica fragrans*): studies on prostaglandins, the intestine and other tissues. Phytother Res 2:124–130

Beyer J, Ehlers D, Maurer HH (2006) Abuse of nutmeg (*Myristica fragrans* Houtt.): studies on the metabolism and the toxicologic detection of its ingredients elemicin, myristicin, and safrole in rat and human urine using gas chromatography/mass spectrometry. Ther Drug Monit 28(4):568–575

Bhamarapravati S, Pendland SL, Mahady GB (2003) Extracts of spice and food plants from Thai traditional medicine inhibit the growth of the human carcinogen *Helicobacter pylori*. In Vivo 17(6):541–544

Burkill IH (1966) A dictionary of the economic products of the Malay Peninsula. Revised reprint, 2 volumes. Ministry of Agriculture and Co-operatives, Kuala Lumpur, vol 1 (A-H), p 1–1240, vol 2 (I-Z), pp 1241–2444

Chatterjee S, Niaz Z, Gautam S, Adhikari S, Variyar PS, Sharma A (2007) Antioxidant activity of some phenolic constituents from green pepper (*Piper nigrum* L.) and fresh nutmeg mace (*Myristica fragrans*). Food Chem 101(2):515–523

Chaubey MK (2008) Fumigant toxicity of essential oils from some common spices against pulse beetle, *Callosobruchus chinensis* (Coleoptera: Bruchidae). J Oleo Sci 57(3):171–179

Checker R, Chatterjee S, Sharma D, Gupta S, Variyar P, Sharma A, Poduval TB (2008) Immunomodulatory and radioprotective effects of lignans derived from fresh nutmeg mace (*Myristica fragrans*) in mammalian splenocytes. Int Immunopharmacol 8(5):661–669

Chhabra SK, Rao AR (1994) Transmammary modulation of xenobiotic metabolizing enzymes in liver of mouse pups by mace (*Myristica fragrans* Houtt.). J Ethnopharmacol 42(3):169–177

Chirathaworn C, Kongcharoensuntorn W, Dechdoungchan T, Lowanitchapat A, Sa-nguanmoo P, Poovorawan Y

(2007) *Myristica fragrans* Houtt. methanolic extract induces apoptosis in a human leukemia cell line through SIRT1 mRNA downregulation. J Med Assoc Thai 90(11):2422–2428

Cho JY, Choi GJ, Son SW, Jang KS, Lim HK, Lee SO, Sung ND, Cho KY, Kim JC (2007) Isolation and antifungal activity of lignans from *Myristica fragrans* against various plant pathogenic fungi. Pest Manag Sci 63(9):935–940

Cho Y, Kim KH, Shim JS, Hwang JK (2008) Inhibitory effects of macelignan isolated from *Myristica fragrans* Houtt. on melanin biosynthesis. Biol Pharm Bull 31(5):986–989

Choo LC, Wong SM, Liew KY (1999) Essential oil of nutmeg pericarp. J Sci Food Agric 79(13):1954–1957

Chung JY, Choo JH, Lee MH, Hwang JK (2006) Anticariogenic activity of macelignan isolated from *Myristica fragrans* (nutmeg) against *Streptococcus mutans*. Phytomedicine 13(4):261–266

Cui CA, Jin DQ, Hwang YK, Lee IS, Hwang JK, Ha I, Han JS (2008) Macelignan attenuates LPS-induced inflammation and reduces LPS-induced spatial learning impairments in rats. Neurosci Lett 448(1):110–114

Daniel D (1994) Nutmeg and derivatives. Food and Agriculture Organization of the United Nations, Rome, Working Paper MISC/94/7

Davis DV, Cooks RG (1982) Direct characterization of nutmeg constituents by mass spectrometry-mass spectrometry. J Agric Food Chem 30(3):495–504

Dhingra D, Sharma A (2006) Antidepressant-like activity of n-hexane extract of nutmeg (*Myristica fragrans*) seeds in mice. J Med Food 9(1):84–89

Dhingra D, Parle M, Kulkarni SK (2006) Comparative brain cholinesterase-inhibiting activity of *Glycyrrhiza glabra, Myristica fragrans*, ascorbic acid, and metrifonate in mice. J Med Food 9(2):281–283

Duan L, Tao HW, Hao XJ, Gu QQ, Zhu WM (2009) Cytotoxic and antioxidative phenolic compounds from the traditional Chinese medicinal plant, *Myristica fragrans*. Planta Med 75(11):1241–1245

Duke JA, Bogenschutz-Godwin MJ, DuCellier J, Duke PA (2002) CRC handbook of medicinal plants, 2nd edn. CRC, Boca Raton, 936 pp

El-Alfy AT, Wilson L, ElSohly MA, Abourashed EA (2009) Towards a better understanding of the psychopharmacology of nutmeg: activities in the mouse tetrad assay. J Ethnopharmacol 126(2):280–286

Facciola S (1990) Cornucopia: a source book of edible plants. Kampong Publications, Vista, 677 pp

Firouzi R, Shekarforoush SS, Nazer AH, Borumand Z, Jooyandeh AR (2007) Effects of essential oils of oregano and nutmeg on growth and survival of *Yersinia enterocolitica* and *Listeria monocytogenes* in barbecued chicken. J Food Prot 70(11):2626–2630

Flach M, Tjeenk Willink M (1999) *Myristica fragrans* Houtt. In: de Guzman CC, Siemonsma JS (eds) Plant resources of South-East Asia no. 13: Spices. Prosea Foundation, Bogor, pp 143–148

Foundation for Revitalisation of Local Health Traditions (2008) FRLHT Database. htttp://envis.frlht.org

Fras I, Friedman JJ (1969) Hallucinogenic effects of nutmeg in adolescent. NY State J Med 69(3):463–465

Gonçalves JL, Lopes RC, Oliveira DB, Costa SS, Miranda MM, Romanos MT, Santos NS, Wigg MD (2005) In vitro anti-rotavirus activity of some medicinal plants used in Brazil against diarrhea. J Ethnopharmacol 99(3):403–407

Grover JK, Khandkar S, Vats V, Dhunnoon Y, Das D (2002) Pharmacological studies on *Myristica fragrans* – antidiarrheal, hypnotic, analgesic and hemodynamic (blood pressure) parameters. Methods Find Exp Clin Pharmacol 24(10):675–680

Gupta S, Yadava JNS, Mehrotra R, Tandon JS (1992) Anti-diarrhoea l profile of an extract and some fractions from *Myristica fragrans* (Nut-meg) on *Escherichia coli* enterotoxin-induced secretory response. Pharm Biol 30(3):179–183

Hada S, Hattori M, Tezuka Y, Kikuchi T, Namba T (1988) New neolignans and lignans from the aril of *Myristica fragrans*. Phytochem 27(2):563–8

Hallström H, Thuvander A (1997) Toxicological evaluation of myristicin. Nat Toxins 5(5):186–192

Han KL, Choi JS, Lee JY, Song J, Joe MK, Jung MH, Hwang JK (2008) Therapeutic potential of peroxisome proliferators–activated receptor-alpha/gamma dual agonist with alleviation of endoplasmic reticulum stress for the treatment of diabetes. Diabetes 7(3):737–745

Hang X, Yang XW (2007) GC-MS analysis of essential oil from nutmeg processed by different traditional methods. Zhongguo Zhong Yao Za Zhi 32(16): 1669–1675

Hattori M, Yang XW, Shu YZ, Kakiuchi N, Tezuka Y, Kikuchi T, Namba T (1988) New constituents of the aril of *Myristica fragrans*. Chem Pharm Bull 36(2): 648–653

Hussain SP, Rao AR (1991) Chemopreventive action of mace (*Myristica fragrans*, Houtt) on methylcholanthrene-induced carcinogenesis in the uterine cervix in mice. Cancer Lett 56(3):231–234

Im YB, Ha I, Kang KW, Lee MY, Han HK (2009) Macelignan: a new modulator of p-glycoprotein in multidrug-resistant cancer cells. Nutr Cancer 61(4): 538–543

Jaiswal P, Kumar P, Singh VK, Singh DK (2009) Biological effects of *Myristica fragrans*. Ann Rev Biomed Sci 11:21–29

Jaiswal P, Kumar P, Singh VK, Singh DK (2010) Enzyme inhibition by molluscicidal components of *Myristica fragrans* Houtt. in the nervous tissue of snail *Lymnaea acuminata*. Enzyme Res, article 478746

Jan M, Faqir F, Qureshi H, Malik SA, Mughal MA (2004) Evaluation of effects of extract from seeds of *Myristica fragrans* on volume and acidity of stimulated gastric secretion, liver and kidney function. J Postgrad Med Inst 18:644–650

Jan M, Faqir F, Hamida, Mughal MA (2005) Comparison of effects of extract of *Myristica fragrans* and verapamil on the volume and acidity of carbachol induced gastric secretion in fasting rabbits. J Ayub Med Coll Abbottabad 17(2):69–71

Jannu LN, Hussain SP, Rao AR (1991) Chemopreventive action of mace (*Myristica fragrans*, Houtt) on DMBA-induced papillomagenesis in the skin of mice. Cancer Lett 56(1):59–63

Janssen J, Laeckman GM (1990) Nutmeg oil: identification and quantification of its most active of platelet constituents as inhibitors aggregation. J Ethnopharmacol 29:179–188

Javaregowda PK, Hong YK, Kim SM, Hong YG (2010) In vitro and in vivo analysis of macelignan induced toxicity on heart. FASEB J 24:lb557

Jin DQ, Lim CS, Hwang JK, Ha I, Han JS (2005) Anti-oxidant and anti-inflammatory activities of macelignan in murine hippocampal cell line and primary culture of rat microglial cells. Biochem Biophys Res Commun 331(4):1264–1269

Joseph J (1980) The nutmeg – its botany, agronomy, production, composition, and uses. J Plant Crops 8(2):61–72

Juhász L, Kürti L, Antus S (2000) Simple synthesis of benzofuranoid neolignans from *Myristica fragrans*. J Nat Prod 63(6):866–870

Jukić M, Politeo O, Miloš M (2006) Chemical composition and antioxidant effect of free volatile aglycones from nutmeg (*Myristica fragrans* Houtt.) compared to its essential oil. Croatica Chem Acta 79(2):209–214

Jung WC, Jang YS, Hieu TT, Lee CK, Ahn YJ (2007) Toxicity of *Myristica fagrans* seed compounds against *Blattella germanica* (Dictyoptera: Blattellidae). J Med Entomol 44(3):524–529

Kasahara H, Miyazawa M, Kameoka H (1995) Absolute configuration of 8-O-4′-neolignans from *Myristica fragrans*. Phytochemistry 40(5):1515–1517

Kim YB, Park IY, Shin KH (1991) The crystal structure of licarin-B, (C20H20O4), a component of the seeds of *Myristica fragrans*. Arch Pharm Res 14(1):1–6

Kim HJ, Chen F, Wang X, Wang Y, McGregor J, Jiang YM (2010) Characterization of antioxidants in nutmeg (*Myristica fragrans* Houttuyn) oil. In: Qian MC, Rimando AM (eds) Flavour and health benefits of small fruits, vol 1035., pp 239–252, Chapter 15

Kimura Y, Ito H, Hatano T (2010) Effects of mace and nutmeg on human cytochrome P450 3A4 and 2C9 activity. Biol Pharm Bull 33(12):1977–1982

Kumari MV, Rao AR (1989) Effects of mace (*Myristica fragrans*, Houtt.) on cytosolic glutathione S-transferase activity and acid soluble sulfhydryl level in mouse liver. Cancer Lett 46(2):87–91

Kwon HS, Kim MJ, Jeong HJ, Yang MS, Park KH, Jeong TS, Lee WS (2008) Low-density lipoprotein (LDL)-antioxidant lignans from *Myristica fragrans* seeds. Bioorg Med Chem Lett 18:194–198

Lee KT, Kim BJ, Kim JH, Heo MY, Kim HP (1997) Biological screening of 100 plant extracts for cosmetic use (I): inhibitory activities of tyrosinase and DOPA auto-oxidation. Int J Cosmet Sci 19(6):291–298

Lee KK, Kim JH, Cho JJ, Choi JD (1999) Effects of 150 plant extracts on elastase activity, and their anti-inflammatory effects. Int J Cosmet Sci 21(2):71–82

Lee BK, Kim JH, Jung JW, Choi JW, Han ES, Lee SH, Ko KH, Ryu JH (2005) Myristicin-induced neurotoxicity in human neuroblastoma SK-N-SH cells. Toxicol Lett 157:49–56

Lee SU, Shim KS, Ryu SY, Min YK, Kim SH (2009) Machilin A isolated from *Myristica fragrans* stimulates osteoblast differentiation. Planta Med 75(2):152–157

Li F, Yang XW (2007) Three new neolignans from the aril of *Myristica fragrans*. Helv Chim Acta 90:1491–1496

Li F, Yang XW (2008a) Biotransformation of myrislignan by rat liver microsomes in vitro. Phytochemistry 69(3):765–771

Li F, Yang XW (2008b) Simultaneous determination of diastereomers (+)-licarin A and isolicarin A from *Myristica fragrans* in rat plasma by HPLC and its application to their pharmacokinetics. Planta Med 74(8):880–884

Ma J, Hwang YK, Cho WH, Han SH, Hwang JK, Han JS (2009) Macelignan attenuates activations of mitogen-activated protein kinases and nuclear factor kappa B induced by lipopolysaccharide in microglial cells. Biol Pharm Bull 32(6):1085–1090

Maeda A, Shin'ichi T, Tomo A, Hisayuki T, Masato N (2006) Physiological activities of *Myristica fragrans* Houtt. Koryo, Terupen oyobi Seiyu Kagaku ni kansuru Toronkai Koen Yoshishu 50:337–339, In Japanese

Maeda A, Tanimoto S, Abe T, Kazama S, Tanizawa H, Nomura M (2008) Chemical constituents of *Myristica fragrans* Houttuyn seed and their physiological activities. Yakugaku Zasshi 128(1):129–133, In Japanese

Mahady GB, Pendland SL, Stoia A, Hamill FA, Fabricant D, Dietz BM, Chadwick LR (2005) In vitro susceptibility of *Helicobacter pylori* to botanical extracts used traditionally for the treatment of gastrointestinal disorders. Phytother Res 19(11):988–991

Malti JE, Bourhim N, Amarouch H (2008) Ttoxicity and antibacterial effect of mace of *Myristica fragrans* used in Moroccan gastronomy: biochemical and histological impact. J Food Safety 28(3):422–441

Morita T, Jinno K, Kawagishi H, Arimoto Y, Suganuma H, Inakuma T, Sugiyama K (2003) Hepatoprotective effect of myristicin from nutmeg (*Myristica fragrans*) on lipopolysaccharide/D-galactosamine-induced liver injury. J Agric Food Chem 51(6):1560–1565

Muchtaridi, Subarnas A, Apriyantono A, Mustarichie R (2010). Identification of compounds in the essential oil of nutmeg seeds (*Myristica fragrans* Houtt.) that inhibit locomotor activity in mice. Int J Mol Sci 11:4771–4781

Narasimhan B, Dhake AS (2006) Antibacterial principles from *Myristica fragrans* seeds. J Med Food 9(3): 395–399

Nguyen PH, Le TV, Kang HW, Chae J, Kim SK, Kwon KI, Seo DB, Lee SJ, Oh WK (2010) AMP-activated protein kinase (AMPK) activators from *Myristica fragrans* (nutmeg) and their anti-obesity effect. Bioorg Med Chem Lett 20(14):4128–4131

Ochse JJ, Bakhuizen van den Brink RC (1980) Vegetables of the Dutch Indies, 3rd edn. Ascher, Amsterdam, 1016 pp

Olajide OA, Ajayi FF, Ekhelar AI, Awe SO, Makinde JM, Alada ARA (1999) Biological effects of *Myristica fragrans* (nutmeg) extract. Phytother Res 13(4): 344–345

Olaleye MT, Akinmoladun AC, Akindahunsi AA (2006) Antioxidant properties of *Myristica fragrans* (Houtt) and its effect on selected organs of albino rats. African J Biotechnol 5(13):1274–1278

Orabi KY, Mossa JS, el-Feraly FS (1991) Isolation and characterization of two antimicrobial agents from mace (*Myristica fragrans*). J Nat Prod 54(3):856–859

Ozaki Y, Soedigdo S, Wattimena YR, Suganda AG (1989) Antiinflammatory effect of mace, aril of *Myristica fragrans* Houtt. and its active principles. Japan J Pharmco 49(2):155–163

Park S, Lee DK, Yang CH (1998) Inhibition of fos-jun-DNA complex formation by dihydroguaiaretic acid and in vitro cytotoxic effects on cancer cells. Cancer Lett 127(1–2):23–28

Parle M, Dhingra D, Kulkarni SK (2004) Improvement of mouse memory by *Myristica fragrans* seeds. J Med Food 7(2):156–161

Porcher MH et al (1995–2020) Searchable World Wide Web multilingual multiscript plant name database. The University of Melbourne, Australia. http://www.plantnames.unimelb.edu.au/Sorting/Frontpage.html

Purseglove JW (1968) Tropical crops: dicotyledons 1 & 2. Longman, London, 719 pp

Purseglove JW, Brown EG, Green CL, Robbins SRJ (1981) Spices (Tropical agriculture series), vol 2. Longman, London, 813 pp

Qiu Q, Zhang G, Sun X, Liu X (2004) Study on chemical constituents of the essential oil from *Myristica fragrans* Houtt. by supercritical fluid extraction and steam distillation. Zhong Yao Cai 27(11):823–826, In Chinese

Ram A, Lauria P, Gupta R, Sharma VN (1996) Hypolipidaemic effect of *Myristica fragrans* fruit extract in rabbits. J Ethnopharmacol 55(1):49–53

Senthilkumar N, Varma P, Gurusubramanian G (2009) Larvicidal and adulticidal activities of some medicinal plants against the malarial vector, *Anopheles stephensi* (Liston). Parasitol Res 104(2):237–244

Sharma A, Mathur R, Dixit VP (1995) Prevention of hypercholesterolemia and atherosclerosis in rabbits after supplementation of *Myristica fragrans* seed extract. Indian J Physiol Pharmacol 39(4):407–410

Sharma M, Kumar M (2007) Radioprotection of Swiss albino mice by *Myristica fragrans* Houtt. J Radiat Res (Tokyo) 48(2):135–41

Sherry CJ, Ray LE, Herron RE (1982) The pharmacological effects of the ligroin extract of nutmeg (*Myristica fragrans*). J Ethnopharmacol 6(1):661–666

Shukla J, Tripathi SP, Chaubey MK (2008) Toxicity of *Myristica fragrans* and *Ilicium verum* essential oils against flour beetle *Tribolium castaneum* Herbst (Coleoptera: Tenebrionidae) *Electr*. J Environ Agric Food Chem 7(7):3059–3064

Sinclair J (1958) A revision of the Malayan Myristicaceae. Gard Bul Singapore 16:205–472

Sinclair J (1968) The genus *Myristica* in Malesia and outside Malesia. Gard Bul Singapore 23:1–540

Singh G, Marimuthu P, de Heluani CS, Catalan C (2005) Antimicrobial and antioxidant potentials of essential oil and acetone extract of *Myristica fragrans* Houtt. (aril part). J Food Sci 70(2):M141–M148

Smith-Palmer A, Stewartt J, Fyfe L (2002) Inhibition of listeriolysin O and phosphatidylcholine-specific production in *Listeria monocytogenes* by subinhibitory concentrations of plant essential oils. J Med Microbiol 51(7):567–574

Sohn JH, Han KL, Choo JH, Hwang JK (2007) Macelignan protects HepG2 cells against tert-butylhydroperoxide-induced oxidative damage. Biofactors 29(1):1–10

Sohn JH, Han KL, Kim JH, Rukayadi Y, Hwang JK (2008) Protective effects of macelignan on cisplatin-induced hepatotoxicity is associated with JNK activation. Biol Pharm Bull 31(2):273–277

Sonavane GS, Palekar RC, Kasture VS, Kasture SB (2002a) Anticonvulsant and behavioural actions of *Myristica fragrans* seeds. Indian J Pharm 34(5):332–338

Sonavane GS, Sarveiya VP, Kasture VS, Kasture SB (2002b) Anxiogenic activity of *Myristica fragrans* seeds. Pharmacol Biochem Behav 71(1):239–244

Spricigo CB, Pinto LT, Bolzan A, Novais AF (1999) Extraction of essential oil and lipids from nutmeg by liquid carbon dioxide. J Supercritical Fluids 15:253–259

Tajuddin AS, Latif A, Qasmi IA (2003) Aphrodisiac activity of 50% ethanolic extracts of *Myristica fragrans* Houtt. (nutmeg) and *Syzygium aromaticum* (L.) Merr. & Perry. (clove) in male mice: a comparative study. BMC Complement Altern Med 3:6

Tajuddin AS, Latif A, Qasmi IA, Amin KM (2005) An experimental study of sexual function improving effect of *Myristica fragrans* Houtt. (nutmeg). BMC Complement Altern Med 5:16

Takikawa A, Abe K, Yamamoto M, Ishimaru S, Yasui M, Okubo Y, Yokoigawa K (2002) Antimicrobial activity of nutmeg against *Escherichia coli* O157. J Biosci Bioeng 94:315–320

Tomaino A, Cimino F, Zimbalatti V, Venuti V, Sulfaro V, De Pasquale A, Saija A (2005) Influence of heating on antioxidant activity and the chemical composition of some spice essential oils. Food Chem 89(4):549–554

Truitt EB Jr, Duritz G, Ebersberger EM (1963) Evidence of monoamine oxidase inhibition by myristicin and nutmeg. Proc Soc Exp Biol Med 112:647–650

U.S. Department of Agriculture, Agricultural Research Service (2010) USDA National Nutrient Database for Standard Reference, Release 23. Nutrient Data Laboratory Home Page. http://www.ars.usda.gov/ba/bhnrc/ndl

Valente VMM, Jham GN, Dhingra OD, Ghiviriga I (2011) Composition and antifungal activity of the Brazilian *Myristica fragrans* Houtt essential oil. J Food Safety 31(2):197–202

Van Gils C, Cox PA (1994) Ethnobotany of nutmeg in the Spice Islands. J Ethnopharmacol 42:117–124

Variyar PS, Chatterjee S, Sajilata MG, Singhal RS, Sharma A (2008) Natural existence of 2-alkylcyclobutanones. J Agric Food Chem 56(24):11817–11823

Voelker TA, Jones A, Cranmer AM, Davies HM, Knutzon DS (1997) Broad-range and binary-range acyl-acyl-carrier protein thioesterases suggest an alternative mechanism for medium-chain production in seeds. Plant Physiol 114(2):669–677

Wahab A, Ul Haq R, Ahmed A, Khan RA, Raza M (2009) Anticonvulsant activities of nutmeg oil of *Myristica fragrans*. Phytother Res 23(2):153–158

Wang Y, Yang XW, Tao HY, Liu HX (2004) GC-MS analysis of essential oils from seeds of *Myristica fragrans* in Chinese market. Zhongguo Zhong Yao Za Zhi 29:339–342, In Chinese

Weiss G (1960) Hallucinogenic and narcotic-like effects of powdered *Myristica* (nutmeg). Psychiatr Q 34(1):346–356

Weiss EA (1997) Myristicaceae. *Essential oil crops*. CAB International, Wallingford, pp 214–234, Chapter 7

Yadav AS, Bhatnagar D (2007a) Free radical scavenging activity, metal chelation and antioxidant power of some of the Indian spices. Biofactors 31(3–4):219–227

Yadav AS, Bhatnagar D (2007b) Modulatory effect of spice extracts on iron-induced lipid peroxidation in rat liver. Biofactors 29(2–3):147–157

Yang S, Na MK, Jang JP, Kim KA, Kim BY, Sung NJ, Oh WK, Ahn JS (2006) Inhibition of protein tyrosine phosphatase 1B by lignans from *Myristica fragrans*. Phytother Res 20(8):680–682

Yang XW, Huang X, Ahmat M (2008) New neolignan from seed of *Myristica fragrans*. Zhongguo Zhong Yao Za Zhi 33(4):397–402, In Chinese

Yang XW, Huang X, Ma L, Wu Q, Xu W (2010) The intestinal permeability of neolignans from the seeds of *Myristica fragrans* in the Caco-2 cell monolayer model. Planta Med 76(14):1587–1591

Yanti, Rukayadi Y, Kim KH, Hwang JK (2008) In vitro anti-biofilm activity of macelignan isolated from *Myristica fragrans* Houtt. against oral primary colonizer bacteria. Phytother Res 22(3):308–312

Yuan ZM, Wang J, Lv J, Jia TZ (2006) Comparing analysis of components in volatile oils of nutmeg and prepared nutmeg by GC-MS. Zhongguo Zhong Yao Za Zhi 31(9):737–739, In Chinese

Acca sellowiana

Scientific Name

Acca sellowiana (O. Berg) Burret.

Synonyms

Acca sellowiana var. *rugosa* (Mattos) Mattos, *Feijoa obovata* (O. Berg) O. Berg, *Feijoa schenkiana* Kiareskou, *Feijoa sellowiana* (O. Berg) O. Berg, *Feijoa sellowiana* f. *elongata* Voronova, *Feijoa sellowiana* var. *rugosa* Mattos, *Feijoa sellowiana* var. *sellowiana*, *Orthostemon obovatus* O. Berg, *Orthostemon sellowianus* O. Berg basionym.

Family

Myrtaceae

Common/English Names

Brazilian Guava, Feijoa, Fig Guava, Guava Brazilera, Guavasteen, Pineapple Guava

Vernacular Names

Argentina: Falsa Guayaba;
Brazil: Goiaba Serrana, Goiaba Verde, Goiaba Abacaxí, Araçá Do Rio Grande, Goiaba-Feijoa, Goiaba Do Campo, Goiaba Silvestre, Goiabeira-Serrana (Portuguese);
Chinese: Fei Hou;
Czech: Fejchoa Sellowova;
Danish: Feijoa, Ananasguava;
Eastonian: Brasiilia Feihoapuu, Vili: Feihoa;
French: Faux Goyavier, Feijoa, Goyave Ananas, Goyave Du Brésil, Goyavier Ananas, Goyavier Du Brésil, Goyavier De Montevideo;
German: Feijoya;
Hawaii: Guavasteen;
Italian: Feijoa;
Japanese: Akka Serowiana, Feijoa Serowiana;
Mexico: Feijoa, Guayaba Chilena;
Portuguese: Araçá Do Rio Grande, Feijoo, Goiaba Do Campo, Goiaba-Feijoa, Goiaba Serrana, Goiaba Silvestre;
Slovašcina: Guava;
Spanish: Feijoo, Guayaba, Guayaba Chilena, Guayaba Sumina, Guayabo (Tree), Guayabo Del Brasil;
Swedish: Feijoa;
Uruguay: Guayaba Del País, Guayabo Grande, Guayabo Chico.

Origin/Distribution

Feijoa originates from the highlands of southern Brazil, parts of Colombia, Uruguay, southern Paraguay, and northern Argentina. It is cultivated in the highlands of Chile and other southern American countries, around the Mediterranean

T.K. Lim, *Edible Medicinal And Non-Medicinal Plants: Volume 3, Fruits*,
DOI 10.1007/978-94-007-2534-8_81, © Springer Science+Business Media B.V. 2012

area (southern France, Portugal, southern Italy), Israel, Algeria, Libya, the Caucasian region around the Black Sea, in India, Australia, South Africa, New Zealand, the Caribbean area and the USA (California, Florida). In Australia, feijoa planting has been officially discouraged, because the fruit is the main host of a fruit fly.

Agroecology

The feijoa likes a warm temperate or subtropical climate with low humidity and a cool season with annual mean rainfall of 760–1,500 mm. It can withstand winter temperatures down to −10°C but the flower buds will be killed and sudden autumn frosts can damage the fruit. It is drought tolerant but requires adequate water for good fruit production. It tolerates partial shade (though fruiting will be reduced) and some exposure to salt spray. It does best on well-drained, fertile, organic or acidic soils.

Edible Plant Parts and Uses

The ripe fruits are consumed fresh. They are aromatic and the thick, translucent flesh is juicy, delicious, sweet to subacid with a characteristic flavour and aroma, which are similar to pineapple and guava or pineapple and strawberry. The flesh is also eaten in salads, or is cooked in puddings, stews, pastries, cakes, pies or tarts and is a popular ingredient in chutney. The fruit is also made into jams, paste, jellies, sauces, crystallized fruits, preserves in syrup and liqueur. The flesh can be used in smoothies, yoghurt, ice-cream, soft drinks, and is also made into cider, wine and a feijoa infused vodka. The bitter fruit peel can also be eaten. The crisp, moist, fleshy petals of its beautiful flowers are also edible and are appreciated raw or as an addition to salads.

Botany

Feijoa is an evergreen much branched shrub or small, much branched tree, 1–5 m high (Plate 1) with a cylindrical trunk and reddish-grey bark that exfoliates and appeared rough and flaky

Plate 1 Young and mature foliage of feijoa shrub

Plate 2 Developing feijoa fruits and leaves

(Plate 3). Leaves are opposite, shortly petiolate, elliptical to ovate, 3–8 cm long, 2–4 cm wide, margin entire, apex obtuse, usually rounded at the base, coriaceous, upper surface glossy dark green, venation prominent, under surface silvery white, tomentose (Plates 1, 2 and 3). Flowers are attractive, pinkish red (Plate 2), solitary in the lowest two or four leaf axils of the current year's

Plate 3 Feijoa fruit and rough, flaky branches

Plate 4 Ripe feijoa fruits

growth, 3–4 cm across. Sepals 4 persistent; petals 4 fleshy, broad elliptic-spoon shaped, reflexed, whitish with red centre; stamens numerous, to 2.5 cm long, carmine with yellow anthers. Fruit is oblong to ovoid, 5–8 cm long by 3–6 cm diameter berry, with remnants of the calyx on the tip, green or yellow-green, fragrant (Plates 3, 4 and 5). Skin texture varies from smooth to rough and pebbly, enclosing a thick, white, granular, juicy, sweet to subacid flesh and the translucent central pulp enclosing the seeds (Plate 5). Seeds numerous, 20–40, very small and oblong.

Nutritive/Medicinal Properties

Nutrient composition of fresh, raw feijoa fruit per 100 g edible portion excluding refuse 43% of the peel is as follows: water 84.94 g, energy 55 kcal (230 kJ), protein 0.98 g, total lipid (fat) 0.0.60 g, ash 0.56 g, carbohydrate 12.92 g, total dietary fibre 6.4 g, total sugars 8.20 g, sucrose 2.93 g, glucose 2.32 g, fructose 2.95 g, Ca 17 mg, Fe 0.14 mg, Mg 9 mg, P 19 mg, K 172 g, Na 3 mg, Zn 0.06 mg, Cu 0.036 mg, Mn 0.084 mg; vitamin C (ascorbic acid) 32.9 mg, thiamin 0.006 mg, riboflavin 0.018 mg, niacin 0.295 mg, vitamin B-6 0.067 mg, pantothenic acid 0.233 mg, folate (total) 23 μg, β-carotene 2 μg, β-cryptoxanthin 3 μg, vitamin A 6 IU, lycopene 5 μg, luetin + zeaxanthine 27 μg, vitamin E (α-tocopherol) 0.16 mg, vitamin K (phylloquinone) 3.5 μg; total saturated fatty acids 0.148 g, 4:0 (butyric) 0.006 g, 10:0 (capric) 0.004 g, 12:0 (lauric) 0.002 g, 14:0 (myristic) 0.004 g, 16:0 (palmitic) 0.107 g, 18:0 (stearic) 0.020 g, 20:0 (arachidic) 0.004 g; total monounsaturated fatty acids 0.081 g, 18:1 c (oleic) 0.081 g; total polyunsaturated fatty acids 0.194 g, 18:2 undifferentiated 0.153 g, 18: n-6, c,c (linoleic) 0.153 g, 18:3 n-3 c,c,c (α-linoleic, ALA) 0.041 g; tryptophan 0.010 g, threonine 0.026 g, isoleucine 0.026 g, leucine 0.039 g, lysine 0.052 g, methionine 0.010 g, Phenylalanine 0.026 g, tyrosine 0.013 g, valine 0.026 g, arginine 0.026 g, histidine 0.013 g, alanine 0.052 g, aspartic acid 0.0104 g, glutamic acid 0.182 g, glycine 0.039 g, proline 0.026 g and serine 0.039 g (U.S. Department of Agriculture, Agricultural Research Service 2010).

Feijoa fruit contained less moisture and ascorbic acid and more protein (amino acids), lipid and ash than other tropical fruits. The ratio of reducing sugars to total sugars was about 50%. The main minerals in feijoa were potassium and

Plate 5 Sliced feijoa fruit

phosphorus. Feijoa contained more organic acids than other generally cultivated fruits (Isobe et al. 2002), the main organic acids being citric acid, tartaric acid, propionic acid and succinic acid; provitamin A, such as α-carotene, β-carotene and β-cryptoxanthin; watersoluble and water-insoluble fibers and flavonoids and no cholesterol. The fruit also contained a small amount of iodine 3 mg/100 g of fresh fruit not dangerous for human health (Ferrara and Montesan 2001). Fruits were also high in pectin (up to 20%).

Methyl benzoate and ethyl benzoate together comprised over 90% of the total volatile oil of feijoa fruit (Hardy and Michael 1970). Other compounds identified were 3-octanone, ethyl butanoate, ethyl cinnamate, ethyl acetate, hexenyl acetate, 2-hexenal, 2-heptanone, 2-nonanone, 2-undecanone, methyl p-anisate, and ethyl p-anisate. Hexenyl propionate and hexenyl benzoate were tentatively identified. A predominant role in the aroma of feijoa fruits was ascribed to methyl and ethyl benzoate. In another study, major volatile constituents of *Feijoa* were reported as germacene D, bicyclogermacene, methyl benzoate, B-caryophyllene, (Z)-3-hexenyl benzoate, linalool, humulene and 3-octane (Binder and Flath 1989). Eighty –five compounds were identified, 47 of which were not previously reported. Another earlier analysis reported 11 components which included myrcene, ethyl hexanoate, trans β-ocimene, 2-heptyl butanoate, cis hex-3-enyl butanoate and cis hex-3-enyl hexanoate (Shaw et al. 1983). Ethyl butanoate, methyl benzoate and ethyl benzoate were important in the aroma of intact feijoa fruit. Subsequent report listed 15 more constituents of which (Z)-hex-3-enal and isopropyl benzoate were reported for the first time (Shaw et al. 1990). Methyl benzoate constituted 82% of the volatile flavour extract.

Feijoa was also found to contain high amounts of bioactive polyphenols, such as catechin, leucoanthocyanins, flavonols, proanthocyanidins, naphthoquinones, terpenes quinones, steroidal saponins and tannins (Nakashima 2001; Vuotto et al. 2000; Bontempo et al. 2007). The flower part of feijoa contained polyphenols such as anthocyanin-3-glucoside (Lowry 1976). Leaves of feijoa also contained catechins, such as (+)-catechin, (−)-epicatechin, (+)-gallocatechin and (−)-epigallocatechin (Vanidze et al. 1991).

About 30 groups of lipid substances have been identified in the flesh and peel of feijoa fruit (Kolesnik et al. 1991). The main groups of lipids in the flesh were triacylglycerols, sterols, cerebrosides, ceramide phosphate inositol oligosides, sulfoquinovosyldiacylglycerols, phosphatidic acids, phosphotidylglycerols and phosphatidylcholines. In the peel, hydrocarbons, sterols, esters of fatty acids and lower alcohols, cerebrosides, digalactosyldiacylglycerols, ceramide oligosides, phosphatidylglycerols predominated. The fatty acids of the flesh were found to include 15 representatives (C12:0–C28:0), and those of the peel 11 representatives (C12:0–C18:3).

Of the 34 constituents identified in the volatile flavour constituents in the feijoa fruit peel oil, (Z)-3-Hexen-1-ol, linalool and methyl benzoate was found to comprise 53% of the oil (Shaw et al. 1989). Essential oil from the peel of *Feijoa sellowiana* contained an average of 68 constituents, of which 67 were identified representing more than 96.4% of the oil (Fernandez et al. 2004). Chemical analysis revealed that feijoa peel oil contained 11 esters (9.1%), two inonoterpene hydrocarbons (0.5%), two oxygenated monoterpenes (1%), 29 sesquiterpene hydrocarbons (64.8%), and 14 oxygenated sesquiterpenes (17.8%). The major constituents were β-caryophyllene (12%), ledene (9.6%), α-humulene (6.3%), β-elemene (4.9%) and δ-cadlnene (4.8%).

The lipid extract of feijoa leaves yielded in addition to the widespread secondary metabolites: α-tocopherol, flavone, stigmasterol and β-carotene, an inseparable mixture of tyrosol esters of lignoceric (1a), cerotic (1b) and montanic (1c) acids, and a novel galactolipid identified as (2 S)-1,2,6'-tri-O-(9Z,12Z,15Z)-octadeca-9,12,15-trienoyl-3-O-β-D-galactopyranosyl glycerol (2) (Ruberto and Tringali 2004).

Feijoa sellowiana has antioxidant, anticancer, anti-inflammatory, antimicrobial, antiviral, hepatoprotective, anti-osteoporotic, anti-hyperthyroidism and immunomodulatory properties imparted by an array of bioactive compounds.

Antioxidant Activity

The fruit and leaf of feijoa were found to have antioxidant capacity but the activity of the leaf was greater than the fruit and was attributed to differences in the content of phenolic compounds (Beyhan et al. 2010). The percentage inhibitions of 40 mg/ml concentration of fresh fruit (MEFF), dry fruit (MEDF) and leaf of feijoa on peroxidation in linoleic acid system were 74%, 75% and 87% respectively, and the same concentration (40 mg/ml) of α-tocopherol showed similar activity. MEFF, MEDF and leaf exhibited effective reducing power, free radical scavenging, and metal chelating activities at equivalent low concentrations (40 μg/ml).

Anticancer Activity

The acetonic extract of the fruit exerted anti-cancer activities on haematological cancer cells (Bontempo et al. 2007). Use of ex-vivo myeloid leukemia patient blast cells confirmed that both the full acetonic feijoa extract and its derived flavone were able to induce apoptosis. In both cell lines and myeloid leukemia patients, the apoptotic activity of feijoa extract and flavone was accompanied by increase of histone and non-histone acetylation levels and by HDAC (Histone deacetylases) inhibition. The findings demonstrated that the feijoa apoptotic active principle could be attributed to the flavone and that this activity was associated with the induction of HDAC inhibition, supporting the hypothesis of its epigenetic, pro-apoptotic regulation in cancer systems.

Antimicrobial and Antiviral Activities

Feijoa fruit peel extracts exhibited therapeutic effects as potential anti-tumour and anti-microbial organism agents (Nakashima 2001). Studies conducted on its anti-tumour activity, 50% cell cytotoxicity (CC50), anti-human immunodeficiency virus activity and anti-bacterial activity showed that two active fractions [A3] of acetone extract and [M2] of methanol extract of feijoa fruit peels had potent inhibitory activity against Gram-positive bacterium, *Staphylococcus epidermidis* and Gram-negative bacteria *Escherichia coli* and *Pseudomonas aeruginosa* as well as fungi, *Candida albicans* and *Candida glabrata*. However, all feijoa extract fractions did not exhibit potent anti-*Helicobacter pylori* (primary causative organism for acute gastritis activity) (MIC_{50} >100 μg/ml). Fraction [A3] displayed significant potency and was relatively cytotoxic to two tumour cell lines (HSC-2 and HSG) and the healthy cell line (HGF). Most fractions did not significantly inhibit HIV-induced cyopathic effects on MT-4 cells. However, fraction [70 M2] showed some anti-HIV activity. The fraction [70 M2] with relatively high water-solubility had the highest anti-HIV activity.

In separate studies, aqueous feijoa fruit extract was also reported to inhibit bacterial growth *Pseudomonas aeruginosa, Enterobacter aerogenes* and *Enterobacter cloacae* (Vuotto et al. 2000). The antibacterial activity of extracts from vegetative plant parts (leaves and stems) was generally less active than that from fruit extracts (skin, pulp and seeds) (Basile et al. 1997). The antibiotic activity of fruit resided essentially in the seeds. Findings again confirmed that *F. sellowiana* showed both antibacterial and antioxidant properties and therefore its extract might be used as a new multifaceted drug.

Hepatoprotective Activity

Methanolic extract of feijoa fruit was found to significantly decrease the activity of aminotransferase enzymes SGOT (serum glutamic oxaloacetic transaminase) and SGPT (serum glutamic pyruvic transaminase) and to increase the level of GSH (glutathione) in mouse with hepatotoxicity induced by MDMA, (3, 4-methylenedioxymethamphetamine, or ecstasy) (Karami et al. 2008). MDMA is a ring-substituted amphetamine derivative that has attracted a great deal of media attention in recent years due to its widespread abuse as a recreational drug by the youth generation. Necrosis and other cellular lesions in liver tissue were decreased. The findings showed that *Feijoa sellowiana* had hepatoprotective activity through its antioxidant property.

Antiinflammatory Activity

Fejoa fruit especially the acetonic extract was shown to possess antiinflammatory activity, utilising J774 cell line model, which expressed inducible nitric oxide synthase (iNOS) following stimulation with lipopolysaccharide (Rossi et al. 2007). Studies demonstrated that at least some part of the antiinflammatory activity of the acetonic extract of feijoa fruit was due to the suppression of nitric oxide production by flavones and stearic acid. The mechanism of this inhibition seemed to be related to an action on the expression of the enzyme iNOS through the attenuation of nuclear factor kappaB (NF-kappaB) and/or mitogen-activated protein kinase (MAPK) activation.

Immunomodulatory Activity

The acetonic extract from *Feijoa sellowiana* fruit was found to contain flavonids which displayed several biological activities, including immunomodulating and antioxidant activities (Ielpo et al. 2000). Research demonstrated that both the raw extracts and flavonoids significantly inhibited chemiluminescence emission by human whole blood phagocytes and isolated polymorphonuclear leukocytes especially when these cells were activated by phorbol miristate acetate(PMA). The antioxidant activity of flavonoids could be increased by changing the chemical structure of the native molecule.

Fejoa fruit was found to have intestinal immunomodulatory activity (Manabe and Isobe 2005). The aqueous extract of feijoa fruit and the in vitro-digested feijoa were found to suppress the secretion of TGF-β (transforming growth factor β) by Caco-2 cells (Human colonic adenocarcinoma cells). The findings indicated that the continued intake of feijoa induced a decrease in TGF-β concentrations in intestinal lamina propria, which in turn caused suppressions of oral tolerance and disorders of mucosal homeostasis. The polyphenols from feijoa were postulated to be involved in the suppressing effect of feijoa on TGF-β secretion by intestinal epithelium.

Antiosteoporotic Activity

The aqueous methanol feijoa leaf extract was investigated on its possible prevention and treatment of osteoporosis by evaluating its stimulating effect on the two human osteoblastic cell lines HOS58 and SaOS-2 (Ayoub et al. 2009). The extract was found to increase significantly the mineralization of cultivated human bone cell. The extract was found to have high phenolic content containing 23 such compounds, among which the new 3-methoxyellagic acid 4-O-β-glucopyranoside was fully identified. The new compound was found to cause a significant increase of mineralized area at 20 μg/ml, while at lower concentrations the effect was not significant. However, an increase of the number of mineralized spots (nodules) at all tested concentrations of the compound was observed.

Antihyperthyroidism Activity

Feijo fruit juice was found to be beneficial in mild cases of thyrotoxicosis (Alijew and Rachimowa 1965). Thyrotoxicosi is a toxic condition that is caused by an excess production of thyroid hormones by the thyroid gland.

Traditional Medicinal Uses

In its native countries, feijoa has been used for medicinal –prophylactic nutrition in diseases of the thyroid gland as it is rich in iodine (Roberts 2001). Feijoa fruit has been reported to contain about 0.6 mg/100 g of Iodine (Kolesnik et al. 1991). The flowers and fruit are made into a tea for this purpose (Roberts 2001). The same brew is used to treat dysentery and diarrhoea but with extra flowers added. In Paraguay, fresh crushed flowers are applied to rashes, mild burns, insect bites and stings and itchy, inflamed areas.A lotion made from the flowers is used to soothe sunburnt skin. Slices of feijoa fruit are used as poultices. Feijoa fruits are also used to treat atherosclerosis, pyelonephritis and diseases of the cardiovascular system (Kolesnik et al. 1991).

Other Uses

In New Zealand, growers sometimes plant feijoa as a windbreak around other wind-sensitive crops. The wood is hard, dense and brittle.

Comments

Feijoa is propagated by seeds and grafting.

Selected References

Alijew RK, Rachimowa AC (1965) Sucfejsel – the juice of the fruits of *Feijoa sellowiana* Berg – in the therapy of thyrotoxicosis. Pharm Zentralhalle Dtschl 104:164–166

Ayoub NA, Hussein SA, Hashim AN, Hegazi NM, Linscheid M, Harms M, Wende K, Lindequist U, Nawwar MA (2009) Bone mineralization enhancing activity of a methoxyellagic acid glucoside from a *Feijoa sellowiana* leaf extract. Pharmazie 64(2):137–141

Basile A, Vuotto ML, Violante U, Sorbo S, Martone G, Castaldo-Cobianchi R (1997) Antibacterial activity in *Actinidia chinensis*, *Feijoa sellowiana* and *Aberia caffra*. Int J Antimicrob Agents 8(3):199–203

Beyhan O, Elmastaş M, Gedikli F (2010) Total phenolic compounds and antioxidant capacity of leaf, dry fruit and fresh fruit of feijoa (*Acca sellowiana*, Myrtaceae). J Med Plants Res 4(11):1065–1072

Binder RG, Flath RA (1989) Volatile components of pineapple guava. J Agric Food Chem 37:734–736

Bontempo P, Mita L, Miceli M, Doto A, Nebbioso A, De Bellis F, Conte M, Minichiello A, Manzo F, Carafa V, Basile A, Rigano D, Sorbo S, Castaldo Cobianchi R, Schiavone EM, Ferrara F, De Simone M, Vietri M, Cioffi M, Sica V, Bresciani F, de Lera AR, Altucci L, Molinari AM (2007) *Feijoa sellowiana* derived natural flavone exerts anti-cancer action displaying HDAC inhibitory activities. Int J Biochem Cell Biol 39(10):1902–1914

Facciola S (1990) Cornucopia. A source book of edible plants. Kampong Publ, Vista, 677pp

Fernandez X, André-Michel L, Poulain S, Lizzani-Cuvelier L, Monnier Y (2004) Chemical composition of the essential oil from feijoa (*Feijoa sellowiana* Berg.) peel. J Essent Oil Res 16(3):274–275

Ferrara L, Montesan D (2001) Nutritional characteristics of *Feijoa sellowiana* fruit: the iodine content. Riv Sci Dell'Alim 30(4):353–356

Govaerts R, Sobral M, Ashton P, Barrie F, Holst BK, Landrum LL, Matsumoto K, Fernanda Mazine F, Nic Lughadha E, Proenca C, Soares-Silva LH, Wilson PG, Lucas E (2010) World checklist of Myrtaceae. The board of trustees of the Royal Botanic Gardens, Kew. Published on the Internet, http://www. kew.org/wcsp/. Accessed 22 Apr 2010

Hardy PJ, Michael BJ (1970) Volatile components of feijoa fruits. Phytochemistry 9(6):1355–1357

Ielpo MT, Basile A, Miranda R, Moscatiello V, Nappo C, Sorbo S, Laghi E, Ricciardi MM, Ricciardi L, Vuotto ML (2000) Immunopharmacological properties of flavonoids. Fitoterapia 71(1):S101–S109

Isobe Y, Kikukawa M, Narita M (2002) Chemical composition of *Feijoa sellowiana* Berg. J Home Econ Japan 53(3):279–283

Karami M, Hassan Salehi H, Naghshvar F (2008) Study of histopathology and antioxidant activity of methanolic extract of *Feijoa sellowiana* against dosage induced by MDMA in mouse liver. Pharmacologyonline 3:315–321

Kolesnik AA, Golubev VN, Gadzhieva AA (1991) Lipids of the fruit of *Feijoa sellowiana*. Chem Nat Comp 27(4):404–407

Landrum LR (1986) *Campomanesia*, *Pimenta*, *Blepharocalyx*, *Legrandia*, *Acca*, *Myrrhinium*, and *Luma* (Myrtaceae). Fl Neotrop 45:1–178

Ledin RB (1957) Tropical and subtropical fruits in Florida (other than *Citrus*). Econ Bot 11:349–376

Lowry J (1976) Anthocyanins of the *Melastomataceae*, *Myrtaceae* and some allied families. Phytochemistry 15:513–516

Manabe M, Isobe Y (2005) Suppressing effects of *Feijoa sellowiana* Berg (Feijoa) on cytokine secretion by intestinal epithelium. Food Sci Technol Res 11(1):71–76

Morton JF (1987) Feijoa. In: Morton JF (ed) Fruits of warm climates. Florida Flair Books, Miami, pp 367–370

Nakashima H (2001) Biological activity of Feijoa peel extracts. Kagoshima Univ Res Cent Pac Islands, Occas Pap 34:169–175

Popenoe W (1974) Manual of tropical and subtropical fruits. Hafner Press, New York, Facsimile of the 1920 edition

Roberts MJ (2001) Edible and medicinal flowers. David Philip Publishers, Cape Town, p 166

Rossi A, Rigano D, Pergola C, Formisano C, Basile A, Bramanti P, Senatore F, Sautebin L (2007) Inhibition of inducible nitric oxide synthase expression by an acetonic extract from *Feijoa sellowiana* Berg. fruits. J Agric Food Chem 55(13):5053–5061

Ruberto G, Tringali C (2004) Secondary metabolites from the leaves of *Feijoa sellowiana* Berg. Phytochemistry 65(2):2947–2951

Shaw GJ, Allen JM, Yates MK (1989) Volatile flavour constituents in the skin oil from *Feijoa sellowiana*. Phytochemistry 28(5):1529–1530

Shaw GJ, Allen JM, Yates MK, Franich RA (1990) Volatile flavour constituents of feijoa (*Feijoa sellowiana*) – analysis of fruit flesh. J Sci Food Agric 50(3):357–361

Shaw GJ, Ellingham PJ, Birch EJ (1983) Volatile constituents of feijoa – headspace analysis of intact fruit. J Sci Food Agric 34(7):743–747

U.S. Department of Agriculture, Agricultural Research Service (2010) USDA National nutrient database for standard reference, release 23, Nutrient Data Laboratory Home Page, http://www.ars.usda.gov/ba/bhnrc/ndl

Vanidze MR, Shalashvili AG, Chkhikvishvili ID, Kvesitadze EG (1991) Catechins of feijoa leaves. Subtrop Kul't 6:87–90

Vuotto ML, Basile A, Moscatiello V, De Sole P, Castaldo-Cobianchi R, Laghi E, Ielpo MT (2000) Antimicrobial and antioxidant activities of *Feijoa sellowiana* fruit. Int J Antimicrob Agents 13(3):197–201

Weston RJ (2020) Bioactive products from fruit of the feijoa (*Feijoa sellowiana*, Myrtaceae): a review. Food Chem 121(4):923–926

Eugenia brasiliensis

Scientific Name

Eugenia brasiliensis **Lam.**

Synonyms

Eugenia bracteolaris Lam. ex DC., *Eugenia brasiliensis* var. *erythrocarpa* Cambess., *Eugenia brasiliensis* var. *leucocarpa* Cambess., *Eugenia dombeyi* Skeels nom. illeg, *Eugenia filipes* Baill., *Eugenia ubensis* Cambess., *Eugenia dombeyi* Skeels nom. illeg., *Myrtus grumixama* Vell., *Stenocalyx brasiliensis* (Lam.) O.Berg, *Stenocalyx brasiliensis* var. *erythrocarpa* (Cambess.) O. Berg, *Stenocalyx brasiliensis* var. *iocarpa* O. Berg, *Stenocalyx brasiliensis* var. *leucocarpa* (Cambess.) O.Berg, *Stenocalyx brasiliensis* var. *silvestris* O.Berg, *Stenocalyx ubensis* (Cambess.) O.Berg.

Family

Myrtaceae

Common/English Names

Brazil Cherry, Brazilian Cherry, Grumichama, Red-Fleshed Grumichama, Spanish Cherry

Vernacular Names

Brazil: Cumbixaba, Grumixama, Grumichameira, Grumixameira (Portuguese);
Czech: Hřebíčkovec Brazilský;
Danish: Brasiliansk Kirsebær;
French: Bois Dè Nefle, Cerisier Du Brésil, Jambosier Du Brésil;
German: Brasilianische Kirschmyrte;
Spanish: Grumichama, Pomarosa Forastera.

Origin/Distribution

The grumichama is indigenous to Brazil; it is found wild in eastern and coastal southern Brazil, especially in the states of Parana and Santa Catarina. The crop is cultivated in Brazil and Paraguay. The plant has been introduced into Australia, Florida, California, Hawaii, Honduras, Cuba, Angola and East Malaysia (Sabah).

Agroecology

The grumichama is hardy to most conditions; once established it can survive light frost (−3.33°C) as experienced in Brazil. When young it needs protection from frosts. The tree is quite drought tolerant but crop quality and development deteriorates during long, dry season unless

T.K. Lim, *Edible Medicinal And Non-Medicinal Plants: Volume 3, Fruits*,
DOI 10.1007/978-94-007-2534-8_82, © Springer Science+Business Media B.V. 2012

Plate 1 Young leaf flush – pink-bronze-coloured leaves

Plate 3 Young solitary fruit and flower cluster

Plate 2 Terminal cluster of flower buds

Plate 4 Close-up of immature green fruit with persistent sepals

adequate water is provided. It does best in full sun but will tolerate partial shade. It thrives on acid soils such as deep fertile, sandy loams but will also grow on rich clays.

Edible Plant Parts and Uses

The cherry like fruits are eaten fresh, candied or stewed. They are also utilised for the production of jelly, jam and pies.

Botany

A slender, erect, evergreen tree reaching 7–10 m high with a short trunk and dense crown. Leaves, opposite, oval-oblong, 9–16 cm long by 5–6 cm wide, simple with recurved margin, minutely pitted on both surface, glossy deep green when mature, pinkish-bronze when young. (Plates 1 and 2) Inflorescence in terminal clusters on branches, or flowers solitary (Plates 2 and 3). Flower 2.5 cm across, with four green sepals, four white petals and numerous stamens (about 100) with plae yellow anthers. Fruit oblate, 1.5–2 cm across, green (Plates 3 and 4) turning to bright-red and finally dark-purple to nearly black as it ripens, and bears the persistent, purple- or red-tinted sepals, (1.25 cm) long, at its apex. The skin thin, firm and exudes dark-red juice. The red or white pulp is juicy subacid to sweet with a touch of aromatic resin. Seeds 1–3, pale tan to gray, hard, hemispherical, 1.25 cm diameter.

Nutritive/Medicinal Properties

The nutritive value per 100 g of grumichama (*Eugenia brasiliensis*) fruit was reported as: energy 55Cal, water 83.5 g, protein 0.35 g, fat 0 g, carbohydrate 13.40 g, ash 0.43 g, Ca 39.5 mg, P 13.6 mg, Fe 0.45 mg, vitamin A 0.02 mg, vitamin B1 0.04 mg, vitamin B2 0.03 mg, vitamin C 18.8 mg and niacin 0.33 mg (Lorenzi et al. 2006).

Eight phenolic compounds: (cyanidin 3-glucoside, delphinidin 3-glucoside, ellagic acid, kaempferol, myricetin, quercetin, quercitrin, and rutin) were found in grumichama fruits (Reynertson et al. 2008). In addition, total phenolic content (TPC) 24.80 mg gallic acid equivalents per g dry weight, total anthocyanin content (TAC) 8.37 cyanidin 3-glucoside mg/g dry weight and antiradical activity, measured as DPPH IC_{50} was 42.73 μg/ml. Cyanidin 3-glucoside was the most abundant compound in grumichama which was also rich in delphinidin 3-glucoside.

The main constituents found in the leaf oil from two different samples of *Eugenia brasiliensis* were α- and β-selinene and β-caryophyllene (Fischer et al. 2005). The specimen collected at Jaboticabal contained β-selinene (17.3%) as the major component, while the specimen from Martinho Prado contained α-selinene (14.8%) as the major compound. Additionally, the specimen from Martinho Prado produced relatively high amounts of *a*- and β-pinene (6.6% and 3.6%, respectively). Altogether 38 compounds were identified, accounting for the range 93.8–99.7%; aliphatic compounds 0.3–0.4%, monterpene hydrocarbons 1.5–11.9%, oxygenated monoterpenes 5.5–14.9%, sesquiterpene hydrocarbons 63.5–66.5%, and oxygenated sesquiterpenes 16.1–23%. Apel et al. (2004) found α- and β-pinene, spathulenol and T-cadinol as the major components of *E. brasiliensis* essential oil.

Clear differences in the pattern of terpenes were observed for both purple and yellow fruit colour varieties of *E. brasiliensis* (Moreno et al. 2007). Although the major components in the leaf oil were the same (α-pinene, β-pinene and 1,8-cineol) for both varieties, in the leaf oil from the purple fruit variety a higher level of oxygenated sesquiterpenes (33.9%) was observed as compared to the yellow fruit variety, in which the majority of the identified compounds were monoterpene hydrocarbons (55.6%) and oxygenated monoterpenes (32.6%) and only 3.8% of oxygenated sesquiterpenes. In the fruit oils, this difference was more evident (Moreno et al. 2007). The purple fruit oil contained almost exclusively sesquiterpenes (57.3% oxygenated and 34.1% hydrocarbons) while the yellow fruit oil was composed of mainly monoterpenes (42.9% hydrocarbons and 18.5% oxygenated). The essential oil composition of the oil from fruits and leaves of the yellow fruit variety were very similar while in the purple fruit variety the fruit oil composition was quite different from that observed in the leaves. In the purple fruit variety, the major fruit oil components were caryophyllene oxide (22.2%) and α-cadinol (10.4%), these compounds were not observed in the leaves oil. In the leaf oil, the major component was the sequiterpene globulol (6.7%), not found in the fruit oil, and very low amount of monoterpenes were detected.

Antiinflammatory Activity

Studies showed that *E. brasiliensis* possessed topical anti-inflammatory activity (Pietrovski et al. 2008). Topical application of hydroalcoholic extract, fractions and isolated compounds from *E. brasiliensis* caused an inhibition of ear oedema in response to topical application of croton oil on the mouse ear. For oedema inhibition, the estimated ID_{50} values for hydroalcoholic extract and fractions (hexane, ethyl acetate and dichloromethane) were 0.17, 0.29, 0.13 and 0.14 mg/ear, respectively, with inhibition of 79%, 87%, 88% and 96%, respectively. Isolated phenolic compounds (quercetin, catechin and gallocatechin) were also effective in inhibiting the oedema (inhibition of 61%, 66% and 37%, respectively). Additionally, both extract and isolated compounds caused inhibition of polymorphonuclear cells influx (inhibition of 85%, 81%, 73% and 76%, respectively). Moreover, hydroalcoholic extract was also effective in inhibiting the arachidonic acid-mediated mouse ear oedema (ID_{50} value was 1.94 mg/ear and inhibition of 60%).

Antibacterial Activity

The major compounds found in the leaf essential oil of *E. brasiliensis* were spathulenol (12.6%) and tau-cadinol (8.7%) (Magina et al. 2009). The oil exhibited antibacterial activity against *Staphylococcus aureus*, *Pseudomonas aeruginosa*, and *Escherichia coli*. It strongly inhibited the growth of *S. aureus* with MIC of 156.2 μg/ml.

Traditional Medicinal Uses

The bark and leaves contain 1.5% of essential oil. In Brazil, the leaf or bark infusion in water is aromatic, astringent, diuretic and employed as a treatment for rheumatism and gastrointestinal disorders.

Other Uses

The small tree size favours its use in urban ornamental landscaping. Its wood is utilised in carpentry, woodworking and turnery. The bark contains about 43% tannin.

Comments

Its slow growth and low rate of dispersion make it rare, and it's generally considered as an endangered species. There is also a variety with white fruits (*E. brasiliensis. leucocarpus* O. Berg) in Brazil.

Selected References

Apel MA, Sobral M, Menut C, Bessiere JM, Schapoval EES, Henriques AT (2004) Essential oils from *Eugenia* species. Part VII. Sections phyllocalyx and stenocalyx. J Essent Oil Res 16:135–138

Burkill IH (1966) A dictionary of the economic products of the Malay Peninsula. Revised reprint. 2 volumes. Ministry of Agriculture and Co-operatives. Kuala Lumpur. Vol. 1 (A–H), pp 1–1240; vol. 2 (I–Z), pp 1241–2444

Fischer DCH, Limberger RP, Amrelia T, Henriques AT, Moreno PRH (2005) Essential oils from leaves of two *Eugenia brasiliensis* specimens from southeastern Brazil. J Essent Oil Res 17:499–500

Govaerts R, Sobral M, Ashton P, Barrie F, Holst BK, Landrum LL, Matsumoto K, Fernanda Mazine F, Nic Lughadha E, Proenca C, Soares-Silva LH, Wilson PG, Lucas E (2010) World checklist of Myrtaceae. The board of trustees of the Royal Botanic Gardens, Kew. Published on the Internet, http://www. kew.org/wcsp/. Accessed 22 April 2010

Kennard WC, Winters HF (1960) Some fruits and nuts for the tropics. USDA Agric Res Serv Misc Publ 801: 1–135

Liberty Hyde Bailey Hortorium (1976) Hortus third. A concise dictionary of plants cultivated in the United States and Canada. Liberty Hyde Bailey Hortorium/Cornell University/Wiley, New York, 1312pp

Lorenzi H, Bacher L, Lacerda M, Sartori S (2006) Brazilian fruits & cultivated exotics (for consuming in Natura). Instituto Plantarum de Etodos da Flora Ltda, Brazil, 740pp

Magina MD, Dalmarco EM, Wisniewski A Jr, Simionatto EL, Dalmarco JB, Pizzolatti MG, Brighente IM (2009) Chemical composition and antibacterial activity of essential oils of *Eugenia* species. J Nat Med 63(3): 345–350

Moreno PRH, Lima MEL, Sobral M, Young MCM, Cordeiro I, Apel MA, Limberger RP, Henriques AT (2007) Essential oil composition of fruit colour varieties of *Eugenia brasiliensis* Lam. Sci Agric (Piracicaba, Brazil) 64(4):428–432

Morton JF (1987) Grumichama. In: Morton JF (ed) Fruits of warm climates. Florida Flair Books, Miami, pp 390–391

Pietrovski EF, Magina MD, Gomig F, Pietrovski CF, Micke GA, Barcellos M, Pizzolatti MG, Cabrini DA, Brighente IM, Otuki MF (2008) Topical anti-inflammatory activity of *Eugenia brasiliensis* Lam. (Myrtaceae) leaves. J Pharm Pharmacol 60(4): 479–487

Reynertson KA, Yang H, Jiang B, Basile MJ, Kennelly EJ (2008) Quantitative analysis of antiradical phenolic constituents from fourteen edible Myrtaceae fruits. Food Chem 109(4):883–890

Verheij EWM, Coronel RE (eds) (1991) Plant resources of south-east Asia, No. 2. Edible fruits and nuts. Prosea Foundation, Bogor, 446pp

Eugenia coronata

Scientific Name

Eugenia coronata **Vahl ex DC.**

Synonyms

Eugenia coronata var. *macrophylla* A.Chev., *Eugenia littorea* Engl. & Brehmer, *Eugenia myrtoides* G.Don.

Family

Myrtaceae

Common/English Name

Crown Eugenia

Vernacular Names

Ghana: Amamɛ (Adangme), Kraka, Kraku (Akan-Brong), ʃWuσtaami (Ga), Amamɛ (Gbe-Vhe).

Origin/Distribution

The species is indigenous to tropical West Africa from Ivory Coast to Benin.

Agroecology

In its native range, it occurs in savannah and fringing forest of the coastal plains. The tree prefers acidic soils rich in humus. It grows in full sun to partial shade and is hardy and drought tolerant.

Edible Plant Parts and Uses

Fruits are edible, sub-acid, eaten raw or more frequently used in jellies, jams and preserves.

Botany

A woody evergreen shrub or small tree, up to 6 m high with ascending branches and an elongate crown. Leaves lime green when young, glossy deep green and coriaceous when mature, lamina simple, obovate to broadly elliptic, up to 7.6 cm long, apex obtuse to acute, base obtuse to broadly cuneate, margin entire with distinct midrib (Plates 1, 2 and 3), aromatic when crushed. Flowers in 1–3 flowered axillary clusters, small with four white petals and four green sepals and numerous long stamens (Plate 2). Fruit ovoid, 1.25–1.8 cm long, green, turning to purplish-black to black when ripe with persistent sepals (Plates 3, 4 and 5) and translucent whitish pulp (Plate 6).

T.K. Lim, *Edible Medicinal And Non-Medicinal Plants: Volume 3, Fruits*,
DOI 10.1007/978-94-007-2534-8_83, © Springer Science+Business Media B.V. 2012

Plate 1 Young ovate lime-green leaves

Plate 4 Ripening fruits

Plate 2 Flowers and mature dark green leaves

Plate 5 Ripe black and immature green fruit

Plate 3 Cluster of young fruits

Plate 6 Ripe purple-black fruit with translucent, whitish flesh

Nutritive/Medicinal Properties

No published information is available on its nutritive value or medicinal uses.

Other Uses

The shrub is used as an ornamental in landscaping and for hedges and boundary markers. Other products from the plant include dyes, stains, inks, tattoos and mordants. The twigs are used as chewsticks (toothbrush) in West Africa.

Comments

The tree is propagated by seeds and cuttings.

Selected References

Burkill HM (1998) Useful plants of west tropical Africa, vol 4, Families M-R. Royal Bot. Gardens, Kew, 969pp

Govaerts R, Sobral M, Ashton P, Barrie F, Holst BK, Landrum LL, Matsumoto K, Fernanda Mazine F, Nic Lughadha E, Proenca C, Soares-Silva LH, Wilson PG, Lucas E (2010) World checklist of Myrtaceae. The board of trustees of the Royal Botanic Gardens, Kew. Published on the Internet, Accessed http://www.kew.org/wcsp/. Accessed 22 Apr 2010

Keay RWJ, Hepper FN (1953–1972) Flora of west tropical Africa, 2nd edn. Crown Agents for Overseas Governments and Administrations, London

Eugenia stipitata

Scientific Name

Eugenia stipitata **McVaugh**.

Synonyms

None

Family

Myrtaceae

Common/English Names

Arazá, Sour Guava, Yellow Araza

Vernacular Names

Brazil: Arazá, *Araça-Boi*, Araçá-Boi (Portuguese);
Costa Rica: Arazá;
Peru: Arazá, Arazá-Buey, Pichi;
Portuguese: Araçá-Boi;
Spanish: Arazá.

Origin/Distribution

The species is believed to have its origin in the extreme west of the Amazon basin, perhaps in the Peruvian Amazon. It is only found in the western Amazon and does not appear to have been widely spread by the Indians although some of the best varieties appear to have been selected by the Peruvian Indians in Iquitos. Its native range includes Bolivia, Brazil, Colombia and Peru. It is exotic to other tropical areas in south, central America and Florida. Specimens have also been introduced elsewhere in the tropics, for example in Tenom, Sabah, Malaysia.

Agroecology

The arazá is a species of the humid, tropical high forest, growing from sea level to 650 m altitude. Most of the wild populations are found on old, non-flooded terraces in highly leached, light, tropical podzolic soils, which are found specifically within the area between the Marañón and Ucayali Rivers and where the Amazon begins and as far as Iquitos. In its natural habitat, the rainfall averages 2,800 mm, with a mean annual temperature of 26°C. In some areas where it has been introduced it grows well enough to withstand a drought of 2 months in an area with only 2,000 mm annual rainfall. It grows well in semi-exposed or exposed areas. The arazá prefers well drained, fertile loamy soils but will tolerate poorer clayey oxisol provided they are well-drained and acid (low pH) soils.

T.K. Lim, *Edible Medicinal And Non-Medicinal Plants: Volume 3, Fruits*,
DOI 10.1007/978-94-007-2534-8_84,

Edible Plant Parts and Uses

Arazá is used to make juices, soft drinks, ice-cream, smoothies, sherbets, jelly, preserves and desserts. The fruit is rarely eaten raw on its own because of its high acidity (pH 2.4 in the case of the juice). It can be eaten fresh sprinkled with sugar.

Botany

The arazá is a shrub or small tree 2.5–15 m, densely branched with brown to reddish brown bark. The leaves are simple, opposite, elliptical to slightly oval, 6–18 by 3.5–9.5 cm, apex acuminate, the base rounded to subcordate, margin entire, dull dark green, with distinct primary and 6–10 pairs of secondary veins on short 3 mm long petioles (Plate 2). Flowers long-pedicellate in 2–5 flowered axillary racemose inflorescences. Flowers with linear bracteoles, calyx 4 rounded, overlapping sepals; petals 5, white, obovate, 7–10 mm by 4 mm, ciliate; stamens 70 and long; ovary with 3–4 locules, each with 5–8 ovules, style 5–8 mm long. Fruit globose to oblate, pale green turning to bright yellow when ripe, reaching 12 cm across (Plates 1, 2 and 3) and weighing up to 750 g, with a thin, velvety skin enclosing a juicy, aromatic, acid, thick pulp enclosing about 12 seeds.

Nutritive/Medicinal Properties

Pinedo et al. (1981) reported that the fruit (on a dry matter basis) contained 8–10.75% protein, 5–6.5 56 fibre, 69–12% other carbohydrates,

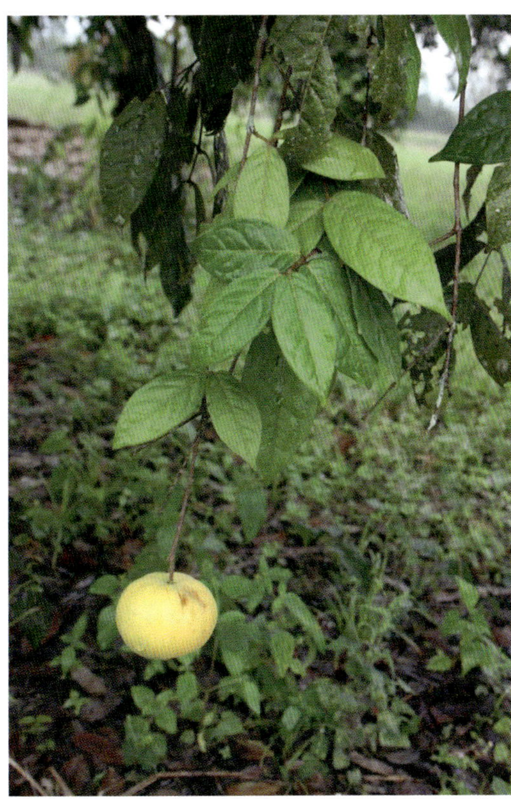

Plate 2 Leaves and ripe fruit

Plate 1 Unripe fruit

Plate 3 Close-up of ripe fruit

0.16–0.21 calcium, 10–12 ppm of zinc and some phosphorus, potassium and magnesium. They found per 100 g fresh weight775 mg vitamin A; 9.84 mg vitamin BI and 768 mg of vitamin C which is double that of oranges.

The nutrient composition of araça-boi fruit pulp was found to contain on a %DM basis (Rogez et al. 2004): 4% dry matter, 11.9% protein, ash 4%, sugars 49.2%, glucose 3.1% fructose 33.9%, sucrose 17.2%, total dietary fibre 39%; amino acid (g/100 g protein) – glycine 4.17 g, alanine 6.84 g, valine 4.77 g, leucine 6.05 g, isoleucine 3.79 g, proline 3.46 g, phenylalanine 4.1 g, tyrosine 2.94 g, serine 4.11 g, threonine 3.64 g, cysteine 1.78 g, methionine 1.84 g, asparagine + aspartic acid 8.57 g, glutamine + glutamic acid 31.86 g, lysine 5.53 g, arginine 4.14 g, histidine 2.43 g; minerals (mg/100 g FW) – Na 1.64 mg, K 27.84 mg, Ca 5.72 mg, Mg 2.52 mg, P 7.4 mg, Fe 0.155 mg and Zn 0.18 mg. The pulp amounted to 82% of the fruit mass, has a creamy colour, with some fine fibres and has a fine, sweet and acid flavour with pH of 2.6. The fruit has a high protein content (1.9%). The amino acid profile was found to be relatively close to the ideal protein profile postulated for humans even for lysine and for the sulphur-containing amino acids which are usually limiting in proteins of vegetal origin. However, the fruit was found to be limiting in methionine and arginine.

Hernández et al. (2007) found the respiratory pattern of arazá fruit to be climacteric, with maximum respiration rates of around 200 mg CO_2/kg/h, preceded by a peak of ethylene production (20 μL C4H4/kg/h), a change in skin colour from green to yellow, a total soluble solids value of 5°Brix, an increase in the sucrose and fructose content up to 2.8 μmol/g, a pH which increased to 3 units, and a decrease in titratable acidity to 400 mmol H+/L. Malic acid was the main organic acid in the edible pulp and ascorbic acid was present in a concentration of 17.8 μmol/g. Ascorbic acid degradations (13.1%) and total carotenoids (79.1%) were the most important factor for loss of quality of the araçá-boi pulp during freezing storage at −12°C during 200 days (Andrade and Caldas 1997). Ascorbic acid and total carotenoids decreased from 24.61 to 23.12 and 0.93 to 0.19 mg 100/g respectively. The pH remained constant. Little variations were found at the Brix/acidity ratio. Dry matter and non reducing sugars increased. Soluble solids and moisture decreased. Reducing sugars showed variations during storage time. García-Reyes and Narváez-Cuenca (2010) found that pasteurization of arazá pulp at 80 C with a holding time of 1 min before frozen storage at −20°C reduced loss of vitamin C and the sensory characteristics of degradation. This technique resulted in a more stable arazá pulp storage product for at least 2 months.

Franco and Shibamoto (2000) found that sesquiterpenes were the main constituents of the oil of the fruits of *E. stipitata*, with germacrene B being the main component (38.3%).

The fruit of *E. stipitata* was found to have an antioxidant activity expressed as a DPPH IC_{50} of 79 μg/ml (Reynertson et al. 2005)

Thirty five of the 37 compounds were identified in leaf oil of *E. stipitata*, with the two other minor components being tentatively identified as epi-zonarene and cadala-1(10),3,8-triene (Medeiros et al. 2003). Together these 37 substances accounted for 94.5% of the oil in the leaves. The leaf oil essentially contained only monoterpenes and sesquiterpenes, of which the main components were sesquiterpenes (69.5%) with about 52% of these being oxygenated. The major sesquiterpenes were caryophyllene derivatives, such as β-caryophyllene (22.7%) and caryophyllene oxide (15.4%). Of the monoterpenes (24.9%), only ~2% were oxygen-containing compounds, with the main constituents being α-pinene (14.1%), myrcene (2.9%) and limonene (2.7%), respectively. The leaf oil of *E. stipitata* displayed promising antimicrobial activity when tested against *Staphylococcus aureus*, *Pseudomonas aeruginosa* and *Listeria monocytogenes* (Medeiros et al. 2003).

Other Uses

In its native range, the species is used for rehabilitating exhausted land. It is commonly cultivated as an ornamental species for recreational purpose.

Comments

The seeds are recalcitrant as that do not survive drying and freezing during ex-situ conservation.

References

Andrade JS, Caldas MLM (1997) Quality of araçá-boi (*Eugenia stiptata* McVaugh) pulp during freezing storage. Acta Hortic ISHS 452:19–22

Bermejo JEH, León J (eds.) (1994) Neglected crops: 1492 from a different perspective. Plant production and protection series No. 26. FAO, Rome, pp 205–209

FAO (1986) Food and fruit bearing forest species. 3: examples from Latin America. FAO forestry paper 44/3. FAO, Rome, 327 pp

Fouqué A (1972). Espèces Fruitières d'Amérique Tropicale. Institut Français de Recherches Fruitierès Outre-Mer (IFAC) Paris

Franco MRB, Shibamoto T (2000) Volatile composition of some Brazilian fruits: umbu-caja (*Spondias cytherea*), camu-camu (*Myrciaria dubia*), araca-boi (*Eugenia stipitata*), and cupuacu (*Theobroma grandiflorum*). J Agric Food Chem 48(4):1263–1265

García-Reyes R-H, Narváez-Cuenca C-E (2010) The effect of pasteurization on the quality of frozen arazá (*Eugenia stipitata* McVaugh) pulp. J Food Qual 33(5):632–645

Gentil DFO, Ferreira SAN (1999) Viability and overcoming dormancy in seeds of araca-boi (*Eugenia stipitata* ssp. *sororia*). Acta Amazonica 29:21–31

Govaerts R, Sobral M, Ashton P, Barrie F, Holst BK, Landrum LL, Matsumoto K, Fernanda Mazine F, Nic Lughadha E, Proenca C, Soares-Silva LH, Wilson PG, Lucas E (2010) World checklist of Myrtaceae. The board of trustees of the Royal Botanic Gardens, Kew. Published on the Internet, http://www. kew.org/wcsp/. Accessed 22 Apr 2010

Hernández MS, Martínez O, Fernández-Trujillo JP (2007) Behavior of arazá (*Eugenia stipitata* Mc Vaugh) fruit quality traits during growth, development and ripening. Sci Hortic 111(3):220–227

Lorenzi H, Bacher L, Lacerda M, Sartori S (2006) Brazilian fruits & cultivated exotics (for consuming in natura). Instito Plantarum de Etodos da Flora Ltda, Nova Odessa, 740 pp

McVaugh R (1969) Botany of the Guyana Highland. Part VIII. Mem N Y Bot Gard 18(2):55–287

Medeiros JR, Medeiros N, Medeiros H, Davin LB, Lewis NG (2003) Composition of the bioactive essential oils from the leaves of *Eugenia stipitata* McVaugh ssp. *sororia* from the Azores. J Essent Oil Res 15(4):293–295

Orwa C, Mutua A, Kindt R, Jamnadass R, Simons A (2009) Agroforestree database: a tree reference and selection guide version 4.0. http://www.worldagroforestry.org/af/treedb/

Pinhedo PM, Ramirez NF, Blasco LM (1981) Notas Preliminares sobre el Araza (*E. stipitata*). Prutal Native de la Amazonia Peruana. INIA/IICA. Pub. Bis. No. 229, Lima

Reynertson KA, Basile MJ, Kennelly EJ (2005) Antioxidant potential of seven myrtaceous fruits. Ethnobot Res Appl 3:25–35

Rogez H, Buxant R, Mignolet E, Souza JNS, Silva EM, Larondelle Y (2004) Chemical composition of the pulp of three typical Amazonian fruits: araça-boi (*Eugenia stipitata*), bacuri (*Platonia insignis*) and cupuaçu (*Theobroma grandiflorum*) *Eur*. Food Res Technol 218(4):380–384

Eugenia uniflora

Scientific Name

Eugenia uniflora **L.**

Synonyms

Eugenia arechavaletae Herter, *Eugenia brasiliana* (L.) Aubl., *Eugenia costata* Cambess., *Eugenia dasyblasta* (O.Berg) Nied., *Eugenia decidua* Merr., *Eugenia indica* Nicheli, *Eugenia lacustris* Barb. Rodr., *Eugenia michelii* Lam., *Eugenia microphylla* Barb. Rodr., *Eugenia myrtifolia* Salisb., *Eugenia oblongifolia* (O.Berg) Arechav., *Eugenia oblongifolia* (O.Berg) Nied. nom. illeg., *Eugenia parkeriana* DC., *Eugenia strigosa* (O. Berg) Arechav., *Eugenia uniflora* var. *atropurpurea* Mattos, *Eugenia willdenowii* (Spreng.) DC., *Eugenia zeylanica* Willd., *Luma arechavaletae* (Herter) Herter, *Luma costata* (Cambess.) Herter, *Luma dasyblasta* (O.Berg) Herter, *Luma strigosa* (O.Berg) Herter, *Myrtus brasiliana* L., *Myrtus brasiliana* var. *diversifolia* Kuntze, *Myrtus brasiliana* var. *lanceolata* Kuntze, *Myrtus brasiliana* var. *lucida* (O.Berg) Kuntze, *Myrtus brasiliana* var. *normalis* Kuntze nom. inval., *Myrtus willdenowii* Spreng., *Myrtus willdenowii* var. *portoriccensis* Spreng. ex DC., *Plinia pedunculata* L. f., *Plinia petiolata* L. nom. illeg., *Plinia rubra* L., *Plinia tetrapetala* L., *Stenocalyx affinis* O. Berg, *Stenocalyx brasiliensis* Berg, *Stenocalyx brunneus* O. Berg, *Stenocalyx costatus* (Cambess.) O.Berg, *Stenocalyx dasyblastus* O. Berg, *Stenocalyx glaber* O. Berg, *Stenocalyx impunctatus* O. Berg, *Stenocalyx lucidus* O. Berg, *Stenocalyx michelii* (Lam.) O. Berg, *Stenocalyx michelii* var. *membranacea* O.Berg, *Stenocalyx michelii* var. *rigida* O.Berg, *Stenocalyx oblongifolius* O.Berg, *Stenocalyx rhampiri* Barb.Rodr., *Stenocalyx ruber* (L.) Kausel, *Stenocalyx strigosus* O. Berg, *Stenocalyx uniflorus* (L.) Kausel, *Syzygium michelii* (Lam.) Duthie.

Family

Myrtaceae

Common/English Names

Barbados Cherry, Brazilian Cherry, Cayenne Cherry, French Cherry, Pitanga, Red Brazil Cherry, Surinam Cherry

Vernacular Names

Argentina: Arrayán, Ñangapiré;
Brazil: Cerisier De Cayenne, Cerisa Carre, Ginga, Pitanga, Pitangueira (Portuguese);
Chamorro: Pitanga;
Chinese: Hong Guo Zi;
Colombia: Cereza Cuadrada;
Cook Islands: Menemene, Menemene, Venevene (Maori);
Cuba: Cerezo De Cayenne;

T.K. Lim, *Edible Medicinal And Non-Medicinal Plants: Volume 3, Fruits*,
DOI 10.1007/978-94-007-2534-8_85, © Springer Science+Business Media B.V. 2012

Czech: Hřebíčkovec Jednokvětý;
Danish: Pitanga;
Dutch: Soete Kers, Surinaamsche Kers;
Eastonian: Pitanga Kirssmürt, Vili Pitanga;
El Salvador: Guinda;
French: Cerisier Carré, Cerisier De Cayenne, Cerise De Pays;
French Antilles: Cerise A Cotes, Cersie Cotes;
French Guiana: Cerise Caree;
German: Cayennekirsche, Surinamkirsche;
Guadeloupe: Cerises-Cotes;
Guyana: Barka Tree, Surianm cherry;
Martinique: Cerises-Cotes;
Niuean: Kafika, Kafika Palangi, Kafika Papalangi;
Portuguese: Pitanga;
Spanish: Cerezo De Cayena;
Sri Lanka: Goraka Jambu;
Surinam: Geribde Kers, Kerseboom, Surinaamse Kers, Surinaamsche Kers, Switie Kersie Wiwiri;
Surinam Sranan: Monkimonkikersi, Monki Monki Kersi;
Swedish: Körsbärsmyrten;
Venezuela: Pendanga.

Origin/Distribution

Pitanga is indigenous to the Amazon rainforest in south America. The plant is native from Surinam, Guyana and French Guiana to eastern and southern Brazil and to northern, eastern and central Uruguay.

It is naturalized in Argentina, Venezuela and Colombia, along the Atlantic coast of Central America, in the Caribbean islands and also in Florida. It is cultivated in all tropical and subtropical regions as fruit and ornamental tree. It has become invasive in Bermuda and some parts of Australia.

Agroecology

Pitanga grows in areas with tropical or subtropical climates from sea level to 1,800 m elevation as found in Guatemala. Young plants are severely damaged by frost but mature established plants are more frost hardy sustaining light damage. Pitanga thrives in full sun or partial shade and requires moderate rainfall and is drought tolerant as it has a deep root system. It is adaptable on a wide range of soil types – sand, sandy loam, stiff clay, soft limestone but is intolerant of saline soil and can withstand waterlogged conditions.

Edible Plant Parts and Uses

Ripe fruit is rich in vitamin C and is can be eaten fresh out-of -hand or sprinkle with sugar. Its predominant food use is as a flavoring and base for jams, jellies and relish. Surinam Cherry is used in shortcake, fruit cups, salads, puddings, custards , yogurts, ice creams, beverage, syrup or juices . The juice is also fermented into vinegar or wine and sometimes prepared as liquor. The fruit is also consumed stewed.

Botany

A multi-branched, erect shrub or small tree growing to 7.5 m high (Plate 1). Leaves are opposite, bronze in color when young, glossy deep-green when mature, than turning red in winter. Leaf is simple, entire, ovate or elliptic-ovate, 3.2–4.2 by 2.3–3 cm, shortly petiolate (Plates 1 and 2) to 1.5–3 mm long, papery, glabrous, with numerous pellucid glands and with 5 secondary veins on each side of midvein, base rounded, slightly cordate to cuneate, apex acuminate, mucronate, or obtuse. Flowers are usually axillary, white, slightly fragrant. Calyx has 4 long elliptic and reflexed lobes. Petals 4 white and obovate. Fruits are succulent , juicy, acid-sweet berries, red to very dark red when ripe, depressed- globose, 1.5 cm long, 2 cm wide , 8-ridged, 1- or 2-seeded and crowned by calyx-limb (Plates 1, 2 and 3).

Nutritive/Medicinal Properties

Pitanga cherries were found to contain per 100 g edible portion: moisture 17.21%, protein 14.71%, carbohydrates 38.55%, fat 15.62%, fibre 9.77%, total ash 4.94%, magnesium 1246.41 mg/kg, zinc 273.34 mg/kg, sodium 997.32 mg/kg, potassium 4271.30 mg/kg (Amoo et al. 2006). Pitanga cherries had the highest composition (g/100 g) of sugar with 7.88 g of hydrated maltose, 6.31 g of anhydrous lactose and 6.64 g of hydrated lactose, dextrose 4.83 g and fructose 5.13 g. Another report based on composite of analyses made in Hawaii, Africa, Florida (Morton 1987) stated the food value per 100 g edible portion to be energy 43–51 cal, moisture 85.4–90.70 g, protein 0.84–1.01 g, fat 0.4–0.88 g, carbohydrates 7.93–12.5 g, fibre 0.34–0.6 g, ash 0.34–0.5 g, Ca 9 mg, P 11 mg, Fe 0.2 mg, carotene (vitamin A) 1,200–2,000 I.U., thiamin 0.03 mg, riboflavin 0.04 mg, niacin 0.03 mg and ascorbic acid 20–30 mg. Lorenzi et al., (2006) reported the following nutrient value for Brazilain cherry: energy 50.4 cal., water 88.05 g, protein 0.97 g, fat 0.64 g, carbohydrate 10.2 g, ahs 0.42 g, Ca 9.0 mg, P 11.0 mg, Fe 0.20 mg, vitamin A 0.0003 mg, vitamin B-1 0.03 mg, vitamin B-2 0.05 mg, niacin 0.16 mg, vitamin C 19.5 mg.

Plate 1 Fruiting branches

Plate 2 Pitanga fruits and leaves

The fruits are very juicy with a high content of vitamin C, calcium and phosphorus, iron, vitamin A, riboflavin and niacin.

Chemical composition results revealed pitanga seeds to be a good source of insoluble dietary fiber, with low protein and fat levels, and no relevant differences were found among pitanga seeds from different flesh colors (Bagetti et al. 2009). The nutrient composition of the seed of pitanga fruits of 3 different coloured variety was reported in grams per 100 g fresh weight as: moisture 57–58.6 g, ash 0.6–0.8 g, protein 3.3–3.7 g, fat 0.5–0.7 g, total carbohydrate 36.4–38.4 g, total fibre 23.0–24.7 g and insoluble fibre

Plate 3 (**a**) Top view of pitanga fruits, (**b**) side and bottom view of fruits

23.0–23.7 g. Pitanga seeds had a high proportion of unsaturated fatty acids (60–70%) being 13–16% monounsaturated fatty acids (MUFA) and 45–47% polyunsaturated fatty acids (PUFAS). PUFAS, especially the n-3 fatty acids, are considered desirable compounds in the human diet because of their effect in reducing the incidence of cardiovascular disease. In seeds from purple and red pitanga the predominant unsaturated fatty acid was linoleic acid (C18:2n6c), followed by oleic acid (C18:1n9c), while seeds from orange pitanga had α-linolenic acid (C18:3n3) as the second most abundant unsaturated fatty acid. Palmitoleic acid was found only in seeds from purple and red pitanga. Seeds from purple and red pitanga had higher linoleic acid and lower α-linolenic acid content than seeds from orange pitanga. Seeds from red pitanga had higher oleic acid (C18:1n9c) followed by seeds from orange and purple pitanga. The only saturated fatty acid found was palmitic acid, which was found in higher proportion in seeds from orange pitanga when compared to the other samples. The researchers suggested that this low value waste of pitanga processing, could be used as a source of natural antioxidants and dietary fibre for animal and/or human nutrition.

Other Phytochemicals

Fifty-four compounds were detected in the volatile constituents of pitanga fruits, and 29 of those were identified (Oliveira et al. 2006). Monoterpenes (75.3% in mass) were found to comprise the largest class of the pitanga fruit volatiles, including trans-β-ocimene (36.2%), cis-ocimene (13.4%), the isomeric β-ocimene (15.4%) and β-pinene (10.3%). Several known therapeutic constituents of pitanga leaf extract, such as selina-1,3,7(11)-trien-8-one (the major constituent) were also found to be present in the fruit volatile extract. Another analysis conducted in Cuba reported that pitanga fruit had 130.6 mg/kg of total volatile compounds (Pino et al. 2003). Thirty-six compounds were identified, of which curzenene (38.9%) and bergaptene (16.2%) were the major ones.

Eugeniflorins D1 and D2 new hydrolysable tannin dimers, were isolated, together with four known polyphenols from *Eugenia uniflora* leaves (Lee et al. 1997). Thirty-nine compounds were found in the pitanga leaf extracts and 26 were identified (Peixoto et al. 2010). The main components identified in the extracts in decreasing quantitative order were: curzerene, germacrene B, $C_{15}H_{20}O_2$ and β-elemene for hydrodistillation; $C_{15}H_{20}O_2$ and curzerene for SC-CO_2 (supercritical carbon dioxide) extracts and 3-hexen-1-ol, curzerene, $C_{15}H_{20}O_2$, β-elemene and germacrene B for SC-CO_2 extracts captured in Porapak-Q polymer.

The main compounds identified in the leaf essential oil were seline-1,3,7(11)-trien-8-one and its oxide, germacrone, furanodiene and curzercne (Weyerstahl et al. 1988). Furanodiene and its rearrangement product, furanoelemene (or curzerene, 50.2%), β-elemene (5.9%) and α-cadinol (4.7%) were identified as the most abundant compounds in pitanga leaf essential oil (Melo et al. 2007). Of nine active aroma compounds identified, furanodiene (along with furanoelemene), β-elemene and (E,E)-germacrene were characterized as the main impact aroma compounds in the odor of this essential oil. Forty-one compounds representing 93.1% of the total volatiles were identified in *E. uniflora* leaf essential oil (Gallucci et al. 2010). The oil comprised 2.3% monoterpene hydrocarbons, 0.2% oxygenated monoterpenes, 26.0% sesquiterpenes hydrocarbons and 64.6% oxygenated sesquiterpenes. The main components were β-elemene (2.8%), β-caryophyllene (5.0%), curzerene (5.5%), germacrene B (10.5%), sehna-l,3,7(ll)-trien-8-one (34.0%), germacrone (4.3%), and selina-l,3,7(ll)-trien-8-one epoxide (17.0%).

Some of the pharmacological properties reported include the following:

Antioxidant Activity

Eugenia uniflora fruit exhibited high antioxidant capacity using the 1,1-diphenyl-2-picrylhydrazyl (DPPH) assay. The aqueous fraction of pitanga fruits contained antioxidant anthocyanins (cyanidin-3-O-β-glucopyranoside and delphinidin-3-O-β-glucopyranoside (Einbond et al. 2004). The antioxidant compounds in pitanga fruits as determined by the DPPH assay was 19.6 μg/ml which compared favourably with known antioxidants ascorbic acid (18.3 μg/ml) and α-tocopherol (53.3 μg/ml) (Reynertson et al. 2005). Carotenoids were found known to be important antioxidant compounds in pitanga pulp as assayed by 2,2-diphenyl-1-picrylhydrazyl radical (DPPH·) scavenging activity and superoxide dismutase (SOD)- and catalase (CAT)-like activities (Spada et al. 2008). Pitanga was found to be a rich fruit source of carotenoids that included: lycopene 73 μg/g, γ-carotene 53 μg/g, β-cryptoxanthin 47 μg/g and rubixanthin 23 μg/g; and it also had phytofluene 13 μg/g, β-carotene 9.5 μg/g, ζ-carotene 4.7 μg/g and unidentified compound 3.4 μg/g (Cavalcante and Rodrigues-Amaya 1992; Rodriguez-Amaya 1999). Fresh *Eugenia uniflora* fruit was reported as one of the richest fruit sources of carotenoids especially lycopene (Porcu and Rodriguez-Amaya 2008). Studies conducted in Brazil showed that ripe pitanga from Campinas had significantly higher (all-E)-lycopene (14.0 versus 71.1 μg/g), (13Z)-lycopene (1.1 versus 5.0 μg/g) and (all-E)-γ-carotene (1.6 vs. 3.8 μg/g) levels as compared to those from Medianeira, Paraná. Significant increases in most of the carotenoids occurred from the partially ripe to the ripe fruits, with (all-E)-lycopene doubling its concentration in fruits from both states. However, the processed products had much lower lycopene content. The mean (all-E)-lycopene concentration was 16.6 μg/g for frozen pulp brand A, 23.0 μg/g for bottled juice brand B and 25.6 μg/g for bottled juice brand C.

The essential oil from pitanga fruits was found to contain α-ocimene (7.4%), β-selinene (7.2%), β-selinene (5.2%), germacrene B (7.2%) and hexadecanoic acid (11.7%) and was demonstrated to have an interesting antioxidant activity (Marin et al. 2008). Pitanga seed extracts was found to have powerful antioxidant capacity that was partially correlated to their high phenolic content and showed some variation according to the pitanga flesh colours (Bagetti et al. 2009).

Studies showed that while ethanolic extract of air dried pitanga fruits (PCA) significantly inhibited the formation of TBARS (thiobarbituric acid reactive species)) induced by prooxidant agents such as iron (II) and sodium nitroprusside (SNP) in both liver and brain tissues homogenates, the ethanolic extract of sun dried pitanga fruits (PCS) did not (Kade et al. 2008). Further investigations revealed that the phenolic content of the PCS was significantly lower compared to PCA. Since phenolics in plants largely contributed to the antioxidative potency of plants, it was concluded that air-drying should be employed in the preparation of extracts of Pitanga cherry leaves before it is administered empirically as a traditional medicament by traditional medical practitioners.

The methanol extracts of six medicinal herbs used in the traditional Paraguayan medicine, *Aristolochia giberti*, *Cecropia pachystachya*, *Eugenia uniflora*, *Piper fulvescens*, *Schinus weinmannifolia* and *Schinus terebinthifolia* were found to protect against enzymatic and non-enzymatic lipid peroxidation in microsomal membranes of the rat (Velázquez et al. 2003). *C. pachystachya*, *E. uniflora*, *S. weinmannifolia* and *S. terebinthifolia* showed the highest scavenging activity on the superoxide and DPPH radicals.

Antimicrobial Activity

The aqueous, organic, and volatile oil extracts of leaves of *Eugenia uniflora* were found to inhibit Gram positive *Staphylococcus aureus* and *Bacillus subtilis*, and Gram negative *Escherichia coli* and *Shigella dysenteriae* (Fadeyit and Akpan 1989). *Pseudomonas aeruginosa*, *Klebsiella pneumoniae*, and *Salmonella typhi* were not inhibited. The aqueous extract was the most active against the organisms compared to the organic and volatile oil extracts. *Eugenia uniflora* was one of 10 Brazilian medicinal plants that exhibited antibacterial activity; it presented moderate activity on both *Staphylococus aureus* and *E. coli* (Holetz et al. 2002). It also displayed anti-candidal activity. The ethanol extract of *E. uniflora* showed antibacterial activity against an *E. coli* strain when exposed to UV-A light (Coutinho et al. 2010b).

The volatile oils of *Eugenia uniflora* were characterized by the occurrence of an unusual sesquiterpene as the major compound. There was abundance of curzerene (19.7%), selina-1,3,7(11)-trien-8-one (17.8%), atractylone (16.9%) and furanodiene (9.6%) in the leaves; and germacrone (27.5%), selina-1,3,7(11)-trien-8-one (19.2%) curzerene (11.3%) and oxidoselina-1,3,7(11)-trien-8-one (11.0%) in the fruits (Ogunwande et al. 2005). The two oils exhibited potent cytotoxic activity and varying antibacterial effects.

The *E. uniflora* lectin (EuniSL) isolated from the seed extract and purified showed a single band on denaturing electrophoresis, with a molecular mass of 67 kDa (Oliveira et al. 2008) . EuniSL agglutinated rabbit and human erythrocytes with a higher specificity for rabbit erythrocytes. The haemagglutination was not inhibited by the tested carbohydrates but glycoproteins exerted a strong inhibitory action. The lectin proved to be thermo-resistant with the highest stability at pH 6.5 and divalent ions did not affect its activity. EuniSL demonstrated a remarkable non-selective antibacterial activity. EuniSL strongly inhibited the growth of *Staphylococcus aureus*, *Pseudomonas aeruginosa* and *Klebsiella* sp. with a minimum inhibitory concentration (MIC) of 1.5 μg/ml, and moderately inhibited the growth of *Bacillus subtilis*, *Streptococcus* sp. and *Escherichia coli* with a MIC of 16.5 μg/ml. The strong antibacterial activity of the studied lectin indicated a high potential for clinical microbiology and therapeutic applications.

Oenothein B extracted from *Eugenia uniflora* effectively inhibited the fungus *Paracoccidioides brasiliensis*, the causative agent of paracoccidioidomycosis (PCM), the most prevalent human systemic mycosis in Latin America (Santos et al. 2007). The oenothein B dosage that most effectively inhibited the development (74%) of *P. brasiliensis* yeast cells in-vitro was 500 μg/ml. The results indicated that oenothein B interfered with the cell morphology of *P. brasiliensis*, probably by inhibiting the transcription of 1,3-β-glucan synthase gene involved in the cell wall synthesis.

Of the 13 tested essential oils, those obtained from *Cinnamomum zeylanicum*, *Ocimum gratissimum*, *Cymbopogon citratus*, *Eugenia uniflora* and *Alpinia speciosa* were found to be the most active, inhibiting 80% of the dermatophyte strains isolated from patients with dermatophytosis and producing inhibition zones of more than 10 mm. in diameter (Lima et al. 1993).

Antihyperglycemic/ Antihypertriglyceridemic Activity

Ethanol leaf extracts of *E. uniflora* were found to have anti hyperglycaemic and antihypertriglyceridemic effects. EtOH (70%) extracts from the leaves were separated into six fractions with different polarity and molecular size, i.e. NP-1-NP-6 (Arai et al. 1999). In an oral glucose tolerance test, NP-1 and 4 inhibited the increase in plasma glucose level. However, in an intraperitoneal glucose tolerance test, such an inhibitory effect was not seen. Thus, the effects of NP-1 and 4 were apparently due to the inhibition of glucose absorption from the intestine. In a sucrose tolerance test, all fractions inhibited the increase in plasma glucose level. In an oral corn oil tolerance test, NP-3 and 4 showed an inhibitory effect on the increase in plasma triglycerides level. On the other hand, NP-3, 4, 5 and 6 inhibited maltase and sucrase activities and all fractions except for NP-1 showed an inhibitory effect on lipase activity dose-dependently. The inhibition of the increase in plasma glucose level by NP-3, 4, 5 and 6 in the oral sucrose tolerance test and the inhibition of the increase in plasma triglycerides by NP-3 and 4 in the oral corn oil tolerance test were apparently due to the inhibition of the decomposition of carbohydrates and fats in the intestine, respectively. Aqueous leaf extract was found to be slightly active on lipid metabolism, and to exert a protective effect on triglycerides and very low-density lipoprotein levels (Ferro et al. 1988).

Hypotensive/Vasorelaxant Activity

Intraperitoneal administration of the aqueous crude extract was found to decrease blood pressure (BP) of normotensive rats dose-dependently until 47.1% of control (Consolini et al. 1999; Consolini and Sarubbio 2002). The ED 50 was 3.1 mg dried leaves/kg. The dose-response curve for phenylephrine on blood pressure was inhibited non-competitively until 80% of its maximal effect. Perfusion pressure (PP) of rat hindquarters (previously vasoconstricted by high-K^+) was decreased by the extract in a concentration-dependent manner. In addition, the extract demonstrated diuretic activity at a dose higher than the hypotensive one. It was almost as potent as amiloride, but while amiloride induced loss of Na^+ and saving of K^+, the extract induced decrease in Na^+ excretion. The results suggested that the empirical use of *Eugenia uniflora* was mostly due to a hypotensive effect mediated by a direct vasodilating activity, and to a weak diuretic effect that could be related to an increase in renal blood flow.

The addition of an increasing cumulative concentration of hydroalcoholic extract from *E. uniflora* (1–300 μg/ml) caused a concentration-dependent relaxation response in intact endothelium-thoracic aorta rings pre-contracted with noradrenaline (30–100 nM) (Wazlawik et al. 1997). The IC_{50} value and the maximum relaxation (Rmax) were 7.02 μg/ml and 83.94%, respectively. The removal of the endothelium completely abolished these responses. The nitric oxide synthase inhibitors N omega-nitro-L-arginine (L-NOARG, 30 μM) and N omega-nitro-L-arginine methyl ester (L-NAME, 30 μM), inhibited the relaxation. These data indicated that in the rat thoracic aorta the hydroalcoholic extract, and its fractions, from the leaves of *E. uniflora* had graded and endothelium-dependent vasorelaxant effects.

Antiviral Activity

Four tannins isolated from the active fractions of *Eugenia uniflora* were found to have inhibitory effect on Epstein Barr virus (EBV) DNA polymerase (Lee et al. 2000). Epstein-Barr virus (EBV) is a human B lymphotropic herpes virus which is known to be closely associated with nasopharyngeal carcinoma (NPC) which is one of the high population malignant tumours among Chinese in southern China and southeast Asia. The results showed the 50% inhibitory concentration (IC_{50}) values of gallocatechin, oenothein B, eugeniflorins D1 and D2 were 26.5 62.3, 3.0 and 3.5 μM, respectively. Furthermore, when compared with the positive control (phosphonoacetic acid), an inhibitor of EBV replication, the IC_{50} value was 16.4 μM. The results confirmed that eugeniflorins D1 and D2 were the potency principles in the inhibition of EBV DNA polymerase from *E. uniflora*.

Antinociceptive and Hypothermic Activity

The chemical composition essential oil of pitangueira leaves was characterized, by a mixture of atractylone (1) and 3-furanoeudesmene (2) as the main constituents of the oil (Amorim et al. 2009). The essential oil, its pentane fraction and the isolated mixture of sesquiterpenes (1 and 2), given orally, significantly inhibited the acetic acid-induced abdominal constrictions, increased the latency time in hot plate test and showed a hypothermic effect. The results suggested that the isolated furanosesquiterpenes were responsible for the antinociceptive and hypothermic effect.

Central Nervous System Activity

A hydroalcoholic extract of *E. uniflora* leaves showed some central nervous system activity in hippocratic screening when given intraperitoneally, but little to no acute or subacute toxicity in doses up to 4,200 mg/kg orally in BALB c mice (Schmeda-Hirschmann et al. 1987). The LD50 of the extract was 220 mg/kg i.p. in mice. A decoction or infusion of the leaves was recommended for treating gout by native herbalists in Paraguay. The known flavonoids quercetin, myricitrin and myricetin were found to be responsible for the xanthine oxidase inhibitory action of the plant extract.

Diuretic Activity

Statistically significant diuretic action in relation to controls was found one hour after administration of the crude extract of pitanga leaves (0.24 mg/kg) (Amat et al. 1999). The major fraction of the extract exhibited diuretic activity 2 hours after administration (0.24 mg/kg). Statistically significant diuretic action in relation to controls was found 1 hour after administration of the crude extract of pitanga leaves (0.24 mg/kg) (Amat et al. 1999). The major fraction of the extract exhibited diuretic activity 2 hours after administration.

Antiinflammatory Activity

The infusion of fresh pitanga leaves had a highly significant antiinflammatory effect when administered p.o. to rats 1 hour before subplantar injection of carrageenin (Schapoval et al. 1994). The infusion increased the pentobarbital sleeping time and also had an effect on intestinal transit, and had no acute toxic effect.

Antidiarrhoeal Activity

Studies showed that aqueous extracts of leaves of *Achyrocline satureioides*, barks of *Eugenia uniflora*, aerial parts of *Foeniculum vulgare*, and barks of *Psidium guajava* exhibited in-vitro cytoxicity against trophozoites of *Giardia lamblia* that causes diarrhoea (Brandelli et al. 2009). These plants are traditionally used for the treatment of diarrhea by the indigenous population Mbyá-Guarani, located at the Lomba do Pinheiro, Porto Alegre, Rio Grande do Sul, Brazil. Results revealed the minimal inhibitory concentrations: 0.313 mg/ml for *A. satureioides* and *E. uniflora*, 0.02 mg/ml for *P. guajava*, and *F. vulgare* did not present any cytotoxic effect. Quantitative assays of viable trophozoites, showed that *A. satureioides* presented the highest cytotoxic effect (93.5%), followed by *P. guajava* (82.2%), and *E. uniflora* (67.3%).

Crude aqueous extract of *E. uniflora* leaves exhibited anti-diarrhoeic activity (Almeida et al. 1995). The extract increased the absorption of water in one or more intestinal portion and reduced gastrointestinal propulsion in relation to the control group.

Muscle Contractile Activity

Of the organic solvents and volatile tested, only the ethyl acetate extract of *E. uniflora* leaf exhibited higher contractile responses on smooth muscles than the reference drug (acetylcholine) (Gbolade et al. 1996). The contractile activity of the isolated rat duodenum demonstrated by the leaf extracts was absent in the volatile oil.

Trypanocidal Activity

E. uniflora extract displayed trypanocidal activity against *Trypanosoma congolense* and *T. brucei*, bloodstream form parasites (Adewunmi et al. 2001). Its extract and other plant extracts had LC 50 of 13–69 μg/ml on both the drug sensitive and multi drug resistant strain of *Trypanosoma congolense*. The LC 50 doses of the active plant extracts on the calf aorta endothelial cells varied between 112 and 13750 μg/ml while the calculated selective indices ranged between 0.71 and 246.8 indicating bright prospects for the development of some of these extracts as potential trypanocidal agents.

Antibiotic Potentiating Activity

Egenia uniflora extract was found to have antibiotic potentiating activity (Coutinho et al. 2010a). The growth of the two strains of *E. coli* bacteria tested was not inhibited in a clinically relevant form by the ethanol extract. Synergism between the extract and gentamicin was demonstrated. In the same extract synergism was observed between chlorpromazine and kanamycin and between amikacin and tobramycin, indicating the involvement of an efflux system in the resistance to these aminoglycosides.

Toxicity Effect

Using the hepatopancreas of *Oreochromis niloticus* as an experimental model, the crude extract of *E. uniflora* leaves and the ethyl, chloroform and hexane fractions were found to induce vasodilation, vascular congestion and toxicity due to the presence of eosinophilic granular cells, rodlet cells, some leukocytic infiltrate and rare focal necroses (Fiuza et al. 2009).

Traditional Medicinal Uses

There is an extensive amount of literature documenting the ethnomedicinal uses of the leaves of Surinam cherry (Consolini et al. 1999; Schapoval et al. 1994; Schmeda-Hirschmann et al. 1987, Weyerstahl et al. 1988).

In Madeira, fruits of *E. uniflora* are eaten for intestinal troubles (Rivera and Obón 1995). Fruits and leaves are also used for their astringent qualities, and are active against high blood pressure (Bandoni et al. 1972). Water decoctions of *E. uniflora* leaves are used in Paraguay to lower cholesterol and blood pressure (Ferro et al. 1988), and have a highly significant anti-inflammatory action (Schapoval, et al. 1994). In Brazil the leaf infusion is taken as a febrifuge and astringent and for stomach problems (Morton 1987). In Surinam, the leaf decoction is drunk as a cold remedy and, in combination with lemongrass, as a febrifuge. In South America the leaf infusion is used as stomachic, febrifuge and astringent. This plant is used in South American traditional medicine as an antihypertensive (Consolini and Sabburio 2002), treatment of diarrhoea (Brandelli et al. 2009) and many other disease (Oliviera et al. 2008). The plant is used by native herbalists in Paraguay for treatment of gout (Schmeda-Hirschmann et al. 1987).

Other Uses

Pitanga is also used for bonsai and in landscaping as hedges. The bark contains tannin and is used for treating leather. The flowers provide a rich source of pollen for honeybees. In Brazil, the leaves are spread over and when trampled upon, they release a pungent oil which repels flies.

Pitanga essential oil is a very interesting ingredient for use in perfumery attributed to its olfactory characteristic (Gallucci et al. 2010). It provides blooming, freshness and a distinctive character (wet wood, green, spice). Skin sensitization and dermal irritation – human patch tests carried out with 1.5% essential oil in ethanol gave no reactions under the current conditions of use as a fragrance ingredient. Photoirritation and photoallergy-UV spectrum studies revealed that the oil did not absorb UV light at wavelengths in the range of 290–400 nm and therefore would have no potential to elicit photoirritation or photoallergy under the current conditions of use as a fragrance ingredient.

Comments

The species is deemed invasive in Bermuda and Australia (New South Wales).

Selected References

Adewunmi CO, Agbedahunsi JM, Adebajo AC, Aladesanmi AJ, Murphy N, Wando J (2001) Ethnoveterinary medicine: screening of Nigerian medicinal plants for trypanocidal properties. J Ethnopharmacol 77(1):19–24

Almeida CE, Karnikowski MG, Foleto R, Baldisserotto B (1995) Analysis of antidiarrhoeic effect of plants used in popular medicine. Rev Saude Publica 29(6):428–433

Amat AG, De Battista GA, Uliana RF (1999) Diuretic activity of *Eugenia uniflora* L. (Myrtaceae) aqueous extract. Acta Hortic ISHS 501:155–158

Amoo IA, Debayo OT, Oyeleye AO (2006) Chemical evaluation of winged beans (*Psophocarpus tetragonolobus*), pitanga cherries (*Eugenia uniflora*) and orchid fruit (orchid fruit *Myristica*). Afr J Food Agric Nutr Dev 6((2):3–12

Amorim ACL, Lima CKF, Hovell AMC, Miranda ALP, Rezende CM (2009) Antinociceptive and hypothermic evaluation of the leaf essential oil and isolated terpenoids from *Eugenia uniflora* L. (Brazilian Pitanga). Phytomedicine 16(10):923–928

Arai I, Amagaya S, Komatsu Y, Okada M, Hayashi T, Kasai M, Arisawa M, Momose Y (1999) Improving effects of the extracts from *Eugenia uniflora* on hyperglycemia and hypertriglyceridemia in mice. J Ethnopharmacol 68(1–3):307–314

Bagetti M, Facco EMP, Rodrigues DB, Vizzotto M, Emanuelli T (2009) Antioxidant capacity and composition of pitanga seeds. Cien Rural 39(8):2504–2510

Bandoni AL, Mendiondo ME, Rondina RVD, Coussio JD (1972) Survey of Argentine medicinal plants. I. Folklore and phytochemical screening. Lloydia 35:69–80

Brandelli CL, Giordani RB, De Carli GA, Tasca T (2009) Indigenous traditional medicine: in vitro anti-giardial activity of plants used in the treatment of diarrhea. Parasitol Res 104(6):1345–1349

Cavalcante ML, Rodriguez-Amaya DB (1992) Carotenoid composition of the tropical fruits *Eugenia uniflora* and *Malpighia glabra*. In: Charalambous G (ed) Food science and human nutrition. Elsevier Scientific Publications, Amsterdam, pp 643–650

Chen J, Craven LA (2007) Myrtaceae. In: Wu ZY, Raven PH, Hong DY (eds) Flora of China, vol 13, Clusiaceae through Araliaceae. Science Press, Beijing, and Missouri Botanical Garden Press, St. Louis

Consolini AE, Sarubbio MG (2002) Pharmacological effects of *Eugenia uniflora* (Myrtaceae) aqueous crude extract on rat's heart. J Ethnopharmacol 81(1):57–63

Consolini AE, Baldini OA, Amat AG (1999) Pharmacological basis for the empirical use of *Eugenia uniflora* L. (Myrtaceae) as antihypertensive. J Ethnopharmacol 66(1):33–39

Coutinho HD, Costa JG, Falcão-Silva VS, Siqueira-Júnior JP, Lima EO (2010a) Potentiation of antibiotic activity by *Eugenia uniflora* and *Eugenia jambolanum*. J Med Food 13(4):1024–1026

Coutinho HD, Costa JGM, Siqueira JP Jr, Lima EO (2010b) In vitro screening by phototoxic properties of *Eugenia uniflora* L., *Momordica charantia* L., *Mentha arvensis* L. and *Turnera ulmifolia* L. R Bras Bioci Porto Alegre 8(3):299–301

Einbond LS, Reynertson KA, Luo X-D, Basile MJ, Kennelly EJ (2004) Anthocyanin antioxidants from edible fruits. Food Chem 84(1):23–28

Fadeyit MO, Akpan UE (1989) Antibacterial activities of the leaf extracts of *Eugenia uniflora* Linn. (Synonym *Stenocalyx michelli* Linn.) Myrtaceae. Phytother Res 3:154–155

Ferro E, Schinini A, Maldonado M, Rosner J, Hirschman GS (1988) *Eugenia uniflora* leaf extract and lipid metabolism in *Cebus apella* monkeys. J Ethnopharmacol 24:321–325

Fiuza TS, Silva PC, De Paula JR, Tresvenzol LM, Sabóia-Morais SM (2009) Bioactivity of crude ethanol extract and fractions of *Eugenia uniflora* (Myrtaceae) in the hepatopancreas of *Oreochromis niloticus* L. Biol Res 42(4):401–414

Gallucci S, Neto AP, Porto C, Barbizan D, Costa I, Marques K, Benevides P, Figueiredo R (2010) Essential oil of *Eugenia uniflora* L.: an industrial perfumery approach. J Essent Oil Res 22(2):176–179

Gbolade AA, Ilesanmi OR, Aladesanmi AJ (1996) The contractile effects of the extracts of *Eugenia uniflora* on isolated rat duodenum. Phytother Res 10:613–615

Govaerts R, Sobral M, Ashton P, Barrie F, Holst BK, Landrum LL, Matsumoto K, Fernanda Mazine F, Nic Lughadha E, Proenca C, Soares-Silva LH, Wilson PG, Lucas E (2010) World checklist of Myrtaceae. The board of trustees of the Royal Botanic Gardens, Kew. Published on the Internet, http://www.kew.org/wcsp/. Accessed 22 Apr 2010

Holetz FB, Pessini GL, Sanches NR, Cortez DA, Nakamura CV, Filho BP (2002) Screening of some plants used in the Brazilian folk medicine for the treatment of infectious diseases. Mem Inst Oswaldo Cruz 97(7):1027–1031

Kade IJ, Ibukun EO, Nogueira CW, da Rocha JB (2008) Sun-drying diminishes the antioxidative potentials of leaves of *Eugenia uniflora* against formation of thiobarbituric acid reactive substances induced in homogenates of rat brain and liver. Exp Toxicol Pathol 60(4–5):365–371

Lee MH, Nishimoto S, Yang LL, Yen KY, Hatano T, Yoshida T, Okuda T (1997) Two macrocyclic hydrolysable tannin dimers from *Eugenia uniflora*. Phytochemistry 44(7):1343–1349

Lee M-H, Chiou J-F, Yen K-Y, Yang L-L (2000) EBV DNA polymerase inhibition of tannins from *Eugenia uniflora*. Cancer Lett 154(2):131–136

Lima EO, Gompertz OF, Giesbrecht AM, Paulo MQ (1993) In vitro antifungal activity of essential oils obtained from officinal plants against dermatophytes. Mycoses 36(9–10):333–336

Lorenzi H, Bacher L, Lacerda M, Sartori S (2006) Brazilian fruits & cultivated exotics (for consuming in natura). Instito Plantarum de Etodos da Flora Ltda, Nova Odessa, Brazil, 740 pp

Marin R, Apel MA, Limberger RP, Raseira MCB, Pereira JFM, Zuanazzi JAS, Henriques AT (2008) Volatile components and antioxidant activity from some Myrtaceous fruits cultivated in Southern Brazil. Lat Am J Pharm 27:172–177

Melo RM, Corrêa VFS, Amorim ACL, Miranda ALP, Rezende CM (2007) Identification of impact aroma compounds in *Eugenia uniflora* L. (Brazilian Pitanga) leaf essential oil. J Braz Chem Soc 18:179–183

Morton J (1987) Surinam Cherry. In: Julia F (ed) Fruits of warm climates. Morton, Miami, pp 386–388

Ogunwande IA, Olawore NO, Ekundayo O, Walker TM, Schmidt JM, Setzer WN (2005) Studies on the essential oils composition, antibacterial and cytotoxicity of *Eugenia uniflora* L. Int J Aromather 15(3):147–152

Oliveira AL, Lopes RB, Cabral FA, Eberlin MN (2006) Volatile compounds from pitanga fruit (*Eugenia uniflora* L.). Food Chem 99(1):1–5

Oliveira MD, Andrade CA, Santos-Magalhães NS, Coelho LC, Teixeira JA, Carneiro-da-Cunha MG, Correia MT (2008) Purification of a lectin from *Eugenia uniflora* L. seeds and its potential antibacterial activity. Lett Appl Microbiol 46(3):371–376

Peixoto CA, Oliveira AL, Cabral FA (2010) Composition of supercritical carbon dioxide extracts of pitanga (*Eugenia uniflora* L.) leaves. J Food Process Eng 33(5):848–860

Pino JA, Bello A, Urquiola A, Aguero J, Marbot R (2003) Fruit volatiles of Cayena cherry (*Eugenia uniflora* L.) from Cuba. J Essent Oil Res 15:70–71

Popenoe W (1974) Manual of tropical and subtropical fruits. Hafner Press, New York, Facsimile of the 1920 edition

Porcu OM, Rodriguez-Amaya DB (2008) Variation in the carotenoid composition of the lycopene-rich Brazilian fruit *Eugenia uniflora* L. Plant Foods Hum Nutr 63(4):195–199

Rehm S (1994) Multilingual dictionary of agronomic plants. Kluwer Academic Publishers, Dordrecht/Boston/London, 286 pp

Reynertson KA, Basile MJ, Kennelly EJ (2005) Antioxidant potential of seven myrtaceous fruits. Ethnobot Res Appl 3:25–35

Rifai MA (1992) *Eugenia uniflora* L. In: Verheij EWM, Coronel RE (eds) Plant resources of South-East Asia No 2. Edible fruits and nuts. Prosea, Bogor, pp 165–167

Rivera D, Obón C (1995) The ethnopharmacology of Madeira and Porto Santo islands, a review. J Ethnopharmacol 46:73–93

Rodriguez-Amaya DB (1999) Latin American food sources of carotenoids. Arch Latinoam Nutr 49(1–5): 74S–84S

Santos GD, Ferri PH, Santos SC, Bao SN, Soares CM, Pereira M (2007) Oenothein B inhibits the expression of PbFKS1 transcript and induces morphological changes in *Paracoccidioides brasiliensis*. Med Mycol 45(7):609–618

Schapoval EE, Silveira SM, Miranda ML, Alice CB, Henriques AT (1994) Evaluation of some pharmacological activities of *Eugenia uniflora* L. J Ethnopharmacol 44(3):137–142

Schmeda-Hirschmann G, Theoduloz C, Franco L, Ferro EB, De Arias AR (1987) Preliminary pharmacological studies on *Eugenia uniflora* leaves: Xanthine oxidase inhibitory activity. J Ethnopharmacol 21(2):183–186

Spada PDS, de Souza GGN, Bortolini GV, Henriques JAP, Salvador M (2008) Antioxidant, mutagenic and antimutagenic activity of frozen fruits. J Med Food 11(1):144–151

Tropicos Org. (2010) Nomenclatural and specimen database of the Missouri Botanical Garden. http://www.tropicos.org/Home.aspx

Velázquez E, Tournier HA, Mordujovich de Buschiazzo P, Saavedra G, Schinella GR (2003) Antioxidant activity of Paraguayan plant extracts. Fitoterapia 74(1–2): 91–97

Wazlawik E, Da Silva MA, Peters RR, Correia JF, Farias MR, Calixto JB, Ribeiro-Do-Valle RM (1997) Analysis of the role of nitric oxide in the relaxant effect of the crude extract and fractions from *Eugenia uniflora* in the rat thoracic aorta. J Pharm Pharmacol 49(4):433–437

Weyerstahl P, Marschall-Weyerstahl H, Christiansen C, Oguntimein BO, Adeoye AO (1988) Volatile constituents of Eugenia uniflora leaf oil. Planta Med 54(6):546–9

Myrciaria dubia

Scientific Name

Myrciaria dubia **(Kunth) McVaugh**.

Synonyms

Eugenia divaricata Benth., *Eugenia grandiglandulosa* Kiaersk., *Marlierea macedoi* D. Legrand, *Myrciaria caurensis* Steyerm., *Myrciaria divaricata* (Benth.) O. Berg, *Myrciaria lanceolata* O. Berg, *Myrciaria lanceolata* var. *angustifolia* O. Berg, *Myrciaria lanceolata* var. *glomerata* O. Berg, *Myrciaria lanceolata* var. *laxa* O. Berg, *Myrciaria obscura* O. Berg, *Myrciaria paraensis* O. Berg, *Myrciaria phillyraeoides* O. Berg, *Myrciaria riedeliana* O. Berg, *Myrciaria spruceana* O. Berg, *Myrtus phillyraeoides* (O. Berg) Willd. ex O. Berg, *Psidium dubium* Kunth (basionym).

Family

Myrtaceae

Common/English Names

Camu-Camu, Guavaberry, Rumberry

Vernacular Names

Brazil: Camu-Camu, Caçari, Arazá De Agua;
Columbia: Guayabo;
Peru: Camo-Camo,Camu-Camu;
Spanish: Camo Camo, Camu-Camu, Guayabito, Guapuro Blanco;
Venezeula: Guaiabito, Guayabato.

Origin/Distribution

The camu-camu is regarded to have originated in the western Amazon basin. Its current native range covers the Amazonian wet lowlands of Colombia, Ecuador, Peru, Bolivia and Brazil. The distribution of Camu camu extends from the center of Para state, Brazil, along the mid and upper Amazon River to the eastern part of Peru; in the north it appears in the Casiquiare

T.K. Lim, *Edible Medicinal And Non-Medicinal Plants: Volume 3, Fruits*,
DOI 10.1007/978-94-007-2534-8_86, © Springer Science+Business Media B.V. 2012

and the upper and middle Orinoco River. The greatest concentration of wild population and varieties is found in the Peruvian Amazon. In Brazil, it is found in Rondônia along the Maçangana and Urupa Rivers and in Amazonas, in the municipalities of Manaus and Manacapuru and along the Javarí, Madeira and Negro.

Agroecology

M. dubia thrives in a hot and humid tropical environment but will also survive in frost free subtropical areas. It is found in areas where the temperature rarely dip below 20°C and with mean annual rainfall greater than 1,200 mm. It occurs in the lowland from sea level to 300 m elevation. In it native range, it occurs in the semi-exposed, bushy areas along the banks of rivers and lakes where it is adapted to periodic flooding for up to 4–5 months . In some areas of Peru and along the Peru-Brazilian border it frequently forms extensive thickets on the river floodplain. It thrives best on well-drained clay oxisols and rich clay loams of the Amazon river floodplains. It also occurs on poorer sandy sites along black water rivers in its native region.

Edible Plant Parts and Uses

Camu-camu fruit is edible but is less frequently consumed raw because of its high acidity which dominates its flavour. The fruit find excellent use as a refreshing juice with added water, sugar or honey and is more popularly used for making jellies, jams, drinks, ice creams, pickles, liqueurs, wines and it can be used in the enrichment of other foods or fruits. In Iquitos (Peru), the juice is sold as a popular bottled carbonated drink. The juice is also exported to the USA for use in "organic" vitamin C tablets and other international markets. In United States, Japan and France markets, products can be found as candies of vitamin C produced with camu-camu. The fruit has only recently come into large-scale cultivation and sale to the world market with Japan being the major buyer. Camu-camu is extremely rich in vitmin C only second to the Australian native, *Terminalia ferdinandiana*. Compared with oranges, the fruits of camu-camu contain 30 times more vitamin C. Camu camu most likely provides other nutritional benefits such as antioxidants that includes phenolics, etc., but these are less communicated to consumers.

Botany

A large, much branched, bushy shrub or small tree growing to 8 m high. Stem is smooth with thin, pale to bronzy-brown bark and the glabrous branches arising low down on the main stem. Leaves are simple, opposite, on 3–9 mm long petioles (Plates 1 and 2). Lamina is narrowly ovate to elliptic, 5–12 cm × 2–4.5 cm, gland-dotted, apex acuminate, base sub-cuneate to rounded, margins entire, dull, dark green above and paler

Plate 1 Immature camu-camu fruits and leaves

Plate 2 Close-view of immature camu-camu fruits

Plate 3 Mature green and early-ripening camu-camu fruit

Plate 4 Ripe camu-camu fruits

green abaxially, with a prominent mid rib and about 20 pairs of obscure lateral veins. Flowers are tiny, white, sub-sessile in axillary racemose clusters of 4 flowers. bracteoles are persistent, broadly ovate and rounded at apex, united by their basal margins into a cuplike involucre. The hypanthium is sessile, glabrous, broadly obconic (as an inverted cone). The hypanthium is adnate to the ovary and prolonged. The bracteoles, hypanthium, calyx-lobes, and corolla have dark raised glands. The clayx has 4 imbriacte, broadly rounded lobes, bright rufous-pubescent on the inner surface, petals 4 white, ovate and ciliate, stamens are numerous, >120, up to 10 mm long, ovary is inferior with a simple style. Fruit is a globose berry 1–3 cm in diameter, with a circular, hypanthial scax at the apex and thin skinned, green turning to reddish-brown to purple-black (Plates 1, 2, 3 and 4), pulp fleshy, soft at maturity and enclosing 2–3 seeds.

Nutritive/Medicinal Properties

Villachica (1996) reported the following nutrient composition of camu-camu fruits: water 94.4 g, energy 17 cal, protein 0.5 g, CHO 4.7 g, fibre 0.6 g, ash 0.2 g, fats 0.2 g, thiamine 0.01 mg, riboflavin 0.04 mg, niacin 0.062 mg, vitamin C 2,994 mg, Ca 27 mg, P 17 mg, Fe 0.5 mg.

Nutrient composition of camu-camu pulp (g/100 g) was reported by Justi et al. (2000) as: moisture 94.1 g, protein 0.4 g, ash 0.3 g, crude fibre 0.1 g, lipids 0.2 g, carbohydrate 3.52 g, vitamin C 1,410 mg; minerals (mg/kg) Na 111.3 mg, K 838.8 mg, Ca 157.3 mg, Fe 5.3 mg, Mg 123.8 mg, Mn 21.1 mg, Zn 3.6 mg, Cu 2 mg, Co 0.1 mg, Ca 0.01 mg, Pb 0.2 mg; fatty acids: C13:0 (tridecanoic) 7.2%, C16:0 (palmitic) 6.6%, C18:0 (stearic) 10%, C18:1w9 (oleic) 11.8%, C18:2w6 (linoleic) 9.7%, C18:3w6 (g-linolenic) 9.3%, C18:3w3 (α-linolenic) 16%, C20:2w6 (eicosadienoic) 10.5%, C20:5w3 (EPA, eicosapentaenoic acid) 7.0%, and C23:0 (tricosanoic) 11.9%.

The nutrient value of ripe camu-camu fruits per 100 g edible portion was reported by Lorenzi et al. (2006) as: energy 20.9 cal., water 94.4 g, protein 0.50 g, fat 0.01 g, carbohydrate 4.7 g, ash 0.20 g, Ca 27 mg, P 7 mg, Fe 0.5 mg, vitamin B-1 0.01 mg, vitamin B-2 0.01 mg, niacin 0.62 mg, and vitamin C 2,994 mg.

Ripe fruit of camu-camu was found to contain g/kg of the following nutrients (Zapata and Dufour 1993): ascorbic acid 9.39 g, dehydroascorbic acid 0.31 g, glucose 8.16 g, fructose 9.51 g, citric acid 19.81 g, isocitric acid 0.15 g, malic acid 5.98 g, acidity (citric acid) 30.8 g, pH 2.56, brix 6.8%, total solids 81 g, total nitrogen 0.735 g; amino acids (in mg/kg) serine 637 mg, valine 316 mg, leucine289mg, glutamate

119 mg, 4-aminobutanoate 108 mg, proline 82 mg, phenylalanine 43 mg, threonine 36 mg, alanine 34 mg; minerals (in mg/kg) K117mg, Ca 65, Mg 51 mg, Na 27 mg, P 295 mg, S 132 mg, Al 2.1 mg, B 0.5 mg, Cu 0.8 mg, Fe 1.8 mg, Mn 2.1 mg, Zn 1.3 mg and Cl 116 mg.

Camu-camu fruits had higher contents of sodium, calcium, potassium, manganese, zinc and copper than acerola but were lower in iron, magnesium, cadmium and lead contents (Justi et al. 2000). Most of the fatty acids detected were polyunsaturated (PUFA) with 52.5%, followed by the saturated (SFA) with 35.7%, and the monounsaturated (MUFA) with 11.8%. Tricosanoic and stearic acid were the predominant saturated acids, with 11.9% and 10.0%, respectively. PUFA α- and γ-linolenic acids with contents of 16.0% and 9.3%, respectively. The α-linolenic acid comprised the greatest proportion of fatty acids detected. The eicosapentaenoic acid (EPA), was foundin a considerable content (7.0%), considering that it is an acid found commonly in fish oils and sea animals. The analysis of the fatty acids obtained from camu-camu showed that they were of higher quality that obtained from acerola. Camu-camu was also one of the richest sources of vitamin C (2.4–3.0 g/100 g in the pulp) found in Brazil (Justi et al. 2000). When stored at −18°C after 28 days, the vitamin C decreased 23% from 1.57 to 1.21 g/100 g. After 335 days of storage, the content found was 1.16 g/100 g of pulp, the ascorbic acid losses amounted to 26%. This content was still higher than the one found for most fruits considered good sources of vitamin C. Dib taxi et al. (2003) developed a process for the microencapsulation of camu-camu juice, optimizing the operational conditions. The optimum conditions for juice yield and vitamin C retention were established as 15% wall material and an air entry temperature of 150°C.

As camu-camu fruit matured, levels of ascorbic and dehydroascorbic acids, reducing sugars (fructose and glucose were the major sugars), amino acids (serine, valine and leucine) and soluble solids were found to increase (Zapata and Dufour 1993). Citric acid was the major acid (from 19.8 up to 29.8 g/kg) and was responsible for the fruit's sour taste. Unlike citric acid, malic acid increased with maturation. Among the macronutrients, potassium was the most abundant mineral (711 mg/kg) and could be considered, like vitamin C, nutritionally significant. During maturation, the fruit pulp colour turned from yellow-green to pink, presumably due to the migration of anthocyanin pigments from the peel.

Camu-camu fruit were harvested and analysed in three stages of maturation according to colour – green, three-quarter green and three-quarter red (Alves et al. 2002). Ascorbic acid content was higher in more mature fruit, varying from 1,791 to 2,061 mg/100 g. The tart flavour of camu-camu was due to its high titrable acidity, even in three-quarter red fruits – 2.6% as citric acid, and low soluble sugar content – 1.5% as glucose in three-quarter red fruit. The contents of starch and pectin, varied respectively from 0.43% to 0.34% and 0.17% to 0.11% as fruit matured from green to three-quarter red, suggesting that the extraction of juice or pulp may be relatively easy and will benefit little from the use of processing enzymes. In contrast, phenolic content may indicate a factor of restriction to palatability, because they were associated with astringency. All the extracted fractions, 50% methanol-soluble and water-soluble, gave high phenolic contents, even though they decreased as fruit matured from green to three-quarter red.

Other Phytochemicals

The following phenolic compounds (mg/g d.w) were found to be present in the methanol extract of *M. dubia* fruits: cyanidin 3-glucoside (0.02×10^{-3} mg), delphinidin 3-glucoside (traces), ellagic acid (0.45 mg), myricetin (n.d.), quercetin (0.24 mg), quercitrin (0.06×10^{-3} mg), and rutin (0.13×10^{-3} mg) (Reynertson et al. 2008). Total phenolic content (TPC), for *M. dubia* was reported as 101.17 mg GAE/g. Traces (<0.1 mg/g) of total anthocyanin content (TAC) was found. Among 14 underutilised Myrtaceous fruits, *Myrciaria cauliflora, M. dubia and M. vexator*, were found to have the highest antiradical activity, measured as DPPH˙ IC_{50}, (*M. cauliflora*, 19.40; *M. dubia*, 57.19; and *M. vexator*, 38.64 μg/ml). The results demonstrated edible fruits in the Myrtaceae to be rich sources of biologically active phenolic compounds.

The major anthocyanins in camu-camu fruits in two different regions of the São Paulo state, Brazil were found to be cyanidin-3-glucoside (88–89.5%), followed by delphinidin-3-glucoside (4.2–5.1%) (Zanatta et al. 2005). Higher total anthocyanin contents were detected in the fruits from Iguape (54.0 mg/100 g) compared to those from Mirandópolis (30.3 mg/100 g), most likely because of the lower temperatures in the Iguape region.

Twenty-one volatile compounds were identified in camu-camu fruit (Franco and Shibamoto 2000). Terpenic compounds predominated among the volatile compounds with α-pinene and d-limonene being the most abundant volatile compounds.

A total of 111, 114 and 138 compounds were identified in the liquid-liquid extraction (LLE), simultaneous distillation-solvent extraction (SDE) and headspace solid-phase microextraction (HSSPME) of camu-camu fruit, respectively (Quijano and Pino 2007). Limonene was the major component of these concentrates, followed by α-pinene. The relative proportions of the major compounds are shown to be dependent upon the isolation methods. The volatile compounds isolated from camu-camu included: acetaldehyde, ethanol, acetic acid; ethyl acetate; 2-pentanone; pentanal; ethyl propionate; methyl butyrate; ethyl isobutyrate; toluene; methyl isovalerate; methyl 2-methylbutyrate; 2,4-dimethyl-3-pentanone; hexanal; ethyl butyrate; butyl acetate; 2-furfural; 3-methylbutanoic acid; ethyl (E)-crotonate; ethyl 2-methylbutyrate; (E)-2-hexenal; (Z)-3-hexenol; hexanol; isoamyl acetate; 2-heptanone; 5-hepten-2-one; 2-heptanol; methyl hexanoate; tricyclene; α-thujene; α-pinene; camphene; thuja-2,4(10)-diene; 2,4-dimethyl-3-pentanone; benzaldehyde; 5-methyl-2-furfural; sabinene; β-pinene; methyl 2-furoate; p-mentha-3-one; benzoate; isopentyl isovalerate; (Z)-3-hexenyl 2-methylbutyrate; 2-methylbutyrate; benzyl isovalerate; myrcene; α-phellandrene; α-terpinene; p-cymene; limonene; benzyl alcohol; (Z)-β-ocimene; 2-phenylacetaldehyde; (E)-β-ocimene; ethyl 2-furoate; γ-hexalactone; amyl isobutyrate; γ-terpinene, acetophenone; trans-linalool oxide (furanoid); heptanoic acid; cis-linalool oxide (furanoid); p-mentha-2,4(8)-diene; terpinolene; 2-nonanone; p-cymenene; 6-camphenone; linalool; nonanal; isoamyl isovalerate; 2-phenylethanol; p-mentha-1,3,8-triene; α-fenchol; cis-p-mentha-2-en-1-ol; 2-ethylhexanoic acid; trans-p-mentha-2,8-dien-1-ol; α-campholenal; allo-ocimene; terpinen-1-ol; trans-pinocarveol; trans-p-menth-2-en-1-ol; cis-β-terpineol; (E,E)-allo-ocimene; camphor; trans-verbenol; (Z)-3-hexenyl isobutyrate; camphene hydrate,2,5-dimethyl-4-hydroxy–3(2H)-furanone; 3-methylbut-2-enyl valerate; benzyl acetate; isoborneol; pinocarvone; borneol; mentha-1,5-dien-8-ol; octanoic acid; ethyl benzoate; tr cis-linalool oxide (pyranoid); terpinen-4-ol; p-cymen-8-ol; p-methylacetophenone; cryptone; dill ether; (Z)-3-hexenyl butyrate; and α-terpineol.

One hundred and fifteen components were identified in camu-camu leaf oil, of which α-pinene (74.3%) and limonene (10.8%) were the major constituents (Pino and Quijano 2008). The composition of the leaf oil was characterized by its richness in monoterpene and sesquiterpene hydrocarbons and the presence of many esters and unsaturated aldehydes.

Other phytochemicals associated with various pharmacological activities of the various plant parts are presented below.

Antioxidant Activity

Camu-camu fruit juice exhibited potent antioxidant activity as determined by the Total Oxidant Scavenging Capacity assay based upon the ethylene yielding reaction of α-keto-γ-methiolbutyric acid with three reactive oxygen species (peroxyl radicals, hydroxyl radicals, and peroxynitrite) (Rodrigues and Marx 2006; Rodrigues et al. 2006). Compared to other fruits camu-camu exhibited outstanding antioxidant activity against peroxyl radicals and peroxynitrite and less so against hydroxyl radicals as shown by the sequences:

peroxyl-radicals: camu-camu > açaí > blueberry > cashew > orange > apple;

peroxynitrite: camu-camu > blueberry > cashew > açaí > orange > apple; and

hydroxyl radicals: cashew = blueberry > apple > camu-camu > orange > açaí.

Beside other Brazilian fruits, camu-camu frozen pulp was found to provide the highest antioxidant. Quercetin and kaempferol derivatives were the main flavonoids present (De Souza et al. 2010). Ellagic acid was also detected.

Ascorbic acid decreased, and anthocyanin, flavonol and flavanol contents, and DPPH antioxidant capacity increased during ripening of camu-camu (Chiniros et al. 2010). Antioxidant compounds from camu camu were fractionated in two fractions: an ascorbic acid-rich fraction (F-I) and a phenolics-rich fraction (F-II). F-I was the major contributor to the DPPH antioxidant capacity (67.5–79.3%) and F-II played a minor role (20.7–32.5%). A total of 30 different phenolic compounds were detected including catechin, delphinidin 3-glucoside, cyanidin 3-glucoside, ellagic acid and rutin. Other phenolic compounds, such as flavan-3-ol, flavonol, flavanone and ellagic acid derivatives, were also present. For the three ripening stages the flavan-3-ols and ellagic acid group were the most representative phenolic compounds in this fruit. Acid hydrolysis of F-II revealed the presence mainly of gallic and ellagic acids, suggesting that camu camu fruit had important quantities of hydrolysed tannins (gallo- and/or ellagitannins). These results confirmed camu camu fruit to be a promising source of antioxidant phenolics.

Antiinflammatory Activity

Camu-camu juice was found to have powerful anti-oxidative and anti-inflammatory properties, compared to vitamin C tablets containing equivalent vitamin C content (Inoue et al. 2008). In the study, 20 male smoking volunteers, considered to have an accelerated oxidative stress state, were recruited and randomly assigned to take daily 70 ml of 100% camu-camu juice, corresponding to 1,050 mg of vitamin C (camu-camu group; n = 10) or 1,050 mg of vitamin C tablets (vitamin C group; n = 10) for 7 days. After 7 days of treatment, oxidative stress markers such as the levels of urinary 8-hydroxy-deoxyguanosine and total reactive oxygen species and inflammatory markers such as serum levels of high sensitivity C reactive protein, interleukin (IL)-6, and IL-8 decreased significantly in the camu-camu group, while there was no change in the vitamin C group. These effects were postulated to be due to the existence of unknown anti-oxidant substances besides vitamin C or unknown substances modulating in vivo vitamin C kinetics in camu-camu.

Hepatoprotective Activity

The juice of camu-camu (*Myrciaria dubia*) was found to significantly suppress D-galactosamine (GalN)-induced liver injury in rats when the magnitude of liver injury was assessed by plasma alanine aminotransferase and aspartate aminotransferase activities, although some other juices (acerola, dragon fruit, shekwasha, and star fruit) also tended to have suppressive effects (Akachi et al. 2010). An active compound was isolated from camu-camu juice and determined to be 1-methylmalate. On the other hand, malate, 1,4-dimethylmalate, citrate, and tartrate had no significant effect on GalN-induced liver injury. It was suggested that 1-methylmalate might be a rather specific compound among organic acids and their derivatives in fruit juices in suppressing GalN-induced liver injury.

Aldose Reductase Inhibition Activity

Ellagic acid (1) and its two derivatives, 4-O-methylellagic acid (2) and 4-(α-rhamnopyranosyl) ellagic acid (3) were isolated from *Myrciaria dubia* as inhibitors of aldose reductase (AR) (Ueda et al. 2004). Compound 3 showed the strongest inhibition against human recombinant AR (HRAR) and rat lens AR (RLAR). Inhibitory activity of compound 3 against HRAR (IC_{50} value = 4.1×10^{-8} M) was 60 times more than that of quercetin (2.5×10^{-6} M). The type of inhibition against HRAR was uncompetitive. Aldose reductase (AR) is an enzyme that catalyses the reduction of glucose to sorbitol in the polyol pathway in insulin-insensitive tissues such as nerve, lens, retina and kidney causing diabetic complications, such as neuropathy, cataract, retinopathy and nephropathy (Haraguchi et al. 1998).

Antiplasmodial Activity

The aqueous and ethanol extract of *Myrciaria dubia* fruit peel exhibited antiplasmodial activity against the chloroquine resistant strain of *Plasmodium falciparum* with IC_{50} of 3 and 6 µg/ml respectively.

Other Uses

Camu-camu has no other documented uses except for its edible fruit. It has never been documented as a traditional herbal remedy for any condition in Amazonia.

Comments

Other closely related, edible species include *M. yciaria baporeti* Le Grand, *M. floribunda* Berg., *M. ibarrae* Lundell., *M. tenella* Berg., *M. trunciflora* Berg., *M. vismeifolia* Berg. and *Plinia cauliflora* (DC.) Kuasel (*Myciaria cauliflora* Berg.).

Selected References

Akachi T, Shiina Y, Kawaguchi T, Kawagishi H, Morita T, Sugiyama K (2010) 1-methylmalate from camu-camu (*Myrciaria dubia*) suppressed D-galactosamine-induced liver injury in rats. Biosci Biotechnol Biochem 74(3):573–578

Alves RE, Filgueiras HAC, Moura CFH, Araújo NCC, Almeida AS (2002) Camu-camu (*Myrciaria dubia* Mc Vaugh): a rich natural source of vitamin C. Proc Interamer Soc Trop Hort 46:11–13

Brako L, Zarucchi JL (1993) Catalogue of theflowering plants and gymnosperms of Peru. Monogr Syst Bot Missouri Bot Gard 45:1–1286

Chiniros R, Galarza J, Betalleluz-Pallardel I, Pedreschi R, Campos D (2010) Antioxidant compounds and antioxidant capacity of Peruvian camu camu (*Myrciaria dubia* (H.B.K.) McVaugh) fruit at different maturity stages. Food Chem 120(4):1019–1024

De Souza SGAE, Lajolo FM, Genovese MI (2010) Chemical composition and antioxidant/antidiabetic potential of Brazilian native fruits and commercial frozen pulps. J Agric Food Chem 58(8):4666–4674

Dib Taxi CM, de Menezes HC, Santos AB, Grosso CR (2003) Study of the microencapsulation of camu-camu (*Myrciaria dubia*) juice. J Microencapsul 20(4):443–448

Franco MRB, Shibamoto T (2000) Volatile composition of some Brazilian fruits: umbu-caja (*Spondias citherea*), camu-camu (*Myrciaria dubia*), araça-boi (*Eugenia stipitata*), and cupuaçu (*Theobroma grandiflorum*). J Agric Food Chem 48(4):1263–1265

Govaerts R, Sobral M, Ashton P, Barrie F, Holst BK, Landrum LL, Matsumoto K, Fernanda Mazine F, Nic Lughadha E, Proenca C, Soares-Silva LH, Wilson PG, Lucas E (2010) World checklist of Myrtaceae. The board of trustees of the Royal Botanic Gardens, Kew. Published on the Internet, http://www. kew.org/wcsp/. Accessed 22 Apr 2010

Haraguchi H, Kanada M, Fukuda A, Naruse K, Okamura N, Yagi A (1998) An inhibitor of aldose reductase and sorbitol accumulation from *Anthocepharus chinensis*. Planta Med 64:68–69

Inoue T, Komoda H, Uchida T, Node K (2008) Tropical fruit camu-camu (*Myrciaria dubia*) has anti-oxidative and anti-inflammatory properties. J Cardiol 52(2):127–132

Instituto Nacional de Pesquisas da Amazonia (INPA) (1986). Food and Fruit Bearing Forest Species 3: Examples from Latin America. Forestry Paper 44–3, Food and Agriculture Organization of the United Nations, Rome, 332 pp

Justi KC, Visentainer JV, Evelázio de Souza N, Matsushita M (2000) Nutritional composition and vitamin C stability in stored camu-camu (*Myrciaria dubia*) pulp. Arch Latinoam Nutr 50(4):405–408

Lorenzi H, Bacher L, Lacerda M, Sartori S (2006) Brazilian fruits & cultivated exotics (for consuming in natura). Instito Plantarum de Etodos da Flora Ltda, Nova Odessa, 740 pp

Macbride JF (1936–1971) Flora of Peru, Botanical series, 6 parts. Field Museum of Natural History, Chicago

McVaugh R (1969) The botany of the Guayana highland – part VIII. Mem N Y Bot Gard 18(2):231

Pino JA, Quijano CE (2008) Volatile constituents of Camu-camu (*Myrciaria dubia* (HBK) McVaugh) leaves. J Essent Oil Res 20(3):205–207

Quijano CE, Pino JA (2007) Analysis of volatile compounds of camu-camu (*Myrciaria dubia* (HBK) Mcvaugh) fruit isolated by different methods. J Essent Oil Res 19:527–533

Reynertson KA, Yang H, Jiang B, Basile MJ, Kennelly EJ (2008) Quantitative analysis of antiradical phenolic constituents from fourteen edible Myrtaceae fruits. Food Chem 109(4):883–890

Rodrigues RB, Marx F (2006) Camu Camu [*Myrciaria dubia* (H.B.K.) McVaugh]: a promising fruit from the Amazon basin. Ernährung (Nutrition) 30(9): 376–381

Rodrigues RB, Papagiannopoulos M, Maia JGS, Yuyama K, Marx F (2006) Antioxidant capacity of camu camu [*Myrciaria dubia* (H.B.K.) Mc Vaugh] pulp. Ernährung (Nutrition) 30(9):357–362

Ueda H, Kuroiwa E, Tachibana Y, Kawanishi K, Ayala F, Moriyasu M (2004) Aldose reductase inhibitors from the leaves of *Myrciaria dubia* (H. B. & K.) McVaugh. Phytomedicine 11(7–8):652–656

Villachica LH (1996) El cultivo del camu-camu *Myrciaria dubia* (H. B. K.) McVaugh en la Amazônia Peruana. Secretaria Pro Tempore del Tratado de Cooperación Amazónica. 46. 95 pp (in Spanish)

Yapu DG, Mozombite DS, Salgado ER, Turba AG (2008) Evaluación de la actividad antiplasmódica in vitro de extractos de *Euterpe oleracea*, *Myrciaria dubia* y *Croton lechleri*. Biofarbo 16(1):16–20 (in Spanish)

Zanatta CF, Cuevas E, Bobbio FO, Winterhalter P, Mercadante AZ (2005) Determination of anthocyanins from camu-camu (*Myrciaria dubia*) by HPLC-PDA, HPLC-MS, and NMR. J Agric Food Chem 53(24):9531–9535

Zapata SM, Dufour J (1993) Camu camu *Myrciaria dubia* (HBK) McVaugh: chemical composition of fruit. J Sci Food Agric 61:349–351

Myrciaria vexator

Scientific Name

Myrciaria vexator McVaugh.

Synonyms

Eugenia palmarum Standl. & L.O. Williams ex P.H. Allen, *Myrciaria pittieri* Burret ex Badillo.

Family

Myrtaceae

Common/English Names

Blue Grape, False Jaboticaba, Vexator

Vernacular Names

Spanish: Jaboticaba Azul.

Origin/Distribution

M. vexator is a native of tropical America – Panama, Costa Rica, Venzuela and Ecuador. It has been introduced to other countries in central and south America and elsewhere in the tropics.

Agroecology

M. vexator is tropical in it climatic requirement but will survive in warm sub-tropical areas. It prefers well-drained, moist soils rich in organic matter and requires copious watering but not flooding.

Edible Plant Parts and Uses

M. vexator has dark purple, almost bluish fruits with excellent, grape-berry flavoured sweet pulp that are usually eaten fresh or used in drinks and jellies.

Botany

A large shrub (2–3 m) or small, much-branched perennial, tree growing to 6–10 m high. The stem and branches are smooth which brown bark that peels off to reveal a gray-white coloured lower layer (Plate 1). The tree is generally glabrous except for the ciliate-margined bracts and bracteoles, and the inner faces of the perianth-parts (sepals and petals). Leaves are simple, elliptic-oblong or elliptic-ovate 2–4.5 cm wide, 5.5–13.5 cm long, acuminate and acutely attenuate pointed, broadly rounded at base, with inconspicuous glands, a prominent mid rib and many obscure lateral veins and borne on stout

T.K. Lim, *Edible Medicinal And Non-Medicinal Plants: Volume 3, Fruits*,
DOI 10.1007/978-94-007-2534-8_87, © Springer Science+Business Media B.V. 2012

Plate 1 Characteristic whitish-gray trunk and branches with flaky brown bark

Plate 2 Lime green leaves

terete petioles 4.5–8 mm long (Plates 2 and 3). Bracts are membranous, ovate, sometimes nearly concealing the buds; bracteoles membranous, distinct, orbicular or sometimes pointed. Flowers are white and borne in 4-8-flowered axillary clusters (Plate 1). Hypanthium short 2 mm and glabrous, calyx lobes and petals oval 2 mm long, stamens numerous around 50 with long protruding, white filaments and anthers. Fruit is globose, 2–3 cm diameter, dark purple to almost blue (Plate 4), the sweetish pulp encloses two seeds.

Plate 3 Axillary inflorescence and open flower

Plate 4 Ripe blue grape fruit

Nutritive/Medicinal Properties

The following phenolic compounds (mg/g d.w) were found to be present in the methanol extract of *M. vexator* fruits: cyanidin 3-glucoside (13.13 mg), delphinidin 3-glucoside (0.29 mg), ellagic acid (0.64 mg), myricetin (0.03×10^{-3} mg), quercetin (0.08 mg), quercitrin (0.05 mg), and rutin (0.11 mg) (Reynertson et al. 2008). Total phenolic content (TPC), for *M. vexator* was reported as 101.17 mg GAE/g. Total anthocyanin content (TAC) was found at 6.84 mg C3G/g dry weight. Among 14 underutilised Myrtaceous fruits, *Myrciaria cauliflora, M. dubia and M. vexator*, were found to have the highest antiradical activity, measured as DPPH˙ IC_{50}, (*M. cauliflora*, 19.40; *M. dubia*, 57.19; and *M. vexator*, 38.64 μg/ml). The results demonstrated edible fruits in the Myrtaceae to be a rich sources of biologically active phenolic compounds.

Other Uses

An interesting tree for ornamental landscaping because of its distinctive brown flaky bark and white-gray trunk and branches, and its small stature.

Comments

Blue grape can be propagated by seeds as well as cuttings.

Selected References

Barrie FR (2007) Myrtaceae. In: Hammel BE, Grayum MH, Herrera C, Zamora N (eds) Manual de Plantas de Costa Rica, vol 6. Monogragh System Bot Missouri Botanical Garden, vol 111. Missouri Botanical Garden Press, St Louis, pp 28–784

Govaerts R, Sobral M, Ashton P, Barrie F, Holst BK, Landrum LL, Matsumoto K, Fernanda Mazine F, Nic Lughadha E, Proenca C, Soares-Silva LH, Wilson PG, Lucas E (2010) World checklist of Myrtaceae. The board of trustees of the Royal Botanic Gardens, Kew. Published on the Internet, http://www. kew.org/wcsp/. Accessed 22 Apr 2010

McVaugh R (1969) The botany of the Guyana highland. Part VIII. Mem N Y Bot Gard 18(2):231

Reynertson KA, Yang H, Jiang B, Basile MJ, Kennelly EJ (2008) Quantitative analysis of antiradical phenolic constituents from fourteen edible Myrtaceae fruits. Food Chem 109(4):883–890

Myrtus communis

Scientific Name

Myrtus communis L.

Comprised of: *Myrtus communis* L. subsp. *communis and Myrtus communis* subsp. *tarentina* (L.) Nyman

Synonyms

(a) For *Myrtus communis* L. subsp. *communis*

Myrtus acuta Mill., *Myrtus acutifolia* (L.) Sennen & Teodoro, *Myrtus angustifolia* Raf. nom. illeg., *Myrtus augustinii* Sennen & Teodoro, *Myrtus baetica* (L.) Mill., *Myrtus baetica* var. *vidalii* Sennen & Teodoro, *Myrtus baui* Sennen & Teodoro, *Myrtus belgica* (L.) Mill., *Myrtus borbonis* Sennen, *Myrtus briquetii* (Sennen & Teodoro) Sennen & Teodoro, *Myrtus buxifolia* Raf. nom. illeg., *Myrtus christinae* (Sennen & Teodoro) Sennen & Teodoro, *Myrtus communis* subsp. *mucronata* Pers., *Myrtus communis* var. *acuminata* Rouy & E.G.Camus, *Myrtus communis* var. *acutifolia* L., *Myrtus communis* var. *christinae* Sennen & Teodoro, *Myrtus communis* var. *angustifolia* L., *Myrtus communis* var. *baetica* L., *Myrtus communis* var. *balearica* Sennen & Teodoro, *Myrtus communis* var. *belgica* L., *Myrtus communis* var. *eusebii* Sennen & Teodoro, *Myrtus communis* var. *foucaudii* Sennen & Teodoro, *Myrtus communis* var. *gervasii* Sennen & Teodoro, *Myrtus communis* var. *grandifolia* Sennen & Teodoro, *Myrtus communis* var. *italica* (Mill.) Rouy & E.G.Camus, *Myrtus communis* var. *joussetii* Sennen & Teodoro, *Myrtus communis* var. *lusitanica* Rouy, *Myrtus communis* var. *mucronata* L., *Myrtus communis* var. *neapolitana* Sennen & Teodoro, *Myrtus communis* var. *romana* L., *Myrtus eusebii* (Sennen & Teodoro) Sennen & Teodoro, *Myrtus gervasii* (Sennen & Teodoro) Sennen & Teodoro, *Myrtus italica* Mill., *Myrtus italica* var. *briquetii* Sennen & Teodoro, *Myrtus italica* var. *petri-ludovici* Sennen & Teodoro, *Myrtus josephi* Sennen & Teodoro, *Myrtus lanceolata* Raf. nom. illeg., *Myrtus latifolia* Raf. nom. illeg., *Myrtus littoralis* Salisb., *Myrtus macrophylla* J.St.-Hil., *Myrtus major* Garsault, *Myrtus media* Hoffmanns., *Myrtus microphylla* J.St.-Hil., *Myrtus minima* Mill., *Myrtus minor* Garsault, *Myrtus mirifolia* Sennen & Teodoro, *Myrtus oerstedeana* O.Berg, *Myrtus petri-ludovici* (Sennen & Teodoro) Sennen & Teodoro, *Myrtus rodesi* Sennen & Teodoro, *Myrtus romana* (L.) Hoffmanns., *Myrtus romanifolia* J.St.-Hil., *Myrtus sparsifolia* O.Berg, *Myrtus theodori* Sennen, *Myrtus veneris* Bubani, *Myrtus vidalii* (Sennen & Teodoro) Sennen & Teodoro.

(b) For *Myrtus communis* subsp. *tarentina* (L.) Nyman

Myrtus communis var. *tarentina* L., *Myrtus tarentina* (L.) Mill.

T.K. Lim, *Edible Medicinal And Non-Medicinal Plants: Volume 3, Fruits*,
DOI 10.1007/978-94-007-2534-8_88, © Springer Science+Business Media B.V. 2012

Family

Myrtaceae

Common/English Names

Common Myrtle, European Myrtle, Greek Myrtle, Myrtle, Sweet Myrtle, True Myrtle

Vernacular Names

Albanian: Mersinë E Rëndomtë, Cimartë, Mërçelë, Mërsina;
Amharic: Addus;
Arabic: Ahmam, Ar-Raihan, Arrihane, As, Aselmûn, Hadass, Halmuch, Houmblass, Raihan, Rayhdn, Rihân;
Armenian: Mrdeni, Mrdi, Mrteni, Mrti, Murt;
Belarusian: Mirt;
Bulgarian: Mirta;
Brazil: Murta (Portuguese);
Catalan: Herba De Poll, Mata Poll, Murta, Murter, Murtera, Murtiñera, Murtó, Murton, Murtons, Murtra, Murtrer, Murtrera, Tei, Tell, Tintorell;
Chinese: Heong Tou Muhk (Cantonese), Xiang Tao Mu, Hsiang Tao Mu;
Croatian: Mirta, Mrtvina;
Czech: Myrta, Myrta Obecná;
Danish: Almindelig Myrte, Myrte;
Dutch: Mirt, Mirte, Myrte;
Eastonian: Harilik Mürt;
Esperanto: Mirto, Ordinara Mirto;
Euskara: Arrayana, Mitre;
Farsi: Moord, Mourd;
Finnish: Myrtti;
French: Eau D'ange, Myrte, Myrte Commun, Myrte Commune, Vrai Myrte;
Polish: Mirt;
Gaelic: Miortal;
Galician: Gorreiro, Matapulgas, Milteira, Miltra, Mirta, Mirteira, Mirto, Mirto Femea, Murta, Murtinho, Murtiños, Murtra;
Georgian: Mirti;
German: Brautmyrte, Echte Myrte, Gewöhnliche Myrte, Myrte;
Greek: Mirtia, Myrsine, Myrtia, Myrtos;
Hebrew: Hadas;
Hungarian: Közönséges Mirtusz, Mirtusz;
India: Bilatimehedi (Bengali), Vilayatimehndi (Hindu), Bola (Sanskrit), Kulinaval, Kuzhinaval (Tamil), Habulas, Vilaiti Mehandi (Urdu);
Italian: Mirto, Mortella;
Japanese: Gimbaika, Ginbaika, Iwai No Ki, Maatoru;
Korean: Meotul, Motul;
Latvian: Mirtes;
Lithuanian: Tokroji Mirta;
Macedonian: Mirta;
Maltese: Rihan;
Marjocan: Murta, Murtonera;
Norwegian: Brudemyrt, Myrte, Myrtel, Vanlig Myrt;
Persian: Mourd;
Polish: Mirt;
Portuguese: Gorreiro, Mastruços, Mata-Pulgas, Miltra, Mirta, Mirto, Mitra, Murta, Murta De Hespanha, Murta Dos Jardins, Murta-Ordinária, Murteira, Murtinheira, Murtinheiro, Murtinho, Murtinhos, Murtra, Myrta, Trovisco, Trovisco-Fêmea, Trovisqueiro;
Provençal: Nerto;
Romanian: Mirt;
Russian: Mirt, Myrt;
Serbian: Mirta, Mrča;
Slovenian: Mirta, Myrta Obyčajná;
Slovašcina: Mirta, Mirta Navadna, Navadna Mirta;
Spanish: Abriján, Arraiana, Arraigan, Arraigán, Arraiganeras, Arraigran, Arraihan, Arraiján, Arrajian, Arrayán Blanco, Arrayán, Arrayán Común, Arrayán De Andalucía, Arrayán Granadino, Arrayán Morisco, Arrayán Poblado Andaluz, Arrayán, Arrayhan Cultivado, Arrayhan Salvaje, Arrejanes, Arrian, Arrijan, Arriján, Astruc, Harrajian, Mata Gallinas, Mirta, Mirtilo, Mirto, Mirto Común, Mortera, Murta, Murta Menuda, Murta Remendada De Granada, Murtal, Murtas, Murtera, Murtiñera, Murto, Murtón, Murtonera, Murtones, Murtra, Murtrón, Mytra;
Swedish: Myrten;

Turkish: Mersin, Nersin, Murt (Wild Form), Ham Beles (Cultivated);
Ukrainian: Myrt;
Valencian: Mortonera, Murta, Murtera, Murtonera, Myrtó;
Yiddish: Hodes, Mirt.

Origin/Distribution

The exact location of its origin is uncertain but is accepted to be in the Mediterranean region and the Middle East – Macaronesia to Pakistan. Wild forms do exist in Turkey. It is widely cultivated for its edible fruits in the coastal regions of Turkey and in the Mediterranean. It has also been introduced into north west Himalaya.

Agroecology

Myrtle does best in full sun, in moist, well-drained, neutral to alkaline, loamy soil. It is drought tolerant and can withstand frost down to –5°C.

Edible Plant Parts and Uses

Ripe berries are eaten raw. The berries and leaves are used to produce an aromatic liqueur, called *Mirto* in the island of Corsica and Sardinia. *Mirto Ross* (red) is produced by macerating the ripe berries in alcohol and *Mirto Blanco* (white) is likewise produced using the leaves. Fresh or dried leaves are used as spice and can be used as a substitute for bay leaves. Myrtle leaf essential oil is used in sauces and in the confectionery and beverage industries besides being used in medicine and the cosmetic and perfume industry.

Botany

Evergreen, bushy, much-branched, strongly-scented, upright erect shrub or small tree, 1–3 m high. Leaves are opposite, usually in 3's, simple, ovate-lanceolate, 2.5–5 cm long, entire, acute,

Plate 1 Pale greenish white myrtle fruits and leaves

Plate 2 Purple myrtle fruits and leaves

base rounded, glossy, dark green, pinnately veined, short petioled, glabrous (Plates 1 and 2), punctuate, aromatic when bruised. Petiole short 1–2 mm with 2–3 glandular setae on the adaxial side of the base. Flowers are axillary, solitary, white or pinkish-tinged, to 2 cm across, hermaphrodite, actinomorphic, epigynous, fragrant, with 2 cm long slender pedicel terminated by 2 oblong prophylls, hypanthia turbinate, densely punctate; calyx of 4–5 deltoid sepals, sepals enlarged in fruit; corolla with 4–5, spreading, obovate petals, numerous free , longer than petals with versatile anthers, in many series forming a central tuft and an inferior, ellipsoid ovary. Fruit is ellipsoid to oblong-ellipsoid, greenish-white (Plate 1) to purplish-blue black when ripe, 12–14 mm long by

7.0–8 mm wide (Plate 2). Each fruit has numerous reinform, whitish seeds.

Nutritive/Medicinal Properties

Nutritional values of myrtle berries were determined by Aydin and Özcan (2007) as: crude oil 2.37%, crude protein 4.17%, crude fibre 17.41%, crude energy 11.21 kcal/g, reducing sugar 8.64%, tannin 76.11 mg/199 g, ash 0.725%, water soluble extract 52.94% and essential oil 0.01%. The average length, width, thickness, the geometric mean diameter of myrtle fruits were 13.75, 8.11, 7.57, 10.53 mm at a moisture content of 8.32% d.b., respectively. The myrtle fruit is composed of pericarp and approximately nine seeds which constituted 63.5% and 36.5% of the whole ripe fruit, respectively (Wannes et al. 2010). The latter presented a weight of 8.8 g% fruits while seed had only 0.5 g% seeds. The moisture contents were 80.1% in pericarp, 72% in whole fruit and 39.7% in seed. The oil yield of seed (11.7%) was significantly higher than that of whole fruit (5.9%) and pericarp (2.1%). Total lipid amounts were 61.26 mg/g in seed, 28.97 mg/g in whole fruit and 4.14 mg/g in pericarp. The amounts of polar glycerolipids were lower than those of neutral glycerolipids in all samples. Triacylglycerol constituted the main neutral glycerolipid with 57.47 mg/g in seed, 25.68 mg/g in whole fruit and 1.67 mg/g in pericarp. The predominant fatty acids of total lipids and different glycerolipid classes were linoleic, palmitic, oleic and α-linolenic acids. Whole fruit, seed and pericarp provided low yields of oil but they were a rich source of essential fatty acids which will be important as an indication of the potentially nutraceutical and industrial utility of myrtle fruit. In separate studies, it was found that reducing sugars increased in two cultivars of myrtle 'Barbara' and 'Daniela' approximately sevenfold from fruit set to complete maturation (Fadda and Mulas 2010). Total sugar content increased similarly ranging from 1.43% to 1.41% at fruit set to 8.28% and 7.56% at maturation for 'Barbara' and 'Daniela', respectively. Titratable acidity decreased during maturation, with significant differences due to cultivar and year of observation. Total phenols and tannins occurred at high levels after fruit set and declined during development. Anthocyanin levels increased, in both cultivars, according to a sigmoid curve. Small increases in ethylene production have been detected during fruit development ranging from 130.57 to 269.14 μl/kg/h measured at the onset of development to 13.04 and 19.36 μl/ kg/h measured at harvest for 'Barbara' and 'Daniela', respectively.

The polyphenol composition of Corsican myrtle berries was characterized by two phenolic acids, four flavanols, three flavonols and five flavonol glycosides. (Barboni et al. 2010a) The major compounds were myricetin-3-O-arabinoside and myricetin-3-O-galactoside. The polyphenolic compositions of the *M. communis* berry extract and the liqueur comprised high concentrations of flavonol glycosides, flavonols and flavanols were reported (Barboni et al. 2010b). The volatile compositions of Corsican myrtle alcoholic products were characterised by high amounts of monoterpene hydrocarbons and oxygenated monoterpenes with α-pinene and 1,8-cineole as major components.

A total of 24 volatile compounds were detected in the alcoholic extracts of myrtle leaves and berries (Tuberoso et al. 2006). The volatile fraction was characterized by the terpene fraction corresponding to that of the essential oils and by a fatty acid ethyl esters fraction. Essential oils were obtained by hydrodistillation, and the yields were on average 0.52% (v/w dried weight) and 0.02% for myrtle leaves and berries, respectively (Tuberoso et al. 2006). A total of 27 components were detected in the essential oils, accounting for 90.6–98.7% of the total essential oil composition. The major compounds in the essential oils were α-pinene (30.0% and 28.5%), 1,8-cineole (28.8% and 15.3%), and limonene (17.5% and 24.1%) in leaves and berries, respectively, and were characterized by the lack of myrtenyl acetate. Chemical analysis of myrtle leaf oil revealed the presence of 70 components, representing 99.23% of the total oil. 1,8-cineole (36.1%), α-pinene (22.5%), linalool (8.4%), bornyl acetate (5.2%), α-terpineol (4.4%), linalyl acetate (4.2%) and limonene (3.8%) were found to be the major components of the oil (Mahboubi and Bidgoli 2010).

Essential oil composition of *Myrtus communis* varied with plant parts and varieties (Wannes et al. 2007). Essential oil yield varied in leaves, fruits and stems. So, in leaves, it was 0.5% for var. italica and 0.3% for var. baetica and was higher than in fruits and stems with respectively 0.1% and 0.04% for italica and 0.07% and 0.03% for baetica. Essential oil composition was characterized by a high percentage of monoterpene hydrocarbons in leaves, largely due to α-pinene with 51.3% for italica and 27.7% for baetica; 1,8-cineole, the alone compound of ether class, was predominant in fruits and stems with respectively 31.6% and 34.7% for italica and 19.8% and 25.8% for baetica.

Chemical composition (%) of *Myrtus communis* essential oil collected from two localities of the Montenegro coastline: Ulcinj (*S1*) and Herceg Novi (*S2*) were determined to be as follows (Mimica-Dukić et al. 2010): isobutyl isobutyrate 1.95%, (S1) 0.7%(S2); α-thujene 0.4%, 0.4%; α-pinene 14.7%, 35.9%; β-pinene 0.2%, 0.3%; β-myrcene 0.3%, 0.2%: α-phellandrene 0.4%, 0.3%; δ-3-carene 0.5%, 0.4%; α-terpinene 0.3%, 0.2%; *p*-cymene 0.9%, 1.2%; limonene 4.15, 4.5%;1,8-cineole 25.7%, 23.9%;(*E*)-β-ocimene 0.4%, 0.3%: γ-terpinene 0.9%, 0.7%; α-terpinolene 1.1%, 0.8%; linalool 10.1%, 10.9%; 4-terpineol 0.3%, 0.3%; cryptone <0.2%, 0.2%; α-terpineol 3.1%, 2.8%; myrtenol 0.8%, 0.6%; geraniol 2.6%, 1.6%; myrtenyl acetate 21.6%, 5.4%; α-terpinyl acetate 1.4%, 0.5%; neryl acetate 0.3%, 0.2%; geranyl acetate 3.4%, 2.3%; methyl eugenol 0.8%, 1.0%; (*e*)-β-caryophyllene 0.6%, 0.5%; α-humulene 1.5%, 1.4%; bicyclogermacrene <0.2%, 0.2%; spathulenol <0.2%, 0.8%. The classes of compounds comprised monoterpene hydrocarbons 24.2%, 45.3%; oxygenated monoterpenes 70.1%, 49.5%; sesquiterpene hydrocarbons 2.1%, 2.1%: oxygenated sesquiterpenes -0, 0.8% respectively for two locations. In both of the samples monoterpenes were found to be the predominant compounds. Among them α-pinene, linalool, 1,8-cineole, and myrtenyl acetate were the major compounds. Significant differences between the samples were found in the ranges of α-pinene (14.7–35.9%) and myrtenyl acetate (5.4–21.6%).

Twenty samples of myrtle leaf oils collected in three locations in north-eastern Algeria ranged between 0.2% and 1.2% (w/w) (Bouzabata et al. 2010). The chemical composition of the oils was largely dominated by monoterpene hydrocarbons, with α-pinene (40.5–64.0%), 1,8-cineole (10.9–29.1%) and limonene (6.7–8.2%) being the major compounds. In all the samples, 3,3,5,5,8,8-hexamethyl-7-oxabicyclo[4.3.0]non-1(6)-ene-2,4-dione was identified (0.8–1.5%). The composition is similar to that reported for myrtle oils from Corsica, Sardinia and Tunisia, but differed from that of Moroccan and Spanish myrtle oils.

Some of the reported pharmacological properties of myrtle are presented below.

Antioxidant Activity

Myrtle leaves contains myrtucommulone A and semimyrtucommulone unique oligomeric non-prenylated acylphloroglucinols that exhibited potent antioxidant properties protecting linoleic acid against free radical attack in simple in vitro systems, inhibiting its autoxidation and its FeC13- and EDTA-mediated oxidation (Rosa et al. 2003). While both compounds lacked pro-oxidant activity, semimyrtucommulone was more powerful than myrtucommulone A. Semimyrtucommulone further exhibited antioxidant activity against lipid peroxidation induced by ferric-nitrilotriacetate, and in cell cultures for cytotoxicity and the inhibition of TBH- or FeC13-induced oxidation. The results of these studies established semimyrtucommulone as a novel dietary antioxidant lead. Myrtle leaves contained the structurally unique oligomeric non-prenylated acylphloroglucinols, semimyrtucommulone and myrtucommulone A, whose antioxidant activity was investigated during the oxidative modification of lipid molecules implicated in the onset of cardiovascular diseases (Rosa et al. 2008). Both acylphloroglucinols showed powerful antioxidant properties during the thermal (140°C) solvent-free degradation of cholesterol. Moreover, the pre-treatment with semimyrtucommulone and myrtucommulone A significantly protected LDL from oxidative damage induced by Cu^{2+} ions at 2 hours of oxidation, and showed remarkable protective effect on the reduction of polyunsaturated fatty acids and cholesterol, inhibiting the increase of their oxidative products (conjugated dienes fatty acids hydroperoxides, 7β-hydroxycholesterol, and 7-ketocholesterol). In view of the widespread culinary use of myrtle leaves, the results indicated the natural compounds semimyrtucommulone and myrtucommulone A to be interesting dietary antioxidants with potential antiatherogenicity.

All extracts prepared from myrtle leaves liquid-liquid extraction (LLE) with different solvents were found to be very rich in polyphenols and to have high antioxidant activity (Roamni et al. 2004). In particular, hydroalcoholic extracts contain galloyl-glucosides, ellagitannins, galloyl-quinic acids and flavonol glycosides; ethylacetate extract and aqueous residues after LLE are enriched in flavonol glycosides and hydrolysable tannins (galloyl-glucosides, ellagitannins, galloyl-quinic acids), respectively. Addition of these extracts did not affect the basal oxidation of human LDL but dose-dependently decreased the oxidation induced by copper ions. Moreover, the myrtle extracts reduce the formation of conjugated dienes. The antioxidant effect of three myrtle extracts decreased in the following order: hydroalcoholic extracts, ethylacetate and aqueous residues after LLE. The extracts had the following IC_{50}=0.36, 2.27 and 2.88 μM, when the sum of total phenolic compounds was considered after the correction of molecular weight based on pure compounds. These results suggested the myrtle extracts to have a potent antioxidant activity mainly due to the presence of galloyl derivatives.

Four hydrolyzable tannins [oenothein B (1), eugeniflorin D(2) (2), and tellimagrandins I (3) and II (4)], two related polyphenolic compounds [gallic acid (5) and quinic acid 3,5-di-O-gallate (6)], and four myricetin glycosides [myricetins 3-O-β-D-xyloside (7), 3-O-β-D-galactoside (8), 3-O-β-D-galactoside 6″-O-gallate (9), and 3-O-α-L-rhamnoside (10)] were isolated from the leaves of *Myrtus communis* (Yoshimura et al. 2008) and their antioxidant activities evaluated by 1,1-diphenyl-2-picrylhydrazyl (DPPH) radical scavenging assay.

The methanol extracts of myrtle fruits exhibited a high level of free radical scavenging activity using 2,2-diphenyl-1-picrylhydrazyl (DPPH) (Serce et al. 2010). There was a wide range (74.51–91.65%) of antioxidant activity among the accessions in the β-carotene-linoleic acid assay. The amount of total phenolics (TP) was determined to be between 44.41 and 74.44 μg Gallic acid equivalent (GAE)/mg, on a dry weight basis. Oleic acid was the dominant fatty acid (67.07%), followed by palmitic (10.24%), and stearic acid (8.19%), respectively. Ethanol and water extracts of myrtle berries showed the highest amount of extracted compounds, but the highest antiradical and antioxidant activities were found in ethanol and ethyl acetate extracts (Tuberoso et al. 2010). These extracts were also the ones with the highest content of phenolic compounds. In addition, the results showed a highly significant correlation between the amount of total phenols and antiradical (R2=0.9993) or antioxidant activities (R2=0.9985) in these extracts. The ethyl acetate extract had the highest protective effect in assays of thermal (140°C) cholesterol degradation and Cu^{2+}-mediated LDL oxidation, inhibiting the reduction of polyunsaturated fatty acids and cholesterol, and the increase of their oxidative products. The results suggested that because of these properties, myrtle berries could be used in dietary supplements preparations or as food additives.

The total phenol content of myrtle leaf and berry extracts ranged between 9.0 and 35.6 mg GAE per g extract (Amensour et al. 2009). For each solvent (methanol, ethanol, water), leaf extracts contained significantly higher amount of total phenolic compounds than berry extracts. Generally, leaf extracts showed higher antioxidant activities than berry extracts, while the overall antioxidant strength was in the order methanol > water > ethanol in leaf extracts and methanol > ethanol > water in berry extracts. The phenolic content exhibited a positive correlation with the antioxidant activity: DPPH assay showed the highest correlation (r=0.949), followed by the reducing power assay (r=0.914) and the lowest for the β-carotene linoleic acid assay (r=0.722).

Myrtle leaf and flower were the important sources for the essential oil production representing a yield of 0.61% and 0.30% (w/w), respectively (Wannes et al. 2010b). The essential oil composition of myrtle leaf and flower was characterized by high concentrations of α-pinene, the main compound of monoterpene hydrocarbon class, with 58.05% for leaf and 17.53% for flower. Stem was rich in oxygenated monoterpenes, largely due to 1,8-cineole with 32.84%. The leaf extract had higher total phenol content (33.67 mg GAE/g) than flower (15.70 mg GAE/g) and stem (11.11 mg GAE/g) extracts. Significant differences were also found in total tannin contents among different myrtle plant parts. The leaf had higher tannin content than the flower; 26.55 mg GAE/g in the leaf, 11.95 mg GAE/g in the flower and 3.33 mg GAE/g in the stem. The highest contents of total flavonoids and condensed tannins were found in the stem (5.17 and 1.99 mg CE/g, respectively) and the leaf (3 and 1.22 mg CE/g, respectively) extracts. The main phenolic class was hydrolysable tannins (gallotannins) in the leaf (79.39%, 8.90 mg/g) and the flower (60.00%, 3.50 mg/g) while the stem was characterized by the predominance of flavonoid class (61.38%, 1.86 mg/g) due to the high presence of catechin (36.91%, 1.12 mg/g). The methanolic extracts of different myrtle plant parts showed better antioxidant activity than essential oils as evaluated by using DPPH radical scavenging, β-carotene-linoleic acid bleaching, reducing power and metal chelating activity assays.

Antiulcerogenic Activity

Two doses of aqueous extracts of *M. communis* 105 and 175 mg/kg significantly reduced the ulcer index in all models of ulcers (Sumbul et al. 2010). Low dose of aqueous extract and high dose of methanolic extract of *M. communis* exhibited more significant effect in comparison to omeprazole (standard drug) in ethanol-induced ulcer model. Both the doses of aqueous and methanolic extracts also reduced the gastric juice volume, total acidity and increased the gastric pH and gastric wall mucus content in all the models of ulcers used in the present study. Histopathological examinations of gastric tissues of rats treated with the aqueous and methanolic extracts in indomethacin-induced ulcer exhibited significant ulcer-protective effect at both the dose levels.

Myrtle oral paste was found to have therapeutic effect on recurrent aphthous stomatitis (RAS) is a common, painful, and ulcerative disorder of the oral cavity in a randomized, double-blind, controlled before-after clinical trial involving 45 patients (Babee et al. 2010). The myrtle oral paste was found to be effective in decreasing the size of ulcers, pain severity and the level of erythema and exudation, and improving the quality of life in patients who suffer from RAS No side effects were reported.

Antiinflammatory Activity

Myrtucommulone (MC) and semimyrtucommulone (S-MC) natural nonprenylated acylphloroglucinol from myrtle leaves were shown to potently suppress the biosynthesis of eicosanoids by direct inhibiting cyclooxygenase-1 and 5-lipoxygenase in-vitro and in-vivo at IC_{50} values in the range of 1.8–29 μM (Feisst et al. 2005). They demonstrated that MC and S-MC prevented the mobilization of Ca^{2+} in polymorphonuclear leukocytes, mediated by G protein signaling pathways at IC_{50} values of 0.55 and 4.5 μM, respectively, and suppressed the formation of reactive oxygen species and the release of elastase at comparable concentrations. The researchers concluded that in view of the ability to suppress typical proinflammatory cellular responses, the unique acylphloroglucinols MC and S-MC from myrtle may possess an anti inflammatory potential, suggesting their therapeutic use for the treatment of diseases related to inflammation and allergy.

Myrtucommulone, was found to inhibit microsomal prostaglandin E2 synthase (mPGES)-1 that efficiently suppressed PGE(2) formation without significant inhibition of the cyclooxygenase (COX)-1 and 2 enzymes (Koeberle et al. 2009). The data indicated the potential use of myrtucommulone in interventions in inflammatory disorders, without the typical side effects of coxibs and non-steroidal antiinflammatory drugs interventions in inflammatory disorders, In another study, Rossi et al. (2009) investigated the effects of myrtucommulone in in-vivo models of inflammation. Myrtucommulone (0.5, 1.5, and 4.5 mg/kg i.p.) was found to reduce the development of mouse carrageenan-induced paw edema in a dose-dependent manner. Moreover, myrtucommulone also exerted antiinflammatory effects in the pleurisy model. In particular, 4 hours after carrageenan injection in the pleurisy model, myrtucommulone reduced: (1) the exudate volume and leukocyte numbers; (2) lung injury (histological analysis) and neutrophil infiltration (myeloperoxidase activity); (3) the lung intercellular adhesion molecule-1 and P-selectin immunohistochemical localization; (4) the cytokine levels (tumour necrosis factor-α and interleukin-β) in the pleural exudate and their immunohistochemical localization in the lung; (5) the leukotriene B(4), but not prostaglandin E(2), levels in the pleural exudates; and (6) lung peroxidation (thiobarbituric acid-reactant substance) and nitrotyrosine and poly (ADP-ribose) immunostaining. The results demonstrated that myrtucommulone exerted potent anti-inflammatory effects in-vivo and offer a novel therapeutic approach for the management of acute inflammation.

Anticancer/Antimutagenic Activity

Tretiakova et al. (2008) found that myrtucommulone, a nonprenylated acylphloroglucinol from myrtle leaves induced apoptosis in cancer cell lines (EC_{50} 3–8 μM), with marginal cytotoxicity for non-transformed human peripheral blood mononuclear cells (PBMC) or foreskin fibroblasts (EC_{50} cell death = 20–50 μM), via the mitochondrial cytochrome c/Apaf-1/caspase-9 pathway.

Antioxidant activity of myricetin-3-o-galactoside and myricetin-3-o-rhamnoside, isolated from the leaves of *Myrtus communis* showed antioxidant activity (Hayder et al. 2008a). The IC_{50} values of lipid peroxidation inhibition by myricetin-3-o-galactoside and myricetin-3-o-rhamnoside were respectively 160 μg/ml and 220 μg/ml. At a concentration of 100 μg/ml, the two compounds showed the most potent inhibitory effect of xanthine oxidase activity by 57% and 59% respectively Myricetin-3-o-rhamnoside was a very potent radical scavenger with an IC_{50} value of 1.4 μg/ml. Moreover, these two compounds induced an inhibitory activity against nifuroxazide, aflatoxin B1 and H2O2 induced mutagenicity using the SOS chromotest and the Comet assay. The protective effect exhibited by these molecules was also determined by analysis of gene expression as response to an oxidative stress using a cDNA μ-array. Myricetin-3-o-galactoside and myricetin-3-o-rhamnoside modulated the expression patterns of cellular genes involved in oxidative stress, DNA damaging repair and in apoptosis.

The hexane, chloroform, ethyl acetate and methanol extracts from leaves of *Myrtus communis*, showed no mutagenicity when tested with *Salmonella typhimurium* strains TA98 and TA100 either with or without metabolic system (S9) (Hayder et al. 2008b). On the other hand, each of the tested extracts exhibited a significant protective effect against the mutagenicity induced by aflatoxin B1 (AFB1) in *Salmonella typhimurium* TA100 and TA98 assay systems, and against the mutagenicity induced by sodium azide in TA100 and TA1535 assay system. Ethyl acetate and methanol extracts showed the highest level of protection towards the direct mutagen, sodium azide, and indirect mutagen AFB1. Recently Mitić-Ćulafic et al. (2009) reported the protective activity of volatile monoterpenes (linalool, myrcene and 1,8-cineole) against the oxidant induced genotoxicity in bacterial cells (*Escherichia coli* WP2 IC185 strain and its oxyR mutant IC202) and cultured human cells (hepatoma HepG2 and human B lymphoid NC–NC cells). They found that myrcene and linalool strongly suppressed t-butyl hydroperoxide (t-BOOH) induced mutagenesis. Further, they reduced DNA damage induced by t-BOOH. In NC–NC cells linalool and myrcene reduced t-BOOH induced DNA damage by about 50% at 0.01 μg/ml. In HepG2 cells linalool reduced DNA damage by 30%, while myrcene was ineffective. All these compounds were identified in myrtle oils, among which 1,8-cineole and linalool were present in substantial amounts.

In both myrtle leaf oil samples from Ulcinj and Herceg Novi, Montenegro monoterpenes were found to be the predominant compounds

(Mimica-Dukić et al. 2010). Among them α-pinene, linalool, 1,8-cineole, and myrtenyl acetate were the major compounds. Significant differences between the samples were found in the ranges of α-pinene (14.7–35.9%) and myrtenyl acetate (5.4–21.6%). Both oils exhibited moderate DPPH scavenging activity, with IC_{50} values of 6.24 and 5.99 mg/ml. Reduction of the spontaneous mutagenesis in presence of myrtle essential oil was only slight, up to 13% at the highest concentration tested. When the oxidative mutagen *Escherichia coli* oxyR mutant IC202 was used, the essential oil expressed higher reduction of mutagenesis, in a concentration dependent manner, with statistical significance for effect at the highest concentration tested (28%). Suppression t-BOOH (t-butyl hydroperoxide) induced mutagenesis was correlated with the observed scavenging activity.

Antigenotoxic Activity

M. communis extracts and oil exhibited antigenotoxic activity. Aqueous extract, the total flavonoids oligomer fraction (TOF), hexane, chloroform, ethyl acetate and methanol extracts and essential oil obtained from *M. communis* significantly decreased the SOS response induced by aflatoxin B1 (10 μg/assay) and Nifuroxazide (20 μg/assay) (Hayder et al. 2004). Ethyl acetate and methanol extracts showed the strongest inhibition of the induction of the SOS response by the indirectly genotoxic aflatoxin B1. The methanol and aqueous extracts exhibited the highest level of protection towards the SOS-induced response by the directly genotoxic Nifuroxazide. SOS response is defined as the DNA repair systems (recA; uvr) induced by the presence of single-stranded DNA that usually occurs from post-replicative gaps caused by various types of DNA damage. The RecA protein, stimulated by single-stranded DNA, is involved in the inactivation of the LexA repressor thereby inducing the response. In addition to anti-genotoxic activity, the aqueous extract, the TOF, the ethyl acetate and methanol extracts showed an important free-radical scavenging activity towards the 1,1-diphenyl-2-picrylhydrazyl (DPPH) radical. These results suggested the potential for future utilization of these extracts as additives in chemoprevention studies.

Antimicrobial Activity

The crude methanol extract of *Myrtus communis* inhibited the growth of 6 Gram positive (*Staphylococcus aureus*, *Micrococcus luteus*, *Streptococcus pneumoniae*, *Streptococcus pyogenes*, *Streptococcus agalactiae*, *Listeria monocytogenes*) and 3 Gram negative bacteria (*Escherichia coli*, *Proteus vulgaris*, *Pseudomonas aeruginosa*) (Mansouri et al. 2001). Further extraction of the crude extract with diethyl ether, ethyl acetate, and ethanol resulted in six different fractions (M1-M6) which were screened for antibacterial activity against the non-fastidious bacteria (*S. aureus*, *M. luteus*, *E. coli*, *P. vulgaris*, and *P. aeruginosa*). The diethyl ether fraction (fraction M1) showed the highest level of activity in comparison to the crude extract and other fractions. The MIC for *S. aureus* and *M. luteus* were reduced from 0.1 in the crude extract to 0.025 mg/ml in fraction M1 and for *E. coli* and *P. aeruginosa* was reduced from over 1 mg/ml in the crude extract to 0.1 mg/ml in fraction M1. Essential oil was also active against the tested bacteria, and *M. luteus* showed the highest level of sensitivity (MIC 1:1600).

Myrtucommulone A, a trimeric acylphloroglucinol isolated from myrtle leaves showed significant antibacterial activity against multi-drug-resistant (MDR) clinically relevant bacteria, while semimyrtucommulone was less active (Appendino et al. 2002). Two acylphloroglucinols, myrtucommulone-A and myrtucommulone-B, were isolated from *Myrtus communis* leaves (Rotstein et al. 1974). Myrtucommulone-A was highly antibacterial against Gram-positive bacteria but was not active against Gram negatives. Methanolic extracts of *Myrtus communis* seeds were reported inhibitory against *Staphylococcus aureus*, *Bacillus cereus* and *Bordetella bronchiseptica* (Bonjar 2004).

Extract of *Myrtus communis* caused death of *Trichomonas vaginalis* at pH 4.65, but failed to do so at pH 6 in in-vitro study (Mahdi et al. 2006). The polar glycosidic fraction from the leaves of myrtle afforded four galloylated nonprenylated phloroglucinol glucosides (3a-d) related to the endoperoxide hormone G3 (4) in terms of structure and biogenesis (Appendino et al. 2006). Despite their close similarity, significant antibacterial activity was shown only by one of these compounds (3b, gallomyrtucommulone B), while the G3 hormone (4) was inactive.

A crude preparation of myrtle was fodun to have antibacterial activity agsint some common human pathogens (Alem et al. 2008). The minimum bactericidal concentration of Myrtle for most tested microorganisms was similar to the minimum inhibitory concentration. i.e. 0.5 mg/ml. for *Staphylococcus aureus*, 2.5 mg/ml for *Proteus mirabilis* and *Proteus vulgaris*, 15 mg/ ml for *Klebsiella aerogenes* and *Salmonella typhi*, 20 mg/ml for *Pseudomonas aeruginosa*. The MBC of Myrtle for the two relatively least sensitive species, *Shigella* and *Escherichia coli* was 40 and 45 mg/ml of media, respectively. The antibacterial activity of Myrtle was markedly increased by 18 times after it has been autoclaved at 121°C for 15 min.

Myrtle leaf oil exhibited good antifungal activity against *Candida albicans* (eight clinical isolates and one ATCC type strains) and different species of *Aspergillus* spp. (*A. niger*, *A. parasiticus*, six isolates of *Aspergillus flavus*) using broth micro dilution assay (Mahboubi and Bidgoli 2010). Myrtle oil showed also significant antifungal activity when combined with amphotericin B. Myrtle oil was found to show good activity towards all strains of *Mycobacterium tuberculosis* but not toward *M. paratuberculosis* (Zanetti et al. 2010). The MIC registered against *M. tuberculosis* was 0.17% (v/v) in comparison with an MIC of 2% (v/v) observed toward *M. paratuberculosis*.

Hypoglycaemic/Antidiabetic Activity

Ethanol-water extract of *Myrtus communis* was found to have anti-hyperglycaemic effect (Elfellah et al. 1984). An ethanol-water extract (2 g/kg) administered intragastrically 30 minutes before streptozotocin abolished the initial hyperglycaemic without affecting the second phase. Myrtle extract given prior to streptozotocin and repeated at 24 hours and 30 hours, did not allow hyperglycaemia to develop until after 48 hours. Administration of myrtle extract 48 hours after streptozotocin significantly reduced the hyperglycaemia and this effect was maintained by its repeated administration. Myrtle extract had no effect on the blood glucose level of normal mice. These studies confirmed the "folk-medicine" indication of myrtle extract as potentially useful in the treatment of diabetes mellitus.

Of several plants tested for α-glucosidase inhibitory activity for control of hyperglycemia in patients with type 2; noninsulin-dependent, diabetes mellitus (NIDDM), *Myrtus communis* was found to potently inhibit the enzyme (IC_{50} = 38 μg/ml) (Onal et al. 2005). *M. communis* oil was found to exert hypoglycaemic as well as mild hypotriglyceridemic activity in alloxan-diabetic rabbits but not in normoglycaemic rabbits (Sepici et al. 2004). The reduction in blood glucose level (51%) was postulated to be due to the reversible inhibition of α-glucosidases present in the brush-border of the small intestinal mucosa, higher rate of glycolysis as envisaged by the higher activity of glucokinase, as one of the key enzymes of glycolysis, and enhanced rate of glycogenesis as evidenced by the higher amount of liver glycogen present after myrtle oil administration.

Traditional Medicinal Uses

The plant including the fruit, leaves and essential oil have been used in traditional folk medicine (Chiej 1984; Chopra et al. 1986; Bown 1995; Mansour et al. 2001; Cakir 2004; Elfellah et al. 1984). The plant is taken internally in the treatment of urinary infections, digestive problems, vaginal discharge, bronchial congestion, sinusitis and dry coughs. In India, it is considered to be useful in the treatment of cerebral affections, especially epilepsy. The essential oil is used as antiseptic, for treatment of acne, wounds, gum infections and haemorrhoids. The oil is also used as a local application for rheumatism. The fruit is carminative. It is used in the treatment of various infectious diseases, including diarrhoea, dysentery haemorrhoids, internal ulceration, rheumatism and eaten in winter as appetiser. The fruit and leaves are traditionally used as antiseptic, disinfectant, and hypoglycemic agents. The leaves are aromatic, balsamic, haemostatic and tonic. The leaves are used as anti-inflammatory, anti-hypertensive, anticarcinogen and haemostatic agents, as a mouthwash, for treatments of candidiasis, for healing wounds, as well as in the therapy of urinary diseases.

Other Uses

Myrtle is cultivated as an ornamental garden shrub, particularly for its numerous flowers in later summer as well as a hedge. In the Mediterranean region myrtle is regarded as a symbol of love and peace and is much prized for use in wedding bouquets. Myrtle is steeped in Greek mythology and Roman culture. It is deemed sacred to the Greek goddesses Aphrodite and Demeter; Roman women wear crowns of woven myrtle branches and myrtle is used in weddings. Myrtle branches are given to the bridegroom as he enters the nuptial chamber after the wedding. In Jewish liturgy, myrtle is one of four scared plants of sukkot, the feast of the tabernacles. Myrtle is also used in pagan rituals.

The essential oils from the bark, leaves and flowers are used in perfumery, soaps and skincare, cosmetic products. A perfumed water, "eau d'ange", is obtained from the flowers. Its wood is used to make walking sticks, tool handles and furniture. Myrtle makes a perfect firewood, transmitting a spicy, aromatic taste to any meat grilled thereafter. Foods flavoured with the smoke of myrtle are common in rural areas of Italy or Sardinia.

M. communis has insecticidal property. Of several plants tested, extracts of *M. communis* were found to be the most toxic against the fourth-in-star larvae of the mosquito *Culex pipiens molestus* (Diptera:Culicidae) with LC_{50} of 16 mg/l (Traboulsi et al. 2002). Over 20 major components were identified in extracts from each plant species. Eight pure components (1,8-cineole, menthone, linalool, terpineol, carvacrol, thymol, (1S)-(−)-α-pinene and (1R)-(+)-α-pinene) were tested against the larvae. Thymol, carvacrol, (1R)-(+)-α-pinene and (1S)-(−)-α-pinene were the most toxic (LC_{50} = 36–49 mg/1), while menthone, 1,8-cineole, linalool and terpineol (LC_{50} = 156–194 mg/l) were less toxic.

Myrtle oil exhibited pronounced insecticidal activity against the stored-product pest, bean weevil *Acanthoscelides obtectus* Say (Coleoptera: Bruchidae) (Ayaz et al. 2010).

The, common hop (*Humulus lupulus*) and myrtle (*Myrtus communis*) methanolic-dichloromethane extracts were found to exhibit the highest growth-inhibitory against effect three strains of *Paenibacillus* larvae, the causal agent of American Foulbrood Disease of honey bees (AFB) with MICs ranging from 2 to 8 μg/ml (Flesar et al. 2010). Acute oral toxicity of the most active natural products was determined on adult honey bees, showing them as non-toxic at concentrations as high as 100 μg per bee.

Comments

Myrtle is propagated by woody stem cuttings and seeds.

Selected References

Alem G, Mekonnen Y, Tiruneh M, Mulu A (2008) In vitro antibacterial activity of crude preparation of myrtle (*Myrtus communis*) on common human pathogens. Ethiop Med J 46(1):63–69

Amensour M, Sendra E, Abrini J, Bouhdid S, Pérez-Alvarez JA, Fernández-López J (2009) Total phenolic content and antioxidant activity of myrtle (*Myrtus communis*) extracts. Nat Prod Commun 4(6):819–824

Appendino G, Bianchi F, Minassi A, Sterner O, Ballero M, Gibbons S (2002) Oligomeric acylphloroglucinols from myrtle (*Myrtus communis*). J Nat Prod 65(3):334–338

Appendino G, Maxia L, Bettoni P, Locatelli M, Valdivia C, Ballero M, Stavri M, Gibbons S, Sterner O (2006) Antibacterial galloylated alkylphloroglucinol glucosides from myrtle (*Myrtus communis*). J Nat Prod 69(2):251–254

Aydın C, Özcan MM (2007) Determination of nutritional and physical properties of myrtle

(*Myrtus communis* L.) fruits growing wild in Turkey. J Food Eng 79(2):453–458

Ayvaz A, Sagdic O, Karaborklu S, Ozturk I (2010) Insecticidal activity of the essential oils from different plants against three stored-product insects. J Insect Sci 10:Article 21

Babaee N, Mansourian A, Momen-Heravi F, Moghadamnia A, Momen-Beitollahi J (2010) The efficacy of a paste containing *Myrtus communis* (Myrtle) in the management of recurrent aphthous stomatitis: a randomized controlled trial. Clin Oral Investig 14(1):65–70

Bailey LH (1949) Manual of cultivated plants most commonly grown in the continental United States and Canada. (Revised edition). The Macmillan Co, New York, 1116 pp

Barboni T, Cannac M, Massi L, Perez-Ramirez Y, Chiaramonti N (2010a) Variability of polyphenol compounds in *Myrtus communis* L. (Myrtaceae) berries from Corsica. Molecules 15(11):7849–7860

Barboni T, Venturini N, Paolini J, Desjobert J-M, Chiaramonti N, Costa J (2010b) Characterisation of volatiles and polyphenols for quality assessment of alcoholic beverages prepared from Corsican *Myrtus communis* berries. Food Chem 122(4):1304–1312

Bonjar GH (2004) Antibacterial screening of plants used in Iranian folkloric medicine. Fitoterapia 75(2):231–235

Bouzabata A, Boussaha F, Casanova J, Tomi F (2010) Composition and chemical variability of leaf oil of *Myrtus communis* from north-eastern Algeria. Nat Prod Commun 5(10):1659–1662

Bown D (1995) Encyclopaedia of herbs and their uses. Dorling Kindersley, London, 424 pp

Cakir A (2004) Essential oil and fatty acid composition of *Hippophae rhamnoides* L., (sea buckthorn) and *Myrtus communis* L. from Turkey. Biochem Syst Ecol 3:809–816

Chiej R (1984) The Macdonald encyclopaedia of medicinal plants. Macdonald, London, 447 pp

Chopra RN, Nayar SL, Chopra IC (1986) Glossary of Indian medicinal plants. (Including the supplement). Council Scientific Industrial Research, New Delhi, 330 pp

Elfellah MS, Akhter MH, Khan MT (1984) Anti-hyperglycemic effect of an extract of *Myrtus communis* in streptozotocin-induced diabetes in mice. J Ethnopharmacol 11:275–281

Facciola S (1990) Cornucopia. A source book of edible plants. Kampong Publ, Vista, 677 pp

Fadda A, Mulas M (2010) Chemical changes during myrtle (*Myrtus communis* L.) fruit development and ripening. Sci Hortic 125(3):477–485

Feisst C, Franke L, Appendino G, Werz O (2005) Identification of molecular targets of the oligomeric nonprenylated acylphloroglucinols from *Myrtus communis* and their implication as anti-inflammatory compounds. J Pharmacol Exp Ther 315(1):389–396

Flesar J, Havlik J, Kloucek P, Rada V, Titera D, Bednar M, Stropnicky M, Kokoska L (2010) In vitro growth-inhibitory effect of plant-derived extracts and compounds against *Paenibacillus* larvae and their acute oral toxicity to adult honey bees. Vet Microbiol 145(1–2):129–133

Govaerts R, Sobral M, Ashton P, Barrie F, Holst BK, Landrum LL, Matsumoto K, Fernanda Mazine F, Nic Lughadha E, Proenca C, Soares-Silva LH, Wilson PG, Lucas E (2010) World checklist of Myrtaceae. The board of trustees of the Royal Botanic Gardens, Kew. Published on the Internet, http://www. kew.org/wcsp/. Accessed 22 Apr 2010

Hayder N, Abdelwahed A, Kilani S, Ben Ammar R, Mahmoud A, Ghedira K, Chekir-Ghedira L (2004) Anti-genotoxic and free radical scavenging activity of extracts from (Tunisian) *Myrtus communis*. Mutat Res 564:89–95

Hayder N, Bouhlel I, Skandrani I, Kadri M, Steiman R, Guiraud P, Mariotte AM, Ghedir K, Dijoux-Franca MG, Chekir-Ghedira L (2008a) In vitro antioxidant and antigenotoxic potential of myricetin-3-O-galactosyde and myricetin-3-o-rhamnoside from *Myrtus communis*: modulation of genes involved in cell defense using cDNA microarray. Toxicol In Vitro 22:567–581

Hayder N, Skandrani I, Kilani S, Bouhlel I, Abdelwahed A, Ammar RB, Mahmoud A, Ghedira K, Chekir-Ghedira L (2008b) Antimutagenic activity of *Myrtus communis* L. using the *Salmonella* microsome assay. S Afr J Bot 74(1):121–125

Koeberle A, Pollastro F, Northoff H, Werz O (2009) Myrtucommulone, a natural acylphloroglucinol, inhibits microsomal prostaglandin E(2) synthase-1. Br J Pharmacol 156(6):952–961

Mahboubi M, Ghazian Bidgoli FB (2010) In vitro synergistic efficacy of combination of amphotericin B with *Myrtus communis* essential oil against clinical isolates of *Candida albicans*. Phytomedicine 17(10):771–774

Mahdi NK, Gany ZH, Sharief M (2006) Alternative drugs against *Trichomonas vaginalis*. East Mediterr Health J 12(5):679–684

Mansouri S, Foroumadi A, Ghaneie T, Najar AG (2001) Antibacterial activity of the crude extracts and fractionated constituents of *Myrtus communis*. Pharm Biol 39(5):399–401

Mimica-Dukić N, Bugarin D, Grbović S, Mitić-Culafić D, Vuković-Gacić B, Orcić D, Jovin E, Couladis M (2010) Essential oil of *Myrtus communis* L. as a potential antioxidant and antimutagenic agents. Molecules 15(4):2759–2770

Mitić-Ćulafić D, Žegura B, Nikolić B, Vuković-Gačić B, Knežević-Vukčević J, Filipič M (2009) Protective effect of linalool, myrcene and eucalyptol against t-butyl hydroperoxide induced genotoxicity in bacteria and cultured human cells. Food Chem Toxicol 47(1):260–266

Onal S, Timur S, Okutucu B, Zihnioğlu F (2005) Inhibition of α-glucosidase by aqueous extracts of some potent antidiabetic medicinal herbs. Prep Biochem Biotechnol 35(1):29–36

Romani A, Coinu R, Carta S, Pinelli P, Galardi C, Vincieri FF, Franconi F (2004) Evaluation of antioxidant effect of different extracts of *Myrtus communis* L. Free Radic Res 38(1):97–103

Rosa A, Deiana M, Casu V, Corona G, Appendino G, Bianchi F, Ballero M, Dessì MA (2003) Antioxidant activity of oligomeric acylphloroglucinols from *Myrtus communis* L. Free Radic Res 37(9):1013–1019

Rosa A, Melis MP, Deiana M, Atzeri A, Appendino G, Corona G, Incani A, Loru D, Dessì MA (2008) Protective effect of the oligomeric acylphloroglucinols from *Myrtus communis* on cholesterol and human low density lipoprotein oxidation. Chem Phys Lipids 155(1):16–23

Rossi A, Di Paola R, Mazzon E, Genovese T, Caminiti R, Bramanti P, Pergola C, Koeberle A, Werz O, Sautebin L, Cuzzocrea S (2009) Myrtucommulone from *Myrtus communis* exhibits potent anti-inflammatory effectiveness in vivo. J Pharmacol Exp Ther 329(1):76–86

Rotstein A, Lifshitz A, Kashman Y (1974) Isolation and antibacterial activity of acylphloroglucinols from *Myrtus communis*. Antimicrob Agents Chemother 6(5):539–542

Sepici A, Gürbüz I, Cevik C, Yesilada E (2004) Hypoglycaemic effects of myrtle oil in normal and alloxan-diabetic rabbits. J Ethnopharmacol 93(2–3):311–318

Serce S, Ercisli S, Sengul M, Gunduz K, Orhan E (2010) Antioxidant activities and fatty acid composition of wild grown myrtle (*Myrtus communis* L.) fruits. Pharmacogn Mag 6:9–12

Sumbul S, Ahmad MA, Asif M, Saud I, Akhtar M (2010) Evaluation of *Myrtus communis* Linn. berries (common myrtle) in experimental ulcer models in rats. Hum Exp Toxicol 29(11):935–944

Traboulsi AF, Taoubi K, el-Haj S, Bessiere JM, Rammal S (2002) Insecticidal properties of essential plant oils against the mosquito *Culex pipiens molestus* (Diptera:Culicidae). Pest Manag Sci 58(5):491–495

Tretiakova I, Blaesius D, Maxia L, Wesselborg S, Schulze-Osthoff K, Cinatl J Jr, Michaelis M, Werz O (2008) Myrtucommulone from *Myrtus communis* induces apoptosis in cancer cells via the mitochondrial pathway involving caspase-9. Apoptosis 13(1):119–131

Tuberoso CI, Barra A, Angioni A, Sarritzu E, Pirisi FM (2006) Chemical composition of volatiles in Sardinian myrtle (*Myrtus communis* L.) alcoholic extracts and essential oils. J Agric Food Chem 54(4):1420–1426

Tuberoso CIG, Rosa A, Bifulco E, Melis MP, Atzeri A, Pirisi FM, Dessì MA (2010) Chemical composition and antioxidant activities of *Myrtus communis* L. berries extracts. Food Chem 123(4):1242–1251

Wannes WA, Mhamdi B, Marzouk B (2007) Essential oil composition of two *Myrtus communis* L. varieties grown in North Tunisia. Ital J Biochem 56(2):180–186

Wannes WA, Mhamdi B, Sriti J, Marzouk B (2010a) Glycerolipid and fatty acid distribution in pericarp, seed and whole fruit oils of *Myrtus communis* var. ital Ind Crops Prod 31(1):77–83

Wannes WA, Mhamdi B, Sriti J, Ben Jemia M, Ouchikh O, Hamdaoui G, Kchouk ME, Marzouk B (2010b) Antioxidant activities of the essential oils and methanol extracts from myrtle (*Myrtus communis* var. *italica* L.) leaf, stem and flower. Food Chem Toxicol 48(5):1362–1370

Yoshimura M, Amakura Y, Tokuhara M, Yoshida T (2008) Polyphenolic compounds isolated from the leaves of *Myrtus communis*. J Nat Med 62(3):366–368

Zanetti S, Cannas S, Molicotti P, Bua A, Cubeddu M, Porcedda S, Marongiu B, Sechi LA (2010) Evaluation of the antimicrobial properties of the essential oil of *Myrtus communis* L. against clinical strains of *Mycobacterium* spp. Interdiscip Perspect Infect Dis 2010:article ID931530

Pimenta dioica

Scientific Name

Pimenta dioica **(L.) Merr.**

Synonyms

Caryophyllus pimenta (L.) Mill., *Eugenia micrantha* Bertol., *Eugenia pimenta* (L.) DC., *Eugenia pimenta* var. *longifolia* DC., *Eugenia pimenta* var. *ovalifolia* DC., *Evanesca crassifolia* Raf. nom. illeg., *Myrtus aromatica* Salisb. nom. superfl., *Myrtus aromatica* Poir. nom. illeg., *Myrtus dioica* L. (basionym), *Myrtus pimenta* L., *Myrtus pimenta* var. *brevifolia* Hayne, *Myrtus pimenta* var. *longifolia* Sims, *Myrtus piperita* Sessé & Moc., *Myrtus tabasco* Willd. ex Schltdl. & Cham., *Pimenta aromatica* Kostel., *Pimenta communis* Benth. & Hook.f., *Pimenta dioica* var. *tabasco* (Willd. ex Schltdl. & Cham.) Standl., *Pimenta officinalis* Lindl., *Pimenta officinalis* var. *cumanensis* Schiede & Deppe, *Pimenta officinalis* var. *longifolia* (Sims) O.Berg, *Pimenta officinalis* var. *ovalifolia* (DC.) O.Berg, *Pimenta officinalis* var. *tabasco* (Willd. ex Schltdl. & Cham.) O.Berg, *Pimenta pimenta* (L.) H.Karst., nom. inval. *Pimenta vulgaris* Lindl., *Pimenta vulgaris* Bello, *Pimentus aromatica* Raf. nom. superfl., *Pimentus geminata* Raf., *Pimentus vera* Raf. nom. superfl.

Family

Myrtaceae

Common/English Names

Allspice, Clover Pepper, Jamaica Pepper, Jamaican Pepper, Pimenta, Pimento

Vernacular Names

Albanian: O:Ispais, Piper Xhamaike;
Algeria: Fulful Meksyk (Arabic);
Arabic: Bahar, Bhar Hub Wa Na'im, Bahar Halu, Fulful Ifranji, Fulful Ifranji Halu, Fulful Tabil, Tawabil Halua;
Bangladesh: Kababchini;
Basque: Jamaikako Piperbeltz;
Belarusian: Angelskae Zelle;
Brazil: Pimenta Da Jamaica, Pimenta Síria (Portuguese);
Bulgarian: Bahhar, Bakhar;
Catalan: Pebre De Jamaica;
Chinese: Bi Wei Hu Jiao, Duo Xiang Guo;
Croatian: Najgvirc, Piment, Začin;
Czech: Hřebíčkový Pepř, Nové Koření, Pimentovník Dvoudomý, Pimentovník Lékařský;
Danish: Allehaande, Jamaicapeber, Piment;

DOI 10.1007/978-94-007-2534-8_89,

Dominican Republic: Clavo De Jamaica, Malagueta;
Dutch: Alles Kruid, Engelse Peper, Jamaicapeper, Piment, Piment Sort, Pimentboom;
Eastonian: Harilik Pimendipuu, Vürts Pipar;
Ecuador: Pimineta Dulce;
Farsi: Felfel Farangi;
Finnish: Maustepippuri;
French: Grand Piment, Piment De La Jamaïque, Piment Des Anglais, Poivre De La Jamaique, Poivrier De La Jamaïque, Quatre-Épices , Toute Épice;
French Guiana: Quatre-Épices, Toutes-Épices, Tout-Épice;
Galician: Pimenta Da Xamaica;
German: Allgewürz, Englisches Gewürz, Jamaicapfeffer, Nelkenpfeffer, Neugewürz, Piment, Pimentbaum, Wunderpfeffer;
Greek: Aromatopeperi, Arōmatopéperi, Bahari, Mpachári, Piménta, Piménto, Pimenta, Pimento, Pipéri Iamaïkḗs, Piperi Iamaïkis;
Guyana: Allspice;
Hawaian: Pimeka Kik;
Haitian: Bwa Pwav (Creole);
Hebrew: Pilpel Angli;
Hungarian: Borsmenta, Jamaicaibors, Szegfűbors, Szegfűbors;
India: Kaabaab Chini (Bengali), Kebab Cheeni (Hindu), Gandamenasu (Kannada), Sarvasugandhi (Malayalam), Sarvasukanthi (Tamil);
Iceland: Allrahanda;
Italian: Pepe Della Giamaica, Pimenta Dioica, Pimento, Pimento Inglese;
Japanese: Hyakumikosho, Oorusupaisu, Orsupaisu;
Korean: Oluspaisu;
Latvian: Jamaikas Pipari, Piments, Smaržīgie Pipari, Virces;
Lithuanian: Gvazdikiniai Medis, Kekinis Pimentas, Kvapusis Pipiras, Pimenta, Vaisiai Pimentas;
Macedonian: Pimento;
Malaysia: Allspice;
Mexico: Malaqueta, Pimienta De Jamaica, Pimienta De Tabasco, Pimineta Gorda;
Norwegian: Allehånde;
Polish: Korzennik Lekarski, Pimenta Lekarska, Ziele Angielskie;
Portuguese: Pimenta- Da- Jamaica, Pimenta Inglesa, Pimenta Síria;
Romanian: Cuisoare, Enibahar, Lenibahar, Mirodenii, Piper De Jamaica;
Russian: Perets Dushistyj, Pimenta Dvudomnaya, Perets Gvozdichnyj, Perets Iamáiskii, Yamajskij Perets;
Serbian: Biber Sa Jamajke;
Slovak: Nové Korenie;
Slovenian: Piment;
Spanish: Especerias, Pimenta Dioica, Pimienta Blanca, Pimienta De Jamaica, Pimienta Gorda, Pimienta Inglesa;
Swedish: Kryddpeppar;
Thai: Olspais;
Tonga: Sipaisi;
Turkish: Jamaika Biberi, Yenibahar;
Ukrainian: Perets Dukhmyaniy, Piment;
Venezuela: Pimienta Guayabita;
Vietnamese: Danh Từ, Hạt Tiêu Gia-Mai-Ca;
Wales: Pupur Jamiaca;
Yiddish: English Gewirts.

Origin/Distribution

Allspice is a native of West Indies – greater Antilles, and possibly parts of Central America – from southern Mexico to Guatemala, Belize, El Salvador, Honduras and Nicaragua. It is now widely cultivated in Jamaica, Guatemala and Mexico (Yucatán), also in northern South America, India and Réunion.

Agroecology

Pimento grows well in semitropical lowland forests with mean minimum of 15°C and maximum of 32°C, mane range of 18–25°C, from sea level to 600 m elevation, with mean annual rainfall of 1,500–2,500 mm evenly distributed throughout the year. In Jamaica, its natural habitat is in the forest in limestone hills near the sea. It is adaptable to a wide range of soil types but thrives best in well drained, moist, fertile, loamy calcareous soil.

Edible Plant Parts and Uses

The allspice or pimento berries are used as a spice and condiment in cooking meats, vegetables, and in desserts. Allspice as a name is attributed to the taste and flavour of a blend of cinnamon, cloves and nutmeg it imparts. The fruit is harvested while immature, as it is then most strongly flavoured. The unripe dried fruits contain 2–5% essential oil. The whole dried fruit is ground to produce the allspice powder of commerce In the West Indies, an allspice liqueur called "pimento dram" is produced. Allspice is one of the most important ingredients in Caribbean gastronomy. It is used in jerk seasoning, mole sauces (homogenous, thick sauces for poultry) and to flavour pickles; it is an important ingredient in commercial sausages and curry powders. Allspice is also indispensable in Middle Eastern food where it is used to spice a variety of stews and meat dishes. In Germany, allspice is used in large quantities by the commercial sausage industry. In America and Great Britain, it is widely used in all sorts of desserts – cakes, confectionary, cookies etc. All over the world, Allspice powder is widely used in soups, stew, pot roast, meat loaf, spaghetti sauce, catsup, as a coating for ham, barbecue sauce, salad dressing marinade, pickles and pickled beets, fish, sweet potatoes, squash, cakes, cookies, candy, frosting, fruit pie and mincemeat. Pimento berry essential oil is also similarly used. The essential oil of the berries and leaves are widely utilised in the meat and canning industries, where the oils replace the ground spices to great advantage by having more uniform quality and can be dosed more easily and with greater accuracy. An oleoresin from the pimento berries is also produced in small quantities.

Botany

A small, evergreen, branched, tree, 6–12 m high with brownish-gay, erect trunk and a bushy, rounded canopy (Plate 1). Leaves are opposite, simple, oval-oblong to elliptical, 6–16 cm long by 3–6 cm wide, obtuse rounded apex, cuneate base, glabrous, entire, leathery, pellucid-dotted on lower surface with 12–16 pairs of prominent main veins, mid-rib depressed on upper surface, glossy deep green when mature, pale green when juvenile (Plates 2 and 3). Inflorescence of axillary pedunculate racemose cymes, the peduncles mainly 3–7 cm long (Plate 4). Flowers, bisexual, greenish white, small, 6–10 mm across; calyx 4-lobed, corolla of four, tiny petals; one style, ovary with 2 locules each with an ovule, and numerous stamens. Fruit globose 4–7 mm in diameter, green turning dark purple when ripe with a thick, woody, brittle pericarp (Plates 3, 4 and 5). Each berry has two hard, dark-brown, reniform seeds.

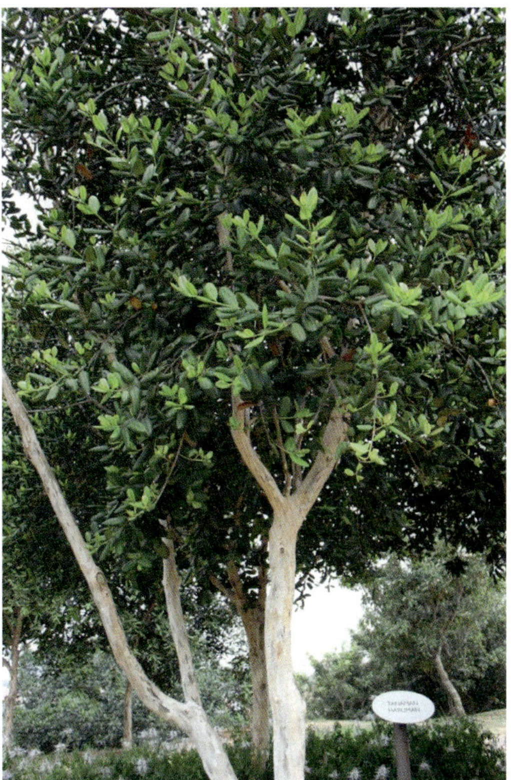

Plate 1 Allspice tree habit

Nutritive/Medicinal Properties

The nutrient composition of ground allspice (*Pimenta dioica*) per 100 g edible portion was reported as follows (U.S. Department of Agriculture, Agricultural Research Service 2010): water 8.46 g, energy 263 kcal (1099 kJ),

Plate 2 Branchlets of allspice

Plate 3 Young fruit and leaves

protein 6.09 g, fat 8.69 g, ash 4.65 g, carbohydrate 72.12 g, total dietary fibre 21.6 g, CA 661 mg, Fe 7.06 mg, Mg 135 mg, P 113 mg, K1044mg, Na 77 mg, Zn 1.01 mg, Cu 0.523 mg, Mn 2.943 mg, Se 2 μg, vitamin C 39.3 mg, thiamine 0.101 mg, riboflavin 0.063 mg, niacin 2.86 mg, vitamin B-6 0.210 mg, total folate 36 μg, vitamin A 540 IU, vitamin A 27 μg RAE,

Plate 4 Clusters of young fruits

Plate 5 Dried allspice berries

total saturated fatty acids 2.255 g, 14:0 (myristic acid) 0.020 g, 16:0 (palmitic acid) 0.490 g, 18:0 (stearic acid) 1.990 g; total monounsaturated fatty acids 0.660 g, 18:1 undifferentiated (oleic acid) 0.660 g; total polyunsaturated fatty acids 2.360 g, 18:2 undifferentiated (linoleic acid) 2.290 g, 18:3 undifferentiated (linolenic acid) 0.070 g and phytosterols 61 mg.

The following chemicals were identified in the leaf essential oil of *P. dioica* from Jamaica (Jirovetz et al. 2007): cis-3-hexenol 0.01%, α-thujene 0.08%, α-pinene 0.15%, β-pinene 0.01%, myrcene 0.11%, α-phellandrene 0.42%, p-cymene 0.51%, limonene 0.12%, β-phellandrene 0.28%, 1,8-cineole 0.31%, trans-β-ocimene 0.21%, γ-terpinene 0.15%, terpinolene 0.47%, linalool 0.06%, terpinen-4-ol 0.19%, α-terpineol 0.09%, chavicol 0.12%, eugenol 76.02%, β-elemene 0.24%, methyl eugenol 7.14%, isoeugenol 0.01%, α-urjunene 0.05%, β-caryophyllene 6.47%, γ-elemene 0.02%, romadendrene 0.26%, α-humulene 1.41%, γ-gurjunene 0.07%, alloaromadendrene 0.26%, β-selinene 0.58%, α-selinene 1.04%, α-(E,E)-farnesene 0.14%, γ-cadinene 0.18%, δ-cadinene 0.37%, α-cadinene 0.08%, trans-nerolidol 0.11%, viridiflorol 0.04%. The major compounds identified in the pimento leaf essential oil were eugenol (76.02%), methyl eugenol (7.14%), β-caryophyllene (6.47%) and -humulene 1.41%. Using supercritical CO2 extraction, Marongui et al. (2005) identified eugenol (77.9%), β-caryophyllene (5.1%), squalene (4.1%) and α-humulene (2.3%) as the major compounds in pimento leaf essential oil. The essential oil of pimento berries has similar chemical composition. Other bioactive chemicals are listed in various sections below.

Allspice has potent antioxidant activity as well as anti inflammatory, anticancer, antimicrobial, hypotensive, antivenom, insecticidal and nematicidal activities.

Antioxidant Activity

Three new galloylglucosides, (4S)-α-terpineol 8-O-β-d-(6-O-galloyl)glucopyranoside (1); (4R)-α-terpineol 8-O-β-d-(6-O-galloyl)glucopyranoside (2), and 3-(4-hydroxy-3-methoxyphenyl)propane-1,2-diol 2-O-β-d-(6-O-galloyl) glucopyranoside (3), were isolated from the berries of *Pimenta dioica* together with three known compounds, gallic acid (4), pimentol (5), and eugenol 4-O-β-d-(6-O-galloyl)glucopyranoside (6) (kikuzaki et al. 2000). These galloylglucosides (1–3, 5, and 6) showed radical-scavenging activity nearly equivalent to that of gallic acid (4) against 1,1-diphenyl-2-picrylhydrazyl radical.

The ethyl acetate-soluble part of allspice, berries of *Pimenta dioica*, showed strong antioxidant activity and radical-scavenging activity against 1,1-diphenyl-2-picrylhydrazyl (DPPH) radical (Miyajima et al. 2004). From the ethyl acetate-soluble part, two new compounds, 5-galloyloxy-3,4-dihydroxypentanoic acid and 5-(5-carboxymethyl-2-oxocyclopentyl)-3Z-pentenyl 6-O-galloyl-β-D-glucoside were isolated together with 11 known polyphenols. Quercetin and its glycosides showed remarkable activity for scavenging DPPH radical and inhibiting peroxidation of liposome induced by 2,2'-azobis-(2-amidinopropane) dihydrocloride (AAPH). Two new compounds also exhibited strong DPPH radical-scavenging activity and inhibitory effect on the peroxidation of liposome as myricetin. The kaempferol, epicatechin and the proanthocyanidin fraction of allspice fruit exhibited antioxidant activity as assayed by the stable radical diphenyl-p-picrylhydrazyl (DPPH) and exhibited IC_{50} values of 7.83, 4.27 and 2.92 g/ml, respectively (Son et al. 2005).

Four new phenolic glycosides, (2-hydroxy-3-methoxy-5-allyl)phenyl β-d-(6-O-E-sinapoyl) glucopyranoside (1), (1' R,5' R)-5-(5-carboxymethyl-2-oxocyclopentyl)-3 Z-pentenyl β-D-(6-O-galloyl)glucopyranoside (2), (S)-α-terpinyl [α-L-(2-O-galloyl)arabinofuranosyl]- (1→6)-β-D-glucopyranoside (3), and (R)-α-terpinyl [α-L-(2-O-galloyl)arabinofuranosyl]- (1→6)-β-D-glucopyranoside (4), were isolated from the berries of *Pimenta dioica* together with eight known flavonoids (Kikizaki et al. 2008). All four glycosides showed radical-scavenging activity against 1,1-diphenyl-2-picrylhydrazyl (DPPH) radicals.

Forty-five constituents were identified in the essential oil of pimento berries (Padmakumari et al. 2011). The major compound identified was eugenol (74.71%, 73.35%), followed by methyl eugenol (4.08%, 9.54%) and caryophyllene (4.90%, 3.30%). The antioxidant assays (free-radical-scavenging activity against 1,1-diphenyl-2-picrylhydrazyl (DPPH), 2,2'-azinobis (3-ethylbenzothiazoline-6-sulphonic acid) diammonium salt (ABTS) radical cation and superoxide anion) showed that the two oil samples possessed very high radical scavenging activities (DPPH IC_{50} 4.82, 5.14 μg/ml, ABTS IC_{50} 2.27, 2.94 μg/ml, superoxide IC_{50} 17.78, 20.65 μg/ml). The metal chelating capacities (IC_{50} 83.62, 101.77 μg/ml) and reducing power were also very high. The results showed the essential oils to have significant antioxidant activity comparable to that of pure eugenol and can be utilised as natural antioxidant with good flavour as well as health benefits.

The essential leaf oil of *Pimenta dioica* exhibited potent antioxidant property (Jirovetz et al. 2007). The scavenging capacity of the pimento leaf oil was strongest in the case of OH• – its IC_{50} value was determined to be 0.29 μg/ml compared to an IC_{50} value for DPPH of 1.79 μg/ml. Xanthine oxidase activity was inhibited 74.83% by the essential *P. dioica* oil, while superoxide scavenging was 95.93%, at 50 μg/ml concentration. The essential pimento leaf oil demonstrated antioxidant activity in a linoleic acid emulsion model system, where at an concentration of 0.005% the sample inhibited conjugated dienes formation by 65.47% and the generation of secondary linoleic acid oxidation products by 72.98%. The antioxidant activity of pimento oil and it main component eugenol were compared with standard antioxidants that included qscorbi acid, rutin, butylated hydroxytoluene (BHT) and butylated hydroxyanisole (BHA) (Jirovetz et al. 2007). Results of the DPPH radical-scavenging activity were as follows: pimento oil IC_{50} 1.79 μg/ml, eugenol 1.26 μg/ml, rutin 14.65 μg/ml, ascorbic acid 4.20 μg/ml, BHT 4.47 μg/ml, BHA 1.12 μg/ml. The results of xanthine oxidase activity were as follows: pimento oil (50 μg/ml) 74.83% inhibition, eugenol (50 μg/ml) 61.29% , BHT (50 μg/ml) 69.03%. The results of the superoxide anion scavenging activity were as follows: pimento oil (50 μg/ml) 95.93% inhibition, eugenol (50 μg/ml) 82.71, BHT (50 μg/ml) 72.20.

Anticancer/Antimutagenic Activities

Two galloylglucosides, 6-hydroxy-eugenol 4-O-(6'-O-galloyl)-β-d-4C1-glucopyranoside (4) and 3-(4-hydroxy-3-methoxyphenyl)-propane-1,2-diol-2-O-(2'6'-di-O-galloyl)-β-d-4C1-glucopyranoside (7), and two C-glycosidic tannins, vascalaginone (10) and grandininol (14), together with 14 known metabolites, gallic acid (1), methyl gallate (2), nilocitin (3), 1-O-galloyl-4,6-(S)-hexahydroxydiphenoyl-(α/β)-d-glucopyranose (5), 4,6-(S)-hexahydroxydiphenoyl-(α/β)-d-glucopyranose (6), 3,4,6-valoneoyl-(α/β)-d-glucopyranose (8), pedunculagin (9), casuariin (11), castalagin (12), vascalagin (13), casuarinin (15), grandinin (16), methylflavogallonate (17) and ellagic acid (18), were identified from the leaves of *Pimenta dioica* (Marzouk et al. 2007). It was found that compound 9 was the most cytotoxic compound against solid tumour cancer cells, the most potent scavenger against the artificial radical DPPH and physiological radicals including ROO*, OH*, and O^{2-}, and strongly inhibited the NO generation and induced the proliferation of T-lymphocytes and macrophages. In contrast, compound 3 was the strongest NO inhibitor and compound 16 the highest stimulator for the proliferation of T-lymphocytes, while compound 10 was the most active inducer of macrophage proliferation.

Five species used as medicinal plants in Cuba that included *Tamarindus indica, Lippia alba, Pimenta dioica, Rheedia aristata* and *Curcuma longa* displayed IC_{50} <30 μg/ml in the DPPH radical reduction assay and IC_{50} <32 μg/ml in lipid peroxidation inhibition testing (Ramos et al. 2003). *Pimenta dioica* and *Curcuma longa* showed also a 20% inhibition of the in-vitro induced z.rad:OH attack to deoxyglucose. Further antimutagenesis assay in *Escherichia coli* IC 188 showed that only *Pimenta dioica* prevented DNA damage by ter-butyl hydroperoxide (TBH). Eugenol, the main constituent of the essential oil of *Pimenta dioica,* also inhibited oxidative mutagenesis by TBH in *Escherichia coli,* at concentrations ranging from 150 to 400 μg/plate.

Antiinflammatory Activity

An aqueous suspension of allspice, produced significant inhibition of carrageenan-induced paw edema, cotton pellet granuloma in rats, a significant inhibition of acetic acid-induced writhing and tail flick reaction time and reduction of yeast-induced hyperpyrexia in mice (Al-Rehaily et al. 2002). The suspension also showed antiulcer and cytoprotective activity by protecting gastric mucosa against indomethacin and various necrotizing agents including 80% ethanol, 0.2 M NaOH and 25% NaCl in rats. The allspice suspension also increased the gastric wall mucus in rats. Acute toxicity studies showed neither mortality nor adverse effects up to a dose of 7.5 g/kg in mice.

Estrogenic Activity

Six of the plant extracts including *P. dioica* exhibited estrogenic effect and were found bound to the estrogen receptors (ER) (Doyle et al. 2009). Four of the six extracts stimulated reporter gene expression in the ER-β-Chemically Activated Luciferase Expression assay. All six extracts modulated expression of endogenous genes in MCF-7 breast cancer cells, with four extracts acting as estrogen agonists and two extracts, *Pimenta dioica* and *Smilax domingensis*, acting as partial agonist/antagonists by enhancing estradiol-stimulated pS2 mRNA expression but reducing estradiol-stimulated PR (Progesterone receptor) and PTGES (prostaglandin E synthase) mRNA expression. Both *P. dioica* and *S. domingensis* induced a 2ERE-luciferase reporter gene (an estrogen responsive luciferase reporter gene plasmid) in transient transfected MCF-7 cells, which was inhibited by the ER antagonist ICI 182,780.

Hypotensive Activity

The intravenous (i.v.) administration of the aqueous extract of *Pimenta dioica* (30, 70, 100 mg/kg) produced a dose-related significant fall in mean arterial blood pressure (MAP). The ED50 was 53.94 mg/kg (Suárez et al. 1997a). The hypotensive effect of identical doses (100 mg/kg) of the aqueous extract (95% decrease) was significantly greater ($P<0.05$) than the effect of the ethanolic extract (67% decrease). The final aqueous fraction produced the greatest hypotensive activity compared to the other fractions of the total aqueous extract. There were no significant changes in the heart rate and no abnormalities were observed in the electrocardiography. The intraperitoneal administration of different extracts of *Pimenta dioica* to conscious normotensive and hypertensive rats caused a depression of the central nervous system (CNS) (Suárez et al. 1997b). The intensity of this depression depended on the dose. Analgesic and hypothermic effects were also observed. The total aqueous extract was more effective than the ethanolic extract and the final aqueous fraction was the most effective. The peritoneal irritation caused by the extract explained only partially the depressive effect over the CNS. When the final aqueous fraction was given orally to SDN (sexually dimorphic nucleus) and SHR (spontaneously hypertensive) rats during 14 days there was no observed change on the sistolic blood pressure, heart rate and weight of the animals. In subsequent studies, the aqueous fraction of *Pimenta dioica* administered intravenously (i.v.) produced a hypotensive action in spontaneously hypertensive rats (SHR) (Suarez et al. 2000). It produced a dose dependent decrease in blood pressure and the ED_{50} was 45 mg/kg. Atropine, propranolol and phentolamine did not affect the hypotensive effect of the final aqueous fraction. With hexamethonium (autonomic ganglion blocker) the hypotensive response was diminished in a significant way. The hypotensive action of the final aqueous extract was not mediated through cholinergic, α or β adrenergic receptors. The extract may possess vasorelaxing activity which could not be evident after autonomic ganglion blockade due to extreme vasodilation present prior to extract administration.

Antimicrobial Activity

Based on the minimum inhibitory concentration (MIC) values, the essential oil of *P. dioica* was found to be superior to the other 24 species of medicinal plants tested in inhibitory activity against six important pathogenic and toxinogenic fungal species that included *Fusarium oxysporum, Fusarium verticillioides, Penicillium expansum, Penicillium brevicompactum, Aspergillus flavus* and *Aspergillus fumigatus* (Zabka et al. 2009). Du et al. (2009) showed that apple-based films with allspice, cinnamon, or clove bud oils were active against 3 food-borne pathogens, *Escherichia coli* O157:H7, *Salmonella enterica*, and *Listeria monocytogenes* by both direct contact with the bacteria and indirectly by vapours emanating from the films. The antimicrobial activities against the 3 pathogens were in the following order: cinnamon oil>clove bud oil>allspice oil. The antimicrobial films were more effective against *L. monocytogenes* than against *S. enterica*.

Antivenom Activity

Ethanolic, ethyl acetate and aqueous extracts of several plants including *P. dioica* was observe to totally inhibit hemorrhagic activity induced by the venom of the snake *Bothrops asper* (Castro et al. 1999). Chemical analysis of these extracts identified catequines, flavones, anthocyanines and condensated tannins, which may be responsible for the inhibitory effect observed, probably owing to the chelation of the zinc required for the catalytic activity of venom's hemorrhagic metalloproteinases.

Traditional Medicinal Uses

Pru is a traditional refreshment and medicinal drink produced by the decoction of three species *Gouania polygamy* (Jacq.) Urban, *Smilax domingensis* Willd., and *Pimento dioica* and fermentation with sugar (Volpato and Godines 2004). Pru is claimed to have hypotensive, stomachic, depurative, and diuretic properties. Pru has long been confined to a number of traditional villages in eastern Cuba, and its origin may be traced back to the ethnobotanical knowledge of French-Haitian people that migrated to Cuba from the end of the 1700 s.

The therapeutic properties of the essential allspice oils and powder are anaesthetic, analgesic, antimicrobial, antioxidant, antiseptic, acaricidal, carminative, muscle relaxant, rubefacient, aromatic stimulant and tonic. Pimento oil can be helpful for the digestive system, for cramp, rheumatism, flatulence, indigestion, colic, diarrhoea, dyspepsia and nausea. Further, the essential oils can help in cases of depression, nervous exhaustion, tension, neuralgia and stress and is used as natural repellent (Duke et al. 2002; Sharma 2003; Seidemann 2005; Jirovetz et al. 2007).

Other Uses

The essential oils of *P. dioica* leaf and fruit are also used in perfumery (especially in aftershaves and deodorants), cosmetics and medicine besides the food industry. Pimento was used as an ointment or a bath additive and sometimes added to commercial medicines to improve their flavour. During the pre-Columbian Mayan civilisation, the Indians used it to embalm their dead and departed. The bark and leaves contain tannin and can be used for tanning purposes. The wood which is very firm and hard with close texture, smooth surface and dark to light salmon colour is used for making walking sticks and umbrellas. Its wood was once in such demand for the making of walking sticks that the tree became endangered and was nearly driven to extinction in the West Indies.

Allspice also has insecticidal and nematicidal activity and has potential to be used as a termiticide and nematicide. Plant essential oils of six (including allspice) out of 26 plant species tested exhibited strong insecticidal activity against Japanese termite (*Reticulitermes speratus* Kolbe) in a fumigation bioassay (Seo et al. 2009). Phenol compounds exhibited the strongest insecticidal activity among the test compounds; furthermore, alcohol and aldehyde groups were more toxic than hydrocarbons. Park et al. (2007) reported good nematicidal activity was achieved against the pinewood nematode, *Bursaphelenchus xylophilus* with essential oils of ajowan (*Trachyspermum ammi*), allspice (*Pimenta dioica*) and litsea (*Litsea cubeba*). These compounds from three plant essential oils were tested individually for their nematicidal activities against the pinewood nematode. LC_{50} values of geranial, isoeugenol, methyl isoeugenol, eugenol, methyl eugenol and neral against pine wood nematodes were 0.120, 0.200, 0.210, 0.480, 0.517 and 0.525 mg/ml, respectively.

Studies showed that *P. dioica* essential oil can be used as an effective alternative acaricide for the control of the cattle tick, *Rhipicephalus (Boophilus) microplus* (Martinez-Velazquez et al. 2011) The essential oil produced 100% mortality at all concentrations.

Comments

Jamaica is also the main exporter of allspice. Several other Central American states (e.g., México, Honduras) produce this spice, but their quality is considered lower than Jamaica.

Selected References

Al-Rehaily AJ, Al-Said MS, Al-Yahya MA, Mossa JS, Rafatullah S (2002) Ethnopharmacological studies on allspice (*Pimenta dioica*) in laboratory animals. Pharm Biol 40(3):200–205

Castro O, Gutiérrez JM, Barrios M, Castro I, Romero M, Umaña E (1999) Neutralization of the hemorrhagic effect induced by *Bothrops asper* (Serpentes: Viperidae) venom with tropical plant extracts. Rev Biol Trop 47(3):605–616, in Spanish

Doyle BJ, Frasor J, Bellows LE, Locklear TD, Perez A, Gomez-Laurito J, Mahady GB (2009) Estrogenic effects of herbal medicines from Costa Rica used for the management of menopausal symptoms. Menopause 16(4):748–755

Du WX, Olsen CW, Avena-Bustillos RJ, McHugh TH, Levin CE, Friedman M (2009) Effects of allspice, cinnamon, and clove bud essential oils in edible apple films on physical properties and antimicrobial activities. J Food Sci 74(7):M372–M378

Duke JA, Bogenschutz-Godwin MJ, DuCellier J, Duke PA (2002) CRC handbook of medicinal plants, 2nd edn. CRC, Boca Raton, 936 pp

Govaerts R, Sobral M, Ashton P, Barrie F, Holst BK, Landrum LL, Matsumoto K, Fernanda Mazine F, Nic Lughadha E, Proenca C, Soares-Silva LH, Wilson PG, Lucas E (2010) World checklist of Myrtaceae. The board of trustees of the Royal Botanic Gardens, Kew. Published on the Internet, http://www. kew.org/wcsp/. Accessed 22 Apr 2010

Jirovetz L, Buchbauer G, Stoilova I, Krastanov A, Stoyanova A, Schmidt E (2007) Spice plants: Chemical composition and antioxidant properties of *Pimenta* Lindl. essential oils, part 1: *Pimenta dioica* (L.) Merr. leaf oil from Jamaica. Nutr Vienna 31(2):55–62

Kikuzaki H, Sato A, Yoko Mayahara Y, Nakatani N (2000) Galloylglucosides from berries of *Pimenta dioica*. J Nat Prod 63(6):749–752

Kikuzaki H, Miyajima Y, Nakatani N (2008) Phenolic glycosides from berries of *Pimenta dioica*. J Nat Prod 71(5):861–865

Landrum LR (1986) *Campomanesia*, *Pimenta*, *Blepharocalyx*, *Legrandia*, *Acca*, *Myrrhinium*, and *Luma* (Myrtaceae). Flora Neotrop Monogr 45:1–178

Marongiu B, Piras A, Porcedda S, Casu R, Pierucci P (2005) Comparative analysis of supercritical CO_2 extract and oil of *Pimenta dioica* leaves. J Essent Oil Res 17:530–532

Martinez-Velazquez M, Castillo-Herrera GA, Rosario-Cruz R, Flores-Fernandez JM, Lopez-Ramirez J, Hernandez-Gutierrez R, Lugo-Cervantes Edel C (2011) Acaricidal effect and chemical composition of essential oils extracted from *Cuminum cyminum*, *Pimenta dioica* and *Ocimum basilicum* against the cattle tick *Rhipicephalus (Boophilus) microplus* (Acari: Ixodidae). Parasitol Res 108(2):481–487

Marzouk MS, Moharram FA, Mohamed MA, Gamal-Eldeen AM, Aboutabl EA (2007) Anticancer and antioxidant tannins from *Pimenta dioica* leaves. Z Naturforsch C 62(7–8):526–536

Miyajima Y, Kikuzaki H, Hisamoto M, Nikatani N (2004) Antioxidative polyphenols from berries of *Pimenta dioica*. Biofactors 22(1–4):301–303

Padmakumaria KP, Sasidharana I, Sreekumar MM (2011) Composition and antioxidant activity of essential oil of pimento (*Pimenta dioica* (L) Merr.) from Jamaica. Nat Prod Res 25(2):152–160

Park IK, Kim J, Lee SG, Shin SC (2007) Nematicidal activity of plant essential oils and components from ajowan (*Trachyspermum ammi*), allspice (*Pimenta dioica*) and Litsea (*Litsea cubeba*) essential oils against pine wood nematode (*Bursaphelenchus Xylophilus*). J Nematol 39(3):275–279

Porcher MH et al (1995–2020) Searchable world wide web multilingual multiscript plant name database. The University of Melbourne, Melbourne. http://www. plantnames.unimelb.edu.au/Sorting/Frontpage.html

Purseglove JW (1968) Tropical crops: dicotyledons. 1 & 2. Longman, London, 719 pp

Ramos A, Visozo A, Piloto J, García A, Rodríguez CA, Rivero R (2003) Screening of antimutagenicity via antioxidant activity in Cuban medicinal plants. J Ethnopharmacol 87(2–3):241–246

Seidemann J (2005) *Pimenta* Lindl. – Allspice – Myrtaceae. World spice plants. Springer, Heidelberg, pp 286–287

Seo SM, Kim J, Lee SG, Shin CH, Shin SC, Park IK (2009) Fumigant antitermitic activity of plant essential oils and components from Ajowan (Trachyspermum ammi, Allspice (*Pimenta dioica*), caraway (*Carum carvi*), dill (*Anethum graveolens*), Geranium (*Pelargonium graveolens*), and Litsea (*Litsea cubeba*) oils against Japanese termite (Reticulitermes speratus Kolbe). J Agric Food Chem 57(15):6596–6602

Sharma R (2003) *Pimenta officinalis*. Medicinal plants of India – an encyclopaedia. Data Publishing House Delhi, New Delhi, 333 pp

Son Y-K, Song T-H, Woo I-A, Ryu H-S (2005) Antioxidative activity of phenolic compounds of allspice (*Pimenta dioica*). J Food Sci Nutr 10:92–94

Suarez A, Ulate G, Ciccio JF (2000) Hypotensive action of an aqueous extract of *Pimenta dioica* (Myrtaceae) in rats. Rev Biol Trop 48(1):53–58

Suárez A, Ulate G, Ciccio JF (1997a) Cardiovascular effects of ethanolic and aqueous extracts of *Pimenta dioica* in Sprague-Dawley rats. J Ethnopharmacol 55(2):107–111

Suárez AU, Ulate GM, Ciccio JF (1997b) Effects of acute and subacute administration of *Pimenta dioica* (Myrtaceae) extracts on normal and hypertensive albino rats. Rev Biol Trop 44–45:39–45, in Spanish

U.S. Department of Agriculture, Agricultural Research Service (2010) USDA national nutrient database for standard reference, release 23. Nutrient Data Laboratory Home Page, Beltsville. http://www.ars.usda.gov/ba/bhnrc/ndl

Volpato G, Godinez D (2004) Ethnobotany of Pru, a traditional Cuban refreshment. Econ Bot 58(3):381–395

Wagner WL, Herbst DR, Sohmer SH (1999) Manual of the flowering plants of Hawaii. Revised edition: Bernice P. Bishop Museum special publication, University of Hawai'i Press/Bishop Museum Press, Honolulu, 1919 pp (two vols)

Ward JF (1961) Pimento. Government Printer, Kingston, 20 pp

Zabka M, Pavela R, Slezakova L (2009) Antifungal effect of *Pimenta dioica* essential oil against dangerous pathogenic and toxinogenic fungi. Ind Crops Prod 30(2):250–253

Plinia cauliflora

Scientific Name

Plinia cauliflora **(DC.) Kuasel.**

Synonyms

Eugenia cauliflora (Mart.) DC., *Eugenia jaboticaba* (Vell.) Kiaersk., *Myrcia jaboticaba* (Vell.) Ball., *Myrciaria cauliflora* (Mart.) O. Berg, *Myrciaria jaboticaba* (Vell.) O.Berg, *Myrtus cauliflora* Mart., *Myrtus jaboticaba* Vell., *Plinia jaboticaba* (Vell.) Kausel.

Family

Myrtaceae

Common/English Names

Brazilian Grape, Brazilian Grapetree, Jaboticaba, Jabotica

Vernacular Names

Argentina: Iba-Purú (Spanish);
Bolivia: Guapurú (Spanish);
Brazil: Jaboticaba, Jaboticaba-De-Sabará, Jaboticaba Sabará, Jaboticabeira, Jabuticaba, Jabuticaba-Açu, Jabuticaba De Campinas, Jabuticaba-Murta, Jabuticaba-Paulista, Jabutica-Ponhema, Yabuticaba, (Portuguese);
Paraguay: Iba-Purú (Spanish);
Spanish: Guapuru, Hivapuru, Iba-Puru, Jaboticaba, Sabara, Ybapuru, Yabuticaba.

Origin/Distribution

The fruit is indigenous to South America – native to the hilly region around Rio de Janeiro and Minas Gerais, Brazil, also around Santa Cruz, Bolivia, Asunción, Paraguay, and northeastern Argentina.

Agroecology

In its native range, jaboticabas grow from sea-level to elevations of 1,000 m in areas with more than 1,200 mm mean annual rainfall and mean annual temperatures of 22–25°C. It is rather frost tolerant as it can withstand brief periods of sub zero temperatures. It thrives best on deep, rich, moist, well-drained, lightly acidic soil with a pH of 5.5–6.5 in full sun or partial shade, but can grow on calcareous (oolitic limestone) and sandy soils. The tree is not tolerant of salty or water-logged soils.

Edible Plant Parts and Uses

Ripe jaboticaba fruits are highly nutritious and are mostly eaten as fresh fruit but are also processed to make juice, dry sweet wine, liquor,

T.K. Lim, *Edible Medicinal And Non-Medicinal Plants: Volume 3, Fruits*,
DOI 10.1007/978-94-007-2534-8_90, © Springer Science+Business Media B.V. 2012

vinegar, sherbets, compotes, jam, jelly, tarts and marmalade. The fruits are often used for making jelly and marmalade, with the addition of pectin. They can be preserved by freezing. The skin is edible but should be avoided as it is tough and high in tannin.

Botany

The jaboticaba is a slow growing large, evergreen shrub or small, much branched, evergreen tree (Plate 1) with a short trunk with diameter of 27–40 cm and dense rounded bushy canopy of 4.5–7 m, growing to 4–12 m high. Young foliage is salmon-pink in colour. Young foliage and branchlets are hairy. Leaves are opposite, on very short, downy petioles, lanceolate to elliptic, 3–6.0 cm long by 1.5–2.5 cm wide, rounded at the base, sharply or bluntly pointed at the apex, leathery, dark-green, and glossy (Plate 2). Flowers small, yellow-white, emerge from the multiple trunks, limbs and large branches in racemose clusters (Plates 1, 2 and 3) of four, on very short, thick pedicels, the flowers have 4 hairy, white petals and about 60 stamens to 4 mm long. Fruit globose to slightly oblate berries (2–3.5 cm

Plate 2 Jaboticaba leaves

Plate 1 Mottled multiple trunks with cauliflorous fascicles of flower buds

Plate 3 Unripe and ripe fruits

Plate 4 Ripe cauliflorous jaboticaba fruits

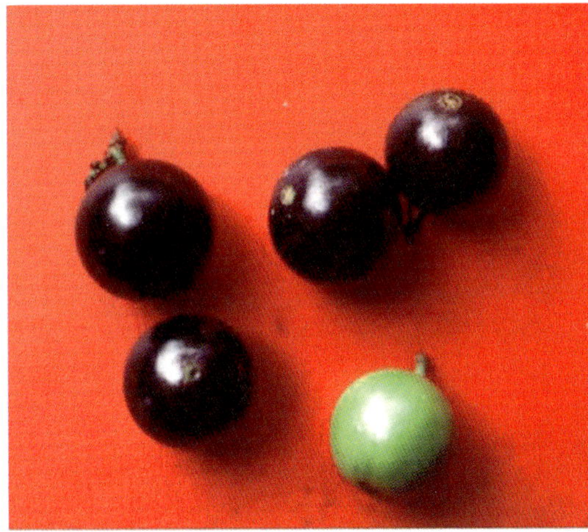

Plate 5 Close-view of jaboticaba fruits

across) and turn from green to dark purple to black at maturity (Plates 3, 4 and 5). They contain 1–3 small, slightly flattened seeds. The pulp is soft, juicy, white, sweet to subacid and covered by the astringent, thick fruit skin.

Nutritive/Medicinal Properties

The food composition of fresh, raw jaboticaba fruit per 100 g edible portion based on analyses made in Cuba was reported as: energy 45.7Cal., moisture 87.1 g, protein 0.11 g, fat 0.01 g, carbohydrates 12.58 g, fibre 0.08 g, ash 0.20 g, Ca 6.3 mg, P 9.2 mg, Fe 0.49 g, thiamin 0.02 mg, riboflavin 0.02 mg, niacin 0.21 mg, ascorbic acid 22.7 mg, tryptophan 1 mg, lysine 7 mg (Morton 1987). The dark-purple fruits were rich in ascorbic acid (22.7 mg per 100 g fresh weight). The nutrient value of ripe jaboticaba fruits per 100 g was reported by Lorenzi et al. (2006) as: energy 51.7 cal., water 87.1 g, protein 0.32 g, fat 0.0 g, carbohydrate 12.58 g, ash 0.20 g, Ca 7.6 mg, P 34.6 mg, Fe 0.87 mg, K 13.2 mg, vitamin B-1 0.04 mg, vitamin B-20.09 mg, niacin 1.30 mg and vitamin C 17.7 mg.

A new depside, jaboticabin (1), and 16 known phenolic compounds: depside 2-O-(3,4-dihydroxybenzoyl)-2,4,6-trihydroxyphenylacetic acid (2), cyanidin 3-glucoside (3), delphinidin 3-glucoside (4), protocatechuic acid (5), methyl protocatechuate (6), gallic acid (7), cinnamic acid (8), o-coumaric acid (9), quercetin (10), quercitrin (11) ,isoquercitrin (12), rutin (13), myricitrin (14), quercimeritrin (15), ellagic acid (16) and pyranocyanin B (17) were identified from jaboticaba fruits (Reynertson et al. 2006). Isoamyl alcohols (2-methyl-1-butanol and 3-methyl-1-butanol) were the most abundant volatile compounds identified in jabuticaba spirit processed by fermentation of jabutica fruits (Duarte et al. 2011).

Foliar contents of total phenols and tannins and the essential oil composition of *Myrciaria cauliflora* populations were shown to be influenced by soil types and foliar nutrients (Duarte et al. 2010). A total of 28 compounds were identified in the essential oil which was predominated by sesquiterpene hydrocarbon compounds 47.50% on sandy-loam soils to 57.22% on clay soils and oxygenated hydrocarbons 40.86% on clay soils to 49.04% on sandy loam soils. Oil yields ranged from 0.26% (sandy loam soils) to 0.48% (clay soils). The oils were predominated by germacrene D (20.58–27.20%), β-eudesmol (15.19–19.20%, α-eudesmol 10.66–12.94%, γ-eudesmol 7.81–11.55%, β-caryophyllene 7.55–8.60%, bicyclogermacrene 6.29–7.82%, elemol 2.12–4.61%, δ-cadinene 2.77–3.32%, and α-copaene 1.99–2.70%. Cluster I included samples which originated from sandy loam soil (S1) with the highest and lowest percentages of γ-eudesmol (11.55%) and germacrene D (20.48%), respectively, as well as high total phenol (136.68 mg g-1) and tannin (60.72 mg g-1) contents. Cluster II, rich in elemol (4.61%), included all the samples cultivated from clay sand loam soils (S2, S3, and S6), whereas clay soils S4 (cluster III) and S5 (cluster IV) had the highest amounts of germacrene D (III: 27.20%; IV: 26.83%) and the lowest levels of elemol (2.12–2.55%). Total phenols ranged from 79.69 mg/g (clay soils) to 136.68 mg/g (sandy loam) and tannins 34.04 mg/g (clay soils) to 60.72 mg/g (sandy-loam). S1 samples had the lowest percentage of germacrene D, bicyclogermacrene and sesquiterpene hydrocarbons, although it showed the highest percentages of oxygenated sesquiterpenes, α-, β- and γ-eudesmol. On the other hand, samples from fertilized soils S4 and S5 showed the lowest amount in elemol, total phenols and tannins, despite showing the highest sesquiterpene content. Despite the high percentage of β-caryophyllene, this constituent did not vary significantly between samples growing on different sites.

Santos et al. (2010) developed a combined extraction process involving ultrasound treatment and agitated solvent extraction to maximise the extraction of phenolic compounds with acceptable degradation of anthocyanin pigments (cyanidin-3-glucoside and delphinidin-3-glucoside) from Brazilian jaboticaba (*Myrciaria cauliflora*) skins. The procedure was found to be cost-effective as it used shorter ultrasonic irradiation and resulted in high antioxidant activity extracts. The HPCD assisted-extraction was effective in extracting high anthocyanins (2.2 mg cyanidin-3-glucoside/g dry skins) and phenolic compounds(13 mg gallic acid equivalents/g dry skins), from jaboticaba fruit skins (Santos and Meirles 2011). The best conditions were obtained at 117 bar extraction pressure, 80°C extraction temperature and 20% volume ratio of solid–liquid mixture/pressurized CO^2.

Phytochemicals in various plant parts and associated pharmacological activities such antiinflammatory, anticancer, antimicrobial and antioxidant activities had been reported. Several potent antioxidant and antiinflammatory anticancer compounds were isolated from the fruit (Reynertson et al. 2006).

Antiinflammatory Activity

Reynertson et al. (2006) reported that jaboticabin (1) and the related depside 2-O-(3,4-dihydroxybenzoyl)-2,4,6-trihydroxyphenylacetic acid (2) from jaboticaba fruit, significantly inhibited chemokine interleukin (IL)-8 production in human small airway epithelial (SAE) cells before and after cigarette smoke treatment of cells. Protocatechuic acid (5), structurally similar to one moiety of the depsides, also inhibited IL-8 production, but not to the same degree as the depsides. The anthocyanins cyanidin 3-glucoside (3) and delphinidin 3-glucoside (4), major constituents of jaboticaba fruits, also displayed significant activity against IL-8 production in SAE cells. Compounds 1–5 were more effective at blocking IL-8 production in 34 untreated SAE cells than catechin, and compounds 2 and 4 were more effective than catechin at blocking cigarette smoke-induced inflammation. Interleukin IL-8 is a cytokine implicated in a wide range of chronic inflammatory conditions, including cancers, rheumatoid arthritis, heart and lung diseases, and these depsides may lead to a promising line of therapy for these conditions. Depsides are known to be potent non-steroidal anti-inflammatories which inhibit prostaglandin and leukotriene B4 biosynthesis.

Anticancer Activity

The cytotoxicity of jaboticabin (1) and the related depside 2-O-(3,4-dihydroxybenzoyl)-2,4,6-trihydroxyphenylacetic acid (2) from jaboticaba fruit and delphinidin 3-glucoside (4) was found to be comparable to 5-fluorouracil (5-FU), a drug used for colon cancer treatment, epigallocatechin gallate (EGCG), and Polyphenon E (Poly E), a standardized decaffeinated green tea extract (Reynertson et al. 2006). Compound 1 was cytotoxic in the HT29 colon cancer cell line (IC_{50} = 65 μM), and 2 was active against HCT116 colon cancer cells (IC_{50} = 30 μM). Compound 4 showed good activity against both the HCT116 and SW480 (human colon adenocarcinoma) cell lines (IC_{50} = 12 and 20 μM, respectively), while cyanidin 3-glucoside (3) inhibited 50% cell growth only at the 100 μM range.

Antioxidant Activity

Compounds from jaboticaba fruits: jaboticabin (1), 2-O-(3,4-dihydroxybenzoyl)-2,4,6-trihydroxyphenylacetic acid (2), cyanidin 3-glucoside (3), delphinidin 3-glucoside (4) exhibited good antioxidant activity . Compounds 1 and 2 exhibited antiradical activity in the 1,1-diphenyl-2-picrylhydrazyl (DPPH) assay (IC_{50}=51.4 and 61.8 μM, respectively). The anthocyanins, cyanidin 3-glucoside (3) and delphinidin 3-glucoside (4), also showed good activity in these assays. In addition, total phenolic content (TPC) 31.63 mg gallic acid equivalents per g dry weight, total anthocyanin content (TAC) 2.78 mg/g, and antiradical activity, measured as DPPH IC_{50} was 19.40 μg/ml (Reynertson et al. 2006).

Cyanidin-3-O-β-glucopyranoside, another anthocyanin antioxidant, was identified from semi-purified aqueous fractions of jaboticaba fruits and exhibited high antioxidant activity (Einbond et al. 2004). Its methanol aqueous extract had an IC_{50} value of 6.2 μg/ml in the DPPH assay.

Antimicrobial Activity

Leaf extract of jaboticaba exhibited in-vitro antibacterial against the following oral pathogens: *Streptococcus mitis, Streptococcus mutans, Streptococcus sanguinis, Streptococcus oralis, Streptococcus salivarius* and *Lactobacillus casei* (Macedo-Costa et al. 2009). Monhanty and Cock (2009) reported that *M. cauliflora* leaf extract inhibited the growth of 9 of the 14 bacteria tested (64%) whilst the fruit extract inhibited the growth of 11 of the 14 bacteria tested (79%). Both Gram-positive and Gram-negative bacterial growth were inhibited by *M. cauliflora* leaf and flower extracts. 7 of the 10 Gram-negative bacteria (70%) and 2 of the 4 Gram-positive bacteria (50%) tested had their growth inhibited by *M. cauliflora* leaf extract whereas the fruit extract inhibited 7 of the 10 Gram-negative bacteria (70%) and 100% of the Gram-positive bacteria tested. *M. cauliflora* leaf extract proved to be toxic in the *Artemia fransiscana* bioassay making it more toxic than Mevinphos but less toxic than potassium dichromate. *M. cauliflora* fruit extract was non-toxic in the *Artemia fransiscana* bioassay indicating its potential as an antibacterial agent for medicinal use.

Traditional Medicinal Uses

Traditionally in Brazil, an astringent decoction of the sun-dried skins has been used as a treatment for hemoptysis, asthma, diarrhoea, dysentery and gargle for chronic inflammation of the tonsils.

Other Uses

Jaboticaba has become poular as bonsai plants in Taiwan and the Caribbean.

Comments

The plant is propagated by seeds and graftings.

Selected References

Asquieri ER, Damiani C, Candido MA, Assis EM (2004) Vino de jabuticaba (*Myrciaria cauliflora* Berg): Estudio de las caracteristicas fisico-químicas y sensoriales de los vinos tinto seco y dulce, fabricados con la fruta integral. Alimentaria 355:111–122, in Spanish

de Jesus N, Martins AB, de Almeida EJ, Vieira JB, Devos R, Scaloppi EJ, Aparecida R, Cunha RF (2004) Caracterizacão de quatro grupos de jaboticabeira, nas condicões de Jaboticabal. SP Rev Bras Frutic 26(3):1–8 (in Portuguese)

Duarte AR, Santos SC, Seraphin JC, Ferri PH (2010) Environmental influence on phenols and essential oils of *Myrciaria cauliflora* leaves. J Braz Chem Soc 21(9):1672–1680

Duarte WF, Amorim JC, Lago LA, Dias DR, Schwan RF (2011) Optimization of fermentation conditions for production of the jabuticaba (*Myrciaria cauliflora*) spirit using the response surface methodology. J Food Sci 76(5):C782–C790

Einbond LS, Reynertson KA, Luo X-D, Basile MJ, Kennelly EJ (2004) Anthocyanin antioxidants from edible fruits. Food Chem 84(1):23–28

Govaerts R, Sobral M, Ashton P, Barrie F, Holst BK, Landrum LL, Matsumoto K, Fernanda Mazine F, Nic Lughadha E, Proenca C, Soares-Silva LH, Wilson PG, Lucas E (2010) World checklist of Myrtaceae. The board of trustees of the Royal Botanic Gardens, Kew. Published on the Internet http://www. kew.org/wcsp/ Accessed 22 Apr 2010

Govaerts R (2006) World checklist of Myrtaceae. The Board of Trustees of the Royal Botanic Gardens, Kew. http://www.kew.org/wcsp/

Kennard WC, Winters HF (1960) Some fruits and nuts for the tropics. USDA Agric Res Serv Misc Publ 801:1–135

Lorenzi H, Bacher L, Lacerda M, Sartori S (2006) Brazilian fruits & cultivated exotics (for consuming in natura). Instito Plantarum de Etodos da Flora Ltda, Brazil, 740 pp

Macedo-Costa MR, Diniz DN, Carvalho CM, do Pereira MSV, Pereira JV, Higino JS (2009) Effectiveness of the *Myrciaria cauliflora* (Mart.) O. Berg. extract on oral bacteria. Rev Bras Farmacogn 19(2b):565–571

Mohanty S, Cock I (2009) Evaluation of the antibacterial activity and toxicity of *Myrciaria caulifloria* methanolic leaf and fruit extracts. Internet J Microbiol 7(Article 2)

Morton JF (1987) Jaboticabas. In: Fruits of warm climates. Julia F. Morton, Miami, pp 371–374

Popenoe W (1974) Manual of tropical and subtropical fruits. Hafner Press, New York, Facsimile of the 1920 edition

Reynertson KA, Wallace AM, Adachi S, Gil RR, Yang H, Basile MJ, D'Armiento J, Weinstein IB, Kennelly EJ (2006) Bioactive depsides and anthocyanins from jaboticaba (*Myrciaria cauliflora*). J Nat Prod 69(8):1228–1230

Reynertson KA, Yang H, Jiang B, Basile MJ, Kennelly EJ (2008) Quantitative analysis of antiradical phenolic constituents from fourteen edible Myrtaceae fruits. Food Chem 109(4):883–890

Santos DT, Meireles MAA (2011) Optimization of bioactive compounds extraction from jabuticaba (*Myrciaria cauliflora*) skins assisted by high pressure CO2. Innov Food Sci Emerg Technol 12(3):398–406

Santos DT, Veggi PC, Meireles MAA (2010) Extraction of antioxidant compounds from Jabuticaba (*Myrciaria cauliflora*) skins: yield, composition and economical evaluation. J Food Eng 101(1):23–31

Wiltbank WJ, Chalfun NNJ, Andersen O (1983) The jaboticaba in Brazil. Proc Amer Soc Hortic Sci Trop Reg 27(Part A):57–69

Psidium acutangulum

Scientific Name

Psidium acutangulum Mart. ex DC.

Synonyms

Britoa acida (DC.) O. Berg, *Guajava acutangula* (Mart. ex DC.) Kuntze, *Psidium acidum* Mart. ex O.Berg nom invalid., *Psidium acutangulum* DC. var. *acidum* C. Mart. ex DC., *Psidium acutangulum* var. *crassirame* O.Berg, *Psidium acutangulum* var. *oblongatum* Mattos, *Psidium acutangulum* var. *tenuirame* O. Berg, *Psidium grandiflorum* Ruiz & Pav. nom. illeg., *Psidium persoonii* McVaugh.

Family

Myrtaceae

Common/English Names

Araçá-pera, Para Guava

Vernacular Names

Bolivia: Guabira (Spanish);
Brazil: Araçá Comum Do Pará, Araçá Do Pará, Araçá-Piranga, Araçá Pomba, Araçándiva, Araçanduba, Goia-Ba-Do-Pará, (Portuguese);
German: Para-Guave;
Peru: Ampiyacu, Guayaba De Agua, Puca Yacu (Spanish).

Origin/Distribution

The species is native to tropical south America (French Guiana, Guyana, Suriname, Venezuela, Bolivia, Ecuador, Peru, Belize and Brazil).

Agroecology

The tree occurs wild and is cultivated at low and medium elevations throughout Amazonia and the coastal areas in its native range.

Edible Plant Parts and Uses

Ripe fruit has translucent, pale yellowish-white acidic pulp with a strong flavour. The fruit is occasionally eaten raw but is more popularly made into drinks because of the high acidity. The para guava is often combined with honey or sugar to make a lemonade-like drink, juice or sherberts. It is also used to make jellies and jams.

T.K. Lim, *Edible Medicinal And Non-Medicinal Plants: Volume 3, Fruits*,
DOI 10.1007/978-94-007-2534-8_91, © Springer Science+Business Media B.V. 2012

Botany

Evergreen shrub or small tree 3-8(−12) m high with winged, quadrangular branchlets. New growth is finely pubescent. The leaves are borne on very short 1 cm petioles, broadly elliptical, 10–14 cm long by 4–6 cm wide, rounded at the base, apex acuminate with distinct mid rib and 6–9 pairs of secondary veins conspicuous on the underside (Plates 1 and 2). Flowers solitary or in fascicles of 2–3 flowers in leaf axils. Flowers on long peduncle with green sepals, 5 white –petals and numerous (> 300) white, long, thin stamens. The fruit is round, pear-shaped or ellipsoid, 3–8 cm across, glabrous, pale green turning to pale-yellow to yellow when ripe, with persistent calyx remnants at the apical end (Plates 1, 2 and 3)

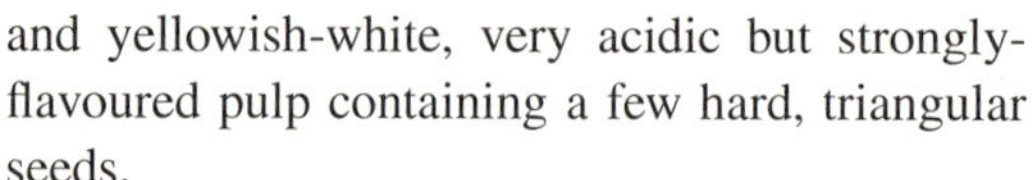

and yellowish-white, very acidic but strongly-flavoured pulp containing a few hard, triangular seeds.

Plate 1 Solitary fruit developed from solitary axillary flower

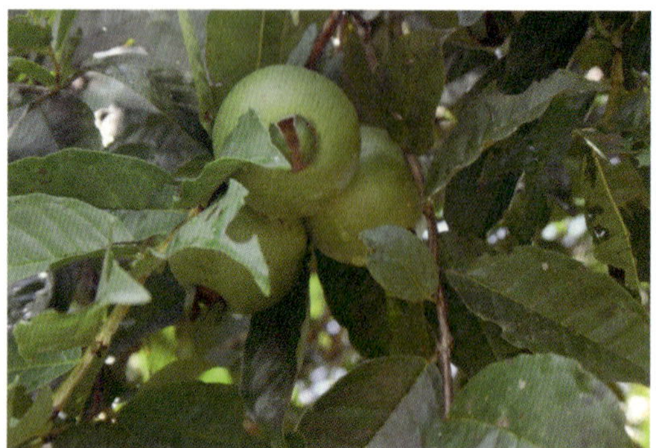

Plate 2 Fruits developed from 3-flowered axillary fascicle

Nutritive/Medicinal Properties

Psidium acutangulum fruits were found to be very acidic with a low pH 2.77 brix/acidity ratio of 3.31, high vitamin C and dietary fibre contents (Stertz et al. 2003). Average fruit weight of 59.19 g to 250.69 g and high yield of pulp (76.27%) were recorded. The fruit was meaty with a stronger and more pleasant flavour than guava. The nutrient composition per 100 g edible portion was reported as follows: moisture 82.4%, energy 34 kcal, carbohydrates including dietary fibre (15.77%), ash (0.49%), protein (0.39%) and lipids (0.86%), dietary fibre 9.66% minerals in ppm Ca- 143.93; Fe- 2.71; P- 1.99; Mg- 93.85; Mn- 1.87; K- 15988.38; Na- 5.43; Zn- 2.28, citric acid 2.67% and vitamin C- 60.98 mg/100 g. *Psidium acutangulum* fruits were found to have high fibre content and higher 2,2-diphenyl-1-picrylhydrazyl radical scavenging activity than guava, *Psidium guajava* (Rincon et al. 2000). Araçá-pera fruit was found to have the following functional properties: total phenolic content of 1,851.38 mg cholorogenic acid equivalent, antioxidant activity 20,324.82 μg/g trolox equivalent, anthocyanins 10.41 mg cyanidin-3-glucoside equivalent and total carotenoid 0.59 mg β-carotene equivalent (Fetter et al. 2010).

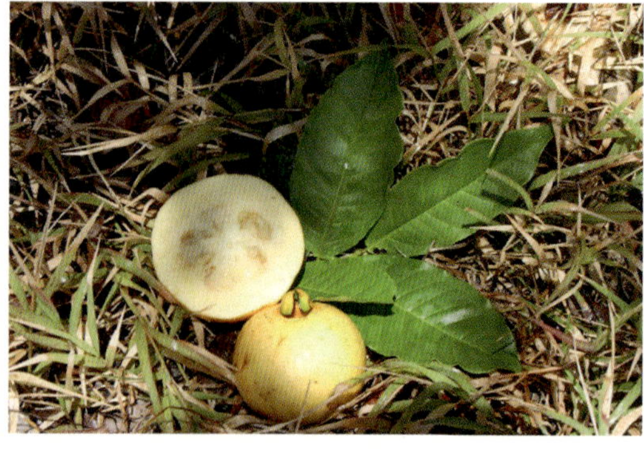

Plate 3 Close-up of whole, halved fruit and leaves

The main compounds identified in the oil of *P. acutangulum* fine stems and leaves were α-pinene (14.8%), 1,8-cineole (12.9%) and β-pinene (10.1%) (da Silva et al. 2003).

Other Uses

P. acutangulum has antimicrobial and insecticidal properties and has potential to be used as such. Twig and leaf extracts of *Psidium acutangulum* were shown to have antifungal properties against the fungi *Rhizoctonia solani, Helminthosporium teres* and *Pythium ultimum* (Miles et al. 1991). The active compound isolated 3′-formyl-2′,4′,6′-trihydroxychalcone, demonstrated activity against the fungi *R. solani* and *H. teres*. A chalcone compound isolated from dichloromethane extracts of *P. acutangulum* leaves showed antimicrobial activity against *Rhizoctonia solani* and *Xanthomonas campestris* and also antifeedant activity in tests with larvae of *Heliothis virescens* (Miles et al. 1990).

Comments

The plant is propagated from seeds, cuttings and grafting.

Selected References

da Silva JD, Luz AIR, da Silva MHL, Andrade EHA, Zoghbi MdGB, Maia JGS (2003) Essential oils of the leaves and stems of four *Psidium* spp. Flav Fragr J 18:240–243

Fetter MDR, Vizzotto M, Corbenini DD, Gonzalez TN (2010) Propriedades funcionais de araçá-amarelo, araçá-vermelho (*Psidium cattleyanum* Sabine) e araçá-pera (*P. acutangulum* DC.) cultivados em Pelotas/RS. Braz J Food Technol III SSA, Novembro 2010 (in Portuguese)

Govaerts R, Sobral M, Ashton P, Barrie F, Holst BK, Landrum LL, Matsumoto K, Fernanda Mazine F, Nic Lughadha E, Proenca C, Soares-Silva LH, Wilson PG, Lucas E (2010) World checklist of Myrtaceae. The board of trustees of the Royal Botanic Gardens, Kew. Published on the Internet http://www. kew.org/wcsp/ Accessed 22 Apr 2010

Lorenzi H, Bacher L, Lacerda M, Sartori S (2006) Brazilian fruits & cultivated exotics (for consuming in natura). Instituto Plantarum de Etodos da Flora Ltda, Nova Odessa, 740 pp

Miles DH, de Medeiros JMR, Chittawong V, Swithenbank C, Lidert Z, Weeks JA, Atwood JL, Hedin PA (1990) 3′-Formyl-2′,4′,6′-trihydroxy-5′-methyl-dihydrochalcone, a prospective new agrochemical from *Psidium acutangulum*. J Nat Prod 53(6):1548–1551

Miles DH, de Medeiros JMR, Chittawong V, Hedin PA, Swithenbank C, Lidert Z (1991) 3′-formyl-2′,4′,6′-trihydroxydihydrochalcone from *Psidium acutangulum*. Phytochemistry 30(4):1131–1132

Morton JF (1987) Brazilian Guava. In: Morton JF (ed) Fruits of warm climates. Julia F. Morton, Miami, pp 365–367

Rincon AM, Jimenez-Escrig A, Padilla FC, Saura-Calixto F (2000) Characterization of the peels of two varieties of guava as fiber source with antioxidant activity. Paper presented at the American Association of Cereal Chemists (AACC) annual meeting, Kansas City, 5–9 Nov 2000

Stertz SC, Neto RC, Wille GMFC, Lima JM, Macedo REF, Freitas RJS, Masson ML (2003) Characterization of Araça Pera (*Psidium acutangulum*, DC.). Paper presented at the 12th World Food Congress, Chicago, 16–19 July 2003

Psidium cattleianum 'Red Strawberry Guava'

Scientific Name

Psidium cattleianum **Afzel. ex Sabine** (for both red and yellow strawberry guava).

Synonyms

Episyzygium oahuense Suess. & A. Ludw., *Eugenia ferruginea* Sieber ex C.Presl, *Eugenia oxygona* Koidz., *Eugenia pseudovenosa* H.Perrier, *Eugenia urceolata* Cordem., *Guajava cattleiana* (Afzel. ex Sabine) Kuntze, *Guajava obovata* (Mart. ex DC.) Kuntze, *Psidium cattleianum* f. *lucidum* O.Deg., *Psidium cattleianum* var. *littorale* (Raddi) Fosberg, *Psidium cattleianum* var. *littorale* (Raddi) Mattos, nom. illeg., *Psidium cattleianum* var. *purpureum* Mattos, *Psidium cattleianum* var. *pyriformis* Mattos, *Psidium coriaceum* var. *grandifolium* O.Berg, *Psidium coriaceum* var. *longipes* O.Berg, *Psidium coriaceum* var. *obovatum* O.Berg, *Psidium ferrugineum* C.Presl, *Psidium indicum* Bojer nom. inval., *Psidium littorale Raddi, Psidium littorale* var. *longipes* (O.Berg) Fosberg, *Psidium obovatum* Mart. ex DC., *Psidium variabile* O.Berg.

Family

Myrtaceae

Common/English Names

Cattley Guava, Cherry Guava, China Guava, Porpay, Purple Guava, Red Strawberry Guava; Strawberry Guava. Lemon guava, Yellow Cattley guava, Yellow strawberry guava (See next chapter)

Vernacular Names

Brazil: Araça-Amarelo, Araçá (Yellow Strawberry Guava), Araçá-Da-Praia, Araçá-De-Comer, Araçá-De-Coroa (Yellow Strawberry Guava), Araçá-Do-Campo, Araçá-Do-Mato, Araçá-Doce, Araçá-Manteiga, Araçá-Pera, Araçá Rosa, Araçá Vermelho, Araçazeiro (Yellow Strawberry Guava) (Portuguese);
Chinese: Cao Mei Fan Shi Liu;
Cook Islands: Tū'Ava Papa'ā, Tuava Papa'ā, Tuava Papa'ā, Tūava Papa'ā, Tūava Papa'ā, Tūvava Papa'ā (Maori);
Costa Rica: Arazá, Cas Dulce (Spanish);
Danish: Jordbærguava;
Eastonian: Maasik-Guajaavipuu (Yellow Strawberry Guava);
French: Goyave De Chine Jaune, Goyave De Chine Rouge, Goyavier, Goyavier De Chine, Gouyave Fraise, Goyavier-Fraise;
Fijian: Ngguava, Quwawa Ni Vavalagi, Waiawichinese (Yellow Strawberry Guava);

T.K. Lim, *Edible Medicinal And Non-Medicinal Plants: Volume 3, Fruits*,
DOI 10.1007/978-94-007-2534-8_92, © Springer Science+Business Media B.V. 2012

German: Erdbeerguave (Yellow Strawberry Guava);
Guatemala: Guayaba De Fresa, Guayaba Japonesa (Spanish);
Hawaiian: Waiawī (Yellow Strawberry Guava), Waiawī 'Ula'Ula;
India: Paayaaraa, Pahaarii (Bengali), Kooyayaa, Samaaii, Simaaii (Tamil);
Japanese: Koba No Banjirou, Teriha Banjirou;
Russian: Psidium Pribrezhnyi;
Samoan: Ku'Ava;
Spanish: Arazá, Cas Dulce, Guayaba, Guayaba De Fresa, Guayaba Japonesa, Guayaba Peruana;
Tahitian: Tuava Popa'a, Tuava Tinito, Tuvava Tinito;
Venezuela: Guayaba Pequeña, Guayaba Peruana.

Origin/Distribution

The strawberry cattley guava is believed to be native to lowlands of eastern and southern Brazil to north-east Uruguay. It is cultivated to a limited extent and has naturalised in various areas of South America and Central America and in the West Indies, Bermuda, the Bahamas, southern and central Florida and southern California. It has been introduced into the Mediterranean region, Asia (highlands of India, Sri Lanka, Malaysia, the Philippines, Vietnam), South Africa, China, Australia and the south Pacific Islands.

Agroecology

The purple strawberry guava is adaptable to a subtropical climate from 150 to 1,300 m elevation, but can be grown in the tropics at higher elevations. It is cold hardier than the yellow strawberry as it can survive subzero temperatures. In its native range is occurs commonly in the pluvial Atlantic forest. It grows mainly in the sandy, coastal restingas (broad-leaved forest) and humid localities in the capoeiras (scrubs) and humid lowlands in Brazil. It is also found in abandoned fields and capoeiras in the humid altitude regions. In some countries it has become a weed. The tree grows well in full sun and need plenty of water for good fruit set and development. It can withstand short periods of drought. It does best in well drained fertile sandy or loamy soils but will grow in calcareous and poor soils.

Edible Plant Parts and Uses

Purple cattley guavas are eaten fresh, out-of-hand and also processed into jams, jellies, tarts, paste, sherbet, ice-creams, butter, beverages, tarts and other desserts. Half-ripe or full-ripe cattleys can be sliced, boiled, and the juice strained to make ade or punch by adding sugar. The seeds are can be roasted as a substitute for coffee and Its leaves may be brewed for tea.

Botany

Psidium cattleianum is an evergreen shrub or small tree 3–7 m high with slender grey-brown slender trunk and cylindrical, glabrous or puberulous branchlets. Leaves opposite, simple, entire, glabrous, glossy dark green (pink–bronze when young (Plate 2)), thick, broadly oblanceolate to obovate, 4–8(−10)cm long, 2.5–4 cm broad, base cuneate, apex acute, 6–7 pairs of lateral veins, pellucid dotted below (Plate 1), ciliolate when young becoming glabrous with age, petioles 4–15 mm long. Flowers axillary, usually solitary, white, fragrant; sepals 4–5 lobes, 3–4 mm long, glabrous, gland-dotted; petals obovate-elliptic,

Plate 1 Fruiting branch and leaves

Plate 2 Young leaves and ripe fruits

Plate 3 Ripe red strawberry guava

10 mm long, slightly concave, white; stamens prominent, free, shorter than petals; ovary 4-loculed; style slender; stigma peltate. Fruit globose or broadly pyriform, 2.5–3.5 cm long, green turning to purplish red (Plates 1, 2, 3 and 4), or yellow (yellow strawberry guava, see next chapter) when ripe; with white, sometimes tinged with pink (Plate 4), translucent, juicy, aromatic flesh, mildly sweet-tasting when ripe; seeds numerous, hard, flattened-triangular, 2.5 mm long.

Nutritive/Medicinal Properties

Analyses carried out in the United States reported that raw, strawberry guava fruit, (minus 15% seeds, stem and blossom ends) had the following proximate composition (per 100 g edible portion): water 80.66 g, energy 69 kcal (289 kJ), protein 0.58 g, total lipid 0.60 g, ash 0.80 g,

Plate 4 White fleshed tinged with pink

carbohydrates 17.36 g, total dietary fibre 5.4 g, Ca 21 mg, Fe 0.22 mg, Mg 17 mg, P 27 mg, K 292 mg, Na 37 mg, vitamin C 37 mg, thiamine 0.030 mg, riboflavin 0.03 mg, niacin 0.60 mg, vitamin A 90 IU, total saturated fatty acids 0.172 g, total monounsaturated fatty acids 0.055 g, total polyunsaturated fatty acids 0.253 g, tryptophan 0.005 g, threonine 0.002 g, isoleucine 0.021 g, leucine 0.039 g, lysine 0.016 g, methionine 0.004 g, phenylalanine 0.001 g, tyrosine 0.007 g, valine 0.020 g, arginine 0.015 g, histidine 0.005 g, alanine 0.029 g, aspartic acid 0.037 g, glutamic acid 0.076 g, glycine 0.029 g, proline 0.018 g, and serine 0.017 g (U.S. Department of Agriculture, Agricultural Research Service 2010).

Analyses of ripe fruits in the Philippines, Hawaii and Florida reported the following constituents (Morton 1987):

Red: seeds, 6%; water, 81.73–84.9%; ash, 0.74–1.50%; crude fibre, 6.14%; protein, 0.75-1-03%; fat, 0.55%; total sugar, 4.42–4.46%.

Yellow: seeds, 10.3%; water, 84.2%; ash, 0.63–0.75%; crude fibre, 3.87%; protein, 0.80%; fat, 0.42%; total sugar, 4.32–10.01%.

Red or *Yellow*: ascorbic acid, 22–50 mg/100 g. Calories per 2.2 lbs (1 kg), 268.

Volatile aroma compounds identified in red (*Psidium cattleianum*, Sabine) and yellow (*Psidium cattleianum* Sabine var. *lucidum* Hort.) guava fruits included: 31 hydrocarbons, 9 acetals, ethers and oxides, 13 aldehydes, 13 ketones, 30 esters, 48 alcohols, 2 acids, 2 sulfur-containing compounds, 4 phenol derivatives, menthofuran and coumarin (Vernin et al. 1998). The following compounds were thought to contribute to the aroma of the red fruit. Fruity notes due to ethyl esters (C4-C16), tiglates, cinnamates, while floral notes could be attributed mainly to terpenic alcohols, 2-phenylethyl alcohol, β-ionone and 1-phenylpropane-1,2-dione. Spicy notes may be due to cinnamaldehyde, eugenol, methyl isoeugenols, while burnt notes were attributed to furfural and 2-acetylfuran. The herbaceous, slightly spicy-like odour could be attributed to 2-tridecanone and the sweet and balsamic notes to benzyl benzoate. The guava aroma was characterized by the quasi absence of lactones.

Two hundred and four compounds were identified in the aroma concentrate of strawberry guava fruit of which ethanol, α-pinene, (Z)-3-hexenol, (E)-β-caryophyllene, and hexadecanoic acid were found to be the major constituents (Pino et al. 2001) The presence of many aliphatic esters and terpenic compounds is thought to contribute to the unique flavour of the strawberry guava fruit.

Pectin fractions from the fruits of *Psidium cattleianum* fractions were found to have high proportions of uronic acids (20–42.6%) and high content of neutral sugars, mainly arabinose and galactose, suggesting the presence of arabinans and galactans as side chains (Vriesmann et al. 2009). A fraction yielded arabinose content of (50.35%). The highest ratio of rhamnose: uronic acid was also observed for these fractions, indicating the presence of rhamnose rich zones. Hemicelluloses (called cross-linking glycans) were extracted with 2 M KOH, yielding a fraction composed mainly of xylose and uronic acid.

Studies reported that aqueous-ethanolic leaf extracts of *Acca sellowiana* (Berg) Burret, *Psidium guajava* L. and *Psidium littorale* Raddi contained aglycones as well as glycosides namely daidzin, genistin, daidzein, genistein, formononetin, biochanin A, prunetin, and several incompletely characterized isoflavones (Lapcik et al. 2005). Among these the main immunoreactive isoflavones were glycitein, glycitin, ononin, sissotrin, including the malonylated and acetylated glucosides.

In a study of exotic Mauritian fruits, strong correlations between antioxidant activity (assessed by both TEAC and FRAP) and total phenolics and proanthocyanidins were observed (Luximon-Ramma et al. 2003). Flavonoids seemed to contribute less to the antioxidant potential of the fruits, while very poor correlations were observed between ascorbate content and antioxidant activity. The highest antioxidant capacities were observed in red and yellow *Psidium cattleianum* Sabine 'Chinese guava', sweet and acid *Averrhoa carambola* L 'starfruit', *Syzygium cumini* L Skeels 'jambolan' and white *Psidium guajava* L 'guava'. These fruits were also characterised by high levels of total phenolics. Maximum levels of total phenolics wee found in red (2,561 μg/g) and yellow (2,409 μg/g) *Psidium cattleianum.*

Research showed that *Psidium cattleianum* leaf extract contained secondary metabolites that had antimicrobial property (Brighenti et al. 2008). The leaf extract killed *Streptococcus mutans* grown in biofilms when applied at high concentrations. At low concentrations it inhibited *S. mutans* acid production and reduced the expression of proteins involved in general metabolism, glycolysis and lactic acid production. Aqueous extract of *Psidium cattleainum* produced a significant reduction on *Streptococcus mutans* counts and decreased the enamel demineralization in Wistar rats subjected to a cariogenic challenge (De Menezes et al. 2010). The extract tested had a significant effect on *S. mutans* in oral biofilm of the rats, decreasing *S. mutans* accumulation and enamel demineralization.

Other Uses

The wood is compact, heavy, durable and resistant. It is used for lathe work, tool handles, charcoal and firewood. The tree is indispensable for mixed planting in reforestation of reclaimed and protected areas in Brazil.

Comments

Erstwhile, the yellow strawberry guava was classified as *Psidium cattleianum* Sabine forma *lucidum* O. Deg., *Psidium cattleianum* Sabine var. *littorale* (Raddi) Fosb., *Psidium littorale* var. *lucidum* (Degener) Fosb.

Selected References

Brighenti FL, Luppens SBI, Delbem ACB, Deng DM, Hoogenkamp MA, Gaetti-Jardim E Jr, Dekker HL, Crielaard W, Ten Cate JM (2008) Effect of *Psidium cattleianum* leaf extract on *Streptococcus mutans* viability, protein expression and acid production. Caries Res 42:148–154

de Menezes TEC, Delbem ACB, Brighenti FL, Okamoto AC, Gaetti-Jardim E Jr (2010) Protective efficacy of *Psidium cattleianum* and *Myracrodruon urundeuva* aqueous extracts against caries development in rats. Pharm Biol 48(3):300–305

Govaerts R, Sobral M, Ashton P, Barrie F, Holst BK, Landrum LL, Matsumoto K, Fernanda Mazine F, Nic Lughadha E, Proenca C, Soares-Silva LH, Wilson PG, Lucas E (2010) World checklist of Myrtaceae. The board of trustees of the Royal Botanic Gardens, Kew. Published on the Internet http://www. kew.org/wcsp/ Accessed 22 Apr 2010

Jansen PCM, Jukema J, Oyen LPA, van Lingen TG (1991) *Psidium littorale* Raddi. In: Verheij EWM, Coronel RE (eds) Plant resources of South-East Asia no 2: edible fruits and nuts. Prosea Foundation, Bogor, p 354

Lapcik O, Klejdus B, Kokoska L, Davidova M, AfandI K, Kuban V, Hampl R (2005) Identification of isoflavones in *Acca sellowiana* and two *Psidium* species (Myrtaceae). Biochem Syst Ecol 33(10):983–992

Lorenzi H (2002) Brazilian trees a guide to the identification and cultivation of Brazilian native trees, vol 1, 4th edn. Instituto Plantarum De Estudos Da lora Ltda, Nova Odessa, 384 pp

Luximon-Ramma A, Bahorun T, Crozier A (2003) Antioxidant actions and phenolic and vitamin C contents of common Mauritian exotic fruits. J Sci Food Agric 85(5):496–502

Martin FW, Campbell CW, Ruberte R (1987) Perennial edible fruits of the tropics: an inventory, agriculture handbook no 642. U.S. Department of Agriculture, Washington, DC

Morton J (1987) Cattley Guava. In: Morton JF (ed) Fruits of warm climates. Julia F. Morton, Miami, pp 363–364

Pino JA, Marbot R, Vázquez C (2001) Characterization of volatiles in strawberry guava (*Psidium cattleianum* Sabine) fruit. J Agric Food Chem 49(12):5883–5887

U.S. Department of Agriculture, Agricultural Research Service (2010) USDA national nutrient database for standard reference, release 23. Nutrient Data Laboratory Home Pages, Beltsville. http://www.ars.usda.gov/ba/bhnrc/ndl

Vernin G, Vernin C, Pieribattesti JC, Roque C (1998) Analysis of the volatile compounds of *Psidium cattleianum* Sabine fruit from Reunion Island. J Essent Oil Res 10(4):353–362

Vriesmann LC, Petkowicz CLO, Carneiro PIB, Costa ME, Beleski-Carneiro E (2009) Acidic polysaccharides from *Psidium cattleianum* (Araçá). Braz Arch Biol Technol 52(2):259–264

Psidium cattleianum 'Yellow Strawberry Guava'

Scientific Name

Psidium cattleianum **Afzel. ex Sabine 'Yellow Strawberry Guava'**.

Synonyms

See *Psidium cattleianum*.

Family

Myrtaceae

Common/English Names

Lemon Guava, Yellow Strawberry Guava, Yellow Cattley Guava

Vernacular Names

Brazil: Araçá, Araçá-De-Coroa, Araçazeiro (Portuguese);
Eastonian: Maasik-Guajaavipuu;
Fijian: Waiawichinese;
Hawaiian: Waiawi.

Origin/Distribution

As for red strawberry guava.

Agroecology

As for the strawberry guava but is rather frost tender and prefers comparatively warmer temperatures than the purple strawberry guava. It will grow on a wide range of soils but need good drainage and is quite drought tolerant.

Edible Plant Parts and Uses

Fruit is eaten as strawberry guava – raw or in jellies, ice-creams, beverages and jams. This species has an agreeable acid-sweet flavour and has high pectin and is suitable for mixing with high-acid, low-pectin fruits for making jellies This species is reportedly more superior in flavour to strawberry guava.

Botany

The plant's habit and morphology is similar to purple strawberry guava but the tree is taller, growing to 10 m high, the shoots are glabrous, leaf petiole are longer 12–15 mm long, lamina is larger 7–10 cm long, obovate and not ciliolate when young (Plates 1 and 2). The fruit is slightly larger 4 cm diameter, globose green when young and yellow when mature, solitary or in clusters of 2–3 (Plates 1 and 2). The flesh is yellow, highly fragrant, sweet with a lemon-guava flavour.

T.K. Lim, *Edible Medicinal And Non-Medicinal Plants: Volume 3, Fruits*,
DOI 10.1007/978-94-007-2534-8_93, © Springer Science+Business Media B.V. 2012

Plate 1 Leaves and fruit

Plate 2 Close-up of leaves and fruits

Nutritive/Medicinal Properties

Analyses of ripe fruits of yellow strawberry guava in the Philippines, Hawaii and Florida reported the following constituents (Morton 1987): seeds, 10.3%; water, 84.2%; ash, 0.63–0.75%; crude fibre, 3.87%; protein, 0.80%; fat, 0.42%; total sugar, 4.32–10.01%, ascorbic acid, 22–50 mg/100 g, energy 268 calories per kg.

Refer also to nutrient values for red strawberry guava.

Volatile aroma compounds identified in red (*Psidium cattleianum*, Sabine) and yellow (*Psidium cattleianum* Sabine var. *lucidum* Hort.) guava fruits included: 31 hydrocarbons, 9 acetals, ethers and oxides, 13 aldehydes, 13 ketones, 30 esters, 48 alcohols, 2 acids, 2 sulfur-containing compounds, 4 phenol derivatives, menthofuran and coumarin (Vernin et al. 1998).

Other Uses

As for red strawberry guava.

Comments

Previously, some authors classified the yellow strawberry guava as *Psidium cattleianum* Sabine forma *lucidum* O. Deg., *Psidium cattleianum* Sabine var. *littorale* (Raddi) Fosb., *Psidium littorale* var. *lucidum* (Degener) Fosb.

Selected References

Jansen PCM, Jukema J, Oyen LPA, van Lingen TG (1991) *Psidium littorale* Raddi. In: Verheij EWM, Coronel RE (eds) Plant resources of South-East Asia no 2: edible fruits and nuts. Prosea Foundation, Bogor, p 354

Morton J (1987) Cattley Guava. In: Morton JF (ed) Fruits of warm climates. Julia F. Morton, Miami, pp 363–364

Vernin G, Vernin C, Pieribattesti JC, Roque C (1998) Analysis of the volatile compounds of *Psidium cattleianum* Sabine fruit from Reunion Island. J Essent Oil Res 10(4):353–362

Wunderlin RP, Hansen BF (2008) Atlas of Florida vascular plants. Institute for Systematic Botany, University of South Florida, Tampa. In: Landry SM, Campbell KN (application development), Florida Center for Community Design and Research. http://www.plantatlas.usf.edu/

Psidium friedrichsthalianum

Scientific Name

***Psidium friedrichsthalianum* (O. Berg) Niedenzu.**

Synonyms

Calyptropsidium friedrichsthalianum O.Berg.

Family

Myrtaceae

Common/English Names

Cas, Cas Acida, Costa Rican Guava (Plate 4), Wild Guava, Wild Guavo

Vernacular Names

Columbia: Guayaba Agria;
Costa Rica: Cas, Cas Ácida, Cas Acido;
Ecuador: Guayaba Del Choco;
El Salvador: Arrayán, Arrayan;
French: Goyavier De Costa Rica, Goyavier Du Costa Rica;
German: Cas, Kostarika-Guave;
Guatemala: Guayaba Ácida;
Honduras: Guayaba Agria, Guayaba De Choco, Guayaba De Costa Rica, Guayaba De Danto;
Mexico: Guayaba Montes;
Nicaragua: Guayaba;
Panama: Guayaba Agria, Guayaba De Agua.

Origin/Distribution

The species is distributed from southern Mexico to Eastern Venezuela. This tree grows naturally in Colombia (especially in the Cauca and Magdalena valleys), throughout Mexico to Panama. The tree is widely grown in Costa Rica. In Panama it is found in the wet tropical forest in the Canal zone, Bocas del Toro and Chiriqui.

Agroecology

The species is grown from sea level to 800 m altitude. In its native range, it is usually found along streams or in swampy woods along the coast and inland and also grown in home gardens. The plant is cold sensitive and is killed by frost.

Edible Plant Parts and Uses

The fruit is mostly used for drinks, ade, jelly, preserves and jam because of its acidity. It is also used as filling for pies. In Costa Rica, the juice is used to make refreshing drinks.

T.K. Lim, *Edible Medicinal And Non-Medicinal Plants: Volume 3, Fruits*,
DOI 10.1007/978-94-007-2534-8_94, © Springer Science+Business Media B.V. 2012

Botany

A small, attractive tree, 6–12 m high with quadrangular branches and brown bark peeling off in sheets exposing the smooth, inner gray bark. Leaves are simple, opposite to sub-opposite, glabrous, elliptic or oblong-elliptic, 3.8–12 cm long, 2.5–5 cm wide, acuminate at the apex, cuneate at the base with entire margins, petioles 3–6 mm long. Leaves bronze coloured when young tuning pale green to green (Plates 1 and 2). Flowers usually borne solitary or in 1-3-flowered axillary cymes, fragrant, white, 2.5 cm across, with 5 white, waxy petals and numerous stamens up to 300. Fruit globose to ellipsoid, green ripening to yellow, 3–6 cm with soft, white acid pulp (Plates 1, 2 and 3). Seeds numerous, tan, irregularly obovoid or reniform, 6–8 mm long, embedded in the pulp.

Nutritive/Medicinal Properties

Analyses carried out in Guatemala reported that the fruit had the following nutrient composition: moisture, 83.15%, protein, 0.78–0.88%, carbohydrates, 5.75–6.75%, fat, 0.39–0.52%, fibre, 7.90% and ash, 0.80% (Morton 1987). The fruit was found to be rich in pectin even when fully ripe.

One hundred and seventy-three volatile components were identified in the aroma concentrate of Costa Rican guava fruit, from which (E)-β-caryophyllene, α-terpineol, α-pinene, α-selinene, β-selinene, δ-cadinene, 4,11-selinadiene, and α-copaene were found to be the major components (Pino et al. 2002). Aliphatic esters and terpenic compounds were thought to contribute to the unique flavour of this fruit.

Plate 1 Immature fruits and young bronze-coloured and mature green leaves (*upper* surface)

Plate 3 Whole and halved fruits

Plate 2 Immature fruits and leaves (*lower* surface)

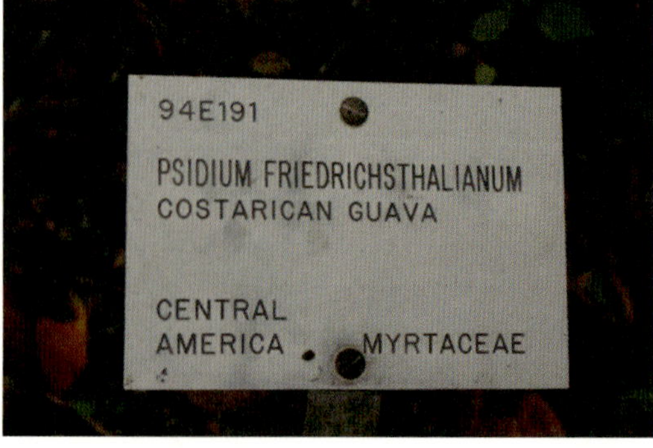

Plate 4 Tree label

Other Uses

The wood is fine-grained and durable.

Comments

Propagation of the plant is usually from seeds.

Selected References

Facciola S (1990) Cornucopia. A source book of edible plants. Kampong Publications, Vista, 677 pp

Govaerts R, Sobral M, Ashton P, Barrie F, Holst BK, Landrum LL, Matsumoto K, Fernanda Mazine F, Nic Lughadha E, Proenca C, Soares-Silva LH, Wilson PG, Lucas E (2010) World checklist of Myrtaceae. The board of trustees of the Royal Botanic Gardens, Kew. Published on the Internet http://www. kew.org/wcsp/ Accessed 22 Apr 2010

McVaugh R (1969) The botany of the Guayana Highland – part VIII. Mem NY Bot Gard 18:252

Morton JF (1987) Costa Rican Guava. In: Morton JF (ed) Fruits of warm climates. Julia F. Morton, Miami, p 365

Pino JA, Marbot R, Vázquez C (2002) Characterization of volatiles in Costa Rican guava [*Psidium friedrichsthalianum* (Berg) Niedenzu] fruit. J Agric Food Chem 50(21):6023–6026

Popenoe W (1974) Manual of tropical and subtropical fruits. Hafner Press, New York, Facsimile of the 1920 edition

Purseglove JW (1968) Tropical crops: dicotyledons. 1 & 2. Longman, London, 719 pp

Psidium guajava

Scientific Name

Psidium guajava L.

Synonyms

Guajava pumila (Vahl) Kuntze, *Guajava pyrifera* (L.) Kuntze, *Myrtus guajava* (L.) Kuntze, *Myrtus guajava* var. *pyrifera* (L.) Kuntze, *Psidium angustifolium* Lam., *Psidium aromaticum* Blanco nom. illeg., *Psidium cujavillus* Burm.f., *Psidium cujavus* L., *Psidium fragrans* Macfad., *Psidium guajava* var. *cujavillum* (Burm.f.) Krug & Urb., *Psidium guajava* var. *minor* Mattos, *Psidium igatemyense* Barb.Rodr., *Psidium intermedium* Zipp. ex Blume, *Psidium pomiferum* L., *Psidium prostratum* O.Berg, *Psidium pumilum* Vahl, *Psidium pumilum* var. *guadalupense* DC., *Psidium pyriferum* L., *Psidium pyriferum* var. *glabrum* Benth., *Psidium sapidissimum* Jacq., *Psidium vulgare* Rich., *Syzygium ellipticum* K.Schum. & Lauterb,

Family

Myrtaceae

Common/English Names

Apple Guava, Guava, Pear Guava, Round Guava, Tropical Guava

Vernacular Names

Afrikaans: Koejawel;
Arabic: Guwâfah Baydâ´, Guwâfah Hhamrâ´, Guwâfah Safrâ';
Argentina: Arazá, Araza-Puita, Arazá-Puitá, Arazapuita, Luma;
Aztec: Xoxococuahuitl;
Banaban: Te Kuao;
Belize: Coloc, Pata, Pa-Ta'h, Piche, Pichi, Pu-Tá, Putah (Maya), Guajava, Guayaba;
Bolivia: Chuará-Catoco, Guayaba Injerta, Guayabo, Guyaba Agria, Sahuinto;
Brazil: Araca, Araca Goiaba, Araçá-Goiaba, Araçá-Guaçú, Araça-Guaiaba, Goiaba, Goiaba-Branca, Goiaba-Pera, Goiaba-Vermelha, Goiabeira, Goiabeira-Branca, Guaiaba, Guaiava, Guava (Portuguese);
Brunei: Biyabas, Jambu Batu;
Bulgaria: Гуава;
Burmese: Malakapen;
Canary Island: Piac (Cacchi);
Cacchi: Ikíec;
Carolinian: Abwas;
Chamorro: Abas, Abas Guayaba, Apas;
Chile: Hurapo;
Chinese: Fan Shi Liu, Fan Tao;
Cook Islands: Tū'Ava, Tuava, Tūava, Tūvava, Tuava 'Enua, (Maori);
Chuukese: Kuafa;
Columbia: Cuayabo Dulce, Guaiaba Dulce, Guaya Dulce, Guayabo;
Costa Rica: Guayaba, Guayabo;
Croatian: Guava;

DOI 10.1007/978-94-007-2534-8_95,

Cuba: Guaba, Guayaba, Guayaba Del Perú, Guayabo, Guayabo Dulce, Gouyabo;
Czech: Kvajava Hruškovitá;
Danish: Almindelig Guava, Guajavatræ;
Dominican Republic: Guayaba Común;
Dutch: Goeajaaba, Guyaba;
Eastonian: Harilik Guajaavipuu, Vili: Guajaav;
Ecuador: Sampi (Shuar),Kuma (Secoya), Dorquila, Guajaba, Guayabo, Guyabo (Spanish);
Fijian: Ngguava, Ngguava Ni India, Quwawa;
Finnish: Guava;
French: Gouyave, Goyavier, Goyavier Commun;
French Polynesia: Tu'avu, Tumu Tuava, Tuvava;
German: Echte Guave, Grosse Gelbe Guajave, Guajave, Guave, Guavenbaum, Guayave;
Ghana: Aduaba, Eguaba, Gua, Gouwa, Oguawa;
Greek: Guava, Gouava;
Guam: Abas;
Guatemala: Cac (Poconchí);
Guinea: Kùávɛilin^G (Kissi);
Guinea-Bissau: Guaiaba (Crioulo);
Hawaiian: Kuawa, Kuawa Ke'Oke'O, Kuawa Lemi, Kuawa Momona, Pauwa;
Haiti: Gwayav (Creole), Pye Gwayav (Kreyòl Ayisyen);
Hebrew: Guyava;
Huasa: Gûway Bâ;
Hungarian: Guáva;
Icelandic: Gúavaber;
I-Kiribati: Te Kuava, Te Kuawa, Te Kuwawa;
India: Madhuram, Madhuriam (Assamese), Peyaaraa (Bengali), Piyaaraa (Gujarati), Amaruud, Amrood, Jamphal (Hindu), Soh Pri Am (Khasi), Malacka-Pela, Pela, Pera (Malayalam), Pington (Manipuri), Jamba, Perunjaam, Tupkel (Marathi), Kawlthei, Kawi-Am (Mizoram), Amruta-Phalam, Aprithaktvacha, Bahu-Bija-Phalam, Dridhabija, Madhuramla, Mansala, Mrduphalam, Mridu, Perala, Peruka, Perukah, Perukam, Pita, Tuvara, Vastula (Sanskrit), Amirtapala, Ampalakkani, Avakacitam, Avakacitamaram, Cenkoyyamaram, Cikappu, Cikappukkoyya, Irattakkoyyamaram, Irattakoyya, Irattamatappal, Irattamatappalmaram, Jaram, Kalarkacikam, Kalarkacikamaram, Kalippacitam, Kalippacitamaram, Koorayaa, Koyya, Palaccaram, Palaccaramaram, Perunkoyya, Tavitatikam, Tavitatikamaram, Uyyakkontan (Tamil); Errajama, Gova, Goyya, Jaama Pandu, Jaamachettu, Thellajaama (Telugu);
Indonesia: Boyawat, Kowayas, Koyabas, Koyawas, Laine Hatu, Lutu Hatu, Wayamas (Alfurese, Sulwaesi), Glima Brih (Aceh), Sotong (Bali), Jambu (Bari), Anta Jau, Attajaan, Galiman, Jambu Horsik (Batak), Biabuto (Boeol), Jambu Paratugala, Jambu Paratukaka, Jambu Putih, Jambu Tella (Bugis), Kayase, Koyawase (W Ceram), Kojawasu, Kujawase, Koyafate (S Ceram), Libu, Njebu (Dyak), Goihawas, Guawa, Nggoi Awa (Flores), Glime Beru (Gajo), Dambu (Gorontalo), Bahaiti, Gawaya, Gowaya (N Halmaheira), Gawaya (S Halmaheira), Jambu Biji, Jambu Piraweh (Jambi), Bayawas, Jambu Klutuk, Jambu Krutuk, Petokal, Tokal (Javanese), Jambu Biji, Jamu Depo, Jambu Klutuk, Jambu Landa (Lampong), Jambu Biawas (Lingga), Jhamhubighi, Jhambhu Bhender (Madurese), Jambu Paratugala (Makassar), Jambu Batu, Jambu Biji, Jambu Susu, Prawas (Malay), Jambu Pertukal (Manado), Masiambu (Nias), Jambu Rutuno (Oelias), Kujabas (Roti), Wo Po Jawa (Sawoe), Jambu Batu (Singkep), Ago, Gejawa, Gewawas (Solor), Jambu Kulutuk, Jambu Siki (Sundanese), Kejawas, Koyabas, Kujawas (Timor), Gawaya (Ternate);
Italian: Guava, Guiava, Guiavo, Guaia Giallo, Pero Dell' Indie, Psidio;
Ixil: Ch'amxuy;
Japanese: Banjirou;
Khmer: Tokal, Trapaek Sruk;
Kosraean: Kuhfahfah;
Laotian: Si Da;
Lithuanian: Gvajava;
Malaysia: Biyawas, Jambu Batu, Jambu Berasa Jambua Bereksa, Jambu Biji, Jambu Biyawas, Jambu Burong, Jambu Padang, Jambu Pelawas, Jambu Kampuchea, Jambu Melekut, Jambu Portugal;
Mali: Biaki, Buyaki, Goyaki (Bambara), Biaki, Buyaki, Goyaki (Malinke), Byaghe, Goyaki Gbyaghe (Senoufo), Goyaghe;
Mexico: Ñi-Joh (Chinanteco), Al-Pil-Ca (Chontal), Ca'aru (Cora), Mo'i (Cuicateco), Bec (Huasteco), Vayeva-Vaxi-Te (Huichol), Chac-Pichi, Gua-Ibasim, Pata (Maya), Posh, Posh-Keip (Mixe), Pojosh (Popoluca), Arrayana, Guayabales, Guayabillo, Guayabo (Spanish), Enandi (Tarasco), A'sihui't (Totonaco), Pata (Tzotzil), Bjui, Pehui, Yagú-Hui (Zapoteco), Pocs-Cuy, Sambadán (Zoque);

Nāhuatl: Chalxócotl, Xālxocotl;
Nauruan: Kuwawa;
Nepalese: Amaruud, Ambaa, Ambaka;
Netherlands Antilles: Goeajaaba, Goejaba, Guajaba, Guava, Guyaba, Yaba (Dutch);
Nicaragua: Guayaba, Guayaba Común, Guayaba De Gusano, Guayaba Perulera, Guayabo, Guyaba Dulce, Guyaba Perulera;
Nigeria: Woba (Efik), Ugwoba (Igbo);
Niuean: Kautoga, Kautoga Tāne, Kautonga, Kautonga, Kautonga Tāne, Lala;
Norwegian: Guava;
Pakistan: Amrood Ka Beech (Urdu);
Palauan: Guabang, Guyab, Kuabang;
Panama: Mulu (Kuna), Guava, Guayaba, Guayabo, Guayaha, Guayava, Guayava Peluda, Guayaya Peluda (Spanish);
Papiamento: Guyaba;
Peru: Bimpish, Guayaba, Guayaba Blanca, Guayabales, Guayabillo, Matos;
Philippines: Bayauas, Bayaua, Bayawas, Bayaya (Bikol), Bayabas (Bisaya), Gayabas, Getabas (Bontok), Bagabas (Cebu-Bisaya), Bayabas, Bayabo (Ibanag), Gaiyabat (Ifugao), Bagabas (Igorot), Bayabas, Guyabas (Iloko), Bayauas (Pangasingan), Biabas (Sulu), Guava, Bayabas, Biyabas, Guayabas, Kalimbahin, Tayabas (Tagalog);
Pohnpeian: Guahva, Kuahpa;
Polish: Gujawa;
Portuguese: Gayaba (Creole), Araçá, Araçá-Uaçu, Goiaba, Goiabeira, Guaiaba;
Puerto Rico: Guaba, Guayaba, Guayabe Silvestre;
Pukapukan: Tuava;
Quecchí: Patá, Pataj;
Rakahanga-Manihiki: Tuava, Tuava;
Romanian: Guava;
Rotuman: Kuava;
Russian: Guaiava, Guava, Psidium Gvaiava;
Saipan: Abas, Abwas;
Samoan: Ku'Ava, Ku'Ava, Kuava, Kuava;
Senegal: Guayaba (Crioulo), Guyab (Fula-Pulaar), Goyap, Guab, Guyaab (Wolof);
Serbia: Gua;
Seychelles: Gouyav (Creole);
Slovak: Guava;
Slovaščina: Guava;
Spanish: Apas, Guayaba, Guayabo;
Sri Lanka: Koiya, Pera;
Suriname: Guava, Guave, Goejaba;
Swedish: Guava;
Tahitian: Tūava, Tuava, Tumu Tuava, Tūvava, Tuvava;
Tanzania: Mpera;
Thai: Farang (Central), Ma Kuai, Ma Man (North), Yamu (South);
Tongan: Kuava;
Tongarevan: Tūava;
Turkish: Guava;
Tuvaluan: Ku'Ava;
Venezuela: Guayaba, Guayabo;
Vietnamese: Oi;
West Africa: Guyab (Fulfulde);
West Indies: Guayaba (Dutch);
Yapese: Abas, Abas Guayaba;
Yoruba: Guafa.

Origin/Distribution

The exact area of origin of *Psidium guajava* is uncertain but is believed to be in the area extending from southern Mexico into Central America. Guava is now common and naturalised in tropical and subtropical America, the Caribbean, Asia, Africa and the Pacific islands.

Agroecology

The guava is robust and hardy and grows in both humid and dry climates in the tropics and subtropics. It thrives in areas with a tropical or equatorial climate where mean temperatures ranges from 27°C to 30°C, with annual rainfall of over 2,500 mm and a high relative humidity of over 80%. It also grows well in the warm, sub-arid, savannah areas as it is quite drought tolerant but adequate irrigation is required for good growth and high yields. Guava can tolerate high temperatures of over 40°C. It can survive only a few degrees of frost. Young trees have been damaged or killed in cold spells at Allahabad, India, in California and in Florida. It can be grown from near sea-level to over 2,000 m elevations. In India, it flourishes up to an altitude of 1,000 m; in

Jamaica, up to 1,200 m; in Costa Rica, to 1,400 m; in Ecuador, to 2,300 m. Guava is not fastidious of soil types; it can grow on heavy clay, marl, light sand, gravel bars near streams, or on limestone; and tolerates a pH range from 4.5 to 9.4. The tree is somewhat salt-tolerant. In Malaysia, guava is grown on very poor sandy soils such as the mixed tin-tailings and Bris soils in the east coast of Peninsular Malaysia to the more fertile sedentary upland soils and alluvial soils; it is also grown in acid sulphate soils. In many tropical countries, guava has naturalised, often forming dense thickets in waste places, disturbed sites, along roadsides, open secondary forests and in pastoral, arable, and plantation land.

Edible Plant Parts and Uses

Raw fully mature and ripe, aromatic guavas are eaten out-of-hand, but are preferred deseeded and served sliced as dessert or in fruit salads. The fruits are also made into puree, juice, canned, stewed, baked or utilised in pies, cakes, puddings, sauce, ice cream, jam, jellies, guava butter, guava cheese, marmalade, beverages, wine, chutney, relish, catsup, and other products. More commonly, the fruit is cooked and cooking eliminates the strong odor. A standard dessert throughout Latin America and the Spanish-speaking islands of the West Indies is stewed guava shells (*cascos de guayaba*). In South Africa, guavas are mixed with cornmeal and other ingredients to make breakfast-food flakes. In the Pacific islands some popular recipes are guava dumplings, guava sauce and stew guava slices. The ingredients for guava dumplings are ripe guava, lemon juice, ground cinnamon (optional), margarine, flour, baking powder, sugar, salt and margarine. The special guava sauce comprises guava pulp, onion, chopped chilli, or ground pepper, garlic, vinegar, ground allspice (optional) ground cinnamon (optional), ground cloves, sugar and salt. Stewed guava slices are boiled with guava juice and served hot or cold with coconut cream. In the Philippines, ripe guava fruit is also used as a vegetable or seasoning for the Filipino sour stew or soup called "sinigang". In Malaysia, the matured or firm ripe fruits are sliced and eaten chilled with a sprinkle of salt or more popularly with a sprinkle of finely ground, preserved dried plum called "Assam Boi".

Guava fruit can be preserved whole or in slices in vinegar and also used for making chutney. Dried guava slices are also dried and preserved with salt or sugar and consumed as snacks. Ready-to-use dehydrated guava products such as dehydrated guava slices and leather, can be prepared from firm and ripe guava fruits (Sagar and Suresh Kumar 2007). The osmo-dried guava slices were found to be acceptable up to 9 months when it was stored in 200 g polyethylene bags at 17–34°C. In Malaysia, the "Kampuchea" cultivar is processed into a much-relished, fresh, chilled guava juice drink. In Taiwan, ripe guava fruits are placed into porcelain jars ad allowed to ferment into a beverage akin to a light wine. In India, ripe guava fruits are similarly processed into guava wine and brandy. A guava extract prepared from small and overripe fruits is used as an ascorbic-acid enrichment for soft drinks and various foods.

The most economically important guava food products are the processed juice or puree products which are canned or aseptically packaged, chilled or frozen, canned guavas and dehydrated guavas. Since 1975, Brazil has been exporting large quantities of guava paste, concentrated guava pulp, and guava shells not only to the United States but to Europe, the Middle East, Africa and Japan. Guava puree or pulp is the starting ingredient for a host of guava food products. It can be made into a nectar drink, fruit punch, syrup or used directly with commercial mixes for making ice-cream, sherbet, yoghurt, smoothies, guava cheese or guava paste. Guava syrup can be used on waffles, ice-cream, milkshakes and puddings. The neat guava puree or sweetened puree can be used for stews, puddings, bakery products such as pastries and the unsweetened puree or pulp used as baby food. In South Africa, a baby-food manufacturer markets a guava-tapioca product. By freeze drying or vacuum puff drying guava puree can be processed into dehydrated powder, a convenient source of vitamin C and pectin a thickening agent, Dried guava powder was used to fortify Allied troops during the second World War. Guava nectar makes a delicious fruit drink. Clarified guava juice is prepared from puree or ripe fruits by removal of the pink colour and insoluble solids. The clarified juice may be blended with other fruit juice, made into jelly or clarified nectar, heated and stored for future use. Guava juice can be processed into carbonated guava beverage.

Guava jams and jellies are usually made from whole fruit ingredients but those made from puree give better quality products. In India and some countries in Central and South America, the principal value of guava lies in the production of guava jelly. The best guava jelly are made from ripe acidic fruits with a pH of 3.3–3.5. The attributes of a good jelly are: a deep red wine colour, firm with a strong musky flavour. Green mature guavas can be utilized as a source of pectin, yielding somewhat more and higher quality pectin than ripe fruits.

In Florida and the West indies, a product similar to guava jam known as guava cheese or paste is made. One recipe comprises butter, sugar and puree, acidified with citric acid and the mixture is heated till it thickens and allow to cool and set. Guava cheese can also be made from ripe guava fruits. Low sugar guava spread is made from 59% puree by weight, 39.6% sugar, 1% low methoxyl pectin and 0.4% anhydrous calcium chloride.

Dehydrated guavas may be reduced to a powder which can be used to flavour ice cream, confections and fruit juices, or boiled with sugar to make jelly, or utilized as pectin to make jelly of low-pectin fruits. India finds it practical to dehydrate guavas during the seasonal glut for jelly-manufacture in the off-season.

Full-size and firm ripe guava fruit not ripe enough for table use are considered best for fruit canning. The fruits are peeled or lye-peeled, deseeded, halved, sliced or diced into cubes and packed in sugar syrup. Canned guavas with higher acidity and total soluble solids have more ascorbic acid retention than other packs.

The oil from guava seeds can be used in salad dressings. In Japan, guava leaf tea sold under the registered name of Bansoureicha (R), by Yakult Honsha, Tokyo, has been approved as one of the Foods for Specified Health Uses and is now commercially available as a health drink. Guava leaf tea consist primarily of the aqueous extract of guava leaves and also to a lesser extent the extract of peels of unripe guava fruit.

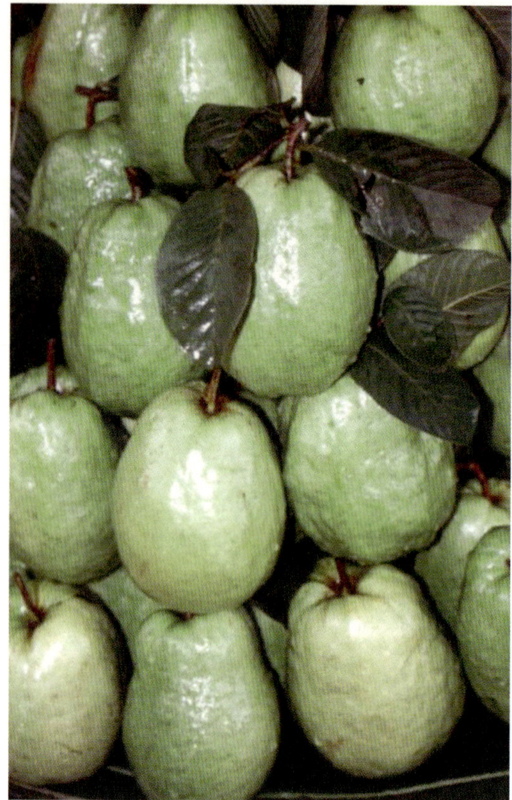

Plate 1 Kampuchean white-fleshed, seeded guava cultivar

Botany

A small tree to 8–10 m high, with smooth grayish brown bark that peels off in strips, spreading branches and quadrangular, pubescent branchlets. Leaves are opposite, ovate-elliptic or oblong-elliptic, acute-acuminate, pubescent beneath, rough adaxially, prominent midrib impressed, lateral nerves 10–20 pairs; blades mostly 7–15 cm long and 3–5 cm wide, rounded at base, apex acute to obtuse, dull green (Plates 2, 6–7). Flowers fragrant, white, large, 2.5 cm across, solitary or 2 or 3 in axillary cymes. Peduncle 1–2 cm long,

Plate 2 Whitefleshed seeded Khao Boon Soom cultivar

Plate 3 Red-fleshed seeded guava

Plate 4 Crystal seedless guava cultivar

Plate 5 Crystal seedless white-fleshed guava

Plate 6 Jade seedless guava and leaves

Plate 7 Maroon seeded cultivar

pubescent. calyx 4–5-lobed, 6–8 mm long, persistent on fruit; petals white, 10–15 mm long, fugacious, usually 4 or 5, obovate, slightly concave; stamens numerous (200–250), white, about as long as petals with pale yellowish anthers; style 10–12 mm long, stigma peltate (Plate 9). Fruit globose, ovoid, or pyriform, 3–10 cm long, green turning to whitish-yellow or faintly pink when ripe, with yellowish-white or pink pulpy, aromatic juicy sweet to sub-sweet flesh, many-seeded or seedless (triploids) (Plates 1–7, 10); seeds numerous (>100), yellowish, reniform. There is a completely maroon-coloured cultivar with maroon banchlets and leaves, crimson flower with crimson stamens and maroon fruit with maroon-coloured flesh (Plates 7 and 8) and also a

Plate 8 Red flower and foliage of the maroon cultivar

Plate 9 White flower of the Kampuchean white-fleshed cultivar

Plate 10 Falangka cultivar with lyrate, crinkled-margin leaves

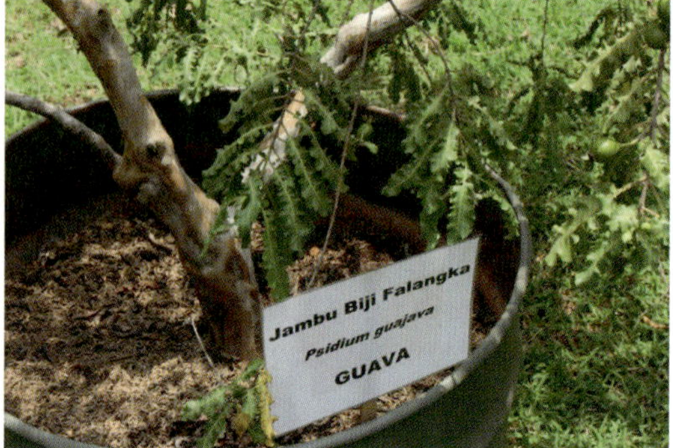

Plate 11 Falangka cultivar

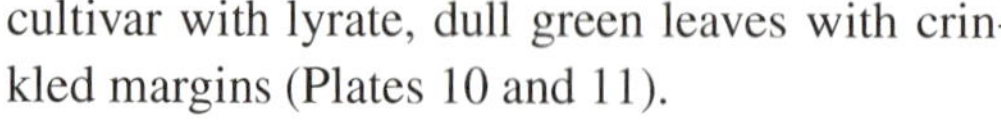

cultivar with lyrate, dull green leaves with crinkled margins (Plates 10 and 11).

Nutritive/Medicinal Properties

Psidium guajava is an important edible food crop and medicinal plant in many tropical and subtropical countries. Its fruit is a rich source of nutrients such as fibre, protein, vitamins and minerals that are beneficial to health. The fruit also contain medicinally important phytochemicals. The leaves also contain many medicinally and biologically important phytochemicals. Many of these medicinally important phytochemicals such as tannins and other phenolics, flavonoids, carotenoids, terpenoids, sesquiterpenes and triterpenes possess innumerable pharmacological properties. Many in-vitro and in-vivo scientific studies have been carried out demonstrating these pharmacological and biological activities (Gutiérrez et al. 2008) that include antioxidant, hepatoprotection, anti-allergy, antimicrobial, antigenotoxic, antiplasmodial, cytotoxic, antispasmodic, cardioprotective, anticough, antidiabetic, inotropic, anti-inflammatory, antidiarrheal, central nervous system (CNS) active, anticestodal, analgesic spermatoprotective, anitmutagenic, spasmolytic, adaptogenic, anti-angiogenesis, anticancer and antinociceptive activities. Many of these in turn supporting its traditional uses. Various aerial parts of the plant have been used in traditional folkloric medicine. A survey of the literature (Gutiérrez et al. 2008) showed *P. guajava* is mainly known for its antispasmodic and antimicrobial properties in the treatment of diarrhoea and dysentery. The plant has also been used extensively as a hypoglycaemic agent. From published literature a wide range of clinical applications have been suggested that include the treatment of infantile rotaviral enteritis, diarrhoea and diabetes, nuclear medicine use in intracellular radio-labelling of blood, and treatment of primary dysmenorrhea.

Nutrients and Phytochemicals in Guava Fruit

Analyses carried out in the United States reported that raw guava (excludes 22% refuse of the skin) had the following nutrient composition (per 100 g edible portion): water 80.80 g, energy 68 kcal (285 kJ), protein 2.55 g, total lipid 0.95 g, ash 1.39 g, carbohydrates 14.32 g, total dietary fibre 5.4 g, total sugars 8.92 g, Ca 18 mg, Fe 0.26 mg, Mg 22 mg, P 40 mg, K 417 mg, Na 2 mg, Zn 0.23 mg, Cu 0.230 mg, Mn 0.150 mg, Se 0.6 μg, vitamin C 228.3 mg, thiamine 0.067 mg, riboflavin 0.040 mg, niacin 1.084 mg, pantothenic acid 0.451 mg, vitamin B-6 0.110 mg, total folate 49 μg, choline 7.6 mg, vitamin A 624 IU, vitamin A 31 μg RAE, vitamin E (α-tocopherol) 0.73 mg, vitamin K (phylloquinone) 2.6 μg, total saturated fatty acids 0.272 g, 14:0 (myristic acid) 0.019 g, 16:0 (palmitic acid) 0.228 g, 18:0 (stearic acid) 0.025 g; total monounsaturated fatty acids 0.087 g, 16:1 undifferentiated (palmitoleic acid) 0.005 g, 18:1 undifferentiated (oleic acid) 0.082; total polyunsaturated fatty acids 0.401 g, 18:2 undifferentiated (linoleic acid) 0.288 g, 18:3 undifferentiated (linolenic acid) 0.112 g; tryptophan 0.022 g, threonine 0.096 g, isoleucine 0.093 g, leucine 0.171 g, lysine 0.072 g, methionine 0.016 g, cystine 0.014 g, phenylalanine 0.006 g, tyrosine 0.031 g, valine 0.087 g, arginine 0.065 g, histidine 0.022 g, alanine 0.128 g, aspartic acid 0.162 g, glutamic acid 0.333 g, glycine 0.128 g, proline 0.078 g, serine 0.075 g, β-carotene 374 μg and lycopene 5,204 μg (U.S. Department of Agriculture, Agricultural Research Service 2010).

Sixteen carotenoids were isolated from the flesh of Brazilian red guavas (*Psidium guajava*) (Mercadante et al. 1999).The carotenoids were identified as phytofluene, (all-E)-, (9Z)-, (13Z)-, and (15Z)-β-carotene, (all-E)-γ-carotene, (all-E)-, (9Z)-, (13Z)-, and (15Z)-lycopene, (all-E,3R)-β-cryptoxanthin, (all-E, 3R)-rubixanthin, (all-E,3 S, 5R,8 S)-cryptoflavin, (all-E,3R,3′R, 6′R)-lutein, (all-E,3 S,5R,6R,3′S,5′R,8′R)-, and (all-E,3 S,5R, 6R,3′S, 5′R,8′S)-neochrome. Chandrika et al. 2009 found that guava 'Horana red' variety contained almost exclusively lycopene (45.3 μg/g fresh weight (FW)), with a small amount of lutein (2.1 μg/g FW), β-carotene (2.0 μg/g FW) and β-cryptoxanthin. The studies showed that guava contained more lycopene (45.3 μg/g FW) than watermelon (37.2 μg/g FW), and that the in- vitro accessibility of lycopene in guava (73%) was more than that in watermelon (25.8%). The authors concluded that guava could be used as a better lycopene source than watermelon.

Total sugars (fructose, glucose and sucrose), ascorbic acid and total soluble solid contents were found to significantly increase with fruit maturity in four guava cultivars studied (El Bulk et al. 1997). The maximum total sugars level varied from 13.7 to 30·6 mg per 100 ml of juice. The maximum level varied from 5.64 to 7.67, 1.90 to 8.00 and 6.20 to 7.78 mg per 100 ml of juice for fructose, glucose and sucrose, respectively, in all cultivars. The maximum ascorbic acid level varied from 88.2 to 113.3 mg per 100 g. Total soluble solids gradually increased with fruit development in all cultivars, which differed in their final value (11.1–13.2 °Brix). Polyphenols significantly decreased with fruit growth and development in all cultivars, which differed in their final value (0.20–0.30%). Total pectin for Shambati and Shendi cultivars significantly increased with fruit development, while for Pakistani and Ganib it reached its maximum when the fruits were 106 days old, and thereafter it declined rapidly. The maximum pectin level varied from 0.62% to 1.00%.

The white- and pink-fleshed guava fruits exhibited a typical climacteric pattern of respiration during fruit ripening (Bashir and Abu-Goukh 2003). Fruit tissue firmness decreased progressively, in a similar manner, in both guava fruit types. Total soluble solids (TSS) and total sugars increased in pulp and peel of both guava types with decrease in flesh firmness. More increase in total sugars was observed after the climacteric peak of respiration. Reducing sugars and titratable acidity increased up to the full-ripe stage and then decreased. Ascorbic acid and phenolic compounds decreased continuously during ripening of the two types. The peel showed higher values of ascorbic acid, total protein and phenolic compounds than the pulp. The white-fleshed guavas had higher levels of TSS, total sugars, reducing sugars, titratable acidity, phenolic compounds and ascorbic acid content then the pink-fleshed fruits.

Wilson and Shaw (1978) identified terpene hydrocarbons (two monoterpenes and nine sesquiterpenes) in guava puree and reported that β-caryophyllene played an important role in the aroma. The hydrocarbons were also dominated by β-caryophyllene (95%) in guavas from United Sates. MacLeod and Troconis (1982), were one of the earliest who identified 40 volatile compounds in guavas from Venezuela. They described that the mixture of 2-methylpropyl acetate, hexyl acetate, benzaldehyde, ethyl decanoate, β-caryophyllene and α-selinene had a guava-like aroma. The essential oil constituents of guava fruit peeling afforded a number of sesquiterpenes and monoterpenes which included α-pinene, aromadendrene, limonene, β-bisabolene, α-copaene, α-humulene, δ-cadinene, ar-curcumene, 1,8-cineole, γ-muurolene, calamenene, camphene, β-pinene, myrcene, p-cymene, α-terpineol, cis-β-ocimene (Oliveros-Belardo et al. 1986).

Chyau et al. (1992) investigated the differences of volatile constituents between mature and ripe guava fruits and identified 34 components. The major components identified in mature fruits were 1,8-cineole, (E)-2-hexenal and (E)-3-hexenal. Ethyl hexanoate and (z)-3-hexenyl acetate were the major volatile components of the ripe guava fruit. β-caryophyllene was present in the highest content among the hydrocarbons. Nishimura et al. (1989) identified a total of 122 volatile components in guava fruits: 13 aldehydes, 17 ketones, 31 alcohols, 10 acids, 28 esters, 10 hydrocarbons and 13 miscellaneous compounds. The major constituents of fresh fruits were C_6 compounds. The total amount of C_6 aldehydes, alcohols and acids comprised 20% of the essence of the fresh white and 44% of the essence of fresh pink fruits. The canned puree contained acetoin which comprised 81% of the essence as the major component. Out of 160 different compounds recorded from guava fruit in Egypt, 132 were identified that included 41 hydrocarbons (alkanes, cycloalkanes, alkenes), 9 aromatics, 3 monoterpenes, 15 sesquiterpene derivatives, 12 carbonyl compounds, 25 esters, 9 lactones, 13 alcohols and 5 miscellaneous compounds (Vernin et al. 1991). The major components were (Z)-hex-3-enl-yl acetate (11%), and the corresponding alcohol (7.5%), pentan-2-one (9.1%), cinnamyl alcohol (10.2%), 3-phenylpropyl acetate (5%) and the corresponding alcohol (3.5%). 3-phenylpropyl acetate, cinnamyl alcohol, gamma lactone and delta lactones (especially jasmine lactone) and ethyl esters were suggested to play an important role in the characteristic sweet and pleasant flavour of guava fruit in Egypt. A total of 25 compounds accounting for 80% of the oil were identified in guava fruits in Nigeria (Ekundayo et al. 1991). Free fatty acids (mainly lauric and myristic acids) were the most abundant group of constituents (34%). Large amounts of β-caryophyllene and oxygen-containing sesquiterpenes (25%) were also typical for Nigerian guava. Caryophylladienols were reported as guava volatiles for the first time. Clery and Hammond (1998) identified new sulfur volatile compounds namely, dimethyl disulfide, dimethyl trisulfide, benzothiazole, 3-mercaptohexanol and 3-mercaptohexyl acetate form pink-flesh guava fruit. These compounds imparted a cassis-like odor and were thought to make an important contribution to the odour of guava.

A total of 73 compounds were identified in the essential oil of white flesh guava fruits grown in Réunion island, 61 by hydrodistillation and 24 by headspace solid-phase microextraction (Paniandy et al. 2000). In the headspace, the major constituents were: hexanal (65.9%), γ-butyrolactone (7.6%), (*E*)-2-hexenal (7.4%), (*E,E*)-2,4-hexadienal (2.2%), (*Z*)-3-hexenal (2%), (*Z*)-2-hexenal (1%), (*Z*)-3-hexenyl acetate (1.3%) and phenol (1.6%). The major volatile constituents present in the hydrodistilled essential oil were β-caryophyllene (24.1%), nerolidol (17.3%), 3-phenylpropyl acetate (5.3%) and caryophyllene oxide (5.1%) were the. Many compounds were identified for the first time in fruits such as γ-butyrolactone (7.6%) in the headspace SPME and nerolidol (17.6%) in the essential oil.

Additionally, some compounds such as (Z)-3-hexenal, (*E,E*)-2,4-hexadienal, γ-butyrolactone, borneol, phenol, cuminyl alcohol were identified only by the headspace method.

Characterization of the aromatic profile in commercial guava essence and fresh fruit puree yielded a total of 51 components (Jordán et al. 2003). Commercial essence was characterized by a volatile profile rich in components with low molecular weight, especially alcohols, esters, and aldehydes, whereas in the fresh fruit puree terpenic hydrocarbons and 3-hydroxy-2-butanone were the most abundant components. New components (3-penten-2-ol and 2-butenyl acetate) were described for the first time as active aromatic constituents in pink guava fruit. Principal differences between the aroma of the commercial guava essence and the fresh fruit puree could be related to acetic acid, 3-hydroxy-2-butanone, 3-methyl-1-butanol, 2,3-butanediol, 3-methylbutanoic acid, (Z)-3-hexen-1-ol, 6-methyl-5-hepten-2-one, limonene, octanol, ethyl octanoate, 3-phenylpropanol, cinnamyl alcohol, α-copaene, and an unknown component. (E)-2-Hexenal appeared to be more significant to the aroma of the commercial essence than of the fresh fruit puree.

Chen et al. (2006) identified a total of 65 compounds from guava fruits. The major constituents identified in the guava fruits were: α-pinene, 1,8-cineole, β-caryophyllene, nerolidol, globulol, C6 aldehydes, C6 alcohols, ethyl hexanoate and (Z)-3-hexenyl acetate. The presence of C6 aldehydes, C6 alcohols, ethyl hexanoate, (Z)-3-hexenyl acetate, terpenes and 1,8-cineole is thought to contribute to the unique flavor of the guava fruit. The following compounds were thought to contribute to the complexity of the guava flavor. In green notes, major constituents were C6 aldehydes, C6 alcohols and (Z)-3-hexenyl acetate, especially hexanal and 2-hexenal which provided strong green aroma. Fruity notes were due to the presence of many aliphatic esters, especially ethyl hexanoate, (Z)-3-hexenyl acetate and hexyl acetate. These esters provided a pleasant odor. Linalool, β-ionone, nerolidol, and β-selinene provided floral note. The odor of 1,8-cineole was fresh, pungent, spicy, minty, fruity and eucalyptus. Although α-pinene was detected as the major volatile in the fruit, it had a weak note. All sesquiterpenes and alcohols provided the same odor as woody, sweet, and citrus-like odor. In the analysis, sesquiterpene alcohols amounted to 6,250 ± 433 μg/kg, nearly 21% of the total volatiles, and included nerolidol, globulol, veridiflorol, ledol, t-cadinol and α-cadinol. Chen et al. (2008) in another recent paper, identified a total of 35 volatile compounds in mature fruits of six Taiwan cultivars, that included 24 terpene hydrocarbons, 2 terpene alcohols, and minor constituents including 1 alcohol, 2 aldehydes, 3 esters, 1 terpene ester and 2 terpene oxides. Although the volatile constituents of the six cultivars were similar, with β-caryophyllene (47.74–58.28%) and aromadendrene (7.11–14.58%) as the major constituents in all cultivars, quantitative differences in the composition of some constituents were observed. *P. guajava* L. cv. Chan-Shan Bar contained higher percentages of 3-hexenyl acetate, 1,8-cineole, and allo-ocimene than other cultivars. Soares et al. 2007 reported that in white guava, the titratable acidity and sugars decreased with maturity while the pH level and amount of vitamin C increased throughout progress of maturation. The behavior of volatile compounds of fruits in the three stages of maturation was: in immature fruits and those in their intermediate stage of maturation, were predominantly the aldehydes such as (E)-2-hexenal and (Z)-3-hexenal. In mature fruits, esters like Z-3-hexenyl acetate and E-3-hexenyl acetate and sesquiterpenes caryophyllene, α-humulene and β-bisabollene were dominant. The major volatile constituents identified in white and red guava fruits were cinnamyl alcohol, ethyl benzoate, ß-caryophyllene, (E)-3-hexenyl acetate and α-bisabolene (Thuaytong and Anprung 2011).

Seventeen aroma-active volatiles previously identified in fresh, pink Colombian guavas were further characterised for their odour activity (Steinhaus et al. 2009). High odor activity values were determined for the green, grassy smelling (Z)-3-hexenal and the grapefruit-like smelling 3-sulfanyl-1-hexanol followed by 3-sulfanylhexyl acetate (black currant-like), hexanal (green, grassy), ethyl butanoate (fruity), acetaldehyde (fresh, pungent), trans-4,5-epoxy-(E)-2-decenal (metallic), 4-hydroxy-2,5-dimethyl-3 (2 H)-furanone (caramel, sweet), cinnamyl alcohol (floral), methyl (2 S,3 S)-2-hydroxy-3-methylpentanoate (fruity), cinnamyl acetate (floral), methional (cooked potato-like), and 3-hy droxy-4,5-dimethyl-2(5 H)-furanone (seasoning-like). Of the aromatic volatiles, (Z)-3-hexenal, 3-sulfanyl-1-hexanol, 4-hydroxy-2,5-dimethyl-3(2 H)-furanone, 3-sulfanylhexyl acetate, hexanal, ethyl butanoate, cinnamyl acetate, and methional were identified as the key aroma compounds of pink guavas. Das and Majumder (2010) found that the water and oxalate-soluble pectic polysaccharide fractions increased, while acid and alkali-soluble pectic fractions had a decreasing trend at the ripening phase of guava fruit. The extent of pectic solubilization was more pronounced in inner pericarp (IP) region as compared to outer pericarp and middle pericarp zone of the fruit tissue. The differential degradation of pectic polymers revealed that ripening as well as tissue softening of guava fruits was centrifugally expressed as evidenced by higher accumulation of sugar and soluble pectic polysaccharides in IP zone of the fruits.

Marcelin et al. (1993) reported that the mesocarp of guava fruit contained about 90% of the total cell wall materials of the edible part (excluding skin and seeds) of guava; about 74% of which were stone cells, while endocarp was relatively richer in parenchymatous tissue. Parenchymatous cell walls had similar composition in both endocarp and mesocarp, and were made up of about 55–60% of neutral polysaccharides (mainly cellulose, xyloglucan, xylan, arabinan, and arabinogalactans of both types I and II) associated with weakly methyl-esterified galacturonan. Stone cell walls were strongly lignified, secondarised elements, about 50% consisting of equivalent amounts of cellulose and acetylated xylan.

Shu et al. (2009) isolated 9 triterpenoids, ursolic acid (1), 1β, 3β-dihydroxyurs-12-en-28-oic acid (2), 2α,3β-dihydroxyurs-12-en-28-oic acid (3), 3β,19α-dihydroxyurs-12en-28-oic acid (4), 19a-hydroxylurs-12-en-28-oic acid-3-O-α-L-arabinopyranoside (5), 3β, 23-dihydroxy urs-12-en-28-oic acid (6), 3β, 19α, 23β-trihydroxylurs-12-en-28-oic acid (7), 2α, 3β,19α, 23β-tetrahydroxyurs-12-en-28-oic acid (8), 3α,19α,23,24-tetrahydroxyurs –12-en-28-oic acid (9) from guava fruits. They isolated also three benzophenone glycosides, viz. 2, 6-dihydroxy-3, 5-dimethyl-4-O-β-D-glucopyranosyl-benzophenone (1), 2, 6-dihydroxy-3-methyl-4-O-(6″-O-galloyl-β-D-glucopyranosyl)-benzophenone (2), 2, 6-dihydroxy-3, 5-dimethyl-4-O-(6″-O-galloyl-β-D-glucopyranosyl)-benzophenone (3) were isolated from ripe guava fruit (Shu et al. 2010a). Pectin methyl esterase (PME) was extracted from guava fruit (Leite et al. 2006). This enzyme catalyses the hydrolysis of methylester groups of cell wall pectins.

Phytochemicals in Guava Seeds

The lipids in guava seeds comprised exclusively of neutral lipids, mostly triglycerides, accounting for 9.4% of the dry weight of the seeds (Opute 1978). Guava seed fat was linoleic acid-rich (79%); palmitic, oleic and stearic acids being the other minor acids present. Ten phenolic and flavonoid compounds including one new acylated flavonol glycoside identified as quercetin-3-O-β-D-(2″-O-galloyl glucoside)-4′-O-vinylpropionat were isolated from *Psidium guajava* seeds (Michael et al. 2002). Salib and Michael (2004) also isolated five known flavonoid glycosides, two phenolic glycosides and two new cytotoxic phenylethanoid glycosides which have been identified as 1-*O*-3,4-dimethoxyphenylethyl-4-*O*-3,4-dimethoxy cinnamoyl-6-*O*-cinnamoyl-β-D-glucopyranose and 1-*O*-3,4-dimethoxyphenylethyl-4-*O*-3,4-dimethoxy cinnamoyl-β-D-glucopyranose from guava seeds. Guava seeds also have glutelins (Bernardino-Nicanor et al. 2005). The functional properties of the five guava seed glutelin extracts were determined. Glut. BoSDS, Glut. BoSDS2-ME, and Glut.Na showed high values for several properties, including surface hydrophobicity, solubility at pH 10, water-holding capacity at pH 3.6, emulsifying activity index and foaming properties.

Phytochemicals in Guava Leaves

Polyphenol compounds obtained from guava leaves included: quercetin, and its 3-arabinopyranoside, guaijaverin; leucocyanidin, ellagic acid and amritoside (Seshadri and Vasishta 1965); avicularin besides quercetin and guajaverin (El Khadem and Mohamed 1958). Three new tannins named guavin A, guavin C and guavin D, each consisting of a hydrolyzable tannin part and a flavan unit, were isolated from the leaves of *Psidium guajava* (Okuda et al. 1987) and Guavin B, an ellagitannin of novel type with a hydroxybenzophenone moiety, was isolated from *Psidium guajava* leaves (Okuda et al. 1984). Other ellagitannin polyphenol compounds isolated from guava leaves included pedunculadgin, casuarinin, casuarictin, strictinin, and isostrictinin (Okuda et al. 1982).

A total of 17 compounds accounting for 86.1% of the aroma were identified in guava leaves (Sagrero-Nieves et al. 1994). α-Selinene (23.7%), α-caryophyllene (18.8%) and δ-selinene (18.3%) were the major compounds identified. The volatile fraction was rich in sesquiterpene compounds. Sixty compounds of the essential oils were identified at rate 90.56% from guava leaves (Li et al. 1999). The major components were caryophyllene (18.81%), copaene (11.80%), [1aR-(1a α, 4a α, 7 α, 7a β, 7b α)]-decahydro-1,1,7-trimethyl-4-methylene-1 H-cycloprop[e] azulene(10.27%), eucalyptol (7.36%).

Silva et al. (2003) found that the oil of *P. guajava* leaf was dominated by α-pinene (23.9%), 1,8-cineole (21.4%) and β-bisabolol (9.2%). Fifty-seven components including 27 terpenes (or sesquiterpenes) along with 14 alcohols and 4 esters were identified obtained from a hydrodistillation of the leaves (Pino et al. 2001). Among these were γ-bisabolene and zingiberene nerolidiol, β-sitosterol, ursolic, crategolic, guayavolic acids, guajavolide and guavenoic acid along with one known triterpene oleanolic acid. Forty-two compounds, representing more than 90% of the volatile mixture, were identified from the volatile oil of Nigerian guava leaves (Ogunwande et al. 2003). The significant compounds were limonene (42.1%) and β-caryophyllene (21.3%). A total of 50 compounds were identified in the guava leaf essential oil from Taiwan (Chen et al. 2007). The major constituents identified in the oil were: β-caryophyllene (27.7%), α-pinene (14.7%) and 1,8-cineole (12.4%).

From the leaves of *Psidium guajava* were isolated two triterpenoids, 20β-acetoxy-2α, 3β-dihydroxyurs-12-en-28-oic acid (guavanoic acid), and 2α,3β-dihydroxy-24-p-z-coumaroyloxyurs-12-en-28-oic acid (guavacoumaric acid), along with six known compounds 2α-hydroxyursolic acid, jacoumaric acid, isoneriucoumaric acid, asiatic acid, ilelatifol D and β-sitosterol-3-O-β-D-glucopyranoside (Begum et al. 2002b); two new triterpenoids, guajavolide (2 α,3 β,6 β,23-tetrahydroxyurs-12-en-28,20 β-olide), and guavenoic acid (2 α,3 β,6 β,23-tetrahydroxyurs-12,20(30)-dien-28-oic acid), along with one known triterpene oleanolic acid (Begum et al. 2002a); three pentacyclic triterpenoids including one new guajavanoic acid and two known obtusinin and goreishic acid I (Begum et al. 2002c). The new constituent guajavanoic acid was characterized as 2α-hydroxy-3β-p-E-coumaroyloxyurs-12, 18-dien-28-oic acid. Further from guava leaves were isolated: five constituents including one new pentacyclic triterpenoid guajanoic acid and four known compounds β-sitosterol, uvaol, oleanolic acid, and ursolic acid (Begum et al. 2004); one new pentacyclic triterpenoid psidiumoic acid (5) along with four known compounds β-sitosterol, obtusol, oleanolic acid, and ursolic acid were isolated from the leaves of *Psidium guajava* (Begum et al. 2007). The new constituent 5 was characterized as 2 α-glycolyl-3β-hydroxyolean-12-en-28-oic acid. Matsuzaki et al. (2010) isolated new benzophenone and flavonol galloyl glycosides from an 80% MeOH extract of *Psidium guajava* leaves together with five known quercetin glycosides. The structures of the novel glycosides were elucidated to be 2,4,6-trihydroxybenzophenone 4-*O*-(6″-*O*-galloyl)-β-D-glucopyranoside (guavinoside A), 2,4,6-trihydroxy-3,5-dimethylbenzophenone 4-*O*-(6″-*O*-galloyl)-β-D-glucopyranoside (guavinoside B), and quercetin 3-*O*-(5″-*O*-galloyl)-α-L-arabinofuranoside (guavinoside C). Two triterpenoids betulinic acid and lupeol were isolated from guava leaf (Ghosh et al. 2010).

Guajadial, a novel caryophyllene-based meroterpenoid, was isolated from the leaves of *Psidium guajava* (Yang et al. 2007). Diguajadial, a new meroterpenoid, which is a symmetric homodimeric ether with two guajadial units was isolated from guava leaves (Yang et al. 2008). Fu et al. (2009) isolated nine compounds from guava leaves which were identified as ursolic acid (1), 2α-hydroxyursolic acid (2), 2α-hydroxyoleanolic acid (3), morin-3-O-α-L-arabopyranoside (4), quercetin (5), hyperin (6), myricetin-3-O-β-D-glucoside (7), quercetin-3-O-β-D-glucuronopyranoside (8), 1-O-galloyl-β-D-glucose (9). Three novel sesquiterpenoid-based meroterpenoids of psidials A–C were isolated from the leaves of *Psidium guajava* (Fu et al. 2010). Psidial B and psidial C represented the new skeleton of the 3,5-diformylbenzyl phloroglucinol-coupled sesquiterpenoid. Psiguadials A and B, two novel sesquiterpenoid-diphenylmethane meroterpenoids with unusual skeletons, along with a pair of known epimers, psidial A and guajadial, were isolated from the leaves of *Psidium guajava* (Shao et al. 2010). Four new compounds were identified from guava leaf essential oil: flavanone-2 2-ene, prenol, dihydro benzophenanrhridine and cryptonine. (Joseph and Priya 2010). Five constituents with galloyl moiety isolated from guava leaves were identified as 1-O-(1, 2-propanediol)-6-O-galloyl-β-D-glucopyranoside (1), gallic acid (2), ellagic acid (3), ellagic acid-4-O-β-D-glucopyranoside (4) and quercetin-3-O-(6″-galloyl) β-D-galactopyranoside (5) (Shu et al. 2010b).

De Lima et al. (2010) demonstrated that the three essential oils of leaves of three domestic Brazilian guava varieties contained many common substances with a prevalence of 1,8-cineole. The essential oil of the Paluma variety contained 1,8-cineole (42.68%) as the major constituent, as well as α-terpineol (38.68%). The principal components of the essential oil of the Século XXI variety were 1,8-cineole (18.83%), *trans*-caryophyllene (12.08%), and selin-11-en-4-αol (20.98%), while those of the Pedro Sato variety and of the wild plant were 1,8-cineole (17.68%) and (12.83%), caryophyllene oxide (9.34%) and (9.09%), and selin-11-en-4-α-ol (21.46%) and (22.19%), respectively.

Phytochemicals in Guava Bark

Six new complex tannins, guajavins A (5) and B (1), psidinins A (9), B (11) and C (13), and psiguavin (15), together with a variety of condensed, hydrolyzable and complex tannins, were isolated from guava bark (Tanaka et al. 1992). The structures of guajavins and psidinins were established to consist of a (+)-gallocatechin unit and a hydrolyzable tannin moiety linked C-glycosidically. Psiguavin was found to be a novel metabolite probably derived from eugenigrandin A (7) through successive oxidation, benzylic acid-type rearrangement, decarboxylation and oxidative coupling of the gallocatechin B-ring and one of the aromatic rings in the hydrolyzable tannin moiety.

Pharmacological properties of various plant parts reported are elaborated below.

Antioxidant Activity

Guava fruit has antioxidant property. Pulp and peel fractions of guava fruit showed high content of dietary fiber (48.55–49.42%) and extractable polyphenols (2.62–7.79%) (Jiménez-Escrig et al. 2001). All fractions tested showed a remarkable antioxidant capacity as studied, using three complementary methods: (i) free radical DPPH• scavenging, (ii) ferric reducing antioxidant power assay (FRAP), and (iii) inhibition of copper-catalyzed in vitro human low-density lipoprotein (LDL) oxidation. The antioxidant activity was correlated with the corresponding total phenolic content. A 1-g (dry matter) portion of peel contained DPPH• activity, FRAP activity, and inhibition of copper-induced in-vitro LDL oxidation, equivalent to 43, 116, and 176 mg of Trolox, respectively. These results indicated that guava could be a suitable source of natural antioxidants. Peel and pulp could also be used to obtain antioxidant dietary fiber (AODF), a new item combining in a single natural product the properties of dietary fiber and antioxidant compounds. Marquina et al. (2008) reported that the antioxidant capacity of the fruit skin was ten times higher than that of the pulp, and the jam was twice that of the core. The highest phenolic content was found in the guava skin (10.36/100 g skin) and the lowest in the jam (1.47/100 g jam), in dry weight. Thaipong et al. (2005) found that the white flesh clone, 'Allahabad Safeda,' showed higher levels of both hydrophilic antioxidant activity (AOAH) [33.3 μM Trolox equivalents (TE)/g fresh weight (FW)] and the lipophilic antioxidant activity (AOAL) (0.25 μM TE/g FW) than the pink flesh clones ('Fan Retief', 'Ruby Supreme,' and an advanced selection) that ranged from 15.5 to 30.4 and from 0.12 to 0.13 μM TE/g FW for AOAH and AOAL, respectively. The AOAH was positively correlated with vitamin C and total phenolic but was negatively correlated with β-carotene. The AOAL was not correlated with these antioxidants. Antioxidant activities of pink guava puree in water and ethanol extracts, based on 2,2-diphenyl-1-picrylhydrazyl (DPPH) assay, were found to be 1.43 mg/gfm and 0.28 mg/gfm, respectively (Ayub et al. 2010). The antioxidant activity values determined by 2-diphenyl-1-picryhydrazyl (DPPH) free radical scavenging and ferric reducing antioxidant power (FRAP) assays were 10.28 μg fresh weight (fw)/μg DPPH and 78.56 μg Trolox equivalent (TE)/g fw for white guava fruit and 7.82 μg/μg DPPH, fw and 111.06 μM TE/g fw for red guava fruit (Thuaytong and Anprung 2011). Ascorbic acid contents were 130 and 112 mg/100 g fw total phenolics contents 145.52 and 163.36 mg gallic acid equivalents (GAE)/100 g fw and total flavonoids contents 19.06 and 35.85 mg catechin equivalents (CE)/100 g fw, in white and red guava, respectively.

Corral-Aguayo et al. (2008) measured the antioxidant capacity (AOC) of eight different fruits namely, guava, avocado, black sapote, mango, papaya, prickly pear fruit, cladodes, and strawberry using six different assays: 2,2′-diphenyl-1-picrylhydrazyl (DPPH), N,N-dimethyl-p-phenylenediamine (DMPD), ferric-ion-reducing antioxidant power (FRAP), oxygen radical absorbance capacity (ORAC), Trolox equivalent antioxidant capacity (TEAC), and total oxidant scavenging capacity (TOSC). The contents of antioxidant nutritional compounds, total soluble phenolics (TSP), vitamin C, vitamin E, β-carotene, and total carotenoids (TC), were correlated with the total antioxidant capacity (AOC) of hydrophilic (HPE) and lipophilic extracts (LPE). HPE of guava had the highest AOC value when evaluated with DMPD, DPPH, FRAP, TEAC, and TOSC assays, whereas with ORAC assay, black sapote had the highest value. HPE of papaya and prickly pear fruit presented the lowest AOC values with all assays. From HPE, vitamin C and TSP contents were highly correlated with AOC for all assays, while from LPE, TC and β-carotene contents possessed a high correlation with AOC only in the DMPD assay.

Kong and Ismail (2010) employed two assays to determine the lipophilic antioxidant capacities namely lycopene equivalent antioxidant capacity (LEAC) and β-carotene bleaching assays of pink guava fruit and by-products of its puree production industry: refiner, siever and decanter. Lycopene content and antioxidant capacity were in the order of fruits > decanter > siever > refiner. Decanter exhibited the highest lycopene content and antioxidant among the studied by-products. It also gave a significant higher lycopene content than pink guava fruit based on the wet basis. There was a significant correlation between lycopene content and LEAC. All samples had a good antioxidant activity in β-carotene bleaching assay but negatively correlated to lycopene content. Decanter was found to be the highest in lycopene content (17 mg/100 g dry basis) and antioxidant capacity (22 μmol LE/100 g dry basis) among the by-products. This by-product of pink guava puree industry can be a potential source of lycopene and antioxidant compounds.

Recent studies showed pink guava puree supplementation increased antioxidant enzyme activity in spontaneous hypertensive rats's blood (Md Nor and Yatim 2011). The specific activities of glutathione peroxidase (GPx) was significantly higher in the low dosage group (LDG, 0.5 g/kg body weight) (2332.5U/L), medium dosage group (MDG, 1.0 g/kg body weight) (2424.8U/L) and high dosage group (HDG, 2.0 g/kg body weight)

(2594.6U/L) respectively, as compared to the control group (2171.8U/L). Significant differences were also seen in glutathione reductase (GR) activities among all treated groups (LDG (132.5U/L), MDG (141.5U/L), HDG (148.8U/L) compared to control group (126.1U/L).

Methanol extract of *P. guajava* was found to contain the highest amount of total phenolics (380.08 4.40 mg/L gallic acid equivalents) of ten selected Nigerian medicinal plants (Akinmoladun et al. 2010). It was also high in total flavonoids (269.72 µg/ml Quercetin Equivalent). Percentage 2,2-diphenyl-1-picrylhydrazyl (DPPH) radical scavenging activity was highest in *Spondias mombin* (88.58%) and *P. guajava* (82.79%) when compared with values obtained for ascorbic acid and gallic acid. All the extracts, generally, had low nitric oxide radical scavenging activities. The extracts in general demonstrated high lipid peroxidation inhibitory activity. The reductive potential was highest in *P. guajava* (0.79). DPPH assay correlated well with total phenolic contents ($r^2=0.76$) and reductive potential ($r^2=0.81$) and fairly with lipid peroxidation inhibitory activity ($r^2=0.51$). There was a good correlation between total phenolic contents and reductive potential ($r^2=0.79$) and a fair correlation between total phenolic contents and lipid peroxidation inhibitory activity ($r^2=0.55$).

The ethanolic extracts of *Psidium guajava* and other Malaysian fruits were found to be better free radical scavengers than the aqueous extracts (Ling et al. 2010). Similar results were seen in the lipid peroxidation inhibition studies. The findings also showed a strong correlation of antioxidant activity with the total phenolic content. These extracts when tested for its heavy metals content, were found to be below permissible value for nutraceutical application. In addition, most of the extracts were found not cytotoxic to 3 T3 and 4 T1 cells at concentrations as high as 100 µg/ml.

Psdium guajava leaf extracts exhibited antioxidant activity which were concentration dependent (Qian and Nihorimbere 2004). The commercial guava leaf extracts and ethanol guava leaf extracts showed almost the same antioxidant power whereas water guava leaf extracts showed lower antioxidant activity as evaluated using 2,2-diphenyl-1-picrylhydryzyl (DPPH*) assay. Remarkably high total phenolic content 575.3 and 511.6 mg of GAE/g of dried weight material (for ethanol guava leaf extracts and water guava leaf extracts, respectively) were obtained. Of 24 plant species assayed for antioxidant activity by ABTS, the ethanol extract from the leaves of guava (*Psidium guajava*) showed the highest antioxidant capacity with the TEAC value of 4.908 mM/mg, followed by the fruit peels of rambutan (*Nephelium lappaceum*) and mangosteen (*Garcinia mangostana*) with the TEAC values of 3.074 and 3.001 mM/mg, respectively (Tachakittirungrod et al. 2007a). Further investigation indicated that the methanol fraction of guava leaf extract possessed the highest antioxidant activity, followed by the butanol and ethyl acetate fractions, respectively. The hexane fraction showed the lowest antioxidant activity. The results demonstrated that the mechanism of antioxidant action of guava leaf extracts was free radical scavenging and reducing of oxidized intermediates. The phenolic content in guava leaf fraction played a significant role on the antioxidant activity via reducing mechanisms. The researchers isolated three compounds from the methanol crude extract of guava leaves that contributed significantly to the antioxidant activity (Tachakittirungrod et al. 2007b). The most active compound was found to be quercetin along with two flavonoid compounds, quercetin-3-O-glucopyranoside and morin. The isolated quercetin, quercetin-3-O-glucopyranoside and morin showed significant scavenging activity with IC_{50} of 1.20, 3.58 and 5.41 µg/ml, respectively. The methanol extract of guava leaves showed concentration- dependent scavenging activity on all reactive oxygen species used (Ogunlana and Ogunlana 2008). Scavenging activity of the extract on hydrogen peroxide and superoxide was more than that of the synthetic antioxidant, buthylated hydroxyanisole (BHA). However, BHA showed greater DPPH scavenging activity than the extract.

Studies by Chen and Yen (2007) reported that 94.4–96.2% of linoleic acid oxidation was inhibited by the addition of guava leaf and guava tea extracts at a concentration of 100 µg/ml. The guava dried fruit extracts exhibited weaker antioxidant effects than did the leaf extracts. The results also demonstrated that the scavenging effects of guava leaf extracts on ABTS.+radicals and superoxide anion increased with increasing concentrations. The guava leaf extracts displayed a significant scavenging ability on the peroxyl radicals. However, the scavenging effects were decreased when the extract concentration was greater than 10 μg/ml. The extracts from leaves of various guava cultivars exhibited more scavenging effects on free radicals than did commercial guava tea extracts and dried fruit extracts. The chromatogram data indicated that guava extracts contained phenolic acids, such as ferulic acid, which appeared to be responsible for their antioxidant activity. Correlation analysis indicated that there was a linear relationship between antioxidant potency, free radical-scavenging ability and the content of phenolic compounds of guava leaf extracts. Studies in Korea found that free radical scavenging activity (FRSA) levels of the guava leaves harvested during May and August were high, and those leaves contained higher amounts of 3-hydroxybutyric acid, acetic acid, glutamic acid, asparagine, citric acid, malonic acid, trans-aconitic acid, ascorbic acid, maleic acid, cis-aconitic acid, epicatechin, protocatechuic acid, and xanthine than the leaves harvested during October and December (Kim et al. 2011). Epicatechin and protocatechuic acid among those compounds seem to have enhanced FRSA of the guava leaf samples harvested in May and August.

Akanji et al. (2009) demonstrated that the guava leaf aqueous extract was able to reduce the trypanosomosis associated lipid peroxidation as well as raise the level of glutathione in the *Trypanosoma brucei brucei* infected but treated animals significantly. Also, the leaf extract was found to lower the malondialdehyde concentrations in the extract treated animals and this may be attributed to its antioxidant properties.

Hypoglycaemic/Antidiabetic Activity

Guava fruit

Cheng and Yang (1983) showed that acute i.p. treatment with 1 g/kg guava juice produced a marked hypoglycemic action in normal and alloxan-treated diabetic mice. Although effective duration of guava was more transient and less potent than chlorpropamide and metformin, blood glucose lowering effect of guava also could be obtained by oral administration in maturity-onset diabetic and healthy volunteers. Thus, it is suggested that guava may be employed to improve and/or prevent the disease of diabetes mellitus. Aqueous extract of *P. guajava* unripe fruit peels exhibited hypoglycaemic as well as antidiabetic effect in normal and streptozotocin induced mild and severely diabetic rats (Rai et al. 2009). At a dose of 400 mg/kg it produced a maximum fall of 21.2% and 26.9% of blood glucose level in normal and mild diabetic rats respectively. In severely diabetic rats the maximum fall of 20.8% and 17.5% in fasting blood glucose and post prandial glucose levels, and 50% in urine sugar levels was observed with the same dose. Haemoglobin level increased by 5.2% and body weight by 2.5% after 21 days treatment.

Raw guava fruit peel extract showed significant hypolipidaemic activity in addition to its hypoglycaemic and antidiabetic activity (Rai et al. 2010). A significant decrease in triglyceride, total cholesterol, high density lipoprotein, very low density lipoprotein and low density lipoprotein, alkaline phosphatase, asparate amino transferase, alanine amino transferase and creatanine levels were observed after 21 days treatment of aqueous extract of raw guava fruit peel in streptozotocin (STZ) induced severely diabetic rats compared to pre-treatment levels. In view of its relative non-toxic nature *P. guajava* raw fruit peel may be a potential antidiabetic agent. After 4 weeks of guava fruit supplementation (125 and 250 mg/kg), guava fruit significantly restored the loss of body weight caused by in streptozotocin (STZ) and reduced blood glucose levels in a dose-dependent manner compared with that in diabetic control rats (Huang et al. 2011). Mechanistically, guava fruit protected pancreatic tissues, including islet β-cells, against lipid peroxidation and DNA strand breaks induced by STZ, and thus reduced the loss of insulin-positive β-cells and insulin secretion. Guava also markedly inhibited pancreatic nuclear factor-kappa B protein expression induced by STZ and restored the activities of antioxidant enzymes, including superoxide dismutase, catalase, and glutathione peroxidase. The findings revealed that guava had a significant antihyperglycemic effect, and that this effect was associated with its antioxidative activity.

Guava leaves

The aqueous guava leaf extract was effective in lowering blood glucose level (Maruyuma et al. 1985). The active constituents were identified as flavonoid glycosides such as strictinin, isostrictinin and pedunculagin which have been used in clinical treatment of diabetes to improve sensitivity of insulin. Guava leaf water extract exhibited potent activity in reducing blood glucose in streptozocin-diabetic rats (Basnet et al. 1995). The main active components were the glycoproteins with molecular weight between 50,000 and 100,000. However, when combined with insulin, there was no additive effect, indicating that the site of action was in the peripheral tissues but not in the pancreas itself. Ojewole (2005) found that the aqueous leaf extract of *P. guajava* possessed hypoglycemic and hypotensive properties. Acute oral administrations of the guava leaf aqueous extract (50–800 mg/kg, p.o.) caused dose-related, significant hypoglycemia in normal (normoglycemic) and streptozotocin-treated, diabetic rats. Additionally, acute intravenous administrations of the guava leaf extract (PGE, 50–800 mg/kg i.v.) produced dose-dependent, significant reductions in systemic arterial blood pressures and heart rates of hypertensive, Dahl salt-sensitive rats. The numerous tannins, polyphenolic compounds, flavonoids, pentacyclic triterpenoids, guiajaverin, quercetin, and other chemical compounds present in the plant were postulated to account for the observed hypoglycemic and hypotensive effects of the plant's leaf extract. The results supported the folkloric, ethnomedical uses of the plant in the management or control of adult-onset, type 2 diabetes mellitus and hypertension in some rural African communities. Mukhtar et al. (2004) showed that in both acute and sub-acute tests, the water extract of guava leaves, at an oral dose of 250 mg/kg, showed statistically significant hypoglycemic activity in alloxan induced diabetic rats. Wang et al. (2005) showed that guava leaf extracts resisted the rise of blood glucose level induced by exogenous glucose and adrenaline to various degrees. The extracts of water, 650 ml/l alcohol and 950 ml/l alcohol significantly decreased the blood glucose level in streptozotocin-induced diabetic mice by 36.3%, 33.5% and 31.3% respectively. Further, among the three extracts, water-soluble extract showed little influence on the growth of mice.

The water-soluble guava leaf extract was found to significantly inhibit, in the dose-dependent manner, the activities of α-glucosidase from small intestinal mucosa of streptozotocin induced diabetic mice (Wang et al. 2007). The extract inhibition concentration (IC_{50}) to sucrase or maltase was 1.0 or 3.0 g/L respectively. The mixed inhibition type was showed to be the competitive and non-competitive inhibition. The results of acute and long-term feeding tests showed a significant reduction in the blood sugar level in diabetic rats fed with either the aqueous or ethanol extract of guava leaves (Shen et al. 2008). Long-term administration of guava leaf extracts increased the plasma insulin level and glucose utilization in diabetic rats. The results also indicated that the activities of hepatic hexokinase, phosphofructokinase and glucose-6-phosphate dehydrogenase in diabetic rats fed with aqueous extracts were higher than in the normal diabetic group. On the other hand, diabetic rats treated with the ethanol extract raised the activities of hepatic hexokinase and glucose-6-phosphate dehydrogenase only. The experiments provided evidence to support the antihyperglycemic effect of guava leaf extract and the health function of guava leaves against type 2 diabetes. Cheng et al. (2009) found that high polarity fractions of the guava leaf extract enhanced glucose uptake in rat clone 9 hepatocytes, and the phenolic, quercetin was identified as the major active compound. The results suggested that quercetin in the aqueous extract of guava leaves promoted glucose uptake in liver cells, and contributed to the alleviation of hypoglycemia in diabetes as a consequence. People in oriental countries, including Japan and Taiwan, boiled guava leaves in water and drank the extract as a folk medicine for diabetes.

During a screening of medicinal plants for inhibition of protein tyrosine phosphatase1B (PTP1B), an extract from *Psidium guajava* leaves was found to exhibit significant inhibitory effect on PTP1B (Oh et al. 2005). PTP-1B is a negative regulator of insulin signalling. Significant blood glucose lowering effects of the extract were observed after intraperitoneal injection of the extract at a dose of 10 mg/kg in both 1- and 3-month-old Lepr(db)/Lepr(db) mice. In addition, histological analysis of the liver from the butanol-soluble fraction treated Lepr(db)/Lepr(db) mice revealed a significant decrease in the number of lipid droplets compared to the control mice. Taken together, it was suggested that the extract from *Psidium guajava* leaves exhibited antidiabetic effect in type 2 diabetic mice model and these effect was, at least in part, mediated via the inhibition of PTP1B. Psidial B and C, novel sesquiterpenoid-based meroterpenoids from guava leaves, showed activity to enzyme protein tyrosine phosphatase-1B (PTP1B) in 10 μM (Fu et al. 2010).

Shen et al. (2008) showed that long term feeding of the guava leaf aqueous extract to diabetic rats significantly reduced blood glucose level, increased plasma insulin level in an oral glucose tolerance test and stimulated activities of some glucose metabolism enzymes. Additionally single feeding of the ertqct also significantly reduced blood glucose level in the oral glucose tolerance test. The findings indicated the potential of the extract in alleviating diabetes symptoms such as hyperglycemia and insulin resistance in diabetic animal models.

Oral administration of *P. guajava* leaf extract (300 mg/kg body weight/day) for 30 days to streptozotocin-induced diabetes rats significantly decreased the levels of blood glucose, glycosylated hemoglobin and improved the levels of plasma insulin and haemoglobin (Subramanian et al. 2009). The levels of protein, urea, creatinine, non-enzymatic antioxidants, and the activities of enzymatic antioxidants such as superoxide dismutase, catalase, glutathione peroxidase, and glutathione *S*-transferase were markedly altered in liver of STZ-induced diabetic rats. Oral administration of *P. guajava* for 30 days restored all these biochemical parameters to near control levels. The present study reveals the efficacy of *P. guajava* leaf extract in the amelioration of diabetes, which may be attributed to its hypoglycemic nature along with its antioxidant potential.

Wu et al. (2009) showed that the inhibitory effects of guava leaf extracts on the formation of α-dicarbonyl compounds were over 95% at 50 μg/ml. Phenolic compounds present, namely gallic acid, catechin and quercetin exhibited over 80% inhibitory effects, but ferulic acid showed no activity. The guava leaf extracts also showed strong inhibitory effects on the production of Amadori products and advanced glycation end products (AGEs) from albumin in the presence of glucose. The phenolic compounds also showed strong inhibitory effects on the glycation of albumin; especially quercetin exhibited over 95% inhibitory effects at 100 μg/ml. According to the results obtained, guava leaf extracts are potent antiglycation agents, which can be of great value in the preventive glycation-associated complications in diabetes. Hyperglycaemia causes increased protein glycation and the formation of early glycation products and advanced glycation end products (AGEs) which are major factors responsible for the complications of diabetes. Guava budding leaf extract showed a potentially active antiglycative effect in an LDL (low density lipoprotein) mimic biomodel, which could be attributed to its large content of polyphenolics (Chen et al. 2010a). The glycation and antiglycative reactions showed characteristic distinct four-phase kinetic patterns. Computer simulation confirmed the dose-dependent inhibition model.

Deguchi et al. (1998) demonstrated that GvEx (hot, aqueous, guava leaf extract), inhibited the in-vitro activities of maltase, sucrase, and α-amylase in a dose-dependent manner. The 50% inhibitory concentration (IC_{50}) of GvEx was 0.6 mg/ml for α-amylase, 2.1 mg/ml for maltase, and 3.6 mg/ml for sucrase, indicating the higher inhibitory activity of α-amylase than the other two enzymes. In an experiment, the oral administration of maltose, sucrose, or soluble starch to GvEx administered normal ICR mice prevented a rise in on postprandial blood glucose elevation. Compared with control, the single ingestion of GvEx significantly reduced the area under the curve (AUC) of postprandial blood glucose levels by 37.8% after loading soluble starch at 250 mg/kg and by 31.0% and 29.6% after loading sucrose and maltose, respectively at 500 mg/kg each. In a separate experiment, GvEx was administered to genetically diabetic model mice (i.e., C57BL/Ksj, db/db, and $Lepr^{db}/Lepr^{db}$) which develop wide spread pathologic abnormalities including not only diabetes and obesity but also well-defined nephropathy. In contrast to control mice without GvEx, the hemoglobin Alc% in blood and providing index of thickening of glomerular mesangial matrix significantly decreased in GvEx-fed mice. Compared with drinking water (control), the GvEx (250 mg/kg/day) solution significantly reduced blood HbA_{1c}% after ingestion for 5 and 7 weeks. GvEx also improved nephropathy with a significant reduction in the thickening index of the glomerular mesangial matrix in the kidney observed at 7 weeks. In contrast, there were no significant effects on weight gain, food and water intakes of the diabetic mice.

The hot water guava leaf extract (GvEx) was found to contain polyphenol with molecular weight of 5,000–30,000 whose fraction has inhibitory activity of carbohydrate digestive enzyme such as α-amylase, maltase and sucrose (Deguchi 2006). In order to clarify the active component of the guava leaf extract, GvEx solution was fractionated in dialysis tubes of 5,000 and 30,000 MW pore size. The inhibitory activity of α-amylase was detected in the fraction with a MW between 5,000 and 30,000. This fraction reacted with ferrous tartrate, indicating that a component with a phenolic hydroxyl group was part of the fraction It has been reported that guava leaf contains polyphenols – tannins like huavin, A, C, D, and ellagitannins like guavin B, pedunculadgin, casuarinin strictnin, and isostrictinin (Okuda et al. 1982; 1984; 1987). However, high-performance liquid chromatography (HPLC) analysis demonstrated that these elementary polyphenols were present in the ethyl acetate extract of guava leaf but not in GvEx. Several instrumental analyses, such as H-nuclear magnetic resonance (NMR), infrared absorption spectrum, and solid C-NMR, suggested that the active component of GvEx was a polymerized polyphenol named guava leaf polyphenol (GvPP), which was composed of ellagic acid, cyanidin and other low-molecular-weight polyphenols (Deguchi 2006).

Deguchi et al. (2003) found that GvEx had synergistic effects with acarbose or voglibose. Acarbose and voglibose were used at concentrations that inhibited the activity of α-amylase and sucrase by approximately 50%, respectively. Addition of GvEx to the reaction system in combination with the respective enzymes-sugars increased the degree of inhibition of each enzyme activity. The in-vitro synergistic effect were confirmed in an in-vivo study using normal ICR mice loaded with either cooked starch (1 g/kg) and sucrose (2 g/kg). The GvEx (250 mg/kg) was administrated together with acarbose or voglibose at a dose that prevented the increase in the postprandial blood glucose level. The GvEx had no effect on the blood glucose-lowering effect of the drugs. However, when GvEx (250 mg/kg) was administered to mice loaded with a high concentration cooked starch (without sucrose) together with acarbose or voglibose at a dose that failed to prevent the increase of the postprandial blood glucose level, there was a significant reduction in the rise of the blood glucose level. These results suggested GvEx to be useful for preventing the increase in the blood glucose level in combination with both acarbose and voglibose without having to increase the dose of these drugs. Further the use of GvEx jointly with acarbose and voglibose had no adverse toxic effects.

Diabetes mellitus (DM) related Advanced Glycation End products (AGEs) are considered to induce functional impairment of cavernosal smooth muscle relaxation and cause erectile dysfunction (ED). Liu et al. (2010) found that administration of *P. guava* budding leaf extract to diabetic animals for 8 weeks reversed the expected impaired relaxation response and nitric oxide production in cavernosal smooth muscle exposed to acetylcholine or electrical field stimulation. The administration of the extract to rats with 8 weeks of uncontrolled diabetes reversed diabetes mellitus-induced harmful effects on vascular smooth muscle. After 8 weeks, the mean glycosylated haemoglobin (HbA1c), serum cholesterol and triglyceride concentrations were significantly higher in the non-extract diabetic than in the age-matched control animals. In diabetic animals fed with the extract, serum cholesterol and triglyceride levels were significantly lower than in the rats given a standard diet. In another recent study, oral administration of ethyl acetate guava leaf extract at different doses showed a significant decrease in blood glucose level (Soman et al. 2010). It also showed an improved antioxidant potential as evidenced by decreased lipid peroxidation and a significant increase in the activity of various antioxidant enzymes such as catalase, superoxide dismutase, glutathione peroxidase and glutathione reductase. Glycated hemoglobin as well as fructosamine indicators of glycation were also reduced significantly in treated groups when compared to diabetic control.

Guava leaf tea

In Japan, Guava Leaf Tea (Bansoureicha (R), Yakult Honsha, Tokyo, Japan) containing the aqueous leaf extract from guava has been approved as one of the Foods for Specified Health Uses and is now commercially available (Deguchi and Miyazaki 2010). In their recent review they described the active component of the aqueous guava leaf extract (GvEx) and its inhibition of α-glucosidase enzymes in-vitro, safety of the extract and Guava Leaf Tea reduction of postprandial blood glucose elevation, and improvement of hyperglycemia, hyperinsulinemia, hypoadiponectinemia, hypertriglycemia and hypercholesterolemia in murine models and several clinical trials.

Oral administration of guava tea prepared from guava leaves, was found to suppress the postprandial blood glucose level of human subjects whose age and BMI index were over 40 and 22.0, respectively Deguchi et al. (1998). To further examine the effects of drinking excessive amounts of Guava Leaf Tea, human healthy subjects in a previous study ingested a three-fold volume (600 ml) of the tea (Deguchi et al. 2000). Notably, neither diarrhea nor hypoglycemia was observed. Single ingestion and the consecutive ingestion of Guava Leaf Tea for 8 or 12 weeks with or without antidiabetic and antihyperlipidemia drugs in human clinical trials demonstrated no side effects or abnormal changes, as described earlier. After 12 weeks of administration of guava tea, the level of fasting blood glucose (FBG) exhibited a significant reduction rate. The levels and reduction rate of triglyceride and total cholesterol in subjects whose levels exceeded beyond normal limits significantly after the administration. Guava tea intake raised no changes in parameters of iron metabolism, liver and kidney functions and of blood chemical data throughout the entire period of the experiment. Neither guava tea nor water as control caused diarrhea. The findings indicated that chronic suppression of postprandial blood glucose by administration of guava tea would be useful for treatment as an alimentotherapy.

Evidence indicated that the possibility of drug interaction with guava tea is low. Deguchi et al. (2003) investigated the effects of GvEx in combination with typical α-GIs acarbose or voglibose on α-amylase activity in-vitro and postprandial blood glucose elevation in mice. GvEx inhibited α-amylase dose-dependently when combined with the low active dose of acarbose or voglibose. When concomitantly administered with acarbose or voglibose to normal mice, acarbose and voglibose each at the active dose suppressed postprandial blood glucose elevation following loading of sugars with no effect of GvEx (250 mg/kg). In contrast, at the inactive dose, acarbose and voglibose did not affect the activity of GvEx (250 mg/kg). The findings indicated that the combined ingestion of GvEx and an α-GI did not induce hypoglycemia in an animal model. In a crossover clinical trial involving 20 hospitalized patients with T2DM, administration of Guava Leaf Tea and voglibose were found to reduce postprandial blood glucose elevation (Ishibashi et al. 2004). The elevated level was significantly reduced with the single administration of Guava Leaf Tea and voglibose to 143 mg/daily and 133 mg/daily, respectively from 160 mg/daily in the patients in the control. The reducing potential was significantly milder with Guava Leaf Tea than with voglibose. There were no side effects, such as hypoglycaemia, due to abnormal interaction in the combined administration of each standard treatment and voglibose or Guava Leaf Tea. Kaneko et al. (2005) found that quercetin and grapefruit juice exhibited higher inhibitory effects on CYP2C8, CYP2C9 and CYP3A4 than GvEx (more than ten-fold) and Guava Leaf Tea (more than two- to ten-fold), respectively. A subsequent histopathological study showed the absence of response to the induction of P450 isoforms in the liver of rats with 1-month repeated oral administration of GvEx (2,000 mg/kg/day). From these findings, it would appear unlikely that Guava Leaf Tea can cause drug interactions based on either inhibition or induction of cytochrome P450 isoforms.

A second long-term clinical trial investigated the effects of consecutive ingestion of Guava Leaf Tea for 2 months on the parameters of diabetes symptoms and safety in 22 diabetic patients receiving therapy, that is, antidiabetic medication with or without an inhibitor of HMG-CoA reductase (Asano et al. 2005). Ingestion of guava tea significantly decreased blood HbA1c% in diabetic patients who had initial values of >6.5% and were assessed to have abnormal control of blood glucose level. Additionally, the ingestion of the tea significantly reduced serum insulin level in diabetic patients with hyperinsulinemia whose serum insulin level was >17 μU/ml before intake. The ingestion of the tea also decreased the parameter values of lipid metabolism, that is, triglycerides (for 4 weeks), nonesterified fatty acids (for 4 weeks), RLP-C (remnant-like particle-cholesterol) (for 4 weeks) and phospholipids (for 8 weeks), in the subjects with values higher than the reference values in patients without fluvastatin treatment. In contrast, neither side effects resulting from alterations in the parameter values of liver and kidney functions or blood chemistry nor changes in doctor's health interviews were observed during the entire clinical trial period. Also, there was no hypoglycemia due to the abnormal interaction between Guava Leaf Tea and antidiabetic drugs with or without an HMG-CoA reductase inhibitor (Asano et al. 2005).

In single-dose and 1-month repeated dose toxicity studies, Kobayashi et al. (2005) demonstrated that the oral administration of GvEx (200 and 2,000 mg/kg/day) and unripe guava fruit (20, 200 and 2,000 mg/kg/day) caused no abnormal effects in rats, indicating that there was neither acute nor chronic toxicity. There was no death through the administration period, and the extract from dry leaves and unripe fruit of Guava did not affect clinical signs, body weight, food intake, water intake, opthalmology, urinalysis, hematology, blood chemistry, organ weight, necropsy and histopathology in all the treated and control groups. Oyama et al. (2005) investigated the mutagenic activity of both GvEx and Guava Leaf Tea and unripe guava fruit. Guava leaf extract (GvEx) showed low mutagenic activities of reverse mutation and DNA damage, while unripe fruit extract showed no activities. They found that Guava Leaf Tea had a lower mutagenic activity than commercial green tea and black tea in a DNA repair test (Rec-assay); however, these teas showed no mutagenic activity in a bacterial reverse mutation test (Ames test). Moreover, GvEx did not induce chromosomal aberrations in a micronuclear test using peripheral blood erythrocytes, which were prepared from mice by a single oral administration of GvEx (2,000 mg/kg). From these findings, it is suggested that Guava Leaf tea prototype product "Bansoreicha (commercial name)" containing Guava leaves and unripe fruit and these commercial teas have no genotoxicity. These findings indicated that Guava Leaf Tea and GvEx induced neither toxicity, mutagenicity, nor abnormal interaction with antidiabetic and anti-hyperlipidemia drugs, and had a lower potential for drug interactions based on either inhibition or induction of cytochrome P450 isoforms. Thus, Guava Leaf Tea and GvEx can be deemed a safe food material, respectively.

Guava tea was also found to improve hypercholesterolemia and hypoadiponectinemia.

To verify the antihyperlipidemic activity of Guava Leaf Tea, a third long-term clinical trial investigated the effects of consecutive intake for

8 weeks on the parameters of hyperlipidemia, diabetes and safety in 23 subjects with borderline or mild hyperlipidemia with or without T2DM (Asano et al. 2007). During the trial, seven subjects were administered fluvastatin, pravastatin, pitavastatin, colestimide (an inhibitor of cholesterol absorption) or ethyl icosapentate (a TG reducer). The consecutive ingestion of guava tea for 8 weeks reduced the serum levels of triglyceride cholesterol (T-CHO), LDL-cholesterol (LDL-CHO) and phospholipid in these subjects. A significant reduction in T-CHO level was also observed in the same subjects receiving no medicinal treatment. On the other hand, the levels of high-density lipoprotein cholesterol (HDL-CHO), TG, NEFA and lipid peroxide were not significantly changed in the same subjects. In contrast, the consecutive ingestion decreased the serum level of TG (week 4) in subjects with hypertriglycemia (initial TG level: >150 mg/dL) and that of phospholipid (weeks 4 and 8) in subjects with hyperphospholipidemia (initial phospholipid level: >250 mg/dL). Additionally, the ingestion of Guava Leaf Tea significantly reduced blood HbA1c % in diabetic subjects (initial HbA1c%: >6.5%), and significantly increased serum adiponectin level in each subject with hypoadiponectinemia and hyperglycemia. This suggested that the trial findings were due to the effects of ingestion of Guava Leaf Tea and not from nutritional intake. There were no abnormal changes in the parameters of liver and kidney function, blood chemistry and doctor's health interviews during the entire trial period. Also, side effects such as hypoglycemia due to the abnormal interaction between Guava Leaf Tea and an HMG-CoA reductase inhibitor, colestimide (an inhibitor of cholesterol absorption) or ethyl icosapentate were not observed. (Asano et al. 2007). Overall, the results indicate that the consecutive ingestion of Guava Leaf Tea together with every meal improved not only hyperglycemia but also hypoadiponectinemia, hypercholesterolemia and hyperlipidemia in pre-diabetic and diabetic patients with or without hyperlipidemia. The consecutive ingestion also ameliorated high blood cholesterol level in subjects with hypercholesterolemia or borderline hypercholesterolemia.

Antidiarrhoeal/Spasmolytic Activities

Thanangkol and Chaichangptipayut (1987) found that guava leaves were more efficient than oxytetracycline in the treatment of acute diarrhea in humans in a double blind study. The spasmolytic effects of *Psidium guajava* leaf methanol, hexane and water extracts were demonstrated in guinea-pig isolated ileum suggesting the existence of two different types of active components in the extracts (Lozoya et al. 1990). The results showed this in-vitro method as a useful model to reproduce some of the characteristics of the oral way of administration of plant extracts. The fraction containing flavonols from the methanol extract of guava leaves, was found to inhibit peristalsis of guinea pig ileum in-vitro (Lozoya et al. 1994). A trace of quercetin aglycone together with five glycosides were isolated from this active fraction and identified as quercetin 3-O-α-L-arabinoside (guajavarin); quercetin 3-O-β-D-glucoside (isoquercetin); quercetin 3-O-β-D-galactoside (hyperin); quercetin 3-O-β-L-rhamnoside (quercitrin) and quercetin 3-O-gentobioside. The results suggested that the spasmolytic activity of the *Psidium guajava* leaf remedy was mainly due to the aglycone quercetin, present in the leaf and in the extract mainly in the form of five flavonols, and whose effect was produced when these products were hydrolyzed by gastrointestinal fluid. The results supported the traditional use of guava leaves as a treatment of acute diarrhea in Mexico. The researchers found that quercetin, a flavonoid contained in guava leaf elicited intestinal smooth muscle relaxation on isolated guinea pig ileum previously contracted by a depolarizing KCl solution (Morales et al. 1994). Quercetin also inhibited intestinal contraction induced by different concentrations of calcium, shifting the contraction curve to the right showing a clear calcium-antagonistic effect. The ileum was more sensitive than aortic smooth muscles to quercetin. The calcium-antagonist property of quercetin was suggested to contribute to the spasmolytic effect of this popular guava leaf herbal remedy. In subsequent studies, Lozoya et al. (2002) found that a phytodrug (QG-5®) developed from guava leaves, standardized in its content of quercetin, exhibited antispasmodic effect in adult patients with acute diarrheic disease. The study involved 50 patients in a randomized, double-blinded, clinical trial. Capsules containing 500 mg of the product were administered to 50 patients every 8 hours for 3 days. Results obtained showed that the used guava product decreased the duration of abdominal pain in these patients. Asiatic acid from guava leaves showed dose-dependent (10–500 μg/ml) spasmolytic activity in spontaneously contracting isolated rabbit jejunum preparations (Begum et al. 2002b). Oral administration of the guava leaf methanol extract reduced intestinal transit time and prevented castor oil-induced diarrhoea in mice (Olajide et al. 1999).

The alcoholic extract of guava leaf showed a morphine-like inhibition of acetylcholine release in the coaxially stimulated guinea-pig ileum, together with an initial increase in muscular tone, followed by a gradual decrease (Lutterodt 1989). The morphine-like inhibition was found to be due to the flavonoid, quercetin (extracted from guava leaf), starting at concentrations of 1.6 μg/ml. The glycoside, quercetin-3-arabinoside (extracted from guava leaf) did not show any such action at concentrations of up to 1.28 mg/ml. The extract inhibited spontaneous contractions in the unstimulated ileum with a concentration-response relationship. Narcotic-like activity of *Psidium guajava* leaf extract was found to have an antidiarrhoel effect in Sprague-Dawley rats (Lutterodt 1992). In experimental groups of rats pretreated with enteral administration of either morphine or aqueous extracts, 1 hour before the challenge with Microlax, the percentage inhibition to the hyperpropulsive rate (antidiarrhoeal activity) was calculated. Both morphine and the extracts produced a dose-response relationship in their antidiarrhoeal effects. A dose of 0.2 ml/kg fresh leaf extract produced 65% inhibition of propulsion. This dose is equiactive with 0.2 mg/kg of morphine sulphate. The antidiarrhoeal action of the extract may be due, in part, to the inhibition of the increased watery secretions that occur commonly in all acute diarrhoeal diseases and cholera. The methanol extract of unripe guava fruit significantly inhibited the growth of *Shigella dysenteriae* and *Vibrio cholera* that cause diarrhea with MIC of 100–200 ug/ml. The extract also reduced gastrointestinal motility of the rat, and inhibited the release of acetylcholine from guinea pig ileum (Ghosh et al. 1993). Quercetin extracted from *Psidium guajava* was found to inhibit the contraction of guinea pig ileum in vitro and the peristaltic motion of mouse small intestine, and reduced the permeability of abdominal capillaries (Zhang et al. 2003). Quercetin was postulated to be responsible for the antidiarrheal mechanism of *Psidium guajava* extract. Studies in mice with diarrhea induced by senna, revealed guava leaf fraction containing moderately-polar quercetin glucosides to be the effective antidiarrhea fraction (Lu et al. 2010). Determination of quercetin glucosides can be used for quality control of guava leaf and its extracts.

Gonçalves et al. (2005) reported that the extract of *Psidium guajava* leaves (8 μg/ml) showed inhibitory activity of 93.8% inhibition against simian rotavirus (SA-11). Rotaviruses have been recognized as the major agents of diarrhea in infants and young children in developed as well as developing countries. In Brazil, diarrhea is one of the principal causes of death, mainly in the infant population. Guava leaf aqueous extract inhibited the replication of rotavirus, reduced toxicity of rotavirus, and weakened its infecting ability in the mouse (Chen and Chen 2002). The main active components included volatile oil, ursolic acid, and quercetin. In addition, guava leaf aqueous extract promoted small intestinal absorption of Na^+ and glucose in mice infected with rotavirus, promoted small intestinal secretion of SIgA, and had protective effects on the mucosa of the small intestine (Chen et al. 2003). Cheng et al. (2005) found that aqueous ethanol extract of dried pulverised guava leaves inhibited the growth of six species of intestinal bacteria at the concentrations of 1 g/ml and decreased spontaneous contractions of rabbit small intestine at the concentrations of 10 mg/ml.

Psidium guajava leaf aqueous extract (PGE) (50–400 mg/kg p.o.) produced dose-dependent and significant protection of rats and mice against castor oil-induced diarrhoea, inhibited intestinal transit, and delayed gastric emptying (Ojewole et al. 2008). Like atropine (1 mg/kg, p.o.), PGE produced dose-dependent and significant antimotility effect, and caused dose-related inhibition of castor oil-induced enteropooling in the animals. Like loperamide (10 mg/kg, p.o.), PGE dose-dependently and significantly delayed the onset of castor oil-induced diarrhoea, decreased the frequency of defaecation, and reduced the severity of diarrhoea in the rodents. Compared with control animals, PGE dose-dependently and significantly decreased the volume of castor oil-induced intestinal fluid secretion, and reduced the number, weight and wetness of faecal droppings. PGE also produced concentration-related and significant inhibitions of the spontaneous, rhythmic, pendular contractions of the rabbit isolated duodenum. The findings of this study indicate that PGE possesses antidiarrhoeal activity, and thus lend pharmacological credence to the suggested folkloric use of the plant as a natural remedy for the treatment, management and/or control of diarrhoea in some rural communities of southern Africa.

Psidium guajava leaf aqueous extract (PGE) exhibited spasmolytic activity ion rat isolated uterine horns (Chiwororo and Ojewole 2009). Graded, escalated concentrations of PGE (0.5–4.0 mg/ml) produced concentration-dependent and significant inhibitions of the amplitude of spontaneous phasic contractions of the isolated rat uterine horn preparations In a concentration-related manner, PGE also significantly inhibited or abolished contractions produced by acetylcholine (0.5–8.0 μg/ml), oxytocin (0.5–4.0 μU), bradykinin (2.5–10 ng/ml), carbachol (0.5–8.0 μg/ml) or potassium chloride (K+, 10–80 mM) in quiescent uterine horn preparations isolated from the oestrogen-dominated rats. The spasmolytic effect of PGE observed in the present study lends pharmacological support to the traditional use of guava leaves in the management, control and/or treatment of primary dysmenorrhoea in some rural African communities.

In another recent study, the hot aqueous decoction of *P. guajava* leaves showed antibacterial activity towards *Shigella flexneri* and *Vibrio cholera* (Birdi et al. 2010). It decreased production of both *Escherichia coli* heat labile toxin (LT) and cholera toxin (CT) and their binding to ganglioside monosialic acid. However, it had no effect on production and action of *E. coli* heat stable toxin (ST). The decoction also inhibited the adherence of enteropathogenic *Escherichia coli* and invasion by both enteroinvasive *E. coli* and *S. flexneri* to HEp-2 cells. Quercetin, on the other hand, had no antibacterial activity at the concentrations used nor did it affect any of the enterotoxins. Although it did not affect adherence of enteropathogenic *Escherichia coli*, it inhibited the invasion of both enteroinvasive *E. coli* and *S. flexneri* to HEp-2 cells. Collectively, the results indicated that the decoction of *P. guajava* leaves to be an effective antidiarrhoeal agent and that the entire spectrum of its antidiarrhoeal activity was not due to quercetin alone.

Anticancer/Antimutagenic Activities

Manosroi et al. (2006) reported that in human mouth epidermal carcinoma (KB) cell line, guava leaf oil showed the highest anti-proliferative activity of 17 Thai medicinal plants, with the IC_{50} value of 0.0379 mg/ml (4.37 times more potent than vincristine). *P. guajava* extracts were found to be efficacious for the prevention of tumour development by depressing Tr cells and subsequently shifting to Th1 cells (Seo et al. 2005). The addition of anti-allergic *P. guajava* extracts blocked interleukin IL-10-mediated, in-vitro induction of T regulatory (Tr) cells from CD4+ splenocytes of C57BL/6 mice, whereas the extracts exerted only a weak or no effect on the development of Th1 (T helper 1) and Th2 cells. Additionally, *P. guajava* extracts shifted the Th1/Th2 balance to a Th1 dominant status by directly attenuating Tr cell activity. Th1 polarization is one of the mechanisms underlying the therapeutic effects of herbal medicine. In a study of tumour immunity, mice pretreated with the extracts exhibited retarded growth of sub-cutaneous inoculated B16 melanoma cells. Psiguadials A and B, and guajadial isolated from guava leaves exhibited potent inhibitory effects on the growth of human hepatoma cells (Shao et al. 2010).

Aqueous extract of *Psidium guajava* budding leaves (PE) was found to possess anti-prostate cancer activity in a cell line model (Chen et al. 2010a, b). The extract was shown to inhibit LNCaP prostate cancer cell proliferation and to down-regulate expressions of androgen receptor (AR) and prostate specific antigen (PSA). The cytotoxicity of the extract was indicated by enhanced LDH release in LNCaP cells. The flow cytometry analysis revealed cell cycle arrests at G(0)/G(1) phase with huge amount of apoptotic LNCaP cells after treatment with the extract for 48 hours in a dose-responsive manner. The molecular action mechanism of the extract to induce apoptosis in LNCaP cells was elucidated by the decreased Bcl-2/Bax ratio, inactivation of phosphor-Akt, activation of phosphor-p38, phospho-Erk1/phospho-Erk2. Compatible with the in vitro study findings, treatment with the extract (1.5 mg/mouse/day) significantly diminished both the PSA serum levels and tumour size in a xenograft mouse tumour model. The data indicate that guava leaf extract is a promising anti-androgen-sensitive prostate cancer agent.

The aqueous soluble polyphenolic fraction of *Psidium guajava* leaves exhibited potent anti-angiogenesis and anti-migration actions on human prostate cancer DU145 cells (Peng et al. 2011). The IC_{50} of the extract for DU145 cells was ~0.57 mg/ml. In addition, the extract effectively inhibited the expressions of vascular endothelial growth factor (VGEF), interleukins IL-6 and IL-8 cytokines, and matrix metallopeptidase MMP-2 and MMP-9, and simultaneously activated TIMP-2 (tissue inhibitor of metalloproteinases-2) and suppressed the cell migration and the angiogenesis. The results showed that the extract possessed a strong anti-DU145 effect and had the potential to be used as an effective adjuvant anti-cancer chemopreventive. Budding leaves of *P. guajava* exhibited potent anticancer activity and were shown to contain huge amounts huge amounts of soluble polyphenolics (SP) including (in mg/g) gallic acid, catechin, epicatechin, rutin and quercetin (100) (Chen et al. 2009). However, reconstitution of these polyphenolics recovered only 40% of the original bioactivity, and the soluble carbohydrate (SC) portion in PE was suspected to contribute the remaining. PE contained a novel rhamnoallosan, which had a carbohydrate/protein (w/w) ratio = 29.06%/10.27% (=2.83, average molecular mass of 5,029 kDa), characteristically with a peptidoglycan, consisting of a composition (%) of rhamnose 36.05%, allose 24.24%, arabinose 8.76%, tallose 7.95%, xylose 7.37%/, fucose 5.90%, glucose 3.69%, mannose 3.19% and galactose 2.85; and of amino acid (in wt %) glycine 37.12%, leucine 12.68%, proline 10.05%, alanine 8.97%, methionine 5.99%, isoleucine 4.89%, valine 4.83%, histidine 4.25%, tyrosine 4.05%, phenylalanine 2.78%, cysteine 1.86%, aspartic acid 1.10%, lysine 0.73% and glutamic acid 0.70%. Kinetic analysis showed comparable apparent cell-killing rate coefficients (k(app)) to be 4.03×103 and 2.92×10^3 cells/mg/h, respectively, by SP and SC, characterising the complementary anti-DU-145 (human prostate cancer cell lines) bioactivity in nature.

The guava leaf extract was found to inhibit the cyclooxygenase reaction of recombinant human prostaglandin endoperoxide H synthases PGHS-1 and PGHS-2 as assessed by conversion of linoleic acid to 9- and 13-hydroxyoctadecadienoic acids (Kawakami et al. 2009). The guava leaf extract also inhibited the prostaglandin PG hydroperoxidase activity of PGHS-1, which was not affected by nonsteroidal anti-inflammatory drugs (NSAIDs). Quercetin which was one of the major components not only inhibited the cyclooxygenase activity of both isoforms but also partially inhibited the PG hydroperoxidase activity. Overexpression of human PGHS-1 and PGHS-2 in the human colon carcinoma cells increased the DNA synthesis rate as compared with mock-transfected cells which did not express any isoforms. The guava leaf extract not only inhibited the PGE(2) synthesis but also suppressed the DNA synthesis rate in the PGHS-1- and PGHS-2-expressing cells to the same level as mock-transfected cells. These results demonstrated the antiproliferative activity of the guava leaf extract which was postulated at least in part to be caused by inhibition of the catalytic activity of PGHS isoforms.

The acetone extracts of guava (*Psidium guajava*) branch were found to have cytotoxic effects on HT-29 human colon cancer cells (Lee and Park 2010). The extract showed highly cytotoxic effects via the MTT [3-(4,5-dimethylthiazol-2-yl)-2,5-diphenyltetrazolium bromide] reduction assay, LDH (Lactate Dehydrogenase) release assay, and colony formation assay. The extract at 250 μg/ml showed 35.5% inhibition against growth of HT-29 cells. As expected, the extract induced characteristic apoptotic effects in HT-29 cells, including chromatin condensation and sharking that occurred 24 hours after the cells had been treated at a concentration level of 250 μg/ml. Guava leaf essential oil exhibited in-vitro anticancer activity when tested with human cervical carcinoma cells (HeLa) (Joseph et al. 2010). Cell treated with the essential oil showed degeneration of cytoplasmic organelles, reflective reduction, increased shrinkage of the HeLa cell lines and apoptotic characteristics.

The water extract of guava was found to be effective in inactivating the mutagenicity of direct-acting mutagens, e.g., 4-nitro-o-phenylenediamine, sodium azide, and the S9-dependent mutagen, 2-aminofluorene, in the tester strains of *Salmonella typhimurium* (Grover and Bala 1993). The chloroform extract was inactive. Autoclaving of the water extract for 15 minutes did not reduce its activity appreciably. The enhanced inhibitory activity of the extracts on pre-incubation suggested the possibility of desmutagens in the extracts.

Antimicrobial Activity

Cáceres et al. (1993) tested guava leaf extracts obtained with three solvents of different polarities (n-hexane, acetone and ethanol) and discovered that the ethanol extract was the most efficient against the pathogenic enterobacteria *Escherichia coli, Salmonella enteritidis* and *Shigella flexneri* tested. Gnan and Demello (1999) reported a complete inhibition of all nine strains of *Staphylococcus aureus* at a concentration of 6.5 mg/ml. Lutterodt et al. (1999) found that guava leaf methanol extract inhibited the causative agents for (i) enteric fever (*Salmonella typhi, Salmonella paratyphi* A, *Salmonella paratyphi* B and *Salmonella paratyphi* C), (ii) food poisoning (*Salmonella typhimurium* and *Staphylococcus aureus*), (iii) dysentery (*Shigella dysenteriae, Shigella flexneri* and *Shigella sonnei*), and (iv) cholera (*Vibrio cholerae*). The growth of all these organisms was inhibited at the MIC of 10 mg/ml of the extract, which is equivalent to 2.5 μg/ml of active extractable flavonoids. The most sensitive organisms (MIC = 1 mg/ml) were *Staphylococcus aureus, Vibrio cholerae* and *Shigella flexneri*. Four antibacterial compounds comprising two new flavonoid glycosides, morin-3-*O*-α-L-lyxopyranoside and morin-3-*O*-α-L-arabopyranoside, and two known flavonoids, guaijavarin and quercetin, were identified from guava leaves (Arima and Danno 2002). The minimum inhibition concentration of morin-3-*O*-α-L-lyxopyranoside and morin-3-*O*-α-L-arabopyranoside was 200 μg/ml for each against *Salmonella enteritidis*, and 250 μg/ml and 300 μg/ml against *Bacillus cereus*, respectively.

Prabu et al. (2006) found that quercetin-3-O-α-l-arabinopyranoside (guaijaverin), from crude methanol extract of *P. guajava* inhibited the growth of cariogenic *Streptococcus mutans*. The anti-*Streptococcus mutans* activity of the guaijaverin was found to be bacteriostatic, both heat and acid stable and alkali labile with the minimum inhibitory concentration (MIC) of 4 mg/ml for MTCC 1943 and 2 mg/ml for CLSM 001. The sub-MIC concentrations (0.0078–2 mg/ml) of the guaijaverin were evaluated for its cariogenic properties such as acid production, cell-surface hydrophobicity, sucrose-dependent adherence to glass surface and sucrose-induced aggregation of *Streptococcus mutans*. The results showed that guaijaverin demonstrated high potential anti-plaque agent by inhibiting the growth of the *Streptococcus mutans*. Fathilah et al. (2009) found that *P. guajava* extract exhibited bacteriostatic effect on selected early dental plaque bacteria: *Streptococcus sanguinis, Streptococcus mitis* and *Actinomyces* sp. At 4 mg/ml, the extract increased the doubling time of *Streptococcus sanguinis* and *Streptococcus mitis* by 1.8- and 2.6-fold, respectively. The effect on *Actinomyces* sp. was observed at a much lower magnitude. It appeared that *P. guajava* extract had bacteriostatic effect on the plaque bacteria by creating a stressed environment that suppressed the growth and propagation of the bacterial cells. Guava extract was shown to be effective in inhibiting the growth of bacteria of the oral biofilm and fungi of oral candidiasis, thus suggesting that the extract can be used as alternative means of dental therapy (Alves et al. 2009). The aqueous extracts of *Piper betle* and *Psidium guajava* exhibited antimicrobial activities against plaque colonisers, *Streptococcus sanguinis, Streptococcus mitis* and *Actinomyces* sp. with MIC values in the range of 2.61–4.69 mg/ml and toxicity values (LC_{50} and EC_{50}) well above their toxic concentrations (Fathilah 2011). *P. betle* and *P. guajava* extracts contained 9.25 and 11.5 ppm fluoride, respectively. Both extracts exhibited positive anti-adherence activity and reduced the cell-surface hydrophobicity of the bacteria which might have rendered them less adherent and hence, minimising their adhesion to the tooth surface during the early stage of plaque development. Both extracts also suppressed the growth of these bacteria. Such an activity was reaffirmed and confirmed by SEM micrographs whereby the bacterial cells were unable to divide or grow successfully and hence suggesting bacteriostatic effect of the extracts.

Psidium guajava and *Juglans regia* leaf extracts exhibited in-vitro inhibitory effect on the main causal agent of acne lesions, *Propionibacterium acnes* and other bacteria isolated from acne lesions (Qa'dan et al. 2005). The zones of inhibition due to the *Psidium guajava* and *Juglans regia* leaf extracts ranged from 15.8 to 17.6 mm against *P. acnes*, 11.3–15.7 mm against *Staphylococcus aureus* and 12.9–15.5 mm against *Staphylococcus epidermidis*, respectively. These zones of inhibition were significantly higher than those of tea tree oil and equivalent in case of Staphylococci spp., but less in case of *P. acnes*, to those obtained from doxycycline or clindamycin. The results indicated that both leaf extracts may be beneficial in treating acne especially when they are known to have anti-inflammatory activities. The methanolic guava leaf extract exhibited antibacterial activity against *E. coli* with minimum inhibitory concentration, 0.78 μg/ml and minimum bactericidal concentration of 50 μg/ml (Dhiman et al. 2011).

Guava (*Psidium guajava*) and neem (*Azadirachta indica*) extracts exhibited antibacterial against strains of foodborne pathogens (Mahfuzul Hoque et al. 2007). Both extracts showed higher antimicrobial activity against Gram-positive bacteria compared to Gram-negative bacteria except for *Vibrio parahaemolyticus, Pseudomonas aeruginosa,* and *Aeromonas hydrophila*. None of the extracts showed antimicrobial activity against *Escherichia coli* O157:H7 and *Salmonella enteritidis*. The minimum inhibitory concentration (MIC) of ethanol extracts of guava showed the highest inhibition for *Listeria monocytogenes* JCM 7676 (0.1 mg/ml), *Staphylococcus aureus* JCM 2151 (0.1 mg/ml), *Staphylococcus aureus* JCM 2179 (0.1 mg/ml), and *Vibrio parahaemolyticus* IFO 12711 (0.1 mg/ml) and the lowest inhibition for *Alcaligenes faecalis* IFO 12669, *Aeromonas hydrophila* NFRI 8282 (4.0 mg/ml), and *Aeromonas hydrophila* NFRI 8283 (4.0 mg/ml). The MIC of chloroform extracts of neem showed similar inhibition for *Listeria monocytogenes* ATCC 43256 (4.0 mg/ml) and *Listeria monocytogenes* ATCC 49594 (5.0 mg/ml). However, ethanol extracts of neem showed higher inhibition for *Staphylococcus aureus* JCM 2151 (4.5 mg/ml) and *Staphylococcus aureus* IFO 13276 (4.5 mg/ml) and the lower inhibition for other microorganisms (6.5 mg/ml).

P. guajava leaf extracts extracted in methanol, acetone and N, N-dimethylformamide exhibited in-vitro antimicrobial activity (Nair and Chanda 2007). The methanol extract was active against 70% of the total Gram-positive bacteria (*Staphylococcus aureus, Staphylococcus epidermidis, Staphylococcus subfava, Staphylococcus* spp., *Bacillus cereus, Bacillus megaterium, Bacillus subtilis, Micrococcus flavus*) studied, while the acetone extract and dimethylformamide extract were active against 80% and 50% of the studied gram-positive bacteria respectively. All the three extracts showed similar activity profiles against gram-negative bacterial strains studied.

They were active against 76.36% of the total gram-negative bacteria studied which included 73.68% *Pseudomonas spp.* (*Pseudomoas aeruginosa, Pseudomonas fluorescens, Pseudomonas testosteronii, Pseudomonas pseudoalcaligenes, Pseudomonas* spp.), 93.75% *Escherichia coli*, 83.33% of *Klebsiella* spp. (*Klebsiella aerogenes, Klebsiella pneumoniae, Klebsiella spp.*) and 66.66% of *Proteus spp.* (*Proteus mirabilis, Proteus vulgaris, Proteus morganii, Proteus* spp.). All of the extracts were inactive against one of the three *Citrobacter species (Citrobacter fruendii, Citrobacter spp.)* and A*lcaligenes fecalis*, while they were active against *Salmonella typhimurium*. The three extracts showed varying results against the fungal strains (*Candida albicans, Candida glabrata, Candida tropicalis, Candida apicola, Candida spp., Cryptococcus neoformans, Cryptococcus luteolus, Trichosporan beigelii)*. The methanol extract was active against 37.5%, acetone extract was active against 56.25% and dimethylformamide extract was active against 31.25% of the total fungal strains studied. All the extracts were inactive against the three *Aspergillus spp.* (*Aspergillus flavus, Aspergillus candidus, Aspergillus niger*) studied.

The essential oils and methanol, hexane, ethyl acetate extracts from guava leaves were found to exhibit inhibitory activity against diarrhea-causing bacteria: *Staphylococcus aureus, Salmonella* spp. and *Escherichia coli,* including strains isolated from seabob shrimp, *Xiphopenaeus kroyeri* (Heller) and laboratory-type strains (Gonçalves et al. 2008). Of the bacteria tested, *Staphylococcus aureus* strains were most inhibited by the extracts. The methanol extract showed greatest bacterial inhibition. No statistically significant differences were observed between the tested extract concentrations and their effect. The essential oil extract showed inhibitory activity against *S. aureus, Bacillus cereus, Enterobactor aerogenes, Pseudomonas fluorescens* and *Salmonella* spp. The strains isolated from the shrimp showed some resistance to commercially available antibiotics. The data supported the use of guava leaf-made medicines in diarrhea cases where access to commercial antibiotics was restricted. Guava leaf essential oil exhibited inhibitory activity in-vitro against *Staphylococcus aureus* and *Salmonella* spp. (Joseph and Priya 2010; Joseph et al. 2010). In Brazil, guava leaf tea is commonly used as a medicine against gastroenteritis and child diarrhea by those who cannot afford or do not have access to antibiotics. Guava leaf sprout extract exhibited inhibitory effect on diarrhoea causing bacteria, *Escherichia coli* and *Staphylococcus aureus* isolated from fish (Vieira et al. 2001). Guava sprout extracts in 50% diluted ethanol were most effective against *E. coli* (EPEC), while those in 50% acetone were less effective. The ethanol, acetone and water-based guava sprout extracts were inhibitory to all four strains of *S. aureus*. However, the extracts prepared with water and 60% acetone were most inhibitory. The scientists concluded that guava sprout extracts constitute a feasible treatment option for diarrhea caused by *E. coli* or by *S. aureus*-produced toxins, due to their quick curative action, easy availability in tropical countries and low cost to the consumer.

Guava extract exhibited antimicrobial activity against pathogenic fish viruses and bacteria (Direkbusarakom et al. 1997). The efficacy of guava extract for the prevention of viral disease and bacterial disease in aquatic animals was estimated using yellow-head virus (YHV), infection in black tiger shrimp and *Aeromonas hydrophila* infection in catfish, respectively. The extract of guava demonstrated anti-viral activity against infectious haematopoietic necrosis virus (IHNV), Oncorhynchus masou virus (OMV) and YHV but was not effective for infectious pancreatic necrosis virus (IPNV). Additionally, the MIC of the extract ranged from 625 to 5,000 μg/ml against all 24 pathogenic bacterial strains tested that included: *Vibrio harveyi* (9 strains), *Vibrio splendidus* (7 strains), *Vibrio parahaemolyticus* (2 strains) and 1 strain of each *Vibrio mimicus, Vibrio vulnificus, Vibrio fluvialis, Vibrio chorelae, Vibrio alginolyticus* and *Aeromonas hydrophila*. The 50% cytotoxicity of the extract to CHSE-214 cell lines was 1,923 μg/ml while the LD_{50} of the extract to black tiger shrimp post larvae was 2,968 μg/ml. These results show that guava extract has low toxicity to salmon cell lines and black tiger shrimp. Moreover, the extract was found effective for prevention of bacterial infection in catfish (*Clarias macrocephalus*) while not suitable for prevention of yellow-head virus infection in black tiger shrimp. From these results, guava can be recommended for treatment of bacterial disease in fish.

The flavonoids (morin, morin-3-O-lyxoside, morin-3-O-arabinoside, quercetin, and quercetin-3-O-arabinoside) isolated from the leaves of *Psidium guajava* were shown to have bacteriostatic effect on all of the tested fish bacterial pathogens (Rattanachaikunsopon and Phumkhachorn 2007).

Abdelrahim et al. (2002) reported a complete inhibition of *Bacillus subtilis, Staphylococcus aureus, Escherichia coli*, and *Pseudomonas aeruginosa* with a methanol extract of the guava bark. Sanches et al. (2005) reported that the aqueous extracts of *P. guajava* leaves, roots and stem bark were active against the Gram-positive bacteria *Staphylococcus aureus* (MICs = 500, 125 and 250 mg/ml, respectively) and *Bacillus subtilis* (MICs = 500 mg/ml), but virtually inactive against the Gram-negative bacteria *Escherichia coli* and *Pseudomonas aeruginosa* (MICs > 1,000 mg/ml). The ethanol:water extracts showed higher antimicrobial activity as compared to aqueous extracts. Fractionation of the ethanol:water extract of *P. guajava* leaves yielded a flavonoid mixture, triterpenes (α- and β-amyrin) and sterol (β-sitosterol). The flavonoid mixture showed good activity on *S. aureus* with MIC of 25 mg/ml. β-sitosterol was inactive for all the bacteria tested. Five flavonoidal compounds were isolated from guava leaves namely quercetin, quercetin-3-O-α-L-arabinofuranoside, quercetin-3-O-β-D-arabinopyranoside, quercetin-3-O-β-D-glucoside and quercetin-3-O-β-D-galactoside (Metwally et al. 2010). Fractions together with the isolates showed good antimicrobial activities.

Crude aqueous mixture and water soluble methanol extract from leaf and bark of *Psidium guajava*, showed strong antibacterial activity against multidrug-resistant *Vibrio cholerae* (Rahim et al. 2010). The in vitro minimum inhibitory concentration of the crude aqueous mixture and water soluble methanol extract, which was bactericidal against 10^7 CFU/ml of *V. cholerae* was determined to be 1,250 and 850 μg/ml, respectively. The antibacterial activity of *P. guajava* was stable at 100°C for 15–20 min, suggesting nonprotein nature of the active component. The growth of *V. cholerae* in rice oral rehydration saline (ORS) was completely inhibited when 10 mg/ml (wt/vol) of crude aqueous mixture was premixed with the ORS in a ratio of 1:7 (vol. extract/vol. ORS). The methanol and dichloromethane (1:1) solvent extract of dried guava plant materials at 0.8 mg/ml showed the highest inhibitory activity against the urogenital sexually transmitted bacterium, *Ureaplasma urealyticum* while the aqueous extract showed poor anti-STI (sexually trasmitted infection) activity (Van Vuuren and Naidoo 2010).

Prebiotic Activity

Prebiotic activity scores for *Lactobacillus acidophilus* LA-5 and *Bifidobacterium lactis* BB-12 were 0.12 and 0.28 in white guava fruit, respectively, and 0.13 and 0.29 in red guava fruit, respectively (Thuaytong and Anprung 2011)

Antiviral Activity

The saponin fraction from guava leaves was found to inhibit HIV-1 mediated cell-cell fusion with an IC_{50} of 7.33 μg/ml (Mao et al. 2010). It obstructed HIV-1 gp41 six helical bundle (6-HB) formation with an activity of 95.93% at 25 μg/ml.

Antiinfl ammatory/Antiarthritic Activities

The methanolic fraction of *Psidium guajava* fruit extract was found to possess significant inhibitory activity against carragenin, kaolin and turpentine-induced oedema formation (Sen et al. 1995). The fraction significantly inhibited protein exudation. The proliferative form of inflammation was significantly counteracted following cotton pellet-induced granuloma formation in rats. Potent antiarthritic activity was observed with the fraction against formaldehyde-induced chronic arthritis in rats. The methanol extract of the leaves of *Psidium guajava* was found to inhibit paw oedema induced by carrageenan in rats and pain induced by acetic acid in mice (Olajide et al. 1999). *Psidium guajava* leaf aqueous extract displayed antiinflammatory and analgesic effects in rats and mice (Ojewole 2006). *P. guajava* leaf aqueous extract (PGE, 50–800 mg/kg, i.p.) produced dose-dependent and significant inhibition of fresh egg albumin-induced acute inflammation (edema) in rats. The numerous tannins, polyphenolic compounds, flavonoids, ellagic acid, triterpenoids, guiajaverin, quercetin, and other chemical compounds present in the plant were postulated to account for the observed antiinflammatory and analgesic effects of the plant's leaf extract. The results provided pharmacological credence to the ethnomedical, folkloric uses of the plant in the management and/or control of painful, arthritic and other inflammatory conditions in some rural communities of Africa. Guava leaf water extract exhibited a restorative effect on the damage due to colonitis caused by trinitro-benzenesulfonic acid through immunity modulation and anti-lipid peroxidation in the rat, suppressing the occurrence of inflammatory responses (Liao et al. 2007). Studies showed ethanolic guava leaf extract to have significant anti-inflammatory activity (Dutta and Das 2011). In rats with acute inflammation induced by carrageenan, there was significant inhibition of paw edema in Groups B and C rats administered the extract (250 and 500 mg/kg body weight, those in Group D asprin given 100 mg/kg body weight in comparison with Group A (control). In subacute inflammation by Granuloma pouch method, there was significant inhibition of exudate formation in Groups B, C, and D in comparison to Group A. In chronic inflammation by Freund's adjuvant-induced arthritis method, there was significant inhibition of paw edema and inhibition of weight reduction in Groups B, C, and D compared with Group A. Downregulation of arthritis index was also significant in Groups B, C, and D in comparison with Group A.

Matsuzaki et al. (2010), found that guavinoside C (3), a benzophenone and the following quercetin glycosides from guava leaves, quercetin 3-O-α-l-arabinofuranoside (4), quercetin 3-O-α-l-arabinopyranoside (5), quercetin 3-O-β-d-xylopyranoside (6), quercetin 3-O-β-d-galactopyranoside (7), and quercetin 3-O-β-d-glucopyranoside (8) (at 100 μg/ml) inhibited histamine release from mast cells with inhibition ratios of 94.4%, 21.9%, 30.5%, 23.9%, 100%, and 93.5%, respectively. Guavinoside A and B did not show inhibitory activity against histamine release at this concentration. Compounds 3–8 (at 100 μg/ml) inhibited NO production by RAW 264.7 cells stimulated with lipopolysaccharide and interferon gamma with inhibition ratios of 50.0%, 33.2%, 32.4%, 65.1%, 55.3%, and 52.1%, respectively. The isolated compounds therefore inhibited chemical mediators, such as histamine and NO, and increased interleukin IL-12 release from RAW 264.7 cells. The results indicated that phenolic compounds isolated from *P. guajava* might be valuable candidates for treating various inflammatory diseases.

Han et al. (2010) investigated the effects of *P. guajava* ethyl acetate extract (PGEA) on IgE-mediated allergic responses in rat mast RBL-2H3 cells. PGEA reduced antigen (DNP-BSA)-induced release of β-hexosaminidase and histamine in IgE-sensitized RBL-2H3 cells. It also inhibited antigen-induced IL-4 and TNF-α mRNA expression and protein production in IgE-sensitized RBL-2H3 cells. PGEA also suppressed antigen-induced COX-2 mRNA and protein expression in these cells, as well as antigen-induced activation of NFAT and reactive oxygen species. Moreover, it inhibited antigen-induced activation of NF-κB and degradation of IκB-α. Additionally, PGEA suppressed antigen-induced phosphorylation of Syk, LAT, Gab2, and PLCγ2 but not Lyn, and inhibited antigen-induced phosphorylation of downstream signaling intermediates including MAP kinases and Akt. Collectively, the anti-allergic effects of PGEA in-vitro suggested its possible therapeutic application to inflammatory allergic diseases, in which its inhibition of inflammatory cytokine production and FcεRI-dependent signaling events in mast cells may be hugely beneficial.

Cardioprotective and Cardiovascular Activities

Aqueous *P. guajava* extract, quercetin and gallic acid (major antioxidative components of guava) were found to have cardioprotective effects against myocardial ischemia-reperfusion injury in isolated rat hearts, primarily through their radical-scavenging actions (Yamashiro et al. 2003). The extract significantly attenuated ischemic contracture during ischemia and improved myocardial dysfunction after reperfusion. Decreases in high-energy phosphates and increases in malondialdehyde in the reperfused hearts were significantly lessened with the plant extract. Quercetin and gallic acid also exerted similar beneficial effects.

The guava budding leaf aqueous extract was found to exert a protective against endothelial cell damages using the human umbilical vein endothelial cell (HUVEC) model (Hsieh et al. 2007a). Chronic cardiovascular and neurodegenerative complications induced by hyperglycemia have been considered to be associated most relevantly with endothelial cell damages. The protective effect of the extract could be ascribed to its high plant polyphenolic (PPP) contents, the latter being potent ROS (reactive oxidation species) inhibitors capable of blocking the glycation of proteins caused by glyoxal (GO) and methylglyoxal (MGO). Results revealed that glyoxal (GO) and methylglyoxal (MGO) resulting from the glycative and autoxidative reactions of the high blood sugar glucose (G) evoked a huge production of ROS and NO, which in turn increased the production of peroxynitrite, combined with the activation of the nuclear factor kappaB (NFkappaB), leading to cell apoptosis.

Studies by Chiwororo and Ojewole (2008) indicated that *Psidium guajava* leaf aqueous extract (PGE) possessed a biphasic effect on rat isolated vascular smooth muscles. Graded concentrations of PGE (0.25–4.0 mg/ml) caused concentration-dependent, initial brief but significant ($P<0.05$) rises of the basal tones and amplitudes of pendular, rhythmic contractions, followed by secondary pronounced, longer-lasting and significant inhibitions of contractile amplitudes of the isolated portal veins. Relatively low concentrations of PGE (<1.0 mg/ml) always contracted freshly-mounted, naïve, endothelium-intact aortic ring preparations. However, relatively high concentrations of PGE (1.0–4.0 mg/ml) always produced initial brief contractions/augmentations of noradrenaline (NA, 10^{-7} M)-induced contractions of endothelium-intact and endothelium-denuded aortic ring preparations, followed by secondary, pronounced relaxations of the aortic ring muscles. Moreover, relatively high concentrations of PGE (1.0–4.0 mg/kg) always relaxed NA-induced contractions of the aortic ring preparations in a concentration-related manner. The arterial-relaxing effects of PGE were more pronounced in endothelium-intact aortic rings than in endothelium-denuded aortic ring preparations. The relaxant effects of PGE on endothelium-intact aortic rings were only partially inhibited by N(G)-nitro-L-arginine methyl ester (L-NAME, 100 μM), a nitric oxide synthase inhibitor, suggesting that the vasorelaxant effect of PGE on aortic rings was probably mediated via both endothelium-derived relaxing factor (EDRF)-dependent and EDRF-independent mechanisms.

Antipyretic Activity

Significant antipyretic activity of the methanolic fraction of *Psidium guajava* fruit extract was observed following yeast-induced pyrexia in rats (Sen et al. 1995). The methanol extract of the leaves of *Psidium guajava* was found to exhibit an antipyretic effect in mice (Olajide et al. 1999).

Ionotropic Activity

The crude extract of *P. guajava* (water-alcohol extract obtained by macerating dry leaves) was found to depress the guinea pig atrial contractility in a concentration-dependent fashion (Conde Garcia et al. 2003). The extract of *Psidium guajava* could block the L-type calcium membrane channels. In the isolated guinea pig left atrium the acetic acid fraction was 20 times more potent in it inotropic effect than the crude extract (EC_{50} = 1.4 g/l). The results showed that extracts from *P. guajava* leaves depressed myocardial inotropism.

Antitussive Activity

Studies showed that water extract of the guava leaf at doses of 2 and 5 g/kg, p.o. decreased the frequency of cough induced by capsaicin aerosol by 35% and 54%, respectively, as compared to the control, within 10 minutes after injection of the extract (Jaiarj et al. 1999). However, the anticough activity was less potent than that of 3 mg/kg dextromethorphan which decreased frequency of cough by 78% ($P<0.01$). An experiment on isolated rat tracheal muscle showed that the extract directly stimulated muscle contraction and also synergized with the stimulatory effect of pilocarpine. This effect was antagonized by an atropine. Additionally, growth of *Staphylococcus aureus* and β-*Streptococcus* group A, was inhibited by water, methanol and chloroform extract of dry guava leaves. The LD_{50} of guava leaf extract was more than 5 g/kg, p.o. These results suggested that guava leaf extract could be recommended as a cough remedy.

Analgesic/Sedative Activity

Studies showed that shortly after intraperitoneal administration of a non-polar fraction from a methanol extract of the dried leaves of *Psidium guajava* in the mouse, typical narcotic-like effects were observed, including catalepsy, analgesia, Straub tail, shallow respiratory movements and exophthalmos (Lutterodt and Maleque 1988). The dose for 90% suppression of exploratory activity was between 3.3 and 6.6 mg/kg intraperitoneally and the onset of action was 6–8 minutes. The duration of activity was dose-dependent and, for a dose of 13.2 mg/kg given intraperitoneally, it was found to be more than 6 hours. Qualitatively similar results on exploratory activity were obtained when the extract was administered orally. Doses of 3.3–6.6 mg/kg i.p. depressed spontaneous locomotor activity and tunnel running was curtailed. Higher doses abolished the spontaneous locomotor reflex action. A flavonoid compound or compounds appeared to account for the activity observed. The guava leaf extract (PGE, 50–800 mg/kg, i.p.) also produced dose-dependent and significant analgesic effects against thermally and chemically induced nociceptive pain in mice (Ojewole 2006). The numerous tannins, polyphenolic compounds, flavonoids, ellagic acid, triterpenoids, guiajaverin, quercetin, and other chemical compounds present in the plant were postulated to account for the observed analgesic effects of the plant's leaf extract.

In a randomised clinical trial involving patients with average age of 19 years, the standardized phyto-drug (*Psidium guajava* folium extract) at a dose of 6 mg/day, was found to reduce severe menstrual pain (dysmenorrhea) significantly compared with conventional treatment and placebo (Doubova et al. 2007). This effect was maintained in cycles 2 and 3, although the reduction in the mean of pain intensity was lower. The group receiving the 3 mg/day extract did not show a consistent effect throughout the three cycles.

Antinociceptive/CNS Activity

Acetic acid-induced algesia (writhing) in mice was significantly inhibited by the methanolic fraction of *Psidium guajava* fruit extract (Sen et al. 1995). Oral administration of 100, 200 and 400 mg/kg of essential oil of guava leaves produced a significant antinociceptive effect in the formalin test and at 200 and 400 mg/kg in the acetic acid- induced writhing test in male albino mice (Santos et al. 1998). Of the major components only α-pinene, but not the β-caryophyllene, demonstrated significant antinociception in the formalin test. Neither the essential oil nor the major components could exert any significant effect in the hot-plate test. Pretreatment of mice with caffeine (20 mg/kg, i.p.), significantly inhibited the antinociceptive effect of essential oil in the formalin test. These results suggested that the antinociceptive effect of *P. guajava* essential oil was probably mediated by endogenously released adenosine. A CNS depressant activity was exhibited by the guava leaf methanol extract by potentiating the phenobarbitone sleeping time in mice (Olajide et al. 1999).

The hexane, ethyl acetate and methanol extracts of *Psidium guajava* leaves (20,100,500 and 1,250 mg/kg) were found to exhibit antinociceptive effects on the central nervous system in mice (Shaheen et al. 2000). The three extracts exhibited mostly dose-dependent antinociceptive effects in chemical and thermal tests of analgesia. The extracts also produced dose-dependent prolongation of pentobarbitone-induced sleeping time. However, they had variable and mostly non-significant effects on locomotor coordination, locomotor activity or exploration. In the pharmacological tests used, the ethyl acetate extract seemed to be the most active, followed by the hexane and then the methanol extracts.

A bioguided fractionation of the hexane extract obtained from *Psidium guajava* leaves led to the isolation of sesquiterpenes with depressant activities on the central nervous system (Meckes et al. 1996). The results demonstrated that the already reported relaxant properties of *Psidium guajava* hexane extract were largely due to the presence of terpenes, especially caryophyllene-oxide and β-selinene, which were by far the largest single components and potentiated pentobarbital sleeping time and the latency of convulsions induced by leptazol in mice. Calcium concentration-response curves showed a rightward displacement when the active fraction was added to isolated guinea-pig ileum depolarized with K^+ (60 mm) and cumulative concentrations of $CaCl_2$, suggesting that caryophyllene-oxide, a known Ca^{2+} antagonist agent could be responsible for the blockade of extracellular Ca^{2+} observed with the active fraction.

Antihypertensive Activity

Studies showed that pink guava puree had antihypertensive properties (Ayub et al. 2010). Final systolic blood pressure values from the beginning and the end of the experiment in spontaneous hypertensive rats (SHR) were much lower in medium dose group (MDG) (231–179 mmHg) given 1.0 g puree/kg b.w/day and high dose group (HDG) (246–169 mmHg) given 2 g/kg b.w/day, compared with the control rats (241–223 mmHg). Final body weights for treatment dosage groups were lower in the MDG (313.01 g), HDG (318.56 g) and LDG (low dose group) (0.5 g/kg b.w./day) (320.01 g) compared to the control group (331.08 g). In a subsequent study, the researchers found that the specific activities of glutathione peroxidase (GPx) was significantly higher in LDG (2332.5U/L), MDG (2424.8U/L) and HDG (2594.6U/L) respectively, as compared to the control group (2171.8U/L). Significant differences were also observed in glutathione reductase (GR) activities among all treated groups (LDG (132.5U/L), MDG (141.5 U/L), HDG (148.8U/L) compared to CG (126.1U/L)). Liver function tests for total antioxidant status (TAS), alanine aminotransferase (ALT), aspartate aminotransferase (AST), lactate dehydrogenase (LDH) and γ-glutamyl transpeptidase (GGT) showed significant differences in the treated group compared to control group. The study showed that pink guava puree supplementation increased antioxidant enzyme activity in spontaneous hypertensive rats' blood concentration.

Aqueous leaves extract of *Psidium guajava* significantly and dose-dependently (0.25–2 mg/ml) contracted aorta rings (Olatunji-Bello et al. 2007). The sensitivity of the aortic rings to cumulative doses of *P. guajava* was significantly enhanced in the presence of phentolamine (antihypertensive agent) suggesting that the effect of *P. guajava* was to a large extent mediated by activation of α-adrenoceptor and to a lesser extent by acting via calcium ion channel. Belemtougri et al. (2006) reported that *Psidium guajava* leaf and *Diospyros mespiliformis* extracts were dose-dependently effective in the inhibition of calcium release induced by caffeine. The guava leaf extracts were more active than extracts of *Diospyros mespiliformis* in the inhibition of calcium release from sarcoplasmic reticulum of rat skeletal muscle cells induced by caffeine.

Crude decoctions show better inhibitory activity. The observed results could explain their use as antihypertensive and antidiarrhoeal agents in traditional medicine in Burkina Faso, by inhibiting intracellular calcium release. It was also reported that extracts of *Psidium guajava* could block the L-type calcium membrane channels (Conde Garcia et al. 2003). According to Re et al. (1999), quercetin isolated from *Psidium guajava* leaf induced a reduction in acetylcholine evoke release and a reduction of presynaptic molecular activity by modulating the cytosolic calcium concentration in mouse neuromuscular junction.

Hepatoprotective Activity

P. guajava leaf extract exhibited hepatoprotective activity (Roy et al. 2006; Roy and Das 2010). In the acute liver damage induced by different hepatotoxins, *P. guajava* leaf extracts (250 and 500 mg/kg, po) significantly reduced the elevated serum levels of aspartate aminotransferase, alanine aminotransferase, alkaline phosphatase and bilirubin (Roy et al. 2006). The higher dose of the extract (500 mg/kg, po) prevented the increase in liver weight when compared to hepatoxin treated control, while the lower dose was ineffective except in the paracetamol induced liver damage. In the chronic liver injury induced by carbon tetrachloride, the higher dose (500 mg/kg, po) of *P. guajava* leaf extract was found to be more effective than the lower dose (250 mg/kg, po). Histological examination of the liver tissues supported the hepatoprotection. In subsequent studies they compared the effect of different solvent extracts (Roy and Das 2010). In the acute liver damage induced by different hepatotoxins, *P. guajava* methanolic leaf extract (200 mg/kg, p.o.) significantly reduced the elevated serum levels of aspartate aminotransferase, alanine aminotransferase, alkaline phosphatase and bilirubin in carbon tetrachloride and paracetamol induced hepatotoxicity. *P. guajava* ethyl acetate leaf extract (200 mg/kg, p.o.) significantly reduced the elevated serum levels of aspartate aminotransferase, alanine aminotransferase and bilirubin in carbon tetrachloride induced hepatotoxicity whereas *P. guajava* aqueous leaf extract (200 mg/kg, p.o.) significantly reduced the elevated serum levels of alkaline phosphatase, alanine aminotransferase and bilirubin in carbon tetrachloride induced hepatotoxicity. *P. guajava* ethyl acetate and aqueous leaf extracts (200 mg/kg, p.o.) significantly reduced the elevated serum levels of aspartate aminotransferase in paracetamol induced hepatotoxicity. They concluded that the methanolic extract of leaves of *Psidium guajava* plant possessed better hepatoprotective activity compared to other extracts.

Studies by Sambo et al. (2009) showed that the aqueous extract of *Psidium guajava* leaf possessed hepatoprotective property at lower dose and a hepatotoxic property at higher dose. Pretreatment with 150 mg/kg of *Psidium guajava* extract showed a slight degree of protection against the induced hepatic injury caused by 100 mg/kg of erythromycin stearate. Biochemical analysis of the serum obtained revealed a significant increase in serum levels of hepatic enzymes measured in the groups administered with 100 mg/kg of erythromycin stearate and 300/450 mg/kg of *Psidium guajava* extract compared to the control groups and those pretreated with 150 mg/kg of *Psidium guajava* extract.

Studies by Uboh et al. (2010) suggested that aqueous extract of *Psidium guajava* leaves may be hepatoprotective, and not hepatotoxic, with hematopoietic potentials in both male and female rats. Liver function tests revealed that the serum ALT, AST and ALP, as well as the concentrations of total protein and albumin in male and female rats were not significantly affected by the oral administration of the extract. Histopathological study also did not show any adverse alteration in the morphological architecture of the liver tissues in both sexes of the animal model. However, red blood cell counts, hemotocrit and hemoglobin concentrations increased significantly on administration of the extract in both male and female rats. Phytochemical analysis of the plant leaves showed the presence of alkaloids, flavonoids, glycosides, polyphenols, reducing compounds, saponins and tannins.

Antigenotoxic Activity

Guava extract has antigenotoxic activity. Bartolome et al. (2006) showed that except for tannic acid (TA), co-treatment of the genotoxicant-activated bacteria with ascorbic acid (AA) and aqueous plant extracts (*Mangifera indica*, *Psidium guajava* and *Syzygium cumini*) afforded protection against all three genotoxicants [mitomycin C (MMC), nalidixic acid (NA) and hydrogen peroxide (HP)]. TA was effective in suppressing the genotoxic effect of MMC and HP. The IC_{50} of the plant extracts and AA varied with the genotoxicant used. Folin-Ciocalteu test, FeCl3 test and DPPH assay confirmed the presence of polyphenolic compounds and hydrolyzable tannins in the plant extracts and the antioxidant capacity of the plant samples.

Anticoagulant Activity

A water extract of guava leaves was found to show ambiguous effects on the haemostatic system (Jaiarj et al. 2000). Guava leaf extract did not affect bleeding times, it stimulated vasoconstriction and platelet aggregation but it inhibited blood coagulation. On this basis, guava leaf extract would not be recommended as a haemostatic agent.

Guava leaf aqueous extracts exhibited significant inhibition of thrombin clotting time shortening induced by methylglyoxal (Hsieh et al. 2007b). Methylglyoxal inhibited antithrombin III activity and over 80% of the activity was lost at 1.2 mM methylglyoxal. The fibrinogen contents in plasma were decreased slightly with increasing concentrations of glyoxal and methylglyoxal. Guava leaf extracts and its active phenolic compounds including ferulic acid, gallic acid and quercetin also displayed a protective effect against methylglyoxal-induced loss of activity of antithrombin III. The results showed that guava leaf extracts can be a potent antiglycative agent and anticoagulant, which can be of great value in the preventive glycation-associated cardiovascular diseases in diabetes.

Adaptogenic Activity

Studies by Lakshmi and Sudhakar (2009) indicated that ethanolic extract of *Psidium guajava leaves* had significant adaptogenic activity against a variety of biochemical and physiological perturbations in different stress models in Swiss mice. Pretreatment with the guava leaf extract significantly ameliorated the stress-induced variations in these biochemical (plasma glucose, triglyceride, cholesterol, BUN and corticosterone) levels and blood cell counts in both acute and chronic stress models. The extract treated animals showed increase in swimming endurance time and increase in anoxia tolerance time in physical and anoxia stress models respectively. Treatment groups also reverted back increase in liver, adrenal gland weights and atrophy of spleen caused by cold chronic stress and swimming endurance stress models in Wistar rats.

Spermatoprotective Activity

Guava leaf extract was found to have beneficial activity effect on on gossypol-associated sperm toxicity in Wistar rats (Akinola et al. 2007a, b). In animals treated with 50 mmg/kg/day of guava leaf extract, significant increase in sperm count from 56.2×10^6 in the control to 72.3×10^6 in treated extract and sperm motility occurred. The findings suggest that the extracts of the leaves of *Psidium guajava* possess beneficial effects on sperm production and quality, and may thus improve the sperm parameters of infertile males with oligospermia and nonobstructive azoospermia. The enhancement of male fertility, may be related to its rich constituents of natural antioxidants.

Urease Inhibition Activity

Quercetin and two quercetin glycosides, avicularin and guaijaverin, isolated from *Psidium guajava* were found to act as urease inhibitors with respective IC_{50} values of 80, 140, and 120 μM respectively (Shabana et al. 2010).

Nuclear Medicine Application

Abreu et al. (2006) found that that aqueous guava extract could present antioxidant action and/or alter the membrane structures involved in ion transport into cells, thus decreasing the radiolabelling of blood constituents with technetium-99 m (^{99m}Tc). The data showed significant alteration of incorporated radioactivity (ATI) in blood constituents from blood incubated with guava extract. Red blood cells labelled with ^{99m}Tc (^{99m}Tc-RBC) are radiobiocomplexes widely used in clinical nuclear medicine for several important applications. Technetium-99 m (^{99m}Tc) has been the most used radionuclide in nuclear medicine diagnosis procedures (single photon emission computed tomography, SPECT) and in labelling of molecular and cellular structures that can be used as radiopharmaceuticals (radiobiocomplexes).

Anticesdotal Activity

Guava leaf extract was shown to possess anticesdotal activity against *Hymenolepis diminuta* infection in rats (Tangpu and Yadav 2006). The leaf extract showed reduction in parasite parasite eggs/g of faeces count in a dose-dependent manner. It further showed comparatively low recovery of worms including scolices in the small intestine and host clearance of parasite in a dose dependent manner. In all the experimental models the anticestodal efficacy of leaf extract was significantly comparable with that of the standard anticestodal drug, praziquantel. The results supported the folk medicinal use of guava leaf extract in the treatment of intestinal-worm infections in northeastern part of India. In the northeastern part of India, various Naga tribes use fresh leaves water decoction of *P. guajava* (locally known as "motiram") as a common remedy for intestinal-worm infections.

Antidrepanocytosis Activity

Chikezie (2011) found that of three medicinal plants (*Anacardium occidentale, Psidium guajava*, and *Terminalia catappa*), aqueous extract of *P. guajava* exhibited the highest capacity to reduced polymerization of deoxygenated sickle cell haemoglobin. In another related in-vitro study, *P. guajava* and *T. catappa* protected the erythrocytes against osmotic stress, as evidenced by decreases in the values of mean corpuscular fragility (MCF) compared with the control sample (Chikezie and Uwakwe 2011). In contrast, *A. occidentale* (800 mg/dL) promoted significant distabilization of sickle erythrocytes and offered no membrane protective effect.

Antiplasmodial Activity

Of ten plants investigated for antiplasmodial against a chloroquine-sensitive strain of malarial parasite, *Plasmodium falciparum*, the two most active extracts were *Psidium guajava* stem-bark extract and *Vangueria infausta* leaf extract, both of which showed IC_{50} values of 10–20 μg/ml (Nundkumar and Ojewole 2002). Phytochemical analysis of the two active plant extracts revealed the presence of anthraquinones, flavonoids, seccoirridoids and terpenoids.

Traditional Medicinal Uses

The roots are official in the Mexican pharmacopoeia; and the leaves in the Dutch and Mexican Pharmacopoeias. The roots, bark, leaves and immature fruits are used in folk medicine because of their astringency, are commonly employed to halt gastroenteritis, diarrhea and dysentery, throughout the tropics. In Japan, guava leaf tea has been approved as one of the Foods for Specified Health Uses and is commonly used for diabetes. In Peninsular Malaysia, leaves have been used for for diarrhoea, stomache-ache and gastroenteritis and as vermifuge. A decotion of leaves is used for skin complaints. A combined decoction of leaves and bark is given to expel the placenta after childbirth and as emmenagogue. In Indonesia, the leaf decoction is used for leucorrhoea, bark used for hystero-epilepsy. In Thailand, the leaves have been used as an antidiarrheal and antidysenteric; externally, they have been used as a deodorant of mouth odor. In the Philippines the astringent, unripe fruit, the leaves, the cortex of the bark and roots – through more often the leaves only – in the form of a decoction, are used for washing ulcers and wounds. The bark and leaves are reported to be astringent, vulnerary, and when decocted, antidiarhetic.

In India, the bark decotion is used in chronic diarrhea of children and sometimes adults; a decoction of the root-bark is recommended as a mouthwash for swollen gums and as local application in prolapsus. Guava jelly is deemed as tonic for the heart and good for constipation and ripe fruit is good aperients. The leaves are chewed as remedy for toothache. The leaf infusion is prescribed in India in cerebral ailments, nephritis and cachexia. An extract is given in epilepsy and chorea and a tincture is rubbed on the spine of children in convulsions. Guava leaf is an important ingredient in many Ayurvedic preparations, for diabetes and other ailments. The leaf decoction is taken as a remedy for coughs, throat and chest ailments, gargled to relieve oral ulcers and inflamed gums. In Costa Rica, a decoction of the flower buds is considered an effective remedy for diarrhoea and flow of blood. In Mexico, the fruit is considered anthelmintic and the leaves used as remedy for itches and decocted leaves are employed for cleansing ulcers. In Uruguay, a decoction of the leaves is used as a vaginal and uterine wash, especially in leucorrhoea. A decoction of the young leaves and shoots is prescribed in the West Indies for febrifuge and antispasmodic baths, and an infusion of the leaves for cerebral affections, nephritis, and cachexia; the pounded leaves are applied locally for rheumatism; an extract is used for epilepsy and chorea; and the tincture is rubbed into the spine of children suffering from convulsions. Leaf extract have been used for diarrhoea and for diabetes in Brazil.

Other Uses

Guava tree provides a yellow to reddish, fine-grained, compact, moderately strong wood that is much used for carpentry and turnery. In Peninsular Malaysia, the wood is used for spear handles, wood engravings and instruments, household and agricultural implements, tree nails, post fro small houses, fence posts, fuelwood and charcoal. In India, it is valued for engravings. Guatemalans use guava wood to make spinning tops, and in El Salvador it is fashioned into hair combs. The leaves and bark are rich in tannin. The bark is used in Central America for tanning hides. Leaves have been used by dyers in Pekan, Peninsular Malaysia to dye silk black; and in Indo-China, to dye cotton black and for dyeing mattings in Indonesia.

Comments

Guava is deemed a weed in many areas and is a natural and preferred host of fruit flies. A major invasive species in the Galapagos Islands, Hawai'i, New Zealand, and southern Africa (Cronk and Fuller 2001). A problem in the Marquesas (French Polynesia), New Caledonia and Fiji. Very invasive in Tonga, especially on 'Eua. Common, and in the future may well become even more widespread in the Cook Islands.

Selected References

Abdelrahim SI, Almadboul AZ, Omer MEA, Elegami A (2002) Antimicrobial activity of *Psidium guajava* L. Fitoterapia 73(7–8):713–715

Abreu PR, Almeida MC, Bernardo RM, Bernardo LC, Brito LC, Garcia EA, Fonseca AS, Bernardo-Filho M (2006) Guava extract (*Psidium guajava*) alters the labelling of blood constituents with technetium-99 m. J Zhejiang Univ Sci B 7(6):429–435

Akanji MA, Adeyemi OS, Oguntoye SO, Sulyman F (2009) *Psidium guajava* extract reduces trypanosomosis associated lipid peroxidation and raises glutathione concentrations in infected animals. EXCLI J 8:148–154

Akinmoladun AC, Obuotor EM, Farombi EO (2010) Evaluation of antioxidant and free radical scavenging capacities of some Nigerian indigenous medicinal plants. J Med Food 13(2):444–451

Akinola OB, Oladosu OS, Dosumu OO (2007a) Ethanol extract of the leaves of *Psidium guajava* Linn enhances sperm output in healthy Wistar rats. Afr J Med Med Sci 36(2):137–140

Akinola OB, Oladosu OS, Dosumu OO (2007b) Spermatoprotective activity of the leaf extract of *Psidium guajava* Linn. Niger Postgrad Med J 14(4):273–276

Alves PM, Queiroz LM, Pereira JV, Pereira Mdo S (2009) In vitro antimicrobial, antiadherent and antifungal activity of Brazilian medicinal plants on oral biofilm microorganisms and strains of the genus *Candida*. Rev Soc Bras Med Trop 42(2):222–224 (In Portuguese)

Arima H, Danno G (2002) Isolation of antimicrobial compounds from guava (*Psidium guajava* L.) and their structural elucidation. Biosci Biotechnol Biochem 66(8):1727–1730

Asano T, Deguchi Y, Tuji A, Makino K, Takamizawa K (2007) Effects and safety of guava tea (Bansoureicha®) on borderline and mild hyperlipidemia. Jpn J Nutr Assess 24:599–605, in Japanese

Asano T, Tuji A, Deguchi Y, Makino K (2005) Clinical effect of guava tea (Bansoureicha®) on diabetes patient. Jpn J Nutr Assess 11:81–85, in Japanese

Ayub MY, Norazmir MN, Mamot S, Jeeven K, Hadijah H (2010) Anti-hypertensive effect of pink guava (*Psidium guajava*) puree on spontaneous hypertensive rats. Int Food Res J 17:89–96

Backer CA, van den Brink RCB Jr (1963) Flora of Java, (spermatophytes only), vol 1. Noordhoff, Groningen, 648 pp

Bartolome A, Mandap K, David KJ, Sevilla F 3rd, Villanueva J (2006) SOS-red fluorescent protein (RFP) bioassay system for monitoring of antigenotoxic activity in plant extracts. Biosens Bioelectron 21(11):2114–2120

Bashir HA, Abu-Goukh A-BA (2003) Compositional changes during guava fruit ripening. Food Chem 80(4):557–563

Basnet P, Kadota S, Pandey RR, Takahashi T, Kojima Y, Shimizu M, Takata Y, Kobayashi M, Namba T (1995) Screening of traditional medicines for their hypoglycemic activity in streptozotocin (STZ)-induced diabetic rats and a detailed study on *Psidium guajava*. Wakan Iyakugaku Zasshi 12(2):109–117

Begum S, Ali SN, Hassan SI, Siddiqui BS (2007) A new ethylene glycol triterpenoid from the leaves of *Psidium guajava*. Nat Prod Res 21(8):742–748

Begum S, Hassan SI, Ali SN, Siddiqui BS (2004) Chemical constituents from the leaves of *Psidium guajava*. Nat Prod Res 18(2):135–140

Begum S, Hassan SI, Siddiqui BS (2002a) Two new triterpenoids from the fresh leaves of *Psidium guajava*. Planta Med 68(12):1149–1152

Begum S, Hassan SI, Siddiqui BS, Shaheen F, Ghayur MN, Gilani AH (2002b) Triterpenoids from the leaves of *Psidium guajava*. Phytochemistry 61(4):399–403

Begum S, Siddiqui BS, Hassan SI (2002c) Triterpenoids from *Psidium guajava* leaves. Nat Prod Lett 16(3): 173–177

Belemtougri RG, Constantin B, Cognard C, Raymond G, Sawadogo L (2006) Effects of two medicinal plants *Psidium guajava* L. (Myrtaceae) and *Diospyros mespiliformis* L. (Ebenaceae) leaf extracts on rat skeletal muscle cells in primary culture. J Zhejiang Univ Sci B 7(1):56–63

Bernardino-Nicanor A, Añón MC, Scilingo AA, Dávila-Ortíz G (2005) Functional properties of guava seed glutelins. J Agric Food Chem 53(9):3613–3617

Birdi T, Daswani P, Brijesh S, Tetali P, Natu A, Antia N (2010) Newer insights into the mechanism of action of *Psidium guajava* L. leaves in infectious diarrhoea. BMC Compl Altern Med 10(1):33

Boyle FP, Seagrave-Smith H, Sakata S, Sherman GD (1957) Commercial guava processing in Hawaii. Hawaii Agric Exp Sta Bull 111:5–30

Burkill IH (1966) A dictionary of the economic products of the Malay Peninsula. Revised reprint, 2 volumes, vol 1 (A–H) pp 1–1240, vol 2 (I–Z) pp 1241–2444. Ministry of Agriculture and Co-operatives, Kuala Lumpur

Cáceres A, Fletes L, Aguilar L, Ramirez O, Figueroa L, Taracena AM, Samayoa B (1993) Plants used in Guatemala for treatment of gastrointestinal disorders. 3. Confirmation of activity against enterobacteria of 16 plants. J Ethnopharmacol 38:31–38

Chandrika UG, Fernando KS, Ranaweera KK (2009) Carotenoid content and in vitro bioaccessibility of lycopene from guava (*Psidium guajava*) and watermelon (Citrullus *lanatus*) by high-performance liquid chromatography diode array detection. Int J Food Sci Nutr 60(7):558–566

Chen GB, Chen BT (2002) Experimental study on anti-rotavirus effect of extract of *Psidiium guajava* leaves in vitro. China J Trad Chin Med Pharm 17(8):502–504, in Chinese

Chen GB, Chen BT, Wang SG, Zhang WJ (2003) An experimental study in vivo anti-rotavirus action of guava leaf. New J Trad Chin Med 35(12):65–67, in Chinese

Chen H-C, Sheu M-J, Wu C-M (2006) Characterization of volatiles in guava (*Psidium guajava* L. cv. Chung-Shan-Yueh-Pa) fruit from Taiwan. J Food Drug Anal 14(4):398–402

Chen J, Craven LA (2007) Myrtaceae. In: Wu ZY, Raven PH, Hong DY (eds) Flora of China. Vol. 13 (Clusiaceae through Araliaceae). Science Press/Missouri Botanical Garden Press, Beijing/St. Louis

Chen H-C, Sheu M-J, Lin L-Y, Wu C-M (2007) Chemical composition of the leaf essential oil of *Psidium guajava* L. from Taiwan. J Essent Oil Res 19:345–347

Chen H-C, Sheu M-J, Lin L-Y, Wu C-M (2008) Volatile constituents of six cultivars of mature guava (*Psidium guajava* L.) fruits from Taiwan. Acta Hortic (ISHS) 765:273–278

Chen H-Y, Yen G-C (2007) Antioxidant activity and free radical-scavenging capacity of extracts from guava (*Psidium guajava* L.) leaves. Food Chem 101(2):686–694

Chen KC, Hsieh CL, Huang KD, Ker YB, Chyau CC, Peng RY (2009) Anticancer activity of rhamnoallosan against DU-145 cells is kinetically complementary to coexisting polyphenolics in *Psidium guajava* budding leaves. J Agric Food Chem 57(14):6114–6122

Chen KC, Chuang CM, Lin LY, Chiu WT, Wang HE, Hsieh CL, Tsai T, Peng RY (2010a) The polyphenolics in the aqueous extract of *Psidium guajava* kinetically reveal an inhibition model on LDL glycation. Pharm Biol 48(1):23–31

Chen KC, Peng CC, Chiu WT, Cheng YT, Huang GT, Hsieh CL, Peng RY (2010b) Action mechanism and signal pathways of *Psidium guajava* L. aqueous extract in killing prostate cancer LNCaP cells. Nutr Cancer 62(2):260–270

Cheng FC, Shen SC, Wu JS (2009) Effect of guava (*Psidium guajava* L.) leaf extract on glucose uptake in rat hepatocytes. J Food Sci 74(5):H132–H138

Cheng JT, Yang RS (1983) Hypoglycemic effect of guava juice in mice and human subjects. Am J Chin Med 11:74–76

Cheng TY, Zhu SH, Wei XK, Chen J (2005) Preliminary studies on antidiarrheic mechanism of *Psidium guajava* L. leaves. Anim Husb Vet Med 37(2):13–15, in Chinese

Chikezie PC (2011) Sodium metabisulfite-induced polymerization of sickle cell hemoglobin incubated in the extracts of three medicinal plants (*Anacardium occidentale, Psidium guajava,* and *Terminalia catappa*). Pharmacogn Mag 7(26):126–132

Chikezie PC, Uwakwe AA (2011) Membrane stability of sickle erythrocytes incubated in extracts of three medicinal plants: *Anacardium occidentale, Psidium guajava,* and *Terminalia catappa*. Pharmacogn Mag 7(26):121–125

Chiwororo WD, Ojewole JA (2008) Biphasic effect of *Psidium guajava* Linn. (Myrtaceae) leaf aqueous extract on rat isolated vascular smooth muscles. J Smooth Muscle Res 44(6):217–229

Chiwororo WD, Ojewole JA (2009) Spasmolytic effect of *Psidium guajava* Linn. (Myrtaceae) leaf aqueous extract on rat isolated uterine horns. J Smooth Muscle Res 45(1):31–38

Chyau CC, Chen SY, Wu CM (1992) Differences of volatile and nonvolatile constituents between mature and ripe guava (*Psidium guajava* Linn) fruits. J Agric Food Chem 40:846–849

Clery RA, Hammond CJ (1998) New sulfur components of pink guava fruit (*Psidium guajava* L.). J Essent Oil Res 20:315–317

Conde Garcia EA, Nascimento VT, Santiago Santos AB (2003) Inotropic effects of extracts of *Psidium guajava* L. (guava) leaves on the guinea pig atrium. Braz J Med Biol Res 36(5):661–668

Corral-Aguayo RD, Yahia EM, Carrillo-Lopez A, González-Aguilar G (2008) Correlation between some nutritional components and the total antioxidant capacity measured with six different assays in eight horticultural crops. J Agric Food Chem 56(22): 10498–10504

Cronk QCB, Fuller JL (2001) Plant Invaders: the Threat to Natural Ecosystems. Earthscan Publications, London, UK. 241 pp

Das A, Majumder K (2010) Fractional changes of pectic polysaccharides in different tissue zones of developing guava (*Psidium guajava* L.) fruits. Sci Hortic 125(3): 406–410

De Lima RK, Cardoso MDG, Andrade MA, Nascimento EA, de Morais SAL, Nelson DL (2010) Composition of the essential oil from the leaves of tree domestic varieties and one wild variety of the guava plant (*Psidium guajava* L., Myrtaceae). Rev Bras Farmacogn 20(1):41–44

Deguchi Y (2006) Effect of guava tea on postprandial blood glucose and diabetes. Assoc J Jpn Soc Med Use Funct Food 3:439–445 (In Japanese)

Deguchi Y, Miyazaki K (2010) Anti-hyperglycemic and anti-hyperlipidemic effects of guava leaf extract. Nutr Metab (Lond) 7:9

Deguchi Y, Osada K, Chonan O, Kobayashi K, Oohashi A, Kitukawa T, Watanuki M, Ooni M, Nakajima K, Hata Y (2000) Effectiveness of consecutive ingestion and excess intake of guava leaves tea in human volunteers. J Jpn Counc Adv Food Ingred Res 1:19–28, in Japanese

Deguchi Y, Osada K, Uchida K, Kimura H, Yoshikawa M, Kudo T, Yasui H, Watanuki M (1998) Effects of extract of guava leaves on the development of diabetes in the db/db mouse and on the postprandial blood glucose of human subjects. Nippon Nogeikagaku Kaishi 72:923–932, in Japanese

Deguchi Y, Osada K, Watanuki M (2003) Effect of guava leaf extract in combination with acarbose or voglibose on increased blood glucose level in sugar-loaded normal mice. J Jpn Soc Nutr Food Sci 56:207–212, in Japanese

Dhawan SSJ, Kainsa RL, Gupta OP (1983) Screening of guava cultivars for wine and brandy making. Haryana Agric Univ J Res 13(3):420–423

Dhiman A, Nanda A, Ahmad S, Narasimhan B (2011) *In vitro* antimicrobial activity of methanolic leaf extract of *Psidium guajava* L. J Pharm Bioall Sci 3:226–229

Direkbusarakom S, Herunsalee A, Yoshimizu M, Ezura Y, Kimura T (1997) Efficacy of guava (*Psidium guajava*) extracts against some fish and shrimp pathogenic agents. In: Flegel TW, MacRae IH (eds) Diseases in Asian aquaculture III. Asian Fisheries Society, Manila, p 363

Doubova SV, Morales HR, Hernández SF, Martínez-García MDC, de Cossío Ortiz MG, Soto MAC, Arce ER, Lozoya X (2007) Effect of a *Psidii guajavae* folium extract in the treatment of primary dysmenorrhea: a randomized clinical trial. J Ethnopharmacol 110(2):305–310

Dutta S, Das S (2011) A study of the anti-inflammatory effect of the leaves of *Psidium guajava* Linn. on experimental animal models. Pharmacogn Res 2(5):313–317

Ekundayo O, Ajani F, Seppaenen-Laakso T, Laakso I (1991) Volatile constituents of *Psidium guajava* L. (guava) fruit. Flavor Fragr J 6:233–236

El Bulk RE, Babiker EFE, El Tinay AH (1997) Changes in chemical composition of guava fruits during development and ripening. Food Chem 59(3):395–399

El Khadem H, Mohamed YS (1958) Constituents of the leaves of *Psidium guajava,* L. Part 11. Quercetin, avicularin, and guajaverin. J Chem Soc 32:3320–3323

Fathilah AR (2011) *Piper betle* L. and *Psidium guajava* L. in oral health maintenance. J Med Plant Res 5(2): 156–163

Fathilah AR, Rahim ZHA, Othman Y, Yusoff M (2009) Bacteriostatic effect of *Piper betle* and *Psidium guajava* extracts on dental plaque bacteria. Pak J Biol Sci 12:518–552

Foundation for Revitalisation of Local Health Traditions (2008) FRLHT Database. htttp://envis.frlht.org

Fu HZ, Luo YM, Zhang DM (2009) Studies on chemical constituents of leaves of *Psidium guajava*. Zhongguo Zhong Yao Za Zhi 34(5):577–579, in Chinese

Fu HZ, Luo YM, Li CJ, Yang JZ, Zhang DM (2010) Psidials A-C, three unusual meroterpenoids from the leaves of *Psidium guajava* L. Org Lett 12(4): 656–659

Ghosh P, Mandal A, Chakraborty P, Rasul MG, Chakraborty M, Saha A (2010) Triterpenoids from *Psidium guajava* with biocidal activity. Indian J Pharm Sci 72(4):504–507

Ghosh TK, Sen T, Das A, Dutta AS, Chaudhuri AKN (1993) Antidiarrhoeal activity of the methanolic fraction of the extract of unripe fruits of *Psidium guajava* Linn. Phytother Res 7(6):431–433

Gnan SO, Demello MT (1999) Inhibition of *Staphylococcus aureus* by aqueous goiaba extracts. J Ethnopharmacol 68(1–3):103–108

Gonçalves FA, Andrade Neto M, Bezerra JN, Macrae A, Sousa OV, Fonteles-Filho AA, Vieira RH (2008) Antibacterial activity of guava, *Psidium guajava* Linnaeus, leaf extracts on diarrhea-causing enteric bacteria isolated from Seabob shrimp, *Xiphopenaeus kroyeri* (Heller). Rev Inst Med Trop Sao Paulo 50(1): 11–15

Gonçalves JLS, Lopes RC, Oliveira DB, Costa SS, Miranda MMFS, Romano MTV, Santos NSO, Wigg MD (2005) In vitro anti-rotavirus activity of some medicinal plants used in Brazil against diarrhea. J Ethnopharmacol 99(3):403–407

Govaerts R, Sobral M, Ashton P, Barrie F, Holst BK, Landrum LL, Matsumoto K, Fernanda Mazine F, Nic Lughadha E, Proenca C, Soares-Silva LH, Wilson PG, Lucas E (2010) World checklist of Myrtaceae . The board of trustees of the Royal Botanic Gardens, Kew. Published on the Internet http://www. kew.org/wcsp/. Accessed 22 Apr 2010

Grover IS, Bala S (1993) Studies on antimutagenic effects of guava (*Psidium guajava*) in *Salmonella typhimurium*. Mutat Res 300(1):1–3

Gutiérrez RM, Mitchell S, Solis RV (2008) *Psidium guajava*: a review of its traditional uses, phytochemistry and pharmacology. J Ethnopharmacol 117(1):1–27

Han EH, Hwang YP, Kim HG, Park JH, Choi JH, Im JH, Khanal T, Park BH, Yang JH, Choi JM, Chun SS, Seo JK, Chung YC, Jeong HG (2010) Ethyl acetate extract of *Psidium guajava* inhibits IgE-mediated allergic responses by blocking FcεRI signaling. Food Chem Toxicol 49(1):100–108

Hsieh CL, Huang CN, Lin YC, Peng RY (2007a) Molecular action mechanism against apoptosis by aqueous extract from guava budding leaves elucidated with human umbilical vein endothelial cell (HUVEC) model. J Agric Food Chem 55(21):8523–8533

Hsieh C-L, Lin Y-C, Yen GC, Chen HY (2007b) Preventive effects of guava (*Psidium guajava* L.) leaves and its active compounds against α-dicarbonyl compounds-induced blood coagulation. Food Chem 103(2):528–535

Huang CS, Yin MC, Chiu LC (2011) Antihyperglycemic and antioxidative potential of *Psidium guajava* fruit in streptozotocin-induced diabetic rats. Food Chem Toxicol 49(9):2189–2195

Ishibashi K, Oka M, Hachiya M, Maeda T, Tajima N (2004) Comparison of voglibose and guava tea (Bansoureicha®) on postprandial blood glucose level. J Pract Diabetes 21:455–458, in Japanese

Jaiarj P, Khoohaswan P, Wongkrajang Y, Peungvicha P, Suriyawong P, Saraya MLS, Ruangsomboon O (1999) Anticough and antimicrobial activities of *Psidium guajava* Linn. leaf extract. J Ethnopharmacol 67(2): 203–212

Jaiarj P, Wongkrajang Y, Thongpraditchote S, Peungvicha P, Bunyapraphatsara N, Opartkiattikul N (2000) Guava leaf extract and topical haemostasis. Phytother Res 14(5):388–391

Jiménez-Escrig A, Rincón M, Pulido R, Fulgencio Saura-Calixto F (2001) Guava fruit (*Psidium guajava* L.) as a new source of antioxidant dietary fiber. J Agric Food Chem 49(11):548–5493

Jordán MJ, Margaría CA, Shaw PE, Goodner KL (2003) Volatile components and aroma active compounds in aqueous essence and fresh pink guava fruit puree (*Psidium guajava* L.) by GC-MS and multidimensional GC/GC-O. J Agric Food Chem 51(5):1421–1426

Joseph B, Priya RM (2010) *In vitro* antimicrobial activity of *Psidium guajava* L. leaf essential oil and extracts using agar well diffusion method. Int J Curr Pharm Res 2:28–32

Joseph B, Priya RM, Helen PAM, Sujatha S (2010) Bio-active compounds in essential oil and its effects of antimicrobial, cytotoxic activity from the *Psidium guajava* (L.) leaf. J Adv Biotechnol 9:10–14

Kaneko K, Iwadate E, Uchida K, Kato I, Deguchi Y, Onoue M (2005) Studies on interaction between Bansoreicha and medicines. Annu Rep Yakult Cent Inst Microbiol Res 24:73–80, in Japanese

Kawakami Y, Nakamura T, Hosokawa T, Suzuki-Yamamoto T, Yamashita H, Kimoto M, Tsuji H, Yoshida H, Hada T, Takahashi Y (2009) Antiproliferative activity of guava leaf extract via inhibition of prostaglandin endoperoxide H synthase isoforms. Prostaglandins Leukot Essent Fatty Acids 80(5–6):239–245

Khurdiya DS, Islam OM, Verma P (1996) Processing and storage of carbonated guava beverage. J Food Process Preserv 20(1):79–86

Kim SH, Cho SK, Hyun SH, Park HE, Kim YS, Choi HK (2011) Metabolic profiling and predicting the free radical scavenging activity of guava (*Psidium guajava* L.) leaves according to harvest time by (1)H-Nuclear Magnetic Resonance Spectroscopy. Biosci Biotechnol Biochem 75(6):1090–1097

Kobayashi T, Kaneko K, Takahashi M, Onoue M (2005) Safety evaluations of guava leaves and unripe fruit of guava: single dose and 1 month-repeated dose oral toxicity studies in rats. Annu Rep Yakult Cent Inst Microbiol Res 24:81–100, in Japanese

Kong KW, Ismail A (2010) Lycopene content and lipophilic antioxidant capacity of by-products from *Psidium guajava* fruits produced during puree production industry. Food Bioprod Process 89(1):53–61

Lakshmi B, Sudhakar M (2009) Screening of *Psidium guajava* leaf extracts for antistress activity in different experimental animal models. Pharmacogn Res 1: 359–366

Lee SB, Park HR (2010) Anticancer activity of guava (*Psidium guajava* L.) branch extracts against HT-29 human colon cancer cells. J Med Plant Res 4(10): 891–896

Leite KMSC, Tadiotti AC, Baldochi D, Oliveira OMMF (2006) Partial purification, heat stability and kinetic characterization of the pectin methylesterase from Brazilian guava, Paluma cultivars. Food Chem 94(4):565–572

Li J, Chen F, Luo J (1999) GC-MS analysis of essential oil from the leaves of *Psidium guajava*. Zhong Yao Cai 22:78–80, in Chinese

Liao ZY, Li YS, Jiang JL (2007) Protective effect of guava leaf extract on colonic tissues with ulcerative colonitis induced by trinitrobenzene sulfonic acid in rats. World Chin J Digestol 15(1):69–71

Lim TK, Khoo KC (1990) Guava in Malaysia production, pests and diseases. Tropical Press, Kuala Lumpur, 260 pp

Ling LT, Radhakrishnan AK, Subramaniam T, Cheng HM, Palanisamy UD (2010) Assessment of antioxidant capacity and cytotoxicity of selected Malaysian plants. Molecules 15:2139–2151

Liu CH, Peng CC, Peng CH, Hsieh CL, Chen KC, Peng RY (2010) Polyphenolics-rich *Psidium guajava* budding leaf extract canreverse diabetes-induced functional impairment of cavernosal smooth muscle relaxation in rats. Res J Med Sci 4:25–32

Lozoya X, Becerril G, Martínez M (1990) Model of intraluminal perfusion of the guinea pig ileum in vitro in the study of the antidiarrheal properties of the guava (*Psidium guajava*). Arch Invest Med (Mex) 21(2):155–162 (In Spanish)

Lozoya X, Meckes M, Aboud-Zaid M, Tortoriello J, Nozolillo C, Arnason JT (1994) Quercetine glycosides in *Psidium guajava* L. leaves and determination of a spasmolytic principle. Arch Med Res 25:11–15

Lozoya X, Reyes-Morales H, Chávez-Soto MA, Martínez-García MC, Soto-González Y, Doubova SV (2002) Intestinal anti-spasmodic effect of a phytodrug of *Psidium guajava* folia in the treatment of acute diarrheic disease. J Ethnopharmacol 83(1–2):19–24

Lu WB, Zhang BE, Wang J, Lu RX, Li RL, Chen WW (2010) Screening of anti-diarrhea effective fractions from guava leaf. Zhong Yao Cai 33(5):732–735, in Chinese

Lutterodt GD (1989) Inhibition of gastrointestinal release of acetylcholine by quercetin as a possible mode of action of *Psidium guajava* leaf extracts in the treatment of acute diarrhoeal disease. J Ethnopharmacol 25(3):235–247

Lutterodt GD (1992) Inhibition of Microlax-induced experimental diarrhea with narcotic-like extracts of *Psidium guajava* leaf in rats. J Ethnopharmacol 37(2):151–157

Lutterodt GD, Ismail A, Basheer RH, Baharudin HM (1999) Antimicrobial effects of *Psidium guajava* extract as one mechanism of its antidiarhoeal action. Malays J Med Sci 6(2):17–20

Lutterodt GD, Maleque A (1988) Effects on mice locomotor activity of a narcotic-like principle from *Psidium guajava* leaves. J Ethnopharmacol 24(2–3): 219–231

MacLeod AJ, Troconis NG (1982) Volatile flavor components of guava. Phytochemistry 21:1339–1342

Mahfuzul Hoque MD, Bari ML, Inatsu Y, Juneja VK, Kawamoto S (2007) Antibacterial activity of guava (*Psidium guajava* L.) and neem (*Azadirachta indica* A. Juss.) extracts against foodborne pathogens and spoilage bacteria. Foodborne Pathog Dis 4(4):481–488

Manosroi J, Dhumtanom P, Manosroi A (2006) Antiproliferative activity of essential oil extracted from Thai medicinal plants on KB and P388 cell lines. Cancer Lett 235(1):114–120

Mao QC, Zhou YC, Li RM, Hu YP, Liu SW, Li XJ (2010) Inhibition of HIV-1 mediated cell-cell fusion by saponin fraction from *Psidium guajava* leaf. Zhong Yao Cai 33(11):1751–1754, in Chinese

Marcelin O, Williams P, Brillouet J-M (1993) Isolation and characterisation of the two main cell-wall types from guava (*Psidium guajava* L.) pulp. Carbohydr Res 240:233–243

Marquina V, Araujo L, Ruíz J, Rodríguez-Malaver A, Vit P (2008) Composition and antioxidant capacity of the guava (*Psidium guajava* L.) fruit, pulp and jam. Arch Latinoam Nutr 58(1):98–102, in Spanish

Maruyuma Y, Matsuda H, Matsuda R, Kubo M, Hatano T, Okuda T (1985) Study on *Psidium guajava* L. (I). Antidiabetic effect and effective components of the leaf of *Psidium guajava* L (Part I). Shoyakugaku Zasshi 39:261–269

Matsuzaki K, Ishii R, Kobiyama K, Kitanaka S (2010) New benzophenone and quercetin galloyl glycosides from *Psidium guajava* L. J Nat Med 64(3):252–256

Md Nor N, Yatim AM (2011) Effects of pink guava (*Psidium guajava*) puree supplementation on antioxidant enzyme activities and organ function of spontaneous hypertensive rat. Sains Malays 40(4):369–372

Meckes M, Calzada F, Tortoriello J, González JL, Martinez M (1996) Terpenoids isolated from *Psidium guajava* with depressant activity on central nervous system. Phytother Res 10(7):600–603

Mercadante AZ, Steck A, Pfander H (1999) Carotenoids from guava (*Psidium guajava* L.): isolation and structure elucidation. J Agric Food Chem 47(1):145–151

Metwally AM, Omar AA, Harraz FM, El Sohafy SM (2010) Phytochemical investigation and antimicrobial activity of *Psidium guajava* L. leaves. Pharmacogn Mag 6(23):212–218

Michael HN, Salib JY, Ishak MS (2002) Acylated flavonol glycoside from *Psidium gauijava* L. seeds. Pharmazie 57(12):859–860

Morales MA, Tortoriello J, Meckes M, Paz D, Lozoya X (1994) Calcium-antagonist effect of quercetin and its relation with the spasmolytic properties of *Psidium guajava* L. Arch Med Res 25(1):17–21

Morton JF (1987) Guava. In: Morton JF (ed) Fruits of warm climates. Julia F. Morton, Miami, pp 356–363

Mukhtar HM, Ansari SH, Ali M, Naved T, Bhat ZA (2004) Effect of water extract of *Psidium guajava* leaves on alloxan-induced diabetic rats. Pharmazie 59:734–735

Nair R, Chanda S (2007) In-vitro antimicrobial activity of *Psidium guajava* L. leaf extracts against clinically important pathogenic microbial strains. Braz J Microbiol 38(3):452–458

Nishimura O, Yamaguchi K, Mihara S, Shibamoto T (1989) Volatile constituents of guava fruits (*Psidium guajava* L.) and canned puree. J Agric Food Chem 37:139–142

Norazmir MN, Ayub MY (2011) Effects of pink guava (*Psidium guajava*) puree supplementation on antioxidant enzyme activities and organ function of spontaneous hypertensive rat. Sains Malays 40(4):369–372

Nundkumar N, Ojewole JAO (2002) Studies on the antiplasmodial properties of some South African medicinal plants used as antimalarial remedies in Zulu folk medicine. Methods Find Exp Clin Pharmacol 24(7): 397–401

Ochse JJ, van den Brink RCB (1931) Fruits and fruitculture in the Dutch East Indies. G. Kolff, Batavia-C, 180 pp

Ogunlana OE, Ogunlana OO (2008) In vitro assessment of the free radical scavenging activity of *Psidium guajava*. Res J Agric Biol Sci 4(6):666–671

Ogunwande IA, Olawore NO, Adeleke KA, Ekundayo O, Koenig WA (2003) Chemical composition of the leaf volatile oil of *Psidium guajava* L. growing in Nigeria. Flavor Fragr J 18(2):136–138

Oh WK, Lee CH, Lee MS, Bae EY, Sohn CB, Oh H, Kim BY, Ahn JS (2005) Antidiabetic effect of extracts from *Psidium guajava*. J Ethnopharmacol 96:411–415

Ojewole JA (2005) Hypoglycemic and hypotensive effects of *Psidium guajava* Linn. (Myrtaceae) leaf aqueous extract. Methods Find Exp Clin Pharmacol 27(10): 689–695

Ojewole JA (2006) Antiinflammatory and analgesic effects of *Psidium guajava* Linn. (Myrtaceae) leaf aqueous extract in rats and mice. Methods Find Exp Clin Pharmacol 28(7):441–446

Ojewole JA, Awe EO, Chiwororo WD (2008) Antidiarrhoeal activity of *Psidium guajava* Linn. (Myrtaceae) leaf aqueous extract in rodents. J Smooth Muscle Res 44(6):195–207

Okuda T, Hatano T, Yazaki K (1984) Guavin B, an ellagitannin of novel type. Chem Pharm Bull 32:3787–3788

Okuda T, Yoshida T, Hatano T, Yazaki K, Ashida M (1982) Ellagitannins of the Casuarinaceae Stachyuraceae and Myrtaceae. Phytochemistry 21:2871–2874

Okuda T, Yoshida T, Hatano T, Yazaki K, Ikegami Y, Shingu T (1987) Guavins A, C, D and complex tannins from *Psidium guajava*. Chem Pharm Bull 35:443–446

Olajide OA, Awe SO, Makinde JM (1999) Pharmacological studies on the leaf of *Psidium guajava*. Fitoterapia 70(1):25–31

Olatunji-Bello II, Odusanya AJ, Raji I, Ladipo CO (2007) Contractile effect of the aqueous extract of *Psidium guajava* leaves on aortic rings in rat. Fitoterapia 78(3):241–243

Oliveros-Belardo L, Smith RM, Robinson JM, Albano V (1986) A chemical study of the essential oil from the fruit peeling of *Psidium guajava* L. Philipp J Sci 115:1–9

Opute FI (1978) The component fatty acids of *Psidium guajava* seed fats. J Sci Food Agric 29:737–738

Oyama W, Urakawa M, Gonda M, Ohsawa T, Yasutake N, Onoue M (2005) Safety of extracts of guava leaves and unripe fruit from *Psidium guajava* L.: bacterial reverse mutation, DNA repair and micronucleus tests. Annu Rep Yakult Cent Inst Microbiol Res 24:113–125, in Japanese

Paniandy JC, Chane-Ming J, Pieribattesti JC (2000) Chemical composition of the essential oil and headspace solid-phase microextraction of the guava fruit (*Psidium guajava* L.). J Essent Oil Res 12(2):153–158

Peng C-C, Peng C-H, Chen KC, Hsieh C-L, Peng RY (2011) The aqueous soluble polyphenolic fraction of *Psidium guajava* leaves exhibits potent anti-angiogenesis and anti-migration actions on DU145 cells. Evid-Based Complim Altern Med 2011:1–8, Article ID 219069

Pino JA, Aguero J, Marbot R, Fuentes V (2001) Leaf oil of *Psidium guajava* L. from Cuba. J Essent Oil Res 13:61–62

Porcher MH et al (1995–2020) Searchable world wide web multilingual multiscript plant name database. The University of Melbourne, Melbourne. http://www.plantnames.unimelb.edu.au/Sorting/Frontpage.html

Prabu GR, Gnanamani A, Sadulla S (2006) Guaijaverin – a plant flavonoid as potential antiplaque agent against *Streptococcus mutans*. J Appl Microbiol 101(2): 487–495

Qa'dan F, Thewaini AJ, Ali DA, Affi R, Elkhawad A, Matalka KZ (2005) The antimicrobial activities of *Psidium guajava* and *Juglans regia* leaf extracts to acne-developing organisms. Am J Chin Med 33(2):197–204

Qian H, Nihorimbere V (2004) Antioxidant power of phytochemicals from *Psidium guajava* leaf. J Zhejiang Univ Sci 5(6):676–683

Rahim N, Gomes DJ, Watanabe H, Rahman SR, Chomvarin C, Endtz HP, Alam M (2010) Antibacterial activity of *Psidium guajava* leaf and bark against multidrug-resistant *Vibrio cholerae*: implication for cholera control. Jpn J Infect Dis 63(4):271–274

Rai PK, Jaiswal D, Mehta S, Watal G (2009) Anti-hyperglycaemic potential of *Psidium guajava* raw fruit peel. Indian J Med Res 129:561–565

Rai PK, Mehta S, Watal G (2010) Hypolipidaemic & hepatoprotective effects of *Psidium guajava* raw fruit peel in experimental diabetes. Indian J Med Res 131:820–824

Rattanachaikunsopon P, Phumkhachorn P (2007) Bacteriostatic effect of flavonoids isolated from leaves of *Psidium guajava* on fish pathogens. Fitoterapia 78(6):434–436

Re L, Barocci S, Capitani C, Vivani C, Ricci M, Rinaldi L, Paolucci G, Scarpantonio A, Leon-Fernandez OS, Morales MA (1999) Effects of some natural extracts on the acetylcholine release at the mouse neuromuscular junction. Pharmacol Res 39:239–245

Roy CK, Das AK (2010) Comparative evaluation of different extracts of leaves of *Psidium guajava* Linn. for hepatoprotective activity. Pak J Pharm Sci 23(1):15–20

Roy CK, Kamath JV, Asad M (2006) Hepatoprotective activity of *Psidium guajava* Linn. leaf extract. Indian J Exp Biol 44(4):305–311

Sagar VR, Suresh Kumar P (2007) Processing of guava in the form of dehydrated slices and leather. Acta Hortic (ISHS) 735:579–589

Sagrero-Nieves L, Bartley JP, Provis-Schwede A (1994) Supercritical fluid extraction of the volatile components from the leaves of *Psidium guajava* L. (guava). Flavor Fragr J 9(3):135–137

Salib JY, Michael HN (2004) Cytotoxic phenylethanol glycosides from *Psidium guajava* seeds. Phytochemistry 65(14):2091–2093

Sambo N, Garba SH, Timothy HA (2009) Effect of the aqueous extract of *Psidium guajava* on erythromycin-induced liver damage in rats. Niger J Physiol Sci 24(2):171–176

Sanches NR, Cortez DAG, Schiavini MS, Nakamura CV, Filho BPD (2005) An evaluation of antibacterial activities of *Psidium guajava* (L.). Braz Arch Biol Technol 48(3):429–436

Santos FA, Rao VSN, Silveira ER (1998) Investigations on the antinociceptive effects of *Psidium guajava* leaf essential oil and its major constituents. Phytother Res 12:24–27

Sen T, Nasralla HSH, Chaudhuri AKN (1995) Studies on the antiinflammatory and related pharmacological activities of *Psidium guajava*: a preliminary report. Phytother Res 9(2):118–122

Seo N, Ito T, Wang NL, Yao ZS, Tokura Y, Furukawa F, Kitanaka S (2005) Anti-allergic *Psidium guajava* extracts exert an antitumor effect by inhibition of T regulatory cells and resultant augmentation of Th1 Cells. Anticancer Res 25:3763–3770

Seshadri TR, Vasishta K (1965) Polyphenols of the leaves of *Psidium guajava* – quercetin, guaijaverin, leucocyanidin and amritoside. Phytochemistry 4(6):989–992

Shabana S, Kawai A, Kai K, Akiyama K, Hayashi H (2010) Inhibitory activity against urease of quercetin glycosides isolated from *Allium cepa* and *Psidium guajava*. Biosci Biotechnol Biochem 74(4):878–880

Shaheen HM, Ali BH, Alqarawi AA, Bashir AK (2000) Effect of *Psidium guajava* leaves on some aspects of the central nervous system in mice. Phytother Res 14(2):107–111

Shao M, Wang Y, Liu Z, Zhang DM, Cao HH, Jiang RW, Fan CL, Zhang XQ, Chen HR, Yao XS, Ye WC (2010) Psiguadials A and B, two novel meroterpenoids with unusual skeletons from the leaves of *Psidium guajava*. Org Lett 12(21):5040–5043

Shen SC, Cheng FC, Wu NJ (2008) Effect of guava (*Psidium guajava* Linn.) leaf soluble solids on glucose metabolism in type 2 diabetic rats. Phytother Res 2622(11):1458–1464

Shu J, Chou G, Wang Z (2009) Triterpenoid constituents in fruits of *Psidum guajava*. Zhongguo Zhong Yao Za Zhi 34(23):3047–3050, in Chinese

Shu J, Chou G, Wang Z (2010a) Two new benzophenone glycosides from the fruit of *Psidium guajava* L. Fitoterapia 81(6):532–535

Shu JC, Chou GX, Wang ZT (2010b) One new galloyl glycoside from fresh leaves of *Psidium guajava* L. Yao Xue Xue Bao 45(3):334–337

Silva JD, Luz AIR, Silva MHL, Andrade EHA, Zoghbi MGB, Maia JGS (2003) Essential oils of the leaves and stems of four *Psidium* spp. Flavor Fragr J 18:240–243

Soares FD, Pereira T, Marques MOM, Monteiro AR (2007) Volatile and non-volatile chemical composition of the white guava fruit (*Psidium guajava*) at different stages of maturity. Food Chem 100(1):15–21

Soetopo L (1992) *Psidium guajava*. In: Verheij EWM, Coronel RE (eds) Plant resources of South-East Asia, no. 2. Edible fruits and nuts. Prosea Foundation, Bogor, pp 266–270

Soman S, Rauf AA, Indira M, Rajamanickam C (2010) Antioxidant and antiglycative potential of ethyl acetate fraction of *Psidium guajava* leaf extract in streptozotocin-induced diabetic rats. Plant Foods Hum Nutr 65(4):386–391

Steinhaus M, Sinuco D, Polster J, Osorio C, Schieberle P (2009) Characterization of the key aroma compounds in pink guava (*Psidium guajava* L.) by means of aroma re-engineering experiments and omission tests. J Agric Food Chem 57(7):2882–2888

Subramanian S, Banu HH, Bai RMR, Shanmugavalli R (2009) Biochemical evaluation of antihyperglycemic and antioxidant nature of *Psidium guajava* leaves extract in streptozotocin-induced experimental diabetes in rats. Pharm Biol 47(4):298–303

Tachakittirungrod S, Ikegami F, Okonogi S (2007a) Antioxidant active principles isolated from *Psidium guajava* grown in Thailand. Sci Pharm 75(4):179–193

Tachakittirungrod S, Okonogi S, Chowwanapoonpohn S (2007b) Study on antioxidant activity of certain plants in Thailand: mechanism of antioxidant action of guava leaf extract. Food Chem 103:381–388

Tanaka T, Ishida N, Ishimatsu M, Nonaka G, Nishioka I (1992) Tannins and related compounds. CXVI. Six new complex tannins, guajavins, psidinins and psiguavin from the bark of *Psidium guajava* L. Chem Pharm Bull 40(8):2092–2098

Tangpu TV, Yadav AK (2006) Anticestodal efficacy of *Psidium guajava* against experimental *Hymenolepis diminuta* infection in rats. Indian J Pharmacol 38:29–32

Thaipong K, Boonprakob U, Cisneros-Zevallos L, Byrne DH (2005) Hydrophilic and lipophilic antioxidant activities of guava fruits. Southeast Asian J Trop Med Public Health 36(Suppl 4):254–257

Thanangkol P, Chaichangptipayut C (1987) Double-blind study of *Psidium guajava* L. and tetracycline in acute diarrhoea. Siriraj Hosp Gaz 39:253–267

Thuaytong W, Anprung P (2011) Bioactive compounds and prebiotic activity in Thailand-grown red and white guava fruit (*Psidium guajava* L.). Food Sci Technol Int 17(3):205–212

U.S. Department of Agriculture, Agricultural Research Service (2010) USDA national nutrient database for standard reference, release 23. Nutrient data laboratory home page. http://www.ars.usda.gov/ba/bhnrc/ndl

Uboh FE, Okon IE, Ekong MB (2010) Effect of aqueous extract of *Psidium guajava* leaves on liver enzymes, histological integrity and hematological indices in rats. Gastroenterol Res 3(1):32–38

van Vuuren SF, Naidoo D (2010) An antimicrobial investigation of plants used traditionally in southern Africa to treat sexually transmitted infections. J Ethnopharmacol 130(3):552–558

Vernin G, Vernin E, Vernin C, Metzger J (1991) Extraction and GC-MS-SPECMA data bank analysis of the aroma of *Psidium guajava* L. fruit from Egypt. Flavour Fragr J 6:143–148

Vieira RHSF, Rodrigues DP, Gonçalves FA, Menezes FGR, Aragão JS, Sousa OV (2001) Microbicidal effect of medicinal plant extracts (*Psidium guajava* Linn. and *Carica papaya* Linn.) upon bacteria isolated from fish muscle and known to induce diarrhea in children. Rev Inst Med Trop S Paulo 43(3):145–148

Wang B, Liu HC, Hong JR, Li HG, Huang CY (2007) Effect of *Psidium guajava* leaf extract on α-glucosidase activity in small intestine of diabetic mouse. Sichuan Da Xue Xue Bao Yi Xue Ban 38(2):298–301, in Chinese

Wang B, Liu HC, Ju CY (2005) Study on the hypoglycemic activity of different extracts of wild *Psidium guajava* leaves in Panzhihua area. Sichuan Da Xue Xue Bao Yi Xue Ban 36:858–861, in Chinese

Wilson CW (1980) Guava. In: Nagy S, Shaw PE (eds) Tropical and sub-tropical fruits: composition, properties, and uses. AVI, Westport, pp 279–299, 270 pp

Wilson CW III, Shaw PE (1978) Terpene hydrocarbons from *Psidium guajava*. Phytochemistry 17(8):1435–1436

Wu J-W, Hsieh C-L, Wang HY, Chen H-Y (2009) Inhibitory effects of guava (*Psidium guajava* L.) leaf extracts and its active compounds on the glycation process of protein. Food Chem 113(1):78–84

Yamashiro S, Noguchi K, Matsuzaki T, Miyagi K, Nakasone J, Sakanashi M, Sakanashi M, Kukita I, Aniya Y, Sakanashi M (2003) Cardioprotective effects of extracts from *Psidium guajava* L and *Limonium wrightii*, Okinawan medicinal plants, against ischemia-reperfusion injury in perfused rat hearts. Pharmacology 67(3):128–135

Yang XL, Hsieh KL, Liu JK (2007) Guajadial: an unusual meroterpenoid from guava leaves *Psidium guajava*. Org Lett 9(24):5135–5138

Yang XL, Hsieh KL, Liu JK (2008) Diguajadial: a dimer of the meroterpenoid from the leaves of *Psidium guajava* (guava). Chin J Nat Med 6(5):333–335

Zhang WJ, Chen BT, Wang CY, Zhu QH, Mo ZX (2003) Mechanism of quercetin as an antidiarrheal agent. Di Yi Jun Yi Da Xue Xue Bao 23(10):1029–1031, in Chinese

Psidium guineense

Scientific Name

Psidium guineense Sw.

Synonyms

Campomanesia multiflora (Cambess.) O.Berg, *Campomanesia tomentosa* Kunth, *Eugenia hauthalii* (Kuntze) K.Schum., *Guajava albida* (Cambess.) Kuntze, *Guajava benthamiana* (O.Berg) Kuntze, *Guajava costa-ricensis* (O. Berg) Kuntze, *Guajava guineensis* (Sw.) Kuntze, *Guajava laurifolia* (O.Berg) Kuntze, *Guajava mollis* (Bertol.) Kuntze, *Guajava multiflora* (Cambess.) Kuntze, *Guajava ooidea* (O.Berg) Kuntze, *Guajava polycarpa* (Lamb.) Kuntze, *Guajava schiedeana* (O.Berg) Kuntze, *Guajava ypanemensis* (O.Berg) Kuntze, *Mosiera guineensis* (Sw.) Bisse, *Myrcianthes irregularis* McVaugh, *Myrtus guineensis* (Sw.) Kuntze, *Myrtus hauthalii* Kuntze, *Psidium albidum* Cambess., *Psidium araca* Raddi, *Psidium araca* var. *sampaionis* Herter, *Psidium benthamianum* O.Berg, *Psidium campicolum* Barb.Rodr., *Psidium chrysobalanoides* Standl., *Psidium costa-ricense* O.Berg, *Psidium dichotomum* Weinm., *Psidium laurifolium* O.Berg, *Psidium lehmannii* Diels, *Psidium minus* Mart. ex DC. nom. inval., *Psidium molle* Bertol., *Psidium molle* var. *gracile* O.Berg, *Psidium molle* var. *robustum* O.Berg, *Psidium monticola* O.Berg, *Psidium monticola* var. *gracile* O.Berg, *Psidium monticola* var. *robustum* O.Berg, *Psidium multiflorum* Cambess., *Psidium ooideum* O.Berg, *Psidium ooideum* var. *grandifolium* O.Berg, *Psidium ooideum* var. *intermedium* O.Berg, *Psidium ooideum* var. *longipedunculatum* Rusby, *Psidium ooideum* var. *parvifolium* O.Berg, *Psidium polycarpon* Lamb., *Psidium popenoei* Standl., *Psidium rotundifolium* Standl., *Psidium rufinervum* Barb.Rodr., *Psidium schiedeanum* O.Berg, *Psidium schippii* Standl., *Psidium sericiflorum* Benth., *Psidium ypanemense* O.Berg.

Family

Myrtaceae

Common/English Names

Brazilian Guava, Castilian Guava, Guinea Guava, Güisara Guava, Guisaro Sour Guava, Wild Guava

Vernacular Names

Afrikaans: Brasiliaanse Koejawel;
Bolivia: Allpa Guayaba, Araza Del Brazil, Arrayán, Chobo, Diondan, Guayaba, Guayaba Hedionda, Guayaba Raijana, Guayaba Silvestre, Guayabilla, Guyaba, Guyabilla;
Brazil: Araçá, Araçá Do Campo, Araçá Verdaleiro, Araçá-Í, Araça-Iba, Araçá-Mirim, Araçá-Pedra, Araçazinho, Arafa-Verdaleiro,

T.K. Lim, *Edible Medicinal And Non-Medicinal Plants: Volume 3, Fruits*,
DOI 10.1007/978-94-007-2534-8_96, © Springer Science+Business Media B.V. 2012

Araja, Araja-Iba, Arajazinho, Awi, Araya Do Campo (Portuguese);
Costa Rica: Cas Extranjero, Dionda, Guisaro, Pichippul;
Cuba: Choba;
Ecuador: Allpa Guayaba, Guayabilla;
El Salvador: Guayabillo, Guayabillo De Tierra Fria;
French: Gouyave Acide, Gouyava De Afrique, Gouyave De L'afrique, Gouyave Du Brésil, Goyavier Acide, Goyavier Du Brésil, Goyavier-Fraise;
German: Stachelbeerguave;
Guadeloupe: Goyave France;
Guatemala: Arrayan, Pataj, Pichippul (Maya), Chamach, Chamacch (Quecchi), Dionda, Guayaba, Guayaba Acida, Guayaba Hedionda, Guisaro (Spanish);
Guianas: Wild Guava, Wilde Guave;
Mexico: Guayaba Ágria;
Panama: Guayabita, Guayaba Arraijan, Guayabita De Sabana;
Peru: Guabillo, Guayaba Brava, Huayava, Sacha Guayaba;
Portuguese: Araçá-Azedo, Araçá Do Campo;
Spanish: Guayaba Ágria, Guayabo Sabanero, Guisaro, Orobua;
Venezeula: Guayaba Ácida, Guayaba Agria, Guayaba Cimarrón; Guayaba De Sabana, Guayaba Sabanera, Guayabite De Cerro, Guayabito, Guayabito Aseyajan, Guayabo, Guayabo De Sabana.

Origin/Distribution

P. guineense is indigenous to the American tropics and sub-tropics, occurring naturally from northern Argentina and Peru to southern Mexico, and in the Caribbean – Trinidad, Martinique, Jamaica and Cuba. It is cultivated to a limited extent in Martinique, Guadeloupe, the Dominican Republic and southern California. At Agartala in Tripura, northeast India, this plant has become thoroughly naturalized and runs wild.

Agroecology

The species is found in the humid to semi arid tropics and subtropics in areas receiving between 1,000 and 2,400 mm annual rainfall. It is occasionally very abundant in areas of sparse vegetation, and appears very occasionally in disturbed areas, such as old pastures, abandoned agricultural areas and along roadsides, being distributed by birds or animals. It thrives in full sun on a wide range of soil types but does best on well-drained, moist, fertile loamy soils up to medium elevations. It tolerates poor soils such as light-textured sandy soils but abhors water-logged and saline soils.

Edible Plant Parts and Uses

This fruit is subacid and is suitable for baking, jams, preserves, juices and sweets although the sweeter varieties may be eaten out of hand. It can be processed into a distinctive jelly which is more flavoursome than guava jelly.

Botany

A small, evergreen erect shrub or small tree, 1.5–6 m high with smooth, greyish bark, hairy young shoots and cylindrical branchlets. Leaves are opposite, leathery, greyish-green, ovate to oblong-elliptic, 3.5–14 cm long by 2.5–8 cm wide, cuneate base, rusty-pubescent below, with prominent midrib and 8–12 pairs of lateral nerves, and pellucid-dotted (Plates 1 and 2). Flowers white, regular, bisexual, borne singly or in clusters of three in the axils of leaf on 1–2.5 cm long terete pedicels. Calyx splitting into irregular, pubescent, pale green lobes, corolla with five white spreading concave-obovate lobes, stamens numerous 150- with bilocular anthers, ovary ellipsoid to globose, three to five locular with numerous ovules and style subulate. Fruit ellipsoid to globose, 1.5–2.5 cm wide with persistent calyx, green

Plate 1 Immature, juvenile fruits and leaves

Plate 2 More mature fruits

(Plates 1 and 2) turning to yellow when mature with pale-yellowish-white flesh and numerous small, hard, kidney-shaped, flattened seeds 5–8 mm diameter.

Nutritive/Medicinal Properties

The proximate nutrient composition of araçá fruit per 100 g edible portion has been reported by Caldeira et al. (2004) as: moisture 85.12%, energy 44.5 kcal, ash 0.85 g, reducing sugars (glucose) 4.74, non reducing sugars (sucrose) 0.29 g, lipid 1.02 g, protein 1.00 g, fibre 4.28 g, Ca 26.8 mg, Mg 17.86 mg, P 17.86 mg, K 212.78 mg, Fe 0.36 mg, Zn, 0.16 mg, Na 0.38 mg, Mn 0.3 mg, Cu 0.12 mg, Araçá (*Psidium guineense*) fruit was reported to be a good source of hydrolyzable tannins and/or flavonols (Gordon et al. 2011).

The major constituents found in the oil of *P. guineense* (leaves and tender stems) were β-bisabolol (17.4%), limonene (6.8%) and epi-α-bisabolol (6.7%) (da Silva et al. 2003).

Hot aqueous extract of *P. guineense* fruit perciarp had in-vitro inhibitory activity against *Staphylococcus aureus, Escherichia coli* and *Aspergillus niger* (Anesini and Perez 1993). Ethanol extracts from the peel and pulp of *P. guineense* and its various factions exhibited antibacterial activity against *Streptococcus mutans* strains (Neira Gonzalez et al. 2005). The fractions with the greatest activity were the so-called M1FA (fraction of ethyl acetate of choba fruit's dry green peel) and M3FA (fraction of ethyl acetate of choba fruit's dry green pulp). The antimicrobial activity of the *P. guineense* species may be attributed to secondary metabolites, taninns, flavonoids, terpenes and aldehydes present in the fruit.

Psidium guineense is used in treating diarrhoeas in Argentinean traditional medicine (Anesini and Perez 1993). In the interior of Brazil, a decoction of the bark or roots is employed to treat urinary diseases, diarrhoea and dysentery and is considered to be diuretic. In Costa Rica, the decoction is said to reduce varicose veins and ulcers on the legs. A leaf decoction is taken to relieve colds and bronchitis.

Other Uses

The wood is strong and resistant and is used for tool handles, beams, posts, planks, utensil and agricultural instruments. It is also used for firewood and charcoal. The bark, rich in tannin, is used for curing hides.

Comments

The plant is propagated from cuttings, air-layering or seeds.

Selected References

Anesini C, Perez C (1993) Screening of plants used in Argentine folk medicine for antimicrobial activity. J Ethnopharmacol 39(2):119–128

Caldeira SD, Hiane PA, Ramos MIL, Ramos Filho MM (2004) Caracterização físico-química do araçá (*Psidium guineense* SW) e do tarumã (*Vitex cymosa* Bert) do estado de Mato Grosso do Sul. Physical-chemical characterization of araça (*Psidium guineense* SW) and Tarumã (*Vitex cymosa* Bert) of Mato Grosso do Sul State (Brazil). Bol Centro Pesqui Process Aliment 22(1):145–154, In Portuguese

da Silva JD, Luz AIR, da Silva MHL, Andrade EHA, Zoghbi MDGB, Maia JGS (2003) Essential oils of the leaves and stems of four *Psidium* spp. Flav Frag J 18(3):240–243

FAO (1986) Food and fruit bearing forest species 3. Examples from Latin America, vol 44/3, FAO forestry paper. FAO, Rome, 308 pp

Gordon A, Jungfer E, da Silva BA, Maia JG, Marx F (2011) Phenolic constituents and antioxidant capacity of four underutilized fruits from the Amazon region. J Agric Food Chem 59(14):7688–7699

Govaerts R, Sobral M, Ashton P, Barrie F, Holst BK, Landrum LL, Matsumoto K, Fernanda Mazine F, Nic Lughadha E, Proenca C, Soares-Silva LH, Wilson PG, Lucas E (2010) World checklist of Myrtaceae. The board of trustees of the Royal Botanic Gardens, Kew. Published on the Internet, http://www. kew.org/wcsp/. Accessed 22 Apr 2010

Jansen PCM, Jukema J, Oyen LPA, van Lingen TG (1992) *Psidium guineense* Swartz. In: Verheij EWM, Coronel RE (eds) Plant resources of South-East Asia No. 2: edible fruits and nuts. Prosea Foundation, Bogor, pp 353–354

Lim TK, Khoo KC (1990) Guava in Malaysia production, pests and diseases. Tropical Press, Kuala Lumpur, 260 pp

McVaugh R (1963) Flora of Guatemala. Myrtaceae. Fieldiana Bot 24 7(3):283–405

Morton JF (1987) Brazilian guava. In: Fruits of warm climates. Julia F. Morton, Miami, pp 365–367

Neira Gonzalez AM, Ramirez Gonzalez MB, Sanchez Pinto NL (2005) Phytochemical study and antibacterial activity of *Psidium guíneense* Sw (choba) against *Streptococcus mutans*, causal agent of dental caries. Rev Cubana Plant Med 10:3–4

Sánchez-Vindas PE (2001) *Calycolpus*, *Eugenia*, *Myrcia*, *Myrcianthes*, *Myrciaria*, *Pimenta*, *Plinia*, *Psidium*, *Syzygium*, *Ugni*. In: Stevens WD, Ulloa C, Pool A, Montiel OM (eds) Flora de Nicaragua. Monogr Syst Bot Missouri Bot Gard 85(2):1566, 1570–1574, 1575–1580

Rhodomyrtus tomentosa

Scientific Name

Rhodomyrtus tomentosa (Aiton) Hassk.

Synonyms

Cynomyrtus tomentosa (Aiton) Scriv., *Myrtus tomentosa* Aiton (basionym).

Family

Myrtaceae

Common/English Names

Australia Murta, Ceylon Hill Cherry, Ceylon Hill Gooseberry, Downy Myrtle, Downy Rose Myrtle, Fluffy Blueberry, Hill Guava, Isenberg Bush, Rhodomyrtus, Rose Myrtle, Tomentose Rose Myrtle

Vernacular Names

Australia: Australia Murta, Fluffy Blueberry;
Brunei: Karamunting, Keramunin;
Chinese: Tao Jun Liang, Tao Chun Liang, Tao Jin Niang, Tao Chin Niang, Gang Shen (Hong Kong);
French: Feijoa, Feijoarte-Groseille, Myrte-Groseille;
Hawaii: Isenberg Bush;
India: Tavute-Gida (Kannada) Taviṭṭu-Kkoyyā, Taviṭṭu-Cceṭi, Taviṭṭu-Ppaẓam (Tamil);
Indonesia: Kemunting (Java), Harendong Sabrang (Sundanese);
Kampuchea: Sragan;
Malaysia: Kemunting (Peninsular), Dundorok, Dunduok, Karamunting (Sabah), Karamunting, Lidah Katak (Sarawak);
Philippines: Rose Myrtle;
Singapore: Kemunting;
Spanish: Guayabillo Forastero;
Sri Lanka: Sitha Pera (Sinhalese);
Thai: Kamunting, Phruat, To, Trat;
Vietnamese: Sim, Hồng Sim, Dương Lê, Co Min (Thái), Mác Nim (Tày), Piêu Nim (Dao).

Origin/Distribution

Downy Rose Myrtle is native to tropical and subtropical Asia – southern Asia and southeast Asia, from India, east to southern China, Taiwan and the Philippines, and south to Malaysia and Sulawesi.

Agroecology

Downy rose myrtle is adaptable to a wide range of elevation and environmental conditions, including slight freezes and salt spray. It can tolerate temperatures down to –7°C. It thrives in environments where annual rainfall exceeds

T.K. Lim, *Edible Medicinal And Non-Medicinal Plants: Volume 3, Fruits*,
DOI 10.1007/978-94-007-2534-8_97, © Springer Science+Business Media B.V. 2012

1,200 mm. In its native range, it occurs in beach forest, shorelines, natural forest, belukars (cleared land that has reverted back to jungle), riparian zones, wetlands, moist and wet forests, bog margins, sometimes forming thick dense thickets, from sea level up to 2,700 m elevation. It is also found as a weed of pasture, rangeland and untended areas. It produces profuse quantities of seeds that germinates readily. In addition, downy rose myrtle is fire adapted and can resprout prolifically after fires.

Edible Plant Parts and Uses

The sweet ripe fruit are consumed fresh or made into pies, tarts, jellies, preserves and jams, or used in salads. In Vietnam, the fruits are used to produce a wine called *rượu sim.*

Botany

A small evergreen shrub 1–2 m high with tomentose cylindrical branchlets (Plate 1). Leaves opposite, sub-coriaceous, elliptic to ovate-elliptic or obovate-elliptic, 5–8 cm long, 1.5–4 cm wide, 3(–5)-nerved, apex obtuse, base obtuse, upper surface glossy and glabrate, lower surface densely tomentose, greyish, petioles 0.4–1 cm long (Plates 4 and 5). Flowers, bisexual, pink 1–3(–5) in axillary cymes, each one subtended by 2 small bracts; sepals 4–5 orbicular, tomentose lobes, 3–4 mm long, petals four–five tomentose rounded petals rose pink, ca.10–13 mm long; stamens numerous, ovary inferior, 3-locular, cell with two rows of ovules, style filiform, stigma capitates (Plates 2 and 3). Fruit sub-globose, green (Plates 4 and 5) turning purple-violet when ripe, 12–14 mm diameter capped by persistent sepals and has many (40–45) tiny, deltoid seeds embedded in an edible flesh.

Two varieties are distinguished (Latiff 1992): – *R. tomentosa* var. *tomentosa* (synonym *Myrtus canescens* Lour.), occurring in South-East Asia, southern China and Indo-China; with whitish-tomentose leaves, apex rounded or obtuse not apiculate, veins not reticulate, pedicels 1–2.5 cm long; – *R. tomentosa* var. *parviflora* (Alston) A.J. Scott (synonym *Rhodomyrtus parviflora* Alston), occurring in India and Sri Lanka; with cream- or yellowish-tomentose leaves, apex apiculate, veins reticulate, pedicels less than 1 cm long.

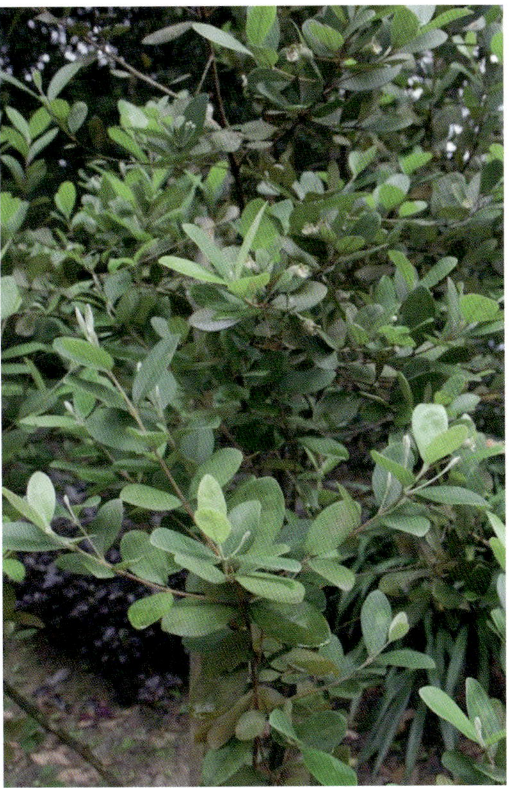

Plate 1 Shrub habit of rose myrtle

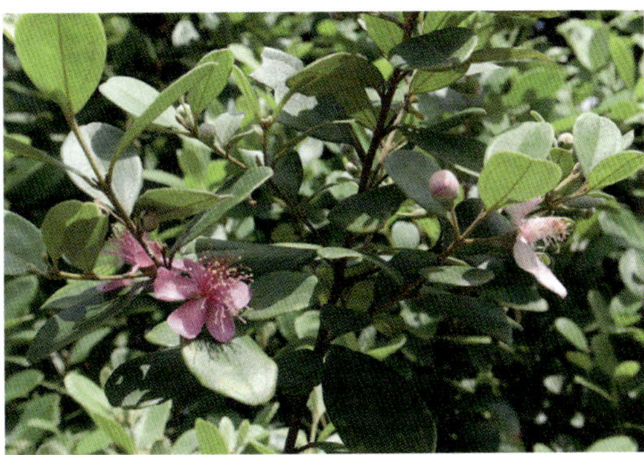

Plate 2 Rose-pink flowers of rose myrtle

Plate 3 Close-up of flower

Plate 4 Young fruits and leaves (upper surface)

Plate 5 Young fruit and leaves (lower surface)

Nutritive/Medicinal Properties

According to the National Institute of Materia Medica, Hanoi, Vietnam (1999), the fruit of *Rhodomyrtus tomentosa* contained 82.5% water, 0.6% proteins, 0.2% fat, 10.7% carbohydrates, 40 mg% Ca, 15 mg% P, 0.9 mg% Fe, 24 IU/100 g vitamin A, 0.07 mg% vitamin B 12, 0.04 mg% riboflavin and 0.3 mg% niacin. The sugar composition of the fruit was reported as 16.3% glucose, 18.1% sucrose dry pulp weight (Ko et al. 1998).

The petroleum extracts of *Rhodomyrtus tomentosa* yielded the following triterpenoids, R4 from the leaves and R5 from the stems besides R1, R2, R3 and the other known compounds already reported (Hui and Li 1976). R1 and R4 were found to be 21αH -hop-22(29)en-3β, 30-diol and 3β-hydroxy-21αH -hop-22(29)-en-30-al respectively, and R2, R3 and R5 were 3β-acetoxy-11α,12α-epoxyoleanan-28,13β-olide, 3β-acetoxy-12α-hydroxyoleanan-28, 13β-olide and 3β-acetoxy-12-oxo-oleanan-28,13β- olide respectively. The ethanol extract of the leaves contained betulinic, ursolic and aliphitolic acids and that of the stems betulonic, betulinic and oleanolic acids. Acylphloroglucinols named rhodomyrtosones A–C and a lepstospermone derivative named rhodomyrtosone D were isolated from the acetone extract of the leaves of *Rhodomyrtus tomentosa* together with six known compounds (Hiranrat and Mahabusarakam 2008).

Aerial parts of the plant contain bioactive phytochemicals with osteogenic and antimicrobial activities.

Osteogenic Activity

Two new anthracene glycosides (1, 2) were isolated from aerial parts of *Rhodomyrtus tomentosa*, along with three known compounds (3–5)

(Tung et al. 2009). The structures of two new compounds were established to be 4,8,9,10-tetrahydroxy-2,3,7-trimethoxyanthracene-6-O-β-D-glucopyranoside (1) and 2,4,7,8,9,10-hexahydroxy-3-methoxyanthracene-6-O-α-L-rhamnopyranoside (2). Among them, compound 1, 2, and 5 significantly increased the alkaline phosphatase activity, collagen synthesis, and mineralization of the nodules of MC3T3-E1 osteoblastic cells compared to those of the control, respectively. Osteoblastic differentiation is an important step of bone formation.

Antimicrobial Activity

The ethyl acetate extract of the leaves of *Rhodomyrtus tomentosa* yielded rhodomyrtone [6,8-dihydroxy-2,2,4,4-tetramethyl-7-(3-methyl-1-oxobutyl)-9-(2-methylpropyl)-4,9-dihydro-1 H-xanthene-1,3(2 H)-di-one] which was found to have significant antibacterial activity against *Escherichia coli* and *Staphylococcus aureus* (Salni et al. 2002). Rhodomyrtone was found to display significant antibacterial activities against Gram-positive bacteria including *Bacillus cereus, Bacillus subtilis, Enterococcus faecalis, Staphylococcus aureus*, methicillin-resistant *S. aureus (MRSA), Staphylococcus epidermidis, Streptococcus gordonii, Streptococcus mutans, Streptococcus pneumoniae, Streptococcus pyogenes*, and *Streptococcus salivarius* (Limsuwan et al. 2009). Especially noteworthy was the activity against MRSA with a minimum inhibitory concentration (MIC) and a minimum bactericidal concentration (MBC) ranging from 0.39 to 0.78 μg/ml. As shown for *S. pyogenes*, no surviving cells were detected within 5 and 6 h after treatment with the compound at 8MBC and 4MBC concentrations, respectively. Rhodomyrtone displayed no bacteriolytic activity. A rhodomyrtone killing test with *S. mutans* showed that this compound caused a few morphological changes as the treated cells were slightly changed in color and bigger than the control when they were killed. The results supported the view that rhodomyrtone had a strong bactericidal activity on Gram-positive bacteria, including major pathogens. Studies showed that sub-inhibitory concentrations [1/32–1/2 minimal inhibitory concentration (MIC)] *R. tomentosa* (0.24–7.81 μg/ml) extracts significantly prevented biofilm formation by *Streptococcus pyogenes* (Limsuwan and Voravuthikunchai 2008). Biofilm formation had been demonstrated as a potentially important mechanism contributing to antibiotic treatment failure on *Streptococcus pyogenes*. It could play a significant role in recurrent and chronic infections. Recent studies showed that Rhodomyrtone had both antimicrobial and anti-infective activities (Limsuwan et al. 2011). Cells of *Streptococcus pyogenes* grown in the presence of rhodomyrtone produced reduced amounts of known virulence factors, such as the glyceraldehyde-3-phosphate dehydrogenase, the CAMP factor, and the streptococcal pyrogenic exotoxin C.

A correlation between antiquorum-sensing and antibiofilm-producing activities was demonstrated. Strong inhibition on quorum sensing was displayed with the extract of *R. tomentosa.* Rhodomyrtone demonstrated pronounced antibacterial activity on staphylococci isolated from acne lesions (Saising et al. 2008). The average inhibition zones of 64 coagulase-positive and 85 coagulase-negative isolates were 12 and 14 mm, respectively. The MIC_{50} and MIC_{90} ranged from 64–512 μg/ml. Rhodomyrtone was very effective against *Staphylococcus aureus* ATCC 25923 with the MIC value at 0.5 μg/ml which is very closed to that of vancomycin. The results indicated the potential use of rhodomyrtone as an alternative agent for staphylococcal cutaneous infections.

Ethanol extract from *Rhodomyrtus tomentosa* leaves was found to be active against *Bacillus cereus* (Voravuthikunchai et al. 2010), with MIC (minimum inhibitory concentration) and MBC (minimum bactericidal concentration) of 16–64 and 32–256 μg/ml, respectively. Rhodomyrtone, a purified compound from the extract, exhibited MIC and MBC at 0.5 and at 2–8 μg/ml, respectively. At two MICs and four MICs, a reduction in the viability of the bacterial cells and endospores was observed within 6–8 h and 2 h after incubation, respectively. Application of the extract in precooked rice and tuna steak demonstrated that after exposure to 16 MICs and 32 MICs, the numbers of viable cells and endospores in both model systems were reduced by at least 2 log within 12 and 6 hours, respectively. The researchers asserted that since the extract consistently produced remarkable activity against both cells and endospores, it could be used as an alternative food additive for controlling *B. cereus* without compromising food safety.

Traditional Medicinal Uses

Roots, leaves and fruits have been utilised in folk medicine. In Peninsular Malaysia, the fruits have been recorded as treatment for dysentery and diarrhoea. The roots and leaves were also used to treat diarrhoea, stomach aches and as a tonic after childbirth. An infusion of the roots was recommended for squirting into the eyes to treat scars on the cornea while a poultice of the leaves was applied on the forehead. In Singapore, the Chinese have used the leaves as a pain killer, the roots to treat heartburn and seeds in a tonic for digestion, and to treat snake bites. In Indonesia, the leaves have been used to treat wounds. Root decoctions are drunk as tonic in Sabah and Sarawak. In Sarawak, pounded leaves with betel leaves and betel nut have been used as poultice on abdomen to mitigate diarrhoea in babies. In Vietnam, buds, young leaves in the form of an extracted juice or dried are used for use as a powder or decoction top treat in colic, diarrhoea, dysentery, abscesses, furunculosis and haemorrhage. Concentrated decoction of leaves is used as an antiseptic wash for wounds, impetigo and abscesses. A combination of the ripe fruit and *Caesalpinia sappan* is also similarly prescribed.

Other Uses

Downy rose myrtle is a popular ornamental plant in gardens in tropical and subtropical areas, grown for its abundant and much prized flowers and sweet, edible fruit despite its invasive habit. It has shown promise as a fire retardant species for use in fire breaks in the Himalayas. The red wood is fine-grained and used for making small objects. The wood with coconut shell is used for making wood tar for blackening teeth and eyebrow. In Peninsular Malaysia, the leaves are used in magic ceremonies when epidemics are being exorcised.

Comments

In many areas like Hawaii and Florida, downy rose myrtle is considered an invasive noxious weed.

Selected References

Ahmad FB, Holdwoldsworth DK (1994) Medicinal plants of Sarawak, Malaysia, Part I. The Kedayans. Pharm Biol 32(4):384–387

Burkill IH (1966) A dictionary of the economic products of the Malay Peninsula. Revised reprint. 2 volumes. Ministry of Agriculture and Co-operatives, Kuala Lumpur, Vol 1 (A-H), pp 1–1240, Vol 2 (I-Z), pp 1241–2444

Chai PPK (2006) Medicinal plants of Sarawak. Lee Ming Press, Kuching, 212 pp

Govaerts R, Sobral M, Ashton P, Barrie F, Holst BK, Landrum LL, Matsumoto K, Fernanda Mazine F, Nic Lughadha E, Proenca C, Soares-Silva LH, Wilson PG, Lucas E (2010) World checklist of Myrtaceae . The board of trustees of the Royal Botanic Gardens, Kew. Published on the Internet, http://www. kew.org/wcsp/. Accessed 22 Apr 2010

Hiranrat A, Mahabusarakam W (2008) New acylphloroglucinols from the leaves of *Rhodomyrtus tomentosa*. Tetrahedron 64(49):11193–11197

Hu SY (2005) Food plants of China. The Chinese University Press, Hong Kong, 844 pp

Hui WH, Li MM (1976) Two new triterpenoids from *Rhodomyrtus tomentosa*. Phytochem 15(11):1741–1743

Keng H, Chin SC, Tan HTW (1990) The concise flora of Singapore: gymnosperms and dicotyledons. Singapore University Press, Singapore, 222 pp

Ko IWP, Corlett RT, Xu RJ (1998) Sugar composition of wild fruits in Hong Kong, China. J Trop Ecol 14:381–387

Langeland KA, Burks KC (eds) (1998) Identification and biology of non-native plants in Florida's natural areas. University Press of Florida, Gainesville, 165 pp

Latiff AM (1992) *Rhodomyrtus tomentosa* (Aiton) Hassk. In: Verheij EWM, Coronel RE (eds) Plant resources of South-East Asia No 2. Edible fruits and nuts. Prosea, Bogor, pp 276–277

Ledin RB (1957) Tropical and subtropical fruits in Florida (other than *Citrus*). Econ Bot 11:349–376

Limsuwan S, Voravuthikunchai SP (2008) *Boesenbergia pandurata* (Roxb.) Schltr., *Eleutherine americana* Merr. and *Rhodomyrtus tomentosa* (Aiton) Hassk. as antibiofilm producing and antiquorum sensing in *Streptococcus pyogenes*. FEMS Immunol Med Microbiol 53(3):429–436

Limsuwan S, Trip EN, Kouwen TR, Piersma S, Hiranrat A, Mahabusarakam W, Voravuthikunchai SP, van Dijl JM, Kayser O (2009) Rhodomyrtone: a new candidate as natural antibacterial drug from *Rhodomyrtus tomentosa*. Phytomed 16(6–7):645–651

Limsuwan S, Hesseling-Meinders A, Voravuthikunchai SP, van Dijl JM, Kayser O (2011) Potential antibiotic and anti-infective effects of rhodomyrtone from *Rhodomyrtus tomentosa* (Aiton) Hassk. on *Streptococcus pyogenes* as revealed by proteomics. Phytomed 18(11):934–940

National Institute of Materia Medica (1999) Selected medicinal plants in Vietnam, vol 2. Science and Technology Publishing House, Hanoi, 460 pp

Saising J, Hiranrat A, Mahabusarakam W, Ongsakul M, Voravuthikunchai SP (2008) Rhodomyrtone from *Rhodomyrtus tomentosa* (Aiton) Hassk. as a natural antibiotic for staphylococcal cutaneous infections. J Health Sci 54(5):589–595

Salni D, Sargent MV, Skelton BW, Soediro I, Sutisna M, White AH, Yulinah E (2002) Rhodomyrtone, an antibotic from *Rhodomyrtus tomentose*. Aust J Chem 55(3):229–232

Scott AJ (1978) A review of *Rhodomyrtus* (Myrtaceae). Kew Bull 33:311–329

Tung NH, Ding Y, Choi EM, Van Kiem P, Van Minh C, Kim YH (2009) New anthracene glycosides from *Rhodomyrtus tomentosa* stimulate osteoblastic differentiation of MC3T3-E1 cells. Arch Pharm Res 32(4):515–520

Voravuthikunchai SP, Dolah S, Charernjiratrakul W (2010) Control of *Bacillus cereus* in foods by *Rhodomyrtus tomentosa* (Ait.) Hassk. leaf extract and its purified compound. J Food Prot 73(10): 1907–1912

Wagner WL, Herbst DR, Sohmer SH (1999) Manual of the flowering plants of Hawaii. Revised edition. Bernice P. Bishop Museum special publication. University of Hawai'i Press/Bishop Museum Press, Honolulu, 1919 pp (two volumes)

Waterhouse BM, Mitchell AA (1998) Northern Australia Quarantine Strategy: Weeds Target List, 2nd edn. Australian Quarantine & Inspection Service, Miscellaneous Publication No. 6/98, 110 pp

Wee YC (1992) A guide to medicinal plants. The Singapore Science Centre, Singapore, 160 pp

Syzygium aqueum

Scientific Name

Syzygium aqueum (Burm. f.) Alston.

Synonyms

Cerocarpus aqueus Hassk., *Eugenia alba* Roxb., *Eugenia aquea* Burm. f., *Eugenia callophylla* (Miq.) Reinw. ex de Vriese, *Eugenia malaccensis* Lour. nom. illeg., *Eugenia mindanaensis* C.B. Robinson, *Eugenia nodiflora* Aubl., *Eugenia observa* Miq., *Eugenia stipularis* (Blume) Miq., *Gelpkea stipularis* Blume, *Jambosa alba* (Roxb.) G.Don, *Jambosa ambigua* Blume, *Jambosa aquea* (Burm. f.) DC., *Jambosa calophylla* Miq., *Jambosa madagascariensis* Blume, *Jambosa obtusissima* (Blume) DC., *Jambosa subsessilis* Miq., *Jambosa timorensis* Blume, *Malidra aquea* Raf., *Myrtus obtusissima* Blume, *Myrtus timorensis* Zipp. ex Span., *Syzygium obversum* (Miq.) Masam.

Family

Myrtaceae

Common/English Names

Bell Apple, Bell Fruit, Water Apple, Water Cherry, Watery Rose Apple

Vernacular Names

Brazil: Jambeiro Aguado, Jambo Branco, Jambo D'agua (Portuguese);
Chinese: Shui Lian Wu;
Dominican Republic: Cajuilito Solimán (Spanish);
Dutch: Djamboe Aer;
French: Jambosier D'eau, Jambolanier D'eau, Pomme D'eau, Pomme De Java;
German: Wachsjambuse, Wasserjambuse;
Indonesia: Jambi Iye, Jambi Pira, Jambi Raya (Aceh), Jambu Er, Njamu Er (Bali), Jambu Aek, Jambu Air, Jambu Erang (Batak), Jambu Ayik (Besemah), Jambu Jene (Bimba), Jambu Salo, Jambu Saio (Bugis), Takaw (Boeol, Manado, Sulawesi), Kepet, Lutune Waele, O'uno, Popte, Tepete (Ceram, Ambon, Moluccas), Kubal (Dyak, Kalimantan), Jambu Airjemer (Gajo), Omuto, Upo (Gorontalo), Rowane, Yarem (Halmaheira), Jambu Pingping (Jambi), Jambu Air, Jambu Wer, Jambu Uwer (Java), Jambu Air, Jambu Ayor, Jambu Kelinga, Jambu Wai (Lampong), Jambu Bertih (Lingga), Jambhu Wir (Madurese), Jambu Jene (Makassar), Jambu Keling (Malay, Bengkoelen), Gora (Manado), Jambu Aye (Minangkabau), Samba (Nias), Jambu Waelo, Kuputol Waelo, Rutu Putio (Oelias), Ansahmoh, Purori (Papua), Compose (Sangir), Wo Luba Kume (Sawoe), Jambu Aer (Singkep), Jambu Ayak (Serawai), Bluwa, Bluwo (Solor), Kebes, Kembes, Kouoa, Kombas, Kumpas, Kumpasa, Mangkoa (Sulawesi, Moluccas) Jambu Air

T.K. Lim, *Edible Medicinal And Non-Medicinal Plants: Volume 3, Fruits*,
DOI 10.1007/978-94-007-2534-8_98,

(Sundanese), Gora Yadi (Ternate), Yadi (Tidore), Jambu, Gambu, Wua Usa (Toraja);
Japanese: Mizu Renbu;
Malaysia: Jambu Chili, Jambu Ayer, Jambu Ayer Mawar, Jambu Penawar;
Papua New Guinea: Lal Lau;
Philippines: Tambis (Bisaya);
Spanish: Perita Costena, Manzana De Agua;
Suriname: Pommerak;
Thai: Chom Pu Pa, Machomphu-Pa.

Origin/Distribution

The species has its origin in tropical Asia to north Queensland. It is commonly cultivated in India, southeast Asia and in the Pacific Islands. In the Philippines, it grows as though wild in the Provinces of Mindanao, Basilan, Dinagat and Samar. It is occasionally grown in Trinidad and Hawaii.

Agroecology

Water apple is tropical in its growth requirement, thriving in warm, wet and humid areas with well distributed annual rainfall. It is found in fairly moist tropical lowlands up to 1,200 m elevation. Water apple grows best in areas with a dry season but with supplemental water supply. The trees prefer heavy soils with- easy access to water instead of having to search for water in light deep soils.

Plate 1 Leaves and fruits of water apple

Plate 2 Clusters of ripe bell-shaped fruits

Edible Plant Parts and Uses

Fruits are eaten fresh out of hand or eaten in *rujak* (fruit salad) in Malaysia and Indonesia. In Indonesia, the fruits are also preserved by pickling (*asinan*). The young leaves are edible. Indonesians use young water apple leaves to wrap snacks of fermented sticky rice.

Botany

A low, evergreen tree, 3–10 m tall, with short and crooked trunk, 30–50 cm diameter, brown flaky bark, often branching near the base and with irregular crown and terete, flaky branches. Leaves opposite, elliptic-cordate to obovate-oblong, 7–25 cm × 2.5–16 cm, base cordate, rounded or obtuse, apex obtuse or obtusely acuminate, entire, glabrous on both surfaces, pellucid, faintly fragrant when bruised, pinkish when young turning light green petiole 0.5–1.5 mm long (Plates 1 and 2). Cymes terminal and axillary, 3–7-flowered; flowers 2.5–3.5 cm in diameter, white; calyx tube 5–7 mm long, 4-fid; petals four, spathulate, reflexed up to 7 mm long, yellowish-white; stamens numerous with white, filiform filaments 0.75–2 cm long, anthers orbicular-ovate

Plate 3 Water apple on sale in the market

yellowish-white; style up to 17 mm long, yellowish white. Fruit a berry, turbinate (bell-shaped), 1.5–2 cm × 2.5–3.5 cm, with a shortly tapered base and a much expanded apex, crowned by the fleshy calyx segments which are curved over the concave apex of the fruit, leaving a small uncovered area, white, pink to red (Plates 1, 2, and 3), glossy; flesh white, very juicy, crisp or spongy, faintly aromatic. Seeds 1–2(–6), rounded, small.

Nutritive/Medicinal Properties

Nutrient composition of water-apple fruit per 100 g of edible portion was reported by Dignan et al. (1994) as energy 68 kJ (17 kcal), moisture 95%, protein 0.8 g, fat 0.1 g, available carbohydrate 3 g, fibre 1.3 g, ash 0.7 g, Ca 2 mg, P 13 mg, Fe 0.2 mg, Na 1 mg, K 48 mg, total vitamin A equivalent 1 μg, β-carotene equivalent 7 μg, thiamin 0.044 mg, vitamin C 16.7 mg and vitamin E traces.

Another analysis of water-apple fruit reported per 100 g of edible portion conducted in Malaysia by Tee et al. (1997) is reported as: energy 17 Kcal, water 95 g, protein 0.8 g, fat 0.1 g carbohydrate 3.1 g, fibre 0.8 g, ash 0.2 g, Ca 2 mg, P 5 mg, Fe 0.2 mg, Na 1 mg, K 48 mg, carotenes 7 μg, RE 1 μg, vitamin B1 0.04 mg and vitamin C 16.7 mg.

A total of 42 volatile constituents were identified in water apple fruit, the largest number and proportion (41.4%) were terpenoids, among which γ-terpinene was dominant (Wong and Lai 1996). Water apple fruit also contained samarangenins A and B, novel proanthocyanidins with doubly bonded structures (Nonaka et al. 1992).

The fruit and leaves have been reported to have antioxidant property and the leaves cosmeceutical properties.

Antioxidant Activity

Recent research reported that that guava, papaya and star fruit had higher primary antioxidant potential, as measured by scavenging DPPH and iron(III) reducing assays. Banana, star fruit, water apple, langsat and papaya were found to have higher secondary antioxidant potential as measured by the iron(II) chelating assay. (Lim et al. 2007). The methanol extracts of *S. aquem* fruits showed that the antioxidant activity increased gradually during postharvest ripening (Tehrani et al. 2011). The total phenol content determined by the Folin-Ciocalteau method revealed a high concentration of phenol content in the fruits, with values around 344.25 mg gallic acid equivalent (GAE)/100 g fresh fruit. Similarly, the flavonoid content showed an increasing trend over the same period.

The phenolic content of *S. aquem* leaf was found to be 186 mg/g gallic acid (GA) equivalent in the aqueous extract and 534 mg/g GA eq in the ethanol extract (Ling et al. 2010b). The free radical scavenging activity of *S. aquem* leaf ethanol extract as measured by DPPH assay was IC_{50} of 4.65 mg/ml, 11.98 mg/ml as measured by the Galvinoxyl assay and 34.52 mg/ml by ABTS assay. The free radical scavenging activity of *S. aquem* leaf aqueous extract as measured by DPPH assay was IC_{50} of 3.07 mg/ml, 6.68 mg/ml as measured by the Galvinoxyl assay and 5.22 mg/ml by ABTS assay. The IC_{50} for inhibition of lipid peroxidation for the aqueous extract and ethanol extract of *S. aqueum* leaf was 1.2 and 2.4 mg/g. The ethanolic extract of *Syzygium aqueum* leaf was found to have a lower Pro-Antidex than the commercially available Emblica™ extract, an antioxidant agent with very low pro-oxidant activity (Ling et al. 2010a). The ProAntidex helps in the identification of extracts with high net free radical scavenging activity potential and is beneficial as a screening parameter to the food industries and healthcare. Palanisamy et al. (2011) reported *S. aqueum* leaf extracts to have a significant composition of phenolic compounds, protective activity against free radicals as well as low pro-oxidant capability. Its ethanolic extract, in particular, was characterized by its excellent radical scavenging activity of EC_{50} of 133 μg/ml 1,1-diphenyl-2-picryl-hydrazyl (DPPH), 65 μg ml – 1 2,2′-azino-bis(3-ethylbenzthiazoline-6-sulphonic acid) (ABTS) and 71 μg/ml (Galvinoxyl), low pro-oxidant capabilities and a phenolic content of 585–670 mg GAE/g extract.

Another study reported that *Paederia foetida* and *Syzygium aqueum* leaves could be significant sources of natural antioxidant compounds that may have potent beneficial health effects (Osman et al. 2009). The total antioxidant activities of the crude extracts of DL-α-tocopherol, fresh *P. foetida*, fresh *S. aqueum*, dried *P. foetida*, dried *S. aqueum*, and quercetin after 160 h reaction time were 79.69%, 78.13%, 73.77%, 66.67%, 55.73% and 42.37%, respectively. The fresh samples had higher antioxidant activity than did the dried samples. In this study, the order of antioxidant activity towards β-carotene oxidation was DL-α-tocopherol > fresh *P. foetida* > fresh *S. aqueum* > dried *P. foetida* > dried *S. aqueum* > quercetin. The antioxidant activity of fresh *P. foetida* was the highest and was comparable to DL-α-tocopherol. All of the tested samples more efficiently slowed the bleaching of β-carotene than did quercetin. The fresh *P. foetida* and *S. aqueum* leaf extracts had 70–76% antioxidant activity and the dried samples had 65–68% antioxidant activity which was higher than the activity of the standard commercial antioxidant, quercetin. Scavenging activity increased with the extracts concentration. At 0.035 mg/ml, the order of scavenging activity extracts was: fresh *P. foetida* > fresh *S. aqueum* > dried *P. foetida* > dried *S. aqueum*. A plateau was reached at 0.055 mg/ml with the scavenging activity > 90% for all extracts.

Cosmeceutical Activity

S. aqueum leaf extract was reported to have cosmeceutical properties (Palanisamy et al. 2011). The extract demonstrated substantial tyrosinase inhibition activity with an IC_{50} of about 60 μg/ml. the extract was also found to have anti-cellulite activity tested for its ability to cause 98% activation of lipolysis of adipocytes (fat cells) at a concentration of 25 μg/ml. In addition, the extract was not cytotoxic to Vero cell lines up to a concentration of 600 μg/ml.

Traditional Medicinal Uses

In Malaysian folkloric medicine, a decoction of the astringent bark is a local application for thrush. A water apple salad used to be served at the ceremony after childbirth.

Other Uses

The wood is hard and is used for small handicraft and tools.

Comments

Water apple can be successfully propagated from stem cuttings which gives a good strike rate.

Selected References

Backer CA, Bakhuizen van den Brink RC Jr (1963) Flora of java (Spermatophytes only), vol 1. Noordhoff, Groningen, 648 pp

Burkill IH (1966) A dictionary of the economic products of the Malay Peninsula. Revised reprint. 2 volumes. Ministry of Agriculture and Co-operatives, Kuala Lumpur, Vol. 1 (A—H) pp. 1—1240, Vol. 2 (I—Z) pp. 1241—2444

Chen J, Craven LA (2007) Myrtaceae. In: Wu ZY, Raven PH, Hong DY (eds) Flora of China (Clusiaceae through Araliaceae), vol 13. Science Press/Missouri Botanical Garden Press, Beijing/St. Louis

Dignan CA, Burlingame, Arthur JM, Quigley RJ, Milligan GC (1994) The Pacific Islands food composition tables. South Pacific Commission, Noumea, pp. 147

Govaerts R, Sobral M, Ashton P, Barrie F, Holst BK, Landrum LL, Matsumoto K, Fernanda Mazine F, Nic Lughadha E, Proenca C, Soares-Silva LH, Wilson PG, Lucas E (2010) World checklist of Myrtaceae. The board of trustees of the Royal Botanic Gardens, Kew. Published on the Internet, http://www. kew.org/ wcsp/. Accessed 22 Apr 2010

Lim YY, Lim TT, Tee JJ (2007) Antioxidant properties of several tropical fruits: a comparative study. Food Chem 103(3):1003–1008

Ling LT, Palanisamy UD, Cheng HM (2010a) Prooxidant/antioxidant ratio (ProAntidex) as a better index of net free radical scavenging potential. Molecules 15(11):7884–7892

Ling LT, Radhakrishnan AK, Subramaniam T, Cheng HM, Palanisamy UD (2010b) Assessment of antioxidant capacity and cytotoxicity of selected Malaysian plants. Molecules 15:2139–2151

Morton J (1987) Water apple. In: Fruits of warm climates. Julia F. Morton, Miami, pp 382–383

Nonaka G, Aiko Y, Aritake K, Nishioka I (1992) Tannins and related compounds CXIX. Samarangenins A and B, novel proanthocyanidins with doubly bonded structures, from *Syzygium samarangense* and *S. aqueum*. Chem Pharm Bull 40:2671–2673

Ochse JJ, Bakhuizen van den Brink RC (1931) Fruits and fruitculture in the Dutch East Indies. G. Kolff & Co, Batavia-C, 180 pp

Osman H, Rahim AA, Isa NM, Bakhir NM (2009) Antioxidant activity and phenolic content of *Paederia foetida* and *Syzygium aqueum*. Molecules 14(3):970–978

Palanisamy UD, Ling LT, Manaharan T, Sivapalan V, Subramaniam T, Helme MH, Masilamani T (2011) Standardized extract of *Syzygium aqueum*: a safe cosmetic ingredient. Int J Cosmet Sci 33(3): 269–275

Panggabean G (1992) *Syzygium aqueum* (Burm.f.) Alst., *Syzygium malaccense* (L.) Merr. & Perry, and *Syzygium samarangense* (Blume) Merr. & Perry. In: Verheij EWM, Coronel RE (eds) Plant resources of South-East Asia. No. 2: Edible fruits and nuts. Prosea Foundation, Bogor, pp 292–294

Tee ES, Noor MI, Azudin MN, Idris K (1997) Nutrient composition of Malaysian foods, 4th edn. Institute for Medical Research, Kuala Lumpur, p 299

Tehrani M, Sharif Hossain ABM, Nasrulhaq-Boyce A (2011) Postharvest physico-chemical and mechanical changes in 'jambu air' (*Syzygium aqueum* Alston) fruits. Aust J Crop Sci 5(1):32–38

Wong KC, Lai FY (1996) Volatile constituents from the fruits of four *Syzygium* species grown in Malaysia. Flavour Frag J 11(1):61–66

Syzygium australe

Scientific Name

***Syzygium australe* (J.C.Wendl. ex Link) B. Hyland.**

Synonyms

Eugenia australis J.C.Wendl. ex Link (basionym), *Eugenia myrtifolia* Sims nom. illeg., *Eugenia simmondsiae* F.M.Bailey, *Jambosa australis* (J.C.Wendl. ex Link) DC., *Jambosa myrtifolia* Heynh., *Jambosa thozetiana* F.Muell., *Myrtus australis* (J.C.Wendl. ex Link) Spreng.

Family

Myrtaceae

Common/English Names

Brush Cherry, Creek Lilly Pilly, Creek Satinash, Magenta Cherry, Purple Monkey Apple, Scrub Cherry

Vernacular Names

None

Origin/Distribution

The species is native to eastern Australia. It occurs in the coastal areas from the South Coast of New South Wales to Central Queensland.

Agroecology

A warm temperate species. It grows in all types of warmer coastal and highland rainforest, often along streams, coastal districts in subtemperate and littoral rainforest on sandy soils or stabilised dunes near the sea. It is found in small numbers in widely separated localities in widely separated localities between Bulahdelah and Jervis Bay on the New South Wales coast.

Edible Plant Parts and Uses

A very versatile fruit. The ripe fruit can be eaten raw or made into jams, jellies, relishes, desserts, tarts, pies, cakes and ice-cream. It is also use

T.K. Lim, *Edible Medicinal And Non-Medicinal Plants: Volume 3, Fruits*,
DOI 10.1007/978-94-007-2534-8_99, © Springer Science+Business Media B.V. 2012

fresh in a salad, fruit salad or muesli, or use as a glaze for cooked meat.

Botany

An erect, evergreen shrub or small tree, 3–5 m tall, with flaky bark; young leafy twigs 4-angled to shortly 4-winged, wings joining above each node to produce a small pocket. Leaves are opposite elliptic to oval, 3–10 cm long by 1–3 cm wide, apex short-acuminate, base cuneate, glabrous and glossy (Plate 1), lower surface paler; lateral and intramarginal veins generally visible; oil glands scattered, often faint; petiole 3–10 mm long. Inflorescences are mostly botryoids, terminal and in the upper leaf axils. Petals 4–6 mm long, white, free and spreading, stamens 15–20 mm long, white, numerous. Fruit in bunches of 3–6, obovoid 15–25 mm long by 15 mm across, reddish pink to red (Plate 1); pulp white, juicy and crisp; seed usually solitary, embryo solitary with smooth cotyledons.

Plate 1 Ripe fruit and leaves of brush cherry

Nutritive/Medicinal Properties

Brush cherry has similar antioxidant levels to blueberries. Its anthocyanins were malvidin based glucosides and malvidin 3, 5-diglucoside was the main anthocyanin pigment (Netzel et al.. 2007). Brush cherry has been suggested as a potential source of bioactive phytochemicals for application in health promoting foods.

Other Uses

The plant is used in ornamental landscaping and as hedges.

Comments

Brush cherry is best propagated from stem cuttings as seed germination is slow and sporadic.

Selected References

Carolin R, Tindale M (1993) Flora of the Sydney region. Reed, Sydney

Floyd AG (1989) Rainforest trees of mainland South-eastern Australia. Inkata Press, Melbourne

Govaerts R, Sobral M, Ashton P, Barrie F, Holst BK, Landrum LL, Matsumoto K, Fernanda Mazine F, Nic Lughadha E, Proenca C, Soares-Silva LH, Wilson PG, Lucas E (2010) World checklist of Myrtaceae . The board of trustees of the Royal Botanic Gardens, Kew. Published on the Internet, http://www.kew.org/wcsp/. Accessed 22 Apr 2010

Netzel M, Netzel G, Tian Q, Schwartz S, Konczak I (2007) Native Australian fruits – a novel source of antioxidants for food. Innov Food Sci Emerg Technol 8(3):339–346

Wrigley J, Fagg M (1996) Australian native plants. New Holland Publishers (Australia) Pty Ltd, Chatswood, 696 pp

Syzygium cumini

Scientific Name

Syzygium cumini (L.) Skeels

Synonyms

Calyptranthes capitellata Buch.-Ham. ex Wall. nom. nud., *Calyptranthes caryophyllifolia* Willd., *Calyptranthes cumini* (L.) Pers., *Calyptranthes cuminodora* Stokes, *Calyptranthes jambolana* (Lam.) Willd., *Calyptranthes jambolifera* Stokes, *Calyptranthes oneillii* Lundell, *Caryophyllus corticosus* Stokes, *Caryophyllus jambos* Stokes, *Eugenia calyptrata* Roxb. ex Wight & Arn., *Eugenia caryophyllifolia* Lam., *Eugenia cumini* (L.) Druce, *Eugenia jambolana* Lam., *Eugenia jambolana* var. *caryophyllifolia* (Lam.) Duthie, *Eugenia jambolana* var. *obtusifolia* Duthie, *Eugenia jambolifera* Roxb. ex Wight & Arn., *Eugenia obovata* Poir., *Eugenia obtusifolia* Roxb., *Eugenia tsoi* Merr. & Chun, *Jambolifera chinensis* Spreng., *Jambolifera coromandelica* Houtt., *Jambolifera pedunculata* Houtt., *Myrtus corticosa* Spreng., *Myrtus cumini* L. basionym, *Myrtus obovata* (Poir.) Spreng., *Syzygium caryophyllifolium* (Lam.) DC., *Syzygium cumini* var. *caryophyllifolium* (Lam.) K.K. Khanna, *Syzygium cumini* var. *obtusifolium* (Roxb.) K.K. Khanna, *Syzygium cumini* var. *tsoi* (Merr. & Chun) H.T. Chang & R.H. Miao, *Syzygium jambolanum* (Lam.) DC., *Syzygium jambolanum* var. *acuminata* O.Berg, *Syzygium jambolanum* var. *elliptica* O.Berg, *Syzygium jambolanum* var. *obovata* O.Berg, *Syzygium obovatum* (Poir.) DC., *Syzygium obtusifolium* (Roxb.) Kostel.

Family

Myrtaceae

Common/English Names

Black Plum, Damson Plum, Indian Blackberry, Jambolan, Java Plum, Jambolan Plum, Malabar Plum, Portuguese Plum, Purple Plum

Vernacular Names

Brazil: Jambolão, Jalão, Jamelão, Azeitona Da Terra, Jambol, Murta (Portuguese);
Burmese: Thabyang Hpyoo;
Chamorro: Duhat;
Chinese: Hai Nan Pu Tao, Hei Mo Shu, Wu Kou Shu, Wu Mo;
Cook Islands: Ka'Ika, Kaika, Paramu, Pisat, Pistāita, Pistati, Pītāti (Maori);
Costa Rica: Ciruelo De Java, Jambolán;
Czech: Hřebíčkovec Kmínový;
Danish: Jambolan;

T.K. Lim, *Edible Medicinal And Non-Medicinal Plants: Volume 3, Fruits*,
DOI 10.1007/978-94-007-2534-8_100, © Springer Science+Business Media B.V. 2012

Fijian: Duhat, Kavika Ni Idia;
French: Faux Pistachier, Jambolanier, Jamélongue, Jamelonguier, Jamelonier, Prune De Java;
German: Jambolanpflaume, Wachsjambuse, Rosenapfel;
Hungarian: Dzsambu (Fa);
India: Kala Jamu, Kothia Jam, Lohajam (Assamese), Jam, Kalajam (Bengali), Chambi, Bor Jamuk, Khimkhol (Garo), Jaman, Jamba, Jambava, Jambhal, Jambu, Jambua, Jambus, Jamnoa, Jamun, Mokni, Phalenda (Hindu), Dieng Sohthongum (Jaintia), Dulle Nerale, Goujalau Mara, Jambu Naerale, Jambuva, Naayinaerale, Naerale, Naeralu, Naerilu Nerale, Narala, Neeraala Mara, Neeram, Nerale Mara, Neralu, Nerula (Kannada), Dieng Ramai (Khasi), Nara, Naval, Njara, Njaval, Perin-Njara (Malayalam), Gulamchat, Jam (Manipuri), Jam, Jaman, Jambul, Rajale, Rajjambula, Thorajambula (Marathi), Hmuipui (Mizoram), Jamkoli (Oriya), Jambava, Jambu, Jambuh, Jambula, Mahaphala, Phalendra, Raja-Jambuh (Sanskrit), Kottainaval, Nagai, Naval, Naaval, Nava-Mara, Neredom (Tamil), Ala Naredu, All Neredu, Jambu, Jambu Naredu, Jambuvu, Jinna, Naredu, Neredu, Pedda Naeredu, Raacahnaeredu, Raasanaeredu (Telugu), Jamun, Poast Jamu (Urdu);
Indonesia: Juwet, Djoowet, Doowe, Juwet Manting, Juwet Sapi (Javanese), Jamblang (Sundanese);
Italian: Iambul, Aceituna Dulce;
Japanese: Janboran, Murasaki Futo Momo; Khmer Pring Bai, Pring Das Krebey;
Laotian: Va;
Malaysia: Jiwat, Juwat-Juwat, Jambelang;
Nepal: Jaambu, Jaamun, Kaalo Jaamun, Phaniir;
Palauan: Mesegerak, Mesekerrák, Mesekerrak, Mesigerak;
Philippines: Lumboi (Bikol), Lomboi (Cebu Bisaya), Lumboi (Ibanag), Dungboi (Igorot), Lungboi, Longboi, Lumboi (Iloko), Duat-Nasi, Lomboi (Pampangan), Duhat, Lumboi (Panay Bisaya), Duhat, Lomboy, Lomboi,Lunaboy (Tagalog);
Samoan: Nonu Fi'Afi'A;
Spanish: Guayabo Pesgua, Yambolana;
Surinam: Koeli, Jamoen, Druif;
Tahitian: Pistas;
Thai: Wa, Wa-Pa, Ma-Ha, Hakhiphae, Look Hwa;
Tibetan: Dza Mbu, Dzam-Bu, Ka Ka Dz Mbu;
Vietnamese: Voi Rung;
Venezuela: Jambolana, Pésjua Extranjera, Guayabo Pésjua.

Origin/Distribution

This species is distributed in tropical and subtropical Asia to Queensland. It is believed to be indigenous to India, Myanmar, Sri Lanka and the Andaman islands. It was introduced long time ago to Malaysia, Indonesia and the Philippines and has become naturalized.

Agroecology

Jambolan requires a tropical to sub tropical conditions for optimum growth. It grows well from sea-level to 1,200 m but, above 2,000 m it does not fruit but can be grown for its timber. It develops most luxuriantly in regions of heavy rainfall, 2,500–3,500 mm annually. It occurs in secondary forests on level areas, wasteland, river banks and has been known to withstand prolonged flooding. Yet it is tolerant of drought after it has made some growth. Despite its ability to thrive in low, wet areas, the tree does well on higher, well-drained sites on loam, marl, sand or oolitic limestone. Dry weather is desirable during the flowering and fruiting periods. It is sensitive to frost when young but mature trees have been undamaged by brief sub-freezing temperatures as occurred in southern Florida. The tree is resistant to coastal high winds and abhors in highly saline, or sodic soils.

Edible Plant Parts and Uses

Jambolan fruits are eaten ripe, fresh or made into tarts, sauces, jams, wine and vinegar. The fruit is astringent, acid to acid-sweet. The fruit is among the most popular fruit in the Philippines. The ripe ones are eaten outright. Good quality jambolan juice is excellent for sherbet, syrup and "squash".

Plate 1 Leaves and developing jambolan fruits

Plate 2 Close-up of developing jambolan fruits

In India, the latter is a bottled drink prepared by cooking the crushed fruits, pressing out the juice, combining it with sugar and water and adding citric acid and sodium benzoate as a preservative. The juice also makes a delicious and excellent red wine, "tinto dulce". Vinegar is made from the juice of the unripe, mature fruit.

Botany

A large, evergreen, much branched tree 5–20 m tall with a trunk diameter of 60–90 cm and scaly bark, terete branches and a canopy spread of 10 m. Leaves are opposite, coriaceous, elliptic-oblong, 5–15 cm long by 2.5–8 cm wide, acuminate, narrow; pinkish when young; when mature, leathery, glossy, dark-green above, lighter beneath, with conspicuous, yellowish midrib and numerous parallel lateral veins; petioles 1–2.5 cm long (Plates 1 and 2). Flowers are fragrant, in cymose clusters, axillary on flowering branches, occasionally terminal, white or pinkish, 12 mm wide; calyx cupular, 4–6 mm wide, up to 8 mm long and toothed; petals coherent, and slightly rounded, caduceous, white to rose-pink; stamens numerous (about 50), exserted, white or pinkish, to 7 mm long, style as long as stamens. Fruit is ovoid to ellipsoid, 2–2.5 cm long, green turning to plae greenish-white to deep purplish-black when ripe (Plates 1, 2, 3 and 4), smooth, with persistent calyx, with white to pale lavender, astringent, juicy pulp and one large, green seed.

Plate 3 Cluster of ripe and unripe jambolan fruits

Plate 4 Ripe jambolan fruits

Nutritive/Medicinal Properties

Analyses showed the fruit to be a good source of vitamin A and vitamin C, calcium and fair one of iron. Food value per 100 g of edible portion was reported as: moisture 83.7–85.8 g, protein 0.7–0.129 g, fat 0.15–0.3 g, crude fiber 0.3–0.9 g, carbohydrates 14.0 g, ash 0.32–0.4 g, calcium 8.3–15 mg, magnesium 35 mg, phosphorus 15–16.2 mg, iron 1.2–1.62 mg, sodium 26.2 mg, potassium 55 mg, copper 0.23 mg, sulfur 13 mg, chlorine 8 mg, vitamin A 80 I.U., thiamine 0.008–0.03 mg, riboflavin 0.009–0.01 mg, niacin 0.2–0.29 mg, ascorbic acid 5.7–18 mg, choline 7 mg and folic acid 3 μg (Morton 1987). Another analysis of the fruit per 100 g of edible portion reported was: energy 60 cal., moisture 82.7%, protein 0.78 g, fat 0.1 g, total carbohydrate 15.8 g, fibre 0.3 g, ash 0.7 g, Ca 8 mg, P 13 mg, Fe 0.2 mg, Na 9 mg, K 116 mg, β-carotene equivalent traces, thiamin traces, riboflavin 0.01 mg, niacin 0.02 mg and ascorbic acid 23 mg (Leung et al. 1972).

Other Phytochemicals

Some reported constituents of the seeds were: protein, 6.3–8.5%; fat, 1.18%; crude fibre, 16.9%; ash, 21.72%; calcium, 0.41%; phosphorus, 0.17%; fatty acids (palmitic, stearic, oleic and linoleic); starch, 41%; dextrin, 6.1%; a trace of phytosterol; and 6–19% tannin (Quisumbing 1978; Morton 1963). Ethanolic extracts of black-plum seeds were found to contain gallic acid, ellagic acid, chebulic acid, corilagin and related ellagitannins, 3,6-hexahydroxydiphenoylglucose and its two isomeric forms, galloylglucoseand quercetin, by chemical and enzymic studies with tannase (Bhatia and Bajaj 1972). From acetone extracts 3,3′,4-tri-O-methylellagic acid was isolated. A possible conversion of quercetin into toxifolin by the use of the antioxidant sodium metabisulphite was observed. Aqueous alcoholic of jambolan bark extract was found to contain bergenin, gallic acid and ethyl gallate (Kopanski and Schnelle 1988).

Recent studies (Mir et al. 2009) identified four new lignan derivatives from the stem bark of *Syzygium cumini*: (7α,8α,2′α)-3,4,5-trimethoxy-7,3′,1′,9′-diepoxylignan (cuminiresinol); (7α,7′α,8α,8′α)-3,4-dioxymethylene-3′,4′-dimethoxy-7,9′,7′,9-diepoxylignan-5′-ol (5′-hydroxy-methyl-piperitol); (7α,7′α,8α,8′α)-3′-methoxy-9-oxo-7,9′,7′,9-diepoxylignan-3,4,4′-triol or 3-demethyl-9-oxo-pinoresinol (syzygiresinol A); (7α,7′α,8α,8′α)-9-oxo-7,9′,7′,9-diepoxylignan-3,4,3′,4′,5′-pentaol or 3,3′-didemethyl-9-oxo-pinoresinol (syzygiresinol B) along with the known lignans di-demethyl-5-hydroxypinoresinol,dimethylpinoresinol, didemethoxypinoresinol, pinoresinol and 4′-methyl-5′-hydroxypinoresinol.

Various parts of the plant have a range of pharmacological activities.

Antioxidant Activity

A significant correlation existed between concentration of the jamun fruit skin extract and percentage inhibition of free radicals or percentage inhibition of lipid peroxidation as determined by using different assays, such as hydroxyl radical-scavenging assay, based on the benzoic acid hydroxylation method, superoxide radical-scavenging assay, based on photochemical reduction of nitroblue tetrazolium (NBT) in the presence of a riboflavin-light-NBT system, DPPH radical-scavenging assay, and lipid peroxidation assay, using egg yolk as the lipid-rich source (Banerjee et al. 2005). It was suggested that the antioxidant property of the fruit skin may come in part from the antioxidant vitamins, phenolics or tannins and anthocyanins present in the fruit.

The whole jamun fruit consisted of 666.0 g/kg pulp, 290.0 g/kg kernel and 50.0 g/kg seed coat (Benherlal and Arumughan 2007). Fresh pulp was rich in carbohydrates, protein and minerals. Total phenolics, anthocyanins and flavonoid contents of jamun pulp were 3.9, 1.34 and 0.07 g/kg, respectively. Kernel and seed coat contained 9.0 and 8.1 g/kg total phenolics respectively. DPPH radical scavenging activity of the samples and standards in descending order was: gallic acid > quercetin > Trolox > kernel ethanol extract > BHT > seed coat ethanol extract > pulp ethanol extract. Superoxide radical scavenging activity (IC_{50}) of kernel ethanol extract was six times higher (85.0 μg/ml) compared to Trolox (540.0 μg/ml) and three times compared to catechin (296.0 μg/ml). Hydroxyl radical scavenging activity (IC_{50}) of kernel ethanol extract was 151.0 μg/ml which was comparable with catechin (188.0 μg/ml). Inhibition of lipid peroxidation of the extracts was also investigated and their activity against peroxide radicals were lower than that of standard compounds BHT, 79.0 μg/ml; quercetin, 166.0 μg/ml; Trolox, 175.0 μg/ml; pulp ethanol extract, 342.0 μg/ml; kernel ethanol extract, 202.0 μg/ml and seed coat ethanol extract, 268.0 μg/ml.

Studies in Thailand found high levels of antioxidants in the jambolan fruit with extracts from the seeds having greater antioxidant capacity and total phenolic content than flesh (Kheaw-on et al. 2009). Acidified methanolic extracts gave consistently higher levels of all parameters studied. Anthocyanin was detected only in the flesh of margenta and dark purple fruits. The high yield of antioxidants found in the fruit strongly suggested that *S. cumini* can be a potentially rich source of natural antioxidant. Another study found that *S. cumini* fruit has low antioxidant activity expressed as antiradical DPPH activity with IC_{50} of 388.69 μg/ml, 6.33 mg C3G/g dry weight of TAC (Total anthocyanins content) and 9.95 mg GAE (gallic acid equivalent)/g dry weight of total phenolic content (TPC) (Reynertson et al. 2008). With regard to anthocyanin compounds, *S. cumini* was found to have more delphinidin 3-glucoside than cyanidin 3-glucoside. It also contained flavonoids quercitin, rutin traces of quercitrin and ellagic acid.

The jambolan fruit also contained tannins which showed a very good DPPH (1,1-diphenyl-2-picrylhydrazyl) radical scavenging activity and ferric reducing/antioxidant power. Hydrolysable tannins were identified as ellagitannins and condensed tannins were identified as B-type oligomers of epiafzelechin (propelargonidin) (Zhang and Lin 2009). The results indicated the promising potential for utilization of the fruit of *S. cumini* as a significant source of natural antioxidants.

Recent studies showed that ethyl acetate fraction of methanolic extract of *Syzygium cumini* leaf had stronger antioxidant activity than the water, chloroform, and *n*-hexane fractions (Ruan et al. 2008). The leaf extract was found to contain phenolic compounds, such as ferulic acid and catechin, A significant linear relationship between antioxidant potency, free radical-scavenging ability and the content of phenolic compounds of leaf extracts was observed.

Antidiabetic Activity

Jambolan seeds

Jambolan seeds have come into prominence in recent years on account of their suggested value in the treatment of diabetes mellitus especially in India, Brazil and Philippines. In the Philippines the fruit juice or pulverised powdered dried seeds have been used in decoction for treating diabetes. Studies reported that the seeds contained an alkaloid, jambosine, and a glycoside, jambolin or antimellin, which halted the diastatic conversion of starch into sugar (Morton 1963; Quisumbing 1978). The seed extract was found to lower blood pressure by 34.6% and this action was attributed to the ellagic acid content. This and 34 other polyphenols in the seeds and bark were isolated and identified.

Oral administration of aqueous jamun seed extract resulted in a significant reduction in blood glucose and resulted in a significant reduction in blood glucose and an increase in total haemoglobin and an increase in total haemoglobin in alloxan diabetic rats (Prince et al. 1998). It prevented decrease in body weight. It also decreased thiobarbituric acid reactive substances (TBARS) and increased in reduced glutathione (GSH), superoxide dismutase (SOD) and catalase (CAT), reflecting its antioxidant property. In another study, oral administration of an aqueous jamun seed extract for 6 weeks caused a significant decrease in lipids, thiobarbituric acid reactive substances (TBARS) and an increase in catalase and superoxide dismutase in the brain of alloxan induced diabetic rats (Prince et al. 2003). Oral administration of an alcoholic jamun seed extract for 6 weeks brought back all the parameters to near normal. The effect of alcoholic seed extract (100 mg/kg) was better than aqueous seed extract(5 g/kg). The effect of both these extracts was better than glibenclamide (600 μg/kg).

The study showed that *S. cumini* seed extracts reduced tissue damage in diabetic rat brain. The alcoholic extract of *Syzygium cumini* seeds was found to have both antidiabetic and antihyperlipidaemic effects (Prince et al. 2004). Oral administration of the extract to alloxan diabetic rats at a dose of 100 mg/kg body weight resulted in a significant reduction in blood glucose and urine sugar and lipids in serum and tissues in alloxan diabetic rats. The extract also increased total haemoglobinand restored all the parameters to normal levels. The effect of the extract was similar to that of insulin. Cold and hot water extracts from jamun seeds exhibited strong human pancreatic amylase (HPA) inhibitory in a concentration-dependent manner and with a IC_{50} values of 42.1 and 4.1 μg/ml (Ponnusamy et al. 2011). This inhibitory action reduced the rate of starch hydrolysis leading to lowered glucose levels.

Another study demonstrated that feeding rats for 21 days of the diets containing 15% powdered un-extracted (intact) seeds containing water soluble gummy fibre, 15% powdered defatted seeds from which lipid and saponins were removed only and 6% water soluble gummy fibre isolated from *S. cumini* seeds significantly lowered blood glucose levels and improved oral glucose tolerance (Pandey and Khan 2002) whereas feeding of the diets containing 15% powdered degummed *S. cumini* seeds from which water soluble gummy fibre was removed but which contained neutral detergent fibre (NDF) and 2.25% water insoluble neutral detergent fibre (NDF) isolated from *S. cumini* seeds neither lowered blood glucose levels nor improved oral glucose tolerance in both normal and diabetic rats. These observations indicated that the hypoglycaemic effect of *S. cumini* seeds was due to water soluble gummy fibre and also that water insoluble neutral detergent fibre (NDF) and other constituents of the seeds had no significant hypoglycaemic effects. Another study reported that the ethanolic extract of seeds of *S. cumini* was reported to increase body weight and decrease blood sugar level in alloxan diabetic albino rats (Singh and Gupta 2007). Level of significance for decrease in blood sugar after feeding alcoholic extract of *S. cumini* seeds in various doses was highly significant. The blood sugar level, which once dropped to normal levels after extract feeding was not elevated when extract feeding was discontinued for 15 days. Studies also showed the bark possessed a significant beneficial effect on glycoproteins in addition to its antidiabetic action. Another earlier study reported on the that hypoglycemic efficacy of the inorganic part using the glucose tolerance test on streptozotocin-induced diabetes (Ravi et al. 2004). Jambolan seed ash-treated diabetic rats exhibited normoglycemia and better glucose tolerance. The study concluded that the inorganic constituents might play an important role in the antidiabetic nature of jambolan seeds.

Administration of petroleum ether, chloroform, acetone, methanol and water extracts of *Syzygium cumini* seeds (100 mg/kg, p.o.) for 21 days caused a decrease in fasting blood sugar in diabetic rats (Farswan et al 2009). Among all the extracts methanol extract was found to lower the fasting blood sugar significantly in diabetic rats. The active principle was identified as cuminoside. Cuminoside caused a significant decrease in fasting blood sugar level, lipidperoxidation level, and improvement in the levels of antioxidant enzymes (reduced glutathione, superoxide dismutase, and catalase) in diabetic rats. A considerable decrease in lipid peroxidation and improvement in the antioxidant enzymes level in non-insulin dependent diabetes mellitus (NIDDM) rats indicated that cuminoside had antioxidant potential with antidiabetic activity and provided a scientific rationale for the use of cuminoside as an antidiabetic agent.

In-vitro studies using the mammalian α-glucosidase from rat intestine revealed the jambolan seed extracts to be more effective in inhibiting maltase when compared to the acarbose control (Shinde et al. 2008). All extracts were found to be more potent against α-glucosidase derived from *Bacillus stearothermophilus* than that against the enzymes from either baker's yeast (*Saccharomyces cerevisiae*) or rat intestine. In an in-vivo study using Goto-Kakizaki (GK) rats, the acetone extract was found to be a potent inhibitor of α-glucosidase hydrolysis of maltose when compared to untreated control animals. Taken together, these results suggested the inhibition of α-glucosidase as a possible mechanism by which this herb acted as an antidiabetic agent. A comparison was made between the antidiabetic activities of methanolic extracts of leaves of *Abroma augusta* and seeds of *Syzygium cumini* in alloxan induced diabetic rats (Nahar et al. 2010). Both extracts caused a significant decrease in serum glucose level, increase in body weight and changes in normal cells in diabetic rats as compared to the standard.

Jambolan fruit pulp

Water extract of jambolan fruit pulp was found to be more effective than the ethanolic extract in reducing fasting blood glucose and improving blood glucose in glucose tolerance test in diabetic rabbits (Sharma et al. 2006). Chromatographic purification of the water extract yielded two hypoglycaemic fractions (F-III more active than F-IV) and also detected the presence of hyperglycemic compounds (F-I and F-II) in the water extract. After treatment of diabetic and severely diabetic rabbits daily once with 25 mg/kg body weight with F-III for 7 and 15 days, respectively, there was a decline in fasting blood glucose (38% diabetic; 48% severely diabetic) and amelioration in blood glucose during glucose tolerance test (48%) in diabetic rabbits. Further, there was an increase in the plasma insulin levels in both diabetic (24.4%) and severely diabetic rabbits (26.3%). The in-vitro studies with pancreatic islets demonstrated that the insulin release was nearly two and half times more than that in untreated diabetic rabbits. The mechanism of action of FIII fraction appeared to be both pancreatic by stimulating release of insulin and extra pancreatic by directly acting on the tissues. Administration of the composite extract aqueous extracts of pulp of *Syzygium cumini* and bark of *Cinnamon zeylanicum* to diabetic rats resulted in a significant reduction on blood glucose levels (Rekha et al. 2010). In addition, it significantly recovered serum insulin levels and prevented the decrease in body weight than a single administration of the extract observed in untreated diabetic rats. Treatment with the combined extract significantly reversed conditions of hyperlipidemia, marked increase in lipid peroxide levels and concomitant decrease in antioxidant enzymes to near normal levels. The results justified the use of a combination of aqueous extracts of pulp of *Syzygium cumini* and bark of *Cinnamon zeylanicum* for the remedial effects against streptozotocin induced diabetic state.

Compared to untreated controls, streptozotocin-induced diabetic rats treated with the lyophilised jambolan fruit-pulp (50 mg/day) for 41 days exhibited no observable difference in body weight, food or water intake, urine volume, glycaemia, urinary urea and glucose, hepatic glycogen, or on serum levels of total cholesterol, HDL cholesterol or triglycerides (Pepato et al. 2005). No change was observed in the masses of epididymal or retroperitoneal adipose tissue or of soleus or extensor digitorum longus muscles. The authors suggested that the lack of any apparent effect on the diabetes may be attributable to the regional ecosystem where the fruit was collected and/or to the severity of the induced diabetes.

Jambolan bark

Studies in India reported that oral administration of stem bark extract exhibited antidiabetic activity by significantly lowering blood glucose and urine sugar levels in diabetic rats (Saravanan and Pari 2007). Additionally, diabetic rats treated with the extract had significantly elevated levels of plasma insulin and C-peptide. The findings of this study indicated that the antidiabetic activity of the extract may involve pancreatic and the extra-pancreatic mechanisms; such apparent dual actions of stem bark extract would be more advantageous to the existing oral antidiabetic monotherapy.

Jambolan leaves

The aqueous extract of *S. cumini* and *Psidium guajava* leaves exhibited higher inhibition against the porcine pancreatic α-amylase among the medicinal plants studied (Karthic et al. 2008). The compounds identified from the seed extract of *S. cumini* were betulinic acid and 3,5,7,4′-tetrahydroxy flavanone; the inhibition was found to be non-competitive in nature. α-Amylase is an enzyme capable of hydrolyzing starch and is the major form of amylase found in humans and other mammals. α-amylase inhibitors are heat-labile proteins that are active against salivary, pancreatic and bacterial α-amylases. Bopp et al. (2009) found that aqueous leaf extracts of *S. cumini* had an effect on adenosine deaminase activity of hyperglycemic subjects. The extract (60–1,000 μg/ml) caused a concentration-dependent inhibition of total adenosine deaminase activity in-vitro and a decrease in the blood glucose level in serum. It also reduced adenosine deaminase levels both in erythrocytes and in hyperglycemic serum. Adenosine deaminase is an important enzyme for modulating the bioactivity of insulin. The results suggested that the decrease of adenosine deaminase activity elicited by jambolan leaf extract may contribute to the control of adenosine levels and the antioxidant defense system of red cells. Another recent study demonstrated that *Syzygium cumini* aqueous leaves extract was able to scavenge oxidant species generated in diabetic conditions and modulated adensosine levels (De Bona et al. 2010). The extract prevented the increase in adenosine deaminase and 5′-Nucleotidase activities and thiobarbituric acid reactive substances (TBARS) levels. They suggested that the extract may promote a compensatory response in platelet function, improving the susceptibility-induced by the diabetes mellitus. Studies by De Bona et al. (2011) showed that *Syzygium cumini* leaf extract reduced the elevated activities of adenosine deaminase and acetylcholinesterase and thiobarbituric acid-reactive substances (TBARS) levels in erythrocytes of Type 2 diabetes mellitus (DM). The extract elevated the low SOD activity and NP-SH levels in erythrocytes of Type 2 DM. The results suggested that the extract was able to promote the reduction of inflammation and oxidative stress parameters, and acted against biochemical changes occurring in diabetes mellitus (DM).

However, there are conflicting reports on the antihyperglycemic/hypoglycaemic activity of jambolan leaves. A series of studies carried out in Brazil strongly refuted the hypoglycemic value of jambolan extracts (Teixeira et al. 1990, 1997, 2000, 2004). None of the herbal tea made from extracts of leaves or seeds elicited any detectable antihyperglycemic effect in both models of normal rats and rats with streptozotocin-induced diabetes mellitus. Studies conducted in Brazil demonstrated that the tea and extracts from different parts of the plant had no effect in normal rats, rats with streptozotocin-induced diabetes, and normal volunteers. In one study of a randomized, parallel, placebo controlled trial, tea prepared from leaves of *S. cumini* did not present any antihyperglycernic effect in 30 non-diabetic young volunteers submitted to a glucose blood tolerance test. In the animal experiments, the crude leaf extract administrated for 2 weeks to streptozotocin-induced diabetes mellitus did not produce any antihyperglycernic effect. In further studies, using double-blind, double-dummy clinical trial of randomized patients with type 2 diabetes, fasting blood glucose levels decreased significantly in participants treated with glyburide and did not change in those treated with the *Syzygium cumini* tea and in the participants who received placebos from tea and glyburide. Body mass index, creatinine, γ-glutamyl transferase, alkaline phosphatase, aspartate aminotransferase (SGOT), alanine aminotransferase (SGPT), 24-h glicosuria, 24-h proteinuria, triglycerides, total, low-density lipoprotein and high-density lipoprotein cholesterol did not vary significantly between the different groups. The study concluded that tea prepared from leaves of *S. cumini* had no hypoglycaemic effect. Therefore the use of such herbal teas in Brazil although very common and widespread, was not recommended by the researchers as a treatment for treating diabetes. Oliveira et al. (2005) found that seven-day treatment with crude ethanolic and aqueous and butanolic fractions (200–2,000 mg/kg, twice daily, per os) of *Syzygium cumini* leaves reduced glycaemia of non-diabetic mice. However, this effect was associated with a reduction of food intake and body weight, indicating that this may not be a genuine hypoglycaemic effect.

Neuropsychopharmacological Activity

Recently, the seed extracts were also reported to exhibit activity on the Central Nervous System (CNS). The seed extracts produced alteration in general behaviour pattern, reduction in spontaneous motility, hypothermia, potentiation of pentobaritone hypnosis, analgesia, reduction in exploratory behaviour pattern, muscle relaxant action, and suppression of aggressive behaviour (Chakraborty et al. 1986). The extract also caused suppression of conditioned avoidance response and showed antagonism to amphetamine group toxicity. These observations suggested that the extract of the seeds of *S. cumini* possessed a potent CNS depressant action.

Radioprotective Activity

Jamun seeds also exhibited radioprotective activity. Mice treated with hydroalcoholic extract of jamun seeds intraperitoneally before exposure to γ radiation showed reduction in the symptoms of radiation sickness and mortality at all exposure doses and caused a significant increase in the animal survival (Jagetia and Baliga 2002). The treatment of mice with different doses of jamun leaf extract, consecutively for five days before irradation, delayed the onset of mortality and reduced the symptoms of radiation sickness when compared with the nondrug-treated irradiated controls (Jagetia and Baliga 2003). All doses of the extract provided protection against the gastrointestinal death increasing the survival by 66.66% after treatment with 20, 30, and 40 mg/kg seed extract versus a 12% survival in the irradiated control group (oil + irradiation). Similarly, the jamun extract provided protection against the radiation-induced bone marrow death in mice treated with 10–60 mg/kg body weight of the extract. However, the best protection was obtained for 30 mg/kg b.wt. jamun extract, with highest the number of survivors (41.66%) after 30 days post-irradiation when compared with the other doses of the jamun extract. In another study, mice treated with 80 mg/kg body weight jamun seed extract intraperitoneally before exposure to 6, 7, 8, 9, 10 and 11 Gy of γ radiation showed reduction in the symptoms of radiation sickness and mortality at all exposure doses (Jagetia et al. 2005). It caused a significant increase in the animal survival when compared with the concurrent double distilled water + irradiation group. The jamun seed extract treatment protected mice against the gastrointestinal as well as bone marrow deaths. Studies by Jagetia et al. (2011) demonstrated that oral treatment of mice with 50 mg/kg body weight of jamun leaf extract protected mice against the radiation-induced DNA damage. The extract also inhibited lipid peroxidation in mice brain. Studies in a cell free system revealed that jamun extract inhibited the formation of OH, O(2)-, DPPH, and ABTS(+) free radicals in a concentration dependent manner. The inhibition of radiation-induced free radical formation may be one of the mechanisms of radioprotection.

Antiinflammatory Activity

The seed extracts also possessed anti-inflammatory activity. Both the ethyl acetate and methanol extracts of the seeds exhibited significant anti-inflammatory activity in carrageenan induced paw oedema in wistar rats at the dose level of 200 and 400 mg/kg administrated orally (Kumar et al. 2008). The methanol extract at the dose of 400 mg/kg showed high significant anti-inflammatory activity at 4 hours, where it caused 62.6% inhibition, as compared to that of 5 mg/kg of diclofenac sodium. The extract did not induce any gastric lesion in both acute and chronic ulcerogenic tests in rats. The results supported the traditional medicinal utilization of the plant. Another study reported that the chloroform fraction of *Syzygium cuminii* seeds caused significant inhibition of carrageenin, kaolin and other mediator-induced oedema (Mahapatra et al. 1986). The extract inhibited exudation of protein, leakage of dye in peritoneal inflammation and migration of leucocytes. The extract also caused inhibition of granuloma formation, experimental arthritis and also turpentine-induced joint oedema. Significant anti-pyretic action of the extract was also observed against yeast-induced pyrexia. These observations confirmed the anti-inflammatory effect of *S. cuminii* seed extract in exudative, proliferative and chronic stages of inflammation along with an anti-pyretic action.

The bark also exhibited anti-inflammatory activity. One study demonstrated that *S. cumini* bark extract has a potent antiinflammatory action against different phases of inflammation without any side effect on gastric mucosa (Muruganandan et al. 2001). Significant anti-inflammatory activity was observed in carrageenin (acute), kaolin-carrageenin (subacute), formaldehyde (subacute)-induced paw oedema and cotton pellet granuloma.

Gastroprotective Activity

Findings of studies suggested that tannins extracted from *S. cumini* had gastroprotective and anti-ulcerogenic effects (Ramirez and Roa 2003). Microscopic examination using Best's Ulcer Staging Index showed that tannins had a very significant decrease in gastric mucosal damage. Average lymphocyte populations showed no significant difference, although both the tannins and Omeprazole group had fewer lymphocytes than the control. A dose comprising of 20.0 g tannins/kg rat weight resulted in significantly lower stomach free radical concentrations.

S. cumini seed also exhibited gastro-protective property. Ethanolic extract of seeds at a dose of 200 mg/kg, when administered orally for 10 days in rats was found to reduce the ulcer index in all gastric ulcer induced by 2 hours cold restraint stress (CRS), aspirin, 95% ethanol and 4 hours pylorus ligation (PL) in rats (Chaturvedi et al. 2007). It tended to decrease acid-pepsin secretion, enhanced mucin and mucosal glycoprotein and decreased cell shedding but had no effect on cell proliferation. It showed antioxidant properties indicated by decrease in LPO and increase in glutathione (GSH) levels in the gastric mucosa of rats. Acute toxicity study indicated LD_{50} to be more than ten times (>2,000 mg/kg) of the effective ulcer protective dose while subacute toxicity study (>1,000 mg/kg) indicated no significant change in the general physiological and haematological parameters, liver and renal function tests. The result of the study indicated that *S. cumnin* seed displayed gastro-protective properties mainly through promotion of mucosal defensive factors and antioxidant status and decreasing lipid peroxidation.

Hepatoprotective Activity

Studies showed that aqueous *Syzygium cumini* leaf extract, may be useful for liver protection but needed to be given over a significant period and prior to liver injury (Moresco et al. 2007). Blood samples of rats with hepatoxicty induced by carbon tetrachloride revealed a significant increase in the aspartate aminotransferase (AST) and alanine aminotransferase (ALT) activities after carbon tetrachloride administration alone, which was significantly lowered by pre-administration with the aqueous extract of *Syzygium cumini*, after repeated doses.

Antiallergic Activity

Recently anti-allergic effect of *jambolan* leaf extract was reported (Brito et al. 2007). HPLC analysis revealed that hydrolyzable tannins and flavonoids are the major components of the extract. Oral administration of jambolan leaf extract in Swiss mice inhibited paw edema induced by compound 48/80 (50% inhibition, 100 mg/kg); and, to a lesser extent, the allergic paw-edema (23% inhibition, 100 mg/k). The extract treatment also inhibited the edema induced by histamine (58% inhibition) and 5/HT (52% inhibition) but had no effect on platelet/aggregating factor/induced paw edema. The extract prevented mast cell degranulation and the consequent histamine release in Wistar rat peritoneal mast cells (50% inhibition, 1 μg/ml) induced by compound 48/80. Pre/treatment of BALB/c mice with 100 mg/kg of the extract significantly inhibited eosinophil accumulation in allergic pleurisy. The studies indicated that its anti-edematogenic effect is due to the inhibition of mast cell degranulation and of histamine and serotonin effects whereas the inhibition of eosinophil accumulation in the allergic pleurisy model.

Antimicrobial Activity

The leaves, stems, flower-buds, opened blossoms, and bark also possessed antibiotic activity. *S. cumini* leaf essential oil was found to have good antibacterial activity better than that from

S. travancoricum (Shafi et al. 2002) The crude hydroalcoholic extract of *S. cumini* was active against *Candida krusei* (MIC = 70 μg/ml), and against multi–resistant strains of *Pseudomonas aeruginosa, Klebsiella pneumoniae* and *Staphylococcus aureus* (de Oliveira et al. 2007). Overall, *Candida krusei*, both multi-resistant and standard strains of *Pseudomonas aeruginosa, Candida albicans* and *Neisseria gonorrhoeae* were the most susceptible to this extract, displaying MIC values from 70 to 90 μg/ml. Aqueous and organic extracts of *Syzygium cumini* was found to have inhibitory effect aginst *Vibrio cholerae* and *Vibrio parahaemolyticus* with MIC values of 2.5–20 mg/ml (Sharma et al. 2009).

The crude methanol and aqueous extracts of jambolan leaves showed inhibitory activity against clinical isolates of Gram negative bacteria such *as Salmonella enteritidis, Salmonella typhi, Salmonella typhi A, Salmonella paratyphi A, Salmonella paratyphi B, Pseudomonas aeruginosa* and *Escherichia coli* and Gram positive bacteria *Bacillus subtilis*, and *Staphylococcus aureus* (Gowri and Vasantha 2010) The methanol extracts was more potent than the aqueous extracts. Flavonoids, alkaloids, glycosides, steroids, phenols, saponins, terpenoid, cardiac glycosides and tannins were detected in the extracts. The oil from jambolan leaves showed activity against all the test organisms: *Pseudomonas aeruginosa, Escherichia coli, Staphylococcus aureus, Candida albicans, Bacillus subtilis* and *Salmonella typhii* (Ugbabe et al. 2010). The leaf extract showed activity against all test organisms except *P. aeruginosa* while the stem bark extract showed no activity on any of the test organisms used. The oil had a saponification value of 363 and an acid value of 4.21. The lethal dose 50% (LD50) in mice was found to be >5,000 mg/kg for the stem bark and at 3,873 mg/kg for the leaf extracts.

Antiviral Activity

The bark of *Eugenia jambolana*, the bark of *Saraca indica* and the stem bark of *Terminalia arjuna* were found to inhibit the HIV type 1 protease activity by more than 70% at a concentration of 0.2 mg/ml (Kusumoto et al. 1995).

Anticancer Activity

Syzygium cumini extract was found to inhibit the growth and induces apoptosis in HeLa and SiHa cervical cancer cell lines in a dose and time dependent manner (Barh and Viswanathan 2008). Studies found that *Syzygium cummini* extract exhibited anti-tumour and anti-oxidative potential against benzo-α-pyrene induced stomach carcinogenesis in mice (Goyal et al. 2010). The extract significantly reduced tumour incidence, tumour burden and cumulative number of gastric carcinomas along with a significant elevation of phase II detoxifying enzymes, and inhibition of lipid peroxidation in the stomach. *S. cumini* seed extract exhibited protective efficacy against DMBA (7,12-dimethyl benz(a)anthracene)-induced skin papillomagenesis in mice (Parmar et al. 2010). The extract significantly reduced the cumulative numbers of skin papillomas and tumour incidence (75%). The results indicated that the anticarcinogenic activity of the extract during DMBA-induced skin papillomagenesis was mediated through alteration of antioxidant status.

Chemoprotective Activity

The administration of aqueous *S. cumini* seed extract reverted the toxic effects (elevated N-acetyl-β-D: -glucosaminidase (NAG) activity in the kidney and urine, the lipid peroxidation levels in the liver and kidney, adenosine deaminase (ADA) activity in the hippocampus, kidney and liver) of methylmercury in neonatal rats (Abdalla et al. 2011). The main compounds present in the extract gallic acid (the major component), chlorogenic acid and rutin, could be responsible for such benefit, since they were found to display antioxidant properties.

Aqueous and ethanolic extracts of jamun seed extract showed significant protective effects against hydroxyl radical induced strand breaks in pBR322 DNA (Arun et al. 2011). In-vivo experiments with aqueous extract showed significant protective effects against chromosomal damage induced by the genotoxic carcinogens urethane and 7,12-dimethyl benz(a)anthracene. Biochemical assays registered significant inhibition of hepatic lipid peroxidation and increase in GSH level and activity of GST, SOD and CAT. The findings suggested that jamun seed extract could possibly play an important role as a chemopreventive agent against oxidative stress and genomic damage.

Antimalarial Activity

Jambolan bark decoction was found to have antiplasmodial activity (Simões-Pires et al. 2009). Among the ellagic acid derivatives isolated ellagic acid, ellagic acid 4-O-α-L-2″-acetylrhamnopyranoside, 3-O-methylellagic acid 3′-O-α-L-rhamnopyranoside and 3-O-methylellagic acid 3′-O-β-D-glucopyranoside, only ellagic acid was able to reduce *Plasmodium falciparum* parasitaemia in-vitro and inhibit β-hematin formation (a known mechanism of action of some antimalarial drugs), suggesting that free hydroxyl groups were necessary for activity within this class of compounds.

Traditional Medicinal Uses

Different parts of jambolan such as seeds, bark, fruit, and leaves have been used medicinally and it has a long tradition in alternative medicine. The jambolan has received far more recognition in traditional folkloric medicine and in the pharmaceutical trade than in any other field. The seeds and the bark are much used in tropical medicine and are shipped from India, Malaysia and Polynesia, and, to a small extent, from the West Indies, to pharmaceutical companies in Europe and England. In India, it is highly esteemed in Ayurvedic and Unani alternative medicine for a range of therapeutic properties. It has been used in traditional medicine as a remedy for *diabetes mellitus* in many countries. Pharmacologically, the fruit is stated to be astringent, stomachic, carminative, antiscorbutic and diuretic. The fruit is a useful astringent in bilious diarrhoea, dysentry and water-diluted juice makes a good gargle for sore throat or a lotion for ringworm of the scalp. In India, the juice of the ripe fruit, or a decoction of the fruit, or jambolan vinegar, may be administered in cases of enlargement of the spleen, chronic diarrhoea and urine retention. Jambolan vinegar is an agreeable stomachic and carminative; it is also used as a diuretic. The seeds are reported to possess hypoglycemic, antibacterial, anti-HIV activity and anti-diarrheal effects. Jambolan is also featured in Chinese medicine where the seeds are employed for digestive ailments. The fruit and seeds are sweet, acrid, sour, tonic, and cooling, and are used in diabetes, diarrhoea and ringworm. The leaves are also used to strengthen the teeth and gums, to treat leucorrhoea, stomachache, fever, strangury, skin diseases, constipation, and to inhibit blood discharges in the faeces. The expressed juice of the leaves has been used alone or in combination with other astringents for gingivitis (bleeding gums). In India, the leaves are used for poulticing in skin complaints. The leaves, steeped in alcohol, are prescribed in diabetes and for dysentery. The bark is astringent, sweet sour, diuretic, digestive and anthelmintic.

In the Philippines, a decoction of the powdered bark is given internally in dyspepsia, dysentery and diarrhoea. It also given as an enema. A decoction made of the bark after scraping off the surface is also used to cleanse ulcers. In India, the bark is used in the treatment of diarrhoea, anaemia, diabetes, dysentery and spongy gum. Bark decoctions are taken in cases of asthma and bronchitis and are gargled or used as mouthwash for the astringent effect on mouth ulcerations, spongy gums, and stomatitis. Ashes of the bark, mixed with water, are spread over local inflammations, or, blended with oil, applied to bums. In modern therapy, tannin is no longer approved on burned tissue because it is absorbed and can cause cancer. Excessive oral intake of tannin-rich plant products can also be dangerous to health.

Other Uses

Flowers have profuse nectar and provide a good foraging source for bees, producing a fine quality honey. In India, the leaves provide food source for tassar silkworms and fodder for livestock. The essential oil from the leaves is used to odorise soap and is blended with other materials in making inexpensive perfume. Brown dyes can be obtained from the bark. The bark contains 8–19% tannin and is much used in tanning leather and preserving fishing nets. The wood is reported to be durable in water and resistant to borers and termites and is used for beams and rafters, posts, bridges, boats, oars, masts, troughs, well-lining, agricultural implements, carts, solid cart wheels, railway sleepers and the bottoms of railroad cars. The wood also provides good fuel-wood.

In India, the tree is venerated by Buddhists and considered to be sacred to Krishna and planted near Hinduistic temples.

Comments

Jambolan is propagated by seeds, root cuttings, air-layering, inarching, grafting, stump planting and budding.

Selected References

Abdalla FH, Bellé LP, Bitencourt PE, De Bona KS, Zanette RA, Boligon AA, Athayde ML, Pigatto AS, Moretto MB (2011) Protective effects of *Syzygium cumini* seed extract against methylmercury-induced sistemic toxicity in neonatal rats. Biometals 24(2):349–356

Arun R, Prakash MV, Abraham SK, Premkumar K (2011) Role of *Syzygium cumini* seed extract in the chemoprevention of in vivo genomic damage and oxidative stress. J Ethnopharmacol 134(2):329–333

Backer CA, Bakhuizen van den Brink RC Jr (1963) Flora of java (Spermatophytes only). (Spermatophytes only), vol 1. Noordhoff, Groningen, 648 pp

Banerjee A, Dasgupta N, De B (2005) In vitro study of antioxidant activity of *Syzygium cumini* fruit. Food Chem 90(4):727–733

Barh D, Viswanathan G (2008) *Syzygium cumini* inhibits growth and induces apoptosis in cervical cancer cell lines: a primary study. ecancermedical Sci 2:83

Benherlal PS, Arumughan C (2007) Chemical composition and in vitro antioxidant studies on *Syzygium cumini* fruit. J Sci Food Agric 87(14):2560–2569

Bhatia IS, Bajaj KL (1972) Tannins in black-plum (*Syzygium cumini* L.) seeds. Biochem J 128:56–60

Bhuiyan MA, Mia MY, Rashid MA (1996) Antibacterial principles of the seed of *Eugenia jambolana*. Bangladesh J Bot 25:239–241

Bopp A, De Bona KS, Bellé LP, Moresco RN, Moretto MB (2009) *Syzygium cumini* inhibits adenosine deaminase activity and reduces glucose levels in hyperglycemic patients. Fund Clin Pharmacol 23(4):501–507

Brito FA, Lima LA, Ramos MFS, Nakamura MJ, Cavalher-Machado SC, Siani AC, Henriques MGMO, Sampaio ALF (2007) Pharmacological study of anti-allergic activity of *Syzygium cumini* (L.) Skeels. Braz J Med Biol Res 40(1):105–115

Burkill IH (1966) A dictionary of the economic products of the Malay Peninsula. Revised reprint, 2 vols. Ministry of Agriculture and Co-operatives, Kuala Lumpur, vol 1 (A–H), pp 1–1240, vol 2 (I–Z), pp 1241–2444

Chakraborty D, Mahapatra PK, Chaudhuri AK (1986) A neuropsychopharmacological study of *Syzygium cumini*. Planta Med 52(2):139–143

Chaturvedi A, Kumar MM, Bhawani G, Chaturvedi H, Kumar M, Goel RK (2007) Effect of ethanolic extract of *Eugenia jambolana* seeds on gastric ulceration and secretion in rats. Indian J Physiol Pharmacol 51(2):131–140

Chen J, Craven LA (2007) Myrtaceae. In: Wu ZY, Raven PH, Hong DY (eds) Flora of China, vol 13 (Clusiaceae through Araliaceae). Science Press, Beijing

Chopra RN, Chopra IC, Handa KL (1958) Indigenous drugs of India, 2nd edn. Dhar and Sons, Calcutta, pp 686–689

Coronel RE (1992) *Syzygium cumini* (L.) Skeels. In: Verheij EWM, Coronel RE (eds) Plant resources of South-East Asia. No. 2. Edible fruits and nuts. Prosea Foundation, Bogor, pp 294–296

De Bona KS, Bellé LP, Sari MH, Thomé G, Schetinger MRC, Morsch VM, Boligon A, Athayde ML, Pigatto AS, Moretto MB (2010) *Syzygium cumini* extract decrease adenosine deaminase, 5′nucleotidase activities and oxidative damage in platelets of diabetic patients. Cell Physiol Biochem 26:729–738

De Bona KS, Bellé LP, Bittencourt PE, Bonfanti G, Cargnelluti LO, Pimentel VC, Ruviaro AR, Schetinger MR, Emanuelli T, Moretto MB (2011) Erythrocytic enzymes and antioxidant status in patients with type 2 diabetes: beneficial effect of *Syzygium cumini* leaf extract in vitro. Diabetes Res Clin Pract 94(1):84–90

de Oliveira GF, Furtado NAJC, da Silva Filho AA, Martins CHG, Bastos JK, Cunha WR, de Andrade e Silva ML (2007) Antimicrobial activity of *Syzygium cumini* (Myrtaceae) leaves extract. Braz J Microbiol 38:381–4

Farswan M, Mazumder PM, Parcha V (2009) Modulatory effect of an isolated compound from *Syzygium cumini* seeds on biochemical parameters of diabetes in rats. Int J Green Pharm 3(2):28–132

Foundation for Revitalisation of Local Health Traditions (2008) FRLHT database. htttp://envis.frlht.org

Govaerts R, Sobral M, Ashton P, Barrie F, Holst BK, Landrum LL, Matsumoto K, Fernanda Mazine F, Nic Lughadha E, Proenca C, Soares-Silva LH, Wilson PG, Lucas E (2010) World checklist of Myrtaceae. The board of trustees of the Royal Botanic Gardens, Kew. Published on the Internet, http://www. kew.org/wcsp/. Accessed 22 Apr 2010

Gowri SS, Vasantha K (2010) Phytochemical screening and antibacterial activity of *Syzygium cumini* (L.) (Myrtaceae) leaves extracts. Int J PharmTech Res 2(2):1569–1573

Goyal P, Verma P, Sharma P, Parmar J, Agarwal A (2010) Evaluation of anti-cancer and anti-oxidative potential of *Syzygium cumini* against benzo[a]pyrene (bap) induced gastric carcinogenesis in mice. Asian Pac J Cancer Prev 11(3):753–758

Jagetia GC, Baliga MS (2002) *Syzygium cumini* (Jamun) reduces the radiation-induced DNA damage in the cultured human peripheral blood lymphocytes: a preliminary study. Toxicol Lett 132(1):19–25

Jagetia GC, Baliga MS (2003) Evaluation of the radioprotective effect of the leaf extract of *Syzygium cumini* (Jamun) in mice exposed to a lethal dose of γ-irradiation. Nahrung 47(3):181–185

Jagetia GC, Baliga MS, Venkatesh P (2005) Influence of seed extract of *Syzygium cumini* (jamun) on mice exposed to different doses of γ-radiation. J Radiat Res (Tokyo) 46(1):59–65

Jagetia GC, Shetty PC, Vidyasagar MS (2011) Inhibition of radiation-induced dna damage by jamun, *Syzygium cumini*, in the cultured splenocytes of mice exposed to different doses of {γ}-radiation. Integr Cancer Ther (in press)

Karthic K, Kirthiram KS, Sadasivam S, Thayumanavan B, Palvannan T (2008) Identification of α- amylase inhibitors from *Syzygium cumini* Linn seeds. Indian J Exp Biol 46(9):677–680

Kheaw-on N, Chaisuksant R, Suntornwat O (2009) Antioxidant capacity of flesh and seed from *Syzygium cumini* fruits. Acta Hort (ISHS) 837:73–78

Kopanski L, Schnelle G (1988) Isolation of bergenin from barks of *Syzygium cumini*. Planta Med 54(6):572

Kumar A, Padmanabhan N, Krishnan MRV (2007) Central nervous system activity of *Syzygium cumini* seed. Pak J Nutr 6(6):698–700

Kumar A, Ilavarasan R, Jayachandran T, Deecaraman M, Mohan Kumar R, Aravindan P, Padmanabhan N, Krishan MRV (2008) Anti-inflammatory activity of *Syzygium cumini* seed. Afri J Biotechnol 7(8): 941–943

Kusumoto IT, Nakabayashi T, Kida H (1995) Screening of various plant extracts used in Ayurvedic medicine for inhibitory effects on human immunodeficience virus type I (HIV-I) protease. Phytother Res 12:488–493

Leung W-TW, Butrum RR, Huang Chang F, Narayana Rao M, Polacchi W (1972) Food composition table for use in East Asia. FAO, Rome, 347 pp

Mahapatra PK, Chakraborty D, Chaudhuri AK (1986) Anti-inflammatory and antipyretic activities of *Syzygium cuminii*. Planta Med 52(6):540

Mir Q, Ali M, Alam P (2009) Lignan derivatives from the stem bark of *Syzygium cumini* (L.) Skeels. Nat Prod Res 23(5):422–430

Moresco RN, Sperotto RL, Bernardi AS, Cardoso RF, Gomes P (2007) Effect of the aqueous extract of *Syzygium cumini* on carbon tetrachloride-induced hepatotoxicity in rats. Phytother Res 21(8):793–795

Morton JF (1963) The jambolan (*Syzygium cumini* Skeels) – its food, medicinal, and ornamental and other uses. Proc Fla State Hort Soc 76:328–338

Morton JF (1987) Jambolan. In: Fruits of warm climates. Julia F. Morton, Miami. pp 375–378

Muruganandan S, Srinivasan K, Chandra S, Tandan SK, Lal J, Raviprakash V (2001) Anti-inflammatory activity of *Syzygium cumini* bark. Fitoterapia 72(4):369–375

Nahar L, Ripa FA, Zulfiker AHM, Rokonuzzaman M, Haque M, Islam KMS (2010) Comparative study of antidiabetic effect of *Abroma augusta* and *Syzygium cumini* on alloxan induced diabetic rat. Agric Biol J N Am 1(6):1268–1272

National Academy of Sciences (1980) Firewood crops: shrub and tree species for energy production. National Academy of Sciences, Washington, DC, 236 pp

Oliveira AC, Endringer DC, Amorim LA, das Graças LBM, Coelho MM (2005) Effect of the extracts and fractions of *Baccharis trimera* and *Syzygium cumini* on glycaemia of diabetic and non-diabetic mice. J Ethnopharmacol 102(3):465–469

Pacific Island Ecosystems at Risk (PIER) (1999) *Syzygium cumini* (L.) Skeels, Myrtaceae. http://www.hear.org/pier/species/syzygium_cumini.htm

Pandey M, Khan A (2002) Hypoglycaemic effect of defatted seeds and water soluble fibre from the seeds of *Syzygium cumini* (Linn.) skeels in alloxan diabetic rats. Indian J Exp Biol 40(10):1178–1182

Parmar J, Sharma P, Verma P, Goyal PK (2010) Chemopreventive action of *Syzygium cumini* on DMBA-induced skin papillomagenesis in mice. Asian Pac J Cancer Prev 11(1):261–265

Pepato MT, Mori DM, Baviera AM, Harami JB, Vendramini RC, Brunetti IL (2005) Fruit of the jambolan tree (*Eugenia jambolana* Lam.) and experimental diabetes. J Ethnopharmacol 96(1–2):43–48

Ponnusamy S, Ravindran R, Zinjarde S, Bhargava S, Ravi Kumar A (2011) Evaluation of traditional Indian antidiabetic medicinal plants for human pancreatic amylase inhibitory effect in vitro. Evid Based Complement Alternat Med ii:515647

Prince PS, Menon VP, Pari L (1998) Hypoglycaemic activity of *Syzygium cumini* seeds: effect on lipid peroxidation in alloxan diabetic rats. J Ethnopharmacol 61(1):1–7

Prince PS, Kamalakkannan N, Menon VP (2003) *Syzygium cumini* seed extracts reduce tissue damage in diabetic rat brain. J Ethnopharmacol 84(2–3):205–209

Prince PS, Kamalakkannan N, Menon VP (2004) Antidiabetic and antihyperlipidaemic effect of alcoholic *Syzygium cumini* seeds in alloxan induced diabetic albino rats. J Ethnopharmacol 91(2–3):209–213

Quisumbing E (1978) Medicinal plants of the Philippines. Katha Publishing Co, Quezon City, 1262 pp

Ramirez RO, Roa CC Jr (2003) The gastroprotective effect of tannins extracted from duhat (*Syzygium cumini* Skeels) bark on HCl/ethanol induced gastric mucosal injury in Sprague-Dawley rats. Clin Hemorheol Microcirc 29(3–4):253–261

Ravi K, Satish Sekar D, Subramanian S (2004) Hypoglycemic activity of inorganic constituents of *Eugenia jambolana* seed on streptozocin induced diabetes in rats. Biol Trace Element Res 99:145–155

Rekha N, Balaji R, Deecaraman M (2010) Antihyperglycemic and antihyperlipidemic effects of extracts of the pulp of *Syzygium cumini* and bark of *Cinnamon zeylanicum* in streptozotocin-induced diabetic rats. J Appl Biosci 28:1718–1730

Reynertson KA, Yang H, Jiang B, Basile MJ, Kennelly EJ (2008) Quantitative analysis of antiradical phenolic constituents from fourteen edible Myrtaceae fruits. Food Chem 109(4):883–890

Ruan ZP, Zhang LL, Lin YM (2008) Evaluation of the antioxidant activity of *Syzygium cumini* leaves. Molecules 13(10):2545–2556

Saravanan G, Pari L (2007) Effect of *Syzygium cumini* bark extract on plasma and tissue glycoproteins in streptozotocin induced diabetic rats. J Cell Tiss Res 7(1):881–887

Shafi PM, Rosamma MK, Jamil K, Reddy PS (2002) Antibacterial activity of *Syzygium cumini* and *Syzygium travancoricum* leaf essential oils. Fitoterapia 73(5):414–416

Sharma SB, Nasir A, Prabhu KM, Murthy PS (2006) Antihyperglycemic effect of the fruit-pulp of *Eugenia jambolana* in experimental diabetes mellitus. J Ethnopharmacol 104(3):367–373

Sharma A, Patel VK, Chaturvedi AN (2009) Vibriocidal activity of certain medicinal plants used in Indian folklore medicine by tribals of Mahakoshal region of central India. Indian J Pharmacol 41(3):129–133

Shinde J, Taldone T, Barletta M, Kunaparaju N, Hu B, Kumar S, Placido J, Zito SW (2008) α-glucosidase inhibitory activity of *Syzygium cumini* (Linn.) Skeels seed kernel in vitro and in Goto-Kakizaki (GK) rats. Carbohydr Res 343(7):1278–1281

Simões-Pires CA, Vargas S, Marston A, Ioset JR, Paulo MQ, Matheeussen A, Maes L (2009) Ellagic acid derivatives from *Syzygium cumini* stem bark: investigation of their antiplasmodial activity. Nat Prod Commun 4(10):1371–1376

Singh N, Gupta M (2007) Effects of ethanolic extract of *Syzygium cumini* (Linn) seed powder on pancreatic islets of alloxan diabetic rats. Indian J Exp Biol 45(10):861–867

Teixeira CC, Knijnik J, Pereira MV, Fuchs FD (1989) The effect of tea prepared from leaves of "jambolão" (*Syzygium cumini*) on the blood glucose levels of normal rats: an exploratory study. In: Proceedings of the Brazilian-sino symposium on chemistry and pharmacology of natural products, Rio de Janeiro, p 191

Teixeira CC, Fuchs FD, Blotta RM, Knijnik J, Delgado I, Netto M, Ferreira E, Costa AP, Müssnich DG, Ranquetat GG, Gastaldo GJ (1990) Effect of tea prepared from leaves of *Syzygium jambos* on glucose tolerance in nondiabetes subjects. Diabetes Care 13:907–908

Teixeira CC, Pinto LP, Kessler FHP, Knijnik L, Pinto CP, Gastaldo GJ, Fuchs FD (1997) The effect of *Sygyzium cumini* (L.) Skeels on post-prandial blood glucose levels in non-diabetic rats and rats with streptozotocin-induced diabetes mellitus. J Ethnopharmacol 56:209–213

Teixeira CC, Rava CA, Mallman da Silva P, Melchior R, Argenta R, Anselmi F, Almeida CR, Fuchs FD (2000) Absence of antihyperglycemic effect of jambolan in experimental and clinical models. J Ethnopharmacol 71(1–2):343–347

Teixeira CC, Weinert LS, Barbosa DC, Ricken C, Esteves JF, Fuchs FD (2004) *Syzygium cumini* (L.) Skeels in the treatment of type 2 diabetes: results of a randomized, double-blind, double-dummy, controlled trial. Diabetes Care 27:3019–3020

Ugbabe GE, Ezeunala MN, Edmond IN, Apev J, Salawu OA (2010) Preliminary phytochemical, antimicrobial and acute toxicity studies of the stem, bark and the leaves of a cultivated *Syzygium cumini* Linn. (Family: Myrtaceae) in Nigeria. Afri J Biotechnol 9(41):6943–6947

Zhang LL, Lin YM (2009) Antioxidant tannins from *Syzygium cumini* fruit. Afri J Biotechnol 8(10):2301–2309

Syzygium jambos

Scientific Name

Syzygium jambos (L.) Alston.

Synonyms

Eugenia decora Salisb., *Eugenia jamboides* Wender., *Eugenia jambos* L. basionym, *Eugenia jambos* var. *sylvatica* Gagnep., *Eugenia jambosa* Crantz, *Eugenia malaccensis* Blanco nom. illeg., *Eugenia malaccensis* f. *cericarpa* (O. Deg.) H.St.John, *Eugenia monantha* Merr., *Eugenia vulgaris* Baill., *Jambosa jambos* (L.) Millsp., *Jambosa malaccensis* f. *cericarpa* O. Deg., *Jambosa palembanica* Blume, *Jambosa vulgaris* DC. nom illeg., *Myrtus jambos* (L.) Kunth, *Plinia jambos* (L.) M.Gómez, *Syzygium jambos* var. *linearilimbum* H.T. Chang & R.H. Miao, *Syzygium jambos* var. *sylvaticum* (Gagnep.) Merr. & L.M. Perry, *Syzygium merrillii* Masam. nom. superfl., *Syzygium monanthum* (Merr.) Merr. & L.M. Perry.

Family

Myrtaceae

Common/English Names

Jambos, Malabar Plum, Plum Rose, Rose-apple

Vernacular Names

Afrikaans: Jamboes;
Argentina: Yambo;
Brazil: Jambo Rosa, Jambeiro, Jambo Amarelo (Portuguese);
Burmese: Thabyu Thabye;
Chinese: Pu Ta;
Columbia: Manzana Rosa;
Cook Islands: Ka'Ika, Ka'Ika Papa'Ā, Ka'Ika Takataka, Ka'Ika Varāni (Maori);
Cuba: Pomarrosa Manzana Rosa;
Czech: Hřebíčkovec Molucký;
Danish: Jambo;
Dominican Republic: Pomarrosa Pomo;
Eastonian: Jambu-Nelgipuu;
Fijian: Kavika, Kavika Ni India, Kavika Ni Vavalangi;
French: Pome Rose, Pommier Rose, Jambosier;
Germany: Rosenapfelbaum, Jambubaum, Rosenapfel;
Guatemala: Manzana Rosa;
Hawaiaan: 'Ohi'A Loke;
India: Golapi-Jamuk (Assamese), Gulab Jamun (Hindu), Pannerale (Kannada), Malakkacampa, Yamu Panawa (Malayalam), Jamb (Marathi), Champai (Tamil), Jambuneredu (Telugu);
Indonesia: Jambu Mawar, Jambu Air Mawar, Jambu Kraton;
Japanese: Futo Momo;
Khmer: Châm-Puu;
Laos: Chièng, Kièng;
Malaysia: Jambu Mawar, Jambu Kelampok;

T.K. Lim, *Edible Medicinal And Non-Medicinal Plants: Volume 3, Fruits*,
DOI 10.1007/978-94-007-2534-8_101, © Springer Science+Business Media B.V. 2012

Nepale: Gulaav Jaamun, Thulo Jamun;
Philippines: Tampoy, Tampoi (Bikol), Bunlaun, Yampoi (Bisaya), Tanpul (Ibanag), Balobar (Pampangan), Bunlauan (Panay Bisaya), Tampoy, Tampoi (Tagalog);
Pohnpeian: Apel En Wai, Iouen Wai, Youenwai;
Portuguese: Jambo Amarelo, Jambo Branco, Jambo Rosa, Jambeiro;
Puerto Rico: Jambo Amarillo, Manzana Rosa, Manzanita Rosa, Pomarrosa;
Russian: Sitsigiui Dzhamboza;
Samoan: Seasea Pālagi, Seasea Papalagi;
Spanish: Manzana Rosa, Pomarrosa, Manzanita De Rosa, Yambo;
Surinam: Pommeroos, Appelroos;
Swedish: Rosenäpple;
Tahitian: Ahi'A Papa'A;
Thailand: Chompu-Nam Dok Mai (Central), Manom Hom (North), Yamu-Panawa (Malay-Yala);
Tongan: Fekika Papalangi;
Vietnamese: Lý, Bô Dào, Roi.

Origin/Distribution

Rose apple is found from Himalaya to West Malesia, but its exact origin is uncertain. It has been introduced and cultivated in home gardens elsewhere in the tropics and subtropics for its edible fruit, also as ornamental or shade tree and wind-break.

Agroecology

The rose apple flourishes in the tropical and near-tropical climates from near sea level to about 1,200 m elevation. In its native range, it occurs in mixed forests, mountain slopes, riverine localities and river valleys. At higher altitudes or much cooler areas it grows but bears no fruit. It prefers a wet climate but tolerates semi-arid conditions as it is quite drought tolerant to brief dry spells. Prolonged dry spells are detrimental. The tree tolerates strong winds as is often planted as wind-breaks. A deep, well-drained, loamy soil is considered ideal for the rose apple although it flourishes also on sandy and limestone soils with very little organic matter.

Edible Plant Parts and Uses

Ripe fruit is eaten fresh but it is rather insipid and are usually eaten by children. It can be cooked, preserved or stewed as dessert. With lemon juice the fruit can be prepared into jam or jelly and more frequently preserved mixed with other fruits, also prepared into sauces and syrup to flavour cold drinks. In Jamaica, the halved or sliced fruits are candied by stewing them in very heavy sugar syrup with cinnamon. Another recipe is to stuff the cup-like halved fruits with a rice-and-meat mixture, cover them with a tomato sauce seasoned with minced garlic, and baking them for about 20 min. The fruit has high pectin content which makes it suitable for making jelly. The fruit can be distilled to yield rose water that is deemed to be as good as that obtained by rose petals. The flowers are a rich source of nectar for honeybees and the honey is a good amber colour.

Botany

An evergreen tree, which can grow up to 10 m tall with a short greyish brown, smooth, 50 cm diameter trunk (Plate 4). The tree is low branching forming a dense crown of wide-spreading branches (Plate 3). Branchlets are terete or subterete, sometimes apically much compressed, sometimes shallowly grooved. The leaves opposite, lanceolate, ovate-lanceolate, or linear, 8–26 × 2–4.5 cm, leathery or stiffly papery, both surfaces with numerous small pellucid glands, apex long-acuminate, base cuneate, petioles 6–12 mm long (Plate 2), shiny and pink when young turning to pale green. Inflorescences are short terminal or axillary corymbs, with 4–10 flowers.

Inflorescences usually terminal cymes with several flowers, sometimes axillary and solitary; peduncle 1–3.5 cm. Flowers white,

Plate 1 Ripe fruit and seed of rose apple

Plate 3 Dense foliage

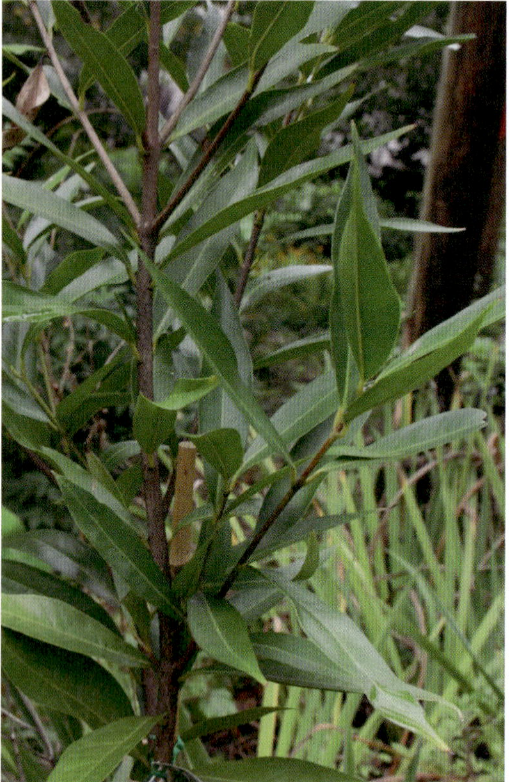

Plate 2 Linear opposite leaves of sapling

Plate 4 Smooth-greyish-brown trunk

showy, fragrant, 4–8 cm across and bisexual. Hypanthium obconic, 0.8–1.5 cm; Calyx lobes 4, semi-orbicular or triangular-ovate, 5–8 × 6–9 mm; Petals distinct, broadly ovate, 1.4–1.5 cm; Stamens numerous about 200 with 1.5–2.8 cm long, creamy-white filaments; anthers 1.5 mm; Style 2–3.5 cm. Fruit a subglobose, ovoid or ellipsoid drupe, with oil glands, pale yellow or flushed with pink when ripe, 2.5–5 cm in diameter with 1 or 2-subglobose seed, pericarp fleshy, white, firm and rose-scented (Plate 1). Embryos numerous.

Nutritive/Medicinal Properties

The nutritional value per 100 g edible portion of the fruit based on analyses made in Central America and elsewhere (Morton 1987) was reported as: energy 56 calories, 84.5–89.1 g water, 0.5–0.7 g protein, 0.2–0.3 g fat, 14.2 g carbohydrates, 101–1.9 g fibre, 0.4–0.44 g ash, 29–45.2 mg Ca, 4 mg Mg, 50 mg K, 0.01 mg Cu, 13 mg S, 4 mg Cl, 123–235 IU carotene, 0.01–0.19 mg thiamin, 0.028–0.05 mg riboflavin, 0.521–0.8 mg niacin and 3–37 mg ascorbic acid. The energy value is 234 kJ/100 g. The pulp contained high pectin content suitable for use as a settling agent. Khoo et al. (2008) reported the fruit to contain 1.41 mg/100 g total carotene content.

Forty-three volatile components were identified in roseapple fruit, 3-phenylpropan-1-ol, (E)-cinnamyl alcohol and other compounds with the C6-C3 skeleton constituted about 60% of roseapple volatiles (Wong and Lai 1996). Another analysis found that the volatile components with potent odour identified in rose apple fruit included hexanal, 3-penten-2-one, hexanol, (*Z*)-3-hexen-1-ol, linalool, isovaleric acid, benzyl alcohol, 2-phenylethylalcohol, (*E*)-cinnamaldehyde, caproic acid, phenylacetic acid, methyl-(*E*)-cinnamate and (*E*)-cinnamyl alcohol (Guedes et al. 2004). Volatiles with lower odour potency included 7-octen-4-ol, acetic acid, 2-octen-1-ol and cinnamic acid.

Scientific studies confirmed that the plant extract had anticancer activity, extracts of the rose apple leaves have antioxidant, analgesic and anti-inflammatory properties and the bark extracts displayed antimicrobial activities.

Antioxidant Activity

Extract of the leaves of *Syzygium jambos* furnished three dihydrochalcones, phloretin 4′-O-methyl ether (2′,6′-dihydroxy-4′-methoxydihydrochalcone), myrigalone G (2′,6′-dihydroxy-4′-methoxy-3′-methyldihydrochalcone), and myrigalone B (2′,6′-dihydroxy-4′-methoxy-3,5′-di-methyldihydrochalcone) which exhibited radical scavenging properties (Jayasinghe et al. 2007). The roseapple fruit exhibited low antioxidant activity expressed as a DPPH (2,2 diphenyl-1-picryl-hydrazyl) with IC_{50} value of 24.7 mg/ml compared to α-tocopherol with an IC_{50} of 53.3 mg/ml (Reynertson et al. 2005).

S. jambos fruit displayed low antioxidant activity with a DPPH IC_{50} value of 247 mg/ml compared to other Myrtaceous fruits and known vitamin antioxidants, ascorbic acid and α-tocopherol with IC_{50} values of 18.3 mg/ml and 53.3 mg/ml, respectively (Reynertson et al. 2005). *S. jambos* had no detectable anthocyanin content and low TPC (total phenolic content) of 8.69 mg GAE/g dry weight (Reynertson et al. 2008). With regard to anthocyanins, it had low levels of quercetin and quercitrin. It also contained ellagic acid.

Antinociceptive/Analgesic Activity

The hydro-alcoholic leaf extract (100–300 mg/kg) significantly reduced pain scores in test rats in all the phases of the formalin test with an analgesic efficacy higher than that shown by diclofenac (Avila-Peña et al. 2007). In the hot plate test, *Syzygium jambos* extract produced a significant increase in the withdrawal response latencies in a dose-dependant manner (10–300 mg/kg i.p.) and with a maximal effect (analgesic efficacy) similar to that of morphine. The leaf extract had remarkable analgesic effects on both cutaneous and deep muscle pain that was not mediated by opioid receptors.

Antiinflammatory Activity

The leaves of *Syzygium jambos* are used in Guatemala for their antiinflammatory and digestive properties. Two flavonol diglycosides quercetin and myricetin 3-O-β-D-xylopyranosyl (1->2) α-L-rhamnopyranosides were isolated from the leaves collected from Guatemala (Slowing et al. 1994b). In another study, orally administered organic leaf extracts of *S. jambos* (hexane, dichloromethane, ethyl acetate and methanol), given at daily doses equivalent to 12.5 g/kg of dried leaf material, inhibited both the acute and chronic phases of experimental model of inflammation (Slowing et al. 1994a). The ethyl acetate and methanol extracts were the most effective and were equal or more effective than 80 mg/kg of phenylbutazone. The methanolic extract was the most active in the chronic phase.

Anticancer Activity

A 70% acetone extract of *S. jambos* exerted the strongest cytotoxic effects on human leukemia cells (HL-60) from a preliminary screening of 15 plants (Yang et al. 2000). The cytotoxic principles were two hydrolyzable tannins:1-O-galloyl castalagin and casuarinin. The compounds significantly inhibited human promyelocytic leukemia cell line HL-60 with IC_{50} were 10.8 and 12.5 μM, respectively. Both showed less cytotoxicity to human adenocarcinoma cell line SK-HEP-1 and normal cell lines of human lymphocytes and Chang liver cells. The apoptosis induced by these two compounds was also demonstrated by DNA fragmentation assay and microscopic observation. These results suggested that the cytotoxic mechanism of both antitumour principle constituents might be the induction of apoptosis in HL-60 cells.

Antimicrobial Activity

Ethanol-aqueous extract of the stem bark of *Syzygium jambos* afforded a number of known triterpenes such as friedelin, β-amyrin acetate, betulinic acid and lupeol (Kuiate et al. 2007). Betulinic acid and friedelolactone were the most active compound with antidermatophytic activity against three dermatophyte species: *Microsporum audouinii, Trichophyton mentagrophytes* and *Trichophyton soudanense*, commonly found in Cameroon, and the most sensitive fungi were *Trichophyton soudanense and Trichophyton mentagrophytes*. Another research reported that acetone and aqueous extracts from the bark displayed antimicrobial activity in vitro (Djipa et al. 2000). The extracts proved to be particularly effective on *Staphylococcus aureus, Yersinia enterocolitica* and coagulase negative staphylococci which included *Staphylococcus hominis, Staphylococcus cohnii* and *Staphylococcus warneri*. The antimicrobial activity was attributed to the high tannin content of the bark extracts.

S. jambos methanol leaf extract inhibited the growth of 4 of the 14 bacteria tested (29%) (Mohanty and cock 2010). Both Gram-positive and Gram-negative bacterial growths were inhibited by *S. jambos* leaf extract, although Gram-positive bacteria appeared more susceptible. Of the 10 g-negative bacteria tested, only *Alcaligenes faecalis* and *Aeromonas hydrophila* were inhibited by *S. jambos* leaf extract. The leaf extract also inhibited the growth of 2 (*Bacillus cereus* and *Staphylococcus aureus*) of the Gram-positive bacteria tested. The leaf extract also proved to be toxic in the *Artemia franciscana* bioassay, with a 48-hours LC_{50} of 387.9 μg/ml, making it slightly more toxic than Mevinphos (505.3 μg/ml) and approximately five fold less toxic than potassium dichromate (80.4 μg/ml). Whilst potassium dichromate's LC_{50} remained constant across the 72-hours test period (24-hours LC_{50}, 86.3; 72-hours LC_{50}, 77.9), the extract and Mevinphos LC_{50} values decreased by 72 hours (87.0 μg/ml and 103.9 μg/ml, respectively), indicating their similar levels of toxicity in the assay.

Absence of Antihyperglecaemic Activity

In Porto Alegre, a southern city of Brazil, the tea prepared from leaves of *Syzygium cumini* or *S. jambos* had been report to be used frequently by diabetic patients (Teixeira et al. 1990). However, studies demonstrated that the tea prepared from leaves of *Syzygium jambos* had no effect on glucose tolerance in non-diabetic (normal) subjects (Teixeira et al. 1990). In subsequent studies, normal rats, rats with streptozotocin-induced diabetes, normal volunteers and patients with diabetes were all found to be negative in regard to an anti-hyperglycemic effect of this plant (Teixeira and Fuchs 2006).

Traditional Medicinal Uses

Eugenia jambos is an antipyretic and antiinfl ammatory herb of Asian folk medicine. All parts of the rose-apple tree have been reported to be used in traditional medicine in the tropics.

Several parts of the tree are used medicinally as a tonic or a diuretic. Bark, leaves and seeds are medicinally used. In Indo-China, all parts of the tree are used as a digestive, a stimulant and a remedy for tooth troubles. In Upper Myanmar, the leaves are boiled and the decoction applied to sore eyes. The leaf decoction also serves as a diuretic and expectorant and treatment for rheumatism. An infusion of the leaves is given for fever in Cambodia. Powdered leaves have been rubbed on the bodies of smallpox patients for the cooling effect. A conserve of the flowers is considered cooling. A sweetened preparation of the flowers is believed to reduce fever. The seeds are employed against diarrhoea, dysentery and catarrh. In El Salvador and Nicaragua, an infusion of roasted, powdered seeds is employed as a remedy for diabetics. In Colombian folk medicine, the seeds are believed to have an anaesthetic property. The bark contains 7–12.4% tannin and has emetic and cathartic properties. The decoction of the bark is administered to relieve asthma, bronchitis and hoarseness. Cuban people believe that the root is an effective remedy for epilepsy.

Other Uses

A yellow coloured essential oil, 26.84% *dl-a*-pinene and 23.84% *l*-limonene derived from the leaves by distillation is important in the perfume industry. The heartwood is heavy and hard, and has been utilised to make beams for construction, furniture, arms for easy chairs, spokes for wheels, knees for all kinds of boats, frames for musical instruments (violins, guitars, etc.), and packing cases. It is also popular for general turnery. It is not durable in the ground and is prone to attack by dry-wood termites. The bark contains 7% tannin on a dry weight basis and is used by local villagers for tanning and dyeing purposes. The flexible branches have been employed in Puerto Rico to make hoops for large sugar casks, and also are valued for weaving large baskets.

The tree grows back rapidly after cutting to a stump and consequently yields a continuous supply of small wood for fuel. Rose apple wood makes very good.

In some countries rose apple is planted more for its ornamental value than as a fruit tree. The showy cream-coloured flowers, dark-green foliage, and moderate size contribute to its popularity. The tree is also planted as boundary, barrier, support tree or as a living fence post or in hedgerows around coffee plantations.

Comments

An important invasive species in French Polynesia and other Pacific Islands, the Galápagos Islands and Mauritius and La Réunion.

Selected References

Avila-Peña D, Peña N, Quintero L, Suárez-Roca H (2007) Antinociceptive activity of *Syzygium jambos* leaves extract on rats. J Ethnopharmacol 112(2):380–385

Backer CA, van den Bakhuizen Brink RC Jr (1963) Flora of java (Spermatophytes only), vol 1. Noordhoff, Groningen, p 648

Burkill IH (1966) A dictionary of the economic products of the malay Peninsula. Revised reprint, 2 vols. Ministry of agriculture and co-operatives. Kuala Lumpur, vol 1 (A–H), pp 1–1240, vol 2 (I–Z), pp 1241–2444

Chen J, Craven LA (2007) Myrtaceae. In: Wu ZY, Raven PH, Hong DY (eds) Flora of China, vol 13 (Clusiaceae through Araliaceae). Science Press, Beijing

Djipa CD, Delmee M, Quetin Leclercq J (2000) Antimicrobial activity of bark extracts of *Syzygium jambos* (L.) Alston (Myrtaceae). J Ethnopharmacol 71(1–2):307–313

Govaerts R, Sobral M, Ashton P, Barrie F, Holst BK, Landrum LL, Matsumoto K, Fernanda Mazine F, Nic Lughadha E, Proenca C, Soares-Silva LH, Wilson PG, Lucas E (2010) World checklist of Myrtaceae. The board of trustees of the Royal Botanic Gardens, Kew. Published on the Internet, http://www. kew.org/wcsp/. Accessed 22 Apr 2010

Guedes C, Pinto A, Moreira R, De Maria C (2004) Study of the aroma compounds of rose apple (*Syzygium jambos* Alston) fruit from Brazil. Eur Food Res Technol A 219(5):460–464

Jayasinghe UL, Ratnayake RM, Medawala MM, Fujimoto Y (2007) Dihydrochalcones with radical scavenging properties from the leaves of *Syzygium jambos*. Nat Prod Res 21(6):551–554

Khoo HE, Ismail A, Mohd.-Esa N, Idris S (2008) Carotenoid content of underutilized tropical fruits. Plant Foods Hum Nutr 63:170–175

Kuiate JR, Mouokeu S, Wabo HK, Tane P (2007) Antidermatophytic triterpenoids from *Syzygium jambos* (L.) Alston (Myrtaceae). Phytother Res 21(2):149–152

Mohanty S, Cock IE (2010) Bioactivity of *Syzygium jambos* methanolic extracts: antibacterial activity and toxicity. Phcog Res 2:4–9

Morton J (1987) Rose apple. In: Morton JF (ed) Fruits of warm climates. Julia F. Morton, Miami, pp 383–386

Reynertson KA, Basile MJ, Kennelly EJ (2005) Antioxidant potential of seven myrtaceous fruits. Ethnobot Res Appl 3:25–35

Reynertson KA, Yang H, Jiang B, Basile MJ, Kennelly EJ (2008) Quantitative analysis of antiradical phenolic constituents from fourteen edible Myrtaceae fruits. Food Chem 109(4):883–890

Slowing K, Carretero E, Villar A (1994a) Anti-inflammatory activity of leaf extracts of *Eugenia jambos* on rats. J Ethnopharmacol 43:9–11

Slowing K, Söllhuber M, Carretero E, Villar A (1994b) Flavonoid glycosides from *Eugenia jambos*. Phytochemistry 37(1):255–258

Teixeira CC, Fuchs FD (2006) The efficacy of herbal medicines in clinical models: the case of jambolan. J Ethnopharmacol 108(1):16–19

Teixeira CC, Fuchs FD, Blotta RM, Knijnik J, Delgado I, Netto M, Ferreira E, Costa AP, Müssnich DG, Ranquetat GG, Gastaldo GJ (1990) Effect of tea prepared from leaves of *Syzygium jambos* on glucose tolerance in non-diabetes subjects. Diabetes Care 13:907–908

Van Lingen TG (1992) *Syzygium jambos* (L.) Alston. In: Verheij EWM, Coronel RE (eds) Plant resources of South-East Asia no 2. Edible Fruits and nuts. Prosea, Bogor, pp 296–298

Wong KC, Lai FY (1996) Volatile constituents from the fruits of four *Syzygium* species grown in Malaysia. Flavour Fragr J 11(1):61–66

Yang L-L, Lee C-Y, Yen K-Y (2000) Induction of apoptosis by hydrolysable tannins from *Eugenia jambos* L. on human leukemia cells. Cancer Lett 157:65–75

Syzygium luehmannii

Scientific Name

Syzygium luehmannii **(F. Muell.) L.A.S. Johnson.**

Synonyms

Austromyrtus exaltata (F.M. Bailey) Burret, *Eugenia leptantha* var. *parvifolia* F.M. Bailey, *Eugenia luehmannii* F. Muell. (basionym), *Eugenia parvifolia* C. Moore, *Myrtus exaltata* F.M. Bailey.

Family

Myrtaceae

Common/English Names

Cherry Alder, Cherry Satinash, Clove Lilli Pilli, Creek Cherry Riberry, Creek Satinash, Red Lily Pilly, Riberry, Small Leaved Lilly Pilly

Vernacular Names

None

Origin/Distribution

The species is indigenous to New Guinea (Goodenough Island) and Australia – sub-tropical New South Wales and Queensland.

Agroecology

In its native range, riberry grows in riverine, littoral and subtropical rainforest.

Edible Plant Parts and Uses

Ripe riberry fruit (Plates 1, 2, and 3) has a juicy, slightly acidic, tart, cranberry-like flavour, that has a hint of cloves. The fruit is most commonly used to make a distinctively flavoured jam, and is also used in sauces served with meat, syrups, ice-cream and confectionery. It has been popular as a gourmet bush-food since the early 1980s, and is commercially cultivated on a small-scale basis.

Botany

Medium-sized to large buttressed tree that can grow to 30 m, with smooth to slightly flaky bark; dense canopy (Plate 3), new foliage growth is pink. Leaves are opposite, ovate to lanceolate, 2.5–7.5 cm long, 1–3 cm wide, glabrous and glossy, with long-acuminate apex, rounded base and paler lower surface (Plates 1 and 2); lateral veins are faint and the intramarginal vein distinct; oil glands numerous and distinct; petiole 2–6 mm long. Inflorescences compact, cymose, paniculate, terminal and in the upper axils; pedicels very short. Flowers creamy-white, mostly 5-merous. Petals 1.5–2.5 mm long, free and spreading. Stamens 3–6 mm long. Fruit obovoid or pyriform, 9–12 mm long, 6–10 mm diameter,

T.K. Lim, *Edible Medicinal And Non-Medicinal Plants: Volume 3, Fruits*,
DOI 10.1007/978-94-007-2534-8_102, © Springer Science+Business Media B.V. 2012

Plate 1 Ripe and unripe fruit and leaves of riberry

Plate 2 Ripe riberry fruit

Plate 3 Dense canopy of riberry

green turning to pink to red when ripe (Plates 1, 2, and 3); seed solitary with a solitary embryo with smooth cotyledons.

Nutritive/Medicinal Properties

The proximate value per 100 g edible portion of raw fruit (Brand Miller et al. 1993) was reported as: water 82 g, energy 84 kJ, protein 0.96 g, fat 0.4 g, ash 0.8 g, total dietary fibre 6.8 g, total sugars 0.42 g, minerals – Ca 100 mg, Fe, 0.9 mg, Mg 48 mg, P, 49 mg, K 250 mg, Na 11 mg, Zn 0.2 mg, Cu 0.3 mg, niacin 0.2 mg.

Riberry is one of many Australian native fruits found to be a rich source of antioxidants, comparable to blueberries (Netzel et al. 2007). It has been suggested as a potential source of bioactive phytochemicals for application in health promoting foods.

Other Uses

Its handsome purple-red growth makes riberry a very popular favourite as a garden ornamental and street tree. It is easily maintained as a smaller tree by light pruning.

Comments

Riberry is easily propagated from seeds or cuttings.

Selected References

Brand Miller J, James KW, Maggiore P (1993) Tables of composition of Australian aboriginal foods. Aboriginal Studies Press, Canberra

Govaerts R, Sobral M, Ashton P, Barrie F, Holst BK, Landrum LL, Matsumoto K, Fernanda Mazine F, Nic Lughadha E, Proenca C, Soares-Silva LH, Wilson PG, Lucas E (2010) World checklist of Myrtaceae . The board of trustees of the Royal Botanic Gardens, Kew. Published on the Internet, http://www. kew.org/wcsp/. Accessed 22 Apr 2010

Hyland BPM (1983) A revision of *Syzygium* and allied genera (Myrtaceae) in Australia. Aust J Bot Suppl Ser 9:1–164

Low T (1991) Wild food plants of Australia. Angus & Robertson, North Ryde, 240 pp

Netzel M, Netzel G, Tian Q, Schwartz S, Konczak I (2007) Native Australian fruits – a novel source of antioxidants for food. Innov Food Sci Emerg Technol 8(3):339–346

Smith K, Smith I (1999) Grow your own bush food. New Holland, Sydney, 139 pp

Syzygium malaccense

Scientific Name

Syzygium malaccense (L.) Merr. & Perry.

Synonyms

Caryophyllus malaccensis (L.) Stokes, *Eugenia domestica* Baill., *Eugenia macrophylla* Lamarck, *Eugenia malaccensis* L. (basionym), *Eugenia pseudomalaccensis* Linden, *Eugenia purpurascens* Baill., *Eugenia purpurea* Roxb., *Jambosa domestica* DC., *Jambosa macrophylla* (Lam.) DC., *Jambosa malaccensis* (L.) DC., *Jambosa purpurascens* DC., *Jambosa purpurea* (Roxb.) Wight & Arn., *Myrtus macrophylla* (Lam.) Spreng. nom. illeg., *Myrtus malaccensis* (L.) Spreng.

Family

Myrtaceae

Common/English Names

Kavika Tree, Large Fruited Rose Apple, Malay Apple, Mountain Apple, Malacca Apple, Malay Apple, Malay Rose Apple, Mountain Apple, Otaheite Cashew, Otaheite Apple, Pomerac

Vernacular Names

Banaban: Te Kabika;
Burmese: Thabyo-Thabyay, Thabyo Thabyang;
Chamorro: Macupa, Makupa;
Chinese: Hong Hua Pu Tao, Ma Lai Pu Tao, Ma Liu Jia Pu Tao, Yang Pu Tao;
Chuukese: Faariyap Faariyap, Fasniyaap Fasniyaap, Feniyap;
Colombia: Cereza Cuadrada, Pomalaca, Pomarosa De Malaca;
Cook Islands: Ka'Ika, Ka'Ika Makatea, Ka'Ika Maori, Ka'Ika Maori, Ka'Ika Papa'Ā, Ka'Ika Tavake, Ka'Ika Tavake, Ka'Ika Tavake, Ka'Ika 'Enua, Kaika Makatea (Maori);
Costa Rica: Mazana De Agua;
Cuba: Ohia, Jamboissier Rouge, Malakel-Apfel, Manzana De Agua, Malayapfel, Mountain Apple, Pomme Malac, Pomarossa De Malaca;
Dominican Republic: Cajualito;
Dutch: Djamboe Bol;
El Salvador: Marañon Japonés;
Fijian: Kavika, Kavika Ndamu, Kavika Ndamundamu, Kavika Vulvula, Yasi Kavika;
French: Jamalac, Jambosier Rouge, Poire De Malaque, Pomme De Malaisie, Pomme De Tahiti, Pomme D'eau, Pomme Malac, Jambose De Malaque, Poirier De Malaque, Pommier-Rose;
German: Groszer Rosenapfel, Malabarischer Rosenapfel, Zahmer Jambusenbaum, Malaka-Apfelbauw, Malakka-Apfel, Malaysia-Apfel;

T.K. Lim, *Edible Medicinal And Non-Medicinal Plants: Volume 3, Fruits*,
DOI 10.1007/978-94-007-2534-8_103, © Springer Science+Business Media B.V. 2012

Guam: Makupa;
Hawaii: ' Ohi'A 'Ai Ai, 'Ohi'A 'Ōhi' Kea, 'Ōhi' Leo, 'Ōhi' 'Ula, 'Ōhi' 'Ai Ke'Oke'O;
India: Burka Jamun;
Indonesia: Jambi Nipoe (Aceh), Jambu Merah, Lutu Kau (Ambon), Njambu Bol (Bali), Maku (Baree), Jambah, Jambol, Jambu Bol (Batak), Jambu Bol, Jambu Boa (Bengkoelen, Jambi), Jambu Djene (Bima), Kuo (Boeol), Jambu Bolu, Jampo Bolu, Jampu Pelo (Bugis), Mutiha, Aloe, Lutune, Nutune, (Ceram), Omtulo (Amahai, Ceram) Rutuno (Sepa, Ceram), Upo (Gorontalo), Goda-Goda, Gogoa, Gorogo (N. Halmaheira), Jambu Tersana, Dersana, Jambu Bol (Java), Jambu Bol, Jambu Kalupa (Lampong), Jambu Keling (Lingga, Singkep), Darsana, Jammbhu Darsana (Madurese), Gora Besar, Jambu Berteh, Jambu Bol (Malay), Gowa Merah (Manado), Jambu Bo, Jambu Jambak, Jambu Gadang (Minangkabau), Wua Sumonda (Mori), Maufa (Nias), Kembes Mea, Kembes Raindang, Kao, Kochuwa, Kombot Moura, Kumpasa Mahendang, Kupa, Kupa Maamu, Kupa Raindang, Mangkoa, Mangkoa Maamu (Northern Sulawesi), Kumkolo, Kopo Kaul, Rutuul (Oelias), Bol, Jambu Bul (Sundanese), Gora Lamo, Gora Tome (Ternate), Suo, Suo Kohori (Tidore), Kau Noel (Timor);
Japanese: Maree Futo Momo;
Khmer: Chompuh Kraham;
Kosrae: Acpuhl;
Malaysia: Jambu Merah, Jambu Bol, Jambu Kling, Jambu Bubul, Jambu Kapal, Jambu Bar, Jambu Melaka, Jambu Susu;
Mangarevan: Ke'Ika;
Niuean: Fekakai;
Northern Marquesas: Kehika Kehika, Kehika Inana;
Palauan: Kidel;
Papua New Guinea: Laulau;
Philippines: Tual (Bagobo), Gubal (Bukidnon), Mangkopa (Panay Bisaya), Tual (Lanao), Makopang-Kalabau, Tamo, Makopa, Yambu (Tagalog), Tersana Rose Apple; ***Pohnpeian***: Apel, Apel En Pohnpei, Paniap;
Portuguese: Jambu, Jambu De Malacca;
Rotuman: Hahi'a;
Russian: Sitsigiui Malakskii;
Samoan: Nonu Fi'Afi'A, Nonu Fiafia;
Societies: 'Ahi'A;
Southern Marquesas: Kehi'A;
Spanish: Pomarosa, Manzana Malaya, Pomarrosa De Malaca, Yambo;
Tahitian: Ahia Tahiti, 'Ahia-Tea, 'Ahia-'Ura;
Thai: Chom-Phu-Daeng, Chompu-Mamieow, Chompu-Saraek;
Tongan: Fekika Kai;
Ulithian: Harafath;
'Uvea, Futuna: Kafika;
Vietnamese: Man Hurong Tau, Cay Dao, Dièu Dò, Cay Roi;
Venezuela: Manzana De Agua, Manzana Malaya, Pera De Agua, Pomagás;
Woleaisan: Faliyap;
Yapese: Arfatlh Arfatlh, Harafath Harafath, Faliap Faliap, Faliyap.

Origin/Distribution

Malay apple is indigenous from southeast Asia to Vanuatu, but its exact origin is uncertain, probably in the lowland rainforests of Peninsular Malaysia, Java and Sumatra. It is now cultivated and naturalized in many countries throughout the tropics, especially in Indo-Malaysia, southeast Asia, East Africa, Central and South America, Melanesia, Polynesia and Micronesia.

Agroecology

In its native range, it occurs in the lowland tropical rainforests up to an altitude of 1,000 m. In the Pacific it has become naturalised since it ancient introduction and occurs in low elevation, moist to wet sites, primarily mesic valley, in thickets or dense or open forest from near sea level to an elevation of about 800 m. Being tropical and subtropical in its requirement it is extremely frost intolerant and grows best in areas with a mean annual temperature 24–30°C and with uniformly distributed rainfall and a mean annual rainfall above 1,500 mm. It does not grow well in areas with seasonal drought unless supplemented by irrigation and it abhors waterlogged conditions. It is adaptable to a wide range of soil type from sand to heavy clays but prefers well drained, medium textured soils like loams, sandy clay loams, sandy clays, clay loams. It grows well in volcanic soils but is intolerant of alkaline or saline soils. The optimum pH range is from 5.5 to 6.8. The tree grows in full sun or partial shade. Strong winds and salt spray are detrimental to the tree.

Edible Plant Parts and Uses

The ripe fruits are eaten fresh, out of hand, in fruit salad, fruit cocktail, stewed with spices as dessert or eaten dipped in sauces. Pomerac fruit can also be processed into candied fruit slices. Half ripe fruits are used for pickles, jelly, and preserves. In Indonesia, the ripe fruit is sliced and eaten as *petjel* (vegetable eaten with *sambal* sauce) and the young pinkish leaves are eaten raw as *lalab* (salad) with rice. Wine is made from the fruit in Puerto Rico. In Indonesia, the flowers are eaten in salads or are preserved in syrup. The flowers are also eaten in Thailand. Young leaves and shoots are consumed as vegetables. Young leaves and flowers also eaten in Papua New Guinea.

Botany

An evergreen tree growing to 15–30 m high with straight trunk, 130 cm diameter and buttresses. The bark is pale-brown, rough, flaky or fissured. Leaves are opposite, petiolate (10 mm long), elliptic-oblong to broadly oblong-lanceolate, 15–30 cm by 7.5–15 cm, entire, acuminate apex, acute base, glabrous and dark green on both sides, venation pinnate, secondary veins closed, not prominent (Plates 1 and 3). The young leaves are pinkish. Inflorescence consisting of short, few-flowered (4–9) cymes borne on the trunk or older branches. Flowers are showy, red, 2.5 cm across, bisexual; calyx turbinate, 1.2–1.8 cm long with 4 pale yellow, rounded lobes; corolla with 4, red or pink rounded petals, 7–11 mm long, early caducous; ovary inferior, with a simple style up to 2.5 cm long; stamens many (about 200), free, red, 1–2 cm long (Plate 2). Fruit is large, fleshy, ovoid, pear-shaped, dumb-bell shaped or oblong berry 4–7.5 cm long, glossy, waxy, white, greenish white, dark pink, red, purple-red or white with red streaks (Plates 3, 4, 5, 6, and 7). Flesh is white, spongy, juicy and mildly sweet to bland. Each fruit usually contains a single large, brown, subglobose, 1.5–2 cm diameter seed (Plate 7).

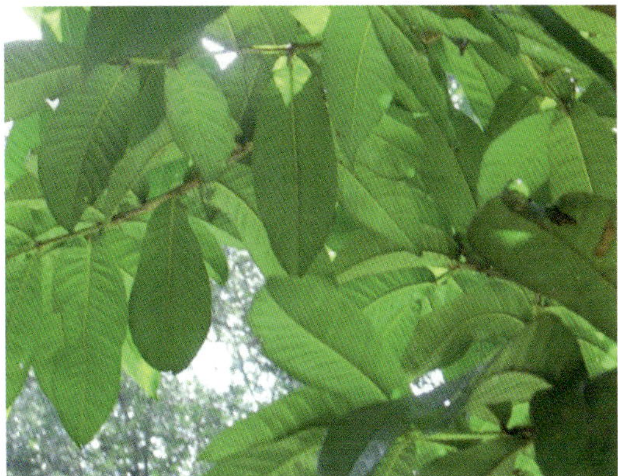

Plate 1 Large, elliptic oblong leaves

Plate 2 Flowers of Malay apple

Plate 3 Fruits and flowers of Malay apple

Plate 4 Close-view of maturing fruits

Plate 5 White dumbbell-shaped cultivar

Plate 6 White Malay apple on sale in a local market

Plate 7 Flesh and seed of red Malay apple cultivar

Nutritive/Medicinal Properties

The nutrient composition of *S. malaccense* fruit per 100 g edible portion was reported by Leung et al. (1972) as: water 91.4 g, energy 30 kcal, protein 0.5 g, fat 0.1 g, carbohydrate 7.62 g, fibre 1.0 g, ash 0.4 g, Ca 18 mg, P 12 mg, Fe 0.4 mg, Na 2 mg, K 114 mg, β-carotene traces, thiamin 0.01 mg, riboflavin 0.01 mg, niacin 0.4 mg and ascorbic acid 17 mg. Another analysis reported the following food value per 100 g edible portion of the fruit (English et al. 1996) as: water 93.8 g, energy 21 kcal (89 kJ), protein 0.7 g, fat 0.2 g, starch <0.1 g, sugars 3.7 g, fibre 1.9 g, ash 0.2 g, Ca 14 mg, Fe 0.9 mg, Na 11 mg, K 75 mg, Zn 0.1 mg, Cu 0.1 mg, Mg 11 mg, Mn 0.1 mg, β-carotene equivalent 16ug, thiamin <0.02 mg, riboflavin

<0.02 mg, niacin 0.2 mg and vitamin C 3 mg. Khoo et al. (2008) reported the fruit to contain 3.95 mg/100 g total carotene content.

Two hundred and three volatile constituents were identified from the essential oil of the fruit (Ismail et al. 2010), with 11 major compounds such as *n*-hexadecanoic acid (18.59%), 9-octadecynoic acid (9.37%), (*Z,Z*)-9-12-octadecadien-1-ol (6.98%), (*E*)-3,7,11- trimethyl-1,6, 10-dodecatrien-3-ol (6.28%), 1-octen-3-ol (2.33%), tetradecanoic acid (1.75%), phytol (1.56%), 2-ethylhexyl *p*-methoxycinnamate (1.43%), (*E*)-2-octen-1-ol (1.26%), 2-phenylethyl accetate (1.25%), and benzyl benzoate (1.00%). The minor compounds consisted of a wide range of compound groups which included fatty acids, aldehydes, ketones, terpenes, alcohols, hydrocarbons, and esters. An earlier study on the essential oil of Malay apple showed 2-phenylethyl acetate, 2-phenylethanol, 1-octen-3-ol and (*E*)-2-octen-1-ol to be the major components (Wong and Lai 1996). The volatiles were characterised by the very low terpenoid content (1.7%). Pino et al. (2004) identified 133 compounds in the aroma concentrate of Malay apple in Cuba, of which 2-phenylethanol and its esters (2-phenylethyl acetate, 2-phenylethyl isopentanoate, 2-phenylethyl benzoate and 2-phenylethyl phenylacetate) were found to be the major constituents. The exotic aroma character of Malay rose apple was found to be the interaction of rose (2-phenylethanol and its esters) and herbaceous (1-octen-3-ol) notes contributing to the complexity of the aroma.

The red-coloured skin of wild-type Malay apple fruit contained five glucose-based anthocyanins (cyanidin-3-O-glucoside, pelargonidin-3-O-glucoside, peonidin-3-O-glucoside, cyanidin-3,5-O-diglucoside and peonidin-3,5-O-diglucoside) (Kotepong et al. 2010). Cyanidin-3-O-glucoside accounted for a large proportion of the total anthocyanin content. The accumulation cyanidin-3-O-glucoside during fruit maturation was correlated with increased activities of phenylalanine ammonia lyase (PAL) and UDPglucose : flavonoid 3-O-glucosyltransferase (UF3GlucT, F3GT). In the wild-type fruit skin, transcripts of seven genes that encode enzymes in the anthocyanin biosynthetic pathway were found. No anthocyanins were detected in the white mutant fruit skin. The skin of the white mutant fruit contained transcripts of all seven genes identified, except F3GT. It also showed no F3GT activity. The data indicated that the lack of anthocyanins in the mutant is due to lack of F3GT expression. The lack of F3GT expression and anthocyanin synthesis during fruit maturation was suggested to be attributed to the lack of MYB transcription factor expression.

In a study conducted in Malaysia, the total carotene content (mg/100 g) of selected underutilized tropical fruit in decreasing order was jentik-jentik (*Baccaurea polyneura*) 19.83 mg>cerapu 2 (*Garcinia prainiana*) 15.81 mg>durian nyekak 2 (*Durio kutejensis*) 14.97 mg>tampoi kuning (*Baccaurea reticulata*) 13.71 mg>durian nyekak (1) 11.16 mg>cerapu 1 6.89 mg>bacang 1 (*Mangifera foetida*) 4.81>kuini (*Mangifera odorata*) 3.95 mg>jambu susu (*Syzygium malaccense*) 3.35 mg>bacang (2) 3.25 mg>durian daun (*Durio lowianus*) 3.04 mg>bacang (3) 2.58 mg>tampoi putih (*Baccaurea macrocarpa*) 1.47 mg>jambu mawar (*Syzygium jambos*) 1.41 mg. Beta carotene content was determined by HPLC to be the highest in jentik-jentik 17.46 mg followed by cerapu (2) 14.59 mg, durian nyekak (2) 10.99 mg, tampoi kuning 10.72 mg, durian nyekak (1) 7.57 mg and cerapu (1) 5.58 mg. These underutilized fruits were found to have acceptable amounts of carotenoids and to have potential as antioxidant fruits.

The essential oil from *S. malaccense* leaves was found to contain 180 compounds, in which 15 (>1.0%) were major and identified as hexanoic acid (12.16%), methyl salicylate (8.27%), 3-hexen-1-ol (7.81%), 1-octen-3-ol (5.89%), *n*-hexadecanoic acid (5.07%), 2-hexenal (4.89%), 3-buten-2-one (3.68%), 1-hexanol (2.96%), phytol (2.95%), acetic acid (2.73%), 3-buten-2-one (2.58%), azulene (1.64%), 2-octen-1-ol (1.34%), α-cadinol (1.11%) and 3-hexen-1-ol (1.10%) (Ismail et al. 2010). The minor compounds of the essential oil from the leaves comprised mostly aldehydes, ketones, sesquiterpenes, alcohols, esters and fatty acids. One common major compound found in both volatile compositions of the leaves and fruit was phytol, which contributed 2.95% and 1.56% of the total oils, respectively. Karioti et al. (2007) identified 38 compounds representing 92.1% of the essential oil from Malay apple leaves in Nigeria. The leaf oil was largely composed of monoterpenes (61.1%) characterized mainly by (+)-α-pinene (7.3%), (−)-β-pinene (8.0%), *p*-cymene (13.5%), and α-terpineol (7.5%). The sesquiterpenes constituted 30.8% of the oil with (−)-β-caryophyllene (9.0%) as the major component.

Some pharmacological properties reported are:

Antioxidant Activity

The percent yield and the amount of total polyphenols in g/100 g calculated as gallic acid in *S. malaccense* dried flowers and crude methanol extract was reported as 22.27%, 0.28 and 3.43 g respectively (Wetwitayaklung et al. 2008). The antioxidant activity in terms of Trolox equivalent antioxidant capacity (TEAC) for the flowers was low (TEAC=0.03 and IC_{50}=3.93.86 μg/50 μl) compared to *Quisqualis indica* flowers (TEAC=0.7 and IC_{50}=13.26 μg/50 μl) but was higher than *Sesbania grandiflora* flowers (TEAC=0.01, IC_{50}=1.32 mg/50 μl). Another study reported that *S. malaccense* fruit had very low antioxidant activity expressed as antiradical DPPH activity with IC_{50} of 269.28 μg/ml, traces of TAC (total anthocyanins content) and 8.58 mg GAE (gallic acid equivalent)/g dry weight of total phenolic content (TPC) (Reynertson et al. 2008). It had the anthocyanin, cyanidin 3-glucoside and the flavonoids, quercitrin and rutin and traces of quercitin and the tannin, ellagic acid.

Xanthine Oxidase Inhibition

Of 19 Myrtaceous plant extracts studied in Puerto Rico, *S. malaccense* plant extract showed the greatest inhibition (64%) of xanthine oxidase with IC_{50}=51 μg/ml and also exhibited in-vitro inhibition of cyclooxgenase-1 and -2 (Guerrero and Guzman 1998). Xanthine oxidase inhibitors are used for prevention of gout and the formation of cataract.

Ichthyotoxic Effects

The hexane fraction of *S. malaccense* leaves was found to be ichthyotoxic against tilapia fish (*Tilapia oreochromis*); however three compounds isolated from the extract namely ursolic acid (1), β-sitosterol (2) and sitost-4-en-3-one (3) exhibited no significant ichthyotoxicity (Ismail et al. 2010).

Traditional Medicinal Uses

In Brazil, various parts of the plant are used as remedies for constipation, diabetes, coughs, pulmonary catarrh, headache, and other ailments. In Cambodia, the root, fruit, leaves, and seeds are used as a febrifuge for fever, and the root is regarded as diuretic. In the Moluccas, the astringent bark is used for making a mouthwash for thrush. A root-bark decoction is used for dysentery, amenorrhea and as abortifacient. In the Marquesas, coconut oil in which fragments of the bark have been soaked is taken as a purgative. The root may enter into an application for itches and a preparation of the root is given to alleviate swelling. An infusion of the bark is used for coughs in Futuna and Niue. In Tonga, an extract of bark scrapings is commonly administered to treat throat infections and stomachache, and is an ingredient in various remedies for a complex of abdominal ailments known as kahi. In Hawaii, the bark was crushed and its juices taken for sore throat, juice extracted from the bark was mixed with salt and applied to cuts. In Malaysia, powder from the dried leaves is reportedly used on a cracked tongue. In Samoa, an infusion of the crushed leaves or scraped bark is very commonly taken as a potion, the leaves are chewed and the juice swallowed, or the leaf juice is dripped into the mouth of an infant to treat mouth infections. In the Cook Islands, a solution of the crushed leaves, or to a lesser extent the grated bark, is widely used to treat thrush.

Other Uses

The tough, reddish timber is employed for construction as house posts, fence posts, rafters, railway sleepers and for carving bowls. In Chuuk (Micronesia), it has been used to make outrigger booms. It is occasionally used for fuel-wood. In Hawai'i, both blossoms and fruit were used to make leis.

Comments

The fruits of *S. malaccense* are bigger than the fruits of *S. aquem* but slightly smaller than *S. samarangense* fruits.

Selected References

Baccus-Taylor GSH, Frederick PD-A, Akingbala JO (2009) Studies on pomerac (*Syzygium malaccense*) candied fruit slices. Acta Hort (ISHS) 806:293–300

Backer CA, Bakhuizen van den Brink RC Jr (1963) Flora of java. (Spermatophytes only), vol 1. Noordhoff, Groningen, 648 pp

Burkill IH (1966) A dictionary of the economic products of the Malay Peninsula. Revised reprint, 2 vols. Ministry of Agriculture and Co-operatives, Kuala Lumpur, vol 1 (A–H), pp 1–1240, vol 2 (I–Z), pp 1241–2444

Cambie RC, Ash JE (1994) Fijian medicinal plants. CSIRO Australia, Collingwood

Chen J, Craven LA (2007) Myrtaceae. In: Wu ZY, Raven PH, Hong DY (eds) Flora of china, vol 13 (Clusiaceae through Araliaceae). Science Press/Missouri Botanical Garden Press, Beijing/St. Louis

English RM, Aalbersberg W, Scheelings P (1996) Pacific island foods – description and nutrient composition of 78 local foods. IAS Technical Report 96/02. Institute of Applied Science, University of the South Pacific, Suva, 94 pp

French BR (1986) Food plants of Papua New Guinea – a compendium. Australia and Pacific Science Foundation, Ashgrove, Brisbane, 408 pp

Govaerts R, Sobral M, Ashton P, Barrie F, Holst BK, Landrum LL, Matsumoto K, Fernanda Mazine F, Nic Lughadha E, Proenca C, Soares-Silva LH, Wilson PG, Lucas E (2010) World checklist of Myrtaceae. The board of trustees of the Royal Botanic Gardens, Kew. Published on the Internet, http://www. kew.org/wcsp/. Accessed 22 Apr 2010

Guerrero RO, Guzman AL (1998) Inhibition of xanthine oxidase by Puerto Rican plant extracts. P R Health Sci J 17(4):359–364

Hancock IR, Henderson CP (1988) Flora of the Solomon Islands. Dodo Creek Research Station, Research Department, Ministry of Agriculture and Lands, Research Bulletin No 7, 204 pp

Ismail IS, Ismail NA, Lajis N (2010) Ichthyotoxic properties and essential oils of *Syzygium malaccense* (Myrtaceae). Pertanika J Sci Technol 18(1):1–6

Karioti A, Skaltsa H, Gbolade AA (2007) Analysis of the leaf oil of *Syzygium malaccense* Merr. et Perry from Nigeria. J Essent Oil Res 19(4):313–315

Khoo HE, Ismail A, Mohd.-Esa N, Idris S (2008) Carotenoid content of underutilized tropical fruits. Plant Foods Hum Nutr 63:170–175

Kotepong P, Ketsa S, van Doorn WG (2010) A white mutant of Malay apple fruit (*Syzygium malaccense*) lacks transcript expression and activity for the last enzyme of anthocyanin synthesis, and the normal expression of a MYB transcription factor. Funct Plant Biol 38(1):75–86

Leung W-TW, Butrum RR, Huang Chang F, Narayana Rao M, Polacchi W (1972) Food composition table for use in east Asia. FAO, Rome, 347 pp

Morton JF (1987) Malay apple. In: Fruits of warm climates. Julia F. Morton, Miami, pp 378–381

Ochse JJ, Bakhuizen van den Brink RC (1931) Fruits and fruitculture in the Dutch East Indies. G. Kolff, Batavia-C, 180 pp

Ochse JJ, Bakhuizen van den Brink RC (1980) Vegetables of the Dutch Indies, 3rd edn. Ascher, Amsterdam, 1016 pp

Panggabean G (1992) *Syzygium aqueum* (Burm.f.) Alst., *Syzygium malaccense* (L.) Merr. & Perry, and *Syzygium samarangense* (Blume) Merr. & Perry. In: Verheij EWM, Coronel RE (eds) Plant resources of south-east Asia. No. 2. Edible fruits and nuts. Prosea Foundation, Bogor, pp 292–294

Pino JA, Marbot R, Rosado CA (2004) Volatile constituents of Malay rose apple [*Syzygium malaccense* (L.) Merr. & Perry]. Flavour Fragr J 19(1):32–35

Reynertson KA, Yang H, Jiang B, Basile MJ, Kennelly EJ (2008) Quantitative analysis of antiradical phenolic constituents from fourteen edible Myrtaceae fruits. Food Chem 109(4):883–890

Walter A, Sam C (2002) Fruits of Oceania. ACIAR monograph No 85. Australian Centre for International Agricultural Research, Canberra, 329 pp

Wetwitayaklung P, Phaechamud T, Limmatvapirat C, Keokitichai S (2008) The study of antioxidant activities of edible flower extracts. Acta Hort (ISHS) 786:185–192

Whistler WA, Elevitch CR (2006) *Syzygium malaccense* (Malay apple), ver. 2.1. In: Elevitch CR (ed) Species profiles for Pacific island agroforestry. Permanent Agriculture Resources (PAR), Holualoa, Hawai'i. http://www traditionaltree.org

Wong KC, Lai FY (1996) Volatile constituents from the fruits of four *Syzygium* species grown in Malaysia. Flavour Fragr J 11:61–66

Syzygium polycephalum

Scientific Name

Syzygium polycephalum **(Miq.) Merr. & L.M. Perry**.

Synonyms

Eugenia polycephala Miq. (basionym), *Jambosa cauliflora* DC., *Jambosa polycephala* (Miq.) Miq., *Syzygium cauliflorum* (DC.) Bennet nom. illeg.

Family

Myrtaceae

Common/English Name

Gowok

Vernacular Names

Indonesia: Gowok (Malay), Gowok, Domjong, Kaliasem, Klesem (Java), Gowok, Kupa, Kupa Benjer, Kupa Gowok, Kupa Manuk (Sundanese); ***Philippines***: Lipote.

Origin/Distribution

Gowok is indigenous to West and Central Malesia. It is common in Java and Kalimantan in Indonesia.

Agroecology

Gowok is strictly a tropical species. It occurs wild in secondary forests, between 200–1,800 m altitude, but is often cultivated in home gardens as a fruit tree.

Edible Plant Parts and Uses

Ripe fruit is eaten fresh or made into *rujak* or candied and used to make a highly sought jelly. Young leafy shoots and sprouts are eaten as vegetables; often eaten raw as lalab with rice.

Botany

Gowok is a low, perennial tree, 8–20 m tall, stem diameter up to 50 cm with a dense crown. Leaves opposite, sessile or shortly petiolate, large, oblong, oblong-lanceolate or subobovate, 17–25 cm × 6–7 cm, base cordate, apex acuminate, obtuse often gland – dotted, dark green

T.K. Lim, *Edible Medicinal And Non-Medicinal Plants: Volume 3, Fruits*,
DOI 10.1007/978-94-007-2534-8_104, © Springer Science+Business Media B.V. 2012

Plate 1 Ripe Gowok fruit clusters with stalks intact

glossy with 12–14 lateral nerves on either side of mid rib. Young leaves purplish. Panicles widely branched, 5–13 cm long on leaves twig. Flowers small, sessile in three or more flowered fascicles on the panicle. Calyx tube, campanulate-turbinate greenish white, with four broadly ovate acute teeth. Petals 4, ovate, yellowish white, early caducous. Stamens white, with 4–6 mm long filaments, numerous around the orbicular disk, style light green, 7 mm long. Fruit a depressed globose berry, 2.5–3.5 cm diameter, fleshy with acid sweet white or pinkish purple pulp, crowned by the calyx limb, dark violet, glossy when ripe (Plates 1 and 2). The fruit is produced in dense clusters on the twigs.

Nutritive/Medicinal Value

No published information is available on the nutritive composition of the fruits.

Syzygium polycephalum is one of many plants used by the Sundanese community in West Java for dysentery.

Other Uses

Its wood is used for house construction.

Plate 2 Ripe, globose, dark purple, glossy Gowok fruit

Comments

Gowok is propagated from seeds and cuttings.

Selected Reference

Backer CA, van den Brink RCB Jr (1963) Flora of java, (spermatophytes only), vol 1. Noordhoff, Groningen, 648 pp

Govaerts R, Sobral M, Ashton P, Barrie F, Holst BK, Landrum LL, Matsumoto K, Fernanda Mazine F, Nic Lughadha E, Proenca C, Soares-Silva LH, Wilson PG, Lucas E (2010) World checklist of Myrtaceae. The board of trustees of the Royal Botanic Gardens, Kew. Published on the Internet, http://www. kew.org/wcsp/. Accessed 22 Apr 2010

Ochse JJ (1927) Indische vruchten [Indonesian fruits]. Volkslectuur, Weltevreden, 330 pp (In Dutch)

Ochse JJ, van den Brink RCB (1980) Vegetables of the Dutch East Indies, 2nd edn. Asher & Co, Amsterdam, 1016 pp

Jansen PCM, Jukema J, Oyen LPA, van Lingen TG (1991) *Syzygium polycephalum* (Miq.) Merr. & Perry. In: Verheij EWM, Coronel RE (eds) Plant resources of South-East Asia No. 2. Edible fruits and nuts. Prosea Foundation, Bogor, p 361

Roosita K, Kusharto CM, Sekiyama M, Fachrurozi Y, Ohtsuka R (2008) Medicinal plants used by the villagers of a sundanese community in west java, Indonesia. J Ethnopharmacol 115(1):72–81

Syzygium samarangense

Scientific Name

Syzygium samarangense **(Blume) Merr. & Perry**.

Synonyms

Eugenia javanica Lam., *Eugenia javanica* var. *parviflora* Craib, *Eugenia javanica* var. *roxburghiana* Duthie, *Eugenia samarangensis (Blume) O. Berg, Jambosa javanica* (Lam.) K.Schum. & Lauterb., *Jambosa samarangense* (Blume) DC., *Myrtus javanica* (Lam.) Blume, *Myrtus samarangensis* Blume (basionym), *Syzygium samarangense* var. *parviflorum* (Craib) Chantaran. & J. Parn.

Family

Myrtaceae

Common/English Names

Java Apple, Java Roseapple, Mountain Apple (Pacific Islands), Samarang Rose Apple, Wax Apple, Wax Jambu

Vernacular Names

Chinese: Jin Shan Pu Tao, Lián Wù, Lián-Bū Nan Yang Pu Tao, Yang Pu Tao;
Costa Rica: Cashu Di Surinam, Makopa;
Curacao: Cashu Di Surinam, Curacaose Appel, Kashu Sürnam;
Dominican Republic: Cajuil De Suliman, Cajuilito;
Dutch: Curacaose Appel, Djamboo Semarang;
Fiji: Kavika;
French: Pomme D'eau De Formose, Pomme De Java;
German: Java-Wachsapfel, Java-Apfel;
Indonesia Kebes, Kembes (Alfurese, North Sulawesi), Jambu Andak, Jambu Mawar (Batak), Jambu Putih (Bengkoelen, Singkep), Libi (Bima), Omuti (Gorontalo), Jambu Bu Buik (Jambi), Jambu Kaget, Jambu Klampok, Jambu Semarang, Jambu Tersana (Java), Jambu Ayoh Mauh, Jambu Mawar, Jambu Ujan (Lampong), Jambu Riou (Lingga), Jambu Selong (Makassar), Jambu Semarang, Jambu Semarang Merah, Jmabu Semarang Putih (Malay), Gora (Manado), Belo, Raun, Tilo (Solor), Jambu Samarang, Jambu Samarang Berem, Jambu Samarang Bodas (Sundanese);
India: Chambekka (Malayalam), Jumrool, Jamrul (Bengali), Amrool;
Japanese: Renbu;
Malaysia: Jambu Ayer Rhio, Jambu Air Mawar;
Mauritius: Zamalac;
Nepalese: Ambake;
Panama: Manzana De Java, Marañon De Curacao;
Papiamentu: Kashu Sürnam ;
Philippines: Makopa, Tambis;
Reunion Islands: Zamalac;

T.K. Lim, *Edible Medicinal And Non-Medicinal Plants: Volume 3, Fruits*,
DOI 10.1007/978-94-007-2534-8_105, © Springer Science+Business Media B.V. 2012

Saint Kitts & Nevis: Morraoca;
Seychelles: Zamalac;
Sri Lanka: Pini Jambu, Jumbu;
Surinam: Curaçaose Appel;
Taiwan: Lián Wù, Lián-Bū;
Thailand: Chom Phûu, Chomphuu Kaem Maem, Chomphuu Kao, Chomphuu Khieo, Chomphuu Nak;
Trinidad & Tobago: Pommerac;
Vietnamese: Man, Roi.

Origin/Distribution

Java apple is indigenous in Bangladesh to the Solomon Islands. It has naturalized in the Philippines since prehistoric times. It is commonly and widely cultivated in Malaysia, Indonesia, Thailand, Cambodia, Laos, Vietnam and Taiwan. It is frequently cultivated in India and in Africa (Zanzibar and Pemba) and also in the Antilles, Suriname and northern Australia.

Agroecology

This species thrives in the warm fairly moist tropical lowlands up to 1,200 m altitude. It is more adaptable than *S. malaccense* to drier conditions and can withstand long dry season with a reliable water supply from streams and ponds. This species prefers heavy, fertile soils.

Edible Plant Parts and Uses

The ripe fruit is eaten fresh out of hand or they are sliced and eaten in fruit cocktails, or dipped in salt, in a sweetened dark soya sauce, in sambal or in a fruit salad "*rujak*" mixed with a spicy peanut sauce. In Indonesia the fruit are also preserved in pickles, "*asinan*". Java apple cal also be stewed like apples. The fruit is frequently used in salads, and used in light sautéed dishes in Indian ocean island cuisine. The ripened fruit varies in hue and can be from yellowish-greenish-white to light pink to red to a dark, maroon purple.

Botany

Evergreen Tree, 5–15 m tall, with short and crooked trunk, 25–50 cm diameter, with pinkish-grey flaky bark, often branched near the base and with wide, irregular crown. Leaves opposite, elliptic to elliptic-oblong, 10–25 cm × 5–12 cm, base cordate, apex obtuse to slightly acuminate, coriaceous with thin margin, pellucid dotted, 14–19 pairs of nerve, rather strongly aromatic when bruised; petiole stout, 3–5 mm long, yellowish green (Plate 2) sometimes tinged with purple. Leaves are pink to dark violet when young turning yellowish green to green with age. Inflorescences are terminal and in axils of fallen leaves, 3–30-flowered; flowers fragrant, 3–4 cm in diameter, calyx-tube 1.5 cm long, 4-lobed, ventricose at apex, lobes 3–5 mm long; petals 4, orbicular to spathulate, 10–15 mm long, yellow-white; stamens numerous, up to 3 cm long; style up to 3 cm long (Plate 1). Fruit a berry, broadly pyriform, crowned by the fleshy calyx with incurved lobes, 3.5–5.5 cm × 4.5–5.5 cm, waxy and glossy, in various hues – white to greenish-white to pink, red, maroon and dark purple (Plates 2–10); flesh white spongy (Plate 3), juicy, aromatic, mildly sweet and crisp. Seeds 0–2, mostly suppressed, globose, up to 8 mm in diameter.

Nutritive/Medicinal Properties

The nutrient composition of *S. samarangense* fruit per 100 g edible portion (Leung et al. 1972) was reported as: water 91.5 g, energy 30 kcal, protein 0.4 g, fat 0.1 g, carbohydrate 7.8 g, fibre 0.8 g, ash 0.2 g, Ca 17 mg, P 9 mg, Fe 0.3 mg, Na 2 mg, K 105 mg, β-carotene 0 μg, thiamin 0.03 mg, riboflavin 0.01 mg, niacin 0.3 mg, ascorbic acid 13 mg. Another analysis conducted in Australia reported that wax jambu had the following food value per 100 g edible portion (Wills 1987): water 90.3%, protein 0.7 g, fat 0.2 g, glucose 2.1 g, fructose 2.4 g, dietary fibre 1.9 g, malic acid 0.10 g, citric acid 0.12 g, oxalic acid 0.02 g, energy 94 kJ, vitamin C 8 mg, thiamin 0.02 mg, riboflavin

Plate 1 Flower buds and opened flowers of Java apple

Plate 2 Leaves and ripe pink fruits of Java apple

Plate 3 Whole and halved Java apple fruits

0.04 mg, niacin 0.5 mg, K 38 mg, Na 1 mg, Ca 13 mg, Mg 5 mg, Fe 0.8 mg and Zn 0.1 mg.

A total of 39 volatile constituents were identified in wax jambu (*S. samarangense*) (Wong and Lai 1996). The volatiles of wax jambu were characterized by the presence of a large number of C9 aldehydes and alcohols.

A triterpene, methyl 3-epi-betulinate in its native form and 4′,6′-dihydroxy-2′-methoxy-3′,5′-dimethyl chalcone along with ursolic acid, jacoumaric acid and arjunolic acid were isolated from the aerial parts of *Syzygium samarangense* (Srivastava et al. 1995).

Various parts of the plant have been reported to have bioactive compounds and to exhibit antioxidant, anticancer, antiviral, antimicrobial, spasmolytic, antihyperglycaemic, protease inhibition, anti-amnesiac and immunomodulatory activities.

Antioxidant Activity

S. samarangse fruit displayed moderate antioxidant activity expressed as antiradical DPPH activity with IC_{50} of 77.51 μg/ml, 0.07 mg C3G/g dry weight of TAC (total anthocyanins content) and 18.04 mg GAE (gallic acid equivalent)/g dry weight of total phenolic content (TPC) (Reynertson et al. 2008). The fruit had traces of the anthocyanins cyanidin 3-glucoside and flavonoids, quercitin and quercitrin, and traces of myricetin and rutin. It also contained ellagic acid. *S. samarangense* also contained samarangenins A and B, novel proanthocyanidins with doubly bonded structures (Nonaka et al. 1992). Another study reported the fruit to have scavenging 1, 1-diphenyl-2-picryl-hydrazyl (DPPH) free radical ability, the ethanolic extract of the fruit exhibited an IC_{50} of 200 μg/ml (Soubir 2007).

Plate 4 Fruits crowned by the fleshy calyx with incurved lobes

The fruit contained two flavonol glycosides as well as epigallocatechin 3-Ogallate, epicatechin3-O-gallate, and samarangenin A and B (Harborne and Baxter 1999) and exhibited an antioxidant activity expressed as a DPPH (2,2 diphenyl−1-picryl-hydrazyl) IC_{50} value of 76.8 mg/ml compared to α-tocopherol with an IC_{50} of 53.3 mg/ml (Reynertson et al. 2005). Eight antioxidants were obtained from the fruit, viz. six quercetin glycosides: reynoutrin, hyperin, myricitrin, quercitrin, quercetin, and guaijaverin (10), one flavanone: (S)-pinocembrin (8), and two phenolic acids: gallic acid and ellagic acid (Simirgiotis et al. 2008).

Anticancer Activity

Ethanolic extracts of the fruit powder showed antioxidant activities indicating java apple fruit to be beneficial to human health. Studies reported the methanolic extracts of the pulp and seeds of

Plate 5 (**a** and **b**) White variety of Java apple

the fruits of *Syzygium samarangense* yielded four cytotoxic flavanoid compounds and eight antioxidant compounds (Simirgiotis et al. 2008). Three C-methylated chalcones, 2′,4′-dihydroxy-3′,5′-dimethyl-6′-methoxychalcone, 2′,4′-dihydroxy-3′-methyl-6′-methoxychalcone (stercurensin), and 2′,4′-dihydroxy-6′-methoxychalcone (cardamonin,) were isolated and displayed cytotoxic activity (IC_{50} = 10, 35, and 35 μM, respectively) against the SW-480 human colon cancer cell line.

Plate 6 Dark red variety of Java apple

Plate 7 Dark maroon variety of Java apple

Among the flavonoids isolated from *Syzygium samarangense*, namely 2′-hydroxy-4′,6′-dimethoxy-3′-methylchalcone (1), 2′,4′-dihydroxy-6′-methoxy-3′,5′-dimethylchalcone (2), 2′,4′-dihydroxy-6′-methoxy-3′-methylchalcone (3), 2′,4′-dihydroxy-6′-methoxy-3′-methyldihydrochalcone (4) and 2′,4′-dihydroxy-6′-methoxy-3′,5′-dimethyldihydrochalcone (5), compound 2 exhibited cytotoxicity testing using the dimethylthiazoldiphenyl tetrazolium (MTT) assay (Amor et al. 2007). Compound 2 exhibited significant differential cytotoxicity against the human mammary adenocarcinoma MCF-7 cell line with an IC_{50} of 0.0015 nM. It was also cytotoxic against the human mammary adenocarcinoma SKBR-3 cell line with an IC_{50} of 0.0128 nM. Doxorubicin, the positive control, had an IC_{50} of 2.60×10^{-4} nM against the MCF-7 cell line and an IC_{50} of 2.76×10^{-5} nM against the SKBR-3 cell line. Compound 2 showed significant selective cytotoxicity against the RAD52 yeast mutant strain. It had an IC_{12} of 0.1482 nM, as compared with the positive control, streptonigrin, which had an IC_{12} of 0.0134 nM. Hence, compound 2 was deemed a cytotoxic natural product with potential anticancer application.

Results of studies suggested that dimethyl cardamonin (2’,4’-dihydroxy-6’-methoxy-3’,5’-

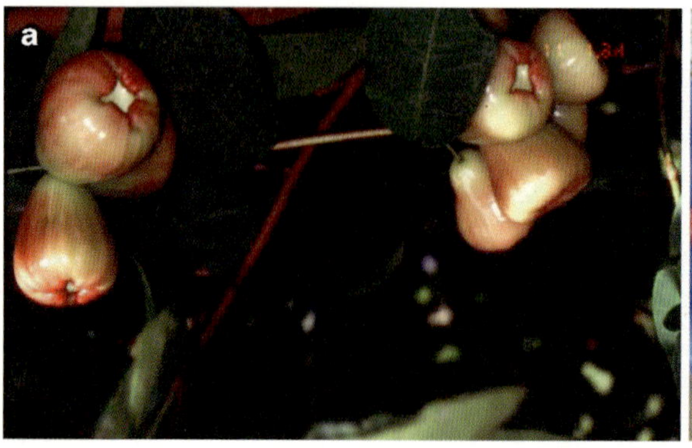

Plate 8 (**a** and **b**) Multi pink-green coloured variety of Java apple

Plate 9 (**a** and **b**) Pale greenish-white variety of Java apple

Plate 10 Red Java apple (*left*) and greenish-white Java apple on sale in the market in Bangkok

dimethylchalcone; DMC), a naturally occurring chalcone, and major compound isolated from *Syzygium samarangense* leaves, possessed anti-proliferative activity (Ko et al. 2011). DMC suppressed colorectal carcinoma HCT116 and LOVO cell proliferation through a G(2) /M phase cell-cycle delay, and induced autophagy, the hallmark of Type II programmed cell death.

Antimicrobial Activity

Scientists found the flower's principal constituent to be tannin. The flowers also had desmethoxy-matteucinol,5-*O*-methyl-4′-desmethoxymatteucinol, oleanic acid and β-sitosterol (Morton 1987). In scientific research the flowers have shown weak antibiotic action against *Staphylococcus aureus, Mycobacterium smegmatis*, and *Candida albicans* (Morton 1987).

Antiviral Activity

Oleanolic acid, an anti-HIV compound and ursolic acid have also been isolated from leaves. Both oleanolic acid and ursolic acid were found effective in protecting against chemically induced liver injury in laboratory animals (Liu 1995). Oleanolic acid has been marketed in China as an oral drug for human liver disorders. Oleanolic acid and ursolic acid have also been long-recognized to have antiinflammatory and antihyperlipidemic properties in laboratory animals. Oleanolic acid and ursolic acid are relatively nontoxic, and have been used in cosmetics and health products (Liu 1995). Oleanolic acid, also possessed anti-HIV activity. Studies showed that oleanolic acid inhibited the human immunodeficiency virus-1 (HIV-1) replication in all the cellular systems (cultures of human peripheral mononuclear cells (PBMC) and of monocyte/macrophages) (Mengoni et al. 2002).

Protease Inhibitory/Antiamnesiac Activity

Compounds isolated from the hexane extract of the leaves of *Syzygium samarangense* exhibited inhibitory activity against the following serine proteases: trypsin, thrombin and prolyl endopeptidase (Amor et al. 2004). The compounds were identified as a mixture of α-carotene and β-carotene (1), lupeol (2), betulin (3), epi-betulinic acid (4), 2′,4′-dihydroxy-6′-methoxy-3′-methylchalcone (5), 2′-hydroxy-4′,6′-dimeth oxy-3′-methylchalcone (6), 2′,4′-dihydroxy-6′-methoxy-3′,5′-dimethylchalcone (7), 2′,4′-dihydroxy-6′-methoxy-3′-methyldihydrochalcone (8) and 7-hydroxy-5-methoxy-6,8-dimethylflavanone (9). Hydrogenation of compounds 5, 6 and 7 yielded compound 8, 2′-hydroxy-4′,6′-dimethoxy -3′-methyldihydrochalcone (10) and 2′,4′-dihydroxy-6′-methoxy-3′,5′-dimethyldihydrochalcone (11), respectively. In addition, β-sitosterol (12) and β-D-sitosterylglucoside (13) were also isolated. Compounds 3–8 and 10 exhibited significant and selective inhibition against prolyl endopeptidase among three serine proteases. Inhibitors of prolyl endopeptidase may improve memory by blocking the metabolism of endogenous neuropeptides and may have possible potential as anti-amnesiac drugs (Yoshimoto et al. 1987). Currently, new drugs are required that can improve memory and learning or delay the neurodegenerative process in conditions such as Alzheimer's disease. Prolyl endopeptidase (PEP) an enzyme with a role in metabolism of proline-containing neuropeptides, such as vasopressin, substance P and thyrotropin-releasing hormone (TRH), were suggested to be involved with learning and memory process (Tezuka et al. 1999).

Spasmolytic Activity

The hexane extract of *Syzygium samarangense* dose-dependently (10–1,000 μg/ml) relaxed the spontaneously contracting isolated rabbit jejunum. Four flavonoids isolated from the hexane extract: 2′-hydroxy-4′,6′-dimethoxy-3′-methylchalcone (SS1), 2′,4′-dihydroxy-6′-methoxy-3′,5′-dimethylchalcone (SS2), 2′, 4′-dihydroxy-6′-methoxy-3′-methylchalcone (SS3) and 7-hydroxy-5-methoxy-6,8-dimethylflavanone (SS4), showed dose-dependent (10–1,000 μg/ml) spasmolytic activity with SS2 being the most potent (Amor et al. 2005; Ghayur et al. 2006). These results indicated that the presence of compounds with spasmolytic and calcium antagonist activity may be responsible for the medicinal use of the plant in diarrhoea. The dihydrochalcone derivative of compound SS1, 2′-hydroxy-4′,6′-dimethoxy-3′-methyldihydrochalcone, when tested for spasmolytic activity, did not significantly relax the smooth muscle relative to the other compounds. The findings indicated the relaxant activity of chalcones, specifically of compounds SS1-3. The flowers were found to be astringent and used in Taiwan to treat fever and stop diarrhoea. Scientists found the flower's principal constituent to be tannin (Morton 1987).

Immunomodulatory Activity

Investigators found that among 16 flavonoids isolated from the acetone extract of the leaves, (−)-strobopinin, myricetin 3-O-(2″-O-galloyl)-α-rhamnopyranoside, (−)-epigallocatechin 3-O-gallate and myricetin 3-O-α-rhamnopyranoside displayed inhibitory potency on human peripheral blood mononuclear cells (PBMC) proliferation activated by phytohemagglutinin (PHA) (Kuo et al. 2004). The inhibitory mechanisms may involve the blocking of interleukin-2 (IL-2) and interferon-γ (IFN-γ) production, since compounds 2, 8, 10 and 11 reduced IL-2 and IFN-γ production in PBMC in a dose-dependent manner.

Antihyperglycaemic Activity

Research reported that the flavonoid, 2′,4′-dihydroxy-3′,5′-dimethyl-6′-methoxychalcone 1, its isomeric flavanone 5-O-methyl-4′-desmethoxymatte-ucinol 2 and 2′4′dihydroxy-6′methoxy-3′-methylchalcone 3 isolated from the leaves significantly lowered the blood glucose levels (BGLs) in glucose-hyperglycaemic mice when administered 15 minutes after a glucose load, indicating their antihyperglycaemic property (Resurreccion-Magno et al. 2005).

Analgesic and Antiinflammatory Activities

Cycloartenyl stearate, lupenyl stearate, sitosteryl stearate, and 24-methylenecycloartanyl stearate (sample 1) from the air-dried leaves of *Syzygium samarangense* exhibited potent analgesic and anti-inflammatory activities at effective doses of 6.25 mg/kg body weight and 12.5 mg/kg body weight, respectively (Raga et al. 2011). Sample 1 also exhibited negligible toxicity on zebrafish embryonic tissues.

A chalcone, 2′,4′-dihydroxy-6′-methoxy-3′-methylchalcone (stercurensin), active compound isolated from the leaves of *Syzygium samarangense* was found to have anti-inflammatory activity (Kim et al. 2011). Pretreatment of isolated mouse peritoneal macrophages with stercurensin reduced lipopolysaccharide-induced iNOS (inducible nitric oxide synthase) and cyclooxygenase-2 expression, thereby inhibiting nitric oxide (NO) and prostaglandin E(2) production, respectively. In addition, an inhibitory effect of stercurensin on NF-κB activation was shown by the recovery of lipopolysaccharide -induced inhibitor of κB (I-κB) degradation after blocking the transforming growth factor-β-activated kinase 1 (TAK1)/I-κB kinase signalling pathways. In mouse models, stercurensin negatively regulated NF-κB-dependent pro-inflammatory mediators and cytokines. These results demonstrated that stercurensin modulated NF-κB-dependent inflammatory pathways through the attenuation of TAK1-TAB1 complex formation. TAB1 is an activator subunit of TAK1.

Other Uses

Java apple is also cultivated as ornamental or wind break. The red hard wood is used for house construction in the Nicobar and Andaman Islands.

Comments

Since the last two decades remarkable success has been achieved in the development of wax jambu varieties which are larger, firmer, crispier, juicier, better flavoured and mildly sweeter and have better handling and keeping qualities. Some of these improved selections such as the dark purplish ones dubbed *Black Pearl* or *Black Diamond* or the yellowish-greenish white cultivar nicknamed *Pearl* are highly sought and highly prized in Taiwan, Thailand, Malaysia, Indonesia and Vietnam.

Selected References

Amor EC, Villasenor IM, Yasin A, Choudhary MI (2004) Prolyl endopeptidase inhibitors from *Syzygium samarangense* (Blume) Merr. & L. M. Perry. Z Naturforsch C 59(1–2):86–92

Amor EC, Villasenor IM, Ghayur MN, Gilani AH, Choudhary MI (2005) Spasmolytic flavonoids from *Syzygium samarangense* (Blume) Merr. & L.M Perry. Z Naturforsch C 60(1–2):67–71

Amor EC, Villasenor IM, Antemano R, Perveen Z, Concepcion G, Choudhary MI (2007) Cytotoxic c-methylated chalcones from *Syzygium samarangense*. Pharm Biol 45(10):777–783

Burkill IH (1966) A dictionary of the economic products of the malay peninsula. Revised reprint, 2 vols, Ministry of Agriculture and Co-operatives, Kuala Lumpur, vol 1 (A–H), pp 1–1240, vol 2 (I–Z), pp 1241–2444

De Nanteuil G, Portevin B, Lepagnol J (1998) Proplyl endopeptidase inhibitors: a new class of memory enhancing drugs. Drugs Future 23:167–179

Ghayur MN, Gilani AH, Khan A, Amor EC, Villaseñor IM, Choudhary MI (2006) Presence of calcium antagonist activity explains the use of *Syzygium samarangense* in diarrhoea. Phytother Res 20(1):49–52

Govaerts R, Sobral M, Ashton P, Barrie F, Holst BK, Landrum LL, Matsumoto K, Fernanda Mazine F, Nic Lughadha E, Proenca C, Soares-Silva LH, Wilson PG, Lucas E (2010) World checklist of Myrtaceae. The board of trustees of the Royal Botanic Gardens, Kew. Published on the Internet, http://www. kew.org/wcsp/. Accessed 22 Apr 2010

Harborne JB, Baxter H (eds) (1999) The handbook of natural flavonoids, vol 2. Wiley, New York, 879 pp

Kim YJ, Kim HC, Ko H, Amor EC, Lee JW, Yang HO (2011) Stercurensin inhibits nuclear factor-κB-dependent inflammatory signals through attenuation of TAK1-TAB1 complex formation. J Cell Biochem 112(2):548–558

Ko H, Kim YJ, Amor EC, Lee JW, Kim HC, Kim HJ, Yang HO (2011) Induction of autophagy by dimethyl cardamonin is associated with proliferative arrest in human colorectal carcinoma HCT116 and LOVO cells. J Cell Biochem 112(9):2471–2479

Kuo YC, Yang LM, Lin LC (2004) Isolation and immunomodulatory effect of flavonoids from *Syzygium samarangense*. Planta Med 70(12):1237–1239

Leung W-TW, Butrum RR, Huang Chang F, Narayana Rao M, Polacchi W (1972) Food composition table for use in East Asia. FAO, Rome, 347 pp

Liu J (1995) Pharmacology of oleanolic acid and ursolic acid. J Ethnopharmacol 49(2):57–68

Mengoni F, Lichtner M, Battinelli L, Marzi M, Mastroianni CM, Vullo V, Mazzanti G (2002) In vitro anti-HIV activity of oleanolic acid on infected human mononuclear cells. Planta Med 68(2):111–114

Morton J (1987) Java apple. In: Fruits of warm climates. Julia F. Morton, Miami, pp 381–382

Nair AGR, Krishnan S, Ravikrishna C, Madhusudanan KP (1999) New and rare flavonol sides from leaves of *Syzygium samarangense*. Fitoterapia 70:148–151

Nonaka G, Aiko Y, Aritake K, Nishioka I (1992) Tannins and related compounds CXIX. Samarangenins A and B, novel proanthocyanidins with doubly bonded structures, from *Syzygium samarangense* and *S. aqueum*. Chem Pharm Bull 40:2671–2673

Ochse JJ, Bakhuizen van den Brink RC (1931) Fruits and fruitculture in the Dutch East Indies. G. Kolff & Co, Batavia, 180 pp

Panggabean G (1992) *Syzygium aqueum* (Burm.f.) Alst., *Syzygium malaccense* (L.) Merr. & Perry, and *Syzygium samarangense* (Blume) Merr. & Perry. In: Verheij EWM, Coronel RE (eds) Plant Resources of South-East Asia. No. 2: Edible fruits and nuts. Prosea Foundation, Bogor, pp 292–294

Raga DD, Cheng CLC, Lee KCLC, Olaziman WZP, De Guzman VJA, Shen CC, Franco FC Jr, Ragasa CY (2011) Bioactivities of triterpenes and a sterol from *Syzygium samarangense*. Z Naturforsch C 66c:235–244

Resurreccion-Magno MHC, Villasenor IM, Harada N, Monde K (2005) Antihyperglycaemic flavonoids from *Syzygium samarangense* (Blume) Merr. and Perry. Phytother Res 19(3):246–251

Reynertson KA, Basile MJ, Kennelly EJ (2005) Antioxidant potential of seven myrtaceous fruits. Ethnobot Res Appl 3:25–35

Reynertson KA, Yang H, Jiang B, Basile MJ, Kennelly EJ (2008) Quantitative analysis of antiradical phenolic constituents from fourteen edible Myrtaceae fruits. Food Chem 109(4):883–890

Simirgiotis MJ, Adachi S, To S, Yang H, Reynertson KA, Basile MJ, Gil RC, Weinstein IB, Kennelly EJ (2008) Cytotoxic chalcones and antioxidants from the fruits of *Syzygium samarangense* (Wax Jambu). Food Chem 107(2):813–819

Soubir T (2007) Antioxidant activities of some local Bangladeshi fruits (*Artocarpus heterophyllus, Annona squamosa, Terminalia bellirica, Syzygium samarangense, Averrhoa carambola* and *Olea europa*). Sheng Wu Gong Cheng Xue Bao 23(2):257–261

Srivastava R, Shaw AK, Kulshreshtha K (1995) Triterpenoids and chalcone from *Syzygium samarangense*. Phytochemistry 38:687–689

Tezuka Y, Fan W, Kasimu R, Kadota SH (1999) Screening of crude drug extracts for prolyl endopeptidase-inhibitory activity. Phytomedicine 6:197–203

Wills RBH (1987) Composition of Australian fresh fruit and vegetables. Food Technol Aust 39(11):523–6

Wong KC, Lai FY (1996) Volatile constituents from the fruits of four *Syzygium* species grown in Malaysia. Flav Fragr J 11:61–66

Yoshimoto T, Kado K, Matsubara F, Koriyama N, Kaneto H, Tsuru D (1987) Specific inhibitors for prolyl endopeptidase and their anti-amnesic effect. J Pharmacobio-Dyn 10:730–735

Syzygium smithii

Scientific Name

***Syzygium smithii* (Poir.) Nied.**

Synonyms

Acmena elliptica G.Benn., *Acmena elliptica* G.Don ex Steud., *Acmena floribunda* DC., *Acmena floribunda* var. *elliptica* A.Cunn. ex DC., *Acmena kingii* G.Don, *Acmena pendula* G.Benn., *Acmena smithii* (Poir.) Merr. & L.M.Perry, *Acmena smithii* var. *minor* (Maiden) Merr. & L.M.Perry, *Eugenia elliptica* Sm. nom. illeg., *Eugenia smithii* Poir., *Eugenia smithii* var. *coriacea* Domin, *Eugenia smithii* var. *minor* Maiden, *Lomastelma elliptica* Raf., *Lomastelma smithii* (Poir.) J.H.Willis, *Myrtus smithii* (Poir.) Spreng., *Syzygium brachynemum* F.Muell.

Family

Myrtaceae

Common/English Name

Lilly Pilly, Lillipilli, Monkey Apple (New Zealand), Narrow Leaved Lilly Pilly

Vernacular Name

Australia: Midjuburi (Aboriginal, Cadigal).

Origin/Distribution

Lilly Pilly is native to eastern Australia, main distribution occurs in New South Wales, Queensland and eastern Victoria.

Agroecology

A very hardy plant adapting to a range of climates from sub-tropical to temperate with summer temperatures of 26–32°C and winter temperatures of 5–15°C and annual mean rainfall of 700–200 mm from sea level to 1,200 m latitudes. It is found along banks of streams and rivers on a wide range of soils that include exposed, stabilised coastal sands, sandstone, basaltic red clay loams and sandy loams derived from granite. The species is resistant to at least moderate frosts and tolerates extended dry periods once established, it will grow in full sun or fairly heavy shade as it is extremely shade tolerant; it regenerates readily under dense forest canopy and in gaps. In an open position it usually retains foliage to near ground level.

T.K. Lim, *Edible Medicinal And Non-Medicinal Plants: Volume 3, Fruits*,
DOI 10.1007/978-94-007-2534-8_106, © Springer Science+Business Media B.V. 2012

Plate 1 Fruit and leaves of Lillypilly

Edible Plant Parts and Uses

Ripe fruit can be eaten raw or made into jams, jellies, drinks, relishes, desserts, tarts, pies, cakes and ice-cream.

Botany

An erect, evergreen, much branched, medium to large tree reaching 8–20 m in height with smooth to slightly flaky reddish-brown bark and dense crown; upper trunk and branches smooth and twigs often 4-angled or shortly 4-winged. Leaves are opposite, petiolate to 3–9 mm long, simple, entire, broadly elliptical to ovate or broadly oblanceolate, 3–11 cm long and 1–5 cm wide, with a drawn–out blunt tip, cuneate base, glabrous, dark green glossy, discolourous, mid vein raised beneath and depressed above, lateral veins numerous and distinct; oil glands usually numerous and conspicuous (Plate 1). Inflorescences are terminal panicles with numerous small flowers. The flowers are creamy-white, with minute, deciduous bracts, four deciduous sepals, four cream petals united to a small, deciduous calyptra, numerous short stamens, ovary inferior, 2-loculed with several ovules per locule and a simple style. Fruit is a berry, 6–15 mm diameter, subglobose, white, pale mauve to lavender, crowned by a circular calyx rim (Plate 1), containing succulent, slightly acid pulp surrounding the seed.

Nutritive/Medicinal Properties

Approximate nutrient composition of *Acmena smithii* fruit per 100 g/edible portion based on analyses conducted by Brand-Miller et al. (1993) was reported as follows: energy 18 kJ, moisture, 93.4 g, total dietary fibre 1.1 g, nitrogen 0.05 g, protein 0.3 g, fat 0.1 g, ash 0.2 g, calcium 8 mg, copper 0.2 mg, iron 0.1 mg, magnesium 3 mg, potassium 35 mg, sodium 2 mg, zinc 0.1 mg, niacin equivalents 0.1 mg and vitamin C 1 mg.

Other Uses

Small bushy or dwarf varieties are used as topiary specimens, hedges, windbreak and ornamental plantings. The timber can be used for general construction, internal fittings, frames and flooring.

Comments

Of several liliy pilly species, *S. smithii* is one of the more resistant to the insect psyllid pest.

Selected References

Ashton DH, Frankenbery J (1976) Ecological studies of *Acmena smithii* (Poir.) Merrill & Perry with special reference to Wilson's Promontory. Aust J Bot 24(4):453–487

Boland DJ, Brooker MIH, Chippendale GM, Hall N, Hyland BPM, Johnston RD, Kleinig DA, McDonald MW, Turner JD (2006) Forest trees of Australia, 5th edn. CISRO Publishing, Collingwood, 768 pp

Brand Miller J, James KW, Maggiore P (1993) Tables of composition of Australian aboriginal foods. Aboriginal Studies Press, Canberra

Carolin R, Tindale M (1993) Flora of the Sydney region. Reed, Chatswood

Govaerts R, Sobral M, Ashton P, Barrie F, Holst BK, Landrum LL, Matsumoto K, Fernanda Mazine F, Nic Lughadha E, Proenca C, Soares-Silva LH, Wilson PG, Lucas E (2010) World checklist of Myrtaceae. The board of trustees of the Royal Botanic Gardens, Kew. Published on the Internet, http://www. kew.org/wcsp/. Accessed 22 Apr 2010

Wrigley J, Fagg M (1996) Australian native plants. New Holland Publishers, Sydney, 696 pp

Syzygium suborbiculare

Scientific Name

Syzygium suborbiculare **(Benth.) T.G. Hartley & Perry**.

Synonyms

Careya jambosoides Lauterb., *Eugenia jambosoides* (Lauterb.) O. Schwarz, *Eugenia suborbicularis* Benth. (basionym), *Syzygium jambosoides* (Lauterb.) Merr. & L.M. Perry

Family

Myrtaceae

Common/English Names

Forest Satinash, Lady Apple, Red Bush Apple

Vernacular Names

None

Origin/Distribution

Lady apple occurs naturally in northern Australia on Cape York Peninsula, Queensland, the north of the Northern Territory and the Kimberley region of Western Australia. It is also found in southern Papua New Guinea. The latitudinal range in Australia is 10–17°S and the altitudinal range is from near sea level to 220 m. in Queensland, Northern Territory and northern Western Australia.

Agroecology

The distribution of Lady apple is mainly in the hot humid and sub-humid zones where there is a distinct summer maximum with the mean annual rainfall is 880–1,760 mm. It commonly grows near creeks and in gullies of sandstone plateaux, on sandy plains near watercourses and sand dunes near the coast and on nearby islands in low rainforest, open-forest and woodlands. Soils are often shallow, sandy and of low fertility. It is a very fire-tolerant species, drought tolerant and tolerates salt spray but frost intolerant.

T.K. Lim, *Edible Medicinal And Non-Medicinal Plants: Volume 3, Fruits*,
DOI 10.1007/978-94-007-2534-8_107, © Springer Science+Business Media B.V. 2012

Edible Plant Parts and Uses

Lady apple is edible. It has a firm crunchy white flesh that has a sharp delicate tang and is revered by all bush travellers and aborigines.

Botany

Usually a small tree up to 12 m tall but also occurs as a low gnarled shrub on sand dunes. The bark is flaky on the trunk though initially smooth, branchlets become scaly with age. Mature leaves are opposite, large, oval to broadly ovate, 10–19 cm long and 5–13 cm wide glossy green above, dull beneath with numerous visible oil dots on the surface. Flowers are large, conspicuous, cream coloured with numerous long white stamens up to 5 cm long, and produced in dense, terminal clusters. The fruit is orbicular, dark red when ripe (Plate 1) and fleshy, 3–7 cm long by 3.5–9 cm across with conspicuous ribs containing one large, globose seed.

Nutritive/Medicinal Properties

The proximate nutrient composition of the fruit based on analyses made in Australia (Brand Miller et al. 1993) is per 100 g edible portion: energy 68 kJ, moisture 89.3 g, , nitrogen 0.1 g, protein 0.6 g, fat 0.5 g, ash 0.6 g, available carbohydrate 0 g, total dietary fibre 4.9 g, Ca 67 mg, Cu 0.1 mg, Mg 32 mg, Fe 1.2 mg, K 85 mg, Na 48 mg, Zn 0.2 mg, thiamine 0.04 mg, riboflavin 0.01 mg, niacin (derived form tryptophan or protein) 0.1 mg, vitamin C 8 mg.

Plate 1 Ripe lady apple fruit

Aborigines are known to use the fruit for colds and juice from the fruit for chest congestion and squeeze the juice and pulp into their ears to relieve earache. An infusion of crushed leaves was used by Aboriginals to treat diarrhoea. Heated leaves are applied to wounds to stop bleeding and reduce swelling. An infusion from the bark is used for stomach pains and to bathe sores.

Other Uses

Australian Aboriginals have used the trunk to make canoes. This species has great potential for use as a shade tree and for amenity planting. The tree is also a nectar source for bees.

Comments

The species is easily propagated from soft tip cuttings.

Selected References

Aboriginal Communities of the Northern Territory (1993) Traditional aboriginal medicines in the Northern Territory of Australia. Conservation Commission of the Northern Territory of Australia, Darwin, 651 pp

Brand Miller J, James KW, Maggiore P (1993) Tables of composition of Australian aboriginal foods. Aboriginal Studies Press, Canberra

Brock J (1988) Top end native plants. John Brock, Darwin, 354 pp

Doran JC, Turnbull JW et al (1997) Australian trees and shrubs; species for land rehabilitation and farm planting in the tropics, ACIAR monograph no. 24. Centre for International Agricultural Research, Canberra, 384 pp

Govaerts R, Sobral M, Ashton P, Barrie F, Holst BK, Landrum LL, Matsumoto K, Fernanda Mazine F, Nic Lughadha E, Proenca C, Soares-Silva LH, Wilson PG, Lucas E (2010) World checklist of Myrtaceae . The board of trustees of the Royal Botanic Gardens, Kew. Published on the Internet, http://www. kew.org/wcsp/. Accessed 22 Apr 2010

Isaacs J (2002) Bush food: aboriginal food and herbal medicine. New Holland, Sydney

Low T (1991) Wild food plants of Australia. Angus & Robertson, North Ryde, 240 pp

Wightman G, Andrews M (1991) Bush tucker identikit. Conservation Commission of The northern Territory, Darwin, 65 pp

Medical Glossary

AAD Allergic airway disease, an inflammatory disorder of the airways caused by allergens.

AAPH 2,2′-Azobis(2-amidinopropane) dihydrochloride, a water-soluble azo compound used extensively as a free radical generator, often in the study of lipid peroxidation and the characterization of antioxidants.

Abeta aggregation Amyloid β protein (Abeta) aggregation is associated with Alzheimer's disease (AD); it is a major component of the extracellular plaque found in AD brains.

Abdominal distension Referring to generalised distension of most or all of the abdomen. Also referred to as stomach bloating often caused by a sudden increase in fibre from consumption of vegetables, fruits and beans.

Ablation therapy The destruction of small areas of myocardial tissue, usually by application of electrical or chemical energy, in the treatment of some tachyarrhythmias.

Abortifacient A substance that causes or induces abortion.

Abortivum A substance inducing abortion.

Abscess A swollen infected, inflamed area filled with pus in body tissues.

ABTS 2,2 Azinobis-3-ethylhenthiazoline-6-sulfonic acid, a type of mediator in chemical reaction kinetics of specific enzymes.

ACAT Acyl CoA: cholesterol acyltransferase.

ACE See angiotensin-converting enzyme.

ACTH (Adrenocorticotropic hormone) Also known as 'corticotropin', is a polypeptide tropic hormone produced and secreted by the anterior pituitary gland.

Acetogenins Natural products from the plants of the family Annonaceae, are very potent inhibitors of the NADH-ubiquinone reductase (Complex I) activity of mammalian mitochondria.

Acetylcholinesterase (AChE) Is an enzyme that degrades (through its hydrolytic activity) the neurotransmitter acetylcholine, producing choline.

Acne vulgaris Also known as chronic acne, usually occurring in adolescence, with comedones (blackheads), papules (red pimples), nodules (inflamed acne spots), and pustules (small inflamed pus-filled lesions) on the face, neck, and upper part of the trunk.

Acidosis Increased acidity.

Acquired immunodeficiency syndrome (AIDS) An epidemic disease caused by an infection by human immunodeficiency virus (HIV-1, HIV-2), retrovirus that causes immune system failure and debilitation and is often accompanied by infections such as tuberculosis.

Acridone An organic compound based on the acridine skeleton, with a carbonyl group at the 9 position.

ACTH Adrenocorticotropic hormone (or corticotropin), a polypeptide tropic hormone produced and secreted by the anterior pituitary gland. It plays a role in the synthesis and secretion of gluco- and mineralo-corticosteroids and androgenic steroids.

Activating transcription factor (ATF) A protein (gene) that binds to specific DNA sequences regulating the transfer or transcription of information from DNA to mRNA.

Activator protein-1 (AP-1) A heterodimeric protein transcription factor that regulates gene expression in response to a variety of stimuli, including cytokines, growth factors, stress,

T.K. Lim, *Edible Medicinal And Non-Medicinal Plants: Volume 3, Fruits*,
DOI 10.1007/978-94-007-2534-8, © Springer Science+Business Media B.V. 2012

and bacterial and viral infections. AP-1 in turn regulates a number of cellular processes including differentiation, proliferation, and apoptosis.

Acyl-CoA dehydrogenases A group of enzymes that catalyzes the initial step in each cycle of fatty acid β -oxidation in the mitochondria of cells.

Adaptogen A term used by herbalists to refer to a natural herb product that increases the body's resistance to stresses such as trauma, stress and fatigue.

Adaptogenic Increasing the resistance of the body to stress.

Addison's disease Is a rare endocrine disorder. It occurs when the adrenal glands cannot produce sufficient hormones (corticosteroids). It is also known as chronic adrenal insufficiency, hypocortisolism or hypocorticism.

Adenocarcinoma A cancer originating in glandular tissue.

Adenoma A benign tumour from a glandular origin.

Adenopathy Abnormal enlargement or swelling of the lymph node.

Adenosine receptors A class of purinergic, G-protein coupled receptors with adenosine as endogenous ligand. In humans, there are four adenosine receptors. A1 receptors and A2A play roles in the heart, regulating myocardial oxygen consumption and coronary blood flow, while the A2A receptor also has broader anti-inflammatory effects throughout the body. These two receptors also have important roles in the brain, regulating the release of other neurotransmitters such as dopamine and glutamate, while the A2B and A3 receptors are located mainly peripherally and are involved in inflammation and immune responses.

ADH See alcohol dehydrogenase.

Adipocyte A fat cell involved in the synthesis and storage of fats.

Adipocytokine Bioactive cytokines produced by adipose tissues

Adiponectin A protein in humans that modulates several physiological processes, such as metabolism of glucose and fatty acids, and immune responses.

Adipose tissues Body fat, loose connective tissue composed of adipocytes (fat cells).

Adoptogen Containing smooth pro-stressors which reduce reactivity of host defense systems and decrease damaging effects of various stressors due to increased basal level of mediators involved in the stress response.

Adrenal glands Star-shaped endocrine glands that sit on top of the kidneys.

Adrenalectomized Having had the adrenal glands surgically removed.

Adrenergic Having to de with adrenaline (epinephrine) and/or noradrenaline (norepinephrine).

Adrenergic receptors A class of G protein-coupled receptors that are targets of the noradrenaline (norepinephrine) and adrenaline (epinephrine).

Adulterant An impure ingredient added into a preparation.

Advanced Glycation End products (AGEs) Resultant products of a chain of chemical reactions after an initial glycation reaction. AGEs may play an important adverse role in process of atherosclerosis, diabetes, aging and chronic renal failure.

Aegilops An ulcer or fistula in the inner corner of the eye.

Afferent Something that so conducts or carries towards, such as a blood vessel, fibre, or nerve.

Agammaglobulinaemia An inherited disorder in which there are very low levels of protective immune proteins called immunoglobulins. Cf. x-linked agammaglobulinaemia.

Agalactia Lack of milk after parturition (birth).

Agglutinin A protein substance, such as an antibody, that is capable of causing agglutination (clumping) of a particular antigen.

Agglutination Clumping of particles.

Agonist A drug that binds to a receptor of a cell and triggers a response by the cell.

Ague A fever (such as from malaria) that is marked by paroxysms of chills, fever, and sweating that recurs with regular intervals.

AHR AhR, aryl hydrocarbon receptor, a cytosolic protein transcription factor.

AIDS See Acquired Immunodeficiency Syndrome.

Akathisia A movement disorder in which there is an urge or need to move the legs to stop unpleasant sensations. Also called restless leg syndrome, the disorder is often caused by long-term use of antipsychotic medications.

Akt signaling pathway Akt are protein kinases involved in mammalian cellular signaling, inhibits apoptotic processes.

Akt/FoxO pathway Cellular processes involving Akt and FoxO transcription factors that play a role in angiogenesis and vasculogenesis.

Alanine transaminase (ALT) Also called Serum Glutamic Pyruvate Transaminase (SGPT) or Alanine aminotransferase (ALAT), an enzyme present in hepatocytes (liver cells). When a cell is damaged, it leaks this enzyme into the blood.

ALAT (Alanine aminotransferase) See Alanine transaminase.

Albumin Water soluble proteins found in egg white, blood serum, milk, various animal tissues and plant juices and tissues.

Albuminaria Excessive amount of albumin in the urine, a symptom of severe kidney disease.

Aldose reductase, aldehyde reductase An enzyme in carbohydrate metabolism that converts glucose to sorbitol.

Alexipharmic An antidote, remedy for poison.

Alexiteric A preservative against contagious and infectious diseases, and the effects of poisons.

Alcohol dehydrogenase (ADH) An enzyme involved in the break-down of alcohol.

Algesic Endogenous substances involved in the production of pain that is associated with inflammation, e.g. serotonin, bradykinin and prostaglandins.

Alkaline phosphatase (ALP) An enzyme in the cells lining the biliary ducts of the liver. ALP levels in plasma will rise with large bile duct obstruction, intrahepatic cholestasis or infiltrative diseases of the liver. ALP is also present in bone and placental tissues.

Allergenic Having the properties of an antigen (allergen), immunogenic.

Allergic Pertaining to, caused, affected with, or the nature of the allergy.

Allergic conjunctivitis Inflammation of the tissue lining the eyelids (conjunctiva) due to allergy.

Allergy A hypersensitivity state induced by exposure to a particular antigen (allergen) resulting in harmful immunologic reactions on subsequent exposures. The term is usually used to refer to hypersensitivity to an environmental antigen (atopic allergy or contact dermatitis) or to drug allergy.

Allogeneic Cells or tissues which are genetically different because they are derived from separate individuals of the same species. Also refers to a type of immunological reaction that occurs when cells are transplanted into a genetically different recipient.

Allografts Or homografts, a graft between individuals of the same species, but of different genotypes.

Alloknesis Itch produced by innocuous mechanical stimulation.

Allostasis The process of achieving stability, or homeostasis, through physiological or behavioral change.

Alopecia Is the loss of hair on the body.

Alopecia areata Is a particular disorder affecting hair growth (loss of hair) in the scalp and elsewhere.

ALP See Alkaline phosphatase.

Alpha-adrenoceptor Receptors postulated to exist on nerve cell membranes of the sympathetic nervous system in order to explain the specificity of certain agents that affect only some sympathetic activities (such as vasoconstriction and relaxation of intestinal muscles and contraction of smooth muscles).

Alpha amylase α-amylase A major form of amylase found in humans and other mammals that cleaves α-bonds of large sugar molecules.

ALT See Alanine transaminase.

Alterative A medication or treatment which gradually induces a change, and restores healthy functions without sensible evacuations.

Alveolar macrophage A vigorously phagocytic macrophage on the epithelial surface of lung alveoli that ingests carbon and other inhaled particulate matter. Also called coniophage or dust cell.

Alzheimer's disease A degenerative, organic, mental disease characterized by progressive brain deterioration and dementia, usually occurring after the age of 50.

Amastigote Refers to a cell that does not have any flagella, used mainly to describe a certain phase in the life-cycle of trypanosome protozoans.

Amenorrhea The condition when a woman fails to have menstrual periods.

Amidolytic Cleavage of the amide structure.

Amoebiasis State of being infected by amoeba such as *Entamoeba histolytica*.

Amoebicidal Lethal to amoeba.

AMPK (5′ AMP-activated protein kinase) Or 5′ adenosine monophosphate-activated protein kinase, enzyme that plays a role in cellular energy homeostasis.

Amyloid beta (Aβ or Abeta) A peptide of 39–43 amino acids that appear to be the main constituent of amyloid plaques in the brains of Alzheimer's disease patients.

Amyotrophic lateral sclerosis Or ALS, is a disease of the motor neurons in the brain and spinal cord that control voluntary muscle movement.

Amyotrophy Progressive wasting of muscle tissues. *adj.* amyotrophic.

Anaemia A blood disorder in which the blood is deficient in red blood cells and in haemoglobin.

Anaesthesia Condition of having sansation temporarily supressed.

Anaesthetic A substance that decreases partially or totally nerve the sense of pain.

Analeptic A central nervous system (CNS) stimulant medication.

Analgesia Term describing relief, reduction or suppression of pain. *adj.* analgetic.

Analgesic A substance that relieves or reduces pain.

Anaphoretic An antiperspirant.

Anaphylaxis A severe, life-threatening allergic response that may be characterized by symptoms such as reduced blood pressure, wheezing, vomiting or diarrhea.

Anaphylactic *adj.* see anaphylaxis.

Anaphylotoxins Are fragments (C3a, C4a or C5a) that are produced during the pathways of the complement system. They can trigger release of substances of endothelial cells, mast cells or phagocytes, which produce a local inflammatory response.

Anaplasia A reversion of differentiation in cells and is characteristic of malignant neoplasms (tumours).

Anaplastic *adj.* see anaplasia.

Anasarca Accumulation of great quantity of fluid in body tissues.

Androgen Male sex hormone in vertebrates. Androgens may be used in patients with breast cancer to treat recurrence of the disease.

Android adiposity Centric fat distribution patterns with increased disposition towards the abdominal area, visceral fat – apple shaped cf gynoid adiposity.

Angina pectoris, Angina Chest pain or chest discomfort that occurs when the heart muscle does not get enough blood.

Angiogenesis A physiological process involving the growth of new blood vessels from pre-existing vessels.

Angiogenic *adj.* see angiogenesis.

Angiotensin An oligopeptide hormone in the blood that causes blood vessels to constrict, and drives blood pressure up. It is part of the renin-angiotensin system.

Angiotensin-converting enzyme (ACE) An exopeptidase, a circulating enzyme that participates in the body's renin-angiotensin system (RAS) which mediates extracellular volume (i.e. that of the blood plasma, lymph and interstitial fluid), and arterial vasoconstriction.

Angioplasty Medical procedure used to open obstructed or narrowed blood vessel resulting usually from atherosclerosis.

Anisonucleosis A morphological manifestation of nuclear injury characterized by variation in the size of the cell nuclei.

Ankylosing spondylitis (AS) Is a type of inflammatory arthritis that targets the joints of the spine.

Annexitis Also called adnexitis, a pelvic inflammatory disease involving the inflammation of the ovaries or fallopian tubes.

Anodyne A substance that relieves or soothes pain by lessening the sensitivity of the brain or nervous system. Also called an analgesic.

Anoikis Apoptosis that is induced by inadequate or inappropriate cell-matrix interactions.

Anorectal Relating to the rectum and anus.

Anorectics Appetite suppressants, substances which reduce the desire to eat. Used on a short term basis clinically to treat obesity. Also called anorexigenics.

Anorexia Lack or loss of desire to eat.

Anorexic Having no appetite to eat.

Anorexigenics See anorectics.

Antagonist A substance that acts against and blocks an action.

Antalgic A substance used to relive a painful condition.

Antecubital vein This vein is located in the antecubital fossa – the area of the arm in front of the elbow.

Anterior uveitis Is the most common form of ocular inflammation that often causes a painful red eye.

Anthelmintic An agent or substance that is destructive to worms and used for expulsion of internal parasitic worms in animals and humans.

Anthocyanins A subgroup of antioxidant flavonoids, are glucosides of anthocyanidins. Which are beneficial to health. They occur as water-soluble vacuolar pigments that may appear red, purple, or blue according to pH in plants.

Anthrax A bacterial disease of cattle and ship that can be transmitted to man though unprocessed wool.

Anthropometric Pertaining to the study of human body measurements.

Antiamoebic A substance that destroys or suppresses parasitic amoebae.

Antiamyloidogenic Compounds that inhibit the formation of Alzheimer's β-amyloid fibrils (fAβ) from amyloid β-peptide (Aβ) and destabilize fAβ.

Antianaphylactic Agent that can prevent the occurrence of anaphylaxis (life threatening allergic response).

Antiangiogenic A drug or substance used to stop the growth of tumours and progression of cancers by limiting the pathologic formation of new blood vessels (angiogenesis).

Antiarrhythmic A substance to correct irregular heartbeats and restore the normal rhythm.

Antiasmathic Drug that treats or ameliorates asthma.

Antiatherogenic That protects against atherogenesis, the formation of atheromas (plaques) in arteries.

Antibacterial Substance that kills or inhibits bacteria.

Antibilious An agent or substance which helps remove excess bile from the body.

Antibiotic A chemical substance produced by a microorganism which has the capacity to inhibit the growth of or to kill other microorganisms.

Antiblennorrhagic A substance that treats blennorrhagia a conjunctival inflammation resulting in mucus discharge.

Antibody A gamma globulin protein produced by a kind of white blood cell called the plasma cell in the blood used by the immune system to identify and neutralize foreign objects (antigen).

Anticarcinomic A substance that kills or inhibits carcinomas (any cancer that arises in epithelium/tissue cells).

Anticephalalgic Headache-relieving or preventing.

Anticestodal A chemical destructive to tapeworms.

Anticholesterolemic A substance that can prevent the build up of cholesterol.

Anticlastogenic Having a suppressing effect of chromosomal aberrations.

Anticoagulant A substance that thins the blood and acts to inhibit blood platelets from sticking together.

Antidepressant A substance that suppresses depression or sadness.

Antidiabetic A substance that prevents or alleviates diabetes. Also called antidiabetogenic.

Antidiarrhoeal Having the property of stopping or correcting diarrhoea, an agent having such action.

Antidote A remedy for counteracting a poison.

Antidopaminergic A term for a chemical that prevents or counteracts the effects of dopamine.

Antidrepanocytary Anti-sickle cell anaemia.

Antidysenteric An agent used to reduce or treat dysentery and diarrhea.

Antidyslipidemic Agent that will reduce the abnormal amount of lipids and lipoproteins in the blood.

Anti-edematous Reduces or suppresses edema.

Anti-emetic An agent that stops vomiting.

Anti-epileptic A drug used to treat or prevent convulsions, anticonvulsant.

Antifebrile A substance that reduces fever, also called antipyretic.

Antifeedant Preventing something from being eaten.

Antifertility Agent that inhibits formation of ova and sperm and disrupts the process of fertilization (antizygotic).

Antifilarial Effective against human filarial worms.

Antifungal An agent that kills or inhibits the growth of fungi.

Antigen A substance that prompts the production of antibodies and can cause an immune response. *adj.* antigenic.

Antigenotoxic An agent that inhibits DNA adduct formation, stimulates DNA repair mechanisms, and possesses antioxidant functions.

Antiganacratia Anti-menstruation.

Antigastralgic Preventing or alleviating gastric colic.

Antihematic Agent that stops vomiting.

Antihemorrhagic An agent which stops or prevents bleeding.

Antihepatotoxic Counteracting injuries to the liver.

Antiherpetic Having activity against Herpes Simplex Virus (HSV).

Antihistamine An agent used to counteract the effects of histamine production in allergic reactions.

Antihyperalgesia The ability to block enhanced sensitivity to pain, usually produced by nerve injury or inflammation, to nociceptive stimuli. *adj.* antihyperalgesic.

Antihypercholesterolemia Term to describe lowering of cholesterol level in the blood or blood serum.

Antihypercholesterolemic Agent that lowers cholesterol level in the blood or blood serum.

Antihyperlidemic Promoting a reduction of lipid levels in the blood, or an agent that has this action.

Antihypersensitive A substance used to treat excessive reactivity to any stimuli.

Antihypertensive A drug used in medicine and pharmacology to treat hypertension (high blood pressure).

Antiinflammatory A substance used to reduce or prevent inflammation.

Antileishmanial Inhibiting the growth and proliferation of *Leishmania* a genus of flagellate protozoans that are parasitic in the tissues of vertebrates.

Antileprotic Therapeutically effective against leprosy.

Antilithiatic An agent that reduces or suppresses urinary calculi (stones) and acts to dissolve those already present.

Antileukaemic Anticancer drugs that are used to treat leukemia.

Antilithogenic Inhibiting the formation of calculi (stones).

Antimalarial An agent used to treat malaria and/or kill the malaria-causing organism, *Plasmodium* spp.

Antimelanogenesis Obstruct production of melanin.

Antimicrobial A substance that destroys or inhibits growth of disease-causing bacteria, viruses, fungi and other microorganisms.

Antimitotic Inhibiting or preventing mitosis.

Antimutagenic An agent that inhibits mutations.

Antimycotic Antifungal.

Antineoplastic Said of a drug intended to inhibit or prevent the maturation and proliferation of neoplasms that may become malignant, by targeting the DNA.

Antineuralgic A substance that stops intense intermittent pain, usually of the head or face, caused by neuralgia.

Antinociception Reduction in pain: a reduction in pain sensitivity produced within neurons when an endorphin or similar opium-containing substance opioid combines with a receptor.

Antinociceptive Having an analgesic effect.

Antinutrient Are natural or synthetic compounds that interfere with the absorption of nutrients and are commonly found in food sources and beverages.

Antioestrogen A substance that inhibits the biological effects of female sex hormones.

Antiophidian Anti venoms of snake.

Antiosteoporotic Substance that can prevent osteoporosis.

Antiovulatory Substance suppressing ovulation.

Antioxidant A chemical compound or substance that inhibits oxidation and protects against free radical activity and lipid oxidation such as vitamin E, vitamin C, or β-carotene (converted to vitamin B), carotenoids and flavonoids which are thought to protect body cells from the damaging effects of oxidation. Many foods including fruit and vegetables contain compounds with antioxidant properties. Antioxidants may also reduce the risks of cancer and age-related macular degeneration(AMD).

Antipaludic Antimalarial.

Antiperiodic Substance that prevents the recurrence of symptoms of a disease e.g. malaria.

Antiperspirant A substance that inhibits sweating. Also called antisudorific, anaphoretic.

Antiphlogistic A traditional term for a substance used against inflammation, an anti-inflammatory.

Antiplatelet agent Drug that decreases platelet aggregation and inhibits thrombus formation.

Antiplasmodial Suppressing or destroying plasmodia.

Antiproliferative Preventing or inhibiting the reproduction of similar cells.

Antiprostatic Drug to treat the prostate.

Antiprotozoal Suppressing the growth or reproduction of protozoa.

Antipruritic Alleviating or preventing itching.

Antipyretic A substance that reduces fever or quells it. Also known as antithermic.

Antirheumatic Relieving or preventing rheumatism.

Antiscorbutic A substance or plant rich in vitamin C that is used to counteract scurvy.

Antisecretory Inhibiting or diminishing secretion.

Antisense Refers to antisense RNA strand because its sequence of nucleotides is the complement of message sense. When mRNA forms a duplex with a complementary antisense RNA sequence, translation of the mRNA into the protein is blocked. This may slow or halt the growth of cancer cells.

Antiseptic Preventing decay or putrefaction, a substance inhibiting the growth and development of microorganisms.

Anti-sickling agent An agent used to prevent or reverse the pathological events leading to sickling of erythrocytes in sickle cell conditions.

Antispasmodic A substance that relieves spasms or inhibits the contraction of smooth muscles; smooth muscle relaxant, muscle-relaxer.

Antispermatogenic Preventing or suppressing the production of semen or spermatozoa.

Antisudorific See antiperspirant.

Antisyphilitic A drug (or other chemical agent) that is effective against syphilis.

Antithermic A substance that reduces fever and temperature. Also known as antipyretic.

Antithrombotic Preventing or interfering with the formation of thrombi.

Antitoxin An antibody with the ability to neutralize a specific toxin.

Antitumoral Substance that acts against the growth, development or spread of a tumour.

Antitussive A substance that depresses coughing.

Antiulcerogenic An agent used to protect against the formation of ulcers, or is used for the treatment of ulcers.

Antivenin An agent used against the venom of a snake, spider, or other venomous animal or insect.

Antivinous An agent or substance that treats addiction to alcohol.

Antiviral Substance that destroys or inhibits the growth and viability of infectious viruses.

Antivomitive A substance that reduces or suppresses vomiting.

Antizygotic See antifertility.

Anuria Absence of urine production and excretion. *adj.* anuric.

Anxiolytic A drug prescribed for the treatment of symptoms of anxiety.

APAF-1 Apoptotic protease activating factor 1.

Apelin Also known as APLN, a peptide which in humans is encoded by the APLN gene.

Aperient A substance that acts as a mild laxative by increasing fluids in the bowel.

Aperitif An appetite stimulant.

Aphonia Loss of the voice resulting from disease, injury to the vocal cords, or various psychological causes, such as hysteria.

Aphrodisiac An agent that increases sexual activity and libido and/or improves sexual performance.

Apnoea Suspension of external breathing.

Apolipoprotein B (APOB) Primary apolipoprotein of low-density lipoproteins which is responsible for carrying cholesterol to tissues.

Apoplexy A condition in which the brain's function stops with loss of voluntary motion and sense.

Apoprotein The protein moiety of a molecule or complex, as of a lipoprotein.

Appendicitis Is a condition characterized by inflammation of the appendix. Also called epityphlitis.

Appetite stimulant A substance to increase or stimulate the appetite. Also called aperitif.

Aphthae White, painful oral ulcer of unknown cause.

Apthous ulcer Canker sore in the lining of the mouth.

Aphthous stomatitis A canker sore, a type of painful oral ulcer or sore inside the mouth or upper throat, caused by a break in the mucous membrane. Also called aphthous ulcer.

Apolipoprotein A-I (APOA1) A major protein component of high density lipoprotein (HDL) in plasma. The protein promotes cholesterol efflux from tissues to the liver for excretion.

Apolipoprotein B (APOB) Is the primary apolipoprotein of low-density lipoproteins (LDL or "bad cholesterol"), which is responsible for carrying cholesterol to tissues.

Apolipoprotein E (APOE) The apolipoprotein found on intermediate density lipoprotein and chylomicron that binds to a specific receptor on liver and peripheral cells.

Apoptogenic Ability to cause death of cells.

Apoptosis Death of cells.

Aphthous ulcer Also known as a canker sore, is a type of oral ulcer, which presents as a painful open sore inside the mouth or upper throat.

Apurinic lyase A DNA enzyme that catalyses a chemical reaction.

Arachidonate cascade Includes the cyclooxygenase (COX) pathway to form prostanoids and the lipoxygenase (LOX) pathway to generate several oxygenated fatty acids, collectively called eicosanoids.

Ariboflavinosis A condition caused by the dietary deficiency of riboflavin that is characterized by mouth lesions, seborrhea, and vascularization.

Aromatase An enzyme involved in the production of estrogen that acts by catalyzing the conversion of testosterone (an androgen) to estradiol (an estrogen). Aromatase is located in estrogen-producing cells in the adrenal glands, ovaries, placenta, testicles, adipose (fat) tissue, and brain.

Aromatic Having a pleasant, fragrant odour.

Aromatherapy A form of alternative medicine that uses volatile liquid plant materials, such as essential oils and other scented compounds from plants for the purpose of affecting a person's mood or health.

Arrhythmias Abnormal heart rhythms that can cause the heart to pump less effectively. Also called dysrhythmias.

Arsenicosis See arsenism.

Arsenism An incommunicable disease resulting from the ingestion of ground water containing unsafe levels of arsenic, also known as arsenicosis.

Arteriosclerosis Imprecise term for various disorders of arteries, particularly hardening due to fibrosis or calcium deposition, often used as a synonym for atherosclerosis.

Arthralgia Is pain in the joints from many possible causes.

Arthritis Inflammation of the joints of the body.

Aryl hydrocarbon receptor (AhR) A ligand-activated transcription factor best known for mediating the toxicity of dioxin and other exogenous contaminants and is responsible for their toxic effects, including immunosuppression.

ASATor AST Aspartate aminotransferase, see aspartate transaminase,

Ascaris A genus of parasitic intestinal round worms.

Ascites Abnormal accumulation of fluid within the abdominal or peritoneal cavity.

Ascorbic acid See vitamin C.

Aspartate transaminase (AST) Also called Serum Glutamic Oxaloacetic Transaminase (SGOT) or aspartate aminotransferase (ASAT) is similar to ALT in that it is another enzyme associated with liver parenchymal cells. It is increased in acute liver damage, but is also present in red blood cells, and cardiac and skeletal muscle and is therefore not specific to the liver.

Asphyxia Failure or suppression of the respiratory process due to obstruction of air flow to the lungs or to the lack of oxygen in inspired air.

Asphyxiation The process of undergoing asphyxia.

Asthenia A nonspecific symptom characterized by loss of energy, strength and feeling of weakness.

Asthenopia Weakness or fatigue of the eyes, usually accompanied by headache and dimming of vision. *adj.* asthenopic.

Asthma A chronic illness involving the respiratory system in which the airway occasionally constricts, becomes inflamed, and is lined with excessive amounts of mucus, often in response to one or more triggers.

Astringent A substance that contracts blood vessels and certain body tissues (such as mucous membranes) with the effect of reducing secretion and excretion of fluids and/or has a drying effect.

Astrocytes Collectively called astroglia, are characteristic star-shaped glial cells in the brain and spinal cord.

Ataxia (loss of co-ordination) results from the degeneration of nerve tissue in the spinal cord and of nerves that control muscle movement in the arms and legs.

Ataxia telangiectasia and Rad3-related protein (ATR) Also known as Serine/threonine-protein kinase ATR, FRAP-related protein 1 (FRP1), is an enzyme encoded by the ATR gene. It is involved in sensing DNA damage and activating the DNA damage checkpoint, leading to cell cycle arrest

ATF-2 Activating transcription factor 2.

Athlete's foot A contagious skin disease caused by parasitic fungi affecting the foot, hands, causing itching, blisters and cracking. Also called dermatophytosis.

Atherogenic Having the capacity to start or accelerate the process of atherogenesis.

Atherogenesis The formation of lipid deposits in the arteries.

Atheroma A deposit or degenerative accumulation of lipid-containing plaques on the innermost layer of the wall of an artery.

Atherosclerosis The condition in which an artery wall thickens as the result of a build-up of fatty materials such as cholesterol.

Atherothrombosis Medical condition characterized by an unpredictable, sudden disruption (rupture or erosion/fissure) of an atherosclerotic plaque, which leads to platelet activation and thrombus formation.

Athymic mice Laboratory mice lacking a thymus gland.

Atonic Lacking normal tone or strength.

Atony Insufficient muscular tone.

Atopic dermatitis An inflammatory, non-contagious, pruritic skin disorder of unknown etiology; often called eczema.

Atresia A congenital medical condition in which a body orifice or passage in the body is abnormally closed or absent.

Atretic ovarian follicles An involuted or closed ovarian follicle.

Atrial fibrillation Is the most common cardiac arrhythmia (abnormal heart rhythm) and involves the two upper chambers (atria) of the heart.

Attention-deficit hyperactivity disorder (ADHD, ADD or AD/HD) Is a neurobehavioral developmental disorder, primarily characterized by "the co-existence of attentional problems and hyperactivity.

Auditory brainstem response (ABR) Also called brainstem evoked response (BSER) is an electrical signal evoked from the brainstem of a human by the presentation of a sound such as a click.

Augmerosen A drug that may kill cancer cells by blocking the production of a protein that makes cancer cells live longer. Also called bcl-2 antisense oligonucleotide.

Auricular Of or relating to the auricle or the ear in general.

Aurones [2-Benzylidenebenzofuran-3(2H)-ones] are the secondary plant metabolites and is a subgroup of flavonoids. See flavonoids.

Autoantibodies Antibodies manufactured by the immune system that mistakenly target and damage specific tissues and organs of the body.

Autolysin An enzyme that hydrolyzes and destroys the components of a biological cell or a tissue in which it is produced.

Autophagy Digestion of the cell contents by enzymes in the same cell.

Autopsy Examination of a cadaver to determine or confirm the cause of death.

Avidity Index Describes the collective interactions between antibodies and a multivalent antigen.

Avulsed teeth Is tooth that has been knocked out.

Ayurvedic Traditional Hindu system of medicine based largely on homeopathy and naturopathy.

Azoospermia Is the medical condition of a male not having any measurable level of sperm in his semen.

Azotaemia A higher than normal blood level of urea or other nitrogen containing compounds in the blood.

Babesia A protozoan parasite (malaria–like) of the blood that causes a hemolytic disease known as Babesiosis.

Babesiosis Malaria-like parasitic disease caused by Babesia, a genus of protozoal piroplasms.

Bactericidal Lethal to bacteria.

Balanitis Is an inflammation of the glans (head) of the penis.

BALB/c mice Balb/c mouse was developed in 1923 by McDowell. It is a popular strain and is used in many different research disciplines, but most often in the production of monoclonal antibodies.

Balm Aromatic oily resin from certain trees and shrubs used in medicine.

Baroreceptor A type of interoceptor that is stimulated by pressure changes, as those in blood vessel wall.

Barrett's esophagus (Barrett esophagitis) A disorder in which the lining of the esophagus is damaged by stomach acid.

Basophil A type of white blood cell with coarse granules within the cytoplasm and a bilobate (two-lobed) nucleus.

BCL-2 A family of apoptosis regulator proteins in humans encoded by the B-cell lymphoma 2 (BCL-2) gene.

BCL-2 antisense oligonucleotide See augmereson.

BCR/ABL A chimeric oncogene, from fusion of BCR and ABL cancer genes associated with chronic myelogenous leukemia.

Bechic A remedy or treatment of cough.

Bed nucleus of the stria terminalis (BNST) Act as a relay site within the hypothalamic-pituitary-adrenal axis and regulate its activity in response to acute stress.

Belching, or burping Refers to the noisy release of air or gas from the stomach through the mouth.

Beri-beri Is a disease caused by a deficiency of thiamine (vitamin B1) that affects many systems of the body, including the muscles, heart, nerves, and digestive system.

Beta-carotene Naturally-occurring retinol (vitamin A) precursor obtained from certain fruits and vegetables with potential antineoplastic and chemopreventive activities. As an antioxidant, β carotene inhibits free-radical damage to DNA. This agent also induces cell differentiation and apoptosis of some tumour cell types, particularly in early stages of tumorigenesis, and enhances immune system activity by stimulating the release of natural killer cells, lymphocytes, and monocytes.

Beta-catenin Is a multifunctional oncogenic protein that contributes fundamentally to cell development and biology, it has been implicated as an integral component in the Wnt signaling pathway.

Beta cells A type of cell in the pancreas in areas called the islets of Langerhans.

Beta-thalassemia An inherited blood disorder that reduces the production of hemoglobin.

Beta-lactamase Enzymes produced by some bacteria that are responsible for their resistance to β-lactam antibiotics like penicillins.

BHT Butylated hydroxytoluene (phenolic compound), an antioxidant used in foods, cosmetics, pharmaceuticals, and petroleum products.

Bifidobacterium Is a genus of Gram-positive, non-motile, often branched anaerobic bacteria. Bifidobacteria are one of the major genera of bacteria that make up the gut flora. Bifidobacteria aid in digestion, are associated with a lower incidence of allergies and also prevent some forms of tumour growth. Some bifidobacteria are being used as probiotics.

Bifidogenic Promoting the growth of (beneficial) bifidobacteria in the intestinal tract.

Bile Fluid secreted by the liver and discharged into the duodenum where it is integral in the digestion and absorption of fats.

Bilharzia, bilharziosis See Schistosomiasis.

Biliary Relating to the bile or the organs in which the bile is contained or transported.

Biliary infections Infection of organ(s) associated with bile, comprise: (a) acute cholecystitis: an acute inflammation of the gallbladder wall; (b) cholangitis: inflammation of the bile ducts.

Biliousness Old term used in the eighteenth and nineteenth centuries pertaining to bad digestion, stomach pains, constipation, and excessive flatulence.

Bilirubin A breakdown product of heme (a part of haemoglobin in red blood cells) produced by the liver that is excreted in bile which causes a yellow discoloration of the skin and eyes when it accumulates in those organs.

Biotin Also known as vitamin B7. See vitamin B7.

Bitter A medicinal agent with a bitter taste and used as a tonic, alterative or appetizer.

Blackhead See comedone.

Blackwater fever Dangerous complication of malarial whereby the red blood cells burst in the blood stream (haemolysis) releasing haemoglobin directly into the blood.

Blain See chilblain.

Blastocyst Blastocyst is an embryonic structure formed in the early embryogenesis of mammals, after the formation of the morula, but before implantation.

Blastocystotoxic Agent that suppresses further development of the blastocyst through to the ovum stage.

Blebbing Bulging e.g. membrane blebbing also called membrane bulging or ballooning.

Bleeding diathesis Is an unusual susceptibility to bleeding (hemorrhage) due to a defect in the system of coagulation.

Blennorrhagia Gonorrhea.

Blennorrhea Inordinate discharge of mucus, especially a gonorrheal discharge from the urethra or vagina.

Blepharitis Inflammation of the eyelids.

Blister Thin vesicle on the skin containing serum and caused by rubbing, friction or burn.

Blood brain barrier (BBB) Is a separation of circulating blood and cerebrospinal fluid (CSF) in the central nervous system (CNS). It allows essential metabolites, such as oxygen and glucose, to pass from the blood to the brain and central nervous system (CNS) but blocks most molecules that are more massive than about 500 Da.

Boil Localized pyrogenic, painful infection, originating in a hair follicle.

Borborygmus Rumbling noise caused by the muscular contractions of peristalsis, the process that moves the contents of the stomach and intestines downward.

Bowman Birk inhibitors Type of serine proteinase inhibitor.

Bouillon A broth in French cuisine.

Bradicardia As applied to adult medicine, is defined as a resting heart rate of under 60 beats per minute.

Bradyphrenia Referring to the slowness of thought common to many disorders of the brain.

Brain derived neutrophic factor (BDNF) A protein member of the neutrophin family that plays an important role in the growth, maintenance, function and survival of neurons. The protein molecule is involved in the modulation of cognitive and emotional functions and in the treatment of a variety of mental disorders.

Bright's disease Chronic nephritis.

Bronchial inflammation See bronchitis.

Bronchiectasis A condition in which the airways within the lungs (bronchial tubes) become damaged and widened.

Bronchitis Is an inflammation of the main air passages (bronchi) to your lungs.

Bronchoalveolar lavage (BAL) A medical procedure in which a bronchoscope is passed through the mouth or nose into the lungs and fluid is squirted into a small part of the lung and then recollected for examination.

Bronchopneumonia Or bronchial pneumonia; inflammation of the lungs beginning in the terminal bronchioles.

Broncho-pulmonary Relating to the bronchi and lungs.

Bronchospasm Is a difficulty in breathing caused by a sudden constriction of the muscles in the walls of the bronchioles as occurs in asthma.

Brown fat Brown adipose tissue (BAT) in mammals, its primary function is to generate body heat in animals or newborns that do not shiver.

Bubo Inflamed, swollen lymph node in the neck or groin.

Buccal Of or relating to the cheeks or the mouth cavity.

Bullae Blisters; circumscribed, fluid-containing, elevated lesions of the skin, usually more than 5 mm in diameter.

Bursitis Condition characterized by inflammation of one or more bursae (small sacs) of synovial fluid in the body.

C-jun NH(2)-terminal kinase Enzymes that belong to the family of the MAPK superfamily of protein kinases. These kinases mediate a plethora of cellular responses to such stressful stimuli, including apoptosis and production of inflammatory and immunoregulatory cytokines in diverse cell systems. *cf:* MAPK.

c-FOS A cellular proto-oncogene belonging to the immediate early gene family of transcription factors.

C-reactive protein A protein found in the blood the levels of which rise in response to inflammation.

c-Src A cellular non-receptor tyrosine kinase.

CAAT element-binding proteins-α (c/EBP-akpha) Regulates gene expression in adipocytes in the liver.

Cachexia Physical wasting with loss of weight, muscle atrophy, fatigue, weakness caused by disease.

Caco-2 cell line A continuous line of heterogeneous human epithelial colorectal adenocarcinoma cells.

Cadaver A dead body, corpse.

Ca2+ ATPase (PMCA) is a transport protein in the plasma membrane of cells that serves to remove calcium (Ca2+) from the cell.

Calcium (Ca) Is the most abundant mineral in the body found mainly in bones and teeth. It is required for muscle contraction, blood vessel expansion and contraction, secretion of hormones and enzymes, and transmitting impulses throughout the nervous system. Dietary sources include milk, yoghurt, cheese, Chinese cabbage, kale, broccoli, some green leafy vegetables, fortified cereals, beverages and soybean products.

Calcium ATPase Is a form of P-ATPase which transfers calcium after a muscle has contracted.

Calcium channel blockers (CCBs) A class of drugs and natural substances that disrupt the calcium (Ca2+) conduction of calcium channels.

Calculus (calculi) Hardened, mineral deposits that can form a blockage in the urinary system.

Calculi infection Most calculi arise in the kidney when urine becomes supersaturated with a salt that is capable of forming solid crystals. Symptoms arise as these calculi become impacted within the ureter as they pass toward the urinary bladder.

Caligo Dimness or obscurity of sight, dependent upon a speck on the cornea.

Calmodulin Is a Calcium Modulated protein that can bind to and regulate a multitude of different protein targets, thereby affecting many different cellular functions.

cAMP dependent pathway Cyclic adenosine monophosphate is a G protein-coupled receptor triggered signaling cascade used in cell communication in living organisms.

CAMP factor Diffusible, heat-stable, extracellular protein produced by Group B *Streptococcus t*hat enhances the hemolysis of sheep erythrocytes by *Staphylococcus aureus.* It is named after Christie, Atkins, and Munch-Peterson, who described it in 1944.

Cancer A malignant neoplasm or tumour in nay part of the body.

Candidiasis Infections caused by members of the fungus genus *Candida* that range from superficial, such as oral thrush and vaginitis, to systemic and potentially life-threatening diseases.

Canker See chancre.

Carboxypeptidase An enzyme that hydrolyzes the carboxy-terminal (C-terminal) end of a peptide bond. It is synthesized in the pancreas and secreted into the small intestine.

Carbuncle Is an abscess larger than a boil, usually with one or more openings draining pus onto the skin.

Carcinogenesis Production of carcinomas. *adj.* carcinogenic.

Carcinoma Any malignant cancer that arises from epithelial cells.

Carcinosarcoma A rare tumour containing carcinomatous and sarcomatous components.

Cardiac Relating to, situated near or affecting the heart.

Cardiac asthma Acute attack of dyspnoea with wheezing resulting from a cardiac disorder.

Cardialgia Heartburn.

Cardinolides Cardiac glycosides with a 5-membered lactone ring in the side chain of the steroid aglycone.

Cardinolide glycoside Cardenolides that contain structural groups derived from sugars.

Cardioactive Having an effect on the heart.

Cardiogenic shock Is characterized by a decreased pumping ability of the heart that causes a shock like state associated with an inadequate circulation of blood due to primary failure of the ventricles of the heart to function effectively.

Cardiomyocytes Cardiac muscle cells.

Cardiomyopathy Heart muscle disease.

Cardiopathy Disease or disorder of the heart.

Cardioplegia Stopping the heart so that surgical procedures can proceed in a still and bloodless field.

Cardiotonic Something which strengthens, tones, or regulates heart functions without overt stimulation or depression.

Cardiovascular Pertaining to the heart and blood vessels.

Caries Tooth decay, commonly called cavities.

Cariogenic Leading to the production of caries.

Carminative Substance that stops the formation of intestinal gas and helps expel gas that has already formed, relieving flatulence: relieving flatulence or colic by expelling gas.

Carnitine palmitoyltransferase I (CPT1) Also known as carnitine acyltransferase I or CAT1 is a mitochondrial enzyme, involved in converting long chain fatty acid into energy.

Carotenes Are a large group of intense red and yellow pigments found in all plants ; these are hydrocarbon carotenoids (subclass of tetraterpenes) and the principal carotene is β-carotene which is a precursor of vitamin A.

Carotenoids A class of natural fat-soluble pigments found principally in plants, belonging to a subgroup of terpenoids containing 8 isoprene units forming a C40 polyene chain. Carotenoids play an important potential role in human health by acting as biological antioxidants. See also carotenes.

Carotenodermia Yellow skin discoloration caused by excess blood carotene.

Carpopedal spasm Spasm of the hand or foot, or of the thumbs and great toes.

Capases Cysteine-aspartic acid proteases, are a family of cysteine proteases, which play essential roles in apoptosis (programmed cell death).

Catalase (CAT) Enzyme in living organism that catalyses the decomposition of hydrogen peroxide to water and oxygen.

Catalepsy Indefinitely prolonged maintenance of a fixed body posture; seen in severe cases of catatonic schizophrenia.

Catamenia Menstruation.

Cataplasia Degenerative reversion of cells or tissue to a less differentiated form.

Cataplasm A medicated poultice or plaster. A soft moist mass, often warm and medicated, that is spread over the skin to treat an inflamed, aching or painful area, to improve the circulation.

Cataractogenesis Formation of cataracts.

Catarrh, Catarrhal Inflammation of the mucous membranes especially of the nose and throat.

Catechins Are polyphenolic antioxidant plant metabolites. They belong to the family of flavonoids; tea is a rich source of catechins. See flavonoids.

Catecholamines Hormones that are released by the adrenal glands in response to stress.

Cathartic Is a substance which accelerates defecation.

Cathepsin K A cysteine protease that plays an essential role in osteoclast function in bone remodelling and resorption in diseases such as osteoporosis, osteolytic bone metastasis and rheumatoid arthritis.

Caustic Having a corrosive or burning effect.

Cauterization A medical term describing the burning of the body to remove or close a part of it.

cdc2 Kinase A member of the cyclin-dependent protein kinases (CDKs).

CDKs Cyclin-dependent protein kinases, a family of serine/threonine kinases that mediate many stages in mitosis.

CD 28 Is one of the molecules expressed on T cells that provide co-stimulatory signals, which are required for T cell (lymphocytes) activation.

CD31 Also known as PECAM-1 (Platelet Endothelial Cell Adhesion Molecule-1), a member of the immunoglobulin superfamily, that mediates cell-to-cell adhesion.

CD36 An integral membrane protein found on the surface of many cell types in vertebrate animals.

CD40 An integral membrane protein found on the surface of B lymphocytes, dendritic cells, follicular dendritic cells, hematopoietic progenitor cells, epithelial cells, and carcinomas.

CD68 A glycoprotein expressed on monocytes/macrophages which binds to low density lipoprotein.

Cecal ligation Tying up the cecam.

Cell adhesion molecules (CAM) Glycoproteins located on the surface of cell membranes involved with binding of other cells or with the extra-cellular matrix.

Cellular respiration Is the set of the metabolic reactions and processes that take place in organisms' cells to convert biochemical energy from nutrients into adenosine triphosphate (ATP), and then release waste products. The reactions involved in respiration are catabolic reactions that involve the oxidation of one molecule and the reduction of another.

Cellulitis A bacterial infection of the skin that tends to occur in areas that have been damaged or inflamed.

Central nervous system Part of the vertebrate nervous system comprising the brain and spinal cord.

Central venous catheter A catheter placed into the large vein in the neck, chest or groin.

Cephalagia Pain in the head, a headache.

Cephalic Relating to the head.

Ceramide oligosides Oligosides with an N-acetyl-sphingosine moiety.

Cerebral embolism A blockage of blood flow through a vessel in the brain by a blood clot that formed elsewhere in the body and traveled to the brain.

Cerebral ischemia Is the localized reduction of blood flow to the brain or parts of the brain due to arterial obstruction or systematic hyperfusion.

Cerebral infarction Is the ischemic kind of stroke due to a disturbance in the blood vessels supplying blood to the brain.

Cerebral tonic Substance that can alleviate poor concentration and memory, restlessness, uneasiness, and insomnia.

Cerebrosides Are glycosphingolipids which are important components in animal muscle and nerve cell membranes.

Cerebrovascular disease Is a group of brain dysfunctions related to disease of the blood vessels supplying the brain.

Cerumen Ear wax, a yellowish waxy substance secreted in the ear canal of humans and other mammals.

cGMP Cyclic guanosine monophosphate is a cyclic nucleotide derived from guanosine triphosphate (GTP). cGMP is a common regulator of ion channel conductance, glycogenolysis, and cellular apoptosis. It also relaxes smooth muscle tissues.

Chalcones A subgroup of flavonoids.

Chancre A painless lesion formed during the primary stage of syphilis.

Chemoembolization A procedure in which the blood supply to the tumour is blocked surgically or mechanically and anticancer drugs are administered directly into the tumour.

Chemokines Are chemotactic cytokines, which stimulate migration of inflammatory cells towards tissue sites of inflammation.

Chemosensitizer A drug that makes tumour cells more sensitive to the effects of chemotherapy.

Chemosis Edema of the conjunctiva of the eye.

Chickenpox Is also known as varicella, is a highly contagious illness caused by primary infection with varicella zoster virus (VZV). The virus causes red, itchy bumps on the body.

Chilblains Small, itchy, painful lumps that develop on the skin. They develop as an abnormal response to cold. Also called perniosis or blain.

Chlorosis Iron deficiency anemia characterized by greenish yellow colour.

Cholagogue Is a medicinal agent which promotes the discharge of bile from the system.

Cholecalcifereol A form of vitamin D, also called vitamin D3. See vitamin D.

Cholecyst Gall bladder.

Cholecystitis Inflammation of the gall bladder.

Cholecystokinin A peptide hormone that plays a key role in facilitating digestion in the small intestine.

Cholera An infectious gastroenteritis caused by enterotoxin-producing strains of the bacterium *Vibrio cholerae* and characterized by severe, watery diarrhea.

Choleretic Stimulation of the production of bile by the liver.

Cholestasis A condition caused by rapidly developing (acute) or long-term (chronic) interruption in the excretion of bile.

Cholesterol A soft, waxy, steroid substance found among the lipids (fats) in the bloodstream and in all our body's cells.

Cholethiasis Presence of gall stones (calculi) in the gall bladder.

Choline A water soluble, organic compound, usually grouped within the Vitamin B complex. It is an essential nutrient and is needed for physiological functions such as structural integrity and signaling roles for cell membranes, cholinergic neuro-transmission (acetylcholine synthesis).

Cholinergic Activated by or capable of liberating acetylcholine, especially in the parasympathetic nervous system.

Cholinergic system A system of nerve cells that uses acetylcholine in transmitting nerve impulses.

Cholinomimetic Having an action similar to that of acetylcholine; called also parasympathomimetic.

Chonotropic Affecting the time or rate, as the rate of contraction of the heart.

Choriocarcinoma A quick-growing malignant, trophoblastic, aggressive cancer that occurs in a woman's uterus (womb).

Chromium (Cr) Is required in trace amounts in humans for sugar and lipid metabolism. Its deficiency may cause a disease called chromium deficiency. It is found in cereals, legumes, nuts and animal sources.

Chromosome Long pieces of DNA found in the center (nucleus) of cells.

Chronic Persisting over extended periods.

Chyle A milky bodily fluid consisting of lymph and emulsified fats, or free fatty acids.

Chylomicrons Are large lipoprotein particles that transport dietary lipids from the intestines to other locations in the body. Chylomicrons are one of the five major groups of lipoproteins (chylomicrons, VLDL, IDL, LDL, HDL) that enable fats and cholesterol to move within the water-based solution of the bloodstream.

Chylorus Milky (having fat emulsion).

Chyluria Also called chylous urine, is a medical condition involving the presence of chyle (emulsified fat) in the urine stream, which results in urine appearing milky.

Chymase Member of the family of serine proteases found primarily in mast cell.

Chymopapain An enzyme derived from papaya, used in medicine and to tenderize meat.

Cicatrizant The term used to describe a product that promotes healing through the formation of scar tissue.

Cirrhosis Chronic liver disease characterized by replacement of liver tissue by fibrous scar tissue and regenerative nodules/lumps leading progressively to loss of liver function.

C-Kit Receptor A protein-tyrosine kinase receptor that is specific for stem cell factor. this interaction is crucial for the development of hematopoietic, gonadal, and pigment stem cells.

Clastogen Is an agent that can cause one of two types of structural changes, breaks in chromosomes that result in the gain, loss, or rearrangements of chromosomal segments. *adj.* clastogenic.

Claudication Limping, impairment in walking.

Climacterium Refers to menopause and the bodily and mental changes associated with it.

Clonic seizures Consist of rhythmic jerking movements of the arms and legs, sometimes on both sides of the body.

Clyster Enema.

C-myc Codes for a protein that binds to the DNA of other genes and is therefore a transcription factor.

CNS Depressant Anything that depresses, or slows, the sympathetic impulses of the central nervous system (i.e., respiratory rate, heart rate).

Coagulopathy A defect in the body's mechanism for blood clotting, causing susceptibility to bleeding.

Cobalamin Vitamin B12. See vitamin B12.

Co-carcinogen A chemical that promotes the effects of a carcinogen in the production of cancer.

Cold An acute inflammation of the mucous membrane of the respiratory tract especially of the nose and throat caused by a virus and accompanied by sneezing and coughing.

Collagen Protein that is the major constituent of cartilage and other connective tissue; comprises the amino acids hydroxyproline, proline, glycine, and hydroxylysine.

Collagenases Enzymes that break the peptide bonds in collagen.

Colic A broad term which refers to episodes of uncontrollable, extended crying in a baby who is otherwise healthy and well fed.

Colitis Inflammatory bowel disease affecting the tissue that lines the gastrointestinal system.

Collyrium A lotion or liquid wash used as a cleanser for the eyes, particularly in diseases of the eye.

Colorectal Relating to the colon or rectum.

Coma A state of unconsciousness from which a patient cannot be aroused.

Comedone A blocked, open sebaceous gland where the secretions oxidize, turning black. Also called blackhead.

Comitogen Agent that is considered not to induce cell growth alone but to promote the effect of the mitogen.

Concoction A combination of crude ingredients that is prepared or cooked together.

Condyloma, Condylomata acuminata Genital warts, venereal warts, anal wart or anogenital wart, a highly contagious sexually transmitted infection caused by epidermotropic human papillomavirus (HPV).

Conglutination Becoming stuck together.

Conjunctival hyperemia Enlarged blood vessels in the eyes.

Conjunctivitis Sore, red and sticky eyes caused by eye infection.

Constipation A very common gastrointestinal disorder characterised by the passing of hard, dry bowel motions (stools) and difficulty of bowel motion.

Constitutive androstane receptor (CAR, NR113) Is a nuclear receptor transcription factor that regulates drug metabolism and homoeostasis.

Consumption Term used to describe wasting of tissues including but not limited to tuberculosis.

Consumptive Afflicted with or associated with pulmonary tuberculosis.

Contraceptive An agent that reduces the likelihood of or prevents conception.

Contraindication A condition which makes a particular treatment or procedure inadvisable.

Contralateral muscle Muscle of opposite limb (leg or arm).

Contralateral rotation Rotation occurring or originating in a corresponding part on an opposite side.

Contusion Another term for a bruise. A bruise, or contusion, is caused when blood vessels are damaged or broken as the result of a blow to the skin.

Convulsant A drug or physical disturbance that induces convulsion.

Convulsion Rapid and uncontrollable shaking of the body.

Coolant That which reduces body temperature.

Copper (Cu) Is essential in all plants and animals. It is found in a variety of enzymes, including the copper centers of cytochrome C oxidase and the enzyme superoxide dismutase (containing copper and zinc). In addition to its enzymatic roles, copper is used for biological electron transport. Because of its role in facilitating iron uptake, copper deficiency can often produce anemia-like symptoms. Dietary sources include curry powder, mushroom, nuts, seeds, wheat germ, whole grains and animal meat.

Copulation To engage in coitus or sexual intercourse. *adj.* copulatory.

Cordial A preparation that is stimulating to the heart.

Corn Or callus is a patch of hard, thickened skin on the foot that is formed in response to pressure or friction.

Corticosteroids A class of steroid hormones that are produced in the adrenal cortex, used clinically for hormone replacement therapy, for suppressing ACTH secretion, for suppression of immune response and as antineoplastic, anti-allergic and anti-inflammatory agents.

Corticosterone A 21-carbon steroid hormone of the corticosteroid type produced in the cortex of the adrenal glands.

Cortisol Is a corticosteroid hormone made by the adrenal glands.

Cornification Is the process of forming an epidermal barrier in stratified squamous epithelial tissue.

Coryza A word describing the symptoms of a head cold. It describes the inflammation of the mucus membranes lining the nasal cavity which usually gives rise to the symptoms of nasal congestion and loss of smell, among other symptoms.

COX-1 See cyclooxygenase-1.

COX-2 See cyclooxygenase-2.

CpG islands Genomic regions that contain a high frequency of CpG sites.

CpG sites The cytosine-phosphate-guanine nucleotide that links two nucleosides together in DNA.

cPLA(2) Cytosolic phospholipases A2, these phospholipases are involved in cell signaling processes, such as inflammatory response.

CPY1B1, CPY1A1 A member of the cytochrome P450 superfamily of heme-thiolate monooxygenase enzymes.

Corticosterone A 21-carbon corticosteroid hormone produced in the cortex of the adrenal glands that functions in the metabolism of carbohydrates and proteins.

Creatin A nitrogenous organic acid that occurs naturally in vertebrates and helps to supply energy to muscle.

Creatine phosphokinase (CPK, CK) Enzyme that catalyses the conversion of creatine and consumes adenosine triphosphate (ATP) to create phosphocreatine and adenosine diphosphate (ADP).

CREB cAMP response element-binding, a protein that is a transcription factor that binds to certain DNA sequences called cAMP response elements.

Crohn Disease An inflammatory disease of the intestines that affect any part of the gastrointestinal tract.

Crossover study A longitudinal, balance study in which participants receive a sequence of different treatments or exposures.

Croup Is an infection of the throat (larynx) and windpipe (trachea) that is caused by a virus (also called laryngotracheobronchitis).

Crytochidism (cryptochism) A developmental defect characterized by the failure of one or both testes to move into the scrotum as the male fetus develops.

Curettage Surgical procedure in which a body cavity or tissue is scraped with a sharp instrument or aspirated with a cannula.

Cutaneous Pertaining to the skin.

CXC8 Also known as interleukin 8, IL-8.

Cyanogenesis Generation of cyanide. *adj.* cyanogenetic.

Cyclooxygenase (COX) An enzyme that is responsible for the formation of prostanoids – prostaglandins, prostacyclins, and thromboxanes that are each involved in the inflammatory response. Two different COX enzymes existed, now known as COX-1 and COX-2.

Cyclooxygenase-1 (COX-1) Is known to be present in most tissues. In the gastrointestinal tract, COX-1 maintains the normal lining of the stomach. The enzyme is also involved in kidney and platelet function.

Cyclooxygenase-2 (COX-2) Is primarily present at sites of inflammation.

Cysteine proteases Are enzymes that degrade polypeptides possessing a common catalytic mechanism that involves a nucleophilic cysteine thiol in a catalytic triad. They are found in fruits like papaya, pineapple, and kiwifruit.

Cystitis A common urinary tract infection that occurs when bacteria travel up the urethra, infect the urine and inflame the bladder lining.

Cystorrhea Discharge of mucus from the bladder.

Cytochrome bc-1 complex Ubihydroquinone: cytochrome c oxidoreductase.

Cytochrome P450 3A CYP3A A very large and diverse superfamily of heme-thiolate proteins found in all domains of life. This group of enzymes catalyzes many reactions involved in drug metabolism and synthesis of cholesterol, steroids and other lipids.

Cytokine Non-antibody proteins secreted by certain cells of the immune system which carry signals locally between cells. They are a category of signaling molecules that are used extensively in cellular communication.

Cytopathic Any detectable, degenerative changes in the host cell due to infection.

Cytoprotective Protecting cells from noxious chemicals or other stimuli.

Cytosolic Relates to the fluid of the cytoplasm in cells.

Cytostatic Preventing the growth and proliferation of cells.

Cytotoxic Of or relating to substances that are toxic to cells; cell-killing.

D-galactosamine An amino sugar with unique hepatotoxic properties in animals.

Dandruff Scurf, dead, scaly skin among the hair.

Dartre Condition of dry, scaly skin.

Debility Weakness, relaxation of muscular fibre.

Debridement Is the process of removing nonliving tissue from pressure ulcers, burns, and other wounds.

Debriding agent Substance that cleans and treats certain types of wounds, burns, ulcers.

Deciduogenic Relating to the uterus lining that is shed off at childbirth.

Decidual stromal cells Like endometrial glands and endothelium, express integrins that bind basement components.

Decoction A medical preparation made by boiling the ingredients.

Decongestant A substance that relieves or reduces nasal or bronchial congestion.

Defibrinated plasma Blood whose plasma component has had fibrinogen and fibrin removed.

Degranulation Cellular process that releases antimicrobial cytotoxic molecules from secretory vesicles called granules found inside some cells.

Delayed afterdepolarizations (DADs) Abnormal depolrization that begins during phase 4 – after repolarization is completed, but before another action potential would normally occur.

Delirium Is common, sudden severe confusion and rapid changes in brain function that occur with physical or mental illness; it is reversible and temporary.

Demulcent An agent that soothes internal membranes. Also called emollient.

Dendritic cells Are immune cells and form part of the mammalian immune system, functioning as antigen presenting cells.

Dentition A term that describes all of the upper and lower teeth collectively.

Deobstruent A medicine which removes obstructions; also called an aperient.

Deoxypyridinoline (Dpd) A crosslink product of collagen molecules found in bone and excreted in urine during bone degradation.

Depilatory An agent for removing or destroying hair.

Depressant A substance that diminish functional activity, usually by depressing the nervous system.

Depurative An agent used to cleanse or purify the blood, it eliminates toxins and purifies the system.

Dermatitis Inflammation of the skin causing discomfort such as eczema.

Dermatophyte A fungus parasitic on the skin.

Dermatosis Is a broad term that refers to any disease of the skin, especially one that is not accompanied by inflammation.

Dermonecrotic Pertaining to or causing necrosis of the skin.

Desquamation The shedding of the outer layers of the skin.

Detoxifier A substance that promotes the removal of toxins from a system or organ.

Diabetes A metabolic disorder associated with inadequate secretion or utilization of insulin and characterized by frequent urination and persistent thirst. See diabetes mellitus.

Diabetes mellitus (DM) (sometimes called "sugar diabetes") is a set of chronic, metabolic disease conditions characterized by high blood sugar (glucose) levels that result from defects in insulin secretion, or action, or both. Diabetes mellitus appears in two forms.

Diabetes mellitus type I (formerly known as juvenile onset diabetes), caused by deficiency of the pancreatic hormone insulin as a result of destruction of insulin-producing β cells of the pancreas. Lack of insulin causes an increase of fasting blood glucose that begins to appear in the urine above the renal threshold.

Diabetes mellitus type II (formerly called non-insulin-dependent diabetes mellitus or adult-onset diabetes), the disorder is characterized by high blood glucose in the context of insulin resistance and relative insulin deficiency in which insulin is available but cannot be properly utilized.

Diads Two adjacent structural units in a polymer molecule.

Dialysis Is a method of removing toxic substances (impurities or wastes) from the blood when the kidneys are unable to do so.

Diaphoresis Is profuse sweating commonly associated with shock and other medical emergency conditions.

Diaphoretic A substance that induces perspiration. Also called sudorific.

Diaphyseal Pertaining to or affecting the shaft of a long bone (diaphysis).

Diaphysis The main or mid section (shaft) of a long bone.

Diarrhoea A profuse, frequent and loose discharge from the bowels.

Diastolic Referring to the time when the heart is in a period of relaxation and dilatation (expansion). *cf.* systolic.

Dieresis Surgical separation of parts.

Dietary fibre Is a term that refers to a group of food components that pass through the stomach and small intestine undigested and reach the large intestine virtually unchanged. Scientific evidence suggest that a diet high in dietary fibre can be of value for treating or preventing such disorders as constipation, irritable bowel syndrome, diverticular disease, hiatus hernia and haemorrhoids. Some components of dietary fibre may also be of value in reducing the level of cholesterol in blood and thereby decreasing a risk factor for coronary heart disease and the development of gallstones. Dietary fibre is beneficial in the treatment of some diabetics.

Digalactosyl diglycerides Are the major lipid components of chloroplasts.

Diosgenin A steroid-like substance that is involved in the production of the hormone progesterone, extracted from roots of *Dioscorea* yam.

Dipsomania Pathological use of alcohol.

Discutient An agent (as a medicinal application) which serves to disperse morbid matter.

Disinfectant An agent that prevents the spread of infection, bacteria or communicable disease.

Diuresis Increased urination.

Diuretic A substance that increases urination (diuresis).

Diverticular disease Is a condition affecting the large bowel or colon and is thought to be caused by eating too little fibre.

DMBA 7,12-Dimethylbenzanthracene. A polycyclic aromatic hydrocarbon found in tobacco smoke that is a potent carcinogen.

DNA Deoxyribonucleic acid, a nucleic acid that contains the genetic instructions used in the development and functioning of all known living organisms.

DOCA Desoxycorticosterone acetate – a steroid chemical used as replacement therapy in Addison's disease.

Dopamine A catecholamine neurotransmitter that occurs in a wide variety of animals, including both vertebrates and invertebrates.

Dopaminergic Relating to, or activated by the neurotransmitter, dopamine.

Double blind Refer to a clinical trial or experiment in which neither the subject nor the researcher knows which treatment any particular subject is receiving.

Douche A localised spray of liquid directed into a body cavity or onto a part.

DPPH 2,2 Diphenyl-1-picryl-hydrazyl – a crystalline, stable free radical used as an inhibitor of free radical reactions.

Dracunculiasis Also called guinea worm disease (GWD), is a parasitic infection caused by the nematode, *Dracunculus medinensis*.

Dropsy An old term for the swelling of soft tissues due to the accumulation of excess water. *adj.* dropsical.

Dysentery (formerly known as flux or the bloody flux) is a disorder of the digestive system that results in severe diarrhea containing mucus and blood in the feces. It is caused usually by a bacterium called *Shigella*.

Dysesthesia An unpleasant abnormal sensation produced by normal stimuli.

Dysgeusia Distortion of the sense of taste.

Dyskinesia The impairment of the power of voluntary movement, resulting in fragmentary or incomplete movements. *adj.* dyskinetic.

Dyslipidemia Abnormality in or abnormal amount of lipids and lipoproteins in the blood.

Dysmenorrhea Is a menstrual condition characterized by severe and frequent menstrual cramps and pain associated with menstruation.

Dysmotility syndrome A vague, descriptive term used to describe diseases of the muscles of the gastrointestinal tract (esophagus, stomach, small and large intestines).

Dyspedia Indigestion followed by nausea.

Dyspepsia Refers to a symptom complex of epigastric pain or discomfort. It is often defined as chronic or recurrent discomfort centered in the upper abdomen and can be caused by a variety of conditions.

Dysphagia Swallowing disorder.

Dysphonia A voice disorder, an impairment in the ability to produce voice sounds using the vocal organs.

Dysplasia Refers to abnormality in development.

Dyspnoea Shortness of breath, difficulty in breathing.

Dysrhythmias See arrhythmias.

Dystocia Abnormal or difficult child birth or labour.

Dystonia A neurological movement disorder characterized by prolonged, repetitive muscle contractions that may cause twisting or jerking movements of muscles.

Dysuria Refers to difficult and painful urination.

E- Selectin Also known as endothelial leukocyte adhesion molecule-1 (ELAM-1), CD62E, a member of the selectin family. It is transiently expressed on vascular endothelial cells in response to IL-1 β and TNF-α.

EC 50 Median effective concentration that produces desired effects in 50% of the test population.

Ecbolic A drug (as an ergot alkaloid) that tends to increase uterine contractions and that is used especially to facilitate delivery.

Ecchymosis Skin discoloration caused by the escape of blood into the tissues from ruptured blood vessels.

ECG See electrocardiography.

EC-SOD Extracellular superoxide dismutase, a tissue enzyme mainly found in the extracellular matrix of tissues. It participates in the detoxification of reactive oxygen species by catalyzing the dismutation of superoxide radicals.

Eczema Is broadly applied to a range of persistent skin conditions. These include dryness and recurring skin rashes which are characterized by one or more of these symptoms: redness, skin edema, itching and dryness, crusting, flaking, blistering, cracking, oozing, or bleeding.

Eczematous rash Dry, scaly, itchy rash.

ED 50 Is defined as the dose producing a response that is 50% of the maximum obtainable.

Edema Formerly known as dropsy or hydropsy, is characterized swelling caused by abnormal accumulation of fluid beneath the skin, or in one or more cavities of the body. It usually occurs in the feet, ankles and legs, but it can involve the entire body.

Edematogenic Producing or causing edema.

EGFR proteins Epidermal growth factor receptor (EGFR) proteins – Protein kinases are enzymes that transfer a phosphate group from a phosphate donor onto an acceptor amino acid in a substrate protein.

EGR-1 Early growth response 1, a human gene.

Eicosanoids Are signaling molecules made by oxygenation of arachidonic acid, a 20-carbon essential fatty acid, includes prostaglandins and related compounds.

Elastase A serine protease that also hydrolyses amides and esters.

Electrocardiography Or ECG, is a transthoracic interpretation of the electrical activity of the heart over time captured and externally recorded by skin electrodes.

Electromyogram (EMG) A test used to record the electrical activity of muscles. An electromyogram (EMG) is also called a myogram.

Electuary A medicinal paste composed of powders, or other medical ingredients, incorporated with sweeteners to hide the taste, suitable for oral administration.

Elephantiasis A disorder characterized by chronic thickened and edematous tissue on the genitals and legs due to various causes.

Embolism Obstruction or occlusion of a blood vessel by a blood clot, air bubble or other foreign matter.

Embrocation Lotion or liniment that relieves muscle or joint pains.

Embryotoxic Term that describes any chemical which is harmful to an embryo.

Emesis Vomiting, throwing up.

Emetic An agent that induces vomiting, *cf*: antiemetic.

Emetocathartic Causing vomiting and purging.

Emmenagogue A substance that stimulates, initiates, and/or promotes menstrual flow. Emmenagogues are used in herbal medicine to balance and restore the normal function of the female reproductive system.

Emollient An agent that has a protective and soothing action on the surfaces of the skin and membranes.

Emulsion A preparation formed by the suspension of very finely divided oily or resinous liquid in another liquid.

Encephalitis Inflammation of the brain.

Encephalopathy A disorder or disease of the brain.

Endocrine *adj.* of or relating to endocrine glands or the hormones secreted by them.

Endocytosis Is the process by which cells absorb material (molecules such as proteins) from outside the cell by engulfing it with their cell membrane.

Endometriosis Is a common and often painful disorder of the female reproductive system. The two most common symptoms of endometriosis are pain and infertility.

Endometritis Refers to inflammation of the endometrium, the inner lining of the uterus.

Endometrium The inner lining of the uterus.

Endoplasmic reticulum Is a network of tubules, vesicles and sacs around the nucleus that are interconnected.

Endostatin A naturally-occurring 20-kDa C-terminal protein fragment derived from type XVIII collagen. It is reported to serve as an anti-angiogenic agent that inhibits the formation of the blood vessels that feed cancer tumours.

Endosteum The thin layer of cells lining the medullary cavity of a bone.

Endosteul Pertaining to the endosteum.

Endothelial progenitor cells Population of rare cells that circulate in the blood with the ability to differentiate into endothelial cells, the cells that make up the lining of blood vessels.

Endothelin Any of a group of vasoconstrictive peptides produced by endothelial cells.

Endotoxemia The presence of endotoxins in the blood, which may result in shock. *adj.* endotoxemic.

Endotoxin Toxins associated with certain bacteria, unlike an 'exotoxin' that is not secreted in soluble form by live bacteria, but is a structural component in the bacteria which is released mainly when bacteria are lysed.

Enema Liquid injected into the rectum either as a purgative or medicine, Also called clyster.

Enteral Term used to describe the intestines or other parts of the digestive tract.

Enteral administration Involves the esophagus, stomach, and small and large intestines (i.e., the gastrointestinal tract).

Enteritis Refers to inflammation of the small intestine.

Enterocolic disorder Inflamed bowel disease.

Enterocytes Tall columnar cells in the small intestinal mucosa that are responsible for the final digestion and absorption of nutrients.

Enterohemorrhagic Causing bloody diarrhea and colitis, said of pathogenic microorganisms.

Enterolactone A lignin formed by the action of intestinal bacteria on lignan precursors found in plants; acts as a phytoestrogen.

Enteropooling Increased fluids and electrolytes within the lumen of the intestines due to increased levels of prostaglandins.

Enterotoxin Is a protein toxin released by a microorganism in the intestine.

Enterotoxigenic Of or being an organism containing or producing an enterotoxin.

Entheogen A substance taken to induce a spiritual experience.

Enuresis Bed-wetting, a disorder of elimination that involves the voluntary or involuntary release of urine into bedding, clothing, or other inappropriate places.

Enophthalmos A condition in which the eye falls back into the socket and inhibits proper eyelid function.

Envenomation Is the entry of venom into a person's body, and it may cause localised or systemic poisoning.

Eosinophilia The state of having a high concentration of eosinophils (eosinophil granulocytes) in the blood.

Eosinophils (or, less commonly, acidophils), are white blood cells that are one of the immune system components.

Epididymis A structure within the scrotum attached to the backside of the testis and whose coiled duct provides storage, transit and maturation of spermatozoa.

Epididymitis A medical condition in which there is inflammation of the epididymis.

Epigastralgia Pain in the epigastric region.

Epigastric discomfort Bloated abdomen, swelling of abdomen, abdominal ditension.

Epilepsy A common chronic neurological disorder that is characterized by recurrent unprovoked seizures.

Epileptiform Resembling epilepsy or its manifestations. *adj.* epileptiformic.

Epileptogenesis A process by which a normal brain develops epilepsy, a chronic condition in which seizures occur. *adj.* epileptogenic.

Episiotomy A surgical incision through the perineum made to enlarge the vagina and assist childbirth.

Epithelioma A usually benign skin disease most commonly occurring on the face, around the eyelids and on the scalp.

Epitrochlearis The superficial-most muscle of the arm anterior surface.

Epistaxis Acute hemorrhage from the nostril, nasal cavity, or nasopharynx (nose-bleed).

Epstein Barr Virus Herpes virus that is the causative agent of infectious mononucleosis. It is also associated with various types of human cancers.

ERbeta Estrogen receptor beta, a nuclear receptor which is activated by the sex hormone, estrogen.

Ergocalciferol A form of vitamin D, also called vitamin D2. See vitamin D.

Ergonic Increasing capacity for bodily or mental labor especially by eliminating fatigue symptoms.

ERK (extracellular signal regulated kinases) Widely expressed protein kinase intracellular signaling molecules which are involved in functions including the regulation of meiosis, mitosis, and post mitotic functions in differentiated cells.

Eructation The act of belching or of casting up wind from the stomach through the mouth.

Eruption A visible rash or cutaneous disruption.

Erysipelas Is an intensely red *Streptococcus* bacterial infection that occurs on the face and lower extremities.

Erythema Abnormal redness and inflammation of the skin, due to vasodilation.

Erythematous Characterized by erythema.

Erythroleukoplakia An abnormal patch of red and white tissue that forms on mucous membranes in the mouth and may become cancer.

Tobacco (smoking and chewing) and alcohol may increase the risk of erythroleukoplakia.

Erythropoietin (EPO) A hormone produced by the kidney that promotes the formation of red blood cells (erythrocytes) in the bone marrow.

Eschar A slough or piece of dead tissue that is cast off from the surface of the skin.

Escharotic Capable of producing an eschar; a caustic or corrosive agent.

Estradiol Is the predominant sex hormone present in females, also called oestradiol.

Estrogen Female hormone produced by the ovaries that play an important role in the estrous cycle in women.

Estrogen receptor (ER) Is a protein found in high concentrations in the cytoplasm of breast, uterus, hypothalamus, and anterior hypophysis cells; ER levels are measured to determine a breast CA's potential for response to hormonal manipulation.

Estrogen receptor positive (ER+) Means that estrogen is causing the tumour to grow, and that the breast cancer should respond well to hormone suppression treatments.

Estrogen receptor negative (ER−) Tumour is not driven by estrogen and need another test to determine the most effective treatment.

Estrogenic Relating to estrogen or producing estrus.

Estrus Sexual excitement or heat of female; or period of this characterized by changes in the sex organs.

Euglycaemia Normal blood glucose concentration.

Exanthematous Characterized by or of the nature of an eruption or rash.

Excitotoxicity Is the pathological process by which neurons are damaged and killed by glutamate and similar substances.

Excipient A pharmacologically inert substance used as a diluent or vehicle for the active ingredients of a medication.

Exocytosis The cellular process by which cells excrete waste products or chemical transmitters.

Exophthalmos or exophthalmia or proptosis Is a bulging of the eye anteriorly out of the orbit. *adj.* exophthalmic.

Exotoxin A toxin secreted by a microorganism and released into the medium in which it grows.

Expectorant An agent that increases bronchial mucous secretion by promoting liquefaction of the sticky mucous and expelling it from the body.

Exteroceptive Responsiveness to stimuli that are external to an organism.

Extrapyramidal side effects Are a group of symptoms (tremor, slurred speech, akathisia, dystonia, anxiety, paranoia and bradyphrenia) that can occur in persons taking antipsychotic medications.

Extravasation Discharge or escape, as of blood from the vein into the surrounding tissues.

FADD Fas-associated protein with death domain, the protein encoded by this gene is an adaptor molecule which interacts with other death cell surface receptors and mediates apoptotic signals.

Familial amyloid polyneuropathy (FAP) Also called Corino de Andrade's disease, a neurodegenerative autosomal dominant genetically transmitted, fatal, incurable disease.

Familial adenomatous polyposis (FAP) Is an inherited condition in which numerous polyps form mainly in the epithelium of the large intestine.

Familial dysautonomia A genetic disorder that affects the development and survival of autonomic and sensory nerve cells.

FasL or CD95L Fas ligand is a type-II transmembrane protein that belongs to the tumour necrosis factor (TNF) family.

FAS: fatty acid synthase (FAS) A multi-enzyme that plays a key role in fatty acid synthesis.

Fas molecule A member of the Tumour Necrosis Factor Receptors, that mediates apoptotic signal in many cell types.

Fauces The passage leading from the back of the mouth into the pharynx.

Favus A chronic skin infection, usually of the scalp, caused by the fungus, *Trichophyton schoenleinii* and characterized by the development of thick, yellow crusts over the hair follicles. Also termed tinea favosa.

Febrifuge An agent that reduces fever. Also called an antipyretic.

Febrile Pertaining to or characterized by fever.

Fetotoxic Toxic to the fetus.

Fibrates Hypolipidemic agents primarily used for decreasing serum triglycerides, while increasing High density lipoprotein (HDL).

Fibril A small slender fibre or filament.

Fibrin Insoluble protein that forms the essential portion of the blood clot.

Fibrinolysis A normal ongoing process that dissolves fibrin and results in the removal of small blood clots.

Fribinolytic Causing the dissolution of fibrin by enzymatic action.

Fibroblast Type of cell that synthesizes the extracellular matrix and collagen, the structural framework (stroma) for animal tissues, and play a critical role in wound healing.

Fibrogenic Promoting the development of fibres.

Fibromyalgia A common and complex chronic pain disorder that affects people physically, mentally and socially. Symptoms include debilitating fatigue, sleep disturbance, and joint stiffness. Also referred to as FM or FMS.

Fibrosarcoma A malignant tumour derived from fibrous connective tissue and characterized by immature proliferating fibroblasts or undifferentiated anaplastic spindle cells.

Fibrosis The formation of fibrous tissue as a reparative or reactive process.

Filarial Pertaining to a thread-like nematode worm.

Filariasis A parasitic and infectious tropical disease that is caused by thread-like filarial nematode worms in the superfamily Filarioidea.

Fistula An abnormal connection between two parts inside of the body.

Fistula-in-ano A track connecting the internal anal canal to the skin surrounding the anal orifice.

5′-Nucleotidase (5′-ribonucleotide phosphohydrolase), an intrinsic membrane glycoprotein present as an ectoenzyme in a wide variety of mammalian cells, hydrolyzes 5′-nucleotides to their corresponding nucleosides.

Flatulence Is the presence of a mixture of gases known as flatus in the digestive tract of mammals expelled from the rectum. Excessive flatulence can be caused by lactose intolerance, certain foods or a sudden switch to a high fibre.

Flavans A subgroup of flavonoids. See flavonoids.

Flavanols A subgroup of flavonoids, are a class of flavonoids that use the 2-phenyl-3,4-dihydro-2H-chromen-3-ol skeleton. These compounds include the catechins and the catechin gallates. They are found in chocolate, fruits and vegetables. See flavonoids.

Flavanones A subgroup of flavonoids, constitute >90% of total flavonoids in citrus. The major dietary flavanones are hesperetin, naringenin and eriodictyol.

Flavivirus A family of viruses transmitted by mosquitoes and ticks that cause some important diseases, including dengue, yellow fever, tick-borne encephalitis and West Nile fever.

Flavones A subgroup of flavonoids based on the backbone of 2-phenylchromen-4-one (2-phenyl-1-benzopyran-4-one). Flavones are mainly found in cereals and herbs.

Flavonoids (or bioflavonoids) are a group of polyphenolic antioxidant compounds in that are occur in plant as secondary metabolites. They are responsible for the colour of fruit and vegetables. Twelve basic classes (chemical types) of flavonoids have been recognized: flavones, isoflavones, flavans, flavanones, flavanols, flavanolols, anthocyanidins, catechins (including proanthocyanidins), leukoanthocyanidins, chalcones, dihydrochalcones, and aurones. Apart from their antioxidant activity, flavonoids are known for their ability to strengthen capillary walls, thus assisting circulation and helping to prevent and treat bruising, varicose veins, bleeding gums and nosebleeds, heavy menstrual bleeding and are also anti-inflammatory.

Flourine F is an essential chemical element that is required for maintenance of healthy bones and teeth and to reduce tooth decay. It is found in sea weeds, tea, water, seafood and dairy products.

Fluorosis A dental health condition caused by a child receiving too much fluoride during tooth development.

Flux An excessive discharge of fluid.

FMD (Flow Mediated Dilation) A measure of endothelial dysfunction which is used to evaluate cardiovascular risk.

Follicle stimulating hormone (FSH) A hormone produced by the pituitary gland. In women, it helps control the menstrual cycle and the production of eggs by the ovaries.

Follicular atresia The break-down of the ovarian follicles.

Fomentation Treatment by the application of war, moist substance.

Fontanelle Soft spot on an infant's skull.

Framboesia See yaws.

FRAP Ferric reducing ability of plasma, an assay used to assess antioxidant property.

Friedreich's ataxia Is a genetic inherited disorder that causes progressive damage to the nervous system resulting in symptoms ranging from muscle weakness and speech problems to heart disease. *cf.* ataxia.

Fulminant hepatitis Acute liver failure.

Functional food Is any fresh or processed food claimed to have a health-promoting or disease-preventing property beyond the basic function of supplying nutrients. Also called medicinal food.

Furuncle Is a skin disease caused by the infection of hair follicles usually caused by *Staphylococcus aureus,* resulting in the localized accumulation of pus and dead tissue.

Furunculosis Skin condition characterized by persistent, recurring boils.

G2-M cell cycle The phase where the cell prepare for mitosis and where chromatids and daughter cells separate.

GABA Gamma aminobutyric acid, required as an inhibitory neurotransmitter to block the transmission of an impulse from one cell to another in the central nervous system, which prevents over-firing of the nerve cells. It is used to treat both epilepsy and hypertension.

GADD 152 A pro-apoptotic gene.

Galctifuge Or lactifuge, causing the arrest of milk secretion.

Galactogogue A substance that promotes the flow of milk.

Galactophoritis Inflammation of the milk ducts.

Galactopoietic Increasing the flow of milk; milk-producing.

Gall bladder A small, pear-shaped muscular sac, located under the right lobe of the liver, in which bile secreted by the liver is stored until needed by the body for digestion. Also called cholecyst, cholecystis.

Gallic Acid Equivalent (GAE) Measures the total phenol content in terms of the standard Gallic acid by the Folin-Ciocalteau assay.

Gamma GT (GGT) γ-glutamyl transpeptidase, a liver enzyme.

Gastralgia (Heart burn) – pain in the stomach or abdominal region. It is caused by excess of acid, or an accumulation of gas, in the stomach.

Gastric Pertaining to or affecting the stomach.

Gastric emptying Refers to the speed at which food and drink leave the stomach.

Gastritis Inflammation of the stomach.

Gastrocnemius muscle The big calf muscle at the rear of the lower leg.

Gastrotonic (Gastroprotective) Substance that strengthens, tones, or regulates gastric functions (or protects from injury) without overt stimulation or depression.

Gavage Forced feeding.

Gene silencing Suppression of the expression of a gene.

Genotoxin A chemical or other agent that damages cellular DNA, resulting in mutations or cancer.

Genotoxic Describes a poisonous substance which harms an organism by damaging its DNA thereby capable of causing mutations or cancer.

Geriatrics Is a sub-specialty of internal medicine that focuses on health care of elderly people.

Gestational hypertension Development of arterial hypertension in a pregnant woman after 20 weeks gestation.

Ghrelin A gastrointestinal peptide hormone secreted by epithelial cells in the stomach lining, it stimulates appetite, gastric emptying, and increases cardiac output.

Gingival Index An index describing the clinical severity of gingival inflammation as well as its location.

Gingivitis Refers to gingival inflammation induced by bacterial biofilms (also called plaque) adherent to tooth surfaces.

Gin-nan sitotoxism Toxicity caused by ingestion of ginkgotoxin and characterised mainly by epileptic convulsions, paralysis of the legs and loss of consciousness.

Glaucoma A group of eye diseases in which the optic nerve at the back of the eye is slowly destroyed, leading to impaired vision and blindness.

Gleet A chronic inflammation (as gonorrhea) of a bodily orifice usually accompanied by an abnormal discharge.

Glial cells Support, non-neuronal cells in the central nervous system that maintain homeostasis, form myelin and provide protection for the brain's neurons.

Glioma Is a type of tumour that starts in the brain or spine. It is called a glioma because it arises from glial cells.

Glioblastoma multiforme Most common and most aggressive type of primary brain tumour in humans, involving glial cells.

Glomerulonephritis (GN) A renal disease characterized by inflammation of the glomeruli, or small blood vessels in the kidneys. Also known as glomerular nephritis. *adj.* glomerulonephritic.

Glomerulosclerosis A hardening of the glomerulus in the kidney.

Glossal Pertaining to the tongue.

GLP-1 Glucagon-like peptide-1 is derived from the transcription product of the proglucagon gene, associate with type 2-diabetes therapy.

Gluconeogenesis A metabolic pathway that results in the generation of glucose from non-carbohydrate carbon substrates such as lactate. *adj.* gluconeogenic.

Glucose transporters (GLUT or SLC2A family) are a family of membrane proteins found in most mammalian cells.

Glucosyltranferase An enzyme that enable the transfer of glucose.

Glucuronidation A phase II detoxification pathway occurring in the liver in which glucuronic acid is conjugated with toxins.

Glutamic Oxaloacetate Transaminase (GOT) Catalyzes the transfer of an amino group from an amino acid (Glu) to a 2-keto-acid to generate a new amino acid and the residual 2-keto-acid of the donor amino acid.

Glutamic pyruvate transaminase (GPT) See Alanine aminotransferase.

Glutathione (GSH) A tripeptide produced in the human liver and plays a key role in intermediary metabolism, immune response and health. It plays an important role in scavenging free radicals and protects cells against several toxic oxygen-derived chemical species.

Glutathione peroxidase (GPX) The general name of an enzyme family with peroxidase activity whose main biological role is to protect the organism from oxidative damage.

Glutathione S-transferase (GST) A major group of detoxification enzymes that participate in the detoxification of reactive electrophilic compounds by catalysing their conjugation to glutathione.

Glycaemic index (GI) Measures carbohydrates according to how quickly they are absorbed and raise the glucose level of the blood.

Glycaemic load (GL) Is a ranking system for carbohydrate content in food portions based on their glycaemic index and the amount of available carbohydrate, i.e. GI x available carbohydrate divided by 100. Glycemic load combines both the quality and quantity of carbohydrate in one 'number'. It's the best way to predict blood glucose values of different types and amounts of food.

Glycation or glycosylation A chemical reaction in which glycosyl groups are added to a protein to produce a glycoprotein.

Glycogenolysis Is the catabolism of glycogen by removal of a glucose monomer through cleavage with inorganic phosphate to produce glucose-1-phosphate.

Glycometabolism Metabolism (oxidation) of glucose to produce energy.

Glycosuria Or glucosuria is an abnormal condition of osmotic diuresis due to excretion of glucose by the kidneys into the urine.

Glycosylases A family of enzymes involved in base excision repair.

Goitre An enlargement of the thyroid gland leading to swelling of the neck or larynx.

Goitrogen Substance that suppresses the function of the thyroid gland by interfering with iodine uptake, causing enlargement of the thyroid, i.e. goiter.

Goitrogenic *adj.* causing goiter.

Gonadotroph A basophilic cell of the anterior pituitary specialized to secrete follicle-stimulating hormone or luteinizing hormone.

Gonatropins Protein hormones secreted by gonadotrope cells of the pituitary gland of vertebrates.

Gonorrhoea A common sexually transmitted bacterial infection caused by the bacterium *Neisseria gonorrhoeae*.

Gout A disorder caused by a build-up of a waste product, uric acid, in the bloodstream. Excess uric acid settles in joints causing inflammation, pain and swelling.

G-protein-coupled receptors (GPCRs) Comprise a large and diverse family of proteins whose primary function is to transduce extracellular stimuli into cells.

Granulation The condition or appearance of being granulated (becoming grain-like).

Gravel Sand-like concretions of uric acid, calcium oxalate, and mineral salts formed in the passages of the biliary and urinary tracts.

Gripe water Is a home remedy for babies with colic, gas, teething pain or other stomach ailments. Its ingredients vary, and may include alcohol, bicarbonate, ginger, dill, fennel and chamomile.

Grippe An epidemic catarrh; older term for influenza.

GSH See Glutathione.

GSH-Px Glutathione peroxidase, general name of an enzyme family with peroxidase activity whose main biological role is to protect the organism from oxidative damage.

GSSG Glutathione disulfides are biologically important intracellular thiols, and alterations in the GSH/GSSG ratio are often used to assess exposure of cells to oxidative stress.

GSTM Glutathione S transferase M1, a major group of detoxification enzymes.

GSTM 2 Glutathione S transferase M2, a major group of detoxification enzymes.

Gynecopathy Any or various diseases specific to women.

Gynoid adiposity Fat distribution mainly to the hips and thighs, pear shaped.

Haemagogic Promoting a flow of blood.

Haematemesis, Hematemesis Is the vomiting of blood.

Haematinic Improving the quality of the blood, its haemoglobin level and the number of erythrocytes.

Haematochezia Passage of stools containing blood.

Haematochyluria, hematochyluria The discharge of blood and chyle (emulsified fat) in the urine, see also chyluria.

Haematoma, hematoma A localized accumulation of blood in a tissue or space composed of clotted blood.

Haematometra, hematometra A medical condition involving bleeding of or near the uterus.

Haematopoiesis, hematopoiesis Formation of blood cellular components from the haematopoietic stem cells.

Haematopoietic *adj.* relating to the formation and development of blood cells.

Haematuria, Hematuria Is the presence of blood in the urine. Hematuria is a sign that something is causing abnormal bleeding in a person's genitourinary tract.

Haeme oxygenase (HO-1, encoded by Hmox1) is an inducible protein activated in systemic inflammatory conditions by oxidant stress, an enzyme that catalyzes degradation of heme.

Haemochromatosis Is a condition in which the body takes in too much iron.

Haemodialysis, Hemodialysis A method for removing waste products such as potassium and urea, as well as free water from the blood when the kidneys are in renal failure.

Haemolyis Lysis of red blood cells and the release of haemoglobin into the surrounding fluid (plasma). *adj.* haemolytic.

Haemoptysis, hemoptysis Is the coughing up of blood from the respiratory tract. The blood can come from the nose, mouth, throat, and the airway passages leading to the lungs.

Haemorrhage, hemaorrhage Bleeding, discharge of blood from blood vessels.

Haemorrhoids, Hemorrhoids A painful condition in which the veins around the anus or lower rectum are enlarged, swollen and inflamed. Also called piles.

Haemostasis, hemostasis A complex process which causes the bleeding process to stop.

Haemostatic, hemostatic Something that stops bleeding.

Halitosis (bad breath) a common condition caused by sulfur-producing bacteria that live within the surface of the tongue and in the throat.

Hallucinogen Drug that produces hallucinogen.

Hallucinogenic Inducing hallucinations.

Haplotype A set of alleles of closely linked loci on a chromosome that tend to be inherited together.

Hapten A small molecule that can elicit an immune response only when attached to a large carrier such as a protein.

HBeAg Hepatitis B e antigen.

HBsAg Hepatitis B s antigen.

Heartburn Burning sensation in the stomach and esophagus caused by excessive acidity of the stomach fluids.

Heat rash Any condition aggravated by heat or hot weather such as intertrigo.

Heat Shock Chaperones (HSC) Ubiquitous molecules involved in the modulation of protein conformational and complexation states, associated with heat stress or other cellular stress response.

Heat Shock Proteins (HSP) A group of functionally related proteins the expression of which is increased when the cells are exposed to elevated temperatures or other cellular stresses.

Helminthiasis A disease in which a part of the body is infested with worms such as pinworm, roundworm or tapeworm.

Hemagglutination A specific form of agglutination that involves red blood cells.

Hemagglutination–inhibition test Measures of the ability of soluble antigen to inhibit the agglutination of antigen-coated red blood cells by antibodies.

Hemagglutinin Refers to a substance that causes red blood cells to agglutinate.

Hemangioma Blood vessel.

Hematocrit Is a blood test that measures the percentage of the volume of whole blood that is made up of red blood cells.

Hematopoietic Pertaining to the formation of blood or blood cells.

Hematopoietic stem cell Is a cell isolated from the blood or bone marrow that can renew itself, and can differentiate to a variety of specialized cells.

Heme oxygenase-1 (HO-1) An enzyme that catalyses the degrdation of heme; an inducible stress protein, confers cytoprotection against oxidative stress in-vitro and in-vivo.

Hemoglobinopathies Genetic defects that produce abnormal hemoglobins and anemia.

Hemolytic anemia Anemia due to hemolysis, the breakdown of red blood cells in the blood vessels or elsewhere in the body.

Hemorheology Study of blood flow and its elements in the circulatory system. *adj.* hemorheological.

Hemorrhagic colitis An acute gasteroenteritis characterized by overtly bloody diarrhea that is caused by *Escherichia coli* infection.

Hemolytic-uremic syndrome Is a disease characterized by hemolytic anemia, acute renal failure (uremia) and a low platelet count.

Hepa-1c1c7 A type of hepatoma cells.

Hepatalgia Pain or discomfort in the liver area.

Heptalgia Pain in the liver and spleen.

Hepatectomy The surgical removal of part or all of the liver.

Hepatic Relating to the liver.

Hepatic cirrhosis Affecting the liver, characterize by hepatic fibrosis and regenerative nodules.

Hepatitis Inflammation of the liver.

Hepatitis A (Formerly known as infectious hepatitis) is an acute infectious disease of the liver caused by the hepatovirus hepatitis A virus.

Hepatocarcinogenesis Represents a linear and progressive cancerous process in the liver in which successively more aberrant monoclonal populations of hepatocytes evolve.

Hepatocellular carcinoma (HCC) Also called malignant hepatoma, is a primary malignancy (cancer) of the liver.

Hepatocytolysis Cytotoxicity (dissolution) of liver cells.

Hepatoma Cancer of the liver.

Hepatopathy A disease or disorder of the liver.

Hepatoprotective (liver protector) a substance that helps protect the liver from damage by toxins, chemicals or other disease processes.

Hepatoregenerative A compound that promotes hepatocellular regeneration, repairs and restores liver function to optimum performance.

Hepatotonic (liver tonic) a substance that is tonic to the liver – usually employed to normalize liver enzymes and function.

Hernia Occurs when part of an internal organ bulges through a weak area of muscle.

HER- 2 Human epidermal growth factor receptor 2, a protein giving higher aggressiveness in breast cancer, also known as ErbB-2, ERBB2.

Herpes A chronic inflammation of the skin or mucous membrane characterized by the development of vesicles on an inflammatory base.

Herpes simplex virus 1 and 2 – (HSV-1 and HSV-2) Are two species of the herpes virus family which cause a variety of illnesses/infections in humans such cold sores, chickenpox or varicella, shingles or herpes zoster (VZV), cytomegalovirus (CMV), and various cancers, and can cause brain inflammation (encephalitis). HSV-1 is commonly associated with herpes outbreaks of the face known as cold sores or fever blisters, whereas HSV-2 is more often associated with genital herpes. They are also called Human Herpes Virus 1 and 2 (HHV-1 and HHV-2) and are neurotropic and neuroinvasive viruses; they enter and hide in the human nervous system, accounting for their durability in the human body.

Herpes zoster Or simply zoster, commonly known as shingles and also known as zona, is a viral disease characterized by a painful skin rash with blisters.

Heterophobia Term used to describe irrational fear of, aversion to, or discrimination against heterosexuals.

Heterophoria An eye condition where the motion of the eyes are not parallel to each other.

HDL-C (HDL Cholesterol) High density lipoprotein-cholesterol, also called "good cholesterol". See also high-density lipoprotein.

Hiatus hernia Occurs when the upper part of the stomach pushes its way through a tear in the diaphragm.

High-density lipoprotein (HDL) Is one of the five major groups of lipoproteins which enable cholesterol and triglycerides to be transported within the water based blood stream. HDL can remove cholesterol from atheroma within arteries and transport it back to the liver for excretion or re-utilization – which is the main reason why HDL-bound cholesterol is sometimes called "good cholesterol", or HDL-C. A high level of HDL-C seems to protect against cardiovascular diseases. cf. LDL.

HGPRT, HPRT (hypoxanthine-guanine phosphoribosyl transferase) An enzyme that catalyzes the conversion of 5-phosphoribosyl-1-pyrophosphate and hypoxanthine, guanine, or 6-mercaptopurine to the corresponding 5′-mononucleotides and pyrophosphate. The enzyme is important in purine biosynthesis as well as central nervous system functions.

Hippocampus A ridge in the floor of each lateral ventricle of the brain that consists mainly of gray matter.

Hippocampal Pertaining to the hippocampus.

Histaminergic Liberated or activated by histamine, relating to the effects of histamine at histamine receptors of target tissues.

Histaminergic receptors Are types of G-protein coupled receptors with histamine as their endogenous ligand.

HIV See Human immunodeficiency virus.

Hives (urticaria) is a skin rash characterised by circular wheals of reddened and itching skin.

HMG-CoAr 3-Hydroxy-3-methyl-glutaryl-CoA reductase or (HMGCR) is the rate-controlling enzyme (EC 1.1.1.88) of the mevalonate pathway.

HMG-CoA 3-Hydroxy-3-methylglutaryl-coenzyme A, an intermediate in the mevalonate pathway .

Hodgkin's disease Disease characterized by enlargement of the lymph glands, spleen and anemia.

Homeodomain transcription factor A protein domain encoded by a homeobox. Homeobox genes encode transcription factors which typically switch on cascades of other genes.

Homeostasis The maintenance of a constant internal environment of a cell or an organism, despite fluctuations in the external.

Homeotherapy Treatment or prevention of disease with a substance similar but not identical to the causative agent of the disease.

Homocysteine An amino acid in the blood.

Homograft See allograft.

Hormonal (female) Substance that has a hormone-like effect similar to that of estrogen and/or a substance used to normalize female hormone levels.

Hormonal (male) Substance that has a hormone-like effect similar to that of testosterone and/or a substance used to normalize male hormone levels.

HRT Hormone replacement therapy, the administration of the female hormones, oestrogen and progesterone, and sometimes testosterone.

HSP27 Is an ATP-independent, 27 kDa heat shock protein chaperone that confers protection against apoptosis.

HSP90 A 90 kDa heat shock protein chaperone that has the ability to regulate a specific subset of cellular signaling proteins that have been implicated in disease processes.

hTERT – (TERT) Telomerase reverse transcriptase is a catalytic subunit of the enzyme telomerase in humans. It exerts a novel protective function by binding to mitochondrial DNA, increasing respiratory chain activity and protecting against oxidative stress–induced damage.

HT29 cells Are human intestinal epithelial cells which produce the secretory component of Immunoglobulin A (IgA), and carcinoembryonic antigen (CEA).

Human cytomegalovirus (HCMV) A DNA herpes virus which is the leading cause of congenital viral infection and mental retardation.

Human factor X A coagulation factor also known by the eponym Stuart-Prower factor or as thrombokinase, is an enzyme involved in blood coagulation. It synthesized in the liver and requires vitamin K for its synthesis.

Human immunodeficiency virus (HIV) A retrovirus that can lead to acquired immunodeficiency syndrome (AIDS), a condition in humans in which the immune system begins to fail, leading to life-threatening opportunistic infections.

Humoral immune response (HIR) Is the aspect of immunity that is mediated by secreted antibodies (as opposed to cell-mediated immunity, which involves T lymphocytes) produced in the cells of the B lymphocyte lineage (B cell).

HUVEC Human umbilical vein endothelial cells.

Hyaluronidase Enzymes that catalyse the hydrolysis of certain complex carbohydrates like hyaluronic acid and chondroitin sulfates.

Hydatidiform A rare mass or growth that forms inside the uterus at the beginning of a pregnancy.

Hydrocholeretic An agent that stimulates an increased output of bile of low specific gravity.

Hydrogogue A purgative that causes an abundant watery discharge from the bowel.

Hydronephrosis Is distension and dilation of the renal pelvis and calyces, usually caused by obstruction of the free flow of urine from the kidney.

Hydrophobia A viral neuroinvasive disease that causes acute encephalitis (inflammation of the brain) in warm-blooded animals. Also called rabies.

Hydropsy See dropsy.

Hyperaemia The increase of blood flow to different tissues in the body.

Hyperalgesia An increased sensitivity to pain (enhanced pricking pain), which may be caused by damage to nociceptors or peripheral nerves.

Hyperammonemia, hyperammonaemia A metabolic disturbance characterised by an excess of ammonia in the blood.

Hypercholesterolemia High levels of cholesterol in the blood that increase a person's risk for cardiovascular disease leading to stroke or heart attack.

Hyperemia Is the increased blood flow that occurs when tissue is active.

Hyperemesis Severe and persistent nausea and vomiting (morning sickness) during pregnancy.

Hyperglycemic, hyperglycaemia High blood sugar; is a condition in which an excessive amount of glucose circulates in the blood plasma.

Hyperglycemic A substance that raises blood sugar levels.

Hyperhomocysteinemia Is a medical condition characterized by an abnormally large level of homocysteine in the blood.

Hyperinsulinemia A condition in which there are excess levels of circulating insulin in the blood; also known as pre-diabetes.

Hyperkalemia Is an elevated blood level of the electrolyte potassium.

Hyperknesis Enhanced itch to pricking.

Hyperleptinemia Increased serum leptin level.

Hypermethylation An increase in the inherited methylation of cytosine and adenosine residues in DNA.

Hyperpiesia Persistent and pathological high blood pressure for which no specific cause can be found.

Hyperplasia Increased cell production in a normal tissue or organ.

Hyperpropulsion Using water pressure as a force to move objects; used to dislodge calculi in the urethra.

Hyperpyrexia Is an abnormally high fever.

Hypertension Commonly referred to as "high blood pressure" or HTN, is a medical condition in which the arterial blood pressure is chronically elevated.

Hypertensive Characterized or caused by increased tension or pressure as abnormally high blood pressure.

Hypertriglyceridaemia or hypertriglycemia A disorder that causes high triglycerides in the blood.

Hypertrophy Enlargement or overgrowth of an organ.

Hyperuricemia Is a condition characterized by abnormally high level of uric acid in the blood.

Hypoadiponectinemia Low plasma adiponectin concentrations associated with obesity and type 2 diabetes; that is closely related to the degree of insulin resistance and hyperinsulinemia than to the degree of adiposity and glucose tolerance.

Hypoalbuminemia A medical condition where levels of albumin in blood serum are abnormally low.

Hypocalcemic tetany A disease caused by an abnormally low level of calcium in the blood and characterized by hyperexcitability of the neuromuscular system and results in carpopedal spasms.

Hypochlorhydria Refer to states where the production of gastric acid in the stomach is absent or low.

Hypocholesterolemic (cholesterol-reducer), a substance that lowers blood cholesterol levels.

Hypocorticism See Addison's disease.

Hypocortisolism See Addison's disease.

Hypoglycemic An agent that lowers the concentration of glucose (sugar) in the blood.

Hypoperfusion Decreased blood flow through an organ, characterized by an imbalance of oxygen demand and oxygen delivery to tissues.

Hypophagic Under-eating.

Hypospadias An abnormal birth defect in males in which the urethra opens on the under surface of the penis.

Hypotensive Characterised by or causing diminished tension or pressure, as abnormally low blood pressure.

Hypothermia A condition in which an organism's temperature drops below that required for normal metabolism and body functions.

Hypothermic Relating to hypothermia, with subnormal body temperature.

Hypoxaemia Is the reduction of oxygen specifically in the blood.

Hypoxia A shortage of oxygen in the body. *adj.* hypoxic.

ICAM-1 (Inter-Cellular Adhesion Molecule 1) Also known as CD54 (Cluster of Differentiation 54), is a protein that in humans is encoded by the ICAM1 gene.

IC 50 The median maximal inhibitory concentration; a measure of the effectiveness of a compound in inhibiting biological or biochemical function.

I.C.V. (intra-cerebroventricular) Injection of chemical into the right lateral ventricle of the brain.

Iceterus Jaundice, yellowish pigmentation of the skin.

Ichthyotoxic A substance which is poisonous to fish.

Icteric hepatitis An infectious syndrome of hepatitis characterized by jaundice, nausea, fever, right-upper quadrant pain, enlarged liver and transaminitis (increase in alanine aminotransferase (ALT) and/or aspartate aminotransferase (AST).

Icterus neonatorum Jaundice in newborn infants.

Idiopathic Of no apparent physical cause.

Idiopathic sudden sensorineural hearing loss (ISSHL) Is sudden hearing loss where clinical assessment fails to reveal a cause.

IgE Immunoglobin E – a class of antibody that plays a role in allergy.

IGFs Insulin-like growth factors, polypeptides with high sequence similarity to insulin.

IgG Immunoglobin G – the most abundant immunoglobin (antibody) and is one of the major activators of the complement pathway.

IgM Immunoglobin M – primary antibody against A and B antigens on red blood cells.

IKAP Is a scaffold protein of the IvarKappaBeta kinase complex and a regulator for kinases involved in pro-inflammatory cytokine signaling.

IKappa B Or IkB-β, a protein of the NF-Kappa-B inhibitor family.

Ileus A temporary disruption of intestinal peristalsis due to non-mechanical causes.

Immune modulator A substance that affects or modulates the functioning of the immune system.

Immunodeficiency A state in which the immune system's ability to fight infectious disease is compromised or entirely absent.

Immunogenicity The property enabling a substance to provoke an immune response.

Immunomodulatory Capable of modifying or regulating one or more immune functions.

Immunoreactive Reacting to particular antigens or haptens.

Immunostimulant Agent that stimulates an immune response.

Immunosuppression Involves a process that reduces the activation or efficacy of the immune system.

Immunotoxin A man-made protein that consists of a targeting portion linked to a toxin.

Impetigo A contagious, bacterial skin infection characterized by blisters that may itch, caused by a *Streptoccocus* bacterium or *Staphylococcus aureus* and mostly seen in children.

Impotence A sexual dysfunction characterized by the inability to develop or maintain an erection of the penis.

Incontinence (fecal) The inability to control bowel's movement.

Incontinence (Urine) The inability to control urine excretion.

Index of structural atypia (ISA) Index of structural abnormality.

Induration Hardened, as a soft tissue that becomes extremely firm.

Infarct An area of living tissue that undergoes necrosis as a result of obstruction of local blood supply.

Infarction Is the process of tissue death (necrosis) caused by blockage of the tissue's blood supply.

Inflammation A protective response of the body to infection, irritation or other injury, aimed at destroying or isolating the injuries and characterized by redness, pain, warmth and swelling.

Influenza A viral infection that affects mainly the nose, throat, bronchi and occasionally, lungs.

Infusion A liquid extract obtained by steeping something (e.g. herbs) that are more volatile or dissolve readily in water, to release their active ingredients without boiling.

Inguinal hernia A hernia into the inguinal canal of the groin.

Inhalant A medicinal substance that is administered as a vapor into the upper respiratory passages.

iNOS, inducible nitric oxide synthases Through its product, nitric oxide (NO), may contribute to the induction of germ cell apoptosis. It plays a crucial role in early sepsis-related microcirculatory dysfunction.

Inotropic Affecting the force of muscle contraction.

Insecticide An agent that destroys insects. *adj.* insecticidal.

Insomnia A sleeping disorder characterized by the inability to fall asleep and/or the inability to remain asleep for a reasonable amount of time.

Insulin A peptide hormone composed of 51 amino acids produced in the islets of Langerhans in the pancreas causes cells in the liver, muscle, and fat tissue to take up glucose from the blood, storing it as glycogen in the liver and muscle. Insulin deficiency is often the cause of diabetes and exogenous insulin is used to control diabetes.

Insulin-like growth factors (IGFs) Polypeptides with high sequence similarity to insulin. They are part of a complex system that cells employ to communicate with their physiologic environment.

Insulin-mimetic To act like insulin.

Insulinogenic Associated with or stimulating the production of insulin.

Insulinotropic Changing the action of insulin.

Integrase An enzyme produced by a retrovirus (such as HIV) that enables its genetic material to be integrated into the DNA of the infected cell.

Interferons (IFNs) Are natural cell-signaling glycoproteins known as cytokines produced by the cells of the immune system of most vertebrates in response to challenges such as viruses, parasites and tumour cells.

Interleukins A group of naturally occurring proteins and is a subset of a larger group of cellular messenger molecules called cytokines, which are modulators of cellular behavior.

Interleukin-1 (IL-1) A cytokine that could induce fever, control lymphocytes, increase the number of bone marrow cells and cause degeneration of bone joints. Also called endogenous pyrogen, lymphocyte activating factor, haemopoietin-1 and mononuclear cell factor, amongst others that IL-1 is composed of two distinct proteins, now called IL-1α and IL-1β.

Interleukin 1 Beta (IL-1β) A cytokine protein produced by activated macrophages. cytokine is an important mediator of the inflammatory response, and is involved in a variety of cellular activities, including cell proliferation, differentiation, and apoptosis.

Interleukin 2 (IL-2) A type of cytokine immune system signaling molecule that is instrumental in the body's natural response to microbial infection.

Interleukin-2 receptor (IL-2R) A heterotrimeric protein expressed on the surface of certain immune cells, such as lymphocytes, that binds and responds to a cytokine called IL-2.

Interleukin-6 (IL-6) An interleukin that acts as both a pro-inflammatory and anti-inflammatory cytokine.

Interleukin 8 (I-8) a cytokine produced by macrophages and other cell types such as epithelial cells and is one of the major mediators of the inflammatory response.

Intermediate-density lipoproteins (IDL) Is one of the five major groups of lipoproteins (chylomicrons, VLDL, IDL, LDL, and HDL) that enable fats and cholesterol to move within the water-based solution of the bloodstream. IDL is further degraded to form LDL particles and, like LDL, can also promote the growth of atheroma and increase cardiovascular diseases.

Intermittent claudication An aching, crampy, tired, and sometimes burning pain in the legs that comes and goes, caused by peripheral vascular disease. I t usually occurs with walking and disappears after rest.

Interoceptive Relating to stimuli arising from within the body.

Interstitium The space between cells in a tissue.

Interstitial Pertaining to the interstitium.

Intertrigo An inflammation (rash) caused by microbial infection in skin folds.

Intima Innermost layer of an artery or vein.

Intoxicant Substance that produce drunkenness or intoxication.

Intraperitoneal (i.p.) The term used when a chemical is contained within or administered through the peritoneum (the thin, transparent membrane that lines the walls of the abdomen).

Intrathecal (i.t.) Through the theca of the spinal cord into the subarachnoid space.

Intromission The act of putting one thing into another.

Intubation Refers to the placement of a tube into an external or internal orifice of the body.

Iodine (I) Is an essential chemical element that is important for hormone development in the human body. Lack of iodine can lead to an enlarged thyroid gland (goitre) or other iodine deficiency disorders including mental retardation and stunted growth in babies and children. Iodine is found in dairy products, seafood, kelp, seaweeds, eggs, some vegetables and iodized salt.

IP See Intraperitoneal.

Iron (Fe) Is essential to most life forms and to normal human physiology. In humans, iron is an essential component of proteins involved in oxygen transport and for haemoglobin. It is also essential for the regulation of cell growth and differentiation. A deficiency of iron limits oxygen delivery to cells, resulting in fatigue, poor work performance, and decreased immunity. Conversely, excess amounts of iron can result in toxicity and even death. Dietary sources include, certain cereals, dark green leafy vegetables, dried fruit, legumes, seafood, poultry and meat.

Ischemia An insufficient supply of blood to an organ, usually due to a blocked artery.

Ischuria Retention or suppression of urine.

Isoflavones A subgroup of flavonoids in which the basic structure is a 3-phenyl chromane skeleton. They act as phytoestrogens in mammals. See flavonoids.

Isomers Substances that are composed of the same elements in the same proportions and hence have the same molecular formula but differ in properties because of differences in the arrangement of atoms.

Isoprostanes Unique prostaglandin-like compounds generated in vivo from the free radical-catalysed peroxidation of essential fatty acids.

Jamu Traditional Indonesian herbal medicine.

Jaundice Refers to the yellow color of the skin and whites of the eyes caused by excess bilirubin in the blood.

JNK (Jun N-terminal Kinase), also known as Stress Activated Protein Kinase (SAPK), belongs to the family of MAP kinases.

Jurkat cells A line of T lymphocyte cells that are used to study acute T cell leukemia.

KB cell A cell line derived from a human carcinoma of the nasopharynx, used as an assay for antineoplastic (anti-tumour) agents.

Kallikreins Peptidases (enzymes that cleave peptide bonds in proteins), a subgroup of the serine protease family; they liberate kinins from kininogens. Kallikreins are targets of active investigation by drug researchers as possible biomarkers for cancer.

Kaposi sarcoma A cancerous tumour of the connective tissues caused by the huma herpesvirus 8 and is often associated with AIDS.

Kaposi sarcoma herpes virus (KSHV) Also known as human herpesvirus-8, is a γ 2 herpesvirus or rhadinovirus. It plays an important role in the pathogenesis of Kaposi sarcoma (KS), multicentric Castleman disease (MCD) of the plasma cell type, and primary effusion lymphoma and occurs in HIV patients.

Keratin A sulphur-containing protein which is a major component in skin, hair, nails, hooves, horns, and teeth.

Keratinocyte Is the major constituent of the epidermis, constituting 95% of the cells found there.

Keratinophilic Having an affinity for keratin.

Keratitis Inflammation of the cornea.

Keratomalacia An eye disorder that leads to a dry cornea.

Kidney stones (calculi) are hardened mineral deposits that form in the kidney.

Kinin Is any of various structurally related polypeptides, such as bradykinin, that act locally to induce vasodilation and contraction of smooth muscle.

Kininogen Either of two plasma α2-globulins that are kinin precursors.

Knockout Gene knockout is a genetic technique in which an organism is engineered to carry genes that have been made inoperative.

Kunitz protease inhibitors A type of protein contained in legume seeds which functions as a protease inhibitor.

Kupffer cells Are resident macrophages of the liver and play an important role in its normal physiology and homeostasis as well as participating in the acute and chronic responses of the liver to toxic compounds.

L-Dopa (L-3,4-Dihydroxyphenylalanine) is an amino acid that is formed in the liver and converted into dopamine in the brain.

Labour Process of childbirth involving muscular contractions.

Lacrimation Secretion and discharge of tears.

Lactagogue An agent that increases or stimulates milk flow or production. Also called a galactagogue.

Lactate dehydrogenase (LDH) Enzyme that catalyzes the conversion of lactate to pyruvate.

Lactation Secretion and production of milk.

Lactic acidosis Is a condition caused by the buildup of lactic acid in the body. It leads to acidification of the blood (acidosis), and is considered a distinct form of metabolic acidosis.

LAK cell A lymphokine-activated killer cell i.e. a white blood cell that has been stimulated to kill tumour cells.

Laminin A glycoprotein component of connective tissue basement membrane that promotes cell adhesion.

Laparotomy A surgical procedure involving an incision through the abdominal wall to gain access into the abdominal cavity. *adj.* laparotomized .

Larvacidal An agent which kills insect or parasite larva.

Laryngitis Is an inflammation of the larynx.

Laxation Bowel movement.

Laxatives Substances that are used to promote bowel movement.

LC 50 Median lethal concentration, see LD 50.

LD 50 Median lethal dose – the dose required to kill half the members of a tested population. Also called LC 50 (median lethal concentration).

LDL See low-density lipoprotein.

LDL Cholesterol See low-density lipoprotein.

LDL receptor (LDLr) A low-density lipoprotein receptor gene.

Lectins Are sugar-binding proteins that are highly specific for their sugar moieties, that agglutinate cells and/or precipitate glycoconjugates. They play a role in biological recognition phenomena involving cells and proteins.

Leishmaniasis A disease caused by protozoan parasites that belong to the genus *Leishmania* and is transmitted by the bite of certain species of sand fly.

Lenticular opacity Also known as or related to cataract.

Leprosy A chronic bacterial disease of the skin and nerves in the hands and feet and, in some cases, the lining of the nose. It is caused by the *Mycobacterium leprae.* Also called Hansen's disease.

Leptin Is a 16 kDa protein hormone with important effects in regulating body weight, metabolism and reproductive function.

Lequesne Algofunctional Index Is a widespread international instrument (10 questions survey) and recommended by the World Health Organization (WHO) for outcome measurement in hip and knee diseases such as osteoarthritis.

Leucocyte White blood corpuscles, colourless, without haemoglobin that help to combat infection.

Leucoderma A skin abnormality characterized by white spots, bands and patches on the skin; they can also be caused by fungus and tinea. Also see vitiligo.

Leucorrhoea Commonly known as whites, refers to a whitish discharge from the female genitals

Leukemia, leukaemia A cancer of the blood or bone marrow and is characterized by an abnormal proliferation (production by multiplication) of blood cells, usually white blood cells (leukocytes).

Leukemogenic Relating to leukemia, causing leukemia.

Leukocytopenia Abnormal decrease in the number of leukocytes (white blood cells) in the blood.

Leukomyelopathy Any diseases involving the white matter of the spinal cord.

Leukopenia A decrease in the number of circulating white blood cells.

Leukoplakia Condition characterized by white spots or patches on mucous membranes, especially of the mouth and vulva.

Leukotriene A group of hormones that cause the inflammatory symptoms of hay-fever and asthma.

Luteolysis Degeneration of the corpus luteum and ovarian luteinized tissues. adj. luteolytic.

Levarterenol See Norepinephrine.

LexA repressor Or Repressor LexA is repressor enzyme that represses SOS response genes coding for DNA polymerases required for repairing DNA damage

Libido Sexual urge.

Lichen planus A chronic mucocutaneous disease that affects the skin, tongue, and oral mucosa.

Ligroin A volatile, inflammable fraction of petroleum, obtained by distillation and used as a solvent.

Liniment Liquid preparation rubbed on skin, used to relieve muscular aches and pains.

Linterized starch Starch that has undergone prolonged acid treatment.

Lipodiatic Having lipid and lipoprotein lowering property.

Lipodystrophy A medical condition characterized by abnormal or degenerative conditions of the body's adipose tissue.

Lipogenesis Is the process by which acetyl-CoA is converted to fats.

Lipolysis Is the breakdown of fat stored in fat cells in the body.

Liposomes Artificially prepared vesicles made of lipid bilayer.

Lipotoxicity Refers to tissues diseases that may occur when fatty acids spillover in excess of the oxidative needs of those tissues and enhances metabolic flux into harmful pathways of nonoxidative metabolism.

Lipotropic Refers to compounds that help catalyse the breakdown of fat during metabolism in the body. e.g. chlorine and lecithin.

Lipoxygenase A family of iron-containing enzymes that catalyse the dioxygenation of polyunsaturated fatty acids in lipids containing a cis,cis-1,4- pentadiene structure.

Lithiasis Formation of urinary calculi (stones) in the renal system (kidneys, ureters, urinary bladder, urethra) can be of any one of several compositions.

Lithogenic Promoting the formation of calculi (stones).

Lithontripic Removes stones from kidney, gall bladder.

Liver X receptors Nuclear hormones that function as central transcriptional regulators for lipid homeostasis.

Lotion A liquids suspension or dispersion of chemicals for external application to the body.

Lovo cells Colon cancer cells.

Low-density lipoprotein (LDL) Is a type of lipoprotein that transports cholesterol and triglycerides from the liver to peripheral tissues. High levels of LDL cholesterol can signal medical problems like cardiovascular disease, and it is sometimes called "bad cholesterol".

LRP1 Low-density lipoprotein receptor-related protein-1, plays a role in intracellular signaling functions as well as in lipid metabolism.

LTB4 A type of leukotriene, a major metabolite in neutrophil polymorphonuclear leukocytes. It stimulates polymorphonuclear cell function (degranulation, formation of oxygen-centered free radicals, arachidonic acid release, and metabolism). It induces skin inflammation.

Luciferase Is a generic name for enzymes commonly used in nature for bioluminescence.

Lumbago Is the term used to describe general lower back pain.

Lung abscess Necrosis of the pulmonary tissue and formation of cavities containing necrotic debris or fluid caused by microbial infections.

Lusitropic An agent that affects diastolic relaxation.

Lutein A carotenoid, occurs naturally as yellow or orange pigment in some fruits and leafy vegetables. It is one of the two carotenoids contained within the retina of the eye. Within the central macula, zeaxanthin predominates, whereas in the peripheral retina, lutein predominates. Lutein is necessary for good vision and may also help prevent or slow down atherosclerosis, the thickening of arteries, which is a major risk for cardiovascular disease.

Luteinising hormone (LH) A hormone produced by the anterior pituitary gland. In females, it triggers ovulation. In males, it stimulates the production of testosterone to aid sperm maturation.

Luteolysis Is the structural and functional degradation of the corpus luteum (CL) that occurs at the end of the luteal phase of both the estrous and menstrual cycles in the absence of pregnancy.

Lymphadenitis-cervical Inflammation of the lymph nodes in the neck, usually caused by an infection.

Lymphatitis Inflammation of lymph vessels and nodes.

Lymphadenopathy A term meaning "disease of the lymph nodes – lymph node enlargement.

Lymphoblastic Pertaining to the production of lymphocytes.

Lymphocyte A small white blood cell (leucocyte) that plays a large role in defending the body against disease. Lymphocytes are responsible for immune responses. There are two main types of lymphocytes: B cells and T cells. Lymphocytes secrete products (lymphokines) that modulate the functional activities of many other types of cells and are often present at sites of chronic inflammation.

Lymphocyte B cells The B cells make antibodies that attack bacteria and toxins.

Lymphocyte T cells T cells attack body cells themselves when they have been taken over by viruses or have become cancerous.

Lymphoma A type of cancer involving cells of the immune system, called lymphocytes.

Lymphopenia Abnormally low number of lymphocytes in the blood.

Lysosomes Are small, spherical organelles containing digestive enzymes (acid hydrolases and other proteases (cathepsins).

Maceration Softening or separating of parts by soaking in a liquid.

Macrophage A type of large leukocyte that travels in the blood but can leave the bloodstream and enter tissue; like other leukocytes it protects the body by digesting debris and foreign cells.

Macular degeneration A disease that gradually destroys the macula, the central portion of the retina, reducing central vision.

Macules Small circumscribed changes in the color of skin that are neither raised (elevated) nor depressed.

Maculopapular Describes a rash characterized by raised, spotted lesions.

Magnesium (Mg) Is the fourth most abundant mineral in the body and is essential to good health. It is important for normal muscle and nerve function, steady heart rhythm, immune system, and strong bones. Magnesium also helps regulate blood sugar levels, promotes normal blood pressure, and is known to be involved in energy metabolism and protein synthesis and plays a role in preventing and managing disorders such as hypertension, cardiovascular disease, and diabetes. Dietary sources include legumes (e.g. soya bean and by-products), nuts, whole unrefined grains, fruit (e.g. banana, apricots), okra and green leafy vegetables.

MAK cell Macrophage-activated killer cell, activated macrophage that is much more phagocytic than monocytes.

Malaise A feeling of weakness, lethargy or discomfort as of impending illness.

Malaria Is an infection of the blood by *Plasmodium* parasite that is carried from person to person by mosquitoes. There are four species of malaria parasites that infect man: *Plasmodium falciparum*, so called 'malignant tertian fever', is the most serious disease, *Plasmodium vivax*, causing a relapsing form of the disease, *Plasmodium malariae*, and *Plasmodium ovale*.

Malassezia A fungal genus (previously known as *Pityrosporum*) classified as yeasts, naturally found on the skin surfaces of many animals including humans. It can cause hypopigmentation on the chest or back if it becomes an opportunistic infection.

Mammalian target of rapamycin (mTOR) Pathway that regulates mitochondrial oxygen consumption and oxidative capacity.

Mammogram An x-ray of the breast to detect tumours.

Mandibular Relating to the mandible, the human jaw bone.

Manganese Is an essential element for heath. It is an important constituent of some enzymes and an activator of other enzymes in physiologic processes. Manganese superoxide dismutase (MnSOD) is the principal antioxidant enzyme in the mitochondria. Manganese-activated enzymes play important roles in the metabolism of carbohydrates, amino acids, and cholesterol. Manganese is the preferred cofactor of enzymes called glycosyltransferases which are required for the synthesis of proteoglycans that are needed for the formation of healthy cartilage and bone. Dietary source include whole grains, fruit, legumes (soybean and by-products), green leafy vegetables, beetroot and tea.

MAO activity Monoamine oxidase activity.

MAPK (Mitogen-activated protein kinase) These kinases are strongly activated in cells subjected to osmotic stress, UV radiation, disregulated K+ currents, RNA-damaging agents, and a multitude of other stresses, as well as inflammatory cytokines, endotoxin , and withdrawal of a trophic factor . The stress-responsive MAPKs mediate a plethora of cellular responses to such stressful stimuli, including apoptosis and production of inflammatory and immunoregulatory cytokines in diverse cell systems.

Marasmus Is one of the three forms of serious protein-energy malnutrition.

Mastectomy Surgery to remove a breast.

Masticatory A substance chewed to increase salivation. Also called sialogue.

Mastitis A bacterial infection of the breast which usually occurs in breastfeeding mothers.

Matrix metalloproteinases (MMP) A member of a group of enzymes that can break down proteins, such as collagen, that are normally found in the spaces between cells in tissues (i.e., extracellular matrix proteins). Matrix metalloproteinases are involved in wound healing, angiogenesis, and tumour cell metastasis. See also metalloproteinase.

MBC Minimum bacterial concentration – the lowest concentration of antibiotic required to kill an organism.

MCP-1 Monocyte chemotactic protein-1, plays a role in the recruitment of monocytes to sites of infection and injury. It is a member of small inducible gene (SIG) family.

MDA Malondialdehyde is one of the most frequently used indicators of lipid peroxidation.

Measles An acute, highly communicable rash illness due to a virus transmitted by direct contact with infectious droplets or, less commonly, by airborne spread.

Medial Preoptic Area Is located at the rostral end of the hypothalamus, it is important for the regulation of male sexual behavior.

Megaloblastic anemia An anemia that results from inhibition of DNA synthesis in red blood cell production, often due to a deficiency of vitamin B12 or folate and is characterized by many large immature and dysfunctional red blood cells (megaloblasts) in the bone marrow.

Melaene (melena) Refers to the black, "tarry" feces that are associated with gastrointestinal hemorrhage.

Melanogenesis Production of melanin by living cells.

Melanoma Malignant tumour of melanocytes which are found predominantly in skin but also in the bowel and the eye and appear as pigmented lesions.

Melatonin A hormone produced in the brain by the pineal gland, it is important in the regulation of the circadian rhythms of several biological functions.

Menarche The first menstrual cycle, or first menstrual bleeding, in female human beings.

Menorrhagia Heavy or prolonged menstruation, too-frequent menstrual periods.

Menopausal Refer to permanent cessation of menstruation.

Menses See menstruation.

Menstruation The approximately monthly discharge of blood from the womb in women of childbearing age who are not pregnant. Also called menses. *adj.* menstrual.

Mesangial cells Are specialized cells around blood vessels in the kidneys, at the mesangium.

Metabonome Complete set of metabologically regulated elements in cells.

Metalloproteinase Enzymes that breakdown proteins and requiring zinc or calcium atoms for proper function.

Meta-analysis A statistical procedure that combines the results of several studies that address a set of related research hypotheses.

Metaphysis Is the portion of a long bone between the epiphyses and the diaphysis of the femur.

Metaphyseal Pertaining to the metaphysis.

Metaplasia Transformation of one type of one mature differentiated cell type into another mature differentiated cell type.

Metastasis Is the movement or spreading of cancer cells from one organ or tissue to another.

Metetrus The quiescent period of sexual inactivity between oestrus cycles.

Metroptosis The slipping or falling out of place of an organ (as the uterus)

Metrorrhagia Uterine bleeding at irregular intervals, particularly between the expected menstrual periods.

Mevinolin A potent inhibitor of 3-hydroxy-3-methylglutaryl coenzyme A reductase (HMG-CoA reductase).

MHC Acronym for major histocompatibility complex, a large cluster of genes found on the short arm of chromosome 6 in most vertebrates that encodes MHC molecules. MHC molecules play an important role in the immune system and autoimmunity.

MIC Minimum inhibitory concentration – lowest concentration of an antimicrobial that will inhibit the visible growth of a microorganism.

Micelle A submicroscopic aggregation of molecules.

Micellization Formation process of micelles.

Microangiopathy (or microvascular disease) is an angiopathy affecting small blood vessels in the body

Microfilaria A pre-larval parasitic worm of the family Onchocercidae, found in the vector and in the blood or tissue fluid of human host.

Micronuclei Small particles consisting of acentric fragments of chromosomes or entire chromosomes, which lag behind at anaphase of cell division.

Microsomal PGE2 synthase Is the enzyme that catalyses the final step in prostaglandin E2 (PGE2) biosynthesis.

Microvasculature The finer vessels of the body, as the arterioles, capillaries, and venules.

Micturition Urination, act of urinating.

Migraine A neurological syndrome characterized by altered bodily perceptions, severe, painful headaches, and nausea.

Mimosine Is an alkaloid, β-3-hydroxy-4 pyridone amino acid, it is a toxic non-protein free amino acid and is an antinutrient.

Mineral apposition rate MAR, rate of addition of new layers of mineral on the trabecular surfaces of bones.

Miscarriage Spontaneous abortion.

Mitochondrial complex I The largest enzyme in the mitochondrial respiratory oxidative phosphorylation system.

Mitochondrial permeability transition (MPT) Is an increase in the permeability of the mitochondrial membranes to molecules of less than 1,500 Da in molecular weight. MPT is one of the major causes of cell death in a variety of conditions.

Mitogen An agent that triggers mitosis, elicit all the signals necessary to induce cell proliferation.

Mitogenic Able to induce mitosis or transformation.

Mitogenicity Process of induction of mitosis.

Mitomycin A chemotherapy drug that is given as a treatment for several different types of cancer, including breast, stomach, oesophagus and bladder cancers.

Mitosis Cell division in which the nucleus divides into nuclei containing the same number of chromosomes.

MMP Matrix metalloproteinases, a group of peptidases involved in degradation of the extracellular matrix (ECM).

Mnestic Pertaining to memory.

Molecular docking Is a key tool in structural molecular biology and computer-assisted drug design.

Molluscidal Destroying molluscs like snails.

Molt 4 cells MOLT4 cells are lymphoblast-like in morphology and are used for studies of apoptosis, tumour cytotoxicity, tumorigenicity, as well as for antitumour testing.

Molybdenum (Mo) Is an essential element that forms part of several enzymes such as xanthine oxidase involved in the oxidation of xanthine to uric acid and use of iron. Molybdenum concentrations also affect protein synthesis, metabolism, and growth. Dietary sources include meat, green beans, eggs, sunflower seeds, wheat flour, lentils, and cereal grain.

Monoamine oxidase A (MAOA) Is an isozyme of monoamine oxidase. It preferentially deaminates norepinephrine (noradrenaline), epinephrine (adrenaline), serotonin, and dopamine.

Monoaminergic Of or pertaining to neurons that secrete monoamine neurotransmitters (e.g., dopamine, serotonin).

Monoclonal antibodies Are produced by fusing single antibody-forming cells to tumour cells grown in culture.

Monocyte Large white blood cell that ingest microbes, other cells and foreign matter.

Monogalactosyl diglyceride Are the major lipid components of chloroplasts.

Monorrhagia Is heavy bleeding and that's usually defined as periods lasting longer than 7 days or excessive bleeding.

Morbidity A diseased state or symptom or can refer either to the incidence rate or to the prevalence rate of a disease.

Morelloflavone A biflavonoid extracted from *Garcinia dulcis*, has shown antioxidative, antiviral, and anti-inflammatory properties.

Morphine The major alkaloid of opium and a potent narcotic analgesic.

MTTP Microsomal triglyceride transfer protein that is required for the assembly and secretion of triglyceride -rich lipoproteins from both enterocytes and hepatocytes.

MUC 5AC Mucin 5AC, a secreted gel-forming protein mucin with a high molecular weight of about 641 kDa.

Mucositis Painful inflammation and ulceration of the mucous membranes lining the digestive tract.

Mucous Relating to mucus.

Mucolytic Capable of reducing the viscosity of mucus, or an agent that so acts.

Mucus Viscid secretion of the mucous membrane.

Multidrug resistance (MDR) Ability of a living cell to show resistance to a wide variety of structurally and functionally unrelated compounds.

Muscarinic receptors Are G protein-coupled acetylcholine receptors found in the plasma membranes of certain neurons and other cells.

Mutagen An agent that induces genetic mutation by causing changes in the DNA.

Mutagenic Capable of inducing mutation (used mainly for extracellular factors such as X-rays or chemical pollution).

Myc Codes for a protein that binds to the DNA of other genes and is therefore a transcription facor, found on chromosome 8 in human.

Mycosis An infection or disease caused by a fungus.

Myelocyte Is a young cell of the granulocytic series, occurring normally in bone marrow, but not in circulating blood.

Myeloid leukaemia (Chronic) A type of cancer that affects the blood and bone marrow, characterized by excessive number of white blood cells.

Myeloma Cancer that arise in the plasma cells a type of white blood cells.

Myeloperoxidase (MPO) Is a peroxidase enzyme most abundantly present in neutrophil granulocytes (a subtype of white blood cells). It is an inflammatory enzyme produced by activated leukocytes that predicts risk of coronary heart disease.

Myeloproliferative disorder Disease of the bone marrow in which excess cells are produced.

Myocardial Relating to heart muscles tissues.

Myocardial infarction (MI) Is the rapid development of myocardial necrosis caused by a critical imbalance between oxygen supply and demand of the myocardium.

Myocardial ischemia An intermediate condition in coronary artery disease during which the heart tissue is slowly or suddenly starved of oxygen and other nutrients.

Myogenesis The formation of muscular tissue, especially during embryonic development.

Myopia Near – or short-sightedness.

Myosarcoma A malignant muscle tumour.

Myotonia dystrophica An inherited disorder of the muscles and other body systems characterized by progressive muscle weakness, prolonged muscle contractions (myotonia), clouding of the lens of the eye (cataracts), cardiac abnormalities, balding, and infertility.

Myringosclerosis Also known as tympanosclerosis or intratympanic tympanosclerosis, is a condition caused by calcification of collagen tissues in the tympanic membrane of the middle ear.

Mytonia A symptom of certain neuromuscular disorders characterized by the slow relaxation of the muscles after voluntary contraction or electrical stimulation.

Myotube A developing skeletal muscle fibre with a tubular appearance.

N-nitrosmorpholine A human carcinogen.

N-nitrosoproline An indicator for N-nitrosation of amines.

NADPH The reduced form of nicotinamide adenine dinucleotide phosphate that serves as an electron carrier.

NAFLD Non-alcoholic fatty liver disease.

Narcotic An agent that produces narcosis, in moderate doses it dulls the senses, relieves pain and induces sleep; in excessive dose it cause stupor, coma, convulsions and death.

Nasopharynx Upper part of the alimentary continuous with the nasal passages.

Natriorexia Excessive intake of sodium evoked by sodium depletion. *adj.* natriorexic, natriorexigenic.

Natriuresis The discharge of excessive large amount of sodium through urine. *adj.* natriuretic.

Natural killer cells (NK cells) A type of cytotoxic lymphocyte that constitute a major component of the innate immune system.

Natural killer T (NKT) cells A heterogeneous group of T cells that share properties of both T cells and natural killer (NK) cells.

Nausea Sensation of unease and discomfort in the stomach with an urge to vomit.

Necropsy See autopsy.

Necrosis Morphological changes that follow cell death, usually involving nuclear and cytoplasmic changes.

Neonatal *adj.* of or relating to newborn infants or an infant.

Neoplasia Abnormal growth of cells, which may lead to a neoplasm, or tumour.

Neoplasm Tumour; any new and abnormal growth, specifically one in which cell multiplication is uncontrolled and progressive. Neoplasms may be benign or malignant.

Neoplastic transformation Conversion of a tissue with a normal growth pattern into a malignant tumour.

Neointima A new or thickened layer of arterial intima formed especially on a prosthesis or in atherosclerosis by migration and proliferation of cells from the media.

Neovasculature Formation of new blood vessels.

Nephrectomised Kidneys surgically removed.

Nephrectomy Surgical removal of the kidney.

Nephric Relating to or connected with a kidney.

Nephrin Is a protein necessary for the proper functioning of the renal filtration barrier.

Nephritic syndrome Is a collection of signs (known as a syndrome) associated with disorders affecting the kidneys, more specifically glomerular disorders.

Nephritis Is inflammation of the kidney.

Nephrolithiasis Process of forming a kidney stone in the kidney or lower urinary tract.

Nephropathy A disorder of the kidney.

Nephrotic syndrome Nonspecific disorder in which the kidneys are damaged, causing them to leak large amounts of protein from the blood into the urine.

Nephrotoxicity Poisonous effect of some substances, both toxic chemicals and medication, on the kidney.

Nerve growth factor (NGF) A small protein that induces the differentiation and survival of particular target neurons (nerve cells).

Nervine A nerve tonic that acts therapeutically upon the nerves, particularly in the sense of a sedative that serves to calm ruffled nerves.

Neuralgia Is a sudden, severe painful disorder of the nerves.

Neuraminidase Glycoside hydrolase enzymes that cleaves the glycosidic linkages of neuraminic acids.

Neuraminidase inhibitors A class of antiviral drugs targeted at the influenza viruses whose mode of action consists of blocking the function of the viral neuraminidase protein, thus preventing the virus from reproducing.

Neurasthenia A condition with symptoms of fatigue, anxiety, headache, impotence, neuralgia and impotence.

Neurasthenic A substance used to treat nerve pain and/or weakness (i.e. neuralgia, sciatica, etc.).

Neurite Refers to any projection from the cell body of a neuron.

Neuritis An inflammation of the nerve characterized by pain, sensory disturbances and impairment of reflexes. *adj.* neuritic.

Neuritogenesis The first step of neuronal differentiation, takes place as nascent neurites bud from the immediate postmitotic neuronal soma.

Neuroblastoma A common extracranial cancer that forms in nerve tissues, common in infancy.

Neuroendocrine *adj.* of, relating to, or involving the interaction between the nervous system and the hormones of the endocrine glands.

Neuroleptic Refers to the effects on cognition and behavior of antipsychotic drugs that reduce confusion, delusions, hallucinations, and psychomotor agitation in patients with psychoses.

Neuropharmacological Relating the effects of drugs on the neurosystem.

Neuroradiology Is a subspecialty of radiology focusing on the diagnosis and characterization of abnormalities of the central and peripheral nervous system. *adj.* neuroradiologic.

Neutropenia A disorder of the blood, characterized by abnormally low levels of neutrophils.

Neurotrophic Relating to neutrophy i.e. the nutrition and maintenance of nervous tissue.

Neutrophil A type of white blood cell, specifically a form of granulocyte.

Neutrophin Protein that induce the survival, development and function of neurons.

NF-kappa B (NF-kB) Nuclear factor kappa B, is an ubiquitous rapid response transcription factor in cells involved in immune and inflammatory reactions.

Niacin Vitamin B3. See vitamin B3.

Niacinamide An amide of niacin, also known as nicotinamide. See vitamin B3.

NIH3T3 cells A mouse embryonic fibroblast cell line used in the cultivation of keratinocytes.

Nitrogen (N) Is an essential building block of amino and nucleic acids and proteins and is essential to all living organisms. Protein rich vegetables like legumes are rich food sources of nitrogen.

NK cells Natural killer cells, a type of cytotoxic lymphocyte that constitute a major component of the innate immune system.

NMDA receptor N-methyl-D-aspartate receptor, the predominant molecular device for controlling synaptic plasticity and memory function. A brain receptor activated by the amino acid glutamate, which when excessively stimulated may cause cognitive defects in Alzheimer's disease.

Nociceptive Causing pain, responding to a painful stimulus.

Non-osteogenic fibroma of bone A benign tumour of bone which shows no evidence of ossification.

Nootropics Are substances which are claimed to boost human cognitive abilities (the functions and capacities of the brain). Also popularly referred to as "smart drugs", "smart nutrients", "cognitive enhancers" and "brain enhancers".

Noradrenalin See Norepinephrine.

Norepinephrine A substance, both a hormone and neurotransmitter, secreted by the adrenal medulla and the nerve endings of the sympathetic nervous system to cause vasoconstriction and increases in heart rate, blood pressure, and the sugar level of the blood. Also called levarterenol, noradrenalin.

Normoglycaemic Having the normal amount of glucose in the blood.

Normotensive Having normal blood pressure.

Nosocomial infections Infections which are a result of treatment in a hospital or a healthcare service unit, but secondary to the patient's original condition.

NK1.1+ T (NKT) cells A type of natural killer T (NKT) cells. See natural killer T cells.

Nuclear factor erythroid 2-related factor 2 (Nrf2) A transcription factor that plays a major role in response to oxidative stress by binding to antioxidant-responsive elements that regulate many hepatic phase I and II enzymes as well as hepatic efflux transporters.

Nucleosomes Fundamental repeating subunits of all eukaryotic chromatin, consisting of a DNA chain coiled around a core of histones.

Nulliparous Term used to describe a woman who has never given birth.

Nyctalopia Night blindness, impaired vision in dim light and in the dark, due to impaired function of certain specialized vision cells.

Nycturia Excessive urination at night; especially common in older men.

Occlusion Closure or blockage (as of a blood vessel).

Occlusive peripheral arterial disease (PAOD) Also known as peripheral vascular disease (PVD), or peripheral arterial disease (PAD) refers to the obstruction of large arteries not within the coronary, aortic arch vasculature, or brain. PVD can result from atherosclerosis, inflammatory processes leading to stenosis, an embolism, or thrombus formation.

Oculomotor nerve The third of 12 paired cranial nerves.

Odds ratio A statistical measure of effect size, describing the strength of association or non-independence between two binary data values.

Odontalgia Toothache. *adj.* odontalgic.

Odontopathy Any disease of the teeth.

Oedema See edema.

Oligoarthritis An inflammation of two, three or four joints.

Oligonucleosome A series of nucleosomes.

Oligospermia or oligozoospermia Refers to semen with a low concentration of sperm, commonly associated with male infertility.

Oliguria Decreased production of urine.

Oligoanuria Insufficient urine volume to allow fo administration of necessary fluids, etc.

Omega 3 fatty acids Are essential polyunsaturated fatty acids that have in common a final carbon–carbon double bond in the n–3 position. Dietary sources of omega-3 fatty acids include fish oil and certain plant/nut oils. The three most nutritionally important omega 3 fatty acids are α-linolenic acid, eicosapentaenoic acid (EPA) and docosahexaenoic acid (DHA). Research indicates that omega 3 fatty acids are important in health promotion and disease and can help prevent a wide range of medical problems, including cardiovascular disease, depression, asthma, and rheumatoid arthritis.

Omega 6 fatty acids Are essential polyunsaturated fatty acids that have in common a final carbon–carbon double bond in the n–6 position. Omega-6 fatty acids are considered essential fatty acids (EFAs) found in vegetable oils, nuts and seeds. They are essential to human health but cannot be made in the body. Omega-6 fatty acids – found in vegetable oils, nuts and seeds – are a beneficial part of a heart-healthy eating. Omega-6 and omega-3 PUFA play a crucial role in heart and brain function and in normal growth and development. Linoleic acid (LA) is the main omega-6 fatty acid in foods, accounting for 85–90% of the dietary omega-6 PUFA. Other omega 6 acids include γ-linolenic acid or GLA, sometimes called gamoleic acid, eicosadienoic acid, arachidonic acid and docosadienoic acid.

Omega 9 fatty acids Are not essential polyunsaturated fatty acids that have in common a final carbon–carbon double bond in the n–9 position. Some n–9s are common components of animal fat and vegetable oil. Two n–9 fatty acids important in industry are: oleic acid (18:1, n–9), which is a main component of olive oil and erucic acid (22:1, n–9), which is found in rapeseed, wallflower seed, and mustard seed.

Oncogenes Genes carried by tumour viruses that are directly and solely responsible for the neoplastic (tumorous) transformation of host cells.

Ophthalmia Severe inflammation of eye, or the conjunctiva or deeper structures of the eye . Also called ophthalmitis.

Ophthalmia (Sympathetic) Inflammation of both eyes following trauma to one eye.

Opiate Drug derived from the opium plant.

Opioid receptors A group of G-protein coupled receptors located in the brain and various organs that bind opiates or opioid substances.

Optic placode An ectodermal placode from which the lens of the embryonic eye develops; also called lens placode.

ORAC (Oxygen radical absorbance capacity) A method of measuring antioxidant capacities in biological samples.

Oral submucous fibrosis A chronic debilitating disease of the oral cavity characterized by inflammation and progressive fibrosis of the submucosa tissues.

Oral thrush An infection of yeast fungus, *Candida albicans*, in the mucous membranes of the mouth.

Orchidectomy Surgery to remove one or both testicles.

Orchidectomised With testis removed.

Orchitis An acute painful inflammatory reaction of the testis secondary to infection by different bacteria and viruses.

Orexigenic Increasing or stimulating the appetite.

Orofacial dyskinesia Abnormal involuntary movements involving muscles of the face, mouth, tongue, eyes, and occasionally, the neck – may be unilateral or bilateral, and constant or intermittent.

Oropharyngeal Relating to the oropharynx.

Oropharynx Part of the pharynx between the soft palate and the epiglottis.

Ostalgia, Ostealgia Pain in the bones. Also called osteodynia.

Osteoarthritis Is the deterioration of the joints that becomes more common with age.

Osteoarthrosis Chronic noninflammatory bone disease.

Osteoblast A mononucleate cell that is responsible for bone formation.

Osteoblastic Relating to osteoblasts.

Osteocalcin A noncollagenous protein found in bone and dentin, also refer to as bone γ-carboxyglutamic acid-containing protein.

Osteoclasts A kind of bone cell that removes bone tissue by removing its mineralized matrix.

Osteoclastogenesis The production of osteoclasts.

Osteodynia Pain in the bone.

Osteogenic Derived from or composed of any tissue concerned in bone growth or repair.

Osteomalacia Refers to the softening of the bones due to defective bone mineralization.

Osteomyelofibrosis A myeloproliferative disorder in which fibrosis and sclerosis finally lead to bone marrow obliteration.

Osteopenia Reduction in bone mass, usually caused by a lowered rate of formation of new bone that is insufficient to keep up with the rate of bone destruction.

Osteoporosis A disease of bone that leads to an increased risk of fracture.

Osteoprotegerin Also called osteoclastogenesis inhibitory factor (OCIF), a cytokine, which can inhibit the production of osteoclasts.

Osteosacrcoma A malignant bone tumour. Also called osteogenic sarcoma.

Otalgia Earache, pain in the ear.

Otic placode A thickening of the ectoderm on the outer surface of a developing embryo from which the ear develops.

Otitis Inflammation of the inner or outer parts of the ear.

Otorrhea Running drainage (discharge) exiting the ear.

Ovariectomised With one or two ovaries removed.

Ovariectomy Surgical removal of one or both ovaries.

Oxidation The process of adding oxygen to a compound, dehydrogenation or increasing the electro-negative charge.

Oxidoreductase activity Catalysis of an oxidation-reduction (redox) reaction, a reversible chemical reaction. One substrate acts as a hydrogen or electron donor and becomes oxidized, while the other acts as hydrogen or electron acceptor and becomes reduced.

Oxytocic *adj.* hastening or facilitating childbirth, especially by stimulating contractions of the uterus.

Oxytocin Is a mammalian hormone that also acts as a neurotransmitter in the brain. It is best known for its roles in female reproduction: it is released in large amounts after distension of the cervix and vagina during labor, and after stimulation of the nipples, facilitating birth and breastfeeding, respectively.

Oxygen radical absorbance capacity (ORAC) A method of measuring antioxidant capacities in biological samples.

Oxyuriasis Infestation by pinworms.

Ozoena Discharge of the nostrils caused by chronic inflammation of the nostrils.

p.o. Per os, oral administration.

P21 Also known as cyclin-dependent kinase inhibitor 1 or CDK-interacting protein 1, is a potent cyclin-dependent kinase inhibitor.

P53 Also known as protein 53 or tumour protein 53, is a tumour suppressor protein that in humans is encoded by the TP53 gene.

P-Selectin Also known as CD62P, GMP-140, LLECAM-3, PADGEM, a member of the selectin family. It is expressed by activated platelets and endothelial cells.

P-glycoprotein (P-gp, ABCB1, MDR1) A cell membrane-associated drug-exporting protein that transports a variety of drug substrates from cancer cells.

Palpebral ptosis The abnormal drooping of the upper lid, caused by partial or total reduction in levator muscle function.

Palpitation Rapid pulsation or throbbing of the heart.

Paludism State of having symptoms of malaria characterized by high fever and chills.

Pancreatectomized Having undergone a pancreatectomy.

Pancreatectomy Surgical removal of all or part of the pancreas.

Pancreatitis Inflammation of the pancreas.

Pantothenic acid Vitamin B5. See vitamin B5.

Papain A protein degrading enzyme used medicinally and to tenderize meat.

Papilloma A benign epithelial tumour growing outwardly like in finger-like fronds.

Papule A small, solid, usually inflammatory elevation of the skin that does not contain pus.

Paradontosis Is the inflammation of gums and other deeper structures, including the bone.

Paralytic Person affected with paralysis, pertaining to paralysis.

Parasitemia Presence of parasites in blood. *adj.* parasitemic.

Parasympathetic nervous system Subsystem of the nervous systems that slows the heart rate and increases intestinal and gland activity and relaxes the sphincter muscles.

Parasympathomimetic Having an action resembling that caused by stimulation of the parasympathetic nervous system.

Parenteral administration Administration by intravenous, subcutaneous or intramuscular routes.

Paresis A condition characterised by partial loss of movement, or impaired movement.

Paresthesia Is an abnormal sensation of the skin, such as burning, numbness, itching, hyperesthesia (increased sensitivity) or tingling, with no apparent physical cause.

Parenteral Is a route of administration via the veins that involves piercing the skin or mucous membrane.

Parotitis Inflammation of salivary glands.

Paroxysm A sudden outburst of emotion or action, a sudden attack, recurrence or intensification of a disease.

Paroxystic Relating to an abnormal event of the body with an abrupt onset and an equally sudden return to normal.

PARP See poly (ADP-ribose) polymerase.

Parturition Act of child birth.

PCE/PCN ratio Polychromatic erythrocyte/normochromatic erythrocyte ratio use as a measure of cytotoxic effects.

pCREB Phosphorylated cAMP (adenosine 3′5′ cyclic monophosphate)-response element binding protein.

PDEF Acronym for prostate-derived ETS factor, an ETS (epithelial-specific E26 transforming sequence) family member that has been identified as a potential tumour suppressor.

Pectoral Pertaining to or used for the chest and respiratory tract.

pERK Phosphorylated extracellular signal-regulated kinase, protein kinases involved in many cell functions.

p53 Also known as protein 53 or tumour protein 53, is a tumour suppressor protein that in humans is encoded by the TP53 gene.

Peliosis See purpura.

Pellagra Is a systemic nutritional wasting disease caused by a deficiency of vitamin B3 (niacin).

Pemphigus neonatorum Staphylococcal scalded skin syndrome, a bacterial disease of infants, characterized by elevated vesicles or blebs on a normal or reddened skin .

Peptic ulcer A sore in the lining of the stomach or duodenum, the first part of the small intestine.

Percutaneous Pertains to a medical procedure where access to inner organs or tissues is done via needle puncture of the skin.

Perfusion To force fluid through the lymphatic system or blood vessels to an organ or tissue.

Periapical periodontitis Is the inflammation of the tissue adjacent to the tip of the tooth's root.

Perifuse To flush a fresh supply of bathing fluid around all of the outside surfaces of a small piece of tissue immersed in it.

Perilipins Highly phosphorylated adipocyte proteins that are localized at the surface of the lipid droplet.

Perimenopause Is the phase before menopause actually takes place, when ovarian hormone production is declining and fluctuating. *adj.* perimenopausal.

Periodontal ligament (PDL) Is a group of specialized connective tissue fibres that essentially attach a tooth to the bony socket.

Periodontitis Is a severe form of gingivitis in which the inflammation of the gums extends to the supporting structures of the tooth. Also called pyorrhea.

Peripheral arterial disease (PAD) See peripheral artery occlusive disease.

Peripheral neuropathy Refers to damage to nerves of the peripheral nervous system.

Peripheral vascular disease (PVD) See peripheral artery occlusive disease .

Peristalsis A series of organized, wave-like muscle contractions that occur throughout the digestive tract.

Perlingual Through or by way of the tongue.

Perniosis An abnormal reaction to cold that occurs most frequently in women, children, and the elderly. Also called chilblains.

Per os (P.O.) Oral administration.

Peroxisome proliferator-activated receptors (PPARs) A family of nuclear receptors that are involved in lipid metabolism, differentiation, proliferation, cell death, and inflammation.

Peroxisome proliferator-activated receptor alpha (PPAR-α) A nuclear receptor protein, transcription factor and a major regulator of lipid metabolism in the liver.

Peroxisome proliferator-activated receptor gamma (PPAR-γ) A type II nuclear receptor protein that regulates fatty acid storage and glucose metabolism.

Pertussis Whooping cough, sever cough.

Peyers Patches Patches of lymphoid tissue or lymphoid nodules on the walls of the ileal-small intestine.

PGE-2 Prostaglandin E2, a hormone-like substance that is released by blood vessel walls in response to infection or inflammation that acts on the brain to induce fever.

Phagocytes Are the white blood cells that protect the body by ingesting (phagocytosing) harmful foreign particles, bacteria and dead or dying cells. *adj.* phagocytic.

Phagocytosis Is process the human body uses to destroy dead or foreign cells.

Pharmacognosis The branch of pharmacology that studies the composition, use, and history of drugs.

Pharmacodynamics Branch of pharmacology dealing with the effects of drugs and the mechanism of their action.

Pharmacokinetics Branch of pharmacology concerned with the movement of drugs within the body including processes of absorption, distribution, metabolism and excretion in the body.

Pharmacopoeia Authoritative treatise containing directions for the identification of drug samples and the preparation of compound medicines, and published by the authority of a government or a medical or pharmaceutical society and in a broader sense is a general reference work for pharmaceutical drug specifications.

Pharyngitis, Pharyngolaryngitis Inflammation of the pharynx and the larynx.

Pharyngolaryngeal Pertaining to the pharynx and larynx.

Phenolics Class of chemical compounds consisting of a hydroxyl group (–OH) bonded directly to an aromatic hydrocarbon group.

Pheochromocytoma Is a rare neuroendocrine tumour that usually originates from the adrenal glands' chromaffin cells, causing overproduction of catecholamines, powerful hormones that induce high blood pressure and other symptoms.

Phlebitis Is an inflammation of a vein, usually in the legs.

Phlegm Abnormally viscid mucus secreted by the mucosa of the respiratory passages during certain infectious processes.

Phlegmon A spreading, diffuse inflammation of the soft or connective tissue due to infection by Streptococci bacteria.

Phoroglucinol A white, crystalline compound used as an antispasmodic, analytical reagent, and decalcifier of bone specimens for microscopic examination.

Phosphatidylglycerol Is a glycerophospholipid found in pulmonary active surface lipoprotein and consists of a L-glycerol 3-phosphate backbone ester-bonded to either saturated or unsaturated fatty acids on carbons 1 and 2.

Phosphatidylinositol 3-kinases (PI 3-kinases or PI3Ks) A group of enzymes involved in cellular functions such as cell growth, proliferation, differentiation, motility, survival and intracellular trafficking, which in turn are involved in cancer.

Phosphatidylserine A phosphoglyceride phospholipid that is one of the key building blocks of cellular membranes, particularly in the nervous system. It is derived from soy lecithin

Phosphodiesterases A diverse family of enzymes that hydrolyse cyclic nucleotides and thus play a key role in regulating intracellular levels of the second messengers cAMP and cGMP, and hence cell function.

Phospholipase An enzyme that hydrolyzes phospholipids into fatty acids and other lipophilic substances.

Phospholipase A2 (PLA2) A small lipolytic enzyme that releases fatty acids from the second carbon group of glycerol. Plays an essential role in the synthesis of prostaglandins and leukotrienes.

Phospholipase C Enzymes that cleaves phospholipase.

Phospholipase C gamma (PLC gamma) Enzymes that cleaves phospholipase in cellular proliferation and differentiation, and its enzymatic activity is upregulated by a variety of growth factors and hormones.

Phosphorus (P) Is an essential mineral that makes up 1% of a person's total body weight and is found in the bones and teeth. It plays an important role in the body's utilization of carbohydrates and fats; in the synthesis of protein for the growth, maintenance, and repair of cells and tissues. It is also crucial for the production of ATP, a molecule the body uses to store energy. Main sources are meat and milk; fruits and vegetables provides small amounts.

Photoaging Is the term that describes damage to the skin caused by intense and chronic exposure to sunlight resulting in premature aging of the skin.

Photocarcinogenesis Represents the sum of a complex of simultaneous and sequential biochemical events that ultimately lead to the occurrence of skin cancer.

Photophobia Abnormal visual intolerance to light.

Photopsia An affection of the eye, in which the patient perceives luminous rays, flashes, coruscations, etc.

Photosensitivity Sensitivity toward light.

Phthisis An archaic name for tuberculosis.

Phytohemagglutinin A lectin found in plant that is involved in the stimulation of lymphocyte proliferation.

Phytonutrients Certain organic components of plants, that are thought to promote human health. Fruits, vegetables, grains, legumes, nuts and teas are rich sources of phytonutrients. Phytonutrients are not 'essential' for life. Also called phytochemicals.

Phytosterols A group of steroid alcohols, cholesterol-like phytochemicals naturally occurring in plants like vegetable oils, nuts and legumes.

Piebaldism Rare autosomal dominant disorder of melanocyte development characterized by distinct patches of skin and hair that contain no pigment.

Piles See haemorrhoids.

Pityriasis lichenoides Is a rare skin disorder of unknown aetiology characterised by multiple papules and plaques.

PKC Protein kinase C, a membrane bound enzyme that phosphorylates different intracellular proteins and raised intracellular Ca levels.

PKC Delta inhibitors Protein Kinase C Δ inhibitors that induce apoptosis of haematopoietic cell lines.

Placebo A sham or simulated medical intervention.

Placode A platelike epithelial thickening in the embryo where some organ or structure later develops.

Plasma The yellow-colored liquid component of blood, in which blood cells are suspended.

Plasmalemma Plasma membrane.

Plasma kallikrien A serine protease, synthesized in the liver and circulates in the plasma.

Plasmin A proteinase enzyme that is responsible for digesting fibrin in blood clots.

Plasminogen The proenzyme of plasmin, whose primary role is the degradation of fibrin in the vasculature.

Plaster Poultice.

Platelet activating factor (PAF) Is an acetylated derivative of glycerophosphorylcholine, released by basophils and mast cells in immediate hypersensitive reactions and macrophages and neutrophils in other inflammatory reactions. One of its main effects is to induce platelet aggregation.

PLC gamma Phospholipase C gamma plays a central role in signal transduction.

Pleurisy Is an inflammation of the pleura, the lining of the pleural cavity surrounding the lungs, which can cause painful respiration and other symptoms. Also known as pleuritis.

Pneumonia An inflammatory illness of the lung caused by bacteria or viruses.

Pneumotoxicity Damage to lung tissues.

Poliomyelitis Is a highly infectious viral disease that may attack the central nervous system and is characterized by symptoms that range from a mild non-paralytic infection to total paralysis in a matter of hours; also called polio or infantile paralysis.

Poly (ADP-ribose) polymerase (PARP) A protein involved in a number of cellular processes especially DNA repair and programmed cell death.

Polyarthritis Is any type of arthritis which involves five or more joints.

Polychromatic erythrocyte (PCE) An immature red blood cell containing RNA, that can be differentiated by appropriate staining techniques from a normochromatic erythrocyte (NCE), which lacks RNA.

Polycystic kidney disease Is a kidney disorder passed down through families in which multiple cysts form on the kidneys, causing them to become enlarged.

Polycythaemia A type of blood disorder characterised by the production of too many red blood cells.

Polymorphonuclear Having a lobed nucleus. Used especially of neutrophilic white blood cells.

Polyneuritis Widespread inflammation of the nerves.

Polyneuritis gallinarum A nervous disorder in birds and poultry.

Polyp A growth that protrudes from a mucous membrane.

Polyphagia Medical term for excessive hunger or eating.

Polyuria A condition characterized by the passage of large volumes of urine with an increase in urinary frequency.

Pomade A thick oily dressing.

Porphyrin Any of a class of water-soluble, nitrogenous biological pigments.

Postpartum Depression Depression after pregnancy; also called postnatal depression.

Postprandial After mealtime.

Potassium (K) Is an element that's essential for the body's growth and maintenance. It's necessary to keep a normal water balance between the cells and body fluids, for cellular enzyme activities and plays an essential role in the response of nerves to stimulation and in the contraction of muscles. Potassium is found in many plant foods and fish (tuna, halibut): chard, mushrooms, spinach, fennel, kale, mustard greens, Brussels sprouts, broccoli, cauliflower, cabbage winter squash, eggplant, cantaloupe, tomatoes, parsley, cucumber, bell pepper, turmeric, ginger root, apricots, strawberries, avocado and banana.

Poultice Is a soft moist mass, often heated and medicated, that is spread on cloth over the skin to treat an aching, inflamed, or painful part of the body. Also called cataplasm.

PPARs Peroxisome proliferator-activated receptors – a group of nuclear receptor proteins that function as transcription factors regulating the expression of genes.

Prebiotics A category of functional food, defined as non-digestible food ingredients that beneficially affect the host by selectively stimulating the growth and/or activity of one or a limited number of bacteria in the colon, and thus improve host health. *cf.* probiotics.

Pre-ecamplasia Toxic condition of pregnancy characterized by high blood pressure, abnormal weight gain, proteinuria and edema.

Prepubertal Before puberty; pertaining to the period of accelerated growth preceding gonadal maturity.

Pre-eclampsia See toxemia.

Pregnane X receptor (PXR; NR1I2) is a ligand-activated transcription factor that plays a role not only in drug metabolism and transport but also in various other biological processes.

Pregnenolone A steroid hormone produced by the adrenal glands, involved in the steroidogenesis of other steroid hormones like progesterone, mineralocorticoids, glucocorticoids, androgens, and estrogens.

Prenidatory Referring to the time period between fertilization and implantation.

Prenylated flavones Flavones with an isoprenyl group in the 8-position, has been reported to have good anti-inflammatory properties.

Proangiogenic Promote angiogensis (formation and development of new blood vessels).

Probiotics Are dietary supplements and live microorganisms containing potentially beneficial bacteria or yeasts that are taken into the alimentary system for healthy intestinal functions. *cf.* prebiotics.

Procyanidin Also known as proathocyanidin, oligomeric proathocyanidin, leukocyanidin, leucoanthocyanin, is a class of flavanols found in many plants. It has antioxidant activity and plays a role in the stabilization of collagen and maintenance of elastin.

Progestational Of or relating to the phase of the menstrual cycle immediately following ovulation, characterized by secretion of progesterone.

Proglottid One of the segments of a tapeworm.

Prognosis Medical term to describe the likely outcome of an illness.

Prolactin A hormone produced by the pituitary gland, it stimulates the breasts to produce milk in pregnant women. It is also present in males but its role is not well understood.

Prolapsus To fall or slip out of place.

Prolapus ani Eversion of the lower portion of the rectum, and protruding through the anus, common in infancy and old age.

Proliferating cell nuclear antigen (PCNA) A new marker to study human colonic cell proliferation.

Proliferative vitreoretinopathy (PVR) A most common cause of failure in retinal reattachment surgery, characterised by the formation of cellular membrane on both surfaces of the retina and in the vitreous.

Promastigote The flagellate stage in the development of trypanosomatid protozoa, characterized by a free anterior flagellum.

Promyelocytic leukemia A subtype of acute myelogenous leukemia (AML), a cancer of the blood and bone marrow.

Pro-oxidants Chemicals that induce oxidative stress, either through creating reactive oxygen species or inhibiting antioxidant systems.

Prophylaxis Prevention or protection against disease.

Proptosis See exophthalmos.

Prostacyclin A prostaglandin that is a metabolite of arachidonic acid, inhibits platelet aggregation, and dilates blood vessels.

Prostaglandins A family of C 20 lipid compounds found in various tissues, associated with muscular contraction and the inflammation response such as swelling, pain, stiffness, redness and warmth.

Prostaglandin E2 (PEG-2) One of the prostaglandins, a group of hormone-like substances that participate in a wide range of body functions such as the contraction and relaxation of smooth muscle, the dilation and constriction of blood vessels, control of blood pressure, and modulation of inflammation.

Prostaglandin E synthase An enzyme that in humans is encoded by the glutathione-dependent PTGES gene.

Prostanoids Term used to describe a subclass of eicosanoids (products of COX pathway) consisting of: the prostaglandins (mediators of inflammatory and anaphylactic reactions), the thromboxanes (mediators of vasoconstriction) and the prostacyclins (active in the resolution phase of inflammation.)

Prostate A gland that surround the urethra at the bladder in the male.

Prostate cancer A disease in which cancer develops in the prostate, a gland in the male reproductive system. Symptoms include pain, difficulty in urinating, erectile dysfunction and other symptoms.

Prostate–specific antigen (PSA) A protein produced by the cells of the prostate gland.

Protein kinase C (PKC) A family of enzymes involved in controlling the function of other proteins through the phosphorylation of hydroxyl groups of serine and threonine amino acid residues on these proteins. PKC enzymes play important roles in several signal transduction cascades.

Protein tyrosine phosphatase (PTP) A group of enzymes that remove phosphate groups from phosphorylated tyrosine residues on proteins.

Proteinase A protease (enzyme) involved in the hydrolytic breakdown of proteins, usually by splitting them into polypeptide chains.

Proteinuria Means the presence of an excess of serum proteins in the urine.

Proteolysis Cleavage of the peptide bonds in protein forming smaller polypeptides. *adj.* proteolytic.

Proteomics The large-scale study of proteins, particularly their structures and functions.

Prothrombin Blood-clotting protein that is converted to the active form, factor IIa, or thrombin, by cleavage.

Prothyroid Good for thyroid function.

Protheolithic Proteolytic see proteolysis.

Proto-oncogene A normal gene which, when altered by mutation, becomes an oncogene that can contribute to cancer.

Prurigo A general term used to describe itchy eruptions of the skin.

Pruritis Defined as an unpleasant sensation on the skin that provokes the desire to rub or scratch the area to obtain relief; itch, itching. *adj.* pruritic.

PSA Prostate Specific Antigen, a protein which is secreted into ejaculate fluid by the healthy prostate. One of its functions is to aid sperm movement.

Psoriasis A common chronic, non-contagious autoimmune dermatosis that affects the skin and joints.

Psychoactive Having effects on the mind or behavior.

Psychonautics Exploration of the psyche by means of approaches such as meditation, prayer, lucid dreaming, brain wave entrainment etc.

Psychotomimetic Hallucinogenic.

Psychotropic Capable of affecting the mind, emotions, and behavior.

Ptosis Also known as drooping eyelid; caused by weakness of the eyelid muscle and damage to the nerves that control the muscles or looseness of the skin of the upper eyelid..

P13-K Is a lipid kinase enzyme involved in the regulation of a number of cellular functions such as cell growth, proliferation, differentiation, motility, survival and intracellular trafficking, which in turn are involved in cancer.

P13-K/AKT signaling pathway Shown to be important for an extremely diverse array of cellular activities – most notably cellular proliferation and survival.

Pthysis Silicosis with tuberculosis.

Ptosis Drooping of the upper eye lid.

PTP Protein tyrosine phosphatase.

PTPIB Protein tyrosine phosphatase 1B.

Puerperal Pertaining to child birth.

Pulmonary embolism A blockage (blood clot) of the main artery of the lung.

Purgative A substance used to cleanse or purge, especially causing the immediate evacuation of the bowel.

Purpura Is the appearance of red or purple discolorations on the skin that do not blanch on applying pressure. Also called peliosis.

Purulent Containing pus discharge.

Purulent sputum Sputum containing, or consisting of, pus.

Pustule Small, inflamed, pus-filled lesions.

Pyelonephritis An ascending urinary tract infection that has reached the pyelum (pelvis) of the kidney.

Pyodermatitis Refers to inflammation of the skin.

Pyorrhea See periodontitis.

Pyretic Referring to fever.

Pyrexia Fever of unknown origin.

Pyridoxal A chemical form of vitamin B6. See vitamin B6.

Pyridoxamine A chemical form of vitamin B6. See vitamin B6.

Pyridoxine A chemical form of vitamin B6. See vitamin B6.

Pyrolysis Decomposition or transformation of a compound caused by heat. *adj.* pyrolytic.

PYY Peptide A 36 amino acid peptide secreted by L cells of the distal small intestine and colon that inhibits gastric and pancreatic secretion.

QT interval Is a measure of the time between the start of the Q wave and the end of the T wave in the heart's electrical cycle. A prolonged QT interval is a biomarker for ventricular tachyarrhythmias and a risk factor for sudden death.

Quorum sensing (QS) The control of gene expression in response to cell density, is used by both gram-negative and gram-positive bacteria to regulate a variety of physiological functions.

Radiolysis The dissociation of molecules by radiation.

Radioprotective Serving to protect or aiding in protecting against the injurious effect of radiations.

RAGE Is the receptor for advanced glycation end products, a multiligand receptor that propagates cellular dysfunction in several inflammatory disorders, in tumours and in diabetes.

RAS See renin-angiotensin system or recurrent aphthous stomatitis.

Rash A temporary eruption on the skin, see uticaria.

Reactive oxygen species Species such as superoxide, hydrogen peroxide, and hydroxyl radical. At low levels, these species may function in cell signaling processes. At higher levels, these species may damage cellular macromolecules (such as DNA and RNA) and participate in apoptosis (programmed cell death).

Rec A Is a 38 kDa *Escherichia coli* protein essential for the repair and maintenance of DNA.

Receptor for advanced glycation end products (RAGE) Is a member of the immunoglobulin superfamily of cell surface molecules; mediates neurite outgrowth and cell migration upon stimulation with its ligand, amphoterin.

Recticulocyte Non-nucleated stage in the development of the red blood cell.

Recticulocyte lysate Cell lysate produced from reticulocytes, used as an in-vitro translation system.

Recticuloendothelial system Part of the immune system, consists of the phagocytic cells located in reticular connective tissue, primarily monocytes and macrophages.

Recurrent aphthous stomatitis, or RAS Is a common, painful condition in which recurring ovoid or round ulcers affect the oral mucosa.

Redox homeostasis Is considered as the cumulative action of all free radical reactions and antioxidant defenses in different tissues.

Refrigerant A medicine or an application for allaying heat, fever or its symptoms.

Renal calculi Kidney stones.

Renal interstitial fibrosis Damage sustained by the kidneys' renal tubules and interstitial capillaries due to accumulation of extracellular waste in the wall of the small arteries and arterioles.

Renin Also known as an angiotensinogenase, is an enzyme that participates in the body's renin-angiotensin system (RAS).

Renin-angiotensin system (RAS) Also called the renin-angiotensin-aldosterone system (RAAS) is a hormone system that regulates blood pressure and water (fluid) balance.

Reperfusion The restoration of blood flow to an organ or tissue that has had its blood supply cut off, as after a heart attack.

Reporter gene A transfected gene that produces a signal, such as green fluorescence, when it is expressed.

Resistin A cysteine-rich protein secreted by adipose tissue of mice and rats.

Resolutive A substance that induces subsidence of inflammation.

Resolvent Reduce inflammation or swelling.

Resorb To absorb or assimilate a product of the body such as an exudates or cellular growth.

Restenosis Is the reoccurrence of stenosis, a narrowing of a blood vessel, leading to restricted blood flow.

Resveratrol Is a phytoalexin produced naturally by several plants when under attack by pathogens such as bacteria or fungi. It is a potent antioxidant found in red grapes and other plants.

Retinol A form of vitamin A, see vitamin A.

Retinopathy A general term that refers to some form of non-inflammatory damage to the retina of the eye.

Revulsive Counterirritant, used for swellings.

Rheumatic Pertaining to rheumatism or to abnormalities of the musculoskeletal system.

Rheumatism, Rheumatic disorder, Rheumatic diseases Refers to various painful medical conditions which affect bones, joints, muscles, tendons. Rheumatic diseases are characterized by the signs of inflammation – redness, heat, swelling, and pain.

Rheumatoid arthritis (RA) Is a chronic, systemic autoimmune disorder that most commonly causes inflammation and tissue damage in joints (arthritis) and tendon sheaths, together with anemia.

Rhinitis Irritation and inflammation of some internal areas of the nose and the primary symptom of rhinitis is a runny nose.

Rhinoplasty Is surgery to repair or reshape the nose.

Rhinorrhea Commonly known as a runny nose, characterized by an unusually significant amount of nasal discharge.

Rhinosinusitis Inflammation of the nasal cavity and sinuses.

Rho GTPases Rho-guanosine triphosphate hydrolase enzymes are molecular switches that regulate many essential cellular processes, including actin dynamics, gene transcription, cell-cycle progression and cell adhesion.

Ribosome inactivating proteins Protein that are capable of inactivating ribosomes.

Rickets Is a softening of the bones in children potentially leading to fractures and deformity.

Ringworm Dermatophytosis, a skin infection caused by fungus.

Roborant Restoring strength or vigour, a tonic.

Rotavirus The most common cause of infectious diarrhea (gastroenteritis) in young children and infants, one of several viruses that causes infections called stomach flu.

Rubefacient A substance for external application that produces redness of the skin e.g. by causing dilation of the capillaries and an increase in blood.

Ryanodine receptor Intracellular Ca++ channels in animal tissues like muscles and neurons.

S.C. Abbreviation for sub-cutaneous, beneath the layer of skin.

S-T segment The portion of an electrocardiogram between the end of the QRS complex and the beginning of the T wave. Elevation or depression of the S-T segment is the characteristics of myocardial ischemia or injury and coronary artery disease.

Sapraemia See septicaemia.

Sarcoma Cancer of the connective or supportive tissue (bone, cartilage, fat, muscle, blood vessels) and soft tissues.

Sarcopenia Degenerative loss of skeletal muscle mass and strength associated with aging.

Sarcoplasmic reticulum A special type of smooth endoplasmic reticulum found in smooth and striated muscle.

SARS Severe acute respiratory syndrome, the name of a potentially fatal new respiratory disease in humans which is caused by the SARS coronavirus (SARS-CoV)

Satiety State of feeling satiated, fully satisfied (appetite or desire).

Scabies A transmissible ectoparasite skin infection characterized by superficial burrows, intense pruritus (itching) and secondary infection.

Scarlatina Scarlet fever, an acute, contagious disease caused by infection with group A streptococcal bacteria.

Schwann cells Or neurolemmocytes, are the principal supporting cells of the peripheral nervous system, they form the myelin sheath of a nerve fibre.

Schistosomiasis Is a parasitic disease caused by several species of fluke of the genus *Schistosoma*. Also known as bilharzia, bilharziosis or snail fever.

Schizophrenia A psychotic disorder (or a group of disorders) marked by severely impaired thinking, emotions, and behaviors.

Sciatica A condition characterised by pain deep in the buttock often radiating down the back of the leg along the sciatic nerve.

Scleroderma A disease of the body's connective tissue. The most common symptom is a thickening and hardening of the skin, particularly of the hands and face.

Scrofula A tuberculous infection of the skin on the neck caused by the bacterium *Mycobacterium tuberculosis*.

Scrophulosis See scrofula.

Scurf Abnormal skin condition in which small flakes or sales become detached.

Scurvy A state of dietary deficiency of vitamin C (ascorbic acid) which is required for the synthesis of collagen in humans.

Secretagogue A substance that causes another substance to be secreted.

Sedative Having a soothing, calming, or tranquilizing effect; reducing or relieving stress, irritability, or excitement.

Seizure The physical findings or changes in behavior that occur after an episode of abnormal electrical activity in the brain.

Selectins Are a family of cell adhesion molecules; e.g. selectin-E, selectin –L, selectin P.

Selenium (Se) A trace mineral that is essential to good health but required only in tiny amounts; it is incorporated into proteins to make selenoproteins, which are important antioxidant enzymes. It is found in avocado, brazil nut, lentils, sunflower seeds, tomato, whole grain cereals, seaweed, seafood and meat.

Sensorineural bradyacuasia Hearing impairment of the inner ear resulting from damage to the sensory hair cells or to the nerves that supply the inner ear.

Sepsis A condition in which the body is fighting a severe infection that has spread via the bloodstream.

Sequela An abnormal pathological condition resulting from a disease, injury or trauma.

Serine proteinase Peptide hydrolases which have an active centre histidine and serine involved in the catalytic process.

Serotonergic Liberating, activated by, or involving serotonin in the transmission of nerve impulses.

Serotonin A monoamine neurotransmitter synthesized in serotonergic neurons in the central nervous system.

Sepsis Condition in which the body is fighting a severe infection that has spread via the bloodstream.

Septicaemia A systemic disease associated with the presence and persistence of pathogenic microorganisms or their toxins in the blood.

Sequelae A pathological condition resulting from a prior disease, injury, or attack.

Sexual potentiator Increases sexual activity and potency, enhances sexual performance due to increased blood flow and efficient metabolism.

Sexually transmitted diseases (STD) Infections that are transmitted through sexual activity.

SGOT, Serum glutamic oxaloacetic transaminase An enzyme that is normally present in liver and heart cells. SGOT is released into blood when the liver or heart is damaged. Also called aspartate transaminase (AST).

SGPT, Serum glutamic pyruvic transaminase An enzyme normally present in serum and body tissues, especially in the liver; it is released into the serum as a result of tissue injury, also called Alanine transaminase (ALT),

Shiga–like toxin A toxin produced by the bacterium *Escherichia coli* which disrupts the function of ribosomes, also known as verotoxin.

Shiga toxigenic *Escherichia coli* **(STEC)** Comprises a diverse group of organisms capable of causing severe gastrointestinal disease in humans.

Shiga toxin A toxin produced by the bacterium *Shigella dysenteriae*, which disrupts the function of ribosomes.

Shingles Skin rash caused by the Zoster virus (same virus that causes chicken pox) and is medically termed Herpes zoster.

Sialogogue Salivation-promoter, a substance used to increase or promote the excretion of saliva.

Sialoproteins Glycoproteins that contain sialic acid as one of their carbohydrates.

Sialyation Reaction with sialic acid or its derivatives; used especially with oligosaccharides.

Sialyltransferases Enzymes that transfer sialic acid to nascent oligosaccharide.

Sickle cell disease Is an inherited blood disorder that affects red blood cells. People with sickle cell disease have red blood cells that contain mostly hemoglobin S, an abnormal type of hemoglobin. Sometimes these red blood cells become sickle-shaped (crescent shaped) and have difficulty passing through small blood vessels.

Side stitch Is an intense stabbing pain under the lower edge of the ribcage that occurs while exercising.

Signal transduction cascade Refers to a series of sequential events that transfer a signal through a series of intermediate molecules until final regulatory molecules, such as transcription factors, are modified in response to the signal.

Silicon (Si) Is required in minute amounts by the body and is important for the development of healthy hair and the prevention of nervous disorders. Lettuce is the best natural source of Silicon.

Sinapism Signifies an external application, in the form of a soft plaster, or poultice.

Sinusitis Inflammation of the nasal sinuses.

SIRC cells Statens Seruminstitut Rabbit Cornea (SIRC) cell line.

SIRT 1 Stands for sirtuin (silent mating type information regulation 2 homolog) 1. It is an enzyme that deacetylates proteins that contribute to cellular regulation.

6-Keto-PGF1 alpha A physiologically active and stable hydrolysis product of Epoprostenol, found in nearly all mammalian tissues.

Skp1 (S-phase kinase-associated protein 1) is a core component of SCF ubiquitin ligases and mediates protein degradation.

Smads A family of intracellular proteins that mediate signaling by members of the TGF-β (transforming growth factor β) superfamily.

Smad2/3 A key signaling molecule for TGF-β.

Smad7 A TGFβ type 1 receptor antagonist.

Smallpox Is an acute, contagious and devastating disease in humans caused by *Variola* virus and have resulted in high mortality over the centuries.

Snuff Powder inhaled through the nose.

SOD Superoxide dismutase, is an enzyme that repairs cells and reduces the damage done to them by superoxide, the most common free radical in the body.

Sodium (Na) Is an essential nutrient required for health. Sodium cations are important in neuron (brain and nerve) function, and in influencing osmotic balance between cells and the interstitial fluid and in maintenance of total body fluid homeostasis. Extra intake may cause a harmful effect on health. Sodium is naturally supplied by salt intake with food.

Soleus muscle Smaller calf muscle lower down the leg and under the gastrocnemius muscle.

Somites Mesodermal structures formed during embryonic development that give rise to segmented body parts such as the muscles of the body wall.

Somites Mesodermal structures formed during embryonic development that give rise to segmented body parts such as the muscles of the body wall.

Somnolence Sleepiness or drowsiness.

Soporific A sleep inducing drug.

SOS response A global response to DNA damage in which the cell cycle is arrested and DNA repair and mutagenesis are induced.

Soyasaponins Bioactive saponin compounds found in many legumes.

Soyasapogenins Triterpenoid products obtained from the acid hydrolysis of soyasaponins, designated soyasapogenols A, B, C, D and E.

Spasmolytic Checking spasms, see antispasmodic.

Spermatorrhoea Medically an involuntary ejaculation/drooling of semen usually nocturnal emissions.

Spermidine An important polyamine in DNA synthesis and gene expression.

Sphingolipid A member of a class of lipids derived from the aliphatic amino alcohol, sphingosine.

Spleen Organ that filters blood and prevents infection.

Spleen tyrosine kinase (SYK) Is an enigmatic protein tyrosine kinase functional in a number of diverse cellular processes such as the regulation of immune and inflammatory responses.

Splenitis Inflammation of the spleen.

Splenocyte Is a monocyte, one of the five major types of white blood cell, and is characteristically found in the splenic tissue.

Splenomegaly Is an enlargement of the spleen.

Sprain To twist a ligament or muscle of a joint without dislocating the bone.

Sprue Is a chronic disorder of the small intestine caused by sensitivity to gluten, a protein found in wheat and rye and to a lesser extent oats and barley . It causes poor absorption by the intestine of fat, protein, carbohydrates, iron, water, and vitamins A, D, E, and K.

Sputum Matter coughed up and usually ejected from the mouth, including saliva, foreign material, and substances such as mucus or phlegm, from the respiratory tract.

SREBP-1 See sterol regulatory element-binding protein-1.

Stanch To stop or check the flow of a bodily fluid like blood from a wound.

Statin A type of lipid-lowering drug.

Status epilepticus Refers to a life-threatening condition in which the brain is in a state of persistent seizure.

STD Sexually transmitted disease.

Steatorrhea Is the presence of excess fat in feces which appear frothy, foul smelling and floats because of the high fat content.

Steatohepatitis Liver disease, characterized by inflammation of the liver with fat accumulation in the liver.

Steatosis Refer to the deposition of fat in the interstitial spaces of an organ like the liver, fatty liver disease.

Sterility Inability to produce offspring, also called asepsis.

Steroidogenisis The production of steroids.

Steroidogenic Relating to steroidogenisis.

Sterol regulatory element-binding protein-1 (SREBP1) Is a key regulator of the transcription of numerous genes that function in the metabolism of cholesterol and fatty acids.

Stimulant A substance that promotes the activity of a body system or function.

Stomachic (digestive stimulant), an agent that stimulates or strengthens the activity of the stomach; used as a tonic to improve the appetite and digestive processes.

Stomatitis Oral inflammation and ulcers, may be mild and localized or severe, widespread, and painful.

Stomatology Medical study of the mouth and its diseases.

Stool Faeces.

Strangury Is the painful passage of small quantities of urine which are expelled slowly by straining with severe urgency; it is usually accompanied with the unsatisfying feeling of a remaining volume inside and a desire to pass something that will not pass.

Straub tail Condition in which an animal carries its tail in an erect (vertical or nearly vertical) position.

STREPs Sterol regulatory element binding proteins, a family of transcription factors that regulate lipid homeostasis by controlling the expression of a range of enzymes required for endogenous cholesterol, fatty acid, triacylglycerol and phospholipid synthesis.

Stria terminalis A structure in the brain consisting of a band of fibres running along the lateral margin of the ventricular surface of the thalamus.

Striae gravidarum A cutaneous condition characterized by stretch marks on the abdomen during and following pregnancy.

Stricture An abnormal constriction of the internal passageway within a tubular structure such as a vessel or duct

Strongyloidiasis An intestinal parasitic infection in humans caused by two species of the parasitic nematode *Strongyloides.* The nematode or round worms are also called thread worms.

Styptic A short stick of medication, usually anhydrous aluminum sulfate (a type of alum) or titanium dioxide, which is used for stanching blood by causing blood vessels to contract at the site of the wound. Also called hemostatic pencil. see antihaemorrhagic.

Subarachnoid hemorrhage Is bleeding in the area between the brain and the thin tissues that cover the brain.

Sudatory Medicine that causes or increases sweating. Also see sudorific.

Sudorific A substance that causes sweating.

Sulfur Sulfur is an essential component of all living cells. Sulfur is important for the synthesis of sulfur-containing amino acids, all polypeptides, proteins, and enzymes such as glutathione an important sulfur-containing tripeptide which plays a role in cells as a source of chemical reduction potential. Sulfur is also important for hair formation. Good plant sources are garlic, onion, leeks and other Alliaceous vegetables, Brassicaceous vegetables like cauliflower, cabbages, Brussels sprout, Kale; legumes – beans, green and red gram, soybeans; horse radish, water cress, wheat germ.

Superior mesenteric artery (SMA) Arises from the anterior surface of the abdominal aorta, just inferior to the origin of the celiac trunk, and supplies the intestine from the lower part of the duodenum to the left colic flexure and the pancreas.

Superoxidae mutase (SOD) Antioxidant enzyme.

Suppuration The formation of pus, the act of becoming converted into and discharging pus.

Supraorbital Located above the orbit of the eye.

SYK, Spleen tyrosine kinase Is a human protein and gene. Syk plays a similar role in transmitting signals from a variety of cell surface receptors including CD74, Fc Receptor, and integrins.

Sympathetic nervous system The part of the autonomic nervous system originating in the thoracic and lumbar regions of the spinal cord that in general inhibits or opposes the physiological effects of the parasympathetic nervous system, as in tending to reduce digestive secretions or speed up the heart.

Synaptic plasticity The ability of neurons to change the number and strength of their synapses.

Synaptogenesis The formation of synapses.

Synaptoneurosomes Purified synapses containing the pre- and postsynaptic termini.

Synaptosomes Isolated terminal of a neuron.

Syncope Fainting, sudden loss of consciousness followed by the return of wakefulness.

Syndactyly Webbed toes, a condition where two or more digits are fused together.

Syneresis Expulsion of liquid from a gel, as contraction of a blood clot and expulsion of liquid.

Syngeneic Genetically identical or closely related, so as to allow tissue transplant; immunologically compatible.

Synovial Lubricating fluid secreted by synovial membranes, as those of the joints.

Synoviocyte Located in the synovial membrane, there are two types. Type A cells are more numerous, have phagocytic characteristics and produce degradative enzymes. Type B cells produce synovial fluid, which lubricates the joint and nurtures nourishes the articular cartilage.

Syphilis Is perhaps the best known of all the STD's. Syphilis is transmitted by direct contact with infection sores, called chancres, syphitic skin rashes, or mucous patches on the tongue and mouth during kissing, necking, petting, or sexual intercourse. It can also be transmitted from a pregnant woman to a fetus after the fourth month of pregnancy.

Systolic The blood pressure when the heart is contracting. It is specifically the maximum arterial pressure during contraction of the left ventricle of the heart.

T cells Or T lymphocytes, a type of white blood cell that play a key role in the immune system.

Tachyarrhythmia Any disturbance of the heart rhythm in which the heart rate is abnormally increased.

Tachycardia A false heart rate applied to adults to rates over 100 beats per minute.

Tachyphylaxia A decreased response to a medicine given over a period of time so that larger doses are required to produce the same response.

Tachypnea Abnormally fast breathing.

Taenia A parasitic tapeworm or flatworm of the genus, *Taenia.*

Taeniacide An agent that kills tapeworms.

TBARS See thiobarbituric acid reactive substances.

T-cell A type of white blood cell that attacks virus-infected cells, foreign cells and cancer cells.

TCA cycle See Tricarboxylic acid cycle.

TCID50 Median tissue culture infective dose; that amount of a pathogenic agent that will produce pathological change in 50% of cell cultures.

Telencephalon The cerebral hemispheres, the largest divisions of the human brain.

Telomerase Enzyme that acts on parts of chromosomes known as telomeres.

Temporomandibular joint disorder (TMJD or TMD syndrome) A disorder characterized by acute or chronic inflammation of the temporomandibular joint, that connects the mandible to the skull.

Tendonitis Is inflammation of a tendon.

Tenesmus A strong desire to defaecate.

Teratogen Is an agent that can cause malformations of an embryo or fetus. *adj.* teratogenic.

Testicular torsion Twisting of the spermatic cord, which cuts off the blood supply to the testicle and surrounding structures within the scrotum.

Tetanus An acute, potentially fatal disease caused by tetanus bacilli multiplying at the site of an injury and producing an exotoxin that reaches the central nervous system producing prolonged contraction of skeletal muscle fibres. Also called lockjaw.

Tete Acute dermatitis caused by both bacterial and fungal infection

Tetter Any of a number of skin diseases.

TGF-beta Transforming growth factor beta is a protein that controls proliferation, cellular differentiation, and other functions in most cells.

Th cells or T helper cells A subgroup of lymphocytes that helps other white blood cells in immunologic processes.

Thermogenic Tending to produce heat, applied to drugs or food (fat burning food)

Thiobarbituric acid reactive substances (TBARS) A well-established method for screening and monitoring lipid peroxidation.

Thixotropy The property exhibited by certain gels of becoming fluid when stirred or shaken and returning to the semisolid state upon standing.

Thrombocythaemia A blood condition characterize by a high number of platelets in the blood.

Thrombocytopenia A condition when the bone marrow does not produce enough platelets (thrombocytes) like in leukaemia.

Thromboembolism Formation in a blood vessel of a clot (thrombus) that breaks loose and is carried by the blood stream to plug another vessel.

Thrombogenesis Formation of a thrombus or blood clot.

Thrombophlebitis Occurs when there is inflammation and clot in a surface vein.

Thromboplastin An enzyme liberated from blood platelets that converts prothrombin into thrombin as blood starts to clot, also called thrombokinase.

Thrombosis The formation or presence of a thrombus (clot).

Thromboxanes Any of several compounds, originally derived from prostaglandin precursors in platelets that stimulate aggregation of platelets and constriction of blood vessels.

Thromboxane B2 The inactive product of thromboxane.

Thrombus A fibrinous clot formed in a blood vessel or in a chamber of the heart.

Thrush A common mycotic infection caused by yeast, *Candida albicans*, in the digestive tract or vagina. In children it is characterized by white spots on the tongue.

Thymocytes Are T cell precursors which develop in the thymus.

Thyrotoxicosis Or hyperthyroidism – an overactive thyroid gland, producing excessive circulating free thyroxine and free triiodothyronine, or both.

TIMP-3 A human gene belongs to the tissue inhibitor of matrix metalloproteinases (MMP) gene family. see MMP.

Tincture Solution of a drug in alcohol.

Tinea Ringworm, fungal infection on the skin.

Tinea favosa See favus.

Tinnitus A noise in the ears, as ringing, buzzing, roaring, clicking, etc.

Tisane A herbal infusion used as tea or for medicinal purposes.

Tissue plasminogen activator A serine protease involved in the breakdown of blood clots.

TNF alpha Cachexin or cachectin and formally known as tumour necrosis factor-α, a cytokine involved in systemic inflammation. primary role of TNF is in the regulation of immune cells. TNF is also able to induce apoptotic cell death, to induce inflammation, and to inhibit tumorigenesis and viral replication.

Tocolytics Medications used to suppress premature labor.

Tocopherol Fat soluble organic compounds belonging to vitamin E group. See vitamin E.

Tocotrienol Fat soluble organic compounds belonging to vitamin E group. See vitamin E.

Toll-like receptors (TLRs) A class of proteins that play a key role in the innate immune system.

Tonic Substance that acts to restore, balance, tone, strengthen, or invigorate a body system without overt stimulation or depression

Tonic clonic seizure A type of generalized seizure that affects the entire brain.

Tonsillitis An inflammatory condition of the tonsils due to bacteria, allergies or respiratory problems.

Topoisomerases A class of enzymes involved in the regulation of DNA supercoiling.

Topoiosmerase inhibitors A new class of anticancer agents with a mechanism of action aimed at interrupting DNA replication in cancer cells.

Total parenteral nutrition (TPN) Is a method of feeding that bypasses the gastrointestinal tract.

Toxemia Is the presence of abnormal substances in the blood, but the term is also used for a serious condition in pregnancy that involves hypertension and proteinuria. Also called pre-eclampsia.

Tracheitis Is a bacterial infection of the trachea; also known as bacterial tracheitis or acute bacterial tracheitis.

Trachoma A contagious disease of the conjunctiva and cornea of the eye, producing painful sensitivity to strong light and excessive tearing.

TRAIL Acronym for tumour necrosis factor-related apoptosis-inducing ligand, is a cytokine that preferentially induces apoptosis in tumour cells.

Tranquilizer A substance drug used in calming person suffering from nervous tension or anxiety.

Transaminase Also called aminotransferase is an enzyme that catalyzes a type of reaction between an amino acid and an α-keto acid.

Transaminitis Increase in alanine aminotransferase (ALT) and/or aspartate aminotransferase (AST) to >5 times the upper limit of normal.

Transcatheter arterial chemoembolization (TACE) Is an interventional radiology procedure involving percutaneous access of to the hepatic artery and passing a catheter through the abdominal artery aorta followed by radiology. It is used extensively in the palliative treatment of unresectable hepatocellular carcinoma (HCC)

Transcriptional activators Are proteins that bind to DNA and stimulate transcription of nearby genes.

Transcriptional coactivator PGC-1 A potent transcriptional coactivator that regulates oxidative metabolism in a variety of tissues.

Transcriptome profiling To identify genes involved in peroxisome assembly and function.

Transforming growth factor beta (TGF-β) A protein that controls proliferation , cellular differentiation, and other functions in most cells.

TRAP 6 Thrombin receptor activating peptide with 6 amino acids.

Tremorine A chemical that produces a tremor resembling Parkinsonian tremor.

Tremulous Marked by trembling, quivering or shaking.

Triacylglycerols Or triacylglyceride, is a glyceride in which the glycerol is esterified with three fatty acids.

Tricarboxylic acid cycle (TCA cycle) A series of enzymatic reactions in aerobic organisms involving oxidative metabolism of acetyl units and producing high-energy phosphate compounds, which serve as the main source of cellular energy. Also called citric acid cycle, Krebs cycle.

Trichophytosis Infection by fungi of the genus *Trichophyton*.

Trigeminal neuralgia (TN) Is a neuropathic disorder of one or both of the facial trigeminal nerves, also known as prosopalgia.

Triglycerides A type of fat (lipids) found in the blood stream.

Trismus Continuous contraction of the muscles of the jaw, specifically as a symptom of tetanus, or lockjaw; inability to open mouth fully.

TrKB receptor Also known as TrKB tyrosine kinase , a protein in humans that acts as a catalytic receptor for several neutrophins.

Trolox Equivalent Measures the antioxidant capacity of a given substance, as compared to the standard, Trolox also referred to as TEAC (Trolox equivalent antioxidant capacity).

Trypanocidal Destructive to trypanosomes.

Trypanosomes Protozoan of the genus *Trypanosoma.*

Trypanosomiasis Human disease or an infection caused by a trypanosome.

Trypsin An enzyme of pancreatic juice that hydrolyzes proteins into smaller polypeptide units.

Trypsin inhibitor Small protein synthesized in the exocrine pancreas which prevents conversion of trypsinogen to trypsin, so protecting itself against trypsin digestion.

Tuberculosis (TB) Is a bacterial infection of the lungs caused by a bacterium called *Mycobacterium tuberculosis,* characterized by the formation of lesions (tubercles) and necrosis in the lung tissues and other organs.

Tumorigenesis Formation or production of tumours.

Tumour An abnormal swelling of the body other than those caused by direct injury.

Tussis A cough.

Tympanic membrane Ear drum.

Tympanitis Infection or inflammation of the inner ear.

Tympanophonia Increased resonance of one's own voice, breath sounds, arterial murmurs, etc., noted especially in disease of the middle ear.

Tympanosclerosis See myringoslcerosis.

Tyrosinase A copper containing enzyme found in animals and plants that catalyses the oxidation of phenols (such as tyrosine) and the production of melanin and other pigments from tyrosine by oxidation.

UCP1 An uncoupling protein found in the mitochondria of brown adipose tissue used to generate heat by non-shivering thermogenesis.

UCP2 enzyme Uncoupling protein 2 enzyme, a mitochondrial protein expressed in adipocytes.

Ulcer An open sore on an external or internal body surface usually accompanied by disintegration of tissue and pus.

Ulcerative colitis Is one of two types of inflammatory bowel disease – a condition that causes the bowel to become inflamed and red.

Ulemorrhagia Bleeding of the gums.

Ulitis Inflammation of the gums.

Unguent Ointment.

Unilateral ureteral obstruction Unilateral blockage of urine flow through the ureter of one kidney, resulting in a backup of urine, distension of the renal pelvis and calyces, and hydronephrosis.

Uraemia An excess in the blood of urea, creatinine and other nitrogenous end products of protein and amino acids metabolism, more correctly referred to as azotaemia.

Urethra Tube conveying urine from the bladder to the external urethral orifice.

Urethritis Is an inflammation of the urethra caused by infection.

Uricemia An excess of uric acid or urates in the blood.

Uricosuric Promoting the excretion of uric acid in the urine.

Urinary Pertaining to the passage of urine.

Urinogenital Relating to the genital and urinary organs or functions.

Urodynia Pain on urination.

Urokinase A serine protease enzyme in human urine that catalyzes the conversion of plasminogen to plasmin.

Urolithiasis Formation of stone in the urinary tract (kidney bladder or urethra).

Urticant A substance that causes wheals to form.

Urticaria (or hives) is a skin condition, commonly caused by an allergic reaction, that is characterized by raised red skin welts.

Uterine Relating to the uterus.

Uterine relaxant An agent that relaxes the muscles in the uterus.

Uterine stimulant An agent that stimulates the uterus (and often employed during active childbirth).

Uterotonic Giving muscular tone to the uterus.

Uterotrophic Causing an effect on the uterus.

Uterus Womb.

Vagotomy The surgical cutting of the vagus nerve to reduce acid secretion in the stomach.

Vagus nerve A cranial nerve, that is, a nerve connected to the brain. The vagus nerve has branches to most of the major organs in the body, including the larynx, throat, windpipe, lungs, heart, and most of the digestive system

Variola Or smallpox, a contagious disease unique to humans, caused by either of two virus variants, *Variola major* and *Variola minor*. The disease is characterised by fever, weakness and skin eruption with pustules that form scabs that leave scars.

Varicose veins Are veins that have become enlarged and twisted.

Vasa vasorum Is a network of small blood vessels that supply large blood vessels. *plur.* vasa vasori.

Vascular endothelial growth factor (VEGF) A polypeptide chemical produced by cells that stimulates the growth of new blood vessels.

Vasculogenesis The process of blood vessel formation occurring by a de novo production of endothelial cells.

Vasoconstrictor Drug that causes constriction of blood vessels.

Vasodilator Drug that causes dilation or relaxation of blood vessels.

Vasodilatory Causing the widening of the lumen of blood vessels.

Vasomotor symptoms Menopausal symptoms characterised by hot flushes and night sweats.

Vasospasm Refers to a condition in which blood vessels spasm, leading to vasoconstriction and subsequently to tissue ischemia and death (necrosis).

Vasculogenesis Process of blood vessel formation occurring by a de novo production of endothelial cells.

VCAM-1 (vascular cell adhesion molecule-1) Also known as CD106, contains six or seven immunoglobulin domains and is expressed on both large and small vessels only after the endothelial cells are stimulated by cytokines.

VEGF Vascular endothelial growth factor.

Venereal disease (VD) Term given to the diseases syphilis and gonorrhoea.

Venule A small vein, especially one joining capillaries to larger veins.

Vermifuge A substance used to expel worms from the intestines.

Verotoxin A Shiga-like toxin produced by *Escherichia coli,* which disrupts the function of ribosomes, causing acute renal failure.

Verruca plana Is a reddish-brown or flesh-colored, slightly raised, flat-surfaced, well-demarcated papule on the hand and face, also called flat wart.

Vertigo An illusory, sensory perception that the surroundings or one's own body are revolving; dizziness.

Very-low-density lipoprotein (VLDL) A type of lipoprotein made by the liver. VLDL is one of the five major groups of lipoproteins (chylomicrons, VLDL, intermediate-density lipoprotein, low-density lipoprotein, high-density lipoprotein (HDL)) that enable fats and cholesterol to move within the water-based solution of the bloodstream. VLDL is converted in the bloodstream to low-density lipoprotein (LDL).

Vesical calculus Calculi (stones) in the urinary bladder

Vesicant A substance that causes tissue blistering.

Vestibular Relating to the sense of balance.

Vestibular disorders Includes symptoms of dizziness, vertigo, and imbalance; it can be result from or worsened by genetic or environmental conditions.

Vestibular system Includes parts of the inner ear and brain that process sensory information involved with controlling balance and eye movement.

Vibrissa Stiff hairs that are located especially about the nostrils.

Viremia A medical condition where viruses enter the bloodstream and hence have access to the rest of the body.

Visceral fat Intra-abdominal fat, is located inside the peritoneal cavity, packed in between internal organs and torso.

Vitamin Any complex, organic compound, found in various food or sometimes synthesized in the body, required in tiny amounts and are essential for the regulation of metabolism, normal growth and function of the body.

Vitamin A Retinol, fat-soluble vitamins that play an important role in vision, bone growth, reproduction, cell division, and cell differentiation, helps regulate the immune system in preventing or fighting off infections. Vitamin A that is found in colorful fruits and vegetables is called provitamin A carotenoid. They can be made into retinol in the body. Deficiency of vitamin A results in night blindness and keratomalacia.

Vitamin B1 Also called thiamine, water-soluble vitamins, dissolve easily in water, and in general, are readily excreted from the body they are not readily stored, consistent daily intake is important. It functions as coenzyme in the metabolism of carbohydrates and branched chain amino acids, and other cellular processes. Deficiency results in beri-beri disease.

Vitamin B2 Also called riboflavin, an essential water-soluble vitamin that functions as coenzyme in redox reactions. Deficiency causes ariboflavinosis.

Vitamin B3 Comprises niacin and niacinamide, water-soluble vitamin that function as coenzyme or co-substrate for many redox reactions and is required for energy metabolism. Deficiency causes pellagra.

Vitamin B5 Also called pantothenic acid, a water-soluble vitamin that function as coenzyme in fatty acid metabolism. Deficiency causes paresthesia.

Vitamin B6 Water-soluble vitamin, exists in three major chemical forms: pyridoxine, pyridoxal, and pyridoxamine. Vitamin B6 is needed in enzymes involved in protein metabolism, red blood cell metabolism, efficient functioning of nervous and immune systems and hemoglobin formation. Deficiency causes anaemia and peripheral neuropathy.

Vitamin B7 Also called biotin or vitamin H, an essential water-soluble vitamin, is involved in the synthesis of fatty acids amino acids and glucose, in energy metabolism. Biotin promotes normal health of sweat glands, bone marrow, male gonads, blood cells, nerve tissue, skin and hair, Deficiency causes dermatitis and enteritis.

Vitamin B9 Also called folic acid, an essential water-soluble vitamin. Folate is especially important during periods of rapid cell division and growth such as infancy and pregnancy. Deficiency during pregnancy is associated with birth defects such as neural tube defects. Folate is also important for production of red blood cells and prevent anemia. Folate is needed to make DNA and RNA, the building blocks of cells. It also helps prevent changes to DNA that may lead to cancer.

Vitamin B12 A water-soluble vitamin, also called cobalamin as it contains the metal cobalt. It helps maintain healthy nerve cells and red blood cells, and DNA production. Vitamin B12 is bound to the protein in food. Deficiency causes megaloblastic anaemia.

Vitamin C Also known as ascorbic acid is an essential water-soluble vitamin. It functions as cofactor for reactions requiring reduced copper or iron metallonzyme and as a protective antioxidant. Deficiency of vitamin C causes scurvy.

Vitamin D A group of fat-soluble, prohormone vitamin, the two major forms of which are vitamin D2 (or ergocalciferol) and vitamin D3 (or cholecalciferol). Vitamin D obtained from sun exposure, food, and supplements is biologically inert and must undergo two hydroxylations in the body for activation. Vitamin D is essential for promoting calcium absorption in the gut and maintaining adequate serum calcium and phosphate concentrations to enable normal growth and mineralization of bone and prevent hypocalcemic tetany. Deficiency causes rickets and osteomalacia. Vitamin D has other roles in human health, including modulation of neuromuscular and immune function, reduction of inflammation and modulation of many genes encoding proteins that regulate cell proliferation, differentiation, and apoptosis.

Vitamin E Is the collective name for a group of fat-soluble compounds and exists in eight chemical forms (α-, β-, γ-, and δ-tocopherol and α-, β-, γ-, and δ-tocotrienol). It has pronounced antioxidant activities stopping the formation of Reactive Oxygen Species when fat undergoes oxidation and help prevent or delay the chronic diseases associated with free radicals. Besides its antioxidant activities, vitamin E is involved in immune function, cell signaling, regulation of gene expression, and other metabolic processes. Deficiency is very rare but can cause mild hemolytic anemia in newborn infants.

Vitamin K A group of fat soluble vitamin and consist of vitamin K1 which is also known as phylloquinone or phytomenadione (also called phytonadione) and vitamin K2 (menaquinone, menatetrenone). Vitamin K plays an important role in blood clotting. Deficiency is very rare but can cause bleeding diathesis.

Vitamin P A substance or mixture of substances obtained from various plant sources, identified as citrin or a mixture of bioflavonoids, thought to but not proven to be useful in reducing the extent of hemorrhage.

Vitiligo A chronic skin disease that causes loss of pigment, resulting in irregular pale patches of skin. It occurs when the melanocytes, cells responsible for skin pigmentation, die or are unable to function. Also called leucoderma.

Vitreoretinopathy See proliferative vitreoretinopathy.

VLA-4 Very late antigen-4, expressed by most leucocytes but it is observed on neutrophils under special conditions.

VLDL See very low density lipoproteins.

Vomitive Substance that causes vomiting.

Vulnerary (wound healer), a substance used to heal wounds and promote tissue formation.

Wart An infectious skin tumour caused by a viral infection.

Welt See wheal.

Wheal A firm, elevated swelling of the skin. Also called a weal or welt.

White fat White adipose tissue (WAT) in mammals, store of energy . cf. brown fat.

Whitlow Painful infection of the hand involving one or more fingers that typically affects the terminal phalanx.

Whooping cough Acute infectious disease usually in children caused by a *Bacillus* bacterium and accompanied by catarrh of the respiratory passages and repeated bouts of coughing.

Wnt signaling pathway Is a network of proteins involved in embryogenesis and cancer, and also in normal physiological processes.

X-linked agammaglobulinemia Also known as X-linked hypogammaglobulinemia, XLA, Bruton type agammaglobulinemia, Bruton syndrome, or sex-linked agammaglobulinemia; a rare x-linked genetic disorder that affects the body's ability to fight infection.

Xanthine oxidase A flavoprotein enzyme containing a molybdenum cofactor (Moco) and (Fe2S2) clusters, involved in purine metabolism. In humans, inhibition of xanthine oxidase reduces the production of uric acid, and prevent hyperuricemia and gout.

Xanthones Unique class of biologically active phenol compounds with the molecular formula C13H8O2 possessing antioxidant properties, discovered in the mangosteen fruit.

Xenobiotics A chemical (as a drug, pesticide, or carcinogen) that is foreign to a living organism.

Xenograft A surgical graft of tissue from one species to an unlike species.

Xerophthalmia A medical condition in which the eye fails to produce tears.

Yaws An infectious tropical infection of the skin, bones and joints caused by the spirochete bacterium *Treponema pertenue,* characterized by papules and pappiloma with subsequent deformation of the skins, bone and joints; also called framboesia.

Yellow fever Is a viral disease that is transmitted to humans through the bite of infected mosquitoes. Illness ranges in severity from an influenza-like syndrome to severe hepatitis and hemorrhagic fever. Yellow fever virus (YFV) is maintained in nature by mosquito-borne transmission between nonhuman primates.

Zeaxanthin A common carotenoid, found naturally as coloured pigments in many fruit vegetables and leafy vegetables. It is important for good vision and is one of the two carotenoids contained within the retina of the eye. Within the central macula, zeaxanthin predominates, whereas in the peripheral retina, lutein predominates.

Zinc (Zn) Is an essential mineral for health. It is involved in numerous aspects of cellular metabolism: catalytic activity of enzymes, immune function, protein synthesis, wound healing, DNA synthesis, and cell division. It also supports normal growth and development during pregnancy, childhood, and adolescence and is required for proper sense of taste and smell. Dietary sources include beans, nuts, pumpkin seeds, sunflower seeds, whole wheat bread and animal sources.

Scientific Glossary

Abaxial Facing away from the axis, as of the surface of an organ.

Abscission Shedding of leaves, flowers, or fruits following the formation of the abscission zone.

Acaulescent Lacking a stem, or stem very much reduced.

Accrescent Increasing in size after flowering or with age.

Achene A dry, small, one-seeded, indehiscent one-seeded fruit formed from a superior ovary of one carpel as in sunflower.

Acid soil Soil that maintains a pH of less than 7.0.

Acidulous Acid or sour in taste.

Actinomorphic Having radial symmetry, capable of being divided into symmetrical halves by any plane, refers to a flower, calyx or corolla.

Aculeate Having sharp prickles.

Acuminate Tapering gradually to a sharp point.

Acute (Botany) tapering at an angle of less than 90 degrees before terminating in a point as of leaf apex and base.

Adaxial Side closest to the stem axis.

Adherent Touching without organic fusion as of floral parts of different whorls.

Adnate United with another unlike part as of stamens attached to petals.

Adpressed Lying close to another organ but not fused to it.

Adventitious Arising in abnormal positions, e.g. roots arising from the stem, branches or leaves, buds arising elsewhere than in the axils of leaves.

Adventive Not native to and not fully established in a new habitat or environment; locally or temporarily naturalized. e.g. an adventive weed.

Aestivation Refers to positional arrangement of the floral parts in the bud before it opens.

Akinete A thick-walled dormant cell derived from the enlargement of a vegetative cell. It serves as a survival structure.

Aldephous Having stamens united together by their filaments.

Alfisols Soil with a clay-enriched subsoil and relatively high native fertility, having undergone only moderate leaching, containing aluminium, iron and with at least 35% base saturation, meaning that calcium, magnesium, and potassium are relatively abundant.

Alkaline soil Soil that maintains a pH above 7.0, usually containing large amounts of calcium, sodium, and magnesium, and is less soluble than acidic soils.

Alkaloids Naturally occurring bitter, complex organic-chemical compounds containing basic nitrogen and oxygen atoms and having various pharmacological effects on humans and other animals.

Allomorphic With a shape or form different from the typical.

Alluvial soil A fine-grained fertile soil deposited by water flowing over flood plains or in river beds.

Alluvium Soil or sediments deposited by a river or other running water.

Alternate Leaves or buds that are spaced along opposite sides of stem at different levels.

T.K. Lim, *Edible Medicinal And Non Medicinal Plants: Volume 3, Fruits*,
DOI 10.1007/978-94-007-2534-8, © Springer Science+Business Media B.V. 2012

Amplexicaul Clasping the stem as base of certain leaves.

Anatomizing Interconnecting network as applied to leaf veins.

Anatropous With the ovule completely inverted.

Andisols Are soils formed in volcanic ash and containing high proportions of glass and amorphous colloidal materials.

Androdioecious With male flowers and bisexual flowers on separate plants.

Androecium Male parts of a flower; comprising the stamens of one flower.

Androgynophore A stalk bearing both the androecium and gynoecium above the perianth of the flower.

Androgynous With male and female flowers in distinct parts of the same inflorescence.

Andromonoecious Having male flowers and bisexual flowers on the same plant.

Angiosperm A division of seed plants with the ovules borne in an ovary.

Annual A plant which completes its life cycle within a year.

Annular Shaped like or forming a ring.

Annulus Circle or ring-like structure or marking; the portion of the corolla which forms a fleshy, raised ring.

Anthelate An open, paniculate cyme.

Anther The part of the stamen containing pollen sac which produces the pollen.

Antheriferous Containing anthers.

Anthesis The period between the opening of the bud and the onset of flower withering.

Anthocarp A false fruit consisting of the true fruit and the base of the perianth.

Anthocyanidins Are common plant pigments. They are the sugar-free counterparts of anthocyanins.

Anthocyanins A subgroup of antioxidant flavonoids, are glucosides of anthocyanidins. They occur as water-soluble vacuolar pigments that may appear red, purple, or blue according to pH in plants.

Antipetala Situated opposite petals.

Antisepala Situated opposite sepals.

Antrorse Directed forward upwards.

Apetalous Lacking petals as of flowers with no corolla.

Apical meristem Active growing point. A zone of cell division at the tip of the stem or the root.

Apically Towards the apex or tip of a structure.

Apiculate Ending abruptly in a short, sharp, small point.

Apiculum A short, pointed, flexible tip.

Apocarpous Carpels separate in single individual pistils.

Apopetalous With separate petals, not united to other petals.

Aposepalous With separate sepals, not united to other sepals.

Appressed Pressed closely to another structure but not fused or united.

Aquatic A plant living in or on water for all or a considerable part of its life span.

Arachnoid (Botany) formed of or covered with long, delicate hairs or fibers.

Arborescent Resembling a tree; applied to non-woody plants attaining tree height and to shrubs tending to become tree-like in size.

Arbuscular mycorrhiza (AM) A type of mycorrhiza in which the fungus (of the phylum Glomeromycota) penetrates the cortical cells of the roots of a vascular plant and form unique structures such as arbuscules and vesicles. These fungi help plants to capture nutrients such as phosphorus and micronutrients from the soil.

Archegonium A flask-shaped female reproductive organ in mosses, ferns, and other related plants.

Areolate With areolea.

Areole (Botany) a small, specialized, cushion-like area on a cactus from which hairs, glochids, spines, branches, or flowers may arise; an irregular angular spaces marked out on a surface e.g. fruit surface. *pl.* areolea.

Aril Specialized outgrowth from the funiculus (attachment point of the seed) (or hilum) that encloses or is attached to the seed. *adj.* arillate.

Arillode A false aril; an aril originating from the micropyle instead of from the funicle or chalaza of the ovule, e.g. mace of nutmeg.

Aristate Bristle-like part or appendage, e.g. awns of grains and grasses.

Aristulate Having a small, stiff, bristle-like part or appendage; a diminutive of aristate

Articulate Jointed; usually breaking easily at the nodes or point of articulation into segments.

Ascending Arched upwards in the lower part and becoming erect in the upper part.

Ascospore Spore produced in the ascus in Ascomycete fungi.

Ascus Is the sexual spore-bearing cell produced in Ascomycete fungi. *pl.* asci.

Asperulous Refers to a rough surface with short, hard projections.

Attenuate Tapered or tapering gradually to a point.

Auricle An ear-like appendage that occurs at the base of some leaves or corolla.

Auriculate Having auricles.

Awn A hair-like or bristle-like appendage on a larger structure.

Axil Upper angle between a lateral organ, such as a leaf petiole and the stem that bears it.

Axile Situated along the central axis of an ovary having two or more locules, as in axile placentation.

Axillary Arising or growing in an axil.

Baccate Beery-like, pulpy or fleshy.

Barbate Bearded, having tufts of hairs.

Barbellae Short, stiff, hair-like bristles. *adj.* barbellate.

Bark Is the outermost layers of stems and roots of woody plants.

Basal Relating to, situated at, arising from or forming the base.

Basaltic soil Soil derived from basalt, a common extrusive volcanic rock.

Basidiospore A reproductive spore produced by Basidiomycete fungi.

Basidium A microscopic, spore-producing structure found on the hymenophore of fruiting bodies of Basidiomycete fungi.

Basifixed Attached by the base, as certain anthers are to their filaments.

Basionym The synonym of a scientific name that supplies the epithet for the correct name.

Beak A prominent apical projection, especially of a carpel or fruit. *adj.* beaked.

Bearded Having a tuft of hairs.

Berry A fleshy or pulpy indehiscent fruit from a single ovary with the seed(s) embedded in the fleshy tissue of the pericarp.

Biconvex Convex on both sides.

Biennial Completing the full cycle from germination to fruiting in more than one, but not more than 2 years.

Bifid Forked, divided into two parts.

Bifoliolate Having two leaflets.

Bilabiate Having two lips as of a corolla or calyx with segments fused into an upper and lower lip.

Bipinnate Twice pinnate; the primary leaflets being again divided into secondary leaflets.

Bipinnatisect Refers to a pinnately compound leaf, in which each leaflet is again divided into pinnae.

Biserrate Doubly serrate; with smaller regular, asymmetric teeth on the margins of larger teeth.

Bisexual Having both sexes, as in a flower bearing both stamens and pistil, hermaphrodite or perfect.

Biternate Twice ternate; with three pinnae each divided into three pinnules.

Blade Lamina; part of the leaf above the sheath or petiole.

Blotched See variegated.

Bole Main trunk of tree from the base to the first branch.

Brachyblast A short, axillary, densely crowded branchlet or shoot of limited growth, in which the internodes elongate little or not at all.

Bracket fungus Shelf fungus.

Bract A leaf-like structure, different in form from the foliage leaves, associated with an inflorescence or flower. *adj.* bracteate.

Bracteate Possessing bracts.

Bracteolate Having bracteoles.

Bracteole A small, secondary, bract-like structure borne singly or in a pair on the pedicel or calyx of a flower. *adj.* bracteolate.

Bristle A stiff hair.

Bulb A modified underground axis that is short and crowned by a mass of usually fleshy, imbricate scales. *adj.* bulbous.

Bulbil A small bulb or bulb-shaped body, especially one borne in the leaf axil or an inflorescence, and usually produced for asexual reproduction.

Bullate Puckered, blistered.

Burr Type of seed or fruit with short, stiff bristles or hooks or may refer to a deformed type of wood in which the grain has been misformed.

Bush Low, dense shrub without a pronounced trunk.

Buttress Supporting, projecting outgrowth from base of a tree trunk as in some Rhizophoraceae and Moraceae.

Caducous Shedding or falling early before maturity refers to sepals and petals.

Caespitose Growing densely in tufts or clumps; having short, closely packed stems.

Calcareous Composed of or containing lime or limestone.

Calcrete A hardpan consisting gravel and sand cemented by calcium.

Callus A condition of thickened raised mass of hardened tissue on leaves or other plant parts often formed after an injury but sometimes a normal feature. A callus also can refer to an undifferentiated plant cell mass grown on a culture medium. *n.* callosity. *pl.* calli, callosities. *adj.* callose.

Calyptra The protective cap or hood covering the spore case of a moss or related plant.

Calyptrate Operculate, having a calyptra.

Calyx Outer floral whorl usually consisting of free sepals or fused sepals (calyx tube) and calyx lobes. It encloses the flower while it is still a bud. *adj.* calycine.

Calyx lobe One of the free upper parts of the calyx which may be present when the lower part is united into a tube.

Calyx tube The tubular fused part of the calyx, often cup shaped or bell shaped, when it is free from the corolla.

Campanulate Shaped like a bell refers to calyx or corolla.

Campylotropous With the ovule partially inverted and curved.

Canaliculate Having groove or grooves.

Candelabriform Having the shape of a tall branched candle-stick.

Canescent Covered with short, fine whitish or grayish hairs or down.

Canopy Uppermost leafy stratum of a tree.

Cap See pileus.

Capitate Growing together in a head. Also means enlarged and globular at the tip.

Capitulum A flower head or inflorescence having a dense cluster of sessile, or almost sessile, flowers or florets.

Capsule A dry, dehiscent fruit formed from two or more united carpels and dehiscing at maturity by sections called valves to release the seeds. *adj.* capsular.

Carinate Keeled.

Carpel A simple pistil consisting of ovary, ovules, style and stigma. *adj.* carpellary.

Carpogonium Female reproductive organ in red algae. *pl.* carpogonia.

Carpophore Part of the receptacle which is lengthened between the carpels as a central axis; any fruiting body or fruiting structure of a fungus.

Cartilaginous Sinewy, having a firm, tough, flexible texture (in respect of leaf margins).

Caryopsis A simple dry, indehiscent fruit formed from a single ovary with the seed coat united with the ovary wall as in grasses and cereals.

Cataphyll A reduced or scarcely developed leaf at the start of a plant's life (i.e., cotyledons) or in the early stages of leaf development.

Catkin A slim, cylindrical, pendulous flower spike usually with unisexual flowers.

Caudate Having a narrow, tail-like appendage.

Caudex Thickened, usually underground base of the stem.

Caulescent Having a well developed aerial stem.

Cauliflory Botanical term referring to plants which flower and fruit from their main stems or woody trunks. *adj.* cauliflorus.

Cauline Borne on the aerial part of a stem.

Chaffy Having thin, membranous scales in the inflorescence as in the flower heads of the sunflower family.

Chalaza The basal region of the ovule where the stalk is attached.

Chartaceous Papery, of paper-like texture.

Chasmogamous Describing flowers in which pollination takes place while the flower is open.

Chloroplast A chlorophyll-containing organelle (plastid) that gives the green colour to leaves and stems. Plastids harness light energy that is used to fix carbon dioxide in the process called photosynthesis.

Chromoplast Plastid containing colored pigments apart from chlorophyll.

Chromosomes Thread-shaped structures that occur in pairs in the nucleus of a cell, containing the genetic information of living organisms.

Cilia Hairs along the margin of a leaf or corolla lobe.

Ciliate With a fringe of hairs on the margin as of the corolla lobes or leaf.

Ciliolate Minutely ciliate.

Cilium A straight, usually erect hair on a margin or ridge. *pl.* cilia.

Cincinnus A monochasial cyme in which the lateral branches arise alternately on opposite sides of the false axis.

Circinnate Spirally coiled, with the tip innermost.

Circumscissile Opening by a transverse line around the circumference as of a fruit.

Cladode The modified photosynthetic stem of a plant whose foliage leaves are much reduced or absent. *cf.* cladophyll, phyllode.

Cladophyll A photosynthetic branch or portion of a stem that resembles and functions as a leaf, like in asparagus. *cf.* cladode, phyllode.

Clamp connection In the Basidiomycetes fungi, a lateral connection or outgrowth formed between two adjoining cells of a hypha and arching over the septum between them.

Clavate Club shaped thickened at one end refer to fruit or other organs.

Claw The conspicuously narrowed basal part of a flat structure.

Clay A naturally occurring material composed primarily of fine-grained minerals like kaolinite, montmorrillonite-smectite or illite which exhibit plasticity through a variable range of water content, and which can be hardened when dried and/or fired.

Clayey Resembling or containing a large proportion of clay.

Cleft Incised halfway down.

Cleistogamous Refers to a flower in which fertilization occurs within the bud i.e. without the flower opening. *cf.* chasmogamous.

Climber Growing more or less upwards by leaning or twining around another structure.

Clone All the plants reproduced, vegetatively, from a single parent thus having the same genetic make-up as the parent.

Coccus One of the sections of a distinctly lobed fruit which becomes separate at maturity; sometimes called a mericarp. *pl.* cocci.

Coenocarpium A fleshy, multiple pseudocarp formed from an inflorescence rather than a single flower.

Coherent Touching without organic fusion, referring to parts normally together, e.g. floral parts of the same whorl. *cf.* adherent, adnate, connate.

Collar Boundary between the above- and below ground parts of the plant axis.

Colliculate Having small elevations.

Column A structure formed by the united style, stigma and stamen(s) as in Asclepiadaceae and Orchidaceae.

Comose Tufted with hairs at the ends as of seeds.

Composite Having two types of florets as of the flowers in the sunflower family, Asteraceae.

Compost Organic matter (like leaves, mulch, manure, etc.) that breaks down in soil releasing its nutrients.

Compound Describe a leaf that is further divided into leaflets or pinnae or flower with more than a single floret.

Compressed Flattened in one plane.

Conceptacles Specialised cavities of marine algae that contain the reproductive organs.

Concolorous Uniformly coloured, as in upper and lower surfaces. *cf.* discolorous

Conduplicate Folded together lengthwise.

Cone A reproductive structure composed of an axis (branch) bearing sterile bract-like organs and seed or pollen bearing structures. Applied to Gymnospermae, Lycopodiaceae, Casuarinaceae and also in some members of Proteaceae.

Conic Cone shaped, attached at the broader end.

Conic-capitate A cone-shaped head of flowers.

Connate Fused to another structure of the same kind. *cf.* adherent, adnate, coherent.

Connective The tissue separating two lobes of an anther.

Connivent Converging.

Conspecific Within or belonging to the same species.

Contorted Twisted.

Convolute Refers to an arrangement of petals in a bud where each has one side overlapping the adjacent petal.

Cordate Heart-shaped as of leaves.

Core Central part.

Coriaceous Leathery texture as of leaves.

Corm A short, swollen, fleshy, underground plant stem that serves as a food storage organ used by some plants to survive winter or other adverse conditions

Cormel A miniature, new corm produced on a mature corm.

Corolla The inner floral whorl of a flower, usually consisting of free petals or a petals fused forming a corolla tube and corolla lobes. *adj.* corolline.

Corona A crown-like section of the staminal column, usually with the inner and outer lobes as in the **Stapelieae**.

Coroniform Crown shaped, as in the pappus of Asteraceae.

Cortex The outer of the stem or root of a plant, bounded on the outside by the epidermis and on the inside by the endodermis containing undifferentiated cells.

Corymb A flat-topped, short, broad inflorescence, in which the flowers, through unequal pedicels, are in one horizontal plane and the youngest in the centre. *adj.* corymbose

Costa A thickened, linear ridge or the midrib of the pinna in ferns. *adj.* costate.

Costapalmate Having definite costa (midrib) unlike the typical palmate leaf, but the leaflets are arranged radially like in a palmate leaf.

Cotyledon The primary seed leaf within the embryo of a seed.

Cover crop Crop grown in between trees or in fields primarily to protect the soil from erosion, to improve soil fertility and to keep off weeds.

Crenate Round-toothed or scalloped as of leaf margins.

Crenulate Minutely crenate, very strongly scalloped.

Crisped With a curled or twisted edge.

Cristate Having or forming a crest or crista.

Crozier Shaped like a shepherd's crook.

Crustaceous Like a crust; having a hard crust or shell.

Cucullate Having the shape of a cowl or hood, hooded.

Culm The main aerial stem of the Graminae (grasses, sedges, rushes and other monocots).

Culm sheath The plant casing (similar to a leaf) that protects the young bamboo shoot during growth, attached at each node of culm.

Cultigen Plant species or race known only in cultivation.

Cultivar Cultivated variety; an assemblage of cultivated individuals distinguished by any characters significant for the purposes of agriculture, forestry or horticulture, and which, when reproduced, retains its distinguishing features.

Cuneate Wedge-shaped, obtriangular.

Cupular Cup-shaped, having a cupule.

Cupule A small cup-shaped structure or organ, like the cup at the base of an acorn.

Cusp An elongated, usually rigid, acute point. *cf.* mucro.

Cuspidate Terminating in or tipped with a sharp firm point or cusp. *cf.* mucronate.

Cuspidulate Constricted into a minute cusp. *cf.* cuspidate.

Cyathiform In the form of a cup, a little widened at the top.

Cyathium A specialised type of inflorescence of plants in the genus Euphorbia and Chamaesyce in which the unisexual flowers are clustered together within a bract-like envelope. *pl.* cyathia.

Cylindric Tubular or rod shaped.

Cylindric-acuminate Elongated and tapering to a point.

Cymbiform Boat shaped, elongated and having the upper surface decidedly concave.

Cyme An inflorescence in which the lateral axis grows more strongly than the main axis with the oldest flower in the centre or at the ends. *adj.* cymose

Cymule A small cyme or one or a few flowers.

Cystidium A relatively large cell found on the hymenium of a Basidiomycete, for example, on the surface of a mushroom.

Cystocarp Fruitlike structure (sporocarp) developed after fertilization in the red algae.

Deciduous Falling off or shedding at maturity or a specific season or stage of growth.

Decompound As of a compound leaf; consisting of divisions that are themselves compound.

Decorticate To remove the bark, rind or husk from an organ; to strip of its bark; to come off as a skin.

Decumbent Prostrate, laying or growing on the ground but with ascending tips. *cf.* ascending, procumbent.

Decurrent Having the leaf base tapering down to a narrow wing that extends to the stem.

Decussate Having paired organs with successive pairs at right angles to give four rows as of leaves.

Deflexed Bent downwards.

Dehisce To split open at maturity, as in a capsule.

Dehiscent Splitting open at maturity to release the contents. *cf.* indehiscent.

Deltate Triangular shape.

Deltoid Shaped like an equilateral triangle.

Dendritic Branching from a main stem or axis like the branches of a tree.

Dentate With sharp, rather coarse teeth perpendicular to the margin.

Denticulate Finely toothed.

Diadelphous Having stamens in two bundles as in Papilionaceae flowers.

Diageotropic The tendency of growing parts, such as roots, to grow at right angle to the line of gravity.

Dichasium A cymose inflorescence in which the branches are opposite and approximately equal. *pl.* dichasia. *adj.* dichasial.

Dichotomous Divided into two parts.

Dicotyledon Angiosperm with two cotyledons.

Didymous Arranged or occurring in pairs as of anthers, having two lobes.

Digitate Having digits or fingerlike projections.

Dikaryophyses Or dendrophydia, irregularly, strongly branched terminal hyphae in the Hymenomycetes (class of Basidiomyctes) fungi.

Dimorphic Having or occurring in two forms, as of stamens of two different lengths or a plant having two kinds of leaves.

Dioecious With male and female unisexual flowers on separate plants. *cf.* monoecious.

Diploid A condition in which the chromosomes in the nucleus of a cell exist as pairs, one set being derived from the female parent and the other from the male.

Diplobiontic life cycle Life cycle that exhibits alternation of generations, which features of spore-producing multicellular sporophytes and gamete-producing multicellular gametophytes. mitoses occur in both the diploid and haploid phases.

Diplontic life cycle Or gametic meiosis, wherein instead of immediately dividing meiotically to produce haploid cells, the zygote divides mitotically to produce a multicellular diploid individual or a group of more diploid cells.

Dipterocarpous Trees of the family Dipterocarpaceae, with two-winged fruit found mainly in tropical lowland rainforest.

Disc (Botany) refers to the usually disc shaped receptacle of the flower head in Asteraceae; also the fleshy nectariferous organ usually between the stamens and ovary; also used for the enlarged style-end in Proteaceae.

Disc floret The central, tubular 4 or 5-toothed or lobed floret on the disc of an inflorescence, as of flower head of Asteraceae.

Disciform Flat and rounded in shaped. *cf.* discoid, radiate.

Discoid Resembling a disc; having a flat, circular form; disk-shaped *cf.* disciform, radiate.

Discolorous Having two colours, as of a leaf which has different colors on the two surfaces. *cf.* concolorous.

Dispersal Dissemination of seeds.

Distal Site of any structure farthest from the point of attachment. *cf.* proximal.

Distichous Referring to two rows of upright leaves in the same plane.

Dithecous Having two thecae.

Divaricate Diverging at a wide angle.

Domatium A part of a plant (e.g., a leaf) that has been modified to provide protection for other organisms. *pl.* domatia.

Dormancy A resting period in the life of a plant during which growth slows or appears to stop.

Dorsal Referring to the back surface.

Dorsifixed Attached to the back as of anthers.

Drupaceous Resembling a drupe.

Drupe A fleshy fruit with a single seed enclosed in a hard shell (endocarp) which is tissue embedded in succulent tissue (mesocarp) surrounded by a thin outer skin (epicarp). *adj.* drupaceous.

Drupelet A small drupe.

Ebracteate Without bracts.

Echinate Bearing stiff, stout, bristly, prickly hairs.

Edaphic Refers to plant communities that are distinguished by soil conditions rather than by the climate.

Eglandular Without glands. *cf.* glandular.

Ellipsoid A 3-dimensional shape; elliptic in outline.

Elliptic Having a 2-dimensional shape of an ellipse or flattened circle.

Eongate Extended, stretched out.

Emarginate Refers to leaf with a broad, shallow notch at the apex. *cf.* retuse.

Embryo (Botany) a minute rudimentary plant contained within a seed or an archegonium, composed of the embryonic axis (shoot end and root end).

Endemic Prevalent in or peculiar to a particular geographical locality or region.

Endocarp The hard innermost layer of the pericarp of many fruits.

Endosperm Tissue that surrounds and nourishes the embryo in the angiosperm seed.

Endospermous Refers to seeds having an endosperm.

Endotrophic As of mycorrhiza obtaining nutrients from inside.

Ensilage The process of preserving green food for livestock in an undried condition in airtight conditions. Also called silaging.

Entire Having a smooth, continuous margin without any incisions or teeth as of a leaf.

Entisols Soils that do not show any profile development other than an A horizon.

Ephemeral Transitory, short-lived.

Epicalyx A whorl of bracts, subtending and resembling a calyx.

Epicarp Outermost layer of the pericarp of a fruit.

Epicormic Attached to the corm.

Epicotyl The upper portion of the embryonic axis, above the cotyledons and below the first true leaves.

Epigeal Above grounds with cotyledons raised above ground.

Epiparasite An organism parasitic on another that parasitizes a third.

Epipetalous Borne on the petals, as of stamens.

Epiphyte A plant growing on, but not parasitic on, another plant, deriving its moisture and nutrients from the air and rain e.g. some Orchidaceae. *adj.* epiphytic.

Erect Upright, vertical.

Essential oils Volatile products obtained from a natural source; refers to volatile products obtained by steam or water distillation in a strict sense.

Etiolation To cause (a plant) to develop without chlorophyll by preventing exposure to sunlight.

Eutrophic Having waters rich in mineral and organic nutrients that promote a proliferation of plant life, especially algae, which reduces the dissolved oxygen content and often causes the extinction of other organisms.

Excentric Off the true centre.

Excrescence Abnormal outgrowth.

Excurrent Projecting beyond the tip, as the midrib of a leaf or bract.

Exserted Sticking out, protruding beyond some enclosing organ, as of stamens which project beyond the corolla or perianth.

Exstipulate Without stipules. *cf.* stipulate.

Extra-floral Outside the flower.

Extrose Turned outwards or away from the axis as of anthers. *cf.* introrse, latrorse.

Falcate Sickle shaped, crescent-shaped.

Fascicle A cluster or bundle of stems, flowers, stamens. *adj.* fasciculate.

Fasciclode Staminode bundles.

Fastigiate A tree in which the branches grow almost vertically.

Ferrosols Soils with an iron oxide content of greater than 5%.

Ferruginous Rust coloured, reddish-brown.

Fertile Having functional sexual parts which are capable of fertilisation and seed production. *cf.* sterile.

Filament The stalk of a stamen supporting and subtending the anther.

Filiform Having the form of or resembling a thread or filament.

Fimbriate Fringed.

Fixed oils Non volatile oils, triglycerides of fatty acids.

Flaccid Limp and weak.

Flag leaf The uppermost leaf on the stem.

Flaky In the shape of flakes or scales.

Flexuous Zig-zagging, sinuous, bending, as of a stem.

Floccose Covered with tufts of soft woolly hairs.

Floral tube A flower tube usually formed by the basal fusion of the perianth and stamens.

Floret One of the small individual flowers of sunflower family or the reduced flower of the grasses, including the lemma and palea.

Flower The sexual reproductive organ of flowering plants, typically consisting of gynoecium, androecium and perianth or calyx and/or corolla and the axis bearing these parts.

Fluted As of a trunk with grooves and folds.

Fodder Plant material, fresh or dried fed to animals.

Foliaceous Leaf-like.

Foliar Pertaining to a leaf.

Foliolate Pertaining to leaflets, used with a number prefix to denote the number of leaflets.

Foliose Leaf-like.

Follicle (Botany) a dry fruit, derived from a single carpel and dehiscing along one suture.

Forb Any herb that is not grass or grass-like.

Free central placentation The arrangement of ovules on a central column that is not connected to the ovary wall by partitions, as in the ovaries of the carnation and primrose.

Frond The leaf of a fern or cycad.

Fruit Ripened ovary with adnate parts.

Fugacious Shedding off early.

Fulvous Yellow, tawny.

Funiculus (Botany) short stalk which attaches the ovule to the ovary wall.

Fusiform A 3-dimensional shape; spindle shaped, i.e. broad in the centre and tapering at both ends thick, but tapering at both ends.

Gall-flower Short styled flower that do not develop into a fruit but are adapted for the development of a specific wasp within the fruit e.g. in the fig.

Gamete A reproductive cell that fuses with another gamete to form a zygote. Gametes are haploid, (they contain half the normal (diploid) number of chromosomes); thus when two fuse, the diploid number is restored.

Gametophyte The gamete-producing phase in a plant characterized by alternation of generations.

Gamosepalous With sepals united or partially united.

Geniculate Bent like a knee, refer to awns and filaments.

Geocarpic Where the fruit are pushed into the soil by the gynophore and mature.

Geophyte A plant that stores food in an underground storage organ e.g. a tuber, bulb or rhizome and has subterranean buds which form aerial growth.

Geotextile Are permeable fabrics which, when used in association with soil, have the ability to separate, filter, reinforce, protect, or drain.

Glabrescent Becoming glabrous.

Glabrous Smooth, hairless without pubescence.

Gland A secretory organ, e.g. a nectary, extrafloral nectary or a gland tipped, hair-like or wart-like organ. *adj.* glandular. *cf.* eglandular.

Glaucous Pale blue-green in colour, covered with a whitish bloom that rubs off readily.

Gley soils A hydric soil which exhibits a greenish-blue-grey soil color due to wetland conditions.

Globose Spherical in shape.

Globular A three-dimensional shape; spherical or orbicular; circular in outline.

Glochidiate Having glochids.

Glochidote Plant having gkochids.

Glochids Tiny, finely barbed hair-like spines found on the areoles of some cacti and other plants.

Glume One of the two small, sterile bracts at the base of the grass spikelet, called the lower and upper glumes, due to their position on the rachilla. Also used in Apiaceae, Cyperaceae for the very small bracts on the spikelet in which each flower is subtended by one floral glume. *adj.* glumaceous.

Guttation The appearance of drops of xylem sap on the tips or edges of leaves of some vascular plants, such as grasses and bamboos.

Guttule Small droplet.

Gymnosperm A group of spermatophyte seed-bearing plants with ovules on scales, which are usually arranged in cone-like structures and not borne in an ovary. *cf.* angiosperm.

Gynoecium The female organ of a flower; a collective term for the pistil, carpel or carpels.

Gynomonoecious Having female flowers and bisexual flowers on the same plant. *cf.* andromonoecious.

Gynophore Stalk that bears the pistil/carpel.

Habit The general growth form of a plant, comprising its size, shape, texture and stem orientation, the locality in which the plant grows..

Halophyte A plant adapted to living in highly saline habitats. Also a plant that accumulates high concentrations of salt in its tissues. *adj.* halophytic.

Hapaxanthic Refer to palms which flowers only once and then dies. c.f. pleonanthic.

Haploid Condition where nucleus or cell has a single set of unpaired chromosomes, the haploid number is designated as n.

Haplontic life cycle Or zygotic meiosis wherein meiosis of a zygote immediately after karyogamy, produces haploid cells which produces more or larger haploid cells ending its diploid phase.

Hastate Having the shape of an arrowhead but with the basal lobes pointing outward at right angles as of a leaf.

Hastula A piece of plant material at the junction of the petiole and the leaf blade; the hastula can be found on the top of the leaf, adaxial or the bottom, abaxial or both sides.

Heartwood Wood from the inner portion of a tree.

Heliophilous Sun-loving, tolerates high level of sunlight..

Heliotropic Growing towards sunlight.

Herb A plant which is non-woody or woody at the base only, the above ground stems usually being ephemeral. *adj.* herbaceous.

Herbaceous Resembling a herb, having a habit of a herb.

Hermaphrodite Bisexual, bearing flowers with both androecium and gynoecium in the same flower. *adj.* hermaphroditic.

Heterocyst A differentiated cyanobacterial cell that carries out nitrogen fixation.

Heterogamous Bearing separate male and female flowers, or bisexual and female flowers, or florets in an inflorescence or flower head, e.g. some Asteraceae in which the ray florets may be neuter or unisexual and the disk florets may be bisexual. *cf.* homogamous.

Heteromorphous Having two or more distinct forms. *cf.* homomorphous.

Heterophyllous Having leaves of different form.

Heterosporous Producing spores of 2 sizes, the larger giving rise to megagametophytes (female), the smaller giving rise to microgametophytes (male). Refer to the ferns and fern allies. *cf.* homosporous.

Heterostylous Having styles of two different lengths or forms.

Heterostyly The condition in which flowers on polymorphous plants have styles of different lengths, thereby facilitating cross-pollination.

Hilar Of or relating to a hilum.

Hilum The scar on a seed, indicating the point of attachment to the funiculus.

Hirsute Bearing long coarse hairs.

Hispid Bearing stiff, short, rough hairs or bristles.

Hispidulous Minutely hispid.

Histosol Soil comprising primarily of organic materials, having 40 cm or more of organic soil material in the upper 80 cm.

Hoary Covered with a greyish layer of very short, closely interwoven hairs.

Holdfast An organ or structure of attachment, especially the basal, root-like formation by which certain seaweeds or other algae are attached to a substrate.

Holocarpic Having the entire thallus developed into a fruiting body or sporangium.

Homochromous Having all the florets of the same colour in the same flower head *cf.* heterochromous.

Homogamous Bearing flowers or florets that do not differ sexually *cf.* heterogamous.

Homogenous endosperm Endosperm with even surface that lacks invaginations or infoldings of the surrounding tissue.

Homogonium A part of a filament of a cyanobacterium that detaches and grows by cell division into a new filament. *pl.* homogonia.

Homomorphous Uniform, with only one form. *cf.* heteromorphous.

Homosporous Producing one kind of spores. Refer to the ferns and fern allies. *cf.* heterosporous.

Hurd fibre Long pith fibre of the stem.

Hyaline Colourless, almost transparent.

Hybrid The first generation progeny of the sexual union of plants belonging to different taxa.

Hybridisation The crossing of individuals from different species or taxa.

Hydathode A type of secretory tissue in leaves, usually of Angiosperms, that secretes water through pores in the epidermis or margin of the leaf.

Hydrophilous Water loving; requiring water in order to be fertilized, referring to many aquatic plants.

Hygrochastic Applied to plants in which the opening of the fruits is caused by the absorption of water.

Hygrophilous Living in water or moist places.

Hymenial cystidia The cells of the hymenium develop into basidia or asci, while in others some cells develop into sterile cells called cystidia.

Hymenium Spore-bearing layer of cells in certain fungi containing asci (Ascomycetes) or basidia (Basidiomycetes).

Hypanthium Cup-like receptacles of some dicotyledonous flowers formed by the fusion of the calyx, corolla, and androecium that surrounds the ovary which bears the sepals, petals and stamens.

Hypha Is a long, branching filamentous cell of a fungus, and also of unrelated Actinobacteria. *pl.* hyphae.

Hypocotyl The portion of the stem below the cotyledons.

Hypodermis The cell layer beneath the epidermis of the pericarp.

Hypogeal Below ground as of germination of seed.

Hysteresis Refers to systems that may exhibit path dependence.

Imbricate Closely packed and overlapping. *cf.* valvate.

Imparipinnate Pinnately compound with a single terminal leaflet and hence with an odd number of leaflets. *cf.* paripinnate.

Inceptisols Old soils that have no accumulation of clays, iron, aluminium or organic matter.

Incised Cut jaggedly with very deep teeth.

Included Referring to stamens which do not project beyond the corolla or to valves which do not extend beyond the rim of a capsular fruit. *cf.* exserted.

Incurved Curved inwards; curved towards the base or apex.

Indefinite Numerous and variable in number.

Indehiscent Not opening or splitting to release the contents at maturity as of fruit. *cf.* dehiscent.

Indumentum Covering of fine hairs or bristles commonly found on external parts of plants.

Indurate To become hard, often the hardening developed only at maturity.

Indusium An enclosing membrane, covering the sorus of a fern. Also used for the modified style end or pollen-cup of some Goodeniaceae (including Brunoniaceae). *adj.* indusiate.

Inferior Said of an ovary or fruit that has sepals, petals and stamens above the ovary. *cf.* superior.

Inflated Enlarged and hollow except in the case of a fruit which may contain a seed. *cf.* swollen.

Inflexed Bent or curved inward or downward, as petals or sepals.

Inflorescence A flower cluster or the arrangement of flowers in relation to the axis and to each other on a plant.

Infrafoliar Located below the leaves.

Infraspecific Referring to any taxon below the species rank.

Infructescence The fruiting stage of an inflorescence.

Inrolled Curved inwards.

Integuments Two distinct tissue layers that surround the nucellus of the ovule, forming the testa or seed coat when mature.

Intercalary Of growth, between the apex and the base; of cells, spores, etc., between two cells.

Interfoliar Inter leaf.

Internode Portion of the stem, culm, branch, or rhizome between two nodes or points of attachment of the leaves.

Interpetiolar As of stipules positioned between petioles of opposite leaves.

Intrastaminal Within the stamens.

Intricate Entangled, complex.

Introduced Not indigenous; not native to the area in which it now occurs.

Introrse Turned inwards or towards the axis or pistil as of anthers. *cf.* extrorse, latrorse.

Involucre A whorl of bracts or leaves that surround one to many flowers or an entire inflorescence.

Involute Having the margins rolled inwards, referring to a leaf or other flat organ.

Jugate Of a pinnate leaf; having leaflets in pairs.

Juvenile Young or immature, used here for leaves formed on a young plant which are different in morphology from those formed on an older plant.

Keel A longitudinal ridge, at the back of the leaf. Also the two lower fused petals of a 'pea' flower in the Papilionaceae, which form a boat-like structure around the stamens and styles, also called carina. *adj.* keeled. *cf.* standard, wing.

Labellum The modified lowest of the three petals forming the corolla of an orchid, usually larger than the other two petals, and often spurred.

Laciniate Fringed; having a fringe of slender, narrow, pointed lobes cut into narrow lobes.

Lamella A gill-shaped structure: fine sheets of material held adjacent to one another.

Lamina The blade of the leaf or frond.

Lanate Wooly, covered with long hairs which are loosely curled together like wool.

Lanceolate Lance-shaped in outline, tapering from a broad base to the apex.

Landrace Plants adapted to the natural environment in which they grow, developing naturally with minimal assistance or guidance from humans and usually possess more diverse phenotypes and genotypes. They have not been improved by formal breeding programs.

Laterite Reddish–coloured soils rich in iron oxide, formed by weathering of rocks under oxidizing and leaching conditions, commonly found in tropical and subtropical regions. *adj.* lateritic.

Latex A milky, clear or sometimes coloured sap of diverse composition exuded by some plants.

Latrorse Turned sideways, i.e. not towards or away from the axis as of anthers dehiscing longitudinally on the side. *cf.* extrorse, introse.

Lax Loose or limp, not densely arranged or crowded.

Leaflet One of the ultimate segments of a compound leaf.

Lectotype A specimen chosen after the original description to be the type.

Lemma The lower of two bracts (scales) of a grass floret, usually enclosing the palea, lodicules, stamens and ovary.

Lenticel Is a lens shaped opening that allows gases to be exchanged between air and the inner tissues of a plant, commonly found on young bark, or the surface of the fruit.

Lenticellate Dotted with lenticels.

Lenticular Shaped like a biconvex lens. *cf.* lentiform.

Lentiform Shaped like a biconvex lens, *cf.* lenticular.

Leptomorphic Temperate, running bamboo rhizome; usually thinner then the culms they support and the internodes are long and hollow.

Liane A woody climbing or twining plant.

Lignotuber A woody, usually underground, tuberous rootstock often giving rise to numerous aerial stems.

Ligulate Small and tongue shaped or with a little tongue shaped appendage or ligule, star shaped as of florets of Asteraceae.

Ligule A strap-shaped corolla in the flowers of Asteraceae; also a thin membranous outgrowth from the inner junction of the grass leaf sheath and blade. *cf.* ligulate.

Limb The expanded portion of the calyx tube or the corolla tube, or the large branch of a tree.

Linear A 2-dimensional shape, narrow with nearly parallel sides.

Linguiform Tongue shaped *cf.* ligulate.

Lithosol A kind of shallow soils lacking well-defined horizons and composed of imperfectly weathered fragments of rock.

Littoral Of or on a shore, especially seashore.

Loam A type of soil mad up of sand, silt, and clay in relative concentration of 40–40–20% respectively.

Lobed Divided but not to the base.

Loculicidal Opening into the cells, when a ripe capsule splits along the back.

Loculus Cavity or chamber of an ovary. *pl.* loculi.

Lodicules Two small structures below the ovary which, at flowering, swell up and force open the enclosing bracts, exposing the stamens and carpel.

Lyrate Pinnately lobed, with a large terminal lobe and smaller laterals ones which become progressively smaller towards the base.

Macronutrients Chemical elements which are needed in large quantities for growth and development by plants and include nitrogen, phosphorus, potassium, and magnesium.

Maculate Spotted.

Mallee A growth habit in which several to many woody stems arise separately from a lignotuber; usually applied to certain low-growing species of *Eucalyptus.*

Mangrove A distinctive vegetation type of trees and shrubs with modified roots, often viviparous, occupying the saline coastal habitats that are subject to periodic tidal inundation.

Marcescent Withering or to decay without falling off.

Margin The edge of the leaf blade.

Medulla The pith in the stems or roots of certain plants; or the central portion of a thallus in certain lichens.

Megasporangium The sporangium containing megaspores in fern and fern allies. *cf.* microsporangium.

Megaspore The large spore which may develop into the female gametophyte in heterosporous ferns and fern allies. *cf.* microspore.

Megasporophyll A leaflike structure that bears megasporangia.

Megastrobilus Female cone, seed cone, or ovulate cone) contains ovules within which, when fertilized by pollen, become seeds. The female cone structure varies more markedly between the different conifer families.

Meiosis The process of cell division that results in the formation of haploid cells from diploid cells to produce gametes.

Mericarp A 1-seeded portion of an initially syncarpous fruit (schizocarp) which splits apart at maturity. *Cf.* coccus.

Meristem The region of active cell division in plants, from which permanent tissue is derived. *adj.* meristematic

-merous Used with a number prefix to denote the basic number of the 3 outer floral whorls, e.g. a 5-merous flower may have 5 sepals, 10 petals and 15 stamens.

Mesic Moderately wet.

Mesocarp The middle layer of the fruit wall derived from the middle layer of the carpel wall. *cf.* endocarp, exocarp, pericarp.

Mesophytes Terrestrial plants which are adapted to neither a particularly dry nor particularly wet environment.

Micropyle The small opening in a plant ovule through which the pollen tube passes in order to effect fertilisation.

Microsporangium The sporangium containing microspores in petridophyes. *cf.* megasporangium.

Microspore A small spore which gives rise to the male gametophyte in heterosporous pteridophytes. Also for a pollen grain. *cf.* megaspore.

Midvein The main vascular supply of a simple leaf blade or lamina. Also called mid-rib.

Mitosis Is a process of cell division which results in the production of two daughter cells from a single parent cell.

Mollisols Soils with deep, high organic matter, nutrient-enriched surface soil (A horizon), typically between 60 and 80 cm thick.

Monadelphous Applied to stamens united by their filaments into a single bundle.

Monocarpic Refer to plants that flower, set seeds and then die.

Monochasial A cyme having a single flower on each axis.

Monocotyledon Angiosperm having one cotyledon.

Monoecious Having both male and female unisexual flowers on the same individual plant. *cf.* dioecious.

Monoembryonic seed The seed contains only one embryo, a true sexual (zygotic) embryo. polyembryonic seed.

Monolete A spore that has a simple linear scar.

Monopodial With a main terminal growing point producing many lateral branches progressively. *cf.* sympodial.

Monotypic Of a genus with one species or a family with one genus; in general, applied to any taxon with only one immediately subordinate taxon.

Montane Refers to highland areas located below the subalpine zone.

Mucilage A soft, moist, viscous, sticky secretion. *adj.* mucilaginous.

Mucous (Botany) slimy.

Mucro A sharp, pointed part or organ, especially a sharp terminal point, as of a leaf.

Mucronate Ending with a short, sharp tip or mucro, resembling a spine. cf. cuspidate, muticous.

Mucronulate With a very small mucro; a diminutive of mucronate.

Mulch Protective cover of plant (organic) or non-plant material placed over the soil, primarily to modify and improve the effects of the local microclimate and to control weeds.

Multiple fruit A fruit that is formed from a cluster of flowers.

Muricate Covered with numerous short hard outgrowths. *cf.* papillose.

Muriculate With numerous minute hard outgrowths; a diminutive of muricate.

Muticous Blunt, lacking a sharp point. *cf.* mucronate.

MYB proteins Are a superfamily of transcription factors that play regulatory roles in developmental processes and defense responses in plants.

Mycorrhiza The mutualistic symbiosis (non-pathogenic association) between soil-borne fungi with the roots of higher plants.

Mycorrhiza (vesicular arbuscular) Endomycorrhiza living in the roots of higher plants producing inter-and intracellular fungal growth in root cortex and forming specific fungal structures, referred to as vesicles and arbuscles. *abbrev.* VAM.

Native A plant indigenous to the locality or region.

Naviculate Boat-shaped.

Necrotic Applied to dead tissue.

Nectariferous Having one or more nectaries.

Nectary A nectar secretory gland; commonly in a flower, sometimes on leaves, fronds or stems.

Nervation Venation, a pattern of veins or nerves as of leaf.

Node The joint between segments of a culm, stem, branch, or rhizome; the point of the stem that gives rise to the leaf and bud.

Nodule A small knoblike outgrowth, as those found on the roots of many leguminous, that containing *Rhizobium* bacteria which fixes nitrogen in the soil.

Nomen Dubium An invalid proposed taxonomic name because it is not accompanied by a definition or description of the taxon to which it applies. *abbrev*. nom. dub.

Nucellus Central portion of an ovule in which the embryo sac develops.

Nomen Illegitimum Illegitimate taxon deemed as superfluous at its time of publication either because the taxon to which it was applied already has a name, or because the name has already been applied to another plant. *abbrev*. nom. illeg.

Nomen Nudum The name of a taxon which has never been validated by a description. *abbrev*. nom. nud.

Nucellar embryony A form of seed reproduction in which the nucellar tissue which surrounds the embryo sac can produce additional embryos (polyembryony) which are genetically identical to the parent plant. This is found in many citrus species and in mango.

Nut A dry indehiscent 1-celled fruit with a hard pericarp.

Nutlet A small. 1-seeded, indehiscent lobe of a divided fruit.

Ob- Prefix meaning inversely or opposite to.

Obconic A 3-dimensional shape; inversely conic; cone shaped, conic with the vertex pointing downward.

Obcordate Inversely cordate, broad and notched at the tip; heart shaped but attached at the pointed end.

Obdeltate Inversely deltate; deltate with the broadest part at the apex.

Oblanceolate Inversely lanceolate, lance-shaped but broadest above the middle and tapering toward the base as of leaf.

Oblate Having the shape of a spheroid with the equatorial diameter greater than the polar diameter; being flattened at the poles.

Oblong Longer than broad with sides nearly parallel to each other.

Obovate Inversely ovate, broadest above the middle.

Obpyramidal Resembling a 4-sided pyramid attached at the apex with the square base facing away from the attachment.

Obpyriform Inversely pyriform, resembling a pear which is attached at the narrower end. *cf.* pyriform.

Obspathulate Inversely spathulate; resembling a spoon but attached at the broadest end. *cf.* spathulate.

Obtriangular Inversely triangular; triangular but attached at the apex. *cf.* triangular.

Obtrullate Inversely trullate; resembling a trowel blade with the broadest axis above the middle. *cf.* trullate.

Obtuse With a blunt or rounded tip, the converging edges separated by an angle greater than 90 degrees.

-oid Suffix denoting a 3-dimensional shape, e.g. spheroid.

Ochraceous A dull yellow color.

Ocreate Having a tube-like covering around some stems, formed of the united stipules; sheathed.

Oleaginous Oily.

Oligotrophic Lacking in plant nutrients and having a large amount of dissolved oxygen throughout.

Operculum A lid or cover that becomes detached at maturity by abscission, e.g. in *Eucalyptus*, also a cap or lid covering the bud and formed by fusion or cohesion of sepals and/or petals. *adj.* operculate.

Opposite Describing leaves or other organs which are borne at the same level but on opposite sides of the stem. *cf.* alternate.

Orbicular Of circular outline, disc-like.

Order A taxonomic rank between class and family used in the classification of organisms, i.e. a group of families believed to be closely related.

Orifice An opening or aperture.

Organosols Soils not regularly inundated by marine waters and containing a specific thickness of organic materials within the upper part of the profile.

Ovary The female part of the pistil of a flower which contains the ovules (immature seeds).

Ovate Egg-shaped, usually with reference to two dimensions.

Ovoid Egg-shaped, usually with reference to three dimensions.

Ovule The young, immature seed in the ovary which becomes a seed after fertilisation. *adj.* ovular..

Ovulode A sterile reduced ovule borne on the placenta, commonly occurring in Myrtaceae.

Oxisols Refer to ferralsols.

Pachymorphic Describes the short, thick, rhizomes of clumping bamboos with short, thick and solid internode (except the bud-bearing internodes, which are more elongated). *cf.* sympodial.

Palate (Botany) a raised appendage on the lower lip of a corolla which partially or completely closes the throat.

Palea The upper of the two membraneous bracts of a grass floret, usually enclosing the lodicules, stamens and ovary. *pl.* paleae. *adj.* paleal. *cf.* lemma.

Paleate Having glumes.

Palm heart Refers to soft, tender inner core and growing bud of certain palm trees which are eaten as vegetables. Also called heart of palm, palmito, burglar's thigh, chonta or swamp cabbage.

Palmate Describing a leaf which is divided into several lobes or leaflets which arise from the same point. *adj.* palmately.

Palmito See palm heart.

Palustrial Paludal, swampy, marshy.

Palustrine Marshy, swampy.

Palustrine herb Vegetation that is rooted below water but grows above the surface in wetland system.

Panduriform Fiddle shaped, usually with reference to two dimensions.

Panicle A compound, indeterminate, racemose inflorescence in which the main axis bears lateral racemes or spikes. *adj.* paniculate.

Pantropical Distributed through-out the tropics.

Papilionaceous Butterfly-like, said of the pea flower or flowers of Papilionaceae, flowers which are zygomorphic with imbricate petals, one broad upper one, two narrower lateral ones and two narrower lower ones.

Papilla A small, superficial protuberance on the surface of an organ being an outgrowth of one epidermal cell. *pl.* papillae. *adj.* papillose.

Papillate Having papillae.

Papillose Covered with papillae.

Pappus A tuft (or ring) of hairs, bristles or scales borne above the ovary and outside the corolla as in Asteraceae often persisting as a tuft of hairs on a fruit. *adj.* pappose.

Papyraceous Resembling parchment of paper.

Parenchyma Undifferentiated plant tissue composed of more or less uniform cells.

Parietal Describes the attachment of ovules to the outer walls of the ovaries.

Paripinnate Pinnate with an even number of leaflets and without a terminal leaflet. *cf.* imparipinnate.

-partite Divided almost to the base into segments, the number of segments written as a prefix.

Patelliform Shaped like a limpet shell; cap-shaped and without whorls.

Patent Diverging from the axis almost at right angles.

Peat Is an accumulation of partially decayed vegetation matter.

Pectin A group of water-soluble colloidal carbohydrates of high molecular weight found in certain ripe fruits.

Pectinate Pinnatifid with narrow segments resembling the teeth of a comb.

Pedicel The stalk of the flower or stalk of a spikelet in Poaceae. *adj.* pedicellate.

Pedicellate Having pedicel.

Peduncle A stalk supporting an inflorescence. *adj.* pedunculate

Pellucid Allowing the passage of light; transparent or translucent.

Pellucid-dotted Copiously dotted with immersed, pellucid, resinous glands.

Peltate With the petiole attached to the lower surface of the leaf blade.

Pendant Hanging down.

Pendulous Drooping, as of ovules.

Penniveined or penni-nerved Pinnately veined.

Pentamerous In five parts.

Perennial A plant that completes it life cycle or lives for more than 2 years. *cf.* annual, biennial.

Perfoliate A leaf with the basal lobes united around – and apparently pierced by – the stem.

Pergamentaceous Parchment-like.

Perianth The two outer floral whorls of the Angiosperm flower; commonly used when the calyx and the corolla are not readily distinguishable (as in monocotyledons).

Pericarp (Botany). The wall of a ripened ovary; fruit wall composed of the exocarp, mesocarp and endocarp.

Persistent Remaining attached; not falling off. *cf.* caduceus.

Petal Free segment of the corolla. *adj.* petaline. *cf.* lobe.

Petiolar relating to the petiole.

Petiolate Having petiole.

Petiole Leaf stalk. *adj.* petiolate.

Petiolulate Supported by its own petiolule.

Petiolule The stalk of a leaflet in a compound leaf. adj. petiolulate.

pH Is a measure of the acidity or basicity of a solution. It is defined as the cologarithm of the activity of dissolved hydrogen ions (H+).

Phenology The study of periodic plant life cycle events as influenced by seasonal and interannual variations in climate.

Phyllary A bract of the involucre of a composite plant, term for one of the scale-like bracts beneath the flower-head in Asteraceae.

Phylloclade A flattened, photosynthetic branch or stem that resembles or performs the function of a leaf, with the true leaves represented by scales.

Phyllode A petiole that function as a leaf. *adj.* phyllodineous. *cf.* cladode.

Phyllopodia Refer to the reduced, scale-like leaves found on the outermost portion of the corm where they seem to persist longer than typical sporophylls as in the fern Isoetes.

Phytoremediation Describes the treatment of environmental problems (bioremediation) through the use of plants which mitigate the environmental problem without the need to excavate the contaminant material and dispose of it elsewhere.

Pileus (Botany) cap of mushroom.

Piliferous (Botany) bearing or producing hairs, as of an organ with the apex having long, hair-like extensions.

Pilose Covered with fine soft hairs.

Pinna A primary division of the blade of a compound leaf or frond. *pl.* pinnae.

Pinnate Bearing leaflets on each side of a central axis of a compound leaf; divided into pinnae.

Pinnatifid, pinnatilobed A pinnate leaf parted approximately halfway to midrib; when divided to almost to the mid rib described as deeply pinnatifid or pinnatisect.

Pinnatisect Lobed or divided almost to the midrib.

Pinnule A leaflet of a bipinnate compound leaf.

Pistil Female part of the flower comprising the ovary, style, and stigma.

Pistillate Having one or more pistils; having pistils but no stamens.

Placenta The region within the ovary to which ovules are attached. *pl.* placentae.

Placentation The arrangement of the placentae and ovules in the ovary.

Plano- A prefix meaning level or flat.

Pleonanthic Refer to palms in which the stem does not die after flowering.

Plicate Folded like a fan.

Plumose Feather-like, with fine hairs arising laterally from a central axis; feathery.

Pneumatophore Modified root which allows gaseous exchange in mud-dwelling shrubs, e.g. mangroves.

Pod A dry 1 to many-seeded dehiscent fruit, as applied to the fruit of Fabaceae ie. Caesalpiniaceae, Mimosaceae and Papilionaceae.

Podzol, Podsolic soil Any of a group of acidic, zonal soils having a leached, light-coloured, gray and ashy appearance. Also called spodosol.

Pollen cone Male cone or microstrobilus or pollen cone is structurally similar across all conifers, extending out from a central axis are microsporophylls (modified leaves). Under each microsporophyll is one or several microsporangia (pollen sacs).

Pollinia The paired, waxy pollen masses of flowers of orchids and milkweeds.

Polyandrous (Botany) having an indefinite number of stamens.

Polyembryonic seed Seeds contain many embryos, most of which are asexual (nucellar) in origin and genetically identical to the maternal parent.

Polygamous With unisexual and bisexual flowers on the same or on different individuals of the same species.

Polymorphic With different morphological variants.

Polypetalous (Botany) having a corolla composed of distinct, separable petals.

Pome A fleshy fruit where the succulent tissues are developed from the receptacle.

Pore A tiny opening.

Premorse Abruptly truncated, as though bitten or broken off as of a leaf.

Procumbent Trailing or spreading along the ground but not rooting at the nodes, referring to stems. *cf.* ascending, decumbent, erect.

Prophyll A plant structure that resembles a leaf.

Prostrate Lying flat on the ground.

Protandous Relating to a flower in which the anthers release their pollen before the stigma of the same flower becomes receptive.

Proximal End of any structure closest to the point of attachment. *cf.* distal.

Pruinose Having a thick, waxy, powdery coating or bloom.

Pseudocarp A false fruit, largely made up of tissue that is not derived from the ovary but from floral parts such as the receptacle and calyx.

Pseudostem The false, herbaceous stem of a banana plant composed of overlapping leaf bases.

Pteridophyte A vascular plant which reproduces by spores; the ferns and fern allies.

Puberulent Covered with minute hairs or very fine down; finely pubescent.

Puberulous Covered with a minute down.

Pubescent Covered with short, soft hairs.

Pulvinate Having a swelling, pulvinus at the base as a leaf stalk.

Pulviniform Swelling or bulging.

Pulvinus Swelling at the base of leaf stalk.

Punctate Marked with translucent dots or glands.

Punctiform Marked by or composed of points or dots.

Punctulate Marked with minute dots; a diminutive of punctate.

Pusticulate Characterized by small pustules.

Pyrene The stone or pit of a drupe, consisting of the hardened endocarp and seed.

Pyriform Pear-shaped, a 3-dimensional shape; attached at the broader end. *cf.* obpyriform.

Pyxidium Seed capsule having a circular lid (operculum) which falls off to release the seed.

Raceme An indeterminate inflorescence with a simple, elongated axis and pedicellate flowers, youngest at the top. *adj.* racemose.

Rachilla The main axis of a grass spikelet.

Rachis The main axis of the spike or other inflorescence of grasses or a compound leaf.

Radiate Arranged around a common centre; as of an inflorescence of Asteraceae with marginal, female or neuter, ligulate ray-florets and central, perfect or functionally male, tubular, disc florets. *cf.* disciform, discoid.

Radical Arising from the root or its crown, or the part of a plant embryo that develops into a root.

Ray The marginal portion of the inflorescence of Asteraceae and Apiaceae when distinct from the disc. Also, the spreading branches of a compound umbel.

Receptacle The region at the end of a pedicel or on an axis which bears one or more flowers. *adj.* receptacular.

Recurved Curved downwards or backwards.

Reflexed Bent or turned downward.

Regosol Soil that is young and undeveloped, characterized by medium to fine-textured unconsolidated parent material that maybe alluvial in origin and lacks a significant horizon layer formation.

Reniform Kidney shaped in outline.

Repand With slightly undulate margin.

Replicate Folded back, as in some corolla lobes.

Resinous Producing sticky resin.

Resupinate Twisted through 180 degrees.

Reticulate Having the appearance of a network.

Retrorse Bent or directed downwards or backwards. *cf.* antrorse.

Retuse With a very blunt and slightly notched apex. *cf.* emarginated.

Revolute With the margins inrolled on the lower (abaxial) surface.

Rhizine A root-like filament or hair growing from the stems of mosses or on lichens.

Rhizoid Root-like filaments in a moss, fern, fungus, etc. that attach the plant to the substratum.

Rhizome A prostrate or underground stem consisting of a series of nodes and internodes with adventitious roots and which generally grows horizontally.

Rhizophore A stilt-like outgrowth of the stem which branches into roots on contact with the substrate.

Rhombic Shaped like a rhombus.

Rhomboid Shaped like a rhombus.

Rib A distinct vein or linear marking, often raised as a linear ridge.

Riparian Along the river margins, interface between land and a stream.

Rosette A tuft of leaves or other organs arranged spirally like petals in a rose, ranging in form from a hemispherical tuft to a flat whorl. *adj.* rosetted, rosulate.

Rostrate Beaked; the apex tapered into a slender, usually obtuse point.

Rostrum A beak-like extension.

Rosulate Having a rosette.

Rotate Wheel shaped; refers to a corolla with a very short tube and a broad upper part which is flared at right angles to the tube. *cf.* salverform.

Rotundate Rounded; especially at the end or ends.

Rugae Refers to a series of ridges produced by folding of the wall of an organ.

Rugose Deeply wrinkled.

Rugulose Finely wrinkled.

Ruminate (Animal) chew repeatedly over an extended period.

Ruminate endosperm Uneven endosperm surface that is often highly enlarged by ingrowths or infoldings of the surrounding tissue. cf. homogenous endosperm.

Rz value Is a numerical reference to the mesh/emulsion equalization on the screen.

Saccate Pouched.

Sagittate Shaped like an arrow head.

Saline soils Soils that contain excessive levels of salts that reduce plant growth and vigor by altering water uptake and causing ion-specific toxicities or imbalances.

Salinity Is characterised by high electrical conductivities and low sodium ion concentrations compared to calcium and magnesium

Salverform Applies to a gamopetalous corolla having a slender tube and an abruptly expanded limb.

Samara An indehiscent, winged, dry fruit.

Sand A naturally occurring granular material composed of finely divided rock and mineral particles range in diameter from 0.0625 μm to 2 mm. *adj.* sandy

Saponins Are plant glycosides with a distinctive foaming characteristic. They are found in many plants, but get their name from the soapwort plant (*Saponaria*).

Saprophytic Living on and deriving nourishment from dead organic matter.

Sapwood Outer woody layer of the tree just adjacent to and below the bark.

Sarcotesta Outermost fleshy covering of Cycad seeds below which is the sclerotesta.

Scabrid Scurfy, covered with surface abrasions, irregular projections or delicate scales.

Scabrous Rough to the touch.

Scale Dry bract or leaf.

Scandent Refer to plants, climbing.

Scape Erect flowering stem, usually leafless, rising from the crown or roots of a plant. *adj.* scapose.

Scapigerous With a scape.

Scarious Fry, thin and membranous.

Schizocarp A dry fruit which splits into longitudinally multiple parts called mericarps or cocci. *adj.* schizocarpous.

Sclerotesta The innermost fleshy coating of cycad seeds, usually located directly below the sarcotesta.

Scorpoid Refers to a cymose inflorescence in which the main axis appears to coil.

Scutellum (Botany) any of various parts shaped like a shield.

Secondary venation Arrangement of the lateral veins arising from the midrib in the leaf lamina.

Secund With the flowers all turned in the same direction.

Sedge A plant of the family Apiaceae, Cyperaceae.

Segmented Constricted into divisions.

Seminal root Or seed root originate from the scutellar node located within the seed embryo and are composed of the radicle and lateral seminal roots.

Senescence Refers to the biological changes which take place in plants as they age.

Sensu lato In a broad or wide sense.

Sensu stricto In a narrow or strict sense.

Sepal Free segment of the calyx. *adj.* sepaline.

Septum A partition or cross wall. *pl.* septa. *adj.* septate.

Seriate Arranged in rows.

Sericeous Silky; covered with close-pressed, fine, straight silky hairs.

Serrate Toothed like a saw; with regular, asymmetric teeth pointing forward.

Serrated Toothed margin.

Serratures Serrated margin.

Serrulate With minute teeth on the margin.

Sessile Without a stalk.

Seta A bristle or stiff hair. *pl.* setae. *adj.* setose, setaceous.

Setaceous Bristle-like.

Setate With bristles.

Setiform Bristle shaped.

Setulose With minute bristles.

Sheathing Clasping or enveloping the stem.

Shrub A woody plant usually less than 5 m high and many-branched without a distinct main stem except at ground level.

Silicula A broad, dry, usually dehiscent fruit derived from two or more carpels which usually dehisce along two sutures. *cf.* siliqua.

Siliqua A silicula which is at least twice as long as broad.

Silt Is soil or rock derived granular material of a grain size between sand and clay, grain particles ranging from 0.004 to 0.06 mm in diameter. *adj.* silty.

Simple Refer to a leaf or other structure that is not divided into parts. *cf.* compound.

Sinuate With deep wavy margin.

Sinuous Wavy.

Sinus An opening or groove, as occurs between the bases of two petals.

Sodic soils Contains high levels of sodium salts that affects soil structure, inhibits water movement and causes poor germination and crop establishment and plant toxicity.

Sodicity Is characterised by low electrical conductivities and high sodium ion concentrations compared to calcium and magnesium.

Soil pH Is a measure of the acidity or basicity of the soil. See pH.

Solitary Usually refer to flowers which are borne singly, and not grouped into an inflorescence or clustered.

Sorocarp Fruiting body formed by some cellular slime moulds, has both stalk and spore mass.

Sorophore Stalk bearing the sorocarp.

Sorosis Fleshy multiple fruit formed from flowers that are crowded together on a fleshy stem e.g. pineapple and mulberry.

Sorus A discrete aggregate of sporangia in ferns. *pl.* sori

Spadix Fleshy spike-like inflorescence with an unbranched, usually thickened axis and small embedded flowers often surrounded by a spathe. *pl.* spadices.

Spathe A large bract ensheathing an inflorescence or its peduncle. *adj.* spathaceous.

Spatheate Like or with a spathe.

Spathulate Spatula or spoon shaped; broad at the tip and narrowed towards the base.

Spicate Borne in or forming a spike.

Spiculate Spikelet-bearing.

Spike An unbranched, indeterminate inflorescence with sessile flowers or spiklets. *adj.* spicate, spiciform.

Spikelet A small or secondary spike characteristics of the grasses and sedges and, generally composed of 2 glumes and one or more florets. Also applied to the small spike-like inflorescence or inflorescence units commonly found in Apiaceae.

Spine A stiff, sharp, pointed structure, formed by modification of a plant organ. *adj.* spinose.

Spinescent Ending in a spine; modified to form a spine

Spinulate Covered with small spines.

Spinulose With small spines over the surface.

Spodosol See podsol.

Sporangium A spore bearing structure found in ferns, fern allies and gymnosperms. *pl.* sporangia. *adj.* sporangial.

Sporidia Asexual spores of smut fungi.

Sporocarp A stalked specialized fruiting structure formed from modified sporophylls, containing sporangia or spores as found in ferns and fern allies.

Sporophore A spore-bearing structure, especially in fungi.

Sporophyll A leaf or bract which bears or subtends sporangia in the fern allies, ferns and gymnosperms.

Sporophyte The spore-producing phase in the life cycle of a plant that exhibits alternation of generations.

Spreading Bending or spreading outwards and horizontally.

Spur A tubular or saclike extension of the corolla or calyx of a flower.

Squama Structure shaped like a fish scale. *pl.* squamae.

Squamous Covered in scales.

Squarrose Having rough or spreading scale-like processes.

Stamen The male part of a flower, consisting typically of a stalk (filament) and a pollen-bearing portion (anther). *adj.* staminal, staminate .

Staminate Unisexual flower bearing stamens but no functional pistils.

Staminode A sterile or abortive stamen, often reduced in size and lacking anther. *adj.* staminodial.

Standard Refers to the adaxial petal in the flower of Papilionaceae. cf. keel, wing.

Starch A polysaccharide carbohydrate consisting of a large number of glucose units joined together by glycosidic bonds α-1-4 linkages.

Stellate Star shaped, applies to hairs.

Stem The main axis of a plant, developed from the plumule of the embryo and typically bearing leaves.

Sterile Lacking any functional sexual parts which are capable of fertilisation and seed production.

Stigma The sticky receptive tip of an ovary with or without a style which is receptive to pollen.

Stilt root A supporting root arising from the stem some distance above the ground as in some mangroves, sometimes also known as a prop root.

Stipe A stalk that support some other structure like the frond, ovary or fruit.

Stipel Secondary stipule at the base of a leaflet. *pl.* stipellae. *adj.* stipellate.

Stipitate Having a stalk or stipe, usually of an ovary or fruit.

Stipulated Having stipules.

Stipule Small leaf-like, scale-like or bristle-like appendages at the base of the leaf or on the petiole. *adj.* stipulate.

Stolon A horizontal, creeping stem rooting at the nodes and giving rise to another plant at its tip.

Stoloniferous Bearing stolon or stolons.

Stoma A pore in the epidermis of the leaf or stem for gaseous exchange. *pl.* stomata.

Stone The hard endocarp of a drupe, containing the seed or seeds.

Stramineous Chaffy; straw-liked.

Striae Parallel longitudinal lines or ridges. *adj.* striate.

Striate Marked with fine longitudinal parallel lines or ridges.

Strigose Bearing stiff, straight, closely appressed hair; often the hairs have swollen bases.

Strobilus A cone-like structure formed from sporophylls or sporangiophores. *pl.* strobili

Style The part of the pistil between the stigma and ovary.

Sub- A prefix meaning nearly or almost, as in subglobose or subequal.

Subcarnose Nearly fleshy.

Sub-family Taxonomic rank between the family and tribe.

Subglobose Nearly spherical in shape.

Subretuse Faintly notched at the apex.

Subsessile Nearly stalkless or sessile.

Subshrub Intermediate between a herb and shrub.

Subspecies A taxonomic rank subordinate to species.

Substrate Surface on which a plant or organism grows or attached to.

Subtend Attached below something.

Subulate Narrow and tapering gradually to a fine point, awl-shaped.

Succulent Fleshy, juicy, soft in texture and usually thickened.

Suckers Young plants sprouting from the underground roots of a parent plant and appearing around the base of the parent plant.

Sulcate Grooved longitudinally with deep furrows.

Sulcus A groove or depression running along the internodes of culms or branches.

Superior Refers to the ovary is free and mostly above the level of insertion of the sepals, and petals. *cf.* inferior.

Suture Line of dehiscence.

Swidden Slash-and-burn or shifting cultivation.

Syconium A type of pseudocarp formed from a hollow receptacle with small flowers attached to the inner wall. After fertilization the ovaries of the female flowers develop into one-seeded achenes, e.g. fig.

Symbiosis Describes close and often long-term mutualistic and beneficial interactions between different organisms.

Sympetalous Having petals united.

Sympodial Refers to a specialized lateral growth pattern in which the apical meristem. *cf* monopodial.

Synangium An organ composed of united sporangia, divided internally into cells, each containing spores. *pl.* synangia.

Syncarp An aggregate or multiple fruit formed from two or more united carpels with a single style. *adj.* syncarpous.

Syncarpous Carpels fused forming a compound pistil.

Tannins Group of plant-derived phenolic compounds.

Taxon The taxonomic group of plants of any rank. e.g. a family, genus, species or any infraspecific category. *pl.* taxa.

Tendril A slender, threadlike organ formed from a modified stem, leaf or leaflet which, by coiling around objects, supports a climbing plant.

Tepal A segment of the perianth in a flower in which all the perianth segments are similar in appearance, and are not differentiated into calyx and corolla; a sepal or petal.

Terete Having a circular shape when cross-sectioned or a cylindrical shape that tapers at each end.

Terminal At the apex or distal end.

Ternate In threes as of leaf with 3 leaflets.

Testa A seed coat, outer integument of a seed.

Tetrasporangium A sporangium containing four haploid spores as found in some algae.

Thallus Plant body of algae, fungi, and other lower organisms.

Thyrse A dense, panicle-like inflorescence, as of the lilac, in which the lateral branches terminate in cymes.

Tomentellose Mildly tomentose.

Tomentose Refers to plant hairs that are bent and matted forming a wooly coating.

Torus Receptacle of a flower.

Transpiration Evaporation of water from the plant through leaf and stem pores.

Tree That has many secondary branches supported clear of the ground on a single main stem or trunk.

Triangular Shaped like a triangle, 3-angled and 3-sided.

Tribe A category intermediate in rank between subfamily and genus.

Trichome A hair-like outgrowth of the epidermis.

Trichotomous Divided almost equally into three parts or elements.

Tridentate Three toothed or three pronged.

Trifid Divided or cleft into three parts or lobes.

Trifoliate Having three leaves.

Trifoliolate A leaf having three leaflets.

Trifurcate Having three forks or branches.

Trigonous Obtusely three-angled; triangular in cross-section with plane faces.

Tripartite Consisting of three parts.

Tripinnate Relating to leaves, pinnately divided three times with pinnate pinnules.

Tripliveined Main laterals arising above base of lamina.

Triploid Describing a nucleus or cell that has three times (3n) the haploid number (n) of chromosomes.

Triveined Main laterals arising at the base of lamina.

Triquetrous Three-edged; acutely 3-angled.

Trullate With the widest axis below the middle and with straight margins; ovate but margins straight and angled below middle, trowel-shaped.

Truncate With an abruptly transverse end as if cut off.

Tuber A stem, usually underground, enlarged as a storage organ and with minute scale-like leaves and buds. *adj.* tuberous.

Tubercle A wart-like protuberance. *adj.* tuberculate.

Tuberculate Bearing tubercles; covered with warty lumps.

Tuberization Formation of tubers in the soil.

Tuft A densely packed cluster arising from an axis. *adj.* tufted.

Turbinate Having the shape of a top; cone-shaped, with the apex downward, inversely conic.

Turgid Distended by water or other liquid.

Turion The tender young, scaly shoot such as asparagus, developed from an underground bud without branches or leaves.

Turnery Articles made by the process of turning.

Twining Winding spirally.

Ultisols Mineral soils with no calcareous material, have less than 10% weatherable minerals in the extreme top layer of soil, and with less the 35% base saturation throughout the soil.

Umbel An inflorescence of pedicellate flowers of almost equal length arising from one point on top of the peduncle. *adj.* umbellate.

Umbellet A secondary umbel of a compound umbel. *cf.* umbellule.

Umbellule An, a secondary umbel of a compound umbel.*cf.* umbellet.

Uncinate Bent at the end like a hook; unciform.

Undershrub Subshrub; a small, usually sparsely branched woody shrub less than 1 m high. *cf.* shrub.

Undulate With an edge/margin or edges wavy in a vertical plane; may vary from weakly to strongly undulate or crisped. *cf.* crisped.

Unifoliolate A compound leaf which has been reduced to a single, usually terminal leaflet.

Uniform With one form, e.g. having stamens of a similar length or having one kind of leaf. *cf.* dimorphic.

Uniseriate Arranged in one row or at one level.

Unisexual With one sex only, either bearing the anthers with pollen, or an ovary with ovules, referring to a flower, inflorescence or individual plant. *cf.* bisexual.

Urceolate Shaped like a jug, urn or pitcher.

Utricle A small bladdery pericarp.

Valvate Meeting without overlapping, as of sepals or petals in bud. *cf.* imbricate.

Valve One of the sections or portions into which a capsule separates when ripe.

Variant Any definable individual or group of individuals which may or may not be regarded as representing a formal taxon after examination.

Variegate, variegated Diverse in colour or marked with irregular patches of different colours, blotched.

Variety A taxonomic rank below that of subspecies.

Vein (Botany) a strand of vascular bundle tissue.

Velum A flap of tissue covering the sporangium in the fern, Isoetes.

Velutinous Having the surface covered with a fine and dense silky pubescence of short fine hairs; velvety. *cf.* sericeous

Venation Distribution or arrangement of veins in a leaf.

Veneer Thin sheet of wood.

Ventral (Botany) facing the central axis, opposed to dorsal.

Vernation The arrangement of young leaves or fronds in a bud or at a stem apex. *cf.* circinnate

Verrucose Warty.

Verticil A circular arrangement, as of flowers, leaves, or hairs, growing about a central point; a whorl.

Verticillaster False whorl composed of a pair of opposite cymes as in Lamiaceae.

Verticillate Whorled, arranged in one or more whorls.

Vertisol A soil with a high content of expansive montmorillonite clay that forms deep cracks in drier seasons or years.

Vertosols Soils that both contain more than 35% clay and possess deep cracks wider than 5 mm during most years.

Vesicle A small bladdery sac or cavity filled with air or fluid. *adj.* vesicular.

Vestigial The remaining trace or remnant of an organ which seemingly lost all or most of its original function in a species through evolution.

Vestiture Covering; the type of hairiness, scaliness or other covering commonly found on the external parts of plants. *cf.* indumentums.

Vibratile Capable of to and fro motion.

Villose Covered with long, fine, soft hairs, finer than in pilose.

Villous Covered with soft, shaggy unmatted hairs.

Vine A climbing or trailing plant.

Violaxanthin Is a natural xanthophyll pigment with an orange color found in a variety of plants like pansies.

Viscid Sticky, being of a consistency that resists flow.

Viviparous Describes seeds or fruit which sprout before they fall from the parent plant.

Whorl A ring-like arrangement of leaves, sepals, stamens or other organs around an axis.

Winged Having a flat, often membranous expansion or flange, e.g. on a seed, stem or one of the two lateral petals of a Papilionaceous flower or one of the petal-like sepals of Polygalaceae. *cf.* keel, standard.

Xanthophylls Are yellow, carotenoid pigments found in plants. They are oxidized derivatives of carotenes.

Xeromorphic Plant with special modified structure to help the plant to adapt to dry conditions.

Xerophyte A plant which naturally grows in dry regions and is often structurally modified to withstand dry conditions.

Zygomorphic Having only one plane of symmetry, usually the vertical plane, referring to a flower, calyx or corolla. *cf.* actinomorphic.

Zygote The fist cell formed by the union of two gametes in sexual reproduction. *adj.* zygotic.

Common Name Index

T.K. Lim, *Edible Medicinal And Non-Medicinal Plants: Volume 3, Fruits*,
DOI 10.1007/978-94-007-2534-8, © Springer Science+Business Media B.V. 2012

Scientific Name Index

T.K. Lim, *Edible Medicinal And Non-Medicinal Plants: Volume 3, Fruits*,
DOI 10.1007/978-94-007-2534-8, © Springer Science+Business Media B.V. 2012

Q

R

S

X

Y

Printed by Printforce, the Netherlands